Plasma Physics and Engineering

Plasma Physics and Engineering

Alexander Fridman
Lawrence A. Kennedy

Taylor & Francis
Taylor & Francis Group

NEW YORK • LONDON

Denise T. Schanck, *Vice President*
Robert L. Rogers, *Senior Editor*
Summers Scholl, *Editorial Assistant*
Savita Poornam, *Marketing Manager*
Randy Harinandan, *Marketing Assistant*

Susan Fox, *Project Editor*
Shayna Murry, *Cover Designer*

Published in 2004 by
Taylor & Francis
29 West 35th Street
New York, NY 10001-2299

Published in Great Britain by
Taylor & Francis
4 Park Square
Milton Park
Abingdon OX14 4RN

www.taylorandfrancis.com

Printed in the United States of America on acid-free paper.

10 9 8 7 6 5 4 3 2 1

Library of Congress Cataloging-in-Publication Data

Fridman, Alexander A., 1953–
Plasma physics and engineering / by Alexander A. Fridman and Lawrence A. Kennedy
p. cm.
ISBN 1-56032-848-7 (hardcover : alk. paper)
1. Plasma (Ionized gases). 2. Plasma engineering. I. Kennedy, Lawrence A., 1937– II. Title.

QC718.F77 2004
530.4′4—dc22 2003022820

Preface

Plasma enjoys an important role in a wide variety of industrial processes, including material processing; environmental control; electronic chip manufacturing; light sources; bio-medicine; and space propulsion. It is also central to understanding most of the universe outside the Earth. As such, the focus of this book is to provide a thorough introduction to the subject and to be a valued reference that serves engineers and scientists as well as students. Plasma is not an elementary subject and the reader is expected to have the normal engineering background in thermodynamics, chemistry, physics, and fluid mechanics upon which to build an understanding of this subject.

This text has been organized into two parts. Part I addresses the basic physics of plasma. Chapter 2 examines the elementary processes of charge species in plasma and Chapter 3 provides a thorough introduction to the elementary processes of excited molecules and atoms in plasmas. Chapter 4 and Chapter 5 examine the kinetics of charged/excited particles and Chapter 6 gives a thorough introduction to the electrostatics, electrodynamics, and fluid mechanics of plasma.

Part II addresses the physics and engineering of electric discharges, specifically examining glow and arc discharges (Chapter 7 and Chapter 8); cold atmospheric pressure discharges typically associated with corona, dielectric barrier, and spark discharges (Chapter 9); plasma created in high-frequency electromagnetic fields characterized by radio-frequency, microwave, and optical discharges (Chapter 10); and discharges in aerosols and dusty plasmas (Chapter 11). The second part of Chapter 12 concludes with discussions on electron beam plasmas. The authors have drawn upon extensive work in the Russian literature in addition to the more accessible results from the West. We believe that this will add an important dimension to development of this subject.

This text is adaptable to a wide range of needs. The material has been taught to graduate and senior-level students from most engineering disciplines and physics. For the latter, it can be packaged to focus on the basic physics of plasma with only selections from discharge applications. For graduate courses, a faster pace can be set that covers Part I and Part II. Presently, the text is used for a plasma engineering course sequence (Plasma I, Plasma II) at Drexel University.

We gratefully acknowledge the loving support of our wives, Irene Fridman and Valaree Kennedy; the governmental research support of the National Science Foundation and the U.S. Department of Energy, together with our long-term industrial sponsors, Air Liquide, Texaco, Kodak, Georgia Pacific, and Applied Materials. We especially appreciate John and Chris Nyheim and the Kaplan family for their support of plasma research at Drexel University and University of Illinois at Chicago. Additionally, we gratefully acknowledge the invaluable, stimulating discussions with our colleagues, Professors Young Cho, Gary Friedman, Baki Farouk, Alexei V. Saveliev and Alexander Gutsol, and our students. We thank K. Gutsol, A. Fridman, G. Fridman, and A. Chirokov for help with illustrations.

Alexander Fridman
Lawrence A. Kennedy

Table of Contents

Part I
FUNDAMENTALS OF PLASMA PHYSICS AND PLASMA CHEMISTRY

CHAPTER

1

PLASMA IN NATURE, IN THE LABORATORY, AND IN INDUSTRY

The term *plasma* is often referred to as the fourth state of matter. As temperature increases, molecules become more energetic and transform in the sequence: solid, liquid, gas, and plasma. In the latter stages, molecules in the gas dissociate to form a gas of atoms and then a gas of freely moving charged particles, electrons, and positive ions. This state is called the plasma state, a term attributed to Langmuir to describe the region of a discharge not influenced by walls and/or electrodes. It is characterized by a mixture of electrons, ions, and neutral particles moving in random directions that, on average, is electrically neutral ($n_e \cong n_i$). In addition, plasmas are electrically conducting due to the presence of these free-charge carriers and can attain electrical conductivities larger than those of metals such as gold and copper.

Plasma is not only most energetic, but also most challenging for researchers state of matter. Temperature of the charged particles in plasma can be so high that their collisions can result in thermonuclear reactions! Plasmas occur naturally, but also can be manmade. Plasma generation and stabilization in the laboratory and in industrial devices are not easy, but very promising for many modern applications, including thermonuclear synthesis, electronics, lasers, and many others. Most computer hardware is made based on plasma technologies, as well as the very large and thin TV plasma screens so popular today.

Plasmas offer two main characteristics for practical applications. They can have temperatures and energy densities greater than those produced by ordinary chemical means. Furthermore, plasmas can produce energetic species that can initiate chemical reactions difficult or impossible to obtain using ordinary chemical mechanisms. The energetic species generated cover a wide spectrum, e.g., charge particles including electrons, ions, and radicals; highly reactive neutral species such as reactive atoms (e.g., O, F, etc.), excited atomic states; reactive molecular fragments; and different wavelength photons. Plasmas can also initiate physical changes in material surfaces.

Applications of plasma can provide major benefits over existing methods. Often, processes can be performed that are not possible in any other manner. Plasma can provide an efficiency increase in processing methods and very often can reduce environmental impact in comparison to more conventional processes.

1.1 OCCURRENCE OF PLASMA: NATURAL AND MANMADE PLASMAS

Although somewhat rare on Earth, plasmas occur naturally and comprise the majority of the universe, encompassing among other phenomena, the solar corona, solar wind, nebula, and Earth's ionosphere. In the Earth's atmosphere, plasma is often observed as a transient event in the phenomenon of lightning strokes. Because air is normally nonconducting, large potential differences can be generated between clouds and Earth during storms. Lightning discharges that occur to neutralize the accumulated charge in the clouds take place in two phases. First, an initial leader stroke progresses in steps across the potential gap between clouds or the Earth. This leader stroke creates a low degree of ionization in the path and provides conditions for the second phase, the return stroke, to occur. The return stroke creates a highly conducting plasma path for the large currents to flow and neutralize the charge accumulation in the clouds.

At altitudes of approximately 100 km, the atmosphere no longer remains nonconducting due to its interaction with solar radiation. From the energy absorbed through these solar radiation processes, a significant number of molecules and atoms become ionized and make this region of the atmosphere a plasma. Further into near-space altitudes, the Earth's magnetic field interacts with charge particles streaming from the Sun. These particles are diverted and often become trapped by the Earth's magnetic field. The trapped particles are most dense near the Poles and account for the Aurora Borealis.

Lightning and the Aurora Borealis are the most common natural plasma observed on Earth and one may think that these are exceptions. However, the Earth's atmospheric environment is an anomaly; we live in a bubble of un-ionized gas surrounded by plasma. Lightning occurs at relatively high pressures and the Aurora Borealis at low pressures. Pressure has a strong influence on plasma, influencing its respective luminosity, energy or temperature, and the plasma state of its components. Lightning consists of a narrow, highly luminous channel with numerous branches. The Aurora Borealis is a low-luminosity, diffuse event.

Because plasmas occur over a wide range of pressures, it is customary to classify them in terms of electron temperatures and electron densities. Figure 1.1 provides a representation of the electron temperature in electronvolts and electron densities (in $1/cm^3$) typical of natural and manmade plasmas. Electron temperature is expressed in electronvolts (eV); 1 eV is equal to approximately 11,600 K. Manmade plasma ranges from slightly above room temperature to temperatures comparable to the interior of stars. Electron densities span over 15 orders of magnitudes. However, most plasmas of practical significance have electron temperatures of 1 to 20 eV with electron densities in the range 10^6 to 10^{18} $1/cm^3$.

Not all particles need to be ionized in a plasma; a common condition is for the gases to be partially ionized. Under the latter condition, it is necessary to examine the particle's densities to determine if it is plasma. In a very low density of charge particles, the influence of the neutral particles can swamp the effects of interaction

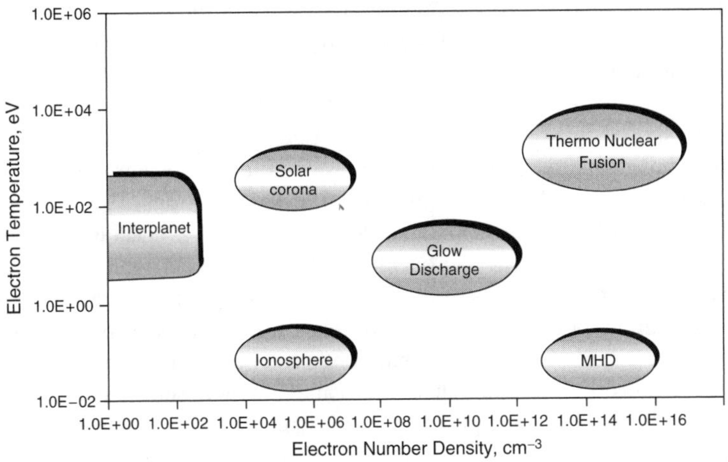

Figure 1.1 Operating regions of natural and manmade plasma

between charged ones. Under these conditions, the neutral particles dominate the collision process.

The force laws appropriate to the specific particles govern collisions. For neutral particles, collisions are governed by forces significant only when the neutrals are in close proximity to each other. Conversely, the longer range Coulomb law governs collisions between charge particles. Thus, the behavior of the collision will change as the degree of ionization increases, with the Coulomb interactions becoming more important. Coulomb's law will govern all particles in fully ionized plasma.

Langmuir was an early pioneer in the field who studied gas discharges and defined plasma to be a region not influenced by its boundaries. The transition zone between the plasma and its boundaries was termed the plasma sheath. The properties of the sheath differ from those of the plasma and these boundaries influence the motion of the charge particles in this sheath. They form an electrical screen for the plasma from influences of the boundary.

As in any gas, the temperature is defined by the average kinetic energy of the particle, molecule, atom, neutral, or charge. Thus, plasmas will typically exhibit multiple temperatures unless sufficient collisions occur between particles to equilibrate. However, because the mass of an electron is much less than the mass of a heavy particle, many collisions are required for this to occur.

The use of an electric discharge is one of the most common ways to create and maintain a plasma. Here, the energy from the electric field is accumulated between collisions by the electron that subsequently transfers a portion of this energy to the heavy neutral particles through collisions. Even with a high collision frequency, the electron temperature and heavy particle temperature normally will be different. Because the collision frequency is pressure dependent, high pressures will increase this frequency and the electron's mean free path between collisions will decrease. One can show that the temperature difference between electrons and heavy neutral particles is proportional to the square of the ratio of the energy an electron receives from the electric field (E) to the pressure (p). Only in the case of small values of E/p do the temperatures of electrons and heavy particles approach each other; thus, this is a basic requirement for *local thermodynamic equilibrium* (LTE) in a plasma. Additionally, LTE conditions require chemical equilibrium as well as restrictions on the gradients.

When these conditions are met, the plasma is termed a *thermal plasma*. Conversely, when there are large departures from these conditions, $T_e > T_n$, the plasma is termed a *nonequilibrium plasma* or *nonthermal plasma*. As an example, for plasma generated by glow discharges, the operating pressures are normally less than 1 kPa and have electron temperatures of the order of 10^4 K with ions and neutral temperatures approaching room temperatures. Physics, engineering aspects, and application areas are quite different for thermal and nonthermal plasmas. Thermal plasmas are usually more powerful, while nonthermal plasmas are more selective. However, these two different types of ionized gases have many more features in common and both are plasmas.

1.2 GAS DISCHARGES

There are many well-established uses of plasma generated by an electric discharge. Historically, the study of such discharges laid the initial foundation of much of our understanding of plasma. In principle, discharges can be simply viewed as two electrodes inserted into a glass tube and connected to a power supply. The tube can be filled with various gases or evacuated. As the voltage applied across the two electrodes increases, the current suddenly increases sharply at a certain voltage characteristic of an electron avalanche.

If the pressure is low (of the order of a few torr) and the external circuit has a large resistance to prohibit a large current flow, a glow discharge develops. This is a low-current, high-voltage device in which the gas is weakly ionized (plasma). Widely used, glow discharges are a very important type of discharge: they form the basis of fluorescent lighting. Such a discharge is present in every fluorescent lamp and their voltage current characteristics are widely employed in constant voltage gas tubes used in electronic circuits. Other common gas vapor discharges used for lighting are mercury vapor and neon.

In power electronics applications, the need for switching and rectification has been addressed with the development of the thyratron and ignitron, which are important constant voltage gas tubes. If the pressure is high (of the order of an atmosphere) and the external circuit resistance is low, a thermal arc discharge can form following breakdown. Thermal arcs usually carry large currents—greater than 1 A at voltages of the order of tens of volts. Furthermore, they release large amounts of thermal energy. These types of arcs are often coupled with a gas flow to form high-temperature plasma jets.

A corona discharge occurs only in regions of sharply nonuniform electric fields. The field near one or both electrodes must be stronger than in the rest of the gas. This occurs near sharp points, edges, or small-diameter wires. These tend to be low-power devices limited by the onset of electrical breakdown of the gas. However, it is possible to circumvent this restriction through the use of pulsating power supplies.

Many other types of discharges are variations or combinations of the preceding descriptions; the gliding arc is one such example. Beginning as a thermal arc located at the shortest distance between the electrodes, it moves with the gas flow at a velocity of about 10 m/s and the length of the arc column increases together with the voltage. When the length of the gliding arc exceeds its critical value, heat losses from the plasma column begin to exceed the energy supplied by the source, so it is not possible to sustain the plasma in a state of thermodynamic equilibrium. As a result, a fast transition into a nonequilibrium phase occurs. The discharge plasma cools rapidly to a gas temperature of about $T_n = 1000$ K and the plasma conductivity is maintained by a high value of the electron temperature $T_e = 1$ eV (about 11,000 K). After this fast transition, the gliding arc continues its evolution, but under nonequilibrium conditions ($T_e >> T_n$). The specific heat losses in this regime are much smaller than in the equilibrium regime (numerically, about three times less). The discharge length increases up to a new critical value of approximately three times its original critical length. The main part of the gliding arc power (up to 75 to 80%) can be dissipated in the nonequilibrium zone.

After decay of the nonequilibrium discharge, the evolution repeats from the initial breakdown. This permits the stimulation of chemical reactions in regimes quite different from conventional combustion and environmental situations. It provides an alternate approach to addressing energy conservation and environmental control. Figure 1.2 illustrates the gliding arc in different configurations, and viewed with different kinds of photography.

Many of the features in these discharges are also typical for discharges in rapidly oscillating fields in which electrodes are not required. As such, a classification of discharges has been developed that does not involve electrode attributes. Such a simplified classification is based upon two properties: (1) state of the ionized gas;

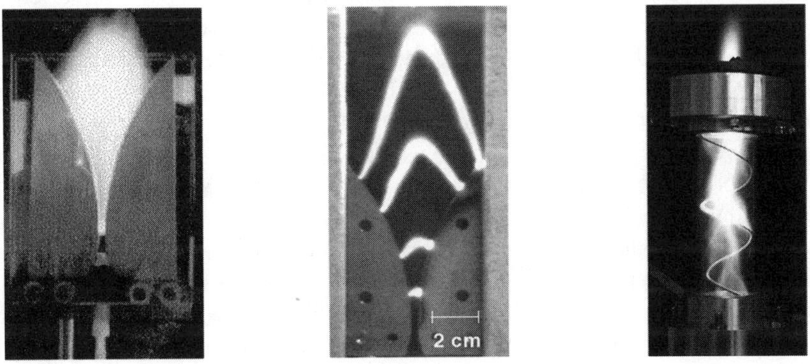

(a) (b) (c)

Figure 1.2 (a) Gliding arc viewed with normal photography; (b) gliding arc viewed with high-speed photography; (c) vortex gliding arc,[622] the so-called gliding arc tornado

Table 1.1 Classification of Discharges

Frequencies	Breakdown	Nonequilibrium Plasma	Equilibrium Plasma
Constant *E* field	Initiation of glow discharge	Glow discharge	High pressure arc
Radio frequencies	Initiation of RF discharge	Capacitively coupled RF discharge in rarefied gas	Inductively coupled plasma
Microwave	Breakdown in waveguides	Microwave discharge in rarefied gas	Microwave plasmatron
Optical range	Gas breakdown by laser radiation	Final stage optical breakdown	Continuous optical discharge

and (2) frequency range of the electric field. The state of the ionized gas distinguishes among (1) breakdown of the gas; (2) sustaining a nonequilibrium plasma by the electric field; and (3) sustaining an equilibrium plasma. The frequency criteria classify the field into: (1) DC, low-frequency, and pulsed E fields; (2) radio-frequency (RF) E fields; (3) microwave fields; and (4) optical fields. These are summarized in Table 1.1 in a classification developed by Raizer.[101]

1.3 PLASMA APPLICATIONS: PLASMAS IN INDUSTRY

Although this text is mainly focused upon the fundamentals of plasma physics and chemistry, ultimately, we are interested in the application of plasmas. The number of industrial applications of plasma technologies is extensive and involves many industries. High-energy efficiency, specific productivity, and selectivity may be achieved in plasmas for a wide range of chemical processes. As an example, for CO_2 dissociation in nonequilibrium plasmas under supersonic flow conditions, it is possible to introduce up to 90% of total discharge power in CO production selectively when the vibrational temperature is about 4000 K and the translational temperature is only around 300 K. The specific productivity of such a supersonic reactor achieves 1,000,000 l/h, with power levels up to 1 MW. This plasma process is also being examined for fuel production on Mars, where the atmosphere primarily consists of CO_2. Here on the Earth, it was applied as a plasma stage in a two-step process for hydrogen production from water.

The key point for practical use of any chemical process in a particular plasma system is to find the proper regime and optimal plasma parameters among the numerous possibilities intrinsic to systems far from equilibrium. In particular, it is desired to provide high operating power for the plasma chemical reactor together with high selectivity of the energy input, while simultaneously maintaining nonequilibrium plasma conditions. Generally, two very different kinds of plasma are used for chemical applications: thermal and nonthermal. Thermal plasma generators have been designed for many diverse industrial applications covering a wide range of operating power levels from less than 1 kW to over 50 MW.

However, in spite of providing sufficient power levels, these are not well adapted to the purposes of plasma chemistry, in which selective treatment of reactants (through, for example, excitation of molecular vibrations or electronic excitation) and high efficiency are required. The main drawbacks of using thermal plasmas for plasma chemical applications are the overheating of reaction media when energy is uniformly consumed by the reagents into all degrees of freedom and, thus, high energy consumption required to provide special quenching of the reagents, etc. Because of these drawbacks, the energy efficiency and selectivity of such systems are rather small (only one of many plasma chemical processes developed in the first decades of the century remains, e.g., the production of acetylene from light hydrocarbons in Germany).

Promising plasma parameters are somewhat achievable in microwave discharges. Skin effect here permits simultaneous achievement of a high level of electron density and a high electric field (and thus a high electronic temperature as well) in the relatively cold gas. Existing super-high frequency discharge technology can be used to generate dense ($n_e = 10^{13}$ electrons/cm^3) nonequilibrium plasmas (T_e = 1 to 2 eV; T_V = 3000 to 5000 K; and T_n = 800 to 1500 K [for supersonic flow, $T_n \leq$ 150 K and less]) at pressures up to 200 to 300 torr and at power levels reaching 1 MW.

An alternative approach for plasma chemical gas processing is the nonthermal one. Silent discharges such as glow, corona, short pulse, microwave, or radiofrequency (RF) electrical discharges are directly produced in the processed gas, mostly under low pressure. The glow discharge in a low-pressure gas is a simple and inexpensive way to achieve a nonthermal plasma. Here, the ionization processes induced by the electric field dominate the thermal ones and give relatively high-energy electrons as well as excited ions, atoms, and molecules that promote selective chemical transitions. However, the power of glow discharges is limited by the glow-to-arc transition. Initially, below 1000 K, gas becomes hot (>6000 K) and the electron temperature, initially sufficiently high (>12,000 K), cools close to the bulk gas temperature. The discharge voltage decreases during such a transition, making it necessary to increase the current in order to keep the power on the same level, which in turn leads to thermalization of the gas. Thus, cold nonequilibrium plasmas created by conventional glow discharges offer good selectivity and efficiency, but at limited pressure and power levels.

In these two general types of plasma discharges, it is impossible simultaneously to keep a high level of nonequilibrium, high electron temperature, and high electron density, whereas most prospective plasma chemical applications simultaneously require high power for high reactor productivity and a high degree of nonequilibrium to support selective chemical process.

Recently, a simpler technique offering similar advantages has been proposed: the Gliding Arc. Such a gliding arc occurs when the plasma is generated between two or more diverging electrodes placed in a fast gas flow. It operates at atmospheric pressure or higher and the dissipated power at nonequilibrium conditions reaches up to 40 kW per electrode pair. The incontestable advantage of the Gliding Arc compared with microwave systems is its cost; it is much less expensive compared to microwave plasma devices.

Although we are not discussing specific plasma applications here, let us go through some applied aspects just to illustrate how different they are from one another. Reviews of various industrial plasma applications are presented in several books (References 9, 13, 90, 431, 264, 154).

1.4 PLASMA APPLICATIONS FOR ENVIRONMENTAL CONTROL

Low-temperature, nonequilibrium plasmas are an emerging technology for abating low volatile organic compound (VOC) emissions and other industrial exhausts. These nonequilibrium plasma processes have been shown to be effective in treating a wide range of emissions including aliphatic hydrocarbons; chlorofluorocarbons; methyl cyanide; phosgene; and formaldehyde, as well as sulfur and organophosphorus compounds, sulfur and nitrogen oxides. Such plasmas may be produced by a variety of electrical discharges or electron beams. The basic feature of nonequilibrium plasma technologies is that they produce plasma in which the majority of the electric energy (more than 99%) goes into production of energetic electrons, instead of heating the entire gas stream. These energetic electrons produce excited species, free radicals, and ions, as well as additional electrons through the electron impact dissociation, excitation, and ionization of the background molecules.

These active species, in turn, oxidize, reduce, or decompose the pollutant molecules. This is in contrast to the mechanism involved in thermal processes that require heating the entire gas stream in order to destroy pollutants. In addition, low-temperature plasma technology is highly selective and has relatively low maintenance requirements. Its high selectivity results in relatively low energy costs for emissions control while low maintenance keeps annual operating expenses low. Furthermore, these plasma discharges normally are very uniform and homogeneous, which results in high process productivity. Although the products of these plasma processes are virtually indistinguishable from incineration products (CO_2, H_2O, SO_2, etc.), chemical reactions occurring in these technologies are substantially different than in incineration.

Large electric fields in discharge reactors create conditions for electric breakdowns during which many electron–ion pairs are formed. Primary electrons are accelerated by the electric field and produce secondary ionization, etc. Created excited species, atoms, radicals, molecular and atomic ions, electrons, and radicals are capable of interacting to a certain degree with VOCs. Some processes are more effective than others; the most important are those leading to formation of OH radicals. Typically, the major mechanisms controlling the chemistry of VOC destruction fall into five major categories.

1. *Radical-induced VOC destruction.* Hydroxyl and other active radicals are formed through a multistep mechanism initiated by energetic electrons. Actually, the higher energy electrons are applied—the higher efficiency can be achieved (see Table 1.2).
2. *Direct electron-induced VOC destruction.* This mechanism usually occurs only in strongly electronegative gases when the electron affinity of a gas specie is comparable to its dissociation energy.

3. *Direct ion-induced VOC destruction.* The direct ion-induced decomposition of VOCs is similar to the hydroxyl-formation mechanism discussed earlier. When the VOC species or their intermediate products produced by another destruction mechanism have a low ionization potential, the charge exchange process can promote further decomposition reactions.
4. *Water droplet- and cluster-enhanced VOC destruction.* Ions produced in an atmospheric-pressure discharge are natural nuclei and will stimulate water condensation or cluster formation. Liquid phase catches positive ions and ozone molecules; captured species induce formation of OH radicals in droplets. The radicals very efficiently destroy VOC molecules when they encounter the droplet.
5. *Ultraviolet VOC destruction.* Ultraviolet (UV) radiation in plasma can also be effective in selective dissociation of VOC. Although this effect is not a primary decomposition mechanism in nonequilibrium plasmas, it plays a very important role as an ancillary VOC destruction process. Typically, UV radiation emitted by the plasma discharge breaks molecular bonds.

Table 1.2 illustrates different plasma approaches to VOC destruction. Electron beams produce plasma with highest average electron energy of 5 to 6 eV; pulsed corona discharges stand very close (3 to 5 eV). However, corona discharges have the advantage of lower capital cost. Generally, both technologies have already found applications for gas treatment from NO_x and SO_2 emissions in coal power plants. Capability of treating large gas streams makes e-beams and pulsed corona discharges the primary candidates considered for VOC abatement. Here, one specific plasma application for environmental control has been discussed in order to introduce some general parameters characterizing such processes. More details on the subject can be found in References 304 and 309.

1.5 PLASMA APPLICATIONS IN ENERGY CONVERSION

Plasma applications in energy conversion are many; however, the most publicized are those related to controlling thermonuclear fusion reactions; magnetohydrodynamic (MHD) power generation; and thermionic energy conversions. Fusion reactions are the source of the Sun's energy as well as the basis of thermonuclear weapons and the more peaceful application goal of power generation. To undergo fusion reactions, the nuclei must interact very close by overcoming the electrostatic repulsive force between particles. Thus, they must approach each other at high velocities.

If one simply employed opposing beams of particles to do this, too many particles would be scattered to make this a practical approach. The scattered particles must be reflected back into the reactor as often as necessary to generate the fusion reaction. For deuterium, the required temperature exceeds 100 million degrees Kelvin to produce a significant amount of fusion. At this temperature, any gas is fully ionized and, therefore, a plasma. The engineering problem for controlled fusion is then to contain this high-temperature gas for an interval sufficiently long for the fusion reactions to occur and then extract the energy. The only way to contain such

Table 1.2 Illustrative Comparison of Nonthermal Plasma Technologies for VOC-Abatement

Plasma Technology	Average Electron Energy	Capital Cost, \$/W	Unit Capacity, kW	Energy Consumption, kWh/m^3 of Pollutant	Known Applications	Technical Limitations
Electron beam technology	5–6 eV	2.0	10–1000	10–70	Removal of SO_2, NOx in thermal power plants	Initial energy of electrons in the beam is limited by 1.5 MeV due to x-ray production; generation of by-products
Gas-phase corona reactor	2–3 eV	2.1	1–20	6–70	Electrostatic precipitators	The voltage cannot be as high as in pulsed corona due to spark formation; packed bed (if applied) limits flow rates
Pulsed corona discharge	3–5 eV	1.0	10	20–150	Removal of polyatomic hydrocarbons in aluminum plant	Voltage is limited by the electrical network coupling and capacitance and inductance of the plasma reactor
Dielectric-barrier discharge	2–3 eV	0.2	0.1–1	20–200	Removal of SO_2, NOx, industrial ozone generators	The voltage cannot be as high as in pulsed corona due to spark formation; low unit capacity
Gliding arc	1–1.5 eV	0.1	120	20–250	Removal of methanol, o-xylene ea.	Discharge is nonuniform

a plasma is to exploit its interactions with magnetic fields. This long-term effort is being pursued at national laboratories in many countries.

The MHD generator exploits the flow of plasma interacting with a magnetic field to generate electrical power. From Maxwell's equations, the interaction of a flowing plasma with a magnetic field (B) will give rise to an induced electric field ($E = v \times B$). Energy can be extracted if electrodes are placed across this flow permitting a current flow through the plasma. This in turn gives rise to a force ($J \times B$) opposite to the flow that decelerates it. A variation of this concept is the plasma accelerator in which the applied fields are configured to create an accelerating force.

Thermionic generators are another example of a plasma energy converter. The simplest configuration is a basic diode where the anode is cooled and a heated cathode causes electrons to boil off. The region between the electrodes is filled with cesium vapor that can be easily ionized through contact with the hot cathode or through multiple steps in the gas. The space charge is neutralized by the positive cesium ions and allows high current to flow. Many different configurations are possible and they provide power generation for space vehicles and mobile power generators. They also have been proposed for miniature mobile power units to replace batteries.

Serious attention is now directed to plasma applications for hydrogen production from different hydrocarbons, including coal and organic wastes. Plasma discharges can be effectively combined in this case with catalytic and different hybrid systems. An extensive review of plasma in energy conversion is given in Reference 9.

1.6 PLASMA APPLICATION FOR MATERIAL PROCESSING

Plasma applications abound in material processing. Initially, thermal plasmas were employed because of the high temperatures they achieved over those obtainable from gaseous flames. Although combustion flame temperatures are limited to approximately 3300 K, a small plasma jet generated by the gas flow through a thermal electric arc provides a working temperature of approximately 6500 K. When one considers the melting and/or vaporization points of many metals and ceramics, it is obvious that these arcs are extremely useful processing tools. More powerful arcs can generate temperatures from 20,000 K to 30,000 K, thus providing environments for activated species surface modifications. The use of electric arcs for welding, melting of metals, and high-purity metal processing is well established.

Plasma is widely employed in the coating industry where its large enthalpy content, high temperatures, and large deposition rates are advantageous for increased throughputs. It provides the ability to mix and blend materials that are otherwise incompatible. Complex alloys, elemental materials, composites, and ceramics can be deposited. Among others, it is the preferred process for the manufacture of electronic components and wear- and heat-resistant coatings. With the advent of nanostructured metals, nanopowder production and *in-situ* coating of these nanoparticles is accomplished with plasma processing.

Surface modification of different classes of material is performed using plasma environments. For example, the choice of a polymer for a specific application is generally driven by the bulk properties of the polymer, but many common polymer surfaces are chemically inert and, therefore, pose challenges for use as substrates for applied layers. Typical applications in which wetting and adhesion are critical issues are printing, painting, and metallization. Surface modification by plasma treatment of polymer surfaces can improve wetting, bonding, and adhesion characteristics without sacrificing desirable bulk properties of the polymer base material. Plasma treatments have proven useful in promoting adhesion of silver to polyethylene terephthalate. In addition, plasma treatments of polyester have been demonstrated to promote adhesion of gelatin-containing layers used in production of photographic film. In both of these examples, low-frequency, capacitively coupled nitrogen plasmas have proven particularly effective. Plasma-assisted material processing in electronics is so widely used and so successful that a few sentences cannot describe it. More details on the subject can be found in an interesting and useful book of Lieberman and Lichtenberg.[13]

CHAPTER

2

ELEMENTARY PROCESSES OF CHARGED SPECIES IN PLASMA

2.1 ELEMENTARY CHARGED PARTICLES IN PLASMA AND THEIR ELASTIC AND INELASTIC COLLISIONS

Because plasma is an ionized medium, the key process in plasma is ionization. This means that a neutral atom or molecule is converted into a positive ion and also an electron, as illustrated in Figure 2.1(a). The main two participants of the ionization process, **electrons** and **positive ions**, are at the same time the most important charged particles in plasma. The principal characteristics and elementary processes of the electrons and positive ions as well as some characteristics of negative and complex ions will be described in this section. Behavior and elementary processes of excited molecules and other active neutral components of plasma will be discussed in Chapter 3.

Normally, concentrations or number densities of electrons and positive ions are equal or near equal in quasi-neutral plasmas, but in "electronegative" gases they can be different. Electronegative gases like O_2, Cl_2, SF_6, UF_6, and $TiCl_4$ consist of atoms or molecules with a high electron affinity, which means they strongly attract electrons. Electrons stick to such molecules or collide with them with the formation of **negative ions**, the third important group of charged particles in plasmas. Concentration of negative ions can exceed that of electrons in electronegative gases. Illustrations of electron attachment and negative ion formation are shown in Figure 2.1(b).

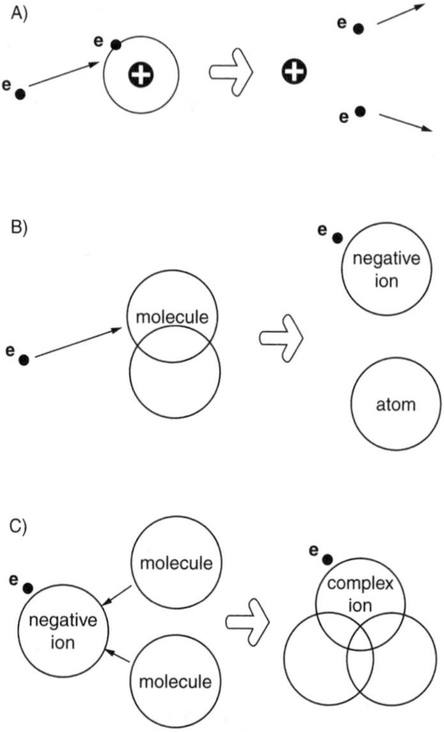

Figure 2.1 Elementary processes of (A) ionization; (B) dissociative attachment and formation of a negative ion; (C) complex ion formation

Charged particles can also appear in plasma in more complicated forms. In high-pressure and low-temperature plasmas, the positive and negative ions attach to neutral atoms or molecules to form quite large **complex ions** or ion clusters, e.g., $N_2^+N_2$ (N_4^+); O^-CO_2 (CO_3^-); and H^+H_2O (H_3O^+), Figure 2.1(c). Ion–molecular bonds in complex ions are usually less strong than regular chemical bonds, but stronger than intermolecular bonds in neutral clusters. In heterogeneous plasmas, ions and electrons can adhere to surfaces and potentially change their behavior. Properties of complex ions and ion clusters are very specific and particularly important for plasma catalysis and plasma surface treatment. However, in general, only three basic types of charged particles are in plasma: electrons, positive ions, and negative ions.

2.1.1 Electrons

Electrons are elementary, negatively charged particles with a mass about three to four orders of magnitude less than the mass of ions and neutral particles (electron charge is an elementary one: $e = 1.6\ 10^{-19}$ C, mass $m_e = 9.11\ 10^{-31}$ kg). Because of their lightness and high mobility, electrons are the first particles in receiving energy from electric fields. Afterward, electrons transmit the energy to all other plasma components, providing ionization, excitation, dissociation, etc.; for this reason, elementary processes with electrons are of major importance in plasma physics and chemistry.

Electrons are energy providers for many plasma chemical processes. The rate of such processes depends on how many electrons have sufficient energy to do the job. This can be described by means of the **electron energy distribution function**, $f(\varepsilon)$, which is a probability density for an electron to have the energy ε (probability for an electron to have energy between ε and $\varepsilon + \Delta$, divided by Δ).

Quite often in a plasma this distribution function strongly depends upon the electric field and gas composition; then it can be very far from equilibrium. The nonequilibrium electron energy distribution functions $f(\varepsilon)$ for different discharge systems and different plasma conditions will be discussed in Chapter 4. Sometimes, however (even in nonequilibrium plasmas of nonthermal discharges), the $f(\varepsilon)$ depends primarily on the electron temperature T_e. Under these circumstances $f(\varepsilon)$ can then be defined by the quasi-equilibrium **Maxwell–Boltzmann distribution function**:

$$f(\varepsilon) = 2\sqrt{\varepsilon/\pi(kT_e)^3}\,\exp(-\varepsilon/kT_e) \tag{2.1}$$

where k is a Boltzmann constant. If temperature is given in energy units, then k = 1 and can be omitted. (This rule will be followed throughout the book.)

For this case, the **mean electron energy**, which is the first moment of the distribution function, is proportional to temperature in the traditional way:

$$\langle\varepsilon\rangle = \int_0^\infty \varepsilon f(\varepsilon)dE = \frac{3}{2}T_e \tag{2.2}$$

Numerically, in most of the plasmas under consideration, the mean electron energy is from 1 to 5 eV. As was mentioned earlier, the electron temperature is measured in energy units (usually eV), and the Boltzmann constant is omitted in Equation 2.2.

2.1.2 Positive Ions

Atoms or molecules lose their electrons in ionization processes and form positive ions (Figure 2.1a). In hot thermonuclear plasmas, the ions are multicharged, but in quasi-cold technological plasmas of interest, their charge is usually equal to +1e ($1.6\ 10^{-19}$ C). Ions are heavy particles, so commonly they cannot receive high energy directly from an electric field because of intensive collisional energy exchange with other plasma components. The collisional nature of the energy transfer results in the ion energy distribution function usually being not far from the quasi-equilibrium Maxwellian one (2.1), with the ion temperature T_i close to the neutral gas temperature T_o. However, in some low pressure discharges and plasma chemical systems, the ion energy can be quite high; this then can be referred to as generating ion beams.

One of the most important parameters of plasma generation is an ionization energy I, which is the energy needed to form a positive ion. Ionization requires quite

Table 2.1 Ionization Energies for Different Atoms, Radicals, and Molecules

$e + N_2 = N_2^+ + e + e$	$I = 15.6$ eV	$e + O_2 = O_2^+ + e + e$	$I = 12.2$ eV
$e + CO_2 = CO_2^+ + e + e$	$I = 13.8$ eV	$e + CO = CO^+ + e + e$	$I = 14.0$ eV
$e + H_2 = H_2^+ + e + e$	$I = 15.4$ eV	$e + OH = OH^+ + e + e$	$I = 13.2$ eV
$e + H_2O = H_2O^+ + e + e$	$I = 12.6$ eV	$e + F_2 = F_2^+ + e + e$	$I = 15.7$ eV
$e + H_2S = H_2S^+ + e + e$	$I = 10.5$ eV	$e + HS = HS^+ + e + e$	$I = 10.4$ eV
$e + SF_6 = SF_6^+ + e + e$	$I = 16.2$ eV	$e + SiH_4 = SiH_4^+ + e + e$	$I = 11.4$ eV
$e + UF_6 = UF_6^+ + e + e$	$I = 14.1$ eV	$e + Cs_2 = Cs_2^+ + e + e$	$I = 3.5$ eV
$e + Li_2 = Li_2^+ + e + e$	$I = 4.9$ eV	$e + K_2 = K_2^+ + e + e$	$I = 3.6$ eV
$e + Na_2 = Na_2^+ + e + e$	$I = 4.9$ eV	$e + Rb_2 = Rb_2^+ + e + e$	$I = 3.5$ eV
$e + CH_4 = CH_4^+ + e + e$	$I = 12.7$ eV	$e + C_2H_2 = C_2H_2^+ + e + e$	$I = 11.4$ eV
$e + CF_4 = CF_4^+ + e + e$	$I = 15.6$ eV	$e + CCl_4 = CCl_4^+ + e + e$	$I = 11.5$ eV
$e + H = H^+ + e + e$	$I = 13.6$ eV	$e + O = O^+ + e + e$	$I = 13.6$ eV
$e + He = He^+ + e + e$	$I = 24.6$ eV	$e + Ne = Ne^+ + e + e$	$I = 21.6$ eV
$e + Ar = Ar^+ + e + e$	$I = 15.8$ eV	$e + Kr = Kr^+ + e + e$	$I = 14.0$ eV
$e + Xe = Xe^+ + e + e$	$I = 12.1$ eV	$e + N = N^+ + e + e$	$I = 14.5$ eV

large energy and as a rule defines the upper limit of microscopic energy transfer in plasma. Detailed information on ionization energies are available.[1] For some gases important in applications, the ionization energies are given in Table 2.1.

As one can see from the table, noble gases like He and Ne have the highest ionization energies I, which is why it is so difficult to ionize them. Alkali metal vapors like Li_2 have the lowest value of I, so even a very small addition of such metals can dramatically stimulate ionization. Also, Table 2.1 shows that, of the main components of air (N_2, O_2, H_2O, and CO_2), water has the lowest ionization energy. This explains the effectiveness in forming H_2O^+ (then $H_2O^+ + H_2O = H_3O^+ + OH$), and thus the high acidity and oxidative ability of humid air after ionization.

2.1.3 Negative Ions

Electron attachment to some atoms or molecules results in formation of negative ions (Figure 2.1b); their charge is equal to –1e ($1.6\ 10^{-19}$ C). Attachment of another electron and formation of multicharged negative ions is actually impossible in the gas phase because of electric repulsion. Negative ions are also heavy particles, so usually their energy balance is not due to the electric field, but rather due to collisional processes. The energy distribution functions for negative ions (the same as for positive ones) are not far from Maxwellian distributions (Equation 2.1), with temperatures also close to those of the neutral gas.

Electron affinity EA can be defined as the energy release during attachment processes, or as the bond energy between the attaching electron and the atom or molecule. EA is usually much less than ionization energy I, even for very electronegative gases. For this reason, destruction of negative ions can be occasionally caused even by thermal collisions with other heavy particles.

Table 2.2 Electron Affinity for Different Atoms, Radicals, and Molecules

$H^- = H + e$	$EA = 0.75$ eV	$O^- = O + e$	$EA = 1.5$ eV
$S^- = S + e$	$EA = 2.1$ eV	$F^- = F + e$	$EA = 3.4$ eV
$Cl^- = Cl + e$	$EA = 3.6$ eV	$O_2^- = O_2 + e$	$EA = 0.44$ eV
$O_3^- = O_3 + e$	$EA = 2.0$ eV	$HO_2^- = HO_2 + e$	$EA = 3.0$ eV
$OH^- = OH + e$	$EA = 1.8$ eV	$Cl_2^- = Cl_2 + e$	$EA = 2.4$ eV
$F_2^- = F_2 + e$	$EA = 3.1$ eV	$SF_5^- = SF_5 + e$	$EA = 3.2$ eV
$SF_6^- = SF_6 + e$	$EA = 1.5$ eV	$SO_2^- = SO_2 + e$	$EA = 1.2$ eV
$NO^- = NO + e$	$EA = 0.024$ eV	$NO_2^- = NO_2 + e$	$EA = 3.1$ eV
$NO_3^- = NO_3 + e$	$EA = 3.9$ eV	$N_2O^- = N_2O + e$	$EA = 0.7$ eV
$HNO_3^- = HNO_3 + e$	$EA = 2.0$ eV	$NF_2^- = NF_2 + e$	$EA = 3.0$ eV
$CH_3^- = CH_3 + e$	$EA = 1.1$ eV	$CH_2^- = CH_2 + e$	$EA = 1.5$ eV
$CF_3^- = CF_3 + e$	$EA = 2.1$ eV	$CF_2^- = CF_2 + e$	$EA = 2.7$ eV
$CCl_4^- = CCl_4 + e$	$EA = 2.1$ eV	$SiH_3^- = SiH_3 + e$	$EA = 2.7$ eV
$UF_6^- = UF_6 + e$	$EA = 2.9$ eV	$PtF_6^- = PtF_6 + e$	$EA = 6.8$ eV
$Fe(CO)_4^- = Fe(CO)_4 + e$	$EA = 1.2$ eV	$TiCl_4^- = TiCl_4 + e$	$EA = 1.6$ eV

Numerical values of the *EA* for many different atoms, radicals, and molecules were collected.[1] Information for some gases of interest is presented in Table 2.2, which shows that the halogens and their compounds have the highest values of electron affinity. One halogen, fluorine, plays an important role in plasma chemical etching that is widely applied in modern microelectronics processing. (Negatively charged particles in such systems are mostly not by generated electrons, but by negative ions.) Oxygen, ozone, and some oxides also are strongly electronegative substances with high electron affinity. For this reason, nonthermal plasma generated in air usually contains a large amount of negative ions.

2.1.4 Elementary Processes of Charged Particles

The ionization of atoms and molecules by electron impact, electron attachment to atoms or molecules, and ion–molecular reactions were discussed previously (Figure 2.1a, b, c). All are examples of elementary plasma chemical processes, e.g., collisions accompanied by transformation of elementary plasma particles. These elementary reactive collisions, as well as others, e.g., electron–ion and ion–ion recombination; excitation and dissociation of neutral species by electron impact; relaxation of excited species; electron detachment and destruction of negative ions; photochemical processes, determine plasma behavior and the properties of a variety of electric discharges.

All the elementary processes can be divided into two classes: elastic and nonelastic. The **elastic collisions** are those in which the internal energies of colliding particles do not change; then the total kinetic energy is conserved as well. Thus, these processes result only in scattering. Alternately, collisions are inelastic. All elementary processes listed in previous paragraphs are inelastic ones. **Inelastic**

collisions result in the transfer of energy from the kinetic energy of colliding partners into internal energy. For example, processes of excitation, dissociation, and ionization of molecules by electron impact are inelastic collisions, including transfer of high kinetic energy of plasma electrons into the internal degrees of freedom of the molecules.

In some instances, the internal energy of excited atoms or molecules can be transferred back into kinetic energy (in particular, into kinetic energy of plasma electrons). Such elementary processes are usually referred as **super elastic collisions**.

Elastic collisions will be discussed at the end of this section. Inelastic processes, which play the most important role in plasma chemistry, will be considered in detail in the balance of this chapter. Super elastic collisions of charged particles are normally important in establishing a nonequilibrium electron energy distribution function, and these will be discussed in Chapter 4.

According to kinetic theory, the elementary processes can be described in terms of six main collision parameters: (1) cross section; (2) probability; (3) mean free path; (4) interaction frequency; (5) reaction rate; and (6) reaction rate coefficient (also termed the reaction rate constant).

2.1.5 Fundamental Parameters of Elementary Processes

The most fundamental characteristic of all elementary processes is their cross section. In a simple way, the **cross section of an elementary process** between two particles can be interpreted as an imaginary circle with area –σ, moving together with one of the collision partners (see Figure 2.2a). If the center of the other collision partner crosses the circle, elementary reaction takes place. The cross sections of elementary processes depend strongly on energy of colliding species.

If two colliding particles can be considered as hard elastic spheres of radii r_1 and r_2, their collisional cross section is equal to $\pi (r_1 + r_2)^2$. Obviously, the interaction radius and cross section can exceed the corresponding geometrical sizes because of the long-distance nature of forces acting between electric charges and dipoles. Alternately, if only a few out of many collisions result in a chemical reaction, the cross section is considered to be less than the geometrical one.

Consider the following as an illustrated example. The typical sizes of atoms and molecules are of the order of 1 to 3 Å and the cross section of simple elastic

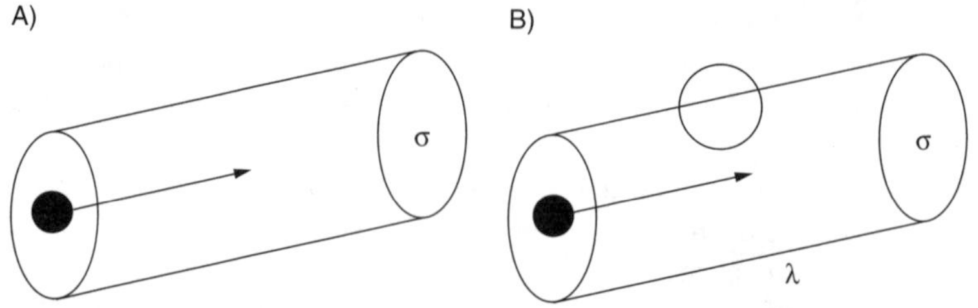

Figure 2.2 (A) Cross section of an elementary process; (B) concept of the mean free path

collisions between plasma electrons (energy 1 to 3 eV) and neutral particles is usually about 10^{-16} to 10^{-15} cm^2. Cross sections of inelastic, endothermic, electron-neutral collisions are normally lower. When the electron energy is very large, the cross section can decrease because of a reduction of the interaction time. The ratio of the inelastic collision cross section to the corresponding cross section of elastic collision under the same condition is often called the dimensionless **probability of the elementary process.**

The **mean free path** λ of one collision partner *A* with respect to the elementary process *A* + *B* with another collision partner *B* can be calculated as:

$$\lambda = 1/(n_B \sigma) \tag{2.3}$$

where n_B is number density (concentration) of the particles *B*. During the mean free path λ, the particle *A* traverses the cylindrical volume $\lambda\sigma$. A reaction occurs if the cylindrical volume transversed by a particle *A* contains at least one *B*-particle, which means $\lambda\sigma n_B = 1$ (see Figure 2.2b). The simplest interpretation of Equation 2.3 is the distance *A* travels before colliding. In the example of electrons colliding with neutrals, the electron mean free path with respect to elastic collisions with neutrals at atmospheric pressure and $n_B = 3{*}10^{19}$ cm^{-3} is approximately $\lambda = 1\,\mu$.

The **interaction frequency** ν of one collision partner *A* (for example, electrons moving with velocity v) with the other collision partner *B* (for example, a heavy neutral particle) can be defined as v/λ or, taking into account Equation 2.3, as:

$$\nu_A = n_B v \sigma \tag{2.4}$$

Actually, this relation should be averaged, taking into account the velocity distribution function f(v) and the dependence of the cross section σ on the collision partners' velocity. In a simple case, when the collision partners' velocity can be attributed mostly to one light particle (e.g. electron), Equation 2.4 can be rewritten in the following integral way:

$$\nu_A = n_B \int f(v)\sigma(v) v\,dv = \langle \sigma v \rangle n_B \tag{2.5}$$

Numerically, for the previously mentioned example of atmospheric pressure, electron neutral, elastic collisions, the interaction frequency is approximately $\nu = 10^{12}$ 1/sec.

2.1.6 Reaction Rate Coefficients

The number of elementary processes w that take place in unit volume per unit time is called the elementary reaction rate. In principal this concept can be used for any kind of reaction: monomolecular, bimolecular, and three-body. For bimolecular processes *A* + *B*, the reaction rate can be calculated by multiplying the interaction frequency of partner *A* with partner *B*—ν_A and number of particles *A* in the unit volume (which is their number density—n_A). Thus: $w = \nu_A n_A$. Taking into account Equation 2.5, this can be rewritten as:

$$w_{A+B} = \langle \sigma v \rangle n_A n_B \tag{2.6}$$

The factor $\langle \sigma v \rangle$ is termed the **reaction rate coefficient** (or reaction rate constant), one of the most useful concepts of plasma chemical and in general chemical kinetics. For bimolecular reactions, it can be calculated as:

$$k_{A+B} = \int \sigma(v) f(v) v dv = \langle \sigma v \rangle \tag{2.7}$$

It is important to emphasize that, in contrast to the reaction cross section σ, which is a function of the partners' energy, the reaction rate coefficient k is an integral factor that includes information on energy distribution functions and depends on temperatures or mean energies of the collision partners. Actually, Equation 2.7 establishes the relation between microkinetics (which is concerned with elementary processes) and macrokinetics (which takes into account real energy distribution functions). Numerically, for the previously considered example of electron neutral, elastic binary collisions, the reaction rate coefficient is approximately $k = 3*10^{-8}$ cm^3/sec.

The concept of reaction rate coefficients can be applied not only for bimolecular processes (Equation 2.7), but also for monomolecular ($A \rightarrow$ products) and three-body processes ($A + B + C \rightarrow$ products). Then the reaction rate is proportional to a product of concentrations of participating particles, and the reaction rate coefficient is just a coefficient of the proportionality:

$$w_A = k_A\, n_A \text{ (a)}; \; w_{A+B+C} = k_{A+B+C}\, n_A\, n_B\, n_C \text{ (b)} \tag{2.8}$$

Obviously, in these cases (Equation 2.8), the reaction rate coefficients cannot be calculated using Equation 2.7); they even have different dimensions. Also, for monomolecular processes (Equation 2.8a), the reaction rate coefficient can be interpreted as the reaction frequency.

2.1.7 Elementary Elastic Collisions of Charged Particles

Elastic collisions (or, elastic scattering) do not result in a change of chemical composition or excitation level of the colliding partners and for this reason they are not able to influence directly plasma chemical processes. On the other hand, due to their high values of cross sections, elastic collisions are responsible for kinetic energy and momentum transfer between colliding partners. This leads to their important role in physical kinetics—in plasma conductivity; drift and diffusion; absorption of electromagnetic energy; and the evolution of energy distribution functions.

These kinetic effects of elastic scattering will be considered in Chapter 4. Here it is only necessary to indicate that the typical value for cross sections of the elastic, electron-neutral collisions for electron energy of 1 to 3 eV is approximately σ = 10^{-16} to 10^{-15} cm^2; the reaction rate coefficient is approximately $k = 3*10^{-8}$ cm^3/sec. For the elastic, ion–neutral collisions at room temperature, the typical value of cross sections is about $\sigma = 10^{-14}$ cm^2 and the reaction rate coefficient is about $k = 10^{-9}$ cm^3/sec.

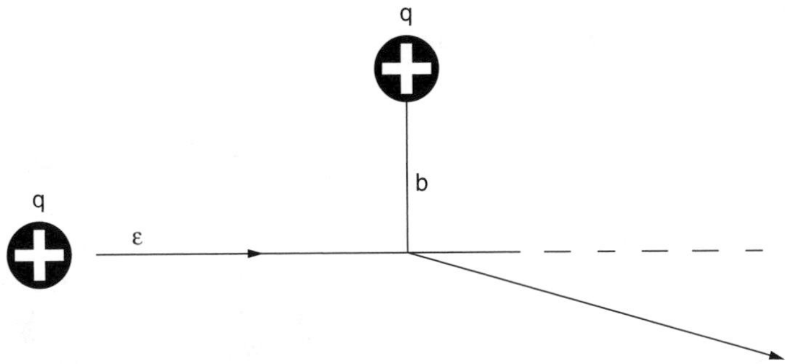

Figure 2.3 Coulomb collisions

Electron–electron, electron–ion, and ion–ion scattering processes are the so-called **Coulomb collisions**. Although their cross sections are quite large with respect to those of collisions with neutral partners, these are relatively infrequent processes in discharges under consideration with a low degree of ionization. An important feature of the Coulomb collisions is a strong dependence of their cross sections on the kinetic energy of colliding particles. This effect can be demonstrated by a simple analysis, illustrated in Figure 2.3. Here two particles have the same charge and for simplicity one collision partner is considered at rest. Scattering takes place if the Coulomb interaction energy (order of $U \sim q^2/b$, where b is the impact parameter) is approximately equal to the kinetic energy ε of a collision partner. Then the impact parameter $b \sim q^2/\varepsilon$, and the reaction cross section σ can be estimated as πb^2 and

$$\sigma(\varepsilon) \sim \pi\, q^4/\varepsilon^2 \tag{2.9}$$

This means, in particular, that the electron–electron scattering cross sections at room temperature are about 1000 times greater than those at the electron temperature of 1 eV typical for electric discharges.

Similar consideration of the charged particle scattering on neutral molecules with a permanent dipole momentum (interaction energy $U \sim 1/r^2$) and an induced dipole momentum (interaction energy $U \sim 1/r^4$) gives, respectively, $\sigma(\varepsilon) \sim 1/\varepsilon$ and $\sigma(\varepsilon) \sim 1/\varepsilon^{1/2}$. This means that the cross section dependence on the partner's energy is not as strong for electron or ion scattering on neutral particles as in the case of the Coulomb collisions (Equation 2.9).

Energy transfer during elastic collisions is possible only as a transfer of kinetic energy. The average fraction γ of kinetic energy, transferred from one particle of mass m to another one of mass M is equal to:

$$\gamma = 2mM/(m+M)^2 \tag{2.10}$$

In elastic collision of electrons with heavy neutrals or ions $m << M$, and thus $\gamma = 2m/M$, which means the fraction of transferred energy is negligible ($\gamma \sim 10^{-4}$).

Detailed fundamental consideration of the elementary elastic collisions of charged species as well as corresponding experimental data can be found in References 2 through 5.

2.2 IONIZATION PROCESSES

The key process in plasma is ionization because it is responsible for plasma generation: birth of new electrons and positive ions. The simplest ionization process—the ionization by electron impact—was illustrated in Figure 2.1(a). In general, all ionization processes can be subdivided into the following five groups.

The first group, **direct ionization by electron impact**, includes the ionization of neutrals: preliminary not-excited atoms, radicals, or molecules by an electron whose energy is sufficiently large to provide the ionization act in one collision. These processes are the most important in cold or nonthermal discharges in which electric fields and thus electron energies are quite high, but the level of excitation of neutral species is relatively moderate.

The second group, **stepwise ionization by electron impact**, includes the ionization of preliminary excited neutral species. These processes are important mainly in thermal or energy-intense discharges, when the degree of ionization (ratio of number densities of electrons and neutrals) and the concentration of highly excited neutral species are quite large.

The third group is the **ionization by collision with heavy particles**. Such processes can take place during ion–molecular or ion–atomic collisions, as well as in collisions of electronically or vibrationally excited species, when the total energy of the collision partners exceeds the ionization potential. The chemical energy of the colliding neutral species can be contributed into ionization also via the so-called processes of associative ionization.

The fourth group is made up of the **photoionization processes,** where neutral collisions with photons result in formation of an electron–ion pair. Photoionization is mainly important in thermal plasmas and some propagation mechanisms of propagation of nonthermal discharges.

These four groups of ionization processes will be considered in this section. The fifth group—surface ionization (electron emission) provided by electron, ion, and photon collisions, or just by surface heating—will be considered in the second part of this book, in discussing different discharges in heterogeneous systems.

2.2.1 Direct Ionization by Electron Impact

This process takes place as a result of interaction of an incident electron, having a high energy ε, with a valence electron of a preliminary neutral atom or molecule. The act of ionization occurs when energy $\Delta\varepsilon$ transferred between them exceeds the ionization potential I (see Section 2.1.2).

Although detailed consideration of the elementary ionization process requires quantum mechanical analysis of the inelastic scattering, a clear physical picture and good quantitative formulas can be derived from the following classical model, first introduced by Thomson.[6,6a,6b] In the framework of the Thomson model, let us suppose, for simplicity, that the valence electron is at rest, and let us neglect the interaction of the two colliding electrons with the rest of the initially neutral particle. The differential cross section of the incident electron scattering with energy transfer $\Delta\varepsilon$ to the valence electron can be defined by the Rutherford formula:

$$d\sigma_i = \frac{1}{(4\pi\varepsilon_0)^2}\frac{\pi e^4}{\varepsilon(\Delta\varepsilon)^2}d(\Delta\varepsilon) \tag{2.11}$$

Integration of the differential expression (Equation 2.11) over $\Delta\varepsilon$, taking into account that for successful ionization acts, the transferred energy should exceed the ionization potential $\Delta\varepsilon \geq I$, gives

$$\sigma_i = \frac{1}{(4\pi\varepsilon_0)^2}\frac{\pi e^4}{\varepsilon}\left(1/I - 1/\varepsilon\right) \tag{2.12}$$

This relation describes the direct ionization cross section σ_i as a function of an incident electron energy ε and it is known as the **Thomson formula**. Obviously, Equation 2.12 should be multiplied, in general, by the number of valence electrons Z_v.

According to the Thomson formula, the direct ionization cross section is growing linearly near the threshold of the elementary process $\varepsilon = I$. In the case of high electron energies $\varepsilon >> I$, the Thomson cross section (Equation 2.12) decreases as $\sigma_i \sim 1/\varepsilon$. Quantum mechanics gives a more accurate but close asymptotic approximation for the high energy electrons $\sigma_i \sim \ln\varepsilon/\varepsilon$. When $\varepsilon = 2I$, the Thomson cross section reaches the maximum value:

$$\sigma_i^{max} = \frac{1}{(4\pi\varepsilon_0)^2}\frac{\pi e^4}{4I^2} \tag{2.13}$$

The Thomson formula can be rewritten more precisely, taking into account kinetic energy ε_v of the valence electron:[2]

$$\sigma_i = \frac{1}{(4\pi\varepsilon_0)^2}\frac{\pi e^4}{\varepsilon}\left(1/\varepsilon - 1/I + \frac{2\varepsilon_v}{3}\left(\frac{1}{I^2} - \frac{1}{\varepsilon^2}\right)\right) \tag{2.14}$$

One can see that the Thomson formula (Equation 2.12) can be derived from Equation 2.14 by supposing that the valence electron is at rest and $\varepsilon_v = 0$.

Useful variation of the Thomson formula can be obtained, taking the valence electron interaction with the rest of the atom as a Coulomb interaction. Then, according to classical mechanics, $\varepsilon_v = I$, and Equation 2.14 gives another relation for the direct ionization cross section, which takes into account motion of valence electrons:

$$\sigma_i = \frac{1}{(4\pi\varepsilon_0)^2}\frac{\pi e^4}{\varepsilon}\left(5/_{3I} - 1/_{\varepsilon} - 2I/_{3\varepsilon^2}\right) \tag{2.15}$$

Equation 2.12 and Equation 2.15 describe the direct ionization in a similar way, but the latter predicts cross section $\sigma_i(\varepsilon)$ maximum ($\sigma_i^{max} \sim \pi e^4/2I^2(4\pi\varepsilon_0)^2$) about two times as large at a slightly lower electron energy, $\varepsilon_{max} = 1.85\ I$.

All the discussed relations for the direct ionization cross section can be generalized in the following form:

$$\sigma_i = \frac{1}{(4\pi\varepsilon_0)^2}\frac{\pi e^2}{I^2} Z_v f\left(\frac{\varepsilon}{I}\right) \tag{2.16}$$

where Z_v is a number of valence electrons, and $f(\varepsilon/I) = f(x)$ is a general function common for all atoms. Thus, for the Thomson formula (Equation 2.12):

$$f(x) = \frac{1}{x} - \frac{1}{x^2} \tag{2.17}$$

Equation 2.16 is in a good agreement with experimental data for different atoms and molecules[7] if:

$$\frac{10(x-1)}{\pi(x+0.5)(x+8)} < f(x) < \frac{10(x-1)}{\pi x(x+8)} \tag{2.18}$$

Formulas for practical estimations are discussed in Barnett.[8,635] The semiempirical formula (Equation 2.18) is quite useful for numerical calculations, if the electron $\sigma_i(\varepsilon)$ is unknown experimentally for some specific atoms and molecules.

2.2.2 Direct Ionization Rate Coefficient

As a general approach, the ionization rate coefficient k_I (Equation 2.7) should be calculated by integration of the cross section $\sigma_i(\varepsilon)$ over the electron energy distribution function (Section 2.1.1). However, because the electron energy distribution is a function of the electron temperature T_e (or mean energy [Equation 2.2]), the ionization rate coefficient $k_I(T_e)$ is also a function of the electron temperature T_e (or mean energy) as well. When the ionization rate coefficient is known, the rate of direct ionization by electron impact w_{ion} can be found as:

$$w_{ion} = k_I(T_e)\, n_e\, n_0 \tag{2.19}$$

In this formula, n_e is the concentration (number density) of electrons, and n_0 is the concentration (number density) of neutral gas atoms or molecules.

The ionization potential I is usually much greater than the mean electron energy. For this reason, the ionization rate coefficient is very sensitive to the electron energy distribution function. Specific distribution functions for nonequilibrium plasmas and

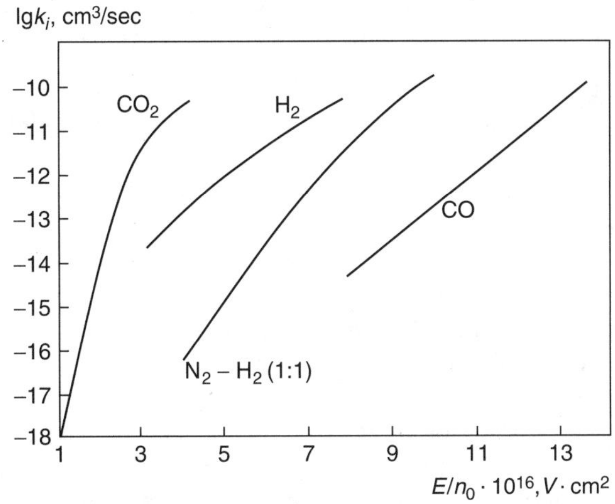

Figure 2.4 Ionization rate coefficient in molecular gases as a function of reduced electric field

their influence on the reaction rates will be discussed in Chapter 4. For the present, for the calculation of the ionization rate as a function of electron temperature, the Maxwellian distribution (Equation 2.1) can be used.

If $T_e << I$, only a small group of electrons can have energy exceeding the ionization potential. For this condition, integrating $\sigma_i(\varepsilon)$ in Equation 2.7, one needs to take into account only the linear part of $\sigma_i(\varepsilon)$ near the threshold of ionization $\varepsilon = I$:

$$\sigma_i^{threshold}(\varepsilon) = \frac{1}{(4\pi\varepsilon_0)^2} Z_v \frac{\pi e^4}{I^2} \left(\varepsilon/I - 1 \right) = \sigma_0 \left(\varepsilon/I - 1 \right) \tag{2.20}$$

Here, $\sigma_0 = Z_v \pi e^4 / I^2 (4\pi\varepsilon_0)^2$ is approximately the geometrical atomic cross section. After integrating Equation 2.7, the direct ionization rate coefficient can be presented as:

$$k_i(T_e) = \sqrt{8T_e/\pi m}\,\sigma_0 \exp\left(- I/T_e \right) \tag{2.21}$$

Equaton 2.21 is convenient for numerical estimations. For such estimations, one can take the cross section parameter σ_0 for molecular nitrogen, $= 10^{-16}$cm^2, and for argon, $= 3*10^{-16}$cm^2. Electron temperature is presented here in energy units, so the Boltzmann coefficient $k = 1$.

Some numerical data on the electron impact direct ionization for different molecular gases, CO_2, H_2, N_2, are presented in Figure 2.4[9] as a function of reduced electric field E/n_0, which is the ratio of electric field over neutral gas concentration. Specific qualitative features of ionization of molecules will be discussed later.

2.2.3 Peculiarities of Dissociation of Molecules by Electron Impact: the Frank–Condon Principle and the Process of Dissociative Ionization

First, consider a process of nondissociative ionization of molecules by direct electron impact, taking as an example the case of ionization of diatomic molecules AB:

$$e + AB \to AB^+ + e + e \tag{2.22}$$

This process takes place predominantly when the electron energy slightly exceeds the ionization potential. It can be described in the first approximation by the Thomson approach, discussed in Section 2.2.1. One can see some peculiarities of ionization of molecules by electron impact, using the illustrative potential energy curves for AB and AB^+, shown in Figure 2.4. It is worthwhile to discuss the potential energy curves in more detail.

The fastest internal motion of atoms inside molecules is molecular vibration. However, even molecular vibrations have typical times of 10^{-14} to 10^{-13} sec, which are much longer than the interaction time between plasma electrons and the molecules $a_0/v_e \sim 10^{-16}$ to 10^{-15} sec (here, a_0 is the atomic unit of length and v_e is the mean electron velocity). This means that any electronic excitation processes under consideration, induced by electron impact (including the molecular ionization; see Figure 2.5), are much faster than all kinds of atomic motions inside the molecules. As a result, all the atoms inside molecules can be considered as frozen during the process of electronic transition, stimulated by electron impact. This fact is known as the **Frank–Condon principle.**[99]

The processes of collisional excitation and ionization of molecules according to the Frank–Condon principle are presented in Figure 2.5 by vertical lines (internuclear distances are constant). This means that the nondissociative ionization process

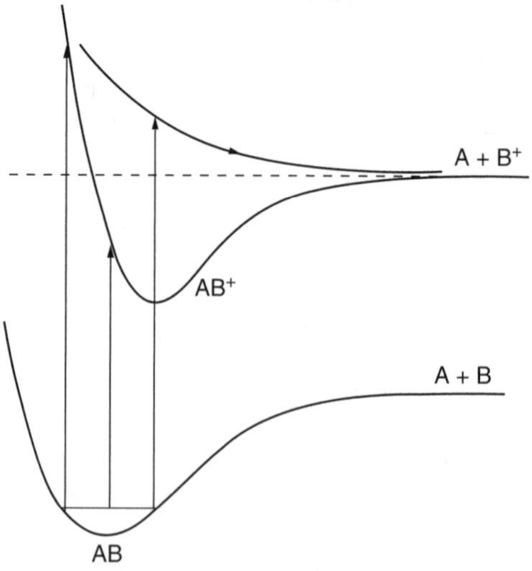

Figure 2.5 Molecular and ionic terms illustrating dissociative ionization

(Equation 2.22) usually results in formation of a vibrationally excited ion $(AB^+)^*$ and requires a little more energy than the corresponding atomic ionization. The phenomena can be described using the **Frank–Condon factors**, which will be discussed later.

When the electron energy is relatively high and essentially exceeds the ionization potential, the dissociative ionization process can take place:

$$e + AB \rightarrow A + B^+ + e + e \quad (2.23)$$

This ionization process corresponds to electronic excitation into a repulsive state of ion $(AB^+)^*$, followed by decay of this molecular ion. This is also illustrated by the vertical line in Figure 2.5. One can see from Figure 2.5 that the energy threshold for the dissociative ionization is essentially greater than for the nondissociative one.

2.2.4 Stepwise Ionization by Electron Impact

If the number density of electrons and thus the concentration of excited neutral species are sufficiently high, the energy I, necessary for ionization, can be provided in two different ways: (1) similar to the case of direct ionization, it can be provided by the energy of plasma electrons; and (2) the high energy of preliminary electronic excitation of neutrals also can be consumed in the ionization act (called stepwise ionization). The total energy needed for ionization in these two processes is the same, but which process is preferable and which contributes the most in the total rate of ionization? The answer is unambiguous: if the level of electronic excitation is sufficiently high, the stepwise ionization is much faster than the direct one, because the statistical weight of electronically excited neutrals is greater than that of free plasma electrons. In other words, when $T_e << I$, the probability of obtaining the high ionization energy I is much lower for free plasma electrons (direct ionization) than for excited atoms and molecules.

In contrast with the direct ionization process, the stepwise process includes several steps, several electron impacts to provide the ionization act. The first electron—neutral collisions result in forming highly excited species, and then a final collision with a relatively low energy electron provides the actual ionization act.

Another illustration of the stepwise ionization can be given from the thermodynamic point of view. In thermodynamic equilibrium, the ionization processes $e + A \rightarrow A^+ + e + e$ are reversed with respect to three-body recombination $A^+ + e + e \rightarrow A^* + e \rightarrow A + e$, which is proceeding through a set of excited states. According to the principle of detailed equilibrium, this means that the ionization process $e + \mathrm{A} \rightarrow \mathrm{A}^+ + e + e$ should go through the set of electronically excited states as well, e.g., the ionization should be a stepwise process. Note that this conclusion can only be referred to quasi-equilibrium or thermal plasmas, as has already been mentioned.

The stepwise ionization rate coefficient k_i^s can be found by summation of the partial rate coefficients $k_i^{s,n}$, corresponding to the nth electronically excited state, over all states of excitation, taking into account their concentrations:

$$k_i^s = \sum_n k_i^{s,n} N_n(\varepsilon_n)/N_0 \tag{2.24}$$

To calculate the maximum stepwise ionization rate, assume that the electronically excited atoms and molecules are in quasi equilibrium with plasma electrons, and that the excited species have an energy distribution function corresponding to the Boltzmann law with the electron temperature T_e:

$$N_n = \left(g_n / g_0 \right) N_0 \exp\left(-\varepsilon_n / T_e \right) \tag{2.25}$$

Here, N_n, g_n, and ε_n are number density, statistical weight, and energy (with respect to ground state) of the electronically excited atoms, radicals, or molecules; the index n is the principal quantum number and actually shows the particle excitation level. From statistical thermodynamics, the statistical weight of an excited particle $g_n = 2g_i n^2$, where g_i is statistical weight of ion. N_0 and g_0 are the concentration and statistical weight of ground state particles. Once more it is noted that the Boltzmann constant can be taken as $k = 1$ and, for this reason, omitted if temperature is given in energy units. This rule will be followed in most cases in this book.

Typical energy transfer from a plasma electron to an electron sitting on an excited atomic level is about T_e. This means that excited particles with energy approximately $\varepsilon_n = I - T_e$ make the most important contribution to the sum (Equation 2.24). Taking into account that $I_n \sim 1/n^2$, the number of states with energy about $\varepsilon_n = I - \mathrm{T_e}$ and ionization potential about $I_n = T_e$ has an order of n. Thus, from Equation 2.24 and Equation 2.25, the following estimation is derived:

$$k_i^s \approx \frac{g_i}{g_0} n^3 \langle \sigma v \rangle \exp\left(-I/T_e \right) \tag{2.26}$$

The cross section σ in Equation 2.26 corresponding to energy transfer (about T_e) between electrons can be estimated as $e^4/T_e^2(4\pi\varepsilon_0)^2$, velocity $v \sim \sqrt{T_e/m}$, and the preferable principal quantum number can be taken from:

$$I_n \approx \frac{1}{(4\pi\varepsilon_0)^2} me^4/\hbar n^2 \approx T_e \tag{2.27}$$

As a result, the stepwise ionization rate finally can be presented based on Equation 2.26 and Equation 2.27:

$$k_i^s \approx \frac{g_i}{g_0} \frac{1}{(4\pi\varepsilon_0)^5} \left(me^{10}/\hbar^3 T_e^3 \right) \exp\left(-I/T_e \right) \tag{2.28}$$

Here, $\hbar$ is the Planck's constant. In general, Equation 2.28 is quite convenient for numerical estimations of the stepwise ionization rate in thermal plasmas, when the electronically excited species are in quasi equilibrium with the gas of plasma electrons. Also, the expression Equation 2.28 for stepwise ionization is in a good agreement (within the framework of the principle of detailed equilibrium) with the rate of the reverse reaction of the three-body recombination $A^+ + e + e \leftrightarrow A^* + e \rightarrow A + e$; this will be discussed in Section 2.3.

Comparing direct ionization (Equation 2.21) with stepwise ionization (Equation 2.28), one can see that the latter can be much faster because of the large statistical weight of excited species involved in the stepwise ionization. The ratio of rate coefficients for these two competing mechanisms of ionization can be easily derived from Equation 2.21 and Equation 2.28:

$$\frac{k_i^s(T_e)}{k_i(T_e)} \approx \frac{g_i a_0^2}{g_0 \sigma_0} \left(\frac{1}{(4\pi\varepsilon_0)^2} me^4 / \hbar T_e \right)^{7/2} \approx \left(\frac{I}{T_e} \right)^{7/2} \tag{2.29}$$

Here, a_0 is the atomic unit of length. This has been taken into account in this relation, the estimations for geometric collisional cross section $\sigma_0 \sim a_0^2$, and for the ionization potential:

$$I \approx \frac{1}{(4\pi\varepsilon_0)^2} me^4 / \hbar^2.$$

Equation 2.29 demonstrates numerically that, for typical discharges with $I/T_e \sim 10$, the stepwise ionization can be 10^3 to 10^4 times faster than the direct one. Remember, this takes place only in the case of high electron concentration, n_e, and high electronic excitation frequency, $k_{en}n_e$, which results in the quasi equilibrium (Equation 2.25) between electronically excited species and plasma electrons. Usually, it can be applied only for quasi-equilibrium plasmas of thermal discharges.

To estimate the contribution of stepwise ionization in nonequilibrium discharges, consider some corrections to Equation 2.29. If deactivation of the excited species because of radiation, collisional relaxation, and losses on the walls has a characteristic time τ_n^* shorter than excitation time $1/k_{en}n_e$, the concentration of excited species will be lower than the equilibrium concentration (Equation 2.25). Equation 2.25 can be corrected by taking into account electron concentration and the nonequilibrium losses of the excited species:

$$N_n \approx \frac{g_n}{g_0} N_0 \frac{k_{en} n_e \tau_n^*}{k_{en} n_e \tau_n^* + 1} \exp\left(-\varepsilon_n / T_e \right) \tag{2.30}$$

Then the stepwise ionization rate in nonequilibrium discharges can be rewritten as a function of electron temperature as well as electron concentration:

$$k_i^s(T_e, n_e) \approx k_i(T_e) \frac{k_{en} n_e \tau_n^*}{k_{en} n_e \tau_n^* + 1} \left(\frac{I}{T_e} \right)^{\frac{7}{2}} \tag{2.31}$$

One can see that when the electron concentration is sufficiently high, this expression corresponds to Equation 2.29 for the stepwise ionization rate in quasi-equilibrium plasmas.

2.2.5 Ionization by High Energy Electron Beams

For many important modern applications, plasma is generated not by electric discharges, but by means of high energy electron beams. The ionization in this case can be more homogeneous and generate uniform nonthermal plasmas even in atmospheric pressure systems. The electron beams also can be effectively combined with the electric field in non-self-sustained discharges, where electron beams providing the ionization and energy consumption are mostly due to the effect of electric field.

Peculiarities of ionization in this case are due to high energy of electrons in the beams. Electron energy in this system usually varies from 50 KeV to 2 MeV. Typical energy losses of the beams in atmospheric pressure air are of about 1 MeV per 1 m. Thus, for generation of high plasma volumes, high electron beam energies are required. The beams with electron energies more than 500 KeV are referred to as relativistic electron beams because these energies exceed the relativistic electron energy at rest ($E = mc^2$).

To describe the direct ionization process induced by the high energy beam electrons with velocities greatly exceeding the velocities of atomic electrons, the Born approximation[6] can be applied. In the framework of this approximation, electron energy losses per unit length dE/dx can be evaluated by the nonrelativistic Bethe–Bloch formula:

$$-\frac{dE}{dx} = \frac{2\pi Z e^4}{\left(4\pi\varepsilon_0\right)^2 m v^2} n_0 \ln \frac{2mEv^2}{I^2} \tag{2.32}$$

Here, Z is an atomic number of neutral particles, providing the beam stopping; n_0 their number density; and v the stopping electron velocity.

In the most interesting case of relativistic electron beams with the electron energies of about 0.5 to 1 MeV, the energy losses going to neutral gas ionization can be numerically calculated by the following relation:

$$-\frac{dE}{dx} = 2 * 10^{-22} n_0 Z \ln \frac{183}{Z^{1/3}} \tag{2.33}$$

where energy losses dE/dx are expressed in MeV/cm, and concentration of neutral particles n_0 is expressed in cm^{-3}.

Equation 2.33 can be rewritten in terms of effective ionization rate coefficient k_i^{eff} for relativistic electrons:

$$k_i^{eff} \approx 3 * 10^{-10} \left(\mathrm{cm}^3/\mathrm{sec}\right) Z \ln \frac{183}{Z^{1/3}} \tag{2.34}$$

Numerically, this ionization rate coefficient is about 10^{-8} to 10^{-7} cm^3/sec. The rate of ionization by relativistic electron beams can be expressed in this case as a function of the electron beam concentration n_b or the electron beam current density j_b:

$$q_e = k_i^{eff} n_b n_0 \approx k_i^{eff} \frac{1}{ec} n_0 j_b \tag{2.35}$$

Here, c is the speed of light.

Energy losses and stopping relations similar to Equation 2.32 through Equation 2.35 will be discussed in Chapter 12 to describe electron beam plasma formation, ionization process, and plasma generation stimulated by multicharged ion stopping, following nuclear decay processes.

2.2.6 Photoionization Processes

Although ionization processes are mostly induced by electron impact, ionization can also be provided by interaction with high energy photons and in collisions of excited heavy particles. The process of photoionization of a neutral particle A with ionization potential I (in eV) by a photon ω with wavelength λ can be illustrated as:

$$\hbar\varpi + A \rightarrow A^+ + e, \ \lambda < \frac{12{,}400}{I(\text{eV})} \tag{2.36}$$

To provide the ionization, the photon wavelength should be quite low. Usually it should be less than 1000 Å, which is ultraviolet radiation. However, to provide effective ionization of preliminary excited atoms and molecules, a photon energy can be less and electromagnetic waves can be longer.

The photoionization cross section increases sharply from zero at the threshold energy (Equation 2.36) to quite high values up to the geometrical cross section. Table 2.3 presents specific numerical values of the photoionization cross sections (near the threshold of the process) for some atoms and molecules.

Although the cross sections shown in the table are quite high, the contribution of the photoionization process is usually not very significant because of low concentration

Table 2.3 Photoionization Cross Sections

Atoms or molecules	Wavelength λ, in A	Cross sections, in cm^2
Ar	787	$3.5{*}10^{-17}$
N_2	798	$2.6{*}10^{-17}$
N	482	$0.9{*}10^{-17}$
He	504	$0.7{*}10^{-17}$
H_2	805	$0.7{*}10^{-17}$
H	912	$0.6{*}10^{-17}$
Ne	575	$0.4{*}10^{-17}$
O	910	$0.3{*}10^{-17}$
O_2	1020	$0.1{*}10^{-17}$
Cs	3185	$2.2{*}10^{-19}$
Na	2412	$1.2{*}10^{-19}$
K	2860	$1.2{*}10^{-20}$

of high energy photons in most situations. However, sometimes photoionization plays a very essential role by rapidly supplying seed electrons for following ionization by electron impact. As examples, consider three important discharge processes:

1. Streamer propagation in nonthermal discharges, where photoionization supplies seed electrons to start electron avalanches
2. Propagation of nonthermal and thermal discharges in fast flow, including supersonic, where other mechanisms of discharge propagation are too slow
3. Preliminary gas ionization by ultraviolet radiation in non-self-sustained discharges, where UV radiation is a kind of replacement of the relativistic electron beams

These preionization processes dominated by photoionization will be considered in detail later during discussion of correspondent discharge systems.

2.2.7 Ionization by Collisions of Heavy Particles: Adiabatic Principle and Massey Parameter

An electron with kinetic energy slightly exceeding the ionization potential is quite effective in performing the ionization act. However, that is not true for ionization by collisions of heavy particles–ions and neutrals. Even when they have enough kinetic energy, they actually cannot provide ionization because their velocities are much less than those of electrons in atoms. Even having enough energy, a heavy particle is very often unable to transfer this energy to an electron inside an atom because the process is far from resonance.

This effect is a reflection of a general principle of particle interaction. A slow motion is "adiabatic," i.e., reluctant to transfer energy to a fast motion. The **adiabatic principle** can be explained in terms of relations between low interaction frequency $\omega_{int} = \alpha v$ (reverse time of interaction between particles) and high frequency of electron transfer in atom $\omega_{tr} = \Delta E\hbar$. Here, $1/\alpha$ is a characteristic size of the interacting neutral particles; v is their velocity; ΔE is a change of electron energy in an atom during the interaction; and $\hbar$ is the Planck's constant.

Only fast Fourier components of the slow interaction potential between particles with frequencies about $\omega_{tr} = \Delta E/\hbar$ provide the energy transfer between the interacting particles. The relative weight or probability of these fast Fourier components is very low if $\omega_{tr} >> \omega_{int}$; numerically, it is approximately: $\exp(-\omega_{tr}/\omega_{int})$. As a result, the probability P_{EnTr} and cross sections of energy transfer processes (including the ionization process under consideration) are usually proportional to the so-called Massey parameter or Massey exponent:

$$P_{EnTr} \propto \exp\left(-\frac{\omega_{tr}}{\omega_{int}}\right) \propto \exp\left(-\frac{\Delta E}{\hbar \alpha v}\right) = \exp\left(-P_{Ma}\right) \tag{2.37}$$

Here, $P_{Ma} = \Delta E/\hbar\alpha v$ is the **adiabatic Massey parameter**. If $P_{Ma} >> 1$, the process of energy transfer is adiabatic, and its probability is exponentially low.

Actually, it takes place during collisions of energetic heavy neutrals and ions. To get the Massey parameter close to one and eliminate the adiabatic prohibition for ionization, the kinetic energy of the colliding heavy particle must be about 10 to 100 KeV, which is about three orders of value more than ionization potential. That is why the kinetic energy of heavy particles is usually ineffective for ionization.

Although kinetic energy of heavy particles in the ground state is ineffective for ionization, the situation can be different if they are electronically excited. If the total electron excitation energy of the colliding heavy particles is close to the ionization potential of one of them, the resonant energy transfer and effective ionization act can take place. Such nonadiabatic ionization processes occurring in collision of heavy particles will be illustrated next by two specific examples: the Penning ionization effect and associative ionization.

2.2.8 The Penning Ionization Effect and Process of Associative Ionization

If electron excitation energy of a metastable atom A^* exceeds the ionization potential of another atom B, their collision can lead to an act of ionization, the so-called **Penning ionization**. The Penning ionization usually takes place through intermediate formation of an unstable excited quasi molecule (in the state of autoionization) and cross sections of the process can be very large.

Thus, the cross sections for the Penning ionization of N_2, CO_2, Xe, and Ar by metastable helium atoms He(2^3S) with excitation energy 19.8 eV reach gas-kinetic values of 10^{-15} cm^2. Similar cross sections can be attained in collisions of metastable neon atoms (excitation energy 16.6 eV) with argon atoms (ionization potential 15.8 eV). An exceptionally high cross section ($1.4*10^{-14}$ cm^2)[7] can be attained in Penning ionization of mercury atoms (ionization potential 10.4 eV) by collisions with the metastable helium atoms He(2^3S, 19.8 eV). In general, a contribution of the Penning ionization in total kinetics of plasma generation can be significant in the presence of the mentioned highly excited metastable species.

If the total electron excitation energy of colliding particles is not sufficient, an ionization process is possible nevertheless when heavy species stick to each other, forming a molecular ion, and thus their bonding energy can also be contributed into the ionization act. Such a process is called *associative ionization*. The process differs from the Penning ionization only by the stability of the molecular ion product. A good example of associative ionization is the process involving collision of two metastable mercury atoms:

$$\mathrm{Hg}\ (6^3P_1,\ E = 4.9\ \mathrm{eV}) + \mathrm{Hg}\ (6^3P_0,\ E = 4.7\ \mathrm{eV}) \rightarrow \mathrm{Hg}_2^+ + e \qquad (2.38)$$

One can see that the total electron excitation energy here, 9.6 eV, is less than the ionization potential of mercury atoms (10.4 eV), but higher than ionization potential for Hg_2- molecule. This is the main "trick" of associative ionization.

Cross sections of the associative ionization (similar to the Penning ionization) can be quite high and close to the gas-kinetic one (10^{-15} cm^2). The associative ionization $A^* + B \rightarrow AB^+ + e$ takes place when there is a crossing of electron term of colliding particles with an electron term of the molecular ion AB^+ and as a result the process is nonadiabatic. Such a situation takes place only for a limited number of excited species.

Actually, the associative ionization is a reverse process with respect to dissociative recombination $e + AB^+ \rightarrow A^* + B$, which is the main recombination mechanism for molecular ions (see the next section). Thus, the associative ionization is effective only for such excited species that can be produced during the dissociative recombination. The relation between the cross sections of associative ionization σ_{ai} and dissociative recombination σ_r^{ei} can be derived from the principle of detailed equilibrium as:[2]

$$\sigma_{ai}(v_{rel}) = \sigma_r^{ei}(v_e)\frac{m^2 v_e^2}{\mu^2 v_{rel}^2}\frac{g_e g_{AB^+}}{g_{A^*} g_B} \tag{2.39}$$

In this relation: v_{rel} and v_e are the relative velocity of heavy particles and the electron velocity, respectively; μ is the reduced mass of heavy particles; and g is statistical weights.

In nonequilibrium plasma chemistry of molecular gases, a major portion of energy can be localized in vibrational excitation. For this reason, ionization (and, in particular, associative ionization) can be very important in the collision of the vibrationally excited molecules. Cross sections and reaction rate coefficients for such processes are very low because of low Frank–Condon factors (see Section 2.2.3); nevertheless, these processes can be very important to describing ionization in the absence of electric field. Here, the reaction of highly vibrationally excited nitrogen molecules can be pointed out:[9]

$$N_2^*\left({}^1\Sigma_g^+, v_1 \approx 32\right) + N_2^*\left({}^1\Sigma_g^+, v_2 \approx 32\right) \rightarrow N_4^+ + e \tag{2.40}$$

Although this associative ionization has enough energy accumulated in the nitrogen molecules with 32 vibrational quanta each, the reaction rate of the process is relatively low: $10^{-15}\exp(-2000\ \text{K}/T)$, cm^3/sec. In general, a good understanding of the phenomenon of ionization in collision of vibrationally excited molecules is still lacking.[12a,20a]

2.3 MECHANISMS OF ELECTRON LOSSES: ELECTRON–ION RECOMBINATION

The ionization processes were considered in Section 2.2 as a source of electrons and positive ions, e.g., as a source of plasma generation. Conversely, the principal loss mechanisms of charged particles, the elementary processes of plasma degradation, will now be examined. Obviously, the losses together with the ionization processes determine a balance of charge particles and plasma density. The variety of channels of charged particle losses can be subdivided into three qualitatively different groups.

The first group includes different types of **electron–ion recombination processes**, in which collisions of the charged particles in a discharge volume lead to their mutual neutralization. These exothermic processes require consuming the large release of recombination energy in some manner. Dissociation of molecules, radiation of excited particles, or three-body collisions can provide the consumption of the recombination energy. These different mechanisms of electron–ion recombination will be discussed in this section.

Electron losses, because of their sticking to neutrals and formation of negative ions (see Figure 2.1b), form the second group of volumetric losses, **electron attachment processes**. These processes are often responsible for the balance of charged particles in such electronegative gases as oxygen (and, for this reason, air); CO_2 (because of formation of O^-); and different halogens and their compounds. Reverse processes of an electron release from a negative ion are called the **electron detachment**. These processes of electron losses related to negative ions will be considered in Section 2.4.

Note that although electron losses in this second group are due to the electron attachment processes, the actual losses of charged particles take place as a consequence following the fast processes of **ion–ion recombination.** The ion–ion recombination process means neutralization during collision of negative and positive ions. These processes usually have very high rate coefficients and will be considered in Section 2.5.

Finally, the third group of charged particle losses is not a volumetric one like all those mentioned previously, but is due to **surface recombination**. These processes of electron losses are the most important in low-pressure plasma systems such as glow discharges. The surface recombination processes are usually kinetically limited not by the elementary act of the electron–ion recombination on the surface, but by transfer (diffusion) of the charged particles to the walls of the discharge chamber. For this reason the surface losses of charged particles will be discussed in Chapter 4, concerning plasma kinetics and transfer phenomena.

2.3.1 Different Mechanisms of Electron–Ion Recombination

The electron–ion recombination is a highly exothermic process like all other recombination processes. The released energy corresponds to the ionization potential and so it is quite high (see Table 2.1). To be effectively realized, such a process should obviously have a channel for accumulation of the energy released during the neutralization of a positive ion and an electron.

Taking into account three main channels of consumption of the recombination energy (dissociation, three-body collisions, and radiation), the electron–ion recombination processes can be subdivided into three principal groups of mechanisms.

In molecular gases, or just in the presence of molecular ions, the fastest electron neutralization mechanism is **dissociative electron–ion recombination**:

$$e + AB^+ \rightarrow (AB)^* \rightarrow A + B^* \tag{2.41}$$

In these processes, the recombination energy usually proceeds via resonance to dissociation of molecular ion and to excitation of the dissociation products. Although these processes principally take place in molecular gases, they can occur even in atomic gases. They can also be important because of preliminary formation of molecular ions in the ion conversion processes: $A^+ + A + A \rightarrow A_2^+ + A$.

In atomic gases, in the absence of molecular ions, the neutralization can be due to **three-body electron–ion recombination**:

$$e + e + A^+ \rightarrow A^* + e \tag{2.42}$$

The excess energy in this case is going to kinetic energy of a free electron, which participates in the recombination act as "a third body partner." It should be noted that heavy particles (ion and neutrals) are unable to accumulate electron recombination energy fast enough in their kinetic energy and are ineffective as the third body partner.

Finally, the recombination energy can be converted into radiation in the process of **radiative electron–ion recombination**:

$$e + A^+ \rightarrow A^* \rightarrow A + \hbar\omega \tag{2.43}$$

The cross section of this process is relatively low and it can be competitive with the three-body recombination only when the plasma density is low.

The electron–ion recombination processes are extremely important in low-temperature plasma kinetics. More detailed data on kinetics of the outlined recombination mechanisms will be discussed in the following four subsections.

2.3.2 Dissociative Electron–Ion Recombination

This resonant recombination process (Equation 2.41) starts with trapping of an electron on a repulsive autoionization level of molecular ion. Then atoms travel apart on the repulsive term and, if the autoionization state is maintained, stable products of dissociation are formed. The described recombination mechanism is quite fast and plays the major role in neutralization and decay of plasma in molecular gases. Reaction rate coefficients for most diatomic and three-atomic ions are on the level of 10^{-7} cm^3/sec; the kinetic information for some important molecular ions can be found in Table 2.4.

This table shows that, in a group of similar ions like molecular ions of noble gases, the recombination reaction rate is growing with a number of internal electrons; recombination of Kr_2^+ and Xe_2^+ is about 100 times faster than in the case of helium.

The rate coefficient of dissociative electron–ion recombination (the highly exothermic process) obviously decreases with temperature, as can be seen from Table 2.4. This process has no activation energy, so its dependencies on electron T_e and gas T_0 temperatures are not very strong. This dependence can be estimated in a semiempirical way:[8,9]

$$k_r^{ei}\left(T_e, T_0\right) \propto \frac{1}{T_0\sqrt{T_e}} \tag{2.44}$$

It should be noted that the recombination reaction rate also depends on the level of vibrational excitation of molecular ions.

Table 2.4 Dissociative Electron–Ion Recombination Reaction Rate Coefficients[a]

Electron–ion dissociative recombination process	Rate coefficient k_r^{ei}, cm^3/sec $T_e = T_0 = 300$ K	Rate coefficient k_r^{ei}, cm^3/sec $T_e = 1$ eV, $T_0 = 300$ K	Electron–ion dissociative recombination process	Rate coefficient k_r^{ei}, cm^3/sec $T_e = T_0 = 300$ K	Rate coefficient k_r^{ei}, cm^3/sec $T_e = 1$ eV, $T_0 = 300$ K
$e + N_2^+ \rightarrow N + N$	$2*10^{-7}$	$3*10^{-8}$	$e + NO^+ \rightarrow N + O$	$4*10^{-7}$	$6*10^{-8}$
$e + O_2^+ \rightarrow O + O$	$2*10^{-7}$	$3*10^{-8}$	$e + H_2^+ \rightarrow H + H$	$3*10^{-8}$	$5*10^{-9}$
$e + H_3^+ \rightarrow H_2 + H$	$2*10^{-7}$	$3*10^{-8}$	$e + CO^+ \rightarrow C + O$	$5*10^{-7}$	$8*10^{-8}$
$e + CO_2^+ \rightarrow CO + O$	$4*10^{-7}$	$6*10^{-8}$	$e + He_2^+ \rightarrow He + He$	10^{-8}	$2*10^{-9}$
$e + Ne_2^+ \rightarrow Ne + Ne$	$2*10^{-7}$	$3*10^{-8}$	$e + Ar_2^+ \rightarrow Ar + Ar$	$7*10^{-7}$	10^{-7}
$e + Kr_2^+ \rightarrow Kr + Kr$	10^{-6}	$2*10^{-7}$	$e + Xe_2^+ \rightarrow Xe + Xe$	10^{-6}	$2*10^{-7}$
$e + N_4^+ \rightarrow N_2 + N_2$	$2*10^{-6}$	$3*10^{-7}$	$e + O_4^+ \rightarrow O_2 + O_2$	$2*10^{-6}$	$3*10^{-7}$
$e + H_3O^+ \rightarrow H_2 + OH$	10^{-6}	$2*10^{-7}$	$e + (NO)_2^+ \rightarrow 2NO$	$2*10^{-6}$	$3*10^{-7}$
$e + HCO^+ \rightarrow CO + H$	$2*10^{-7}$	$3*10^{-8}$			

[a] Room gas temperature $T_0 = 300$ K; electron temperature $T_e = 300$ K and $T_e = 1$ eV.

Table 2.5 Ion Conversion Reaction Rate Coefficients at Room Temperature

Ion conversion process	Reaction rate coefficient	Ion conversion process	Reaction rate coefficient
$N_2^+ + N_2 + N_2 \rightarrow N_4^+ + N_2^+$	8*10^-29 cm^6/sec	$O_2^+ + O_2 + O_2 \rightarrow O_4^+ + O_2^+$	3*10^-30 cm^6/sec
$H^+ + H_2 + H_2 \rightarrow H_3^+ + H_2$	4*10^-29 cm^6/sec	$Cs^+ + Cs + Cs \rightarrow Cs_2^+ + Cs$	1.5*10^-29 cm^6/sec
$He^+ + He + He \rightarrow He_2^+ + H$	9*10^-32 cm^6/sec	$Ne^+ + Ne + Ne \rightarrow Ne_2^+ + Ne$	6*10^-32 cm^6/sec
$Ar^+ + Ar + Ar \rightarrow Ar_2^+ + Ar$	3*10^-31 cm^6/sec	$Kr^+ + Kr + Kr \rightarrow Kr_2^+ + Kr$	2*10^-31 cm^6/sec
$Xe^+ + Xe + Xe \rightarrow Xe_2^+ + Xe$	4*10^-31 cm^6/sec		

2.3.3 Ion Conversion Reactions as a Preliminary Stage of Dissociative Electron–Ion Recombination

It is interesting that, if pressure is sufficiently high, the recombination of atomic ions like Xe^+ is increasing not by means of three-body (Equation 2.42) or radiative (Equation 2.43) mechanisms, but through the preliminary formation of molecular ions through so-called **ion conversion reactions** such as $Xe^+ + Xe + Xe \rightarrow Xe_2^+ + Xe$. Then the molecular ion can be quickly neutralized in the rapid process of dissociative recombination (see Table 2.3).

The ion conversion reaction rate coefficients are quite high;[7,10,11] some of them are included in Table 2.5. When pressure exceeds 10 torr, the ion conversion is usually faster than the following process of dissociative recombination, which becomes a limiting stage in the overall kinetics. An analytical expression for the ion-conversion three-body reaction rate coefficient can be quite easily derived just from dimension analysis:[5]

$$k_{ic} \propto \left(\frac{\beta e^2}{4\pi\varepsilon_0}\right)^{5/4} \frac{1}{M^{1/2} T_0^{3/4}} \tag{2.45}$$

In this relation, M and β are mass and polarization coefficient of colliding atoms; T_0 is the gas temperature.

The ion conversion effect takes place as a preliminary stage of recombination not only for simple atomic ions, as was discussed earlier, but also for some important molecular ions. As is clear from Table 2.5, the polyatomic ions have very high recombination rates, often exceeding 10^{-6} cm^3/sec at room temperature. This results in an interesting fact: the recombination of molecular ions like N_2^+ and O_2^+ at elevated pressures sometimes goes through intermediate formation of dimers like N_4^+ and O_4^+. Thus, the ion conversion mentioned previously can be essential for recombination of these important molecular ions of air plasma as well.

2.3.4 Three-Body Electron–Ion Recombination

This three-body recombination process (Equation 2.42) $e + e + A^+ \rightarrow A^* + e$ is the most important one for high density equilibrium plasma with temperature about

1 eV. In this case the concentration of molecular ions is very low (because of thermal dissociation) due to the fast mechanism of dissociative recombination described earlier, and the three-body reaction is dominant.

The recombination process starts with a three-body capture of an electron by a positive ion and the formation of a very highly excited atom with binding energy of about T_e. The initially formed highly excited atom then loses its energy in step-by-step deactivation through electron impacts. A final relaxation step from the lower excited state to the ground state has a relatively high value of energy transfer and is usually due to radiative transition.

The three-body electron–ion recombination process (Equation 2.42) is a reverse one with respect to the stepwise ionization considered in Section 2.2.4. For this reason, the reaction rate coefficient of this recombination process can be derived from Equation 2.26 for the stepwise ionization rate coefficient k_i^s and from the Saha thermodynamic equation for ionization/recombination balance and equilibrium electron density (see details in Chapter 4):

$$k_r^{eei} = k_i^s \frac{n_0}{n_e n_i} = k_i^s \frac{g_0}{g_e g_i}\left(\frac{2\pi\hbar}{mT_e}\right)^{3/2} \exp\left(\frac{I}{T_e}\right) \approx \frac{e^{10}}{\left(4\pi\varepsilon_0\right)^5 \sqrt{mT_e^9}} \tag{2.46}$$

Here, n_e, n_i, and n_0 are number densities of electrons, ions, and neutrals; g_e, g_i, and g_0 are their statistical weights; e and m are electron charge and mass; electron temperature T_e is taken as usual in energy units; and I is an ionization potential.

It is convenient for practical calculations of the two-body electron–ion recombination to rewrite Equation 2.46 in the following numerical form:

$$k_r^{eei}, \frac{\text{cm}^6}{\text{sec}} = \frac{\sigma_0}{I} 10^{-14} \left(I/T_e \right)^{4.5} \tag{2.47}$$

In this numerical formula, σ_0, cm^2 is the gas-kinetic cross section, the same as one involved in Equation 2.20 and Equation 2.21, I and T_e are the ionization potential and electron temperature, to be taken in eV.

The rate coefficient of the recombination process depends strongly on electron temperature but not exponentially. As can be seen from Equation 2.47, the typical value of k_r^{eei} at room temperature is about 10^{-20} cm^6/sec; at $T_e = 1$ eV this rate coefficient is about 10^{-27} cm^6/sec. At room temperature, the three-body recombination is able to compete with dissociative recombination when electron concentration is quite high and exceeds 10^{13} cm^{-3}. If the electron temperature is about 1 eV, three-body recombination can compete with the dissociative recombination only in the case of exotically high electron density exceeding 10^{20} cm^{-3}.

As was mentioned earlier, the excessive energy in the process under consideration is going to the kinetic energy of a free electron—the "third body." In this case heavy particles, ions, and neutrals are too slow and ineffective as third-body partners. The corespondent rate coefficient of the third order reaction is about 10^8 times lower than Equation 2.46.

2.3.5 Radiative Electron–Ion Recombination

The electron–ion recombination process (Equation 2.43) $e + A^+ \rightarrow A + \hbar\omega$ is a relatively slow one because it requires a photon emission during the short interval of the electron–ion interaction. This type of recombination can play a major role in the balance of charged particles only in the absence of molecular ions and, in addition, if the plasma density is quite low and the three-body mechanisms are suppressed.

Typical values of cross sections of the radiative recombination process are about 10^{-21} cm^2, which is low. The reaction rate coefficients are therefore not high and can be simply estimated as a function of electron temperature by the following numeric formula:[12]

$$k^{ei}_{rad.rec.} \approx 3 \cdot 10^{-13} \left(T_e, \text{eV}\right)^{-3/4}, \ \text{cm}^3/\text{sec} \tag{2.48}$$

Comparing numerical relation (2.47) and relation (2.48), one can see that the reaction rate of radiative recombination exceeds the reaction rate of the three-body process when electron concentration is not high enough:

$$n_e < 3 \cdot 10^{13} \left(T_e, \text{eV}\right)^{-3/4}, \ \text{cm}^{-3} \tag{2.49}$$

Equation 2.47 and Equation 2.48 are obviously good only for estimations; more precise data depend on the specific type of colliding positive ion.

2.4 ELECTRON LOSSES DUE TO FORMATION OF NEGATIVE IONS: ELECTRON ATTACHMENT AND DETACHMENT PROCESSES

A simple illustration of the electron attachment process and negative ion formation was presented in Figure 2.1(b). Electron attachment plays an essential role not only as a channel for electron losses, but in dissociation of electronegative molecules and other plasma chemical and plasma-catalytic processes, which will be discussed later. The electron balance in electronegative gases, which have high electron affinity (oxygen, different halogens, and their compounds; see Table 2.2) can be strongly affected by these processes. In some other gases, like CO_2 and H_2O, the electron attachment can be equally important because of high electron affinity of decomposition products and formation of negative ions like O^- and H^- during dissociation.

The electron attachment processes are essential in the electron balance in weakly ionized plasma with low electron concentration and low degree of ionization. This fact is due to the first kinetic order of this process with respect to concentration of electrons, which means the reaction rate of the dissociative attachment is directly proportional to electron density. In contrast, the recombination processes considered in the previous section are of second or third kinetic order. This means the electron–ion recombination reaction rates are proportional to the square or cube of electron concentration and, therefore, they can be neglected with respect to electron attachment if the electron density and ionization degree are low.

The negative ions formed as a result of electron attachment can be neutralized quite easily and extremely fast in the processes of ion–ion recombination considered in the next section. For this reason, the attachment processes in general can provide significant losses of electrons and charged particles, which can result in prevention of ignition and propagation of electric discharges. Fortunately for plasma generation, the electron attachment processes can be effectively suppressed by the reverse reactions of electron detachment. These processes of decay of negative ions and restoration of free electrons can be provided by collisions with excited or chemically active species, energetic electrons, etc. Now, consider a step-by-step discussion of a variety of the electron attachment–detachment processes, a variety of mechanisms of formation, and decay of the negative ions.

2.4.1 Dissociative Electron Attachment to Molecules

This process is important in molecular gases like CO_2, H_2O, SF_6, and CF_4 when molecular fragments (dissociation products) have positive electron affinities (see Table 2.2):

$$e + AB \rightarrow (AB^-)^* \rightarrow A + B^- \tag{2.50}$$

The mechanism of this process is somewhat similar to the dissociative recombination described previously and proceeds by intermediate formation of an autoionization state $(AB^-)^*$. This excited state is unstable and its decay leads to the reverse process of autodetachment ($AB + e$) or to dissociation ($A + B^-$). During the attachment (Equation 2.50), an electron is captured and not able to provide an energy balance of the elementary process. For this reason, the dissociative attachment is a resonant reaction, which means it requires quite definite values of the electron energy.

The most typical potential energy curves illustrating the dissociative attachment (Equation 2.50) are presented in Figure 2.6(a). The electron attachment process starts in this case with a vertical transition from AB molecular ground state electronic term to a repulsive state of AB^- (obviously following the Frank–Condon principle discussed in Section 2.2.3). During the repulsion, before the $(AB^-)^*$ reaches an intersection point of AB and AB^- electronic terms, the reverse autodetachment reaction ($AB + e$) is very possible. But after passing the intersection, the AB potential energy exceeds that of AB^- and further repulsion results in dissociation ($A + B^-$).

To estimate the cross section of the dissociative attachment, it is necessary to take into account that the repulsion time with possible autodetachment is proportional to the square root of reduced mass of the AB molecule[5] $\sqrt{M_A M_B/(M_A + M_B)}$. The characteristic electron transition time is much shorter and proportional to the square root of its mass m. For this reason, the maximum cross section of dissociative attachment with the described configuration of electronic terms can be roughly estimated by the following formula:[13]

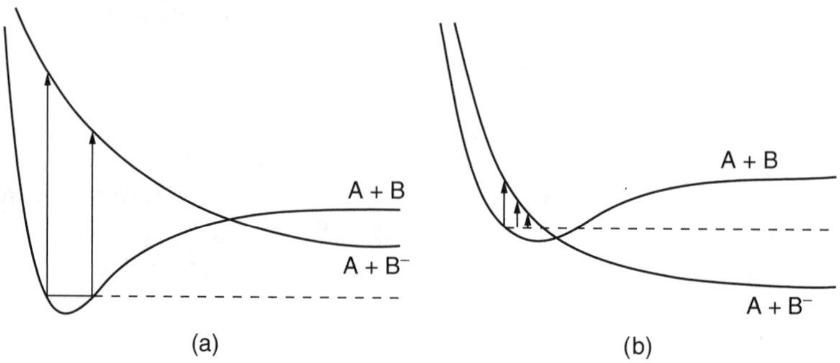

Figure 2.6 Dissociative attachment: (a) low electron affinity; (b) high electron affinity

$$\sigma_{d.a.}^{\max} \approx \sigma_0 \sqrt{\frac{m(M_A + M_B)}{M_A M_B}} \tag{2.51}$$

From this simplified formula, one can see that the maximum cross section is two orders of value less than the gas-kinetic cross section σ_0, and numerically is about 10^{-18} cm^2.

Some change in the reaction cross section (Equation 2.51) occurs if the AB^- electronic term corresponds not to the repulsive, but rather to the attractive, state;[5] however, this is not a common situation. A more interesting process occurs when the electron affinity of a product exceeds the dissociation energy, which takes place for some halogens and their compounds in particular. Corresponding potential energy curves illustrating the dissociative attachment are presented in Figure 2.6b. One can see that in this case (in contrast to the process depicted by Figure 2.6a), even very low energy electrons can effectively provide the dissociative attachment.

Also, if the electron affinity of a product exceeds the dissociation energy, the intersection point of AB and AB^- electronic terms (Figure 2.6b) is actually located inside the so-called geometrical sizes of the dissociating molecules. As a result, during the repulsion of $(AB^-)^*$, probability of the reverse autodetachment reaction ($AB + e$) is very low and the cross section of the dissociative attachment process can in this case reach the gas-kinetic cross section σ_0 of about 10^{-16} cm^2.

Cross sections of the dissociative attachment for several molecular gases are presented in Figure 2.7[9] as a function of electron energy. The figure shows that usually only electrons, having enough energy (more than the difference between dissociation energy and electron affinity), can provide the dissociative attachment. Also, the dissociative attachment cross section as a function of electron energy $\sigma_a(\varepsilon)$ has a resonant structure. It reflects the resonance nature of the process, which can be effective only within a narrow range of electron energy.

The resonant structure of $\sigma_a(\varepsilon)$ permits estimating the dissociative attachment rate coefficient k_a as a function of electron temperature T_e by the following integration of the $\sigma_a(\varepsilon)$ over the Maxwellian distribution (Equation 2.1):

$$k_a(T_e) \approx \sigma_{d.a.}^{\max}(\varepsilon_{\max}) \sqrt{\frac{2\varepsilon_{\max}}{m}} \frac{\Delta\varepsilon}{T_e} \exp\left(-\frac{\varepsilon_{\max}}{T_e}\right) \tag{2.52}$$

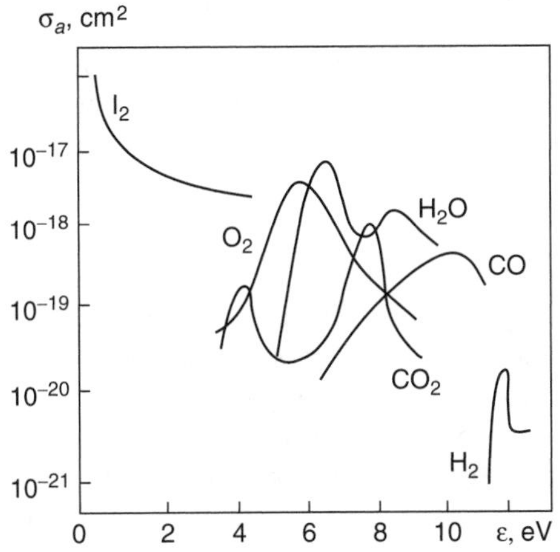

Figure 2.7 Cross sections of dissociative electron attachment to different molecules

Table 2.6 Resonance Parameters for Dissociative Attachment of Electrons to Different Molecules

Dissociative attachment process	ε_{max}, eV	$\sigma^{max}_{d.a.}$, cm^2	$\Delta\varepsilon$, eV
$e + O_2 \rightarrow O^- + O$	6.7	10^{-18}	1
$e + H_2 \rightarrow H^- + H$	3.8	10^{-21}	3.6
$e + NO \rightarrow O^- + N$	8.6	10^{-18}	2.3
$e + CO \rightarrow O^- + C$	10.3	$2*10^{-19}$	1.4
$e + HCl \rightarrow Cl^- + H$	0.8	$7*10^{-18}$	0.3
$e + H_2O \rightarrow H^- + OH$	6.5	$7*10^{-18}$	1
$e + H_2O \rightarrow O^- + H_2$	8.6	10^{-18}	2.1
$e + H_2O \rightarrow H + OH^-$	5	10^{-19}	2
$e + D_2O \rightarrow D^- + OD$	6.5	$5*10^{-18}$	0.8
$e + CO_2 \rightarrow O^- + CO$	4.35	$2*10^{-19}$	0.8

In this relation, the only resonance leading to the dissociative attachment, and thus only the resonance of the $\sigma_a(\varepsilon)$, was taken into account; ε_{max} and $\sigma^{max}_{d.a.}$ are the electron energy and maximum cross section (Equation 2.51) corresponding to the resonance and $\Delta\varepsilon$ is its energy width. Equation 2.52 is quite convenient for numerical calculations; the necessary parameters, ε_{max}, $\sigma^{max}_{d.a.}$, and $\Delta\varepsilon$ are provided in Table 2.6. As was mentioned earlier, the information is given in each case only for one resonance, which is supposed to have the maximum contribution into kinetics of the dissociative attachment.

2.4.2 Three-Body Electron Attachment to Molecules

These attachment processes of formation of negative ions in the collision of an electron with two heavy particles (at least one of which is supposed to have positive electron affinity) can be shown as:

$$e + A + B \rightarrow A^{-} + B \tag{2.53}$$

The three-body electron attachment can be a principal channel of electron losses through formation of negative ions; when electron energies are not high enough for the dissociative attachment; and when pressure is elevated (usually more than 0.1 atm) and the third kinetic order processes are preferable. In contrast to the dissociative attachment, the three-body process does not require consumption of electron energy. For this reason, its rate coefficient does not depend strongly on electron temperature (at least, within the temperature range of interest, about 1 eV).

In this case, it needs to be pointed out that heavy particles are responsible for consumption of energy released during the attachment. Electrons are usually kinetically not effective as a third body B because of the low degree of ionization and low energy release during the attachment (in contrast with the three-body electron–ion recombination [Equation 2.42]).

Atmospheric pressure nonthermal discharges in air are probably the most important systems in which the three-body attachment plays the key role in the balance of charged particles. That is why the three-body electron attachment to molecular oxygen (e + O_2 + M → O_2^- + M) can be taken as a good example. Consider the mechanism of Equation 2.53 in more detail.

The three-body electron attachment to molecular oxygen proceeds by the two-stage *Bloch–Bradbury mechanism.*[14,15] The first stage of the process includes an electron attachment to the molecule with formation of a negative oxygen ion in an unstable autoionization state:

$$e + O_2 \xleftrightarrow{k_{att},\,\tau} \left(O_2^-\right)^* \tag{2.54}$$

Here, k_{att} is a rate coefficient of the intermediate electron trapping and τ is lifetime of the excited unstable ion with respect to collisionless decay into initial state.

The second stage of the Bloch–Bradbury mechanism includes collision with the third-body particle M with either relaxation and stabilization of O_2^- (rate coefficient k_{st}) or collisional decay of the unstable ion into initial state (rate coefficient k_{dec}):

$$\left(O_2^-\right)^* + M \xrightarrow{k_{st}} O_2^- + M \tag{2.55}$$

$$\left(O_2^-\right)^* + M \xrightarrow{k_{dec}} O_2 + e + M \tag{2.56}$$

Taking into account the steady-state conditions for number density of the intermediate excited ions $(O_2^-)^*$, the rate coefficient for the total process of three-body electron attachment to molecular oxygen e + O_2 + M → O_2^- + M can expressed in the following way:

Table 2.7 Reaction Rate Coefficients of Electron Attachment to Oxygen Molecules at Room Temperature with Different Third-Body Partners

Three-body attachment	Rate coefficient	Three-body attachment	Rate coefficient
$e + O_2 + Ar \rightarrow O_2^- + Ar$	$3*10^{-32}$ cm^6/sec	$e + O_2 + Ne \rightarrow O_2^- + Ne$	$3*10^{-32}$ cm^6/sec
$e + O_2 + N_2 \rightarrow O_2^- + N_2$	$1.6*10^{-31}$ cm^6/sec	$e + O_2 + H_2 \rightarrow O_2^- + H_2$	$2*10^{-31}$ cm^6/sec
$e + O_2 + O_2 \rightarrow O_2^- + O_2$	$2.5*10^{-30}$ cm^6/sec	$e + O_2 + CO_2 \rightarrow O_2^- + CO_2$	$3*10^{-30}$ cm^6/sec
$e + O_2 + H_2O \rightarrow O_2^- + H_2O$	$1.4*10^{-29}$ cm^6/sec	$e + O_2 + H_2S \rightarrow O_2^- + H_2S$	10^{-29} cm^6/sec
$e + O_2 + NH_3 \rightarrow O_2^- + NH_3$	10^{-29} cm^6/sec	$e + O_2 + CH_4 \rightarrow O_2^- + CH_4$	$>10^{-29}$ cm^6/sec

$$k_{3M} = \frac{k_{att}k_{st}}{\frac{1}{\tau} + \left(k_{st} + k_{dec}\right)n_0} \tag{2.57}$$

In this relation, n_0 is the concentration of the third-body heavy particles M. When the pressure is not too high, $(k_{st} + k_{dec})n_0 << \tau^{-1}$, and Equation 2.57 for the reaction rate can be significantly simplified:

$$k_{3M} \approx k_{att}k_{st}\tau \tag{2.58}$$

As one can see from Equation 2.58, the three-body attachment process has a third kinetic order. It equally depends on rate coefficient of formation and stabilization of negative ions on a third particle. The latter strongly depends on type of the third particle. In general, the more complicated a role a molecule plays of the third body (M), the easier it stabilizes the $(O_2^-)^*$ and the higher is the total reaction rate coefficient k_{3M}. Numerical values of the total rate coefficients k_{3M} are presented in Table 2.7 for room temperature and different third-body particles.

The three-body attachment rate coefficients are shown in Figure 2.8 as a function of electron temperature (the gas, in general, is taken to be room temperature). For simple estimations, when $T_e = 1$ eV and $T_0 = 300$ K, one can take $k_{3M} = 10^{-30}$ cm^6/sec. The rate of the three-body process is greater than dissociative attachment (k_a) when the gas number density exceeds a critical value $n_0 > k_a(T_e)/k_{3M}$. Numerically, in oxygen with $T_e = 1$ eV, $T_0 = 300$ K, this means $n_0 > 10^{18}$ cm^{-3}, or in pressure units, $p > 30$ torr.

2.4.3 Other Mechanisms of Formation of Negative Ions

Three other mechanisms of formation of negative ions should be mentioned; these can usually be neglected, but in some specific situations should be taken into account. The first process is the *polar dissociation*:

$$e + AB \rightarrow A^+ + B^- + e \tag{2.59}$$

This process actually includes ionization and dissociation; therefore, the threshold energy is quite high in this case. On the other hand, the electron is not captured,

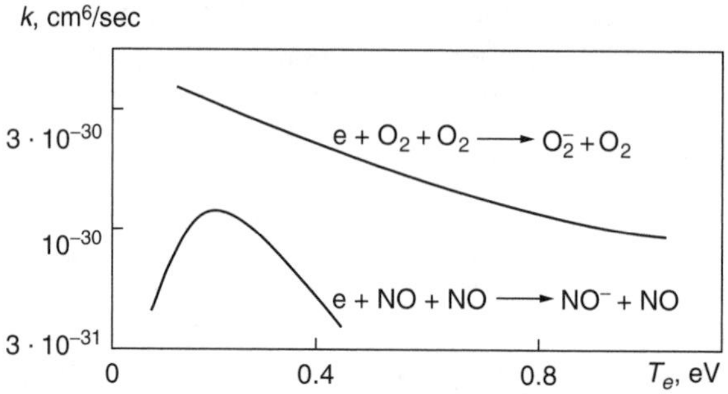

Figure 2.8 Rate coefficients of the three-body electron attachment to molecules

and the process is not a resonant one and can be effective in a wide range of high electron energies. For example, in molecular oxygen, the maximum value of the cross section of the polar dissociation is about $3*10^{-19}$ cm^2 and corresponds to electron energy 35 eV.

To stabilize formation of a negative ion during electron attachment, the excessive energy of an intermediate excited ion particle can be emitted. As a result, a negative ion can be formed in the process of radiative attachment:

$$e + M \rightarrow (M^-)^* \rightarrow M^- + \hbar\omega \tag{2.60}$$

Such an electron capture can take place at low electron energies, but the probability of the radiative process is very low, about 10^{-5} to 10^{-7}. Corresponding values of attachment cross sections are about 10^{-21} to 10^{-23} cm^2.

Finally, it should be noted that some electronegative polyatomic molecules such as SF_6 have a negative ion state very close to a ground state (only 0.1 eV in the case of SF_6^-). As a result, lifetime of such metastable negative ions can be long. Such an attachment process is resonant and for very low electron energies has maximum cross sections of about 10^{-15} cm^2.

2.4.4 Mechanisms of Negative Ion Destruction: Associative Detachment Processes

The negative ions formed as described earlier can then be neutralized in the fast ion–ion recombination or can release an electron by its detachment and ion destruction. Competition between these two processes defines the balance of charged particles, and quite often even defines the type of regime of the discharges. Different mechanisms can provide the destruction of negative ions with an electron release. These mechanisms are discussed in detail in special books and reviews.[3,5,15,16] Consider three detachment mechanisms of the most importance in plasma-chemical systems. In nonthermal discharges, probably the most important one is the **associative detachment**:

$$A^- + B \rightarrow (AB^-)^* \rightarrow AB + e \tag{2.61}$$

Table 2.8 Associative Detachment Rate Coefficients at Room Temperature

Associative detachment process	Reaction enthalpy	Rate coefficient
$H^- + H \rightarrow H_2 + e$	–3.8 eV (exothermic)	$1.3*10^{-9}$ cm^3/sec
$H^- + O_2 \rightarrow HO_2 + e$	–1.25 eV (exothermic)	$1.2*10^{-9}$ cm^3/sec
$O^- + O \rightarrow O_2 + e$	–3.8 eV (exothermic)	$1.3*10^{-9}$ cm^3/sec
$O^- + N \rightarrow NO + e$	–5.1 eV (exothermic)	$2*10^{-10}$ cm^3/sec
$O^- + O_2 \rightarrow O_3 + e$	0.4 eV (endothermic)	10^{-12} cm^3/sec
$O^- + O_2(^1\Delta_g) \rightarrow O_3 + e$	–0.6 eV (exothermic)	$3*10^{-10}$ cm^3/sec
$O^- + N_2 \rightarrow N_2O + e$	–0.15 eV (exothermic)	10^{-11} cm^3/sec
$O^- + NO \rightarrow NO_2 + e$	–1.6 eV (exothermic)	$5*10^{-10}$ cm^3/sec
$O^- + CO \rightarrow CO_2 + e$	–4 eV (exothermic)	$5*10^{-10}$ cm^3/sec
$O^- + H_2 \rightarrow H_2O + e$	–3.5 eV (exothermic)	10^{-9} cm^3/sec
$O^- + CO_2 \rightarrow CO_3 + e$	(endothermic)	10^{-13} cm^3/sec
$C^- + CO_2 \rightarrow CO + CO + e$	–4.3 (exothermic)	$5*10^{-11}$ cm^3/sec
$C^- + CO \rightarrow C_2O + e$	–1.1 (exothermic)	$4*10^{-10}$ cm^3/sec
$Cl^- + O \rightarrow ClO + e$	0.9 (endothermic)	10^{-11} cm^3/sec
$O_2^- + O \rightarrow O_3 + e$	–0.6 eV (exothermic)	$3*10^{-10}$ cm^3/sec
$O_2^- + N \rightarrow + e$	–4.1 eV (exothermic)	$5*10^{-10}$ cm^3/sec
$OH^- + O \rightarrow O_2 + e$	–1 eV (exothermic)	$2*10^{-10}$ cm^3/sec
$OH^- + N \rightarrow HNO + e$	–2.4 eV (exothermic)	10^{-11} cm^3/sec
$OH^- + H \rightarrow H_2O + e$	–3.2 eV (exothermic)	10^{-9} cm^3/sec

This process is a reverse one with respect to the dissociative attachment (Equation 2.50) and so it can also be illustrated by Figure 2.6. The associative detachment is a nonadiabatic process that occurs by intersection of electronic terms of a complex negative ion $A^-–B$ and corresponding molecule AB. For this reason (nonadiabatic reaction), the rate coefficients of such processes are usually quite high. Typical values of the coefficients are about $k_d = 10^{-10}$ to 10^{-9} cm^3/sec and are not far from those of gas-kinetic collisions. One should take into account that because the reaction (Equation 2.61) corresponds to the intersection of electronic terms of $A^-–B$ and AB, it can have an energy barrier even in exothermic conditions. Rate coefficients of some associative detachment processes are presented in Table 2.8 together with enthalpy of the reactions.

As one can see from the table, the associative detachment is quite fast. For example, the reaction $O^- + CO \rightarrow CO_2 + e$ can effectively suppress the dissociative attachment in CO_2 and corespondent electron losses, if CO concentration is high enough. This effect is of importance in nonthermal discharges in CO_2 and in CO_2–laser mixture, which will be discussed later. One should note, however, that fast three-body cluster formation processes are able to stabilize O^- with respect to associative detachment. The following clusterization reactions:

$$O^- + CO_2 + M \rightarrow CO_3^- + M, k = 10^{-27}\,\text{cm}^6/\text{sec} \qquad (2.62)$$

$$O^- + O_2 + M \rightarrow O_3^- + M, k = 10^{-30}\,\text{cm}^6/\text{sec} \quad (2.63)$$

convert O^- ions into CO_3^- and O_3^-, which are more stable with respect to detachment and thus promotes ion–ion recombination and loss of charge particles.

2.4.5 Electron Impact Detachment

This detachment process can be described as $e + A^- \rightarrow A + 2e$ and is an essential one in the balance of negative ions when the ionization degree is high. This process with the negative hydrogen atom also plays an important role in plasma chemical water decomposition in nonequilibrium discharges, which will be discussed later.

The electron impact detachment is somewhat similar to the direct ionization of neutrals by an electron impact (Thomson mechanism; see Section 2.2.1). The main difference in this case is due to repulsive coulomb force acting between the incident electron and the negative ion. For incident electron energies about 10 eV, the cross section of the detachment process can be high, about 10^{-14} cm^2. Typical detachment cross section dependence on energy or on the incident electron velocity v_e can be illustrated by that for the detachment from negative hydrogen ion ($e + H^- \rightarrow H + 2e$):[17]

$$\sigma(v_e) \approx \frac{\sigma_0 e^4}{(4\pi\varepsilon_0)^2 \hbar^2 v_e^2}\left(-7.5\ln\frac{e^2}{4\pi\varepsilon_0 \hbar v_e} + 25\right) \quad (2.64)$$

In this relation, σ_0 is the geometrical atomic cross section, more exactly defined in Equation 2.20. Numerically, the detachment cross section as a function of electron energy is presented in Figure 2.9[9] for hydrogen and oxygen atomic ions. One can see from the figure that the maximum cross sections of about 10^{-15} to 10^{-14} cm^2 corresponds to electron energies about 10 to 50 eV. These energies exceed electron affinities more than ten times in contrast to the Thomson mechanism of the electron impact ionization (which can be explained by the coulomb repulsion).

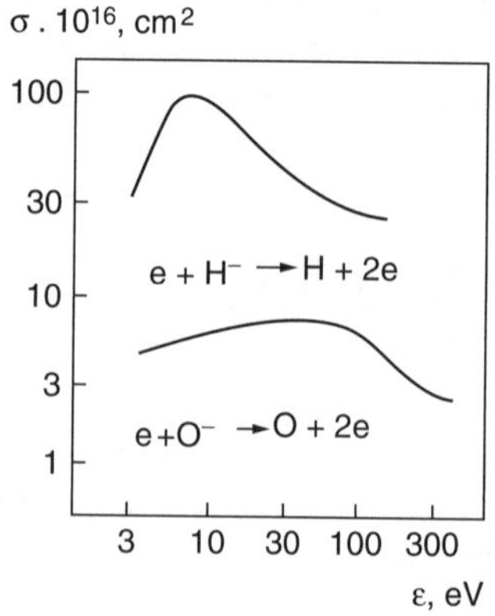

Figure 2.9 Negative ion destruction by an electron impact

Table 2.9 Thermal Destruction of an Oxygen Molecular Ion

Detachment process	Reaction enthalpy	Rate coefficient, 300 K	Rate coefficient, 600 K
$O_2^- + O_2 \rightarrow O_2 + O_2 + e$	0.44 eV	$2.2*10^{-18}$ cm^3/sec	$3*10^{-14}$ cm^3/sec
$O_2^- + N_2 \rightarrow O_2 + N_2 + e$	0.44 eV	$<10^{-20}$ cm^3/sec	$1.8*10^{-16}$ cm^3/sec

2.4.6 Detachment in Collisions with Excited Particles

For the case of using **electronic excitation** energy of particle B, this detachment process ($A^- + B^* \rightarrow A + B + e$) is similar to the Penning ionization (see Section 2.2.8). If the electronic excitation energy of a collision partner B exceeds the electron affinity of another particle A, the detachment process can proceed effectively as an electronically nonadiabatic reaction (without significant energy exchange with translational degrees of freedom of the heavy particles; see Section 2.2.7). As an example, consider the exothermic detachment of an electron from an oxygen molecular ion in collision with a metastable electronically excited oxygen molecule (excitation energy 0.98 eV):

$$O_2^- + O_2\left({}^1\Delta_g\right) \rightarrow O_2 + O_2 + e, \Delta H = -0.6 \text{ eV} \tag{2.65}$$

The rate coefficient of the detachment reaction is very high; even at room temperature, it is about $2*10^{-10}$ cm^3/sec.

Electron detachment also can be effective in collisions with **vibrationally excited molecules**. Consider again as an example the destruction of an oxygen molecular ion, $O_2^- + O_2^*(v > 3) \rightarrow O_2 + O_2 + e$. This process provides an essential contribution into thermal detachment. Rate coefficients of the detachment process in the quasi-equilibrium thermal systems (characterized only by the temperature) are presented in Table 2.9.

From simple calculations, one can see that the first process presented in Table 2.9 has the activation energy close to the reaction enthalpy. Actually, these quasi-equilibrium detachment processes are using mostly vibrational energy of colliding partners because in other cases the reaction is strongly adiabatic (see Section 2.2.7).

The detachment process stimulated by vibrational excitation of molecules proceeds according to the modified Bloch–Bradbury mechanism (Equation 2.54 through Equation 2.57). In this case, the process starts with collisional excitation of O_2^- to the vibrationally excited states with $v > 3$ (these excitation and vibrational energy transfer processes will be considered in detail in Chapter 3 and Chapter 5). The excited ions $(O_2^-)^*(v > 3)$ are then in the state of autoionization, which results in an electron detachment during a period of time shorter than the interval between two collisions.

Excitation of $(O_2^-)^*(v > 3)$ is easier and faster in oxygen than in nitrogen, where the process is less resonant. This explains the significant difference in detachment rate coefficients in collisions with oxygen and nitrogen, shown in Table 2.9. Kinetics of the detachment process can be described in this case in a conventional way for all reactions stimulated by vibrational excitation of molecules (see Chapter 5). The traditional Arrhenius formula, $k_d \propto \exp(-E_a/T_v)$, is applicable here, if molecular vibrations are in internal quasi equilibrium with the only vibrational temperature T_v. The activation energy of the detachment process can be taken in this case equal to the electron affinity to oxygen molecules ($E_a \approx 0.44$ eV) in the same way one can describe NO^-, F^-, and Br^- detachment kinetics.[5]

In more complicated situations of the vibrational excitation typical for nonthermal discharges, the vibrational energy distribution function should be used instead of the Boltzmann function, which results in corresponding modifications in the Arrhenius formula (see Chapter 5).

2.5 ION–ION RECOMBINATION PROCESSES

In electronegative gases, when the attachment processes are involved in the balance of electrons and ions, the actual losses of charged particles are mostly due to ion–ion recombination. This process means mutual neutralization of positive and negative ions in binary or three-body collisions. The ion–ion recombination can proceed by a variety of different mechanisms that dominate in different pressure ranges. However, all of them are characterized by very high rate coefficients.

At high pressures (usually more than 30 torr), three-body mechanisms dominate the recombination. The recombination rate coefficient in this case reaches a maximum of about 1 to $3*10^{-6}$ cm^3/sec near atmospheric pressures for room temperature. Traditionally, the rate coefficient of ion–ion recombination is recalculated with respect to concentrations of positive and negative ions, that is, to the second kinetic order. Due to the three-molecular nature of the recombination mechanism in this pressure range, the recalculated (second kinetic order) recombination rate coefficient depends on pressure. Near atmospheric pressure corresponds to the fastest neutralization. An increase and a decrease of pressure result in proportional reduction of the three-body ion–ion recombination rate coefficient.

Obviously, at low pressures, the three-body mechanism becomes relatively slow. In this pressure range, the binary collisions with transfer of energy into electronic excitation make a major contribution to the ion–ion recombination. The different mechanisms of the ion–ion recombination that dominate the whole process of neutralization in different pressure ranges will be discussed next.

2.5.1 Ion–Ion Recombination in Binary Collisions

This neutralization process proceeds in a binary collision of a negative and positive ion with the released energy going to electronic excitation of a neutral product:

$$A^- + B^+ \rightarrow A + B^* \tag{2.66}$$

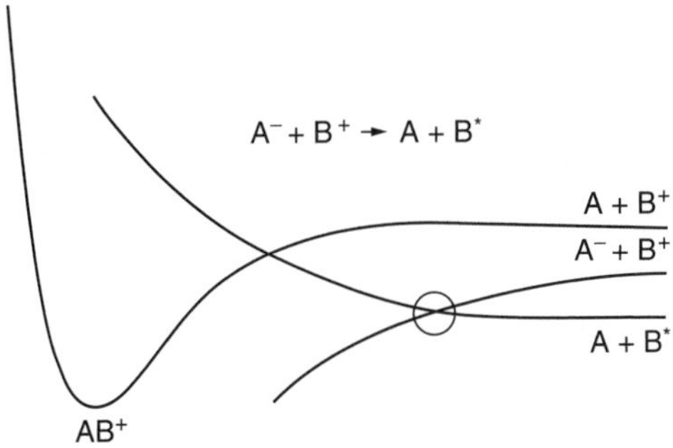

Figure 2.10 Terms illustrating ion–ion recombination

This reaction has a second kinetic order and dominates the ion–ion recombination in low-pressure discharges (p < 10 to 30 torr). Electronic terms illustrating the recombination are presented in Figure 2.10. The ions A^- and B^+ are approaching each other following the attractive A^-–B^+ Coulomb potential curve. When the distance between the heavy particles is large, this potential curve lies below the term of A^-–B^+ (lower value of electron affinity of particle A: EA_A), and above the final A–B^* electronic term on the energy interval:

$$\Delta E \approx {}^{I_B}\!/_{n} - EA_A \tag{2.67}$$

Here, I_B is the ionization potential of particle B. If the principal quantum number of particle B after recombination is quite high, $n = 3$ to 4, then ΔE is low. This means that the electronic terms of A^-–B^+ and A–B^* are relatively close when the ions A^- and B^+ are approaching each other. When the principal quantum number n is not specifically defined, then the energy affinity EA_A can be taken as a reasonable estimation for the energy interval (Equation 2.67), that is, $\Delta E \approx EA_A$.

As can be seen from Figure 2.11,[9] the low value of ΔE results in the possibility of effective transition between the electronic terms (from A^-–B^+ to A–B^*), when distance R_{ii} between ions is still large. Even for this long distance R_{ii} between the ions, the coulomb attraction energy is already sufficient to compensate the initial energy gap ΔE between the terms:

$$R_{ii} \approx \frac{e^2}{4\pi\varepsilon_0} \frac{1}{\left({}^{I_B}\!/_{n^2} - EA_A\right)} \approx \frac{e^2}{4\pi\varepsilon_0 EA_A} \tag{2.68}$$

The high value of R_{ii} results in large cross sections of the ion–ion recombination (Equation 2.66). This can be estimated taking into account conservation of angular momentum during the attractive coulomb collision and estimating the maximum kinetic energy equal to the potential energy EA_A (Equation 4.160). The impact parameter b (see Section 2.1.7) can be found as a function of the ion kinetic energy ε in the center of mass system:

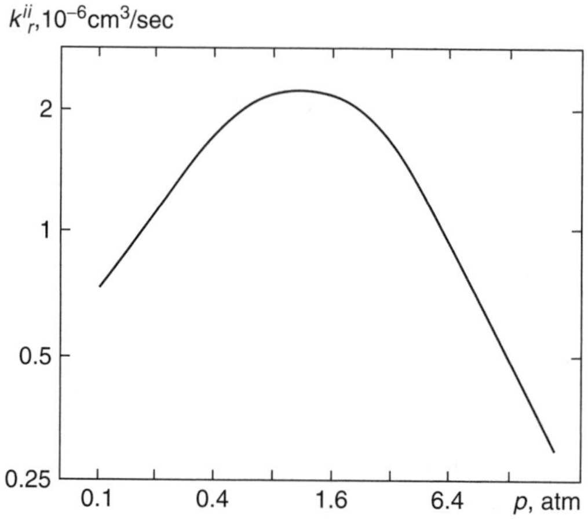

Figure 2.11 Ion–ion recombination rate coefficient in air

$$b \approx R_{ii} \frac{\sqrt{\mathrm{EA_A}}}{\sqrt{\varepsilon}} \tag{2.69}$$

Based on the expressions for the impact parameter b and for the reactive distance R_{ii}, the formula to calculate the cross section of the ion–ion recombination process can be given as:

$$\sigma^{ii}{}_{rec} = \pi b^2 \approx \pi \frac{e^4}{(4\pi\varepsilon_0)^2} \frac{1}{\mathrm{EA_A}} \frac{1}{\varepsilon} \tag{2.70}$$

It should be mentioned that quantum mechanical calculation of the ion–ion recombination cross section should take into account the effect of electron tunneling.[16] As a result, it gives the more precise dependence of the cross section of the ion–ion recombination in binary collisions of the ion kinetic energy as:

$$\sigma^{ii}_{rec} \propto 1/\sqrt{\varepsilon}$$

Equation 2.70 can be rewritten referring to the atomic units,

$$\sigma^{ii}_{rec} \approx \sigma_0 \frac{I}{\mathrm{EA_A}} \frac{I}{\varepsilon}$$

where I is a typical value of ionization energy. This means that the recombination cross section exceeds the gas-kinetic one by several orders of magnitude. The same is true for the reaction rate coefficients, which usually are of the order of 10^{-7} cm^3/sec. Some binary recombination rate coefficients are presented in Table 2.10 together with energy released in the process.

Table 2.10 Reaction Rate Coefficients of Ion–Ion Recombination in Binary Collisions at Room Temperature

Recombination process	Released energy, $I - EA$	Rate coefficient
$H^- + H^+ \rightarrow H + H$	12.8 eV	$3.9*10^{-7}$ cm^3/sec
$O^- + O^+ \rightarrow O + O$	12.1 eV	$2.7*10^{-7}$ cm^3/sec
$O^- + N^+ \rightarrow O + N$	13.1 eV	$2.6*10^{-7}$ cm^3/sec
$O^- + O_2^+ \rightarrow O + O_2$	11.6 eV	10^{-7} cm^3/sec
$O^- + NO \rightarrow O + NO$	7.8 eV	$4.9*\ 10^{-7}$ cm^3/sec
$SF_6 + SF_5^+ \rightarrow SF_6 + SF_5$	15.2 eV	$4*10^{-8}$ cm^3/sec
$NO_2^- + NO^+ \rightarrow NO_2 + NO$	5.7 eV	$3*10^{-7}$ cm^3/sec
$O_2^- + O^+ \rightarrow O_2 + O$	13.2 eV	$2*10^{-7}$ cm^3/sec
$O_2^- + O_2^+ \rightarrow O_2 + O_2$	11.6 eV	$4.2*10^{-7}$ cm^3/sec
$O_2^- + N_2^+ \rightarrow O_2 + N_2$	15.1 eV	$1.6*10^{-7}$ cm^3/sec

2.5.2 Three-Body Ion–Ion Recombination: Thomson's Theory

The neutralization process can effectively proceed in a triple collision of a heavy neutral with a negative and a positive ion:

$$A^- + B^+ + M \rightarrow A + B \tag{2.71}$$

This process dominates the ion–ion recombination at moderate and high pressures ($p > 10$ to 30 torr). The three-body reaction has the third kinetic order only in the moderate pressure range, usually less than 1 atm (for high pressures, the process is limited by ion mobility; see below). This means that only at moderate pressures is the reaction rate proportional to the product of concentrations of all three collision partners, positive and negative ions, and neutrals.

At moderate pressures (about 0.01 to 1 atm), the three-body ion–ion recombination can be described in the framework of the Thomson theory.[6] According to this theory, recombination takes place if negative and positive ions approach each other closer than the critical distance $b \approx e^2/4\pi\varepsilon_0 T_0$. In this case, their coulomb interaction energy reaches the level of thermal energy, which is approximately T_0, the gas temperature or temperature of heavy particles. Then, during the collision, the third body (a heavy neutral particle) can absorb energy approximately equal to the thermal energy T_0 from an ion and provide the act of recombination.

The collision frequency of an ion with a neutral particle leading to energy transfer of about T_0 between them and subsequently resulting in recombination can be found as $n_0\sigma v_t$, where n_0 is the neutral's number density; σ is a typical cross section of ion–neutral elastic scattering; and v_t is an average thermal velocity of the heavy particles. After the collision, the probability P_+ for a positive ion to be closer than $b \approx e^2/4\pi\varepsilon_0 T_0$ to a negative ion and, as a result to be neutralized in the following recombination, can be estimated as:

Table 2.11 Rate Coefficients of Three-Body Ion–Ion Recombination Processes at Room Temperature and Moderate Pressures

Ion–ion recombination	Rate coefficients, 3rd order	Rate coefficients, 2nd order, 1 atm
$O_2^- + O_4^+ + O \rightarrow O_2 + O_2 + O_2 + O_2 + O_2$	$1.55*10^{-25}$ cm^6/sec	$4.2*\ 10^{-6}$ cm^3/sec
$O^- + O_2^+ O_2 \rightarrow O_3 + O_2$	$3.7*10^{-25}$ cm^6/sec	10^{-5} cm^6/sec
$N_2^- + NO^+ + O_2 \rightarrow NO_2 + NO + O_2$	$3.4*10^{-26}$ cm^6/sec	$0.9*10^{-6}$ cm^6/sec
$NO_2^- + NO^+ + N_2 \rightarrow NO_2 + NO + N_2$	10^{-25} cm^6/sec	$2.7*10^{-6}$ cm^6/sec

Note: In the last column, the coefficients are recalculated to the second kinetic order and pressure of 1 atm, that is, multiplied by concentration of neutrals $2.7*10^{19}$ cm^{-3}.

$$P_+ \approx n_- b^3 \approx n_- \frac{e^6}{\left(4\pi\varepsilon_0\right)^3 T_0^3} \tag{2.72}$$

The frequency of the collisions of a positive ion with neutral particles, resulting in ion–ion recombination, can then be found as $\nu_{ii} = (\sigma v_t n_0) P_+$. Consequently, the total three-body ion–ion recombination rate can be presented according to Thomson's theory as:

$$w_{ii} \approx \left(\sigma v_t\right) \frac{e^6}{\left(4\pi\varepsilon_0\right)^3 T_0^3} n_0 n_- n_+ \tag{2.73}$$

As can be seen, the recombination has the third kinetic order with the rate coefficient k_{r3}^{ii} decreasing with gas temperature:

$$k_{r3}^{ii} \approx \left(\sigma \sqrt{\frac{1}{m}}\right) \frac{e^6}{\left(4\pi\varepsilon_0\right)^3} \frac{1}{T_0^{5/2}} \tag{2.74}$$

At room temperature, the recombination coefficients are about 10^{-25} cm^6/sec. For some specific reactions, these third-order kinetic reaction rate coefficients are presented in Table 2.11. Traditionally, the coefficients in this table can be recalculated to the second kinetic order, k_{r2}^{ii}, which means also tabulating $k_{r2}^{ii} = k_{r3}^{ii} n_0$.

Comparing Table 2.10 and Table 2.11, the binary and triple collisions are seen to contribute equally to the ion–ion recombination at pressures of about 10 to 30 torr.

2.5.3 High-Pressure Limit of Three-Body Ion–Ion Recombination: Langevin Model

According to Equation 2.73 and Equation 2.74, the ion–ion recombination rate coefficient recalculated to the second kinetic order, $k_{r2}^{ii} = k_{r3}^{ii} n_0$, grows linearly with pressure at fixed temperatures. This growth is limited in the range of moderate pressures (usually not more than 1 atm) by the frameworks of Thomson's theory. This approach requires the capture distance b to be less than an ion mean free path $(1/n_0\sigma)$. This requirement prevents using Thomson's theory for high concentration of neutrals and high pressures (more than 1 atm):

$$n_0\sigma b \approx n_0 \frac{e^2}{(4\pi\varepsilon_0)\mathrm{T}_0}\sigma < 1 \tag{2.75}$$

In the opposite case of high pressures, $n_0\sigma b > 1$, the recombination is limited by the phase, in which a positive and negative ion approach each other, moving in electric field and overcoming multiple collisions with neutrals. The Langevin model developed in 1903 describes the ion–ion recombination in this pressure range.

This motion of a positive and negative ion to collide in recombination can be considered as their drift in the coulomb field $e/(4\pi\varepsilon_0)r^2$, where r is the distance between them. Use positive and negative ion mobility μ_+ and μ_- (see Chapter 4 for details) as a coefficient of proportionality between ion drift velocity and the strength of electric field. Then the ion drift velocity to meet each other can be expressed as:

$$v_d = \frac{e}{(4\pi\varepsilon_0)^2}(\mu_+ + \mu_-) \tag{2.76}$$

Consider a sphere with radius r surrounding a positive ion and then a flux of negative ions with concentration n_- approaching the positive one. As a result, the recombination frequency v_{r+} for the positive ion can be written as:

$$\nu_{r+} = 4\pi r^2 \nu_d n_- \tag{2.77}$$

Based on Equation 2.77 for neutralization frequency with respect to one positive ion, the total ion–ion recombination reaction rate can be found as $\mathrm{w} = n_+\nu_{r+} = 4\pi^2 v_d n_- n_+$. This case shows the obvious second kinetic order of the process with respect to ion concentrations. The final Langevin expression for the ion–ion recombination rate coefficient $k_r^{ii} = \mathrm{w}/n_+n_-$ in the limit of high pressures (usually more than 1 atm) can then be given, taking into account Equation 2.76 and Equation 2.77, in the following form:

$$k_r^{ii} = 4\pi e(\mu_+ + \mu_-) \tag{2.78}$$

This recombination rate coefficient is proportional to the ion mobility, which decreases proportionally with pressure (see Chapter 4 for details).

Comparing the Thomson and Langevin models, it is seen that at first, recombination rate coefficients grow with pressure $k_{r2}^{ii} = k_{r3}^{ii} n_0$ (see Equation 2.74), and then at high pressures they begin to decrease as 1/p together with ion mobility (Equation 2.78). The highest recombination coefficient, according to Equation 2.75, corresponds to the concentration of neutrals:

$$n_0 \approx \frac{4\pi\varepsilon_0 T_0}{\sigma e^2} \tag{2.79}$$

At room temperature this gas concentration corresponds to atmospheric pressure. The maximum value of the ion–ion recombination rate coefficient can then be found by substituting the number density n_0 (2.79) into $k_{r2}^{ii} = k_{r3}^{ii} n_0$ (2.74):

$$k_{r,\max}^{ii} \approx \frac{e^4}{\left(4\pi\varepsilon_0\right)^2} \frac{v_t}{T_0^2} \tag{2.80}$$

Numerically, the maximum recombination rate coefficient is about 1 to $3*10^{-6}$ cm^3/sec. This coefficient decreases for an increase or a decrease of pressure from atmospheric pressure. The total experimental dependence of the k_r^{ii} on pressure is presented in Figure 2.11 The generalized ion–ion recombination model combining the Thomson and Langevin approaches for moderate and high pressures was developed by Natanson.[18]

2.6 ION–MOLECULAR REACTIONS

Another group of fast processes taking place in collisions of heavy particles are the ion–molecular reactions, some of which have already been discussed. The positive ion conversion $A^+ + B + M \to AB^+ + M$ was considered in Section 2.3.3 as a preliminary stage of the dissociative electron–ion recombination. This process is able to accelerate electron losses in several gases, including air, at high pressures. Another ion–molecular process, clusterization of negative ions, was considered briefly in Section 2.4.4, using important examples of the formation of stable complex ions from a negative oxygen ion O– + $O^- + CO_2 + M \to CO_3^- + M$ and $O^- + O_2 + M \to O_3^- + M$. These processes play a significant role in charge balance, providing "protection" of the O^- negative ions against destruction in the electron detachment process.

The fast ion–molecular reactions, as was already discussed, make an important contribution in the balance of charged particles. These also can provide plasma chemical processes by themselves. Probably the best examples here are the ion-cluster growth in dusty plasmas and ion–molecular chain reactions of pollutant oxidation in air. Corresponding problems of applied plasma chemistry will be discussed later in this book. In this section, the focus will be on the main fundamental aspects of kinetics of different ion–molecular reactions.

2.6.1 Ion–Molecular Polarization Collisions: Langevin Rate Coefficient

The ion–molecular processes can start with scattering in the polarization potential, leading to the so-called Langevin capture of a charged particle and formation of an intermediate ion–molecular complex. If a neutral particle has no permanent dipole moment, the ion–neutral charge–dipole interaction and scattering is due to the dipole moment p_M induced in the neutral particle by the electric field E of an ion:

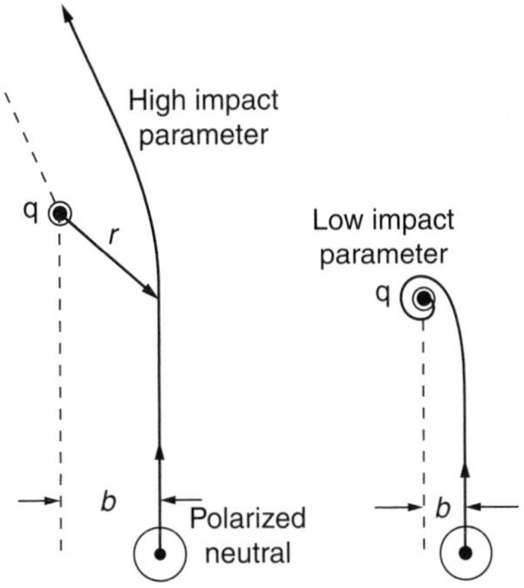

Figure 2.12 Langevin scattering in the polarization potential

$$p_m = \alpha\varepsilon_0 \mathrm{E} = \alpha \frac{e}{4\pi r^2} \tag{2.81}$$

Here, r is the distance between the interacting particles and α is the polarizability of a neutral atom or molecule (see Table 2.12).

The typical orbits of relative ion and neutral motion during polarization scattering are shown in Figure 2.12. As can be seen, when the impact parameter is high, the orbit has a hyperbolic character, but when the impact parameter is sufficiently low, the scattering leads to the Langevin polarization capture. The Langevin capture means that the spiral trajectory results in "closer interaction" and formation of the ion–molecular complex, which then can either spiral out or provide inelastic changes of state and formation of different secondary products.

This ion–molecular capture process based on polarization occurs when the interaction energy between charge and induced dipole

$$\mathrm{P}_m \mathrm{E} = \alpha \frac{e}{4\pi r^2} \frac{e}{4\pi\varepsilon_0 r^2}$$

becomes the order of the kinetic energy of the colliding partners

$$\frac{1}{2}\mathrm{M}\mathrm{v}^2$$

where M is their reduced mass and v their relative velocity. From this qualitative equality of kinetic and interaction energies, the **Langevin cross section** of the ion–molecular polarization capture can be found as $\sigma_L \approx \pi r^2$:

$$\sigma_L = \sqrt{\frac{\pi\alpha e^2}{\varepsilon_0 \mathrm{M}\mathrm{v}^2}} \tag{2.82}$$

Table 2.12 Polarizabilities of Atoms and Molecules

Atom or molecule	A, 10^{-24} cm^3	Atom or molecule	A, 10^{-24} cm^3
Ar	1.64	H	0.67
C	1.78	N	1.11
O	0.8	CO	1.95
Cl_2	4.59	O_2	1.57
CCl_4	10.2	CF_4	1.33
H_2O	1.45	CO_2	2.59
SF_6	4.44	NH_3	2.19

Detailed derivation of the exact formula for the Langevin cross section can be found in the book by Lieberman and Lichtenberg.[13]

According to Equation 2.82, the Langevin capture rate coefficient $k_L \approx \sigma_{L}v$ does not depend on velocity and therefore does not depend on temperature:

$$k_L = \sqrt{\frac{\pi \alpha e^2}{\varepsilon_0 M}} \tag{2.83}$$

It is convenient for calculations to express the polarizability α in 10^{-24} cm^3 (cubic angstroms; see Table 2.11) and reduced mass in atomic mass units (amu), and then to rewrite and use Equation 2.83 in the following numerical form:

$$k_L^{ion/neutral} = 2.3 * 10^{-9}\ \mathrm{cm^3/sec} * \sqrt{\frac{\alpha, 10^{-24}\,\mathrm{cm}^3}{\mathrm{M, amu}}} \tag{2.84}$$

As can be seen, the typical value of the Langevin rate coefficient for ion–molecular reactions is about 10^{-9} cm^3/sec, which is 10 times higher than the typical value of gas-kinetic rate coefficient for binary collisions of neutral particles: $k_0 \approx 10^{-10}$ cm^3/sec. Also, α/M in Equation 2.84 reflects the "specific volume" of an atom or molecule, because the polarizability α actually corresponds to the particle volume. This means that the Langevin capture rate coefficient grows with a decrease of "density" of an atom or molecule.

The relations in Table 2.12 describe interactions of a charge with an induced dipole. If an ion interacts with a molecule having not only an induced dipole but also a permanent dipole moment μ_D, then the Langevin capture cross section becomes larger. The corresponding correction of Equation 2.83 was done by Su and Bowers[19] and can be expressed as:

$$k_L = \sqrt{\frac{\pi e^2}{\varepsilon_0 M}} \left(\sqrt{\alpha} + c\mu_D \sqrt{\frac{2}{\pi T_0}} \right) \tag{2.85}$$

Here T_0 is a gas temperature (in energy units) and parameter $0 < c < 1$ describes the effectiveness of a dipole orientation in electric field of an ion. For molecules like

$H_2 0$ or HF with a large permanent dipole moment, the ratio of the second to the first terms in Equation 2.85 is about $\sqrt{I/T_0}$, where I is an ionization potential. This means that the Langevin cross sections and rate coefficients for the dipole molecules and radicals can exceed by a factor of 10 the numerical values obtained for pure polarization collisions Equation 2.84.

Note that derivation of the Langevin formula (Equation 2.82) did not specify the mass of a charged particle. This means that the Langevin cross section and rate coefficients Equation 2.82 and Equation 2.83 also can be applied for an electron-neutral interaction. Obviously, in this case the reduced mass M is close to an electron mass and the numerical formula for the Langevin rate coefficient for the electron-neutral collision can be presented as:

$$\kappa_L^{electron/neutral} = 10^{-7}\ \mathrm{cm}^3/\mathrm{sec} * \sqrt{\alpha, 10^{-24}\mathrm{cm}^3} \tag{2.86}$$

Numerical value of the Langevin rate coefficient for the electron-neutral collision is about 10^{-7} $\mathrm{cm}^3/\mathrm{sec}$ and also does not depend on temperature.

2.6.2 The Ion–Atom Charge Transfer Processes

During a collision, an electron can be transferred from a neutral particle to a positive ion, or from negative ion to a neutral particle. These processes are referred to as charge transfer or charge exchange. The charge exchange reaction without significant defect of the electronic state energy ΔE during collision is called the **resonant charge transfer**; otherwise, the charge transfer is called **nonresonant**. The resonant charge transfer is a nonadiabatic process and usually has a very large cross section.

Consider a charge exchange between a neutral particle B and a positive ion A^+, assuming the particle B/B^+ is at rest:

$$\mathrm{A}^+ + \mathrm{B} \rightarrow \mathrm{A} + \mathrm{B}^+ \tag{2.87}$$

The energy scheme of this reaction is illustrated in Figure 2.13, in which one can see the coulomb potential energy of an electron in the coulomb field of A^+ and B^+:

$$U(z) = -\frac{e^2}{4\pi\varepsilon_0 z} - \frac{e^2}{4\pi\varepsilon_0 |r_{AB} - z|} \tag{2.88}$$

Here, r_{AB} is a distance between the centers of A and B. The maximum value of the potential energy corresponds to $z = r_{AB}/2$ and is equal to:

$$U_{\max} = -\frac{e^2}{\pi\varepsilon_0 r_{AB}} \tag{2.89}$$

The charge transfer, Equation 2.87, is possible in the framework of classical mechanics (see Figure 2.13) if the maximum of potential energy $U_{\max}$ is lower than the initial energy E_B of an electron to be transferred from level n of particle B:

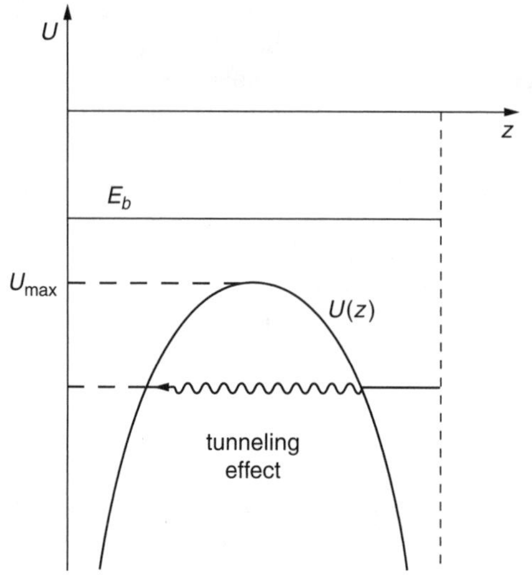

Figure 2.13 Energy terms for resonant charge exchange

$$E_B = -\frac{I_B}{n^2} - \frac{e^2}{4\pi\varepsilon_0 r_{AB}} \geq U_{max} \tag{2.90}$$

Here, I_{AB} is the ionization potential of atom B. From Equation 2.89 and Equation 2.90, the maximum distance between the interacting heavy particles is found when the charge transfer is still permitted by classical mechanics:

$$r_{AB}^{max} = \frac{3e^2 n^2}{4\pi\varepsilon_0 I_B} \tag{2.91}$$

If the charge exchange is resonant and therefore not limited by the defect of energy, the classical reaction cross section can be found from Equation 2.91 as $\pi r_{AB}{}^2$:

$$\sigma_{ch.tr}^{class} = \frac{9e^4 n^4}{16\pi\varepsilon_0^2 I_B^2} \tag{2.92}$$

The classical cross section of charge exchange does not depend on the kinetic energy of the interacting species and its numerical value for ground state transfer ($n = 1$) is approximately the gas-kinetic cross section for collisions of neutrals.

The actual cross section of a resonant charge transfer can be much higher than Equation 2.92, taking into account the quantum mechanical effect of electron tunneling from B to A^+ (see Figure 2.13). This effect can be estimated by calculating the electron tunneling probability P_{tunn} across a potential barrier of height about I_B and width d:

$$P_{tunn} \approx \exp\left(-\frac{2d}{\hbar}\sqrt{2meI_B}\right) \tag{2.93}$$

The frequency of electron oscillation in the ground state atom B is about $I_B/\hbar$, so the frequency of the tunneling is $I_B P_{tunn}/\hbar$. Taking into account Equation 2.93, one finds the maximum barrier width d_{max} when the tunneling frequency still exceeds the reverse ion–neutral collision time: $I_B P_{tunn}/\hbar > v/d$; as a result, the tunneling can take place. Here, v is the relative velocity of the colliding partners. Then the cross section πd_{max}^2 of the electron tunneling from B to A^+ leading to the resonance charge exchange can be estimated as:

$$\sigma_{ch.tr}^{tunn} \approx \frac{1}{I_B}\left(\frac{\pi\hbar^2}{8me}\right)\left(\ln\frac{I_B d}{\hbar} - \ln v\right)^2 \tag{2.94}$$

The more detailed derivation of the electron tunneling cross section[60a] results in more complicated expressions, which nevertheless can be presented in a numerical form very similar to Equation 2.94.

$$\sqrt{\sigma_{ch.tr}^{tunn}, \mathrm{cm}^2} \approx \frac{1}{\sqrt{I_B \mathrm{eV}}}\left(6.5*10^{-7} - 3*10^{-8}\ln v, \frac{\mathrm{cm}}{\mathrm{sec}}\right) \tag{2.95}$$

This numerical formula can be applied to calculate the resonant charge transfer cross section in the velocity range $v = 10^5$ to 10^8 cm/sec. At higher energies, when the velocity v exceeds 10^8 cm/sec, the tunneling cross section decreases to the gas-kinetic cross sections (about $3*10^{16}$ cm^2) and then it is stabilized. The stabilized cross section does not depend on energy and is given by the classical expression, Equation 2.92.

At the lower limit of application of Equation 2.95, $v = 10^5$ cm/sec and the tunneling cross section reaches values as high as 10^{-14} cm^2. Equation 2.94 and Equation 2.95 were derived assuming straight line collision trajectories. Therefore, applying these two equations at lower velocities is impossible because, when

$$v \le 10^5 \frac{\mathrm{cm}}{\mathrm{sec}} \Big/ \sqrt{M}$$

(M is a reduced mass), the collision trajectory is strongly perturbed during the charge exchange.

Actually, in low energy collision, when

$$v \le 10^5 \frac{\mathrm{cm}}{\mathrm{sec}} \Big/ \sqrt{M}$$

the ion and neutral are captured in a complex. In this case, the exchanging electron always has enough time for tunneling and therefore the equal probability 1/2 is found on either particle. Then the resonant charge exchange cross section can be found as half of the Langevin capture cross section (Equation 2.82):

$$\sigma_L = \frac{1}{2}\sqrt{\frac{\pi\alpha e^2}{\varepsilon_0 M}}\frac{1}{v} \tag{2.96}$$

The typical numerical value of this cross section at room temperature is about 10^{-14} cm^2 and higher. It grows because the velocity (and temperature) decreases even faster than in the case of intermediate velocities (Equation 2.95).

2.6.3 Nonresonant Charge Transfer Processes

Consider an electron transfer between oxygen and nitrogen atoms as an example of the nonresonant charge exchange with energy defect close to 1 eV:

$$N^+ + O \to N + O^+, \ \Delta E = -0.9 \text{ eV} \tag{2.97}$$

The principal potential curves illustrating the nonresonant charge transfer (Equation 2.97) are shown in Figure 2.14. From Table 2.1, the ionization potential of oxygen (I = 13.6 eV) is lower than that of nitrogen (I = 14.5 eV). This is the reason that the electron transfer from oxygen to nitrogen is an exothermic process and the separated $N+O^+$ energy level is located 0.9 eV lower than the separated $O+N^+$ energy level.

The reaction (Equation 2.97) begins with N^+ approaching O by following the attractive NO^+ term. Then this term crosses with the repulsive NO^+ term and the system experiences a nonadiabatic transfer, which results in formation of O^+ + N. The excess energy of approximately 1 eV goes into the kinetic energy of the products. The cross section of such exothermic charge exchange reactions at low thermal energies is of the order of the resonant cross sections of tunneling (Equation 2.95) or Langevin capture (Equation 2.96). The endothermic reactions of charge exchange, like the reverse process N + O^+ → N^+ + O, are usually very slow at low gas temperatures with low energies of the colliding ions and heavy neutral particles.

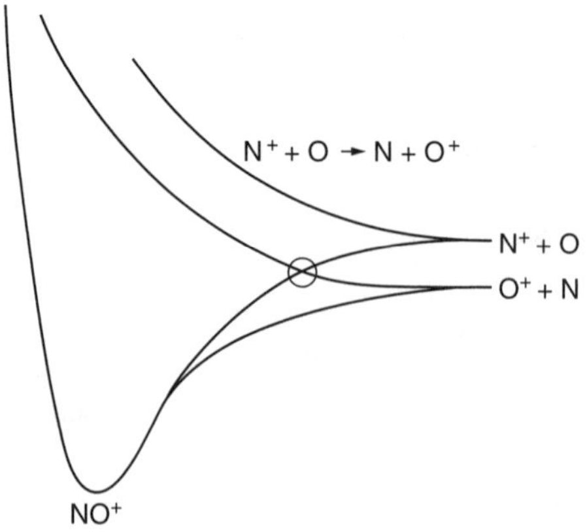

Figure 2.14 Terms illustrating nonresonant charge exchange

In this section, as well as in a previous one, resonant and nonresonant charge exchange processes between a neutral atom and a positive atomic ion were considered. Actually, the kinetic formulas are similar for an electron transfer between neutral molecules and molecular ions. These also can be applied for charge exchange with atomic and molecular negative ions. All these charge exchange processes are very fast (following the tunneling or Langevin kinetics) if they are exothermic and slow if they are endothermic.

The important role of the nonresonant charge exchange processes can be illustrated by the so-called effect of **acidic behavior of nonthermal air plasma**. Ionization of air in nonthermal discharges primarily leads to a large amount of N_2^+ ions (with respect to other positive ions) because of the high molar fraction of nitrogen in air. Later, low ionization potential and high dipole moment of water molecules can result in fast charge exchange:

$$N_2^+ + H_2O \rightarrow N_2 + H_2O^+, \quad k(300\ \text{K}) = 2.2 * 10^{-9}\,\text{cm}^3/\text{sec} \qquad (2.98)$$

The whole ionization process can be significantly focused on formation of water ions H_2O^+ even though the molar fraction of water in air is low. The generated water ions can then react with neutral water molecules in the quite fast ion–molecular reaction:

$$H_2O^+ + H_2O \rightarrow H_3O + OH, \ \Delta H = -12\ \text{kcal/mol}, \quad k(350\ \text{K}) = 0.5 * 10^{-9}\ \text{cm}^3/\text{sec} \qquad (2.99)$$

As a result, the production of H_3O^+ ions and OH radicals that provide the acidic behavior of the air plasma can be observed. The selective generation of OH radicals in nonthermal air discharges is the fundamental basis for employing discharges for purifying air from different pollutants.

2.6.4 Ion–Molecular Reactions with Rearrangement of Chemical Bonds

Reaction 2.99 demonstrates the acidic behavior of nonthermal air plasma. This class of ion–molecular processes consists of very different reactions, which can be subdivided into many groups, including:

- $(A)B^+ + C \rightarrow A + (C)B^+$, reactions with a positive ion transfer
- $A(B^+) + C \rightarrow (B^+) + AC$, reactions with a neutral transfer
- $A(B^+) + (C)D \rightarrow (A)D + (C)B^+$, double exchange reactions
- $A(B^+) + (C)D \rightarrow AC^+ + BD$, reconstruction processes

These groups of processes are shown with positive ions, but similar reactions take place with negative ions.

The special feature of the ion–molecular reactions that make them very distinctive from regular atom-molecular processes between neutral species can be stated as the following statement: **most exothermic ion-molecular reactions have no**

activation energy.[20] This is because quantum mechanical repulsion between molecules, which provides the activation barrier even in the exothermic reactions of neutrals, can be suppressed by the charge–dipole attraction in ion–molecular reactions. Thus, the rate coefficients are very large and can be found based on the Langevin relations (Equation 2.83 and Equation 2.85).

Not every Langevin capture leads to ion–molecular reactions with complicated rearrangement of chemical bonds or the orbital symmetry and spin-forbidden reactions. For this reason, the pre-exponential Arrhenius factors of the complicated exchange reaction, orbital symmetry, and spin forbidden reactions can be lower than the Langevin rate coefficients.[21,615] In general, though, the exothermic ion–molecular reactions with rearrangement of chemical bonds are much faster than the corresponding processes between neutrals. This effect is especially important at low temperatures when even small activation barriers can dramatically slow down exothermic reactions. An extensive listing of the ion–molecular reaction rate coefficients is presented in the monograph of Virin et al.[10]

2.6.5 Ion–Molecular Chain Reactions and Plasma Catalysis

The absence of activation energies in exothermic ion–molecular reactions facilitates organization of chain reactions in ionized media. Thus, for example, the important industrial process of SO_2 oxidation into SO_3 and sulfuric acid is highly exothermic; however, it cannot be arranged as an effective chain process without a catalyst because of the energy barriers of intermediate elementary reactions between neutral species. However, the long length SO_2 chain oxidation becomes effective in ionized media through the negative ion mechanism in water or in droplets in heterogeneous plasma.[22,23]

This chain oxidation process begins with the SO_2 dissolving and forming a negative ion SO_3^{2-} in the case of low acidity (pH > 6.5). Then a charge transfer process with a OH radical results in forming an active SO_3^- ion radical and subsequently initiating a chain reaction:

$$\mathrm{OH} + SO_3^{2-} \rightarrow \mathrm{OH}^- + SO_3^- \tag{2.100}$$

The active SO_3^- ion radical becomes the main chain-carrying particle, initiating the following room-temperature chain mechanism. The first reaction of chain propagation is attachment of an oxygen molecule:

$$SO_3^- + \mathrm{O}_2 \rightarrow SO_5^-, \; k = 2.5 * 10^{-12}\,\mathrm{cm}^3/\mathrm{sec} \tag{2.101}$$

Then the chain propagation can go through intermediate formation of SO_4^- ion radical:

$$SO_5^- + SO_3^{2-} \rightarrow SO_4^{2-} + SO_4^-, \; k = 5 * 10^{-14}\,\mathrm{cm}^3/\mathrm{sec} \tag{2.102}$$

$$SO_4^- + SO_3^{2-} \rightarrow SO_4^{2-} + SO_3^-, \; k = 3.3 * 10^{-12}\,\mathrm{cm}^3/\mathrm{sec} \tag{2.103}$$

or similarly through intermediate formation of SO_5^{2-} ion radical:

$$SO_5^- + SO_3^{2-} \rightarrow SO_5^{2-} + SO_3^-, \quad k = 1.7 * 10^{-14}\,\text{cm}^3/\text{sec} \tag{2.104}$$

$$SO_5^{2-} + SO_3^{2-} \rightarrow SO_4^{2-} + SO_4^{2-}, \quad k = 2 * 10^{-14}\,\text{cm}^3/\text{sec} \tag{2.105}$$

Considering these reactions, one should take into account that the ion SO_4^{2-} is a product of the SO_2 oxidation. This stable ion corresponds to the SO_3^- molecule in the gas phase and to sulfuric acid H_2S in a water solution. Here, the chain termination is due to losses of SO_3^- and SO_5^- and the chain length easily exceeds 10^3.

Details of the ion–molecular chain oxidation of SO_2 in clusters and small droplets in heterogeneous plasma will be considered later in the discussion of plasma application for purifying air from SO_2, NO_x, and volatile organic compounds (VOC). Here, it will just be concluded that plasma as an ionized medium permits avoiding activation energy barriers of elementary exothermic reactions and, as a result, can operate similar to traditional catalysts. This concept of **plasma catalysis**[24,25] is useful not only in processes of environment control, but also in organic fuel conversion, hydrogen production, and other applied plasma chemical technologies.

2.6.6 Ion–Molecular Processes of Cluster Growth: the Winchester Mechanism

From Section 2.6.1 it is clear that ion–molecular reactions are very favorable to clusterization due to the effective long-distance charge–dipole and polarization interaction leading to Langevin capturing. Indeed, both positive and negative ion–molecular reactions can make a fundamental contribution in the nucleation and cluster growth phases of dusty plasma generation in electric discharges[481,482] and even soot formation in combustion.

Specific details of cluster growth in nonthermal discharges will be discussed during consideration of the physics and chemistry of the dusty plasmas; here, the fundamental effects of ion–molecular processes of cluster growth will be pointed out. Besides the effect of Langevin capture (Section 2.6.1), the less known but very important **Winchester mechanism** of ion–molecular cluster growth needs to be considered.

A key point of the Winchester mechanism is thermodynamic advantage of the ion-cluster growth processes. If a sequence of negative ions defining the cluster growth is considered:

$$A_1^- \rightarrow A_2^- \rightarrow A_3^- \rightarrow \ldots \rightarrow A_n^- \rightarrow \ldots \tag{2.106}$$

the corresponding electron affinities EA_1, EA_2, EA_3, … EA_n, … are usually increasing, ultimately reaching the value of the work function (electron extraction energy), which is generally larger than the electron affinity for small molecules. As a result, each elementary step reaction of the cluster growth $A_n^- \rightarrow A_{n+1}^-$ has an *a priori* tendency to be exothermic. Exothermic ion–molecular reactions have no activation

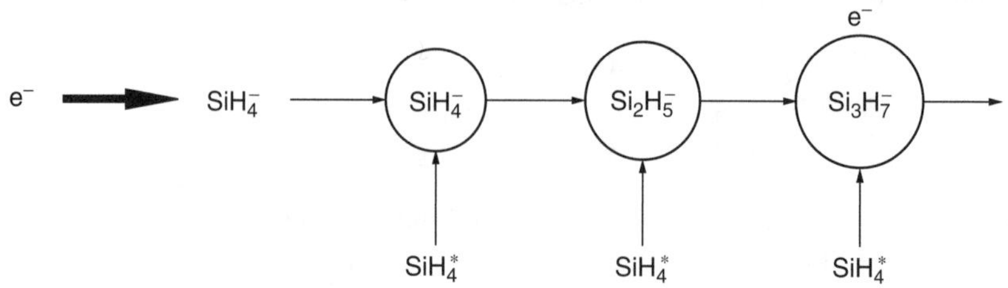

Figure 2.15 Ion–molecular reactions of cluster growth

barrier and are usually very fast; thus the Winchester mechanism explains effective cluster growth based on ion–molecular processes. The phenomenon is important until the cluster becomes too large and the difference in electron affinities $EA^-_{n+1} - EA_n$ becomes negligible.

Each elementary step $A^-_n \rightarrow A^-_{n+1}$ includes the cluster rearrangements with an electron usually going to the furthest end of the complex. This explains the origin of the term "Winchester," which sticks to the described mechanism of cluster growth. The Winchester mechanism is valid not only for negative but also for positive ion clusters. The thermodynamic advantage of ion-cluster growth for positive ions also holds. If a sequence of positive ions defining the cluster growth is considered:

$$A_1^+ \rightarrow A_2^+ \rightarrow A_3^+ \rightarrow \ldots \rightarrow A_n^+ \rightarrow \ldots \quad (2.107)$$

the corresponding ionization energies $I(A_1)$, $I(A_2)$, $I(A_3)$,…, $I(A_n)$,… are usually decreasing to ultimately reach the value of the work function (electron extraction energy), which is generally lower than ionization energy for small molecules. As a result, each elementary step reaction of the cluster growth $A^+_n \rightarrow A^+_{n+1}$ has an *a priori* tendency to be exothermic for positive ions as well. Thus, the Winchester mechanism explains effective cluster growth for positive and for negative ions.

As a specific example, one can mention a sequence of ion–molecular reactions defining an initial nucleation stage and cluster growth during the dusty plasma formation in low-pressure silane SiH_4 and silane–argon SiH_4–Ar discharges:[26–28]

$$Si_nH^-_{2n+1} + SiH_4 \rightarrow Si_{n+1}H^-_{2n+3} + H_2 \quad (2.108)$$

The nucleation process can be initiated by a dissociative attachment to a silane molecule: $e + SiH_4 + SiH_3^- + H$ ($k = 10^{-12}$ cm^3/sec at electron temperature $T_e = 2$ eV) and then continues by the sequence (Equation 2.108) of exothermic, thermoneutral, and slightly endothermic reactions as is illustrated in Figure 2.15.

Next, the thermal effects of the first four reactions of Equation 2.108 demonstrating the Winchester mechanism in silane plasma are given:

$$SiH_3^- + SiH_4 \rightarrow Si_2H_5^- + H_2 - 0.07 \text{ eV} \quad (2.109)$$

$$Si_2H_5^- + SiH_4 \rightarrow Si_3H_7^- + H_2 + 0.07 \text{ eV} \quad (2.110)$$

$$Si_3H_7^- + SiH_4 \rightarrow Si_4H_9^- + H_2 + 0.07 \text{ eV} \quad (2.111)$$

$$Si_4H_9^- + SiH_4 \rightarrow Si_5H_{11}^- + H_2 + 0.00 \text{ eV} \quad (2.112)$$

As can be seen, the Winchester mechanism shows only the tendency of energy effects on ion-cluster growth. Not all reactions of the sequence are exothermic; thermoneutral and even endothermic stages can be found. This results in "bottle neck" phenomena and other peculiarities of nucleation kinetics in low-pressure silane discharges, which will be discussed later.

PROBLEMS AND CONCEPT QUESTIONS

2.1 **Electron Energy Distribution Functions.** If electron energy distribution function is Maxwellian and the electron temperature is T_e = 1 eV, (1) what is the mean velocity $\langle v \rangle$ of the electrons in this case? (2) What is the mean electron energy? (3) Which electron energy $m\langle v \rangle^2/2$ corresponds to the mean velocity? (4) How can you explain the difference between the two "average electron energies" in terms of standard deviation?

2.2 **Ionization Potentials and Electron Affinities.** Why are most ionization potentials (Table 2.1) greater than electron affinities (Table 2.2)?

2.3 **Positive and Negative Ions.** Why is it possible to produce the double or multicharged positive ions such as A^{++} or A^{+++} in plasma, but impossible to generate in practical gas discharge plasmas the double or multicharged negative ions such as A^{--} or A^{---}?

2.4 **Mean Free Path of Electrons.** In which gases are the mean free paths of electrons with electron temperature T_e = 1 eV longer and why (helium, nitrogen, or water vapor at the similar pressure conditions)?

2.5 **Reaction Rate Coefficients.** Recalculate the Maxwellian electron energy distribution function into the electron velocity distribution function $f(v)$ and then find an expression for reaction rate coefficient (Equation 2.7) in the case when $\sigma = \sigma_0$ = const.

2.6 **Elastic Scattering.** Charged particle scattering on neutral molecules with permanent dipole momentum can be characterized by the following cross section dependence on energy: $\sigma(\varepsilon)$ = const/ε. How does the rate coefficient of the scattering depend on temperature in this case?

2.7 **Direct Ionization by Electron Impact.** Using Equation 2.18 for the general ionization function $f(x)$, find the electron energy corresponding to maximum value of direct ionization cross section. Compare this energy with the electron energy optimal for direct ionization according to the Thomson formula.

2.8 **Comparison of Direct and Stepwise Ionization.** Why does direct ionization make a dominant contribution in nonthermal electric discharges while in thermal plasmas, stepwise ionization is more important?

2.9 **Stepwise Ionization.** Estimate a stepwise ionization reaction rate in Ar at electron temperature 1 eV, assuming quasi equilibrium between plasma electrons and electronic excitation of atoms, and using Equation 2.29.

2.10 **Electron Beam Propagation in Gases.** How does the propagation length of a high energy electron beam depend on pressure at fixed temperature? How does it depend on temperature at a fixed pressure?

2.11 **Photoionization.** Why can the photoionization effect play the dominant role in propagation of nonthermal and thermal discharges in fast flows, including supersonic ones? What is the contribution of photoionization in the propagation of slow discharges?

2.12 **Massey Parameter.** Estimate the adiabatic Massey parameter for ionization of Ar atom at rest in collision with Ar^+ ion, having kinetic energy twice exceeding the ionization potential. Discuss the probability of ionization in this case.

2.13 **Ionization in an Ion–Neutral Collision.** How large is an H^+ atom energy supposed to be for this ion to reach a velocity the same as an electron velocity in a hydrogen atom? Compare this energy with the H-atom ionization potential. What is your conclusion about ionization process initiated by ion impact?

2.14 **Ionization in Collision of Excited Heavy Particles.** Why is ionization in the collision of vibrationally excited molecules usually much less effective than in a similar collision of electronically excited molecules?

2.15 **Ionization in Collision of Vibrationally Excited Molecules.** In principle, ionization can take place in collisions of nitrogen molecules having 32 vibrational quanta each. Is it possible to increase the probability of such ionization by increasing the vibrational energy of the collision partners? Is this increase possible by increasing electronic energy of the collision partners?

2.16 **Dissociative Electron–Ion Recombination.** According to Equation 2.44, a rate coefficient of dissociative recombination is reversibly proportional to gas temperature. However, in reality, when the electric field is fixed, this rate coefficient is decreasing faster than the reverse proportionality. How can you explain this?

2.17 **Ion Conversion Preceding Electron–Ion Recombination.** How large is the relative concentration of complex ions such as $N_2^+(N_2)$ supposed to be to compete with conventional electron recombination with N_2^+ ions in the same discharge conditions?

2.18 **Three-Body Electron–Ion Recombination.** Derive the reaction rate coefficient of the three-body recombination, using Equation 2.26 for the stepwise ionization rate coefficient and the Saha thermodynamic equation for ionization/recombination balance.

2.19 **Radiative Electron–Ion Recombination.** Give an example of discharge parameters (type of gas, range of pressures, plasma densities, etc.) when the radiative electron–ion recombination dominates between other mechanisms of losses of charged particles.

2.20 **Dissociative Attachment.** Calculate the dissociative attachment rate coefficients for molecular hydrogen and molecular oxygen at electron temperatures of 1 and 5 eV. Compare the results and give your comments.

2.21 **Negative Ions in Oxygen.** What are the discharge and gas parameters supposed to be if the negative oxygen ions are mostly generated by three-body attachment processes and not by dissociative attachment?

2.22 **Associative Detachment.** Calculate the relative concentration of CO in CO_2 discharge plasma if the main associative detachment process in this discharge dominates over dissociative attachment. Suppose that the electron temperature is 1 eV and the concentration of negative ions is 10 times less than positive ions.

2.23 **Detachment by Electron Impact.** Detachment of an electron from negative ion by electron impact can be considered "ionization" of the negative ion by electron impact. Why can the Thomson formula not be used in this case instead of Equation 2.64?

2.24 **Negative Ions in Thermal Plasma.** Why is concentration of negative ions in thermal plasmas usually very low even in electronegative gases? Which specific features of thermal plasmas with respect to nonthermal plasmas are responsible for this?

2.25 **Ion–Ion Recombination in Binary Collisions.** Determine how the ion–ion recombination rate coefficient depends on temperature. Use Equation 2.70 for the cross section of the binary collision recombination rate and assume the Maxwellian distribution function with temperature T_0 for positive and negative ions.

2.26 **Three-Body Ion–Ion Recombination.** Comparing Thomson's and Langevin's approaches for the three-body ion–ion recombination and correspondent rate coefficients (Equation 2.74 and Equation 2.62), find a typical value of pressure when the recombination reaches the maximum rate. Consider oxygen plasma with heavy particle temperature of 1 eV as an example.

2.27 **Langevin Cross Section.** Calculate the Langevin cross section for Ar^+ ion collision with an argon atom and with a water molecule. Comment on your result. Compare the contribution of the polarization term and the charge–dipole interaction term in the case of Ar^+ ion collision with a water molecule.

2.28 **Resonant Charge Transfer Process.** Is it possible to observe the resonant charge transfer process during an ion–neutral interaction with only a slight perturbation of trajectories of the collision partners?

2.29 **Tunneling Effect in Charge Transfer.** Estimate how strong the tunneling effect can be in resonant charge transfer in ion–neutral collision. For two different velocities, compare the cross section of charge exchange between an argon atom and argon positive ion calculated with and without taking into account the tunneling effect.

2.30 **Nonresonant Charge Exchange.** Most negative ions in nonthermal atmospheric pressure air discharges are in the form of O_2^-. What will happen to the negative ion distribution if even a very small amount of fluorine compounds is added to air?

2.31 **Plasma Catalytic Effect.** Compare the catalytic effect of plasma provided by negative ions, positive ions, active atoms, and radicals. All these species are usually quite active, but which one is the most generally relevant to the plasma catalysis?

2.32 **Winchester Mechanism.** Why is the ion mechanism of cluster growth usually less effective in combustion plasma during soot formation than in a similar situation in electrical discharge plasma?

CHAPTER

3

ELEMENTARY PROCESSES OF EXCITED MOLECULES AND ATOMS IN PLASMA

3.1 ELECTRONICALLY EXCITED ATOMS AND MOLECULES IN PLASMA

The extremely high chemical activity of a plasma is based on high and quite often superequilibrium concentrations of active species. The active species generated in a plasma include: chemically aggressive atoms and radicals; charged particles—electron and ions; and excited atoms and molecules. Elementary processes of atoms and radicals are traditionally considered in a framework of chemical kinetics;[29,30] elementary processes of charged particles were considered in Chapter 2 and elementary processes of excited species will be considered in this chapter.

Excited species, in particular vibrationally excited molecules, are of special importance in plasma chemical kinetics. This is primarily due to the fact that most of the discharge energy in molecular gases can be focused on vibrational excitation of molecules by electron impact. Often more than 95% of electron gas energy can be transferred to the vibrational excitation. For that reason, the vibrational excitation, relaxation, and reactions of vibrationally excited molecules are considered in the most careful way in this book and in plasma chemistry in general.

Excited species can be subdivided into three groups: electronically excited atoms and molecules; vibrationally excited molecules; and rotationally excited molecules. In the first two sections of this chapter, general features will be considered; then discussions will address excitation processes in plasma; and, in the final sections, elementary chemical reactions of excited and high-energy particles will be discussed.

3.1.1 Electronically Excited Particles: Resonance and Metastable States

High electron temperatures and thus high values of electron energies in electric discharges provide a high excitation rate of different electronically excited states of atoms and molecules by electron impact. Energy of the electronically excited particles usually is relatively high (about 5 to 10 eV), but their lifetime is generally very short (usually about 10^{-8} to 10^{-6} sec). If radiative transition to the ground state is not forbidden by quantum mechanical selection rules, such a state is called the **resonance excited state**. The resonance states have the shortest lifetime (about 10^{-8} sec) with respect to radiation and, therefore, their direct contribution in kinetics of chemical reactions in plasma is usually small.

If the radiative transition is forbidden by selection rules, the lifetime of the excited particles can be much longer because of absence of spontaneous transition. Such states are called the **metastable excited states**. The energy of the first resonance excited states, as well as the energy of the low-energy metastable states and their radiative lifetime, is given in Table 3.1 for some atoms of interest for plasma chemical applications.

Because of their long lifetime with respect to radiation, the metastable electronically excited atoms and molecules are able to accumulate the necessary discharge energy and significantly contribute to the kinetics of different chemical reactions in plasma. It should be noted that the metastable excited particles can lose their energy not only by radiation, but also by means of different collisional relaxation processes that will be discussed later in this chapter. As can be seen from the table, the radiative lifetime of metastable atoms can be very long (up to $1.4*10^5$ sec) and can effectively stimulate plasma chemical processes. This table also shows that the energy of excitation of the metastable states can be quite low (sometimes less than 1 eV); therefore, their concentration in electric discharges can be high.

3.1.2 Electronically Excited Atoms

Collision of a high-energy plasma electron with a neutral atom in a ground state can result in energy transfer from the free plasma electron to a bound electron in the atom. This is the main source of electronically excited atoms in plasma. The most important energy growth of a bound electron during excitation is due to an increase in the **principal quantum number *n***, but it also usually grows with the value of **angular momentum quantum number *l***.

This effect can be illustrated quantitatively in a case of strong excitation of any atom to a relatively high principal quantum number n. In this case, one electron is moving quite far from the nucleus and, therefore, moving in an almost coulomb potential. Such an excited atom is somewhat similar to a hydrogen atom. Its energy can be described by the **Bohr formula with the Rydberg correction** term Δ_l, which takes into account the short-term deviation from the coulomb potential when the electron approaches the nucleus:[31]

$$E = -\frac{me^4}{2\hbar^2(4\pi\varepsilon_0)^2}\frac{1}{(n+\Delta_l)^2} \tag{3.1}$$

Table 3.1 Lowest Electronically Excited States and Lowest Metastable States for Excited Atoms with Their Radiative Lifetimes

Atom and its ground state	First resonance excited states	Resonance energy	Low energy metastable states	Metastable energy	Metastable lifetime
He ($1s^2\,{}^1S_0$)	$2p\ {}^1P_1{}^0$	21.2 eV	$2s\ {}^3S_1$	19.8 eV	$2*10^{-2}$ sec
He($1s^2\,{}^1S_0$)			$2s\ {}^1S_0$	20.6 eV	$9*10^3$ sec
Ne ($2s^2p^6\,{}^1S_0$)	$3s\ {}^1P_1{}^0$	16.8 eV	$4s\ {}^3P_2$	16.6 eV	$4*10^2$ sec
Ne ($2s^2p^6\,{}^1S_0$)			$4s\ {}^3P_0$	16.7 eV	20 sec
Ar ($3s^2p^6\,{}^1S_0$)	$4s\ {}^2P_1{}^0$	11.6 eV	$4s\ {}^2P_{0,2}{}^0$	11.6 eV	40 sec
Kr ($4s^2p^6\,{}^1S_0$)	$5s\ {}^3P_1{}^0$	10.0 eV	$5s\ {}^3P_2$	9.9 eV	2 sec
Kr ($4s^2p^6\,{}^1S_0$)			$5s\ {}^3P_0$	10.6 eV	1 sec
H ($1s\,{}^2S_{1/2}$)	$2p\ {}^2P_{1/2,3/2}{}^0$	10.2 eV	$2s\ {}^2S_{1/2}$	10.2 eV	0.1 sec
N ($2s^2p^3\,{}^4S_{3/2}$)	$3s\ {}^4P_{1/2,3/2,5/2}{}^0$	10.3 eV	$2p^3\ {}^2D_{3/2}$	2.4 eV	$6*10^4$ sec
N ($2s^2p^3\,{}^4S_{3/2}$)			$2p^3\ {}^2D_{5/2}$	2.4 eV	$1.4*10^5$ sec
N ($2s^2p^3\,{}^4S_{3/2}$)			$2p^3\ {}^2P_{1/2}{}^0$	3.6 eV	40 sec
N ($2s^2p^3\,{}^4S_{3/2}$)			$2p^3\ {}^2P_{3/2}{}^0$	3.6 eV	$1.7*10^2$ sec
O ($2s^2p^4\,{}^3P_2$)	$3s\ {}^3S_1{}^0$	9.5 eV	$2p^4\ {}^3P_1$	0.02 eV	—
O ($2s^2p^4\,{}^3P_2$)			$2p^4\ {}^3P_0$	0.03 eV	—
O ($2s^2p^4\,{}^3P_2$)			$2p^4\ {}^1D_2$	2.0 eV	10^2 sec
O ($2s^2p^4\,{}^3P_2$)			$2p^4\ {}^1S_0$	4.2 eV	1 sec
Cl ($3s^2p^5\,{}^2P_{3/2}$)	$4s\ {}^2P_{1/2,3/2}{}^0$	9.2 eV	$3p^5\ {}^3P_{1/2}{}^0$	0.1 eV	10^2 sec

The Rydberg correction term $\Delta_l = 0$ in hydrogen atoms and Equation 3.1 become identical to the conventional Bohr formula. In general, Δ_l does not depend on the principal quantum number n, but depends on the angular momentum quantum number l of the excited electron as well as on the L and S momentum of the whole atom. If the L and S quantum numbers are fixed, the Rydberg correction term Δ_l decreases fast with increasing angular momentum quantum number l. When the quantum number l is growing, the excited electron spends less time in the vicinity of nucleus and the energy levels are closer to those of hydrogen. This effect can be demonstrated by numerical values of Δ_l for excited helium in the cases of $S = 0$ and $S = 1$ (see Table 3.2).

Table 3.2 Rydberg Correction Terms Δ_l for Excited Helium Atoms

	He ($S = 0$)	He ($S = 1$)
He ($l = 0$)	$\Delta_0 = -0.140$	$\Delta_0 = -0.296$
He ($l = 1$)	$\Delta_1 = +0.012$	$\Delta_1 = -0.068$
He ($l = 2$)	$\Delta_2 = -0.0022$	$\Delta_2 = -0.0029$

In general, energy levels of excited atoms depend not only on the principal quantum number n of the excited electrons and the **total angular orbital momentum** L, but also on the **total spin number** S and **total momentum quantum number** J. Thus, the triplet terms ($S = 1$) typically lie below the correspondent singlet terms ($S = 0$) and, therefore, the energy levels in corresponding excited atomic states are lower for the higher total spin numbers.

The total orbital angular momentum (corresponding quantum number L) and the total spin momentum (quantum number S) are coupled by weak magnetic forces. This results in splitting of a level with fixed values of L and S into groups of levels with different total momentum quantum numbers J from a maximum $J = L + S$ to a minimum $J = |L - S|$ (together, there are $2S + 1$ energy levels in the multiplet, if $S \leq L$). Also, the energy difference inside the multiplet between two levels $J + 1$ and J (with the same L and S) is proportional to $J + 1$.

It should be mentioned that the preceeding discussion of the energy hierarchy of electronically excited levels implies the usual case of the so-called ***L–S* coupling**. The orbital angular momenta of individual electrons in this case are strongly coupled among themselves; then they can be described by only the quantum number L. Similarly, individual electron spin momenta are presented together by total quantum number S. The L–S coupling cannot be applied to noble gases and heavy (and, therefore, larger) atoms when coupling between electrons decreases and electron spin–orbit interaction becomes predominant, resulting in the so-called ***j–j* coupling**.

Before discussing the selection rules for dipole radiation, it would be helpful to review the standard designation of atomic levels. As an example, take the ground state of atomic nitrogen from Table 3.1: N ($2s^2p^3\ {}^4S_{3/2}$). This designation includes information about individual outer electrons: $2s^2p^3$ means that, on the outer shell with the principal quantum number $n = 2$, there are two s-electrons ($l = 0$) and three p-electrons ($l = 1$). The next part of the designation, ${}^4S_{3/2}$, describes the outer electrons of the ground state atomic nitrogen collectively rather than individually. The left-hand superscript "4" in ${}^4S_{3/2}$ denotes the **multiplicity 2*S* + 1**, which corresponds, as mentioned before, to the number of energy levels in a multiplet, if $S \leq L$. From the multiplicity one can obviously determine the total spin number of the ground state atomic nitrogen $S = 3/2$. The upper-case letter S in ${}^4S_{3/2}$ denotes the total angular orbital momentum $L = 0$ (other values of $L = 1, 2, 3, 4$ correspond to upper-case letters P, D, F, and G). The right-hand subscript "3/2" in ${}^4S_{3/2}$ denotes the total momentum quantum number $J = 3/2$.

One can see from the Table 3.1 that some term symbols, such as the case of the first excited state of helium 2p ${}^1P_1{}^0$, also contain a right-hand superscript 0. This superscript is the designation of **parity**, which can be odd (superscript 0) or even (no right-hand superscript), depending on the odd or even value of the sum of angular momentum quantum numbers for individual electrons in the atom. For completed shells, the parity is obviously even.

To determine which excited states are metastable and which can be easily deactivated by dipole radiation, it is convenient to use **selection rules**. Actually, these rules indicate whether an electric dipole transition (and thus radiation emission or absorption) is allowed or forbidden. The selection rules allowing the dipole radiation in the case of *L–S* coupling are:

1. *The parity must change.* Thus, as seen from Table 3.1, ground states and first resonance excited states have different parity.
2. *The multiplicity must remain unchanged.* This rule can also be observed from Table 3.1, in which spin–orbit interaction is predominant, resulting not in *L–S* but *j–j* coupling. Note that this rule cannot be applied to noble gases.
3. *Quantum numbers* J *and* L *must change by +1, –1, or 0 (however, transitions 0 → 0 are forbidden).* This can also be seen in the correspondence between the ground and first resonance excited states presented in Table 3.1.

It is also interesting to check how these selection rules prohibit radiative transfers from metastable to ground states for the excited atoms considered in Table 3.1

3.1.3 Electronic States of Molecules and Their Classification

High-energy plasma electrons provide electronic excitation of molecules as well as atoms in electric discharges. In this case, energy transfer from the free plasma electron to a bound electron in a ground-state molecule results in excitation. The electronically excited molecules can be metastable and have long lifetimes and thus make an important contribution in plasma chemical kinetics.

Classification of electronically excited states of diatomic and linear polyatomic molecules is somewhat similar to that of atoms. In this case, molecular terms can be specified by the quantum number $\Lambda = 0, 1, 2, 3$ (corresponding Greek symbols $\Sigma, \Pi, \Delta, \Phi$), which describes the absolute value of the component of the total orbital angular momentum along the internuclear axis. If $\Lambda \neq 0$, the states are doubly degenerate because of two possible directions of the angular momentum component. Molecular terms are then specified by a quantum number S, which designates the total electron spin angular momentum and defines the multiplicity $2S + 1$. Similar with atomic terms, the multiplicity is written here as a prefixed superscript. Thus, the designation $^2\Pi$ means $\Lambda = 1, S = 1/2$. Note that the description and classification of nonlinear polyatomic molecules are more complicated, in particular because of absence of a single internuclear axis.

In the case of Σ-states (e.g., $\Lambda = 0$), to designate whether the wave function is symmetric or antisymmetric with respect to reflection at any plane, including the internuclear axis, the right-hand superscripts "+" and "–" are used. Furthermore, to designate whether the wave function is symmetric or antisymmetric with respect to the interchange of nuclei in homonuclear molecules such as N_2, H_2, O_2, F_2, etc., the right-hand subscripts *g* or *u* are written. Remember, this type of symmetry can be applied only for diatomic molecules. Thus, the molecular term designation $^1\Sigma_g^+$ means $\Lambda = 0$, $S = 0$, and the wave function is symmetric with respect to reflection at any plane including the internuclear axis and to interchange of nuclei.

Finally, to denote the normal ground-state electronic term, the upper-case *X* is usually written before the symbol of the term symbol. Upper-case letters *A*, *B*, *C*,

Table 3.3 Electronic Terms of Nonexcited Diatomic Molecules, Radicals, and Products of Their Dissociation

Molecules and radicals	Ground state term	Electronic states of dissociation products
C_2	$X^1\,\Sigma_g^+$	C(^{3}P) + C(^{3}P)
CF	$X^2\,\Pi$	C(^{3}P) + F(^{2}P)
CH	$X^2\,\Pi$	C(^{3}P) + H(^{2}S)
CN	$X^2\,\Sigma^+$	C(^{3}P) + N(^{4}S)
CO	$X^1\,\Sigma^+$	C(^{3}P) + O(^{3}P)
BF	$X^1\,\Sigma^+$	B(^{2}P) + F(^{2}P)
F_2	$X^1\,\Sigma_g^+$	F(^{2}P) + F(^{2}P)
H_2	$X^1\,\Sigma_g^+$	F(^{2}S) + F(^{2}S)
HCl	$X^1\,\Sigma^+$	H(^{2}S) + Cl(^{2}P)
F	$X^1\,\Sigma^+$	H(^{2}S) + F(^{2}P)
Li_2	$X^1\,\Sigma_g^+$	Li(^{2}S) + Li(^{2}S)
N_2	$X^1\,\Sigma_g^+$	N(^{4}S^0) + N(^{4}S^0)
NF	$X^3\,\Sigma^-$	N(^{4}S^0) +F(^{2}P)
NH	$X^3\,\Sigma^-$	N(^{4}S^0) +H(^{2}S)
NO	$X^2\,\Pi$	N(^{4}S^0) +O(^{3}P)
O_2	$X^3\,\Sigma_g^-$	O(^{3}P) + O(^{3}P)
OH	$X^2\,\Pi$	O(^{3}P) +H(^{2}S)
SO	$X^3\,\Sigma^-$	S(^{3}P) +O(^{3}P)

etc. before the main symbol denote the consequence of excited states having the same multiplicity as a ground state. Lower-case letters *a*, *b*, *c*, etc. before the main symbol denote, conversely, the consequence of excited states having multiplicity different from that of a ground state.

Electronic terms of ground states of some diatomic molecules and radicals are given in the Table 3.3, together with specification of electronic states of atoms, corresponding to dissociation of the diatomic molecules. From this table, the majority of chemically stable (saturated) diatomic molecules are seen to have completely symmetrical normal electronic state with $S = 0$. In other words, a ground electronic state for the majority of diatomic molecules is $X^1\Sigma^+$ or $X^1\Sigma_g^+$ for the case of homonuclear molecules. Exceptions to this rule include O_2 (normal term $X^3\Sigma_g^-$) and NO (normal term $X^2\Pi$). Obviously, this rule cannot be applied to radicals.

3.1.4 Electronically Excited Molecules and Metastable Molecules

Properties of electronically excited molecules and their contribution to plasma chemical kinetics depend upon whether these molecules are stable or not stable with respect to radiative and collisional relaxation (deactivation) processes. The collisional relaxation processes strongly depend on the degree of resonance during the collision and the corresponding Massey parameter considered in Section 2.2.7. However, the energy transfer for such a deactivation and the corresponding Massey parameter are usually very large and the probability of relaxation is low. Nevertheless, the deactivation of molecules can be effective in collision with chemically active atoms and radicals, as well as in surface collisions.

The most important factor defining the stability of excited molecules is radiation, which will be considered in more detail. Electronically excited molecules, as well as electronically excited atoms, can easily decay to a lower energy state by a photon emission if the corresponding transition is not forbidden by the selection rules. The selection rules for electric dipole radiation of excited molecules require:

$$\Delta\Lambda = 0, \pm 1; \quad \Delta S = 0; \tag{3.2}$$

For transitions between Σ-states, and for transitions in the case of homonuclear molecules, additional selection rules require:

$$\Sigma^{+} \rightarrow \Sigma^{+} \text{ or } \Sigma^{-} \rightarrow \Sigma^{-}, \text{ and } g \rightarrow u \text{ or } u \rightarrow g \tag{3.3}$$

If radiation is allowed, its frequency can be as high as 10^9 sec^{-1}; thus, in this case, the lifetime of electronically excited species is short. Some data on such lifetimes for diatomic molecules and radicals are given in Table 3.4 with the excitation energy of the corresponding states (this energy is taken between the lowest vibrational levels).

Table 3.4 Lifetimes and Energies of Electronically Excited Diatomic Molecules and Radicals on Lowest Vibrational Levels

Molecule or radical	Electronic state	Energy of the state	Radiative lifetime
CO	$A^1\Pi$	7.9 eV	$9.5*10^{-9}$ sec
C_2	$d^3\Pi_g$	2.5 eV	$1.2*10^{-7}$ sec
CN	$A^2\Pi$	1 eV	$8*10^{-6}$ sec
CH	$A^2\Delta$	2.9 eV	$5*10^{-7}$ sec
N_2	$b^3\Pi_g$	7.2 eV	$6*10^{-6}$ sec
NH	$b^1\Sigma^+$	2.7 eV	$2*10^{-2}$ sec
NO	$b^2\Pi$	5.6 eV	$3*10^{-6}$ sec
O_2	$A^3\Sigma_u^+$	4.3 eV	$2*10^{-5}$ sec
OH	$A^2\Sigma^+$	4.0 eV	$8*10^{-7}$ sec
H_2	$B^1\Sigma_u^+$	11 eV	$8*10^{-10}$ sec
Cl_2	$A^3\Pi$	2.1 eV	$2*10^{-5}$ sec
S_2	$B^3\Sigma_u^-$	3.9 eV	$2*10^{-8}$ sec
HS	$A^2\Sigma^+$	—	$3*10^{-7}$ sec
SO	$A^3\Pi$	4.7 eV	$1.3*10^{-8}$ sec

Table 3.5 Lifetimes and Energies of Metastable Diatomic Molecules[a]

Metastable molecule	Electronic state	Energy of the state	Radiative lifetime
N_2	$A^3\Sigma_u^+$	6.2 eV	13 sec
N_2	$a'^1\Sigma_u^-$	8.4 eV	0.7 sec
N_2	$a^1\Pi_g$	8.55 eV	$2*10^{-4}$ sec
N_2	$E^3\Sigma_g^+$	11.9 eV	300 sec
O_2	$a^1\Delta_g$	0.98 eV	$3*10^3$ sec
O_2	$b^1\Sigma_g^+$	1.6 eV	7 sec
NO	$a^4\Pi$	4.7 eV	0.2 sec

[a] On the lowest vibrational level.

In contrast to the resonance states, the electronically excited metastable molecules have very long lifetimes: seconds, minutes, and sometimes even hours, resulting in the very important role of the metastable molecules in plasma chemical kinetics. These very chemically reactive species can be generated in a discharge zone and then transported to a quite distant reaction zone. Information on excitation energies of some metastable molecules (with respect to the ground electronic state) and their radiative lifetimes is presented in Table 3.5. It should be noted that the discussed particles are metastable with respect to radiation, but in principle they can lose their excitation energy during different intermolecular relaxation processes and in relaxation collisions with different surfaces.

As seen from Table 3.5, the oxygen molecules have two very low-lying (energy about 1eV) metastable singlet states with long lifetimes. Such electronic excitation energy is already close to the typical energy of vibrational excitation and therefore the relative concentration of such metastable particles in electric discharges can be very high. Their contribution in plasma chemical kinetics is especially important, taking into account the very fast relaxation rate of vibrationally excited oxygen molecules.

The energy of each electronic state depends on the instantaneous configuration of nuclei in a molecule. In the simplest case of diatomic molecules, the energy depends only on the distance between two atoms. This dependence can be clearly presented by **potential energy curves**. These are very convenient to illustrate different elementary molecular processes like ionization, excitation, dissociation, relaxation, chemical reactions, etc.

The potential energy curves were already used in Chapter 2 to describe electron–molecular and other fundamental collisional processes (see, for example, Figure 2.5, Figure 2.6, Figure 2.10, Figure 2.13, and Figure 2.14). These figures are simplified to illustrate the curves of potential energy and to explain a process; real curves are much more sophisticated. Examples of real potential energy curves are given in Figure 3.1 for some diatomic molecules important in plasma chemical applications, e.g., H_2 (3.1a), N_2(3.1b), CO(3.1c), NO(3.1d), and O_2(3.1e). Their approximation by the Morse curve is presented in Figure 3.2, which will be discussed in Section 3.2.1.

Potential energy curves for different electronic states of polyatomic molecules are much more complicated and depend on many parameters describing the configuration of these molecules. Some, e.g., CO_2 molecular terms, will be discussed later during consideration of specific plasma chemical processes.

3.2 VIBRATIONALLY AND ROTATIONALLY EXCITED MOLECULES

When the electronic state of a molecule is fixed, the potential energy curve is fixed as well and determines the interaction between atoms of the molecules and their possible vibrations. The molecules also participate as a multibody system in rotational and translational motion. Quantum mechanics of the molecules shows that only discrete energy levels are permitted in molecular vibration and rotation. The structure of these levels for diatomic and polyatomic molecules will be considered later in connection with the potential energy curves.

It should be pointed out here that the vibrational excitation of molecules plays an essential and extremely important role in plasma physics and chemistry of molecular gases. First, this is because the largest portion of the discharge energy usually transfers from the plasma electrons primarily to excitation of molecular vibrations, and only after that to different channels of relaxation and chemical reactions. Several molecules, such as N_2, CO, H_2, and CO_2, can maintain vibrational energy without relaxation for a relatively long time, keeping vibrational energy without relaxation, and resulting in accumulating fairly large amounts of the energy that then can be selectively used in chemical reactions. Such vibrationally excited molecules are the most energy effective in the stimulation of dissociation and other endothermic chemical reactions. This emphasizes the special importance of vibrationally excited molecules in plasma chemistry and the attention paid to the physics and kinetics of these active species.

3.2.1 Potential Energy Curves for Diatomic Molecules: Morse Potential

The potential curve $U(r)$ corresponds closely to the actual behavior of the diatomic molecule if it is represented by the so-called **Morse potential**:

$$U(r) = D_0\left[1 - \exp\left(-\alpha\left(r - r_0\right)\right)\right]^2 \tag{3.4}$$

In this expression, r is a the distance between atoms in a diatomic molecule. Parameters r_0, α, and D_0 are usually referred to as the Morse potential parameters. The distance r_0 is the equilibrium distance between the nuclei in the molecule; the

(a) (b) (c) (d)

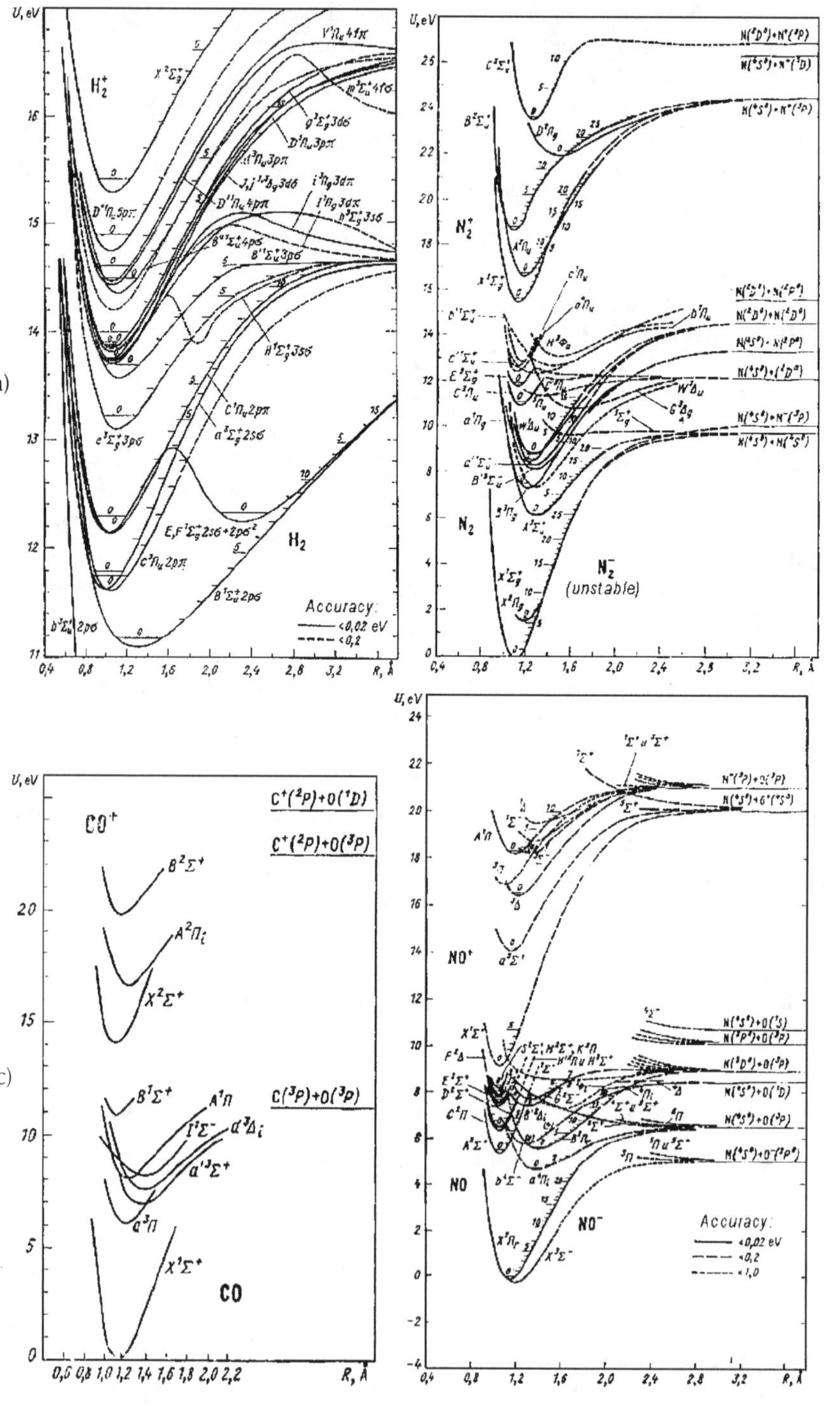

Figure 3.1 Potential energy curves of diatomic molecules: H_2 (a), N_2 (b), CO (c), NO (d), O_2 (e)

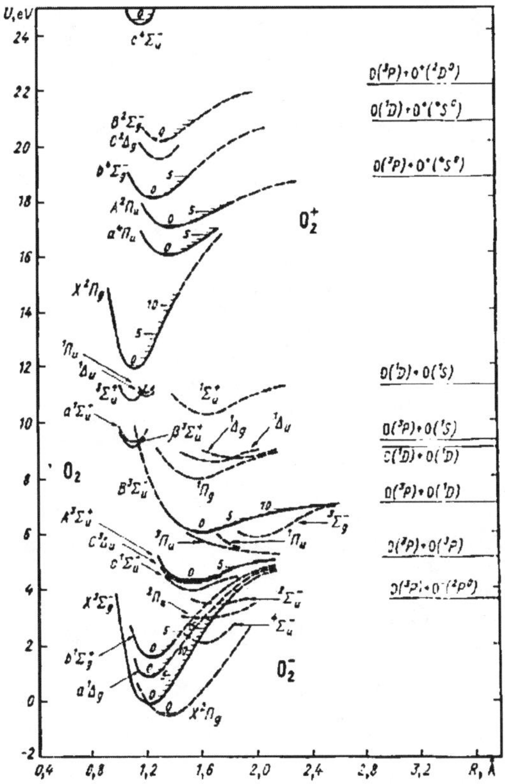

(e)

Figure 3.1 (continued) Potential energy curves of diatomic molecules: H_2 (a), N_2 (b), CO (c), NO (d), O_2 (e)

parameter α is a force coefficient of interaction between the nuclei; and the energy parameter D_0 is the difference between the energy of the equilibrium molecular configuration ($r = r_0$) and that corresponding to the free atoms ($r \to \infty$).

The Morse parameter D_0 is called the dissociation energy of a diatomic molecule with respect to the minimum energy (see Figure 3.2). As can be seen from this figure, the real dissociation energy D is less than the Morse parameter on the so-called value of "zero-vibrations," which is one half of a vibrational quantum $1/2\ \hbar\omega$:

$$D = D_0 - \frac{1}{2}\hbar\omega \tag{3.5}$$

This difference between D and D_0 is not large and often can be neglected. Nevertheless, it can be very important sometimes, in particular for isotopic effects in plasma chemical and general chemical kinetics, which will be discussed later. The Morse parameter D_0 is determined by the electronic structure of the molecule, and as a result it is the same for different isotopes. However, the vibrational quantum $\hbar\omega$ depends on the mass of the oscillator and thus is different for different isotopes. The difference in vibrational quantum results in differences in the real dissociation

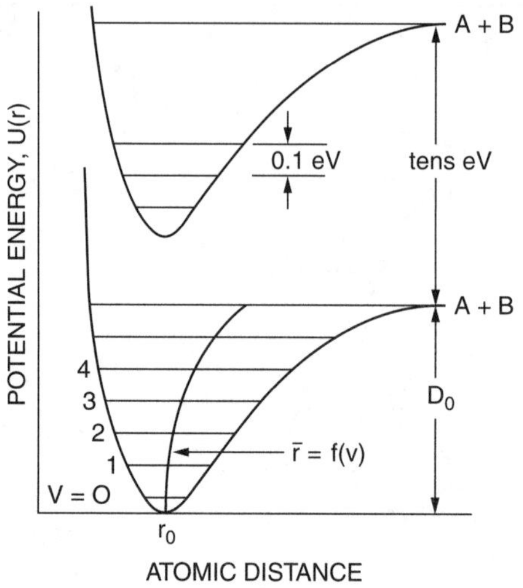

Figure 3.2 Potential energy diagram of a diatomic molecule *AB* in the ground and electronically excited states

energy D and therefore differences in the chemical kinetics of dissociation and reaction for different isotopes.

The part of the energy curve when $r > r_0$ corresponds to an attractive potential and that of $r < r_0$ corresponds to repulsion between nuclei. Near a $U(r)$ minimum point $r = r_0$, the potential curve is close to a parabolic one, $U = D_0\alpha^2(r - r_0)^2$, which corresponds to harmonic oscillations of the molecule. With energy growth, as is shown in Figure 3.2, the potential energy curve becomes quite asymmetric; the central line demonstrates the increase of an average distance between atoms, and the molecular vibration becomes anharmonic.

The Morse potential can be applied to describe the potential energy curve of diatomic molecules in many different specific problems. It is especially important in consideration of molecular vibration because it permits analytical description of energy levels of highly vibrationally excited molecules when the harmonic approximation can no longer be applied.

3.2.2 Vibration of Diatomic Molecules: Model of Harmonic Oscillator

Consider at first the potential curve of interaction between atoms in a diatomic molecule as a parabolic one: $U = D_0\alpha^2(r - r_0)^2$. This is an approximation of a diatomic molecule as a harmonic oscillator. As was previously mentioned, such an approximation is quite accurate for low-amplitude molecular oscillations. In this case, a quantum mechanical consideration predicts the following sequence of discrete vibrational energy levels for diatomic molecules:

$$E_v = \hbar\omega\left(v + \frac{1}{2}\right), \quad \omega = \alpha r_0\sqrt{\frac{2D_0}{I}} \tag{3.6}$$

Here, the vibrational energy E_v is taken with respect to the bottom of the potential curve; $\hbar$ is the Planck's constant;

$$I = \frac{M_1 M_2}{M_1 + M_2} r_0^2$$

is the momentum of inertia of the diatomic molecule with mass of nuclei M_1 and M_2; ω is the vibration frequency; and v is the vibrational quantum number of the molecule ($v = 0, 1, 2, 3, \ldots$).

The sequence of vibrational levels is shown in Figure 3.2. These levels are equidistant in the framework of the harmonic approximation; that is, the energy distance between them is constant and equals to the **vibrational quantum** $\hbar\omega$. The vibrational quantum is actually the smallest portion of vibrational energy that can be used during an energy transfer or exchange. The value of the quantum is fixed in the framework of harmonic approximation, but it changes (see Figure 3.2) from level to level in the case of an anharmonic oscillator.

One should also note that even the lowest vibrational level, $v = 0$, is located above the bottom of the potential curve. "No oscillation" maintains some energy there anyway. The level of the "zero vibrations" corresponds to $1/2\ \hbar\omega$ and defines, as mentioned earlier, the quantum mechanical effect on dissociation energy (Equation 3.5).

Now compare typical frequencies and energies (which are obviously proportional to the frequencies) for electronic excitation and vibrational excitation of molecules. Electronic excitation does not depend on the mass of heavy particles M (but only on electron mass m; see Equation 3.1). However, the second one—vibrational excitation (Equation 3.6)—is proportional to

$$\frac{1}{\sqrt{M}} \left(\omega \propto \frac{1}{\sqrt{I}} \propto \frac{1}{\sqrt{M}} \right)$$

(Equation 3.6). This means that a vibrational quantum is typically $\sqrt{m/M} \approx 100$ less than characteristic electronic energy (ionization potential I~10 to 20 eV). As a result, the typical value of a vibrational quantum, which is usually about 0.1 to 0.2 eV, can be easily calculated.

This typical value of vibrational quantum (about 0.1 to 0.2eV) occurs in a very interesting energy interval. On the one hand, this energy is relatively low with respect to typical electron energies in electric discharges (1 to 3 eV) so, for this reason, vibrational excitation by electron impact is very effective. On the other hand, the vibrational quantum energy is large enough to provide the high values of the Massey parameter (Equation 2.37) $P_{Ma} = \Delta E/\hbar\alpha v = \omega/\alpha v >> 1$ at relatively low gas temperatures, and to make vibrational deactivation (relaxation) in collision of heavy particles a slow, adiabatic process (see Section 2.2.7). As a result (at least in nonthermal discharges), the molecular vibrations are easy to activate and difficult to deactivate, making vibrationally excited molecules very special in different applications of plasma chemistry.

Table 3.6 Vibrational Quantum and Coefficient of Anharmonicity for Diatomic Molecules in Their Ground Electronic States

Molecule	Vibrational quantum	Coefficient of anharmonicity	Molecule	Vibrational quantum	Coefficient of anharmonicity
CO	0.27 eV	$6*10^{-3}$	Cl_2	0.07 eV	$5*10^{-3}$
F_2	0.11 eV	$1.2*10^{-2}$	H_2	0.55 eV	$2.7*10^{-2}$
HCl	0.37 eV	$1.8*10^{-2}$	HF	0.51 eV	$2.2*10^{-2}$
N_2	0.29 eV	$6*10^{-3}$	NO	0.24 eV	$7*10^{-3}$
O_2	0.20 eV	$7.6*10^{-3}$	S_2	0.09 eV	$4*10^{-3}$
I_2	0.03 eV	$3*10^{-3}$	B_2	0.13 eV	$9*10^{-3}$
SO	0.14 eV	$5*10^{-3}$	Li_2	0.04 eV	$5*10^{-3}$

Actual values of vibrational quantum for some diatomic molecules are presented in Table 3.6. From this table and in accordance with Equation 3.6, the lightest molecule H_2 is seen to have the highest oscillation frequency and thus the highest value of vibrational quantum $\hbar\omega = 0.55$ eV. Detailed information about all parameters characterizing the oscillations of diatomic molecules, including the vibration frequency and anharmonicity, can be found in Huber and Herzbert[33] and Radzig and Smirnov.[34]

3.2.3 Vibration of Diatomic Molecules: Model of Anharmonic Oscillator

The parabolic potential $U = D_0\alpha^2(r - r_0)^2$ and harmonic approximation (Equation 3.6) for vibrational levels are possible to use only for low vibrational quantum numbers sufficiently far from the level of dissociation. Equation 3.6 is even unable to explain the molecular dissociation; vibrational quantum number and vibrational energy can grow according to this formula without any limits. To solve this problem, the quantum mechanical consideration of oscillations of diatomic molecules should be done based on the Morse potential (Equation 3.4). This is an approximation of a diatomic molecule as an anharmonic oscillator. Treating it as an anharmonic oscillator, the discrete vibrational levels of the diatomic molecules, according to exact quantum mechanical considerations, have the following energies:

$$E_v = \hbar\omega\left(v + \frac{1}{2}\right) - \hbar\omega x_e\left(v + \frac{1}{2}\right)^2, \quad x_e = \frac{\hbar\omega}{4D_0} \tag{3.7}$$

These energies of anharmonic vibrational levels are taken in the same manner as previously with respect to the bottom of the potential curve. Equation 3.7 introduces a new important parameter of a diatomic molecule, x_e, which is a dimentionless coefficient of anharmonicity. As can be seen from Equation 3.7, the typical value of anharmonicity is $x_e \sim 0.01$. The actual values of anharmonicity for some diatomic molecules are also presented in Table 3.6.

Equation 3.7 for harmonic oscillators obviously coincides with Equation 3.6 for anharmonic oscillators; if the coefficient of anharmonicity is equal to zero, $x_e = 0$. Also, the energy distance between zero level and the first level of an anharmonic oscillator ($\Delta E_0 = \hbar\omega(1 - 2x_e)$) is fairly close to that of a harmonic oscillator, which has a vibrational quantum $\hbar\omega$.

Vibrational levels are equidistant for harmonic oscillators $\Delta E_v = \hbar\omega$, but this is not the case if anharmonicity is taken into account. For anharmonic oscillators, according to Equation 3.7, the energy distance $\Delta E_v(v, v + 1)$ between vibrational levels v and $v + 1$ becomes less and less with an increase of vibrational quantum number v:

$$\Delta E_v = E_{v+1} - E_v = \hbar\omega - 2x_e\hbar\omega(v+1) \tag{3.8}$$

Thus, taking into account the effect of anharmonicity, only a finite number of vibrational levels exists in a diatomic molecule. Equation 3.8 provides the possibility of determining the maximum value of the vibrational quantum number $v = v_{max}$, which corresponds to $\Delta E_v(v, v + 1) = 0$ and thus to dissociation of the diatomic molecule:

$$v_{max} = \frac{1}{2x_e} - 1 \tag{3.9}$$

Calculating the maximum possible vibrational energy $E_v(v = v_{max})$ based on Equation 3.7 also leads to a relation similar to Equation 3.7 among the vibrational quantum, coefficient of anharmonicity, and dissociation energy of a diatomic molecule. The distance between vibrational levels (Equation 3.8) can also be referred to as a "vibrational quantum," which obviously is not a constant, but a function of the vibrational quantum number. In this case, the smallest vibrational quantum corresponds to the energy difference between the last two vibrational levels, $v = v_{max} - 1$ and $v = v_{max}$:

$$\hbar\omega_{min} = \Delta E_v\left(v_{max} - 1, v_{max}\right) = 2x_e * \hbar\omega \tag{3.10}$$

This last vibrational quantum before the level of dissociation of a molecule is the smallest one, with a typical numerical value of about 0.003 eV. Corresponding Massey parameter (Equation 2.37) $P_{Ma} = \Delta E/\hbar\alpha v = \omega/\alpha v$ is also relatively low. This means that transition between high vibrational levels during collision of heavy particles is a fast, nonadiabatic process in contrast to adiabatic transitions between low vibrational levels (see Section 2.2.7). Thus, relaxation (deactivation) of highly vibrationally excited molecules is much faster than relaxation of molecules with the only quantum. This will be discussed in more detail later in the chapter.

Thus, the vibrationally excited molecules can be quite stable with respect to collisional deactivation. On the other hand, their lifetime with respect to spontaneous radiation is also relatively long. Corresponding to a transition between vibrational levels of the same electronic state, the electric dipole radiation is permitted for molecules having permanent dipole moments p_m. In the framework of the model of harmonic oscillator, in this case the selection rule requires $\Delta v = \pm 1$. However, the other transitions, $\Delta v = \pm 2, \pm 3, \pm 4, \ldots$, are also possible in the case of the anharmonic oscillator, although with a much lower probability.

The transitions allowed by the selection rule $\Delta v = \pm 1$ provide spontaneous infrared (IR) radiation. The radiative lifetime of vibrationally excited molecule (as well as electronically excited ones) can then be found according to the classical formula for an electric dipole, p_m, oscillating with frequency ω (here c is the speed of light):

$$\tau_R = 12\pi\varepsilon_0 \frac{\hbar c^3}{p_m^2} \frac{1}{\omega^3} \tag{3.11}$$

As can be seen from this relation, the radiative lifetime strongly depends on the oscillation frequency. As shown earlier in Section 3.2.2, the ratio of frequencies corresponding to vibrational excitation and electronic excitation is about $\sqrt{m/M} \approx$ 100. Then, taking into account Equation 3.11, the radiative lifetime of vibrationally excited molecules should be approximately $(M/m)^{\bar{2}} \approx 10^6$ times longer than that of electronically excited particles. Numerically, this means that even for transitions allowed by the selection rule, the radiative lifetime of vibrationally excited molecules is about 10^{-3} to 10^{-2} sec—quite long with respect to typical time of resonant vibrational energy exchange and some chemical reactions stimulated by vibrational excitation.

3.2.4 Vibrationally Excited Polyatomic Molecules: the Case of Discrete Vibrational Levels

The vibration of polyatomic molecules is much more complicated than that of diatomic molecules because of possible strong interactions between different vibrational modes inside multibody systems. This strong interaction takes place even at low excitation levels because of degenerate vibrations and intramolecular resonances. These questions attract special attention in molecular spectroscopy. At high excitation levels, most important for plasma chemistry, this interaction is due to anharmonicity and leads to a quasi continuum of vibrational states. Details about this sophisticated subject can be found in Herzenberg[35] and Rusanov et al.[36] Here, the discussion will only address the most important aspects concerning energy characteristics of vibrationally excited, triatomic molecules.

The nonlinear triatomic molecules have three vibrational modes (normal vibrations) with three frequencies, ω_1, ω_2, ω_3 (that, in general, are different), and without any degenerate vibrations. When the energy of vibrational excitation is relatively low, the interaction between the vibrational modes is not strong and the structure of vibrational levels is discrete. The relation for vibrational energy of such triatomic molecules at relatively low excitation levels is a generalization of the similar Equation 3.7 for a diatomic, anharmonic oscillator:

$$E_v(v_1, v_2, v_3) = \hbar\omega_1\left(v_1 + \frac{1}{2}\right) + \hbar\omega_2\left(v_2 + \frac{1}{2}\right) + \hbar\omega_3\left(v_3 + \frac{1}{2}\right) + x_{11}\left(v_1 + \frac{1}{2}\right)^2 +$$

$$x_{22}\left(v_2 + \frac{1}{2}\right)^2 + x_{33}\left(v_3 + \frac{1}{2}\right)^2 + x_{12}\left(v_1 + \frac{1}{2}\right)\left(v_2 + \frac{1}{2}\right) + \qquad (3.12)$$

$$x_{13}\left(v_1 + \frac{1}{2}\right)\left(v_3 + \frac{1}{2}\right) + x_{23}\left(v_2 + \frac{1}{2}\right)\left(v_3 + \frac{1}{2}\right)$$

The six coefficients of anharmonicity have energy units in this relation in contrast with the coefficients for diatomic molecules (Equation 3.7). Table 3.7 gives information about vibrations of some triatomic molecules, including their vibrational quanta and coefficients of anharmonicity, as well as type of symmetry, that describes the types of normal vibrations and, in particular, indicates presence of degenerate modes.

The types of molecular symmetry clarify peculiarities of vibrational modes of the triatomic molecules. Symbols of molecular symmetry groups, given in the second column of Table 3.7, indicate transformations of coordinates, rotations, and reflections that keep the Schroedinger equation unchanged for a triatomic molecule. The major groups of symmetry are explained below.

1. **The group C_{nv}** includes an nth-order axis of symmetry (symmetry with respect to rotation on angle $2\pi/n$ around the axis) and n planes of symmetry passing through the axis with an angle π/n between them. The subscript v in the symbol C_{nv} of the group reflects that these planes of symmetry are vertical. In particular, a group $C_{\infty v}$ means complete axial symmetry of a molecule and symmetry with respect to reflection in any plane passing through the axis. This group of symmetry ($C_{\infty v}$) describes linear triatomic molecules like COS or N_2O with different atoms from both sides of a central one. The group of symmetry C_{2v} describes different nonlinear molecules such as H_2O or NO_2 with identical atoms on both sides of a central one.
2. **The group C_{nh}** includes an nth-order axis of symmetry (symmetry with respect to rotation on angle $2\pi/n$ around the axis) and a plane of symmetry perpendicular to the axis. Similar to the preceding list item, a subscript h in the symbol C_{nh} of the group is a reminder that the plane of symmetry is horizontal. In the simplest case of $n = 1$, the group C_{nh} is usually denoted as C_s. The symmetry group C_s includes only symmetry with respect to a plane of a molecule. This group describes nonlinear molecules like HDO that are actually not symmetrical.
3. **The group D_{nh}** includes an nth-order principal axis of symmetry (symmetry with respect to rotation on angle $2\pi/n$ around the axis) and n second-order axes crossing at angle Π/n and perpendicular to the principal one. It also includes n vertical symmetry planes passing through the principal axis and a second-order one, and a horizontal symmetry plane containing all the n second-order axes (this horizontal plane is the reason for h in the subscript). In particular, a group $D_{\infty h}$ means complete axial symmetry of a molecule and symmetry with respect

Table 3.7 Parameters of Oscillations of Triatomic Molecules

Molecules and symmetry		Normal vibrations and their quanta, eV			Coefficients of anharmonicity, 10^{-3} eV					
Molec.	Sym.	ν_1	ν_2	ν_3	x_{11}	x_{22}	x_{33}	x_{12}	x_{13}	x_{23}
NO_2	C_{2v}	0.17	0.09	0.21	–1.1	–0.06	–2.0	–1.2	–3.6	–0.33
H_2S	C_{2v}	0.34	0.15	0.34	–3.1	–0.71	–3.0	–2.4	–11.7	–2.6
SO_2	C_{2v}	0.14	0.07	0.17	–0.49	–0.37	–0.64	–0.25	–1.7	–0.48
H_2O	C_{2v}	0.48	0.20	0.49	–5.6	–2.1	–5.5	–1.9	–20.5	–2.5
D_2O	C_{2v}	0.34	0.15	0.36	–2.7	–1.2	–3.1	–1.1	–10.6	–1.3
T_2O	C_{2v}	0.285	0.13	0.30	–1.9	–0.83	–2.2	–0.76	–7.5	–0.90
HDO	$C_{1h} = C_s$	0.35	0.18	0.48	–5.1	–1.5	–10.2	–2.1	–1.6	–2.5
HTO	$C_{1h} = C_s$	0.29	0.17	0.48	–3.6	–1.3	–10.2	–1.7	–1.3	–2.4
DTO	$C_{1h} = C_s$	0.29	0.14	0.35	–3.6	–0.88	–5.4	1.4	–0.97	–1.4
CO_2	$D_{\infty h}$	0.17	0.085	0.30	–0.47	–0.08	–1.6	0.45	–2.4	–1.6
CS_2	$D_{\infty h}$	0.08	0.05	0.19	–0.13	0.02	–0.64	0.1	–0.61	–0.83
COS	$C_{\infty v}$	0.11	0.06	0.26	–0.50	0.02	–1.4	–0.1	–0.28	–0.91
HCN	$C_{\infty v}$	0.26	0.09	0.43	–0.88	–0.33	–6.5	–0.31	–1.3	–2.4
N_2O	$C_{\infty v}$	0.28	0.07	0.16	–0.65	–0.02	–1.9	–0.06	–3.4	–1.8

to reflection in any plane passing through the axis, as well as symmetry with respect to reflection in a plane perpendicular to the axis. This group ($D_{\infty h}$) describes the most symmetrical linear triatomic molecules like CO_2 or CS_2 with identical atoms from both sides of a central one.

Triatomic molecules have altogether nine degrees of freedom related to different motions of their nuclei. In the general case of nonlinear molecules, these include three translational, three rotational, and three vibrational degrees of freedom. Linear triatomic molecules, however, have only two rotational degrees of freedom because of their axial symmetry (groups $C_{\infty v}$ and $D_{\infty h}$). For this reason, they have four vibrational degrees of freedom (four vibrational modes), but two of them are actually degenerated. These degenerated vibrational modes have the same frequency, but two different polarization planes passing through the molecular axis and perpendicular to each other. In other words, the linear triatomic molecules have three vibrational modes, but one mode is doubly degenerate.

As an example, a linear CO_2 molecule (as well as all other molecules of symmetry group $D_{\infty h}$) has three normal vibrational modes (illustrated in Figure 3.3): asymmetric valence vibration ν_3 (see Table 3.7); symmetric valence vibration ν_1; and a doubly degenerate symmetric deformation vibration ν_2. It should be noted that a resonance occurs in CO_2 molecules $\nu_1 \sim 2\nu_2$ between the two different types of symmetric vibrations (see Table 3.7). For this reason, symmetric modes are sometimes taken in plasma chemical calculations for simplicity as one triple degenerate vibration.

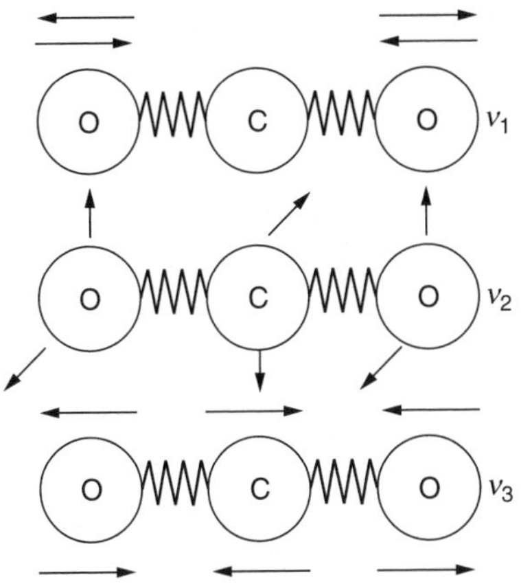

Figure 3.3 CO_2 vibrational modes

The degenerated symmetric deformation vibrations ν_2 can be polarized in two perpendicular planes (see Figure 3.3), which can result after summation in quasi rotation of the linear molecule around its principal axis. Angular momentum of the quasi rotation is characterized by the special quantum number l_2 that assumes the values:

$$l_2 = v_2,\ v_2 - 2,\ v_2 - 4,\ \ldots,\ 1 \text{ or } 0 \tag{3.13}$$

where v_2 is number of quanta on the degenerate mode.

A level of vibrational excitation of the linear triatomic molecule can then be denoted as $CO_2(v_1, v_2^{l_2}, v_3)$. For example, in the molecule $CO_2(0,2^2,0)$, only symmetric deformation vibrations are excited with effective "circular" polarization. The relation for vibrational energy of the linear triatomic molecules at relatively low excitation levels is somewhat similar to formula Equation 3.12 for nonlinear molecules. However, it obviously includes also the quantum number l_2 and correspondent coefficient g_{22}, which describe the polarization of the doubly degenerate symmetric deformation vibrations:

$$E_v(v_1, v_2, v_3) = \hbar\omega_1\left(v_1 + \frac{1}{2}\right) + \hbar\omega_2(v_2 + 1) + \hbar\omega_3\left(v_3 + \frac{1}{2}\right) + x_{11}\left(v_1 + \frac{1}{2}\right)^2 +$$
$$x_{22}(v_2 + 1)^2 + g_{22}l_2^2 + x_{33}\left(v_3 + \frac{1}{2}\right)^2 + x_{12}\left(v_1 + \frac{1}{2}\right)(v_2 + 1) + \tag{3.14}$$
$$x_{13}\left(v_1 + \frac{1}{2}\right)\left(v_3 + \frac{1}{2}\right) + x_{23}(v_2 + 1)\left(v_3 + \frac{1}{2}\right)$$

Numerically, the difference in energies between Equation 3.12 and Equation 3.14 is not so important in plasma chemical kinetics as the related effect of interaction between degenerate vibrations and rotations, which can significantly influence vibrational relaxation of polyatomic molecules.

3.2.5 Highly Vibrationally Excited Polyatomic Molecules: Vibrational Quasi Continuum

A good example of the vibrational quasi continuum is the vibrational excitation of CO_2 molecules to the level of their dissociation in nonthermal discharges.[37] Plasma electrons at an electron temperature of 1 to 2 eV excite mostly the lower vibrational levels of CO_2 and predominantly the asymmetric valence mode of the vibrations. The population of highly excited states usually involved in plasma chemical reactions occurs in the course of VV relaxation (resonance or close-to-resonance intermolecular exchange of vibrational energy).

At low levels of vibrational excitation, the VV exchange occurs independently along the different modes. The asymmetric modes of CO_2 are, however, better populated in plasma because they have higher rates of excitation by electron impact, higher rates of VV exchange, and lower rates of VT relaxation (quantum losses to the translational degrees of freedom). However, as the level of excitation increases, the vibrations of different types are collisionlessly mixed due to the intermode anharmonicity (Equation 3.12) and the Coriolis interaction.

The intramolecular VV′ quantum exchange results in the **vibrational quasi continuum** of the highly excited states of CO_2 molecules. Shchuryak[616] proposed the classical description of the transition in terms of beating. Energy of the asymmetric mode changes in the course of beating, corresponding to its interaction with symmetric modes ($\nu_3 \rightarrow \nu_1 + \nu_2$) and, as a result, the effective frequency of this oscillation mode changes as well:

$$\Delta\omega_3 \approx \left(x_{33}\omega_3\right)^{\frac{1}{3}}\left(A_0\omega_3 n_{sym}\right)^{\frac{2}{3}} \tag{3.15}$$

Here, $A_0 \sim 0.03$ is a dimensionless characteristic of the interaction between modes ($\nu_3 \rightarrow \nu_1 + \nu_2$) and n_{sym} is the total number of quanta on the symmetric modes (ν_1 and ν_2), assuming quasi equilibrium between modes inside a CO_2 molecule. As level of excitation n_{sym} increases, the value of $\Delta\omega_3$ grows and, at a certain critical number of quanta, $n_{sym}{}^{cr}$, it covers the defect of resonance $\Delta\omega$ of the asymmetric-to-symmetric quantum transition:

$$n_{sym}^{cr} \approx \frac{1}{\sqrt{x_{33}A_0^2}}\left(\frac{\Delta\omega}{\omega_3}\right)^{3/2} \tag{3.16}$$

When the number of quanta exceeds the critical value (Equation 3.14; fulfillment of the so-called Chirikov stochasticity criterion[38]), the motion becomes quasi random and the modes become mixed in the vibrational quasi continuum. For the general case of polyatomic molecules, the critical excitation level n^{cr} decreases with the number N of vibrational modes: $n^{cr} \approx 100/N^3$. This means that for molecules with four or more atoms, the transition to the vibrational quasi continuum takes place at a quite low level of excitation.

The polyatomic molecules in the state of vibrational quasi continuum can be considered as an infinitely high number of small oscillators. These oscillators can be characterized by the distribution function of the squares of their amplitudes $I(\omega)$, which is actually the vibrational Fourier spectrum of the system.[36] In this case:

$$I(\omega)\hbar d\omega = \frac{dE}{\omega} \tag{3.17}$$

is an adiabatic invariant (shortened action) for oscillators with energy dE in the frequency range from ω to $\omega+\Delta\omega$ and, according to the correspondence principle, is an analogue of the number of quanta for these oscillators.

If a polyatomic molecule is simulated by a set of harmonic oscillators with vibrational frequencies ω_{0i} and quantum numbers n_i, then the vibrational Fourier spectrum $I(\omega)$ of the system can be presented as a sum of δ-functions:

$$I(\omega) = \sum_i n_i \delta(\omega - \omega_{0i}) \tag{3.18}$$

Anharmonicity and interaction between modes causes the vibrational spectrum $I(\omega)$ to differ from Equation 3.16 in two basic ways. First, the anharmonicity can be considered as a shift of the fundamental vibration frequencies:

$$\omega_i(n_i) = \omega_{0i} - 2\omega_{0i}x_{0i}n_i \tag{3.19}$$

$$x_{0i} = \frac{1}{2}\sum_{j=1}^{j=N} x_{ij}(1+\delta_{ij})\frac{n_j}{n_i} \tag{3.20}$$

In these relations, x_{ij} are the anharmonicity coefficients; N is the number of vibrational modes; and δ_{ij} is the Kronecker δ-symbol. Second, the energy exchange between modes with characteristic frequency δ_i leads to a broadening of the vibrational spectrum lines[39,40] that can be described by the Lorentz profile:

$$I(\omega) = \frac{1}{\pi}\sum_i \frac{n_i\delta_i}{[\omega - \omega_i(n_i)]^2 + \delta_i^2} \tag{3.21}$$

When the third-order resonances prevail in the intermode exchange of vibrational energy (similar to the earlier case of CO_2 molecules), the frequency δ_i is growing with growth of vibrational energy as $\delta_i \sim E^{3/2}$. In the case of CO_2, the frequency δ_i can be simply estimated by the intermode anharmonicity $\delta \approx x_{23}n_3n_{sym}/h$.

With the growth of the number of atoms in a molecule, the vibrational energy of transition to the quasi continuum decreases (see discussion after Equation 3.14) and the line width at a fixed energy increases. It is also important to point out that the energy exchange frequency δ (at least for molecules like CO_2 and not the highest excitation levels) is less than the difference in fundamental frequencies. This means that, in spite of rapid mixing, the modes in the vibrational quasi continuum still keep their individuality.

3.2.6 Rotationally Excited Molecules

The rotational energy of a diatomic molecule with a fixed distance r_0 between nuclei can be found from the Schroedinger equation as a function of the rotational quantum number:

$$E_r = \frac{\hbar^2}{2I} J(J+1) = BJ(J+1) \tag{3.22}$$

Here, B is the rotational constant and

$$I = \frac{M_1 M_2}{M_1 + M_2} r_0^2$$

is the momentum of inertia of the diatomic molecule with mass of nuclei M_1 and M_2.

This relation obviously can be generalized to the case of polyatomic molecules, taking into consideration rotation around different principal axes according to corresponding type of symmetry (see Table 3.7). For example, the linear triatomic molecules with group of symmetry $D_{\infty h}$ or $C_{\infty v}$ (e.g., CO_2 or N_2O) have only one rotational constant B (because they have two rotational degrees of freedom with the same momentum of inertia). Alternately, the nonlinear triatomic molecules such as H_2O or H_2S have three rotational degrees of freedom with different values of momentum of inertia and the rotational constants (see Table 3.8).

Note also that the momentum of inertia and thus the correct rotational constant B are sensitive to a change of the distance r_0 between nuclei during vibration of a molecule. As a result, the rotational constant B (which must be used in Equation 3.22), for diatomic molecules, decreases with a growth of the vibrational quantum number:

$$B = B_e - \alpha_e \left(v + \frac{1}{2} \right) \tag{3.23}$$

In this relation, B_e is the rotational constant corresponding to the zero-vibration level and usually is presented in tables and handbooks (the subscript e stands here for "equilibrium"). Another molecular constant, α_e, is very small compared to B_e and describes the influence of molecular vibration on the momentum of inertia and rotational constant.

Estimate a typical numerical value of rotational constant and thus a characteristic value of the rotational energy. As discussed in Section 3.2.2, value of a vibrational quantum is typically $\sqrt{m/M} \approx 100$ less than characteristic electronic energy because a vibrational quantum is proportional to

$$\frac{1}{\sqrt{M}} \left(\omega \propto \frac{1}{\sqrt{I}} \propto \frac{1}{\sqrt{M}} \right),$$

Table 3.8 Rotational Constants B_e and α_e for a Ground State of Different Molecules

Molecule	B_e, 10^{-4} eV	α_e, 10^{-7} eV	Molecule	B_e, 10^{-4} eV	α_e, 10^{-7} eV
CO	2.39	21.7	HF	26.0	987
H_2	75.5	3796	HCl	12.9	381
F_2	1.09	—	N_2	2.48	21.5
NO	2.11	22.1	O_2	1.79	19.7
OH	23.4	885	NH	20.7	804
CO_2	0.484	—	CS_2	0.135	—
HCN	1.84	—	N_2O	0.520	—
COS	0.252	—	H_2O	A:34.6; B:18.0; C:11.5	—
H_2S	A:12.9; B:11.1; C:5.87	—	NO_2	A:9.92; B:0.53; C:0.51	—
O_3	A:4.40; B:0.55; C:0.48	—	SO_2	A:2.52; B:0.42; C:0.36	—

Note: Symbols A, B, and C denote different rotational modes in nonlinear triatomic molecules.

where M is a typical mass of heavy particles and m is an electron mass. In a similar way, the values of rotational constant B and rotational energy are proportional to

$$\frac{1}{M}\left(B \propto \frac{1}{I} \propto \frac{1}{M}\right).$$

This means that the value of the rotational constant is typically $\sqrt{m/M} \approx 100$ less than value of a vibrational quantum (which is about 0.1 eV) and numerically is about 10^{-3} eV (or even 10^{-4} eV). These rotational energies in temperature units (1 eV = 11,600 K) correspond to about 1 to 10 K; that is the reason why molecular rotation levels are already well populated even at room temperature (in contrast to molecular vibrations). The actual values of the rotational constants B_e and α_e for some diatomic and triatomic molecules are presented in Table 3.8.

The selection rules for rotational transition in molecular spectra require $\Delta J = -1, 0,$ or $+1$, taking into account that transitions from $J = 0$ to $J = 0$ are not allowed (the different possibilities of coupling, usually under detailed consideration in molecular spectroscopy, are not discussed here). Because collisional processes also prefer a transfer of the only rotational quantum (if any), this will be examined in some more detail.

As one can see from relation Equation 3.22, the energy levels in the rotational spectrum of a molecule are not equidistant. For this reason, the rotational quantum—an energy distance between the consequent rotational levels—is not a constant. In contrast to the case of molecular vibrations, the rotational quantum is growing with the increase of quantum number J and thus with the growth of rotational energy of a molecule. The value of the rotational quantum (in the simplest case of fixed distance between nuclei) can be easily found from Equation 3.22:

$$E_r(J+1) - E_r(J) = \frac{\hbar^2}{2I} 2(J+1) = 2B(J+1) \tag{3.24}$$

As shown earlier, the typical value of the rotational constant is 10^{-3} to 10^{-4} eV, which means that, at room temperatures, the quantum number J is about 10. As a result, in this case even the largest rotational quantum is relatively small, about $5*10^{-3}$eV. In contrast to the vibrational quantum, the rotational one corresponds to low values of the Massey parameter (Equation 2.37) $P_{Ma} = \Delta E/\hbar\alpha v = \omega/\alpha v$, even at low gas temperatures. This means that the energy exchange between rotational and translational degrees of freedom is a fast nonadiabatic process (see Section 2.2.7). Consequently, the rotational temperature of molecular gas in a plasma is usually very close to the translational temperature, even in nonequilibrium discharges, while vibrational temperatures can be significantly higher.

3.3 ELEMENTARY PROCESSES OF VIBRATIONAL, ROTATIONAL, AND ELECTRONIC EXCITATION OF MOLECULES IN PLASMA

Generation of excited particles in plasma is mostly due to the direct electron impact to be considered in this section. Although other mechanisms generating excited species, such as recombination of charged particles or neutral species and exothermic chemical reactions, normally are less important in plasma chemistry, some of these processes will be discussed later in this chapter. Photoexcitation also can be quite significant and will be considered in detail in Chapter 5 with other radiation processes.

3.3.1 Vibrational Excitation of Molecules by Electron Impact

As has been already mentioned, vibrational excitation is probably the most important elementary process in nonthermal molecular plasma—responsible for the major portion of energy exchange between electrons and molecules and also contributing significantly in the kinetics of nonequilibrium chemical processes in plasma. The elastic collisions of electrons and molecules are not effective in the process of vibrational excitation because of the significant difference in their masses ($m/M << 1$). This can be easily evaluated. Typical energy transfer from an electron with kinetic energy ε to a molecule in the elastic collision is about $\delta(m/M)$; vibrational quantum is as discussed in the previous section: $\hbar\omega \approx I\sqrt{m/M}$. For this reason, the classical cross section of vibrational excitation in an elastic electron–molecular collision can be estimated as:

$$\sigma_{vib}^{elastic} \approx \sigma_0 \frac{\varepsilon}{I}\sqrt{\frac{m}{M}} \tag{3.25}$$

In this relation, $\sigma_0 \sim 10^{-16}\,cm^2$ is the gas-kinetic cross section and I is an ionization potential. This gives numerical values of the cross section of vibrational excitation $\sim 10^{-19}\,cm^2$ for electron energies of about 1 eV. Quantum mechanical calculations of the vibrational excitation cross section during elastic scattering result in a similar expression with the selection rule $\Delta v = 1$.

However, numerous experiments show that vibrational excitation cross sections can be much larger, e.g., about the same as atomic ones ($10^{-16} cm^2$). Also, these cross sections are nonmonotonic functions of energy and the probability of multiquantum excitation is not very low. Schultz presented a detailed review of these cross sections in which information indicated that vibrational excitation of a molecule AB from vibrational level v_1 to v_2 is usually not a direct elastic process, but rather a resonance one, proceeding through formation of an intermediate nonstable negative ion:[41]

$$AB(v_1) + e \xleftrightarrow{\Gamma_{1i},\Gamma_{i1}} AB^-(v_i) \xrightarrow{\Gamma_{i2}} AB(v_2) + e \tag{3.26}$$

In this relation, v_i is the vibrational quantum number of the nonstable negative ion and $\Gamma_{\alpha\beta}$ (measured in sec^{-1}) are the probabilities of corresponding transitions between different vibrational states. Cross sections of the resonance vibrational excitation process (Equation 3.26) can be found in the quasi steady-state approximation using the Breit–Wigner formula:

$$\sigma_{12}(v_i, \varepsilon) = \frac{\pi\hbar^2}{2m\varepsilon}\frac{g_{AB^-}}{g_{AB}g_e}\frac{\Gamma_{1i}\Gamma_{i2}}{\frac{1}{\hbar^2}(\varepsilon - \Delta E_{1i})^2 + \Gamma_i^2} \tag{3.27}$$

Here, ε is the electron energy; ΔE_{1i} is energy of transition to the intermediate state $AB(v_1) \rightarrow AB^-(v_i)$; g_{AB^-}, g_{AB}, and g_e are corresponding statistical weights; and Γ_i is the total probability of $AB^-(v_i)$ decay through all channels.

Equation 3.27 illustrates the resonance structure of vibrational excitation cross section dependence upon electron energy. The energy width of the resonance spikes in this function is approximately $\hbar\Gamma_i$, which is related to the lifetime of the nonstable, intermediate, negative ion $AB^-(v_i)$. The maximum value of the cross section (Equation 3.27) is approximately the atomic value ($10^{-16} cm^2$). To analyze the vibrational excitation cross section further as a function of energy, one should consider separately the three cases corresponding to different lifetimes of the intermediate ionic states (the so-called "resonances").

3.3.2 Lifetime of Intermediate Ionic States during Vibrational Excitation

First, consider the so-called **short-lifetime resonance**. In this case, the lifetime of the autoionization states $AB^-(v_i)$ is much shorter than the period of oscillation of a molecule's nuclei ($\tau \ll 10^{-14}$ sec). The energy width of the autoionization level $\sim\hbar\Gamma_i$ is very large for the short-lifetime resonances in accordance with the uncertainty principle. This results in relatively wide maximum spikes (usually several electronvolts) in

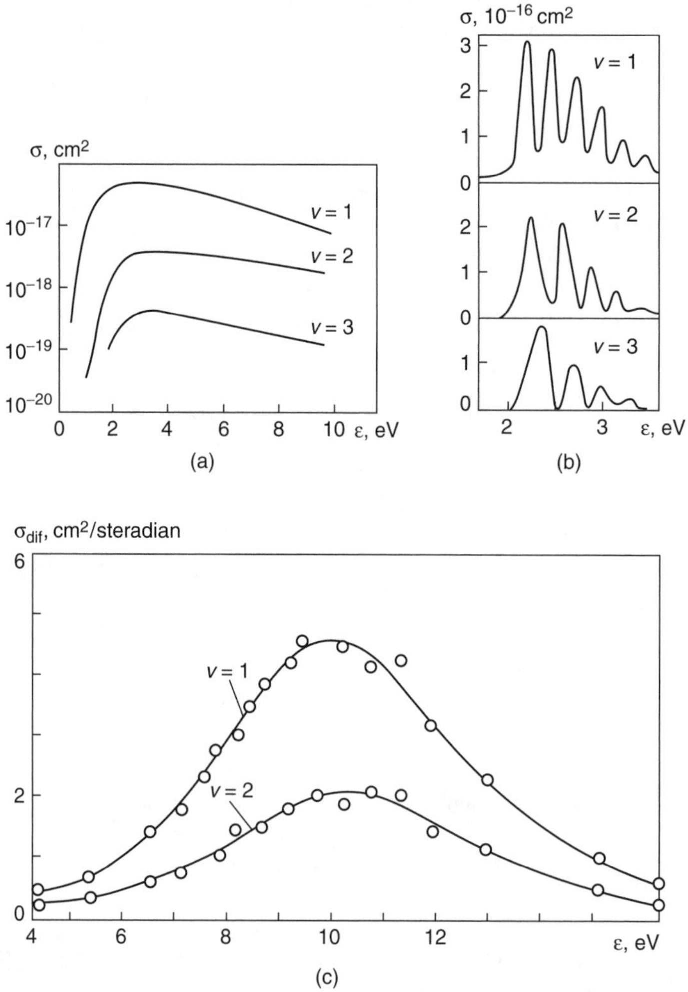

Figure 3.4 Cross sections of vibrational excitation as a function of electron energy in the case of (a) short-lifetime resonances (H_2, v = 1, 2, 3); (b) boomerang resonances (N_2, v = 1, 2, 3); (c) long-lifetime resonances (O_2, v = 1, 2, 3).[41]

the dependence of the vibrational excitation cross section on electron energy. There is no fine energy structure of $\sigma_{12}(\varepsilon)$ in this case, as can be seen in Figure 3.4.

Because of the short lifetime of the autoionization state $AB^-(v_i)$, displacement of nuclei during the lifetime period is small. As a result, decay of the nonstable, negative ion leads to excitation mostly to low vibrational levels. Vibrational excitation through intermediate formation of short-lifetime resonances is observed in particular in such molecules as H_2, N_2O, and H_2O. Electron energies corresponding to the most effective vibrational excitation of these molecules, as well as the maximum

Table 3.9 Cross Sections of Vibrational Excitation of Molecules by Electron Impact

Molecule	Most effective electron energy	Maximum cross section	Molecule	Most effective electron energy	Maximum cross section
N_2	1.7–3.5 eV	$3*10^{-16}cm^2$	NO	0–1 eV	$10^{-17}cm^2$
CO	1.2–3.0 eV	$3.5*10^{-16}cm^2$	NO_2	0–1 eV	—
CO_2	3–5 eV	$2*10^{-16}cm^2$	SO_2	3–4 eV	—
C_2H_4	1.5–2.3 eV	$2*10^{-16}cm^2$	C_6H_6	1.0–1.6 eV	—
H_2	~3 eV	$4*10^{-17}cm^2$	CH_4	Thresh. 0.1 eV	$10^{-16}cm^2$
N_2O	2–3 eV	$10^{-17}cm^2$	C_2H_6	Thresh. 0.1 eV	$2*10^{-16}cm^2$
H_2O	5–10 eV	$6*10^{-17}cm^2$	C_3H_8	Thresh. 0.1 eV	$3*10^{-16}cm^2$
H_2S	2–3 eV	—	Cyclo-Propane	Thresh. 0.1 eV	$2*10^{-16}cm^2$
O_2	0.1–1.5 eV	$10^{-17}cm^2$	HCl	2–4 eV	$10^{-15}cm^2$

cross sections of the process, are presented in Table 3.9. Cross sections of the lowest resonances are presented in this table; after integration over the electron energy distribution function, these usually contribute most to the vibrational excitation rate coefficient.

If an intermediate ion lifetime is approximately a molecular oscillation period ($\sim 10^{-14}$sec), such a resonance is usually referred to as the **boomerang resonance.** In particular, this boomerang type of vibrational excitation takes place for low-energy resonances in N_2, CO, and CO_2. The boomerang model, developed by Herzenberg,[35] considers formation and decay of the negative ion during one oscillation as interference only of the coming and reflected waves. The interference of the nuclear wave packages results in an oscillating dependence of excitation cross section on the electron energy with typical spikes period about 0.3 eV (see Figure 3.4 and Table 3.9).

Boomerang resonances usually require larger electron energies for excitation of higher vibrational levels. For example, the excitation threshold of N_2 ($v = 1$) is 1.9 eV, and that of N_2 ($v = 10$) from ground state is about 3 eV. Excitation of CO ($v = 1$) requires a minimal electron energy of 1.6 eV, while the threshold for CO ($v = 10$) excitation from the ground state is about 2.5 eV. The maximum value of the vibrational excitation cross section for boomerang resonances usually decreases with growth in the vibrational quantum number v.

The **long-lifetime resonance** corresponds to autoionization states $AB^-(v_i)$ with much longer lifetimes ($\tau = 10^{-14}$ to 10^{-10} sec) than a period of oscillation of the nuclei in a molecule. In particular, this type of vibrational excitation takes place for low-energy resonances in such molecules as O_2, NO, and C_6H_6. The long-lifetime resonances usually result in quite narrow isolated spikes (about 0.1 eV) in cross-section dependence on electron energy (see Figure 3.4 and Table 3.9). In contrast to the boomerang resonances, the maximum value of the vibrational excitation cross section remains the same for different vibrational quantum numbers.

Table 3.10 Rate Coefficients of Vibrational Excitation of Molecules by Electron Impact

Molecule	$T_e = 0.5$ eV	$T_e = 1$ eV	$T_e = 2$ eV
H_2	$2.2*10^{-10}$ cm^3/sec	$2.5*10^{-10}$ cm^3/sec	$0.7*10^{-9}$ cm^3/sec
D_2	—	—	10^{-9} cm^3/sec
N_2	$2*10^{-11}$ cm^3/sec	$4*10^{-9}$ cm^3/sec	$3*10^{-8}$ cm^3/sec
O_2	—	—	10^{-10}–10^{-9} cm^3/sec
CO	—	—	10^{-7} cm^3/sec
NO	—	$3*10^{-10}$ cm^3/sec	—
CO_2	$3*10^{-9}$ cm^3/sec	10^{-8} cm^3/sec	$3*10^{-8}$ cm^3/sec
NO_2	—	—	10^{-10}–10^{-9} cm^3/sec
H_2O	—	—	10^{-10} cm^3/sec
C_2H_4	—	10^{-8} cm^3/sec	—

3.3.3 Rate Coefficients of Vibrational Excitation by Electron Impact: Semiempirical Fridman's Approximation

Table 3.9 shows that the electron energies most effective in vibrational excitation are from 1 to 3 eV, which usually corresponds to the maximum in the electron energy distribution function. The vibrational excitation rate coefficients, which are the results of integration of the cross sections over the electron energy distribution function, are obviously very large and reach 10^{-7} cm^3/sec in this case. Such rate coefficients are given in Table 3.10 for different diatomic or polyatomic molecules and different electron temperatures T_e. As one can see from this table, the excitation rate coefficients are quite large. For such molecules as N_2, CO, CO_2, almost each electron–molecular collision leads to vibrational excitation at $T_e = 2$ eV. This explains why such a large fraction of electron energy from nonthermal discharges in several gases goes mostly into the vibrational excitation at electron temperatures of $T_e = 1$ to 3 eV

Vibrational excitation by electron impact is preferably a one-quantum process. Nevertheless, excitation rates of multiquantum vibrational excitation can also be important. Detailed kinetic information about electron impact excitation rate coefficients $k_{eV}(v_1, v_2)$ for excitation of molecules from an initial vibrational level v_1 to a final level v_2 is very important in numerous plasma chemical and laser problems. Unfortunately, this information is quite limited because it is complicated to measure the rate coefficients $k_{eV}(v_1, v_2)$ experimentally as well as to calculate them theoretically.

For this reason, the following semiempirical **Fridman's approximation for multiquantum vibrational excitation** can be very useful. It was first applied for different diatomic and polyatomic gases.[9,37] This semiempirical approximation determines the relationship between the one-quantum and the multiquantum vibrational excitation stimulated by electron impact. The approach permits finding the excitation rate $k_{eV}(v_1, v_2)$ based on the much better known vibrational excitation rate coefficient

Table 3.11 Parameters of Multiquantum Vibrational Excitation by Electron Impact

Molecule	a	b	Molecule	a	b
N_2	0.7	0.05	H_2	3	—
CO	0.6	—	O_2	0.7	—
$CO_2(v_3)$	0.5	—	NO	0.7	—

$k_{eV}(0, 1)$ corresponding to the excitation from ground state to the first vibrational level (see Table 3.10). According to the Fridman's approximation:

$$k_{eV}(v_1, v_2) = k_{eV}(0,1)\frac{\exp(-\alpha(v_2 - v_1))}{1+\beta v_1} \tag{3.28}$$

Parameters α and β for different gases, which are necessary for numerical calculations based on Equation 3.28, are summarized in Table 3.11. The approximative nature of the approach must be stressed. Thus, sometimes in practical plasma chemical modeling, it is helpful to consider the parameters α and β as functions of electron temperature. Also, in nonthermal discharges, the non-Maxwellian behavior of the electron energy distribution function should be taken into account to modify Equation 3.28.

In general, the Fridman's approximation of the multiquantum vibrational excitation by electron impact and Equation 3.28 shows the following physical statements:

- Plasma electrons are more effective in excitation of low vibrational levels.
- The one-quantum vibrational excitation process is the most probable one.
- The probability of multiquantum vibrational excitation by electron impact decreases exponentially with the number of quanta transferred.

3.3.4 Rotational Excitation of Molecules by Electron Impact

If electron energies exceed values of about 1 eV (see Table 3.9), the rotational excitation can proceed resonantly through the autoionization state of a negative ion as in the case of vibrational excitation. However, the relative contribution of this multistage rotational excitation is small, taking into account the low value of rotational quantum with respect to the vibrational one. The rotational excitation can be relatively important in electron energy balances at lower electron energies when the resonant vibrational excitation is already ineffective, but rotational transitions are still possible by long-distance electron–molecular interaction.

The nonresonant rotational excitation of molecules by electron impact can be illustrated using the classical approach. Typical energy transfer from an electron with kinetic energy ε to a molecule in an elastic collision, inducing rotational excitation, is about $\varepsilon(m/M)$. Typical distance between rotational levels, rotational quantum, is $I(m/M)$, where I is the ionization potential. Thus, the classical cross section of the nonresonant rotational excitation can be related to the gas-kinetic collisional cross section with $\sigma_0 \sim 10^{-16}$ cm^2 in the following manner:

$$\sigma^{elastic}_{rotational} \approx \sigma_0 \frac{\varepsilon}{I} \tag{3.29}$$

Comparing this relation with Equation 3.25, it can be seen that the cross section of the nonresonant rotational excitation can exceed that of the nonresonant vibrational excitation by a factor of 100.

A quantum mechanical approach leads to similar results and conclusions. An electron collision with a dipole molecule induces rotational transitions with a change of the rotational quantum number $\Delta J = 1$. Quantum mechanical cross sections of rotational excitation of a linear dipole molecule by a low-energy electron can then be calculated as:[42]

$$\sigma(J \to J+1, \varepsilon) = \frac{d^2}{3\varepsilon_0 a_0 \varepsilon} \frac{J+1}{2J+1} \frac{\sqrt{\varepsilon}+\sqrt{\varepsilon'}}{\sqrt{\varepsilon}-\sqrt{\varepsilon'}} \tag{3.30}$$

In this relation, d is a dipole moment; a_0 is the Bohr radius; $\varepsilon' = \varepsilon - 2B(J+1)$ is the electron energy after collision; and B is the rotational constant (see Table 3.8). Numerically, this cross section is approximately 1 to $3{*}10^{-16}$ cm^2 when the electron energy is about 0.1 eV.

Homonuclear molecules such as N_2 or H_2 have no dipole moment, and any rotational excitation of such molecules is due to electron interaction with their quadrupole moment Q. In this case, the rotational transition takes place with the change of rotational quantum number $\Delta J = 2$. The cross section of the rotational excitation by a low-energy electron is obviously lower in this case and can be calculated as:[43]

$$\sigma(J \to J+2, \varepsilon) = \frac{8\pi Q^2}{15e^2 a_0^2} \frac{(J+1)(J+2)}{(2J+1)(2J+3)} \ln\sqrt{\frac{\varepsilon}{\varepsilon'}} \tag{3.31}$$

Numerically, this cross section of rotational excitation of the homonuclear molecules by electron impact is usually in the order of 1 to $3{*}10^{-17}$ cm^2 when the electron energy is about 0.1 eV.

In general, the rotationally excited molecules are usually much less important in nonthermal plasma chemistry than the vibrationally excited species. This is primarily due to the low value of rotational quantum and thus very fast "nonadiabatic" energy exchange leading to equilibrium of molecular rotations with the translational degrees of freedom. The rotational and translational degrees of freedom are quite often considered together in plasma chemical kinetics. Thus, to evaluate the elastic or quasi-elastic energy transfer from the electron gas to neutral molecules, the rotational excitation can be taken into account combined with the elastic collisions. The process is then characterized by the gas-kinetic rate coefficient $k_{e0} \approx \sigma_0 \langle v_e \rangle \approx$

$3 \cdot 10^{-8}$ cm³/sec ($\langle v_e \rangle$ is the average thermal velocity of electrons) and each collision is considered a loss of about $\varepsilon(m/M)$ of electron energy.

3.3.5 Electronic Excitation of Atoms and Molecules by Electron Impact

In contrast to the vibrational and rotational excitation processes discussed previously, electronic excitation by electron impact usually needs large electron energies ($\varepsilon >$ 10 eV). The **Born approximation** can be applied to calculate the processes' cross sections when electron energies are large enough. For excitation of optically permitted transitions from an atomic state i to another state k, the Born approximation gives the following process cross section:

$$\sigma_{ik}(\varepsilon) = 4\pi a_0^2 f_{ik} \left(\frac{Ry}{\Delta E_{ik}} \right)^2 \frac{\Delta E_{ik}}{\varepsilon} \ln \frac{\varepsilon}{\Delta E_{ik}} \tag{3.32}$$

In this relation Ry is the Rydberg constant; a_0 is the Bohr radius; f_{ik} is the force of oscillator for the transition $i \rightarrow k$; and ΔE_{ik} is energy of the transition. The relation is valid in the framework of the Born approximation, e.g., for high electron energies $\varepsilon >> \Delta E_{ik}$.

The cross sections of electronic excitation of molecules by electron impact should be known for practical calculations and modeling in a wide range of electron energies, starting from the threshold of the process. In this case, semiempirical formulas can be very useful; two important ones were introduced by Drawin:[44,45]

$$\sigma_{ik}(\varepsilon) = 4\pi a_0^2 f_{ik} \left(\frac{Ry}{\Delta E_{ik}} \right)^2 \frac{x-1}{x^2} \ln(2.5x) \tag{3.33}$$

and by Smirnov:[7a]

$$\sigma_{ik}(\varepsilon) = 4\pi a_0^2 f_{ik} \left(\frac{Ry}{\Delta E_{ik}} \right)^2 \frac{\ln(0.1x + 0.9)}{x - 0.7} \tag{3.34}$$

In both relations, $x = \varepsilon/\Delta E_{ik}$; obviously, the semiempirical formulas correspond to the Born approximation at high electron energies ($x >> 1$).

Electronic excitation of molecules to the optically permitted states follows relations similar to Equation 3.32 through Equation 3.34, but additionally include the Frank–Condon factors (see Section 2.2.3) and internuclear distances. Numerically, the maximum values of cross sections for excitation of all optically permitted transitions are about the size of atomic gas-kinetic cross sections, $\sigma_0 \sim 10^{-16}$ cm². To reach the maximum cross section of the electronic excitation, the electron energy should be greater than the transition energy ΔE_{ik} by a factor of two.

The dependence $\sigma_{ik}(\varepsilon)$ is quite different for the excitation of electronic terms from which optical transitions (radiation) are forbidden by selection rules (see Section 3.1). In this case, the exchange interaction and details of electron shell structure become important. As a result, the maximum cross section, which is also

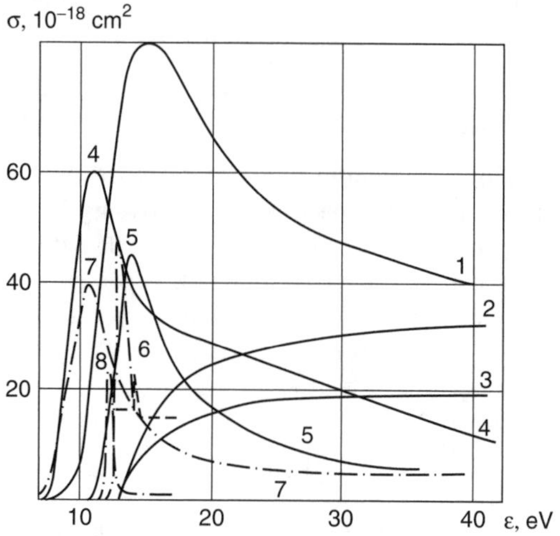

Figure 3.5 Cross section of excitation of different electronic states of N_2 ($X^1\Sigma_q^+$, v = 0) by electron impact: (1) $a^1\Pi_g$; (2) $b^1\Pi_u$ ($v_k = 0-4$); (3) transitions 12, 96 eV; (4) $B^3\Pi_g$; (5) $C^3\Pi_u$; (6) $a''^1\Sigma_g^+$; (7) $A^3\Sigma_u^+$; (8) $E^3\Sigma_g^+$.[94]

about the size of the atomic one, $\sigma_0 \sim 10^{-16}$ cm^2, can be reached at much lower electron energies $\varepsilon/\Delta E_{ik} \approx 1.2-1.6$. This leads to an interesting effect of predominant excitation of the optically forbidden and metastable states by electron impact in nonthermal discharges, where electron temperature T_e is usually much less than the transition energy ΔE_{ik}. Obviously, this effect requires availability of the optically forbidden and metastable states at relatively low energies. Some cross sections of electronic excitation by electron impact are presented in Figure 3.5; detailed information on the subject can be found in the monograph of Slovetsky.[94]

3.3.6 Rate Coefficients of Electronic Excitation in Plasma by Electron Impact

The rate coefficients of electronic excitation can be calculated by integration of the cross sections $\sigma_{ik}(\varepsilon)$ over the electron energy distribution functions, which, in the simplest case, are Maxwellian (see Section 2.1.1). Because electron temperature is usually much less than the electronic transition energy ($T_e << \Delta E_{ik}$), the rate coefficient of the process is exponential in the same manner as in the case of ionization by direct electron impact (see Section 2.2):

$$k_{el.excit.} \propto \exp\left(-\frac{\Delta E_{ik}}{T_e}\right) \tag{3.35}$$

For numerical calculations and modeling, it is convenient to use the following semiempirical relation between the electronic excitation rate coefficients and the

Table 3.12 Parameters for Semiempirical Approximation of Rate Coefficients of Electronic Excitation and Ionization of CO_2 and N_2 by Electron Impact

Molecule	Excitation level or ionization	C_1	C_2*10^{16}, $V*cm^2$	Molecule	Excitation level or ionization	C_1	C_2*10^{16}, $V*cm^2$
N_2	$A^3\Sigma_u^+$	8.04	16.87	N_2	$c^1\Pi_u$	8.85	34.0
N_2	$B^3\Pi_g$	8.00	17.35	N_2	$a^1\Pi_u$	9.65	35.2
N_2	$W^3\Delta_u$	8.21	19.2	N_2	$b'^1\Sigma_u^+$	8.44	33.4
N_2	$B'^3\Sigma_u^-$	8.69	20.1	N_2	$c^3\Pi_u$	8.60	35.4
N_2	$a'^1\Sigma_u^-$	8.65	20.87	N_2	$F^3\Pi_u$	9.30	32.9
N_2	$a^1\Pi_g$	8.29	21.2	N_2	Ionization	8.12	40.6
N_2	$W^1\Delta_u$	8.67	20.85	CO_2	$^3\Sigma_u^+$	8.50	10.7
N_2	$C^3\Pi_u$	8.09	25.5	CO_2	$^1\Delta_u$	8.68	13.2
N_2	$E^3\Sigma_g^+$	9.65	23.53	CO_2	$^1\Pi_g$	8.84	14.8
N_2	$a''^1\Sigma_g^+$	8.88	26.5	CO_2	$^1\Sigma_g^+$	8.23	18.9
N_2	$b^1\Pi_u$	8.50	31.88	CO_2	Other levels	8.34	20.9
N_2	$c'^1\Sigma_u^+$	8.56	35.6	CO_2	Ionization	8.38	25.5

reduced electric field E/n_0 proposed by Kochetov and Pevgov.[47] The same formula can be applied to calculate the ionization rate coefficient as well:

$$\log k_{el.excit.} = -C_1 - \frac{C_2}{E/n_0} \tag{3.36}$$

In this numeric relation, the rate coefficient $k_{el.excit.}$ is expressed in cm^3/sec; E is an electric field strength measured in V/cm; and n_0 is number density measured in 1/cm^3. The numerical values of the parameters C_1 and C_2 for different electronically excited states (and also ionization) of CO_2 and N_2 are presented in Table 3.12.

Equation 3.36 implies that the electron energy distribution function is not disturbed by vibrational excitation of the molecule in plasma. If the level of vibrational excitation of molecules is quite high, superelastic collisions provide higher electron energies and higher electronic excitation rate coefficients for the same value of the reduced electric field E/n_0. This effect will be discussed in detail in Section 4.3, considering the influence of vibrational excitation on the electron energy distribution functions in nonthermal plasmas. It can be taken into account for calculating the electronic excitation rate coefficients by including two special numerical terms related to vibrational temperature T_v in Equation 3.36:

$$\log k_{el.excit.} = -C_1 - \frac{C_2}{E/n_0} + \frac{40z + 13z^2}{\left[(E/n_0)\cdot 10^{16}\right]^2} - 0.02\left(\frac{T_v}{1000}\right)^{2/3} \tag{3.37}$$

In this numeric relation, the vibrational temperature T_v is to be expressed in degrees Kelvin. The Boltzman factor z can be found as:

$$z = \exp\left(-\frac{\hbar\omega}{T_v}\right) \tag{3.38}$$

Equation 3.37 and Equation 3.38 give a good approximation (accuracy about 20%) for the electronic excitation rate coefficients with reduced electric fields $5*10^{-16}$ V*cm^2 < E/n_0 < $30*10^{-16}$ V*cm^2 and vibrational temperatures less than 9000 K. At low vibrational temperatures, $z << 1$ and Equation 3.37 obviously coincides with the simplified numerical formula (Equation 3.36).

3.3.7 Dissociation of Molecules by Direct Electron Impact

Electron impacts are able to stimulate dissociation of molecules by vibrational and electronic excitation processes. Vibrational excitation by electron impact, however, usually results in initial formation of molecules with only one quantum or few quanta. Dissociation takes place in this case as a nondirect multistep process, including energy exchange (VV relaxation) between molecules to collect an amount of vibrational energy sufficient for dissociation. Such processes are effective only for limited (but very important) groups of gases such as N_2, CO_2, H_2, and CO and will be discussed later. In contrast, the dissociation through electronic excitation of molecules can proceed in just one collision; therefore, it is referred to as stimulated by the direct electron impact. The dissociation through electronic excitation can be observed with any kind of diatomic and polyatomic molecules making this process specifically important in many plasma chemical applications.

The dissociation through electronic excitation can proceed as an elementary process by different mechanisms or through different intermediate steps of intramolecular transitions. These mechanisms are illustrated in Figure 3.6.

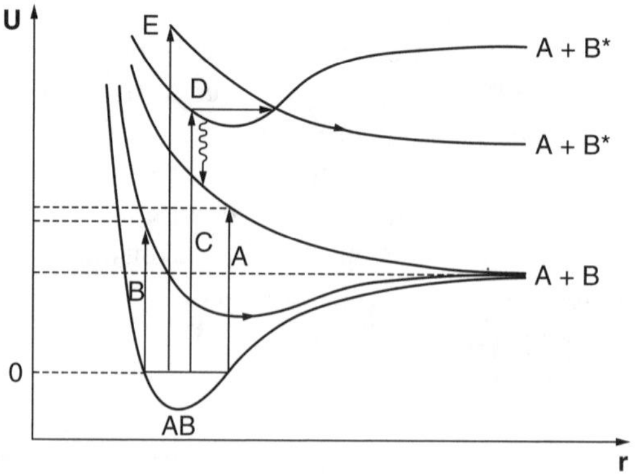

Figure 3.6 Mechanisms of dissociation of molecules through electronic excitation

- *Mechanism A* begins with the direct electronic excitation of a molecule from ground state to a repulsive state, followed by dissociation. In this case, the required electron energy can significantly (few electronvolts) exceed the dissociation energy. This mechanism generates hot (high-energy) neutral fragments that, for example, could significantly affect surface chemistry in low-pressure nonthermal discharges.
- *Mechanism B* includes the direct electronic excitation of a molecule from ground state to an attractive state with energy exceeding the dissociation threshold. The excitation of the state then results in the following dissociation. As one can see from Figure 3.6, the energy of the dissociation fragments is lower in this case.
- *Mechanism C* consists in the direct electronic excitation of a molecule from ground state to an attractive state corresponding to electronically excited dissociation products. The excitation of the state can lead to radiative transition to a low-energy repulsive state (see Figure 3.6) followed by dissociation. Energy of the dissociation fragments in this case is similar to that of mechanism A.
- *Mechanism D* (similar to mechanism *C*) starts with the direct electronic excitation of a molecule from ground state to an attractive state corresponding to electronically excited dissociation products. In contrast to the previous case, excitation of the state leads to nonradiative transfer to a highly excited repulsive state (see Figure 3.6) followed by dissociation into electronically excited fragments. This mechanism is usually referred to as predissociation.
- *Mechanism E* is similar to mechanism A and consists of direct electronic excitation of a molecule from ground state to a repulsive state, but with dissociation into electronically excited fragments. This mechanism requires the largest values of electron energies and, therefore, the corresponding rate coefficients are relatively low.

Cross sections of the dissociation of different molecules by direct electron impact are presented as a function of electron energy in Figure 3.7. Some rate coefficients of the process are given in Figure 3.8 as a function of electron temperature.

3.3.8 Distribution of Electron Energy in Nonthermal Discharges between Different Channels of Excitation and Ionization

Previously, the principal channels of energy transfer from electrons to neutrals during inelastic collisions were considered. In thermal plasmas, these energy exchange processes can be quite close to local equilibrium. In nonthermal discharges, contributions of superelastic collisions are relatively low. As a result, most electron energy initially received from an electric field can be simply distributed between elastic energy losses and different channels of excitation and ionization. The corresponding distribution function shows the relative role of different active species; therefore, it is very important in nonequilibrium plasma chemistry. Such distributions of electron energy in nonthermal discharges as a function of reduced electric field E/n_0 in different atomic and molecular gases[9] are presented in Figure 3.9 through Figure 3.16.

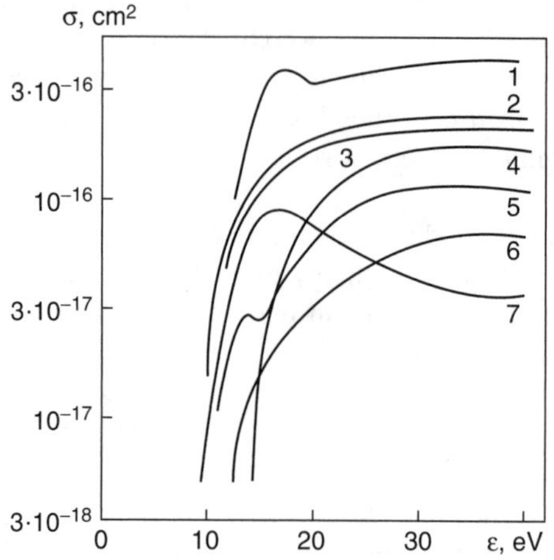

Figure 3.7 Cross sections of dissociation of molecules through electronic excitation as a function of electron energy: (1) CH_4; (2) O_2; (3) NO; (4) N_2; (5) CO_2; (6) CO; (7) H_2

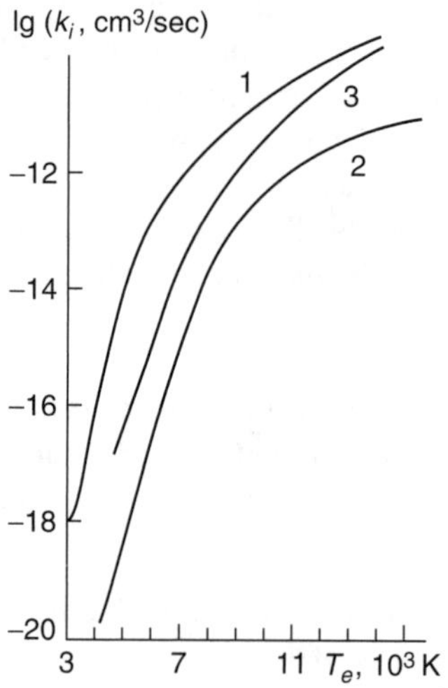

Figure 3.8 (1) Rate coefficients of stepwise N_2 electronic state excitation; (2) direct dissociation of initially nonexcited molecules N_2; (3) N_2 dissociation in stepwise electronic excitation sequence

It should be noted that all the energy distributions between different excitation channels presented in these figures were calculated numerically. Such self-consistent calculations should take into account the influence of different elastic, inelastic, and superelastic collisions on electron energy distribution function $f(\varepsilon)$. On the other hand, this electron energy distribution function $f(\varepsilon)$ should then be applied to calculate the rate coefficients of all the mentioned elastic, inelastic, and superelastic processes. For example, electron excitation rate coefficients in Figure 3.15 and Figure 3.16 were calculated in this quite sophisticated, self-consistent way.

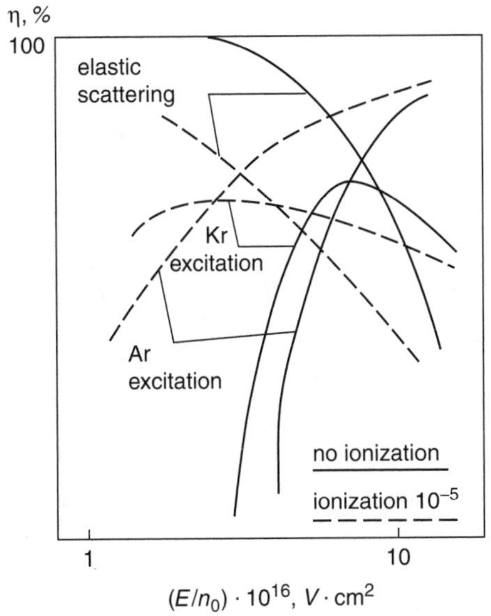

Figure 3.9 Electron energy distribution between excitation channels in Kr (5%)–Ar (95%) mixture

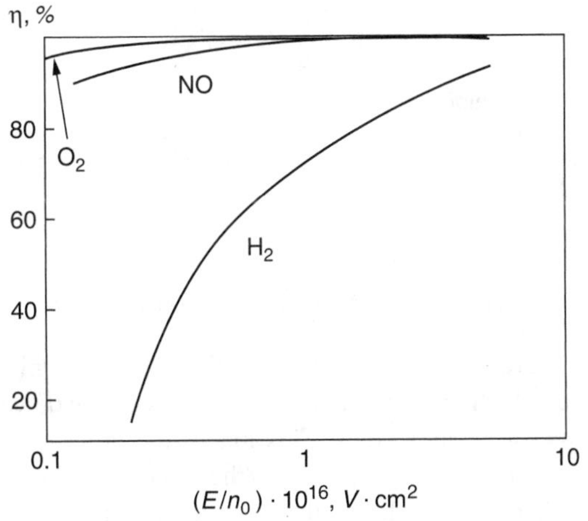

Figure 3.10 Fraction of electron energy spent on vibrational excitation of O_2, NO, and H_2

As can be seen from Figure 3.9 through Figure 3.16, all the energy distributions among different excitation channels have similar general features. For example, contribution of rotational excitation of molecules and elastic energy losses is significant only at low values of the reduced electric field E/n_0, and thus at low electron temperatures. This is expected since these processes are nonresonant and take place at low electron energies (<<1 eV).

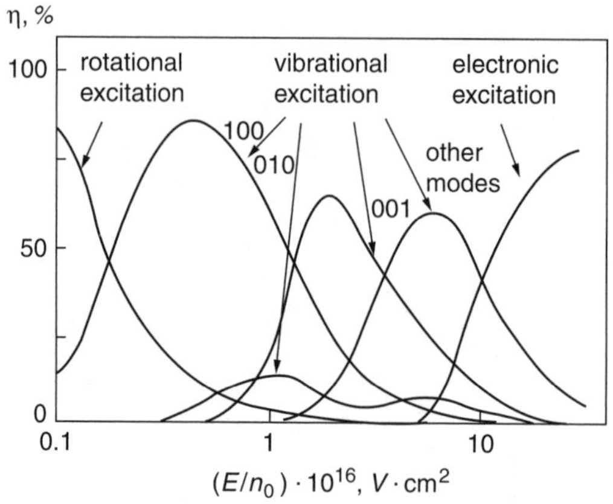

Figure 3.11 Electron energy distribution between excitation channels in CO_2

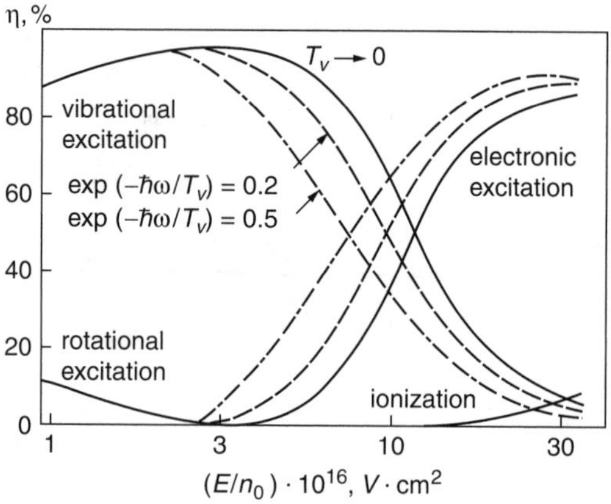

Figure 3.12 Electron energy distribution between excitation channels in nitrogen

At electron temperatures of about 1 eV, which are very typical for nonthermal discharges, almost all electron energy and thus most discharge power can be localized on vibrational excitation of molecules. This makes the process of vibrational excitation exceptionally important and special in the nonequilibrium plasma chemistry of molecular gases. Obviously, the contribution of electron attachment processes, including the dissociative attachment, can effectively compete with vibrational excitation at similar electron temperatures, but only in strongly electronegative gases. Finally, the contribution of electronic excitation and ionization becomes significant at higher values of E/n_0 and higher electron temperatures because of high-energy thresholds of these processes.

The electron energy distributions between different excitation channels strongly depend on gas composition (e.g., Figure 3.11 to Figure 3.16); the addition of CO to CO_2 changes the distribution of electron energy losses quite significantly. The

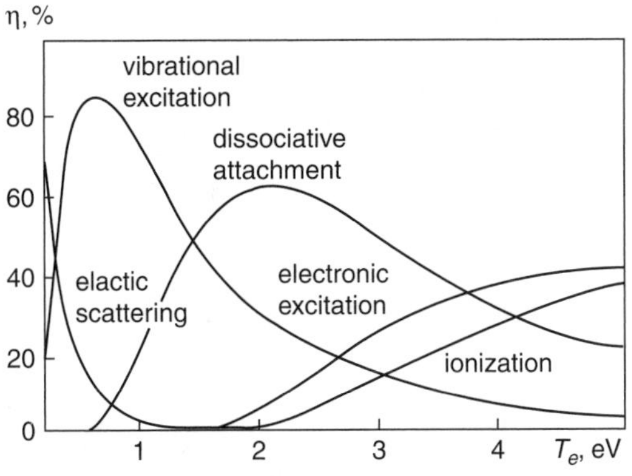

Figure 3.13 Electron energy distribution between excitation channels in water vapor

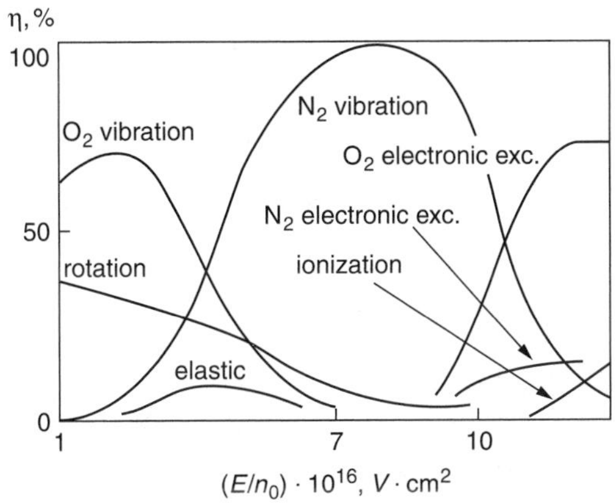

Figure 3.14 Electron energy distribution between excitation channels in air

carbon monoxide CO molecules have higher cross sections of vibrational excitation than those of CO_2 and, therefore, they decrease the fraction of high-energy electrons in the nonthermal discharge. Thus, the addition of CO to CO_2 results in a reduction of the electronic excitation and ionization rate coefficients and in the growth of energy going into vibrational excitation. Interestingly, the CO molecules are products of CO_2 dissociation, which can be effectively stimulated by vibrational excitation of the molecules in plasma. For this reason, the CO_2 dissociation in plasma, stimulated by vibrational excitation, can be considered an autocatalytic process, which can be accelerated by its own products.

The influence of vibrational temperature on the electron energy distribution between different excitation channels can be illustrated using a distribution calculated for a nonthermal discharge in molecular nitrogen. Results of the calculations are presented in Figure 3.12. As this figure shows, for higher values of vibrational temperatures, the concentration of the excited molecules is also high and the efficiency

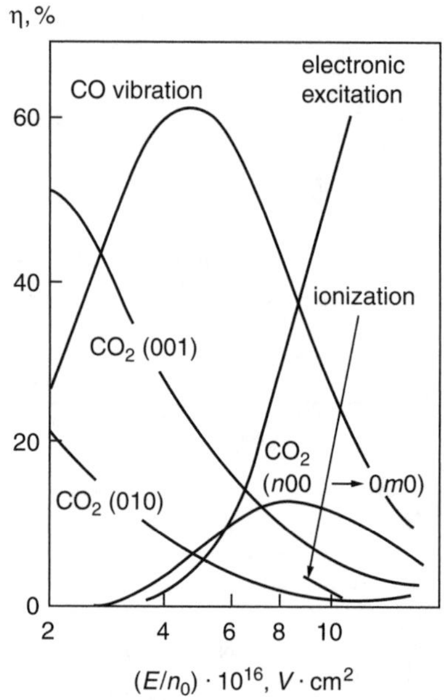

Figure 3.15 Electron energy distribution in CO_2 (50%)–CO (50%) mixture

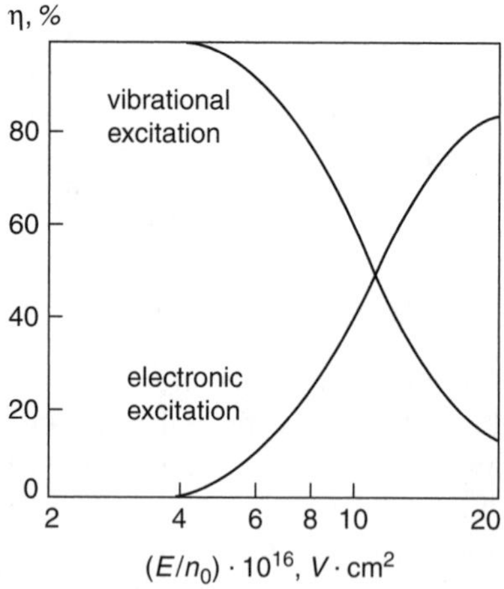

Figure 3.16 Electron energy distribution between excitation channels in CO

of their further vibrational excitation is relatively lowered. This results in an increase in the fraction of high-energy electrons and thus in intensification of electronic excitation and ionization processes. In other words, the fraction of discharge energy going into the vibrational excitation of molecules at the same value of the reduced electric field is relatively larger when the vibrational temperatures are relatively lower. This effect can be easily observed in this figure.

In general, Figure 3.9 through Figure 3.16 demonstrate that, at electron temperatures of about 1 eV (most typical for nonthermal discharges) and at the corresponding values of the reduced electric fields E/n_0, the principal fraction of electron energy can go to vibrational excitation of molecules (see Chapter 4 for the relation between E/n_0 and electron temperature). This means that most of the nonequilibrium, cold discharge power can be selectively focused on only the mechanism of vibrational excitation and the resulting chemical reactions. This makes plasma chemical reactions stimulated by vibrational excitation an interesting and promising method for minimizing the energy cost of different plasma technologies using discharges in molecular gases.

3.4 VIBRATIONAL (VT) RELAXATION; THE LANDAU–TELLER FORMULA

In the previous section, different mechanisms of formation of excited atoms and molecules in plasma by electron impact were discussed. Next, the mechanisms of deactivation and conversion of the excited species are considered. These energy exchange reactions are usually referred to as relaxation processes. As explained earlier, the vibrational excitation of molecules plays a very special role in plasma chemistry. Thus, the consideration of losses of excited particles with vibrational deactivation will begin with processes of vibrational–translational (VT) relaxation.

3.4.1 Vibrational–Translational (VT) Relaxation: Slow Adiabatic Elementary Process

This important process is usually called simply the vibrational relaxation or VT relaxation. The principal qualitative features of vibrational relaxation can be demonstrated in the framework of classical mechanics by considering collision of a classical harmonic oscillator with an atom or molecule. The oscillator is considered under the influence of an external force $F(t)$, which represents the intermolecular collision (see Figure 3.17). The one-dimensional motion of the harmonic oscillator can then be described in the center of mass system by the Newton equation:

$$\frac{d^2y}{dt^2} + \omega^2 y = \frac{1}{\mu_0} F(t) \tag{3.39}$$

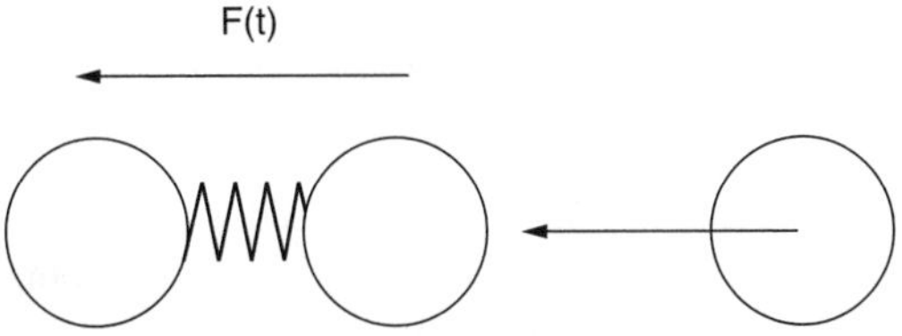

Figure 3.17 The Landau–Teller effect

In this relation, y is the deviation of the vibrational coordinate from equilibrium; ω is the oscillator frequency; and μ_0 is its reduced mass.

Suppose that the oscillator is initially ($t \to -\infty$) not excited, which means $y(t \to -\infty) = 0$, $dy/dt\ (t \to -\infty) = 0$. Then the vibrational energy transferred to the oscillator during the collision can be expressed as:

$$\Delta E_v = \frac{\mu_0}{2}\left[\left(\frac{dy}{dt}\right)^2 + \omega^2 y^2\right]_{t=\infty} \tag{3.40}$$

It is important to note that in the model under consideration, the reverse influence of vibrationally excited molecules on the external force $F(t)$ is neglected. This means that the collision partners are actually taken into account independently, and the same approach and the same expressions for ΔE_v can be applied also to describe the reverse process of energy transfer from the preliminary excited oscillator to the external force, which is actually another atom or molecule.[48,49]

It is convenient to calculate the energy transfer ΔE_v by introducing a new complex variable $\xi(t) = dy/dt + i\omega y$ instead of the vibrational variable y. Then, the oscillator energy transfer (Equation 3.40) can be found as square of the module of the complex variable:

$$\Delta E_v = \frac{\mu_0}{2}\left|\xi(t)\right|^2_{t=\infty} \tag{3.41}$$

To find the complex function $\xi(t)$, the dynamic equation (Equation 3.39) can be rewritten as a first-order equation:

$$\frac{d}{dt}\left(\frac{dy}{dt} + i\omega y\right) - i\omega\left(\frac{dy}{dt} + i\omega y\right) = \frac{1}{\mu_0}F(t); \frac{d\xi}{dt} - i\omega\xi = \frac{1}{\mu_0}F(t) \tag{3.42}$$

Solution of the linear nonuniform differential equation with initial conditions given previously $y(t \to -\infty) = 0$, $dy/dt\ (t \to -\infty) = 0$ can easily be found as:

$$\xi(t) = \exp(i\omega t)\int_{-\infty}^{t} \frac{1}{\mu_0}F(t')\exp(-i\omega t')dt' \tag{3.43}$$

Substituting the expression for $\xi(t)$ into Equation 3.41, the energy transfer during the VT relaxation process can be presented as:

$$\Delta E_v = \frac{1}{2\mu_0}\left|\int_{-\infty}^{+\infty} F(t)\exp(-i\omega t)dt\right|^2 \tag{3.44}$$

This expression shows that the energy transferred from an oscillator (or to an oscillator) is determined by square of module of the Fourier component of the force $F(t)$ on frequency ω corresponding to that of the oscillator. In other words, only the usually small Fourier component of the perturbation force on the oscillator frequency is effective in collisional excitation (or deactivation) of vibrational degrees of freedom of molecules.

A simple estimation of the integral can be found by considering time t as a complex variable (a trick successfully applied by Landau[31]) and supposing that $F(t) \to \infty$ in a singularity point $t = \tau + i\tau_{col}$, where $\tau_{col} = 1/\alpha v$ is time of the collision; α is the reverse radius of interaction between molecules; and v is a relative velocity of the colliding particles. The singularity $t = \tau + i\tau_{col}$ can be considered as the closest one to the real axis. Then, the integration of Equation 3.44 can be accomplished by going around the singularity while shifting the integration line to the upper semi-plane. As a result of the integration, the vibrational energy transfer from (or to) a molecule during an elementary VT relaxation process can be estimated as:

$$\Delta E_v \propto \exp(-2\omega\tau_{col}) \tag{3.45}$$

This interesting and important relation[50] demonstrates the adiabatic behavior of the vibrational relaxation. Usually, the Massey parameter (see Section 2.2.7) at moderate gas temperatures is fairly high for molecular vibration $\omega\tau_{col} >> 1$, which explains the adiabatic behavior and results in the exponentially low vibrational energy transfer during VT relaxation. During the adiabatic collision, a molecule has sufficient time for many vibrations and the oscillator can actually be considered "structureless," which explains the low level of the energy transfer.

It should be pointed out that the exponentially low probability of the adiabatic VT relaxation at low gas temperatures as well as the intensive vibrational excitation by electron impact is responsible for the unique role of vibrational excitation in plasma chemistry. Molecular vibrations for some gases such as N_2 CO_2 are able to "trap" energy of nonthermal discharges; it is easy to activate them and difficult to deactivate them.

3.4.2 Quantitative Relations for Probability of the Elementary Process of Adiabatic VT Relaxation

Equation 3.45 illustrates the adiabatic behavior of VT relaxation. To make this formula quantitative, one can specify an interaction potential between a molecule BC with an atom A in the following exponential form:

$$V(r_{AB}) = C\exp(-\alpha r_{AB}) = C\exp[-\alpha(r - \lambda y)] \tag{3.46}$$

Here, it is supposed that interaction potential is mainly due to interactions between atoms A and B; r is the coordinate of relative motion of mass centers of A and BC; $\lambda = m_C/(m_B + m_C)$; and m_B and m_C are masses of the corresponding atoms. Taking into account this interaction potential in the form Equation 3.46 and integrating Equation 3.44 gives:

$$\Delta E_v = \frac{8\pi\omega^2\mu^2\lambda^2}{\alpha^2\mu_0^2}\exp\left(-\frac{2\pi\omega}{\alpha v}\right) \tag{3.47}$$

In this relation, μ is a reduced mass of A and BC and μ_0 is a reduced mass of the molecule BC. As can be seen, the exact relation (Equation 3.47) for vibrational energy transfer is in good agreement with the qualitative expression (Equation 3.45). Equation 3.47 is actually a classical one; however, if the quantum $\hbar\omega$ is the minimum value of vibrational energy transfer, then the probability of transfer of the one quantum can be found based on Equation 3.47 as:

$$P_{01}^{VT}(v) = \frac{\Delta E_v}{\hbar\omega} = \frac{8\pi^2\omega\mu^2\lambda^2}{\hbar\alpha^2\mu_0^2}\exp\left(-\frac{2\pi\omega}{\alpha v}\right) \tag{3.48}$$

As has been already discussed, reverse effect of the vibrationally excited molecule on the external force $F(t)$ was neglected. For this reason, the probability Equation 3.48 can describe vibrational activation and deactivation processes. Quantum mechanics generalizes Equation 3.44 to describe the probability $P_{mn}^{VT}(v)$ of an oscillator transition from an initial state with vibrational quantum number m to a final state n. First order of perturbation theory gives:

$$P_{mn}^{VT}(v) = \frac{\langle m|y|n\rangle^2}{\hbar^2}\left|\int_{-\infty}^{+\infty} F(t)\exp\left(i\omega_{mn}t\right)dt\right|^2 \tag{3.49}$$

In this relation, $\hbar\omega_{mn} = E_m - E_n$ is the transition energy and $\langle m|y|n\rangle$ is a matrix element corresponding to the eigen functions (m and n) of nonperturbated Hamiltonian of the oscillator. The integral Equation 3.49 can be estimated in the same manner as Equation 3.44, resulting in the following qualitative formula for the transition probability:

$$P_{mn}^{VT}(v) \propto \langle m|y|n\rangle^2 \exp\left(-2|\omega_{mn}|\tau_{col}\right) \tag{3.50}$$

This formula shows as well as Equation 3.45 the adiabatic behavior of the vibrational relaxation. Using the specific potential Equation 3.46, the integration Equation 3.49 gives the following quantum mechanical expression for probability of the transition:

$$P_{mn}^{VT}(v) = \frac{16\pi^2\mu^2\omega_{mn}^2\lambda^2}{\alpha^2\hbar^2}\langle m|y|n\rangle^2 \exp\left(-\frac{2\pi|\omega_{mn}|}{\alpha v}\right) \tag{3.51}$$

As can be seen, the quantum mechanical and classical expressions for probability of vibrational relaxation (Equation 3.48 and Equation 3.51) are similar, but the quantum mechanical expression additionally includes the square of the matrix elements of relaxation transition. The matrix elements for harmonic oscillators $\langle m|y|n\rangle$ are nonzero only for transitions with $n = m \pm 1$. Then, as is known from quantum mechanics:[31]

$$\langle m|y|n\rangle = \sqrt{\frac{\hbar}{2\mu_0\omega}}\left(\sqrt{m}\,\delta_{n,m-1} + \sqrt{m+1}\,\delta_{n,m+1}\right) \tag{3.52}$$

Here, the symbol $\delta_{ij} = 1$ if $i = j$, and $\delta_{ij} = 0$ if $i \neq j$. This means that, according to Equation 3.51 and Equation 3.52, only the one-quantum VT relaxation processes are possible, However, taking into account the higher powers of y in expansion of $V(r - \lambda y$; see Equation 3.46), the multiquantum relaxation processes become possible, but obviously with lower probability. Multiquantum adiabatic VT relaxation is also slow because of much larger values of the Massey parameters in the exponents (Equation 3.50 and Equation 3.51) in this case.

It is interesting to mention that combining the quantum mechanical relations Equation 3.51 and Equation 3.52 gives an expression for the average collisional energy transfer of a quantum oscillator $\Delta E_v = P_{01}^{VT}(v) \bullet \hbar\omega$, which exactly coincides with the corresponding relation for a classical oscillator (Equation 3.47). This demonstrates that vibrational relaxation processes can quite often be described accurately in the framework of classical mechanics.

3.4.3 VT Relaxation Rate Coefficients for Harmonic Oscillators: Landau–Teller Formula

To find the rate coefficient of the vibrational VT relaxation of a harmonic oscillator, rewrite the relaxation probability Equation 3.51 and Equation 3.52 as a function of the relative velocity v of the colliding particles in the following form:

$$P_{n+1,n}^{VT}(v) \propto (n+1)\exp\left(-\frac{2\pi\omega}{\alpha v}\right) \tag{3.53}$$

and then integrate it over the Maxwellian distribution function for relative velocities of colliding partners. After such integration, the expression is derived for the averaged probability of VT relaxation as a function of translational temperature T_0 and vibrational quantum number n, which is actually a number of quanta on a molecule:

$$P_{n+1,n}^{VT}(T_0) = (n+1)P_{1,0}^{VT}(T_0) \tag{3.54}$$

The relaxation rate coefficient obviously can be found based on the expression for the relaxation probability by using the relation $k_{VT}^{n+1,n}(T_0) = P_{n+1,n}^{VT}(T_0) \bullet k_0$, where $k_0 \approx 10^{-10}$ cm^3/sec is the rate coefficient of gas-kinetic collisions.

Equation 3.54 shows the relaxation rate dependence on the number of vibrational quanta. To find the vibrational relaxation temperature dependence (Equation 3.54), take into account that the function under integral of probability (Equation 3.53) over the Maxwellian distribution has a sharp maximum at the relative velocity:

$$v^* = \sqrt[3]{\frac{2\pi T_0}{\mu\alpha}} \tag{3.55}$$

It helps to find the integral of probability (Equation 3.53) over the Maxwellian distribution and the temperature dependence of vibrational relaxation probability:

$$P_{10}^{VT}(T_0) \propto \exp\left[-\frac{2\pi\omega}{\alpha v*} - \frac{\mu v^{*2}}{2T_0}\right] \equiv \exp\left(-\frac{3\pi\omega}{\alpha v*}\right) \equiv \exp\left[-3\left(\frac{\hbar^2\mu\omega^2}{2\alpha^2 T_0}\right)^{1/3}\right] \quad (3.56)$$

As one can see, the probability of VT relaxation exponentially depends on the adiabatic Massey factor calculated for the velocity v*:

$$\frac{3\pi\omega}{\alpha v^*} \propto \frac{1}{T_0^{1/3}}.$$

Numerically, this adiabatic factor is typically about 5 to 15 over a wide range of temperatures. Equation 3.56 obviously can be rewritten for the rate coefficient of the VT relaxation of the vibrational quantum as a function of translational gas temperature T_0:[50]

$$k_{VT}^{10} \propto \exp\left(-B/T_0^{1/3}\right), B = \sqrt[3]{\frac{27\hbar^2\mu\omega^2}{2\alpha^2 T_0}} \quad (3.57)$$

Equation 3.57 shows the exponential growth of the VT relaxation rate with translational temperature T_0 and plays an important role in plasma chemistry, gas laser, and shock-wave kinetics. It is well known as the **Landau–Teller formula**.

Generalization of the Landau–Teller formula for better quantitative description of the adiabatic VT relaxation was developed in the frameworks of Schwartz-Slawsky-Herzfeld (SSH) theory.[51,52] For numerical calculations of the vibrational VT relaxation rate coefficient, it is convenient to use the following semiempiric relation based on the Landau–Teller formula, derived by Lifshitz:[53]

$$k_{VT}^{10} = 3.03 * 10^6 (\hbar\omega)^{2.66} \mu^{2.06} \exp\left[-0.492(\hbar\omega)^{0.681} \mu^{0.302} T_0^{-1/3}\right] \quad (3.58)$$

Here, the vibrational relaxation rate coefficient should be expressed in cm^3/mol*sec; T_0 and $\hbar\omega$ in degrees Kelvin; and μ, the reduced mass of colliding particles, in atomic mass units. Similarly, another semiempiric formula[54] can be applied to calculate the VT relaxation time as a function of pressure:

$$\ln(p\tau_{VT}) = 1.16 * 10^{-3} \mu^{1/2} (\hbar\omega)^{4/3} \left(T_0^{-1/3} - 0.015\mu^{1/4}\right) - 18.42 \quad (3.59)$$

Pressure p is expressed in atmospheres in this relation; τ_{VT} in seconds; T_0 and $\hbar\omega$ in degrees Kelvin; and reduced mass μ in atomic mass units. Table 3.13 gives numerical values of vibrational relaxation rate coefficients at room temperature for some one-component gases of interest for plasma chemical applications. The temperature dependence of rate coefficients of the vibrational VT relaxation for some molecules is given in Table 3.14. Relaxation of most of molecules considered in this table follows the Landau–Teller tendency in the temperature dependence.

Table 3.13 Vibrational VT Relaxation Rate Coefficients for One-Component Gases at Room Temperature

Molecule	$k_{VT}^{10}(T_0 = 300K), \frac{cm^3}{sec}$	Molecule	$k_{VT}^{10}(T_0 = 300K), \frac{cm^3}{sec}$
O_2	$5*10^{-18}$	F_2	$2*10^{-15}$
Cl_2	$3*10^{-15}$	D_2	$3*10^{-17}$
Br_2	10^{-14}	$CO_2(01^10)$	$5*10^{-15}$
J_2	$3*10^{-14}$	$H_2O(010)$	$3*10^{-12}$
N_2	$3*10^{-19}$	N_2O	10^{-14}
CO	10^{-18}	COS	$3*10^{-14}$
H_2	10^{-16}	CS_2	$5*10^{-14}$
HF	$2*10^{-12}$	SO_2	$5*10^{-14}$
DF	$5*10^{-13}$	C_2H_2	10^{-12}
HCl	10^{-14}	CH_2Cl_2	10^{-12}
DCl	$5*10^{-15}$	CH_4	10^{-14}
HBr	$2*10^{-14}$	CH_3Cl	10^{-13}
DBr	$5*10^{-15}$	$CHCl_3$	$5*10^{-13}$
HJ	10^{-13}	CCl_4	$5*10^{-13}$
HD	10^{-16}	NO	10^{-13}

3.4.4 Vibrational VT Relaxation of Anharmonic Oscillators

The most important peculiarity of relaxation of anharmonic oscillators is due to reduction of the transition energy with an increase of vibrational quantum number:

$$\omega_{n,n-1} = \omega(1 - 2x_e n) \tag{3.60}$$

Here, x_e is the coefficient of anharmonicity of a vibrationally excited diatomic molecule and n is the vibrational quantum number.

As can be seen from Equation 3.54, the probability and thus the rate coefficient of the vibrational VT relaxation increase with the vibrational quantum number n even in the case of harmonic oscillators. Furthermore, in the case of anharmonic oscillators, the increase of vibrational quantum number n leads also to a reduction of the Massey parameter, making the VT relaxation less adiabatic, and accelerates the process exponentially:

$$P_{n+1,n}^{VT}(T_0) = (n+1)P_{1,0}^{VT}(T_0)\exp(\delta_{VT} n) \tag{3.61}$$

The temperature dependence of the probability of VT relaxation of anharmonic oscillators (Equation 3.61) is similar to that of harmonic oscillators (Equation 3.54) and corresponds to the Landau–Teller formula. The exponential parameter δ_{VT} can be found in the case of anharmonic oscillators from the following two-part relation:

Table 3.14 Temperature Dependence of Vibrational VT Relaxation Rate Coefficients[a] $k_{VT}^{10}(T_0)$ for One-Component Gases

Molecule	Temperature dependence $k_{VT}^{10}(T_0), \frac{cm^3}{sec}$ [b]
O_2	$10^{-10}\exp(-129*T_0^{-1/3})$
Cl_2	$2*10^{-11}\exp(-58*T_0^{-1/3})$
Br_2	$2*10^{-11}\exp(-48*T_0^{-1/3})$
J_2	$5*10^{-12}\exp(-29*T_0^{-1/3})$
CO	$10^{-12}T_0\exp(-190*T_0^{-1/3}+1410*T_0^{-1})$
NO	$10^{-12}\exp(-14*T_0^{-1/3})$
HF	$5*10^{-10}T_0^{-1}+6*10^{-20}T_0^{2.26}$
DF	$1.6*10^{-5}T_0^{-3}+3.3*10^{-16}T_0$
HCl	$2.6*10^{-7}T_0^{-3}+1.4*10^{-19}T_0^{2}$
F_2	$2*10^{-11}\exp(-65*T_0^{-1/3})$
D_2	$10^{-12}\exp(-67*T_0^{-1/3})$
$CO_2(01^10)$	$10^{-11}\exp(-72*T_0^{-1/3})$

[a] $k_{VT}^{10}(T_0)$.
[b] Temperature T_0 in Kelvin degrees.

$$\delta_{VT} = 4\gamma_n^{2/3}x_e, \quad if\ \gamma_n \geq 27 \tag{3.62a}$$

$$\delta_{VT} = \frac{4}{3}\gamma_n x_e, \quad if\ \gamma_n < 27 \tag{3.62b}$$

The adiabatic factor γ_n in this relation is actually the Massey parameter for the vibrational relaxation transition $n + 1 \rightarrow n$ with the transition energy $E_{n+1} - E_n$. This adiabatic factor can be calculated as:

$$\gamma_n(n+1\rightarrow n) = \frac{\pi(E_{n+1}-E_n)}{\hbar\alpha}\sqrt{\frac{\mu}{2T_0}} \tag{3.63}$$

For numerical calculations of the Massey parameter γ_n, it is convenient to use the following numeric relation:

$$\gamma_n = \frac{0.32}{\alpha}\sqrt{\frac{\mu}{T_0}}\hbar\omega\left(1 - 2x_e(n-1)\right) \tag{3.64}$$

In this relation, the reduced mass μ is expressed in atomic units; reverse radius of interaction between colliding particles α is expressed in $Å^{-1}$; and vibrational quantum $\hbar\omega$ and gas temperature T_0 are given in degrees Kelvin.

3.4.5 Fast Nonadiabatic Mechanisms of VT Relaxation

As discussed earlier, vibrational relaxation is quite slow in adiabatic collisions when no chemical interaction occurs between colliding partners. Thus, the probability of a vibrationally excited N_2 deactivation in collision with another N_2 molecule at room temperature can be as low as 10^{-9}. However, the vibrational relaxation process can be arranged much faster in a nonadiabatic way, with a quantum transfer at almost each collision if the colliding partners interact chemically. This can happen in collisions of vibrationally excited molecules with active atoms and radicals, in surface collisions, etc. Now consider separately the main nonadiabatic mechanisms of VT relaxation.

1. **VT relaxation in molecular collisions with atoms and radicals**. These processes can be illustrated by the relaxation of vibrationally excited N_2 molecules on atomic oxygen, analyzed by Andreev and Nikitin[55] (see Figure 3.18). The energy distance initially between degenerated electronic terms grows as a molecule and an atom approach each other. Finally, when this energy of electronic transitions becomes equal to a vibrational quantum, the nonadiabatic relaxation (the so-called vibronic transition; Figure 3.18) can take place. Temperature dependence of the process is due to the low activation energy and is not actually significant. Typical values of the nonadiabatic VT relaxation rate coefficients are very high, usually about 10^{-13} to 10^{-12} cm^3/sec. Sometimes, as in the case of relaxation of alkaline atoms, the nonadiabatic VT relaxation rate coefficients reach those for gas-kinetic collisions, e.g., about 10^{-10} cm^3/sec (see Equation 3.96 and Figure 3.20).
2. **VT relaxation through intermediate formation of long-life complexes**. Fast nonadiabatic relaxation is also possible if a collision results in forming long-life chemically bonded complexes. This takes place, in particular, in collisions of H_2O^*–H_2O, CO_2^*–H_2O, CH_4^*–CO_2, CH_4^*–H_2O, $C_2H_6^*$–O_2, and NO^*–Cl_2 related to important plasma chemical applications. Interaction between the collision partners is based quite often in this case on hydrogen bonds. In this situation, relaxation rate coefficients also can reach gas-kinetic values. Temperature dependence is not strong and usually is negative so the relaxation rate coefficients decrease with temperature growth. Another important peculiarity of this relaxation mechanism is its multiquantum nature; this means that the probabilities of one-quantum and multiquantum transfer of vibrational energy are relatively close.

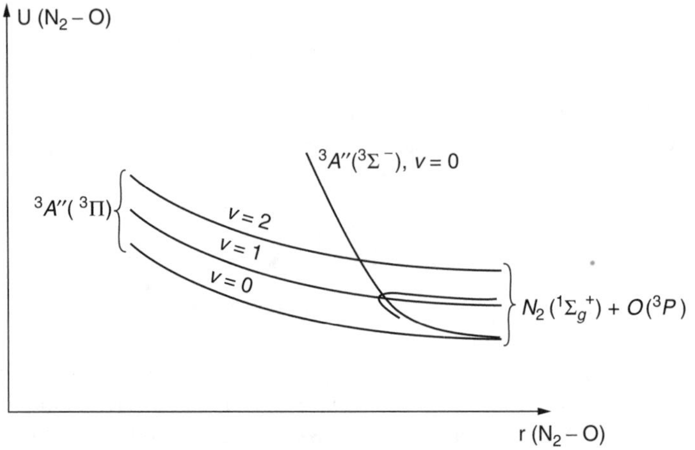

Figure 3.18 Nonadiabatic VT relaxation of N_2 molecules in collisions with O atoms

3. **VT relaxation in symmetrical exchange reactions**. Very fast nonadiabatic relaxation takes place in chemical reactions such as:

$$A' + (BA'')^*(n=1) \to A'' + BA'(n=0) \tag{3.65}$$

These processes are effective when activation energies of corresponding chemical processes are low. A very important example is the oxygen exchange reaction proceeding without any activation energy through the intermediate formation of an excited ozone molecule: $O_2^* + O \to O_3^* \to O + O_2$. Here, the relaxation rate coefficient is approximately 10^{-11} cm^3/sec and practically does not depend on temperature. A similar situation takes place in the fast nonadiabatic VT relaxation processes of H_2 molecules on H-atoms, different halogen molecules on corresponding halogen atoms, and hydrogen halides on the hydrogen atoms.

4. **Heterogeneous VT relaxation, losses of vibrational energy in surface collisions, and accommodation coefficients**. Heterogeneous relaxation is nonadiabatic and a fast process if it proceeds through an adsorption stage and forms intermediate complexes with a surface as is usually the case.[56] The losses of vibrational energy can be determined in this situation from the probability of the vibrational relaxation calculated with respect to one surface collision. This probability is usually referred to as the **accommodation coefficient** and can be found in Table 3.15 for different molecules and different surfaces.

3.4.6 VT Relaxation of Polyatomic Molecules

The peculiar features of VT relaxation of vibrationally excited polyatomic molecules are because, in this case, not one but a set of oscillators interacts with an incident atom or molecule. At a low level of excitation, vibrational modes of a polyatomic molecule can be considered separately (see Equation 3.12 and Equation 3.14) and

Table 3.15 Accommodation Coefficients for Heterogeneous VT Relaxation

Molecule	Vibration mode	Quantum energy, cm^{-1}	Temperature K	Surface	Accommodation coefficient
CO_2	$v_2 = 1$	667	277–373	Platinum	0.3–0.4
CO_2	$v_2 = 1$	667	297	NaCl	0.22
CO_2	$v_3 = 1$	2349	300–350	Pyrex	0.2–0.4
CO_2	$v_3 = 1$	2349	300	Brass, Teflon, Mylar	0.2
CO_2	$v_3 = 1$	2349	300–1000	Quartz	0.05–0.45
CO_2	$v_3 = 1$	2349	300–560	Molybdenum glass	0.3–0.4
CH_4	$v_4 = 1$	1306	273–373	Platinum	0.5–0.9
H_2	$v = 1$	4160	300	Quartz	$5*10^{-4}$
H_2	$v = 1$	4160	300	Molybdenum glass	$1.3*10^{-4}$
H_2	$v = 1$	4160	300	Pyrex	10^{-4}
D_2	$v = 1$	2990	77–275	Molybdenum glass	$(0.2–8)*10^{-4}$
D_2	$v = 1$	2990	300	Quartz	10^{-4}
N_2	$v = 1$	2331	350	Pyrex	$5*10^{-4}$
N_2	$v = 1$	2331	282–603	Molybdenum glass	$(1–3)*10^{-3}$
N_2	$v = 1$	2331	350	Pyrex	$5*10^{-4}$
N_2	$v = 1$	2331	300	Steel, aluminum, copper	$3*10^{-3}$
N_2	$v = 1$	2331	300	Teflon, alumina	10^{-3}
N_2	$v = 1$	2331	295	Silver	$1.4*10^{-2}$
CO	$v = 1$	2143	300	Pyrex	$1.9*10^{-2}$
HF	$v = 1$	3962	300	Molybdenum glass	10^{-2}
HCl	$v = 1$	2886	300	Pyrex	0.45
OH	$v = 9$	—	300	Boron acid	1
N_2O	$v_2 = 1$	589	273–373	Platinum, NaCl	0.3
N_2O	$v_3 = 1$	2224	300–350	Pyrex	0.2
N_2O	$v_3 = 1$	2224	300–1000	Quartz	0.05–0.33
N_2O	$v_3 = 1$	2224	300–560	Molybdenum glass	0.01–0.03
CF_3Cl	$v_3 = 1$	732	273–373	Platinum	0.5–0.6

their relaxation can be calculated for each mode following the Landau–Teller approach or a nonadiabatic model described previously (see Table 3.13 and Table 3.14).

For higher levels of excitation corresponding to the vibrational quasi continuum of polyatomic molecules (see Section 3.2.4 and Equation 3.14), the mean square of the vibrational energy transferred to translational degrees of freedom $\langle \Delta E_{VT}^2 \rangle$ is obtained by averaging the exponential Landau–Teller factors over the vibrational Fourier spectrum of the system $I(\omega)$,[36] which was described in Section 3.2.4:

$$\langle \Delta E_{VT}^2 \rangle = \int_0^\infty \left(\Omega\tau_{col}\right)_{VT}^2 (\hbar\omega)^2 I(\omega) \exp\left(-\frac{\omega}{\alpha v}\right) d\omega \tag{3.66}$$

The factor $(\Omega\tau_{col})_{VT}^2$ characterizes the smallness of the probability of transition and is due to the smallness of the amplitude of vibrations relative to the interaction radius. Numerically, this factor is usually of the order of $(\Omega\tau_{col})^2_{VT} \approx 0.01$. If a polyatomic molecule can be considered as a group of harmonic oscillators with frequencies ω_{0i} and vibrational quantum numbers n_i, then the vibrational Fourier spectrum $I(\omega)$ can be presented as a sum of δ-functions (Equation 3.16). Obviously, in this case (Equation 3.66) the vibrational energy transfer during a collision reduces to the known expression corresponding to the Landau–Teller model:

$$\langle \Delta E^2_{VT} \rangle = \sum_i (\Omega\tau_{col})^2_{VTi} n_i (\hbar\omega_{0i})^2 \exp\left(-\frac{\omega_{0i}}{\alpha v}\right) \quad (3.67)$$

For the case of vibrationally excited molecules in quasi continuum, the vibrational Fourier spectrum $I(\omega)$ can be described by the Lorentz profile (Equation 3.19) as was discussed in Section 3.2.4. Then, the energy transfer from the most rapidly relaxing vibrational mode (with the smallest quantum $\hbar\omega_n$) can be obtained from integration of Equation 3.66:

$$\langle E^2_{VT} \rangle = (\Omega\tau_{col})^2_{VT} n \left[\frac{\alpha v}{\alpha v + \delta} \exp\left(-\frac{\omega_n - \delta}{\alpha v}\right) + \frac{\delta}{3\pi\omega_n}\left(\frac{\alpha v}{\omega_n}\right)^3 \right] (\hbar\omega_n)^2 \quad (3.68)$$

Here, δ is the intermode vibrational energy exchange frequency considered while discussing Equation 3.19. Factor n is the number of quanta on the mode under consideration. If the low-frequency mode is degenerated or is a part of Fermi-resonance modes, then n implies the total number of quanta, taking into account the degeneracy.

As is seen from Equation 3.68, the VT relaxation of polyatomic molecules in quasi continuum is determined by two effects: an adiabatic effect and a quasi-resonant effect.[36] The first term is the adiabatic effect and is somewhat similar to the case of diatomic molecules. However, it is growing faster with n because of the effective reduction of the vibrational frequency and the Massey parameter $(\omega_n - \delta(n))/\alpha v$ due to broadening of the given mode line in the vibrational spectrum $I(\omega)$. The second term in Equation 3.68 corresponds to quasi-resonance (nonadiabatic) relaxation of polyatomic molecules at low frequencies $\omega \approx \alpha v$. Excitation at these low frequencies in the vibrational spectrum of polyatomic molecules becomes possible due to interaction of their fundamental modes.

Comparison of these two relaxation effects shows that the quasi-resonant VT relaxation has no exponentially small factor and can exceed the adiabatic relaxation. This means that VT relaxation of polyatomic molecules actually becomes nonadiabatic and very fast when high levels of their vibrational excitation result in transition to a quasi continuum of vibrational spectrum.

3.4.7 Effect of Rotation on the Vibrational Relaxation of Molecules

Interaction between molecular vibrations and rotations is able to accelerate VT relaxation. This effect is most important for molecules with low momentum of inertia having at the same time a large reduced mass of colliding particles. Vibrational relaxation of methane and hydrogen halides can be taken as an example. In general, molecules containing a hydrogen atom can lose their vibrational energy faster through rotation.[30]

This effect can be explained because the Massey parameter $\omega/\alpha v$ in this case depends, not on relative translational velocity of colliding partners, but rather on the velocity of molecular rotation at the point of minimal distance between them. Molecules that consist of heavy and light atoms have rotational velocity faster than translational, which results in reducing the Massey parameter and accelerating the vibrational relaxation.

Polyatomic molecules have also a specific effect of rotations on vibrational relaxation—the so-called VRT relaxation process. Degenerated vibrational modes can have "circular" polarization (see Section 3.2.3) and angular momentum of the quasi rotations (Equation 3.11). This opens a fast, nonadiabatic channel for energy transfer from vibrational to translational degrees of freedom through intermediate rotations of polyatomic molecules.

3.5 VIBRATIONAL ENERGY TRANSFER BETWEEN MOLECULES: VV RELAXATION PROCESSES

In the two previous sections, main processes of collisional generation and deactivation of vibrationally excited molecules in plasma were considered. These determined the level of vibrational excitation of molecular gases in discharges, which is important to understand the energy balance of the discharges and in practice to stimulate different plasma chemical processes. Kinetics of the plasma chemical reactions of vibrationally excited molecules, however, is determined, not by the concentration of the excited molecules, but mostly by the number density of highly excited species with energies sufficient for dissociation and/or other endothermic chemical reactions.

Generation of these highly vibrationally excited molecules is usually not due to direct electron impact if the gas pressure is not very low. The highly excited particles can be effectively formed during collisional energy exchange processes between molecules. These fundamental processes of great importance in plasma chemistry are usually resonant or close to resonance and called VV relaxation or VV exchange processes.

3.5.1 Resonant VV Relaxation

These processes usually imply vibrational energy exchange between molecules of the same kind, for example: $N_2^*(v = 1) + N_2(v = 0) \rightarrow N_2(v = 0) + N_2^*(v = 1)$. The resonant VV exchange between diatomic molecules can be characterized by the probability $q_{mn}^{sl}(v)$ of a collisional transition, when one oscillator changes its vibrational quantum number from s to l, and the other from m to n. Quantum mechanics gives an expression for the probability $q_{mn}^{sl}(v)$ similar to that obtained for VT relaxation (Equation 3.49). Instead of the ω_{mn} used in the case of VV exchange, the frequency describing the change of vibrational energy during the collision should be employed:

$$\omega_{ms,nl} = \frac{1}{\hbar}\left(E_m + E_s - E_l - E_n\right) = \frac{\Delta E}{\hbar} \tag{3.69}$$

Also, instead of the matrix elements of transition $m \to n$ for VT relaxation (Equation 3.49), the probability of the VV exchange is determined by a product of matrix elements of transitions $m \to n$ and $s \to l$ in the two interacting oscillators:

$$q_{mn}^{sl}(v) = \frac{\langle m|y_1|n\rangle^2 \langle s|y_2|l\rangle^2}{\hbar^2} \left| \int_{-\infty}^{+\infty} F(t)\exp\left(i\omega_{ms,nl}t\right)dt \right|^2 \tag{3.70}$$

The squared module of a Fourier component of interaction force $F(t)$ on a transition frequency (Equation 3.69) characterizes here the level of resonance (of adiabatic behavior) of the process in the same manner as in the case of VT relaxation. As can be seen from Equation 3.52, the matrix elements for harmonic oscillators $\langle m|y|n\rangle$ are nonzero only for transitions with $n = m \pm 1$, which means only the one-quantum VV relaxation processes are possible in this approximation.

Expression for the VV relaxation probability of the one-quantum exchange as a function of translational gas temperature T_0 can be obtained by averaging the probability $q_{mn}^{sl}(v)$ over the Maxwellian distribution:

$$Q_{n+1,n}^{m,m+1}\left(T_0\right) = (m+1)(n+1)Q_{10}^{01}\left(T_0\right) \tag{3.71}$$

Taking into account the higher powers in expansion of intermolecular interaction potential, multiquantum VV exchange processes become possible even for harmonic oscillators, but obviously with lower probability:[48]

$$Q_{0,k}^{m,m-k} \approx \frac{m!}{(m-k)!k!2^{k-1}}\left(\Omega\tau_{col}\right)^{2k} \tag{3.72}$$

In this relation, τ_{col} is time of a collision and Ω is the VV transition frequency during the collision. The factor $\Omega\tau_{col}$ is usually small due to the smallness of the vibrational amplitude with respect to intermolecular interaction radius in the same manner as in case of VT relaxation.

VV relaxation for most molecules is due to the exchange interaction (short-distance forces) that results in a short time of collision τ_{col} and $\Omega\tau_{col} \approx 0.1$ to 0.01. In this case, the probability of only the one-quantum transfer is about $Q_{10}^{01} \approx (\Omega\tau_{col})^2 \approx 10^{-2} - 10^{-4}$ (see Table 3.16). The probability of multiquantum VV exchange in this case, according to Equation 3.72 is very low (e.g., about 10^{-9} even for resonant three-quantum exchanges).

However, for some molecules, such as CO_2 and N_2O, VV relaxation is due to dipole or multipole interaction (long-distance forces). Because of the longer interaction distance, the collision time is much longer and the factor $\Omega\tau_{col}$ and the VV

Table 3.16 Resonant VV Relaxation Rate Coefficients at Room Temperature and Their Ratio to Those of Gas-Kinetic Collisions (k_{VV}/k_0)

Molecule	k_{VV}, cm³/sec	k_{VV}/k_0	Molecule	k_{VV}, cm³/sec	k_{VV}/k_0
CO_2(001)	$5*10^{-10}$	4	HF	$3*10^{-11}$	0.5
CO	$3*10^{-11}$	0.5	HCl	$2*10^{-11}$	0.2
N_2	10^{-13}–10^{-12}	10^{-3}–10^{-2}	HBr	10^{-11}	0.1
H_2	10^{-13}	10^{-3}	DF	$3*10^{-11}$	—
N_2O(001)	$3*10^{-10}$	3	HJ	$2*10^{-12}$	—

relaxation probability Q_{10}^{01} become much larger and approach unity: $Q_{10}^{01} \approx (\Omega\tau_{col})^2 \approx 1$ (see Table 3.16). Taking into account Equation 3.71), it is interesting to note that the cross sections of the resonance VV exchange between highly vibrationally excited molecules can exceed in this case the gas-kinetic cross section. Also, the multiquantum resonant VV exchange (Equation 3.72) is decreasing with the number of transferred quanta, but not as fast as in the case of short-distance exchange interaction between molecules.

Table 3.16 gives numerical values of the resonant VV exchange rate coefficients at room temperature and a quantum exchange between the first and zero vibrational levels. Ratios of the rate coefficient to the correspondent gas-kinetic rate can also be found in the table. In general, the resonant VV exchange is usually much faster at room temperature than VT relaxation (see Table 3.13). This leads to the possibility of efficient generation of highly vibrationally excited, and thus very reactive, molecules in nonthermal discharges.

Temperature dependence of the VV relaxation probability $Q_{10}^{01} \approx (\Omega\tau_{col})^2$ is different for that provided by dipole and exchange interactions. The transition frequency Ω is proportional to the average interaction energy; thus in the case of the short distance exchange interaction, $\Omega \propto T_0$. Taking into account that $\tau_{col} \propto 1/v \propto T_0^{1/2}$, $Q_{10}^{01} \approx (\Omega\tau_{col})^2 \propto T_0$. Thus, the probability of VV relaxation provided by short-distance forces is proportional to temperature. Conversely, in the case of long-distance dipole and multipole interactions, the VV relaxation probability decreases with an increase of the translational gas temperature T_0. Because the transition frequency Ω does not depend on temperature in this case, $Q_{10}^{01} \approx (\Omega\tau_{col})^2 \propto \tau_{col}^2 \propto 1/T_0$.

3.5.2 VV Relaxation of Anharmonic Oscillators

Although the influence of anharmonicity is related to a change of the transition matrix elements, it is primarily related to the increase of the transition energy. VV exchange becomes nonresonant and slightly adiabatic and, as a result, slower than the resonant one. The expression for the transition probability becomes more complicated than Equation 3.71 and includes the exponential adiabatic factors:[48]

$$Q_{n+1,n}^{m,m+1} = (m+1)(n+1)Q_{10}^{01} \exp\left(-\delta_{VV}|n-m|\right)\left[\tfrac{2}{3} - \tfrac{1}{2}\exp\left(-\delta_{VV}|n-m|\right)\right] \quad (3.73)$$

In the same manner as for the case of VT relaxation of anharmonic oscillators (see Equation 3.62 and Equation 3.63), in this relation:

$$\delta_{VV} = \frac{4}{3} x_e \gamma_0 = \frac{4}{3} \frac{\pi \omega x_e}{\alpha} \sqrt{\frac{\mu}{2T_0}} \quad (3.74)$$

Note that the adiabatic factor (Equation 3.63) is much lower for a VV exchange than for VT relaxation. That is why Equation 3.62a can be neglected in the case of VV relaxation, and only Equation 3.62b is required to be taken into account. In general, in one-component gases, $\delta_{VV} = \delta_{VT}$. For numerical calculations of δ_{VV}, it is convenient to use the following formula:

$$\delta_{VV} = \frac{0.427}{\alpha} \sqrt{\frac{\mu}{T_0}}\, x_e \hbar\omega \quad (3.75)$$

In this relation, the reduced mass of colliding molecules μ should be given in a.m.u. (atomic units); reverse intermolecular interaction radius α in A^{-1}; and translational gas temperature T_0 and vibrational quantum $\hbar\omega$ in degrees Kelvin.

The preceding formulas were related to the VV exchange of anharmonic oscillators due to the short-distance forces (the exchange interaction between colliding molecules), which is correct in most cases. Taking into account the long-distance forces leads to modification of Equation 3.73:[57]

$$Q_{n+1,n}^{m,m+1} \approx (m+1)(n+1)\left\{\left(Q_{10}^{01}\right)_S \exp\left(-\delta_{VV}|n-m|\right)\left[\tfrac{3}{2} - \tfrac{1}{2}\exp\left(-\delta_{VV}|n-m|\right)\right] + \left(Q_{10}^{01}\right)_L \exp\left(-\Delta_{VV}(n-m)^2\right)\right\} \quad (3.76)$$

In this relation, $(Q_{10}^{01})_S$ and $(Q_{10}^{01})_L$ represent probabilities of the only quantum transfer due to, respectively, short-distance and long-distance forces, and parameter Δ_{VV} is related to the Massey parameter and determines the adiabatic degree for the VV exchange of anharmonic oscillators provided by the long-distance forces. When the probability $(Q_{10}^{01})_L \ll 1$ is negligible, Equation 3.76 coincides with Equation 3.73.

Note that the long collisional time in the case of the VV exchange provided by long-distance forces results in relatively large Massey parameters $\Delta\omega * \tau_{col}$. As a result, $\Delta_{VV} > \delta_{VV}$ and, according to Equation 3.76, the long-distance forces are able to affect the rate of VV exchange only close to resonance, which means only for collisions of the anharmonic oscillators with close values of vibrational quantum numbers. Figure 3.19[58] illustrates this effect by showing the dependence of the VV exchange rate coefficient in pure CO at room temperature on the vibrational quantum number.

Now compare the rate coefficients of VV exchange and VT relaxation processes, taking into account the anharmonicity. The effect of anharmonicity on the VV

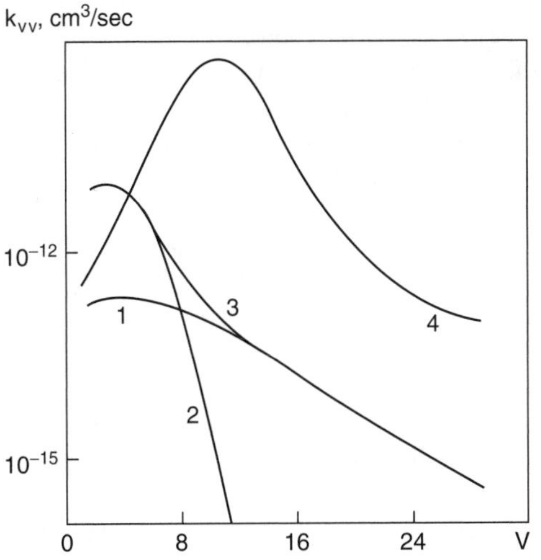

Figure 3.19 VV exchange rate coefficient dependence on vibration quantum number (CO, $T_0 = 300$ K): (1) contribution of short-distance forces; (2) contribution of long-distance forces; (3) total curve for transitions ($v \rightarrow v - 1$, $0 \rightarrow 1$); (4) total curve for transitions ($v \rightarrow v - 1$, $8 \rightarrow 9$)[58]

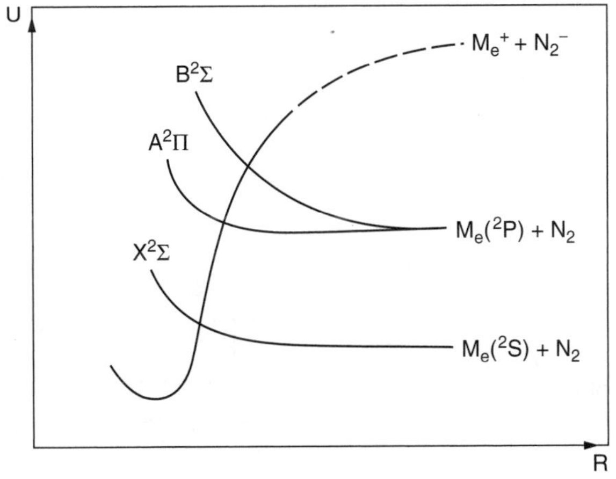

Figure 3.20 Electronic energy relaxation through formation of intermediate ionic complexes; example of alkaline atom relaxation $Me(^2P) \rightarrow Me(^2S)$ with N_2 vibrational excitation

exchange is negligible in resonant collisions, when $|m - n|\,\delta_{VV} << 1$ and Equation 3.73 actually coincides with Equation 3.71 for relaxation of harmonic oscillators. Anharmonicity becomes quite necessary in considerating VV exchange processes occurring in collisions of relatively highly excited molecules and molecules on low vibrational levels. Taking into account the effect of anharmonicity, the rate coefficient of the VV exchange process, $k_{VV}(n) = k_0 Q^{0,1}_{n+1,n}$, decreases with growth of the vibrational quantum number n; this is in contrast to the VT relaxation rate coefficient, $k_{VT}(n) = k_0 P^{VT}_{n+1,n}$, which increases with n ($k_0 \approx 10^{-10}$ cm^3/sec is the rate coefficient of gas-kinetic collisions). It is interesting to compare the rate coefficients of VV exchange and VT relaxation processes for vibrationally excited molecules based on Equation 3.73, Equation 3.61, and Equation 3.62b:

$$\xi(n) = \frac{k_{VT}(n)}{k_{VV}(n)} \approx \frac{P^{VT}_{10}}{Q^{01}_{10}} \exp\left[\left(\delta_{VV} + \delta_{VT}\right)n\right] \tag{3.77}$$

As can be seen from the relation, $\xi << 1$ at low excitation levels and the VT relaxation is much slower than the VV exchange because $P^{VT}_{10} / Q^{01}_{10} << 1$. This means that the population of highly vibrationally excited states can increase in nonthermal plasma much faster than losses of vibrational energy; this also explains the high efficiency of this type of excitation in plasma chemistry.

The ratio $\xi(n)$, however, is growing exponentially with n and the VT relaxation catches up to the VV exchange ($\xi = 1$) at some critical value of the vibrational quantum number (and corresponding vibrational energy $E^*(T_0)$). This critical value of vibrational energy actually shows a maximum level of effective vibrational excitation $E^*(T_0)$ in plasma chemical systems and can be calculated as a function of gas temperature:[59]

$$E^*(T_0) = \hbar\omega\left(\frac{1}{4x_e} - b\sqrt{T_0}\right) \tag{3.78}$$

The first term in this relation obviously corresponds to the dissociation energy D of a diatomic molecule. Parameter b depends only on the molecular gas under consideration. Thus, for CO molecules, $b = 0.90$ K$^{-0.5}$; for N_2, $b = 0.90$ K$^{-0.5}$; and for HCl, $b = 0.90$ K$^{-0.5}$.

Taking into account anharmonicity, the VV relaxation process between molecules of the same kind can include a multiquantum exchange. The simplest example here is the two-quantum resonance that occurs when a molecule with high vibrational energy (e.g., $\approx$ 3/4 D) can resonantly exchange two quanta for the only vibrational quantum of another molecule at low vibrational level. In general, such processes can be considered as intermolecular VV′ exchange (see below), and their small probability can be estimated by Equation 3.72.

3.5.3 Intermolecular VV′ Exchange

Vibrational energy exchange between molecules of a different kind is usually referred to as VV′ exchange. Consider first the VV′ exchange in a mixture of diatomic molecules A and B with only slightly different vibrational quanta $\hbar\omega_A > \hbar\omega_B$. In the same manner as was done in the VV exchange of anharmonic molecules of the same kind, here, also, the adiabatic factors mostly determine the small probability of the process. Then, Equation 3.73 can be generalized to describe the VV′ exchange when a molecule A transfers a quantum ($v_A + 1 \rightarrow v_A$) to a molecule B ($v_B + 1 \rightarrow v_B$):

$$Q_{v_A+1,v_A}^{v_B,v_B+1} = (v_A+1)(v_B+1)Q_{10}^{01}(AB)\exp\left(-\left|\delta_B v_B - \delta_A v_A + \delta_A p\right|\right)\exp(\delta_A p) \quad (3.79)$$

Here, $Q_{10}^{01}(AB)$ is the probability of a quantum transfer from A to B for the lowest levels; parameters δ_A and δ_B are related to the process adiabaticity and can be found based on Equation 3.74 and Equation 3.75, taking for each molecule a separate coefficient of anharmonicity: x_{eA} and x_{eB}. Parameter $p > 0$ is the vibrational level of the oscillator A, corresponding to the exact resonant transition $A(p+1 \to p) - B(0 \to 1)$ of a quantum from molecule A to B:

$$p = \frac{\hbar(\omega_A - \omega_B)}{2x_{eA}\hbar\omega_A} \quad (3.80)$$

The product $Q_{10}^{01}(AB)\exp(\delta_A p)$ in Equation 3.79 corresponds to the exact resonant exchange and does not include the adiabatic smallness; that is, $Q_{10}^{01}(AB)\exp(\delta_A p) \approx (\Omega\tau_{col})^2$, where the factor $(\Omega\tau_{col})^2$ is about 10^{-2} to 10^{-4} and characterizes the resonant transition of a quantum (see Section 3.5.1). Taking this into account, Equation 3.79 for the one-quantum VV′ exchange can be rewritten for simplicity as:

$$Q_{v_A+1,v_A}^{v_B,v_B+1} = (v_A+1)(v_B+1)(\Omega\tau_{col})^2\exp\left(-\left|\delta_B v_B - \delta_A v_A + \delta_A p\right|\right) \quad (3.81)$$

A formula for calculating the nonresonant transition of the only quantum from a molecule A to B can be derived from Equation 3.81:

$$Q_{10}^{01}(AB) = (\Omega\tau_{col})^2\exp(-\delta p) \quad (3.82)$$

For numerical calculations of the nonresonant, one-quantum VV′ exchange between a nitrogen molecule with a similar one, Rapp[60] proposed the following semiempirical formula corresponding to Equation 3.82:

$$Q_{10}^{01}(AB) = 3.7 * 10^{-6} T_0\, ch^{-2}\left(0.174\Delta\omega\hbar/\sqrt{T_0}\right) \quad (3.83)$$

In this relation the defect of resonance $\hbar\Delta\omega$ should be expressed in cm^{-1}, and temperature T_0 in Kelvin degrees.

Equation 3.82 and Equation 3.83 obviously demonstrate that the probability of VV′ exchange decreases with growth of the defect of resonance $\hbar\Delta\omega$. Here, it should be noted that when the defect of resonance is quite large $\hbar\Delta\omega \approx \hbar\omega$, different multiquantum resonant VV′ exchange processes are able to assume importance. The probability of such multiquantum resonant VV′ exchange processes can be calculated by the following formula[48]:

$$Q_{n,n-s}^{m,m+r} = \frac{1}{r!s!}\frac{n!(m+r)!}{(n-s)!m!}Q_{s0}^{0r} \quad (3.84)$$

This formula correlates to Equation 3.71 in the case of transfer of the only quantum $r = s = 1$. To estimate the probability Q_{s0}^{0r}, it is necessary to take into account that this transition in the harmonic approximation is due to a term of expansion of the intermolecular interaction potential, including coordinates of two oscillators in powers r and s. Each power of these coordinates in the interaction potential corresponds to a small factor $\Omega\tau_{col}$ in the expression for probability. As a result:

$$Q_{s0}^{0r} \propto \left(\Omega\tau_{col}\right)^{r+s} \text{ and } Q_{n,n-s}^{m,m+r} = \frac{1}{r!s!}\frac{n!(m+r)!}{(n-s)!m!}\left(\Omega\tau_{col}\right)^{r+s} \tag{3.85}$$

This formula for multiquantum resonant VV′ exchange obviously correlates with that for multiquantum resonant VV exchange (Equation 3.72), if $r = s$ and $m = 0$. Rate coefficients of some nonresonant VV′ exchange processes at room temperature, interesting for plasma chemical applications, are given in Table 3.17.

3.5.4 VV Exchange of Polyatomic Molecules

Vibrational modes of a polyatomic molecule can be considered separately at a low level of excitation (see Equation 3.12 and Equation 3.14) and their VV exchange can be calculated using the same formulas as those described previously (see Table 3.17). For higher levels of excitation corresponding to the vibrational quasi continuum of polyatomic molecules (see Section 3.2.4 and Equation 3.14), the mean square of vibrational energy transferred in the elementary process of VV exchange $\left\langle \Delta E_{VV}^2 \right\rangle$ can be found similarly to Equation 3.66 for VT relaxation:[36]

$$\left\langle \Delta E_{VV}^2 \right\rangle_{12} = \iint_{0,\infty} \left(\Omega\tau_{col}\right)_{VV}^2 I_1(\omega_1) I_2(\omega_2) \exp\left(-\frac{|\omega_1 - \omega_2|}{\alpha v}\right) (\hbar\omega_1)^2 \, d\omega_1 \, d\omega_2 \tag{3.86}$$

In this relation, $I(\omega)$ is the vibrational Fourier spectrum of a polyatomic molecule that can be described by the Lorentz profile (Equation 3.19) and indices 1 and 2 are, respectively, the quantum-transferring and quantum-accepting molecules.

Consider the VV exchange within one type of vibration between a molecule from the discrete spectrum region (in the first excited state, frequency ω_0) and a molecule in the quasi continuum. Taking the vibrational spectra of the two molecules as Equation 3.18 and Equation 3.21 and an average factor $(\Omega\tau_{col})^2$ for the chosen mode of vibrations, Equation 3.86 can be rewritten as:

$$\left\langle \Delta E_{VV}^2 \right\rangle = \left(\Omega\tau_{col}\right)_{VV}^2 \left(\hbar\omega_0\right)^2 \int_0^\infty \frac{1}{\pi} \frac{n\delta}{\left(\omega - \omega_n\right)^2 + \delta^2} \exp\left(-\frac{|\omega_0 - \omega|}{\alpha v}\right) d\omega \tag{3.87}$$

Here, n and ω_n are the number of quanta and corresponding frequency value (Equation 3.19 and Equation 3.20). If both colliding partners are in the discrete vibrational spectrum region, which corresponds to $\delta \to 0$, then the VV exchange is obviously nonresonant. Exponential smallness of the transferred energy (Equation 3.87) is then due to the adiabatic factor $\exp(-|\omega_0 - \omega|/\alpha v)$. If one of the molecules is in quasi continuum, the line width δ usually exceeds the anharmonic shift $|\omega_0 - \omega|$[61] and, as seen from Equation 3.87, the VV exchange is resonant:

Table 3.17 Rate Coefficients of Nonresonant VV′ Relaxation[a]

VV′ exchange process	$k_{VV'}$, cm^3/sec	VV′ exchange process	$k_{VV'}$, cm^3/sec
$H_2(v = 0) + HF(v = 1) \rightarrow H_2(v = 1) + HF(v = 0)$	$3*10^{-12}$	$O_2(v = 0) + CO(v = 1) \rightarrow O_2(v = 1) + CO(v = 0)$	$4*10^{-13}$
$DCl(v = 0) + N_2(v = 1) \rightarrow DCl(v = 1) + N_2(v = 0)$	$5*10^{-14}$	$N_2(v = 0) + CO(v = 1) \rightarrow N_2(v = 1) + CO(v = 0)$	10^{-15}
$N_2(v = 0) + HCl(v = 1) \rightarrow N_2(v = 1) + HCl(v = 0)$	$3*10^{-14}$	$N_2(v = 0) + HF(v = 1) \rightarrow N_2(v = 1) + HF(v = 0)$	$5*10^{-15}$
$NO(v = 0) + CO(v = 1) \rightarrow NO(v = 1) + CO(v = 0)$	$3*10^{-14}$	$O_2(v = 0) + COl(v = 1) \rightarrow O_2(v = 1) + CO(v = 0)$	10^{-16}
$DCl(v = 0) + CO(v = 1) \rightarrow DCl(v = 1) + CO(v = 0)$	10^{-13}	$CO(v = 0) + D_2(v = 1) \rightarrow CO(v = 1) + D_2(v = 0)$	$3*10^{-14}$
$CO(v = 0) + H_2(v = 1) \rightarrow CO(v = 1) + H_2(v = 0)$	10^{-16}	$NO(v = 0) + CO(v = 1) \rightarrow NO(v = 1) + CO(v = 0)$	10^{-14}
$HCl(v = 0) + CO(v = 1) \rightarrow HCl(v = 1) + CO(v = 0)$	10^{-12}	$HBr(v = 0) + CO(v = 1) \rightarrow HBr(v = 1) + CO(v = 0)$	10^{-13}
$HJ(v = 0) + CO(v = 1) \rightarrow HJ(v = 1) + CO(v = 0)$	10^{-14}	$HBr(v = 0) + HCl(v = 1) \rightarrow HBr(v = 1) + HCl(v = 0)$	10^{-12}
$HJ(v = 0) + HCl(v = 1) \rightarrow HJ(v = 1) + HCl(v = 0)$	$2*10^{-13}$	$DCl(v = 0) + HCl(v = 1) \rightarrow DCl(v = 1) + HCl(v = 0)$	10^{-13}
$O_2(v = 0) + CO_2(01^10) \rightarrow O_2(v = 1) + CO_2(00^00)$	$3*10^{-15}$	$H_2O(000) + N_2(v = 1) \rightarrow H_2O(010) + N_2(v = 0)$	10^{-15}
$CO_2(00^00) + N_2(v = 1) \rightarrow CO_2(00^01) + N_2(v = 0)$	10^{-12}	$CO(v = 1) + CH_4 \rightarrow CO(v = 1) + CH_4^*$	10^{-14}
$CO(v = 1) + CF_4 \rightarrow CO(v = 1) + CF_4^*$	$2*10^{-16}$	$CO(v = 1) + SF_6 \rightarrow CO(v = 1) + SF_6^*$	10^{-15}
$CO(v = 1) + SO_2 \rightarrow CO(v = 1) + SO_2^*$	10^{-15}	$CO(v = 1) + CO_2 \rightarrow CO(v = 1) + CO_2^*$	$3*10^{-13}$
$CO_2(00^00) + HCl(v = 1) \rightarrow CO_2(00^01) + HCl(v = 0)$	$3*10^{-13}$	$CO_2(00^00) + HF(v = 1) \rightarrow CO_2(00^01) + HF(v = 0)$	10^{-12}
$CO_2(00^00) + DF(v = 1) \rightarrow CO_2(00^01) + DF(v = 0)$	$3*10^{-13}$	$CO(v = 0) + CO_2(00^01) \rightarrow CO(v = 1) + CO_2(00^00)$	$3*10^{-15}$
$H_2O^* + CO_2 \rightarrow H_2O + CO_2^*$	$10–^{12}$	$O_2(v = 0) + CO_2(10^00) \rightarrow O_2(v = 1) + CO_2(00^00)$	10^{-13}
$CS_2(00^00) + CO(v = 1) \rightarrow CS_2(00^01) + CO(v = 0)$	$3*10^{-13}$	$N_2O(00^00) + CO(v = 1) \rightarrow N_2O(00^01) + CO(v = 0)$	$3*10^{-12}$

[a] $T_0 = 300$ K.

$$\left\langle \Delta E_{VV}^2 \right\rangle = \left(\Omega\tau_{col}\right)_{VV}^2 n \frac{\alpha v}{\delta + \alpha v} \left(\hbar\omega_0\right)^2 \tag{3.88}$$

This leads to an interesting conclusion: in contrast to diatomic molecules (see Equation 3.73 and Equation 3.77), the VV exchange with a polyatomic molecule in the quasi continuum is resonant and the corresponding rate coefficient does not decrease with a growth in the excitation level. Now we can compare the rate coefficients of the VV exchange (with a molecule in the first excited state) and VT relaxation in the same manner as was done for diatomic molecules (Equation 3.77). If the nonresonant term prevails in VT relaxation (Equation 3.68), the ratio of the rate coefficients for polyatomic molecules can be expressed as:

$$\xi(n) = \frac{\left\langle \Delta E_{VT}^2 \right\rangle (n)}{\left\langle \Delta E_{VV}^2 \right\rangle (n)} \approx \frac{\left(\Omega\tau_{col}\right)_{VT}^2}{\left(\Omega\tau_{col}\right)_{VV}^2} \exp\left(-\frac{\omega_n - \delta(n)}{\alpha v} \right) \tag{3.89}$$

As one can see from the relation, the VV exchange proceeds faster than the VT relaxation $\xi(n) << 1$ at low levels of excitation (as in the case of diatomic molecules). Intermode exchange frequency $\delta(n)$ is growing with n faster than $\Delta\omega = \omega_0 - \omega_n$. For this reason, $\xi(n)$ reaches unity and the VT relaxation catches up with the VV exchange at lower levels of vibrational excitation than in the case of diatomic molecules (Equation 3.78). In a plasma, it is much more difficult to sustain the high population of upper vibrational levels of polyatomic molecules in quasi continuum than those of diatomic molecules.

VV quantum transfer from a weakly excited diatomic molecule to a highly excited one results in transfer of the anharmonic defect of energy $\hbar\omega_0 - \hbar\omega_n$ into translational degrees of freedom. This effect, which is so obvious in the case of diatomic molecules, also takes place in the case of polyatomic molecules in the quasi continuum. The average value of vibrational energy $\langle \Delta E_T \rangle$ transferred into translational degrees of freedom in the process of VV quantum transfer from a molecule in the first (discrete) excited state (frequency ω_0) to a molecule in quasi continuum within the framework of this approach can be found as:

$$\left\langle \Delta E_T \right\rangle = \frac{\left(\Omega\tau_{col}\right)_{VV}^2}{\pi} \iint_{0,\infty} \frac{n\delta(n)}{\left(\omega' - \omega_n\right)^2 + \delta^2(n)} \exp\left(-\frac{\left|\omega' - \omega''\right|}{\alpha v} \right) \times \tag{3.90}$$
$$\delta\left(\omega'' - \omega_0\right) \hbar\left(\omega'' - \omega'\right) d\omega' \, d\omega''$$

In the quasi continuum under consideration, the principal frequency is ω_n, with the effective number of quanta n. After integrating Equation 3.90 and taking into account that $\delta > \left|\omega_n - \omega_0\right|$, the expression for vibrational energy $\langle \Delta E_T \rangle$ transferred into translational degrees of freedom during the VV exchange can be written as:

$$\left\langle \Delta E_T \right\rangle = n \left(\Omega\tau_{col}\right)_{VV}^2 \frac{\alpha v}{\alpha v + \delta} \left(\hbar\omega_0 - \hbar\omega_n\right) \tag{3.91}$$

Comparison of Equation 3.91 and Equation 3.88 shows that each act of the VV exchange between a weakly excited diatomic molecule and a highly excited one in quasi continuum is actually nonresonant and leads to the transfer of the anharmonic defect of energy $\hbar\omega_0 - \hbar\omega_n$ into translational motion. This effect can be interpreted in the quasi-classical approximation as conservation in the collision process of the adiabatic invariant-shortened action, which is the ratio of vibrational energy of each oscillator to its frequency:[62]

$$\frac{E_{v1}}{\omega_1} + \frac{E_{v2}}{\omega_2} = const \tag{3.92}$$

The conservation of the adiabatic invariant corresponds in quantum mechanics to conservation of the number of quanta during a collision. As is clear from Equation 3.92, the transfer of vibrational energy E_v from a higher-frequency oscillator to a lower-frequency one results in a decrease of the total vibrational energy and thus losses of energy into translational motion (Equation 3.91). The ratio of rate coefficients of the direct (considered previously) and reverse processes of the VV relaxation energy transfer to highly excited molecules can be found based on the relation (Equation 3.91) and the principle of detailed balance:

$$\frac{k_{VV}^{+}(T_v, T_0)}{k_{VV}^{-}(T_v, T_0)} = \left(1 + \frac{\hbar\omega_0}{E_v}\right)^{s-1} \exp\left(-\frac{\hbar\omega_0}{T_v} + \frac{\hbar(\omega_0 - \omega_n)}{T_0}\right) \tag{3.93}$$

In this relation, s is the effective number of vibrational degrees of freedom of the molecule and T_0 and T_v are the translational and vibrational temperatures. The ratio (Equation 3.93) is indicative of the possibility of super-equilibrium populations of the highly vibrationally excited states at $T_0 < T_v$, which is typical for nonequilibrium discharges in molecular gases. The Treanor effect in the nonequilibrium plasma chemistry related to those super-equilibrium populations of the highly vibrationally excited states will be discussed in following chapters.

3.6 PROCESSES OF ROTATIONAL AND ELECTRONIC RELAXATION OF EXCITED MOLECULES

3.6.1 Rotational Relaxation

Rotational–rotational (RR) and rotational–translational (RT) energy transfer processes (relaxation) are usually nonadiabatic because rotational quanta are generally small (see Section 3.2.5) for not very highly excited molecules and thus the Massey parameter is also small. As a result in this case, collision of a rotator with an atom or another rotator can be considered as a free, classical collision accompanied by a neccessary energy transfer. This means that the probability of RT and RR relaxation is actually very high.

A formula for calculating the number of collisions necessary for the "total" energy exchange between rotational and translational degrees of freedom (RT relax-

Table 3.18 Number of Collisions Z_{rot} Necessary for RT Relaxation in One-Component Gases

Molecule	$Z_{rot}^{\infty}(T_0 \to \infty)$	$Z_{rot}(T_0 = 300K)$	Molecule	$Z_{rot}^{\infty}(T_0 \to \infty)$	$Z_{rot}(T_0 = 300K)$
Cl_2	47.1	4.9	N_2	15.7	4.0
O_2	14.4	3.45	H_2	—	~500
D_2	—	~200	CH_4	—	15
CD_4	—	12	$C(CH_3)_4$	—	7
SF_6	—	7	CCl_4	—	6
CF_4	—	6	$SiBr_4$	—	5
SiH_4	—	28			

ation) was proposed by Parker[63] as a function of translational temperature T_0. The Parker formula was later corrected by Bray and Jonkman:[64]

$$Z_{rot} = \frac{Z_{rot}^{\infty}}{1+\left(\pi/2\right)^{3/2}\sqrt{\frac{T^*}{T_0}}+\left(2+\pi^2/4\right)\frac{T^*}{T_0}} \tag{3.94}$$

In this relation, T^* is the depth (energy) of the potential well corresponding to intermolecular attraction; $Z_{rot}^{\infty} = Z_{rot}(T_0 \to \infty)$ is the number of collisions necessary for RT relaxation at very high temperatures.

The increase of the collision number Z_{rot} with temperature in the Parker formula becomes clear noting that the intermolecular attraction accelerates the RT energy exchange. The attraction effect obviously becomes less effective when the translational temperature increases, thus explaining Equation 394. Numerical values of the collision numbers that are necessary for the RT relaxation are given in Table 3.18 for room temperature and for high temperatures $Z_{rot}^{\infty} = Z_{rot}(T_0 \to \infty)$. These two parameters permit determining the energy depth T^* of a potential well corresponding to the intermolecular attraction in the Parker formula. It actually helps to calculate the number of collisions (and thus the rotational relaxation rate coefficient) at any temperature, using Equation 3.94.

As one can see from Table 3.18, the factor Z_{rot} is usually small; at room temperature, it is approximately 3 to 10 for most molecules. An exception occurs in the case of hydrogen and deuterium, where Z_{rot} is approximately 200 to 500. This is due to high values of rotational quanta and relatively high Massey parameters, which make this process slightly adiabatic.

The preceding discussion was related to the common situation of the RT relaxation at relatively low levels of rotational excitation when the rotational quantum is very small. It is necessary to take into account, however, that the rotational quantum $2B(J + 1)$ grows with the level of rotational excitation and rotational quantum number (see Section 3.2.5). As a result, the RT relaxation of the highly rotationally excited molecules can become much more adiabatic and thus much slower than was discussed before and shown in Table 3.18.

In general, one can conclude that RT relaxation (and, furthermore, RR relaxation) is a fast process usually requiring only several collisions. In most systems under consideration, the rates of the rotational relaxation processes are comparable with the rate of thermalization. Thermalization, or TT relaxation, is a process sustaining equilibrium inside translational degrees of freedom. Thus, under most conditions taking place (even in nonequilibrium discharges), the rotational degrees of freedom can be considered to be in quasi equilibrium with the translational degrees of freedom and can be characterized by the same temperature T_0.

3.6.2 Relaxation of Electronically Excited Atoms and Molecules

Electronically excited atoms and molecules have many different channels of relaxation—channels to transfer energy into other degrees of freedom. For example, super-elastic collisions and energy transfer from electronically excited particles back to plasma electrons can be of great importance here. Also, the spontaneous radiation on the permitted transitions makes a much more important contribution than in the case of vibrational excitation (because of much higher transition frequencies). These processes will be considered later in the chapters related to plasma chemical kinetics. This section will only discuss the fundamental basics of the relaxation of electronically excited atoms and molecules in collisions with other heavy neutral particles.

The relaxation of electronic excitation, which is usually related to transfer of several electronvolts into translational degrees of freedom, is a strongly adiabatic process with very high Massey parameters ($\omega\tau_{col} \sim$ 100 to 1000). In this case, the relaxation is very slow, with low probability and low values of cross section. Thus, for example, the relaxation process:

$$Na(3^2P) + Ar \rightarrow Na + Ar \tag{3.95}$$

has a cross section not exceeding 10^{-19} cm^2, which corresponds to probability 10^{-9} or less.

Sometimes, however, the relaxation processes similar to Equation 3.95 with very high values of the Massey parameters can have pretty high probabilities. Thus, relaxation of electronically excited oxygen atoms $O(^1D)$ on heavy atoms of noble gases requires only several collisions. This effect can be explained by taking into account that the Massey parameter includes transition frequency in a quasi molecule formed during a collision. This frequency can be lower than that directly calculated from energy of the electronically excited atom.

Electronically excited atoms and molecules obviously can transfer their energy, not only into translational, but also into vibrational and rotational degrees of freedom. These processes can proceed with a higher probability because of the lower energy losses from the internal degrees of freedom and thus smaller Massey parameters. Even stronger effects can be achieved by formation of intermediate ionic complexes. For example, the relaxation probability is close to unity in the case of electronic

Table 3.19 Cross Sections of Nonadiabatic Relaxation of Electronically Excited Sodium Atoms

Relaxation processes	Cross sections, cm^2
$Na^* + N_2(v=0) \rightarrow Na + N_2^*(v>0)$	$2*10^{-14} cm^2$
$Na^* + CO_2(v=0) \rightarrow Na + CO_2^*(v>0)$	$10^{-14} cm^2$
$Na^* + Br_2(v=0) \rightarrow Na + Br_2^*(v>0)$	$10^{-13} cm^2$

energy transfer from excited metal atoms Me* to vibrational excitation of nitrogen molecules:

$$Me^* + N_2(v = 0) \rightarrow Me^+N_2^- \rightarrow Me + N_2(v > 0) \quad (3.96)$$

In this case, initial and final energy terms of the interaction Me–N_2 cross the ionic one (see illustration of the process in Figure 3.20), providing the transition between them.[65] Massey parameters of the transitions near crossings of the terms are low, which makes the relaxation processes nonadiabatic and very fast. The maximum cross sections of such processes of relaxation of excited sodium atoms are given in Table 3.19.

3.6.3 Electronic Excitation Energy Transfer Processes

These processes can be effective only very close to resonance, within about 0.1 eV or less. This means they can take place only for a limited set of collision partners. The He–Ne gas laser provides practically important examples of these kinds of highly resonant processes:

$$He(^1S) + Ne \rightarrow He + Ne(5s) \quad (3.97)$$

$$He(^3S) + Ne \rightarrow He + Ne(4s) \quad (3.98)$$

These processes of transfer of excitation from electronically excited helium atoms lead to a population inversion for the 4*s* and 5*s* levels of neon atoms. Subsequently, this effect results in coherent radiation from the He–Ne laser. Interaction radius for collisions of electronically excited atoms and molecules is usually large (up to 1 nm). For this reason, the values of cross sections of the electronic excitation transfer obviously can be very large if these processes are quite close to the resonance. For example, such a resonance electronic excitation transfer process from mercury atoms to atoms of sodium has a large value of cross section reaching 10^{-14} cm^2.

It should be noted that the electronic excitation transfer in collision of heavy particles could also take place as a fast nonadiabatic transfer inside the intermediate quasi molecule formed during the collision. However, this fast nonadiabatic transfer requires crossing of the corresponding potential energy surfaces. Such a crossing of terms becomes possible in interatoms' collisions only at a very close distance between nuclei, which in traditional plasma systems corresponds to nonrealistic interaction energies of about 10 keV.

When the degree of ionization in plasma exceeds values of about 10^{-6}, the electronic excitation transfer can occur faster by interaction with the electron gas than in collisions of heavy particles. Such excitation transfer proceeds through the super-elastic collisions of the excited heavy particles with electrons (deactivation) followed by their excitation through electron impact. These processes can dominate the energy exchange between electronically excited atoms and molecules and will be discussed under electronic excitation kinetics in Chapter 5. In some cases, effective electronic excitation transfer can be provided by radiative transitions.

3.7 ELEMENTARY CHEMICAL REACTIONS OF EXCITED MOLECULES; FRIDMAN–MACHERET α-MODEL

High concentration of excited atoms and molecules in electric discharges leads to high rates of many different plasma chemical reactions. Vibrational and electronic excitation plays the most important role in stimulation of endothermic processes in plasma (although the special effect of translational degrees of freedom will be considered in the next section). The chemical processes stimulated by vibrational excitation provide the highest energy efficiency in plasma (see Chapter 5) and their description is more general as well. For this reason, they will be considered in this section in more detail. Some relations and discussions related to vibrationally excited molecules also can be applied to some extent to electronically excited particles.

3.7.1 Rate Coefficient of Reactions of Excited Molecules

A convenient formula for calculation of the rate coefficient of an elementary reaction of an excited molecule with vibrational energy E_v at translational gas temperature T_0 can be derived using the information theory approach:[66]

$$k_R(E_v, T_0) = k_{R0} \exp\left(-\frac{E_a - \alpha E_v}{T_0}\right) \theta(E_a - \alpha E_v) \tag{3.99}$$

In this relation, E_a is the Arrhenius activation energy of the elementary chemical reaction; coefficient α is the efficiency of vibrational energy in overcoming the activation energy barrier; k_{R0} is the pre-exponential factor of the reaction rate coefficient; and $\theta(x - x_0)$ is the Heaviside function ($\theta(x - x_0) = 1$ when $x > 0$ and $\theta(x - x_0) = 0$ when $x < 0$).

According to Equation 3.99, the reaction rates of vibrationally excited molecules follow the traditional Arrhenius law. The activation energy in the Arrhenius law is reduced by the value of vibrational energy taken with efficiency α. If the vibrational temperature exceeds the translational one $T_v >> T_0$ and the chemical reaction is mostly determined by the vibrationally excited molecules, then Equation 3.99 can be simplified:

$$k_R(E_v) = k_{R0}\,\theta(\alpha E_v - E_a) \tag{3.100}$$

The effective energy barrier of chemical reaction (or effective activation energy) in this case is equal to E_a/α. The pre-exponential factor can be calculated in the framework of the transition state theory,[48] but quite often can be taken simply as the gas-kinetic rate coefficient $k_{R0} \approx k_0$.

Equation 3.99 describes the rate coefficients of reactions of vibrationally excited molecules after averaging over the Maxwellian distribution function for translational degrees of freedom. The probability of these reactions without the averaging can be found based on the LeRoy formula.[67] This formula gives the probability $P_v(E_v,E_t)$ of the elementary reaction as a function of vibrational E_v and translational E_t energies of the colliding partners:

$$P_v(E_v,E_t) = 0, \quad \text{if} \quad E_t < E_a - \alpha E_v,\ E_a \geq \alpha E_v \tag{3.101a}$$

$$P_v(E_v,E_t) = 1 - \frac{E_a - \alpha E_v}{E_t}, \quad \text{if} \quad E_t > E_a - \alpha E_v,\ E_a \geq \alpha E_v \tag{3.101b}$$

$$P_v(E_v,E_t) = 1, \quad \text{if} \quad E_a < \alpha E_v, \quad \text{any } E_t \tag{3.101c}$$

Averaging $P_v(E_v,E_t)$ over the Maxwellian distribution of the translational energies E_t obviously gives Equation 3.99, which is actually the most important relation for calculation of the elementary reaction of excited particles. As one can see from Equation 3.99 through Equation 3.101, the kinetics of chemical reactions of vibrationally excited molecules mostly depends on two parameters: the activation energy E_a and the efficiency of vibrational energy α. Different theoretical methods of determining these two key parameters will be next considered.

3.7.2 Potential Barriers of Elementary Chemical Reactions: Activation Energy

Extensive information on chemical kinetic data, including values of the activation energies, was collected by Baulch[68] and Kondratiev.[69] Today, the easiest access to this information is through the NIST chemical kinetics database on the World Wide Web. However, these databases are usually not sufficient for describing the detailed mechanism of complicated plasma chemical reactions. In these cases, the activation energies can be found theoretically from the potential energy surfaces, calculated using such sophisticated quantum mechanical methods as the LEPS (London–Eyring–Polanyi–Sato) method[70] or the DCM (diatomic complexes in molecules) method.[29] Although an example of such a detailed quantum mechanical calculation will be presented later in this section, several simple semiempirical methods can be successfully applied to find the value of activation energy. The most convenient of them are presented here.

- **Polanyi–Semenov Rule.** According to this rule,[71] the activation energies for a group of similar exothermic chemical reactions can be found as:

$$E_a = \beta + \alpha\,\Delta H \tag{3.102}$$

In this formula, ΔH is a reaction enthalpy (negative for the exothermal reactions), and α and β are constants of the model. According to Semenov,[71] these parameters are equal to $\alpha = 0.25$ to 0.27 and $\beta = 11.5$ kcal/mol for the following large group of exchange reactions:

$H + RH \rightarrow H_2 + R$, $D + RH \rightarrow DH + R$, $H + RCHO \rightarrow$
$H_2 + RCO$, $H + RCl \rightarrow HCl + R$, $H + RBr \rightarrow HBr + R$;
$Na + RCl \rightarrow NaCl + R$, $Na + RBr \rightarrow NaBr + R$;
$CH_3 + RH \rightarrow CH_4 + R$, $CH_3 + RCl \rightarrow CH_3Cl + R$, $CH_3 + RBr \rightarrow CH_3Br + R$;
$OH + RH \rightarrow H_2O + R$.

Endothermic chemical reactions where the reaction enthalpy ΔH is positive can be considered as a reverse process with respect to the exothermic reactions. Thus, Equation 3.102 can be rewritten for them with the same coefficients α and β:[72]

$$E_a = \Delta H + \left(\beta + \alpha(-\Delta H)\right) = \beta + (1-\alpha)\Delta H \tag{3.103}$$

- **Kagija Method.** This method was developed to find the activation energy of exchange reactions $A + BC \rightarrow AB + C$, based on the dissociation energy D of the molecule BC; the reaction enthalpy ΔH; and the one semiempirical parameter γ.[73] The relation for activation energy in this case is similar to that of the Polanyi–Semenov rule:

$$E_a = \frac{D}{(2\gamma D - 1)^2} + \frac{2\gamma D(2\gamma^2 D^2 - 3\gamma D + 2)}{(2\gamma D - 1)^3}\Delta H \tag{3.104}$$

The Kagija parameter γ is fixed for groups of similar reactions:
  - For the reactions of detachment of hydrogen atoms by alkil radicals, $\gamma = 0.019$.
  - For the reactions of detachment of hydrogen atoms by halogen atoms, $\gamma = 0.025$.
  - For the reactions of detachment of halogen atoms from alkil–halides, the parameter γ grows from $\gamma = 0.019$ for chlorides to $\gamma = 0.03$ for iodides.
- **Sabo Method.** According to this approach, the activation energy can be found based on the sum of bond energies of the initial molecules ΣD_i and the final molecules ΣD_f with one semiempirical parameter a:[74]

$$E_a = \sum D_i - a\sum D_f \tag{3.105}$$

For exothermic substitution exchange reactions, the Sabo parameter $a = 0.83$; for endothermic substitution exchange reactions $a = 0.96$; for reactions of disproportioning $a = 0.60$; and for exchange reactions of inversion $a = 0.84$.

- **Alfassi–Benson Method.** This approach is related to the exchange reactions $R + AR' \rightarrow AR + R'$ and claims that the activation of the process depends mainly upon the sum A_e of electron affinities to the particles R and R' and obviously on the reaction enthalpy ΔH.[75] Such analysis of different reaction of atoms H, Na, O, and radicals OH, CH_3, OCH_3, and CF_3 results in the following semiempirical formula:

$$E_a = \frac{14.8 - 3.64 * A_e}{1 - 0.025 * \Delta H} \tag{3.106}$$

In this relation, A_e, ΔH, and E_a are expressed in kilocalories per mole. Several other semiempirical formulas for calculation of activation energies based on electron affinities of the radicals R and R' have been proposed.

3.7.3 Efficiency α of Vibrational Energy in Overcoming Activation Energy Barrier

The coefficient α is the key parameter in Equation 3.99, and thus the key parameter describing the influence of excited molecules in plasma on their chemical reaction rates. The database of these important parameters of nonequilibrium kinetics is relatively limited, however, because methods of their determination are quite sophisticated (some of these are reviewed by Levitsky and Macheret.[76] Numerical values of the coefficient α for different chemical reactions obtained experimentally, as well as by detailed modeling applying the method of classical trajectories, are presented in Table 3.20. A more detailed database concerning the α-coefficients can be found in Rusanov and Fridman.[9]

Information presented in Table 3.20 permits dividing chemical reactions into several classes (endothermic processes; exothermic processes; simple exchange; double exchange, etc.) with the most probable value of the coefficients α in each class. Such a classification[76] can be expressed in the form of Table 3.21. This table is useful for determining the efficiency of vibrational energy α in elementary reactions if it is not known experimentally or from special detailed modeling.

3.7.4 Fridman–Macheret α-Model

This model permits calculating the α-coefficient of the efficiency of vibrational energy in elementary chemical processes based primarily on information about the activation energies of the corresponding direct and reverse reactions.[77] It describes the exchange reaction $A + BC \rightarrow AB + C$ with the reaction path profile shown in Figure 3.21(A, B).

Vibration of the molecule BC can be taken into account using the approximation of the vibronic term;[55,78] this is also shown on the figure by a dash-dotted line. The energy profile, corresponding to the reaction of a vibrationally excited molecule with

Table 3.20 Efficiency α of Vibrational Energy in Overcoming the Activation Energy Barrier

Reaction	α_{exp}[a]	α_{MF}[b]	Reaction	α_{exp}[a]	α_{MF}[b]
$F + HF^* \rightarrow F_2 + H$	0.98	0.98	$F + DF^* \rightarrow F_2 + D$	0.99	0.98
$Cl + HCl^* \rightarrow H + Cl_2$	0.95	0.96	$Cl + DCl^* \rightarrow D + Cl_2$	0.99	0.96
$Br + HBr^* \rightarrow H + Br_2$	1.0	0.98	$F + HCl^* \rightarrow H + ClF$	0.99	0.96
$Cl + HF^* \rightarrow ClF + H$	1.0	0.98	$Br + HF^* \rightarrow BrF + H$	1.0	0.98
$J + HF^* \rightarrow JF + H$	1.0	0.98	$Br + HCl^* \rightarrow BrCl + H$	0.98	0.98
$Cl + HBr^* \rightarrow BrCl + H$	1.0	0.97	$J + HCl^* \rightarrow JCl + H$	1.0	0.98
$SCl + HCl^* \rightarrow SCl_2 + H$	0.96	0.98	$S_2Cl + HCl^* \rightarrow S_2Cl_2 + H$	0.98	0.97
$SOCl + HCl^* \rightarrow SOCl_2 + H$	0.95	0.96	$SO_2Cl + HCl^* \rightarrow SO_2Cl_2 + H$	1.0	0.96
$NO + HCl^* \rightarrow NOCl + H$	1.0	0.98	$FO + HF^* \rightarrow F_2O + H$	1.0	0.98
$O_2 + OH^* \rightarrow O_3 + H$	1.0	1.0	$NO + OH^* \rightarrow NO_2 + H$	1.0	1.0
$ClO + OH^* \rightarrow ClO_2 + H$	1.0	1.0	$CrO_2Cl + HCl^* \rightarrow CrO_2Cl_2 + H$	0.94	0.9
$PBr_2 + HBr^* \rightarrow PBr_3 + H$	1.0	0.97	$SF_5 + HBr^* \rightarrow SF_5Br + H$	1.0	0.98
$SF_3 + HF^* \rightarrow SF_4 + H$	0.89	0.98	$SF_4 + HF^* \rightarrow SF_5 + H$	0.97	0.99
$H + HF^* \rightarrow H_2 + F$	1.0	0.95	$D + HF^* \rightarrow HD + F$	1.0	0.95
$Cl + HF^* \rightarrow HCl + F$	0.96	0.97	$Br + HF^* \rightarrow HBr + F$	1.0	0.98
$J + HF^* \rightarrow HJ + F$	0.99	0.99	$Br + HCl^* \rightarrow HBr + Cl$	1.0	0.95
$J + HCl^* \rightarrow HJ + Cl$	1.0	0.98	$J + HBr^* \rightarrow HJ + Br$	1.0	0.96
$OH + HF^* \rightarrow H_2O + F$	1.0	0.95	$HS + HF^* \rightarrow H_2S + F$	1.0	0.98
$HS + HCl^* \rightarrow H_2S + Cl$	1.0	1.0	$HO_2 + HF^* \rightarrow H_2O_2 + F$	1.0	0.98
$HO_2 + HF^* \rightarrow H_2O_2 + F$	1.0	0.98	$NH_2 + HF^* \rightarrow NH_3 + F$	1.0	0.97
$SiH_3 + HF^* \rightarrow SiH_4 + F$	1.0	1.0	$GeH_3 + HF^* \rightarrow GeH_4 + F$	1.0	1.0
$N_2H_3 + HF^* \rightarrow N_2H_4 + F$	0.97	0.98	$CH_3 + HF^* \rightarrow CH_4 + F$	0.98	0.97
$CH_2F + HF^* \rightarrow CH_3F + F$	1.0	0.97	$CH_2Cl + HF^* \rightarrow CH_3Cl + F$	1.0	0.97
$CCl_3 + HF^* \rightarrow CHCl_3 + F$	0.98	0.98	$CH_2Br + HF \rightarrow CH_3Br + F$	1.0	0.97
$C_2H_5 + HF^* \rightarrow C_2H_6 + F$	1.0	0.99	$CH_2CF_3 + HF^* \rightarrow CH_3CF_3 + F$	1.0	0.98
$C_2H_5O + HF^* \rightarrow (CH_3)_2O + F$	1.0	1.0	$C_2H_5Hg + HF^* \rightarrow (CH_3)_2Hg + F$	0.99	0.97
$HCO + HF^* \rightarrow H_2CO + F$	1.0	0.99	$FCO + HF^* \rightarrow HFCO + F$	1.0	0.99
$C_2H_3 + HF^* \rightarrow C_2H_4 + F$	0.99	0.97	$C_3H_5 + HF^* \rightarrow C_3H_6 + F$	1.0	0.98
$C_6H_5 + HF^* \rightarrow C_6H_6 + F$	1.0	0.97	$CH_3 + JF^* \rightarrow CH_3J + F$	0.81	0.9
$S + CO^* \rightarrow CS + O$	1.0	0.99	$F + CO^* \rightarrow$ cf. $+ O$	1.0	1.0
$CS + CO^* \rightarrow CS_2 + O$	0.83	0.9	$CS + SO^* \rightarrow CS_2 + O$	0.90	0.95
$SO + CO^* \rightarrow COS + O$	0.96	0.92	$F_2 + CO^* \rightarrow CF_2 + O$	1.0	1.0
$C_2H_4 + CO^* \rightarrow (CH_2)_2\,C + O$	0.94	0.98	$CH_2 + CO^* \rightarrow C_2H_2 + O$	0.90	0.94
$H_2 + CO^* \rightarrow CH_2 + O$	0.96	—	$OH + OH^* \rightarrow H_2O + O(^1D_2)$	1.0	0.97
$NO + NO^* \rightarrow N_2O + O(^1D_2)$	1.0	—	$CH_3 + OH^* \rightarrow CH_4 + O(^1D_2)$	1.0	1.0
$O + NO^* \rightarrow O2 + N$	0.94	0.86	$H + BaF^* \rightarrow HF + Ba$	0.99	1.0
$O + AlO^* \rightarrow Al + O_2$	0.67	0.7	$O + BaO^* \rightarrow Ba + O_2$	1.0	0.99
$CO + HF^* \rightarrow CHF + O$	1.0	1.0	$CO_2 + HF^* \rightarrow O_2 + CHF$	1.0	—
$CO^* + HF \rightarrow CHF + O$	1.0	1.0	$CF_2O + HF^* \rightarrow CHF_3 + O(^1D_2)$	1.0	1.0
$H_2CO + HF^* \rightarrow CH_3F + O(^1D_2)$	1.0	1.0	$J + HJ^* \rightarrow J_2 + H$	1.0	1.0
$O + N_2 \rightarrow NO + N$ (nonadiabatic)	1.0	1.0	$O + N_2 \rightarrow NO + N$ (adiabatic)	0.6	1.0
$F + ClF^* \rightarrow F_2 + Cl$	1.0	0.93	$O + (CO^+)^* \rightarrow O_2 + C^+$	0.9	1.0

Table 3.20 (continued) Efficiency α of Vibrational Energy in Overcoming the Activation Energy Barrier

Reaction	α_{exp} [a]	α_{MF} [b]	Reaction	α_{exp} [a]	α_{MF} [b]
$H + HCl^* \rightarrow H_2 + Cl$	0.3	0.4	$NO + O_2^* \rightarrow NO_2 + O$	0.9	1.0
$O + H_2^* \rightarrow OH + H$	0.3	0.5	$O + HCl^* \rightarrow OH + Cl$	0.60	0.54
$H + HCl^* \rightarrow HCl + H$	0.3	0.5	$H + H_2^* \rightarrow H_2 + H$	0.4	0.5
$NO + O_3^* \rightarrow NO_2(^3B_2) + O_2$	0.5	0.3	$SO + O_3^* \rightarrow SO_2(^1B_1) + O_2$	0.25	0.15
$O^+ + N_2^* \rightarrow NO^+ + N$	0.1	—	$N + O_2^* \rightarrow NO + O$	0.24	0.19
$OH + H_2^* \rightarrow H_2O + H$	0.24	0.22	$F + HCl^* \rightarrow HF + Cl$	0.4	0.1
$H + N_2O^* \rightarrow OH + N_2$	0.4	0.2	$O_3 + OH^* \rightarrow O_2 + O + OH$	0.02	0
$H_2 + OH^* \rightarrow H_2O + H$	0.03	0	$Cl_2 + NO^* \rightarrow ClNO + Cl$	0	0
$O_3 + NO^* \rightarrow NO_2(^2A_1) + O_2$	0	0	$O_3 + OH^* \rightarrow HO_2 + O_2$	0.02	0

[a] α_{exp} = coefficient obtained experimentally or in detailed modeling.
[b] α_{MF} = coefficient found from the Fridman–Macheret α-model.

Table 3.21 Classification of Chemical Reactions for Determining Vibrational Energy Efficiency Coefficients α

Reaction type	Simple exchange	Simple exchange	Simple exchange	Double exchange
	Bond break in excited molecule	Bond break in nonexcited molecule (through complex)	Bond break in nonexcited molecule (direct)	
Endothermic	0.9–1.0	0.8	<0.04	0.5–0.9
Exothermic	0.2–0.4	0.2	0	0.1–0.3
Thermoneutral	0.3–0.6	0.3	0	0.3–0.5

energy E_v (the vibronic term), can be obtained in this approach by a parallel shift up of the initial profile $A + BC$ on the value of E_v. The part of the reaction path profile corresponding to products $AB + C$ remains the same, if the products are not excited.

From the simple geometry shown in Figure 3.21(B), it can be seen that the effective decrease of the activation energy related to the vibrational excitation E_v is equal to:

$$\Delta E_a = E_v \frac{F_{A+BC}}{F_{A+BC} + F_{AB+C}} \tag{3.107}$$

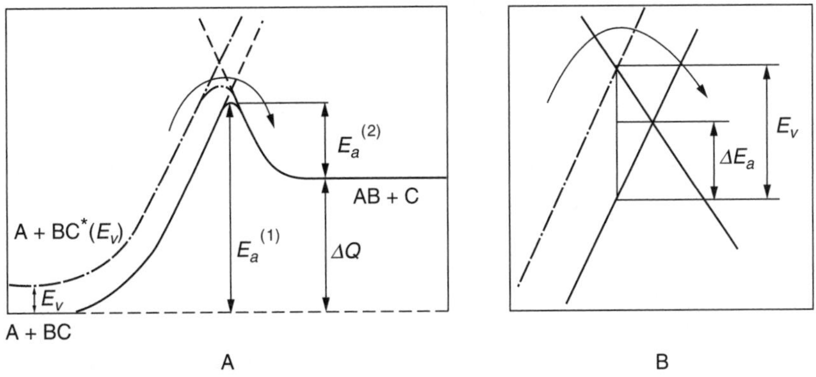

Figure 3.21 Efficiency of vibrational energy in a simple exchange reaction $A + BC \rightarrow AB + C$: (A) solid curve: reaction profile; dashed line: a vibronic term corresponding to A interaction with a vibration-excited molecule $BC^*(E_v)$; (B) part of the reaction profile near the barrier summit

In this relation, F_{A+BC} and F_{AB+C} denote the characteristic slopes of the terms $A + BC$ and $AB + C$ presented in Figure 3.21(B). If the energy terms exponentially depend on the reaction coordinates with the decreasing parameters γ_1 and γ_2 (reverse radii of the corresponding exchange forces), then, as is clear from the figure:

$$\frac{F_{A+BC}}{F_{AB+C}} = \frac{\gamma_1 E_a^{(1)}}{\gamma_2 E_a^{(2)}} \tag{3.108}$$

The subscripts "1" and "2" stand here for direct ($A + BC$) and reverse ($AB + C$) reactions. The decrease of activation energy (Equation 3.107) together with Equation 3.108 not only explains the main kinetic relation (Equation 3.99) for reactions of vibrationally excited molecules, but also determines the value of the coefficient α (which is actually equal to $\alpha = \Delta E_a/E_v$):

$$\alpha = \frac{\gamma_1 E_a^{(1)}}{\gamma_1 E_a^{(1)} + \gamma_2 E_a^{(2)}} = \frac{E_a^{(1)}}{E_a^{(1)} + \gamma_1/\gamma_2 \, E_a^{(2)}} \tag{3.109}$$

Usually, the exchange force parameters for direct and reverse reactions γ_1 and γ_2 are close to each other and their ratio is close to unity $\gamma_1/\gamma_2 \approx 1$, which leads to the approximate but very convenient main formula of the Fridman–Macheret α-model:

$$\alpha \approx \frac{E_a^{(1)}}{E_a^{(1)} + E_a^{(2)}} \tag{3.110}$$

The geometric derivation of this formula is very similar to the corresponding geometric derivation of the Polanyi–Semenov rule (Equation 3.102). Equation 3.110 is in good numerical agreement with the experimental data (see Table 3.20) and reflects the three most important tendencies of the α-coefficient:

- The efficiency α of the vibrational energy is highest, that is, close to unity (100%), for strongly endothermic reactions with activation energies close to the reaction enthalpy.
- The efficiency α of the vibrational energy is lowest, that is, close to zero, for exothermic reactions without activation energies.
- The sum of the α-coefficients for direct and reverse reactions is equal to unity $\alpha^{(1)} + \alpha^{(2)} = 1$.

Also, it should be noted that Equation 3.110 does not actually include any detailed information on the dynamics of the elementary chemical reaction nor the type of excitation. For this reason, one can expect good results applying the relation (Equation 3.110) to the wide range of different chemical reactions of excited species. The similarity of the Fridman–Macheret geometric approach with that of the Polanyi–Semenov approach (Equation 3.102) confirms this statement as well.

3.7.5 Efficiency of Vibrational Energy in Elementary Reactions Proceeding through Intermediate Complexes

As an example of such elementary reactions, consider the synthesis of lithium hydride:

$$Li\left({}^2S\right) + H_2^*\left({}^1\Sigma_g^+, v\right) \rightarrow LiH\left({}^1\Sigma^+\right) + H\left({}^2S\right) \qquad (3.111)$$

The minimum of the potential energy of the system Li–H_2 corresponds to the group of symmetry C_{2v}. The potential energy surface for the reaction with the C_{2v} configuration was calculated using the DCM method[79] and is presented in Figure 3.22. The corresponding reaction path profile is shown in Figure 3.23, where one can see the potential well with energy depth of about 0.4 eV.

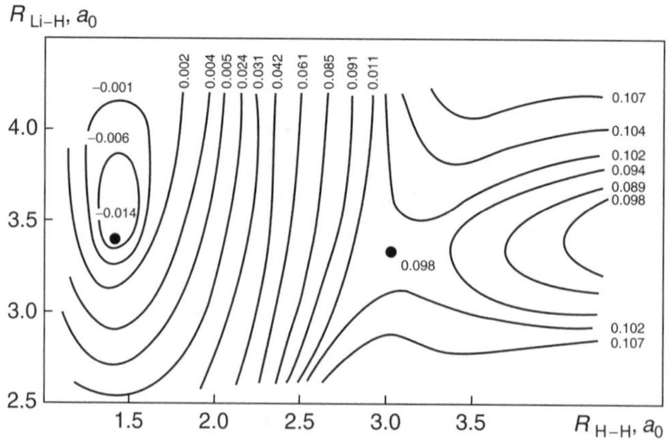

Figure 3.22 Li–H_2 potential energy surface, energy, and distances in atomic units

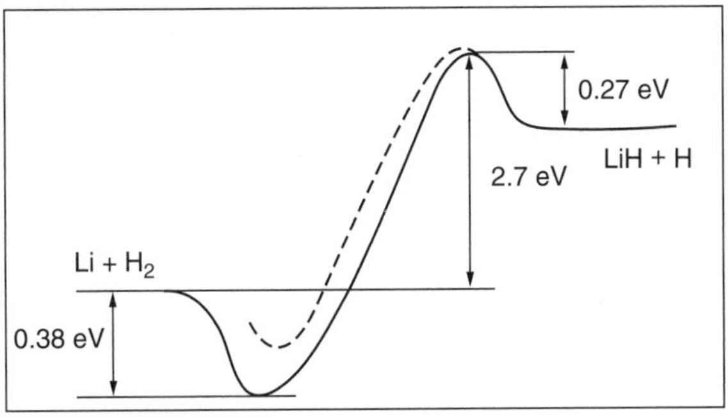

Figure 3.23 Reaction profile Li + H_2 → LiH + H. dashed line illustrates the effect of angular momentum conservation.

The potential well makes the lithium hydride synthesis proceed through an intermediate complex, which can be described using the statistical theory of chemical processes.[80] The total cross section of the reaction (Equation 3.111) is expressed in the framework of this theory as a product of the cross section of the intermediate complex formation σ_a and the probability of its subsequent decay with the formation of LiH + H. The attachment cross section σ_a at the fixed orbital quantum number l for the relative motion of the reactants can be written as:

$$\sigma_a^l = \frac{2\pi}{k^2}(2l+1)\frac{\hbar\Gamma}{\Delta E_k} \tag{3.112}$$

In this relation, ΔE_k is the average distance between energy levels of the complex; $\hbar\Gamma$ is the level energy width of the complex; and k is the wave vector of the relative motion of the reactants. Taking into account that, in Equation 3.112 $\hbar\Gamma/\Delta E_k = 1/2\pi$, and then considering the probabilities of different decay channels of the intermediate complex, the total cross section of the reaction (Equation 3.111) can be expressed in the following way:

$$\sigma_R(E) = \sum_l \sigma_a^l \frac{k_1(E,l)}{k_1(E,l)+k_{-1}(E,l)} = \sum_l \frac{\pi}{k^2}(2l+1)A_\omega\left(\frac{E-E_1^a}{E}\right)^{s-1}\theta\left(E-E_1^a\right) \tag{3.113}$$

Here, $E_1{}^a = 2.7$ eV is activation energy of Equation 3.111; k_1 and k_{-1} are the rate coefficients of the direct and reverse channels of decay of the intermediate complex; s is the number of vibrational degrees of freedom of the complex; $A_\omega \approx 3$ is the frequency factor; and $\theta(E - E_1{}^a)$ is the Heaviside function ($\theta(x - x_0) = 1$ when $x > 0$, and $\theta(x - x_0) = 0$ when $x < 0$). If E_v, E_r, and E_t are, respectively, the vibrational energy of the hydrogen molecule and the rotational and translational energies of the reacting particles, then the total effective energy of the intermediate complex E, calculated with respect to the initial level of potential energy, can be found as the following sum:

$$E = E_v + \alpha_r E_r + \alpha_t E_t \tag{3.114}$$

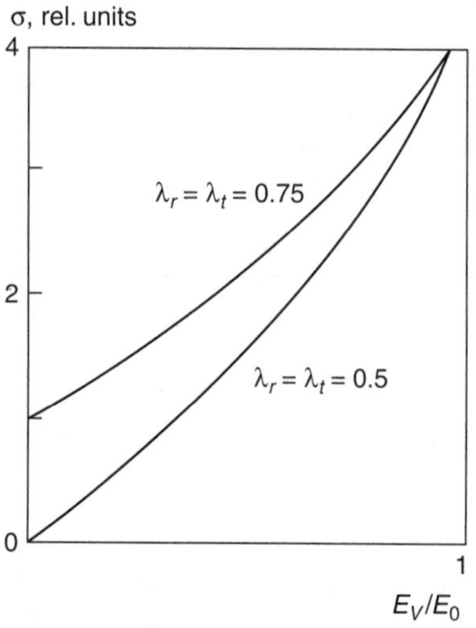

Figure 3.24 Contribution of different degrees of freedom in a cross section of the reaction

The efficiencies α_r and α_t of the rotational and translational energies in overcoming the activation energy barrier are less than 100% and can be estimated as:

$$\alpha_r \approx \alpha_t \approx 1 - \frac{r_e^2}{R_a^2} \tag{3.115}$$

The lower efficiencies of the rotational and translational energies are due to conservation of angular momentum. This results in part of the rotational and translational energies remaining in rotation at the last moment of decay of the complex (see Figure 3.23), when the characteristic distance between products is equal to R_a. In Equation 3.115, r_e is the equilibrium distance between hydrogen atoms in a hydrogen molecule H_2.

The cross section of the reaction (Equation 3.111) is shown in Figure 3.24 as a function of the fraction E_v/E_0 of vibrational energy E_v in the total energy E_0, assuming the fixed value of the total energy $E_0 = 2E_1{}^a = \text{const}$. As the figure shows, the vibrational energy of hydrogen is more efficient than rotational and translational energies in stimulating the typical reaction (Equation 3.111) proceeding through formation of the intermediate complex. According to Equation 3.113 Equation 3.114, the efficiency of the vibrational energy in such reactions is close to 100%.

Using this fact, it is possible to use the general formula (Equation 3.99) with the efficiency coefficient $\alpha \approx 1$ for stimulating the endothermic reactions going through complexes by vibrational excitation of reagents. This has also been confirmed in detailed consideration of the nonadiabatic Zeldovich reaction of NO synthesis:

$$O(^3P) + N_2^*(^1\Sigma_g^+) \Leftrightarrow N_2O^*(^1\Sigma^+) \rightarrow NO(^2P) + N(^4S) \tag{3.116}$$

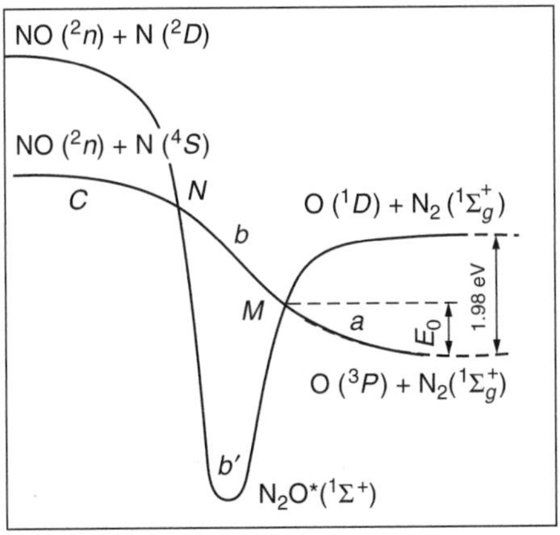

Figure 3.25 Reaction energy profile: $O + N_2 \rightarrow NO + N$

which is proceeding through formation of the intermediate complex (see Figure 3.25) and can be effectively stimulated ($\alpha = 1$) in plasma by vibrational excitation of nitrogen molecules[81] (see also Table 3.20). In this case, it is necessary to take into account that the pre-exponential factor k_{R0} in the relation (Equation 3.99) for such reactions is not constant, but is proportional to the statistical theory factor:[48]

$$k_{R0} \approx k_{R00} P_{L-Z} \left(\frac{E_v - E_a}{E_v} \right)^{s-1} \tag{3.117}$$

This statistical theory factor (in parentheses; see also Equation 3.113) reflects the relative probability of the localization of vibrational energy on the chemical bond, which is supposed to be broken; s is a number of degrees of freedom; the Landau–Zener factor P_{L-Z} shows the probabilities of transitions between electronic terms during formation and decay of the intermediate complex molecule (for the nonadiabatic reactions, see Figure 3.25).

3.7.6 Dissociation of Molecules Stimulated by Vibrational Excitation in Nonequilibrium Plasma

Dissociation of diatomic and polyatomic molecules in nonequilibrium plasma can be very effectively stimulated by their vibrational excitation. Such dissociation of diatomic molecules has a trivial elementary mechanism[81] and is mostly controlled by vibrational kinetics, which will be discussed in Chapter 5. A good and practically important example of the polyatomic molecule dissociation stimulated by vibrational excitation is the process of CO_2 decomposition in nonthermal discharges.[82] Kinetics of such processes will be discussed later. To analyze the elementary process, it is interesting simply to look at the scheme of electronic terms of CO_2 presented in Figure 3.26. From this figure, the adiabatic dissociation of CO_2 molecules (with spin conservation) in step-by-step vibrational excitation:

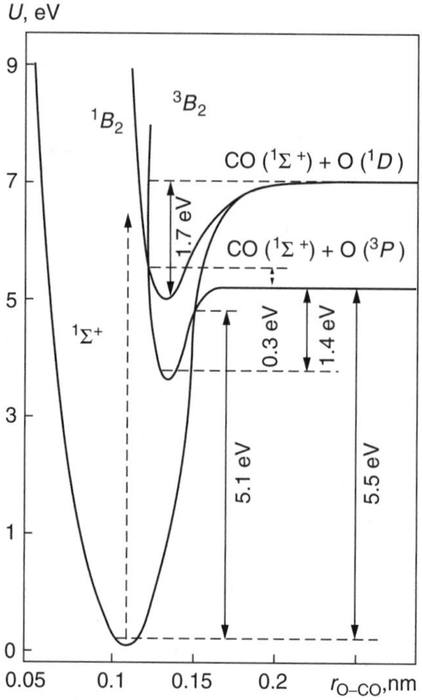

Figure 3.26 Low electronic terms of CO_2

$$CO_2*\left({}^1\Sigma^+\right)\rightarrow CO\left({}^1\Sigma^+\right)+O\left({}^1D\right) \tag{3.118}$$

leads to the formation of an oxygen atom in the electronically excited state and requires more than 7 eV of energy. At the same time, the nonadiabatic transition ${}^1\Sigma^+ \rightarrow {}^3B_2$ in the crossing point of the terms (change of spin during the transition) provides the more effective dissociation process stimulated by the step-by-step vibrational excitation:

$$CO_2*\left({}^1\Sigma^+\right)\rightarrow CO_2*\left({}^3B_2\right)\rightarrow CO\left({}^1\Sigma^+\right)+O\left({}^3D\right) \tag{3.119}$$

This process results in formation of an oxygen atom in the electronically ground state 3^P and requires spending exactly OC = O bond energy, which is equal to 5.5 eV.

Theoretical description of the elementary dissociation process stimulated by vibrational excitation of molecules is similar to the one discussed previously in Section 3.7.5 considering reactions going through complexes. The dissociation processes have high coefficients of efficiency of vibrational energy utilization in overcoming the activation energy barriers (the coefficient $\alpha = 1$ in Equation 3.99). The pre-exponential factors for the dissociation processes stimulated by vibrational excitation can also be described in frameworks of the statistical theory of the monomolecular reactions[48] or simply by using Equation 3.117. The dissociation rate coefficient for diatomic molecules obviously does not include the statistical theory factor because $s = 1$ in Equation 3.117.

3.7.7 Dissociation of Molecules in Nonequilibrium Conditions with Essential Contribution of Translational Energy

Vibrational temperatures in most nonthermal discharges exceed translational temperatures $T_v > T_0$, as discussed earlier. Some practically important phenomena such as dissociation of molecules after a shock wave front, however, are related to the opposite nonequilibrium conditions, when $T_v < T_0$. In this case, vibrational and translational temperatures make contributions into stimulation of dissociation. For these cases, it is convenient to describe the stimulation using the so-called nonequilibrium factor Z:[83]

$$Z(T_0, T_V) = \frac{k_R(T_0, T_v)}{k^0(T_0, T_0 = T_v)} = Z_h(T_0, T_v) + Z_l(T_0, T_v) \tag{3.120}$$

In this formula, $k_R(T_0, T_v)$ is the dissociation rate coefficient; $k_R^0(T_0, T_v = T_o)$ is the corresponding equilibrium rate coefficient for temperature T_0; and the nonequilibrium factors Z_h and Z_l are related to dissociation from high h and low l vibrational levels.

The nonequilibrium factor Z can be found using the **Macheret–Fridman model.**[84] The model is based on the assumption of classical impulsive collisions. Dissociation is considered to occur mainly through an optimum configuration that is a set of collisional parameters minimizing the energy barrier. This optimum configuration defines the threshold kinetic energy for dissociation, and through it, the exponential factor of the rate coefficient. The probability of finding the colliding system near the optimum configuration determines the pre-exponential factor of the rate. In the frameworks of the Macheret–Fridman model, the nonequilibrium factor for dissociation from the high vibrational levels can be expressed as:

$$Z_h(T_0, T_v) = \left\{ \frac{1 - \exp\left(-\frac{\hbar\omega}{T_v}\right)}{1 - \exp\left(-\frac{\hbar\omega}{T_0}\right)} (1 - L) \right\} \exp\left[-D\left(\frac{1}{T_v} - \frac{1}{T_0}\right)\right] \tag{3.121}$$

where D is the dissociation energy and L is the pre-exponential factor related to configuration of collisions. For dissociation from the low vibrational levels, which is applicable in this case, the Macheret–Fridman model gives:

$$Z_l(T_0, T_v) = L \exp\left[-D\left(\frac{1}{T_{eff}} - \frac{1}{T_0}\right)\right] \tag{3.122}$$

where the effective temperature is introduced by the following relation (m is mass of an atom in dissociating molecule and M is mass of an atom in the other colliding partner):

$$T_{eff} = \alpha T_v + (1-\alpha)T_0, \quad \alpha = \left(\frac{m}{M+m}\right)^2 \tag{3.123}$$

In contrast to the general formula (Equation 3.99) showing efficiency of vibrational energy, the relation Equation 3.123 shows efficiency of vibrational energy and efficiency of translational energy in dissociating molecules under nonequilibrium conditions. This formula is convenient for practical calculations because to find the dissociation rate coefficient only requires the conventional equilibrium rate coefficient and recalculated temperature.

Experimental values of the nonequilibrium factor Z are presented in Figure 3.27 together with the theoretical predictions. It is interesting to note that the nonequilibrium factor $Z(T_0)$ at $T_0 > T_v$ taken as a function of temperature T_0 corresponds to a curve with a minimum. When at first T_0 is still close to T_v, the dissociation is mostly due to contributions from high vibrational levels. While the translational temperature is growing in this case, the relative population of the high vibrational levels is decreasing; this results in a reduction of the nonequilibrium factor Z. When the translational temperature becomes very high, the dissociation becomes "direct," e.g., primarily due to strong collisions with vibrationally nonexcited molecules. In this case, the nonequilibrium factor Z starts growing. Thus, the minimum point of the function $Z(T_0)$ corresponds actually to the transition to direct dissociation mainly from vibrationally nonexcited states (see Figure 3.27).

Several semiempirical models have also been proposed to make simple estimations of the nonequilibrium dissociation at $T_0 > T_v$. An examination of some of the most used and comparison with the Macheret–Fridman model follow.

Park Model.[85] Because of its simplicity, this model is probably the most widely used in practical calculations. According to this model, the dissociation rate coefficient can be found using the conventional Arrhenius formula for equilibrium reactions, by simply replacing the temperature with a new effective value:

$$T_{eff} = T_0^s T_v^{1-s} \tag{3.124}$$

Here, the power s is a fitting parameter of the Park model that was finally recommended by the author to be taken as $s = 0.7$ (in first publications, however, the effective temperature was taken approximately as $T_{eff} = \sqrt{T_0 T_v}$). The Macheret–Fridman model permits theoretically finding the Park parameter as:

$$s = 1 - \alpha = 1 - \left(\frac{m}{m+M}\right)^2 \tag{3.125}$$

It can be seen that if two atomic masses are not very different, $m \sim M$, then the parameter $s \approx 0.75$ according to Equation 3.125, which is fairly close to the value $s = 0.7$ recommended by Park.

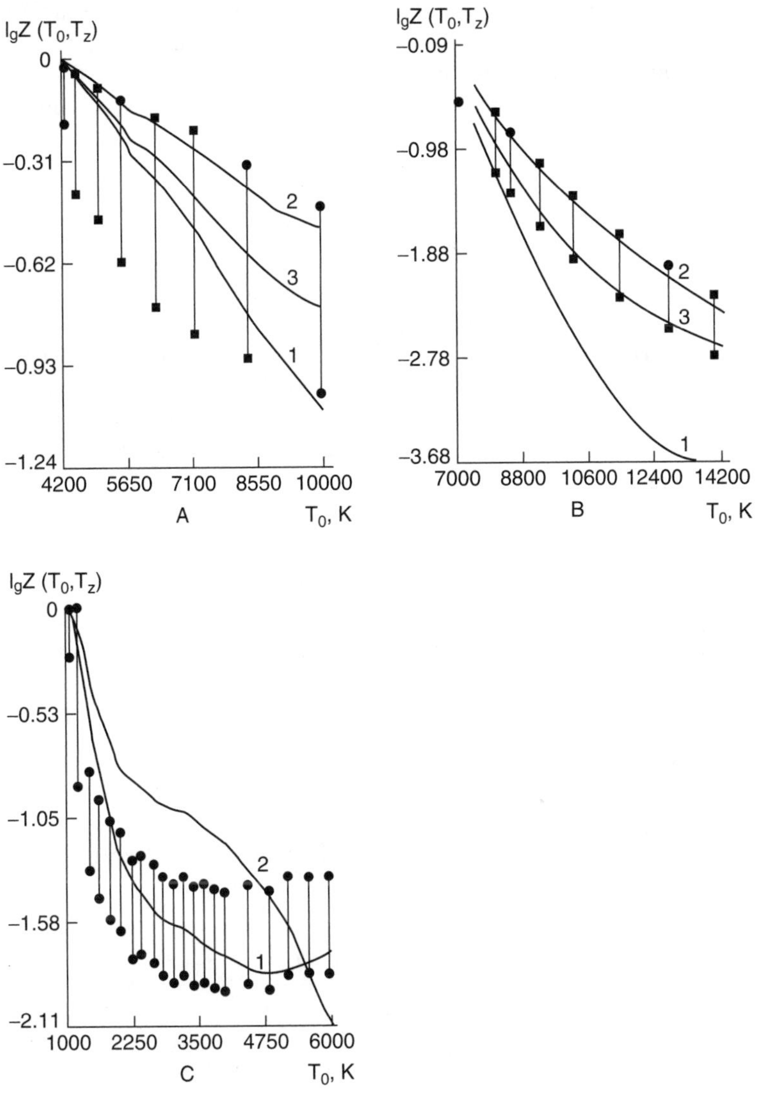

Figure 3.27 Experimental and theoretical values of the nonequilibrium factor Z in (A) oxygen; (B) nitrogen; (C) iodine; (1) Macheret–Fridman model; (2) Kuznetsov model; (3) Losev model[83]

Losev Model.[86] According to this model, the dissociation rate coefficient can be found using the conventional Arrhenius formula for equilibrium reactions with vibrational temperature T_v and effective value D_{eff} of dissociation energy. To find D_{eff}, the actual dissociation energy D should be decreased by the value of translational temperature taken with efficiency β:

$$D_{eff} = D - \beta T_0 \tag{3.126}$$

This relation is somewhat similar to Equation 3.99, with the vibrational and translational energies replacing each other. The coefficient β is the fitting parameter of the Losev model and, for estimations, it should be taken as $\beta \approx 1.0$ to 1.5. The Macheret–Fridman model permits theoretically finding the Losev semiempirical parameter as:

$$\beta \approx \frac{D(T_0 - T_v)}{T_0^2}, \quad T_0 > T_v \tag{3.127}$$

As is clear from Equation 3.127, the Losev parameter β can be considered as a constant only for a narrow range of temperatures.

Marrone–Treanor Model.[87] This model supposes the exponential distribution (with the parameter $-U$) of the probabilities of dissociation from different vibrational levels. If $U \to \infty$, the probabilities of dissociation from all vibrational levels are equal. The nonequilibrium factor Z can be found in the framework of this semiempirical model as:

$$Z(T_0, T_v) = \frac{Q(T_0)Q(T_f)}{Q(T_v)Q(-U)}, \quad T_0 > T_v \tag{3.128}$$

The statistical factors $Q(T_j)$ can be defined for the four "temperatures": translational temperature T_0; vibrational temperature T_v; and two special temperatures: $-U$ and

$$T_f = \frac{1}{\dfrac{1}{T_v} - \dfrac{1}{T_0} - \dfrac{1}{-U}}.$$

All are denoted below in general as T_j, using the following relation:

$$Q(T_j) = \frac{1 - \exp\left(-\dfrac{D}{T_j}\right)}{1 - \exp\left(-\dfrac{\hbar\omega}{T_j}\right)} \tag{3.129}$$

The negative temperature $-U$ is a fitting parameter of the model. Recommended values of this semiempirical coefficient U are $0.6T_0$ to $3T_0$. The Macheret–Fridman model permits theoretically finding the Marrone–Treanor semiempirical parameter as:

$$U = \frac{T_0[\alpha T_0 + (1-\alpha)T_v]}{(T_0 - T_v)(1-\alpha)} \tag{3.130}$$

In the same manner as the Losev parameter β, the Marrone–Treanor parameter U can be considered a constant only for a narrow range of temperatures. More detailed information about the two-temperature kinetics of dissociation for $T_0 > T_v$ can be found in Sergievska, Kovach, and Losev.[88]

3.7.8 Chemical Reactions of Two Vibrationally Excited Molecules in Plasma

As demonstrated in Section 3.7.4, the efficiency of the vibrational energy in overcoming the activation energy barriers of the strongly endothermic reactions is usually close to 100% ($\alpha = 1$). However, it was assumed in this statement that only one of the reacting molecules keeps the vibrational energy (see Table 3.20). In some plasma chemical processes, two molecules involved in reaction can be strongly excited vibrationally. An interesting and important example of such a reaction of two excited molecules is the carbon oxide disproportioning leading to carbon formation:

$$CO^*(v_1) + CO^*(v_2) \rightarrow CO_2 + C \tag{3.131}$$

The disproportioning is a strongly endothermic process, (enthalpy of about 5.5 eV) with activation energy of 6 eV, which can be stimulated by vibrational energy of both CO molecules. This elementary reaction was studied using the vibronic terms approach.[89] According to this approach, the general Equation 3.99 and Equation 3.100 also can be applied to the reaction of two excited molecules. In this case, both excited molecules contribute their energy to decrease the activation energy.

$$\Delta E_a = \alpha_1 E_{v1} + \alpha_2 E_{v2} \tag{3.132}$$

In this relation, E_{v1} and E_{v2} are the vibrational energies of the two excited molecules (subscript "1" corresponds to the molecule losing an atom; subscript "2" is related to another molecule accepting the atom); and α_1 and α_2 are coefficients of efficiency of using vibrational energy from these two sources. Specifically, for the reaction Equation 3.131, it was shown that $\alpha_1 = 1$and $\alpha_2 \approx 0.2$. This reflects the fact that vibrational energy of the donating molecule is much more efficient in the stimulation of endothermic exchange reactions than that of the accepting molecule.

PROBLEMS AND CONCEPT QUESTIONS

3.1 **Lifetime of Metastable Atoms.** Compare the relatively long radiative lifetime of a metastable helium atom He (2s $^3S_1$) with the time interval between collisions of the atom with other neutrals and with a wall of a discharge chamber (radius 3 cm) in a room temperature discharge at low pressure of 10 torr. Comment on the result.

3.2 **Rydberg Correction Terms in the Bohr Model.** Estimate the relative accuracy (in percentage) of applying the Bohr formula (Equation 3.1) without the Rydberg correction for excited helium atoms with the following quantum numbers: (a) $n = 2$, $l = 0$, $S = 1$; and (b) $n = 3$, $l = 2$, $S = 0$. Explain the large difference in accuracy.

3.3 **Metastable Atoms and Selection Rules.** Using Table 3.1, check the change in parity and quantum numbers *S*, *J*, and *L* between excited metastable states and corresponding ground states for different atoms. Compare these changes in parity and quantum numbers with the selection rules for dipole radiation.

3.4 **Molecular Vibration Frequency.** Consider a diatomic homonuclear molecule as a classical oscillator with mass *M* of each atom. Take interaction between the atoms in accordance with the Morse potential with known parameters. Based on that, find a vibration frequency of the diatomic homonuclear molecule. Compare the frequency with an exact quantum mechanical frequency given by Equation 3.6.

3.5 **Anharmonicity of Molecular Vibrations.** Equation 3.7 gives a relation between anharminicity coefficient, vibrational quantum, and the dissociation energy for diatomic molecules. Using this relation and numerical data from Table 3.6, find dissociation energies of molecular nitrogen, oxygen, and hydrogen. Compare the calculated values with actual dissociation energies of these molecules (for N_2, $D = 9.8$ eV; for O_2, $D = 5.1$ eV; and for H_2, $D = 4.48$ eV).

3.6 **Maximum Energy of an Anharmonic Oscillator.** Based on Equation 3.7, calculate the maximum vibrational energy $E_v(v = v_{\max})$ of an anharmonic oscillator corresponding to the maximum vibrational quantum number (Equation 3.9). Compare this energy with dissociation energy of the diatomic molecule calculated, based on Equation 3.7, between vibrational quantum and coefficient of anharmonicity. Comment on the role of Equation 3.5 in this case.

3.7 **Isotopic Effect in Dissociation Energy of Diatomic Molecules.** Dissociation of which diatomic molecule-isotope—light one or heavy one—is a kinetically faster process in the thermal quasi-equilibrium discharges? Make your decision based on Equation 3.5 and Equation 3.6 for dissociation energy and vibration frequency.

3.8 **Parameter Massey for Transition between Vibrational Levels.** Estimate the Massey parameter (Equation 2.37) for a process of loss of one vibrational quantum from an excited nitrogen molecule in a collision with another heavy particle at room temperature. Consider the loss of the highest (Equation 3.10) and the lowest $\hbar\omega$ vibrational quantum. Compare and comment on the results.

3.9 **Anharmonicity of Vibrations of Polyatomic Molecules.** Why can the coefficients of anharmonicity be both positive and negative for polyatomic molecules, while for diatomic molecules they are always effectively negative?

3.10 **Symmetric Modes of CO_2 Vibrations.** In Equation 3.15 and Equation 3.16, n_{sym} is the total number of quanta on the symmetric modes (ν_1 and ν_2) of CO_2. It can be presented as $n_{sym} = 2n_1 + n_2$, taking into account the relation between frequencies of the symmetric valence and deformation vibrations ($\nu_1 \sim 2\nu_2$). Estimate the values of vibrational quantum and anharmonicity for such a combined symmetric vibrational mode.

3.11 **Transition to the Vibrational Quasi Continuum.** Based on Equation 3.14, determine numerically at what level of vibrational energy CO_2 modes can no longer be considered individually.

3.12 **Effect of Vibrational Excitation on Rotational Energy.** Based on Equation 3.23, calculate the maximum possible relative decrease of the rotational constant B of the N_2 molecule due to the effect of vibration.

3.13 **Cross Section of Vibrational Excitation by Electron Impact.** Using the Breit–Wigner formula, estimate the half-width of a resonance pike in the vibrational excitation cross section dependence on electron energy. Discuss the values of the energy half-width for the resonances with a different range of lifetime.

3.14 **Multiquantum Vibrational Excitation by Electron Impact.** Based on the Fridman's approximation and taking into account known values of the basic excitation rate coefficient $k_{eV}(0,1)$, as well as parameters α and β of the approximation, calculate the total rate coefficient of the multiquantum vibrational excitation from an initial state v_1 to all other vibrational levels $v_2 > v_1$. Using Table 3.11, compare the total vibrational excitation by electron impact rate coefficient with the basic one $k_{eV}(0,1)$ for different gases.

3.15 **Rotational Excitation of Molecules by Electron Impact.** Compare qualitatively the dependences of rotational excitation cross section on electron energy in the cases of excitation of dipole and quadrupole molecules (Equation 3.30 and Equation 3.31). Discuss the qualitative difference of the cross sections' behavior near the process threshold.

3.16 **Cross Sections of Electronic Excitation by Electron Impact.** Find electron energies (in units of ionization energy, $x = \varepsilon/I$) corresponding to the maximum value of cross sections of the electronic excitation of permitted transitions. Use semiempirical relations of Drawin (Equation 3.33) and Smirnov (Equation 3.34) and compare the results.

3.17 **Electronic Excitation Rate Coefficients.** Using Table 3.12 for nitrogen molecules, calculate and draw the dependence of rate coefficients of electronic excitation of $A^3\Sigma_u^+$ and ionization in a wide range of reduced electric fields E/n_0 typical for nonthermal discharges. Discuss the difference in behavior of these two curves.

3.18 **Influence of Vibrational Temperature on Electronic Excitation Rate Coefficients.** Calculate the relative increase of the electronic excitation rate coefficient of molecular nitrogen in a nonthermal discharge with reduced electric field $E/n_0 = 3*10^{-16}$ Vcm^2 when the vibrational temperature increases from room temperature to $T_v = 3000$ K.

3.19 **Dissociation of Molecules through Electronic Excitation by Direct Electron Impact.** Explain why the electron energy threshold of the dissociation through electronic excitation almost always exceeds the actual dissociation energy, in contrast to dissociation stimulated by vibrational excitation in which the threshold is equal quite often to the dissociation energy. Use the dissociation of diatomic molecules as an example.

3.20 **Distribution of Electron Energy between Different Channels of Excitation and Ionization.** Analyzing Figures 3.9 through Figure 3.16, find the molecular gases with the highest possibility to localize selectively discharge energy on vibrational excitation. What is the highest percentage of electron energy that can be selectively focused on vibrational excitation? Which values of reduced electric field or electron temperatures are optimal to reach this selective regime?

3.21 **Probability of VT Relaxation in Adiabatic Collisions.** Probability of the adiabatic VT relaxation (for example, in Equation 3.48) exponentially depends on the relative velocity v of colliding partners, which is actually different before and after collision (because of the energy transfer during the collision). Which one from the two translational velocities should be taken to calculate the Massey parameter and relaxation probability?

3.22 **VT Relaxation Rate Coefficient as a Function of Vibrational Quantum Number.** The VT relaxation rate grows with the number of quantum for two reasons (see, for example, Equation 3.53). First, the matrix element increases as $(n + 1)$ and, second, vibrational frequency decreases because of anharmonicity. Show numerically which effect is dominating the acceleration of VT relaxation with level of vibrational excitation.

3.23 **Temperature Dependence of VT Relaxation Rate.** The vibrational relaxation rate coefficient for adiabatic collisions usually decreases with reduction of translational gas temperature in good agreement with the Landau–Teller formula. In practice, however, this temperature dependence function has a minimum, usually at about room temperature, and then the relaxation accelerates with further cooling. Explain the phenomena.

3.24 **Semiempirical Relations for VT Relaxation Rate Coefficients.** Calculate vibrational relaxation rate coefficients for a couple of molecular gases at room temperature, using the semiempirical formula (Equation 3.58). Compare your results with data obtained from Table 3.13; estimate the relative accuracy of using the semiempirical relation.

3.25 **VT Relaxation in Collision with Atoms and Radicals.** How small a concentration of oxygen atoms should be in nonthermal air-plasma at room temperature to neglect the effect of VT relaxation of vibrationally excited nitrogen molecules in collision with the atoms? Make the same estimation for nonequilibrium discharges in pure oxygen and explain difficulties in accumulating essential vibrational energy in such discharges.

3.26 **Surface Relaxation of Molecular Vibrations.** Estimate the pressure at which heterogeneous vibrational relaxation in a room temperature nitrogen discharge becomes dominant with respect to volume relaxation. Take the discharge tube radius about 3 cm, suppose molybdenum glass as the discharge wall material, and estimate the diffusion coefficient of nitrogen molecules in the system by the simplified numerical formula $D \approx 0.3/p(atm)$ cm^3/sec.

3.27 **Vibrational Relaxation of Polyatomic Molecules.** How does relative contribution of resonant and nonresonant (adiabatic) relaxation of polyatomic molecules depend on the level of their vibrational excitation? Consider cases of excitation of independent vibrational modes and also transition to the quasi continuum. Use the relation Equation 3.68 and discussions concerning transition to the quasi continuum in Section 3.2.4.

3.28 **Resonant Multiquantum VV Exchange.** Estimate the smallness of the double-quantum resonance VV exchange A_2^* $(v = 2) + A_2$ $(v = 0) \rightarrow A_2$ $(v = 0) +$ A_2^* $(v = 2)$ and triple-quantum resonance VV exchange A_2^* $(v = 3) + A_2(v = 0) \rightarrow$ A_2 $(v = 0) + A_2^*$ $(v = 3)$ in diatomic molecules, taking the matrix element factor $(\Omega\tau_{col})^2 = 10^{-3}$.

3.29 **Resonant VV Relaxation of Anharmonic Oscillators.** How does the probability, and thus rate coefficient, of the resonant VV exchange in the case of relaxation of identical diatomic molecules A_2^* $(v = n + 2) + A_2(v = n) \rightarrow A_2^*$ $(v = n) + A_2^*$ $(v = n + 1)$ depend on their vibrational quantum number.

3.30 **Comparison of Adiabatic Factors for Anharmonic VV and VT Relaxation Processes.** As shown in the discussion after Equation 3.73 and Equation 3.74, the adiabatic factors for anharmonic VV and VT relaxation processes are equal in one-component gases $\delta_{VV} = \delta_{VT}$. What is the reason for difference of the adiabatic factors in the case of collisions of nonidentical anharmonic diatomic molecules?

3.31 **VV Exchange of Anharmonic Oscillators Provided by Dipole Interaction.** Why should the long-distance forces usually be taken into account in the VV relaxation of anharmonic oscillators only for exchange close to resonance?

3.32 **Nonresonant One-Quantum VV′ Exchange.** Estimate the rate coefficient of the nonresonant VV′ exchange between a nitrogen molecule and another one (presented in Table 3.17) using the semiempirical Rapp approximation. Compare the obtained result with experimental data found in Table 3.17.

3.33 **VV Relaxation of Polyatomic Molecules.** Based on Equation 3.89, explain why excitation of high vibrational levels of polyatomic molecules in nonthermal discharges is much less effective than similar excitation of diatomic molecules.

3.34 **Rotational RT Relaxation.** Using the Parker formula and data from Table 3.18, calculate the energy depth of the intermolecular attraction potential T^* and probability of the RT relaxation in pure molecular nitrogen at translational gas temperature 400 K.

3.35 **LeRoy Formula and α-Model.** Derive the general Equation 3.99 describing stimulation of chemical reactions by vibrational excitation of molecules, averaging the LeRoy formula over the different values of translational energy E_t. Use the Maxwellian function for distribution of molecules over the translational energies.

3.36 **Semiempirical Methods of Determining Activation Energies.** Estimate activation energy of at least one reaction from the group $CH_3 + RH \rightarrow CH_4 + R$ using (*a*) the Polanyi–Semenov rule; (*b*) the Kagija method; and (*c*) the Sabo method. Compare the results and comment on them; which method looks more reliable in this case?

3.37 **Efficiency of Vibrationally Excited Molecules in Stimulation of Endothermic and Exothermic Reactions.** Using the potential energy surfaces of exchange reactions A + BC → AB + C, illustrate the conclusion of the Fridman–Macheret α-model stating that vibrational excitation of molecules is more efficient in stimulation of endothermic (rather than exothermic) reactions.

3.38 **Accuracy of the Fridman–Macheret α-Model.** Using Table 3.20 of the α-efficiency coefficients, determine the typical values of relative accuracy of application the Fridman–Macheret α-model for (*a*) endothermic; (*b*) exothermic; and (*c*) thermoneutral reactions. Which of these three groups of chemical reactions of vibrationally excited molecules can be described by this approach more accurately and why?

3.39 **Contribution of Translational Energy in Dissociation of Molecules under Nonequilibrium Conditions.** Using expression for the nonequilibrium factor Z of the Fridman–Macheret model, derive the formula for dissociation of molecules at translational temperature much exceeding the vibrational one.

3.40 **Efficiency of Translational Energy in Elementary Endothermic Reactions.** According to Equation 3.115 and to common sense (conservation of momentum and angular momentum), efficiency of translational energy in endothermic reactions should be less than 100%. This means that the effective dissociation energy in this case should exceed the real one. On the other hand, the Fridman–Macheret model permits dissociation when total translational energy is equal to D. Explain the "contradiction."

3.41 **The Park Model.** Is it possible to apply the Park model for semiempirical description of dissociation of diatomic molecules in nonequilibrium plasma when $T_v > T_0$? Analyze the effect of vibrational and translational temperatures in this case.

CHAPTER

4

PLASMA STATISTICS AND KINETICS OF CHARGED PARTICLES

4.1 STATISTICS AND THERMODYNAMICS OF EQUILIBRIUM AND NONEQUILIBRIUM PLASMAS: BOLTZMANN, SAHA, AND TREANOR DISTRIBUTIONS

Properties of thermal plasma quite often can be described simply based on the temperature of the system and using quasi-equilibrium approaches. General aspects of statistics and thermodynamics of such systems, including the Boltzmann distribution; chemical and radiation equilibrium; and the Saha equation for ionization equilibrium, will be considered in this section. Elements and examples of the nonequilibrium statistics of charged and neutral particles, such as the Treanor distribution of the vibrationally excited molecules, also will be discussed.

4.1.1 Statistical Distribution of Particles over Different States: Boltzmann Distribution

Consider an isolated system with the constant total energy E consisting of a large fixed number N of particles in different states (i).The number of particles n_i are in the state i, defined by a set of quantum numbers and energy E_i:

$$N = \sum_i n_i, \qquad E = \sum_i E_i n_i \tag{4.1}$$

Collisions of the particles lead to continuous change of populations n_i, but the total group follows the conservation laws (Equation 4.1) and the system is in thermodynamic equilibrium. The objective of the statistical approach in this case is to find the distribution function of particles over the different states i without details of probabilities of transitions between the states. Such statistical function can be derived, in general, because the probability to find n_i particles in the state i is proportional to the number of ways in which this distribution can be arranged.

The so-called **thermodynamic probability** $W(n_1, n_2, \ldots, n_i, \ldots)$ is the probability to have n_1 particles in the state "1," n_2 particles in the state "2," etc. It can be found that N particles can be arranged in $N!$ different ways, but because n_i particles have the same energy, this number should be divided by the relevant factor to exclude repetitions:

$$W(n_1, n_2, \ldots, n_i, \ldots) = A\frac{N!}{N_1!N_2!\ldots N_i!\ldots} = A\frac{N!}{\prod_i n_i!} \tag{4.2}$$

where A is a factor of normalization. One can find the most probable numbers of particles $\overline{n_i}$ when the probability W (Equation 4.2) as well as its logarithm:

$$\ln W(n_1, n_2, \ldots, n_i, \ldots) = \ln(AN!) - \sum_i \ln n_i! \approx \ln(AN!) - \sum_i \int_0^{n_i} \ln x dx \tag{4.3}$$

have a maximum. Achievement of the maximum of the function $\ln W$ of many variables at a point, where n_i are equal to the most probable numbers of particles $\overline{n_i}$, requires:

$$0 = \sum_i \left(\frac{\partial \ln W}{\partial n_i}\right)_{n_i=\overline{n_i}} dn_i = \sum_i \ln \overline{n_i}\, dn_i \tag{4.4}$$

Differentiation of the conservation laws (Equation 4.1) gives:

$$\sum_i dn_i = 0, \quad \sum_i E_i dn_i = 0 \tag{4.5}$$

Multiplying the two equations (Equation 4.5) respectively by parameters $\ln C$ and $1/T$ and then adding Equation 4.4 yields:

$$\sum_i \left(\ln \overline{n_i} - \ln C + \frac{E_i}{T}\right) dn_i = 0 \tag{4.6}$$

This sum (Equation 4.6) is supposed to be equal to zero at any independent values of dn_i. This is possible only if the expression is equal to zero, which leads to the famous expression of the **Boltzmann distribution function:**

$$\overline{n_i} = C \exp\left(-\frac{E_i}{T}\right) \tag{4.7}$$

where C is the normalizing parameter related to the total number of particles and T is the statistical temperature of the system, related to the average energy per one particle. As always in this book, the same units are used for energy and temperature.

Equation 4.7 was derived for the case when the subscript i was related to the only state of a particle. If the state is degenerated, it is necessary to add into Equation 4.7 the statistical weight g, showing the number of states with the given quantum number:

$$\overline{n_j} = C\, g_j \exp\left(-\frac{E_j}{T}\right) \tag{4.8}$$

In this case, the subscript j corresponds to the group of states with the statistical weight g_j and energy E_j. The Boltzmann distribution can be expressed then in terms of the number densities N_j and N_0 of particles in j states and the ground state (0):

$$N_j = N_0 \frac{g_j}{g_0} \exp\left(-\frac{E_j}{T}\right) \tag{4.9}$$

In this relation, g_j and g_0 are the statistical weights of the j and ground states.

The general Boltzmann distribution (Equation 4.7 and Equation 4.9) can be applied to derive many specific distribution functions important in plasma chemistry. One of them is obviously the Maxwell–Boltzmann (or just the Maxwellian) distribution (Equation 2.1) of particles over the translational energies or velocities, which was discussed in Chapter 2. Some other specific quasi-equilibrium statistical distributions important in plasma physics and engineering are considered next.

4.1.2 Equilibrium Statistical Distribution of Diatomic Molecules over Vibrational–Rotational States

Vibrational levels of diatomic molecules are not degenerated and their energy with respect to vibrationally ground state can be taken as $E_v = \hbar\omega v$, when the vibrational quantum numbers are low (see Equation 3.6). Then, according to Equation 4.9, the number density of molecules with v vibrational quanta can be found as:

$$N_v = N_0 \exp\left(-\frac{\hbar\omega v}{T}\right) \tag{4.10}$$

The total number density of molecules N is a sum of densities in different vibrational states:

$$N = \sum_{v=0}^{\infty} N_v = \frac{N_0}{1 - \exp\left(-\frac{\hbar\omega}{T}\right)} \tag{4.10a}$$

Based on Equation 4.10a, the distribution of molecules over the different vibrational levels can be renormalized in a traditional way with respect to the total number density N of molecules:

$$N_v = N\left[1-\exp\left(-\frac{\hbar\omega}{T}\right)\right]\exp\left(-\frac{\hbar\omega v}{T}\right) \tag{4.11}$$

Now it is necessary to find the Boltzmann vibrational–rotational distribution N_{vJ}

$$\sum_J N_{vJ} = N_v.$$

One should take into account Equation 3.22 for rotational energy ($B << T$), and the statistical weight for a rotational state with quantum number J, which is equal to $2J+1$. Based on the general relation (Equation 4.9), this distribution can be presented as:

$$N_{vJ} = N\frac{B}{T}(2J+1)\left[1-\exp\left(-\frac{\hbar\omega}{T}\right)\right]\exp\left(-\frac{\hbar\omega v + BJ(J+1)}{T}\right) \tag{4.12}$$

4.1.3 Saha Equation for Ionization Equilibrium in Thermal Plasma

The Boltzmann distribution (Equation 4.9) can be applied to find the ratio of average number of electrons and ions, $\overline{n_e} = \overline{n_i}$, to average number of atoms, $\overline{n_a}$, in a ground state or, in other words, to describe the ionization equilibrium $A^+ + e \Leftrightarrow A$ in plasma:

$$\frac{\overline{n_i}}{\overline{n_a}} = \frac{g_e g_i}{g_a}\int\left[\frac{d\vec{p}d\vec{r}}{(2\pi\hbar)^3}\exp\left(-\frac{I+\frac{p^2}{2m}}{T}\right)\right] \tag{4.13}$$

In this relation, I is the ionization potential; p is momentum of a free electron, so $I + p^2/2m$ is the energy necessary to produce the electron; g_a, g_i, and g_e are the statistical weights of atoms, ions, and electrons; and

$$\frac{d\vec{p}\,d\vec{r}}{(2\pi\hbar)^3}$$

is the statistical weight corresponding to continuous spectrum, which is the number of states in a given element of the phase volume $d\vec{p}d\vec{r} = dp_x dp_y dp_z\, dx dy dz$.

Integration of Equation 4.13 over the electron momentum gives:

$$\frac{\overline{n_i}}{\overline{n_a}} = \frac{g_e g_i}{g_a}\left(\frac{mT}{2\pi\hbar^2}\right)^{3/2}\exp\left(-\frac{I}{T}\right)\int d\vec{r} \tag{4.14}$$

Here, m is an electron mass; $\int d\vec{r} = V/\overline{n_e}$ is volume corresponding to one electron; and V is total system volume. Introducing in Equation 4.14 the number densities of electrons $N_e = \overline{n_e}/V$, ions $N_i = \overline{n_i}/V$, and atoms $N_a = \overline{n_a}/V$, yields the **Saha equation** describing the ionization equilibrium in thermal plasmas:

$$\frac{N_e N_i}{N_a} = \frac{g_e g_i}{g_a}\left(\frac{mT}{2\pi\hbar^2}\right)^{3/2} \exp\left(-\frac{I}{T}\right) \tag{4.15}$$

It can be estimated from Equation 4.15 that the effective statistical weight of the continuum spectrum is very high. As a result, the Saha equation predicts very high values of the degree of ionization N_e/N_a, which can be close to unity even when temperature is still much less than ionization potential $T << I$.

4.1.4 Dissociation Equilibrium in Molecular Gases

The Saha equation (Equation 4.15) derived before for the ionization equilibrium $A^+ + e \Leftrightarrow A$ can be generalized to describe the dissociation equilibrium $X + Y \Leftrightarrow XY$. The relation between densities N_X of atoms X, N_Y of atoms Y, and N_{XY} of molecules XY in the ground vibrational–rotational state can be written based on Equation 4.15 as:

$$\frac{N_X N_Y}{N_{XY}(v=0, J=0)} = \frac{g_X g_Y}{g_{XY}}\left(\frac{\mu T}{2\pi\hbar^2}\right)^{3/2} \exp\left(-\frac{D}{T}\right) \tag{4.16}$$

In this relation, g_X, g_Y, and g_{XY} are the corresponding statistical weights related to their electronic states; μ is the reduced mass of atoms X and Y; D is dissociation energy of the molecule XY. Most of the molecules are not in a ground state but in an excited one at the high temperatures typical for thermal plasmas.

Substituting the ground state concentration $N_{XY}(v = 0, J = 0)$ in Equation 4.16 by the total N_{XY} concentration and using Equation 4.12:

$$N_{XY}(v=0, J=0) = \left[1-\exp\left(-\frac{\hbar\omega}{T}\right)\right]\frac{B}{T} N_{XY} \tag{4.17}$$

the final statistical relation for the equilibrium dissociation of molecules is obtained.

$$\frac{N_X N_Y}{N_{XY}} = \frac{g_X g_Y}{g_{XY}}\left(\frac{\mu T}{2\pi\hbar^2}\right)^{3/2} \frac{B}{T}\left[1-\exp\left(-\frac{\hbar\omega}{T}\right)\right]\exp\left(-\frac{D}{T}\right) \tag{4.18}$$

4.1.5 Equilibrium Statistical Relations for Radiation: Planck Formula and Stefan–Boltzmann Law

According to the general expression of the Boltzmann distribution function (Equation 4.7), the relative probability to have n photons in a state with frequency ω is equal in equilibrium to $\exp(-\hbar\omega n)$. Then the average number of photons in the state with frequency ω can be found in the form of the so-called **Planck distribution**:

$$\overline{n_\omega} = \frac{\sum_n n \exp\left(-\frac{\hbar\omega n}{T}\right)}{\sum_n \exp\left(-\frac{\hbar\omega n}{T}\right)} = \frac{1}{\exp\frac{\hbar\omega}{T} - 1} \tag{4.19}$$

The electromagnetic energy in the frequency interval (ω, $\omega + d\omega$) and in volume V then can be expressed as energy of the $\overline{n_\omega}$ photons in one state, $\hbar\omega\overline{n_\omega}$, multiplied by the number of states in the given volume of the phase space $2 * V d\vec{k}/(2\pi)^3$ (the factor 2 is related here to the number of polarizations of electromagnetic wave):

$$dE(V, \omega \sim \omega + d\omega) = \hbar\omega\overline{n_\omega} 2 \frac{V d\vec{k}}{(2\pi)^3} \tag{4.20}$$

Using Equation 4.20 and the dispersion relation $\omega = kc$ (c is the speed of light), the Planck distribution (Equation 4.19) can be rewritten in terms of the spectral density of radiation

$$U_\infty = \frac{1}{V} \frac{dE(V, \omega \sim \omega + d\omega)}{d\omega},$$

which is the electromagnetic radiation energy taken per unit volume and unit interval of frequencies:

$$U_\omega = \frac{\hbar\omega^3}{\pi^2 c^3} \overline{n_\omega} = \frac{\hbar\omega^3}{\pi^2 c^3 \left(\exp\frac{\hbar\omega}{T} - 1\right)} \tag{4.21}$$

yielding the **Planck formula** for the spectral density of radiation. The Planck formula at a high temperature limit $\hbar\omega/T \ll 1$ gives the classical **Rayleigh–Jeans formula,** which does not contain the Planck constant $\hbar$:

$$U_\omega = \frac{\omega^2 T}{\pi^2 c^3} \tag{4.22}$$

and, at a low temperature limit $\hbar\omega/T \gg 1$, it gives the **Wien formula** describing the exponential decrease of radiation energy:

$$U_\omega = \frac{\hbar\omega^3}{\pi^2 c^3} \exp\left(-\frac{\hbar\omega}{T}\right) \tag{4.23}$$

The Planck formula (Equation 4.21) permits calculating the total equilibrium radiation flux falling on the surface. In the case of a black body, this is equal to the radiation flux from the surface:

$$J = \frac{c}{4} \int_0^\infty U_\omega d\omega = \frac{\pi^2 T^4}{60 c^2 \hbar^3} = \sigma T^4 \tag{4.24}$$

This is the **Stefan–Boltzmann law**, with the Stefan–Boltzmann coefficient σ determined only by the fundamental physical constants:

$$\sigma = \frac{\pi^2}{60c^2\hbar^3} = 5.67 * 10^{-12} \frac{W}{cm^2K^4} \tag{4.25}$$

4.1.6 Concepts of Complete Thermodynamic Equilibrium (CTE) and Local Thermodynamic Equilibrium (LTE) for Plasma Systems

In plasma systems, **complete thermodynamic equilibrium (CTE)** is related to uniform homogeneous plasma, in which kinetic and chemical equilibrium as well as all the plasma properties are unambiguous functions of temperature. This temperature is supposed to be homogeneous and the same for all degrees of freedom, all the plasma system components, and all their possible reactions. In particular, the following five equilibrium statistical distributions (described previously) obviously should take place with the same temperature T:

1. The Maxwell–Boltzmann velocity or translational energy distributions (Equation 2.1) for all neutral and charged species that exist in the plasma
2. The Boltzmann distribution (Equation 4.9 and Equation 4.12) for the population of excited states for all neutral and charged species that exist in the plasma
3. The Saha equation (Equation 4.15) for ionization equilibrium to relate the number densities of electrons, ions, and neutral species
4. The Boltzman distribution (Equation 4.18) for dissociation or, more generally, the thermodynamic relations for chemical equilibrium
5. The Planck distribution (Equation 4.19) and Planck formula (Equation 4.21) for spectral density of electromagnetic radiation

Plasma in CTE conditions cannot be practically realized in the laboratory. Nevertheless, thermal plasmas sometimes are modeled in this way for simplicity. To imagine a plasma in CTE conditions, consider a very large plasma volume such that the central part is homogeneous and does not sense the boundaries. In this case, the electromagnetic radiation of the plasma can be considered as black body radiation with plasma temperature.

Actual (even thermal) plasmas are quite far from these ideal conditions. Most plasmas are optically thin over a wide range of wavelengths and, as a result, the plasma radiation is much less than that of a black body. Plasma nonuniformity leads to irreversible losses related to conduction, convection, and diffusion, which also disturb the complete thermodynamic equilibrium (CTE). A more realistic approximation is the so-called **local thermodynamic equilibrium (LTE)**.

According to the LTE-approach, a thermal plasma is considered optically thin and thus radiation is not required to be in equilibrium. However, the collisional (not radiative) processes are required to be locally in equilibrium similar to that described earlier for CTE but with a temperature T, which can differ from point to point in space and time. Detailed consideration of LTE conditions can be found, for example, in Boulos et al.[90]

4.1.7 Partition Functions

Determination of the thermodynamic plasma properties, free energy, enthalpy, entropy, etc. at first requires calculation of the partition functions, which are the links between microscopic statistics and macroscopic thermodynamics. Their calculation is an important statistical task (see Mayer and Mayer[91]) that should be accomplished to simulate different plasma properties.

The partition function Q of an equilibrium particle system at the equilibrium temperature T in general can be expressed as a statistical sum over different states s of the particle with energies E_s and statistical weights g_s:

$$Q = \sum_s g_s \exp\left(-\frac{E_s}{T}\right) \tag{4.26}$$

The translational and internal degrees of freedom of the particles can be considered independent. Then, their energy can be expressed as the sum $E_s = E^{tr} + E^{\text{int}}$ and the partition function as a product of translational and internal partition functions. In this case, for particles of a chemical species i and plasma volume V:

$$Q_i = Q_i^{tr} Q_i^{\text{int}} = \left(\frac{m_i T}{\hbar^2}\right)^{3/2} V Q^{\text{int}} \tag{4.27}$$

If the translational partition function in Equation 4.27 implies calculation of the continuous spectrum statistical weight (see the preceding section), the partition functions of internal degrees of freedom (including molecular rotation and vibration) depend on the system characteristics.[62] A particular numerical example of the partition functions of nitrogen atoms and ions is presented in Figure 4.1.

4.1.8 Thermodynamic Functions of Thermal Plasma Systems

If the total partition function Q_{tot} of all the particles in an equilibrium system is known, the different thermodynamic functions can be calculated. For example, the Helmholtz free energy F, related to a reference free energy F_0, can be found as:[62]

$$F = F_0 - T \ln Q_{tot} \tag{4.28}$$

If the system particles are noninteracting (the ideal or perfect gas approximation), then the total partition function can be expressed as a product of partition functions Q_i of a single particle of a chemical component i:

$$Q_{tot} = \frac{\prod_i Q_i^{N_i}}{\prod_i N_i!} \tag{4.29}$$

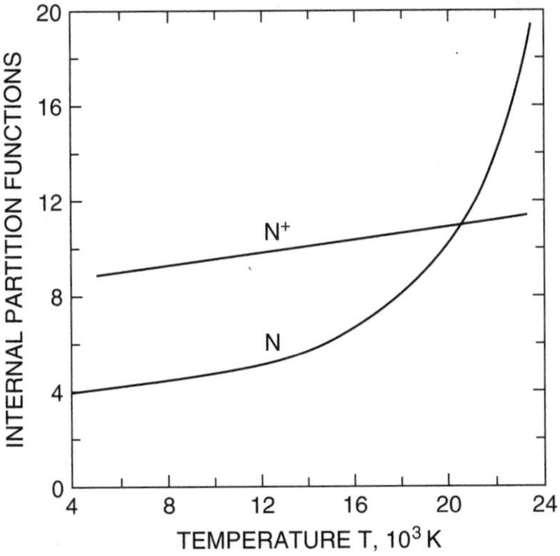

Figure 4.1 Partition function of nitrogen atoms and ions (From Boulos, M. et al. *Thermal Plasma*.[90])

where N_i is the total number of particles of the species i in the system. Using Stirling's formula for factorial ($\ln N! = N \ln N - N$) and substituting the expression for the partition function (Equation 4.29) into Equation 4.28, the final formula for calculation of the Helmholtz free energy of the system of noninteracting particles (e is the base of natural logarithms obtained):

$$F = F_0 - \sum_i N_i T \ln \frac{Q_i e}{N_i} \tag{4.30}$$

Other thermodynamic functions can be easily calculated based on Equation 4.30. For example, the internal energy U can be found as:

$$U = U_0 + T^2 \left(\frac{\partial (F/T)}{\partial T} \right)_{V,N_i} = \sum_i N_i \left[\frac{3}{2} T + T^2 \left(\frac{\partial \ln Q_i^{\text{int}}}{\partial V} \right)_{T,N_i} \right] \tag{4.31}$$

Also, pressure p can be found from Equation 4.30 and the differential thermodynamic relations:

$$p = -\left(\frac{\partial (F/T)}{\partial V} \right)_{T,N} = \sum_i N_i T \left(\frac{\partial \ln Q_i}{\partial V} \right)_{T,N_i} \tag{4.32}$$

The preceding relations do not take into account interactions between particles. Ion and electron number densities grow at higher temperatures of thermal plasmas; the coulomb interaction becomes important and should be added to the thermodynamic functions. The Debye model (see, for example, Drawin[92]) provides the coulomb interaction factor term to Equation 4.30:

$$F = F_0 - \sum_i N_i T \ln \frac{Q_i e}{N_i} - \frac{TV}{12\pi} \left(\frac{e^2}{\varepsilon_0 TV} \sum_i Z_i^2 N_i \right)^{3/2} \tag{4.33}$$

Here, Z_i is the number of charges of species I (for electrons $Z_i = -1$). This **Debye correction** in the Helmholtz free energy F obviously leads to corrections of the other thermodynamic functions. For example, the Gibbs energy G with the Debye correction becomes:

$$G = G_0 - \sum_i N_i T \ln \frac{Q_i}{N_i} - \frac{TV}{8\pi} \left(\frac{e^2}{\varepsilon_0 TV} \sum_i Z_i^2 N_i \right)^{3/2} \tag{4.34}$$

The relation for pressure (Equation 4.32) with the Debye correction should also be modified from the case of ideal gas as:

$$p = \frac{NT}{V} - \frac{T}{24\pi} \left(\frac{e^2}{\varepsilon_0 TV} \sum_i Z_i^2 N_i \right)^{3/2} \tag{4.35}$$

Numerically, the Debye correction becomes relatively important (typically around 3%) at temperatures of thermal plasma of about 14,000 to 15,000 K.

The thermodynamic functions help to calculate chemical and ionization composition of thermal plasmas. Examples of such calculations can be found in Boulos et al.[90] Numerous thermodynamic results related to equilibrium chemical and ionization composition can be found in handbooks[93] or in thermodynamic database Web sites.

4.1.9 Nonequilibrium Statistics of Thermal and Nonthermal Plasmas

Correct description of the nonequilibrium plasma systems and processes generally requires application of detailed kinetic models, which will be discussed later in this and the following two chapters. Application of statistical models in this case can lead to huge deviations from reality, as was demonstrated by Slovetsky.[94] However, statistical approaches can be simple as well as quite successful in description of the nonequilibrium plasma systems. Several examples on the subject will be discussed next.

The first example is related to thermal discharges with small deviation of the electron temperature from the temperature of heavy particles. Such a situation can take place in boundary layers separating quasi-equilibrium plasma from electrodes and walls. In this case, **two-temperature statistics and thermodynamics** can be developed.[90] These models suppose that partition functions depend on electron temperature and the temperature of heavy particles. Electron temperature in this case determines the partition functions related to ionization processes; chemical processes are in turn determined by the temperature of heavy particles.

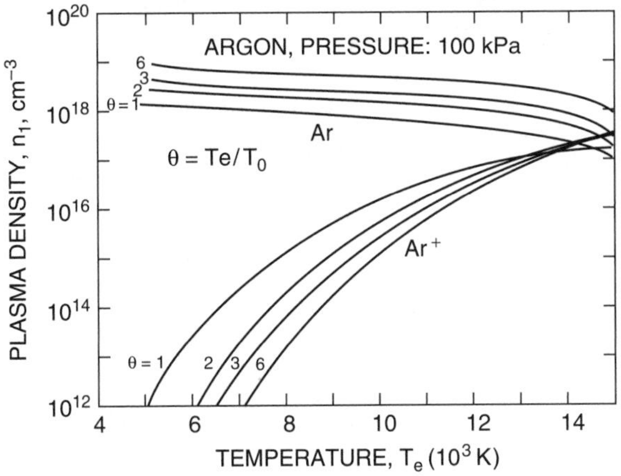

Figure 4.2 Dependence of number densities of Ar and Ar^+ on electron temperature at atmospheric pressure for two-temperature plasma with $\theta = T_e/T_0 = 1, 2, 3, 6$[90]

The partition functions found in such a way can then be used to calculate the free energy, Gibbs potential, and, finally, composition and thermodynamic properties of the two-temperature plasma. An example of such a calculation of composition in two-temperature Ar plasma is given in Figure 4.2. It should be noted that this approach is valid only for small levels of nonequilibrium.

The second example of application of nonequilibrium statistics is related to nonthermal plasma systems and processes having a high level of nonequilibrium. It is concerned with the process of gasification of solid surfaces:

$$A(\text{solid}) + bB^*(\text{gas}) \rightarrow cC(\text{gas}) \tag{4.36}$$

stimulated by particles B excited in nonthermal discharges, predominantly in one specific state.[95,96,96a]

In this case, deviation from the conventional equation for the equilibrium constant

$$K = \frac{(Q_C)^c}{Q_A(Q_B)^b} \tag{4.37}$$

is related only in a more general form of definition to the partition functions. Differences in temperatures—translational T_t, rotational T_r, and vibrational T_v—as well as the nonequilibrium distribution function (population) $f_e^X(\varepsilon_e^X)$ of electronically excited states should be taken into account for each reactant and product $X = A, B, C$:

$$Q_X = \sum_e g_e^X f_e^X(\varepsilon_e^X) \sum_{(t,r,v)} \prod_{k=t,r,v} g_{ke}^X \exp\left(-\frac{\varepsilon_{ke}^X}{T_k^X}\right) \tag{4.38}$$

Here, g and ε are the statistical weights and energies of corresponding states. However, the general introduction of the three temperatures can lead to internal inconsistencies. The statistical approach can be applied in a consistent way for the reactions (Equation 4.36) if the nonthermal plasma stimulation of the process is limited to the electronic excitation of the only state of a particle B with energy E_b. The population of the excited state can then be expressed by δ-function and the Boltzmann factor with an effective electronic temperature T^*:

$$f_e^B\left(\varepsilon_e^B\right) = \delta\left(\varepsilon_e^B - E_B\right)\exp\left(-\frac{E_B}{T^*}\right) \tag{4.39}$$

Other degrees of freedom of all the reaction participants are considered actually in quasi equilibrium with the gas temperature T_0:

$$f_e^{A,C}\left(\varepsilon_e^{A,C}\right) = 1, \quad T_{k=t,r,v}^X = T_0 \tag{4.40}$$

In the framework of the **single excited state approach**, in particular, the quasi-equilibrium constant of the nonequilibrium process (Equation 4.36) can be estimated as:

$$K \approx \exp\left\{-\frac{1}{T_0}\left[\left(\Delta H - \Delta F^A\right) - bE_B\right]\right\} = K_0 \exp\frac{bE_B}{T_0} \tag{4.41}$$

In this relation, ΔH is the reaction enthalpy; ΔF_A is the free energy change corresponding to heating to T_0; and K_0 is the equilibrium constant at temperature T_0. As one can see from Equation 4.41, the equilibrium constant can be significantly increased due to the electronic excitation, and equilibrium significantly shifted toward the reaction products.

Again, note that both examples of statistical description of nonequilibrium plasma systems are very approximate and can be used only for qualitative description. The next example, however, demonstrates the possibility of correct qualitative and quantitative statistical description of a strongly nonequilibrium system.

4.1.10 Nonequilibrium Statistics of Vibrationally Excited Molecules: Treanor Distribution

As shown previously, the equilibrium distribution of diatomic molecules over the different vibrationally excited states follows the Boltzmann formula (Equation 4.10 and Equation 4.11) obviously with the temperature T. As demonstrated in Chapter 3, vibrational excitation of diatomic molecules in nonthermal plasma can be much faster than vibrational VT relaxation rate. As a result, the vibrational temperature T_v of the molecules in plasma can essentially exceed the translational one, T_0. The vibrational temperature in this case is usually defined in accordance with Equation 4.10 by the ratio of populations of ground state and the first excited level:

$$T_v = \frac{\hbar\omega}{\ln\left(N_0/N_1\right)} \tag{4.42}$$

The diatomic molecules can be considered as harmonic oscillators (Equation 3.6) for not very large levels of vibrational excitation. In this case, the VT relaxation can be neglected with respect to the VV exchange of vibrational quanta and the VV exchange is absolutely resonant and not accompanied by energy transfer to translational degrees of freedom. As a result, the vibrational degrees of freedom are independent from the translational one, and the vibrational distribution function follows the same Boltzmann formula (Equation 4.10 and Equation 4.11) but with the vibrational temperature T_v (Equation 4.42) exceeding the translational one, T_0.

An extremely interesting nonequilibrium phenomenon takes place if the anharmonicity of the diatomic molecules is taken into account. In this case, the one-quantum VV exchange is not completely resonant and the translational degrees of freedom become involved in the vibrational distribution, which results in strong deviation from the Boltzmann distribution.

Considering vibrational quanta as quasi particles and using the Gibbs distribution with a variable number of quasi particles v, the relative population of the vibrational levels can be expressed as:[97]

$$N_v = N_0 \exp\left(-\frac{\mu v - E_v}{T_0}\right) \tag{4.43}$$

In this relation, the parameter μ is the chemical potential; E_v is energy of the vibrational level v taken with respect to the zero level. Comparing Equation 4.43 and Equation 4.10, the effective chemical potential of a quasi particle μ is found:

$$\mu = \hbar\omega\left(1 - \frac{T_0}{T_v}\right) \tag{4.44}$$

From the Gibbs distribution (Equation 4.43), chemical potential (Equation 4.44), and definition (Equation 4.42) of the vibrational temperature, the distribution of vibrationally excited molecules over different vibrational levels can be expressed in the form of the famous **Treanor distribution function**:

$$f(v, T_v, T_0) = B\exp\left(-\frac{\hbar\omega v}{T_v} + \frac{x_e \hbar\omega v^2}{T_0}\right) \tag{4.45}$$

Here, x_e is the coefficient of anharmonicity and B is the normalizing factor of the distribution. Comparison of the parabolic-exponential Treanor distribution with the linear-exponential Boltzmann distribution is illustrated in Figure 4.3.

The Treanor distribution function was first derived statistically by Treanor et al.[98] When $T_v > T_0$, it shows that the population of highly vibrationally excited levels can

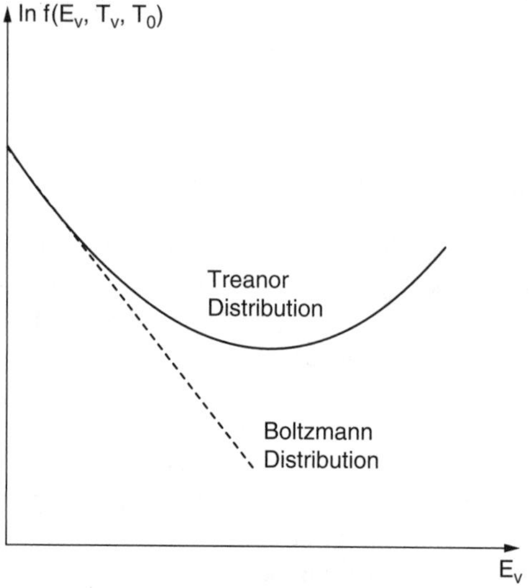

Figure 4.3 Comparison of Treanor and Boltzmann distribution functions

be many orders of value higher than predicted by the Boltzmann distribution, even at vibrational temperature (Figure 4.3). The Treanor distribution (or the Treanor effect) explains, in particular, the very high rates and high energy efficiency of chemical reactions stimulated by vibrational excitation in plasma. This will be discussed in the next chapter.

The Treanor distribution function (Equation 4.45) can be transformed back to the traditional form of the Boltzmann distribution in the following two situations. First, if molecules can be considered as harmonic oscillators and the anharmonicity coefficient is zero, $x_e = 0$, then Equation 4.45 gives the Boltzmann distribution $f(v) = B\exp(-\hbar\omega v/T_v) = B\exp(-E_v/T_v)$ even if the vibrational and translational temperatures are different. Second, if the translational and vibrational temperatures are equal, $T_v = T_0 = T$, the Treanor distribution (Equation 4.45) converts back to the Boltzmann function:

$$f(v,T) = B\exp\left(-\frac{\hbar\omega v}{T} + \frac{x_e\hbar\omega v^2}{T}\right) = B\exp\left(-\frac{\hbar\omega v - x_e\hbar\omega v^2}{T}\right) \propto \exp\left(-\frac{E_v}{T}\right) \quad (4.46)$$

Physical interpretation of the Treanor effect can be illustrated by the collision of two anharmonic oscillators, presented in Figure 4.4. First recall the VV exchange of harmonic molecules. Here, the resonant inverse processes of a quantum transfer from a lower excited harmonic oscillator to a higher excited one and back are not limited by translational energy and, as a result, have the same probability. That is the reason why the population of vibrationally excited, harmonic molecules keeps following the Boltzmann distribution even when the vibrational temperature in nonthermal plasma greatly exceeds the translational one.

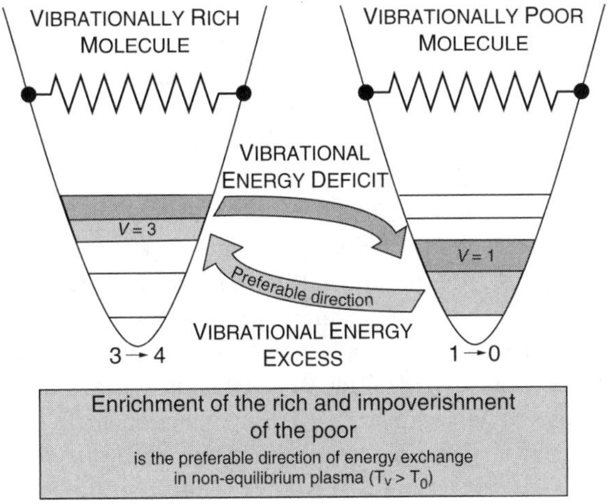

Figure 4.4 Overpopulation of highly vibrationally excited states (Treanor effect) in nonequilibrium plasma chemistry

In contrast, the VV exchange of anharmonic oscillators (see Figure 4.4) under conditions of a deficiency of translational energy ($T_v > T_0$) gives priority to a quantum transfer from a lower excited anharmonic oscillator to a higher excited one. The reverse reaction (a quantum transfer from a higher excited oscillator to a lower excited one) is less favorable because to cover the anharmonic defect of vibrational energy requires translational energy that cannot be provided effectively if $T_v > T_0$. Thus, the molecules with a larger number of vibrational quanta keep receiving more than those with a small number of quanta. In other words, the rich molecules become richer and the poor molecules become poorer. That is why the Treanor effect of overpopulation of highly excited vibrational states in nonequilibrium conditions is sometimes referred to as *capitalism in molecular life*.

As one can see from Figure 4.3 and Equation 4.45, the exponentially parabolic Treanor distribution function has a minimum at a specific number of vibrational quanta:

$$v_{\min}^{Tr} = \frac{1}{2x_e}\frac{T_0}{T_v} \tag{4.47}$$

Obviously, the Treanor minimum corresponds to low levels of vibrational excitation only if the vibrational temperature greatly exceeds the translational one. In equilibrium, $T_v = T_0 = T$ and point of the Treanor minimum goes beyond the physically possible range of the vibrational quantum numbers (Equation 3.9) for diatomic molecules.

Population of the vibrational level, corresponding to the Treanor minimum, can be found from Equation 4.45 and Equation 4.47 as:

$$f_{\min}^{Tr}\left(T_v, T_0\right) = B\exp\left(-\frac{\hbar\omega T_0}{4x_e T_v^2}\right) \tag{4.48}$$

At vibrational levels higher than the Treanor minimum (Equation 4.47), the population of vibrational states according to the Treanor distribution (Equation 4.45) becomes inverse, which means that the vibrational levels with higher energy are more populated than those with lower energy. This interesting effect of inverse population takes place in nonthermal discharges of molecular gases and, in particular, plays an important role in CO lasers.

However, it should be noted that according to the Treanor distribution, the population of states with high vibrational quantum numbers is growing without limitation (see Figure 4.3). This unlimited growth is not physically realistic. Deriving the Treanor distribution, the VT relaxation with respect to the VV exchange was neglected, which is correct only at relatively low levels of vibrational excitation (see Equation 3.77 and Equation 3.78). The VT relaxation prevails at large vibrational energies and causes the vibrational distribution function to fall exponentially in accordance with the Boltzmann distribution function corresponding to the translational temperature T_0. Description of this effect is beyond the statistical approach and requires analysis of the kinetic equations for vibrationally excited molecules; this will be considered in the next chapter.

4.2 BOLTZMANN AND FOKKER–PLANCK KINETIC EQUATIONS: ELECTRON ENERGY DISTRIBUTION FUNCTIONS

Kinetic equations permit description of the time and space evolution of different distribution functions, taking into account numerous energy exchange and chemical processes. In this chapter, the electron distribution functions will be discussed; those for excited atoms and molecules will be considered in the next chapter. Electron energy (or velocity) distribution functions in nonthermal discharges can be very sophisticated and quite different from the quasi-equilibrium Boltzmann distribution discussed previously in the framework of statistical approach. The distribution functions are usually strongly exponential and thus they can significantly influence reaction rates, which makes the analysis of their evolution especially important in plasma chemical kinetics. This section will begin with a general consideration of the kinetic equations and then applying them to describe and analyze specific electron energy distribution functions.

4.2.1 Boltzmann Kinetic Equation

The evolution of a distribution function $f(\vec{r}, \vec{v}\ t)$ can be considered in the six-dimensional **phase space** $(\vec{r}, \vec{v})$ of particle positions and velocities. Quantum numbers related to internal degrees of freedom are not included because primary interest is now in the electron distributions. In the absence of collisions between particles:

$$\frac{df}{dt} = \frac{f\left(\vec{r} + \overrightarrow{dr}, \vec{v} + \overrightarrow{dv}, t + dt\right) - f(\vec{r}, \vec{v}, t)}{dt} \tag{4.49}$$

Taking into account that in the collisionless case, the number of particles in a given state is fixed: $df/dt = 0$, $\overrightarrow{dv}/dt = \vec{F}/m$ and $\overrightarrow{dr}/dt = \vec{v}$, where m is the particle mass and $\vec{F}$ is an external force, Equation 4.49 can be rewritten as:

$$\frac{df}{dt} = \frac{\partial f}{\partial t} + \vec{v}\frac{\partial f}{\partial \vec{r}} + \frac{\vec{F}}{m}\frac{\partial f}{\partial \vec{v}} = 0 \tag{4.50}$$

The external electric E and magnetic B forces for electrons can be introduced into the relation as: $\vec{F} = -e(\vec{E} + \vec{v} \times \vec{B})$ and leads from Equation 4.50 to the **collisionless Vlasov equation** for the electron distribution function:[99]

$$\frac{\partial f}{\partial t} + \vec{v} * \nabla_r f - \frac{e}{m}\left(\vec{E} + \vec{v} \times \vec{B}\right) * \nabla_v f = 0 \tag{4.51}$$

Here, the operators ∇_r and ∇_v denote the electron distribution function gradients related to space and velocity coordinates.

Binary collisions between particles are very important in kinetics. Because they occur over a very short time, velocities can be changed practically instantaneously, and df/dt no longer is zero. The corresponding evolution of the distribution function can be described by adding the special term $I_{col}(f)$, the so-called collisional integral, to the kinetic equation (Equation 4.50):[99]

$$\frac{df}{dt} = \frac{\partial f}{\partial t} + \vec{v} * \nabla_r f + \frac{\vec{F}}{m} * \nabla_v f = I_{col}(f) \tag{4.52}$$

This is the well-known **Boltzmann kinetic equation**, which is used most to describe the evolution of distribution functions and to calculate different macroscopic kinetic quantities. The collisional integral of the Boltzmann kinetic equation (see, for example, Lifshitz and Pitaevsky[99]) is a nonlinear function of $f(\vec{v})$ and can be presented as:

$$I_{col}(f) = \int \left[f(v')f(v_1') - f(v)f(v_1)\right]\left|v - v_1\right|\frac{1}{\varepsilon}\,d\omega\,dv_1 \tag{4.53}$$

where v_1 and v are velocities before collision; v_1' and v' velocities after collision (all the velocities in this expression are vectors); ε is a parameter proportional to the mean free path ratio to the typical system size; and $d\omega$ is the area element in the plane perpendicular to the $(v_1 - v)$ vector.

In equilibrium, the collisional integral (Equation 4.53) should be equal to zero, which requires the following relation between velocities:

$$f(v')f(v_1') = f(v)f(v_1) \text{ or } \ln f(v') + \ln f(v_1') = \ln f(v) + \ln f(v_1) \tag{4.54}$$

This relation corresponds to the conservation of kinetic energy during the elastic collision:

$$\frac{m(v')^2}{2} + \frac{m(v_1')^2}{2} = \frac{mv^2}{2} + \frac{mv_1^2}{2} \tag{4.55}$$

and explains the proportionality $\ln f \propto mv^2/2$, which leads to the equilibrium Maxwell distribution function of particles at equilibrium:

$$f_0(\vec{v}) = B\exp\left(-\frac{mv^2}{2T}\right) \tag{4.56}$$

After calculation of the normalizing factor B, the final form of the equilibrium Maxwellian (or Maxwell–Boltzmann) distribution of particles over translational energies is:

$$f_0(\vec{v}) = \left(\frac{m}{2\pi T}\right)^{3/2} \exp\left(-\frac{mv^2}{2T}\right) \tag{4.57}$$

This relation, obviously, corresponds to the earlier considered Maxwell–Boltzmann distribution of electrons over translational energies (Equation 2.1). It should be noted that the nonequilibrium ideal gas entropy (including, with some restrictions, the case of gas of the plasma electrons) can be expressed by the following integral:

$$S = \int f \ln\frac{e}{f}\, d\vec{v} \tag{4.58}$$

Differentiating the entropy (Equation 4.58) and taking into account the Boltzmann kinetic equation (Equation 4.52), the time evolution of the entropy is:

$$\frac{dS}{dt} = -\int \ln f * I_{col}(f)\, d\vec{v} \tag{4.59}$$

In equilibrium, when the collisional integral $I_{col}(f)$ (Equation 4.53) is equal to zero, the entropy reaches its maximum value, which is known as the **Boltzmann H-Theorem.**

4.2.2 The τ-Approximation of the Boltzmann Kinetic Equation

As can be seen from Equation 4.52 and Equation 4.53, the most informative part of the Boltzmann kinetic equation is related to the collisional integral. Unfortunately, the expression (Equation 4.53) of the collisional integral is quite complicated for mathematical calculations. Several approximations have been developed to simplify the calculations. Probably the simplest one is the so-called **τ-approximation**:

$$I_{col}(f) = -\frac{f - f_0}{\tau} \tag{4.60}$$

The τ-approximation is based on the fact that the collisional integral should be equal to zero when the distribution function is Maxwellian (Equation 4.57 or Equation 2.1). The characteristic time of evolution of the distribution function is the collisional time τ (the reverse collisional frequency Equation 2.4). The Boltzmann equation in the framework of the τ-approximation becomes simple:

$$\frac{\partial f}{\partial t} = -\frac{f - f_0}{\tau} \tag{4.61}$$

with the solution exponentially approaching the Maxwellian distribution function f_0:

$$f(\vec{v},t) = f_0 + \left[f(\vec{v},0) - f_0\right]\exp\left(-\frac{t}{\tau}\right) \tag{4.62}$$

Here, $f(\vec{v},0)$ is the initial distribution function. Obviously, the τ-approximation is a very simplified but useful approach. Details about the Boltzmann equation and its solutions can be found in Lifshitz and Pitaevsky.[99]

4.2.3 Macroscopic Equations Related to Kinetic Boltzmann Equation

Now, the kinetic Boltzmann equation (Equation 4.52) can be integrated over the particle velocity. Then, the right-hand side of the resulting equation (the integral of I_{col} over velocities) represents the total change of the particle number in unit volume per unit time $G - L$ (generation rate minus rate of losses):

$$\int \overrightarrow{dv}\frac{\partial f}{\partial t} + \int \overrightarrow{dv} * \left(\vec{v}\frac{\partial f}{\partial \vec{r}}\right) + \frac{\vec{F}}{m}\int \frac{\partial f}{\partial \vec{v}}\overrightarrow{dv} = G - L \tag{4.63}$$

If a given volume does not include particle formation or loss, the right side of the equation is equal to zero. The distribution function is normalized here on the number of particles; then $\int f\overrightarrow{dv} = n$ (where n is the number density of particles) and the first term in Equation 4.63 corresponds to $\partial n/\partial t$. The particle flux is defined as $\int \vec{v} f \overrightarrow{dv} = n\vec{u}$, where $\vec{u}$ is the particle mean velocity, which is normally referred to as the **drift velocity**. In this case, the second term in Equation 4.63 is equal to $\partial/\partial\vec{r}\ (n\vec{u})$. The third term in this equation is equal to zero because, at infinitely high velocities, the distribution function is equal to zero. As a result, Equation 4.63 can be expressed as:

$$\frac{\partial n}{\partial t} + \nabla_r\left(n\vec{u}\right) = G - L \tag{4.64}$$

This macroscopic relation reflects the particle conservation and is known as the **continuity equation**. Again, the right-hand side of the equation is equal to zero if there are no processes of particle generation and loss.

The **momentum conservation equation** can be derived in a similar manner as the continuity equation. In this case, the Boltzmann equation should be multiplied by $\vec{v}$ and then integrated over velocity, which, for the gas of charged particles, gives as a result:[100]

$$mn\left[\frac{\partial \vec{u}}{\partial t} + \left(\vec{u} * \nabla\right)\vec{u}\right] = qn\left(\vec{E} + \vec{u}\times\vec{B}\right) - \nabla * \Pi + F_{col} \tag{4.65}$$

The second term from the right-hand side is the divergence of the pressure tensor Π_{ij}, which can be replaced in a plasma by pressure gradient ∇p; q is the particle charge; and E and B are the electric and magnetic fields. The last term, F_{col}, represents the collisional force, which is the time rate of momentum transfer per unit volume due to collisions with other species (i). The collisional force for plasma electrons and ions is usually due to collisions with neutrals. The **Krook collision operator** can approximate the collisional force by summation over the other species i:

$$\overrightarrow{F_{col}} = -\sum_i mn\nu_{mi}\left(\vec{u} - \vec{u_i}\right) - m\vec{u}(G - L) \tag{4.66}$$

In this relation, ν_{mi} is the momentum transfer frequency for collisions with the species i; $\vec{u}$ and $\overrightarrow{u_i}$ are the mean or drift velocities. The first term in the Krook collision operator can be interpreted as friction. The second term corresponds to the momentum transfer due to creation or destruction of particles and is generally small.

Equation 4.65 for plasma in electric field can be simplified, considering in the Krook operator only neutral species, assuming they are at rest ($u_i = 0$), and also neglecting the inertial force $(\vec{u} * \nabla)\vec{u} = 0$. This results in the most common form of the momentum conservation (or force) equation:

$$m\frac{\partial \vec{u}}{\partial t} = q\vec{E} - \frac{1}{n}\nabla p - m\nu_m\vec{u} \tag{4.67}$$

Finally, the **energy conservation equation** for the electron and ion fluids can be derived by multiplying the Boltzmann equation by 1/2 mv^2 and integrating over velocity:

$$\frac{\partial}{\partial t}\left(\frac{3}{2}p\right) + \nabla * \frac{3}{2}(p\vec{u}) + p\nabla * \vec{u} + \nabla * \vec{q} = \frac{\partial}{\partial t}\left(\frac{3}{2}p\right)_{col} \tag{4.68}$$

In this relation, $\vec{q}$ is the heat flow vector; 3/2 p corresponds to the energy density; and the term $\partial/\partial t\,(3/2\,p)_{col}$ represents changes of the energy density due to collisional processes, including ohmic heating, excitation, and ionization. The energy balance of the steady-state discharges usually can be described by the simplified form of the equation (Equation 4.68):

$$\nabla * \frac{3}{2}(p\vec{u}) = \frac{\partial}{\partial t}\left(\frac{3}{2}p\right)_{col} \tag{4.69}$$

which balances the macroscopic energy flux with the rate of energy change due to collisional processes.

4.2.4 Fokker–Planck Kinetic Equation for Determination of Electron Energy Distribution Functions

The Boltzmann kinetic equation obviously can be used for determinating the electron energy distribution functions in different nonequilibrium conditions (see, for example,

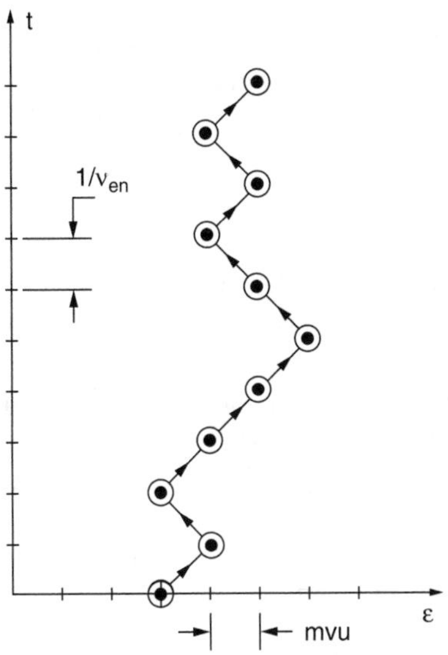

Figure 4.5 Electron diffusion and drift in energy space

the appendix to the book of Lieberman and Lichtenberg).[13] An improved physical interpretation of the electron energy distribution function evolution can be presented using the Fokker–Planck approach. The evolution of the electron energy distribution function is considered in the framework of this kinetic approach as an electron diffusion and drift in the space of electron energy (see, for example, Rusanov and Fridman[9] and Raizer[101]).

To derive the kinetic equation for the electron energy distribution function $f(\varepsilon)$ in the Fokker–Planck approach, consider the dynamics of energy transfer from the electric field to electrons. Suppose that most of the electron energy obtained from the electric field is then transferred to neutrals through elastic and inelastic collisions. Let an electron have a velocity $\vec{v}$ after a collision with a neutral particle. Then, the electron receives an additional velocity during the free motion between collisions (see the illustration in Figure 4.5):

$$\vec{u} = -\frac{e\vec{E}}{m\nu_{en}} \tag{4.70}$$

which corresponds to its drift in the electric field $\vec{E}$. Here, ν_{en} is frequency of the electron–neutral collisions and e and m are an electron charge and mass. The corresponding change of the electron kinetic energy between two collisions is equal to:

$$\Delta\varepsilon = \frac{1}{2}m\left(\vec{v}+\vec{u}\right)^2 - m\vec{v}^2 = m\vec{v}\vec{u} + \frac{1}{2}m\vec{u}^2 \tag{4.71}$$

Usually, the drift velocity is much lower than the thermal velocity $u << v$, and the absolute value of the first term $m\vec{v}\vec{u}$ in Equation 4.71 is much greater than that of the second. However, the thermal velocity vector $\vec{v}$ is isotropic and the value of $m\vec{v}\vec{u}$ can be positive and negative with the same probability. The average contribution of the term into $\Delta\varepsilon$ is equal to zero, and the average electron energy increase between two collisions is related only to the square of drift velocity:

$$\langle \Delta\varepsilon \rangle = \frac{1}{2} m\vec{u}^2 \tag{4.72}$$

Comparing Equation 4.71 and Equation 4.72 leads to the conclusion that the **electron motion along the energy spectrum (or in energy space) can be considered as a diffusion process**. An electron with energy $\varepsilon = 1/2\, mv^2$ receives or loses per one collision an energy portion about mvu, depending on the direction of its motion—along or opposite to the electric field (see Figure 4.5). The energy portion mvu can be considered in this case as the electron mean free path along the energy spectrum (or in the energy space). Taking into account also the possibility of an electron motion across the electric field, the corresponding coefficient of electron diffusion along the energy spectrum can be introduced as:

$$D_\varepsilon = \frac{1}{3}(mvu)^2 \nu_{en} = \frac{2}{3} mu^2\, \varepsilon\, \nu_{en} \tag{4.73}$$

Besides the diffusion along the energy spectrum, drift related to the permanent average energy gains and losses also occurs in the energy space. Such average energy consumption from the electric field is described by Equation 4.72. The average energy losses per one collision are mostly due to the elastic scattering $(2m/M)\,\varepsilon$ and the vibrational excitation $P_{eV}(\varepsilon)\hbar\omega$ in the case of molecular gases. Here, M and m are the neutral particle and electron mass, and $P_{eV}(\varepsilon)$ is the probability of vibrational excitation by electron impact. The mentioned three effects define the electron drift velocity in the energy space neglecting the super-elastic collisions as follows:

$$u_\varepsilon = \left[\frac{mu^2}{2} - \frac{2m}{M}\varepsilon - P_{eV}(\varepsilon)\hbar\omega\right]\nu_{en} \tag{4.74}$$

The electron energy distribution function $f(\varepsilon)$ can be considered as a number density of electrons in the energy space and can be found from the continuity equation (Equation 4.64) also in the energy space. Based upon the expressions for diffusion coefficient (Equation 4.73) and drift velocity (Equation 4.74), such a **continuity equation for an electron motion along the energy spectrum** can be presented as the **Fokker–Planck kinetic equation:**

$$\frac{\partial f(\varepsilon)}{\partial t} = \frac{\partial}{\partial \varepsilon}\left[D_\varepsilon \frac{\partial f(\varepsilon)}{\partial \varepsilon} - f(\varepsilon) u_\varepsilon\right] \tag{4.75}$$

This form of the Fokker–Planck kinetic equation for electrons is obviously much easier to interpret and to solve for different nonequilibrium plasma conditions than the general Boltzmann kinetic equation (Equation 4.52). Next, different specific solutions of the equation will be analyzed.

4.2.5 Different Specific Electron Energy Distribution Functions: Druyvesteyn Distribution

The steady-state solution of the Fokker–Planck equation (Equation 4.75) for the electron energy distribution function in nonequilibrium plasma, corresponding to $f(\varepsilon \to \infty) = 0,\ df/d\varepsilon(\varepsilon \to \infty) = 0$, can be presented as:

$$f(\varepsilon) = B\exp\left\{\int_0^{\varepsilon} \frac{u_\varepsilon}{D_\varepsilon} d\varepsilon'\right\} \tag{4.76}$$

where B is the pre-exponential normalizing factor.

Using in Equation 4.76 the expressions (Equation 4.73 and Equation 4.74) for the diffusion coefficient and drift velocity in the energy space, the electron energy distribution function can be rewritten in the following integral form:

$$f(\varepsilon) = B\exp\left[-\int_0^{\varepsilon} \frac{3m^2}{Me^2E^2} \nu_{en}^2 \left(1 + \frac{M}{2m}\frac{\hbar\omega}{\varepsilon'} P_{eV}(\varepsilon)\right) d\varepsilon'\right] \tag{4.77}$$

This distribution function derived for the constant electric field can be generalized to the alternating electric fields. In this case, the electric field strength E should be replaced in Equation 4.77 by the effective strength:

$$E_{eff}^2 = E^2 \frac{\nu_{en}^2}{\omega^2 + \nu_{en}^2} \tag{4.78}$$

In this relation, E is the average value of the electric field strength and ω is the field frequency. From Equation 4.78, the electric field can be considered as quasi constant at room temperature and pressure 1 torr if frequencies ω << 2000 MHz.

Consider four specific electron energy distribution functions that can be derived after integration of the general relation (Equation 4.77) and correspond to energy dependences $\nu_{en}(\varepsilon)$ and $P_{eV}(\varepsilon)$:

1. **Maxwellian distribution**. For atomic gases, the elastic collisions dominate the electron energy losses,

$$P_{eV} \ll \frac{2m}{M}\frac{\varepsilon}{\hbar\omega},$$

and the electron–neutral collision frequency can be approximated as constant $\nu_{en}(\varepsilon) = const$; then, integration of Equation 4.77 gives the Maxwellian distribution (Equation 2.1) with the electron temperature:

$$T_e = \frac{e^2E^2M}{3m^2\nu_{en}^2} \tag{4.79}$$

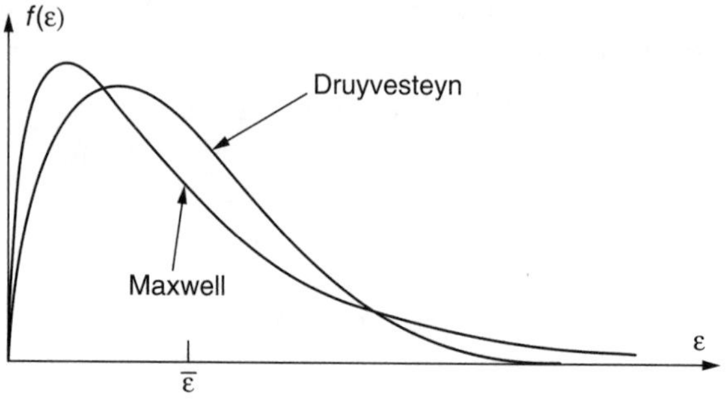

Figure 4.6 Maxwell and Druyvesteyn distribution functions at the same value of mean energy. Statistical weight effect related to the pre-exponential factor B and resulting in $f(0) = 0$ is also taken into account here.

2. **Druyvesteyn distribution**. If, again, the elastic collisions dominate the electron energy losses; but rather than the collision frequency, the electrons' mean free path λ is taken constant, then $\nu_{en} = v/\lambda$ and the integration of Equation 4.77 gives the exponential-parabolic Druyvesteyn distribution function, first derived in 1930:

$$f(\varepsilon) = B\exp\left[-\frac{3m}{M}\frac{\varepsilon^2}{(eE\lambda)^2}\right] \tag{4.80}$$

The Druyvesteyn distribution is decreasing with energy much faster than the Maxwellian one for the same mean energy. It is well illustrated in Figure 4.6. Equation 4.80 also leads to an interesting conclusion. The value $eE\lambda$ in the Druyvesteyn exponent corresponds to the energy that an electron receives from the electric field during one mean free path λ along the field. As can be seen from Equation 4.80, this requires from an electron about $\sqrt{M/m} \approx 100$ of the mean free paths λ along the electric field to get the average electron energy.

3. **Margenau distribution**. This distribution takes place in conditions similar to those of the Druyvesteyn distribution, but in an alternating electric field. Based on Equation 4.77 and Equation 4.78, integration yields the Margenau distribution function for electrons in high-frequency electric fields:

$$f(\varepsilon) = B\exp\left[-\frac{3m}{M}\frac{1}{(eE\lambda)^2}\left(\varepsilon^2 + \varepsilon m\omega^2\lambda^2\right)\right] \tag{4.81}$$

This was first derived in 1946. The Margenau distribution function (Equation 4.81) obviously corresponds to the Druyvesteyn distribution (Equation 4.80) if the electric field frequency is equal to zero $\omega = 0$.

4. **Distributions controlled by vibrational excitation**. The influence of vibrational excitation of molecules by electron impact on the electron energy distribution function is very strong in molecular gases

$$P_{eV} >> \frac{2m}{M}\frac{\varepsilon}{\hbar\omega}.$$

If it is assumed that $P_{eV}(\varepsilon) = \text{const}$ over some energy interval $\varepsilon_1 < \varepsilon < \varepsilon_2$ and, similar to the Druyvesteyn conditions, $\lambda = \text{const}$ ($\nu_{en} = v/\lambda$), then the electron energy distribution function in the energy interval $\varepsilon_1 < \varepsilon < \varepsilon_2$ can be presented as:

$$f(\varepsilon) \propto \exp\left[-\frac{6P_{eV}\hbar\omega}{(eE\lambda)^2}\varepsilon\right] \tag{4.82}$$

Outside this energy interval, the Druyvesteyn distribution (Equation 4.122) takes place. Comparing the relations Equation 4.80 and Equation 4.82, it should be taken into account that the vibrational excitation of molecules provides higher electron energy losses than elastic collisions ($P_{eV}\hbar\omega >> (m/M)\,\varepsilon$). This means that the electron energy distribution function is reduced in the energy interval $\varepsilon_1 < \varepsilon < \varepsilon_2$, related to the vibrational excitation. This sharp decrease of the distribution function can be described by the following small factor:

$$\alpha_v = \frac{f(\varepsilon_2)}{f(\varepsilon_1)} \approx \exp\left(-\frac{2M}{m}P_{eV}\hbar\omega\frac{\varepsilon_2 - \varepsilon_1}{\langle\varepsilon\rangle^2}\right) \tag{4.83}$$

where $\langle\varepsilon\rangle$ is the average electron energy of the Druyvesteyn distribution.

The parameter α_v of the distribution function reduction actually describes the probability for an electron to come through the energy interval of intensive vibrational excitation. It is proportional to probability for the electron to participate in the high-energy processes of electronic excitation and ionization. In other words, the ionization and electronic excitation rate coefficients are proportional to the parameter α_v.

For this reason, the parameter α_v can be used to analyze the influence of the vibrational temperature T_v on the ionization and electronic excitation rate coefficients. The probability P_{eV} is proportional to the number density of vibrationally nonexcited molecules $N(v = 0) \propto 1 - \exp(\hbar\omega/T_v)$ (see Equation 4.11), taking into account the possibility of super-elastic collisions. As a result, the influence of the vibrational temperature T_v on the ionization and electronic excitation rate coefficients can be characterized as the double exponential function:

$$\alpha_v \approx \exp\left[-\frac{2M}{m}P_{eV}^0\hbar\omega\frac{\varepsilon_2 - \varepsilon_1}{\langle\varepsilon\rangle^2}\left(1 - \exp\left(-\frac{\hbar\omega}{T_v}\right)\right)\right] \tag{4.84}$$

Here, P_{eV}^0 is the vibrational excitation probability of nonexcited molecules ($T_v = 0$). Equation 4.84 illustrates the strong increase of electron energy distribution function as well as ionization and electronic excitation rate coefficients with vibrational temperature.

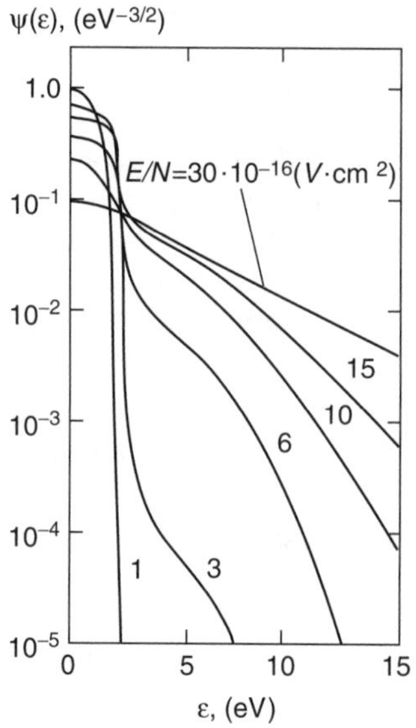

Figure 4.7 Electron distribution function $\psi(\varepsilon) = n(\varepsilon)/(n_e \tilde{A}\varepsilon)$ in nitrogen (From Raizer, Y. P. 1991. *Gas Discharge Physics*. Springer, Berlin)[101]

4.2.6 Electron Energy Distribution Functions in Different Nonequilibrium Discharge Conditions

The analytical distribution functions discussed previously can be used only for qualitative analysis of nonequilibrium plasma systems. The quantitative distributions found experimentally or from detailed kinetic modeling are more sophisticated, as can be seen from some examples presented in Figure 4.7. Some general specific features of such electron energy distribution functions will be discussed in examples of real discharge and gas conditions.

At first, it should be noted that the electron energy distribution function in nonequilibrium discharges in noble gases is usually close to the Druyvesteyn distribution (Equation 4.80) if the ionization degree is not high enough for essential contribution of electron–electron collisions. For the same mean energies, $f(\varepsilon)$ in molecular gases is much closer to the Maxwellian function. Quite strong deviations of $f(\varepsilon)$ in molecular gases from the Maxwellian distribution take place only for high-energy electrons at relatively low mean energies of about 1.5 eV.

Even a small admixture of a molecular gas (about 1%) into a noble gas dramatically changes the electron energy distribution function, strongly decreasing the fraction of high-energy electrons. The influence of such a small molecular gas admixture is strongest at relatively low values of reduced electric field E/n_0 (n_0 is the number density of gas), when the Ramsauer effect is essential. Addition of only 10% of air into argon makes the electron energy distribution function appear similar to that for molecular gas.[94]

The electron–electron collisions were not taken into account earlier in the kinetic equation (Equation 4.75). They can influence the distribution function and make it Maxwellian, when the ionization degree n_e/n_0 is high. This effect is convenient to characterize using the following numerical factor:[47]

$$a = \frac{\nu_{ee}}{\delta \nu_{en}} \approx 10^8 \frac{n_e}{n_0} \times \frac{1}{\langle \varepsilon, eV \rangle^2} \times \frac{10^{-16}\text{cm}^2}{\sigma_{en}, \text{cm}^2} \times \frac{10^{-4}}{\delta} \tag{4.85}$$

In this expression, ν_{ee} and ν_{en} are frequencies of electron–electron and electron–neutral collisions, corresponding to the average electron energy $\langle \varepsilon \rangle$; σ_{en} is the cross section of the electron–neutral collisions at the same electron energy; and δ is the average fraction of electron energy transferred to a neutral particle during their collision.

The electron–electron collisions make the electron energy distribution function Maxwellian at higher values of the factor $a >> 1$. Note that the Maxwellization of high-energy electrons ($\varepsilon >> \langle \varepsilon \rangle$) becomes possible at higher levels of the degree of ionization than Maxwellization of electrons with average energy. The Maxwellization of the high-energy electrons with energy ε by means of the electron–electron collisions requires $a >> \varepsilon^2/T_e^2$.

As one can see from Equation 4.85, Maxwellization of electrons in a plasma of noble gases (smaller δ) requires lower degrees of ionization than in cases of molecular gases (higher δ). For example, effective Maxwellization of the electron energy distribution function in argon plasma at $E/n_0 = 10^{-16}\ V * \text{cm}^2$ begins at the ionization degree $n_e/n_0 = 10^{-7} - 10^{-6}$. In molecular nitrogen at the same value of the reduced electric field and ionization degrees up to 10^{-4}, the deviations of the electron distribution provided by electron–electron collisions are still weak.

4.2.7 Relations between Electron Temperature and Reduced Electric Field

The electron energy distribution functions described earlier permit finding the relation between the reduced electric field and average electron energy, which then can be recalculated to an effective electron temperature: $\langle \varepsilon \rangle = 3/2\ T_e$ even for non-Maxwellian distributions. Such a relation can be derived, for example, from averaging the electron drift velocity in energy space (Equation 4.74) over the entire energy spectrum, taking into account the expression (Equation 4.70):

$$\frac{e^2 E^2}{m \nu_{en}^2} = \delta * \frac{3}{2} T_e \tag{4.86}$$

Here, the factor δ characterizes the average fraction of electron energy lost in a collision with a neutral particle:

$$\delta \approx \frac{2m}{M} + \langle P_{eV} \rangle \frac{\hbar \omega}{\langle \varepsilon \rangle} \tag{4.87}$$

In atomic gases, $\delta = 2m/M$ and Equation 4.87 permits finding the exact value of the factor δ for the formula Equation 4.86. In molecular gases, however, this factor is usually considered a semiempirical parameter. Thus, in typical conditions of nonequilibrium discharges in nitrogen, one can take for numerical calculations: $\delta \approx 3*10^{-3}$.

Further analysis of Equation 4.86 can be done, taking into account the following relations between parameters:

$$\nu_{en} = n_0\langle\sigma_{en}v\rangle, \quad \lambda = \frac{1}{n_0\sigma_{en}}, \quad \langle v\rangle = \sqrt{\frac{8T_e}{\pi m}} \qquad (4.88)$$

Combining these formulas with Equation 4.86 gives the relation between electron temperature, electric field, and the mean free path of electrons:

$$T_e = \frac{eE\lambda}{\sqrt{\delta}}\sqrt{\pi/12} \qquad (4.89)$$

This relation is in good agreement with Maxwell (Equation 4.79) and Druyvesteyn (Equation 4.80) electron energy distribution functions. It is convenient to rewrite Equation 4.89 as a relation between electron temperature and the reduced electric field E/n_0:

$$T_e = \left(\frac{E}{n_0}\right)\frac{e}{\langle\sigma_{en}\rangle}\sqrt{\frac{\pi}{12\delta}} \qquad (4.90)$$

This linear relation between electron temperature and the reduced electric field is obviously only a qualitative one. As can be seen from Figure 4.8, the relation $T_e(E/n_0)$ can be more complicated in reality.

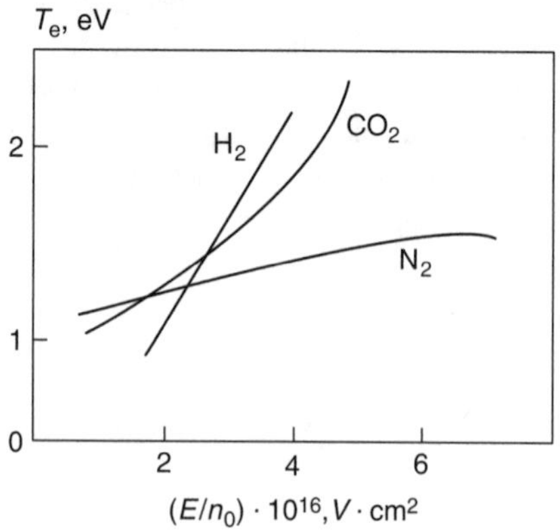

Figure 4.8 Electron temperature as a function of reduced electric field

4.3 ELECTRIC AND THERMAL CONDUCTIVITY IN PLASMA: DIFFUSION OF CHARGED PARTICLES

4.3.1 Isotropic and Anisotropic Parts of Electron Distribution Functions

The previous section concentrated on the electron energy distribution functions, which are related to the isotropic part of general electron velocity distribution $f(\vec{v})$. The electron velocity distribution $f(\vec{v})$ is, obviously, anisotropic in electric field because of the peculiarity of motion in the direction of the field. This anisotropy will be considered in this section; it determines the electric current, plasma conductivity, and all related effects.

To describe the anisotropy of the electron velocity distribution, it is necessary to recall from the previous section that an electron receives an additional velocity $\vec{u}$ during the free motion between collisions (see Equation 4.70 and illustration in Figure 4.5). If the anisotropy is not very strong (which is the normal case) and $u << v$, it can be assumed that the fraction of electrons in the point $\vec{v}$ of the real anisotropic distribution f is directly related to the fraction of electrons in the point $\vec{v}-\vec{u}$ of the correspondent isotropic distribution $f^{(0)}$. In this case, taking into account that directions of $\vec{E}$ and $\vec{u}$ are opposite for electrons, the electron velocity distribution function can be presented as:

$$f(\vec{v})=f^{(0)}(\vec{v}-\vec{u})\approx f^{(0)}(\vec{v})-\vec{u}\frac{\partial}{\partial\vec{v}}f^{(0)}(\vec{v})=f^{(0)}(\vec{v})+u\cos\theta\frac{\partial f^{(0)}(v)}{\partial v} \quad (4.91)$$

Here, θ is the angle between directions of the velocity $\vec{v}$ and the electric field $\vec{E}$. Taking into account the azimuthal symmetry of $f(\vec{v})$ and omitting the dependence on θ for the isotropic function $f^{(0)}$, Equation 4.91 can be rewritten in the following scalar form:

$$f(v,\theta)=f^{(0)}(v)+\cos\theta * f^{(1)}(v) \quad (4.92)$$

In this relation, the function $f^{(1)}(v)$ is the one responsible for the anisotropy of the electron velocity distribution. According to Equation 4.91, this function can be expressed as:

$$f^{(1)}(v)=u\frac{\partial f^{(0)}(v)}{\partial v}=\frac{eE}{m\nu_{en}}\frac{\partial f^{(0)}(v)}{\partial v} \quad (4.93)$$

Equation 4.92 can be interpreted as the first two terms of the series expansion of $f(\vec{v})$ using the orthonormalized system of Legendre polynomals. The first term of the expansion is the isotropic part of the distribution function. It corresponds to the electron energy distribution functions (see Equation 2.1 and other functions considered in the previous section) and related to them as:

$$f(\varepsilon)\,d\varepsilon=f^{(0)}(v)*4\pi v^2\,dv \quad (4.94)$$

The second term of the expansion (Equation 4.92) includes information on the electric field and thus is related to the electric current provided by plasma electrons. Based on that, the electron current density can be presented as:

$$j = -e\int \vec{v} f(\vec{v})\vec{dv} = -\frac{4\pi}{3}e\int_0^\infty v^3 f^{(1)}(v)dv \tag{4.95}$$

Note that the results concerning anisotropy of the electron distribution function could be also generalized for the case of high-frequency fields. Often in this case, it is sufficient to replace the electric field by the effective one (Equation 4.78). Special effects related to high-frequency electromagnetic fields will be discussed in Chapter 6.

Equation 4.93 can be used to compare the anisotropic and isotropic parts of the electron velocity distribution function:

$$f^{(1)} = u\frac{\partial f^{(0)}}{\partial v} \sim \frac{u}{\langle v \rangle} * f^{(0)} \tag{4.96}$$

This shows that the smallness of the anisotropy of the distribution function $f^{(1)} << f^{(0)}$, which determines the framework of presented consideration, is directly related to smallness of the electron drift velocity with respect to the thermal velocity.

4.3.2 Electron Mobility and Plasma Conductivity

Taking into account Equation 4.93 for the anisotropic part of distribution function $f^{(1)}(v)$ and the conventional relation between current density and the strength of electric field $j = \sigma E$, the formula for the **electron conductivity** in plasma follows:

$$\sigma = \frac{4\pi e^2}{3m}\int_0^\infty \frac{v^3}{\nu_{en}(v)}\left[-\frac{\partial f^{(0)}(v)}{\partial v}\right]dv \tag{4.97}$$

In the simple case, when $\nu_{en}(v) = \nu_{en} = const$, the integration (Equation 4.97) by parts gives the well-known and very convenient for application relation between the electric conductivity of the plasma and the electron concentration:

$$\sigma = \frac{n_e e^2}{m\nu_{en}} \tag{4.98}$$

The following numerical formula based on Equation 4.98 can be used for practical calculations of the conductivity of the weakly ionized plasma under consideration:

$$\sigma = 2.82 * 10^{-4}\frac{n_e\left(\text{cm}^{-3}\right)}{\nu_{en}\left(\sec^{-1}\right)}, \text{Ohm}^{-1}\text{cm}^{-1} \tag{4.99}$$

Equation 4.98 and Equation 4.99 imply that electrons mostly provide plasma conductivity; this is correct in most cases. Only when the ions' concentration much exceeds that of electrons does the ions' contribution become important as well. For this case, a corresponding expression for ions similar to Equation 4.99 should be added to calculate the total conductivity.

Plasma conductivity presented in Equation 4.99 depends on the electron concentration n_e and frequency of electron–neutral collisions ν_{en} (calculated with respect to the momentum transfer). The electron concentration for an equilibrium thermal plasma can be taken in this case from the Saha equation (Equation 4.15); for nonequilibrium, nonthermal plasma, it can be found from a balance of charged particles (see, for example, Section 4.5). The frequency of electron–neutral collisions ν_{en} is proportional to the gas number density (or, in other words, to pressure) and can be found numerically for some specific gases in Table 4.1. In general, as can be seen from Equation 4.99 and the preceding discussion, the conductivity is proportional to the degree of ionization n_e/n_0 in a plasma.

Obviously, Equation 4.98 and Equation 4.99 can be applied in particular to calculate the power transferred from an electric field to plasma electrons. This leads to another well-known formula of the so-called **joule heating**:

$$P = \sigma E^2 = \frac{n_e e^2 E^2}{m\nu_{en}} \tag{4.100}$$

Also, Equation 4.98 and Equation 4.99 can be used to determine another important characteristic of an electron motion in plasma—the **electron mobility** μ_e, which is the coefficient of proportionality between the electron drift velocity v_d and electric field:

$$v_d = \mu_e E, \quad \mu_e = \frac{\sigma_e}{en_e} = \frac{e}{m\nu_{en}} \tag{4.101}$$

To calculate the electron mobility, the following numerical relation similar to Equation 4.99 can be used:

$$\mu_e = \frac{1.76 * 10^{15}}{\nu_{en}\left(\sec^{-1}\right)}, \frac{\text{cm}^2}{V\sec} \tag{4.102}$$

Equation 4.101 demonstrates the proportionality of the drift velocity and reduced electric field E/n_0. This is quite close to reality (see Figure 4.9) if the electron–neutral collision cross section is not changing significantly, although this is not always true. This cross section, for example, decreases significantly in noble gases at low electron energies, which is known as the **Ramsauer effect**. As a result, the effective electron mobility in noble gases can be relatively high at low values of the reduced electric field E/n_0 or E/p (in argon it takes place at E/p of about 10^{-3} V/cm Torr).

It should be noted that the reduced electric field is presented in Figure 4.9 as E/p, not as E/n_0. In the same way in Table 4.1, the gas concentration n_0 is also replaced by pressure. Such a way of presentation is historically typical for the description of room temperature nonthermal discharges where n_0 (cm^{-3}) = 3.295 *

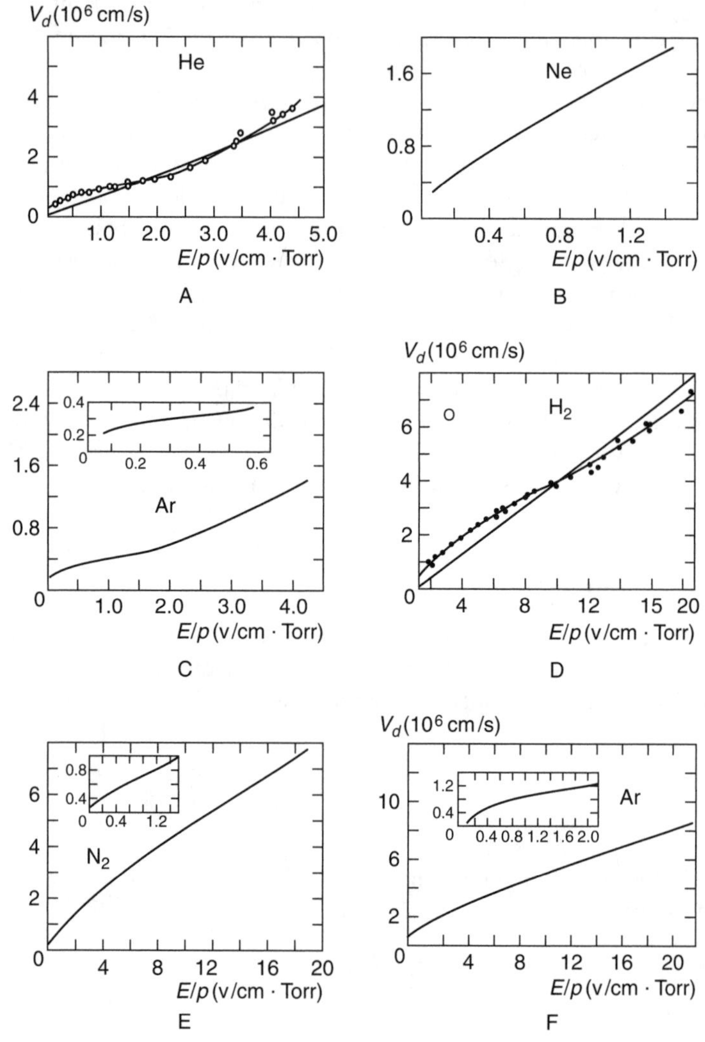

Figure 4.9 Electron drift velocities in different inert and molecular gases (From Raizer, Y. P. 1991. *Gas Discharge Physics*. Springer, Berlin)[101]

$10^{16} * p$(Torr). If the gas temperature T_0 of a nonthermal discharge differs from the room temperatureone T_{00}, the pressure should be recalculated to the gas concentration n_0 or at least replaced by the effective pressure $p_{eff} = p\ T_{00}/T_0$.

4.3.3 Similarity Parameters Describing Electron Motion in Nonthermal Discharges

Taking into account the Equation 4.98 and Equation 4.101 and the preceding remark about effective pressure, it is convenient to construct the so-called **similarity param-**

Table 4.1 Estimated Similarity Parameters Describing Electron–Neutral Collision Frequency, Electron Mean Free Path, Electron Mobility, and Conductivity at $E/p = 1 - 30$ V/cm.Torr

Gas	$\lambda * p$, 10^{-2} cm Torr	ν_{en}/p, 10^9 sec $^{-1}$ Torr^{-1}	$\mu_e * p$, 10^6 cm^2 Torr/V sec	$\frac{\sigma * p}{n_e}$, $10^{-13}\frac{\text{Torr cm}^2}{\text{Ohm}}$
Air	3	4	0.45	0.7
N_2	3	4	0.4	0.7
H_2	2	5	0.4	0.6
CO_2	3	2	1	2
CO	2	6	0.3	0.5
Ar	3	5	0.3	0.5
Ne	12	1	1.5	2.4
He	6	2	0.9	1.4

eters ν_{en}/p, λp, $\mu_e\, p$, $\sigma p/n_e$. These parameters are approximately constant for each gas, at low temperature conditions and $E/p = 1 - 30$ V/cm.Torr. Thus, they can be applied for simple determination of the electron–neutral collision frequency ν_{en}; electron mean-free-path λ: electron mobility μ_e; and conductivity σ. Numerical values of these similarity parameters for several gases are collected in Table 4.1.

4.3.4 Plasma Conductivity in Perpendicular Static Uniform Electric and Magnetic Fields

The direction of current in the presence of magnetic field becomes noncollinear to the electric field and the conductivity should be considered as the tensor σ_{ij}:

$$j_i = \sigma_{ij} E_j \tag{4.103}$$

In general, this motion is quite sophisticated and needs to be discussed in detail, for example, in Uman.[102] To illustrate the physics of the phenomenon, consider at first the motion of a particle with electric charge q in the static uniform perpendicular electric E and magnetic B fields ($E << cB$; here, c is the speed of light) in the absence of collisions. This motion of a charged particle is shown in Figure 4.10 and can be described by the equation:

$$m\frac{\overrightarrow{dv}}{dt} = q\left(\vec{E} + \vec{v}\times\vec{B}\right) \tag{4.104}$$

This motion can be analyzed in a reference frame, moving with respect to the initial laboratory reference frame with some velocity $\overrightarrow{v_{EB}}$. The particle velocities in the new and old reference frames are obviously related as: $\vec{v}' = \vec{v} + \overrightarrow{v_{EB}}$. Then, Equation 4.104 in the new moving reference frame can be rewritten as:

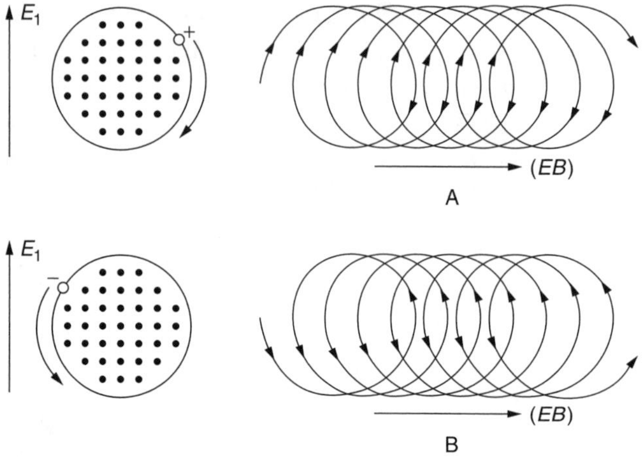

Figure 4.10 Illustration of (A) positive and (B) negative particles' drift in perpendicular electric and magnetic fields

$$m\frac{\overrightarrow{dv'}}{dt} = q\left(\vec{E} + \overrightarrow{v_{EB}} \times \vec{B}\right) + q\vec{v}' \times \vec{B} \tag{4.105}$$

The trick is to find the velocity $\overrightarrow{v_{EB}}$, which makes the first right-hand-side term in Equation 4.105 equal to zero: $\vec{E} + \overrightarrow{v_{EB}} \times \vec{B} = 0$ and reduces the particle motion to $m\,\overrightarrow{dv'}/dt = q\vec{v}' \times \vec{B}$, that is, to a circular motion and rotation around the magnetic field lines (see Figure 4.10). The frequency of this rotation is the so-called cyclotron frequency:

$$\omega_B = \frac{eB}{m} \tag{4.106}$$

Multiplying the requirement $\vec{E} + \overrightarrow{v_{EB}} \times \vec{B} = 0$ by $\vec{B}$, a formula for the velocity $\overrightarrow{v_{EB}}$ can be derived. The "spiral" moves (see Figure 4.3.2) with this velocity in the direction perpendicular to electric and magnetic fields:

$$\overrightarrow{v_{EB}} = \frac{\vec{E} \times \vec{B}}{B^2} \tag{4.107}$$

This motion of a charged particle is usually referred to as the **drift in crossed electric and magnetic fields**. It is interesting that the direction of the drift velocity does not depend on a charge sign and thus is the same for electrons and ions.

Equation 4.106 and Equation 4.107 describe the collisionless drift of charged particles in crossed electric and magnetic fields. Collisions with neutral particles cause the charge particles to move additionally along the electric field. This combined motion can be described in general by the conductivity tensor (Equation 4.103).

This tensor includes two important conductivity components: one representing conductivity along the electric field, $\sigma_{\parallel}$, and the second corresponding to the current perpendicular to electric and magnetic fields $\sigma_{\perp}$:

$$\sigma_{\parallel} = \frac{\sigma_0}{1+\left(\dfrac{\omega_B}{\nu_{en}}\right)^2}, \quad \sigma_{\perp} = \sigma_0 \frac{\left(\dfrac{\omega_B}{\nu_{en}}\right)}{1+\left(\dfrac{\omega_B}{\nu_{en}}\right)^2} \tag{4.108}$$

In these relations, σ_0 is the conventional conductivity (Equation 4.98) in absence of magnetic field. When the magnetic field is low or pressure high, then $\omega_B << \nu_{en}$. In this case, according to Equation 4.108, the transverse conductivity $\sigma_{\perp}$ can be neglected, and the longitudinal conductivity $\sigma_{\parallel}$ actually coincides with the conventional one σ_0. Conversely, in the case of high magnetic fields and low pressures ($\omega_B >> \nu_{en}$), first, the electrons become "trapped" by magnetic field and start drifting across the electric and magnetic fields. The longitudinal conductivity then can be neglected and the transverse conductivity becomes independent of pressure and mass of a charged particle:

$$\sigma_{\perp} \approx \frac{n_e e^2}{m\omega_B} = \frac{n_e e}{B} \tag{4.109}$$

The transverse conductivity (Equation 4.108) obviously corresponds in this case to the collisionless drift velocity (Equation 4.109) in the crossed electric and magnetic fields. Note that this relation for plasma conductivity is valid when the cyclotron velocity slightly exceeds the electron–neutral collision frequency ($\omega_B >> \nu_{en}$) but not too much. If the magnetic field is much larger than is sufficient for $\omega_B >> \nu_{en}$, the ion motion can also become quasi collisionless and, according to Equation 4.107, ions and electrons actually move together without any current.

4.3.5 Conductivity of Strongly Ionized Plasma

The electric conductivity in a weakly ionized plasma discussed previously (see Equation 4.98 and Equation 4.99) was always related to the resistance provided by electron–neutral collisions. In strongly ionized plasma with larger degrees of ionization, the electron–ion scattering also must be taken into account. In this case, the electron–neutral collision frequency ν_{en} in Equation 4.98 and Equation 4.99 as well as further relations must be replaced by the total frequency, including the electron–neutral and electron–ion collisions:

$$v_{\Sigma} = v_{en} + n_e \langle v \rangle \sigma_{Coul} \tag{4.110}$$

In this relation, it is assumed that the ion and electron concentrations are equal (n_e); $\langle v \rangle$ is the average electron velocity; and σ_{coul} is the specially averaged **coulomb cross section** of the electron–ion collisions, which can be numerically calculated as:

$$\sigma_{Coul} = \frac{4\pi}{9}\frac{e^4 \ln\Lambda}{\left(4\pi\varepsilon_0 T_e\right)^2} = \frac{2.87 * 10^{-14} \ln\Lambda}{\left(T_e, eV\right)}, \text{cm}^2 \tag{4.111}$$

Although trajectory deflection in the electron–ion interaction is low when they are relatively far apart one from another, such large-distance collisions make an important contribution to the momentum transfer cross section because of the long-range nature of coulomb forces. This effect is taken into account in Equation 4.111 by multiplication of the natural cross section for interaction of charged particles $e^4/(4\pi\varepsilon_0 T_e)^2$ by the so-called **Coulomb logarithm:**

$$\ln\Lambda = \ln\left[\frac{3}{2\sqrt{\pi}}\frac{\left(4\pi\varepsilon_0 T_e\right)^{3/2}}{e^3 n_e^{1/3}}\right] = 13.57 + 1.5\log\left(T_e, eV\right) - 0.5\log n_e \tag{4.112}$$

Based upon Equation 4.111 and Equation 4.112, it can be calculated that the electron–ion collisions become significant in the total electron–collision frequency (Equation 4.110) and thus in the electric conductivity when the degree of ionization exceeds some critical value of about: $n_e/n_0 \geq 10^{-3}$. When the degree of ionization much exceeds the critical value, the electron–ion collisions become dominant in Equation 4.110. The electric conductivity in this case (see Equation 4.110) becomes almost independent on the electron concentration n_e (only through the Coulomb logarithm $\ln\Lambda$) and reaches its maximum value:

$$\sigma = \frac{9\varepsilon_0 T_e^2}{m\langle v\rangle e^2 \ln\Lambda} = 1.9 * 10^2 \frac{\left(T_e, eV\right)^{3/2}}{\ln\Lambda}, \text{Ohm}^{-1}\text{cm}^{-1} \tag{4.113}$$

4.3.6 Ion Energy and Ion Drift in Electric Field

One can find a relation among ions' average energy $\langle\varepsilon_i\rangle$; gas temperature T_0; and electric field E by analyzing the balance of an ion–molecular collision:[101]

$$\left\langle\varepsilon_i\right\rangle = \frac{3}{2}T_0 + \frac{M}{2M_i}\left(1 + \frac{M_i}{M}\right)^3 \frac{e^2E^2}{M_i\nu_{in}^2} \tag{4.114}$$

Here, M_i and M are masses of an ion and a neutral particle; ν_{in} is frequency of ion–neutral collisions corresponding to momentum transfer. If the reduced electric field in plasma is not too strong (usually $E/p < 10$ V/cm torr), the ion energy (Equation 4.114) only slightly exceeds that of neutrals. For electrons, because of their low mass, such a situation takes place only at very low reduced electric fields $E/p < 0.01$ V/cm torr. When the reduced electric fields are much stronger than mentioned previously ($E/p >> 10$ V/cm torr) and the second term in Equation 4.114 exceeds the first one, then the ion velocity, collision frequency ν_{in}, and ion energy are growing with the electric field:

$$\langle \varepsilon_i \rangle \approx \frac{1}{2}\sqrt{\frac{M}{M_i}}\left(1+\frac{M_i}{M}\right)^{3/2} eE\lambda \tag{4.115}$$

Here, λ is the ion mean free path.

In the case of weak and moderate electric fields ($E/p < 10$ V/cm torr), the ion drift velocity is proportional to the electric field and the ion mobility is constant:

$$\overrightarrow{v_{id}} = \frac{e\vec{E}}{v_{in} * MM_1/(M+M_1)}, \quad \mu_i = \frac{e}{v_{in} * MM_1/(M+M_1)} \tag{4.116}$$

Note that the ion mobility is actually constant as long as the ion–neutral collision frequency ν_{in} is constant, which takes place in cases of polarization-induced collisions (see the Langevin model; Equation 2.82 through Equation 2.84). The following convenient numerical relation can be used for calculations of the ion mobility in the framework of the Langevin model:

$$\mu_i = \frac{2.7 * 10^4 \sqrt{1 + M/M_i}}{p(\text{Torr})\sqrt{A * \left(\alpha/a_0^3\right)}} \tag{4.117}$$

In this relation, α is the polarizability of a neutral particle (see Section 2.6); A is its molecular mass; and a_0 is the Bohr radius.

At relatively strong electric fields (E/p >> 10 V/cm torr), the ion energies are higher and the ion–neutral collision cross section is limited by the gas-kinetic one. Similar to Equation 4.115, λ rather than ν_{in} is better considered as constant. Taking into account Equation 4.115, the drift velocity in the case of strong electric fields can be found as:

$$v_{id} \approx \sqrt[4]{\frac{M_i}{M}\left(1+\frac{M_i}{M}\right)}\sqrt{\frac{eE\lambda}{M_i}} \tag{4.118}$$

Thus, the linear dependence of the ion drift velocity on the electric field for weak fields changes to the square root proportionality at strong electric fields. This effect of decreasing ion mobility is illustrated in Figure 4.11.

4.3.7 Free Diffusion of Electrons and Ions

If electron and ion concentrations are quite low, their diffusion in plasma can be considered independent and described, for example, by the continuity equation (Equation 4.64). The total flux of charged particles ($\vec{\Phi} = n\vec{u}$, which, in Equation 4.64, was referred to as "drift") includes, in this case, the actual drift in the electric field described in Section 4.3.2, and diffusion following the Fick's law:

$$\overrightarrow{\Phi_{e,i}} = \pm n_{e,i}\mu_{e,i}\vec{E} - D_{e,i}\frac{\partial n_{e,i}}{\partial \vec{r}} \tag{4.119}$$

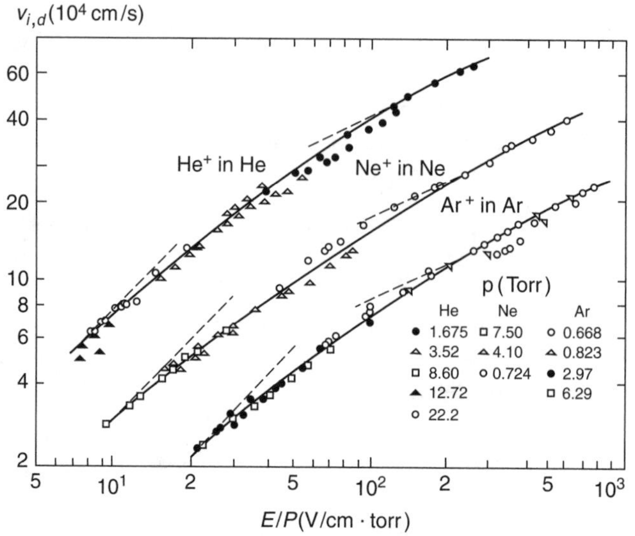

Figure 4.11 Drift velocities of ions in inert gases $T = 300$ K: $v_{id} \propto E$—dashed line on the left; $v_{id} \propto \sqrt{E}$—dashed line on the right (From Raizer, Y. P. 1991. *Gas Discharge Physics*. Springer, Berlin)[101]

The signs "+" and "–" here correspond to the charge of particles. The continuity equation (Equation 4.64) then can be rewritten as:

$$\frac{\partial n_{e,i}}{\partial t} + \frac{\partial}{\partial \vec{r}}\left(\pm n_{e,i}\mu_{e,i}\vec{E} - D_{e,i}\frac{\partial n_{e,i}}{\partial \vec{r}} \right) = G - L \tag{4.120}$$

The characteristic diffusion time corresponding to the diffusion distance R can be estimated based on Equation 4.120 as $\tau_D \approx R^2/D_{e,i}$. This simple relation is widely used to estimate all kinds of diffusion processes. The diffusion coefficients for electrons and ions can be estimated by the following formula, including the corresponding values of thermal velocity, mean free path, and collisional frequency with neutrals ($\langle v_{e,i}\rangle$, $\lambda_{e,i}$, $\nu_{en,in}$):

$$D_{e,i} = \frac{\langle v_{e,i}^2 \rangle}{3\nu_{en,in}} = \frac{\lambda_{e,i}\langle v_{e,i}\rangle}{3} \tag{4.121}$$

It is clear from Equation 4.121 that the diffusion coefficients are inversely proportional to the gas number density or pressure. This means that the similarity parameter $D_{e,i} * p$ (see Section 4.3.3) can be used to calculate easily the coefficients of free diffusion at room temperature. Some numerical values of the similarity parameters for electrons and ions are presented in Table 4.2.

It should be noted that, in accordance to Equation 4.131, the coefficient of free electron diffusion grows with temperature in nonthermal constant pressure discharges as $D_e \propto T_e^{1/2} T_0$. Assuming that ions and neutrals have the same temperature,

Table 4.2 Coefficients of Free Diffusion D of Electrons and Ions at Room Temperature in Different Gases

Diffusion	$D * p$, cm²/sec Torr	Diffusion	$D * p$, cm²/sec Torr	Diffusion	$D * p$, cm²/sec Torr
e in He	$2.1*10^5$	e in Ne	$2.1*10^6$	e in Ar	$6.3*10^5$
e in Kr	$4.4*10^4$	e in Xe	$1.2*10^4$	e in H_2	$1.3*10^5$
e in N_2	$2.9*10^5$	e in O_2	$1.2*10^6$	N_2^+ in N_2	40

the coefficient of free ion diffusion grows with temperature in nonthermal constant pressure discharges as $D_i \propto T_0^{3/2}$.

4.3.8 Einstein Relation among Diffusion Coefficient, Mobility, and Mean Energy

Taking the electron mobility as Equation 4.101 and the electron free diffusion coefficient as Equation 4.121, the relation between them and the average electron energy can be found:

$$\frac{D_e}{\mu_e} = \frac{2/3 \langle \varepsilon_e \rangle}{e} \tag{4.122}$$

A similar relation for ions can be derived from Equation 4.116 and Equation 4.121. In the case of Maxwellian distribution function, the average energy is equal to 3/2 T and formula Equation 4.122 can be expressed in a more general form valid for different particles:

$$\frac{D}{\mu} = \frac{T}{e} \tag{4.123}$$

known as the **Einstein relation.** The Einstein relations (Equation 4.122 and Equation 4.123) are very useful, in particular, for determinating the diffusion coefficients of charged particles, based on experimental measurements of their mobility.

4.3.9 Ambipolar Diffusion

The electron and ion diffusion cannot be considered "free" and independent, as was earlier assumed, when the ionization degree is relatively high. The electrons are moving faster than ions and form the charge separation zone with a strong polarization field that adjusts the electron and ion fluxes. This electric polarization field accelerates ions and slows down electrons, which makes all of them diffuse together as a "team." This phenomenon is known as the **ambipolar diffusion** and is illustrated in Figure 4.12.

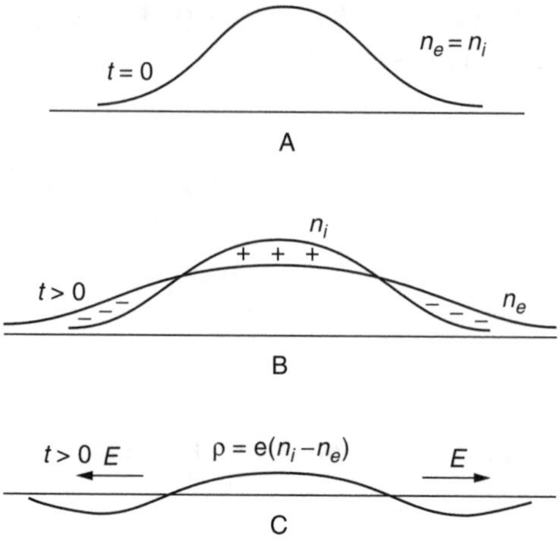

Figure 4.12 Ambipolar diffusion effect and plasma polarization in the presence of electron and ion density gradients

If the separation of charges is small, the electron and positive ion concentrations are approximately equal, $n_e \approx n_i$, as are their fluxes. The equality of the fluxes, including diffusion and drift in the polarization field, can be expressed based on Equation 4.119 as:

$$\overrightarrow{\Phi_e} = -\mu_e \vec{E} n_e - D_e \frac{\partial n_e}{\partial \vec{r}}, \quad \overrightarrow{\Phi_i} = \mu_i \vec{E} n_i - D_i \frac{\partial n_i}{\partial \vec{r}}, \quad \overrightarrow{\Phi_e} = \overrightarrow{\Phi_i} \tag{4.124}$$

Dividing the first relation by μ_e and the second by μ_i and then adding up the results gives the general relation for electron and ion fluxes:

$$\overrightarrow{\Phi_{e,i}} = -\frac{D_i \mu_e + D_e \mu_i}{\mu_e + \mu_i} \frac{\partial n_{e,i}}{\partial \vec{r}} \tag{4.125}$$

This permits introduction of the coefficient of ambipolar diffusion $\overrightarrow{\Phi_{e,i}} = -D_a \frac{\partial n_{e,i}}{\partial \vec{r}}$, which describes the collective diffusive motion of electrons and ions:

$$D_a = \frac{D_i \mu_e + D_e \mu_i}{\mu_e + \mu_i} \tag{4.126}$$

Noting that $\mu_e >> \mu_i$ and $D_e >> D_i$, Equation 4.126 can be rewritten as $D_a = D_i + D_e\ \mu_i/\mu_e$ and lead to the conclusion that ambipolar diffusion is greater than that of ions and less than that of electrons. Thus, the ambipolar diffusion speeds up ions and slows down electrons. Taking into account also the Einstein relation, (Equation 4.123) yields:

$$D_a \approx D_i + \frac{\mu_i}{e} T_e \tag{4.127}$$

It follows that, in equilibrium plasmas with equal electron and ion temperatures, $D_a = 2D_i$. For nonequilibrium plasmas with $T_e >> T_e$, the ambipolar diffusion $D_a = \mu_i/e\, T_e$ corresponds to the temperature of the fast electrons and mobility of the slow ions.

4.3.10 Conditions of Ambipolar Diffusion: Debye Radius

To determine the conditions of ambipolar diffusion with respect to the free diffusion of electrons and ions, first the absolute value of the polarization field from Equation 4.124 must be estimated:

$$E \approx \frac{D_e}{\mu_e}\frac{1}{n_e}\frac{\partial n_e}{\partial r} = \frac{T_e}{e}\frac{\partial \ln n_e}{\partial r} \propto \frac{T_e}{eR} \tag{4.128}$$

Here, R is the characteristic length of change of the electron concentration. From another view, the difference between the ion and electron concentrations $\Delta n = n_i - n_e$ characterizing the space charge is related to the electric field by the Maxwell equation:

$$\frac{\partial}{\partial \vec{r}}\vec{E} = \frac{e\,\Delta n}{\varepsilon_0}.$$

Simplifying the Maxwell equation for estimations as $E/R \propto e\Delta n/\varepsilon_0$ and combining this relation with Equation 4.128 yields the relative deviation from quasi-neutrality:

$$\frac{\Delta n}{n_e} \approx \frac{T_e}{e^2 n_e}\frac{1}{R^2} = \left(\frac{r_D}{R}\right)^2, \quad r_D = \sqrt{\frac{T_e \varepsilon_0}{e^2 n_e}} \tag{4.129}$$

Here, r_D is the so-called **Debye radius**. This important plasma parameter characterizes the quasi neutrality. The Debye radius is the characteristic of strong charge separation and plasma polarization. If the electron concentration is high and the Debye radius is small, $r_D << R$, then, according to Equation 4.129, the deviation from quasi neutrality is small, the electrons and ions move as a group, and the diffusion should be considered ambipolar. Conversely, if the electron concentration is relatively low and the Debye radius is large, $r_D \geq R$, then the plasma is not quasi neutral, the electrons and ions move separately, and their diffusion should be considered free.

For calculations of the Debye radius, it is convenient to use the following numerical formula:

$$r_D = 742\sqrt{\frac{T_e, eV}{n_e, \text{cm}^{-3}}}, \text{cm} \tag{4.130}$$

For example, if the electron temperature is 1 eV and the electron concentration exceeds 10^8 cm^{-3}, then according to Equation 4.130, the Debye radius is less than

0.7 mm and diffusion should be considered ambipolar for gradient scale greater than 3 mm to 1 cm.

A quite large collection of numerical data concerning diffusion and drift of electron and ions can be found in McDaniel and Mason.[103,104]

4.3.11 Thermal Conductivity in Plasma

Heat transfer—in particular, the behavior of thermal conductivity in a plasma—is a quite complex subject of plasma engineering; it is primarily important for thermal plasmas. In-depth discussion of the subject can be found in Boulos et al.[90] and Eletsky et al.[105,141] Here, some fundamental aspects of high temperature, heat transfer, and the peculiarities of the coefficients of thermal conductivity at high temperatures in plasma will be briefly discussed. The equation of heat transfer due to the thermal conductivity in moving one-component gas can be expressed similarly to the continuity equation (see Section 4.2.3 and, specifically, Equation 4.64):

$$\frac{\partial}{\partial t}\left(n_0\langle\varepsilon\rangle\right) + \nabla * \vec{q} = 0 \tag{4.131}$$

Here, n_0 and $\langle\varepsilon\rangle$ are the gas number density and average energy of a molecule; $\vec{q}$ is the heat flux. In Equation 4.131, the average energy of a molecule should be replaced by the equilibrium temperature T_0 and specific heat c_v with respect to one molecule. Also, it is necessary to consider that the heat flux includes two terms: one related to the gas motion with velocity $\vec{u}$ and the other related to the thermal conductivity with coefficient κ: $\vec{q} = \langle\varepsilon\rangle n_0\vec{u} - \kappa\nabla * T_0$. This gives a convenient equation describing the thermal conductivity in moving quasi-equilibrium gas:

$$\frac{\partial}{\partial t}T_0 + \vec{u}\nabla * T_0 = \frac{\kappa}{c_v n}\nabla^2 T_0 \tag{4.132}$$

The thermal conductivity coefficient in a one-component gas without dissociation, ionization, and chemical reactions can be estimated similarly to Equation 4.121 as:

$$\kappa \approx \frac{1}{3}\lambda\langle v\rangle n_0 c_v \propto \frac{c_v}{\sigma}\sqrt{\frac{T_0}{M}} \tag{4.133}$$

Here, σ is a typical cross section of the molecular collisions, and M is the molecular mass. According to Equation 4.133, the thermal conductivity coefficient does not depend on gas density and grows slowly with temperature.

The growth of the value of the thermal conductivity with temperature in plasma at high temperatures, however, can be much faster than was given by Equation 4.133 because of the influence of dissociation, ionization, and chemical reactions. For example, consider the effect of dissociation and recombination $2A \Leftrightarrow A_2$ on acceleration of the temperature dependence of thermal conductivity. Molecules are mostly dissociated into atoms in a zone with higher temperature (see illustration in Figure

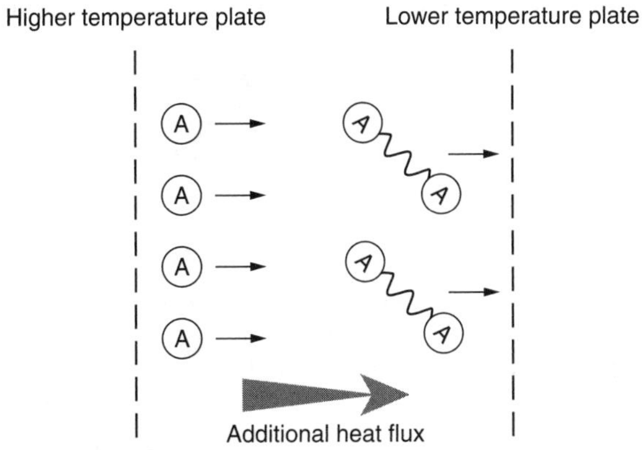

Figure 4.13 Recombination contribution to high-temperature thermal conductivity

4.13) and much less dissociated in lower temperature zones. Then, the quasi-equilibrium diffusion of the molecules (D_m) to the higher temperature zone leads to their intensive dissociation, consumption of dissociation energy E_D, and to the related big heat flux:

$$\overrightarrow{q_D} = -E_D D_m \nabla n_m = -\left(E_D D_m \frac{\partial n_m}{\partial T_0} \right) \nabla T_0 \tag{4.134}$$

which can be interpreted as accelerating thermal conductivity.

When the concentration of molecules is less than that of atoms, $n_m << n_a$ and $T_0 << E_D$, the equilibrium relation (Equation 4.18) permits evaluation of the molecules' density derivative $\partial n_m / \partial T_0 = E_D / T_0^2 \, n_m$. Substituting this derivative into Equation 4.134, the coefficient of thermal conductivity related to the dissociation of molecules is given by:

$$\kappa_D = D_m \left(\frac{E_D}{T_0} \right)^2 n_m \tag{4.135}$$

Compare the temperature dependence of the coefficient of thermal conductivity related to the dissociation of molecules with the one (Equation 4.133) not taking into account any reactions (Equation 4.133). Based on Equation 4.135, Equation 4.121, and Equation 4.133, the ratio of the two coefficients can be expressed as:

$$\frac{\kappa_D}{\kappa} \approx \frac{1}{c_v} \left(\frac{E_D}{T_0} \right)^2 \frac{n_m}{n_a} \tag{4.136}$$

As can be seen, the contribution of dissociation to the temperature dependence of the thermal conductivity related to dissociation can be very significant because $T_0 << E_D$. Obviously, the strongest effect takes place within the relatively narrow range of typical dissociation temperatures, when the concentrations of atoms and molecules are of the same order and very sensitive to temperature.

Relations similar to Equation 4.136 can be derived for showing the contributions of ionization and chemical reactions to the temperature dependence of thermal conductivity. These effects also are able to increase thermal conductivity because they are related with the transfer of relatively large quantities of energy. Ionization energy and typical chemical reaction enthalpy usually greatly exceed temperature. However, the relative influence of these effects depends on the relative concentration of corresponding species, which could be fairly low even in high-temperature thermal plasmas.

In the manner seen from Equation 4.136, the different mechanisms of thermal conductivity make their major contribution in specific temperature ranges. The concentration of corresponding species (e.g., electron and ion concentrations in the case of ionization) in this specific temperature range, on the one hand, is very sensitive to temperature; on the other hand, it could be relatively high (see Equation 4.136). Such requirements make the appropriate temperature range relatively narrow. As a result, the dependence of the total thermal conductivity coefficient on temperature, which grows in quasi-equilibrium systems (first of all, in the case of inert gases), can be sometimes quite sophisticated, particularly in the case of molecular gases.

Specific examples of the temperature dependence of thermal conductivity for different inert gases are presented in Figure 4.14 and, for different molecular gases, in Figure 4.15. The influence of different effects on the total thermal conductivity can be very well illustrated by the "roller-coaster" like $\kappa(T_0)$ dependence in air (see Figure 4.15).

4.4 BREAKDOWN PHENOMENA: TOWNSEND AND SPARK MECHANISMS, AVALANCHES, STREAMERS, AND LEADERS

The first three sections of Chapter 4 have been concerned with the general aspects of plasma statistics and kinetics. In the following two sections, the kinetics of charged particles is to be applied to describe some general fundamental features of starting and sustaining electric discharges. The breakdown phenomena will be considered in this section and the different discharge modes in the next.

4.4.1 Electric Breakdown of Gases: Townsend Mechanism

Electric breakdown is a complicated multistage threshold process that occurs when the electric field exceeds some critical value. During the short breakdown period, usually 0.01 to 100 μsec, the nonconducting gas becomes conductive and, as a result, generates different kinds of plasmas. The breakdown mechanisms can be very sophisticated, but all of them usually start with the electron avalanche, e.g., multiplication of some primary electrons in cascade ionization.

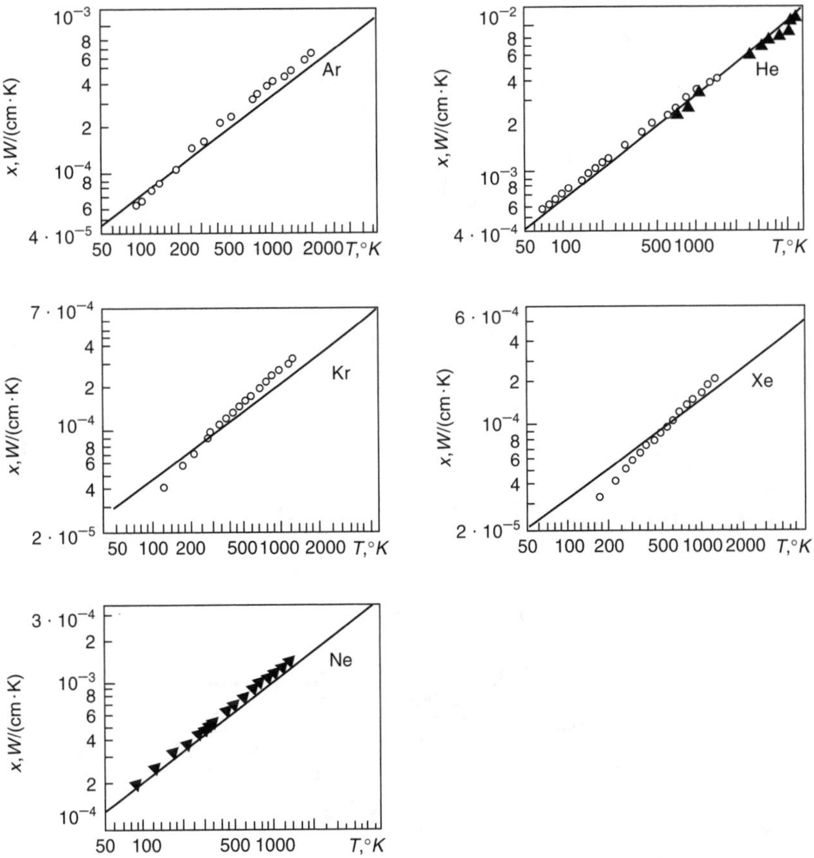

Figure 4.14 Thermal conductivity of inert gases[90]

Consider first the simplest breakdown in a plane gap with electrode separation, d, connected to a DC power supply with voltage V, which provides the homogeneous electric field $E = V/d$ (Figure 4.16). It is apparent that occasional formation of primary electrons near the cathode occurs that provides the very low initial current i_0. Each primary electron drifts to the anode, concurrently ionizing the gas and generating an avalanche. The avalanche evolves in time and in space because multiplication of electrons proceeds along with their drift from the cathode to the anode (Figure 4.16).

It is convenient to describe the ionization in an avalanche, not by the ionization rate coefficient, but rather by the **Townsend ionization coefficient α** that shows electron production per unit length or the multiplication of electrons (initial density n_{e0}) per unit length along the electric field: $dn_e/dx = \alpha n_e$ or the same $n_e(x) = n_{e0} \exp(\alpha x)$. The Townsend ionization coefficient is related to the ionization rate coefficient $k_i(E/n_0)$ (Section 2.2.1) and electron drift velocity v_d as:

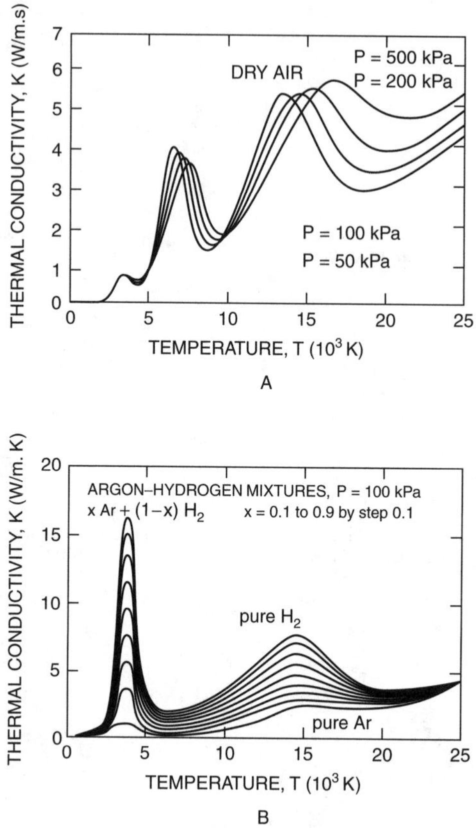

Figure 4.15 Temperature dependence of thermal conductivity of plasmas in air and Ar-H_2 (From Boulos, M. et al. *Thermal Plasma.*)[90]

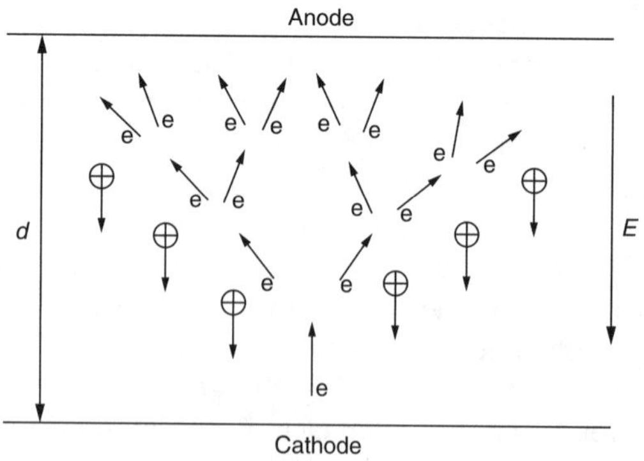

Figure 4.16 Townsend breakdown gap

$$\alpha = \frac{\nu_i}{v_d} = \frac{1}{v_d} k_i(E/n_0) n_0 = \frac{1}{\mu_e} \frac{k_i(E/n_0)}{E/n_0} \tag{4.137}$$

where ν_i is the ionization frequency with respect to one electron and μ_e is the electron mobility. Noting that breakdown begins at room temperature and that electron mobility is inversely proportional to pressure, it is convenient to present the Townsend coefficient α as the similarity parameter α/p depending on the reduced electric field E/p. Such dependence $\alpha/p = f(E/p)$ is presented in Figure 4.17 for different inert and molecular gases.[101]

According to the definition of the Townsend coefficient α, each primary electron generated near the cathode produces $\exp(\alpha d) - 1$ positive ions in the gap (Figure 4.16). Here, the electron losses due to recombination and attachment to electronegative molecules were neglected. The electron–ion recombination was neglected because the degree of ionization is very low during breakdown; attachment processes important in electronegative gases will be discussed.

All the $\exp(\alpha d) - 1$ positive ions produced in the gap per one electron are moving toward the cathode and altogether eliminate $\gamma * [\exp(\alpha d) - 1]$ electrons from the cathode in the process of secondary electron emission. Here, another Townsend coefficient γ is the secondary emission coefficient, characterizing the probability of a secondary electron generation on the cathode by an ion impact. Obviously, the **secondary electron emission coefficient** γ depends on cathode material, state of surface, type of gas, and reduced electric field E/p (defining ion energy). Some relevant numerical data for different conditions are given in Figure 4.18. Typical values of γ in electric discharges are 0.01 to 0.1; the effect of photons and metastable atoms and molecules (produced in avalanche) on the secondary electron emission is usually incorporated in the same "effective" γ coefficient.

Taking into account the current of primary electrons i_0 and electron current due to the secondary electron emission from the cathode, the total electronic part of the cathode current i_{cath} can be found from:

$$i_{cath} = i_0 + \gamma\, i_{cath}\left[\exp(\alpha d) - 1\right] \tag{4.138}$$

The total current in the external circuit is equal to the electronic current at the anode because of the absence of ion current. For this reason the total current can be found as $i = i_{cath} \exp(\alpha d)$. Taking Equation 4.138 into account, it leads us to **the Townsend formula**, first derived in 1902 to describe the ignition of electric discharges:

$$i = \frac{i_0 \exp(\alpha d)}{1 - \gamma\left[\exp(\alpha d) - 1\right]} \tag{4.139}$$

The current in the gap is non-self-sustained as long as the denominator in Equation 4.139 is positive. As soon as the electric field, (and, therefore, the Townsend α coefficient) becomes sufficiently large, the denominator in Equation 4.139 goes to zero and transition to self-sustained current (breakdown!) takes place. Thus, the simplest breakdown condition in the gap can be expressed as:

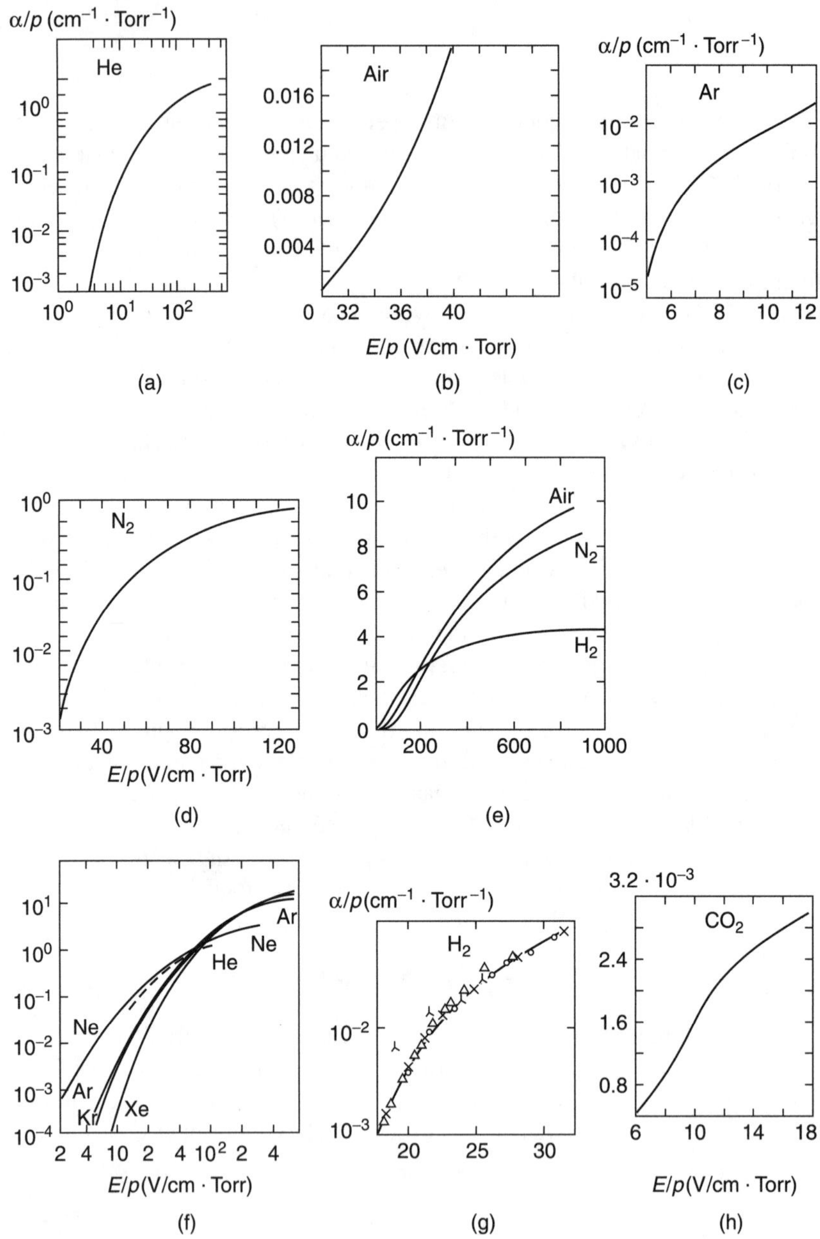

Figure 4.17 Townsend ionization coefficients in different gases

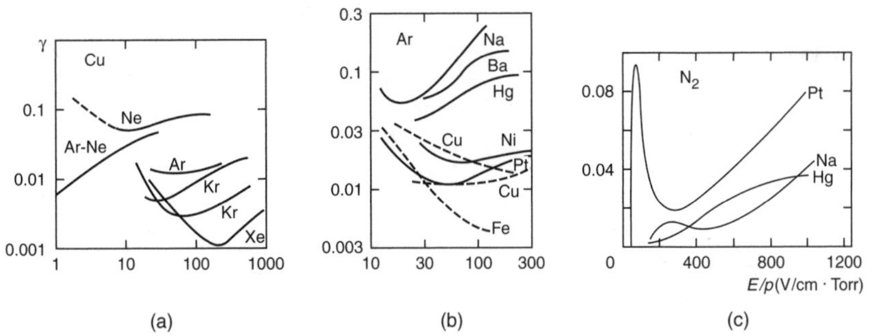

Figure 4.18 Effective secondary emission coefficient: (a) copper cathode in inert gases; (b) various metals in Ar; (c) various metals in N_2 (From Raizer, Y. P. 1991. *Gas Discharge Physics*. Springer, Berlin)[101]

$$\gamma[\exp(\alpha d)-1]=1, \quad \alpha d=\ln\left(\frac{1}{\gamma}+1\right) \tag{4.140}$$

This mechanism of ignition of a self-sustained current in a gap controlled by secondary electron emission from the cathode is usually referred to as the **Townsend breakdown mechanism**.

4.4.2 Critical Breakdown Conditions: Paschen Curves

To derive relations for the breakdown voltage and electric field based on Equation 4.140, it is convenient to rewrite Equation 4.137 for the Townsend coefficient α in the following conventional, semiempirical way; this relates the similarity parameters α/p and E/p, and was initially proposed by Townsend:

$$\frac{\alpha}{p}=A\exp\left(-\frac{B}{E/p}\right) \tag{4.141}$$

The parameters A and B of Equation 4.141 for numerical calculations of α in different gases at E/p = 30 ÷ 500 V/cm torr, are given in Table 4.3.

Table 4.3 Numerical Parameters A and B for Calculation of Townsend Coefficient α

Gas	A, 1/cm Torr	B, V/cm Torr	Gas	A, 1/cm Torr	B, V/cm Torr
Air	15	365	N_2	10	310
CO_2	20	466	H_2O	13	290
H_2	5	130	He	3	34
Ne	4	100	Ar	12	180
Kr	17	240	Xe	26	350

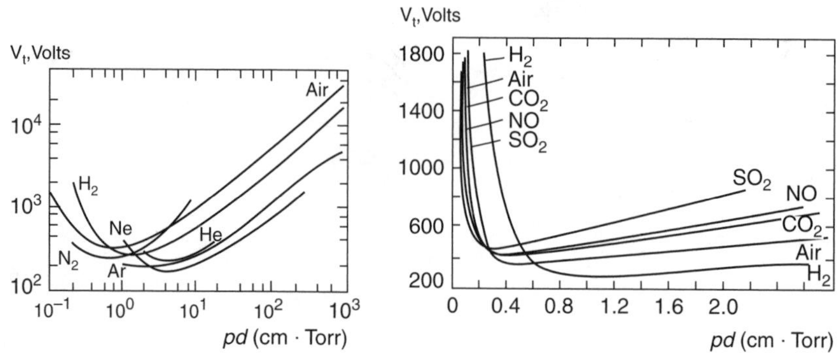

Figure 4.19 Paschen curves for different gases (From Raizer, Y. P. 1991. *Gas Discharge Physics*. Springer, Berlin)[101]

Combining Equation 4.140 and Equation 4.141 gives the following convenient formulas for calculating breakdown voltage and breakdown reduced electric field as functions of the important similarity parameter *pd*:

$$V = \frac{B(pd)}{C + \ln(pd)}, \quad \frac{E}{p} = \frac{B}{C + \ln(pd)} \tag{4.142}$$

In these relations, parameter B is obviously the same as in Equation 4.141 and Table 4.3. However, the parameter A is replaced by $C = \ln A - \ln\ln(1/\gamma + 1)$, taking into account the very weak, double logarithmic influence of secondary electron emission.

The breakdown voltage dependence on the similarity parameter *pd*, which is quite well described by Equation 4.142, is usually referred to as the **Paschen curve**. The experimental Paschen curves for different gases are presented in Figure 4.19. These curves have a minimum voltage point corresponding to the easiest breakdown conditions that can be found from Equation 4.142:

$$V_{\min} = \frac{eB}{A}\ln\left(1 + \frac{1}{\gamma}\right), \left(\frac{E}{p}\right)_{\min} = B, (pd)_{\min} = \frac{e}{A}\ln\left(1 + \frac{1}{\gamma}\right) \tag{4.143}$$

where $e = 2.72$ is the base of natural logarithms.

Numerically, the typical value of minimum voltage is about 300 V, corresponding to reduced electric field of about 300 V/cm torr and the parameter *pd* of about 0.7 cm torr. The right-hand branch of the Paschen curve (pressure greater than 1 torr for a gap of about 1 cm) is related to a case in which the electron avalanche has enough distance and gas pressure to provide intensive ionization even at moderate electric fields. In this case, the reduced electric field is almost fixed; it is slowly reducing logarithmically with *pd* growth. Conversely, the left-hand branch of the Paschen curve is related to the case in which ionization is limited by the avalanche size and gas pressure. Obviously, in this latter case, ionization sufficient for breakdown can be provided only by very high electric fields.

It is interesting to mention that the reduced electric field at the Paschen minimum corresponds to the so-called **Stoletov constant** $(E/p)_{min} = B$, which is the minimum discharge energy necessary to produce one electron–ion pair (minimum price of ionization). Here, the price of ionization can be expressed as $W = eE/\alpha$ (e is the charge of an electron), and its minimum, the Stoletov constant, is equal to

$$W_{min} = \frac{2.72 * eB}{A}.$$

The Stoletov constant usually exceeds the ionization potentials several times because electrons dissipate energy to vibrational and electronic excitation per each act of ionization. A typical numerical estimation for the minimum ionization price in electric discharges with high electron temperatures is about 30 eV.

Analyzing Equation 4.143, it is seen that the reduced electric field at the Paschen minimum $(E/p)_{min} = B$ does not depend on γ and thus on a cathode material in contrast to the minimum voltage V_{min} and the corresponding similarity parameter $(pd)_{min}$.

4.4.3 Townsend Breakdown of Larger Gaps: Specific Behavior of Electronegative Gases

The Townsend mechanism of breakdown is relatively homogeneous and includes development of independent avalanches; it usually occurs at $pd < 4000$ Torr · cm (at atmospheric pressure $d < 5$ cm). For larger gaps (more than 6 cm at atmospheric pressure), the avalanches disturb the electric field and are no longer independent. This leads to the spark mechanism of breakdown. Here, the discussion will focus on relatively large gaps, but still not sufficiently large ($pd < 4000$ Torr · cm) for sparks.

The reduced electric field E/p necessary for breakdown (Equation 4.142) is slightly logarithmically decreasing with pd in the framework of the Townsend breakdown mechanism ($pd < 4000$ Torr · cm). This is illustrated by the $E(d)$ dependence in atmospheric air, presented in Figure 4.20.[101] For larger gaps and larger avalanches, the reduced electric field E/p is less sensitive to the secondary electron emission and cathode material. This explains the E/p decrease with pd.

However, this reduction in electronegative gases is limited by electron attachment processes (usually dissociative attachment; Section 2.4.1, Equation 2.52). The influence of attachment processes can be taken into account in a similar manner as with ionization (Equation 4.137) by introducing an additional Townsend coefficient β:

$$\beta = \frac{\nu_a}{v_d} = \frac{1}{v_d} k_a\left(E/n_0\right) n_0 = \frac{1}{\mu_e} \frac{k_a(E/n_0)}{E/n_0} \tag{4.144}$$

In this relation, $k_a(E/n_0)$ and ν_a are the attachment rate coefficient (for dissociative attachment, see Equation 2.52) and attachment frequency with respect to an electron. Thus, all together, three Townsend coefficients α, β, and γ, describe the Townsend mechanism of electric breakdown. The Townsend coefficient β shows the electron

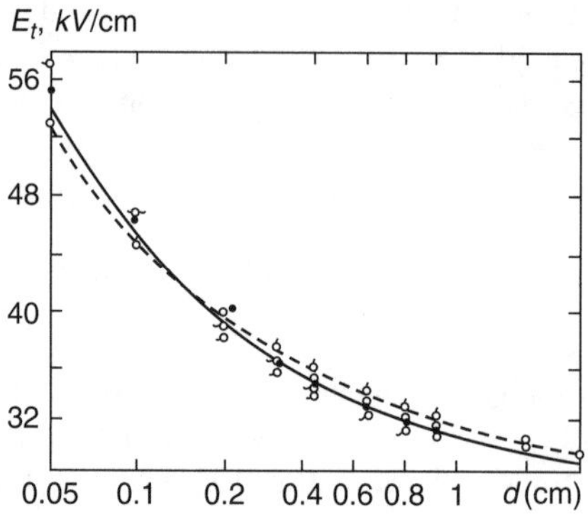

Figure 4.20 Breakdown electric field in atmospheric air

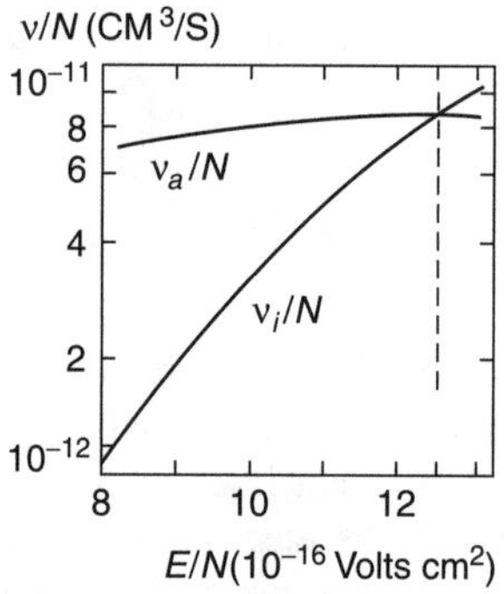

Figure 4.21 Frequencies of ionization (ν_i) and electron attachment (ν_a) in air, N is neutral gas density

losses due to attachment per unit length. Combining α (see Equation 4.137) and β gives:

$$\frac{dn_e}{dx} = (\alpha - \beta)n_e, \quad n_e(x) = n_{e0}\exp[(\alpha - \beta)x] \tag{4.145}$$

The Townsend coefficient β is an exponential function of the reduced electric field in the same way as α, although this function is not as sharp (see Figure 4.21).[101] For this reason, ionization much exceeds attachment at relatively high values of reduced electric fields and, in this case, the β coefficient can be neglected with respect to α.

When the gaps are relatively large (centimeter range at atmospheric pressure), the Townsend breakdown electric field in electronegative gases actually becomes constant and limited by attachment processes. Obviously, in this case, breakdown of electronegative gases requires much higher values of the reduced electric fields. The breakdown electric fields at high pressures for electronegative and nonelectronegative gases are presented in Table 4.4.

The Townsend mechanism of breakdown was discussed earlier in the most general situations. Such specific cases as electric breakdown in microwave; radio-

Table 4.4 Electric Fields Sufficient for Townsend Breakdown of Centimeters-Size Gaps at Atmospheric Pressure

Gas	*E/p*, kV/cm	Gas	*E/p*, kV/cm	Gas	*E/p*, kV/cm
Air	32	O_2	30	N_2	35
H_2	20	Cl_2	76	CCl_2F_2	76
CSF_8	150	CCl_4	180	SF_6	89
He	10	Ne	1.4	Ar	2.7

frequency and low-frequency fields; optical breakdown; and breakdown of vacuum gaps require special consideration. Such a detailed consideration can be found, in particular, in Raizer.[101]

4.4.4 Sparks Vs. Townsend Breakdown Mechanism

As has been mentioned, the Townsend mechanism of quasi-homogeneous breakdown can be applied only for low pressures and short discharge gaps ($pd < 4000$ Torr · cm, at atmospheric pressure $d < 5$ cm). Another breakdown mechanism, the so-called spark, takes place in larger gaps at high pressures. In contrast to the Townsend mechanism, **sparks** provide breakdown in a local narrow channel, without direct relation to electrode phenomena and with very high currents (up to 10^4 to 10^5 A) and current densities.

Sparks as well as Townsend breakdown are primarily related to avalanches. However, in large gaps they cannot be considered as independent and stimulated by electron emission from cathode. The sparks breakdown at high *pd* and considerable overvoltage develops much faster than the time necessary for ions to cross the gap and provide the secondary emission. The high conductivity spark channel can be formed even faster than electron drift time from cathode to anode! In this case, the breakdown voltage is independent of the cathode material, which is also a qualitative difference of Townsend and spark mechanisms of breakdown.

The mechanism of spark breakdown is based on the concept of a **streamer.** This thin ionized channel grows fast along the positively charged trail left by an intensive primary avalanche between electrodes. The avalanche also generates photons, which in turn initiate numerous secondary avalanches in the vicinity of the primary one. Electrons of the secondary avalanches are pulled by the strong electric field into the positively charged trail of the primary avalanche, creating a streamer propagating fast between electrodes. The streamer theory was originally developed by Raether,[106] Loeb,[107] and Meek and Craggs.[108]

If the distance between electrodes is multimeter or even kilometers long, as in the case of lightning, the individual streamers are not sufficient to provide large-scale spark breakdown. In this case, the so-called **leader** is moving from one electrode to another. The leader, as well as the streamer, is also a thin channel but much more conductive than an individual streamer. Leaders actually include the streamers as elements. Prior to considering the physics of streamers and leaders, a more detailed description of an individual avalanche is necessary.

4.4.5 Physics of Electron Avalanches

According to Equation 4.145 and taking into account possible electron attachment, the increase of the total number of electrons N_e, positive N_+ and negative N_- ions in an avalanche moving along the axis x follow the equations:

$$\frac{dN_e}{dx} = (\alpha - \beta)N_e, \quad \frac{dN_+}{dx} = \alpha N_e, \quad \frac{dN_-}{dx} = \beta N_e \tag{4.146}$$

where α and β are the ionization and attachment Townsend coefficients. If the avalanche starts from the only primary electron, the numbers of charged particles, electrons, and positive and negative ions can be found from Equation 4.146 as:

$$N_e = \exp[(\alpha - \beta)x], \quad N_+ = \frac{\alpha}{\alpha - \beta}(N_e - 1), \quad N_- = \frac{\beta}{\alpha - \beta}(N_e - 1) \tag{4.147}$$

The electrons in an avalanche move together along the direction of the nondisturbed electric field E_0 (axis x) with the drift velocity $v_d = \mu_e E_0$. Concurrently, free diffusion (D_e) spreads the group of electrons around the axis x in radial direction r. Taking into account the drift and the diffusion, the electron density in an avalanche $n_e(x,r,t)$ can be found from Equation 4.120 in the following form as:[109]

$$n_e(x,r,t) = \frac{1}{(4\pi D_e t)^{3/2}} \exp\left[-\frac{(x - \mu_e E_0 t)^2 + r^2}{4 D_e t} + (\alpha - \beta)\mu_e E_0 t \right] \tag{4.148}$$

From Equation 4.148, the electron density decreases with distance from the axis x following the Gaussian law. Then, the avalanche radius r_A (where the electron density is e times less than on the axis x) grows with time and the distance x_0 of the avalanche propagation:

$$r_A = \sqrt{4 D_e t} = \sqrt{4 D_e \frac{x_0}{\mu_e E_0}} = \sqrt{\frac{4 T_e}{e E_0} x_0} \tag{4.149}$$

In Equation 4.149, the Einstein relation between electron mobility and free diffusion coefficient (Equation 4.123) was taken into account.

The space distribution of positive and negative ion densities during the short interval of avalanche propagation, when the ions actually remain at rest, can be calculated using Equation 4.148 for n_e

$$n_+(x,r,t) = \int_0^t \alpha \mu_0 E_0 n_e(x,r,t')dt', \quad n_-(x,r,t) = \int_0^t \beta \mu_0 E_0 n_e(x,r,t')dt' \tag{4.150}$$

A simplified expression for the space distribution of positive ion density not too far from the axis x can be derived based on Equation 4.150 and Equation 4.148 in the absence of attachment and in the limit $t \to \infty$ as:[109]

$$n_+(x,r) = \frac{\alpha}{\pi r_A^2(x)} \exp\left[\alpha x - \frac{r^2}{r_A^2(x)}\right] \tag{4.151}$$

where $r_A(x)$ is the avalanche radius. This distribution reflects the fact that the ion concentration in the trail of the avalanche grows along the axis in accordance with the multiplication and exponential increase of the number of electrons (Equation 4.147). Concurrently, the radial distribution is Gaussian with an effective avalanche radius $r_A(x)$ growing with x (Equation 4.149).

An avalanche propagation is illustrated in Figure 4.22. Although the avalanche radius (Equation 4.149) is growing parabolically (proportionally to $\sqrt{x}$), the visible avalanche outline is wedge shaped. This means that the visible avalanche radius is growing linearly (proportionally to x; see the picture in Figure 4.22). This happens because the visible avalanche radiation is determined by the absolute density of excited species that is approximately proportional to the exponential factor Φ in the Equation 4.151) and obviously grows with x. The visible avalanche radius $r(x)$ can then be expressed from Equation 4.151, taking into account the smallness of r at small x, as:

$$\frac{r^2(x)}{r_A^2(x)} = \alpha x - \ln\Phi, \quad r(x) \approx r_A(x)\sqrt{\alpha x} = \sqrt{\frac{4T_e x}{eE_0} * \alpha x} = x\sqrt{\frac{4T_e \alpha}{eE_0}} \tag{4.152}$$

This explains the linearity of $r(t)$ and the wedge-shape form of an avalanche (Figure 4.22).

The qualitative change in avalanche behavior takes place when the charge amplification $\exp(\alpha x)$ is large. In this case, the considerable space charge is created with its own significant electric field $\vec{E}_a$, which should be added to the external field $\vec{E}_0$. The electrons are at the head of the avalanche while the positive ions remain behind, creating a dipole with the characteristic length $1/\alpha$ (distance that the electrons move before ionization) and charge $N_e \approx \exp(\alpha x)$. For breakdown fields about 30 kV/cm in atmospheric pressure air, the α-coefficient is approximately 10 cm^{-1} and the characteristic ionization length can be estimated as $1/\alpha \approx 0.1$ cm.

The external electric field distortion due to the space charge of the dipole is shown in Figure 4.23. In front of the avalanche head (and behind the avalanche), the external $\vec{E}_0$ and internal $\vec{E}_a$ electric fields make a total field stronger, which in turn accelerates ionization. Conversely, in between the separated charges or "inside the avalanche," the total electric field is lower than the external one, which slows down the ionization. The space charge also creates a radial electric field (see Figure 4.23). At a distance of approximately the avalanche radius (Equation 4.149), the electric field of the charge $N_e \approx \exp(\alpha x)$ reaches the value of the external field $\vec{E}_0$ at some critical value of αx. For example, during a 1-cm gap breakdown in air, the avalanche radius is about $r_A = 0.02$ cm and the critical value of αx when the avalanche

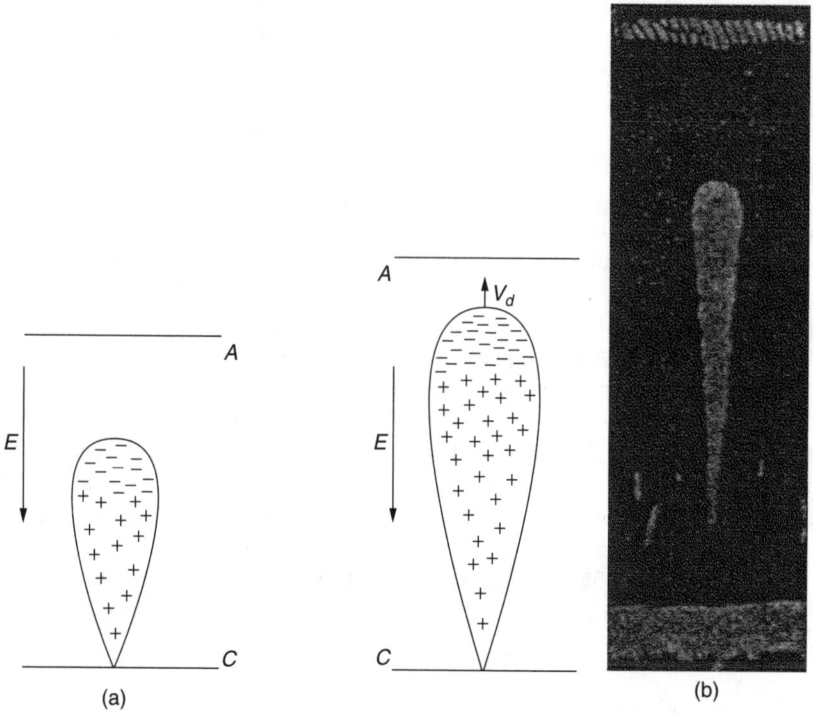

Figure 4.22 Avalanche evolution: (a) schematic; (b) photograph (From Raizer, Y. P. 1991. *Gas Discharge Physics*. Springer, Berlin)

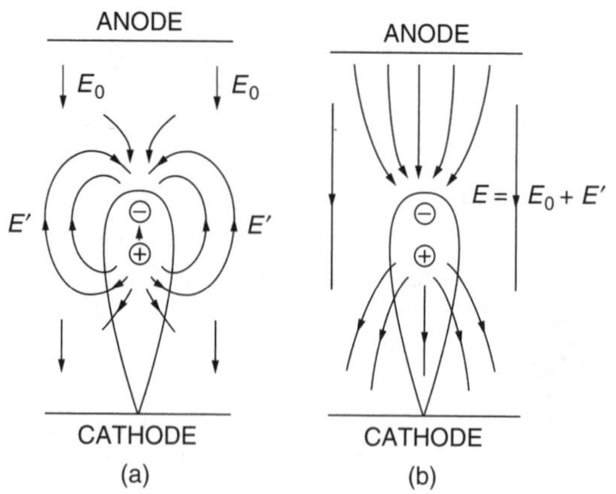

Figure 4.23 Eelctric field distribution in an avalanche. External and space charge fields are shown (a) separately; (b) combined.

electric field becomes comparable with E_0 is $\alpha x = 18$ (note that this corresponds to the Meek criterion of streamer formation discussed later).

When $\alpha x \geq 14$, the radial growth of an avalanche due to repulsion drift of electrons exceeds the diffusion effect and should be taken into account. In this case, the avalanche radius grows with x as (see problem in the problem section):

$$r = \sqrt[3]{\frac{3e}{4\pi\varepsilon_0 \alpha E_0}} \exp\frac{\alpha x}{3} = \frac{3}{\alpha}\frac{E_a}{E_0} \tag{4.153}$$

This fast growth of the transverse avalanche size restricts the electron density in the avalanche to the maximum value (see problem in the problem section):

$$n_e = \frac{\varepsilon_0 \alpha E_0}{e} \tag{4.154}$$

When the transverse avalanche size reaches the characteristic ionization length $1/\alpha \approx 0.1$ cm (the avalanche "dipole" size), the broadening of the avalanche head dramatically slows down. Obviously, the avalanche electric field is about the size of the external one in this case (see Equation 4.153). Typical values of maximum electron density in an avalanche are about 10^{12} to 10^{13} cm^{-3}.

As soon as the avalanche head reaches the anode, the electrons sink into the electrode and it is mostly the ionic trail that remains in the discharge gap. The electric field distortion due to the space charge in this case is shown in Figure 4.24. Because electrons are no longer present in the gap, the total electric field is due to the external field, the ionic trail and also the ionic charge "image" in the anode (see Figure 4.24). The resulting electric field in the ionic trail near the anode is less than the external electric field, but farther away from the electrode; it exceeds E_0. The total electric field reaches maximum value at the characteristic ionization distance, which is about $1/\alpha \approx 0.1$ cm from the anode.

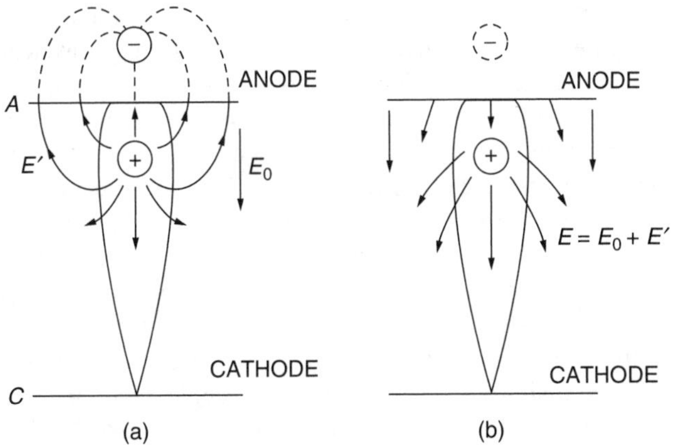

Figure 4.24 Electric field distribution when the avalanche reaches the anode. External and space charge fields are shown (a) separately; (b) combined.

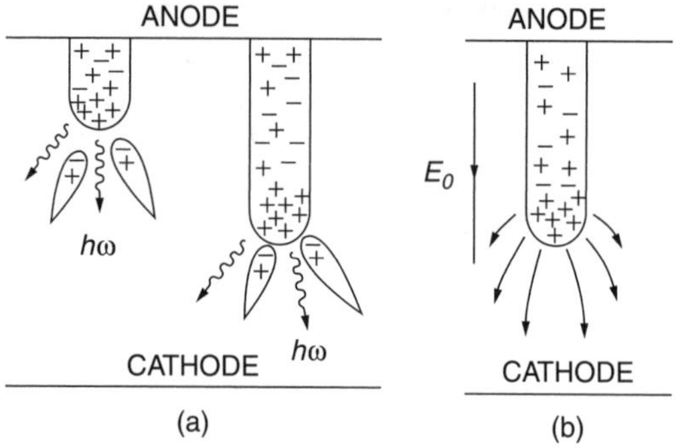

Figure 4.25 Cathode-directed streamer: (a) propagation; (b) electric field near streamer head

4.4.6 Cathode- and Anode-Directed Streamers

A strong primary avalanche is able to amplify the external electric field and form a thin, weakly ionized plasma channel, the so-called streamer, growing very fast between electrodes. When the streamer channel connects the electrodes, the current may be significantly increased to form the spark. The avalanche-to-streamer transformation takes place when the internal field of an avalanche becomes comparable with the external electric field, e.g., when the amplification parameter αd is sufficiently large (see Section 4.4.5).

If the discharge gap is relatively small, the transformation occurs only when the avalanche reaches the anode. Such streamer growth from anode to cathode is known as the **cathode-directed** or **positive streamer**. If the discharge gap and overvoltage are large, the avalanche-to-streamer transformation can take place quite far from the anode. In this case, the so-called **anode-directed** or **negative streamer** is able to grow toward both electrodes.

The mechanism of formation of a cathode-directed streamer is illustrated in Figure 4.25. High-energy photons emitted from the primary avalanche provide photoionization in the vicinity, which initiates secondary avalanches. Electrons of secondary avalanches are pulled into the ionic trail of the primary one (see the electric field distribution in Figure 4.24) and create a quasi-neutral plasma channel. Subsequently, the process repeats, providing growth of the streamer.

The cathode-directed streamer begins near the anode, where the positive charge and electric field of the primary avalanche are the highest. The streamer appears and operates as a thin conductive needle growing from the anode. Obviously, the electric field at the tip of the "anode needle" is very large, which stimulates the fast streamer propagation in the direction of cathode. Usually, streamer propagation is limited by neutralization of the ionic trail near the tip of the needle. Here, the electric field is so high that it provides electron drift with velocities of about 10^8 cm/sec. This explains the high speed of streamer growth, which is also about 10^8 cm/sec and exceeds by a factor of 10 the typical electron drift velocity in the external breakdown field. The latter is usually about 10^7 cm/sec.

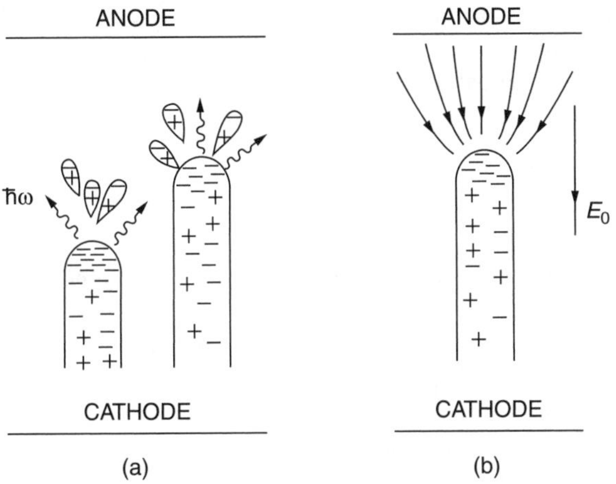

Figure 4.26 Anode-directed streamer: (a) propagation; (b) electric field near streamer head

All streamer parameters are in some way related to those of the primary avalanche. Thus, the diameter of the streamer channel is about 0.01 to 0.1 cm and corresponds to the maximum size of a primary avalanche head, which can be estimated as the ionization length $1/\alpha$ (see Section 4.4.5). Plasma density in the streamer channel also corresponds to the maximum electron concentration in the head of the primary avalanche, which is about 10^{12} to 10^{13} cm^{-3}. It is important to note that the specific energy input (electron energy transferred to one molecule) in a streamer channel is small during the short period (~30 nsec) of the streamer growth between electrodes. In molecular gases, it is about 10^{-3} eV/mol, which in temperature units corresponds to ~10 K.

The anode-directed streamer occurs between electrodes if the primary avalanche becomes sufficiently strong even before reaching the anode. Such a streamer, which is actually growing in two directions, is illustrated in Figure 4.26. The mechanism of the streamer propagation in the direction of the cathode is obviously the same as that of cathode-directed streamers. The mechanism of the streamer growth in the direction of the anode is also similar, but, in this case, electrons from the primary avalanche head neutralize the ionic trail of secondary avalanches. It should be noted that the secondary avalanches could be initiated here by photons as well as by some electrons moving in front of the primary avalanche.

4.4.7 Criterion of Streamer Formation: Meek Breakdown Condition

Formation of a streamer, as shown in Section 4.4.5 and Section 4.4.6, requires the electric field of space charge in the avalanche E_a to be of the order of the external field E_0:

$$E_a = \frac{e}{4\pi\varepsilon_0 r_A^2}\exp\left[\alpha\left(\frac{E_0}{p}\right) * x\right] \approx E_0 \qquad (4.155)$$

Assuming the avalanche head radius to be the ionization length $r_a \approx 1/\alpha$, the criterion of streamer formation in the gap with the distance d between electrodes can be presented as the requirement for the avalanche amplification parameter αd to exceed the critical value:

$$\alpha\left(\frac{E_0}{p}\right) * d = \ln \frac{4\pi\varepsilon_0 E_0}{e\alpha^2} \approx 20, \quad N_e = \exp(\alpha d) \approx 3 \cdot 10^8 \tag{4.156}$$

This convenient and important criterion of the streamer formation is known as the **Meek breakdown condition** ($\alpha d \geq 20$).

Electron attachment processes in electronegative gases slow down the electron multiplication in avalanches and increase the value of the electric field required for a streamer formation. The situation here is similar to the case of a Townsend breakdown mechanism (see Section 4.4.3 and Equation 4.145). Actually, the ionization coefficient α in the Meek breakdown condition (Equation 4.156) should be replaced for electronegative gases by $\alpha - \beta$. However, when the discharge gaps are not large (in air $d \leq 15$ cm), the electric fields required by the Meek criterion are relatively high; then $\alpha >> \beta$ and the attachment can be neglected (see Figure 4.19).

Increasing the distance d between electrodes in electronegative gases does lead to gradual decreases of the electric field necessary for streamer formation, but it is limited by some minimum value. The minimal electric field required for streamer formation can be found from the ionization-attachment balance $\alpha(E_0/p) = \beta(E_0/p)$ (see Figure 4.19). In atmospheric-pressure air, this limit is about 26 kV/cm. At atmospheric pressure in such a strongly electronegative gas as SF_6, which is commonly used for electric insulation, the balance $\alpha(E_0/p) = \beta(E_0/p)$ requires the very high electric field of 117.5 kV/cm.

Electric field nonuniformity has a strong influence on breakdown conditions and an avalanche transformation into a streamer. Quite obviously, the nonuniformity, as in the case of corona discharge (see next chapters), decreases the breakdown voltage for a given distance between electrodes. This is clear because the breakdown condition (Equation 4.156) requires not αd but the integral

$$N_e = \int_{x_1}^{x_2} \alpha(E)\, dx$$

to exceed the certain threshold. The function $\alpha(E)$ is strongly exponential (see Equation 4.141), which results in significantly higher values of the integral in a nonuniform field with respect to a uniform one.

In other words, the voltage applied to the rod electrode due to the nonuniform electric field should provide the intensive electron multiplication only near the electrode to initiate a streamer. Once the plasma channel is initiated, its growth is then controlled mostly by the high electric field of its own streamer tip. Such a situation is considered in the next section as propagation of the quasi-self-sustained streamer. In this case, for very long (≥1 meter) and nonuniform systems, the average electric field can be as low as 2 to 5 kV/cm (see Figure 4.27).[101]

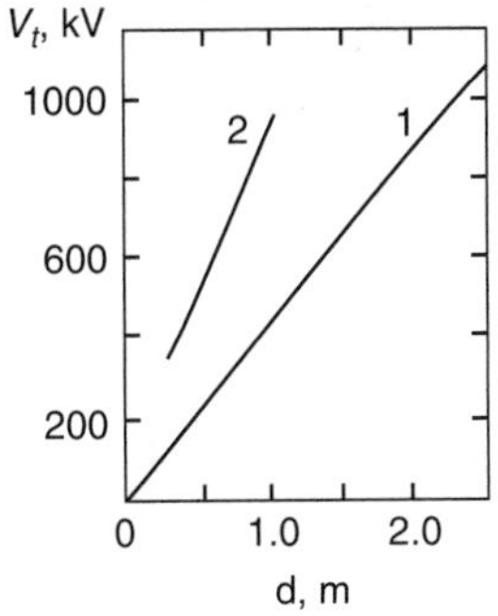

Figure 4.27 Breakdown voltage in air at 50 Hz: (1) rod–rod gap; (2) rod–plane gap

The breakdown threshold in the nonuniform constant electric field also depends on the polarity of the principal electrode, where the electric field is higher. Then, the threshold voltage in a long gap between a negatively charged rod and a plane is about twice as large as in the case of a negatively charged rod (see Figure 4.27). This polarity effect is due to the nonuniformity of the electric field near the electrodes. In the case of the rod anode, the avalanche approaches the anode where the electric field becomes stronger and stronger, which facilitates the avalanche–streamer transition. Also, the avalanche electrons easily sink into the anode in this case, leaving near the electrode the ionic trail enhancing the electric field. In the case of rod cathode, the avalanche moves from the electrode into the low electric field zone, which requires higher voltage for the avalanche–streamer transition.

4.4.8 Streamer Propagation Models

The model of propagation of **quasi-self-sustained streamers** was proposed by Dawson and Winn[110] and further developed by Galimberti.[111] This model, illustrated in Figure 4.28, assumes very low conductivity of a streamer channel and makes the streamer propagation autonomous and independent from the anode. Photons initiate the avalanche at a distance x_1 from center of the positive charge zone of radius r_0. According to this model, the avalanche then develops in the autonomous electric field of the positive space charge $E(x) = eN_+/4\pi\varepsilon_0 x^2$; the number of electrons is increasing by ionization as

$$N_e = \int_{x_1}^{x_2} \alpha(E)\,dx;$$

and the avalanche radius grows by the mechanism of free diffusion as:

$$\frac{dr^2}{dt} \approx 4D_e, \quad r(x_2) = \left[\int_{x_1}^{x_2} \frac{4D_e}{\mu_e E(x)}\,dx\right]^{1/2} \tag{4.157}$$

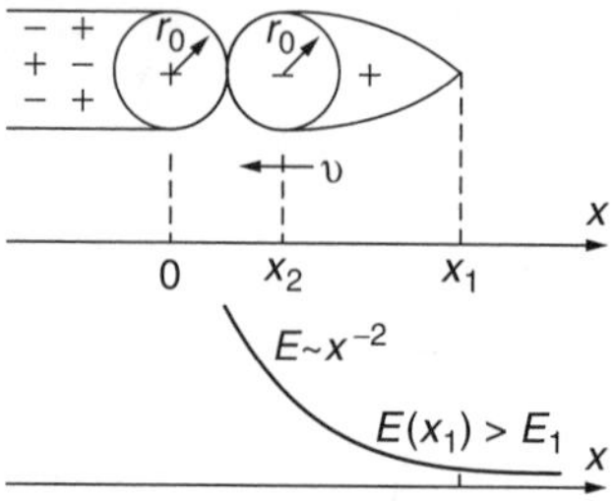

Figure 4.28 Self-sustaining streamer

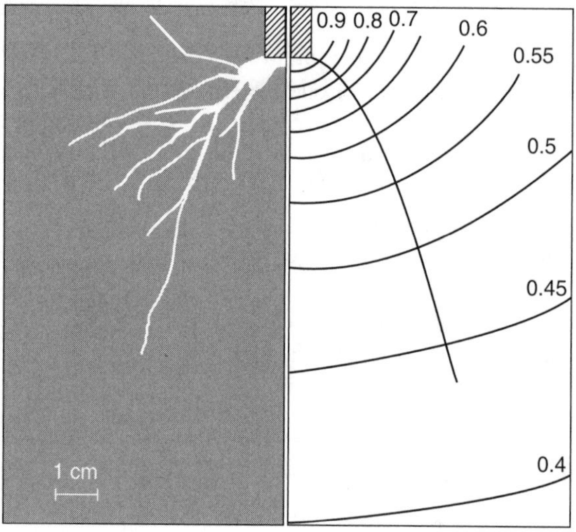

Figure 4.29 Streamer propagation from a positive 2-cm rod to a plane at distance of 150 cm and constant voltage of 125 kV. Equipotential surfaces are shown at right.[101]

To provide continuous and steady propagation of the self-sustained streamer, its positive space charge N_+ should be compensated by the negative charge of avalanche head $N_e = N_+$ at the meeting point of the avalanche and streamer: $x_2 = r_0 + r$. Also, the radii of the avalanche and streamer should be correlated at this point, $r = r_0$. These equations permit describing the streamer parameters including the propagation velocity, which can be found as x_2 divided by time of the avalanche displacement from x_1 to x_2. Comparison of the Gallimberti model of streamers in the strongly nonuniform electric field with an experimental photograph of a streamer corona is presented in Figure 4.29.[101] The model of a quasi-self-sustained streamer is helpful in describing the breakdown of long gaps with high-voltage and low-average electric fields (see previous section and Figure 4.27).

A qualitatively different model of streamer propagation was developed by Klingbeil et.al[112] and Lozansky and Firsov.[109] In contrast to the preceding approach of the self-sustained propagation of streamers, this model considers the streamer channel as an ideal conductor connected to an anode (see Section 4.4.6). In this model, the ideally conducting streamer channel is considered an anode elongation in the direction of external electric field E_0 with the shape of an ellipsoid of revolution.

For ideally conducting streamers, the streamer propagation at each point of the ellipsoid is normal to its surface. The propagation velocity is equal to the electron drift velocity in the appropriate electric field. To calculate the streamer growth velocity, a convenient formula for the maximum electric field E_m on the tip of the streamer with length l and radius r was proposed:[113]

$$\frac{E_m}{E_0} = 3 + \left(\frac{l}{r}\right)^{0.92}, \quad 10 < \frac{l}{r} < 2000 \tag{4.158}$$

The model of the ideally conducting streamer is also in reasonably good agreement with experimental results. Other, more detailed numerical streamer propagation models that take into account the finite conductivity of the streamer channel will be discussed during consideration of the corona and dielectric barrier discharges.

4.4.9 Concept of a Leader: Breakdown of Multimeter and Kilometer Long Gaps

The spark or streamer breakdown mechanism discussed earlir can be summarized as a sequence of three processes: (1) development of an avalanche and the avalanche–streamer transformation; (2) the streamer growth from anode to cathode; and (3) triggering of a return wave of intense ionization, which results in a spark formation. However, this breakdown mechanism cannot be applied directly to very long gaps, particularly in electronegative gases (including air).

The spark cannot be formed by the described mechanism in long gaps because the streamer channel conductivity is not sufficiently large to transfer the anode potential close to the cathode and there stimulate the return wave of intense ionization and spark. This effect is especially strong in electronegative gases in which the streamer channel conductivity is low. Also, in nonuniform electric fields, the streamer head grows from the strong to the weak field region, which delays its propagation. The streamers stop in the long, nonuniform air gaps without reaching the opposite electrode (see Figure 4.29). Usually, the streamer length is about 0.1 to 1 m.

Breakdown of longer gaps, including those with the multimeter- and kilometer-long interelectrode distances, is related to formation and propagation of the leaders. The leader is a highly ionized and highly conductive (with respect to streamer) plasma channel and grows from the active electrode along the path prepared by the preceding streamers. Because of the large conductivity, the leaders are much more effective than streamers in transferring the anode potential close to the cathode and there stimulating the return wave of intense ionization and spark. Historically, leaders were first investigated in connection with the natural phenomenon of lightning.

Leader propagation is illustrated in Figure 4.30. Leaders have large conductivity and are able to make "ideally" conductive channels (see the previous section) much

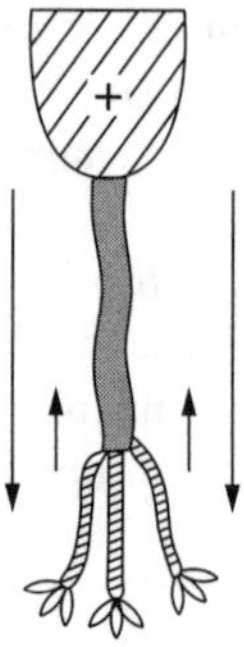

Figure 4.30 Leader propagation

longer than streamers. This results in a stronger electric field (Equation 4.158) on the leader head, which is able to create new streamers growing from the head and preparing the path for further propagation of the leader. The difference between leaders and streamers is more quantitative than qualitative: longer length; higher degree of ionization; higher conductivity; and higher electric field; a streamer absorbs avalanches and a leader absorbs streamers (Figure 4.30).

Heating effects of the relatively short centimeters-long streamers are estimated in Section 4.4.6 as about 10 K. For meters-long channels, at least near the active electrode, the heating effect reaches 3000 K. This heating, together with the corresponding high level of nonequilibrium excitation of atoms and molecules, probably explains transformation of a streamer channel into the leader. A 3000-K temperature is not sufficient for thermal ionization of air, but this temperature together with the elevated nonequilibrium excitation level is sufficiently high for other mechanisms of increase of electric conductivity in plasma channel.

Gallimberti[114] assumed a mechanism for streamer-to-leader transition in air related to thermal detachment of electrons from the negative ions O_2^- that are the main products of electron attachment in electronegative gas. The effective destruction of the ions O_2^- and, as a result, compensation of electron attachment become possible if temperature exceeds 1500 K in dry air and 2000 K in humid air (where the electron affinity of complexes $O_2^-(H_2O)_n$ is higher). Such temperatures are available in the plasma channel and can provide the formation of a high-conductivity leader in electronegative gas. During the evolution of a streamer in air, at first the joule heat is stored in vibrational excitation of N_2 molecules. While temperature of air is increasing, the VT relaxation grows exponentially (Section 3.4, the Landau–Teller formula [Equation 3.57]), providing the explosive heating of the plasma channel.

As an example, typical parameters of a leader growing from an anode are: leader current is about 100 A (for comparison, streamer current is 0.1 to 10 mA); diameter of a plasma channel is about 0.1 cm; quasi-equilibrium plasma temperature is 20,000 to 40,000 K; electric conductivity is about 100 Ohm^{-1} cm^{-1}; and propagation velocity is 2 * 10^6 cm/sec.

In the present section, only the general physical features and kinetics of charged species in the avalanches, streamers, and leaders were discussed. Their role in the specific discharge system, in particular in corona, spark, and dielectric barrier discharges, will be considered in later chapters.

4.5 STEADY-STATE REGIMES OF NONEQUILIBRIUM ELECTRIC DISCHARGES

Nonequilibrium steady-state plasma can be self-sustained in very different ways, depending on the discharge type, its gas-dynamics, electrodynamics, and gas composition. The situation is more complicated in plasma chemical systems in which the gas composition is changing during the processes and influencing the balance of charged particles. The steady-state regimes of specific electric discharges will be discussed later. In this section, some general features of the charge balance in steady-state nonequilibrium discharges and peculiarities of such discharge propagation in plasma chemical systems will be considered.

4.5.1 Steady-State Discharges Controlled by Volume and Surface Recombination Processes

The steady-state regimes of nonequilibrium discharges are provided by a balance of generation and loss of charged particles. The generation of electrons and positive ions is mostly due to volume ionization processes (discussed in Section 2.2). To sustain the steady-state plasma, the ionization should be quite intensive; this usually requires the electron temperature to be on the level of 1/10 of the ionization potential (~1 eV).

The losses of charged particles can be related to volume processes of recombination (Section 2.3) or attachment (Section 2.4), but they can also be provided by diffusion of charged particles to the walls with subsequent surface recombination. These two mechanisms of charge losses define two different regimes of sustaining the steady-state discharge: the first controlled by volume processes and the second controlled by diffusion to the walls. Before discussing these, the conditions of their realization will be established.

If the degree of ionization in a plasma is relatively high and diffusion considered as ambipolar (see Section 4.3.9, Equation 4.126), the frequency of charge losses due to the diffusion to the walls can be described as:

$$\nu_D = \frac{D_a}{\Lambda_D^2} \tag{4.159}$$

In this relation, D_a is the coefficient of ambipolar diffusion (see Equation 4.126 and Equation 4.127), and Λ_D is the characteristic diffusion length, which can be calculated for different shapes of the discharge chambers:

- For a cylindrical discharge chamber of radius R and length L:

$$\frac{1}{\Lambda_D^2} = \left(\frac{2.4}{R}\right)^2 + \left(\frac{\pi}{L}\right)^2 \tag{4.160}$$

- For a parallel-piped discharge with side lengths L_1, L_2, L_3:

$$\frac{1}{\Lambda_D^2} = \left(\frac{\pi}{L_1}\right)^2 + \left(\frac{\pi}{L_2}\right)^2 + \left(\frac{\pi}{L_3}\right)^2 \tag{4.161}$$

- For a spherical discharge chamber with radius R:

$$\frac{1}{\Lambda_D^2} = \left(\frac{\pi}{R}\right)^2 \tag{4.162}$$

Based on Equation 4.159 through Equation 4.162 for diffusion charge losses, a criterion of predominantly volume process-related charge losses and thus a criterion for the volume process-related steady-state regime of sustaining the nonequilibrium discharges is obtained:

$$k_i(T_e)n_0 >> \frac{D_a}{\Lambda_D^2} \tag{4.163}$$

In this relation, $k_i(T_0)$ is the ionization rate coefficient (see Equation 2.21 or Equation 2.28), n_0 is the neutral gas density.

The criterion (Equation 4.163) actually restricts pressure because $D_a \propto 1/p$ and $n_0 \propto p$. When pressure in a discharge chamber exceeds 10 to 30 torr (range of moderate and high pressures), the diffusion usually is relatively slow and the balance of charge particles is due to volume processes. In this case, the kinetics of electrons as well as positive and negative ions can be characterized by the following simplified set of equations:

$$\frac{dn_e}{dt} = k_i n_e n_0 - k_a n_e n_0 + k_d n_0 n_- - k_r^{ei} n_e n_+ \tag{4.164}$$

$$\frac{dn_+}{dt} = k_i n_e n_0 - k_r^{ei} n_e n_+ - k_r^{ii} n_+ n_- \tag{4.165}$$

$$\frac{dn_-}{dt} = k_a n_e n_0 - k_d n_0 n_- - k_r^{ii} n_+ n_- \tag{4.166}$$

In this set of equations, n_+ and n_- are concentrations of positive and negative ions; n_e and n_0 are concentrations of electrons and neutral species; and rate coefficients k_i, k_a, k_d, k_r^{ei}, k_r^{ii} are related to the processes of ionization by electron impact, dissociative or other electron attachment, electron detachment from negative ions, and electron–ion and ion–ion recombination. One should note that the rate coefficients of processes involving neutral particles (k_i, k_a, k_d) are expressed in the system (Equation 4.164 through Equation 4.166) with respect to the total gas density.

If a moderate- or high-pressure gas is not electronegative, then the volume balance of electrons and positive ions in the nonequilibrium discharge obviously can be reduced to the simple ionization–recombination balance. However, for electronegative gases, two qualitatively different self-sustained regimes can be achieved (at different effectiveness of electron detachment): one controlled by recombination and the other by electron attachment.

4.5.2 Discharge Regime Controlled by Electron–Ion Recombination

In some plasma chemical systems the destruction of negative ions by, for example, the associative electron detachment (see Equation 2.61 and Table 2.8) can be faster than ion–ion recombination:

$$k_d n_0 >> k_r^{ii} n_+ \tag{4.167}$$

In this case, the actual losses of charged particles are also due to electron–ion recombination in the same manner as for nonelectronegative gases. Such situations can take place in plasma chemical processes of CO_2 and H_2O dissociation, and NO-synthesis in air. In these systems, the associative electron detachment processes:

$$O^- + CO \rightarrow CO_2 + e, \quad O^- + NO \rightarrow NO_2 + e, \quad O^- + H_2 \rightarrow H_2O + e \tag{4.167a}$$

proceed very fast; these require about 0.1 μsec at concentrations of the CO, NO, and H_2 molecules ~10^{17} cm^{-3}. This electron liberation time (at relatively low degrees of ionization) is shorter than the time of ion–ion recombination and, according to Equation 4.167, the discharge regime controlled by electron–ion recombination can take place.

The electron attachment and detachment processes are in the dynamic quasi equilibrium in the recombination regime (Equation 4.167) during time intervals sufficient for electron detachment ($t >> 1/k_d n_0$). Then, the concentration of negative ions can be considered in dynamic quasi equilibrium with electron concentration:

$$n_- = \frac{k_a}{k_d} n_e = n_e \varsigma \tag{4.168}$$

Using the quasi-constant parameter $\varsigma = k_a/k_d$, the set of Equation 4.164 through Equation 4.166 can be reduced to the kinetic equation for electron concentration. This equation can be derived by taking into account the plasma quasi neutrality, $n_+ = n_e + n_- = n_e(1 + \varsigma)$, and substituting Equation 4.168 into the kinetic equation (Equation 4.165) for positive ions:

$$\frac{dn_e}{dt} = \frac{k_i}{1+\varsigma} n_e n_0 - \left(k_r^{ei} + \varsigma k_r^{ii}\right) n_e^2 \tag{4.169}$$

The parameter $\varsigma = k_a/k_d$ shows the detachment ability to compensate the electron losses due to attachment. If $\varsigma << 1$, the attachment influence on the electron balance is negligible and the kinetic equation (Equation 4.169) becomes equivalent to that for nonelectronegative gases including only ionization and electron–ion recombination.

The kinetic equation includes the effective rate coefficient of ionization $k_i^{eff} = k_i/1+\varsigma$, which interprets the fraction of electrons lost in the attachment process as not generated by ionization at all. In the same manner, the coefficient $k_r^{eff} = k_r^{ei} + \varsigma k_r^{ii}$ in Equation 4.169 can be interpreted as the effective coefficient of recombination. This coefficient describes direct electron losses through the electron–ion recombination and indirect electron losses through attachment and following ion–ion recombination.

Equation 4.169 describes the electron concentration evolution to the steady-state n_e magnitude of the recombination-controlled regime:

$$\frac{n_e}{n_0} = \frac{k_i^{eff}(T_e)}{k_r^{eff}} = \frac{k_i}{\left(k_r^{ei} + \varsigma k_r^{ii}\right)(1+\varsigma)} \tag{4.170}$$

The important peculiarity of the recombination-controlled regime of a nonequilibrium discharge is the fact that there is a steady-state degree of ionization (n_e/n_0) for each value of electron temperature T_e.

It should be noted that the criterion of the recombination-controlled regime (Equation 4.167) can be rewritten using only rate coefficients and taking into account the plasma quasi neutrality $n_+ = n_e + n_- = n_e\ (1 + \varsigma)$ and Equation 4.170 for the degree of ionization:

$$k_d >> \frac{(k_i - k_a)\,k_r^{ii}}{k_r^{ei}} \tag{4.171}$$

Actually, this criterion means that the recombination-controlled regime takes place when the electron detachment rate coefficient k_d (calculated with respect to total gas density n_0) is sufficiently large. In other words, the concentration of particles providing the effective destruction of negative ions should be relatively high.

4.5.3 Discharge Regime Controlled by Electron Attachment

This regime takes place if the balance of charged particles is again due to volume processes and the discharge parameters correspond to inequalities opposite to Equation 4.167 and Equation 4.171. In this case, the negative ions produced by electron attachment go almost instantaneously into ion–ion recombination, and electron losses are mostly due to the attachment process. The steady-state solution of Equation 4.165 for the attachment-control regime can be presented as:

$$k_i(T_e) = k_a(T_e) + k_r^{ei}\frac{n_+}{n_0} \tag{4.172}$$

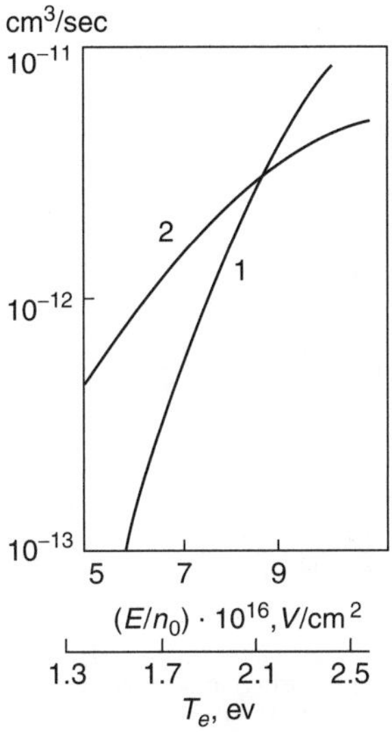

Figure 4.31 (1) Rate coefficients of ionization; (2) dissociative attachment for CO_2

In the attachment-controlled regime, the electron attachment is usually faster than recombination and Equation 4.172 actually requires $k_i(T_e) \approx k_a(T_e)$. The exponential functions $k_i\,(T_e)$ and $k_a\,(T_e)$ usually appear as shown in Figure 4.31 (see also Figure 4.21) and have the only crossing point T_{st}. This crossing point actually determines the steady-state electron temperature in the nonequilibrium discharge—self-sustained in the attachment-controlled regime.

In contrast to the recombination-controlled regime, the steady-state nonequilibrium discharge can be controlled by electron attachment only at high electron temperatures when $T_e \geq T_{st}$. From Equation 4.172, the generation of electrons cannot compensate their losses at lower values of electron temperature without detachment.

4.5.4 Discharge Regime Controlled by Charged Particle Diffusion to Walls: Engel–Steenbeck Relation

When gas pressure is relatively low and the inequality opposite to Equation 4.163 is valid, the balance of charged particles is governed by the competition between ionization in volume and diffusion of charged particles to the walls (Equation 4.159 through Equation 4.162). The balance of direct ionization by electron impact (Equation 2.21) and ambipolar diffusion to the walls of a long discharge chamber of radius R (see Equation 4.159, Equation 4.160, and Equation 4.126) gives the following relation between electron temperature and pressure (or the similarity parameter pR):

$$\left(\frac{T_e}{I}\right)^{1/2} \exp\left(\frac{I}{T_e}\right) = \frac{\sigma_0}{\mu_i p}\left(\frac{8I}{\pi m}\right)^{1/2}\left(\frac{n_0}{p}\right)(2.4)^2(pR)^2 \tag{4.173}$$

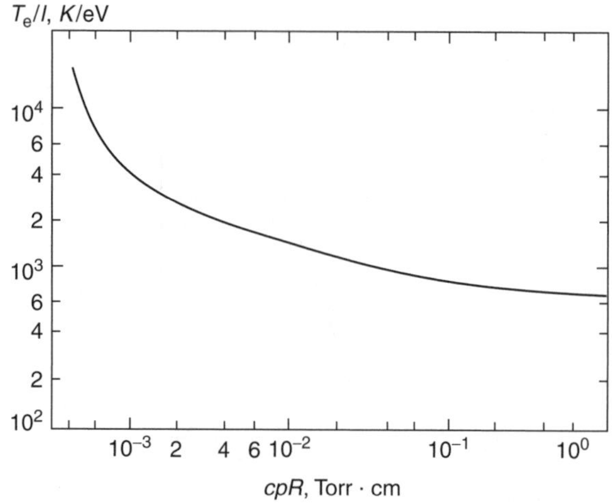

Figure 4.32 Universal relation among electron temperature, pressure, and discharge tube radius

This is the famous **Engel–Steenbeck relation** for the diffusion-controlled regime of nonequilibrium discharges. Here, I is the ionization potential; μ_i is the ion mobility; σ_0 is the electron–neutral gas-kinetic cross section; and m is an electron mass. If the gas temperature is fixed, the parameters $\mu_i p$ and n_0/p in Equation 4.173 are constant and the Engel–Steenbeck relation can be written as:

$$\sqrt{\frac{T_e}{I}}\exp\left(\frac{I}{T_e}\right) = C(pR)^2 \tag{4.174}$$

The constant C depends only on the type of gas. Table 4.5 provides values of C for some gases. The universal relation between T_e/I and the similarity parameter cpR for the diffusion-controlled regime is usually also presented as a graph (see Figure 4.32). The gas type parameters c for this graph are also given in Table 4.5. As can be seen from the Engel–Steenbeck curve, the electron temperature in the diffusion-controlled regimes of nonequilibrium discharges decreases with increases of pressure and radius of the discharge tube.

Table 4.5 Numerical Parameters of the Engel–Steenbeck Relation

Gas	C, Torr^{-2} cm^{-2}	c, Torr^{-1} cm^{-1}	Gas	C, Torr^{-2} cm^{-2}	c, Torr^{-1} cm^{-1}
N_2	$2*10^4$	$4*10^{-2}$	Ar	$2*10^4$	$4*10^{-2}$
He	$2*10^2$	$4*10^{-3}$	Ne	$4.5*10^2$	$6*10^{-3}$
H_2	$1.25*10^3$	10^{-2}			

It is important to mention that, in contrast to the steady-state regimes sustained by volume processes, the diffusion-controlled regimes of nonequilibrium discharges are sensitive to radial density distribution of charged particles. Such distribution for a long cylindrical discharge tube can be described by Bessel function:

$$n_e(r) \propto J_0\left(\frac{r}{R}\right) \tag{4.175}$$

More details about the steady-state regimes will be discussed in following chapters concerned with specific electric discharges.

4.5.5 Propagation of Electric Discharges

The phenomenon of discharge propagation is an intermediate one between breakdown and steady-state discharges. Although it is tempting to view the propagation of a discharge in gas flow as only a continuous breakdown of newer portions of gas coming into a high electric field zone, this is incorrect because the breakdown and steady-state discharge conditions are usually quite different.

The electric fields necessary to initiate a discharge usually exceed those necessary to sustain the nonequilibrium steady-state plasma. This can be explained by the formation of excited and chemically active species in plasma, which facilitates the nonthermal ionization mechanisms; facilitates destruction of negative ions; and finally reduces the requirements on the electric field. For this reason, discharge propagation is not only breakdown of new portions of incoming gas, but also spreading of conditions facilitating effective ionization.

Thermal plasma propagation is related to the heat transfer processes providing high temperatures sufficient for thermal ionization of the incoming gas.[115] The nonthermal plasma propagation at high electric fields can be provided simply by electron diffusion in front of the discharge.[116,117] As mentioned earlier, the most specific mechanism of discharge propagation in plasma chemical systems is related to preliminary propagation of such a discharge facilitating species as excited atoms and molecules, or products of chemical reactions providing the effective electron detachment.[118]

Analyze this mechanism, taking as an example the nonthermal discharge, one-dimensional propagation in CO_2 in a uniform electric field, corresponding to electron temperature of about 1 eV (see Figure 4.33). The breakdown of CO_2 is controlled by the dissociative attachment $e + CO_2 \rightarrow CO + O^-$ and requires quite large electric fields and electron temperatures exceeding 2 eV (see Figure 4.31). However, CO_2 dissociation in plasma produces sufficient CO to provide effective electron detachment (Equation 4.167a) and the recombination-controlled regime corresponding to the lower electric fields under consideration.

Numerical values of the parameters of the CO_2 discharge under consideration, propagating in fast gas flow are $n_e = 10^{13}$ cm^{-3}, $n_0 = 3 \cdot 10^{18}$ cm^{-3}, $T_e = 1$ eV, $T_0 \approx$ 700 K.[119] Based on Equation 4.167 and Equation 4.170 for these nonthermal plasma

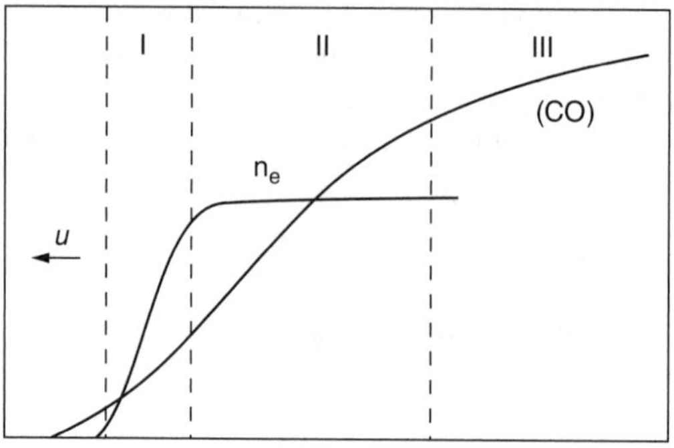

Figure 4.33 Electron and CO density distributions in the front of propagating discharge; I: low electron concentration zone; II: discharge zone where CO diffusion provides effective detachment and sufficient electron density; III: effective CO_2 dissociation zone

conditions, the critical value of CO concentration, separating the attachment and recombination-controlled regimes, is:

$$[\mathrm{CO}]_{cr} = \frac{k_a k_r^{ii}}{k_d k_r^{ei}} n_0 \approx \left(10^{-3} \div 10^{-4}\right) \cdot n_0 \qquad (4.176)$$

Note that the electron detachment rate coefficient $k_d = 7 \cdot 10^{-10}$ cm^3/sec (see Table 2.7) is taken here in the conventional way, e.g., with respect to CO concentration. If the CO concentration exceeds the critical value $[\mathrm{CO}] > [\mathrm{CO}]_{cr}$, the recombination-controlled balance gives the relatively high electron density:

$$n_e(T_e) = \frac{k_i(T_e)}{k_r^{ei}} n_0 \qquad (4.177)$$

Conversely, if the carbon monoxide concentration is below the critical limit, $[\mathrm{CO}] < [\mathrm{CO}]_{cr}$, then the electron concentration is very low, controlled by the dissociative attachment, and also proportional to the CO concentration:

$$n_e = \frac{k_i k_d}{k_a k_r^{ii}} [\mathrm{CO}] \qquad (4.178)$$

Thus, propagation of the electron concentration and of the discharge in general is related to the propagation of the CO concentration. Most of CO production in this system is due to dissociation of vibrationally excited CO_2 molecules and takes place in the main plasma zone III (Figure 4.33). CO diffusion from zone III into zone II provides the sufficiently high CO concentration for sustaining the high electron concentration (Equation 4.177) that subsequently provides the vibrational excitation and CO_2 dissociation in zone III. Further decrease of the CO concentration below the critical value in zone I corresponds to a dramatic fall of the electron concentration (Equation 4.156). Thus, in this case, the discharge propagation can be interpreted as the propagation of a self-sustained ionization wave, supported by CO production after the ionization front, which diffuses ahead and facilitates the ionization conditions.

4.5.6 Propagation of Nonthermal Ionization Waves Self-Sustained by Diffusion of Plasma Chemical Products

To determine the evolution of the profile of electron concentration and the velocity of the ionization wave, the linear one-dimensional propagation of CO concentration along the axis x in the plasma-dissociating CO_2 was described by differential equation with the only variable $\xi = x + ut$:[118]

$$u\frac{\partial[\mathrm{CO}]}{\partial\xi} = D\frac{\partial^2[\mathrm{CO}]}{\partial\xi^2} + g(\xi, T_v, T_0, n_e, n_0) \tag{4.179}$$

In this equation, u is the velocity of the ionization wave (discharge propagation); T_v and T_0 are vibrational and translational gas temperatures; n_e and n_0 are electron and neutral gas concentrations; D is the diffusion coefficient of CO molecules (taken here as a constant); and g is a model source of CO as a result of CO_2 dissociation:

$$g(\xi) = \frac{n_0}{\tau_{eV}}\exp\left\{-\frac{[\mathrm{CO}]_{cr}}{[\mathrm{CO}](\xi - u\tau_{eV})} - \alpha\frac{[\mathrm{CO}](\xi)}{[\mathrm{CO}]_0}\right\} \tag{4.180}$$

Here, $\tau_{eV} = 1/k_{eV}n_e^{II}$ is the vibrational excitation time in zone II; $[\mathrm{CO}]_0 = [\mathrm{CO}](\xi_0)$ is the maximum concentration of CO at the end of zone III; $\xi_0 = u\tau_{chem}$; τ_{chem} is the total chemical reaction time in zone III; and parameter $\alpha \approx 3$ shows the exponential smallness of the dissociation rate at the end of zone III $\xi \to \xi_0, t \to \tau_{chem}$, when the process is actually completed. The boundary conditions for the Equation 4.179 should be taken as: $[\mathrm{CO}](-\infty) = 0, [\mathrm{CO}](\xi_0) = [\mathrm{CO}]_0$.

The source $g(\xi)$ is not powerful at the negative values of ξ, and perturbation theory can be applied to solve the nonlinear equations (Equation 4.179 and Equation 4.180). The nonperturbed equation (g = 0) gives the solution $[\mathrm{CO}] = [\mathrm{CO}]_0 \exp(\xi u/D)$. Contribution of the source $g(\xi)$ in the first order of the perturbation theory leads to the following linear equation:

$$\frac{\partial}{\partial\xi}\left\{u[\mathrm{CO}](\xi) - D\frac{\partial}{\partial\xi}[\mathrm{CO}](\xi)\right\} = \frac{n_0}{\tau_{eV}}\exp\left[-\aleph\exp\left(-\frac{\xi u}{D}\right)\right] \tag{4.181}$$

with the numerical parameter $\aleph$, which is equal to:

$$\aleph = \frac{[\mathrm{CO}]_{cr}}{[\mathrm{CO}]_0}\exp\frac{u^2\tau_{eV}}{D} \tag{4.182}$$

Taking into account the asymptotical decrease of the right-hand part of Equation 4.181 as $\xi \to -\infty$, the solution of this equation can be presented as:

$$[CO](\xi) = [CO]_0 \exp\left(\frac{\xi u}{D}\right)\left\{1 - \frac{D}{u^2\tau_{eV}}\frac{n_0}{[CO]_0}\exp\left(-\frac{\xi u}{D} - \aleph\exp\left(-\frac{\xi u}{D}\right)\right)\right\} \quad (4.183)$$

In a similar manner, first-order perturbation theory permits solving the nonlinear Equation 4.179 and Equation 4.180 in the interval $0 \le \xi \le \xi_0$ because, as $\xi \to \xi_0$, the source $g(\xi)$ also becomes very small. First-order perturbation theory gives:

$$[CO](\xi) = [CO]_0\left\{1 + \frac{n_0}{[CO]_0}\frac{\xi - \xi_0}{u\tau_{eV}}\exp(-\alpha)\right\} \quad (4.184)$$

Equation 4.183 and Equation 4.184 give solutions for the concentration profiles for positive and negative magnitudes of the automodel variable ξ. To find the united solution, Equation 4.183 and Equation 4.184 are matched at the wave front $\xi = 0$. Such cross linking of the solutions is possible only at the similar value of the parameter μ^2, e.g., only at the magnitude of the velocity of the ionization wave:

$$u^2 = \frac{D}{\tau_{eV}}\ln\left[\gamma \ln\left(\frac{De^{\alpha}}{u^2\tau_{chem}}\right)\right] \quad (4.185)$$

where γ $[CO]_0/[CO]_{cr} \approx 10^3 - 10^4$ (see Equation 4.176). Taking into account the value parameter $\alpha \sim 3$, the approximate solution of the transcendent equation (Equation 4.185) for the velocity of the ionization wave can be expressed as:

$$u^2 \approx \frac{D}{\tau_{eV}}\ln\gamma \quad (4.186)$$

This velocity of the ionization wave and nonthermal discharge propagation can be physically interpreted as the velocity of diffusion transfer of the detachment active heavy particles (CO) ahead of the discharge front on a distance necessary for effective vibrational excitation of CO_2 molecules with their further dissociation. For numerical calculations, it is convenient to rewrite Equation 4.186 in terms of speed of sound c_s; Mach number M; and the ionization degree in plasma n_e/n_0:

$$u \approx 30c_s\sqrt{\frac{n_e}{n_0}}, \quad M = 30\sqrt{\frac{n_e}{n_0}} \quad (4.187)$$

From Equation 4.186 and Equation 4.187, the velocity of the nonequilibrium ionization wave propagation actually does not strongly depend on the details of the propagation mechanism. It depends mostly on the degree of ionization in the main plasma zone and also on the critical amount $(1/\gamma)$ of the ionization active species (in this case CO), which should be transported in front of the discharge to facilitate ionization. This means that the final relation for the ionization wave velocity (Equation 4.186 and Equation 4.187) can be used for other similar mechanisms of nonthermal discharge propagation related to diffusion of some active heavy plasma species in front of the discharge to facilitate further propagation of the ionization wave. Note that the degree of ionization in discharges under consideration is moderate and the discharge velocity (Equation 4.187) is always lower than the speed of sound. Discharge propagation in very fast flows, in particular in supersonic flows, requires special consideration.[115]

4.5.7 Nonequilibrium Behavior of Electron Gas: Difference between Electron and Neutral Gas Temperatures

There are two principal aspects of the nonequilibrium behavior of an electron gas: the first is related to temperature differences between electrons and heavy particles; the second is concerned with significant deviation of the degree of ionization from that predicted by the Saha equilibrium.

Ionization in plasma is mostly provided by electron impact and the ionization process should be quite intensive to sustain the steady-state plasma. This usually requires the electron temperature T_e to be at least on the level of 1/10 of the ionization potential (~1 eV). Indeed, the electron temperature in a plasma is usually about 1 to 3 eV for thermal equilibrium and cold nonequilibrium discharges. The gas temperature T_0 determines the equilibrium or nonequilibrium plasma behavior; in thermal discharges, T_0 is high and the system can be close to equilibrium, but in nonthermal discharges T_0 is low and the degree of nonequilibrium T_e/T_0 can be high, sometimes up to 100.

The effect of nonequilibrium $T_e/T_0 >> 1$ is due to the fact that, for some reason, the neutral gas is not heated in the discharge. Thus, in low-pressure discharges, it is related to intensive heat losses to the discharge chamber walls. The difference between the gas temperature in plasma T_0 and room temperature T_{00} in such discharges can be estimated from the simple relation:

$$\frac{T_0 - T_{00}}{T_{00}} = \frac{P}{P_0} \tag{4.188}$$

In this relation: P is the discharge power per unit volume (specific power); P_0 is the critical value of specific power corresponding to the gas temperature increase over T_{00} = 300 K. Taking into account that the thermal conductivity and adiabatic discharge heating per one molecule are reversibly proportional to pressure, practically, the critical specific power does not depend on pressure and numerically is $P_0 = 0.1 \div 0.3$ W/cm^3. It is important from the point of industrial applications to note that the specific power of the described low-pressure nonequilibrium discharges (cooled by thermal conductivity) is limited by P_0 and practically is very low.

In moderate- and high-pressure nonequilibrium discharges (usually more than 20 to 30 torr) where heat losses to the wall are low, neutral gas overheating can be prevented by high velocities and low residence times in the discharge or by short time of discharge pulses in nonsteady-state systems. In both cases, this restricts the specific energy input, E_v, that is the energy transferred from the electric discharge into the neutral gas calculated per one molecule (it can be expressed in eV/mol or in J/cm^3). Similar to Equation 4.188, the relation estimating overheating for moderate- and high-pressure nonequilibrium discharges can be presented as:

$$\frac{T_0 - T_{00}}{T_{00}} = \frac{E_v(1-\eta)}{c_p T_{00}} \tag{4.189}$$

Here, c_p is the specific heat and η is the energy efficiency of the plasma chemical process, which is the fraction of discharge energy spent to perform the chemical reaction. The energy efficiency η in moderate- and high-pressure plasma chemical systems stimulated by vibrational excitation can be large—up to 70 to 90% in the particular case of CO_2 dissociation—which facilitates the conditions necessary for sustaining the nonequilibrium. In this case, Equation 4.189 restricts the specific energy input by a fairly high critical value, $E_v \approx 1.5$ eV/mol. However, the nonequilibrium discharge can be overheated and converted into a quasi-equilibrium discharge even at lower values of specific energy input due to possible plasma instabilities. These will be discussed in Chapter 6.

Another important element of the nonequilibrium behavior of electron gas in electric discharges is the strong deviation of the electron energy distribution function from the Maxwellian distribution. This effect is also primarily due to the low neutral gas temperature. Thus, this leads to low concentration of electronically excited atoms and molecules, which causes electrons to establish their energy balance directly with the electric field. The nonequilibrium behavior of electron energy distribution functions was discussed in Section 4.2.

4.5.8 Nonequilibrium Behavior of Electron Gas: Deviations from the Saha Formula—Degree of Ionization

The quasi-equilibrium electron concentration and degree of ionization can be easily found as the function of one temperature, based on the Saha formula (Equation 4.15). Although the ionization processes (in thermal and nonthermal discharges) are provided by the electron gas, for nonequilibrium discharges the Saha formula with electron temperature T_e gives the ionization degree n_e/n_0 several orders of magnitude higher than the real one. Obviously, the Saha formula assuming the neutral gas temperature gives even many fewer electron concentrations and much worse agreement with reality.

This nonequilibrium effect is due to the presence of additional channels of charged particle losses in cold gas. These are much faster than reverse processes of recombination leading to the Saha equilibrium. The Saha equilibrium actually implies the balance of ionization by electron impact and the double-electron, three-body recombination:

$$e + A \Leftrightarrow A^+ + e + e \tag{4.190}$$

In reality, however, the charged particle losses in moderate- and high-pressure systems are mostly due to the dissociative recombination:

$$e + AB^+ \rightarrow A + B^*$$

This recombination process is much faster than the three-body recombination process, but at the same time it is not compensated by the reverse process of associative ionization because of a relatively low concentration of electronically excited atoms. The low (under-equilibrium) concentration of the excited species in nonthermal plasma is due to intensive energy transfer in other degrees of freedom. Furthermore, in nonequilibrium, low-pressure discharges, the ionization is not compensated by three-body recombinations but rather by the diffusion to the walls, which is also much faster and obviously not balanced by electron emission.

In general, electron concentration and degree of ionization in nonthermal discharges are far from equilibrium. The Saha formula gives results that are several orders of magnitude higher than reality. Even simplified calculations of electron concentration and the degree of ionization in nonthermal discharges require relevant kinetic analysis of the recombination-controlled, attachment-controlled, diffusion-controlled, or other more sophisticated regimes.

PROBLEMS AND CONCEPT QUESTIONS

4.1 **Average Vibrational Energy and Vibrational Specific Heat**. Based on the equilibrium Boltzmann distribution function Equation 4.11, find the average value of vibrational energy and related specific heat of a diatomic molecule. Analyze the result for the case of high ($T >> \hbar\omega$) and low ($T >> \hbar\omega$) temperatures.

4.2 **Ionization Equilibrium, the Saha Equation**. Using the Saha equation, estimate the degree of ionization in the thermal Ar plasma at atmospheric pressure and an equilibrium temperature of T = 20,000 K. Explain why the high ionization degree can be reached at temperatures much less than the ionization potential.

4.3 **Statistics of Plasma Radiation at Complete Thermodynamic Equilibrium (CTE)**. Calculate the total radiation of the Ar arc discharge plasma with a surface area of 10 cm^2 at an equilibrium temperature of T = 20,000 K, assuming CTE conditions. Does such thermal radiation power look reasonable for the arc discharge?

4.4 **The Debye Corrections of Thermodynamic Functions in Plasma**. Estimate the Debye correction for the Gibbs potential per unit volume of a thermal Ar plasma at atmospheric pressure and equilibrium temperature of T = 20,000 K. The ionization degree can be calculated using the Saha equation (see the preceding problem).

4.5 **The Treanor Distribution Function**. Estimate the normalizing factor B for the Treanor distribution function, taking into account only relatively low levels of vibrational excitation. Find a criterion for application of such a normalizing factor (most of the molecules should be located on the vibrational levels lower than the Treanor minimum).

4.6 **Average Vibrational Energy of Molecules Following the Treanor Distribution**. Based on the nonequilibrium Treanor distribution function, find the average value of vibrational energy, taking into account only relatively low vibrational levels. Find an application criterion for the result (most of the molecules should be located on vibrational levels lower than the Treanor minimum).

4.7 **The Boltzmann Kinetic Equation**. Prove that the Maxwell distribution function is an equilibrium solution of the Boltzmann equation. Compare the Maxwell distribution functions for three-dimensional velocities (Equation 4.57), absolute value of velocities, and translational energies (Equation 2.1).

4.8 **Entropy of Electrons in Nonequilibrium Plasma.** Using Equation 4.58, estimate the entropy of plasma electrons, following the Druyvesteyn electron energy distribution function, as a function of reduced electric field in nonthermal discharges.

4.9 **The Krook Collisional Operator for Momentum Conservation Equation**. Compare the frictional and particle generation terms in the Krook collisional operator, and explain why the frictional term usually dominates in nonthermal discharges.

4.10 **The Fokker–Planck Kinetic Equation**. Using the Fokker–Planck kinetic equation in Equation 4.75, estimate the time necessary to reestablish the steady-state electron energy distribution function after fluctuation of the electric field.

4.11 **The Druyvesteyn Electron Energy Distribution Function**. Calculate the average electron energy for the Druyvesteyn distribution. Define the effective electron temperature of the distribution and compare it with that of the Maxwellian distribution function.

4.12 **The Margenau Electron Energy Distribution Function**. Compare the behavior of Margenau and Druyvesteyn distribution functions. Estimate the average electron energy and effective electron temperature for the Margenau electron energy distribution function. Consider the case of very high frequencies when the Margenau distribution becomes exponentially linear with respect to electron energy.

4.13 **Influence of Vibrational Temperature on Electron Energy Distribution Function**. Simplify the general relation (Equation 4.84), describing the influence of vibrational temperature on the electron energy distribution function, for the case of high vibrational temperatures $T_v >> \hbar\omega$. Using this relation, estimate the typical acceleration of ionization rate coefficient corresponding to 10% increase of vibrational temperature.

4.14 **Influence of Molecular Gas Admixture to a Noble Gas on Electron Energy Distribution Function**. Using the general Equation 4.77, estimate the percentage of nitrogen to be added to a nonthermal discharge initially sustained in argon to change behavior of the electron energy distribution function.

4.15 **Electron–Electron Collisions and Maxwellization of Electron Energy Distribution Function**. Calculate the minimum ionization degree n_e/n_0, when the Maxwellization provided by electron–electron collisions becomes essential in establishing the electron energy distribution function in the case of (1) argon–plasma T_e = 1 eV and (2) nitrogen–plasma T_e = 1 eV.

4.16 **Relation between Electron Temperature and Reduced Electric Field**. Calculate the reduced electric field necessary to provide an electron temperature T_e = 1 eV in the case of (1) argon–plasma and (2) nitrogen–plasma. Compare and discuss the result.

4.17 **Electron Conductivity**. Calculate the electron conductivity in plasma using the general Equation 4.97, assuming a constant value of the electron mean free path λ. Compare the result with the conventional expression of conductivity (Equation 4.98) corresponding to a constant value of the electron–neutral collision frequency.

4.18 **Joule Heating**. Using Equation 4.100 for the joule heating, prove the stability of a thermal plasma sustained at a constant value of current density $\vec{j} = \sigma\vec{E}$ with respect to local fluctuation of plasma temperature.

4.19 **Similarity Parameters**. Based on the similarity parameters presented in Table 4.1, find the electron mean free pass, electron–neutral collision frequency, and electron mobility in atmospheric pressure air at (1) room temperature and (2) 600 K.

4.20 **Electron Drift in the Crossed Electric and Magnetic Fields**. Explain why the electron drift in crossed electric and magnetic fields is impossible at fixed values of magnetic field and relatively high values of electric field. Estimate the criteria for the critical electric field. Illustrate the electron motion if the electric field exceeds the critical one.

4.21 **Electric Conductivity in the Crossed Electric and Magnetic Fields**. Calculate the lower and upper application limits of the ratio ω_B/ν_{en} for the transverse plasma conductivity (Equation 4.109) in a strong magnetic field. Discuss the difference in magnetic fields and gas pressures necessary for trapping plasma electrons and ions in the magnetic field.

4.22 **Plasma Rotation in the Crossed Electric and Magnetic Fields, Plasma Centrifuge**. Describe plasma motion in the electric field $\vec{E}(r)$ created by a long charged cylinder and uniform magnetic field $\vec{B}$ parallel to the cylinder. Derive a relation for the maximum operation pressure of such a plasma centrifuge. Is it necessary to trap ions in the magnetic field to provide the plasma rotation as a whole in this system? Is it necessary to trap electrons in the magnetic field?

4.23 **Electric Conductivity of Strongly Ionized Plasma**. In calculating the electric conductivity of a strongly ionized plasma (Equation 4.110), why are the electron–neutral and electron–ion, but not electron–electron, collisions taken into account?

4.24 **Free Diffusion of Electrons**. Using the similarity parameters presented in Table 4.2, estimate the free diffusion time of electrons with temperature of about 1 eV to the wall of the 1-cm diameter cylindrical chamber filled atmospheric pressure nitrogen.

4.25 **Ambipolar Diffusion**. Estimate the ambipolar diffusion coefficient in room temperature atmospheric pressure nitrogen at an electron temperature of about 1 eV. Compare this diffusion coefficient with that for free diffusion of electrons and free diffusion of nitrogen molecular ions (take the ion temperature as room temperature).

4.26 **Debye Radius and Ambipolar Diffusion**. Estimate the Debye radius and thus the typical size scale at which diffusion of charged particles should be considered, not as ambipolar but free, in the nonthermal discharge plasma. Take as an example the electron temperature of 1 eV and electron concentration of 10^{12} cm^{-3}. Which specific plasma physical systems and plasma chemical processes can involve such size scales?

4.27 **Thermal Conductivity in Plasma Related to Dissociation and Recombination of Molecules.** Based on Equation 4.136 for the effect of dissociation on thermal conductivity and Equation 4.18 for equilibrium degree of dissociation, determine a general expression for the temperature range corresponding to the strong influence of the dissociation on thermal conductivity. Compare the result with numerical data presented in Figure 4.15.

4.28 **Thermal Conductivity in Plasma Related to Ionization and Charged Particle Recombination.** Discuss the effect of ionization–recombination and different chemical reactions on the total coefficient of thermal conductivity in high-temperature quasi-equilibrium plasma. In the same way as was done for effect of dissociation (Equation 4.134 and Equation 4.135), derive a formula for calculating the coefficient of thermal conductivity related to the ionization process. Compare the influence of these two effects on the total thermal conductivity.

4.29 **The Townsend Breakdown Mechanism.** Based on Equation 4.140, Equation 4.142, and Table 4.3, compare the Townsend breakdown conditions in molecular gases (take air as an example) and monatomic gas (take argon as an example). How can you explain the large exponential difference between these two cases?

4.30 **The Stoletov Constant and Energy Price of Ionization.** Using the electron energy distribution between ionization and different channels of excitation presented in Figure 3.12 through Figure 3.15, estimate the energy price of ionization in electric discharges as a function of the reduced electric field. Compare this ionization energy price with the Stoletov constant, and explain the similarity and difference between them.

4.31 **Effect of Electron Attachment on Breakdown Conditions, Formation of Streamers and Leaders.** Which attachment mechanism, the dissociative attachment or three-body attachment, is more important in preventing different breakdown-related phenomena? Is there a difference from this perspective in preventing a Townsend breakdown and formation of streamers and leaders?

4.32 **Radial Growth of an Avalanche due to Repulsion of Electrons.** Based on consideration of the radial electron drift in the avalanche head, derive Equation 4.153 for the transverse growth of the avalanche due to repulsion of electrons. Estimate the critical value of αx when this repulsion effect exceeds that of the free electron diffusion.

4.33 **Limitation of Electron Density in Avalanche due to Electron Repulsion.** Analyzing the electric charge of an avalanche head and its size (when it is controlled by electron repulsion), show that the electron density in the head reaches a maximum value that does not depend on x. Estimate and discuss the numerical value of the maximum electron concentration.

4.34 **Energy Input and Temperature in a Streamer Channel.** As was shown, the specific energy input (electron energy transferred to one molecule) in a streamer channel is small during the short period (~30 nsec) of streamer growth between electrodes. In molecular gases, it is about 10^{-3} eV/mol, which in temperature units corresponds to ~10 K. In this case, estimate the vibrational temperature in the streamer channel taking a vibrational quantum as $\hbar\omega = 0.3$ eV.

4.35 **Streamer Propagation Velocity**. In the framework of the model of ideally conducting streamer channel (Equation 4.158), estimate the difference between streamer velocity and electron drift velocity in the external electric field. Explain the streamer propagation velocity's dependence on its length at other parameters fixed.

4.36 **Leader.** Estimate the electric field on the head of a 2-m long leader and calculate the correspondent electron drift velocity. Compare the electron drift velocity with a typical leader propagation velocity and explain the difference (for streamer channel growth, these velocities are usually supposed to be equal).

4.37 **Steady-State Nonthermal Discharge Regime in Nonelectronegative Gas.** Taking into account the absence of negative ions in nonelectronegative gases, simplify the set of Equation 4.164 through Equation 4.166 and find the steady-state concentrations of charged particles in the discharge regime. Compare the results with the concentrations of charged particles in the recombination-controlled regime of a discharge in electronegative gas.

4.38 **Recombination-Controlled Regime of Steady-State Nonthermal Discharge.** Analyzing Equation 4.170, determine how the ionization degree in the recombination-controlled regime depends on concentration of species, providing electron detachment from negative ions. Take as an example the steady-state nonthermal discharge in CO_2, with small additions of CO, providing the effective detachment (Equation 4.167a).

4.39 **Attachment-Controlled Regime of Steady-State Nonthermal Discharge**. In the attachment-controlled regime (Equation 4.172), the electron temperature and reduced electric field are actually fixed by the ionization-attachment balance $k_i(T_e) \approx k_a(T_e)$. What is the restriction on electron concentration and degree of ionization in this case? How can these be calculated? From this point of view, analyze the difference between the nonthermal glow and microwave discharges.

4.40 **The Engel–Steenbeck Model, Diffusion-Controlled Discharges**. Calculate the electron temperature in nonthermal sustained discharge in nitrogen with a pressure of 1 torr and radius of 1 cm, based on the Engel–Steenbeck model. Estimate the reduced electric field corresponding to the electron temperature. Compare this reduced electric field with that required for nitrogen breakdown in similar conditions and discuss the result.

4.41 **Propagation of Nonthermal Discharges**. Imagine a steady-state nonthermal discharge in a gas flow with the ionization wave propagation velocity exceeding the gas flow velocity. Discuss.

4.42 **Ionization Wave Propagation**. According to Equation 4.187, what level of electron concentration and degree of ionization are necessary to provide the velocity of the ionization wave close to the speed of sound? Explain why such discharge conditions are physically nonrealistic for the type of ionization wave under consideration.

CHAPTER

5

KINETICS OF EXCITED PARTICLES IN PLASMA

5.1 VIBRATIONAL DISTRIBUTION FUNCTIONS IN NONEQUILIBRIUM PLASMA: FOKKER–PLANCK KINETIC EQUATION

Excited atoms and molecules play an extremely important role in plasma chemical processes, as discussed in Chapter 3. That chapter was concerned with elementary processes of excited atoms and molecules, with probability analysis of energy transfer and chemical reactions taking place during collision of the excited species. Actually, plasma chemical kinetics is even more sensitive to the concentration of these active particles and to the population of the excited states. Statistics can describe the distribution functions of the excited species only in quasi equilibrium or some very specific nonequilibrium conditions (see Section 4.1). In general, determining the distribution functions of excited species requires the detailed analysis of relevant kinetic equations that will take place in this chapter.

Once the distribution functions of excited atoms and molecules are determined, the nonequilibrium rates of relaxation and chemical reactions in plasma can be found. This is usually referred to as plasma chemical microkinetics, in contrast to the macrokinetic, which includes the macroscopic balance of plasma chemical processes. This will be considered at the end of this chapter. As in Chapter 3, most of the attention will be paid to the vibrationally excited molecules, taking into account their special contribution in the energy balance of nonthermal discharges in molecular gases (see Section 3.3 and Section 3.7).

5.1.1 Nonequilibrium Vibrational Distribution Functions: General Concept of Fokker–Plank Equation

The vibrational distribution functions were first introduced (as the population of vibrationally excited states) in Section 4.1.2 in relation to quasi-equilibrium systems, and then generalized to the case of nonequilibrium statistics (Section 4.1.10). The statistical approaches, however, provide only qualitative analysis of vibrational kinetics. Only kinetic equations permit describing evolution of the vibrational distribution functions, taking into account the variety of relaxation processes and chemical reactions.

The electrons in nonthermal discharges mostly provide excitation of low molecular vibrational levels (see Section 3.3), which determines average vibrational energy or vibrational temperature. Formation of highly excited and chemically active molecules depends on many different processes in plasma, but mostly on the competition of multistep VV exchange processes and vibrational quanta losses in VT relaxation (see Section 3.4 and Section 3.5). This competition should be described to determine the population of the highly excited states.

Different kinetic approaches were developed to describe the evolution of the vibrational distribution functions in nonequilibrium systems (see, for example, Reference 81 and Reference 120 through Reference 122). The vibrational kinetics in plasma chemical systems includes not only the relaxation processes in the collision of heavy particles and chemical reactions, but also the direct energy exchange between the electron gas and excited molecules (the so-called eV processes; see below). The necessity of simultaneous consideration of so many kinetic processes with widely different natures suggests using the Fokker–Plank approach developed for vibrational kinetics by Rusanov, Fridman, and Sholin.[123] In the framework of this approach, **the evolution of vibrational distribution function is considered as diffusion and drift of molecules in the space of vibrational energies (or, in other words, the diffusion and drift of molecules along the vibrational spectrum)**. This approach is actually similar to that applied for determining the electron energy distribution functions (see Section 4.2.4).

The continuous distribution function $f(E)$ of molecules over vibrational energies E (the vibrational distribution function) can be considered in this case as the density in energy space and be determined from the continuity equation along the energy spectrum (compare with the similar Equation 4.75) for the electron energy distribution functions):

$$\frac{\partial f(E)}{\partial t} + \frac{\partial}{\partial E} J(E) = 0 \tag{5.1}$$

In this continuity equation, $J(E)$ is the flux of molecules in the energy space, which additively includes all relaxation and energy exchange processes, while taking into account chemical reactions from different vibrationally excited states. The most important energy space fluxes contributing to Equation 5.1 and the related vibrational distribution functions will be analyzed in the following sections.

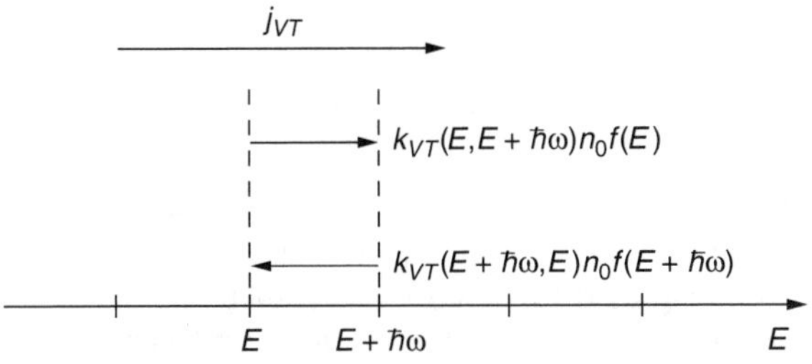

Figure 5.1 VT flux in energy space

5.1.2 Energy–Space Diffusion-Related VT Flux of Excited Molecules

Consider the vibrational kinetics in plasma chemical systems with the case of one-component diatomic molecular gas. The VT relaxation of molecules with vibrational energy E and vibrational quantum $\hbar\omega$ is conceptually viewed as diffusion along the vibrational energy spectrum and is illustrated in Figure 5.1. The VT flux then can be expressed as:

$$j_{VT} = f(E)\,k_{VT}(E, E+\hbar\omega)\,n_0\hbar\omega - f(E+\hbar\omega)\,k_{VT}(E+\hbar\omega, E)\,n_0\hbar\omega \tag{5.2}$$

Here, $k_{VT}(E+\hbar\omega, E)n_0$ and $k_{VT}(E, E+\hbar\omega)n_0$ are the frequencies of the direct and reverse processes of vibrational relaxation, whose ratio is exp $\hbar\omega/T_0$; n_0 is the neutral gas density. Expanding the vibrational distribution function in series, $f(E+\hbar\omega) = f(E) + \hbar\omega\,\partial f(E)/\partial E$, and denoting the relaxation rate coefficient $k_{VT}(E+\hbar\omega, E) \equiv k_{VT}(E)$, Equation 5.2 can be rewritten in the final form of the energy–space diffusion VT flux:

$$j_{VT} = -D_{VT}(E)\left[\frac{\partial f(E)}{\partial E} + \tilde{\beta}_0 f(E)\right] \tag{5.3}$$

In this relation, the diffusion coefficient D_{VT} of excited molecules in the vibrational energy space, related to VT relaxation, was introduced:

$$D_{VT}(E) = k_{VT}(E)n_0 \cdot (\hbar\omega)^2 \tag{5.4}$$

This definition of the diffusion coefficient along the energy spectrum is quite natural and corresponds to that for the conventional diffusion coefficient in coordinate space (see Equation 4.121). Obviously, the mean free path λ in the conventional coordinate space corresponds to the vibrational quantum $\hbar\omega$ in energy space, and the frequency of collisions ν in the coordinate space corresponds to the quantum transfer frequency $k_{VT}(E)n_0$ in energy space. Obviously, the energy–space diffusion coefficient $D_{VT}(E)$ grows exponentially with increasing vibrational energy, according to the Landau–Teller relation (Equation 3.61).

The translational temperature parameter $\tilde{\beta}_0$ in Equation 5.3 is defined by the following relation:

$$\tilde{\beta}_0 = \frac{1 - \exp\left(-\frac{\hbar\omega}{T_0}\right)}{\hbar\omega} \tag{5.5}$$

If the translational gas temperature is relatively high, $T_0 >> \hbar\omega$, then $\tilde{\beta}_0 = \beta_0 = 1/T_0$. The first term in the flux relation (Equation 5.3) can be interpreted as diffusion and the second term as the drift in energy space; then $\tilde{\beta}_0 = \beta_0 = 1/T_0$ is the ratio of diffusion coefficient over mobility completely in accordance with the Einstein relation. However, for lower translational temperatures, $T_0 \le \hbar\omega$, $\tilde{\beta}_0 = \beta_0 = 1/T_0$. This corresponds to the quantum-mechanical effect of a VT flux decrease with respect to the case of the quasi-continuum vibrational spectrum. Note that Equation 5.3 can only be applied for not very abrupt changes of the vibrational distribution function:

$$\left| \hbar\omega \frac{\partial \ln f(E)}{\partial E} \right| << 1 \tag{5.6}$$

that is necessary to make the $f(E)$ series expansion.

When VT relaxation is the dominating process, then the kinetic equation for vibrational distribution function based on Equation 5.1, Equation 5.3, and Equation 5.6 can be presented as:

$$\frac{\partial f(E)}{\partial t} = \frac{\partial}{\partial E}\left\{ D_{VT}(E)\left[\frac{\partial f(E)}{\partial E} + \frac{1}{T_0} f(E)\right]\right\} \tag{5.7}$$

At steady-state conditions ($\partial/\partial t = 0$), after integration, Equation 5.7 gives

$$\frac{\partial f(E)}{\partial E} + \frac{1}{T_0} f(E) = const(E).$$

Taking into account the boundary conditions at $E \to \infty$: $\partial f(E)/\partial E = 0$, $f(E) = 0$ yields $const(E) = 0$. As a result, the solution of Equation 5.7 corresponds to the steady-state domination of the VT relaxation and leads to the quasi-equilibrium exponential Boltzmann distribution with temperature T_0: $f(E) \propto \exp(-E/T_0)$. For comparison, see Equation 4.10 and Equation 4.11.

5.1.3 Energy–Space Diffusion-Related VV Flux of Excited Molecules

The VV exchange in contrast to the case of VT relaxation has two vibrationally excited molecules involved in the process. As a result, the VV flux is nonlinear with respect to the vibrational distribution function:[124]

$$j_{VV} = k_0 n_0 \hbar\omega \int_0^{\infty} \left[Q_{E+\hbar\omega,E}^{E',E'+\hbar\omega} f(E+\hbar\omega) f(E') - Q_{E,E+\hbar\omega}^{E'+\hbar\omega,E'} f(E) f(E'+\hbar\omega) \right] dE' \tag{5.8}$$

The VV exchange probabilities Q in this integral correspond to those presented in Equation 3.73; k_0 is the rate coefficient of neutral–neutral gas-kinetic collisions.

Taking into account the defect of vibrational energy in the VV exchange process $2x_e(E' - E)$ (x_e is the coefficient of anharmonicity), the probabilities Q are related to each other by the detailed equilibrium relation:

$$Q_{E,E+\hbar\omega}^{E'+\hbar\omega,E'} = Q_{E+\hbar\omega,E}^{E',E'+\hbar\omega} \exp\left[-\frac{2x_e(E'-E)}{T_0}\right] \tag{5.9}$$

It is important to point out that the Treanor distribution function (Equation 4.45), derived in Section 4.1.10 as the nonequilibrium statistical distribution of vibrationally excited molecules, makes the VV flux (Equation 5.8) in general equal to zero. This means that the Treanor distribution function (Equation 4.45) is a steady-state solution of the Fokker–Planck kinetic equation (Equation 5.1) if VV exchange is the dominating process and the vibrational temperature T_v exceeds the translational temperature T_0.

Replacing the vibrational quantum numbers by vibrational energy, the Treanor distribution function can be rewritten as:

$$f(E) = B\exp\left(-\frac{E}{T_v} + \frac{x_e E^2}{T_0\hbar\omega}\right), \quad \frac{1}{T_v} = \left.\frac{\partial \ln f(E)}{\partial E}\right|_{E\to 0} \tag{5.10}$$

The exponentially parabolic Treanor distribution function provides significant overpopulation of the highly vibrationally excited states and is illustrated in Figure 4.3 in comparison with the Boltzmann distribution function. The physical nature of the Treanor distribution, related to the predominant quantum transfer from vibrationally poor to vibrationally rich molecules (effect of the "capitalism in molecular life"), was discussed in the Section 4.10 and illustrated in Figure 4.4.

To analyze the complicated VV flux along the vibrational energy spectrum, it is convenient to divide the total VV flux and, hence, the total integral Equation 5.18 into two parts, corresponding to linear $j^{(0)}$ and nonlinear $j^{(1)}$ flux-components:

$$j_{VV}(E) = j_{vv}^{(0)}(E) + j_{VV}^{(1)}(E) \tag{5.11}$$

The linear flux component $j_{VV}^{(0)}(E)$ corresponds to the nonresonant VV exchange collisions of a highly vibrationally excited molecule of energy E with the bulk of low vibrational energy molecules, which is related in Equation 5.8 to the integration domain $0 < E' < T_v$. The nonlinear flux component $j_{VV}^{(1)}(E)$ corresponds to the resonant VV exchange collisions of two highly vibrationally excited molecules. Vibrational energies of the two molecules are mostly in the interval confined by the adiabatic parameter δ_{VV} (3.5.6): $|E - E'| \le \delta_{VV}^{-1}$) and the integral Equation 5.18 domain $T_v < E' < \infty$. Next, the linear and nonlinear VV flux-components will be considered separately.

5.1.4 Linear VV Flux along Vibrational Energy Spectrum

As discussed, the linear flux component $j_{VV}^{(0)}(E)$ corresponds to the nonresonant VV exchange with the bulk of low vibrational energy molecules and the integration domain $0 < E' < T_v$ in Equation 5.8. Taking into account that the distribution function for molecules with relatively low vibrational energy is $f(E') \approx 1/T_v \exp(-E'/T_v)$, assuming in a similar manner as in Section 5.2 that:

$$f(E+\hbar\omega) = f(E) + \hbar\omega \frac{\partial f(E)}{\partial E}, \exp\left(-\frac{\hbar\omega}{T_v} + \frac{2x_e E}{T_0}\right) \approx 1 - \frac{\hbar\omega}{T_v} + \frac{2x_e E}{T_0},$$

and using the integral Equation 5.8 yields the following expression for the linear component of the VV flux along the vibrational energy spectrum:

$$j_{vv}^{(0)} = -D_{VV}(E)\left[\frac{\partial f(E)}{\partial E} + \left(\frac{1}{T_v} - \frac{2x_e E}{T_0 \hbar\omega}\right) f(E)\right] \quad (5.12)$$

Similar to Equation 5.3, the diffusion coefficient D_{VV} of excited molecules in the vibrational energy space, related to the nonresonant VV exchange of a molecule of energy E with the bulk of low vibrational energy molecules, is introduced:

$$D_{vv}(E) = k_{VV}(E) n_0 (\hbar\omega)^2 \quad (5.13)$$

The relevant VV exchange rate coefficient can be expressed based on Equation 3.73 as: $k_{vv}(E) \approx ET_v/(\hbar\omega)^2 Q_{01}^{10} k_0 \exp(-\delta_{vv} E)$, where k_0 is the rate coefficient of neutral–neutral gas-kinetic collisions, and the adiabatic parameter δ_{VV} (Equation 3.5.6) is recalculated with respect to the vibrational energy rather than the vibrational quantum number.

The definition of the linear VV diffusion coefficient along the energy spectrum (Equation 5.13) corresponds to the diffusion coefficient in the coordinate space (Equation 4.121). The mean free path λ in this case is again the vibrational quantum $\hbar\omega$, and collisional frequency ν is the nonresonant quantum exchange frequency $k_{VV}(E)n_0$. Note that the solution of the linear kinetic equation $j_{VV}^{(0)}(E) = 0$ (with flux Equation 5.12) leads to the Treanor distribution function (Equation 5.10).

5.1.5 Nonlinear VV Flux along Vibrational Energy Spectrum

The nonlinear flux component $j_{VV}^{(1)}(E)$ is related to the resonant VV exchange $|E - E'| \leq \delta_{VV}^{-1}$. It is convenient in the domain under consideration, $T_v < E' < \infty$, to rewrite the integral (Equation 5.18) in the following integral-differential form:

$$j_{VV}^{(1)} = k_0 n_0 Q_{10}^{01} \frac{E+\hbar\omega}{\hbar\omega} \int_{T_v}^{\infty} \frac{E'+\hbar\omega}{\hbar\omega} \exp\left(-\delta_{vv}|E'-E|\right) \times$$

$$\left\{\hbar\omega\left[f(E')\frac{\partial f(E)}{\partial E} - f(E)\frac{\partial f(E')}{\partial E'}\right] + \frac{2x_e(E'-E)}{T_0} f(E) f(E')\right\} \hbar\omega \, dE' \quad (5.14)$$

When the contribution of the quasi-resonance VV exchange processes is dominant, the under-integral function has a sharp maximum at $E = E'$. In this case, assume $f(E) \approx f(E')$ and

$$\frac{\partial \ln f(E')}{\partial E'} - \frac{\partial \ln f(E)}{\partial E} \approx \frac{\partial^2 \ln f(E)}{\partial E^2}(E' - E).$$

Then, integration of Equation 5.4 over the energy interval $(E - \delta_{VV}^{-1}, E + \delta_{VV}^{-1})$ gives the final differential expression of the nonlinear VV flux along the vibrational energy spectrum:

$$j_{VV}^{(1)} = -D_{VV}^{(1)} \frac{\partial}{\partial E}\left[f^2(E) E^2 \left(\frac{2x_e}{T_0} - \hbar\omega \frac{\partial^2 \ln f(E)}{\partial E^2} \right) \right] \tag{5.15}$$

The energy-space diffusion coefficient $D_{VV}^{(1)}$ describes the resonance VV exchange processes and can be found as:

$$D_{VV}^{(1)} = 3k_0 n_0 Q_{10}^{01} \left(\delta_{VV} \hbar\omega\right)^{-3} \tag{5.16}$$

In the same manner as in the case of the linear kinetic equation $j_{VV}^{(0)}(E) = 0$ with the flux (Equation 5.12), one solution of the nonlinear kinetic equation $j_{VV}^{(1)}(E) = 0$ with flux (Equation 5.15) is again the Treanor distribution function (Equation 5.10). Indeed, the Treanor distribution satisfies the equality:

$$\frac{2x_e}{T_0} - \hbar\omega \frac{\partial^2 \ln f(E)}{\partial E^2} = 0.$$

However, it is important to point out that the Treanor distribution function is not the only solution of the nonlinear kinetic equation $j_{VV}^{(1)}(E) = 0$; the other plateau-like distribution will be discussed next.

5.1.6 Equation for Steady-State Vibrational Distribution Function Controlled by VV and VT Relaxation Processes

The vibrational distribution functions in nonequilibrium plasma are usually controlled by VV exchange and VT relaxation processes. The vibrational excitation of molecules by electron impact, chemical reactions, radiation, etc. is mainly related to the averaged energy balance and temperatures. However, details of the relative population of vibrationally excited states are mostly controlled by competition of VV and VT relaxation. Equation 5.3, Equation 5.12, and Equation 5.15 for VT and VV fluxes permit determining the vibrational distribution functions provided by these relaxation processes.

At steady-state conditions, the general Fokker–Planck kinetic equation (Equation 5.1) gives $J(E) = \text{const}$. Taking into account that at $E \to \infty$: $\partial f(E)/\partial E = 0, f(E) = 0$ yields $const(E) = 0$. As a result, the steady-state vibrational distribution function provided by VT and VV relaxation processes can be found from the equation:

$$j_{VV}^{(0)}(E) + j_{VV}^{(1)}(E) + j_{VT}(E) = 0 \tag{5.17}$$

Considering the specific relations Equation 5.3, Equation 5.12, and Equation 5.15 for the fluxes, the Fokker–Planck kinetic equation can be written as:

$$D_{VV}(E)\left(\frac{\partial f(E)}{\partial E}+\frac{1}{T_v}f(E)-\frac{2x_eE}{T_0\hbar\omega}f(E)\right)+$$

$$D^{(1)}_{VV}\frac{\partial}{\partial E}\left[f(E)^2E^2\left(\frac{2x_e}{T_0}-\hbar\omega\frac{\partial^2\ln f(E)}{\partial E^2}\right)\right]+ \tag{5.18}$$

$$D_{VT}(E)\left(\frac{\partial f(E)}{\partial E}+\tilde{\beta}_0 f(E)\right)=0$$

The first two terms in kinetic Equation 5.17 and Equation 5.18 are related to VV relaxation and prevail at low vibrational energies; the third term is related to VT relaxation and dominates at higher vibrational energies (see Equation 3.77 and Equation 3.78). The linear part of the equation (Equation 5.18), including the first and third terms, is easy to solve, but the second nonlinear term makes a solution much more complicated. It is helpful to point out three cases of **strong, intermediate,** and **weak excitation,** corresponding to different levels of contribution of the nonlinear term (which is obviously the most important at higher excitation levels). Separation of these three cases is determined by two dimensionless parameters: $\delta_{VV}T_v$ and $x_eT_v^2/\hbar\omega T_v$. The strong excitation case will be considered first.

5.1.7 Vibrational Distribution Functions: Strong Excitation Regime

The regime of **strong excitation** takes place in nonthermal plasma at high values of average vibrational energy and vibrational temperature:

$$\delta_{VV}T_v \geq 1 \tag{5.19}$$

The adiabatic parameter δ_{VV} was discussed in Section 3.5.2. Its typical value at room temperature is approximately $\delta_{VV} \approx (0.2 \div 0.5)/\hbar\omega$ (in this chapter, δ is expressed with respect to vibrational energy, not quantum number). This means, numerically, that the strong excitation regime requires very high vibrational temperatures, usually exceeding 5000 to 10,000 K.

When vibrational temperatures are so high, the resonant VV exchange between highly excited molecules (the nonlinear VV flux) dominates the nonresonant VV exchange of the excited molecules with the bulk of low excited molecules (the linear VV flux). The kinetic equation (Equation 5.18) at moderately high vibrational energies (e.g., when the VT flux can still be neglected) can be simplified to the following form:

$$E^2f^2(E)\left(\frac{2x_e}{T_0}-\hbar\omega\frac{\partial^2\ln f(E)}{\partial E^2}\right)=F \tag{5.20}$$

Here, F is the constant proportional to the quantum flux from low levels (where they appear due to excitation by electron impact) to high levels (where they disappear in VT relaxation and chemical reactions).

More detailed solutions of Equation 5.20 can be found in Likalter and Naidis[121] and Rusanov and Fridman.[125] However, the following simple analysis of Equation 5.20 permits the solution. At low vibrational energies, when the product $Ef(E)$ is large,

$$\frac{2x_e}{T_0} - \hbar\omega \frac{\partial^2 \ln f(E)}{\partial E^2} \approx 0$$

and the vibrational distribution function is close to the Treanor distribution. For larger energies, $Ef(E)$ becomes smaller and

$$\frac{2x_e}{T_0} - \hbar\omega \frac{\partial^2 \ln f(E)}{\partial E^2} = \frac{F}{E^2 f^2(E)}$$

grows and cannot be taken as zero as was previously assumed. As a result, the vibrational distribution deviates from the Treanor distribution, becoming flatter; the derivative

$$\frac{\partial^2 \ln f(E)}{\partial E^2}$$

decreases and finally leads to the so-called **hyperbolic plateau distribution**:

$$\frac{2x_e}{T_0} = \frac{F}{E^2 f^2(E)}, \quad f(E) = \frac{C}{E} \tag{5.21}$$

The "plateau level" C can be found from the power P_{eV} of vibrational excitation in plasma, calculated per molecule:

$$C(P_{eV}) = \frac{1}{\hbar\omega}\sqrt{\frac{P_{eV} T_0 (\delta_{VV}\hbar\omega)^3}{4x_e k_0 n_0 Q_{10}^{01}}} \tag{5.22}$$

In general, the vibrational distribution function in the strong excitation regime first follows the Treanor function at relatively low energies:

$$E < E_{Tr} - \hbar\omega\sqrt{\frac{T_0}{2x_e\hbar\omega}} \tag{5.23}$$

Here, $E_{Tr} = \hbar\omega \cdot v_{\min}^{Tr}$ is the Treanor minimum energy, corresponding to the minimum of the Treanor distribution function (see Equation 4.47). As the vibrational energies exceed the Treanor minimum point ($E > E_{Tr}$), the distribution function becomes the hyperbolic plateau:

$$f(E) = B\frac{E_{Tr}}{E}\exp\left(-\frac{T_0\hbar\omega}{4x_e T_v^2} - \frac{1}{2}\right) \tag{5.24}$$

where B is the Treanor normalization factor. Finally, transition from the plateau (Equation 5.24) to the rapidly decreasing Boltzmann distribution with temperature T_0 takes place at higher vibrational energies that exceed the critical one:

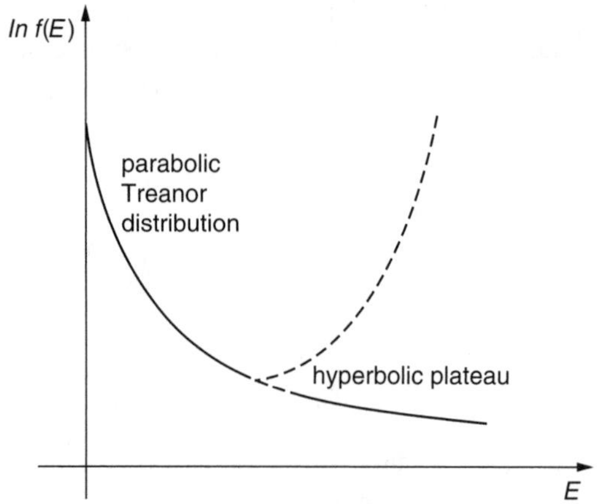

Figure 5.2 Vibrational energy distribution function due to VV exchange, strong and intermediate excitation regimes

$$E(\text{plateau}-VT)=\frac{1}{\delta_{VT}}\left[\ln\frac{k_0Q_{10}^{01}}{k_{VT}(E=0)}\frac{E_{Tr}}{\hbar\omega}-\frac{T_0\hbar\omega}{4x_eT_v^2}-\frac{1}{2}\right] \tag{5.25}$$

This plateau–Boltzmann transitional energy corresponds to the equality of probabilities of the vibrational VT relaxation and the resonance VV exchange. The typical behavior and shape of the vibrational distribution function in the regime of strong excitation in nonthermal molecular plasma is illustrated in Figure 5.2.

5.1.8 Vibrational Distribution Functions: Intermediate Excitation Regime

In the **regime of intermediate excitation**, the vibrational distribution function is similar to that presented in Figure 5.2. It also includes the Treanor function (Equation 5.10) at $E > E_{Tr}$; the hyperbolic plateau (Equation 5.21 and Equation 5.24) at energies exceeding the Treanor minimum ($E > E_{Tr}$); and the sharp plateau Boltzmann fall at higher energies (Equation 5.25). The intermediate excitation regime takes place in nonequilibrium plasma when the vibrational temperature is sufficiently high to satisfy the inequality in Equation 5.19 (it is less than the so-called VV exchange radius: $T_v < \delta_{VV}^{-1}$, which is numerically about 5000 to 10,000 K), but sufficiently high to provide the conditions for the intensive Treanor effect:

$$\frac{x_eT_v^2}{T_0\hbar\omega}\geq 1 \tag{5.26}$$

Under such conditions, the population of the vibrationally excited states at the Treanor minimum $E = E_{Tr}$ is quite high (Equation 4.48); the nonlinear resonance VV exchange dominates and provides the plateau at $E > E_{Tr}$, even though $T_v < \delta_{VV}^{-1}$. At low levels of vibrational energy $E < E_{Tr}$, the linear nonresonant VV exchange dominates the nonlinear one. However, the point is that the vibrational distribution function does not change because both components of VV exchange, the nonresonant Equation 5.12 and resonant Equation 5.15, result in the same Treanor distribution at $E < E_{Tr}$.

In conclusion, for two different reasons and at two different conditions, the strong and intermediate excitation regimes lead to the same vibrational distribution function illustrated in Figure 5.2.

5.1.9 Vibrational Distribution Functions: Regime of Weak Excitation

This regime takes place in nonthermal plasma when vibrational temperature is not very high and in contrast to the regimes of strong and intermediate excitation:

$$\delta_{VV} T_v < 1, \frac{x_e T_v^2}{T_0 \hbar\omega} < 1 \tag{5.27}$$

In this case, the nonlinear term in Equation 5.18 can be neglected and the Fokker–Planck kinetic equation simplified to the form:

$$\frac{\partial f(E)}{\partial E}\left(1+\xi(E)\right)+f(E)\cdot\left(\frac{1}{T_v}-\frac{2x_e E}{T_0\hbar\omega}+\xi(E)\tilde{\beta}_0\right)=0 \tag{5.28}$$

Here, $\xi(E)$ is the ratio of the rate coefficients of VT relaxation and nonresonant VV exchange, determined by the exponential relation (Equation 3.77). The solution of the linear differential Equation 5.28 gives, after integration,

$$f(E) = B\exp\left[-\frac{E}{T_v}+\frac{x_e E^2}{T_0\hbar\omega}-\frac{\tilde{\beta}_0-\dfrac{1}{T_v}}{2\delta_{VV}}\ln\left(1+\xi(E)\right)\right] \tag{5.29}$$

This continuous vibrational distribution function $f(E)$ for the weak excitation regime was derived by Rusanov et al.[123] and corresponds to the discrete distribution $f(v)$ over the vibrational quantum numbers:

$$f(v) = f_{Tr}(v)\prod_{i=0}^{v}\frac{1+\xi_i\exp\dfrac{\hbar\omega}{T_0}}{1+\xi_i} \tag{5.30}$$

This is usually referred to as the **Gordiets vibrational distribution.**[126] In this relation, $f_{Tr}(v)$ is the discrete form of the Treanor distribution. As can be seen from Equation 5.29 and Equation 5.30, at low vibrational energies ($E < E^*(T_0)$; Equation 3.78): $\xi(E) << 1$, so the vibrational distribution is close to the Treanor distribution. Conversely, at high energies $E < E^*(T_0)$: $\xi(E) >> 1$, the vibrational distribution is exponentially decreasing, according to the Boltzmann law, with temperature T_0.

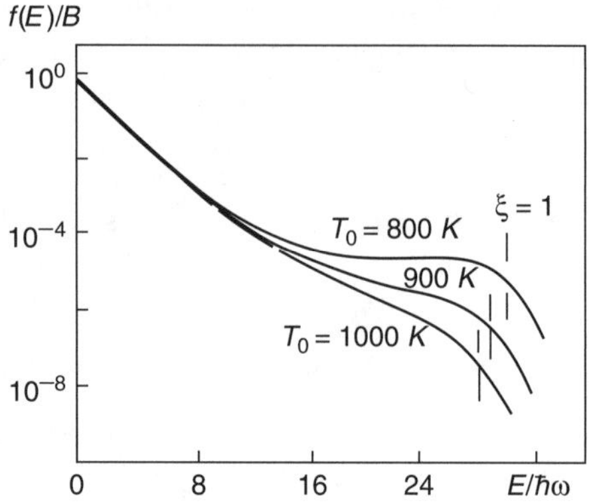

Figure 5.3 Nonequilibrium vibrational distributions in nitrogen, T_v = 3000 K

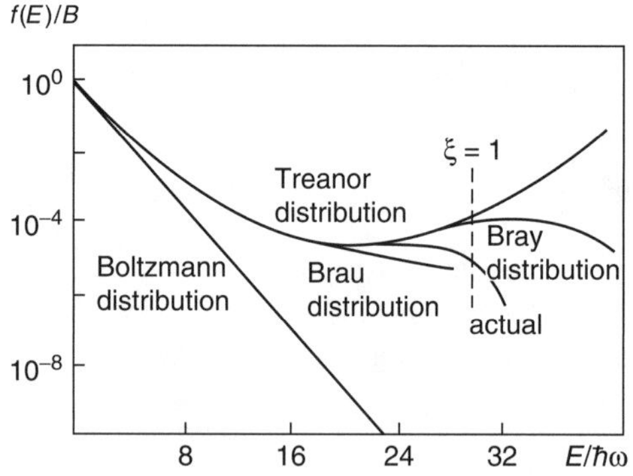

Figure 5.4 Different models of vibrational distribution in nitrogen, T_v = 3000 K; T_0 = 800 K

The vibrational distribution function in nitrogen at vibrational temperature T_v = 3000 K and different translational temperatures is shown in Figure 5.3. Figure 5.4 compares different types of vibrational distribution functions at the same conditions (in both figures, the dashed lines stand for $\xi(E) = 1$). Obviously, Figure 5.4 demonstrates the significant difference between the Treanor and Boltzmann vibrational distribution functions. Also, Figure 5.4 illustrates two less trivial effects:

1. The Bray[9,642] distribution (1968), only taking into account VT relaxation when $\xi(E) \geq 1$, overestimates the actual $f(E)$.
2. Conversely, the completely classical ($\tilde{\beta}_0 = \beta_0$) Brau[9,643] distribution (1972) underestimates the actual distribution function $f(E)$.

Equation 5.29 permits stressing the following qualitative features of the vibrational distribution function $f(E)$ that are controlled by the VV and VT relaxation processes in the regime of weak excitation:

1. The logarithm of the vibrational distribution function ln $f(E)$ always has an **inflection point**

$$\frac{\partial^2 \ln f(E)}{\partial E^2} = 0$$

corresponding to the vibrational energy below $E^*(T_0)$:

$$E_{\inf l} = E^*(T_0) - \frac{1}{2\delta_{VV}} \ln \frac{\delta_{VV} T_0}{x_e} < E^*(T_0) \quad (5.31)$$

Here, $E^*(T_0)$ is the critical vibrational energy when the resonant VV and VT relaxation rate coefficients are equal. Obviously, because $E_{\inf l} < E^*(T_0)$, the VV exchange is still faster than VT relaxation at the inflection point:

$$\xi\left(E_{\inf l}\right) = \frac{x_e}{\delta_{vv} T_0} < 1 \quad (5.32)$$

In particular, for nitrogen at $T_0 = 1000$ K, $\xi(E_{\inf l}) \approx 0.1$. It should be noted that the inflection point on the function ln $f(E)$ exists even if the function is continuously decreasing with vibrational energy (see Figure 5.3).

2. Sometimes the vibrational distribution function includes the **domain of inverse population**

$$\frac{\partial \ln f(E)}{\partial E} > 0$$

(see Figure 5.3). The existence criterion of the inverse population can be presented as the requirement of a positive logarithmic derivative at the inflection point:

$$2\delta_{VV}\left(E^*(T_0) - E_{Tr}\right) > \ln \frac{x_e}{\delta_{VV} T_0} \quad (5.33)$$

3. If the inequality (Equation 5.33) is valid and the function $f(E)$ has the domain of inverse population, then the vibrational distribution function has a maximum point $E_{\max}$, which can be found from the following equation:

$$\frac{2x_e}{T_0}\left(E_{\max} - E_{Tr}\right) = \xi\left(E_{\max}\right) \quad (5.34)$$

As can be seen from Equation 5.34, the parameter $\xi(E_{\max})$ is usually small numerically ($\xi(E_{\max}) < 1$). This means that $E_{\max} < E^*(T_0)$.

It is interesting to note that the qualitative characteristics of the vibrational distribution function $f(E)$, the inflection and maximum energies $E_{\inf l}$ and $E_{\max}$ are related to the influence of the VT relaxation and both are less than $E^*(T_0)$. This leads to the conclusion that VT relaxation already has a significant effect on the vibrational distribution function $f(E)$ when the VT relaxation rate is slower with respect to VV exchange ($\xi < 1$). This effect is well illustrated in Figure 5.3 and Figure 5.4.

5.2 NONEQUILIBRIUM VIBRATIONAL KINETICS: eV PROCESSES, POLYATOMIC MOLECULES, AND NON-STEADY-STATE REGIMES

The Fokker–Planck approach developed in the previous section can be applied to the analysis of different important aspects of vibrational kinetics in nonthermal plasma. Some of these will be considered in the following three sections. The first section considers the direct influence of molecular excitation by electron impact on the vibrational distribution functions (eV relaxation processes)—probably the most specific aspect of vibrational kinetics in plasma.

5.2.1 eV Flux along Vibrational Energy Spectrum

Vibrational excitation of molecules by electron impact (eV relaxation process) was briefly discussed in an indirect way in Section 5.1. There the assumed effective excitation of only the low vibrational levels that determined vibrational temperature T_v of the plasma was assumed. Evolution of the vibrational distribution function $f(E)$ in this approach was controlled by VV and VT relaxations and described by Fokker–Planck kinetic equation with vibrational temperature T_v as a parameter.

Such an approach of the indirect influence of the electron gas on $f(E)$ is relevant when VV exchange is much faster than the vibrational excitation (eV processes), which then can be considered only as boundary conditions. However, at high degrees of ionization in plasma, the frequency of eV processes becomes comparable with that of VV exchange and the eV flux along the vibrational energy spectrum must be taken into account. The eV flux in energy space describes the direct influence of energy exchange of excited molecules with the electron gas on the vibrational distribution function $f(E)$.

In contrast to VV relaxation and VT exchange, eV relaxation can often be effective as a one-quantum, and also multiquantum, process. The probability of the multiquantum eV processes, described by the Fridman's approximation (Equation 3.28) can be large. In general, the eV flux along the vibrational energy spectrum can be subdivided into two terms, corresponding to two mentioned types of eV relaxation processes:[127]

$$j_{eV}(E) = j_{eV}^{(1)}(E) + j_{eV}^{(0)}(E) \tag{5.35}$$

The linear eV flux $j_{eV}^{(1)}$ describes the eV processes with transfer of one quantum or a few quanta. This flux in the energy space can be expressed similarly to the linear VV and VT fluxes in the differential Fokker–Planck form as:

$$j_{eV}^{(1)}(E) = -D_{eV}\left(\frac{\partial f(E)}{\partial E} + \frac{1}{T_e} f(E)\right) \tag{5.36}$$

Here, T_e is the electron temperature and D_{eV} is the effective one-quantum, vibrational excitation diffusion coefficient in energy space:

$$D_{eV} = \lambda k^0_{eV} n_e (\hbar\omega)^2, \quad \lambda \approx \frac{2}{\alpha^3} \tag{5.37}$$

The one-quantum excitation rate coefficient $k^0_{eV} = k_{eV}(0,1)$ and parameter α correspond to the Fridman's approximation (Equation 3.28); numerically, $\alpha \approx 0.5 \div 0.7$. The factor λ accounts for the transfer of a few quanta; numerically, in nitrogen, $\lambda \approx 10$.

The eV flux component $j^{(0)}_{eV}(E)$ in Equation 5.35 is related to the multiquantum excitation of molecules from low levels to energy E and can be expressed in integral form as:

$$j^{(0)}_{eV}(E) = -\int_E^{\infty} k^0_{eV} n_e f(0) \exp\left(-\alpha \frac{E'}{\hbar\omega}\right)\left[1 - \frac{f(E')}{f(0)} \exp\frac{E'}{T_e}\right] dE' \tag{5.38}$$

This integral flux is actually a source (either positive or negative) of vibrationally excited molecules. The multiquantum excitation flux (Equation 5.38), as well as the one-quantum excitation flux, becomes equal to zero when the vibrational distribution $f(E)$ is the Boltzmann function with temperature T_e: $f(E) \propto \exp(-E/T_e)$.

When the eV de-excitation processes (super-elastic collisions) can be neglected, the integral expression for the multiquantum excitation eV flux can be simplified. Such a simplification can be done if $f(E) << f(0)\exp(-E/T_e)$. Then, after integration, the simplified expression for the eV flux component can be written as:

$$j^{(0)}_{eV}(E) = -D_{eV} \frac{\alpha^2}{2\hbar\omega} f(0) \exp\left(-\frac{E\alpha}{\hbar\omega}\right) \tag{5.39}$$

Comparing the eV flux components (Equation 5.36, Equation 5.38, and Equation 5.39) illustrates that the one-quantum eV processes (Equation 5.36) dominate ($j^{(1)}_{eV} >> j^{(0)}_{eV}$) if the vibrational distribution function $f(E)$ decreases with energy slower than $\exp(-\alpha E/\hbar\omega)$. Conversely, in the typical case of $T_v < \hbar\omega/\alpha$, the multiquantum processes dominate the eV relaxation at least at low vibrational energies. Based on the derived expressions for the eV relaxation fluxes, their influence on the vibrational distribution function at high and intermediate degrees of ionization values can be analyzed.

5.2.2 Influence of eV Relaxation on Vibrational Distribution at High Degrees of Ionization

In this case, the criteria of high electron density and high degree of ionization mean the domination of the vibrational excitation frequency over the frequency of VV exchange, even at low vibrational levels:

$$\frac{n_e}{n_0} >> \frac{k^0_{eV}}{k^0_{vv}}, \quad k^0_{vv} = Q^{10}_{01} k_0 \tag{5.40}$$

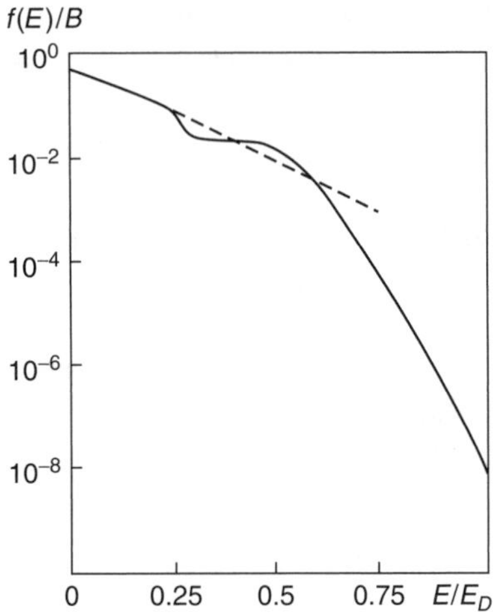

Figure 5.5 Vibrational distribution in N_2:Ar = 1:100 mixture, $T_e = 1$ eV; $T_0 = 750$ K; ionization degree = 10^{-6}

Numerically, this requires a fairly high degree of ionization in nonthermal plasma, typically exceeding 10^{-4} to 10^{-3}. Here, most of the vibrationally excited molecules are in quasi equilibrium with the electron gas, which can be characterized by the electron temperature T_e. The vibrational distribution function $f(E)$ is close to the Boltzmann function with temperature T_e for a wide range of vibrational energies from lowest to highly critical:

$$E_{VT-eV} \approx \frac{1}{\delta_{VT}} \ln \frac{k_{eV} n_e}{k_{VT}(0) n_0} \tag{5.41}$$

At this critical energy, the VT relaxation rate matches the rate of the eV processes and, according to Boltzmann distribution, the vibrational distribution function falls exponentially with temperature T_0. The vibrational distribution functions controlled by eV and VT relaxation processes at high degrees of ionization are shown in Figure 5.5. This figure presents the results of numerical calculations of Sergeev and Slovetsky.[128]

It is interesting to note that the $f(E)$ behavior around the transition energy (Equation 5.41) is not trivial and includes the "microplateau," which can be seen in Figure 5.4. One can find the relevant analysis and explanation of the phenomenon as well as other details of eV processes in vibrational kinetics in Demura et al.[124]

5.2.3 Influence of eV Relaxation on Vibrational Distribution at Intermediate Degrees of Ionization

If, in contrast to Equation 5.40 $n_e/n_0 << k^0_{eV}/k^0_{vv}$ and the vibrational excitation by electron impact is much slower than VV exchange, obviously, the direct influence of eV processes on $f(E)$ can be neglected (the indirect effect on vibrational temperature still exists).

The case of intermediate ionization degree implies $n_e/n_0 \sim k^0_{eV}/k^0_{vv}$. This means that, at low levels of vibrational excitation, VV relaxation is still sufficiently strong to build the Treanor distribution; however, at higher energies, eV processes dominate, thus creating the Boltzmann distribution with temperature T_e. Obviously, at higher values of vibrational energy, VT relaxation prevails (see Equation 5.41), leading to the Boltzmann distribution with temperature T_0.

The direct influence of eV processes on distribution function $f(E)$ is not possible when VV processes are dominated by the resonance exchange (strong excitation regime). This can easily be illustrated in the case of multiquantum eV relaxation, when the kinetic equation, $j^{(1)}_{VV} + j^{(0)}_{eV} = 0$, can be written after integration as:

$$D^{(1)}_{VV}E^2 f^2\left(\frac{2x_e}{T_0} - \hbar\omega\frac{\partial^2 \ln f}{\partial E^2}\right) = -D_{eV}\frac{\alpha f(0)}{2}\left[1-\exp\left(\frac{-\alpha E}{\hbar\omega}\right)\right] \approx -D_{eV}\frac{\alpha f(0)}{2} \quad (5.42)$$

Equation 5.42 is identical to the kinetic equation (Equation 5.20) because $\exp(-\alpha E/\hbar\omega) << 1$. As a result, the solution of Equation 5.42 is the hyperbolic plateau (Equation 5.21), which includes only indirect information on eV processes.

Thus, the direct influence of eV processes on $f(E)$ at intermediate degrees of ionization takes place only in the regime of weak excitation. The vibrational distribution in this case is controlled by eV and nonresonant VV relaxation at relatively low energies and can be found from the linear kinetic equation:

$$D_{VV}(E)\left(\frac{\partial f}{\partial E} + \frac{f}{T_v} - \frac{2x_e E}{T_0\hbar\omega}f\right) + D_{eV}\left(\frac{\partial f}{\partial E} + \frac{f}{T_e}\right) + D_{eV}\frac{\alpha^2}{2\hbar\omega}f(0)\exp\left(-\frac{\alpha E}{\hbar\omega}\right) = 0 \quad (5.43)$$

Solution of this equation can be presented in the following integral form:

$$f(E) = \varphi(E)\left[1 - \int_0^E \frac{\alpha^2 f(0)\eta_{eV}(E')\exp\left(-\frac{\alpha E'}{\hbar\omega}\right)}{2\hbar\omega\varphi(E')\left(1+\eta_{eV}(E')\right)}dE'\right] \quad (5.44)$$

In this expression, the factor $\eta_{eV}(E) = D_{eV}(E)/D_{VV}(E)$ is proportional to the degree of ionization in the plasma and shows the relation between eV and nonresonant VV processes. The special function $\varphi(E)$ is given by the following integral, which is similar to that corresponding to the weak excitation regime controlled by VV and VT relaxation (see Equation 5.28 and Equation 5.29):

$$\varphi(E) = \frac{1}{T_v}\exp\left[-\int_0^E \frac{\frac{1}{T_v} - \frac{2x_e E'}{T_0\hbar\omega} + \frac{1}{T_e}\eta_{eV}(E')}{1+\eta_{eV}(E')}dE'\right] \quad (5.45)$$

If the vibrational temperature is sufficiently high $T_v > \hbar\omega/\alpha$ (this assumption does not change qualitative conclusions), then the integral in Equation 5.44 is small and the vibrational distribution function $f(E) \approx \varphi(E)$ can be found from Equation 5.45.

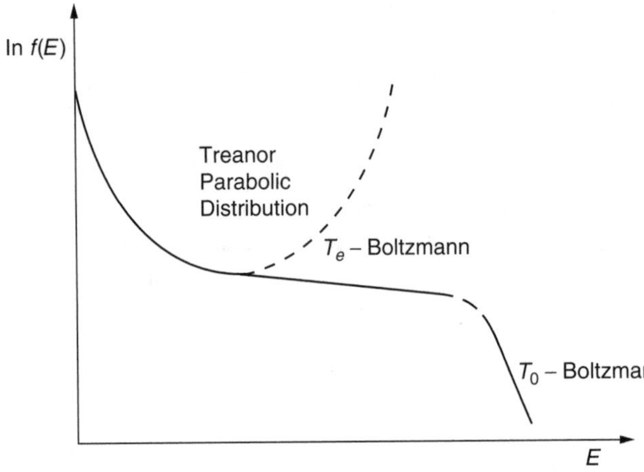

Figure 5.6 eV-processes' influence on vibrational distribution at intermediate ionization degrees

Equation 5.45 permits analyzing the vibrational distribution function $f(E) \approx \varphi(E)$ in the case of intermediate degrees of ionization. At low vibrational energies, when $\eta_{eV}(E) \leq 1$, the vibrational distribution is close to the Treanor function. The factor $\eta_{eV}(E)$ grows with energy and, at some point, the function $f(E) \approx \varphi(E)$ becomes the Boltzmann distribution with temperature T_e. A detailed discussion of this $f(E)$ behavior can be found in Rusanov and Fridman.[9] The relevant graphic illustration is presented in Figure 5.6.

It is interesting to note that the transition from the Treanor to Boltzmann distribution with high electron temperature T_e can be interpreted also as a transition to the plateau (see Figure 5.6). Remember, however, that this plateau has nothing in common with the hyperbolic plateau related to the resonance VV exchange.

5.2.4 Diffusion in Energy Space and Relaxation Fluxes of Polyatomic Molecules in Quasi Continuum

The vibrational distribution functions discussed earlier were related to the case of diatomic molecules. If polyatomic molecules are not strongly excited, their vibrational levels can also be considered discrete rather than continuous (see Section 3.2.3). Vibrational kinetics of polyatomic molecules in this case is quite similar to that of diatomic molecules.[129–131] An interesting example of such discrete vibrational distribution functions is presented in Figure 5.7,[517] where even the Treanor effect can be seen. Vibrational distribution between major groups of levels corresponds to vibrational temperature, while inside of each group it corresponds to rotational tempeature (which is close to translational temperature).

The specific features of polyatomic molecules manifest themselves at higher excitation levels, when interaction between vibrational modes is strong and the molecules are in the state of vibrational quasi continuum (Section 3.2.4). VT and VV relaxation processes for polyatomic molecules in quasi continuum were discussed

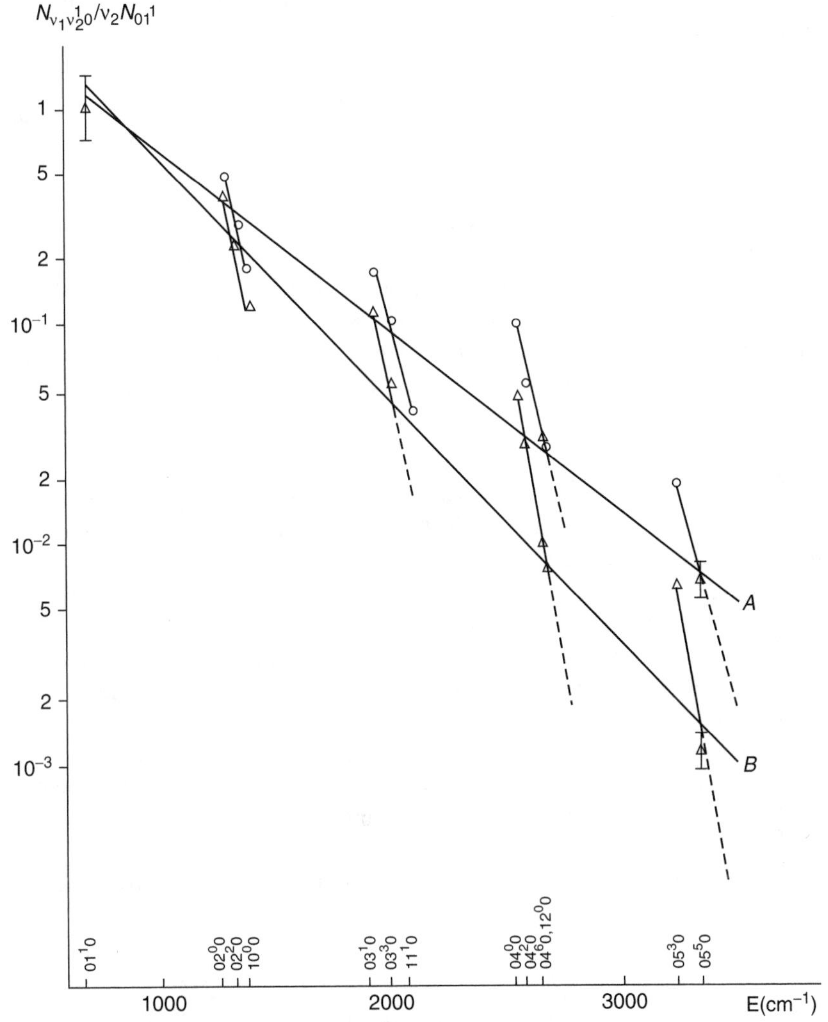

Figure 5.7 Vibrational distribution of CO_2 in symmetric valence and deformation modes: (A) $T_{v1} = 780 \pm 40$ K, $T_{r1} = 150 \pm 15$;K; (B) $T_{v2} = 550 \pm 40$ K, $T_{r2} = 110 \pm 10$ K[517]

in Section 3.4.6 and Section 3.5.4. Based on the analysis of VV and VT relaxation, consider now the vibrational distributions in these conditions.

It should be noted that fluxes related to the intermode energy exchange do not appear directly in the Fokker–Planck equation because they are very fast and compensate each other. The distribution function $f(E)$ of the polyatomic molecules in quasi continuum over the total vibrational energy E can then be found from the kinetic equation:

$$\frac{\partial f(E)}{\partial E} + \frac{\partial}{\partial E}\left(j_{VV}^{poly} + j_{VT}^{poly}\right) = 0 \tag{5.46}$$

For polyatomic molecules, the VT relaxation flux in energy space can be expressed similarly to Equation 5.3 and Equation 5.4 as:[617]

$$j_{VT}^{poly} = -\sum_{i=1}^{N} D_{VT}^{i}(E)\rho(E)\left[\frac{\partial}{\partial E}\left(\frac{f(E)}{\rho(E)}\right)+\frac{1}{T_0}\frac{f(E)}{\rho(E)}\right] \tag{5.47}$$

The main peculiarity of this relation with respect to the similar one for diatomic molecules (Equation 5.3) is the presence of the statistic factor $\rho(E) \propto E^{s-1}$ showing the density of vibrational states and taking into account the effective number s of vibrational degrees of freedom. $\tilde{\beta}_0 \approx 1/T_0$, which reflects the relative smallness of vibrational quantum in quasi continuum with respect to temperature. The summation is taken over all N vibrational modes and the diffusion coefficient in energy space, D_{VT}^{i}, is related to each of the modes i:

$$D_{VT}^{i}(E) = \left\langle (E_{VT}^{i})^2 \right\rangle k_0 n_0 \tag{5.48}$$

Here, k_0 is the rate coefficient of gas-kinetic collisions; n_0 is gas density; and the averaged square of VT energy transfer from a mode in quasi-continuum, $\left\langle (E_{VT}^{i})^2 \right\rangle$, is determined by Equation 4.120. It must be emphasized that the flux (Equation 5.47) is equal to zero for the Boltzmann distribution function with the statistical weight factor: $f(E) \propto \rho(E)\exp(-E/T_0)$.

To introduce the VV flux j_{VV}^{poly} in Equation 5.12 for polyatomic molecules in quasi continuum, note first that this flux, in contrast to diatomic molecules (Equation 5.11), has only the linear component $j_{VV}^{(0)}$. This flux corresponds to the VV exchange between excited molecules of high vibrational energy E with low excited molecules of the "thermal reservoir." The nonlinear VV flux component $j_{VV}^{(1)}$ can be neglected because of the resonant nature of the VV exchange of polyatomic molecules in quasi continuum (Equation 3.88), even if they have different vibrational energies. Then, the VV flux for the polyatomic molecules in quasi continuum can be presented in the following linear differential form:[36]

$$j_{VV}^{poly}(E) = -\sum_{i=1}^{N} D_{VV}^{i}(E)\rho(E)\left[\frac{\partial}{\partial E}\left(\frac{f(E)}{\rho(E)}\right)+\frac{1}{T_v^i}\left(\frac{f(E)}{\rho(E)}\right)-\left(\frac{f(E)}{\rho(E)}\right)\frac{E}{sT_0\hbar\omega_{0i}}\sum_{i=1}^{N} x_{ij}\left(1+\delta_{ij}\right)g_i\frac{\omega_{0i}}{\omega_{0j}}\right] \tag{5.49}$$

It is interesting to compare this sophisticated VV flux (Equation 5.49) for polyatomic molecules with the VT flux for polyatomic molecules (Equation 5.47) and with VV flux for diatomic molecules (Equation 5.12). The flux given by Equation 5.49 looks somewhat like a combination of these two.

In Equation 5.49, T_{vi} and $\hbar\omega_{oi}$ are vibrational temperatures and the first quantum for different modes; s is the number of effective vibrational degrees of freedom; δ_{ij} is the Kronecker delta-symbol; x_{ij} are the anharmonicity coefficients; and g_i is the degree of degeneracy of a vibrational mode. Here, the diffusion coefficient in energy space D^{i}_{VV} is related to the VV exchange between excited molecules with molecules of the thermal reservoir with temperature T_v:

$$D^{i}_{VT}(E) = \left\langle \left(E^{i}_{VV}\right)^2 \right\rangle k_0 n_0 \tag{5.50}$$

where the averaged square of VV energy transfer $\left\langle (E^{i}_{VV})^2 \right\rangle$ is determined by Equation 3.88. The most intriguing part of the VV flux (Equation 5.49) is the third term leading to the Treanor effect. This term arose from the balance of direct and reverse VV exchange processes Equation 3.93). Thus, as was discussed in Section 3.5.4, the Treanor effect is still valid for polyatomic molecules even though vibrations are in quasi continuum.

5.2.5 Vibrational Distribution Functions of Polyatomic Molecules in Nonequilibrium Plasma

The kinetic equation (Equation 5.46) or, in particular, the steady-state equation, $j^{poly}_{VV} + j^{poly}_{VT} = 0$, can be solved with Equation 5.47 and Equation 5.49 for VV and VT fluxes to find the vibrational distribution function $f(E)$. First, the steady-state case controlled only by the VV exchange ($j^{poly}_{VV} = 0$) should be analyzed. Assuming the only mode primarily determining $f(E)$ in quasi continuum, one obtains, after integration of Equation 5.49, the following distribution function:

$$\frac{f(E)}{\rho(E)} = B\exp\left[-\frac{E}{T_{Va}} + \frac{E^2}{2sT_0}\sum_{j=1}^{N}\frac{x_{aj}}{\hbar\omega_{0j}}\left(1+\delta_{aj}\right)g_j\right] \tag{5.51}$$

This is a generalization of the Treanor distribution (Equation 5.10) for the case of polyatomic molecules in quasi continuum. As a reminder, parameter B is the normalization factor. This generalization includes the statistical weight factor $\rho(E) \propto E^{s-1}$, which characterizes the density of the vibrational states and also effective coefficient of anharmonicity:

$$x_m = \frac{1}{2s}\sum_{j=1}^{N}\frac{x_{aj}}{\hbar\omega_{0j}}(1+\delta_{aj})g_j \tag{5.52}$$

These effective anharmonicity coefficients for polyatomic molecules are usually less than those for diatomic molecules. This makes the Treanor effect weaker for polyatomic molecules: the more vibrational degrees of freedom, the weaker is the Treanor effect.

Taking the VT flux given by Equation 5.47 into consideration leads to the following integral form of the vibrational distribution function:

$$f(E) = B\rho(E)\exp\left(-\int_{E_c}^{E}\frac{\dfrac{1}{T_{va}} - \dfrac{2x_m E'}{T_0\hbar\omega_{0a}} + \dfrac{\xi(E')}{T_0}}{1+\xi(E')}\,dE'\right) \tag{5.53}$$

Here, E_c designates the energy at which a polyatomic molecule enters the quasi continuum. Somewhat similar to Equation 5.28, the factor

$$\xi(E) = \sum_i D^i_{VT} / D^a_{VV}$$

characterizes the ratio of VT and VV relaxation rates for polyatomic molecules in quasi continuum. The factor $\xi(E)$ was analyzed in the Section 3.5.4. (see Equation 3.89). It was shown that $\xi(E)$ is much larger for polyatomic molecules than for diatomic molecules. This results in probably the most important peculiarity of the vibrational distribution function in polyatomic molecules. Although the Treanor function can be observed in quasi continuum, transition to Boltzmann distribution with translational temperature T_0 takes place in this case at much lower levels of vibrational excitation.

5.2.6 Non-Steady-State Vibrational Distribution Functions

Analytical solutions of the non-steady-state kinetic equation are known only for a few specific problems (see, for example, Macheret et al.[127] and Zhdanok et al.[132]). As an example, consider the nonsteady VV exchange with a variable diffusion coefficient in energy space: $D_{VV}(E) = D^{(0)} \exp(-\delta_{VV} E)$. Neglecting the Treanor term in the Fokker–Planck kinetic equation gives:

$$\frac{\partial f}{\partial t} = D^{(0)} \exp(-\delta_{VV} E) \left(\frac{\partial f}{\partial E} + \frac{1}{T_v} f \right) \tag{5.54}$$

This partial differential equation can be reduced to an ordinary differential equation by introduction of the following new variable:

$$Z(E,t) = D^{(0)} t \delta_{VV}^2 \exp\left(-\delta_{VV} E\right) \tag{5.55}$$

This new variable has a clear physical meaning: $Z(E,t) = 1$ describes the propagation of the front of the Boltzmann distribution function. In terms of this new variable $Z(E,t)$, the non-steady-state kinetic equation (Equation 5.54) can be rewritten as a second-order ordinary differential equation for the distribution function $f(Z)$:

$$Z^2 f''_{ZZ} + f'_Z \left[1 - \left(2 - \frac{1}{\delta_{VV} T_v} \right) Z \right] - \frac{1}{\delta_{VV} T_v} f(Z) = 0 \tag{5.56}$$

For the case of $Z(E,t) << 1$ corresponding to both low time intervals from the starting point of VV exchange and to high vibrational energies, the asymptotic solution for the non-steady-state vibrational distribution function can be presented as:

$$f(E,Z) \propto \exp\left(-\frac{1}{Z} - \frac{1}{T_v} E \right) \tag{5.57}$$

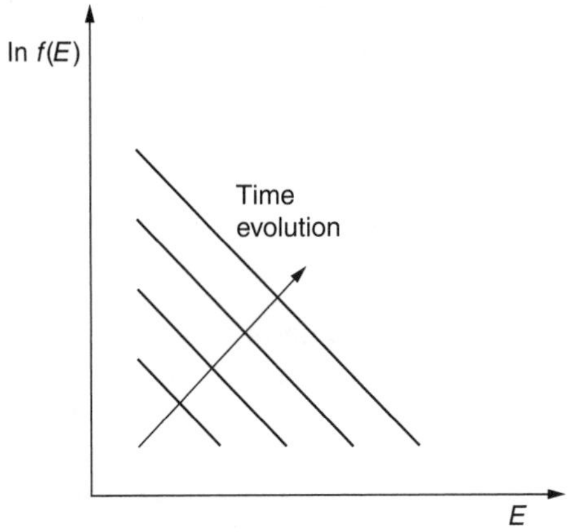

Figure 5.8 Time evolution of a vibrational distribution function

Equation 5.57 can be applied for estimations of the evolution of the Boltzmann function during the VV exchange for not very low values of $Z(E,t) << 1$ also. The non-steady-state distribution (Equation 5.57) is illustrated in Figure 5.8.

5.3 MACROKINETICS OF CHEMICAL REACTIONS AND RELAXATION OF VIBRATIONALLY EXCITED MOLECULES

The elementary processes and microkinetics of chemical reactions and the relaxation of excited species were considered in Chapter 3. Microkinetics implies the probabilities and rates of the processes from a fixed excitation state or with fixed excitation energy. In contrast, macrokinetics of the vibrationally excited molecules means averaging the microkinetic process rates over the vibrational distribution function.

5.3.1 Chemical Reaction Influence on Vibrational Distribution Function: Weak Excitation Regime

The macrokinetic rates of reactions of vibrationally excited molecules are self-consistent with the influence of the reactions on the vibrational distribution functions $f(E)$. This chemical reaction effect on $f(E)$ can be taken into account by introducing into the Fokker–Planck kinetic equation (Equation 5.1) an additional flux related to the reaction:

$$j_R(E) = -\int_E^{\infty} k_R(E')n_0 f(E')dE' = -J_0 + n_0 \int_0^E k_R(E')f(E')dE' \qquad (5.58)$$

Here, $J_0 = -j_R(E=0)$ is the total flux of the molecules in the chemical reaction (this means that the total reaction rate $w_R = n_0 J_0$) and $k_R(E)$ is the microscopic reaction rate coefficient given by Equation 3.99. In the weak excitation regime controlled by nonresonant VV and VT relaxation processes (usually the case in plasma chemistry), this leads to the equation: $j_{VV}^{(0)} + j_{VT} + j_R = 0$. Taking into account the specific expressions for all these fluxes (Equation 5.3, Equation 5.12, and Equation 5.58) results in the nonuniform linear kinetic equation:

$$\frac{\partial f(E)}{\partial E}(1+\xi(E)) + f(E)\cdot\left(\frac{1}{T_v} - \frac{2x_e E}{T_0 \hbar\omega} + \tilde{\beta}_0 \xi(E)\right) = \frac{1}{D_{VV}(E)} j_R(E) \qquad (5.59)$$

The exact solution of this nonuniform linear equation can be found in the form of $f(E) = C(E) * f^{(0)}(E)$, where $f^{(0)}(E)$ is the solution (Equation 5.29) of the corresponding uniform Equation 5.28. In other words, the function $f^{(0)}(E)$ makes the left-hand side of Equation 5.59 equal to zero. The function $C(E)$ can then be found from the equation:

$$D_{VV}(E)(1+\xi(E)) f^{(0)}(E)\frac{\partial C(E)}{\partial E} = j_R(E) \qquad (5.60)$$

After integration of Equation 5.60, the vibrational distribution function perturbed by plasma chemical reaction can be expressed in the following integral equation:

$$f(E) = f^{(0)}(E)\left[1 - \int_0^E \frac{-j_R(E')dE'}{D_{VV}(E')f(E')(1+\xi(E'))}\right] \qquad (5.61)$$

The function $-j_R(E)$ actually determines the flux of molecules along the energy spectrum that are going to participate in chemical reaction when they have sufficient energy ($E \geq E_a$; E_a is activation energy). This means that at relatively low vibrational energies $E < E_a$, where the chemical reaction can be neglected,

$$-j_R(E) = \int_{E_a}^{\infty} k_R(E')n_0 f(E')dE' = J_0 = const.$$

As a result at these energies ($E < E_a$), the perturbation of the vibrational distribution function $f^{(0)}(E)$ by chemical reaction (Equation 5.61) can be given as:

$$f(E) = f^{(0)}(E)\left[1 - J_0\int_0^E \frac{dE'}{D_{VV}(E')f^{(0)}(E')(1+\xi(E'))}\right] \qquad (5.62)$$

The total flux J_0, which is taken here as a constant parameter, is related to the total reaction rate and will be calculated later (Equation 5.67).

At high vibrational energies exceeding the activation energy ($E < E_a$), according to Equation 5.58, $j_R(E) \approx -k_R(E)n_0 f(E)\hbar\omega$. Then the integral equation (Equation 5.61) can be converted at $E \geq E_a$ into:

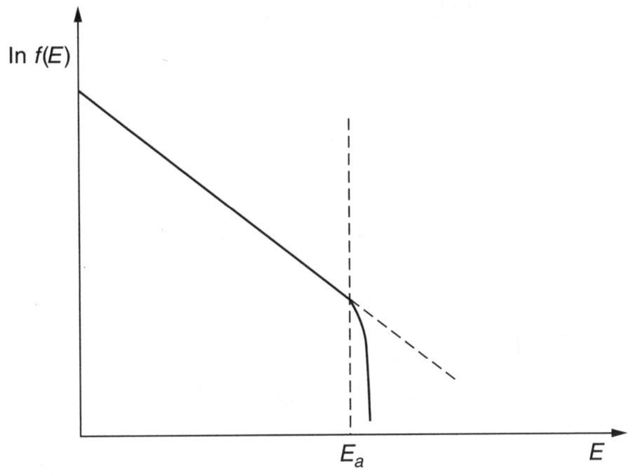

Figure 5.9 Reaction influence on vibrational distribution in weak excitation regime

$$\left\{\frac{f(E)}{f^{(0)}(E)}\right\} = \int_{E_a}^{E} \left\{\frac{f(E)}{f^{(0)}(E)}\right\} \frac{k_R(E')\hbar\omega\, dE'}{D_{VV}(E')(1+\xi(E'))} \tag{5.63}$$

Equation 5.63 can be easily solved:

$$f(E) \propto f^{(0)}(E)\exp\left[-\int_{E_a}^{E} \frac{k_R(E')\hbar\omega\, dE'}{D_{VV}(E')(1+\xi(E'))}\right] \tag{5.64}$$

and determines the decrease of the vibrational distribution function at $E \geq E_a$, e.g., in the domain of fast chemical reactions. Taking into account chemical reactions, the total vibrational distribution function is the combination of Equation 5.62 and Equation 5.64. Such a function is illustrated in Figure 5.9. As one can see from this figure, the significant influence of chemical reaction on vibrational population takes place only at energies at which the reaction is already effective.

5.3.2 Macrokinetics of Reactions of Vibrationally Excited Molecules: Weak Excitation Regime

Equation 5.61 and Equation 5.64 can be applied to calculate rates of the reactions of vibrationally excited molecules in the specific limits of slow and fast reactions:

1. The **fast reaction limit** implies that the chemical reaction is fast for $E \geq E_a$:

$$D_{VV}(E = E_a) << n_0 \cdot k_R(E + \hbar\omega) \cdot (\hbar\omega)^2 \tag{5.65}$$

and the chemical process in general is limited by the VV diffusion along the vibrational spectrum to the threshold $E = E_a$. In this case, according to Equation

5.64, the distribution function $f(E)$ decays very fast at $E > E_a$, and one can take $f(E = E_a) = 0$ in Equation 5.61:

$$1 = \int_0^{E_a} \frac{-j_R(E')dE'}{D_{VV}(E')f^{(0)}(E')(1+\xi(E'))} \tag{5.66}$$

This equation allows finding the total chemical process rate for the fast reaction limit, taking into account that $-j_R(E) = J_0 = const$ at $E < E_a$:

$$w_R = n_0 J_0 = n_0 \left\{ \int_0^{E_a} \frac{dE'}{D_{VV}(E')f^{(0)}(E')(1+\xi(E'))} \right\}^{-1} \tag{5.67}$$

As can be seen from Equation 5.67, the chemical reaction rate in this case is determined by the frequency of the VV relaxation and by the nonperturbed vibrational distribution function $f^{(0)}(E)$. The rate given by Equation 5.67 in the fast reaction limit actually is not sensitive to the detailed characteristics of the elementary chemical reaction once it is sufficiently large. Such a situation in practical plasma chemistry takes place, for example, in CO_2 and H_2O monomolecular dissociation processes, proceeding as second kinetic order reactions.[133]

2. The **slow reaction limit** corresponds to the opposite inequality in Equation 5.65. In this case, the population of the highly reactive states $E > E_a$, provided by VV exchange, takes place faster than the elementary chemical reaction. According to Equation 5.61 and Equation 5.64, the vibrational distribution function is almost nonperturbed by the chemical reaction $f(E) \approx f^{(0)}(E)$, and the total macroscopic reaction rate coefficient can be found as:

$$k_R^{macro} = \int_0^{\infty} k_R(E') f(E') dE' \tag{5.68}$$

The preceding consideration assumed a high efficiency of vibrational energy in the chemical reaction $\alpha = 1$. This can be generalized by using the microscopic rate coefficient $k_R(E)$ in the form of Equation 3.99 with an arbitrary value of α. Then, integrating Equation 5.68 over the distribution function, which is mostly controlled by VV exchange, leads to the following approximation of the macroscopic rate coefficient:[134]

$$k_R(T_v, T_0) = k_R^{(0)} \exp\left(-\frac{E_a}{T_0}\right) + k_R^{(v)} \exp\left(-\frac{E_a}{\alpha T_v} + \frac{x_e E_a^2}{T_0 \hbar\omega \alpha^2}\right) \tag{5.69}$$

In this relation, $k_R^{(0)}$ and $k_R^{(v)}$ are the pre-exponential factors of the reaction rate coefficient. According to Equation 5.69, if $\alpha T_v < T_0$, the chemical reaction proceeds by the quasi-equilibrium mechanism related to the translational temperature $k_R \propto \exp(-E_a/T_0)$. Conversely, high vibrational temperatures ($\alpha T_v > T_0$) correspond to effective stimulation of chemical processes by vibrational excitation, and the macroscopic reaction rate is related to the population of vibrational levels with energy exceeding E_a/α. Actually, the energy threshold E_a/α can be interpreted as the effective activation energy for reactions stimulated by vibrational excitation.

It is important to point out that, in contrast to the fast reaction limit (Equation 5.67), in the slow reaction limit (Equation 5.69) the information about elementary chemical process is explicitly presented in the formula for rate coefficient. Thus, the pre-exponential factors $k_R^{(0)}$ and $k_R^{(v)}$ in Equation 5.69 for chemical reactions proceeding through long-lifetime complexes include the factor $(T_v/E_a)^{s-1}$. This factor corresponds before averaging to the statistical theory factor $(E - E_a/E)^{s-1}$ in the microscopic reaction rate.[76] Similarly, the pre-exponential factors $k_R^{(0)}$ and $k_R^{(v)}$ in Equation 5.69 for electronically nonadiabatic chemical reactions include the relevant Landau–Zener transition factors.[31]

5.3.3 Macrokinetics of Reactions of Vibrationally Excited Molecules in Regimes of Strong and Intermediate Excitation

For this case, it is logical to assume that the distribution function $f(E)$ is controlled at the activation energy ($E = E_a$) by the resonance VV exchange processes. The chemical reaction influence on the plateau distribution ($E_{Tr} < E < E_a < E(\text{plateau} - VT)$) can be described by the Fokker–Planck kinetic equation derived from Equation 5.1, Equation 5.15, and Equation 5.58:

$$\frac{\partial}{\partial E}\left(D_{VV}^{(1)}E^2 f^2 \frac{2x_e}{T_0}\right) = J_0 \tag{5.70}$$

In the slow reaction limit, this means:

$$D_{VV}^{(1)}E_{Tr}^2 f^2\left(E_{Tr}\right)\frac{2x_e}{T_0} >> k_R(E+\hbar\omega)n_0\hbar\omega \tag{5.71}$$

The distribution function is actually not perturbed by the chemical process, and the macroscopic reaction rate can be found by integration (Equation 5.68) as:

$$k_R^{macro} = k_R\left(E \geq E_a\right)E_{Tr}f\left(E_{Tr}\right)\ln\frac{E(\text{plateau} - VT)}{E_a} \tag{5.72}$$

In this relation, *E(plateau–VT)* is the transition energy (Equation 5.25) from the hyperbolic plateau to the Boltzmann distribution with temperature T_0 (Equation 5.25); $k_R(E \geq E_a)$ is the microscopic reaction rate coefficient at vibrational energies exceeding the activation energy; and

$$E_{Tr} = \frac{\hbar\omega}{2x_e}\frac{T_0}{T_v}$$

is the Treanor minimum point.

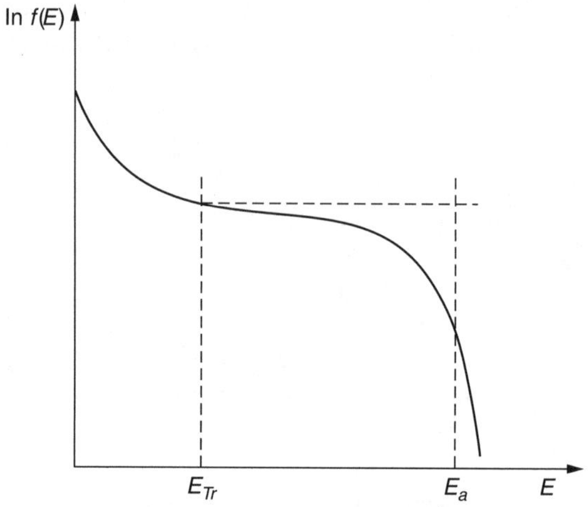

Figure 5.10 Reaction influence on vibrational distribution in strong excitation regime

In the fast reaction limit, when the inequality opposite to Equation 5.71 is valid, the vibrational population at high energies $E \geq E_a$ is negligible. Then, one can take $f(E = E_a) = 0$ as a boundary condition for the kinetic equation (Equation 5.71) to derive the vibrational distribution function strongly affected by the chemical reaction:

$$f(E) = f(E_{Tr})\frac{E_{Tr}}{E}\sqrt{\frac{E_a - E}{E_a - E_{Tr}}} \tag{5.73}$$

The perturbation of vibrational distribution functions in the case of strong excitation is illustrated in Figure 5.10. The macroscopic rate coefficient can then be expressed as:

$$k_R^{macro} = \frac{2x_e(\hbar\omega)^2}{T_0(E_a - E_{Tr})(\delta_{VV}\hbar\omega)^3} Q_{01}^{10} k_0 E_{Tr}^2 f^2(E_{Tr}) \tag{5.74}$$

The preceding relations imply the transition point from the Treanor distribution to the plateau distribution in the Treanor minimum. This is always correct in the strong excitation regime, but could be slightly different in the intermediate one. Detailed analysis on the subject can be found in Rusanov and Fridman.[9]

5.3.4 Macrokinetics of Reactions of Vibrationally Excited Polyatomic Molecules

The kinetic equation describing the reaction influence on the vibrational distribution function of polyatomic molecules can be derived based on Equation 5.46, Equation 5.47, Equation 5.49, and Equation 5.58 (neglecting the anharmonicity) as:[134]

$$D_{VV}\rho(E)\left[\frac{\partial}{\partial E}\left(\frac{f(E)}{\rho(E)}\right)(1+\xi)+\frac{f(E)}{\rho(E)}\left(\frac{1}{T_v}+\frac{\xi(E)}{T_0}\right)\right]=j_R(E) \quad (5.75)$$

The solution $f^{(0)}(E)$ of the uniform equation, related to the linear nonuniform equation (Equation 5.75), is the vibrational distribution function (Equation 5.53) without anharmonicity. Equation 5.61 remains true in this case and permits calculating the distribution function perturbed by the reaction and macroscopic reaction rate coefficient. For example, the plasma chemical dissociation of polyatomic molecules such as CO_2 and H_2O, effectively stimulated by vibrationally excitation of the molecules, corresponds to the fast reaction limit. The macroscopic dissociation rate coefficients of these processes can be derived from Equation 5.75, Equation 5.53, and Equation 5.61[134,134a] as:

$$k_R^{macro}=\frac{k_{VV}^0}{\Gamma(s)}\frac{\hbar\omega}{T_v}\left(\frac{E_a}{T_v}\right)^s\exp\left(-\frac{E_a}{T_v}\right)\sum_{r=0}^{\infty}\frac{(s+r-1)!}{(s-1)!r!}\frac{\gamma(r+1,E_a/T_v)}{(E_a/T_v)^r} \quad (5.76)$$

In this relation, $\Gamma(s)$ is the γ-function; $\gamma(r+1, E_a/T_v)$ is the incomplete γ-function; s is the number of vibrational degrees of freedom; and k_{VV}^0 and $\hbar\omega$ are the lowest vibrational quantum and the corresponding low-energy VV exchange rate coefficient. The sum in Equation 5.76 is not a strong function of E_a/T_v and, for numerical calculations, can be taken approximately as 1.1 to 1.3 if $E_a = 3 \div 5$ eV and $T_v = 1000 \div 4000$ K.

Detailed analysis of plasma chemical reactions of vibrationally excited polyatomic molecules can be found in the review of Rusanov, Fridman, and Sholin.[36]

5.3.5 Macrokinetics of Reactions of Two Vibrationally Excited Molecules

Microkinetics of elementary reactions of two vibrationally excited molecules, for example, $CO^* + CO^* \rightarrow CO_2 + C$ was considered in Section 3.7.8. Assuming that the effective reaction takes place, when the sum of vibrational energies of two partners exceeds the activation energy, $E' + E'' \geq E_a$ (compare with Equation 3.132). Then we can find the total reaction rate ($w_R = n_0 J_0$) for the vibrational distribution function $f^{(0)}(E)$ (Equation 5.29) not perturbed by the chemical process:[134b]

$$J_0=\iint_{E'+E''\geq E_a} f^{(0)}(E')f(E'')\,k_R(E',E'')n_0\,dE'\,dE'' \quad (5.77)$$

If $E_a > 2E^*(T_0)$, the process stimulation by vibrational excitation is ineffective because population of excited states is low due to VT relaxation (here, $E^*(T_0)$ is the critical vibrational energy when rates of VT and VV relaxation become equal [Equation 3.78]).

If $E_a < E^*(T_0)$, the Treanor effect is stronger if all the vibrational energy is located on just one molecule, and the reaction of two excited molecules actually uses effectively only one of them. Indeed, when total vibrational energy is fixed, $E' + E'' = const$, then $(E')^2 + (E'')^2 = (E' + E'')^2 - 2E'E''$and the Treanor effect is the strongest if $E'' = 0$.[134b] As a result, the nontrivial kinetic effect can be obtained from Equation 5.77 only if $E^*(T_0) < E_a < 2E^*(T_0)$.

The integration (Equation 5.77) determines the optimal vibrational energies of reaction partners E_{opt} and $E_a - E_{opt}$, which make the biggest contribution to the total reaction rate:

$$E_{opt} = E^*(T_0) - \frac{1}{2\delta_{VV}} \ln \frac{T_0}{2x_e\left[2E^*(T_0) - E_a\right]} \tag{5.78}$$

The macroscopic rate coefficient for the reaction of two vibrationally excited molecules can be then presented as:

$$k_R^{macro}(T_v, T_0) \propto \exp\left[-\frac{E_a}{T_v} + \frac{x_e}{T_0\hbar\omega}\left(E_{opt}^2 + \left(E_a - E_{opt}\right)^2\right) - \frac{\tilde{\beta}_0 x_e}{\delta_{VV} T_0}\left(2E^*(T_0) - E_a\right)\right] \tag{5.79}$$

Comparison of relations Equation 5.78 and Equation 5.31 shows that the optimal energy exceeds the inflection energy point of $f(E)$: $E_{infl} < E_{opt} < E^*(T_0)$. Thus, although $E_a > E^*(T_0)$, the VT relaxation does not significantly slow down the chemical reaction of two excited molecules because $E_{opt} < E^*(T_0)$.

The previously considered slow reaction limit is most typical for reactions of two vibrationally excited molecules. The distribution function is almost not disturbed by reaction in this case. Consideration of less typical fast reaction limit and influence of reactions of two excited molecules on distribution function can be found in Rusanov and Fridman.[9]

5.3.6 Vibrational Energy Losses Due to VT Relaxation

The elementary processes of vibrational relaxation were discussed in Chapter 3. Now, it is possible to calculate the vibrational relaxation processes averaged over the vibrational distribution function $f(E)$. It is necessary to find out the losses of the average vibrational energy, defined as:

$$\varepsilon_v = \int_0^\infty E f(E)\,dE \tag{5.80}$$

Multiplying Equation 5.1 by E and then integrating, the balance relation for the average vibrational energy is obtained:

$$\frac{d\varepsilon_v}{dt} = \int_0^\infty j_{VV}(E)\,dE + \int_0^\infty j_{VT}(E)\,dE \tag{5.81}$$

Here, only the vibrational energy losses related to VT and VV relaxation processes were taken into account. Beginning with consideration of the losses due to VT relaxation and taking the VV flux from Equation 5.3 and the VT diffusion coefficient in energy space as:

$$D_{VT}(E) = k_{VT}(E)n_0(\hbar\omega)^2 = k_{VT}^0\left(\frac{E}{\hbar\omega}+1\right)\exp\left(\delta_{VT}E\right)n_0(\hbar\omega)^2 \tag{5.82}$$

the formula for vibrational energy losses due to VT relaxation can be derived, based on Equation 5.81:

$$\left(\frac{d\varepsilon_v}{dt}\right)^{VT} = D_{VT}(0)f(0) - \int_0^\infty D_{VT}(0)\left[\frac{\left(\tilde{\beta}_0-\delta_{VT}\right)E-1}{\hbar\omega} - \left(\tilde{\beta}_0-\delta_{VT}\right)\right]\exp\left(\delta_{VT}E\right)f dE \tag{5.83}$$

These vibrational energy losses can be subdivided into two classes: (1) related to low vibrational levels and prevailing in conditions of weak excitation; and (2) related to high vibrational levels and dominating in conditions of strong excitation.

5.3.7 VT Relaxation Losses from Low Vibrational Levels: Losev Formula and Landau–Teller Relation

Based on Equation 5.83, the VT losses related to the low vibrational levels can be expressed in the following form, corresponding to the **Losev formula**:[135]

$$\left(\frac{d\varepsilon_v}{dt}\right)_L = -k_{VT}(0)n_0\left[1-\exp\left(-\frac{\hbar\omega}{T_0}\right)\right]\cdot\left[\frac{1-\exp\left(-\frac{\hbar\omega}{T_v}\right)}{1-\exp\left(-\frac{\hbar\omega}{T_v}+\delta_{VT}\hbar\omega\right)}\right]^2\left(\varepsilon_v-\varepsilon_{v0}\right) \tag{5.84}$$

Here, $\varepsilon_{v0} = \varepsilon_v(T_v = T_0)$. Neglecting the effect of anharmonicity ($\delta_{VT} = 0$), the Losev formula can be rewritten in the well-known **Landau–Teller relation**:

$$\left(\frac{d\varepsilon_v}{dt}\right)_L = -k_{VT}(0)n_0\left[1-\exp\left(-\frac{\hbar\omega}{T_0}\right)\right]\left(\varepsilon_v-\varepsilon_{v0}\right) \tag{5.85}$$

Obviously, in equilibrium between vibrational and translational degrees of freedom, when $T_v = T_0$ and thus $\varepsilon_v = \varepsilon_0$, the losses of vibrational energy become equal to zero.

5.3.8 VT Relaxation Losses from High Vibrational Levels

The contribution of the high levels into the VT losses of vibrational energy is usually related to the highest vibrational levels before the fast fall of the distribution function due to the VT relaxation or chemical reaction. Contribution of vibrational levels, corresponding to the Boltzmann distribution with temperature T_0, is small because of exponential decrease of the product $k_{VT}(E) \cdot f(E)$ with energy ($\delta_{VT}T_0 << 1$). The vibrational population falls at $E > E_a$; the fast reaction limit leads to even faster decrease of $k_{VT}(E) \cdot f(E)$.

The VT losses from the high vibrational levels in the **fast reaction limit**, when the previously mentioned fast fall of $f(E)$ is related to chemical reaction, can be calculated from the relation (Equation 5.83) as ($T_0 << \hbar\omega, \delta_{VT}T << 1$):

$$\left(\frac{d\varepsilon_v}{dt}\right)_H^{VT} \approx -D_{VT}(0)f(0) - \left[\frac{\left(\tilde{\beta}_0 - \delta_{VT}\right)E_a - 1}{\hbar\omega} - \left(\tilde{\beta}_0 - \delta_{VT}\right)\right] \tag{5.86}$$

$$D_{VT}(0)\exp\left(\delta_{VT}E_a\right)f^{(0)}\left(E_a\right)\Delta \approx -k_{VT}\left(E_a\right)n_0\hbar\omega\, f^{(0)}\left(E_a\right)\Delta$$

Here, $f(0)(E)$ is the vibrational distribution function not perturbed by chemical reaction, and the effective integration domain:

$$\Delta = \left|\frac{1}{T_v} - \delta_{VT} - \frac{2x_e E_a}{T_0\hbar\omega}\right|^{-1}$$

Physical interpretation of the relation (Equation 5.86) is fairly clear. The rate of VT energy losses from high levels is the product of the fraction of molecules $f(0)(E_a) \cdot \Delta$ making the key contribution in VT relaxation; the relaxation frequency of these molecules $k_{VT}(E_a)n_0$; and the energy transfer in a relaxation act $\hbar\omega$.

Energy balance of plasma chemical systems becomes more understandable physically when the vibrational energy losses due to the VT relaxation from high levels are presented per one act of chemical reaction: $\Delta\varepsilon_{VT}$. Based on Equation 5.67 and Equation 5.86 and assuming $\xi(E_a) << 1$, the VT losses per one act of fast chemical reaction can be expressed as:

$$\Delta\varepsilon_{VT} \approx \frac{k_{VT}(E_a)}{k_{VV}(E_a)}\Delta \tag{5.87}$$

This relation reflects the fact that fast reaction and VT relaxation from high levels are related to the same excited molecules with energies slightly exceeding E_a. Frequencies of the processes are, however, different and proportional respectively to k_{VV} and k_{VT}. Also, Equation 5.87 shows that, if $\xi(E_a) << 1$, the VT relaxation from high levels does not much affect the energy efficiency of plasma chemical processes stimulated by vibrational excitation. Equation 5.87 was derived for the case of predominantly nonresonance VV exchange; if VV relaxation is mostly resonant, Equation 5.87 can be used as the higher limit of the losses.

The VT losses from the high vibrational levels in the **slow reaction limit**, when the previously mentioned fast fall of $f(E)$ is related to VT relaxation reaction, should be calculated directly from Equation 5.83. These losses, specifically in the strong excitation regime, can be higher than those considered previously.[124]

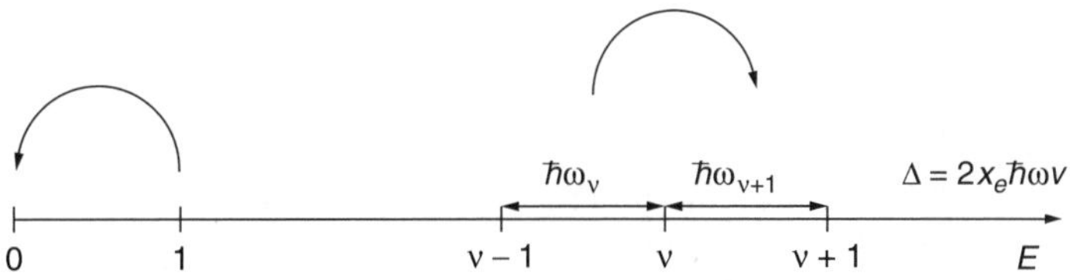

Figure 5.11 Vibrational energy losses in nonresonant VV exchange

5.3.9 Vibrational Energy Losses Due to Nonresonance Nature of VV Exchange

These losses can be calculated based on the general relation (Equation 5.81) in the same way as was done before in the case of VT relaxation. It is physically clearer and more practical in plasma chemistry, however, to analyze these losses per one act of chemical reaction.

The VV losses per one act of chemical reaction can be calculated in frameworks of the following model, illustrated in Figure 5.11. A diatomic molecule is excited by electron impact from the zero level to the first vibrational level, so a quantum comes to the system as a $\hbar\omega$. Further population of the higher excited vibrational levels is due to the one-quantum VV exchange. Quanta become smaller and smaller during the VV exchange due to anharmonicity (Equation 3.8). Thus, each step up on the "vibrational ladder" requires the resonance defect energy transfer $2x_e\hbar\omega\, v$ from vibrational to translational degrees of freedom; the total losses corresponding to excitation of a molecule to the nth vibrational level can be found as the following sum:

$$\Delta\varepsilon^{VV}(n) = \sum_{v=0}^{v=n-1} 2x_e\hbar\omega\, v = x_e\hbar\omega\,(n-1)\,n \tag{5.88}$$

In particular, for the process of dissociation of diatomic molecules stimulated by vibrational excitation, $n = n_{max} \approx 1/2x_e$. In this case (see Equation 3.7), the total losses of vibrational energy per one act of dissociation, associated with anharmonicity of molecular oscillations and nonresonance nature of VV relaxation, are just equal to the dissociation energy:

$$\Delta\varepsilon_D^{VV}(n = n_{max}) \approx \frac{\hbar\omega}{4x_e} = D_0 \tag{5.89}$$

This means that, taking these losses into account, the total vibrational energy necessary for dissociation provided by the VV exchange is equal not to D_0, but to $2D_0$. This amazing fact was first mentioned by Sergeev and Slovetsky[128] and then described by Demura et al.[124] Similarly, the relation Equation 5.88 predicts vibrational energy losses due to VV exchange per one act of chemical reaction of a diatomic molecule. If the reaction is stimulated by vibrational excitation and has activation energy $E_a << D_0$, the VV losses can be calculated as:

$$\Delta\varepsilon_R^{VV}(E_a) \approx \frac{1}{4} D_0 \left(\frac{E_a}{D_0}\right)^2 = x_e \frac{E_a^2}{\hbar\omega} \qquad (5.90)$$

As an example, consider the plasma chemical process of NO synthesis in air, stimulated by vibrational excitation of N_2 molecules. According to the Zeldovich mechanism, this synthesis is limited by the reaction:

$$O + N_2^* \rightarrow NO + N, \quad E_a = 1.3\,eV, \quad D_0 = 10\,eV \qquad (5.91)$$

Energy losses (Equation 5.90) in this reaction are equal to 0.28 eV per NO molecule, which results in 14% decrease of energy efficiency of the total plasma chemical process.[136]

In this section, the most important effects related to vibrational energy losses due to VV relaxation have been considered. Next, some other related phenomena that could be significant in special conditions are noted.[9]

1. Equation 5.88 through Equation 5.90 determine the losses of vibrational energy in steady-state systems. The non-steady-state initial establishment of the vibrational distribution function also includes the conversion of the first "big quanta" $\hbar\omega$ into smaller ones. According to Equation 5.88, this leads to additional VV energy losses of about $x_e T_v^2 / \hbar\omega$.
2. Each act of VT relaxation from the highly excited states with "smaller quantum" $\hbar\omega_s$ corresponds to VV losses of vibrational energy $\hbar\omega - \hbar\omega_s$ into translational degrees of freedom. Obviously, this effect takes place only for population of the highly excited states of diatomic molecules due to the one-quantum VV relaxation.
3. The evolution of the vibrational distribution $f(E)$, due mostly to one-quantum VV exchange, was implied earlier. However, at vibrational quantum numbers about $v \approx 1/2\, v_{max} = 1/4x_e$, the value of vibrational quantum becomes exactly twice less than the initial value $\hbar\omega$, which makes the double-quantum exchange possible (though still with low probability). Obviously, this kind of VV exchange decreases the VV losses of vibrational energy.
4. At higher ionization degrees, when evolution of vibrational distribution function is provided mostly by direct interaction with electron gas (eV processes), the losses of vibrational energy due to VV exchange become much less significant.

5.4 VIBRATIONAL KINETICS IN GAS MIXTURES: ISOTOPIC EFFECT IN PLASMA CHEMISTRY

Population of vibrationally excited molecular states in gas mixtures is provided not only by VV, VT, and eV relaxation processes, but also by the nonresonant vibration–vibration VV′ exchange between different components of the molecular gas. Even if the difference in oscillation frequencies of two components of the gas mixture is small (e.g., isotopes), the VV′ exchange can result in significant differences in the level of their vibrational excitation. Usually, a gas component with a low value of vibrational quantum becomes excited to the higher vibrational levels. Finally, this effect leads to a fundamental difference in the rates of plasma chemical reactions of the gas mixture components, providing in particular the strong isotopic effect in plasma chemical kinetics.[9,137,141,142]

The vibrational kinetics of gas mixtures has been studied and reviewed in numerous works (see, for example, Rich and Bergman[137]). Discussion in the next section will cover only the relevant effects in nonequilibrium plasma of double-component mixture of diatomic molecules with close values of oscillation frequency and the corresponding isotopic effect in nonequilibrium plasma chemical kinetics.

5.4.1 Kinetic Equation and Vibrational Distribution in Gas Mixture

Consider a kinetic equation describing the vibrational distribution functions $f_i(E)$ for a double-component gas mixture: subscripts $i = 1, 2$ correspond to the first (lower oscillation frequency) and second (higher frequency) molecular components:

$$\frac{\partial f_i(E)}{\partial t} + \frac{\partial}{\partial E}\left[j_{VV}^{(i)}(E) + j_{VT}^{(i)}(E) + j_{eV}^{(i)}(E) + j_{VV'}^{(i)}(E)\right] \tag{5.92}$$

The VV, VT, and eV fluxes in energy space $j_{VV}^{(i)}(E), j_{VT}^{(i)}(E), j_{eV}^{(i)}(E)$ are similar to those for a one-component gas and can be taken, respectively, for example, as Equation 5.12, Equation 5.3, and Equation 5.36. The VV′ relaxation flux, which is special for gas mixtures and provides predominant population of a component with lower oscillation frequency ($I = 1$), can be expressed as:[136]

$$j_{VV'}^{(i)} = -D_{VV'}^{(i)}\left(\frac{\partial f_i}{\partial E} + \beta_i f_i - 2x_e^{(i)}\beta_0 \frac{E}{\hbar\omega_i} f_i\right) \tag{5.93}$$

In this relation, $x_e^{(i)}, \omega_i$ are the anharmonicity coefficient and the oscillation frequency for the molecular component i. The correspondent diffusion coefficient in energy space is:

$$D_{VV'}^{(i)} = k_{VV'}^{(i)}\, n_{l\neq i}\,(\hbar\omega_i)^2 \tag{5.94}$$

Here, the rate coefficient $k_{VV'}^{(i)}$ corresponds to the intercomponent VV′ exchange process and is related to the rate coefficient $k_{VV}^{(0)}$ of the resonant VV exchange at the low vibrational levels of the component i as:

$$k_{VV'}^{(i)} = k_{VV}^{(0)} \exp\left[\frac{\delta_{VV}\hbar(\omega_i - \omega_l)}{2x_e^{(i)}} - \delta_{VV}E\right] \tag{5.95}$$

The concentration $n_{l\neq i}$ in Equation 5.94 represents the number density of a component l of the gas mixture, which interacts with the component i ($i,l = 1,2$). The reverse temperature factor β_i in the flux Equation 5.93 can be found from the relation:

$$\beta_i = \frac{\omega_l}{\omega_i}\beta_{vl} + \frac{\omega_i - \omega_l}{\omega_i}\beta_0 \tag{5.96}$$

The other reverse temperature parameters in Equation 5.96 are quite traditional: $\beta_0 = T_0^{-1}, \beta_{vi} = (T_{vi})^{-1}$.

A particularly interesting case is that of isotope mixtures that consist of a large fraction of a light gas component (higher frequency of molecular oscillations, concentration n_2) and only a small fraction of heavy component (lower oscillation frequency, concentration n_1). Usually:

$$\frac{n_2}{n_1} << \exp\left[-\frac{\hbar\delta_{VV}}{2x_e}(\omega_2 - \omega_1)\right] \tag{5.97}$$

In this case, the steady-state solution of the Fokker–Planck kinetic equation (Equation 5.92) with the relaxation fluxes (Equation 5.12, Equation 5.3, Equation 5.36, and Equation 5.93) gives the following vibrational distribution functions for two components of a gas mixture:

$$f_{1,2}(E) = B_{1,2}\exp\left[-\int_0^E \frac{\beta_{1,v2} - 2x_e^{(1,2)}\beta_0 \dfrac{E'}{\hbar\omega_{1,2}} + \tilde{\beta}_0^{(1,2)}\xi_{1,2} + \beta_e\eta_{1,2}}{1 + \xi_{1,2} + \eta_{1,2}}dE'\right] \tag{5.98}$$

Here, $B_{1,2}$ are the normalization factors; reverse electron temperature parameter $\beta_e = T_e^{-1}$; and factors ξ_i and η_l describe the relative contribution of VT and eV processes with respect to VV (*VV′*) exchange (see Section 5.1 and Section 5.2):

$$\xi_i(E) = \xi_i(0)\exp(2\delta_{VV}E), \quad \eta_i(E) = \eta_i(0)\exp(\delta_{VV}E) \tag{5.99}$$

5.4.2 Treanor Isotopic Effect in Vibrational Kinetics

VT and eV relaxation processes can be neglected ($\xi_i << 1, \eta_i << 1$) on the initial low-energy part of the vibrational distribution functions $f_{1,2}(E)$. In this case, Equation 5.98 gives, for both gas components ($i = 1,2$), the Treanor vibrational distribution functions (Equation 5.10) with obviously the same translational temperature T_0, but with different vibrational temperatures T_{v1} and T_{v2}:

$$\frac{\omega_1}{T_{v1}} - \frac{\omega_2}{T_{v2}} = \frac{\omega_1 - \omega_2}{T_0} \tag{5.100}$$

This relation is known as the **Treanor formula for isotopic mixture**. The Treanor formula shows that under nonequilibrium conditions ($T_{v1,v2} > T_0$) of isotopic mixture, the component with the lower oscillation frequency (heavier isotope) has a higher vibrational temperature. In this case, the predominant VV transfer of vibrational energy from the molecules with larger vibrational quanta to those with smaller quanta

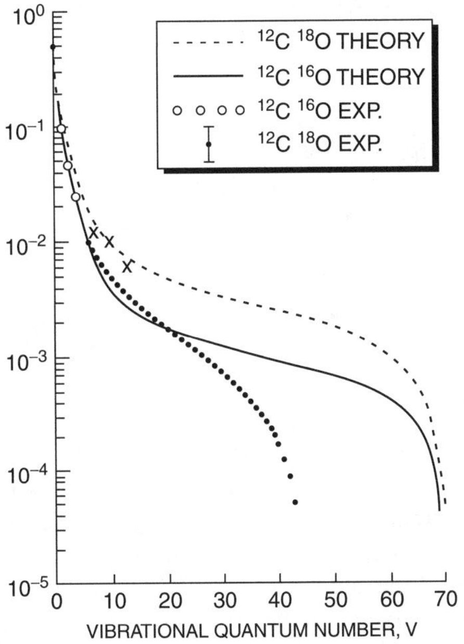

Figure 5.12 Isotopic effect in vibrational population for $^{12}C^{18}O/^{12}C^{16}O$ mixture (From Capitelli, M. 1986. In *Topics in Current Physics*. Capitelli, M., Ed. Berlin: Springer–Verlag)

is similar to the main Treanor effect (Section 4.1.10) providing overpopulation of highly excited states with lower values of vibrational quanta.

Experimental illustration of this Treanor effect in the isotopic gas mixture of $^{12}C^{16}O$ and $^{12}C^{18}O$ is presented in Figure 5.12.[138] This figure shows that the difference in relative population of highly excited vibrational states can be significant. The Treanor effect can be applied for isotope separation in plasma chemical reactions stimulated by vibrational excitation, as first suggested by Belenov et al.[139] The ratio of rate coefficients of chemical reactions of two different vibrationally excited isotopes $\kappa = k_R^{(1)}/k_R^{(2)}$ is proportional to the ratio of the population of vibrational levels $E \geq E_a$ for the two isotopes (see Section 5.3). This so-called coefficient of selectivity in cases of slow and fast reactions can be expressed as:

$$\kappa = \frac{f_1(E_a)\,\Delta E_1}{f_2(E_a)\,\Delta E_2} \tag{5.101}$$

Here, $\Delta E_1 \approx \Delta E_2$ are the parameters of the $f_{1,2}(E)$ exponential decrease at $E = E_a$. Assuming the Treanor functions for vibrational distributions $f_{1,2}(E)$ of both gas components, the coefficient of selectivity can be expressed as:

$$\kappa \approx \exp\left[\frac{\Delta\omega}{\omega} E_a \left(\frac{1}{T_0} - \frac{1}{T_{v2}}\right)\right] \tag{5.102}$$

where the relative defect of resonance

$$\frac{\Delta\omega}{\omega} = \frac{\omega_2 - \omega_1}{\omega_2}.$$

It is very interesting that the coefficient of selectivity does not depend on the vibrational temperature in the nonequilibrium conditions ($T_v > T_0$), but depends only on the translational temperature T_0, the relative defect of resonance $\Delta\omega/\omega$, and activation energy E_a.

Another interesting remark is related to the direction of the isotopic effect in vibrational kinetics, which is opposite to that observed in quasi-equilibrium chemical kinetics. As discussed previously after Equation 3.5, the dissociation energy and, similarly, the activation energy, are sensitive to the "zero-vibration level" $1/2\ \hbar\omega$. Heavy isotopes with the lower value of $1/2\ \hbar\omega$ have higher activation (and dissociation) energies and, as a result, their quasi-equilibrium reactions are slower. This is the conventional isotopic effect. Conversely, in vibrational kinetics, heavy isotopes react faster due to the Treanor effect. Thus, the isotopic effect in vibrational kinetics is stronger and has opposite direction with respect to the conventional one.

Numerical values of the coefficient of selectivity for different plasma chemical processes of isotope separation, stimulated by vibrational excitation of the molecules-isotopes, are presented in Figure 5.13. The coefficients of selectivity are shown in the figure in a convenient form as the function (Equation 5.102) of the defect of mass $\Delta m/m$ and the process activation energy. As can be seen, the selectivity for light molecules obviously can be very high. The related detailed calculations of isotope separation using the Treanor effect were carried out for nitrogen and carbon monoxide isotopes[140] and for hydrogen-isotopes.[141.142]

5.4.3 Influence of VT Relaxation on Vibrational Kinetics of Mixtures: Reverse Isotopic Effect

Next, take into account the contribution of VT relaxation in the isotopic effect, but still neglect the direct influence of the eV processes. The direct influence of eV processes in the general expression (Equation 5.98) for the distribution functions in mixtures at all vibrational energies can be neglected if the ionization degree in plasma is relatively low:

$$\left(\frac{n_e}{n_0}\right)^2 << \frac{k_{VV'}(0)k_{VT}(0)}{k_{eV}^2} \qquad (5.103)$$

Here, $n_0 = n_1 + n_2 \approx n_2$; the total concentration of the gas isotope, k_{eV}, $k_{VT}(0)$, $k_{VV'}(0)$, are, respectively, the rate coefficients of eV relaxation, and VT and VV′ relaxation processes at low vibrational levels. At the not very high electron concentrations (Equation 5.103), the eV factor $\eta_{1,2}(E)$ can be neglected and the integral (Equation 5.98) can be taken:

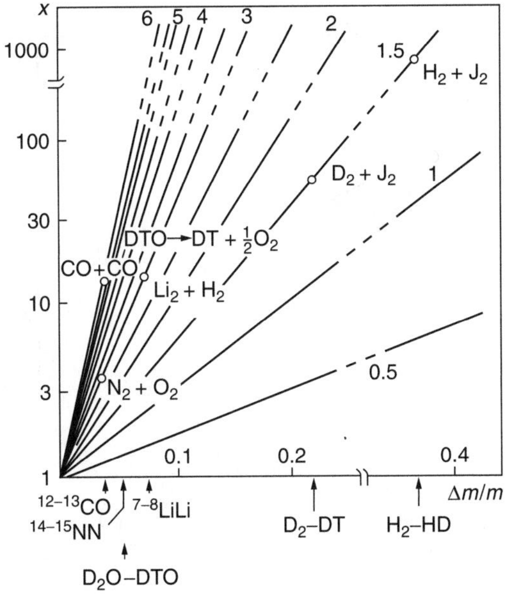

Figure 5.13 Isotopic effect for different molecules. Numbers on curves represent activation energies of specific reactions

$$f_{1,2}(E) = B_{1,2} \exp\left[-\beta_{1,v2} E + \frac{x_e^{(1,2)} \beta_0 E^2}{\hbar w_{1,2}} - \frac{\tilde{\beta}_0^{(1,2)}}{2\delta_{VV}} \ln\left(1 + \xi_{1,2}\right) \right] \tag{5.104}$$

The vibrational distribution functions (Equation 5.104) for two isotopes are illustrated in Figure 5.14. As can be seen from the figure, the population $f_1(E)$ of the same low vibrational levels is higher for the heavier isotope ("1," usually small additive). It obviously corresponds to the Treanor effect and Equation 5.100. However, the situation becomes completely opposite at higher levels of excitation, where the vibrational population of the relatively light isotope exceeds the population of the heavier one. This phenomenon is known as the reverse isotopic effect in vibrational kinetics.[136]

The physical basis of the reverse isotopic effect is quite clear. The vibrational distribution function $f_1(E)$ for the heavier isotope (small additive) is determined by the VV′ exchange, which is slower than VV exchange because of the defect of resonance. As a result, the VT relaxation makes the vibrational distribution $f_1(E)$ start falling at lower energies $E_1(\xi_1 = 1)$ with respect to the distribution function $f_2(E)$ of the main isotope, which is determined by VV exchange and starts falling at the higher vibrational energy $E_2(\xi_2) > E_1$. From Equation 5.104:

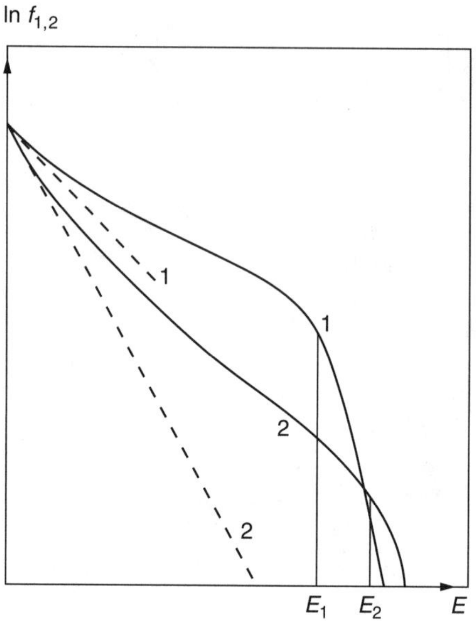

Figure 5.14 Vibrational distribution functions for two isotopes (1, 2) without significant influence of eV processes; dashed lines represent Boltzmann distribution

$$E_1 = \frac{1}{2\delta_{VV}} \ln \frac{k_{VV}(0)}{k_{VT}(0)} - \frac{\hbar\Delta\omega}{4x_e}, \quad E_2 = \frac{1}{2\delta_{VV}} \ln \frac{k_{VV}(0)}{k_{VT}(0)} \tag{5.105}$$

Thus, the reverse isotopic effect takes place if $E_1 < E_a < E_2$ (see Figure 5.14) and the light isotope is excited more strongly and reacts faster than the heavier one. The coefficient of selectivity for the reverse isotopic effect can be calculated from Equation 5.104 and Equation 5.105[136] and expressed as:

$$\kappa \approx \exp\left(-\frac{\Delta\omega}{\omega}\frac{D_0}{T_0}\right) \tag{5.106}$$

Here D_0 is the dissociation energy of the diatomic molecules in the manner as that for the direct effect (Equation 5.102). The selectivity coefficient (Equation 5.106) for the reverse isotopic effect does not depend directly on vibrational temperature. It is even more interesting that, although the coefficient of selectivity was derived for a chemical reaction with activation energy $E_a < D_0$, the activation energy E_a is not explicitly presented in Equation 5.106. Taking into account $E_a < D_0$, it can be seen that the reverse isotopic effect is much stronger than the direct one. The reverse isotopic effect can be achieved in a narrow range of translational temperatures:

$$\frac{\Delta T_0}{T_0} = 2\frac{\Delta\omega}{\omega}\left(1 - \frac{E_a}{D_0}\right)^{-1} \tag{5.107}$$

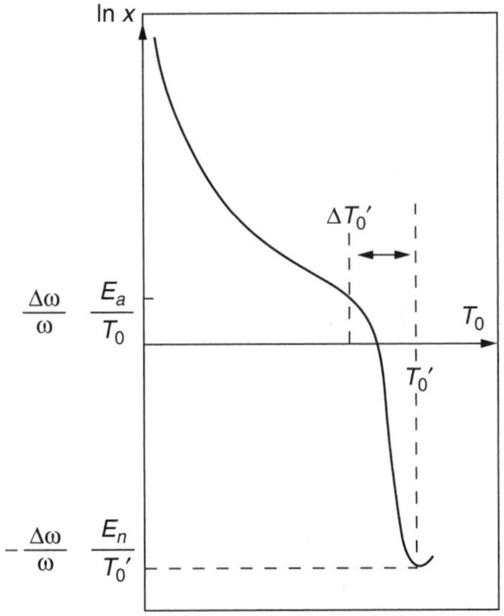

Figure 5.15 Isotopic effect dependence on translational temperature T_0; temperature T_0' corresponds to maximum value of the inverse isotopic effect coefficient

The selectivity coefficient dependence on translational temperature is illustrated in Figure 5.15. As one can see from the figure, the direct effect takes place at relatively low translational temperatures. By increasing the translational temperature, one can find the narrow temperature range (Equation 5.107) of the reverse effect, where the isotopic effect changes direction and becomes much stronger.

5.4.4 Influence of eV Relaxation on Vibrational Kinetics of Mixtures and Isotopic Effect

Consider the vibrational kinetics of the isotopic mixture in plasma with high degrees of ionization, when the inequality opposite to Equation 5.103 is valid. In this case, the vibrational distribution functions (at not very high energies) are controlled by VV (VV′) and eV relaxation processes. These distribution functions for two isotopes can be found by integrating Equation 5.98 as:

$$f_{1,2}(E) = B_{1,2} \exp\left[-\beta_{1,v2} E + \frac{x_e^{(1,2)} \beta_0 E^2}{\hbar w_{1,2}} + \frac{\tilde{\beta}_{1,v2} - \beta_e}{\delta_{VV}} \ln\left(1 + \eta_{1,2}\right) \right] \quad (5.108)$$

The vibrational distribution functions for two isotopes are illustrated in Figure 5.16. As can be seen from the figure, at relatively high vibrational energies, when eV processes are dominating, $f_{1,2}(E) \propto \exp(-\beta_e E)$, which can be considered as the plateau on the vibrational distribution (see Section 5.2). The VV′ exchange processes are slower than VV exchange because of the defect of resonance. For this reason,

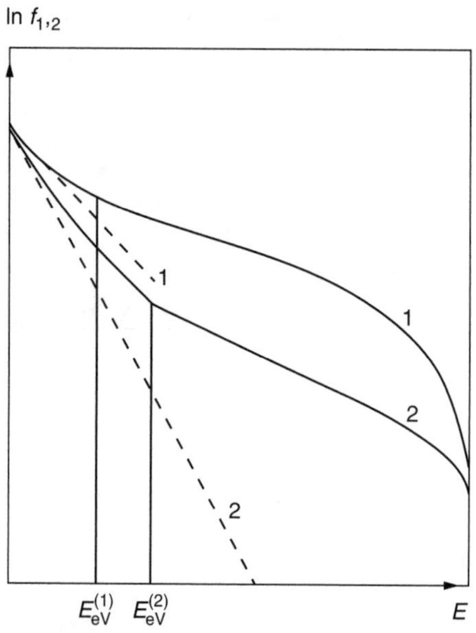

Figure 5.16 Vibrational distribution functions for two isotopes (1, 2) with significant influence of eV processes; dashed lines represent Boltzmann distribution

the transition from the Treanor distribution to the eV plateau (see Figure 5.16) takes place for heavy isotopes at lower vibrational energies:

$$E_{eV}^{(2)} - E_{eV}^{(1)} = \frac{\hbar\,\Delta\omega}{2x_e} \tag{5.109}$$

As a result, the direct isotopic effect can be higher in this case than in Equation 5.102, which does not take into account eV processes.

The coefficient of selectivity (κ) dependence on the plasma degree of ionization is presented in Figure 5.17. At low degrees of ionization, the coefficient of selectivity obviously corresponds to Equation 5.102. Then, because of this effect, κ grows with the degree of ionization in the plasma. The selectivity coefficient reaches the high value:

$$\kappa \approx \exp\left(\frac{\Delta\omega}{\omega}\frac{E_{eV}^{(1)}}{T_0}\right) \tag{5.110}$$

when the activation energy E_a corresponds to the eV plateau of the vibrational distribution functions for both isotopes (see Figure 5.16). In this relation, the vibrational energy $E_{eV}^{(1)}$ corresponds to the vibrational distribution transition from the Treanor function to the eV plateau. Obviously, the vibrational energy $E_{eV}^{(1)}$ is reversibly proportional to n_e/n_0, and thus the selectivity coefficient also decreases with the

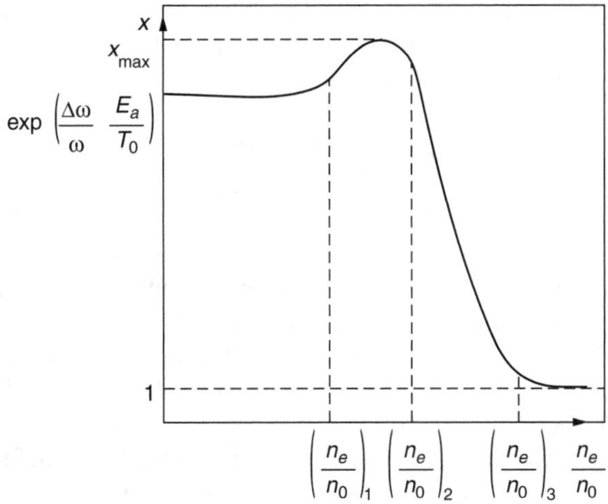

Figure 5.17 Isotopic effect coefficient dependence on ionization degree

degree of ionization. The maximum coefficient of selectivity takes place at the following degree of ionization:

$$\frac{n_e}{n_0} \approx \frac{k_{VV}(0)}{k_{eV}} \exp(-\delta_{VV} E_a) \tag{5.111}$$

At higher electron concentrations, both vibrational distributions follow the Boltzmann functions with temperature T_e from the relatively low vibrational levels; the isotopic effect decreases with the ionization and finally disappears (see Figure 5.17).

5.4.5 Integral Effect of Isotope Separation

The preceding selectivity coefficients described the ratio of rate coefficients of chemical reactions for different molecules–isotopes. Practical calculations require, however, determination of the separation coefficient:

$$R = \frac{n_1/n_2}{(n_1/n_2)_0} \tag{5.112}$$

This integral coefficient of isotope separation determines the change of molar fractions of different isotopes in mixtures and, thus, describes the effect of isotope enrichment in a system. Calculation of this coefficient is more complicated in general and requires all details of the chemical process. However, a relatively simple expression for the separation coefficient can be found if the main channels of VT relaxation and chemical reaction are related to the same molecules.[136] Such a situation takes place in the plasma chemical systems $H_2 - J_2$ and $N_2 - O_2$, considered for separation of hydrogen and nitrogen isotopes in reactions stimulated by vibrational excitation.

The particle balance equation in these mentioned kinetic systems (assuming the small relative difference of isotope mass $\Delta m/m << 2T_0/E_a$) gives the following expression for the integral coefficient of isotope separation:

$$R \approx \exp\left[\frac{1}{P_{VT}(T_0)}\exp\left(-\frac{E_a}{T_{v1}}\right) * \frac{E_a}{2T_0}\frac{\Delta m}{m}\right] \quad (5.113)$$

Here, $P_{VT}(T_0)$ is the averaged probability of VT relaxation. According to Equation 5.113, the coefficient of isotope separation R depends on the vibrational temperature much more strongly than on the mass defect $\Delta m/m$. This provides the possibility to reach high values of the isotope separation coefficient R even at low values of the mass defect $\Delta m/m$, e.g., for separation of heavy molecular isotopes of special importance. However, the strongest isotopic effect of that kind can be reached for the separation of light isotopes. An important example is H_2–HD, where the selectivity in reactions $H_2 + J_2 \rightarrow 2HJ$ and $H_2 + Br_2 \rightarrow 2HBr$ reaches 1000.

5.5 KINETICS OF ELECTRONICALLY AND ROTATIONALLY EXCITED STATES: NONEQUILIBRIUM TRANSLATIONAL DISTRIBUTIONS: RELAXATION AND REACTIONS OF "HOT" ATOMS IN PLASMA

Although the vibrationally excited molecules considered previously play a very important role in plasma chemical kinetics, nonequilibrium excitation of other degrees of freedom can make significant contributions as well. This contribution, however, is less general and usually depends on specific details of the plasma chemical process. In this section, some special features of kinetics of the electronically and rotationally excited species and high translational energy ("hot") atoms in nonequilibrium plasma will be discussed.

5.5.1 Kinetics of Population of Electronically Excited States: Fokker–Planck Approach

Transfer of electronic excitation energy in collisions of heavy particles is effective in contrast to VV exchange only for a limited number of specific electronically excited states (see Chapter 3). Even for high levels of electronic excitation, the transitions between electronic states are mostly due to collisions with plasma electrons at degrees of ionization exceeding 10^{-6}. In relatively low-pressure plasma systems, radiation transition can be significant as well.[94] Models describing the population of electronically excited species in plasma have been developed.[143,144] Description of the highly electronically excited states can be accomplished in the framework of the Fokker–Planck diffusion approach[145] similar to that considered for vibrational excitation.

The possibility to apply the diffusion in the energy space approach is due to: (1) low energy intervals between the highly electronically excited states and (2) transitions occurring mostly between close levels in the relaxation collisions with plasma electrons. The population of these highly electronically excited states in plasma, $n(E)$, due to energy exchange with electron gas, can then be found in the framework of the diffusion approach from the Fokker–Planck kinetic equation similar to the kinetic equation (Equation 5.1) for vibrationally excited states:

$$\frac{\partial n(E)}{\partial t} = \frac{\partial}{\partial E}\left[D(E)\left(\frac{\partial n(E)}{\partial E} - \frac{\partial \ln n^0}{\partial E} n(E)\right)\right] \tag{5.114}$$

In this relation, $n^0(E)$ is the quasi-equilibrium population of the electronically excited states, corresponding to the electron temperature T_e:

$$n^0(E) \propto E^{-\frac{5}{2}} \exp\left(-\frac{E_1 - E}{T_e}\right) \tag{5.115}$$

Here, E is the absolute value of the bonded electron energy; transition to continuum corresponds to the zero electron energy $E = 0$; and E_1 is the ground state energy ($E_1 \geq E$). Note that for $E >> T_e$, $\partial \ln n^0(E)/\partial E = 1/T_e$ in Equation 5.1. The diffusion coefficient $D(E)$ in energy space, related to the energy exchange between bonded electrons of highly electronically excited particles with plasma electrons, can be expressed as:[97a]

$$D(E) = \frac{4\sqrt{2\pi}\, e^4 n_e E}{3\sqrt{mT_e}\,(4\pi\varepsilon_0)^2} \Lambda \tag{5.116}$$

where Λ is the Coulomb logarithm for the electronically excited state with ionization energy E.

5.5.2 Simplest Solutions of Kinetic Equation for Electronically Excited States

To solve the kinetic equation in quasi-steady-state conditions, it is convenient to introduce a new variable, the relative dimensionless population of electronically excited states:

$$y(E) = \frac{n(E)}{n^0(E)} \tag{5.117}$$

Boundary conditions for Equation 5.114 can be taken as $y(E_1) = y_1$, $y(0) = y_e y_i$. Parameters: y_e, y_i are the electron and ion densities in plasma, divided by the corresponding equilibrium values. Then the quasi-steady-state solution of the kinetic equation (Equation 5.114) can be expressed in the following way:

$$y(E) = \frac{y_1 \chi\left(\frac{E}{T_e}\right) + y_e y_i \left[\chi\left(\frac{E_1}{T_e}\right) - \chi\left(\frac{E}{T_e}\right)\right]}{\chi\left(\frac{E_1}{T_e}\right)} \tag{5.118}$$

Here, $\chi(x)$ is a function determined by the integral:

$$\chi(x) = \frac{4}{3\sqrt{\pi}} \int_0^x t^{\frac{3}{2}} \exp(-t)\, dt \tag{5.119}$$

and having the following asymptotic approximations:

$$\chi(x) \approx 1 - \frac{4}{3\sqrt{\pi}} e^{-x} x^{3/2}, \quad if\ x >> 1 \tag{5.120}$$

$$\chi(x) \approx \frac{1}{2\sqrt{\pi}} x^{5/2}, \quad if\ x << 1 \tag{5.121}$$

For electronically excited levels with energies $E << T_e << E_1$ close to continuum, this gives the relative population:

$$y(E) \approx y_e y_i \left[1 - \frac{1}{2\sqrt{\pi}}\left(\frac{E}{T_e}\right)^{5/2}\right] + y_1 \frac{1}{2\sqrt{\pi}}\left(\frac{E}{T_e}\right)^{5/2} \to y_e y_i \tag{5.122}$$

The population of electronically excited states (Equation 5.117 and Equation 5.122) is decreasing exponentially with the effective Boltzmann temperature T_e and the absolute value corresponding to equilibrium with continuum $y(E) \to y_e y_i$. For the opposite case $E >> T_e$, the population of electronically excited states far from continuum can be found as:

$$y(E) = y_1 + y_e y_i \frac{4}{3\sqrt{\pi}} \exp\left(-\frac{E}{T_e}\right) \cdot \left(\frac{E}{T_e}\right)^{3/2} \to y_1 \tag{5.123}$$

In this range of the electronic excitation energy (far from continuum), the population is also exponential with effective temperature T_e, but the absolute value corresponds to equilibrium with the ground state.

Note that, assuming diffusion of neutral particles in plasma along the energy spectrum, the Fokker–Planck approach is much less accurate in describing the population of lower electronic levels, which are quite remote from one another. The more accurate, "modified" diffusion approach, including discrete consideration of the lower levels, was developed by Biberman et al.[146] Practically, the Boltzmann distribution of electronically excited states with the temperature equal to the temperature of plasma electrons requires a very high degree of ionization in plasma $n_e/n_0 \geq 10^{-3}$ (although domination of energy exchange with electron gas requires only $n_e/n_0 \geq 10^{-6}$). This is mostly due to the influence of some resonance transitions and the non-Maxwellian behavior of electron energy distribution function at the lower degree of ionization.

The radiative deactivation of electronically excited particles is required at low pressures, usually when $p < 1 - 10$ torr. Contribution of the radiative processes decreases with the growth of excitation energy, e.g., approaching continuum. This can be explained by the reduction of the intensity of the radiative processes and, conversely, intensification of collisional energy exchange when the electron bonding energy in the atom becomes smaller. It follows that the numerical formula for a critical value of the electron bonding energy applies:

$$E_R, eV = \left(\frac{n_e, cm^{-3}}{4.5 \cdot 10^{13} cm^{-3}} \right)^{1/4} * \left(T_e, eV\right)^{-1/8} \tag{5.124}$$

Collisional energy exchange dominates when the excitation level is higher and electron binding energy is lower ($E < E_R$). When the excitation level is not high and the bonding energy is significant ($E > E_R$), then even though the electronic excitation is collisional, the deactivation is mostly due to radiation.

5.5.3 Kinetics of Rotationally Excited Molecules: Rotational Distribution Functions

The fast elementary processes of RT and RR rotational relaxation were discussed in Section 3.6.1. The conclusion was that even in nonthermal discharges, the rotational and translational degrees of freedom are usually in equilibrium between themselves and can be characterized by the same temperature T_0. Now consider the kinetics and evolution of the rotational energy distribution functions in nonequilibrium plasma conditions.

For simplicity, consider the rotational and translational relaxation kinetics of a small admixture of relatively heavy diatomic molecules (m_{BC}) in a light inert gas (m_A), first described by Safarian and Stupochenko.[147] Low values of energy transferred per one collision permit using the Fokker–Planck kinetic equation in this case:

$$\frac{\partial f(E_t, E_r, t)}{\partial t} = \frac{\partial}{\partial E_t}\left[bE_t \left(\frac{\partial f}{\partial E_t} - f \frac{\partial \ln f^{(0)}}{\partial E_t} \right) \right] + \frac{\partial}{\partial E_r}\left[bE_r \left(\frac{\partial f}{\partial E_r} - f \frac{\partial \ln f^{(0)}}{\partial E_r} \right) \right] \tag{5.125}$$

In this relation, $f(E_t, E_r, t)$ is the distribution function related to translational E_t and rotational E_r energies; $f^{(0)}(E_t, E_r)$ is the equilibrium distribution function corresponding to temperature T_0 of the monatomic gas; and the diffusion coefficient in energy space is characterized by parameter:

$$b = \frac{32}{3} \frac{m_A}{m_{BC}} N_A T_0 \Omega_{col} \tag{5.126}$$

where N_A is the concentration of the monatomic gas particles (which is actually close to the total gas density) and Ω_{col} is the dimensionless collisional integral.

As can be seen, the kinetic equation (Equation 5.125) independently represents the rotational and translational relaxation processes. Then, the averaged rotational distribution function:

$$f(E_r,t)=\int_0^\infty f(E_t,E_r,t)dE_t \tag{5.127}$$

can be described by Equation 5.125, the simplified Fokker–Planck kinetic equation including only rotational energy:

$$\frac{\partial f(E_r,t)}{\partial t}=\frac{\partial}{\partial E_r}\left[bE_r\left(\frac{\partial f(E_r,t)}{\partial E_r}-\frac{1}{T_0}f(E_r,t)\right)\right] \tag{5.128}$$

Multiplication of Equation 5.128 by E_r and following integration from 0 to ∞ results in the macroscopic relation for the total rotational energy E_r^{total}:

$$\frac{dE_r^{total}}{dt}=\frac{E_r^{total}-E_{r,0}^{total}(T_0)}{\tau_{RT}},\quad \tau_{RT}=\frac{T_0}{b} \tag{5.129}$$

Here, τ_{RT} is the RT relaxation time and $E_r^{total}(T_0)$ is the equilibrium value of rotational energy at the inert gas temperature T_0. It is interesting to note that Equation 5.129 is valid for any kind of initial rotational distribution functions $f(E_r,t=0)$. General solution of the kinetic equation (Equation 5.128) at the arbitrary initial rotational distribution functions $f(E_r,t=0)$ can be presented as the series:

$$f(E_r,t)=\sum c_n\exp\left(-\frac{nt}{\tau_{RT}}\right)\cdot L_n\left(\frac{E_r}{T_0}\right)\cdot\exp\left(-\frac{E_r}{T_0}\right) \tag{5.130}$$

where $L_n(x)$ are the Laguerre polynomials.

Now take the initial rotational distribution corresponding to the Boltzmann function:

$$f(E_r,t)=\frac{N_{BC}}{T_R(t)}\exp\left[-\frac{E_r}{T_r(t)}\right] \tag{5.131}$$

with the initial rotational temperature $T_r(t=0)\neq T_0$. Then the solution of kinetic Equation 5.128 shows that the rotational distribution function $f(E_r, t)$ maintains the same form of the Boltzmann distribution (Equation 5.131) during the relaxation to equilibrium, but obviously with changing value of the rotational temperature approaching the translational one:

$$T_r(t)=T_0+\left[T_r(t=0)-T_0\right]\exp\left(-\frac{t}{\tau_{RT}}\right) \tag{5.132}$$

This important property of always maintaining the same Boltzmann distribution function during relaxation to equilibrium (with only temperature change) is usually referred to as **canonical invariance**.[148]

5.5.4 Nonequilibrium Translational Energy Distribution Functions: Effect of "Hot Atoms"

Translational energy distributions of plasma electrons are often very far from the equilibrium Maxwellian function (see Section 4.2.5). This is due to the relatively low electron density and thus the low frequency of electron–electron collisions (necessary for their Maxwellization) with respect to the frequencies of other processes creating the electron energy distribution function.

The situation with neutral particles in weakly ionized plasma is very different because their density is much higher. Relaxation of the translational energy of neutral particles requires only a couple of collisions and usually determines the shortest time scale in plasma chemical systems. As a result, the assumption of a local quasi equilibrium is usually valid for the translational energy subsystem of neutral particles, even in strongly nonequilibrium plasma chemical systems. The translational energy distributions in most discharge conditions are Maxwellian with one local temperature T_0 for all neutral components participating in plasma chemical processes.

However, this general rule has some important exceptions because of the possible formation of high-energy neutral particles (usually atoms) in plasma that strongly perturb the conventional Maxwellian distribution. Generation of the high-energy hot atoms can be related to fast exothermic chemical reaction or to fast vibrational relaxation processes, if their frequencies can somewhat exceed the very high frequency of Maxwellization. Consider separately these two possible sources of hot atoms in plasma chemical systems.

5.5.5 Kinetics of Hot Atoms in Fast VT Relaxation Processes: Energy–Space Diffusion Approximation

Different elementary processes of the fast, nonadiabatic VT relaxation were discussed in Section 3.4.5. Probably the fastest of these are the relaxation processes of molecules, Mo, on alkaline atoms, Me, proceeding through intermediate formation of ionic complexes $[Me^+Mo^-]$. Thus, molecular nitrogen N_2^* vibrational relaxation on atoms of Li, K, and Na actually takes place at each collision, which makes these alkaline atoms hot when they are added to the nonequilibrium nitrogen plasma ($T_v >> T_0$).

Consider the evolution of the translational energy distribution function $f(E)$ of a small admixture of alkaline atoms into a nonequilibrium molecular gas ($T_v >> T_0$)—for simplicity, a diatomic gas. Under these conditions, the translational energy distribution is determined by the kinetic competition of the fast VT relaxation energy exchange between the alkaline atoms and the diatomic molecules and the Maxwellization TT processes in collisions of the same partners. The exponential part of the steady-state distribution function $f(E)$ then can be found from the Fokker–Planck kinetic equation describing the atoms' diffusion along the translational energy spectrum:[149]

$$D_{VT}\left(\frac{\partial f}{\partial E}+\frac{f}{T_v}\right)+D_{TT}\left(\frac{\partial f}{\partial E}+\frac{f}{T_0}\right)=0 \tag{5.133}$$

Obviously, this continuous approach can be applied only for high energies: $E \gg \hbar\omega$. Consideration of lower energies requires more complicated discrete models (see the next section). In Equation 5.133, D_{VT} and D_{TT} are the diffusion coefficients of the alkaline atoms, Me, along the energy spectrum related to VT relaxation and Maxwellization, respectively. The translational energy distribution function of the alkaline atoms $f(E)$ depends upon the ratio of the diffusion coefficients $\mu(E) = D_{TT}(E)/D_{VT}(E)$, which is proportional to ratio of the corresponding relaxation rate coefficients. Integration of the linear differential equation (Equation 5.133) gives:

$$f(E)=B\exp\left[-\int\frac{\frac{1}{T_v}+\frac{\mu(E)}{T_0}}{1+\mu(E)}dE\right] \tag{5.134},$$

where B is the pre-exponential normalization factor.

In general, the Maxwellization is much faster than VT relaxation and $\mu(E) \gg 1$. According to Equation 5.134, this means that the distribution $f(E)$ is Maxwellian with temperature equal to T_0, the translational temperature of the molecular gas. However, the situation can be very different for the admixture of light alkaline atoms (e.g., Li, atomic mass m) into a relatively heavy molecular gas (e.g., N_2 or CO_2, molecular mass $M \gg m$), taking into account that the VT and TT relaxation frequencies are almost equal for this mixture. Because of the large difference in masses, the energy transfer during Maxwellization can be lower than $\hbar\omega$ at low energies $\mu \ll 1$. In accordance with Equation 5.134, this means that the translational temperature of the light alkaline atoms can be equal, not to the translational T_0, but rather to the vibrational T_v (!) temperature of the main molecular gas (see illustration in Figure 5.18).

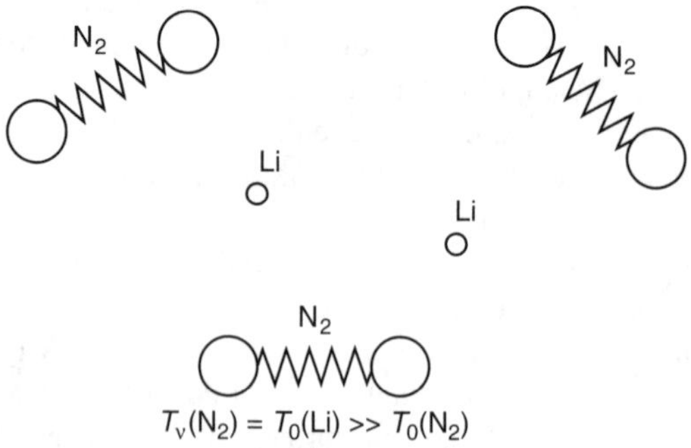

Figure 5.18 Equilibrium between N_2 vibrational degrees of freedom and Li translational degrees of freedom

For further calculation of the distribution function, the factor $\mu(E)$,when $m \ll M$, can be expressed as:

$$\mu(E) \approx \frac{E+T_v}{\hbar\omega}\left(\frac{m}{M}\frac{E}{T_v}\right)^2 \tag{5.135}$$

Taking Equation 5.134 and Equation 5.135 into account, it can be seen that the translation distribution function $f(E)$ can be Maxwellian with vibrational temperature T_v only at relatively low translational energies $E < E^*$. At higher energies $E > E^*$, the factor is always large $\mu \gg 1$ and the exponential decrease of $f(E)$ always corresponds to the temperature T_0. The critical value of the translational energy $E^*(T_v)$ can be found from Equation 5.134 and Equation 5.135 as:

$$E^*(T_v) = \frac{M}{m}\sqrt{T_v\hbar\omega}, \quad if\ T_v > \hbar\omega\left(\frac{M}{m}\right)^2 \tag{5.136a}$$

$$E^*(T_v) = (\hbar\omega)^{1/3}\left(\frac{T_v M}{m}\right)^{2/3}, \quad if\ T_v < \hbar\omega\left(\frac{M}{m}\right)^2 \tag{5.136b}$$

Note that the visible $f(E)$ decrease with temperature T_v is possible only at $E > T_v$, which requires $E^* > T_v$. Considering Equation 5.136, it can be seen that alkaline atoms can have "vibrational" temperature only if this temperature is not very high:

$$T_v < \hbar\omega\left(\frac{M}{m}\right)^2 \tag{5.137}$$

5.5.6 Hot Atoms in Fast VT Relaxation Processes: Discrete Approach and Applications

The Fokker–Planck energy-space diffusion approximation applied before is valid only for energy intervals exceeding $\hbar\omega$. Determination of $f(E)$ on the smaller scale $E < \hbar\omega$ is more complicated and requires discrete consideration. In the most interesting conditions when $\mu \ll 1$ and $f(E)$ are dominated mostly by the process of VT relaxation ($T_0 \ll \hbar\omega$, $m \ll M$), the distribution function is actually almost discrete; the only possible translational energies are $0 \cdot \hbar\omega$, $1 \cdot \hbar\omega$, $2 \cdot \hbar\omega$, …. The Maxwellization process (TT relaxation) and the partial energy transfer into molecular rotation make the translational distribution function more smooth.

Quite sophisticated mathematical consideration of the discrete random motion of the atoms along the translational energy spectrum results in the following analytical distribution function at $E \le \hbar\omega$:[149]

$$f(E) = \frac{2}{\sqrt{\pi\frac{M}{m}\left(1-\frac{E}{\hbar\omega}\right)}}\left\{1-\Phi\left[\frac{\hbar\omega}{2T_v}\sqrt{\frac{M}{m}\left(1-\frac{E}{\hbar\omega}\right)}\right]\right\}\exp\left[\left(\frac{\hbar\omega}{2T_v}\right)^2\frac{M}{m}\left(\frac{E}{\hbar\omega}-1\right)\right] \tag{5.138}$$

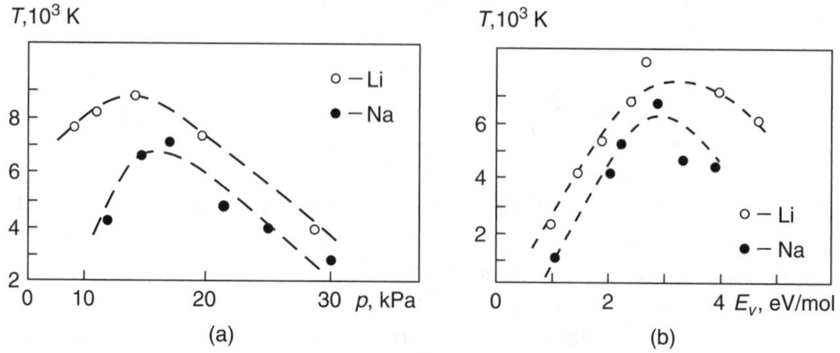

Figure 5.19 Li and Na atoms' temperature in CO_2 microwave discharge as function of (a) pressure (energy input of 3 J/cm^3); (b) energy input (pressure of 15.6 kPa)[134]

Here, $\Phi(x)$ is the normal distribution function. In this case, as one can see from Equation 5.138, the translational energy distribution $f(E)$ at $E \le \hbar\omega$ has nothing in common with any kind of Maxwellian distribution functions. Therefore, if $T_v < \hbar\omega$, the alkaline atoms' distribution cannot be described by the vibrational temperature.

From this statement and restriction (Equation 5.137), one can conclude that temperature of the light alkaline atoms corresponds to the vibrational temperature of molecular gas only if:

$$\hbar\omega \le T_v \le \hbar\omega\left(\frac{M}{m}\right)^2 \tag{5.139}$$

Within this framework, Doppler broadening measurements of the alkaline atoms can be used to determine the vibrational temperature of the molecular gas. According to Equation 5.139, this is possible only if $M/m \gg 1$. Furthermore, the heavier the alkaline atoms, the larger is the difference between their temperature and vibrational temperature of molecules.

This interesting method of diagnostics has been applied to measure the vibrational temperature of CO_2 in a nonequilibrium microwave discharge by adding a small amount of lithium and sodium atoms.[134,134a,134b] The Doppler broadening was observed by Fourier analysis of the Li spectrum line (transition $2^2s_{1/2}$ to $2^2p_{1/2}$, λ = 670.776 nm) and the Na spectrum line (transition $3^2s_{1/2}$ to $3^2p_{1/2}$, λ = 588.995 nm). Some results of these experiments are presented in Figure 5.19 to demonstrate the effect of hot atoms related to fast VT relaxation. Although the gas temperature in the nonthermal discharge was less than 1000 K, the alkaline temperature (Figure 5.19) was up to 10 times larger. Also, as can be seen from the figure, the sodium temperature (M/m = 1.9) was always lower than temperature of lithium atoms (M/m = 6.3). Similar measurements in nonequilibrium nitrogen plasma using in-resonator laser spectroscopy of Li, Na, and Cs additives were accomplished.[150]

It should be noted that the relaxation-induced effect of hot atoms could be observed not only in the case of alkaline atom additives. As discussed in Section 3.4.5, fast nonadiabatic VT relaxation with frequencies comparable to Maxwellization can be generated by symmetrical exchange reactions (Equation 3.65). Even the adiabatic Landau–Teller mechanism of vibrational relaxation can provide the hot atoms in some special conditions. This interesting effect is due to exponential acceleration of VT relaxation with the translational energy of atoms. Accelerated once, a group of atoms can support their high energy in fast Landau–Teller VT relaxation collisions because of their high velocity and thus low values of the Massey parameter.

5.5.7 Hot Atom Formation in Chemical Reactions

If fast atoms generated in exothermic reactions react again before Maxwellization, they are able to perturb the translational distribution function significantly. The formed group of hot atoms is able to accelerate the exothermic chemical reactions. The fast laser-chemical chain reaction stimulated in $H_2 - F_2$ mixtures by initial fluorine dissociation in an electron beam plasma:

$$H + F_2 \rightarrow HF^* + F, \quad F + H_2 \rightarrow HF^* + H \tag{5.140}$$

is a good example of such generation of hot atoms in fast exothermic chemical reactions.[151]

In nonequilibrium plasma chemical systems, the hot atoms can be generated in endothermic chemical reactions as well. For example, as shown in Section 5.3.2, vibrationally excited molecules participate in an endothermic reaction with some excess of energy, which depends on slope of the vibrational distribution function $f(E)$ and in average equals to:

$$\langle \Delta E_v \rangle = \left| \left[\frac{\partial \ln f}{\partial E}(E = E_a) \right]^{-1} \right| \tag{5.141}$$

This energy can be large and greatly exceeds the value of a vibrational quantum. This situation occurs primarily in the slow reaction limit (see Section 5.3.2).

Vibrationally excited molecules in nonthermal discharges are very effective in stimulating endothermic chemical reactions. However, exothermic elementary reactions with activation barriers are not effectively stimulated by molecular vibrations (see Section 3.7.3) and can retard the entire plasma chemical process. In this case, hot atoms can make a difference due to their high efficiency in stimulating exothermic processes. Plasma chemical NO synthesis in nonthermal air plasma is a good relevant example.

The NO synthesis in moderate atmospheric pressure and nonthermal air plasma can be effectively stimulated by vibrational excitation of N_2 molecules. The synthesis proceeds following the Zeldovich mechanism, including the chain reaction:[126]

$$\mathrm{O} + \mathrm{N_2}^* \rightarrow \mathrm{NO} + \mathrm{N}, \quad E_a^{(1)} = 3.2\ \mathrm{eV}, \quad \Delta H = 3.2\ \mathrm{eV} \tag{5.142}$$

$$\mathrm{N} + \mathrm{O_2} \rightarrow \mathrm{NO} + \mathrm{O}, \quad E_a^{(2)} = 0.3\ \mathrm{eV}, \quad \Delta H = -1.2\ \mathrm{eV} \tag{5.143}$$

The fast reverse reaction of NO destruction:

$$N + NO \rightarrow N_2 + O \tag{5.144}$$

has no activation energy and as an alternative reaction of atomic nitrogen strongly competes with the direct synthesis reaction (Equation 5.144). Comparing the rates of Equation 5.143 and Equation 5.144, it can be seen that competition between these reactions restricts the concentration ratio $[NO]/[O_2]$, and thus the degree of conversion of air into NO, because the reaction Equation 5.143 has activation energy and the reaction Equation 5.144 does not. Numerically, this prevents formation of more than approximately 0.5 to 1% of NO in nonequilibrium discharges with $T_v \approx 3000 - 5000$ K and $T_0 \approx 600 - 800$ K.

In many modern applications, the nonthermal discharges are applied to destroy, rather than to produce, NO to eliminate nitrogen oxides from air. It follows that the gas temperature as well as vibrational temperature should be decreased to provide domination of the reaction Equation 5.144. However, if the objective is to produce NO, the exothermic reaction (Equation 5.143) with activation energy 0.3 eV should be accelerated to overcome influence of the reverse process (Equation 5.144).

Vibrational excitation is not effective to stimulate the reaction (Equation 5.143), but the hot nitrogen atoms generated in Equation 5.142 with reasonably high energy (Equation 5.141) are able to accelerate (Equation 5.143) while barely influencing the rate of the reverse reaction.[151] The kinetic equation for the nitrogen atoms in this plasma chemical system can be solved considering their formation in endothermic reactions (Equation 5.142) stimulated by vibrational excitation; Maxwellization in TT relaxation; and their losses in chemical processes. Under the simplest conditions, the following distribution function is obtained:

$$f_N(E) \approx f_N^{(0)}(E) + \frac{\nu_R}{\nu_{\max}} f_{N_2}^{vib}\left(E_a^{(1)} + \frac{E}{\beta}\right) \tag{5.145}$$

Here, $f_N^{(0)}(E)$ is the nitrogen atoms' distribution function not perturbed by the hot atoms' effect; $\nu_R/\nu_{\max}$ is frequency ratio of Equation 5.134 to the Maxwellization; $f_{N_2}^{vib}(E)$ is vibrational distribution function of nitrogen molecules; $E_a^{(1)}$ is activation energy of the reaction; and coefficient β shows the fraction of the excess vibrational energy released in the reaction (Equation 5.134) and going to translational energy of a nitrogen atom.

Interpretation of the distribution function (Equation 5.145) is clear. The hot atoms are represented by the second term and actually copy the vibrational distribution function of nitrogen molecules after adjusting the activation barrier with the scaling factor β. The hot atoms are able to increase the maximum conversion of air into NO. This effect is shown in Figure 5.20 and obviously becomes stronger at higher oxygen fractions in the N_2–O_2 mixtures.[151] It should be noted that the effect of hot atoms is of special interest for plasma chemical processes in supersonic gas flows. In these systems, the gas temperature can be so low that several secondary exothermic processes are stopped even though they have very small activation energy.[152]

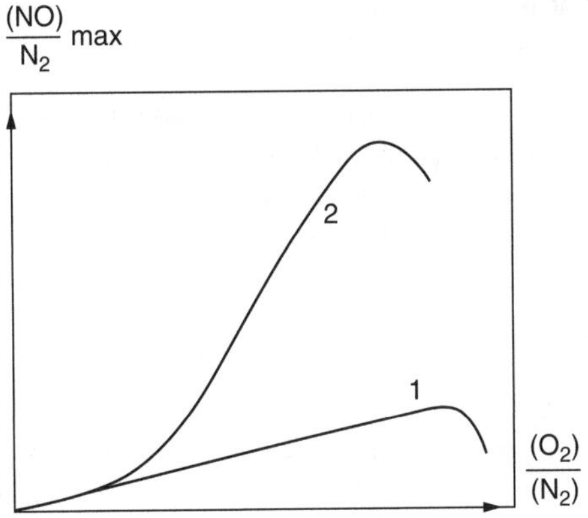

Figure 5.20 Maximum NO yield at low translational temperature ($T_0 < 1500$ K) (1) without and (2) with contribution of hot atom effect

5.6 ENERGY EFFICIENCY, ENERGY BALANCE, AND MACROKINETICS OF PLASMA CHEMICAL PROCESSES

One of the most important problems of applied plasma chemistry is the minimization of energy consumption. Plasma chemical processes mostly consume electricity, a relatively expensive form of energy. This imposes tough requirements on the energy efficiency of the processes in order to be competitive with other technologies using only heat or chemical sources of energy. The problem remains important, but somewhat less restrictive, when plasma is applied to unique technologies related to production of expensive products, e.g., in electronics. Requirements for energy efficiency are especially crucial when plasma technology is applied to such large-scale applications as chemical synthesis, fuel conversion, or emission control of industrial and automotive exhaust gases.

Energy cost and energy efficiency of a plasma chemical process are closely related to its mechanism. The same plasma chemical processes in different discharge systems or under different conditions (corresponding to different mechanisms) result in entirely different expenses of energy. For example, plasma chemical purification of air from small amounts of SO_2 using pulse corona discharge requires 50 to 70 eV/mol. The same process stimulated under special plasma conditions by relativistic electron beams requires about 1 eV/mol; thus, it requires two orders of magnitude less of electrical energy.[153]

The energy efficiency analysis is obviously unique for each specific plasma chemical process. This analysis will be done in relevant chapters related to specific plasma chemical processes. In this section, the general peculiarities of energy efficiency of plasma chemical processes provided through different channels or by different mechanisms will first be discussed. Then some general features of the energy balance of plasma chemical systems in relation with their macrokinetics will be considered.

5.6.1 Energy Efficiency of Quasi-Equilibrium and Nonequilibrium Plasma Chemical Processes

Energy efficiency is one of the most important characteristics of an endothermic plasma chemical process. The **energy efficiency** η is the ratio of the thermodynamically minimal energy cost of the plasma chemical process (which is usually the reaction enthalpy ΔH) to the actual energy consumption of the plasma W_{plasma}:

$$\eta = \Delta H / W_{plasma} \tag{5.146}$$

This definition of energy efficiency is most convenient during optimization of the reaction mechanisms occurring within a plasma. However, for more applied aspects of the problem, it is more convenient to consider the actual energy consumption, taking into account the efficiency of energy conversion from the source of electricity through the power supply into the plasma.[154]

The energy efficiency of the quasi-equilibrium plasma chemical systems organized in thermal discharges is usually relatively low (less than 10 to 20%) due to two major factors:

1. Thermal energy in the quasi-equilibrium plasma is distributed over all components and all degrees of freedom of the system, but most of them are useless in stimulating the plasma chemical reaction under consideration. Obviously, this makes products of chemical processes using thermal plasma relatively more expensive if heat recuperation is not arranged.
2. If the high-temperature gas generated in a thermal plasma and containing the process products is slowly cooled afterwards, the process products will be converted back into the initial substances by the reverse reactions. Conservation of products of the quasi-equilibrium plasma chemical processes requires applying quenching, e.g., very fast cooling of the product to avoid reverse reactions, and an important factor limiting energy efficiency.

Energy efficiency of quasi-equilibrium plasma chemical processes intrinsically depends on the type of quenching employed. Methods for calculating a product's energy cost will be considered for different types of quenching in the next section. Energy efficiency of the nonequilibrium plasma chemical systems can be much higher. If correctly organized, the nonthermal plasma chemical process's discharge power can be selectively focused on the chemical reaction of interest without heating the whole gas. Also, low bulk gas temperatures automatically provide product stability with respect to reverse reactions without any special quenching. However, it should be noted that energy efficiency in this case strongly depends on the specific mechanism of the process. The most important general mechanisms of nonequilibrium plasma chemical processes will be compared later with regard to their energy efficiency.

5.6.2 Energy Efficiency of Plasma Chemical Processes Stimulated by Vibrational Excitation of Molecules

This mechanism can provide the highest energy efficiency of endothermic plasma chemical reactions in nonequilibrium conditions due to four factors:

1. The major fraction (70 to 95%) of the discharge power in most molecular gases (including N_2, H_2, CO, CO_2, etc.) at $T_e \approx 1\,eV$ can be transferred from the plasma electrons to the vibrational excitation of molecules (see Figure 3.9 through Figure 3.16).
2. The rate of VT relaxation is usually low at low gas temperatures. For this reason, the optimal choice of the degree of ionization and the specific energy input permits spending most of the vibrational energy on stimulating chemical reactions.
3. The vibrational energy of molecules is the most effective in stimulating endothermic chemical processes (see Section 3.7.3).
4. The vibrational energy necessary for an endothermic reaction is usually equal to the activation barrier of the reaction and is much less than the energy threshold of the corresponding processes proceeding through electronic excitation. For example, the dissociation of H_2 through vibrational excitation requires 4.4 eV. The same process proceeding through excitation of electronically excited state ${}^3\Sigma_u^+$ requires more than twice the energy: 8.8 eV (see Figure 3.1a).

Some processes, stimulated by vibrational excitation, are able to consume most of the discharge energy. Probably the best example is the dissociation of CO_2, which can be arranged in nonthermal plasma with energy efficiency up to 90%.[155,156,156a,164] In this case, almost all of the discharged power is spent selectively on the chemical process. No other single mechanism can provide such high energy efficiency of a plasma chemical process.

5.6.3 Dissociation and Reactions of Electronically Excited Molecules and Their Energy Efficiency

Some principal channels of the physical organization of this mechanism of plasma chemical processes were considered in Section 3.3.7. None of the four kinetic factors previously listed that positively influence the energy efficiency can be applied to plasma chemical processes proceeding through electronic excitation. As a result, the energy cost of chemical products generated in this latter manner is relatively high. Numerically the energy efficiency of this type of plasma chemical processes usually does not exceed 20 to 30%. Plasma chemical processes stimulated by electronic excitation can be energy effective if they initiate chain reactions. For example, such a situation takes place in NO synthesis, where the Zeldovich mechanism (Equation 5.142 and Equation 5.143) can be effectively initiated by dissociation of molecular oxygen through electronic excitation.

5.6.4 Energy Efficiency of Plasma Chemical Processes Proceeding through Dissociative Attachment

The dissociative attachment was discussed in Section 2.4.1 as a process essentially influencing the balance of charged particles. Now consider the dissociative attachment of electrons as a plasma chemical process. The energy threshold of the dissociative attachment is lower than the threshold of dissociation into neutrals (see Section 2.4.1). When the electron affinity of the products is large (e.g., some halogens and their compounds; see Table 2.2), the dissociative attachment can even be exothermic and has no energy threshold at all. This means not only a low energy price for the reaction, but also the transfer of the important part of electron energy into dissociative attachment (see Section 3.3.8). These are very positive factors in obtaining low energy cost of products through the dissociative attachment.

However, the energy efficiency of the dissociative attachment is strongly limited by the energy cost of producing an electron lost during attachment and following ion–ion recombination. The energy cost of an electron, (e.g., the cost of ionization) can be easily calculated from Figure 3.6 through Figure 3.16 and numerically is approximately 30 to 100 eV (at high electron temperatures typical for strongly electronegative gases, it is about 30 eV).

Consider as an example the plasma chemical dissociation of NF_3. This strongly electronegative gas can be dissociated almost completely into atomic fluorine and molecular nitrogen in low-pressure nonthermal discharges. The process is promoted by dissociative attachment with an energy price of 30 eV/mol, which exactly corresponds to the energy cost of an electron in the NF_3 discharge.[157]

The energy price of dissociative attachment controlled by the cost of an electron in an electric discharge (about 30 to 100 eV) is obviously high and much exceeds the dissociation energy. Thus, the plasma chemical process based on dissociative attachment can only become really energy effective if the same electron, generated in plasma, is able to participate in the reaction many times. In such chain reactions, the detachment process and liberation of the electron from negative ion should be faster than the loss of the charged particle in ion–ion recombination.

As an example of such a chain process, consider the dissociation of water in a nonthermal plasma, which, at high degrees of ionization, can be arranged as a chain reaction:[158]

$$e + H_2O \rightarrow H^- + OH \tag{5.147}$$

$$e + H^- \rightarrow H + e + e \tag{5.148}$$

The first reaction in the chain is the dissociative attachment and the second is the electron impact detachment process (see Section 2.4.5 and Figure 2.9). The rate coefficient of the detachment is fairly large and, at electron temperatures of about 2 eV, can reach 10^{-6} cm^3/sec. The energy efficiency of this chain process of the plasma chemical water dissociation can be relatively high (40 to 50%; see Figure 5.24). However, achieving such energy efficiency requires very high values of the degree of ionization, typically exceeding 10^{-4}.[158] The required high value of the degree of ionization is due to the kinetic competition of the detachment reaction (Equation 5.148) with the fast ion–molecular reaction:

$$H^- + H_2O \rightarrow H_2 + OH^- \tag{5.149}$$

The negative hydroxyl ion OH^- produced in the ion–molecular reaction (Equation 5.149) is stable with respect to detachment and participates in fast ion–ion recombination, which in turns leads to loss of the charged particle and to chain termination.

5.6.5 Methods of Stimulation of Vibrational-Translational Nonequilibrium in Plasma

As shown previously, the most energy-effective possible mechanism of organization of nonequilibrium plasma chemical processes is related to the vibrational excitation of molecules (obviously this mechanism cannot be arranged in all gases nor under all discharged conditions). In the next two sections the principal conditions and approaches necessary for stimulating significant vibrational–translational nonequilibrium in plasma will be discussed.

Vibrational–Translational Nonequilibrium Provided by High Degree of Ionization. This is the most typical manner of nonequilibrium, vibrational excitation in nonthermal plasma. It requires the vibrational excitation frequency (proportional to the electron concentration n_e) to be faster than VT relaxation frequency (proportional to gas concentration n_0 or, in some more specific cases, to the concentration of the most active neutral species in the relaxation). For this reason, significant vibrational–translational nonequilibrium can be reached when the degree of ionization exceeds the critical value:

$$\frac{n_e}{n_0} >> \frac{k_{VT}^{(0)}(T_0)}{k_{eV}(T_e)} \tag{5.150}$$

Here, $k_{eV}(T_e)$ is the rate coefficient of excitation of the lower vibrational levels in plasma by electron impact; $k_{VT}^{(0)}(T_0)$ is the VT relaxation rate coefficient from the low vibrational levels (correlated with the concentration n_0).

Equation 5.150 is obviously an approximation. The more accurate criterion requires details of the neutral gas composition in plasma to take into account relaxation on atoms, radicals, etc. Nevertheless, the criterion (Equation 5.150) is quite convenient and usually gives good agreement with experiments. Thus, to achieve $T_v >> T_0$ in room-temperature nitrogen plasma (with $T_e \approx 1\,eV$), Equation 5.150 correctly requires $n_e/n_0 >> 10^{-9}$; at the same conditions in CO_2 plasma, it requires higher degrees of ionization $n_e/n_0 >> 10^{-6}$.

Vibrational–Translational Nonequilibrium Provided by Fast Gas Cooling. The vibrational–translational nonequilibrium $T_v >> T_0$ can be achieved by an increase of vibrational temperature and also by rapidly decreasing the translational temperature.

Such stimulation of the vibrational–translational nonequilibrium can be organized by using expansion in the supersonic nozzle for rapid cooling of the gas highly preheated in a thermal plasma. This effect plays the key role in providing nonequilibrium and subsequent radiation in gas dynamic lasers.[159]

It should be noted that creation of the vibrational–translational nonequilibrium due to gas cooling is not necessarily restricted to supersonic expansion. It can also take place in plasma chemical systems because of specific gas flow characteristics in a reactor and fast mixing of hot and cold gases (see the specific heat effect in the next section). Another method of stimulating vibrational–translational nonequilibrium is related to the possible **fast transfer of vibrational energy in nonequilibrium systems**. This phenomenon, which is a generalization of the Treanor effect on vibrational energy transfer, is more sophisticated and will be discussed in the next section.

5.6.6 Vibrational–Translational Nonequilibrium Provided by Fast Transfer of Vibrational Energy: Treanor Effect in Vibrational Energy Transfer

Cold gas flowing around the high-temperature thermal plasma zone provides the vibrational–translational nonequilibrium in the area of their contact.[160] In particular, this effect is due to the higher rate of vibrational energy transfer from the quasi-equilibrium high-temperature zone with respect to the rate of translational energy transfer.[156a,618]

The possibility of fast transfer of vibrational energy in nonequilibrium conditions is illustrated in Figure 5.21. The average value of a vibrational quantum is lower because of the anharmonicity at higher vibrational temperatures. The fast VV exchange during the transfer of the vibrational quanta (see Figure 5.21) makes the vibrational quanta motion from high T_v to lower T_v more preferable than in the opposite direction. This effect can be interpreted as domination of the vibrational energy transfer over the transfer of translational energy and can also be interpreted as an additional VV flux of vibrational energy[161,162]:

$$J_{VV} \approx -un_0\hbar\omega \iint Q_{v_2,v_2+1}^{v_1,v_1-1} \frac{2x_e\hbar\omega}{T_0}(v_2 - v_1) f_1(v_1) f_2(v_2)\, dv_1 dv_2 \tag{5.151}$$

In this relation, u is the average thermal velocity of molecules; $f(v)$ is the vibrational distribution function expressed with respect to the number v of vibrational quanta; Q is the probabilities of VV exchange; n_0 is the gas density; and subscripts "1" and "2" are related to two planes (Figure 5.21) with the distance between them equal to the mean free path.

The described phenomenon of preferable transfer of vibrational energy in nonequilibrium conditions is actually a generalization of the Treanor effect (see Figure 4.4) on the case of space transfer of vibrational energy. Integrating Equation 5.151, using Equation 3.73 for the probabilities Q of VV exchange between anharmonic

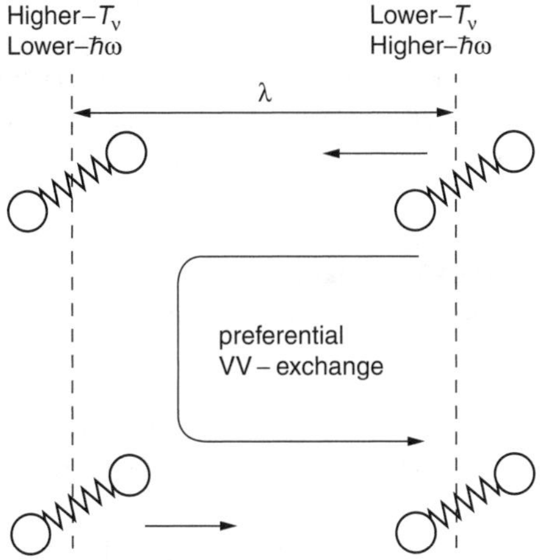

Figure 5.21 Treanor effect in vibrational energy transfer

oscillators and considering only the relaxation processes close to resonance results in the VV vibrational energy transfer flux:

$$J_{VV} = -un_0\hbar\omega \frac{2x_e\hbar\omega}{T_0} Q_{10}^{o1}\left(\langle v_1^2\rangle\langle v_2\rangle - \langle v_1\rangle\langle v_2^2\rangle\right) \tag{5.152}$$

where $\langle v\rangle$ and $\langle v^2\rangle$ are the first and second momentum of the vibrational distribution function (average vibrational quantum number and its average square)

The additional VV vibrational energy transfer flux (Equation 5.152) can be recalculated into the relative increase of the coefficient of vibrational temperature conductivity ΔD_v with respect to the conventional coefficient of translational temperature conductivity D_0. This can be done, based on Equation 5.152 and taking the vibrational distribution as the initial part of the Treanor function:[161,162]

$$\frac{\Delta D_v}{D_0} \approx 4Q_{01}^{10}q\frac{1+30q+72q^2}{(1+2q)^2}, \quad q = \frac{x_eT_v^2}{T_0\hbar\omega} \tag{5.153}$$

As can be seen from Equation 5.153, the effect is strong for molecules such as CO_2, CO, and N_2O with VV exchange provided by long-distance forces, which results in $Q_{01}^{10} \approx 1$. As always with the Treanor effects, anharmonicity $x_e \neq 0$, $q \neq 0$ is necessary for the phenomenon to occur.

Numerically, the effect can be very significant. For example, in room-temperature CO_2 plasma with $T_v = 6000$ K, the transfer of vibrational energy (D_v) according to Equation 5.153 can exceed the transfer of translational energy (D_0) almost 100 times. Such large differences in coefficients of vibrational and translational energy conductivity are obviously able to support a high level of vibrational–translational nonequilibrium in the system.

It should be noted at this point that the Treanor effect (Equation 5.153) is strong in developing nonequilibrium only if the factor q is large beforehand. This means that some nonequilibrium occurs preliminarily in a different way. Such an initial vibrational–translational nonequilibrium can obviously be due to electron impact; however, at relatively low degrees of ionization, this can be be provided also by three effects:

1. **Effect of nonresonant VV exchange**. Vibrational quanta move in a system with a temperature gradient from the higher T_v zone (where average quantum is lower because of anharmonicity) to the lower T_v zone (where average quantum is larger). During the vibrational thermal conductivity, the vibrational quanta have a tendency to grow, taking a defect of energy from the translational degrees of freedom in the nonresonant VV relaxation. Thus, even in quasi-equilibrium systems, the vibrational thermal conductivity leads to some small but notable gas cooling ΔT and thus some vibrational–translational nonequilibrium on the level:

$$\frac{\Delta T_o}{T_0} \approx \frac{T_v - T_0}{T_v} \approx x_e \frac{T}{\hbar\omega} \tag{5.154}$$

2. **Recombination effect**. Diffusion of atoms and radicals from the high-temperature zone of thermal discharges to the low-temperature zone results in recombination and thus an over-equilibrium concentration of vibrationally excited molecules. Deviation of the vibrational temperature from the translational temperature can be quite significant in this case.
3. **Specific heat effect**. Mixing quasi-equilibrium hot and cold gases averages their energies and translational temperatures, but not their vibrational temperatures (because of the nonlinear dependence of vibrational energy on temperature and growth of specific heat with temperature). The growth of the vibrational specific heat with temperature results in vibrational–translational nonequilibrium $T_v > T_0$ with temperature difference of about 1/2 $\hbar\omega$.

5.6.7 Energy Balance and Energy Efficiency of Plasma Chemical Processes Stimulated by Vibrational Excitation of Molecules

As was discussed in Section 5.6.1, the energy efficiency of plasma chemical processes strongly depend on their mechanisms. For example, in Section 5.6.2 it was shown that the highest energy efficiency could be expected (when possible) in the process stimulated by vibrational excitation of molecules. When the main mechanism (for example, the vibrational excitation) is specified, the energy efficiency becomes a strong function of discharge parameters: degree of ionization, specific energy input, etc. Optimization of the discharge parameters is very individualized for different processes, but some general principles can be illustrated. It is necessary now to analyze the general principles of energy efficiency optimization for the nonequilibrium plasma chemical processes stimulated by vibrational excitation of molecules.

The vibrational energy balance in the plasma chemical process can be illustrated by the following simplified, one-component equation, taking into account vibrational excitation by electron impact, VT relaxation, and chemical reaction:

$$\frac{d\varepsilon_v(T_v)}{dt} = k_{eV}(T_e)n_e\hbar\omega\,\theta\left(E_v - \int_0^t k_{eV}n_e\hbar\omega\,dt\right) - \tag{5.155}$$
$$k_{VT}^{(0)}(T_0)n_0\left[\varepsilon_v(T_v) - \varepsilon_{v0}(T_v = T_0)\right] - k_R(T_V, T_0)n_0\Delta Q_\Sigma$$

The left side of this equation represents the change of the average vibrational temperature per molecule, which in the one-mode approximation at not very high levels of nonequilibrium $x_eT_v^2/T_0\hbar\omega < 1$ can be expressed by the Planck formula (see Section 4.1):

$$\varepsilon_v(T_v) = \frac{\hbar\omega}{\exp\left(\frac{\hbar\omega}{T_v}\right) - 1} \tag{5.156}$$

The first term on the right side of the energy balance (Equation 5.155) describes the vibrational excitation of molecules (rate coefficient k_{eV}) by electron impact (density n_e). The Heaviside function in this term, $\theta(x)$: $[\theta = 1, if\ x \geq 0; \theta = 0, if\ x < 0]$, restricts the vibrational energy input in the discharge per one molecule by the value of E_v.

If most discharge power is going into vibrational excitation of molecules, the **specific energy input E_v** per one molecule is equal to the ratio of the discharge power over the gas flow rate through the discharge. This very important and experimentally convenient discharge parameter is related with the energy efficiency (η) and the degree of conversion (χ) of a plasma chemical process by the simple formula:

$$\eta = \frac{\chi \times \Delta H}{E_v} \tag{5.157}$$

where ΔH is the enthalpy of formation of a product molecule in the plasma chemical reaction.

The second term on the right side of Equation 5.155 clearly describes the VT relaxation of the lower vibrational levels; n_0 is the gas density. The third term on the right side is related to the vibrational energy losses in the chemical reaction (reaction rate $k_R(T_v, T_0)$) and the relaxation channels having rates corresponding to those of the chemical reaction. In this term, the total losses of ΔQ_Σ vibrational energy per one act of chemical reaction can be expressed as:

$$\Delta Q_\Sigma = E_a/\alpha + \langle \Delta E_v \rangle + \Delta\varepsilon_{VT} + \Delta\varepsilon_R^{VV} \tag{5.158}$$

Here, E_a/α is the vibrational energy necessary to overcome the activation barrier of chemical reaction (α is the efficiency of vibrational energy; see Section 3.7.3); $\langle \Delta E_v \rangle$ characterizes the excess of vibrational energy in chemical reaction Equation 5.141;

$\Delta\varepsilon_{VT}$ and $\Delta\varepsilon_R^{VV}$ are losses of vibrational energy due, respectively, to VT relaxation from high levels and to nonresonance of VV exchange, expressed with respect to one act of chemical reaction (Equation 5.87 and Equation 5.90). Following the discussion of the vibrational energy balance in the plasma chemical systems (Equation 5.155), some general features of the energy efficiency of processes stimulated by vibrational excitation as a function of discharge parameters can be considered.

5.6.8 Energy Efficiency as Function of Specific Energy Input and Degree of Ionization

The chemical reaction rate coefficient depends more strongly on the vibrational temperature T_v than on all other terms in the balance equation (Equation 5.155). In a simple case of weak vibrational excitation, the rate coefficient can be estimated (see Section 5.3) as:

$$k_R(T_v, T_0) = A(T_0)\exp\left(-\frac{E_a}{T_v}\right) \tag{5.159}$$

The pre-exponential factor $A(T_0)$ differs from the conventional Arrhenius one, A_{Arr}. For example, in the case of the Treanor distribution function,

$$A(T_0) = A_{Arr}\exp\left(\frac{x_e E_a^2}{T_0 \hbar\omega}\right).$$

Based on Equation 5.155, the critical value of vibrational temperature when the reaction and VT relaxation rates are equal is given by:

$$T_v^{\min} = E_a \ln^{-1}\frac{A(T_0)\Delta Q_\Sigma}{k_{VT}^{(0)}(T_0)\hbar\omega} \tag{5.160}$$

An increase of vibrational temperature results in much stronger exponential acceleration of the chemical reaction with respect to vibrational relaxation. This means that, at high vibrational temperatures $T_v > T_v^{\min}$, almost all the vibrational energy is going to chemical reaction according to Equation 5.155. Conversely, at lower vibrational temperatures $T_v < T_v^{\min}$, almost all vibrational energy can be lost in vibrational relaxation.

The critical vibrational temperature $T_v^{\min}$ determines the threshold dependence of the energy efficiency of plasma chemical processes $\eta(E_v)$, stimulated by vibrational excitation, on the specific energy input E_v (see Equation 5.155):

$$\left(E_v\right)_{threshold} = \varepsilon_v\left(T_v^{\min}\right) = \frac{\hbar\omega}{\exp\left(\frac{\hbar\omega}{T_v^{\min}}\right) - 1} \tag{5.161}$$

If the specific energy input is lower than the threshold value $E_v < (E_v)_{threshold}$, the vibrational energy cannot reach the critical value $T_v < T_v^{\min}$ and the energy efficiency

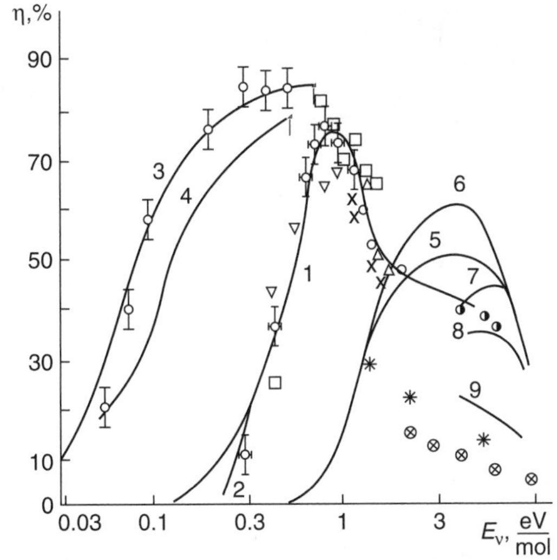

Figure 5.22 Energy efficiency of CO_2 dissociation as a function of energy input. (1, 2) Nonequilibrium calculation in 1-T_v and 2-T_v approximations; nonequilibrium calculations for supersonic flows; (3) $M = 5$; (4) $M = 3.5$; calculations of thermal dissociation with (5) ideal and (6) super-ideal quenching, thermal dissociation with quenching rates of (7) 10^9 K/sec; (8) 10^8 K/sec; (9) 10^7 K/sec. Different experiments in microwave discharges: ⫩, I, Δ, *x*. Experiments in supersonic microwave discharges: Ī. Experiments in different RF CCP discharges: ○, ∇. Experiments in RF ICP discharges: ◑. Experiments in different arc discharges: ⊗, *.

is very low. The typical threshold values of energy input for plasma chemical reactions stimulated by vibrational excitation are approximately 0.1 to 0.2 eV/mol.

The threshold in the dependence $\eta(E_v)$ is a very important qualitative feature of the plasma chemical processes stimulated by vibrational excitation. For example, to avoid intensive dissociation of carbon dioxide in a gas-discharge CO_2 laser, the specific energy input is always limited by the threshold value. The dependences $\eta(E_v)$ for specific plasma chemical processes of dissociation of CO_2, H_2O, and NO synthesis, stimulated by vibrational excitation, are shown in Figure 5.22 through Figure 5.24.[9] All three processes show the essential threshold effect for the energy efficiency discussed.

Another important feature of the plasma chemical processes stimulated by vibrational excitation, which can be observed in Figure 5.22 through Figure 5.24, is the maximum in the dependence $\eta(E_v)$. The optimal value of the specific energy input is usually about 1 eV. At larger values of the specific energy input, the translational temperature increases, which accelerates VT relaxation and decreases the energy efficiency. Also, at large energy inputs E_v, the major fraction of the discharge energy can be spent on useless excitation of products. Even if the plasma chemical process degree of conversion χ approaches 100% at high energy inputs, the energy efficiency (see Equation 5.157) decreases hyperbolically with the energy input $\eta \propto \Delta H / E_v$.

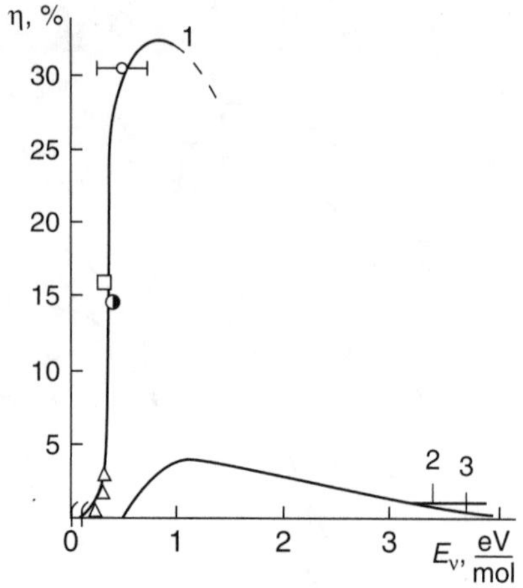

Figure 5.23 Energy efficiency of NO synthesis in air as a function of energy input. 1—Nonequilibrium process stimulated by vibrational excitation; 2, 3—thermal synthesis with (2) ideal and (3) absolute quenching. Experiments with microwave discharges: ⊢○⊣, |; with discharges sustained by electron beams: ◑, Δ, ○.

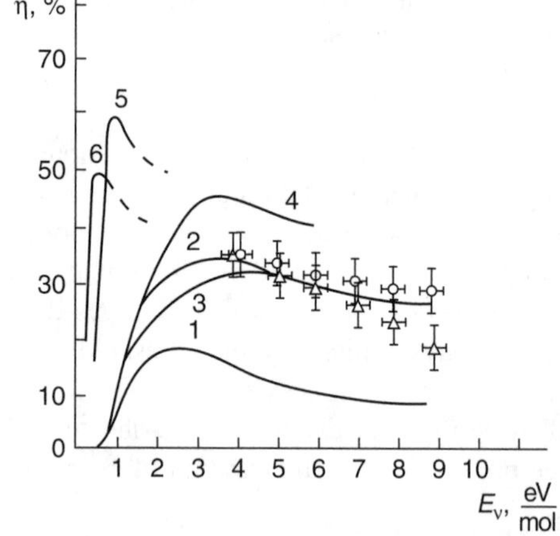

Figure 5.24 Energy efficiency of water vapor dissociation as a funciton of energy input: (1, 2) thermal dissociation with absolute and ideal quenching; (3, 4) super-absolute and super-ideal quenching mechanisms; (5) nonequilibrium decomposition due to dissociative attachment; (6) nonequilibrium dissociation stimulated by vibrational excitation. Different experiments in microwave discharges: ⊹, ⸸.

If the vibrational temperature exceeds the critical one, $T_v > T_v^{min}$, the relaxation losses in Equation 5.155 can be neglected with almost all of the discharge energy going through vibrational excitation into the chemical reaction. The steady-state

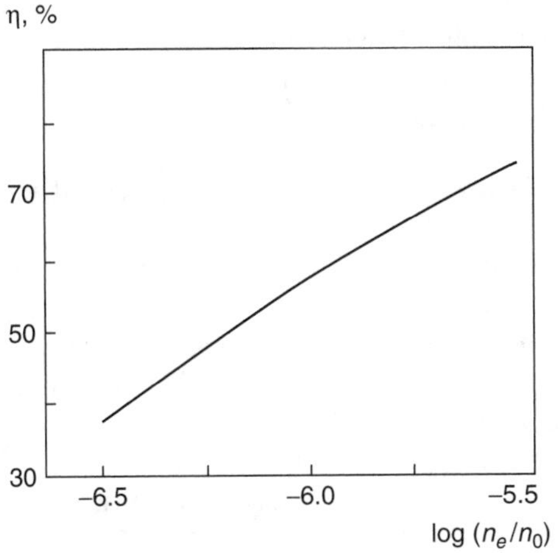

Figure 5.25 Dependence of CO_2 dissociation energy efficiency on ionization degree, E_v = 0.5 eV/mol

value of the vibrational temperature then can be found from the balance equation (Equation 5.155, taking into account Equation 5.159) as:

$$T_v^{st} = E_a \ln^{-1}\left(\frac{A(T_0)\Delta Q_\Sigma}{k_{eV}\hbar\omega} \cdot \frac{n_0}{n_e}\right) \tag{5.162}$$

Obviously, this steady-state vibrational temperature exceeds the critical temperature (Equation 5.160) when the electron concentration exceeds the corresponding critical value (Equation 5.150). In general, the energy efficiency of plasma chemical processes stimulated by vibrational excitation grows with the degree of ionization. An example of the dependence $\eta(n_e/n_0)$ is presented in Figure 5.25 for the plasma chemical process of CO_2 dissociation stimulated by vibrational excitation.

5.6.9 Components of Total Energy Efficiency: Excitation, Relaxation, and Chemical Factors

The total energy efficiency of any plasma chemical process can be subdivided into the three main components: (1) excitation factor (η_{ex}); (2) relaxation factor (η_{rel}); and (3) chemical factor (η_{chem}):

$$\eta = \eta_{ex} \times \eta_{rel} \times \eta_{chem} \tag{5.163}$$

The **excitation factor** shows the fraction of discharge energy directed to producing the principal agent of the plasma chemical reaction. For example, in the plasma chemical process of NF_3 dissociation for cleaning CVD chamber,[157] this factors shows the efficiency of generation of F-atoms responsible for cleaning. In plasma chemical processes stimulated by vibrational excitation, the excitation factor (η_{ex}) illustrates the fraction of electron energy (or discharge energy, which is often the same) going to vibrational excitation of molecules. This factor obviously depends on the electron temperature and gas composition; at $T_e \approx 1\ eV$, it can reach 90% and higher (see Section 3.3.8).

The **relaxation factor** (η_{rel}) is the efficiency related to conservation of the principal active species (such as the F-atoms or vibrationally excited molecules in the preceding examples) with respect to their losses in relaxation, recombination, etc. In plasma chemical processes stimulated by vibrational excitation, the factor (η_{rel}) shows the fraction of vibrational energy that can avoid vibrational relaxation and be spent in the chemical reaction of interest. Taking into account that vibrational energy losses into VT relaxation are small inside the discharge zone if $T_v > T_v^{\min}$, the factor (η_{rel}) can be expressed as a function of specific energy input E_v near the threshold value (Equation 5.161):

$$\eta_{rel} = \frac{E_v - \varepsilon_v\left(T_v^{\min}\right)}{E_v} \tag{5.164}$$

Equation 5.164 includes the threshold of the dependence $\eta(E_v)$ (see Figure 5.22 through Figure 5.24) and again shows that when the vibrational temperature finally becomes lower than the critical one, most of the vibrational energy going to VT relaxation comes from lower levels. The relaxation factor (η_{rel}) strongly depends on temperature in accordance with the Landau–Teller theory (see Section 3.4). For example, at very low translational temperatures ($T_0 \approx 100$ K) in a supersonic microwave discharge, the relaxation factor can be very high, more than 97%.[163]

The **chemical factor** (η_{chem}) is a component of the total energy efficiency (Equation 5.163), which shows the efficiency of application of the principal active discharge species in the chemical reaction of interest.[164] In the example of plasma chemical process of NF_3 dissociation and F-atom generation for cleaning CVD chambers, the chemical factor (η_{chem}) shows the efficiency of atomic fluorine in gasification of silicon and its solid compounds. In plasma chemical reactions stimulated by vibrational excitation, based on the balance equation, this factor can be presented as:

$$\eta_{chem} = \frac{\Delta H}{E_a/\alpha + \left\langle \Delta E_v \right\rangle + \Delta\varepsilon_{VT} + \Delta\varepsilon_R^{VV}} \tag{5.165}$$

The chemical factor restricts energy efficiency mainly because the activation energy quite often exceeds the reaction enthalpy: $E_a > \Delta H$. For example, in the plasma chemical process of NO synthesis, proceeding by the Zeldovich mechanism (Equation 5.142 and Equation 5.143), the second reaction is significantly exothermic. This results in relatively low value of the chemical factor ($\eta_{chem} \approx$ 50%) and restricts the total efficiency of the plasma chemical process.[164]

5.7 ENERGY EFFICIENCY OF QUASI-EQUILIBRIUM PLASMA CHEMICAL SYSTEMS: ABSOLUTE, IDEAL, AND SUPER-IDEAL QUENCHING

In this section, the effects of quenching kinetics on energy efficiency of chemical processes in quasi-equilibrium plasma will be discussed. Special types of quenching, including strongly nonequilibrium quenching, will be considered using examples of CO_2 and H_2O dissociation in thermal plasmas. Finally, the possible influence of transfer phenomena on the energy efficiency of plasma chemical processes will be analyzed.

5.7.1 Concepts of Absolute, Ideal, and Super-Ideal Quenching

The quasi-equilibrium plasma chemical processes can be subdivided into two phases. In the first, reagents are thermally heated to high temperatures necessary to shift chemical equilibrium in the direction of products. In the second phase, called quenching, temperature decreases sufficiently fast to protect the products produced in the first high-temperature phase from reverse reactions.

Absolute quenching means that the cooling process is sufficiently fast to save all the products formed in the high temperature zone. In the high-temperature zone, initial reagents are partially converted into the products of the process, but also into some unstable atoms and radicals. In the case of absolute quenching, stable products are saved during the cooling process, but atoms and radicals are converted back into the initial reagents.

Ideal quenching means that the cooling process is very effective and able to maintain the total degree of conversion on the same level as was reached in the high-temperature zone. Ideal quenching not only saves all the products formed in the high-temperature zone, but also provides conversion of all the relevant atoms and radicals into the process products. It is interesting that during the quenching phase, the total degree of conversion cannot only be saved, but also even increased. Such quenching is usually referred to as super-ideal.

Super-ideal quenching permits increasing the degree of conversion during the cooling stage, using the chemical energy of atoms and radicals as well as the excitation energy accumulated in molecules.[169] In particular, super-ideal quenching can be organized when the gas cooling is faster than VT relaxation, and the vibrational–translational, nonequilibrium $T_v > T_0$ can be achieved during the quenching (see Section 5.6.5). In this case, direct endothermic reactions are stimulated by vibrational excitation, while reverse exothermic reactions related to translational degrees of freedom proceed more slowly. Such an unbalance between direct and reverse reactions provides additional conversion of the initial substances or, in other words, provides super-ideal quenching. Next, kinetic

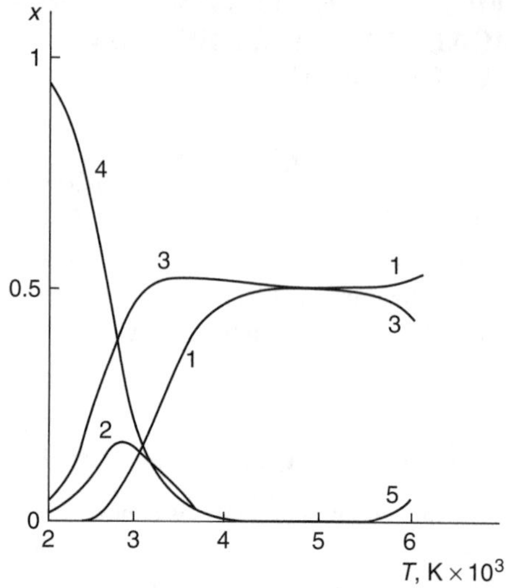

Figure 5.26 Equilibrium composition for CO_2 thermal decomposition (p = 0.16 atm): (1) O; (2) O_2; (3) CO; (4) CO_2; (5) C

details of different quenching mechanisms on specific examples of plasma chemical processes will be discussed.

5.7.2 Ideal Quenching of CO_2 Dissociation Products in Thermal Plasma

Thermodynamically, equilibrium composition of the CO_2 dissociation products in a thermal plasma depends only on temperature and pressure. This temperature dependence, which is much stronger than pressure dependence, is presented in Figure 5.26. The main products of the plasma chemical process under consideration:

$$CO_2 \rightarrow CO + \frac{1}{2}O_2, \quad \Delta H_{CO_2} = 2.9 \text{ eV} \tag{5.166}$$

are the saturated molecules CO and O_2. Also, the high-temperature heating provides significant concentration of atomic oxygen and atomic carbon at very high T_0.

If quenching of the products is not sufficiently fast, the slow cooling could be quasi equilibrium, and reverse reactions would return the composition to the initial one, e.g., pure CO_2. Fast cooling rates ($>10^8$ *K*/sec) permit saving CO in the products.[165] In this case, atomic oxygen recombines into molecular oxygen O + O + M → O_2 + M faster than reacting with carbon monoxide O + CO + M → CO_2 + M, thus maintaining the CO_2 degree of conversion and the conditions for ideal quenching. In this case, this is the same as absolute quenching.

Energy efficiency of the quasi-equilibrium process with ideal quenching can be calculated for this case in the following manner. First, the energy input per one initial CO_2 molecule required to heat the dissociating CO_2 at constant pressure to temperature T can be expressed as:

$$\Delta W_{CO_2} = \frac{\sum x_i I_i(T)}{x_{CO_2} + x_{CO}} - I_{CO_2}(T = 300\ \mathrm{K}) \tag{5.167}$$

In this relation, $x_i(p,T) = n_i/n$ is the quasi-equilibrium relative concentration of i component of the mixture; $I_i(T)$ is the total enthalpy of the component.

If β is the number of CO molecules produced from one initial CO_2 molecule taking into account quenching, then the total energy efficiency of the quasi-equilibrium plasma chemical process can be given as:

$$\eta = \frac{\Delta H_{CO_2}}{\left(\Delta W_{CO_2}/\beta\right)} \tag{5.168}$$

The conversion is equal to

$$\beta^0 = \frac{x_{CO}}{x_{CO_2} + x_{CO}}$$

in the case of ideal quenching, which leads to the final expression of energy efficiency:

$$\eta = \frac{\Delta H_{CO_2} x_{CO}}{\sum x_i I_i(T) - (x_{CO_2} + x_{CO}) I_{CO_2}(T = 300\ \mathrm{K})} \tag{5.169}$$

The energy efficiency of CO_2 dissociation in thermal plasma with ideal quenching is presented in Figure 5.27 as a function of temperature. As this figure shows, a maximum energy efficiency of about 50% can be reached at T = 2900 K. The energy cost of CO production by CO_2 dissociation in quasi-equilibrium plasma chemical systems is shown in Figure 5.28 as a function of the cooling rate.[166] As can be seen from the figure, a high cooling rate of 10^8/sec is required to provide conditions close to ideal quenching.

5.7.3 Nonequilibrium Effects during Product Cooling: Super-Ideal Quenching

As mentioned in Section 5.7.1, vibrational–translational nonequilibrium effects in the process of cooling can provide additional conversion. In the plasma chemical process of CO_2 dissociation, this effect of super-ideal quenching is related to shifting equilibrium of the reaction:

$$O + CO_2 \Leftrightarrow CO + O_2, \quad \Delta H = 0.34\ \mathrm{eV} \tag{5.170}$$

This direct endothermic reaction of CO formation can be effectively stimulated by vibrational excitation of CO_2 molecules, and not balanced by reverse exothermic reaction at $T_v > T_0$. The energy efficiency of the quasi-equilibrium thermal CO_2 dissociation with super-ideal quenching of products can be found in this case as:

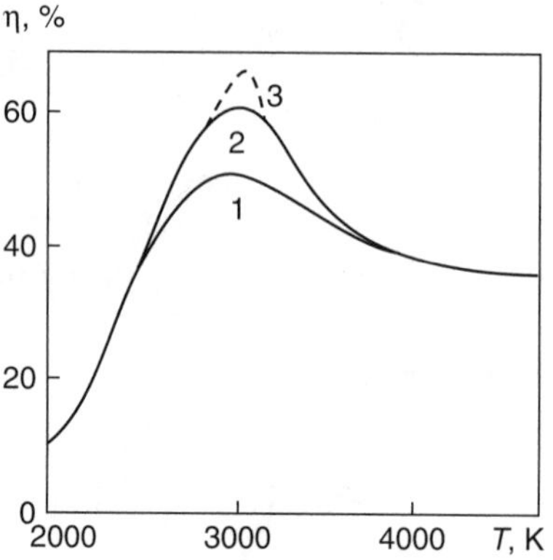

Figure 5.27 Energy efficiency of CO_2 thermal dissociation as a function of heating temperature at (p = 0.16 atm): 1: ideal quenching; 2: super-ideal quenching; 3: upper limit of super-ideal quenching

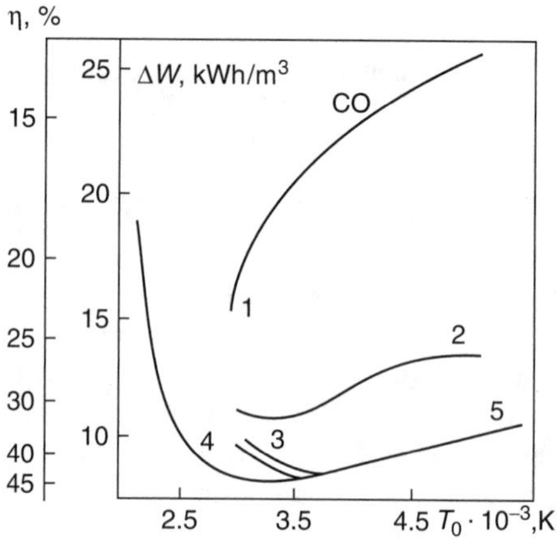

Figure 5.28 Dependence of energy efficiency η and energy cost ΔW of CO_2 thermal dissociation on heating temperature (p = 1 atm) and different quenching rates: 1: 10^6 K/sec; 2: 10^7 K/sec; 3: 10^8 K/sec; 4: 10^9 K/sec; 5: instantaneous cooling

$$\eta = \frac{\Delta H_{CO_2}}{\Delta W_{CO_2}} \frac{x_{CO} + x_O}{x_{CO_2} + x_{CO}}, \quad \text{if } x_{CO_2} > x_O \tag{5.171}$$

$$\eta = \frac{\Delta H_{CO_2}}{\Delta W_{CO_2}}, \quad \text{if } x_{CO_2} < x_O \tag{5.172}$$

Numerically, the energy efficiency of super-ideal quenching $\eta(T)$ as a function of temperature is also shown in Figure 5.27. Thus, it is seen that the maximum efficiency $\eta = 64\%$ can be reached at a temperature in the hot zone of about $T = 3000$ K. This

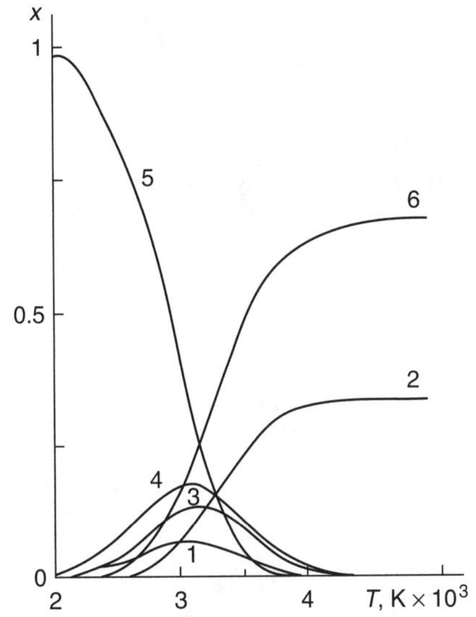

Figure 5.29 Equilibrium composition for H_2O thermal decomposition: 1: O_2; 2: O; 3: OH; 4: H_2; 5: H_2O, 6: H

efficiency is 14% higher than the maximum one for ideal quenching (but still much less than the maximum efficiency of the pure nonequilibrium process; see Figure 5.25). Detailed analysis of super-ideal quenching under different conditions can be found in Potapkin et al.[167,169]

5.7.4 Mechanisms of Absolute and Ideal Quenching for H_2O Dissociation in Thermal Plasma

Equilibrium composition of the H_2O dissociation products in thermal plasma is presented in Figure 5.29 as a function of temperature at fixed pressure. The main products of the plasma chemical process under consideration:

$$H_2O \rightarrow H_2 + \frac{1}{2}O_2, \quad \Delta H_{H_2O} = 2.6 \text{ eV} \tag{5.173}$$

are the saturated molecules H_2 and O_2. In contrast to the case of CO_2, thermal dissociation of water vapor results in a variety of atoms and radicals. As Figure 5.29 shows, concentrations of O, H, and OH are significant in this process. Even when H_2 and O_2 initially formed in the high-temperature zone are saved from reverse reactions, the active species O, H, and OH can be converted into products (H_2 and O_2) or back into H_2O. This qualitatively different behavior of radicals determines the key difference between the absolute and ideal mechanisms of quenching.

In the case of the absolute quenching, when the active species are converted back to H_2O, the energy efficiency of the plasma chemical process (Equation 5.173) of hydrogen production can be found as:

$$\eta = \frac{\Delta H_{H_2O}}{\left(\Delta W_{H_2O} / \beta_{H_2^0}\right)} \tag{5.174}$$

Here, $\beta^0_{H_2}(T)$ is the quasi-equilibrium conversion of water vapor into hydrogen. In other words, $\beta^0_{H_2}(T)$ is the number of hydrogen molecules formed in the quasi-equilibrium phase calculated per one initial H_2O molecule:

$$\beta^0_{H_2} = \frac{x_{H_2}}{x_{H_2O} + x_O + 2x_{O_2} + x_{OH}} \tag{5.175}$$

The discharge energy input per one initial H_2O molecule to heat the dissociating water at constant pressure to temperature T can be expressed similarly to Equation 5.167 as:

$$\Delta W_{H_2O} = \frac{\sum x_i I_i(T)}{x_{H_2O} + x_O + 2x_{O_2} + x_{OH}} - I_{H_2O}(T = 300\text{ K}) \tag{5.176}$$

Based on Equation 5.174 through Equation 5.176, the final formula for calculating the energy efficiency of dissociating water and hydrogen production in thermal plasma with absolute quenching of the process products can be presented as:

$$\eta = \frac{\Delta H_{H_2O} x_{H_2}}{\sum x_i I_i(T) - \left(x_{H_2O} + x_O + 2x_{O_2} + x_{OH}\right) I_{H_2O}(T = 300\text{ K})} \tag{5.177}$$

The energy efficiency of the thermal water dissociation in case of ideal quenching can be calculated in a similar way, taking into account the additional conversion of active species, H, OH, and O into the product of the process (H_2 and O_2):

$$\eta = \Delta H_{H_2O} \left[\frac{\sum x_i I_i(T)}{x_{H_2} + 1/2\left(x_H + x_{OH}\right)} - \frac{x_{H_2O} + x_O + 2x_{O_2} + x_{OH}}{x_{H_2} + 1/2\left(x_H + x_{OH}\right)} I_{H_2O}(T = 300\text{ K}) \right]^{-1} \tag{5.178}$$

Numerical values of the energy efficiencies of water dissociation in thermal plasma for two cases of absolute and ideal quenching are presented in Figure 5.30, which shows that complete usage of atoms and radicals to form products (ideal quenching) permits increasing the energy efficiency by a factor of almost two.

5.7.5 Effect of Cooling Rate on Quenching Efficiency: Super-Ideal Quenching of H_2O Dissociation Products

Slow cooling shifts the equilibrium between direct and reverse reactions in the exothermic direction of destruction of molecular hydrogen:

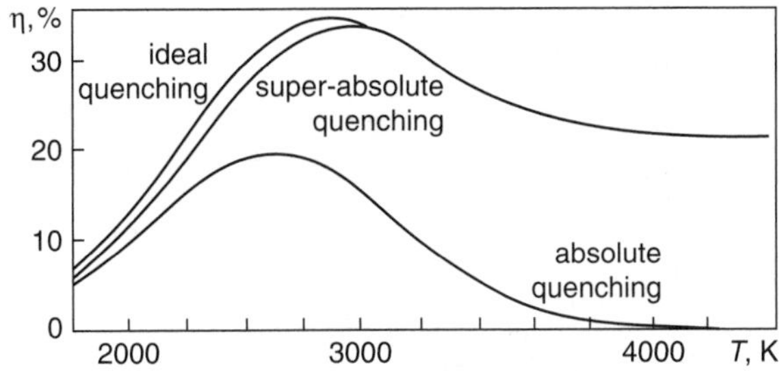

Figure 5.30 Energy efficiency of H_2O thermal dissociation as function of heating temperature, $p = 0.05$ atm

$$O + H_2 \rightarrow OH + H \tag{5.179}$$

$$OH + H_2 \rightarrow H_2O + H \tag{5.180}$$

The following three-body recombination leads to formation of water molecules:

$$H + OH + M \rightarrow H_2O + M \tag{5.181}$$

and its reaction rate coefficient is about 30 times larger than the rate coefficient of the alternative three body recombination with formation of molecular hydrogen:

$$H + H + M \rightarrow H_2 + M \tag{5.182}$$

The reactions in Equation 5.179 through Equation 5.182 explain the destruction of hydrogen and the reformation of water molecules during slow cooling of the water dissociation products.

If the cooling rate is sufficiently high ($>10^7$ K/sec), the reactions of O and OH with saturated molecules very soon become less effective. This provides conditions for absolute quenching. Instead of participating in the processes depicted in Equation 5.179 and Equation 5.180, the active species OH and O react to form mostly atomic hydrogen:

$$OH + OH \rightarrow O + H_2O \tag{5.183}$$

$$O + OH \rightarrow O_2 + H \tag{5.184}$$

The atomic hydrogen (which is in excess with respect to OH) recombines in molecular form (Equation 5.182), at higher cooling rates ($>2 \cdot 10^7$ K/sec) providing the additional conversion of radicals into the stable process products. This means that the super-absolute (almost ideal; see Figure 5.30) quenching can be achieved at these cooling rates. Energy efficiency of thermal water dissociation and hydrogen production as a function of the cooling rate is presented in Figure 5.31.[168]

Vibrational–translational nonequilibrium during fast cooling of the water dissociation products can lead to the surer-ideal quenching effect in a similar way as that described in Section 5.7.3 regarding CO_2 dissociation. The super-ideal quench-

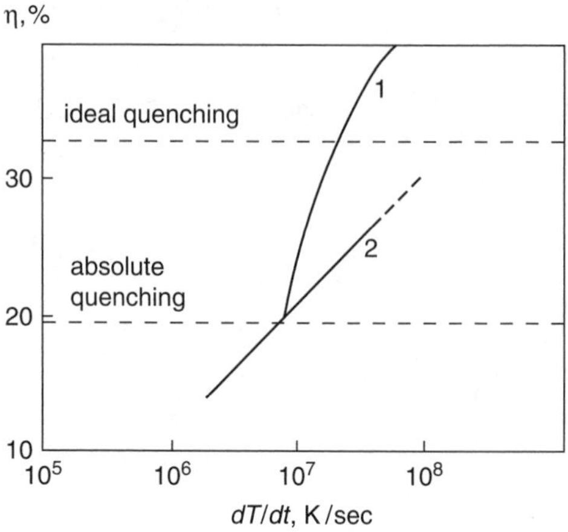

Figure 5.31 Energy efficiency of H_2O thermal dissociation as a function of quenching rate; heating temperature 2800 K; $p = 0.05$ atm: (1) vibrational–translational nonequilibrium; (2) vibrational–translational equilibrium

ing effect can be related in this case to the shift of quasi equilibrium (at $T_v > T_0$) of the reactions:

$$H + H_2O \Leftrightarrow H_2 + OH, \quad \Delta H = 0.6 \text{ eV} \tag{5.185}$$

The direct endothermic reaction of hydrogen production is stimulated here by vibrational excitation, while the reverse reaction stays relatively slow. The energy efficiency of water dissociation in such nonequilibrium quenching conditions is presented in Figure 5.31 as a function of cooling rate.[167,169] From this figure, the super-ideal quenching of the water dissociation products requires cooling rates exceeding $5 \cdot 10^7$ K/sec and provides energy efficiency up to approximately 45%.

5.7.6 Mass and Energy Transfer Equations in Multicomponent Quasi Equilibrium Plasma Chemical Systems

To analyze the influence of transfer phenomena on the energy efficiency of plasma chemical processes, it is necessary to begin with some general aspects of mass and energy transfer. The conservation equations, describing enthalpy (I) and mass transfer in multicomponent (number of components N) quasi-equilibrium reacting gas, are similar in thermal plasma and combustion systems and can be presented as:[170]

- Conservation equation for total enthalpy:

$$\rho \frac{dI}{dt} = -\nabla \vec{q} + \frac{dp}{dt} + \Pi_{ik} \frac{\partial v_i}{\partial x_k} + \rho \sum_{\alpha=1}^{N} Y_\alpha \overrightarrow{v_\alpha} \overrightarrow{f_\alpha} \tag{5.186}$$

- Continuity equation for chemical components:

$$\rho \frac{dY_\alpha}{dt} = -\nabla\left(\rho Y_\alpha \overrightarrow{v_\alpha}\right) + \omega_\alpha \tag{5.187}$$

Neglecting the contribution of radiation heat transfer, the thermal flux $\vec{q}$ in Equation 5.186 can be expressed as:

$$\vec{q} = -\lambda \nabla T + \rho \sum_{\alpha=1}^{N} I_\alpha Y_\alpha \overrightarrow{v_\alpha} \tag{5.188}$$

The diffusion velocity $\overrightarrow{v_\alpha}$ for α-chemical component can be found from the following relation:

$$\nabla x_\alpha = \sum_{\beta=1}^{N} \frac{x_\alpha x_\beta}{D_{\alpha\beta}}\left(\overrightarrow{v_\beta} - \overrightarrow{v_\alpha}\right) + \left(Y_\alpha - x_\alpha\right)\frac{\nabla p}{p} + \frac{\rho}{p}\sum_{\beta=1}^{N} Y_\alpha Y_\beta\left(\overrightarrow{f_\alpha} - \overrightarrow{f_\beta}\right) + \sum_{\beta=1}^{N}\left[\frac{x_\alpha x_\beta}{D_{\alpha\beta}}\frac{1}{\rho}\left(\frac{D_{T,\beta}}{Y_\beta} - \frac{D_{T,\alpha}}{Y_\alpha}\right)\right]\frac{\nabla T}{T} \tag{5.189}$$

In the preceding relations,

$$I = \sum_{\alpha=1}^{N} Y_\alpha I_\alpha$$

is the total enthalpy per unit mass of the mixture including the enthalpy of formation; Π_{ik} is the tensor of viscosity; $\overrightarrow{f_\alpha}$ is external force per unit mass of the component α; ω_α is the rate of mass change of the component α per unit volume as a result of chemical reactions; λ, $D_{\alpha\beta}$, $D_{T,\alpha}$ are the coefficients of thermal conductivity, binary diffusion, and thermodiffusion, respectively; $x_\alpha = n_\alpha/n$, $Y_\alpha - \rho_\alpha/\rho$ are the molar and mass fractions of the component α; v_i is the ith component of hydrodynamic velocity; and p is the pressure of the gas mixture.

To simplify Equation 5.186 through Equation 5.189, it is necessary to neglect the total enthalpy change due to viscosity; assume the forces $\overrightarrow{f_\alpha}$ the same for all components; and take the binary diffusion coefficients as $D_{\alpha\beta} = D(1 + \delta_{\alpha\beta})$, $\delta_{\alpha\beta} < 1$. After such simplification, these equations describing the energy and mass transfer can be rewritten as:

$$\rho \frac{dI}{dt} = -\nabla \vec{q} + \frac{dp}{dt} \tag{5.190}$$

$$\vec{q} = -\frac{\lambda}{c_p}\nabla I + (1 - Le)\sum_{\alpha=1}^{N} I_\alpha D \rho_\alpha \nabla \ln Y_\alpha + \sum_{\alpha=1}^{N} \rho_\alpha I_\alpha \overrightarrow{v_\alpha^c} + \sum_{\alpha=1}^{N} \rho_\alpha I_\alpha \overrightarrow{v_\alpha^g} \tag{5.191}$$

$$\overrightarrow{v_\alpha} = -D\nabla \ln Y_\alpha + \overrightarrow{v_\alpha^c} + \overrightarrow{v_\alpha^g} \tag{5.192}$$

plus the continuity Equation 5.187.

In these equations,

$$Le = \frac{\lambda}{\rho c_p D}$$

is the Lewis number, and

$$c_p = \sum_{\alpha=1}^{N} Y_\alpha \frac{\partial I_\alpha}{\partial T}$$

is the constant-pressure specific heat of the unit mass of the mixture. The component of diffusion velocity in Equation 5.191 and Equation 5.192:

$$\overrightarrow{v_a^c} = D\sum_{\beta=1}^{N} Y_\beta \left(\frac{\overrightarrow{F_\beta}}{x_\beta} - \frac{\overrightarrow{F_\alpha}}{x_\alpha} \right) \tag{5.193}$$

is related to baro- and thermodiffusion effects. Another component of the diffusion velocity in Equation 5.191 and Equation 5.192:

$$\overrightarrow{v_\alpha^g} = \sum_{\beta=1}^{N} x_\beta \delta_{\alpha\beta} \left(\overrightarrow{v_\beta^{(0)}} - \overrightarrow{v_\alpha^{(0)}} \right) + \sum_{\gamma=1}^{N} Y_\gamma \sum_{\beta=1}^{N} \delta_{\beta\gamma} \left(\overrightarrow{v_\beta^{(0)}} - \overrightarrow{v_\gamma^{(0)}} \right) \tag{5.194}$$

is related to difference in the binary diffusion coefficients.

The α-component force in Equation 5.193 and the α-component velocities in Equation 5.194 are given by the expressions:

$$\overrightarrow{F_\alpha} = (Y_\alpha - x_\alpha)\frac{\nabla p}{p} + \sum_{\gamma=1}^{N} \frac{x_\alpha x_\gamma}{\rho D_{\alpha\beta}} \left(\frac{D_{T,\gamma}}{Y_\gamma} - \frac{D_{T,\alpha}}{Y_\alpha} \right) \frac{\nabla T}{T} \tag{5.195}$$

$$\overrightarrow{v_\alpha^{(0)}} = -D\nabla \ln Y_\alpha + D\sum_{\beta=1}^{N} Y_\beta \left(\frac{\overrightarrow{F_\beta}}{x_\beta} - \frac{\overrightarrow{F_\alpha}}{x_\alpha} \right) \tag{5.196}$$

The set of equations introduced here describes the energy and mass transfer in quasi-equilibrium plasma chemical systems. These will now be analyzed to determine the influence of the transfer phenomena on energy efficiency of chemical processes in thermal plasma.

5.7.7 Influence of Transfer Phenomena on Energy Efficiency of Plasma Chemical Processes

Based on these energy and mass transfer equations, it is necessary first to establish the following important rule. The transfer phenomena do not change the limits of energy efficiency of quasi-equilibrium plasma chemical processes (which are absolute and ideal quenching; see Section 5.166 through Section 5.170) if two requirements are satisfied:

1). The binary diffusion coefficients are fixed, equal to each other, and equal to the reduced coefficient of thermal conductivity (the Lewis number $Le = 1$, $\delta_{\alpha\beta} = 0$).
2). The effective pressure in the system is constant, thus implying the absence of external forces and large velocity gradients.

Taking the preceding conditions into account, the enthalpy balance equation (Equation 5.190) and the continuity equation (Equation 5.187) for the steady-state one-dimensional case can be rewritten as:

$$\rho v \frac{\partial}{\partial x} I = \frac{\partial}{\partial x}\left(\rho D \frac{\partial}{\partial x} I\right) \tag{5.197}$$

$$\rho v \frac{\partial}{\partial x} Y_\alpha = \frac{\partial}{\partial x}\left(\rho D \frac{\partial}{\partial x} Y_\alpha\right) + \omega_\alpha \tag{5.198}$$

It is convenient to introduce new dimensionless functions describing reduced enthalpy and reduced mass fractions:

$$\xi = \frac{I - I^r}{I^l - I^r}, \quad \eta = \frac{Y_\alpha - Y_\alpha^r}{Y_\alpha^l - Y_\alpha^r} \tag{5.199}$$

where Y_α^l, I^l and Y_α^r, I^r are the mass fractions and enthalpy, respectively, on the left (l) and right (r) boundaries of the region under consideration. With these new functions, the enthalpy conservation (Equation 5.197) and continuity (Equation 5.198) equations can be rewritten as:

$$\rho v \frac{\partial \xi}{\partial x} = \frac{\partial}{\partial x} \rho D \frac{\partial \xi}{\partial x} \tag{5.200}$$

$$\rho v \frac{\partial \eta_\alpha}{\partial x} = \frac{\partial}{\partial x} \rho D \frac{\partial \eta_\alpha}{\partial x} + \omega_\alpha \left(Y_\alpha^l - Y_\alpha^r\right) \tag{5.201}$$

The boundary conditions for these equations on the left and right sides of the region can be written as:

$$\xi(x = x_l) = \eta(x = x_l) = 1, \quad \xi(x = x_r) = \eta(x = x_r) = 0 \tag{5.202}$$

In general, chemical reactions (ω_α in the continuity equation [Equation 5.201]) during the diffusion process are able to increase or to decrease concentration of the reacting components. However, considering the endothermic quasi-equilibrium processes (which is usually the case in thermal plasma), chemical reactions only decrease the concentration of products. To determine the maximum yield of the products, the reaction rate during diffusion should be neglected, assuming $\omega_\alpha = 0$

in the continuity equation (Equation 5.201). In this case, equations and boundary conditions for η and ξ are completely identical, which means they are equal and that gradients of total enthalpy I and mass fraction Y_α are related by the formula:[171]

$$\left(Y_\alpha^l - Y_\alpha^r\right)^{-1} \frac{\partial}{\partial x} Y_\alpha = \left(I^l - I^r\right)^{-1} \frac{\partial}{\partial x} I \tag{5.203}$$

From Equation 5.203 and taking into account the effect of chemical reactions, it can be concluded that the minimum ratio of the enthalpy flux to the flux of products is equal to:

$$A = \frac{I^l - I^r}{Y_\alpha^l - Y_\alpha^r} \tag{5.204}$$

The ratio (Equation 5.204) gives the energy price of the process product and completely correlates with the expressions for energy efficiency of quasi-equilibrium plasma chemical processes considered in Section 5.7.1, Section 5.7.2, and Section 5.7.4. Therefore, in absence of external forces and when the diffusion and reduced thermal conductivity coefficients are equal (the Lewis number $Le = 1$), the minimum energy cost of products of plasma chemical reaction is determined by the product formation in the quasi-equilibrium high-temperature zone. This minimum energy cost (maximum energy efficiency) obviously corresponds to the discussed limits of absolute and ideal quenching and, considering the nonuniformity of heating, can be calculated as:

$$\langle A \rangle = \frac{\int y(T)\left[I(T) - I_0\right] dT}{\int y(T)\chi(T)\, dT} \tag{5.205}$$

Here, $y(T)$ is the mass fraction of initial substance heated to temperature T; $I(T)$ and I_0 are the total enthalpy of the mixture at temperature T and initial temperature, respectively; and $\chi(T)$ is the degree of conversion of the initial substance (for ideal quenching) or degree of conversion into final product (for absolute quenching) achieved in the high temperature zone.

The possibility of using the thermodynamic relation (Equation 5.205) for calculating the energy cost of chemical reactions in thermal plasma under conditions of intensive heat and mass transfer is a consequence of the **similarity principle of concentrations and temperature fields**. This principle is valid at the Lewis number $Le = 1$ in absence of external forces and is widely applied in combustion theory.[170]

If the Lewis number differs from unity ($Le = \lambda/\rho c_p D \neq 1$), the maximum energy efficiency can be different from that calculated in the framework of ideal quenching. For example, if light and fast hydrogen atoms are products of dissociation, then the correspondent Lewis number is relatively small ($Le \ll 1$). In this case, diffusion of products is faster than the energy transfer, and energy cost of the products can be less with respect to ideal quenching (Equation 5.205).

Significant deviation of energy efficiency from the case of ideal quenching can be achieved due to the influence of strong external forces. For example, if relatively heavy molecular clusters are products of chemical processes in the fast-rotating plasma, centrifugal forces make the product flux much more intensive than the heat flux, resulting in significant increase of energy efficiency. This interesting effect takes place in the plasma chemical process of H_2S dissociation with formation of sulfur clusters and will be discussed in Chapter 11.

The previously discussed statistical approach (taking into account kinetic and transfer process limitations) was applied in Nester et al.[93] to analyze and calculate product composition, degree of conversion and energy efficiency of 156 different endothermic chemical processes stimulated in thermal plasma. Some examples of practically important quasi-equilibrium plasma chemical processes described in this manner are presented in Figure 5.32 through Figure 5.38. Product composition and energy cost are presented on the figures as functions of specific energy input.

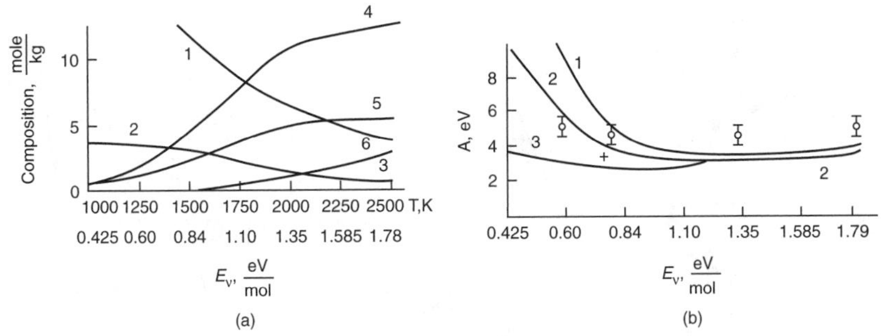

Figure 5.32 Thermal reaction in mixture S + CO_2. (a) Composition: (1) CO_2; (2) S_2; (3) S; (4) CO; (5) SO_2; (6) SO. (b) Energy price of CO formation: (1) ideal quenching; (2) super-ideal quenching; (3) process stimulated by CO_2 vibrational excitation. Experiments with sulfur vapor (⌶) with sulfur powder (+).

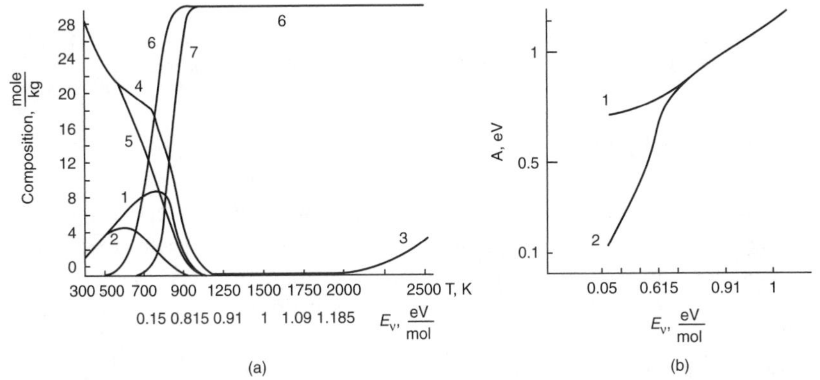

Figure 5.33 Thermal reaction in mixture C + H_2O. (a) Composition: (1) CO_2; (2) CH_4; (3) H; (4) C (solid); (5) H_2O; (6) H_2; (7) CO. (b) Energy price of CO/H_2 formation: (1) absolute quenching; (2) ideal quenching.

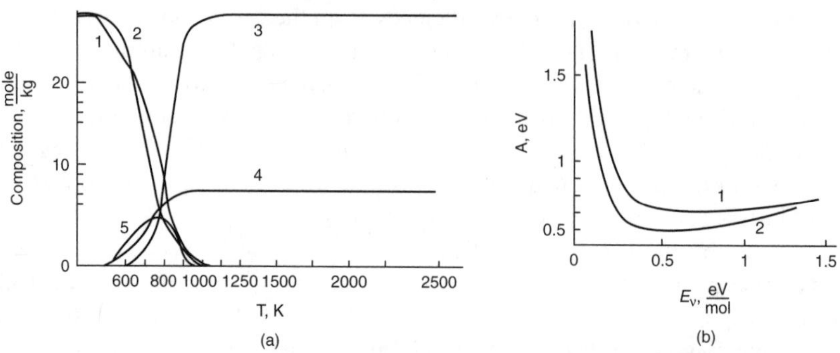

Figure 5.34 Thermal reaction in mixture $CH_4 + H_2O$. (a) Gas phase composition: (1) H_2O; (2) CH_4; (3) CO; (4) $H_2/10$; (5) CO_2. (b) Energy price of CO/H_2 formation: (1) absolute quenching; (2) ideal quenching.

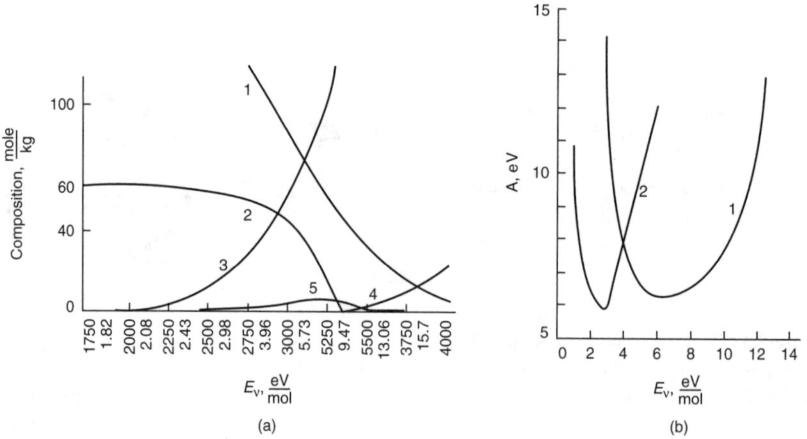

Figure 5.35 Thermal dissociation of CH_4. (a) Composition: (1) H_2; (2) C (solid); (3) H; (4) C; (5) C_2H_2. (b) Energy price of C_2H_2 formation: (1) absolute quenching A/10; (2) ideal quenching.

PROBLEMS AND CONCEPT QUESTIONS

5.1 **Diffusion of Molecules along the Vibrational Energy Spectrum.** Considering VV exchange with frequency $k_{vv}n_0$ as diffusion of molecules along the vibrational energy spectrum, estimate the average time necessary for a molecule to get the vibrational energy corresponding to the Treanor minimum point E_{Tr}.

5.2 **VT Relaxation Flux in Energy Space.** The kinetic equation (Equation 5.7) is often used to describe the vibrational distribution function at relatively low translational temperatures ($T_0 < \hbar\omega$), when the criterion Equation 5.6 is not valid. Analyze the steady-state solution of the kinetic equation (Equation 5.7) in this case and explain the contradiction.

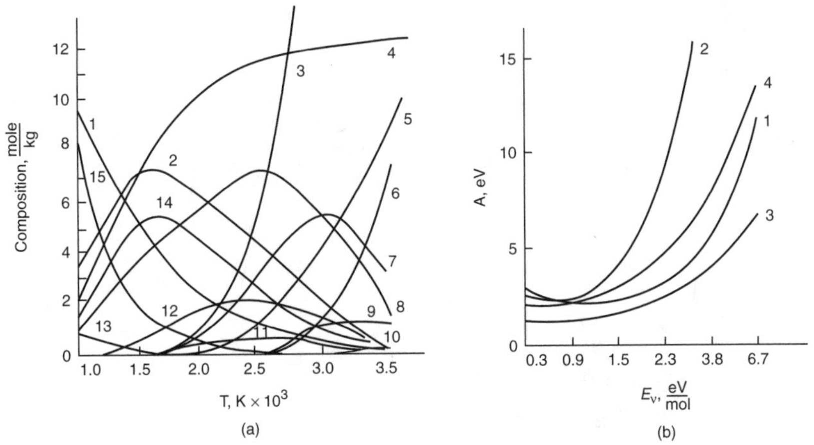

Figure 5.36 Thermal reaction mixture $H_2S + CO_2$. (a) Composition: (1) CO_2; (2) H_2O; (3) H; (4) CO; (5) S; (6) O; (7) SO; (8) H_2; (9) OH; (10) O_2; (11) HS; (12) SO_2; (13) COS; (14) S_2; (15) H_2S. (b) Energy price of CO/H_2 formation: (1) absolute quenching; (2) absolute quenching with respect to atomic sulfur; (3) ideal quenching; (4) ideal quenching with respect to atomic sulfur.

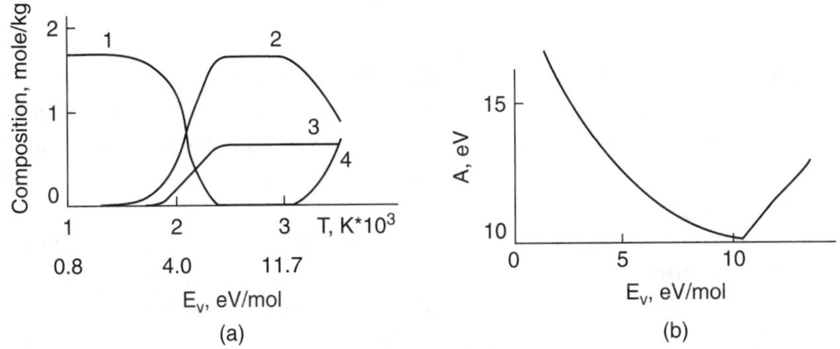

Figure 5.37 Thermal decomposition of ZrI_4. (a) Composition: (1) ZrI_4; (2) Zr (condensed phases); (3) I/10; 4) Zr. (b) Energy price of Zr formation.

5.3 **The Treanor Distribution Function and Criterion of the Fokker–Planck Approach.** Using the criterion (Equation 5.6), determine the conditions for application of the Fokker–Planck approach to the case of the Treanor vibrational distribution function. Are these conditions difficult to achieve in non-equilibrium plasma systems?

5.4 **Flux of Molecules and Flux of Quanta along the Vibrational Energy Spectrum.** Prove that the flux of molecules along the vibrational energy spectrum is equal to the derivative $\partial/\partial v$ of the flux of quanta in the energy space. Based on this, explain why the constant F in the kinetic equation (Equation 5.20) for the strong excitation regime is proportional to the flux of quanta in the energy space. Find the coefficient of proportionality between them.

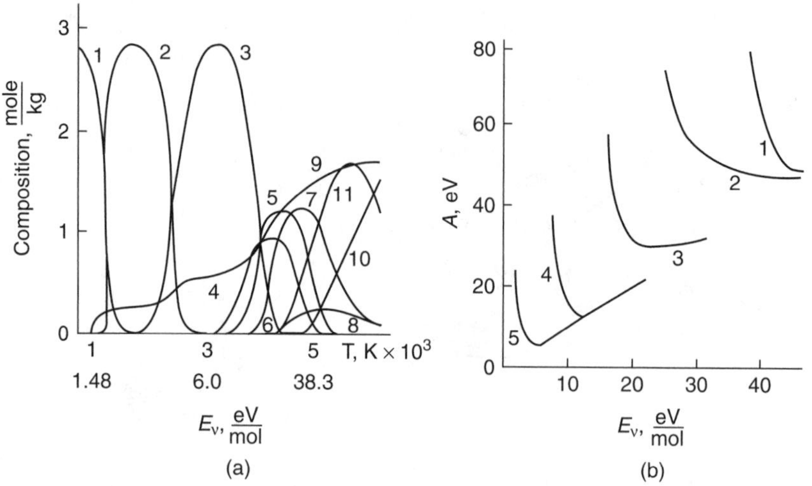

Figure 5.38 Thermal dissociation of UF_6. (a) Composition: (1) UF_6; (2) UF_5; (3) UF_4; (4) UF_3; (5) UF_2; (6) UF_2^+; (7) UF; (8) UF^+; (9) $F/10$; (10) U^+; (11) U. (b) Energy price of uranium atom formation: (1) absolute quenching; (2) ideal quenching—formation of UF_3 from UF_4; (3) ideal quenching—formation of UF_4 from UF_5; (4) ideal quenching—formation of UF_5 from UF_6; (5) ideal quenching.

5.5 **The Hyperbolic Plateau Distribution.** Derive the relation between the plateau coefficient C (Equation 5.22) and the degree of ionization in nonthermal plasma n_e/n_0. Compare the result with the plateau distribution (Equation 5.24) to find a relation between vibrational temperature and the degree of ionization and then analyze the result.

5.6 **Vibrational Distribution Functions in the Strong and Intermediate Excitation Regimes.** Compare numerically the criteria of the strong excitation ($\delta_{VV} T_v \geq 1$) and intermediate excitation ($x_e T_v^2 / T_0 \hbar\omega \geq 1$). Which one requires the larger value of vibrational temperature? Consider different levels of translational temperature T_0.

5.7 **The Gordiets Vibrational Distribution Function.** Compare the discrete Gordiets vibrational distribution with the continuous distribution function (Equation 5.29). Pay special attention to the exponential decrease of the vibrational distribution functions at high vibrational energies in the case of low translational temperatures $T_0 < \hbar\omega$.

5.8 **The eV Flux along the Vibrational Energy Spectrum.** Prove that differential one-quantum (Equation 5.36) and integral multiquantum (Equation 5.38) eV fluxes in energy space becomes equal to zero, if the vibrational distribution $f(E)$ can be presented as the Boltzmann function with temperature T_e: $f(E) \propto \exp(-E/T_e)$.

5.9 **eV Processes at Intermediate Ionization Degrees.** The vibrational distribution function for this case is $f(E) = \varphi(E)$ (Equation 5.45) at the weak excitation regime and $T_v > \hbar\omega/\alpha$. Using a more general equation integral (Equation 5.44), estimate the relevant vibrational distribution function at higher vibrational temperatures $T_v > \hbar\omega/\alpha$.

5.10 **Treanor Effect for Polyatomic Molecules.** Explain how the Treanor effect, which is especially related to the discrete anharmonic structure of vibrational levels (see Section 4.1.10), can be achieved in polyatomic molecules in the quasi continuum of vibrational states, in which there are no clear discrete structure of vibrational levels.

5.11 **The Treanor–Boltzmann Transition in Vibrational Distributions of Polyatomic Molecules.** Explain why the transition from the Treanor or Boltzmann distribution function $f(E)$ with temperature T_v (at low energies) to the Boltzmann distribution function with temperature T_0 (at high energies) takes place in polyatomic molecules at lower vibrational levels than the same transition in the case of diatomic molecules.

5.12 **Non-Steady-State Vibrational Distribution Function.** Using the approximation Equation 5.57, estimate the evolution of the average vibrational energy during the early stages of the VV exchange process, stimulated by the electron impact excitation of the lowest vibrational levels.

5.13 **Reactions of Vibrationally Excited Molecules, Fast Reaction Limit.** Calculate the integral (Equation 5.67), neglecting VT relaxation ($\xi(E) << 1$), and determine the explicit relation for the reaction rate of the chemical process, stimulated by vibrational excitation and limited by VV relaxation to reach the activation energy.

5.14 **Reactions of Vibrationally Excited Molecules, Slow Reaction Limit.** Equation 5.69 describes the reaction rates of excited molecules nonsymmetric with respect to the contribution of vibrational and translational degrees of freedom: vibrationally excited molecules have activation energy E_a/α, while other degrees of freedom have the activation energy E_a. Does this mean that the vibrational energy is less effective than translational energy? Surely not, so explain the "contradiction."

5.15 **Reactions of Vibrationally Excited Polyatomic Molecules.** The rate coefficient of the polyatomic molecule reactions (Equation 5.76) can be presented over a narrow range of vibrational temperatures in the simple Arrhenius manner $k_R^{macro} \propto \exp(-E_a/T_v)$. Determine the effective activation energy in this case as a function of vibrational temperature T_v and number of degrees of freedom s.

5.16 **Macrokinetics of Reactions of Two Vibrationally Excited Molecules.** For the case of the Treanor vibrational distribution function, show that the main contribution into the rate of reaction of two vibrationally excited molecules $E' + E'' = E_a$ is made by such collisions when all the energy is actually located on one molecule.

5.17 **Vibrational Energy Losses Due to VT Relaxation from High Levels.** Estimate the VT losses from high vibrational levels per one act of chemical reaction in the regime of strong excitation, when the reaction is controlled by the resonant VV exchange. Compare the result with Equation 5.87 and discuss the difference.

5.18 **Contribution of High and Low Levels in the Total Rate of VT Relaxation.** The relative contribution of high and low vibrational levels in VT relaxation depends on the degree of ionization in the plasma. Estimate the maximum degree of ionization when the total VT relaxation rate is still controlled by low excitation levels.

5.19 **Vibrational Energy Losses Due to Nonresonant VV Exchange.** Derivation of Equation 5.88 for calculating the VV exchange vibrational energy losses assumed predominantly nonresonant VV relaxation. However, this relation is correct for any kind of one-quantum VV exchange. Derive Equation 5.88 in a general way, comparing the starting "big" quantum after excitation by electron impact and the final "smaller" one after series of one-quantum exchange processes.

5.20 **VV and VT Losses of Vibrational Energy of Highly Excited Molecules.** Calculate the ratio of the vibrational energy losses related to VT relaxation of highly excited diatomic molecules to the correspondent losses during the series of the one-quantum VV relaxation process, which provided the formation of the highly excited molecule. What is the theoretically minimum value of the ratio? At which vibrational quantum number are these losses equal?

5.21 **Vibrational Energy Losses Related to Double-Quantum VV Exchange.** Estimate the vibrational energy losses related to VV exchange per one act of plasma chemical dissociation of a diatomic molecule stimulated by vibrational excitation. Assume domination of the one-quantum VV relaxation at lower vibrational levels until the double quantum resonance becomes possible (assume no VV losses at higher vibrational energies).

5.22 **Treanor Formula for Isotopic Mixtures.** Based on the Treanor formula (Equation 5.100) and assuming for an isotopic mixture that $\Delta\omega/\omega = 1/2\ \Delta m/m$, estimate the difference in vibrational temperatures of nitrogen molecules–isotopes at room temperature and averaged level of vibrational temperature in the system $T_v \approx 3000$ K.

5.23 **Isotopic Effects in Vibrational Kinetics and in Conventional Quasi-Equilibrium Kinetics.** Demonstrate that the isotopic effect in vibrational kinetics is usually stronger than the conventional one. Compare the typical values of the selectivity coefficients for these two isotopic effects. Is it possible for these two effects to take place together and to compensate each other? Give the numerical example.

5.24 **Coefficient of Selectivity for Separation of Heavy Isotopes.** Based on Equation 5.102 and Figure 5.13, estimate the possible level of the selectivity coefficient for separation of uranium isotopes, assuming a chemical process with UF_6 at room temperature. Assume the activation energy of the process is approximately 5eV. Discuss the result.

5.25 **Reverse Isotopic Effect in Vibrational Kinetics.** For the hydrogen–isotope separation (H_2–HD) plasma chemical process stimulated by vibrational excitation at room temperature find: (*a*) the typical value of the selectivity coefficient for the conventional direct isotopic effect in vibrational kinetics (E_a = 2 eV); (*b*) the typical value of the selectivity coefficient for the reverse isotopic effect; and (*c*) the relative temperature interval necessary for achievement of the reverse effect.

5.26 **Effect of eV Processes on Isotope Separation in Plasma.** Based on Equation 5.110, estimate the maximum value of the selectivity coefficient determined by the influence of eV processes in the hydrogen–isotope mixture (H_2–HD) in a plasma at room temperature. Compare the result with the typical values of the separation coefficient for conventional direct and reverse isotopic effects in nonequilibrium vibrational kinetics.

5.27 **Canonical Invariance of Rotational Relaxation Kinetics.** Based on the kinetic equation (Equation 5.128), prove the canonical invariance of the relaxation to equilibrium of the initial Boltzmann distribution function with rotational temperature exceeding the translational one. Show that the rotational distribution function $f(E_r, t)$ always maintains the same form of the Boltzmann distribution (Equation 5.131) during the relaxation to equilibrium, but obviously with changing values of the rotational temperature as it approaches the translational temperature.

5.28 **Hot Atoms Generated by Fast VT Relaxation.** In nonthermal plasma of molecular gases ($T_v >> T_0$) admixture of the light alkaline atoms is able to establish the quasi equilibrium with vibrational (not translational) degrees of freedom of the molecules. This can be explained by the high rate of the VT relaxation in this case and the large difference in masses that slows down the Maxwellization (TT exchange) processes. Conversely, why can the same effect not be achieved with addition of relatively heavy alkaline atoms when the Maxwellization process is also somewhat slower for the same reason of large differences in masses?

5.29 **Relation between Translation Temperature of Alkaline Atoms and Vibrational Temperature of Molecular Gas.** Using Equation 5.134 for the translational energy distribution function of small addition of alkaline atoms in a molecular gas, derive the relation between the corresponding temperatures: T_v, T_0, and T_{Me}, which can be measured experimentally. Estimate the Treanor effect of the molecular gas anharmonicity on the relation among the temperatures.

5.30 **Hot Atoms Generated in Fast Endothermic Plasma Chemical Reactions and Stimulated by Vibrational Excitation.** Compare the average energy of hot atoms generated in the fast endothermic plasma chemical reactions in the cases of strong, intermediate, and weak excitation. Use Equation 5.141 and assume the slow reaction limit in analyzing the type of the vibrational energy distribution function.

5.31 **Energy Efficiency of Quasi-Equilibrium and Nonequilibrium Plasma Chemical Processes.** Explain why the energy efficiency of the nonequilibrium plasma chemical process can be higher than that for the correspondent quasi-equilibrium process. Is it in contradiction with thermodynamic principles?

5.32 **Plasma Chemical Reactions Controlled by Dissociative Attachment.** For the chain reaction of water molecules' dissociation in plasma (Equation 5.147 and Equation 5.148), determine the chain length as a function of the degree of ionization. Estimate the related values of the energy cost for this mechanism as a function of the degree of ionization.

5.33 **The Treanor Effect in Vibrational Energy Transfer.** Estimate, using the relation Equation 5.6.8, how big the parameter $q = x_e T_v^2 / T_0 \hbar\omega$ in nitrogen plasma ($Q_{01}^{10} \approx 3 \cdot 10^{-3}$) should be to observe the Treanor effect in vibrational energy transfer ($\Delta D_v \approx D_0$).

5.34 **Stimulation of Vibrational–Translational Nonequilibrium by the Specific Heat Effect.** Assuming no energy exchange between vibrational and other degrees of freedom, calculate temperatures T_v and T_0 after mixing 1 mol of air at room temperature with an equal amount of air at 1000 K. Explain and discuss difference in the temperatures.

5.35 **Plasma Chemical Processes Stimulated by Vibrational Excitation of Molecules.** Explain why plasma chemical reactions can be effectively stimulated by vibrational excitation of molecules only if the specific energy input in nonthermal discharge exceeds the critical value, while reactions related to electronic excitation or dissociative attachment can proceed effectively at any levels of the specific energy input.

5.36 **Absolute and Ideal Quenching of Products of Chemical Reactions in Thermal Plasma.** Explain why energy efficiencies of the absolute and ideal quenching of the CO_2 dissociation products in a thermal plasma are identical while, in the case of dissociation of water molecules in thermal plasma, they are significantly different.

5.37 **Super-Ideal Quenching Effect Related to Vibrational–Translational Nonequilibrium.** How can vibrational–translational nonequilibrium be achieved during the quenching stage if the degree of ionization is usually not sufficient in this case to provide vibrational excitation faster than vibrational relaxation?

5.38 **Super-Ideal Quenching Effects Related to Selectivity of Transfer Processes.** Compare the conditions necessary for effective super-ideal quenching related to three different causes: (*a*) vibrational–translational nonequilibrium during the cooling process; (*b*) the Lewis number significantly differs from unity during separation and collection of products; and (*c*) cluster products move relatively fast from the rotating high-temperature discharge zone.

CHAPTER

6

ELECTROSTATICS, ELECTRODYNAMICS, AND FLUID MECHANICS OF PLASMA

6.1 ELECTROSTATIC PLASMA PHENOMENA: DEBYE RADIUS AND SHEATHS, PLASMA OSCILLATIONS, AND PLASMA FREQUENCY

6.1.1 Ideal and Nonideal Plasmas

The majority of plasmas are similar to gases from the point of view that, most of the time, electrons and ions are moving straight forward between collisions. This means that the interparticle potential energy $U \propto e^2/4\pi\varepsilon_0 R$, corresponding to the average distance between electrons and ions $R \approx n_e^{-1/3}$, is much less than their relative kinetic energy (which is approximately equal to temperature T_e, expressed in energy units):

$$\frac{n_e e^6}{(4\pi\varepsilon_0)^3 T_e^3} << 1 \tag{6.1}$$

Plasma satisfying this condition is called **ideal plasma.** The nonideal plasma (corresponding to the inverse of inequality Equation 6.1 and very high density of charged particles) is not found in nature. Even creation of such plasma in a laboratory is, so far, problematic. Therefore, only electrostatically ideal plasmas will be considered in this chapter. Nonideal plasma effects will only be considered in Section 11.8 dealing with dusty plasmas.

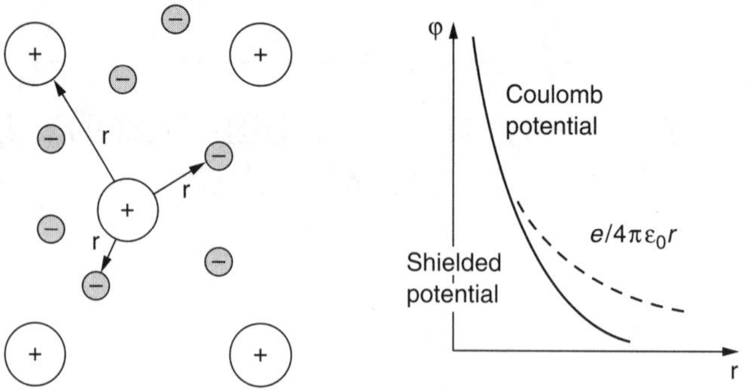

Figure 6.1 Plasma polarization around a charged particle

6.1.2 Plasma Polarization, "Screening" of Electric Charges, and External Electric Fields: Debye Radius in Two-Temperature Plasma

External electric fields induce plasma polarization, thus preventing penetration of the field inside plasma. Plasma "screens" the external electric field, which means this field exponentially decreases (protecting in this way the quasi neutrality of plasma; see Equation 4.129). Such an effect of the electrostatic "screening" of the electric field around a specified charged particle is illustrated in Figure 6.1.

To describe the space evolution of the potential φ induced by the external electric field, Poisson's equation can be used:

$$div\vec{E} = -\Delta\varphi = \frac{e}{\varepsilon_0}(n_i - n_e) \qquad (6.2)$$

Here, n_e and n_i are the number densities of electrons and positive ions and E is the electric field strength. One can assume the Boltzmann distribution for electrons (temperature T_e) and ions (temperature T_i), and the same quasi-neutral plasma concentration n_{e0} for electrons and ions:

$$n_e = n_{e0}\exp\left(+\frac{e\varphi}{T_e}\right), \quad n_i = n_{e0}\exp\left(-\frac{e\varphi}{T_i}\right) \qquad (6.3)$$

Combining the Poisson's equation with the Boltzmann distributions (Equation 6.3) and expanding in a Taylor series ($e\varphi << T_e, T_i$) results in the following equation:

$$\Delta\varphi = \frac{\varphi}{r_D^2}, \quad r_D = \sqrt{\frac{\varepsilon_0}{n_{e0}e^2\left(1/T_e + 1/T_i\right)}} \qquad (6.4)$$

In this equation, r_D is the **Debye radius**, a widely used and important electrostatic plasma parameter discussed earlier (see Section 4.3.10) in relation with the ambipolar diffusion and plasma quasi neutrality.

In the simplest one-dimensional case, Equation 6.4 becomes $d^2 \varphi/d^2x = \varphi/r_D^2$. This describes the penetration and exponential decrease of the external electric field (E_0 on the plasma boundary at $x = 0$) along the axis x ($x > 0$) perpendicular to the plasma boundary:

$$\vec{E} = -\nabla\varphi = \vec{E}_0 \exp\left(-\frac{x}{r_D}\right) \tag{6.5}$$

In a similar manner, reduction of the electric field of a specified charge q located in plasma (see Figure 6.1) can also be described by Equation 6.4, but in spherical symmetry:

$$\Delta\varphi \equiv \frac{1}{r}\frac{d^2}{dr^2}(r\varphi) = \frac{1}{r_D^2}\varphi \tag{6.6}$$

Taking into account the boundary condition $\varphi = q/4\pi\varepsilon_0 r$ at $r \to 0$, the solution of Equation 6.6 again results in the exponential decrease of electric potential (and electric field):

$$\varphi = \frac{q}{4\pi\varepsilon_0 r}\exp\left(-\frac{r}{r_D}\right) \tag{6.7}$$

Thus, the Debye radius gives the characteristic of the plasma size necessary for screening or suppressing the external electric field. Obviously, the same distance is necessary to compensate the electric field of one specified charged particle in plasma (Figure 6.1). It is important to note that the same Debye radius indicates the scale of plasma quasi neutrality (Equation 4.129).

There is some physical discrepancy between one-temperature (T_e) and two-temperature (T_e, T_i) relations for the Debye radius, between Equation 4.129 and Equation 6.4. According to the latter equation, if $T_e >> T_i$, the Debye radius depends mostly on the lower (ion) temperature, while according to Equation 4.129 and to common sense, it should depend on electron temperature. In reality, the heavy ions at low T_i are unable to establish the quasi-equilibrium Boltzmann distribution. It is correct to assume at $T_e >> T_i$ that the ions are at rest and $n_i = n_{e0} = const$. In this case, the term $1/T_i$ is negligible and the Debye radius can be found from Equation 4.129 or, numerically, from Equation 4.130.

6.1.3 Plasmas and Sheaths

Not all ionized gases are plasma. Plasma is supposed to be quasi neutral ($n_e \approx n_i$) and provide the screening of an external electric field and the field around a specified charged particle. To obtain qualities, the typical size of plasma should be much larger than the Debye radius. In Table 6.1, the characteristic parameters of some plasma systems, including comparison of their size and Debye radius, are given. As

Table 6.1 Debye Radius and Typical Size of Different Plasma Systems

Type of plasma	Typical n_e, cm^{-3}	Typical T_e, eV	Debye radius, cm	Typical size, cm
Earth ionosphere	10^5	0.03	0.3	10^6
Flames	10^8	0.2	0.03	10
He–Ne laser	10^{11}	3	0.003	3
Hg lamp	10^{14}	4	$3 \cdot 10^{-5}$	0.3
Solar chromosphere	10^9	10	0.03	10^9
Lightning	10^{17}	3	$3 \cdot 10^{-6}$	100

the table shows, the Debye radius in all examples presented in the table is much less than the typical size of a system, which means that all these types of ionized gases are real plasmas.

Although, in general, plasmas are quasi neutral ($n_e \approx n_i$), they contact the wall surfaces across positively charged thin layers called sheaths rather than across quasi-neutral layers. An example of a sheath between a plasma and zero-potential surfaces is illustrated in Figure 6.2. Formation of the positively charged sheaths is due to the fact that electrons are much faster than ions, and the electron thermal velocity $\sqrt{T_e/m}$ is about 1000 times faster than the ion's thermal velocity $\sqrt{T_i/M}$. The fast electrons are able to stick to the wall surface, leaving the area near the walls for positively charged ions alone.

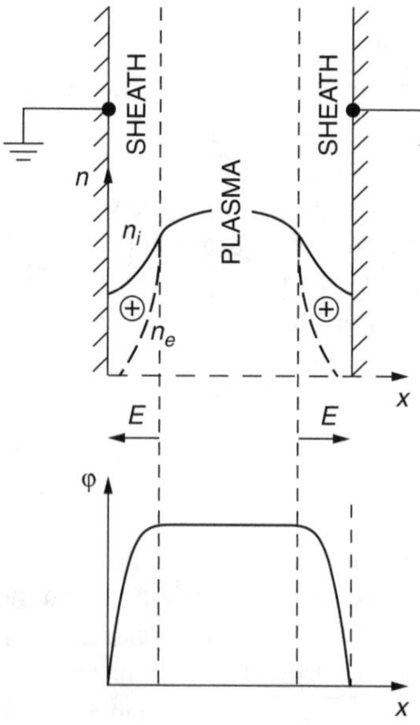

Figure 6.2 Plasma and sheaths

The positively charged sheath results in a typical potential profile for the discharge zone, also presented in Figure 6.2. The bulk of the plasma is quasi neutral and thus it is iso-potential ($\varphi = const$) for the case under consideration, according to the Poisson's equation (Equation 6.2). Near the discharge walls, the positive potential falls sharply, providing a high electric field, accelerating ions, and decelerating electrons, which again sustains the plasma quasi neutrality. It is interesting to note that because of this effect of ion acceleration in the sheath, the energy of ions bombarding the walls does not correspond to the ion temperature, but to the temperature of electrons.

6.1.4 Physics of DC Sheaths

Sheaths play a significant role in various discharge systems engineered for practically important applications related to surface treatment. Most of these systems operate at low gas temperatures ($T_e >> T_0, T_i$) and low pressures, where the sheath can be considered as collisionless. In this case, the basic one-dimensional equation governing the DC sheath potential φ in the direction perpendicular to the wall can be obtained from Poisson's equation, energy conservation for the ions, and Boltzmann distribution for electrons:

$$\frac{d^2\varphi}{dx^2} = \frac{en_s}{\varepsilon_0}\left[\exp\frac{\varphi}{T_e} - \left(1 - \frac{e\varphi}{E_i}\right)^{-1/2}\right] \tag{6.8}$$

In this equation, n_s is plasma density at the sheath edge; $E_{is} = 1/2\ Mu_{is}^2$ is the initial energy of an ion entering the sheath (u_{is} is the corresponding velocity); and the potential is assumed to be zero ($\varphi = 0$) at the sheath edge ($x = 0$). Multiplying Equation 6.8 by $d\varphi/dx$ and then integrating (assuming boundary conditions: $\varphi = 0$, $d\varphi/dx = 0$ at x = 0; see Figure 6.3) permits finding the electric field in the sheath (the potential gradient) as a function of potential:

$$\left(\frac{d\varphi}{dx}\right)^2 = \frac{2en_s}{\varepsilon_0}\left[T_e\exp\left(\frac{e\varphi}{T_e}\right) - T_e + 2E_{is}\left(1 - \frac{e\varphi}{E_{is}}\right)^{1/2} - 2E_{is}\right] \tag{6.9}$$

Equation 6.9 can be solved numerically to determine the potential distribution; however, a simple analysis leads to a very important conclusion. The solution of Equation 6.9 can exist only if its right-hand side is positive. Expanding Equation 6.9 to the second order in a Taylor series, leads to the conclusion that the sheath can exist only if the initial ion velocity exceeds the critical one; this is known as the Bohm velocity u_B:

$$u_{is} \geq u_B = \sqrt{T_e/M} \tag{6.10}$$

As can be seen, the Bohm velocity is equal to the velocity of ions having energy corresponding to the electron temperature. The condition (Equation 6.10) of a sheath existence is usually referred to as the **Bohm sheath criterion**.

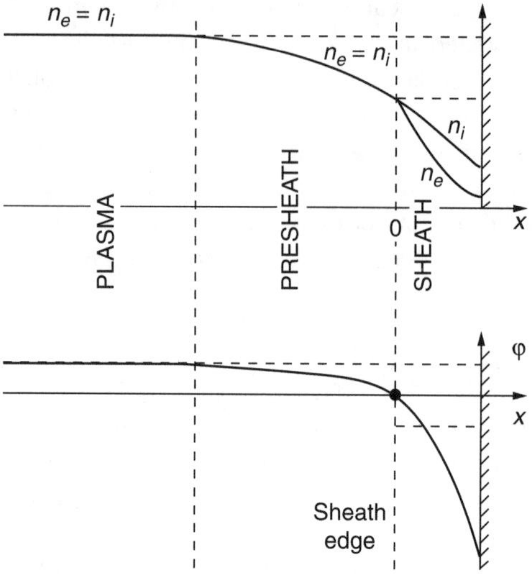

Figure 6.3 Sheath and presheath in contact with a wall

To provide ions with the energy and directed velocity necessary to satisfy the Bohm criterion, there must be a quasi-neutral region (wider than the sheath, e.g., several r_D) with some electric field. This region, illustrated in Figure 6.3, is called the **presheath.** Obviously, the minimum presheath potential (between bulk plasma and sheath) should be equal to:

$$\varphi_{presheath} \approx \frac{1}{2e} M u_B^2 = \frac{T_e}{2e} \tag{6.11}$$

Balancing the ion and electron fluxes to the floating wall yields the expression for the change of potential across the sheath as:

$$\Delta\varphi = \frac{1}{e} T_e \ln \sqrt{M/2\pi m} \tag{6.12}$$

Because the ion-to-electron mass ratio M/m is large, the change of potential across the sheath (even at a floating wall) exceeds five to eight times that across the presheath in this case of a floating potential. Numerical calculations based (Equation 6.9) a typical width of sheath as about a few Debye radii ($s \approx 3r_D$).

6.1.5 High-Voltage Sheaths: Matrix and Child Law Sheath Models

Equation 6.12 was related to the floating potential that exceeds electron temperatures five to eight times. However, in general, the change of potential across a sheath (sheath voltage V_0) is often driven to be very large compared to electron temperature T_e/e. In this case, the electron concentration in the sheath can be neglected and only ions need be taken into account. As an interesting consequence of the electron absence, the sheath region appears dark when observed.

The simplest model of such a high-voltage sheath assumes uniformity of the ion density in the sheath. This sheath is usually referred to as the **matrix sheath**. In the framework of the simple matrix sheath model, the sheath thickness can be expressed in terms of the Debye radius r_D corresponding to plasma concentration at the sheath edge:

$$s = r_D \sqrt{\frac{2V_0}{T_e}} \tag{6.13}$$

When voltage is high enough numerically, the matrix sheath thickness can be large and exceed the Debye radius 10 to 50 times,.

The more accurate approach to describing the sheath should take into account the decrease of the ion density as the ions accelerate across the sheath. This is done in a model of the so-called **Child law sheath**. In the framework of this model, the ion current density $j_0 = n_s e u_B$ is taken equal to that of the well-known **Child law of space-charge-limited current** in a plane diode:

$$j_0 = n_s e u_B = \frac{4\varepsilon_0}{9}\sqrt{\frac{2e}{M}}\frac{1}{s^2}V_o^{3/2} \tag{6.14}$$

Considering the Child law (Equation 6.14) as an equation with respect to s, the relation for the thickness of the Child law sheath is:

$$s = \frac{\sqrt{2}}{3} r_D \left(\frac{2V_0}{T_e}\right)^{3/4} \tag{6.15}$$

Numerically, the Child law sheath can be of an order of 100 Debye lengths in typical low-pressure discharges applied for surface treatment. More details regarding sheaths, including collisional sheaths, sheaths in electronegative gases, radio-frequency plasma sheaths, and pulsed potential sheaths can be found in Lieberman and Lichtenberg.[13]

6.1.6 Electrostatic Plasma Oscillations: Langmuir or Plasma Frequency

It was shown previously that the typical space size characterizing plasma (scale of electroneutrality and screening of external fields) is the Debye radius. In the same manner, the typical plasma time scale and typical time of a plasma response to external fields is determined by the plasma frequency.

Consider the electrostatic plasma oscillations illustrated in Figure 6.4. Assume in a one-dimensional approach that all electrons at $x > 0$ are initially shifted to the right on the distance x_0, while heavy ions are not perturbed and located in the same position. This will result in the electric field pushing the electrons back. If $E = 0$ at $x < 0$, this electric field restoring the plasma quasi neutrality can be found at $x > x_0$ from the one-dimensional Poisson's equation as:

$$\frac{dE}{dx} = \frac{e}{\varepsilon_0}(n_i - n_e), \quad E = -\frac{e}{\varepsilon_0} n_{e0} x_0 \ (at\ x > x_0) \tag{6.16}$$

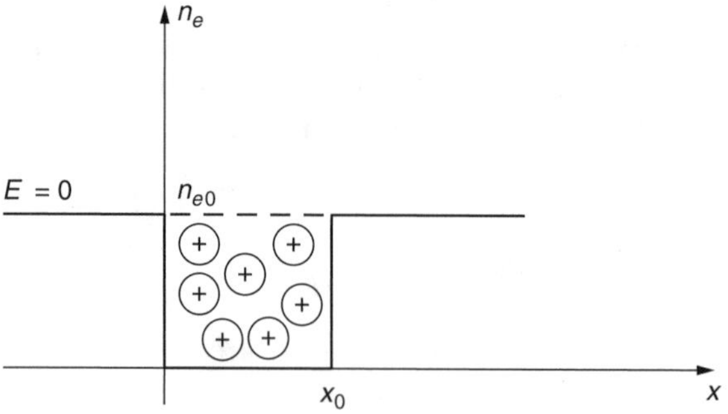

Figure 6.4 Electron density distribution in plasma oscillations

This electric field pushes all the initially shifted electrons to move back to the left ($e < 0$, see Figure 6.4) with their boundary (at $x = x_0$). The resulting electron motion is described by the equation of electrostatic plasma oscillations:

$$\frac{d^2 x_0}{dt^2} = -\omega_p^2 x_0, \quad \omega_p = \sqrt{\frac{e^2 n_e}{\varepsilon_0 m}} \tag{6.17}$$

Here, the parameter ω_p is the **Langmuir frequency** or **plasma frequency**. This is a very important plasma characteristic that determines the time scale of a plasma response to external electric perturbations and will appear repeatedly is analyzing plasma oscillations, waves, and other phenomena.

Comparing Equation 6.7 and Equation 4.129 for plasma frequency and Debye radius, it is interesting to note that:

$$\omega_p \times r_D = \sqrt{2T_e/m} \tag{6.18}$$

The product of the Debye radius and the plasma frequency gives thermal velocity of electrons, a fact that has a very clear physical interpretation. The time of a plasma reaction to an external perturbation ($1/\omega_p$) corresponds to the time required by a thermal electron (with velocity $\sqrt{2T_e/m}$) to travel the distance r_D necessary to provide screening of the external perturbation.

Because the plasma (or Langmuir) frequency depends only on the plasma density, it can conveniently be calculated by the following numerical formula:

$$\omega_p\left(\sec^{-1}\right) = 5.65 \cdot 10^4 \sqrt{n_e\left(cm^{-3}\right)} \tag{6.19}$$

Although different types of electrical discharges have different plasma frequencies, these usually reside in the microwave frequency region of 1 to 100 GHz.

6.1.7 Penetration of Slow Changing Fields into Plasma: Skin Effect

The problem of penetration of oscillating fields into plasmas is not only an electrostatic phenomenon. However, it is helpful to discuss the skin effect because of the general importance of this concept as another space scale of plasma response to external fields. Consider the penetration of a changing field (actually electromagnetic) into a plasma with frequency less than the plasma frequency ($\omega < \omega_p$). In this case, Ohm's law can be applied in the simple form:

$$\vec{j} = \sigma \vec{E} \tag{6.20}$$

where $\vec{j}$ is the current density in plasma; $\vec{E}$ is the electric field strength; and σ is plasma conductivity corresponding to constant electric field (see Section 4.3.2). To describe the evolution of the field (which is actually electromagnetic), penetrating into plasma, the Maxwell equation for curl of magnetic field must be taken into account

$$curl\,\vec{H} = \vec{j} + \varepsilon_0 \frac{\partial \vec{E}}{\partial t} \tag{6.21}$$

Assume now that the frequency of the field under consideration is also low with respect to the plasma conductivity and thus the displacement current (second current term in Equation 6.21) can be neglected. Then, combining Equation 6.20 and Equation 6.21 gives the relations help between the changing electric and magnetic fields:

$$curl\,\vec{H} = \sigma \vec{E} \tag{6.22}$$

Taking the electric field from Equation 6.22 and substituting into the Maxwell equation for the curl of electric field:

$$curl\,\vec{E} = -\mu_0 \frac{\partial \vec{H}}{\partial t} \tag{6.23}$$

yields the final differential relation for the amplitude of electromagnetic field decrease during penetration into a plasma:

$$\frac{\partial \vec{H}}{\partial t} = -\frac{1}{\mu_0 \sigma} curl\,curl\,\vec{H} = -\frac{1}{\mu_0 \sigma} \nabla\left(div\vec{H}\right) + \frac{1}{\mu_0 \sigma} \Delta \vec{H} = \frac{1}{\mu_0 \sigma} \Delta \vec{H} \tag{6.24}$$

Here, the conventional relation for double curl and the Maxwell equation, $div\vec{H} = 0$, was taken into account. It should be noted that a similar equation can be derived for an electric field penetrating into a plasma. Equation 6.24 describes the decrease of amplitude of low-frequency electric and magnetic fields during their penetration into a plasma. The characteristic space scale of this decrease can easily be found from Equation 6.24 as:

$$\delta = \sqrt{\frac{2}{\omega \mu_0 \sigma}} \tag{6.25}$$

If this space-scale δ is small with respect to the plasma size, then the external fields and currents are located only on the plasma surface layer (within the depth δ). This effect is known as the **skin effect**, and the boundary layer where the external fields penetrate and where plasma currents are located is usually referred to as the **skin layer**.

As one can see from Equation 6.25, the depth of skin layer depends on the frequency of the electromagnetic field ($f = \omega/2\pi$) and the plasma conductivity. For calculating the skin layer depth, it is convenient to use the following numeric formula:

$$\delta(\text{cm}) = \frac{5.03}{\sigma^{1/2}(1/\text{Ohm}\cdot\text{cm})\cdot f^{1/2}(\text{MHz})} \tag{6.26}$$

In this section, only the general concept of the skin layer was introduced. More detailed considerations of electrostatic and electromagnetic wave propagation in a plasma will take place later in this chapter.

6.2 MAGNETO-HYDRODYNAMICS OF PLASMA

In this section, the general aspects of plasma behavior in magnetic fields will be discussed and, in particular, magneto-hydrodynamics (MHD) of plasmas. The magneto-hydrodynamic approach is most useful in the case of highly ionized or completely ionized plasma. Initially, this was especially important in problems relating to controlled thermonuclear fusion; it is important now for low-temperature plasma applications, such as the plasma centrifuges, and specialized propulsion systems.

6.2.1 Equations of Magneto-Hydrodynamics

Alfven[172] first pointed out the interesting behavior of a plasma in magnetic fields. Motion of a high-density plasma (the degree of ionization is assumed high in this section) induces electric currents that, together with the magnetic field, influence the motion of the plasma. Such phenomena can be described by the system of magneto-hydrodynamic equations, including:

a) The Navier–Stokes (N.S.) equation neglecting viscosity, but taking into account the magnetic ampere force acting on the plasma current with density $\vec{j}$ (noting that B is magnetic induction; M is the mass of ions; and $n_e = n_i$ is the plasma density, then $Mn_e = \rho$):

$$Mn_e\left[\frac{\partial\vec{v}}{\partial t} + (\vec{v}\nabla)\vec{v}\right] + \nabla p = \left[\vec{j}\vec{B}\right] \tag{6.27}$$

b) The continuity equations for plasma electrons and ions, moving together as a fluid with the same macroscopic velocity, $\vec{v}$:

$$\frac{\partial n_e}{\partial t} + div(n_e\vec{v}) = 0 \tag{6.28}$$

c) The Maxwell equations for magnetic field, neglecting the displacement current because of relatively low velocities:

$$curl\,\vec{H} = \vec{j}, \quad div\,\vec{B} = 0 \tag{6.29}$$

d) The Maxwell equation for electric field $curl\,E = -\partial\vec{B}/\partial t$ together with the first equation of Equation 6.29 and Ohm's law ($\vec{j} = \sigma(\vec{E} + [\vec{v}\vec{B}])$ for a plasma with conductivity σ gives the relation for the magnetic field:

$$\frac{\partial\vec{B}}{\partial t} = curl[\vec{v}\vec{B}] + \frac{1}{\sigma\mu_0}\Delta\vec{B} \tag{6.30}$$

The system of magneto-hydrodynamic equations is useful for describing large-scale plasma motion in magnetic fields when the peculiarities in behavior of different plasma components are not essential. In particular, MHD is able to describe plasma equilibrium and confinement in magnetic fields, which is of great importance in problems related to thermonuclear plasma systems.[173,174,212a]

6.2.2 Magnetic Field "Diffusion" in Plasma: Effect of Magnetic Field Frozen in Plasma

Some interesting physical phenomena of plasma interactions with magnetic field can be analyzed based on magneto-hydrodynamic Equation 6.30. First, if the plasma is at rest ($\vec{v} = 0$), Equation 6.30 can be reduced to the diffusion equation:

$$\frac{\partial\vec{B}}{\partial t} = D_m\Delta\vec{B}, \quad D_m = \frac{1}{\sigma\mu_0} \tag{6.31}$$

The factor D_m in this equation can be interpreted as a coefficient of "diffusion" of the magnetic field into the plasma; sometimes it is also called the **magnetic viscosity**. If the characteristic time of the magnetic field change is $\tau = 1/\omega$, then the characteristic length of magnetic field diffusion according to Equation 6.31 is $\delta \approx \sqrt{2D_m\tau} = \sqrt{2/\sigma\omega\mu_0}$, which again is the skin-layer depth (Equation 6.25).

Also, Equation 6.31 can be applied to describe the damping time τ_m of currents and magnetic fields in a conductor with characteristic size *L*:

$$\tau_m = \frac{L^2}{D_m} = \mu_0\sigma L^2 \tag{6.32}$$

According to Equation 6.32, for example, the damping time is infinitely high for superconductors. The magnetic field damping time is also very long for large special objects; for solar spots, this time exceeds 300 years.

Equation 6.31 and Equation 6.32 can be interpreted in another important way. If the plasma conductivity is high ($\omega \to \infty$), the diffusion coefficient of the magnetic field is small ($D_m \to 0$) and the magnetic field is unable to "move" with respect to

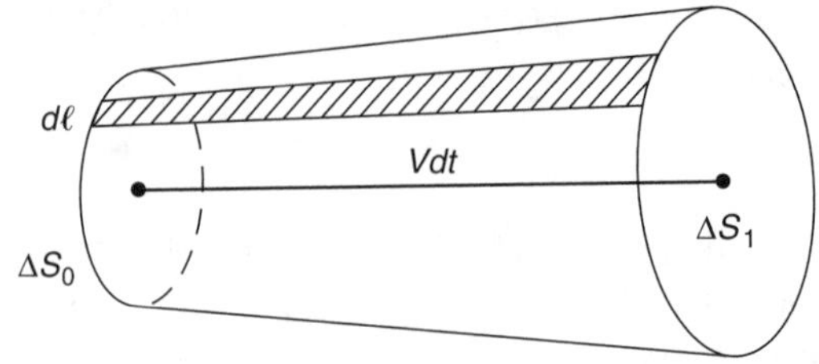

Figure 6.5 Magnetic field frozen in plasma

the plasma. Thus, the magnetic field sticks to plasma or, in other words, the **magnetic field is frozen in the plasma**.

To demonstrate the effect of frozen magnetic field, consider displacement (correspondent to time interval *dt*) of the surface element ΔS, moving with velocity $\vec{v}$ together with plasma (see Figure 6.5). The effect takes place when the plasma conductivity is high ($\sigma \to \infty$) and the second left-side term of Equation 6.30 can be neglected:

$$\frac{\partial \vec{B}}{\partial t} = curl[\vec{v}\vec{B}] \tag{6.33}$$

Because of $div\,\vec{B}=0$, difference in fluxes $\vec{B}\,d\vec{S}/dt$ through surface elements ΔS_0 and ΔS_1 is equal to the flux $\int \vec{B}[\vec{v}\,d\vec{l}\,] = \int [\vec{B}\vec{v}]d\vec{l}$ through the side surface of the fluid element in Figure 6.5. Based on this fact and on Equation 6.33, the time derivative of the magnetic field flux $\int \vec{B}d\vec{S}$ through the moving surface element ΔS can be expressed as:

$$\begin{aligned}\frac{d}{dt}\int \vec{B}d\vec{S} &= \int \frac{\partial \vec{B}}{\partial t}d\vec{S} + \int \vec{B}\frac{d\vec{S}}{dt} = \int \frac{\partial \vec{B}}{\partial t}d\vec{S} + \oint \left[\vec{B}\vec{v}\right]d\vec{l} \\ &= \int curl\left[\vec{v}\vec{B}\right]d\vec{S} + \oint \left[\vec{B}\vec{v}\right]d\vec{l}\end{aligned} \tag{6.34}$$

According to Stokes' theorem, the last sum of the equality (Equation 6.34) is equal to zero, resulting in the conclusion:

$$\frac{d\Phi}{dt} = \frac{d}{dt}\int \vec{B}\,d\vec{S} = 0, \quad \Phi = const \tag{6.35}$$

Equation 6.35 shows that magnetic flux Φ through any surface element moving with the plasma is constant. It again shows that the magnetic field is "frozen" in a moving plasma with high conductivity. This effect significantly simplifies understanding the behavior of magnetic fields related to flows of highly ionized plasma.

6.2.3 Magnetic Pressure: Plasma Equilibrium in Magnetic Field

Under steady-state plasma conditions ($d\vec{v}/dt = 0$), the Navier–Stokes equation 6.27 can be simplified to:

$$grad\, p = [\vec{j}\vec{B}] \tag{6.36}$$

This equation can be interpreted as a balance of hydrostatic pressure p and ampere force exerted on plasma. Taking into account the first part of Equation 6.29, eliminate the current from the balance of forces (Equation 6.36):

$$\nabla p = \left[\vec{j}\vec{B}\right] = \mu_0\left[curl\,\vec{H} \times \vec{H}\right] = -\frac{\mu_0}{2}\nabla H^2 + \mu_0\left(\vec{H}\nabla\right)\vec{H} \tag{6.37}$$

Combination of the gradients in Equation 6.37 leads to the following equation describing plasma equilibrium in magnetic field:

$$\nabla\left(p + \frac{\mu_0 H^2}{2}\right) = \mu_0\left(\vec{H}\nabla\right)\vec{H} = \frac{\mu_0 H^2}{R}\vec{n} \tag{6.38}$$

Here, R is the radius of curvature of the magnetic field line and $\vec{n}$ is the unit normal vector to the line (directed inside to the curvature center). Thus, the force $\mu_0 H/R\,\vec{n}$ is related to bending of the magnetic field lines and can be interpreted as the **tension of magnetic lines**. This tension tends to make the magnetic field lines straight and is equal to zero when the lines are straight. The pressure term $\mu_0 H^2/2$ is usually called the **magnetic pressure.** The sum of the hydrostatic and magnetic pressures $p + \mu_0 H^2/2$ is referred to as the total pressure.

Equation 6.38 for plasma equilibrium in magnetic field can be considered as the dynamic balance of the gradient of total pressure and the tension of magnetic lines. If the magnetic field lines are straight and parallel, then $R \to \infty$ and the "tension" of magnetic lines is equal to zero. In this case, Equation 6.38 gives the equilibrium criterion:

$$p + \frac{\mu_0 H^2}{2} = const \tag{6.39}$$

Thus, plasma equilibrium in a magnetic field actually means a balance of pressure in plasma with the outside magnetic pressure (compare this statement with Equatiobn 6.36). This balance is the most typical condition of the plasma equilibrium in magnetic field. However, plasma equilibrium is also sometimes possible for special configurations of magnetic fields when $p << \mu_0 H^2/2$. In this case, the outside magnetic pressure is compensated by the "tension" of magnetic lines. Such equilibrium configurations of magnetic field are called forceless configurations.

6.2.4 The Pinch-Effect

This effect means self-compression of a plasma in its own magnetic field. Consider the Pinch effect in a long cylindrical discharge plasma with the electric current along the axis of the cylinder (see Figure 6.6). This is the so-called Z-pinch.[175] Equilibrium of the completely ionized Z-pinch plasma is determined by Equation 6.36 or Equation 6.38, combined with the first Maxwell (Equation 6.29) for the self magnetic field of the plasma column:

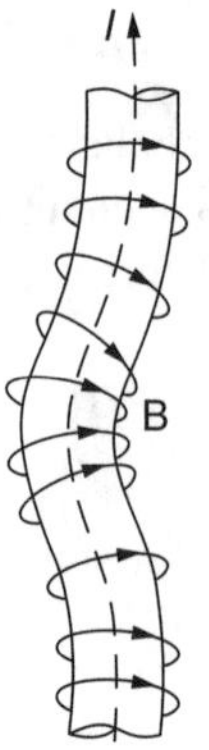

Figure 6.6 Pinch effect when discharge channel is bent

$$N_L T = \frac{\mu_0}{8\pi} I^2 \tag{6.40}$$

This equilibrium criterion is called the **Bennet relation.**[176] In this relation, N_L is the plasma density per unit length of the cylinder and T is the plasma temperature. The Bennett relation shows that plasma temperature should grow proportionally to the square of the current to provide a balance of plasma pressure and magnetic pressure of the current. Thus, to reach thermonuclear temperatures of about 100 keV in a plasma with density of 10^{15} cm^{-3} and cross section of 1 cm^2, the necessary current, according to Equation 6.40, should be about 100 kA.

A current of O(100 KA) can be achieved in Z-pinch discharges, which in the early 1950s stimulated enthusiasm in controlled thermonuclear fusion research. However, hopes for easily controlled fusion in Z-pinch discharges were shattered because very fast instabilities arose to destroy the plasma. Related to plasma bent and nonuniform plasma compression, these Z-pinch instabilities are illustrated in Figure 6.6.

If the discharge column is bent, the magnetic field and magnetic pressure become larger on the concave side of the plasma, which leads to a break of the channel. This instability of Z-pinch is usually referred to as the "wriggle" instability. Similar to this instability, if the discharge channel becomes locally thinner, the magnetic field ($B \propto 1/r$) and magnetic pressure at this point grow, leading to further compression and to a subsequent break of the channel.

6.2.5 Two-Fluid Magneto-Hydrodynamics: Generalized Ohm's Law

In the magneto-hydrodynamic approach, the electron $\vec{v}_e$ and ion $\vec{v}_i$ velocities were considered equal. Actually, this is in contradiction with the presence of electric current $\vec{j} = en_e(\vec{v}_i - \vec{v}_e)$ in the quasi-neutral plasma with density n_e. Thus, it is clear that effects related to separate motion of electrons and ions require consideration under a two-fluid model of magneto-hydrodynamics. In such a model, the

Navier–Stokes equation for electrons is somewhat similar to Equation 6.1, but includes electron mass m, pressure p_e, and velocity and, additionally, takes into account the friction between electrons and ions (which corresponds to the last term in the equation, where ν_e is the frequency of electron collisions):

$$mn_e \frac{d\vec{v}_e}{dt} + \nabla p_e = -en_e\vec{E} - en_e\left[\vec{v}_e\vec{B}\right] - mn_e\nu_e\left(\vec{v}_e - \vec{v}_i\right) \tag{6.41}$$

Obviously, a similar Navier–Stokes equation for ions, including the same friction term but with opposite sign, can be written.

The first term in Equation 6.41 is related to the electron inertia and can be neglected because of very low electron mass. Also, denoting the ion's velocity as $\vec{v}$ and the plasma conductivity as $\sigma = n_e e^2/m\nu_e$, Equation 6.41 can be rewritten in the form known as the **generalized Ohm's law**:

$$\vec{j} = \sigma(\vec{E} + [\vec{v}\vec{B}]) + \frac{\sigma}{en_e}\nabla p_e - \frac{\sigma}{en_e}[\vec{j}\vec{B}] \tag{6.42}$$

The generalized Ohm's law differs from the conventional one because it takes into account two additional terms: electron pressure gradient and the $[\vec{j}x\vec{B}]$ term related to the **Hall effect**. These two additional terms show that current direction is not always correlated straight with the direction of the electric field (even corrected by $[\vec{v}\vec{B}]$).

The Hall effect is related to electron conductivity in the presence of a magnetic field that provides an electric current in the direction perpendicular to electric field. This effect was discussed earlier in Section 4.3.4 (see Equation 4.108) in terms of plasma drift in crossed electric and magnetic fields. The generalized Ohm's law can be applied as well as the equations of motion for individual species to analyze different types of charged particle drifts in magnetic fields (see Thompson[177]).

Solution of Equation 6.42 with respect to electric current is complicated because of the presence of current in two terms of the generalized Ohm's law. This equation can be solved if the plasma conductivity is sufficiently large. In this case ($\sigma \to \infty$), the generalized Ohm's law can be expressed in the form:

$$\vec{E} + [\vec{v}\vec{B}] + \frac{1}{en_e}\nabla p_e = \frac{1}{en_e}[\vec{j}\vec{B}] \tag{6.43}$$

This is more convenient for further analysis, including detailed consideration of the Hall effect in plasma.[178]

If the electron temperature is uniform in space, the electron hydrodynamic velocity $\vec{v}$ can be used and the generalized Ohm's law (Equation 6.43) rewritten as:

$$\vec{E} = -[\vec{v}_e\vec{B}] - \frac{1}{e}\nabla(T_e \ln n_e) \tag{6.44}$$

Using this expression for the electric field in the Maxwell equation *curl* $\vec{E} = \partial\vec{B}/\partial t$ and taking into account that curl of gradient is equal to zero:

$$\frac{\partial \vec{B}}{\partial t} = curl\left[\vec{v}_e \vec{B}\right] \tag{6.45}$$

This form of the two-fluid MHD equation is similar to Equation 6.33 for the one-fluid MHD, which was applied to determine the effect of a magnetic field frozen in plasma (see Section 6.2.2); only the common plasma velocity $\vec{v}$ in Equation 6.33 is replaced in Equation 6.45 by electron gas velocity $\vec{v}_e$. Thus, it follows that the magnetic field in the framework of the two-fluid MHD approximation is frozen in the electron gas. This interesting conclusion also shows that qualitative variations between the one-fluid and two-fluid MHD models can be expected only when significant differences exist between electron and ion fluid velocities.

6.2.6 Plasma Diffusion across Magnetic Field

The generalized Ohm's law (Equation 6.42) generated with the two-fluid magneto-hydrodynamics model is able to describe electron and ion diffusion in a direction perpendicular to the uniform magnetic field. Obviously, this effect is of great importance for a variety of problems related to plasma confinement in a magnetic field, so calculations of the diffusion coefficients for electrons and ions in this direction in the framework of this model gives:[179]

$$D_{\perp,e} = \frac{D_e}{1+\left(\dfrac{\omega_{B,e}}{\nu_e}\right)^2}, \quad D_{\perp,i} = \frac{D_i}{1+\left(\dfrac{\omega_{B,i}}{\nu_i}\right)^2} \tag{6.46}$$

In this relation, D_e and D_i are the coefficients of free diffusion of electrons and ions in a plasma without magnetic field (see Section 4.3.7); ν_e and ν_i are the collisional frequencies of electrons and ions; and $\omega_{B,e}$ and $\omega_{B,i}$ are the electron and ion cyclotron frequencies (see Equation 4.106):

$$\omega_{B,e} = \frac{eB}{m}, \quad \omega_{B,i} = \frac{eB}{M} \tag{6.47}$$

which show the frequencies of electron and ion collisionless rotation in magnetic field. Because ions are much heavier than electrons ($M >> m$), electron cyclotron frequency is much greater than ion cyclotron frequency in the same magnetic field B.

If diffusion is ambipolar, which is actually the case in the highly ionized plasma under consideration, then Equation 4.126 can still be applied to find D_a in a magnetic field. Obviously, the free diffusion coefficients for electrons and ions in this equation should be replaced by those in the magnetic field (Equation 6.46). The regular electron and ion mobilities μ_e, μ_i in Equation 4.126 also should be replaced by those corresponding to the drift perpendicular to the magnetic field (Equation 4.108). This leads to the following coefficient of ambipolar diffusion perpendicular to the magnetic field:

$$D_{\perp} = \frac{D_a}{1 + \frac{\omega_{B,i}^2}{\nu_i^2} + \frac{\mu_i}{\mu_e}\left(1 + \frac{\omega_{B,e}^2}{\nu_e^2}\right)} \tag{6.48}$$

Here, D_a is the regular coefficient of ambipolar diffusion (Equation 4.126) in a plasma without magnetic field. Equation 6.48 for ambipolar diffusion in a plasma in the direction perpendicular to the magnetic field can be simplified when electrons are magnetized, $\omega_{B,e}/\nu_e >> 1$. This means the electrons are trapped in the magnetic field and rotating around the magnetic lines. If at the same time heavy ions are not magnetized ($\omega_{B,i}/\nu_i << 1$), then Equation 6.48 can be rewritten as:

$$D_{\perp} = \frac{D_a}{1 + \frac{\mu_i}{\mu_e}\frac{\omega_{B,e}^2}{\nu_e^2}} \tag{6.49}$$

Under strong magnetic fields, when $\mu_i/\mu_e\ \omega_{B,e}^2/\nu_e^2 >> 1$, the relation for ambipolar diffusion can be further simplified as:

$$D_{\perp} \approx D_a \frac{\mu_e}{\mu_i}\frac{\nu_e^2}{\omega_{B,e}^2} \approx D_e \frac{\nu_e^2}{(e/M)^2}\frac{1}{B^2} \tag{6.50}$$

This shows that, in strong magnetic fields, the coefficient $D_{\perp}$ significantly decreases with the strength of the magnetic field: $D_{\perp} \propto 1/B^2$. The magnetic field is not transparent for the plasma and can be used to prevent the plasma from decay. This, as well as the MHD plasma equilibrium, is very important for plasma confinement in a magnetic field and has been experimentally demonstrated by D'Angelo and Rynn.[180]

The slow ambipolar diffusion across the strong magnetic field can be interpreted using the illustration in Figure 6.7. The magnetized electron is trapped by the magnetic field and rotates along the Larmor circles until a collision pushes the electron to another Larmor circle. The **electron Larmor radius**:

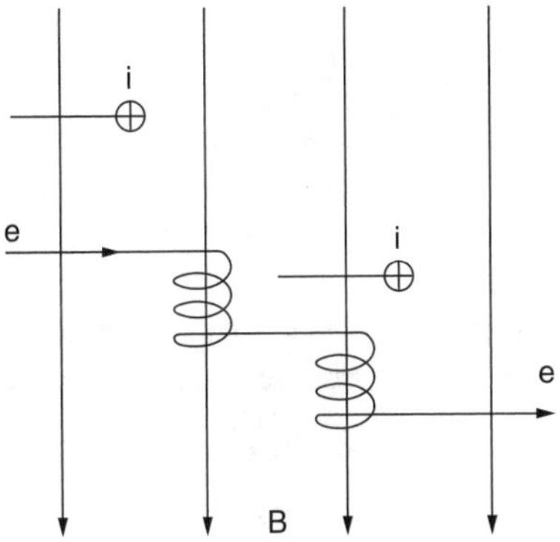

Figure 6.7 Ambipolar diffusion across magnetic field

$$\rho_L = \frac{v_\perp}{\omega_{B,e}} = \frac{1}{eB}\sqrt{2T_e m} \tag{6.51}$$

is the radius of the circular motion of a magnetized electron (see Figure 6.7); here, $v_\perp$ is the component of the electron thermal velocity perpendicular to the magnetic field. In the case of diffusion across the strong magnetic field, the Larmor radius actually plays the same role as the mean free path plays in diffusion without a magnetic field. This can be clearly illustrated by rewriting Equation 6.50 in terms of the electron Larmor radius:

$$D_\perp \approx D_e \frac{\nu_e^2}{\omega_{B,e}^2} \approx \rho_L^2 \nu_e \tag{6.52}$$

As can be seen, Equation 6.52 for the diffusion coefficient across the strong magnetic field is similar to Equation 4.121 for free diffusion without a magnetic field, but with the mean free path replaced by the Larmor radius.

When the magnetic field is high and electrons are magnetized ($\omega_{B,i}/\nu_i << 1$), then the electron Larmor radius is much shorter than the electron mean free path ($\rho_L << \lambda_e$). Plasma diffusion across the strong magnetic field in this case is much slower than without the magnetic field and significantly decreases with the strength of the magnetic field.

6.2.7 Conditions for Magneto-Hydrodynamic Behavior of Plasma: Alfven Velocity and Magnetic Reynolds Number

Next, determine the plasma conditions when the magneto-hydrodynamic effects take place, e.g., when the fluid dynamics are strongly coupled with the magnetic field. First, magneto-hydrodynamics demands the "diffusion" of the magnetic field to be less than the "convection." Mathematically, this means the curl in the right-hand side of Equation 6.30 should much exceed the Laplacian. If the space scale of plasma is L, this requirement can be presented as:

$$\frac{vB}{L} >> \frac{1}{\sigma\mu_0}\frac{B}{L^2}, \quad or \quad v >> \frac{1}{\sigma\mu_0 L} \tag{6.53}$$

As was shown in Section 6.2.2, these requirements (Equation 6.53) of high conductivity and velocity are sufficient to make the magnetic field frozen in the plasma. This is the most important feature of magneto-hydrodynamics.

Using the concept of magnetic viscosity D_m (Equation 6.31), the condition (Equation 6.53) of magneto-hydrodynamic behavior can be rewritten as:

$$\mathrm{Re}_m = \frac{vL}{D_m} >> 1 \tag{6.54}$$

Here, Re_m is the **magnetic Reynolds number**, which is somewhat similar to the conventional Reynolds number but with kinematic viscosity replaced by the magnetic

Table 6.2 Magnetic Reynolds Numbers in Different Laboratory and Nature Plasmas

Type of plasma	B, Tesla	Space scale, m	ρ, kg/m^3	σ, 1/Ohm · cm	R_m
Ionosphere	10^{-5}	10^5	10^{-5}	0.1	10
Solar atmosphere	10^{-2}	10^7	10^{-6}	10	10^8
Solar corona	10^{-9}	10^9	10^{-17}	10^4	10^{11}
Hot interstellar gas	10^{-10}	10 light yr	10^{-21}	10	10^{15}
Arc discharge plasma	0.1	0.1	10^{-5}	10^3	10^3
Hot confined plasma, $n = 10^{15}$ cm^{-3}, T = 10^6	0.1	0.1	10^{-6}	10^3	10^4

viscosity. The physical interpretation of the magnetic Reynolds number can be further clarified taking into account that the plasma velocity v in magneto-hydrodynamic systems usually satisfies the approximate balance of dynamic and magnetic pressures:

$$\frac{\rho v^2}{2} \propto \frac{\mu_0 H^2}{2}, \quad or \quad v \propto v_A = \frac{B}{\sqrt{\rho \mu_0}} \tag{6.55}$$

where $\rho = Mn_e$ is the plasma density. The characteristic plasma velocity v_A corresponding to the equality of the dynamic and magnetic pressures is called the **Alfven velocity**. The criterion of magneto-hydrodynamic behavior can then be written:

$$\mathrm{Re}_m = \frac{v_A L}{D_m} = BL\sigma\sqrt{\frac{\mu_0}{\rho}} >> 1 \tag{6.56}$$

This form of the magnetic Reynolds number and criterion for magneto-hydrodynamic behavior were derived by Lundquist.[181] It shows that magneto-hydrodynamic plasma behavior can take place not only at high values of plasma conductivity and at high magnetic fields, but also at large sizes and low densities. Numerically, the magnetic Reynolds numbers for some laboratory and nature plasmas are presented in Table 6.2.

This short section merely touched a large and very sophisticated branch of plasma physics related to magneto-hydrodynamics and plasma confinement in magnetic fields. More information on the subject can be found, for example, in Chen[182] and Rutherford and Goldston.[183]

6.3 INSTABILITIES OF LOW-TEMPERATURE PLASMA

Instability is one of the most serious problems of plasma physics, applying to hot plasma confined in a magnetic field and to nonequilibrium low-temperature plasma. The instabilities of Z-pinch discussed in Section 6.2.4 are an example of hot plasma instabilities. The variety of these instabilities is incredible, e.g., "gutter" instability, "screw" instability, current-convective, drift, and many others. Decades of attempts to employ hot confined plasma for controlled thermonuclear fusion are actually a half-century struggle by physicists and engineers against different plasma instabilities.

Although the instabilities of hot confined plasmas are not the subject of this book, for information on the subject, the reader is directed to Melrose[184] and Mikhailovskii.[185,186] Here the focus will be on instabilities of low-temperature plasmas, which are also numerous and usually related to a specific type of electric discharge. Thus, specific instabilities related to the glow discharge or plasma-beam discharge will be considered later in chapters relevant to these discharges. In this section, the general physical features of low-temperature plasma instabilities will be discussed as well as their peculiarities in plasma chemical systems.

6.3.1 Types of Instabilities of Low-Temperature Plasmas: Peculiarities of Plasma Chemical Systems

The most serious instabilities of nonthermal, nonequilibrium plasmas are related to the system's tendency to restore thermodynamic quasi equilibrium between different degrees of freedom. Typically in nonequilibrium plasmas, electron, vibrational, and translational temperatures remain in the following hierarchy: $T_e >> T_v >> T_0$. A balance of excitation and relaxation processes (considered in Chapter 3 and Chapter 4) sustains the difference between temperatures. Unfortunately, this balance is stable only at some specific range of discharge parameters. Beyond this range, relaxation processes become unstable; their rates increase in a self-accelerating way that leads to decay of the nonequilibrium system and restoration of thermodynamic quasi equilibrium.

Thus, the homogeneous state of nonequilibrium discharges is possible only in a limited range of parameters. Usually, this means low pressures, low specific energy inputs, and low specific powers (power per unit volume). Small fluctuation of the plasma parameters beyond this stability range can increase exponentially, resulting in the discharge transition into a nonhomogeneous form. This problem is especially serious in nonequilibrium plasma chemistry, where optimal conditions necessary for high efficiency of chemical processes are not always the same as the conditions necessary for stability and homogeneity of relevant nonthermal discharges. Thus, the optimal specific energy input (discharge energy per one molecule) for quite a few nonequilibrium plasma chemical processes is about $E_v \approx 1$ eV/mol (see Section 5.6.8). Unfortunately, this specific energy input for plasma chemical systems exceeds about 10 times the optimal one ($E_v \approx 0.1$ eV/mol) for highly effective nonequilibrium discharges applied for stimulation of gas lasers.

However, it should be noted that the requirements of $E_v \approx 0.1$ eV/mol are related to the final gas heating, while the optimal energy input $E_v \approx 1$ eV/mol is related to the initial energy going mostly into vibrational excitation (see Section 5.6.8). Thus, if the energy efficiency of plasma chemical process reaches 90% (which is quite difficult), the requirement can be met. Remember that plasma chemical systems, in

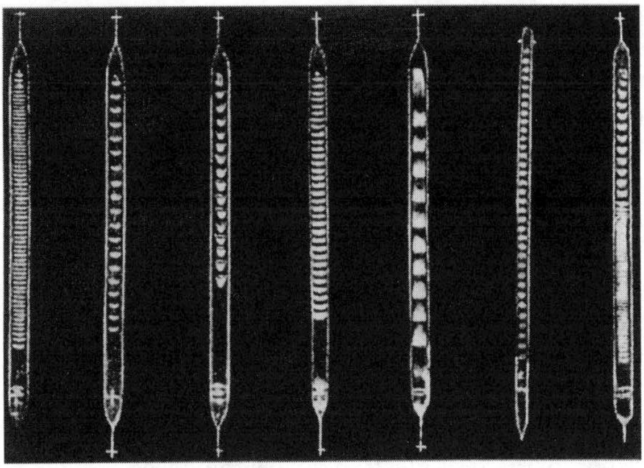

Figure 6.8 General view of striated discharges[101]

contrast to gas lasers, can be effective in nonhomogeneous systems as well, if the required level of the plasma nonequilibrium remains.[187]

The instabilities of nonequilibrium plasma can be subdivided into two types of phenomena: striation and contraction. Both phenomena will be discussed later with reference to a glow discharge. Here, their definitions and the qualitative differences between them will be merely pointed out.

Striation is an instability related to formation of plasma structure that looks like a series of alternating light and dark layers along the discharge current. The general appearance of a striated discharge is shown in Figure 6.8. Normally, the striations are able to move fast with velocities up to 100 m/sec, but they also can be at rest; usually, the conditions of their appearance do not depend on pressure and specific energy input. It is very important that the striations do not significantly affect plasma parameters. The characteristics of plasma chemical processes in nonequilibrium discharges with and without striations are close. Often the striations are not detectable by the naked eye. The physical phenomenon of striation is related to ionization instability (see below) and can be interpreted as ionization oscillations and waves.[188,189]

Contraction is an instability related to plasma "self-compression" into one or several bright current filaments. The contraction takes place when the pressure or specific energy input exceeds some critical values. In contrast to striation, contraction significantly changes plasma parameters. The plasma filament formed as a result of contraction is close to quasi equilibrium. For this reason, contraction is a serious factor limiting power and efficiency of gas lasers and nonequilibrium plasma chemical systems. The physical mechanism of contraction is related to the ionization overheating instability, which is discussed next.

6.3.2 Thermal (Ionization-Overheating) Instability in Monatomic Gases

Consider the ionization-overheating instability (often called "thermal" instability) for a plasma of monatomic gases, where the instability is easier to explain. The instability in this case is due to the strong exponential dependence of the ionization rate coefficient and thus the electron concentration on the reduced electric field E_0/n_0. The thermal instability can be illustrated by the following closed chain of causal links, which can start from fluctuation of the electron concentration:

$$\delta n_e \uparrow \rightarrow \delta T_0 \uparrow \rightarrow \delta n_0 \downarrow \rightarrow \delta\left(\frac{E}{n_0}\right) \uparrow \rightarrow \delta n_e \uparrow \tag{6.57}$$

The local increase of electron concentration δn_e leads to intensification of gas heating by electron impact and thus to an increase of temperature δT_0. Taking into account that pressure $p = n_0 T_0$ is constant, the local increase of temperature δT_0 leads to decrease of gas density δn_0 and to an increase of the reduced electric field $\delta(E/n_0)$ (electric field E = const). Finally, the increase of the reduced electric field $\delta(E/n_0)$ results in the further increase of electron concentration δn_e, which makes the chain (Equation 6.57) closed and determines the positive feedback.

The sequence (Equation 6.57) gives the physical meaning of the thermal instability and contraction in plasma of atomic gases. Because of the strong exponential dependence of the ionization rate on E/n_0, a small local initial overheating δT_{00} according to the mechanism of (Equation 6.57) grows up exponentially:

$$\delta T_0(t) = \delta T_{00} \cdot \exp \Omega t \tag{6.58}$$

The exponential increase in overheating leads to formation of the hot filament or, in other words, to contraction. Here δT_{00} is the initial perturbation of temperature. The parameter of exponential growth of the initial perturbation is called the **instability increment**. If the system is stable, the instability increment is negative ($\Omega > 0$); the positive increment ($\Omega > 0$) means the actual instability.

The increment of thermal instability (Equation 6.57) can be found by linearization of the differential equations of heat and ionization balance. In the case of high- or moderate-pressure discharges when the influence of walls can be neglected, the thermal instability increment is equal to:[190]

$$\Omega = \hat{k}_i \frac{\sigma E^2}{n_o c_p T_0} = \hat{k}_i \frac{\gamma - 1}{\gamma} \frac{\sigma E^2}{p} \tag{6.59}$$

Here, the dimensionless factor $\hat{k}_i = \partial \ln k_i / \partial \ln T_e$ is the logarithmic sensitivity of the ionization rate coefficient to electron temperature (directly related to E/n_0; see Section 4.2.7). Numerically, this factor is usually about 10; σ is plasma conductivity, p is pressure, and c_p and γ are specific heat and specific heat ratio.

Analyzing the increment (Equation 6.59), one should take into account that

$$\nu_{Tp} = \frac{\gamma - 1}{\gamma} \frac{\sigma E^2}{p}$$

is the frequency of gas heating by electric current at constant pressure. This means that the instability increment (Equation 6.59) is mostly determined by the heating frequency, but actually exceeds ν_{Tp} because of the strong sensitivity of ionization to electron temperature.

Equation 6.59 shows that steady-state discharges at high and moderate pressures are always unstable with respect to the considered thermal instability. However, the thermal instability (Equation 6.57) and the related effect of discharge contraction can be suppressed by intense cooling as in the case of heat losses to the walls at low discharge pressures. Also, one can simply avoid thermal instability if the specific energy input is low and does not provide sufficient energy for contraction. In other words, the thermal instability can be neglected if the gas residence time in discharge (or discharge duration) is small with respect to the thermal instability time ($1/\Omega$).

6.3.3 Thermal (Ionization-Overheating) Instability in Molecular Gases with Effective Vibrational Excitation

The key difference of thermal instability in molecular gases is related to the fact that fluctuation of electron density δn_e does not lead directly to heating δT_0, but rather through intermediate vibrational excitation. Vibrational-translational (VT) relaxation is relatively slow ($\tau_{VT}\nu_{Tp} >> 1$, τ_{VT} is the VT relaxation time) in the highly effective laser and plasma chemical systems. In this case, the thermal instability becomes more sensitive to VT relaxation than to excitation (ν_{VT}) as was the case in monatomic gases (Equation 6.59).

In the most interesting case of fast vibrational excitation slow relaxation ($\tau_{VT}\nu_{Tp} >> 1$), the increment of thermal instability in the molecular gas plasma can be expressed as:[190–192]

$$\Omega_T = \frac{b}{2} \pm \sqrt{\frac{b^2}{4} + c} \tag{6.60}$$

where the parameters b and c are:

$$b = \frac{1}{\tau_{VT}}\left(1 - \frac{\partial \ln \tau_{VT}}{\partial \ln \varepsilon_v}\right) + \nu_{Tp}\left(2 + \hat{\tau}_{VT}\right) \tag{6.61}$$

$$c = \frac{\nu_{Tp}}{\tau_{VT}}\left(-\frac{\partial \ln n_e}{\partial \ln n_0}\right)\left(1 - \frac{\partial \ln \tau_{VT}}{\partial \ln \varepsilon_v}\right) \tag{6.62}$$

In these relations the dimensionless factor

$$\hat{\tau}_{VT} = \frac{\partial \ln \tau_{VT}}{\partial \ln T_0}$$

is the logarithmic sensitivity of the vibrational relaxation time to translational temperature. Numerically, this factor is usually about $-3 \div -5$; ε_v is vibrational energy of a molecule.

Analysis of Equation 6.60 through Equation 6.62 shows that thermal instability in molecular gases can proceed in two different ways, thermal and vibrational modes:

1. **The thermal mode of instability** corresponds to the condition: $b < 0$. Taking into account the relatively high rate of vibrational excitation ($\nu_{Tp}\tau_{VT} >> 1$), one can see that $b^2 >> c$. Then Equation 6.60 gives the instability increment for thermal mode:

$$\Omega_T = |b| \approx -\nu_{Tp}(2+\hat{\tau}_{VT}) = k_{VT}n_0 \frac{\hbar\omega}{c_p T_0}(\hat{k}_{VT}-2) \tag{6.63}$$

In this relation, the dimensionless factor

$$\hat{k}_{VT} = \frac{\partial \ln k_{VT}}{\partial \ln T_0}$$

is the logarithmic sensitivity of the vibrational relaxation rate coefficient to translational temperature T_0. Numerically, this sensitivity factor is usually about three divided by five.

As can be seen from Equation 6.63, the increment of instability in the thermal mode corresponds to the frequency of heating due to vibrational VT relaxation, multiplied by $(\hat{k}_{VT} - 2) >> 1$ because of the strong exponential dependence of the Landau–Teller relaxation rate coefficient on gas temperature (see Section 3.4.3). The pure thermal mode does not include any direct characteristics of ionization (only overheating; no factors directly related to n_e). Thus, it is the "overheating" part of the entire ionization-overheating instability.[190–192]

2. **The vibrational mode of instability** corresponds to the condition: $c > 0$ (at any sign of the parameter b). The instability increment at a relatively high rate of vibrational excitation ($\nu_{Tp}\tau_{VT} >> 1$) can be expressed for this mode as:

$$\Omega_T = \frac{c}{|b|} \approx k_{VT}n_0 \frac{\hat{k}_i}{\hat{k}_{VT} - 2 - (\nu_{Tp}\tau_{Tp})^{-1}} \tag{6.64}$$

The instability increment for the vibrational mode is also mostly characterized by the frequency of vibrational relaxation. However, in contrast to the thermal mode, the sensitivity term in Equation 6.64 does include factors related to the electron concentration and the overheating effect on ionization rate. This makes the instability similar to that in monatomic gases (Equation 6.59). Thus, it is the "ionization" part of the entire ionization-overheating instability.

To make the information on the thermal instability complete, one should note that, in the case of slow excitation ($\nu_{Tp}\tau_{VT} << 1$), the instability increment can be expressed as:

$$\Omega_T = \sqrt{\frac{\nu_{Tp}}{\tau_{VT}}\left(-\frac{\partial \ln n_e}{\partial \ln n_0}\right)\left(1 - \frac{\partial \ln \tau_{VT}}{\partial \ln \varepsilon_v}\right)} \tag{6.65}$$

Although important, this case is less interesting for specific laser and plasma chemical applications.

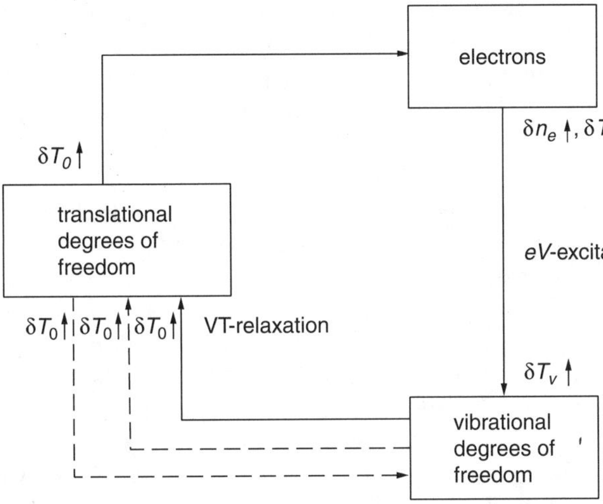

Figure 6.9 Thermal instability in molecular gases: solid lines correspond to vibrational mode; dashed lines correspond to thermal mode

6.3.4 Physical Interpretation of Thermal and Vibrational Instability Modes

The different modes of thermal instability discussed previously are illustrated in Figure 6.9. In the same manner as in the case of monatomic gases (Equation 6.57), local decreases of gas density in molecular gases lead to an increase of the reduced electric field and electron temperature. An increase of electron temperature results in an acceleration of ionization and growth of electron concentration and vibrational temperature. Growth of the vibrational temperature makes VT relaxation and heating more intensive, leading to an increase of the translational temperature and to a decrease of the gas density (because $p = n_0T = \text{const}$). This is the mechanism of the vibrational mode of instability, which is somewhat similar to Equation 6.57.

Thermal mode, which is also shown in Figure 6.9, in general, is not directly related to electron density, temperature, and ionization. This is due to the strong exponential dependence of the VT relaxation rate coefficient on translational temperature. Even small increases of the translational temperature lead to significant acceleration of VT relaxation rate (Landau–Teller mechanism; see Section 3.4.3), intensification of heating, and then to further growth of the translational temperature. This phenomenon is sometimes called the *thermal explosion of vibrational reservoir*. Increase of temperature in the thermal mode, obviously, also stimulates growth of the electron density, but it is not included inside the principal chain of events in this mode of instability.

The instability increment in thermal (Equation 6.63) and vibrational (Equation 6.64) modes usually has close numerical values. Sometimes, however, even the qualitative behavior of these two instability modes is different. For example, vibrational energy losses into translational degrees of freedom at strong vibrational nonequilibrium are mostly provided by VV and VT relaxation from high vibrational levels (see Section 5.3.8 and Section 5.3.9). In this case, an increase of translational temperature T_0 leads to a decrease of the effective VT relaxation and gas heating; the thermal instability mode is impossible under such conditions. At the same time, the increment of the vibrational mode increases.

The opposite situation takes place in nonequilibrium discharges sustained in supersonic flows (see the next section). In this case, gas heating leads, not to a decrease, but to an increase of gas density and to a reduction of electron temperature and reduction of further heating. This means the plasma is stable with respect to the vibrational mode, but the thermal mode of instability is still in place.

Plasma chemical processes have stabilizing and destabilizing effects on discharge instabilities. A significant part of the vibrational energy in plasma chemical systems can be consumed in endothermic reactions instead of heating. Obviously, this provides plasma stabilization with respect to the ionization-overheating instability; in turn, destabilization is due to fast heat release in exothermic reactions. Next, these two opposite plasma chemical effects on discharge stability will be discussed.

6.3.5 Nonequilibrium Plasma Stabilization by Chemical Reactions of Vibrationally Excited Molecules

Both modes (Equation 6.63 and Equation 6.64) of the strong thermal instability that leads to discharge contraction have typical time comparable with the time of VT relaxation ($\tau_{VT} = 1/k_{VT}n_0$). This makes the thermal instability less dangerous for effective plasma chemical processes (see Section 5.6.5) in which the reaction time is approximately the time of vibrational excitation ($\tau_{eV} = 1/k_{eV}n_e$), which is shorter than time of VT relaxation (see Equation 5.150). An effective plasma chemical process can be completed before development of the strong but slow thermal instability.

One of such **fast ionization instability** with a frequency of approximately $\tau_{eV} = 1/k_{eV}n_e$ is due to direct increase of T_e and acceleration of ionization by an increase of T_v.[193] The effect is provided by super-elastic collisions of electrons with vibrationally excited molecules (Section 4.2.5; Equation 4.84).

The scheme of the fast ionization instability is shown in Figure 6.10. The initial small increase of electron concentration leads to intensification of vibrational excitation and growth of the vibrational temperature. Higher vibrational temperatures because of super-elastic collisions results in acceleration of ionization and further increase of the electron concentration. This instability includes neither VT relaxation nor any heating, and it is fast (controlled by vibrational excitation $\tau_{eV} = 1/k_{eV}n_e$). Fortunately, this instability can be stabilized by endothermic chemical reactions consuming vibrational energy and decreasing T_v, which is also illustrated in Figure 6.10.

Increment of this fast ionization instability can be calculated by linearization of the differential equations for the vibrational energy balance and balance of electrons, taking into account the endothermic reactions stimulated by vibrational excitation:[194,195]

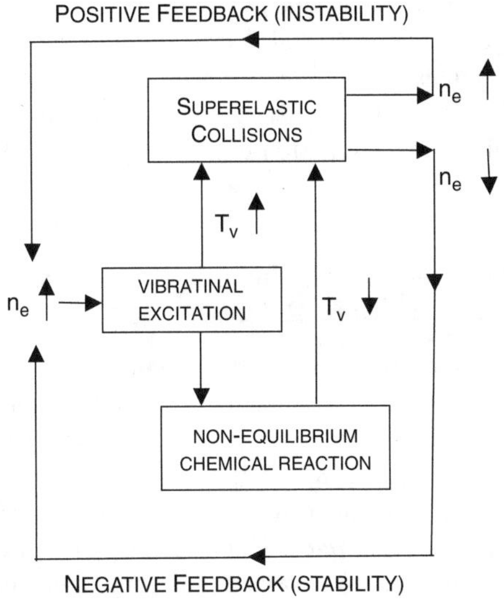

Figure 6.10 Ionization instability of molecular gas in chemically active plasma

$$\Omega_v = k_{eV} n_e \frac{\hbar\omega}{c_v^v T_v}\left(\tilde{k}_i - \tilde{k}_r\right) \tag{6.66}$$

Here, $c_v^v = \partial\varepsilon_v / \partial T_v$ is the vibrational heat capacity; $\tilde{k}_i = \partial \ln k_i / \partial \ln T_v$ is the dimensionless sensitivity of the ionization rate coefficient to the vibrational temperature; and $\tilde{k}_r = \partial \ln k_r / \partial \ln T_v$ is the dimensionless sensitivity of the chemical reaction rate to vibrational temperature (in the case of weak excitation $\tilde{k}_r = E_a / T_v$ and at strong excitation $\tilde{k}_r = T_o \hbar\omega / x_e T_v^2$; see Section 5.3). The factor $\tilde{k}_i = \partial \ln k_i / \partial \ln T_v$ can be found, taking into account the influence of the vibrational temperature and thus super-elastic processes on the ionization rate (see Section 4.2.5 and Equation 4.84), as:

$$\tilde{k}_i = \left(\frac{\hbar\omega}{T_e}\right)^2 \frac{\Delta\varepsilon}{T_v} \tag{6.67}$$

In the preceding equation, $\Delta\varepsilon \approx 1 - 3$ eV is the energy range of effective vibrational excitation (see Table 3.9) and E_a is the activation energy. Based on Equation 6.66 and Equation 6.67, the fast ionization instability increment for the case of not very strong excitation can be expressed as:

$$\Omega_v = k_{eV} n_e \frac{\hbar\omega E_a}{c_v^v T_v^2}\left(\frac{\hbar^2\omega^2}{T_e^2}\frac{\Delta\varepsilon}{E_a} - 1\right) \tag{6.68}$$

The "unity" in parentheses corresponds to the case in which most of vibrational excitation energy is going into chemical reactions. Numerically, for a wide range of typical discharge parameters, this "unity" makes the increment negative, providing plasma stability with respect to the fast self-acceleration of ionization. Obviously, if only part of the vibrational excitation energy is going to chemical reaction, the stabilization effect is less.

6.3.6 Destabilizing Effect of Exothermic Reactions and Fast Mechanisms of Chemical Heat Release

Chemical reactions of vibrationally excited molecules stabilize the perturbations of ionization (Equation 6.68), but unfortunately they are also able to amplify instabilities related to the plasma direct overheating. As was discussed in Section 5.6.7 (see Equation 5.155 and Equation 5.158), each act of chemical reaction stimulated by vibrational excitation is accompanied by transfer of energy, $\xi E_a = \Delta Q_\Sigma - \Delta H$, into heat. This so-called *chemical heat release* includes the effect of exothermic reactions, nonresonant VV exchange, and VT relaxation from high vibrational levels. The chemical heat release causes problems to effective plasma chemical processes stimulated by vibrational excitation because of the very high frequency corresponding to the rate of the reactions, which can much exceed the frequency of VT relaxation from low vibrational levels.

The scheme of ionization-overheating instability is illustrated in Figure 6.11. As can be seen, the instability of chemical heat release operates in the framework of the thermal (ionization-overheating) instability. However, it is much faster because of the presence of the fast heating mechanisms controlled by the fast chemical reactions. The increment of the instability can be presented as:[196]

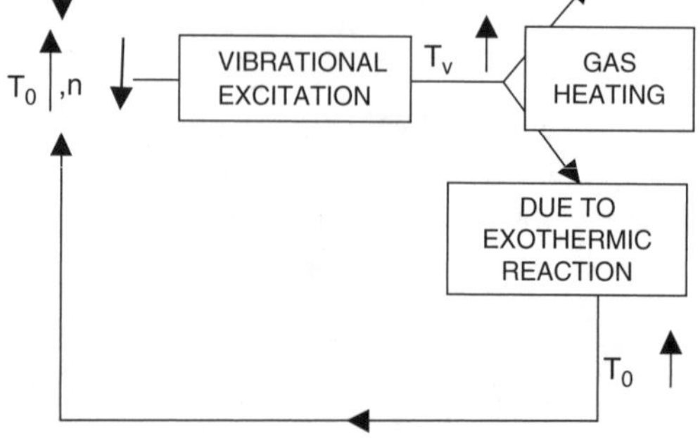

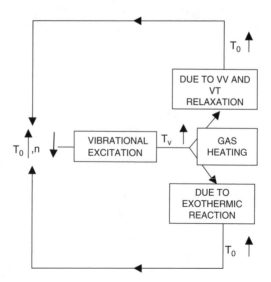

Figure 6.11 Instability related to "chemical" heat release

$$\Omega_T = k_{eV} n_e \frac{\hbar\omega}{c_p T_o} \left\{ \xi\left(\hat{k}_r - 2\right) - \tilde{k}_r \frac{c_p T_0}{c_v^v T_v} \right. \\ \left. + \sqrt{\left[\xi\left(\hat{k}_r - 2\right) + \tilde{k}_r \frac{c_p T_0}{c_v^v T_v} \right]^2 + 4\xi \tilde{k}_r \frac{c_p T_0}{c_v^v T_v} \left(\hat{k}_i - \hat{k}_r\right)} \right\} \tag{6.69}$$

Here, the dimensionless factor $\tilde{k}_r = \partial \ln k_r / \partial \ln T_0$ is the logarithmic sensitivity of chemical reaction rate to translational temperature. Equation 6.69 demonstrates that the time of instability is comparable with the time of vibrational excitation. This establishes a serious limit on the maximum specific energy input into homogeneous nonequilibrium discharge as:

$$E_v^{\max} = k_{eV} \frac{\hbar\omega}{\Omega_T} = c_p T_0 \left\{ \xi\left(\hat{k}_r - 2\right) - \tilde{k}_r \frac{c_p T_0}{c_v^v T_v} + \right. \\ \left. + \sqrt{\left[\xi\left(\hat{k}_r - 2\right) + \tilde{k}_r \frac{c_p T_0}{c_v^v T_v} \right]^2 + 4\xi \tilde{k}_r \frac{c_p T_0}{c_v^v T_v} \left(\hat{k}_i - \hat{k}_r\right)} \right\}^{-1} \tag{6.70}$$

This important and strong restriction of the plasma parameters required for stability of a moderate- or high-pressure discharge (with chemical reactions stimulated by vibrational excitation) is shown numerically in Figure 6.12 in coordinates of vibrational and translational temperatures. As can be seen from the figure, at most of the conditions, the homogeneous, steady-state, nonequilibrium plasma chemical discharges are unstable with respect to overheating if pressure is moderate or high, and there is no discharge stabilization by walls.

This explains why most effective plasma chemical processes, stimulated by vibrational excitation of molecules at high or moderate pressures, are experimentally observed only in space-nonuniform, nonhomogeneous or non-steady-state discharges.[9] However, this ionization-overheating instability does not affect plasma chemical discharges in supersonic flows; in that case, a small increase of temperature leads, not to reduction, but to growth of the gas density (see Figure 6.11), thus providing plasma stability.

6.3.7 Electron Attachment Instability

The attachment instability is not as "dangerous" as the thermal one considered earlier, but it is quite general and a phenomenon observed often. The attachment instability takes place if electron detachment essentially compensates electron attachment (see Section 4.5.2). This can be observed, for example, in glow discharges when the perturbation of n_e does not affect current. In this case, the following sequence of perturbations can illustrate the instability:

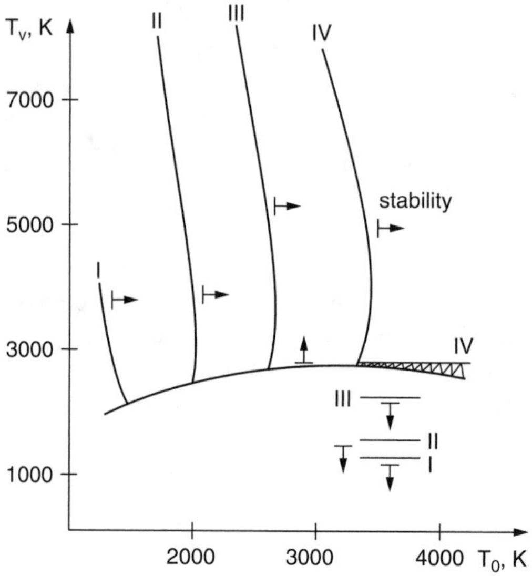

Figure 6.12 Thermal instability limits. Specific energy input: (I) 0.1 eV/mol; (II) 0.3 eV/mol; (III) 0.5 eV/mol; (IV) 0.7 eV/mol. Range of stability parameters is crossed out.

$$\delta n_e \uparrow \rightarrow \delta T_e \downarrow \rightarrow \nu_a \downarrow \rightarrow \delta n_e \uparrow \tag{6.71}$$

Next, analyze the chain of causal links (Equation 6.71). The small increase of electron concentration leads to decrease of local electric field (current is not perturbed) and thus to a decrease of electron temperature. Obviously, this intensifies ionization, but the effect of weakening of attachment may have a stronger effect (because the ionization rate is much less than the rate of attachment in the presence of intensive detachment process; see below). Before the perturbation, attachment and detachment processes were in balance; therefore, weakening the attachment rate at a constant level of detachment results in an increase of electron concentration and, finally, in the instability. Increment of the attachment instability can be expressed in this case as:

$$\Omega_a \approx k_a n_0 \left[\hat{k}_a \left(1 - \frac{k_i}{k_a} \frac{\hat{k}_i}{\hat{k}_a} \right) - \frac{n_+}{n_-} \right] \tag{6.72}$$

In this relation, k_a is the rate coefficient of electron attachment; dimensionless factor $\tilde{k}_a = \partial \ln k_a / \partial \ln T_e$ is the sensitivity of this rate coefficient to electron temperature; and n_+ and n_- are concentrations of positive and negative ions, respectively.

As can be seen from Equation 6.72, the characteristic time of attachment instability is, obviously, the time of electron attachment. It is surprising that the attachment instability does not take place in the discharge regime controlled by electron attachment where $k_a \approx k_i$ (see Section 4.5.3). The instability increment is negative in this case because ionization is more sensitive to electron temperature than attachment ($\hat{k}_i > \hat{k}_a$). The attachment instability can be observed only in the presence of intensive detachment (recombination regime; see Section 4.5.2), where $k_a >> k_i$ and parentheses in Equation 6.69 become positive.

The attachment instability discussed earlier usually leads to formation of **electric field domains**, which are a form of striations. For example, an initial local fluctuation $\delta n_e > 0, \delta T_e < 0$ in the presence of high concentration of negative ions results in their decay, growth of electron concentration, and further decrease of a local electric field. This is called a *weak field domain*; an opposite local fluctuation, $\delta n_e < 0, \delta T_e > 0$, leads to formation of a *strong field domain*. The domains usually move toward an anode much more slowly than the electron drift velocity. It is important to note that the domains do not significantly affect the characteristics of plasma chemical processes in nonequilibrium discharges.

6.3.8 Other Instability Mechanisms in Low-Temperature Plasma

Next, some less general and less "dangerous" instability mechanisms that can take place in nonequilibrium discharges stimulating striation or contraction will be discussed.

a) **Ionization instability controlled by dissociation of molecules**. As shown in Section 4.2.6, the effective electron temperature is significantly higher in monatomic gases than in corresponding molecular gases at the same value of reduced electric field. This is due to significant reduction of electron energy distribution function in an energy interval corresponding to intensive vibrational excitation (see Equation 4.84). This effect explains the nonequilibrium discharge instability related to dissociation of molecular gases, which can be illustrated by the scheme:[197–199]

$$\delta n_e \uparrow \to \delta(\mathit{dissociation}) \uparrow \to \delta T_e \uparrow \to \delta n_e \uparrow \tag{6.73}$$

Increase of electron concentration (or temperature) leads to intensification of dissociation and conversion of molecules into atoms, and then results in further growth of the electron concentration and temperature. Increment of the instability (Equation 6.73) is determined by the dissociation time.

b) **Stepwise ionization instability**. This instability is somewhat similar to the fast ionization instability related to vibrational excitation of molecules and was considered in Section 6.3.5. Here also the increase of electron concentration leads to additional population of excited species (but in this case electronically excited particles), providing faster ionization and a further increase of electron concentration. Increment of this instability is approximately the frequency of electronic excitation. In contrast to the instability related to vibrational excitation (Section 6.3.5), this one cannot be as effectively stabilized by chemical reactions.

c) **Electron maxwellization instability**. Electron energy distribution functions are restricted at high energies by a variety of channels of inelastic electron collisions with atoms and molecules. Maxwellization of electrons at higher electron densities provides larger amounts of high-energy electrons (Equation 4.127) and a related stimulated ionization in this way. Instability can be illustrated by the following sequence of events:

$$\delta n_e \uparrow \rightarrow \delta(Maxwellization) \uparrow \rightarrow \delta f(E) \uparrow \rightarrow \delta n_e \uparrow \tag{6.74}$$

An increase of electron concentration leads to maxwellization; growth of the electron energy distribution function at high energies; intensification of dissociation; and, finally, further growth of electron concentration. This instability mechanism (as well as stepwise ionization) may lead to striation or contraction at sufficiently high electron densities.

d) **Instability in fast oscillating fields**. An example of instability in oscillating fields is the ionization instability of a microwave plasma in low-pressure monatomic gases $\nu_{en} << \omega$.[200] The mechanism of this instability is similar to that related to the modulation instability of hot plasma. An increase of electron density in a layer (perpendicular to electric field) provides growth of the plasma frequency, which approaches the microwave field frequency ω and leads to an increase of the field. The growth of the electric field results in intensification of ionization and further increase of electron density, which determines the instability. At higher pressures ($\nu_{en} >> \omega$), temperature perturbation also plays a significant role in development of the ionization instability.

The instability results in formation of overheated filaments parallel to the electric field E. The maximum increment of the instability corresponds to perturbations with the wave vector:

$$\kappa_m = \left(\frac{\partial \ln k_i}{\partial \ln E} \frac{n_e e^2}{\varepsilon_0 m(\omega^2 + \nu_{en}^2)} \frac{k_i n_e}{D_a} \frac{\omega^2}{c^2} \right)^{1/4} \tag{6.75}$$

Here, D_a is the coefficient of ambipolar diffusion; ν_{en} is the frequency of electron–neutral collisions; and c is the speed of light.

The distance between the filaments formed as a result of the instability can be estimated as $l \approx 2\pi/\kappa_m$. For electron temperatures of about 1eV; electron density close to the critical one (when microwave frequency is close to plasma frequency); air pressure of 200 torr; and wave length of $\lambda = 1$ cm, the distance between filaments is about 0.2 to 0.3 cm.

6.4 NONTHERMAL PLASMA FLUID MECHANICS IN FAST SUBSONIC AND SUPERSONIC FLOWS

Fluid mechanic aspects of generation for thermal and nonthermal plasmas will be discussed in the following chapters concerned with specific types of electric discharges. In this section, general problems of fluid mechanics and relaxation dynamics of nonequilibrium plasmas in the fast subsonic and supersonic flows will be discussed.

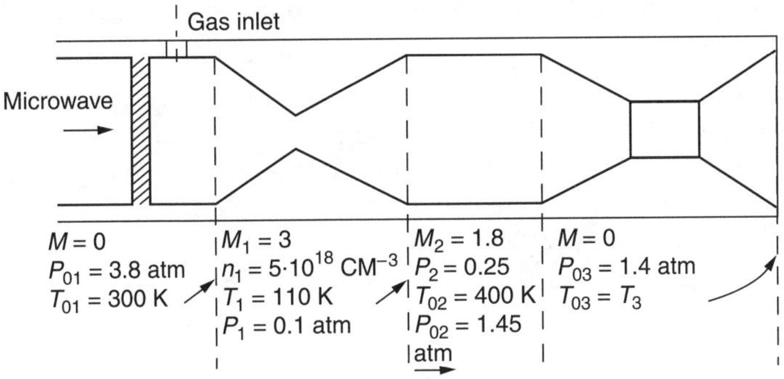

Figure 6.13 Typical parameters of a nozzle system. Subscripts 1 and 2 are related to inlet and exit of a discharge zone; subscript 3 is related to exit from the nozzle system; subscript 0 is related to stagnation pressure and temperature.

6.4.1 Nonequilibrium Supersonic and Fast Subsonic Plasma Chemical Systems

As has been shown in Section 5.6.8, the optimal energy input in energy-effective plasma chemical systems should be about 1 eV/mol at moderate and high pressures. This determines the proportionality between discharge power and gas flow rate in the optimal regimes of plasma chemical processes; high discharge power requires high flow rates through the discharge. However, the gas flow rates through the discharge plasma zone are limited. Instabilities restrict space sizes and pressures of the steady-state discharge systems; velocities are usually limited by speed of sound. As a result, the flow rate, and thus power, are restricted by some critical value. This subject will be considered in depth related to scaling of microwave discharges, where the maximum power in nonthermal conditions is limited to approximately 100 kW.[201]

Higher powers of steady state, nonequilibrium discharges can be reached in supersonic gas flows. Under such conditions, maximum power of a steady-state, nonequilibrium microwave discharge is increased up to 1 MW.[202] A simplified scheme and flow parameters of such a discharge system (power = 500 kW, Mach number $M = 3$) is presented in Figure 6.13. Ignition of the nonequilibrium discharge takes place in this case after a supersonic nozzle in a relatively low-pressure zone. In the after-discharge zone, pressure is restored in a diffuser, so initial and final pressures are above the atmospheric.

Besides the two mentioned technological advantages of nonthermal, supersonic discharges, high values of power and pressure (before and after the discharge), two more fundamental points of interest should be mentioned. Gas temperature in the discharge is much less than room temperature (down to 100 K; see Figure 6.13), which provides low VT relaxation rates and much higher levels of nonequilibrium. Also, the high flow velocities make this discharge more stable with respect to contraction at higher pressures and energy inputs.[203] For example, homogeneous glow discharges can be organized in atmospheric pressure and fast flows at energy input of 0.5 kJ/g.[204]

6.4.2 Gas Dynamic Parameters of Supersonic Discharges: Critical Heat Release

Plasma and therefore the heat release zone in supersonic discharge (Figure 6.13) are located between the nozzle and diffuser. The gas flow beyond the heat release zone (and shock in the diffuser) can be considered as isentropic at high Reynolds numbers and smooth duct profile. Gas pressure p, density n_0, and temperature T are related in the isentropic zones to the Mach number M as:

$$p\left(1+\frac{\gamma-1}{2}M^2\right)^{\frac{\gamma}{\gamma-1}} = const \tag{6.76}$$

$$n_0\left(1+\frac{\gamma-1}{2}M^2\right)^{\frac{1}{\gamma-1}} = const \tag{6.77}$$

$$T\left(1+\frac{\gamma-1}{2}M^2\right) = const \tag{6.78}$$

Here, γ is the specific heat ratio. The Mach number M is determined by variation of the cross section S of the plasma chemical system (duct):

$$SM\left(1+\frac{\gamma-1}{2}M^2\right)^{\frac{\gamma+1}{2(1-\gamma)}} = const \tag{6.79}$$

The preceding equations permit calculating the supersonic gas flow parameters at the beginning of the discharge as a function of gas parameters in the initial tank and Mach number after the nozzle. Data for CO_2 are presented in Figure 6.14. For example, if initial tank pressure is about 5 atm at room temperature and $M = 0.05$, then in the supersonic flow after nozzle and before the discharge, the gas pressure is 0.1 atm, Mach number $M \sim 3$, and gas temperature T ~ 100 K.

The supersonic gas motion in the discharge zone and afterward strongly depends on the plasma heat release q. This plasma heat release leads to an increase of stagnation temperature $\Delta T_0 = q / c_p$, and therefore decreases the velocity according to the relations for the supersonic flow in the duct with the constant cross section:

$$\frac{\sqrt{T_0}}{f(M)} = const, \quad f(M) = \frac{M\sqrt{1+\frac{\gamma-1}{2}M^2}}{1+\gamma M^2} \tag{6.80}$$

Equation 6.4.5 permits determining the **critical heat release** for the supersonic reactor with constant cross section, which corresponds to the decrease of the initial Mach number from $M > 1$ before the discharge to $M = 1$ afterward:

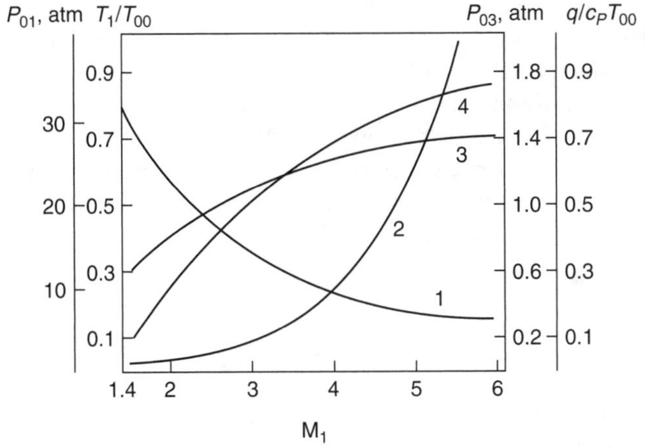

Figure 6.14 Gas dynamic characteristics of a discharge in supersonic flow. (1) Discharge inlet temperature T_1; (2) initial tank pressure p_{01}; (3) exit pressure p_{03} at critical heat release; (4) critical heat release q—as functions of Mach number in front of discharge. Initial gas tank temperature T_{00} = 300 K; static pressure in front of discharge p_1 = 0.1 atm.

$$q_{cr} = c_p T_{00} \left[\frac{\left(1+\gamma M^2\right)^2}{2(\gamma+1)M^2\left(1+\frac{\gamma-1}{2}M^2\right)} - 1 \right] \tag{6.81}$$

Here, T_{00} is the initial gas temperature in the tank and the reactor cross section in the discharge zone is considered constant. As can be seen from Equation 6.81, if the initial Mach number is not very close to unity, the critical heat release can be estimated as $q_{cr} \approx c_p T_{00}$. Numerical values of the critical heat release at different initial Mach numbers for the supersonic CO_2 flow can be found in Figure 6.14.

Further increase of the heat release in plasma over the critical value leads to formation of non-steady-state flow perturbations like shock waves that are detrimental to nonequilibrium plasma chemical systems. Such generation of shock waves will be discussed in the last section of this chapter. Even taking into account the high-energy efficiency of chemical reactions in supersonic flows, the critical heat release seriously restricts the specific energy input and, subsequently, the degree of conversion of the plasma chemical process. Thus, the maximum degree of conversion of CO_2 dissociation in a supersonic microwave discharge (T_{00} = 300 K and M = 3) does not exceed 15 to 20%, even at the extremely high energy efficiency of about 90%.[206]

6.4.3 Supersonic Nozzle and Discharge Zone Profiling

The effect of critical heat release on stability and efficiency of supersonic plasma chemical systems can be mitigated by an increase of initial temperature T_{00} or by reagents' dilution in noble gases. However, the most effective approach for suppressing the critical heat is by special profiling of the supersonic nozzle and discharge zone.[207]

The supersonic reactor can be designed to maintain a constant Mach number during the heat release in the discharge zone. The cross section of the duct should gradually increase to accelerate the supersonic flow and compensate the deceleration effect related to the heat release q. The heat release in this case has no limit. To provide the constant Mach number and conditions of unrestricted heat release, the reactor cross section S should increase with q as:

$$S = S_0\left[1 + \frac{q}{T\left(c_p + \gamma M_0^2/2\right)}\right]^{\frac{\gamma M_0^2+1}{2}} \tag{6.82}$$

Here, S_0, M_0, and T are the reactor cross section, Mach number, and temperature in the beginning of the discharge zone; specific heat is dimensionless because energy, heat, and temperature are considered in the same units. With sufficiently large Mach numbers, Equation 6.82 can be simplified and used for estimations as:

$$S \approx S_0\left(1 + \frac{q}{c_p T_{00}}\right)^{\frac{\gamma M_0^2+1}{2}} \tag{6.83}$$

Although the idea to keep the Mach number high without restrictions of heat release looks very attractive, the required increase of the discharge cross section is too large and not realistic for significant values of $q/c_p T_{00}$. More reasonable increases in the cross section S/S_0 require nonconstant Mach number with constant pressure conditions during the heat release in the supersonic plasma chemical reactor. The heat release in this case is obviously not unlimited, but the critical limit here is not as serious as that for the reactor with constant cross section:

$$q_{cr}(p = const) = c_p T_{00} \frac{M_0^2 - 1}{1 + \frac{\gamma - 1}{2} M_0^2} \tag{6.84}$$

Numerically, at relatively high Mach numbers, this critical heat release at constant pressure is approximately $q_{cr}(p = const) \approx 5c_p T_{00}$, e.g., five times less great than for the case of constant cross section. This is enough to permit relatively high specific energy input in the plasma and degree conversion (see the discussion after Equation 6.81).

To provide conditions for the constant pressure plasma chemical process, the required increase of the reactor cross section is much less than Equation 6.83 and can be expressed as:

$$S = S_0\left(1 + \frac{q}{c_p T}\right) = S_0\left[1 + \frac{q}{c_p T_{00}}\left(1 + \frac{\gamma - 1}{2} M_0^2\right)\right] \tag{6.85}$$

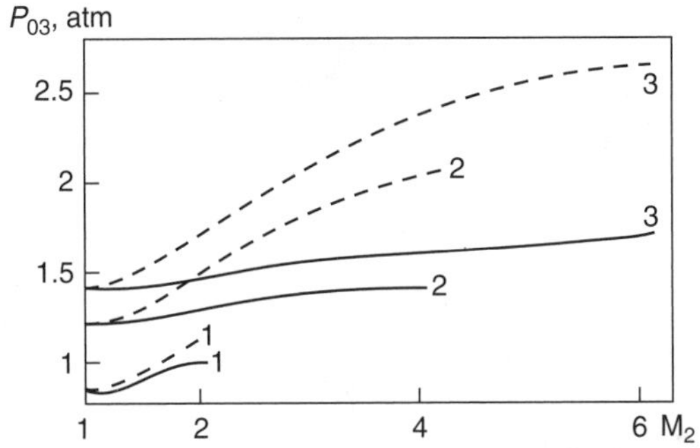

Figure 6.15 Pressure restoration in a diffuser (p_{o3}). M_1 and M_2 are Mach numbers in the discharge inlet and exit; M_1 = (1) 2; (2) 4, and (3) 6. Dashed lines correspond to ideal diffuser; solid lines take into account nonideal shock waves; static pressure at the discharge inlet is 0.1 atm.

For this constant pressure system, in order to obtain heat release values twice that of the critical value at S = const, it is necessary to increase the reactor cross section in the discharge zone by a factor of six.

6.4.4 Pressure Restoration in Supersonic Discharge Systems

As noted in Section 6.4.1, the important technological advantage of the supersonic plasma chemical systems is the possibility of operating a discharge at moderate pressures, while keeping system inlet and exit at high pressures. This requires an effective supersonic diffuser to restore the pressure after the discharge zone. Some relevant information on the pressure restoration after the supersonic discharge with different Mach numbers is presented in Figure 6.15. For this case, the initial pressure before supersonic nozzle was 3.8 atm at room temperature and the heat release was about half of the critical value. As can be seen from the figure, the pressure recovery can be quite decent (up to 1.5 to 2.5 atm), although heating and stopping of the supersonic flow in this way is obviously not optimal for pressure recovery.

Energy efficiency of the plasma chemical process of CO_2 dissociation in a supersonic discharge is presented in Figure 6.16, taking into account the energy spent on gas compression for nonconverted gas recirculation.[206] Comparing Figure 6.16 with Figure 5.22 (pure plasma chemical efficiency of the process) shows that energy spent on compression in this supersonic system is about 10% of the total process energy cost. Also, as can be seen from Figure 6.16, the energy cost of compression makes the process less effective at high Mach numbers ($M > 3$). This occurs even though the plasma chemical efficiency increases with Mach number because of reduction of vibrational relaxation losses.

6.4.5 Fluid Mechanic Equations of Vibrational Relaxation in Fast Subsonic and Supersonic Flows of Nonthermal Reactive Plasma

Most discharge and process characteristics of moderate- to high-pressure, nonequilibrium plasma chemical systems related to vibrational excitation of molecules are

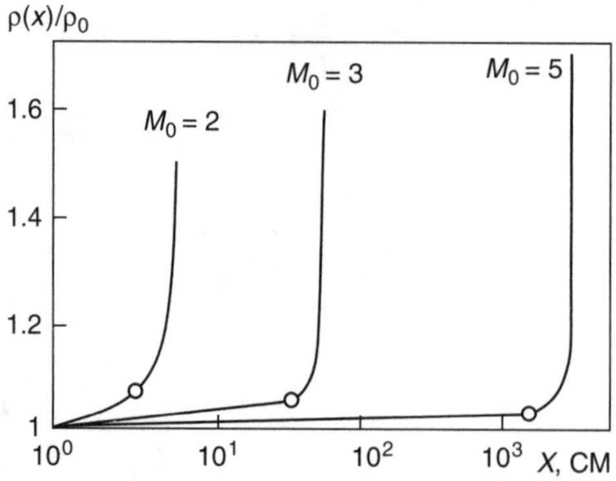

Figure 6.16 Gas density profiles in plasma chemical reaction zone at different Mach numbers

very sensitive to vibrational relaxation. Consider the peculiarities of vibrational relaxation in fast subsonic and supersonic flows, taking into account gas compressibility and the heat release in chemical reactions. Based on the kinetic scheme discussed in Section 5.6.7, the relevant one-dimensional fluid mechanic equations can be taken as follows:

1. Continuity equation:

$$\rho u = const \tag{6.86}$$

2. Momentum conservation equation:

$$p + \rho u^2 = const \tag{6.87}$$

3. Translational energy balance:

$$\rho u \frac{\partial}{\partial x}\left(\frac{R}{\mu}\frac{\gamma}{\gamma - 1}T_0 + \frac{u^2}{2}\right) = \xi P_R(T_v) + P_{VT} \tag{6.88}$$

4. Vibrational energy balance:

$$\rho u \frac{\partial}{\partial x}\left[\varepsilon_v(T_v)\right] = P_{ex} - P_R(T_v) - P_{VT} \tag{6.89}$$

These equations include specific powers of chemical reaction (P_R), VT relaxation (P_{VT}), and vibrational excitation (P_{ex}), which can be expressed as:

$$P_R = k_R(T_v)\rho^2\left(\frac{N_A}{\mu}\right)^2 \Delta Q \tag{6.90}$$

$$P_{VT} = k_{VT}(T_0)\rho^2\left(\frac{N_A}{\mu}\right)^2 \left[\varepsilon_v(T_v) - \varepsilon_v(T_0)\right] \tag{6.91}$$

$$P_{ex} = k_{eV} n_e \rho \frac{N_A}{\mu} \hbar\omega \tag{6.92}$$

In these relations, $k_R(T_v)$, $k_{VT}(T_0)$, and $k_{eV}(T_e)$ are the rate coefficients of chemical reaction, VT relaxation, and vibrational excitation by electron impact; n_e is electron density; $\gamma = c_p/c_v$ is the specific heat ratio, taking into account that the heat capacities include, in this case, only translational and rotational degrees of freedom; p, ρ, u, μ are pressure, density, velocity, and molecular mass of gas; N_A is the Avogadro number; ΔQ is vibrational energy spent per one act of chemical reaction; ξ is the fraction of this vibrational energy that goes into translational degrees of freedom; and T_v, T_0 are vibrational and translational temperatures.

Now introduce a new dimensionless density variable, $y = \rho_0/\rho(x)$, assume $T_v \approx \hbar\omega > T_0$, and also use the ideal gas equation of state:

$$p = \frac{\rho}{\mu} RT_0 \tag{6.93}$$

Then, rewrite the system of relaxation dynamic (Equation 6.86 to Equation 6.89) in the form:

$$u = u_0 y \tag{6.94}$$

$$T_0 = T_{00}\left[\left(1 + \gamma M_0^2\right)y - \gamma M_0^2 y^2\right] \tag{6.95}$$

$$y^2 \frac{\partial}{\partial x}[y(1+\gamma M_0^2) - \frac{\gamma+1}{2} M_0^2 y^2] = \xi Q_R + Q_{VT} \tag{6.96}$$

$$y^2 \frac{\partial}{\partial x} \varepsilon_v(T_v) = \frac{k_{eV} n_e}{u_0} \hbar\omega y - c_p T_{00} Q_R - c_p T_{00} Q_{VT} \tag{6.97}$$

In this system of equations:

$$Q_R = \frac{k_R(T_v)n_0}{u_0}\frac{\Delta Q}{c_p T_{00}}, \quad Q_{VT} = \frac{k_{VT}(T_{00})n_0}{u_0}\frac{\hbar\omega}{c_p T_{00}} \tag{6.98}$$

M is the Mach number; n_0 is gas concentration; and the subscript "0" means that the corresponding parameter is related to the inlet of the plasma chemical reaction zone.

It is convenient to analyze the solution of the system of Equation 6.94 and Equation 6.95 for the cases of strong and weak contribution of the chemical heat release (ξ).

6.4.6 Dynamics of Vibrational Relaxation in Fast Subsonic and Supersonic Flows

If the contribution of the chemical heat release can be neglected ($\xi\Delta Q << Q_{VT}$), then Equation 6.96 can be solved with respect to the reduced gas density $y = \rho(x)/\rho_0$ and analyzed in the following integral form:[208]

$$\int_1^y \frac{(y')^2\left[1+\gamma M_0^2-(\gamma+1)M_0^2 y'\right]dy'}{\exp\left[\frac{B}{T_{00}^{1/3}}\left(y'\left(1+\gamma M_0^2\right)-(y')^2\gamma M_0^2\right)^{1/3}\right]} = \frac{k_0 n_0}{u_0}\frac{\hbar\omega}{c_p T_{00}} \tag{6.99}$$

In the integration of Equation 6.96, the vibrational relaxation rate coefficient was taken as $k_{VT}(T_0) = k_0 \exp(-B/T_0^{1/3})$, based on the Landau–Teller approach (see Section 3.4.3). The dependence of $\rho(x)$, calculated from Equation 6.99, is presented in Figure 6.16 for supersonic flows with different initial Mach numbers. As can be seen from the figure, the growth of density during vibrational relaxation in supersonic flow is "explosive," which reflects the thermal mode of the overheating instability of vibrational relaxation discussed in Section 6.3.4 and illustrated in Figure 6.9. The density during the explosive overheating is a specific feature of supersonic flow.

"Explosions" similar to those shown in Figure 6.16 can be observed in the growth of temperature and pressure and in the decrease of velocity during the vibrational relaxation in supersonic flow. Linearization of the system of Equation 6.94 through Equation 6.96 allows finding the increment Ω (see Section 6.3) describing the common time scale for the explosive temperature growth and corresponding explosive changes of density and velocity:

$$\Omega_{VT} = k_{VT}(T_{00})n_0 \frac{\hbar\omega}{c_p T_{00}} \frac{2+\hat{k}_{VT}\left(\gamma M_0^2 - 1\right)}{M_0^2 - 1} \tag{6.100}$$

Here, T_{00}, n_0, and M_0 are the initial values of translational temperature, gas density, and Mach number at the beginning of the relaxation process; $\tilde{k}_{VT} = \partial \ln k_{VT}/\partial \ln T_0$ is the logarithmic sensitivity of vibrational relaxation rate coefficient to translational temperature (normally $\hat{k}_{VT} >> 2$; see below). Equation 6.100 shows that the dynamics of vibrational relaxation is qualitatively different at different Mach numbers:

1. **For subsonic flows with low Mach numbers**:

$$M_0 < \sqrt{\frac{1}{\gamma}\left(1 - \frac{2}{\hat{k}_{VT}}\right)} \tag{6.101}$$

the increment (Equation 6.100) is positive ($\Omega_{VT} > 0$) and explosive heating takes place. At very low Mach numbers ($M << 1$), Equation 6.100 for the increment can be simplified:

$$\Omega_{VT} = k_{VT}(T_{00})n_0 \frac{\hbar\omega}{c_p T_{00}}\left(\hat{k}_{VT} - 2\right) \tag{6.102}$$

This expression coincides with Equation 6.63 describing the increment of ionization-overheating instability in thermal mode.

The dependence of the overheating increment on Mach number Ω_{VT} (M_0) at fixed values of stagnation temperature and initial gas density is shown in Figure

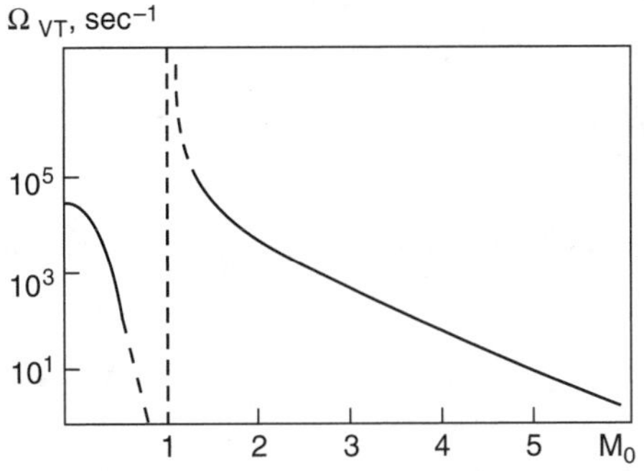

Figure 6.17 Relaxation frequency Ω_{VT} as a function of Mach number; stagnation temperature T_{00} = 300 K and density $n_0 = 3*10^{18}$ cm^{-3}

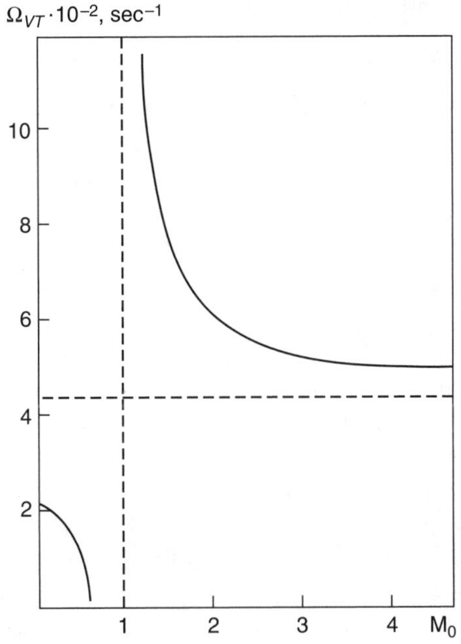

Figure 6.18 Relaxation frequency Ω_{VT} as function of Mach number at fixed thermodynamic temperature T_{00} = 100 K and density $n_0 = 3*10^{18}$ cm^{-3}

6.17. The same dependence Ω_{VT} (M_0), but at fixed values, of thermodynamic temperature and initial gas density, is presented in Figure 6.18. From these figures, the frequency (increment) of explosive vibrational relaxation decreases for subsonic flows as the Mach number increases. This phenomenon is because the relaxation heat release at higher Mach numbers contributes more to flow acceleration than temperature growth.

2. **For transonic flows with Mach numbers**:

$$\sqrt{\frac{1}{\gamma}\left(1-\frac{2}{\hat{k}_{VT}}\right)} < M_0 < 1 \tag{6.103}$$

the increment (Equation 6.100) is negative ($\Omega_{VT} < 0$) and the effect of explosive heating does not take place. The system is stable with respect to VT relaxation overheating.

This can be explained in the following way. The relaxation heat release increasingly contributes more to flow acceleration than temperature growth as the velocity approaches the speed of sound. In transonic flows, this effect becomes dominant. VT relaxation contributes little to increasing the thermodynamic temperature but the density decrease, which actually decreases the relaxation rate. It provides the plasma stability with respect to the relaxation overheating. However, in this case, the total relaxation heating should not exceed the critical value (pretty low in transonic flows), which results in shock wave generation.

3. **For supersonic flows** ($M > 1$), the increment (Equation 6.100) again becomes positive ($\Omega_{VT} > 0$) and the effect of explosive overheating takes place. At high Mach numbers:

$$\Omega_{VT} \approx k_{VT}(T_{00})n_0 \frac{\hbar\omega}{c_p T_{00}} \gamma \hat{k}_{VT} \tag{6.104}$$

which exceeds the increment (Equation 6.102) for subsonic flows. Therefore, the overheating instability is faster in the supersonic flows. This can be explained by noting that heating of the supersonic flow not only increases its temperature directly but also decelerates the flow, leading to additional temperature growth. This effect can be seen in Figure 6.18, in which the initial thermodynamic temperature is fixed. If the stagnation temperature is fixed, then the increase in Mach number results first in significant cooling. According to the Landau–Teller approach, this provides low values of the relaxation rate coefficient k_{VT} and significant reduction of the overheating instability increment (Equation 6.104), which can be seen in Figure 6.17.

6.4.7 Effect of Chemical Heat Release on Dynamics of Vibrational Relaxation in Supersonic Flows

Linearization of the system of equations describing vibrational relaxation in fast flows (see Section 6.4.5), which takes into account the chemical heat release, leads to the following expression for the overheating instability increment:

$$\Omega_{VT} = k_{VT}(T_{00})n_0 \frac{\hbar\omega}{c_p T_{00}} \frac{2+\hat{k}_{VT}\left(\gamma M_0^2 - 1\right)}{M_0^2 - 1} + k_R(T_v)n_0 \frac{2\xi\Delta Q}{\left(M_0^2-1\right)c_p T_{00}} \tag{6.105}$$

This expression for increment coincides with the corresponding Equation 6.100 in absence of the chemical heat release ($\xi = 0$). It is interesting that the chemical heat release stabilizes the overheating in subsonic flows where the second term in Equation 6.105 is negative. This effect has the same explanation as that for stabilizing overheating in transonic flows (see Section 6.4.6). In supersonic flows, the chemical heat release obviously accelerates the overheating. Note that the plasma chemical reaction rate coefficient is a function of vibrational temperature, which was considered in this case as unperturbed.

Equation 6.105 implies that the plasma chemical reaction and thus the chemical heat release take place throughout the relaxation process. This is not the case in plasma chemical processes stimulated by vibrational excitation, where the reaction time is shorter than relaxation (see Section 5.6.7). Taking this fact into account, time averaging of Equation 6.105 gives the following corrected expression for the increment of overheating:

$$\langle \Omega_{VT} \rangle = k_{VT}(T_{00}) n_0 \frac{\hbar\omega}{c_p T_{00}} \frac{2 + \hat{k}_{VT}\left(\gamma M_0^2 - 1\right)}{M_0^2 - 1 - 2\dfrac{q_R}{c_p T_{00}}} \tag{6.106}$$

Here, q_R is the total integral chemical heat release per one molecule, which is total energy transfer into translational degrees of freedom related to VV exchange, VTrelaxation from high vibrational levels, and heating due to exothermic chemical reactions. When the chemical heat release can be neglected, Equation 6.106 also coincides with Equation 6.100.

The values of the overheating increment Ω_{VT} are recalculated often into the length $1/\Lambda$ or reverse length Λ of vibrational relaxation. Typical dependence of the reverse length Λ of vibrational relaxation, taking into account the chemical heat release on the initial Mach number M_0, is presented in Figure 6.19.[208] This figure also permits comparing results of detailed modeling with analytical formulas of the previously considered linear approximation.

6.4.8 Spatial Nonuniformity of Vibrational Relaxation in Chemically Active Plasma

Assume that the one-dimensional space distribution of vibrational $T_v\,(x)$ and translational $T_0\,(x)$ temperatures initially have small fluctuations in the system related to gas:

$$T_0(t=0, x) = T_{00} + g(x), \quad T_v(t=0, x) = T_{v0} + h(x) \tag{6.107}$$

Because of the strong exponential temperature dependence of VT relaxation rate, heat transfer is unable to restore spatial uniformity of the temperatures when the density of the vibrationally excited molecule is relatively high and heterogeneous vibrational relaxation can be neglected. The effect of convective heat transfer on the spatial nonuniformity of vibrational relaxation in chemically active plasma is quite

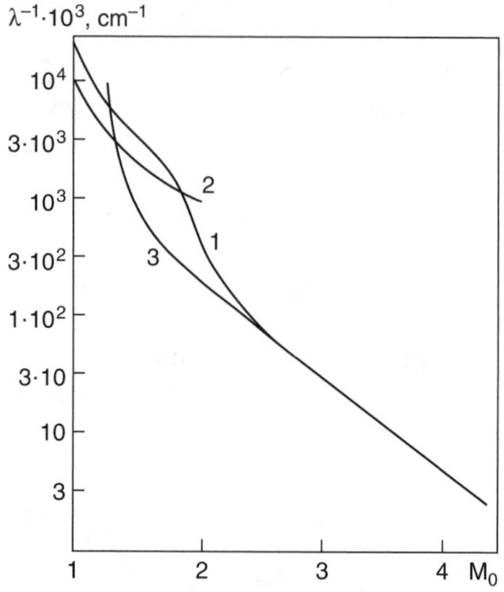

Figure 6.19 Inverse length of vibrational relaxation as function of Mach number at discharge inlet. Stagnation temperature T_{00} = 300 K is fixed; (1) = numerical calculation and (2, 3) = analytical model.

sophisticated.[209] However, at low Peclet numbers, the problem can be simplified taking into consideration only vibrational and translational energy conduction, VT relaxation, and chemical reactions:

$$n_0 c_v \frac{\partial T_0}{\partial t} = \lambda_T \frac{\partial T_0}{\partial x^2} + k_{VT} n_0^2 \left[\varepsilon_v(T_v) - \varepsilon_v(T_0)\right] + \xi k_R n_0^2 \Delta Q \tag{6.108}$$

$$n_0 c_v^v \frac{\partial T_v}{\partial t} = \lambda_V \frac{\partial T_v}{\partial x^2} - k_{VT} n_0^2 \left[\varepsilon_v(T_v) - \varepsilon_v(T_0)\right] - k_R n_0^2 \Delta Q \tag{6.109}$$

In this system, $\lambda_V, \lambda_T, c_v^v, c_v$ are vibrational and translational coefficients of thermal conductivities and specific heats. To analyze time and spatial evolution of small perturbations T_1 (x, t), T_{v1} (x, t) of the translational and vibrational temperatures, the system of Equation 6.108 and Equation 6.109 can be linearized:

$$\frac{\partial T_1}{\partial t} = \omega_{TT} T_1 + \omega_{TV} T_{v1} + D_T \frac{\partial^2 T_1}{\partial x^2} \tag{6.110}$$

$$\frac{\partial T_{v1}}{\partial t} = \omega_{VT} T_1 + \omega_{VV} T_{v1} + D_V \frac{\partial^2 T_{v1}}{\partial x^2} \tag{6.111}$$

Here, the following frequencies have been introduced:

$$\omega_{TT} = k_{VT} n_0 \left[\frac{\varepsilon_v(T_v) - \varepsilon_v(T_0)}{c_v T_0} \hat{k}_{VT} - \frac{c_v^v(T_0)}{c_v}\right] \tag{6.112}$$

$$\omega_{VV} = -k_{VT}n_0 - k_R n_0 \frac{\Delta Q}{c_v^v T_v}\hat{k}_R \tag{6.113}$$

which characterize the changes of translational temperature due to perturbations of T_0 and changes of vibrational temperature to perturbations of T_v. In a similar way, two other frequencies:

$$\omega_{TV} = k_{VT}n_0 \frac{c_v^v(T_v)}{c_v} + \xi k_R n_0 \frac{\Delta Q}{c_v T_v}\hat{k}_R \tag{6.114}$$

$$\omega_{TT} = -k_{VT}n_0 \frac{c_v}{c_v^v(T_v)}\left[\frac{\varepsilon_v(T_v) - \varepsilon_v(T_0)}{c_v T_0}\hat{k}_{VT} - \frac{c_v^v(T_0)}{c_v}\right] \tag{6.115}$$

describe changes of translational temperature due to perturbations of T_v and changes of vibrational temperature due to perturbations of T_0. In the preceding relations, logarithmic sensitivity factors are $\hat{k}_R = \partial \ln k_R / \partial \ln T_v$ and $\hat{k}_{VT} = \partial \ln k_{VT} / \partial \ln T_0$ and $D_T = \lambda_T / n_0 c_v$ and $D_V = \lambda_V / n_0 c_v^v$ are the reduced coefficients of translational and vibrational thermal conductivity.

To obtain the dispersion equation for evolution of fluctuations of vibrational and translational temperatures, consider the temperature perturbations in the linearized system of Equation 6.110 and Equation 6.111 in the exponential form with amplitudes A, B as:

$$T_1(x,t) = A\cos kx \cdot \exp(\Lambda t), \quad T_{v1}(x,t) = B\cos kx \cdot \exp(\Lambda t) \tag{6.116}$$

The system of Equation 6.110 and Equation 6.111 then leads to the following dispersion equation relating increment Λ of amplification of the perturbations with their wave number k:

$$\left(\Lambda - \Omega_T^k\right)\left(\Lambda - \Omega_V^k\right) = \omega_{VT}\omega_{TV} \tag{6.117}$$

In this dispersion equation, Ω_T^k and Ω_V^k are the so-called thermal and vibrational modes, which are determined by the wave number k as:

$$\Omega_T^k = \omega_{TT} - D_T k^2 \tag{6.118}$$

$$\Omega_V^k = \omega_{TT} - D_V k^2 \tag{6.119}$$

One can note, then, that $\omega_{VT}\omega_{TV}$ is the nonlinear coupling parameter between the vibrational and thermal modes. If the product is equal to zero, the modes Equation 6.118 and Equation 6.119 are independent. The solution of Equation 6.117 with respect to the increment $\Lambda(k)$ can be presented as:

$$\Lambda_{\pm}(k) = \frac{\Omega_T^k + \Omega_V^k}{2} \pm \sqrt{\frac{\left(\Omega_T^k - \Omega_V^k\right)^2}{4} + \omega_{VT}\omega_{TV}} \tag{6.120}$$

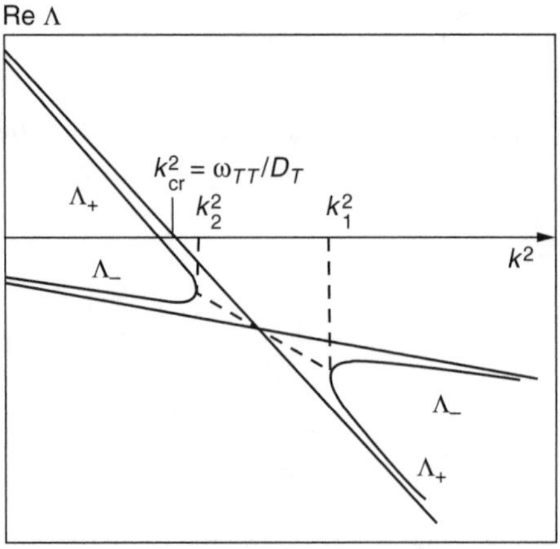

Figure 6.20 *Re* Λ as function of perturbation wave number

This increment describes the initial linear phase of the spatial nonuniform vibrational relaxation in a chemically active plasma, which results in different effects on structure formation in corresponding discharge systems.

6.4.9 Space Structure of Unstable Vibrational Relaxation

Analysis of different spatial, nonuniform VT relaxation instability spectra, expressed by the dispersion Equation 6.120, shows a variety of possible phenomena in the system, including interaction between the instability modes, wave propagation, etc.[209] The following relation between the frequencies can usually characterize plasma chemical processes stimulated by vibrational excitation of molecules:

$$\omega_{TT} > 0, \omega_{VV} < 0, \omega_{VT}\omega_{TV} < 0, \max\{\omega^2_{VV}, \omega^2_{TT}\} > |\omega_{VT}\omega_{TV}|$$

which greatly simplifies the analysis of the dispersion equation. In this case, growth of the initial temperature perturbations (Equation 6.116) is controlled by the thermal mode and takes place under the following condition:

$$\Omega^k_T = \omega_{TT} - D_T k^2 = k_{VT} n_0 \left[\frac{\varepsilon_v(T_v) - \varepsilon_v(T_0)}{c_v T_0} \hat{k}_{VT} - \frac{c^v_v(T_0)}{c_v} \right] - D_T k^2 > 0 \quad (6.121)$$

Stability diagrams or dispersion curves *Re* Λ (k_2), which illustrate the evolution of temperature perturbations at $D_T > D_V$ and frequency relations mentioned earlier, are shown in Figure 6.20. Obviously, $\text{Re}\,\Lambda(k^2) > 0$ means amplification of initial perturbations, and $\text{Re}\,\Lambda(k^2) < 0$ their stabilization. As can be seen from Figure 6.20, the unstable harmonics (Equation 6.116) correspond to the thermal mode (Equation 6.121), which is related to the Λ_+ line on the figure.

Thus, during the linear phase of the relaxation process, short-scale perturbations $k > k_{cr} = \sqrt{\omega_{TT}/D_T}$ decrease and disappear because of thermal conductivity, while

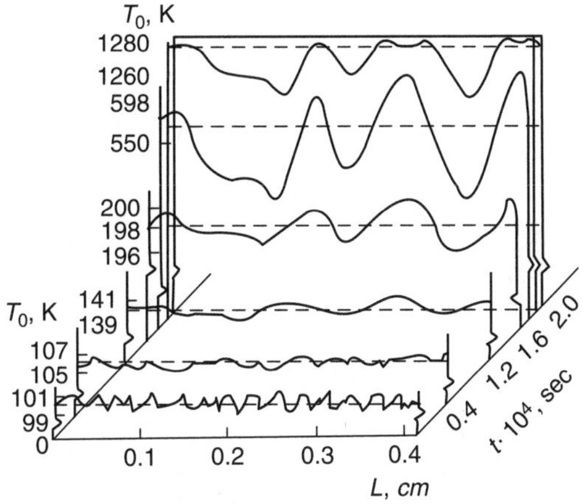

Figure 6.21 Dynamics of space nonuniform relaxation;[209] dashed line corresponds to uniform solution

perturbations with longer wavelength $\lambda > \lambda_{cr} = 2\pi/k_{cr}$ grow exponentially. When amplitudes of the temperature perturbations become sufficiently large, the regime of the vibrational relaxation is nonlinear. This results in the formation of high-gradient spatial structures, which consist of periodical temperature zones with quite different relaxation times. Minimal distance between such zones can be determined by the previously calculated critical wavelength:

$$\lambda_{cr} = 2\pi\sqrt{\frac{D_T c_v T_0}{k_{VT} n_0 \hat{k}_{VT}\left(\varepsilon_v(T_v) - \varepsilon_v(T_0)\right)}} \tag{6.122}$$

Formation of the spatial structure of temperature provided by vibrational relaxation on the nonlinear phase of evolution is shown in Figure 6.21.[209] For CO_2 ($n_0 = 3 \cdot 10^{18}$ cm^{-3}) with the following initial and boundary ($x = 0, L$) conditions:

$$T_0(x,t=0) = 100K,\ T_v(x,t=0) = 2500K,\ \frac{\partial T_0}{\partial x}(x=0,L;t) = \frac{\partial T_v}{\partial x}(x=0,L;t) = 0$$

Initial perturbations of vibrational temperature were taken as "white noise" with the mean-square deviation $\langle \delta T_0^2 \rangle = 0.8 \cdot 10^{-4} K^2$.

As can be seen from Figure 6.21, the evolution of perturbations can be subdivided into two phases: linear ($t < 5 \cdot 10^{-4}$ sec) and nonlinear ($t > 5 \cdot 10^{-4}$ sec). The linear phase can be characterized by damping of the short-scale perturbations and thus by an approximate 100 times decrease of the mean-square level of fluctuations $\langle \delta T_0^2 \rangle$ (see Figure 6.22). In the nonlinear phase, perturbations grow significantly (see Figure 6.22), forming the structures with characteristic sizes of about 0.1 to 0.2 cm, in accordance with Equation 6.122. Experimentally, such structures were observed in the supersonic microwave discharge discussed in Section 6.4.1.

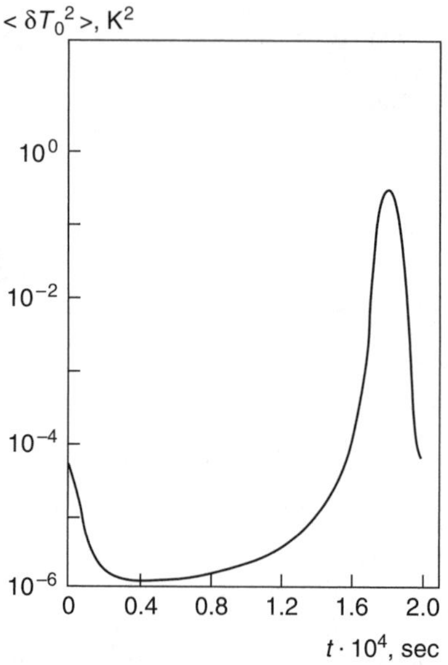

Figure 6.22 Time evolution of translational temperature mean square deviation

6.5 ELECTROSTATIC, MAGNETO-HYDRODYNAMIC, AND ACOUSTIC WAVES IN PLASMA

In general, plasma behaves somewhat similar to a gas interacting between particles. However, oscillations, waves, and noises play a much more significant role in plasma. This is due to the long-distance nature of interaction between charged particles and to variety of different types of oscillations and waves that can be generated because of external fields and plasma nonhomogeneities. Also, the amplitude of plasma oscillations and waves can grow so large that they become a major factor in determining plasma behavior and evolution.

6.5.1 Electrostatic Plasma Waves

The electrostatic plasma oscillations discussed in Section 6.1.6 are able to propagate in plasma as longitudinal waves. Electric fields in these longitudinal waves (in contrast to the transverse electromagnetic wave) are in the direction of the wave propagation, which is the direction of wave vector $\vec{k}$.

To analyze these electrostatic plasma waves, the amplitude A^1 of oscillations of any macroscopic parameter $A(x,t)$ can be considered small ($A^1 << A_0$). Then the oscillations A(x,t) for the linearized relevant equations can be expressed as:

$$A = A_0 + A^1 \exp[i(kx - \omega t)] \tag{6.123}$$

where A_0 is the value of the macroscopic parameter in absence of oscillations; k is the wave number; and ω is the wave frequency.

To obtain the dispersion equation of the electrostatic plasma waves, consider the linearized system of Equation 6.123, including continuity for electrons; momentum conservation for the electron gas without dissipation; the adiabatic relation for the gas; and the Poisson equation:

$$-i\omega n_e^1 + ikn_e u^1 = 0 \tag{6.124}$$

$$-i\omega u^1 + ik\frac{p^1}{mn_e} + \frac{eE^1}{m} = 0 \tag{6.125}$$

$$\frac{p^1}{p_0} = \gamma\frac{n_e^1}{n_e} \tag{6.126}$$

$$ikE^1 = -\frac{1}{\varepsilon_0}en_e^1 \tag{6.127}$$

In the system of equations, n_e^1, u^1, p^1, E^1 are the amplitude of oscillations of electron concentration, velocity, pressure, and electric field; n_e is the unperturbed plasma density; e, m, and γ are charge, mass, and specific heat ratio for an electron gas; $p_0 = n_e m\langle v_x^2\rangle = n_e T_e$ is the electron gas pressure in the absence of oscillations; $\langle v_x^2\rangle$ is the averaged square of electron velocity in direction of oscillations; and T_e is the electron temperature.

From the system of linearized equations, the dispersion equation (relation between frequency and wave number) for the electrostatic plasma waves can be derived as:

$$\omega^2 = \omega_p^2 + \frac{\gamma T_e}{m}k^2, \quad \omega_p = \sqrt{\frac{n_e e^2}{\varepsilon_0 m}} \tag{6.128}$$

Here, ω_p is the plasma frequency, which was introduced and discussed in Section 6.1 (see Equation 6.17 through Equation 6.19).

From the dispersion equation (Equation 6.128), it is seen that the electrostatic wave frequency is close to the plasma frequency, if the wavelength $2\pi/k$ much exceeds the Debye radius (see Equation 6.18). In the opposite limit of very short wavelengths, the phase velocity of the electrostatic plasma waves corresponds to the thermal speed of electrons.

6.5.2 Collisional Damping of Electrostatic Plasma Waves in Weakly Ionized Plasma

In the dispersion Equation 6.128, the electron–neutral collisions (frequency ν_{en}) have been neglected. However, in weakly ionized plasma, electron and neutral collisions can influence the plasma oscillations quite significantly. Plasma oscillations actually do not exist when $\omega < \nu_{en}$. At higher frequencies when $\omega > \nu_{en}$, electron–neutral collisions provide damping of the electrostatic plasma oscillations in accordance with the corrected dispersion equation (Equation 6.128):

$$\omega = \sqrt{\omega_p^2 + \frac{\gamma T_e}{m} k^2} - i\nu_{en} \quad (6.129)$$

The amplitude of plasma oscillations (Equation 6.123) with frequency in the form of Equation 6.129 decays exponentially as a function of time: $\propto \exp(-\nu_{en} t)$. This effect is usually referred to as the collisional damping of the electrostatic plasma waves. Taking into account that the wave frequencies are usually near the Langmuir frequency, the numerical criterion of the existence of electrostatic plasma waves with respect to collisional damping can be expressed based on Equation 6.129 as:

$$\frac{\sqrt{n_e, \mathrm{cm}^{-3}}}{n_0, \mathrm{cm}^{-3}} >> 10^{-12} \mathrm{cm}^{3/2} \quad (6.130)$$

Here, n_0 is the neutral gas density. For example, in atmospheric pressure room temperature discharges, $n_0 = 3 \cdot 10^{19}$ cm^{-3}, and the criterion (Equation 6.130) requires very large values of electron concentrations $n_e >> 10^{15}$ cm^{-3}.

6.5.3 Ionic Sound

Electrostatic plasma oscillations related to motion of ions are called the ionic sound. The ionic sound waves are longitudinal in the same manner as the electrostatic plasma waves, e.g., the direction of electric field in the wave coincides with the direction of the wave vector $\vec{k}$. To derive the dispersion equation of the ionic sound, one should start with the Poisson equation for the potential φ of the plasma oscillations:

$$\frac{\partial^2 \varphi}{\partial x^2} = \frac{e}{\varepsilon_0}(n_e - n_i) \quad (6.131)$$

Then, taking take into account that light electrons quickly correlate their local instantaneous concentration in the wave n_e (x,t) with the potential φ (x,t) of the plasma oscillations in accordance with the Boltzmann distribution:

$$n_e = n_p \exp\left(+\frac{e\varphi}{T_e}\right) \approx n_p \left(1 + \frac{e\varphi}{T_e}\right) \quad (6.132)$$

Here, the unperturbed density of the homogeneous plasma is denoted as n_p. Linearizing (Equation 6.123) permits obtaining from Equation 6.131 and Equation 6.132 the following expression for amplitude of oscillations of ion density:

$$n_i^1 = n_p \frac{e\varphi}{T_e}\left(1 + k^2 \frac{\varepsilon_0 T_e}{n_p e^2}\right) \quad (6.133)$$

Combining Equation 6.133 with the linearized motion equation for ions in the electric field of the electrostatic wave:

$$M\frac{d\vec{u}_i}{dt} = e\vec{E} = -e\nabla\varphi, \quad M\omega u_i^1 = ek\varphi \tag{6.134}$$

and with the linearized continuity equation for ions:

$$\frac{\partial n_i}{\partial t} + \nabla(n_i\vec{u}_i) = 0, \quad \omega n_i^1 = kn_p u_i^1 \tag{6.135}$$

yields the dispersion equation for the ionic sound:

$$\left(\frac{\omega}{k}\right)^2 = c_{si}^2 \frac{1}{1+k^2 r_D^2}, \quad c_{si} = \sqrt{\frac{T_e}{M}} \tag{6.136}$$

Here, c_{si} is the speed of ionic sound; M is mass of an ion; $\vec{u}_i$ and u_i^1 are the ionic velocity and amplitude of its oscillation; and r_D is the Debye radius (Equation 4.129). Note that the speed of ionic sound (Equation 6.136) coincides with the Bohm velocity (Equation 6.10) discussed in Section 6.1.4 regarding the DC sheaths.

As can be seen from the dispersion (Equation 6.136), the ionic sound waves $\omega/k = c_{si}$ propagate in the plasma with wavelengths exceeding the Debye radius ($kr_D << 1$). For wavelengths shorter than the Debye radius ($kr_D >> 1$), Equation 6.136 describes plasma oscillations with the frequency:

$$\omega_{pi} = \frac{c_{si}}{r_D} = \sqrt{\frac{n_p e^2}{M\varepsilon_0}} \tag{6.137}$$

known as the plasma-ion frequency. The expression for the plasma-ion frequency is similar to that describing the plasma-electron (Langmuir) frequency (Equation 6.17), but with the ionic mass M instead of the electron mass m.

6.5.4 Magneto-Hydrodynamic Waves

Special types of waves occur in a plasma located in a magnetic field. Dispersion equations for such waves obviously can be derived by the conventional routine of linearization of magneto-hydrodynamic equations discussed in Section 6.2.1. However, it is interesting, to describe these MHD waves simply by a physical analysis of the oscillations of the magnetic field frozen in plasma.

Consider a plasma with a high magnetic Reynolds number ($\mathrm{Re}_m >> 1$), where the magneto-hydrodynamic approach can be applied and the magnetic field is frozen in plasma (see Section 6.2.7). In this case, any displacement of the magnetic field $\vec{H}$ leads to corresponding plasma displacement, which results in plasma oscillations and propagation of specific magneto-hydrodynamic waves.

The propagation velocity of the elastic oscillations can be determined using the conventional relation for speed of sound, $v_A = \sqrt{\partial p/\partial \rho}$, where p is pressure and ρ =

$Mn_p = Mn_e$ is the plasma's density. The total pressure of the relatively cold plasma is equal to its magnetic pressure $p = \mu_0 H^2/2$ (see Section 6.2.3), which leads to the relation:

$$v_A = \sqrt{\frac{\partial(\mu_0 H^2/2)}{\partial(Mn_p)}} = \sqrt{\frac{\mu_0 H}{M}\frac{\partial H}{\partial n_p}} \tag{6.138}$$

Taking into account that the magnetic field is frozen in the plasma, the derivative in Equation 6.138 is $\partial H/\partial n_p = H/n_p$ and the propagation velocity of the magneto-hydrodynamic waves in plasma is:

$$v_A = \sqrt{\frac{\mu_0 H^2}{\rho}} = \frac{B}{\sqrt{\mu_0 \rho}} \tag{6.139}$$

This wave propagation velocity is known as the **Alfven velocity** and often appears in magneto-hydrodynamics of plasma (see Section 6.2.7 and Equation 6.55).

There are two types of plasma oscillations in magnetic field that have the same Alfven velocity, but different directions of propagation. The plasma oscillations propagating along the magnetic field like a wave propagating along elastic string are usually referred as the **Alfven wave** or magneto-hydrodynamic wave. Also, oscillation of a magnetic line induces oscillation of other nearby magnetic lines. This results in propagation perpendicular to the magnetic field of another wave with the same Alfven velocity. This wave is called the **magnetic sound**. These are the main features of the phenomena; more details regarding magneto-hydrodynamic, electrostatic, and other plasma waves can be found, for example, in the text of Stix.[210]

6.5.5 Collisionless Interaction of Electrostatic Plasma Waves with Electrons

Damping plasma waves, discussed in Section 6.5.2, was related to electron–neutral collisions. Consider now the interesting and practically important effect of energy exchange between electrons and plasma oscillations without any collisions. It is convenient to illustrate the collisionless energy exchange in a reference frame moving with a wave (for where the wave is at rest, see Figure 6.23). As is seen from the figure, an electron can be trapped in a potential well created by the wave, which leads to their effective interaction.

If an electron moves in the reference frame of electrostatic wave with velocity u along the wave and, after reflection, changes to the opposite direction and its velocity to $-u$, then change of electron energy due to the interaction with electrostatic wave is:

$$\Delta\varepsilon = \frac{m(v_{ph} + u)^2}{2} - \frac{m(v_{ph} - u)^2}{2} = 2mv_{ph}u \tag{6.140}$$

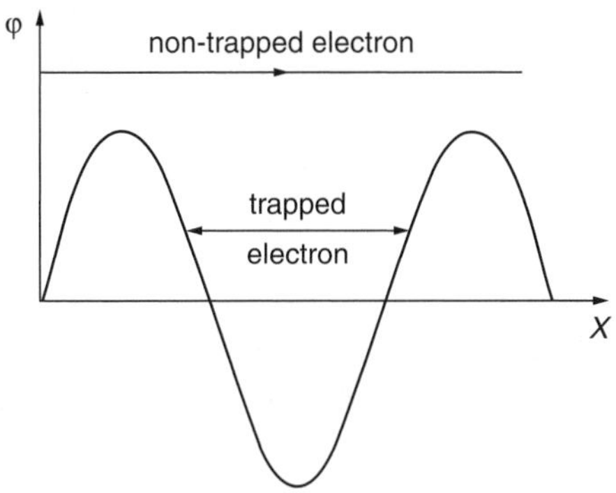

Figure 6.23 Electron interaction with plasma oscillations

Here, $v_{ph} = \omega/k$ is the phase velocity of the wave. If φ is the amplitude of the potential oscillations in the wave, then the typical velocity of a trapped electron in the reference frame of plasma wave is $u \approx \sqrt{e\varphi/m}$ and typical energy exchange between electrostatic wave and a trapped electron can be estimated as:

$$\Delta\varepsilon \approx v_{ph}\sqrt{me\varphi} = e\varphi\sqrt{\frac{mv_{ph}^2}{e\varphi}} \tag{6.141}$$

In contrast to Equation 6.140, typical energy exchange between an electrostatic wave and a nontrapped electron is approximately $e\varphi$. This means that, at low amplitudes of oscillations, the collisionless energy exchange is mostly due to the trapped electrons.

The described collisionless energy exchange tends to equalize the electron distribution function $f(v)$ at the level of electron velocities near to their resonance with the phase velocity of the wave. Actually, the equalization (trend to form plateau on $f(v)$) takes place in the range of electron velocities between $v_{ph} - u$ and $v_{ph} + u$. It has typical frequency, corresponding to the frequency of a trapped electron oscillation in a potential well of the wave:

$$\nu_{ew} \approx uk \approx k\sqrt{\frac{e\varphi}{m}} \approx \sqrt{\frac{eE^1k}{m}} \approx \sqrt{\frac{e^2n_e^1}{\varepsilon_0 m}} = \omega_p\sqrt{\frac{n_e^1}{n_e}} \tag{6.142}$$

Here, k is the wave number; m is the electron mass; E^1 is the amplitude of electric field oscillation; ω_p is the plasma frequency; and n_e, n_e^1 are electron density and amplitude of its oscillation. In Equation 6.142, the linearized Maxwell relation (Equation 6.127) between amplitudes of oscillations of electric field and electron density was taken into account.

Although the electrons' interaction with the Langmuir oscillations tend to form a plateau on $f(v)$ at electron velocities close to $v_p = \omega/k$, the electron–electron maxwellization collisions tend to restore the electron distribution function $f(v)$. The frequency of the electron–electron maxwellization process can be found based on estimations of the thermal electron temperature v_{Te} and the electron–electron coulomb cross section σ_{ee} as:

$$\nu_{ee} = n_e v_{Te} \sigma_{ee} \approx n_e \sqrt{\frac{T_e}{m}} \cdot \frac{e^4}{(4\pi\varepsilon_0)^2 T_e^2} \tag{6.143}$$

If the maxwellization frequency exceeds the electron–wave interaction frequency $\nu_{ee} >> \nu_{ew}$, the electron distribution function is almost nonperturbed by the collisionless interaction with the electrostatic wave. Comparing the frequencies (Equation 6.142 and Equation 6.143) shows that changes in the electron distribution functions $f(E)$ during the collisionless interaction can be neglected, if relative perturbations of electron density are small:

$$\frac{n_e^1}{n_e} << \frac{n_e e^6}{(4\pi\varepsilon_0)^3 T_e^3} \tag{6.144}$$

It is clear from Figure 6.23 and Equation 6.140 that electrons with velocities $v_{ph} + u$ transfer their energy to the electrostatic plasma wave, while electrons with velocities $v_{ph} - u$ receive energy from the plasma wave. Thus, the collisionless damping of electrostatic plasma oscillations takes place when:

$$\frac{\partial f}{\partial v_x}(v_x = v_{ph}) < 0 \tag{6.145}$$

In this relation, v_x is the electron velocity component in the direction of wave propagation; here, the derivative of the electron distribution function is taken at the electron velocity equal to the phase velocity $v_{ph} = \omega/k$ of the wave.

Usually, the electron distribution functions $f(v)$ decreases at high velocities corresponding to v_{ph}, the inequality (Equation 6.145) is satisfied, and the collisionless damping of the electrostatic plasma waves takes place. However, the opposite situation of amplification of the electrostatic plasma waves due to the interaction with electrons is also possible.

Injection of an electron beam in a plasma creates distribution function, where the derivative (Equation 6.145) is positive (see Figure 6.24); this corresponds to energy transfer from the electron beam and exponential amplification of electrostatic plasma oscillations. The electron beam keeps transferring energy to plasma waves until the total electron distribution function becomes always decreasing (see Figure 6.24). This so-called **beam instability** is used for stimulation of nonequilibrium plasma-beam discharge, which will be discussed in Section 11.3.

6.5.6 Landau Damping

If the inequality (Equation 6.145) is valid, the electrostatic plasma oscillations transfer energy to electrons as described earlier. Estimate the rate of this energy transfer from the plasma oscillations per unit time and per unit volume as:

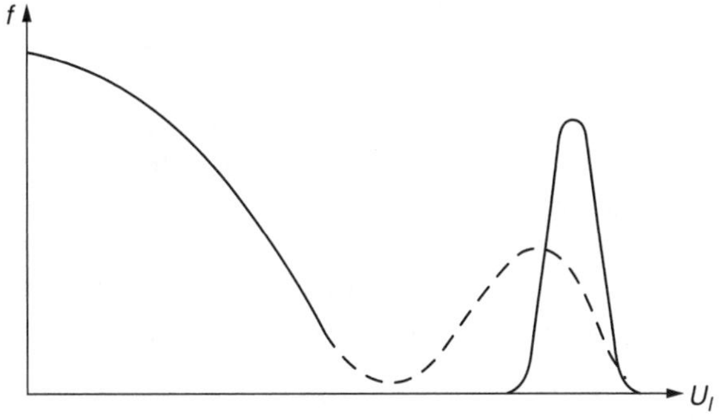

Figure 6.24 Electron energy distribution function for electron beam interaction with plasma: solid curve = initial distribution; dashed curve = resulting distribution

$$\frac{\partial W}{\partial t} \approx [f(v+u) - f(v-u)]un_e \cdot \nu_{ew} \cdot \Delta\varepsilon \approx \frac{\partial f}{\partial v_x}\frac{n_e e^2}{m}\omega\varphi^2 \tag{6.146}$$

Frequency ν_{ew} is determined by Equation 6.142.

Taking into account that $\varphi^2 \approx (E^1)^2/k^2 \approx W/\varepsilon_0 k^2$, rewrite Equation 6.146 in a conventional form for the decreasing specific energy W of damping oscillations:

$$\frac{\partial W}{\partial t} = -2\gamma W, \quad \gamma \approx -\frac{\omega_p^2\omega}{k^2}\frac{\partial f}{\partial v_x}\left(v_x = \frac{\omega}{k}\right) > 0 \tag{6.147}$$

This damping of the electrostatic plasma waves, $W = W_0 \exp(-2\gamma t)$, due to their collisionless interaction with electrons is known as the **Landau damping** (see, for example, O'Neil[211] and Ivanov[212]).

The existence criterion for electrostatic plasma oscillations can be expressed as $\gamma << \omega$. This can be rewritten, based on Equation 6.147 for plasma oscillations ($\omega \sim \omega_p$) and a Maxwellian electron distribution function, as $kr_D << 1$, where r_D is the Debye radius. When this criterion is valid, the phase velocity of the plasma wave greatly exceeds the thermal velocity of electrons. In this case, the plasma wave is able to trap only a small fraction of electrons from the tail of electron distribution function $f(v)$ and, as a result, damping of the wave is not strong.

6.5.7 Beam Instability

Consider dissipation of an electron beam injected into a plasma. Assume that the velocity of the beam electrons greatly exceeds the thermal velocity of plasma electrons, and that the electron beam density is much less than the plasma density ($n_b << n_p$). Dissipation and stopping of the electron beam obviously can be provided by collisions with electrons, ions, or neutral particles. However, the dissipation and stopping of an electron beam sometimes can be faster than that; this is known as **beam instability** or the **Langmuir paradox**.

The effect of beam instability is actually opposite with respect to Landau damping, as was mentioned earlier and is illustrated in Figure 6.24. Electron beam (where the inequality Equation 6.145 has an opposite direction) transfers energy to excitation and amplification of plasma oscillations. This collisionless dissipation of electron beam energy explains the Langmuir paradox. In a similar manner as was done previously, the dispersion equation for plasma waves in the presence of the electron beam can be derived by linearization of continuity equation, momentum conservation, and Poisson equation. This leads to the dispersion equation, in which u is the electron beam velocity and ω_p is the plasma frequency:

$$1 = \frac{\omega_p^2}{\omega^2} + \frac{\omega_p^2}{(\omega - ku)^2}\frac{n_b}{n_p} \tag{6.148}$$

The strongest collisionless interaction of the electron beam is related to plasma waves with phase velocity close to the electron beam velocity $|\omega/k - u| << \omega/k$. Also, the oscillation frequency ω is close to the plasma frequency ω_p because the electron beam density is much less than the plasma density ($n_b << n_p$). Taking into account these two facts and assuming $\omega = \omega_p + \delta$, $|\delta/\omega_p| << 1$, the dispersion equation is:

$$\delta = \omega_p\left(\frac{n_b}{2n_p}\right)^{1/3} e^{2\pi i m/3} \tag{6.149}$$

where m is an integer. The amplification of the specific energy $W \propto (E^1)^2$ of plasma oscillations corresponds, in accordance with Equation 6.123, to the imaginary part of ω (and thus δ). This amplification of the Langmuir oscillation energy W again can be described by Equation 6.147, but in this case with a negative coefficient γ:

$$\frac{\partial W}{\partial t} = -2\gamma W, \quad \gamma \approx -\frac{\sqrt{3}}{2}\left(\frac{n_b}{2n_p}\right)^{1/3}\omega_p = -0.69\left(\frac{n_b}{n_p}\right)^{1/3}\omega_p < 0 \tag{6.150}$$

Thus, an electron beam is able to transfer its energy effectively into electrostatic plasma oscillations, which thereafter can dissipate to other plasma degrees of freedom. More details related to the beam instabilities can be found in Ivanov and Kadomtsev.[178,212a]

6.5.8 Buneman Instability

The beam instability considered previously is an example of the so-called kinetic instabilities, where amplification of plasma oscillations is due to difference in the motion of different groups of charged particles (beam electrons and plasma electrons in this case). Another example of such kinetic instabilities is the Buneman instability, which occurs if the average electron velocity differs from that of ions.

To describe the Buneman instability, assume that all plasma ions are at rest while all plasma electrons are moving with velocity u. As in the beam instability, electron energy dissipates here to generate and amplify electrostatic plasma oscillations. The dispersion equation for the Buneman instability is similar to Equation 6.148; however, in this case, concentrations of two groups of charged particles (electrons and ions, no beam) are equal, but their mass is very different ($m/M << 1$):

$$1 = \frac{m}{M}\frac{\omega_p^2}{\omega^2} + \frac{\omega_p^2}{(\omega - ku)^2} \tag{6.151}$$

Taking into account that $m/M << 1$, it follows from Equation 6.151 that $\omega - ku$ should be quite close to the plasma frequency. Then, assuming $\omega = \omega_p + ku + \delta$, $|\delta/\omega_p| << 1$, rewrite the dispersion (Equation 6.151) as:

$$\frac{2\delta}{\omega_p} = \frac{m}{M}\frac{\omega_p^2}{(\omega_p + ku + \delta)^2} \tag{6.152}$$

The electrons interact most efficiently with a wave having the wave number $k = -\omega_p/u$. For this wave, based on Equation 6.152,

$$\delta = \left(\frac{m}{2M}\right)^{1/3} \omega_p e^{2\pi i m/3} \tag{6.153}$$

where m is an integer (the strongest plasma wave amplification corresponds to $m = 1$). Similar to Equation 6.149 and Equation 6.150, the coefficient of amplification γ of the electrostatic plasma oscillation amplitude can be expressed in this case as:

$$-\gamma = \operatorname{Im}\delta = \frac{\sqrt{3}}{2}\left(\frac{m}{2M}\right)^{1/3} \omega_p = 0.69\left(\frac{m}{M}\right)^{1/3} \omega_p \tag{6.154}$$

Note that, because $k = -\omega_p/u$, the frequency of the amplified plasma oscillations in the case of Buneman instability is close to the coefficient of wave amplification $\omega \approx |\gamma|$. More details regarding kinetic instabilities, generation and amplification of electrostatic plasma wave can be found in Mikhailovskii.[185]

6.5.9 Dispersion and Amplification of Acoustic Waves in Nonequilibrium Weakly Ionized Plasma: General Dispersion Equation

Nonequilibrium plasma of molecular gases is able to amplify acoustic waves significantly due to different relaxation mechanisms. In principle, amplification of sound related to phased heating has been known since 1878.[213] The Rayleigh mechanism of acoustic instability in weakly ionized plasma due to joule heating has been analyzed by different authors (see Jacob and Mani[214]). Consider the dispersion $k(\omega)$ and amplification of acoustic waves in nonequilibrium, chemically active plasma, taking into account a heat release provided by vibrational relaxation and chemical reactions. Chemical reactions are assumed to be stimulated by vibrational excitation of molecules. Linearizing the continuity equation, momentum conservation, and balance of translational and vibrational energies, the dispersion equation for acoustic waves (Equation 6.123) in the vibrationally nonequilibrium chemically active plasma can be given as:[214a]

$$\frac{k^2c_s^2}{(\omega-\vec{k}\vec{v})^2}=1-\frac{i(\omega-\vec{k}\vec{v})[e_T(\gamma-1)+e_n]-(e_v\bar{e}_T-e_T\bar{e}_v)\gamma(\gamma-1)+(e_n\bar{e}_v-e_v\bar{e}_n)\gamma}{(\omega-\vec{k}\vec{v})^2+i(\omega-\vec{k}\vec{v})(e_n-e_T-\gamma\bar{e}_v)+\gamma(e_v\bar{e}_T-e_T\bar{e}_v)+\gamma(e_n\bar{e}_v-e_v\bar{e}_n)} \tag{6.155}$$

In this dispersion equation, $c_s=\sqrt{\gamma T_0/M}$ is the "frozen" speed of sound (which means that vibrational degrees of freedom are "frozen" and do not follow variations of T_0); M is mass of heavy particles; γ is the specific heat ratio; $\vec{v}$ is the gas velocity; and $e_{n,T,v}$ and $\bar{e}_{n,T,v}$ are the translational (e) and vibrational ($\bar{e}$) temperature changes related to perturbations of gas concentration n_0, translational T_0, and vibrational T_v temperatures, respectively:

$$e_n=2(\nu_{VT}+\xi\nu_R) \tag{6.156}$$

$$e_T=\hat{k}_{VT}\nu_{VT}-k_{VT}n_0\frac{c_v^v(T_0)}{c_v} \tag{6.157}$$

$$e_v=k_{VT}n_0\frac{c_v^v(T_v)T_v}{c_vT_0}+\hat{k}_R\xi\nu_R \tag{6.158}$$

$$\bar{e}_n=\nu_{eV}\frac{c_vT_0}{c_v^v(T_v)T_v}\left(1+\frac{\partial\ln n_e}{\partial\ln n_0}\right)-2\nu_{VT}\frac{c_vT_0}{c_v^v(T_v)T_v}-2\xi\nu_R\frac{c_vT_0}{c_v^v(T_v)T_v} \tag{6.159}$$

$$\bar{e}_T=-\hat{k}_{VT}\nu_{VT}\frac{c_vT_0}{c_v^v(T_v)T_v}+k_{VT}n_0\frac{T_0}{T_v} \tag{6.160}$$

$$\bar{e}_v=-k_{VT}n_0-\hat{k}_R\nu_R\frac{c_vT_0}{c_v^v(T_v)T_v} \tag{6.161}$$

The characteristic frequencies of VT relaxation, chemical reaction, and vibrational excitation of molecules by electron impact (Equation 6.156 through Equation 6.161) can be expressed respectively by:

$$\nu_{VT}=\frac{\gamma-1}{\gamma}\cdot\frac{k_{VT}n_0^2\left[\varepsilon_v(T_v)-\varepsilon_v(T_0)\right]}{p} \tag{6.162}$$

$$\nu_R=\frac{\gamma-1}{\gamma}\cdot\frac{k_Rn_0^2\Delta Q}{p} \tag{6.163}$$

$$\nu_{eV} = \frac{\gamma - 1}{\gamma} \cdot \frac{k_{eV} n_e n_0 \hbar\omega}{p} \tag{6.164}$$

Here,

$$\varepsilon_v(T_v) = \frac{\hbar\omega}{\exp(\hbar\omega/T_v) - 1}$$

is the average vibrational energy of molecules; c_v, c_v^v are translational and vibrational heat capacities; $p = n_0 T_0$ is the gas pressure; k_{eV}, k_{VT}, and k_R are rate coefficients of vibrational excitation, vibrational relaxation, and chemical reaction; n_e and n_0 are electron and gas concentrations; ΔQ is the vibrational energy consumption per one act of chemical reaction; ξ is that part of the energy going into translational degrees of freedom (the chemical heat release); and the dimensionless factors

$$\hat{k}_R = \frac{\partial \ln k_R(T_v)}{\partial T_v}, \quad \hat{k}_{VT} = \frac{\partial \ln k_{VT}}{\partial \ln T_0}$$

represent before the logarithmic sensitivities of rate coefficients k_R and k_{VT} to vibrational and translational temperatures, respectively.

Note that the dispersion (Equation 6.155) describes not only propagation of sound but also evolution of aperiodic perturbations. For example, if $\omega = 0$, the dispersion (Equation 6.155) leads to Equation 6.100 for frequency and length of relaxation in a fast flow. Without heat release, the dispersion (Equation 6.155) obviously gives $v_{ph} = v \pm c_s$ and describes sound waves propagating along and against the gas flow.

6.5.10 Analysis of Dispersion Equation for Sound Propagation in Nonequilibrium Chemically Active Plasma

Analyze the dispersion (Equation 6.155) in gas at rest ($\vec{k}\vec{v} = 0$, although some results can be generalized). The frequency ω will be considered a real number, while the wave $k = k_0 - i\delta$ will be considered a complex number. Then k_0 characterizes the acoustic wavelength and δ is related to the space amplification of sound.

1. **Acoustic wave in molecular gas at equilibrium**. There is no special vibrational excitation in this case without sound $T_v = T_0$. If the VT relaxation time is constant ($\tau_{VT} = const$), then Equation 6.155 gives the well-known **dispersion equation of relaxation gas-dynamics**:

$$\frac{k^2 c_s^2}{\omega^2} = 1 + \frac{\frac{c_S^2}{c_e^2} - 1}{1 - i\omega\tau_{VT}} \tag{6.165}$$

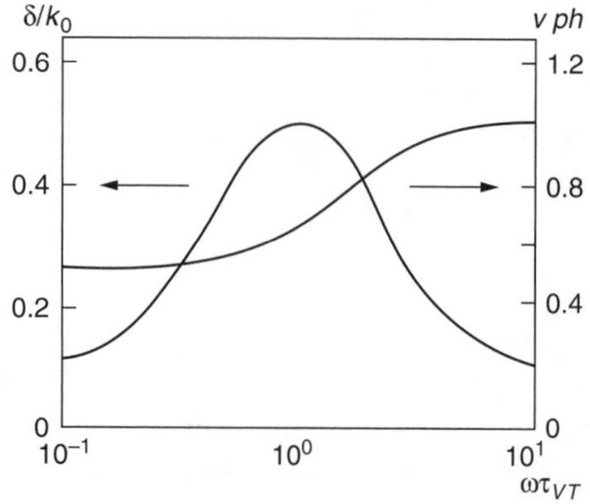

Figure 6.25 Phase velocity (in speed of sound units) and attenuation coefficient as functions of relaxation parameter

The dispersion equation shows that at the high-frequency limit, the acoustic wave propagates with the "frozen" sound velocity c_s (vibrational degrees of freedom are frozen, i.e., not participating in sound propagation). The acoustic wave velocity at low frequencies ($\omega\tau_{VT} << 1$) is the conventional equilibrium speed of sound c_e. The sound waves are always damping (Equation 6.165), but the maximum damping takes place when $\omega\tau_{VT} \approx 1$. The phase velocity v_{ph} and damping δ/k_0 as functions of $\omega\tau_{VT}$ are shown in Figure 6.25.[134a,b]

2. **Acoustic wave in high-frequency limit**. In this case, the gas conditions are initially nonequilibrium $T_v > T_0$, and the sound frequency exceeds all the relaxation frequencies (Equation 6.156 through 6.161): $\omega >> \max\{\bar{e}_n, \bar{e}_T, \ldots, e_v\}$. The dispersion (Equation 6.155) can be simplified at these conditions and expressed as:

$$\frac{k^2 c_s^2}{\omega^2} = 1 - i\frac{(\gamma-1)\hat{k}_{VT}\nu_{VT} + 2(\nu_{VT} + \xi\nu_R)}{\omega} \tag{6.166}$$

From Equation 6.166, the acoustic wave propagates in the high-frequency limit at the "frozen" sound velocity with only slight dispersion:

$$v_{ph} = \frac{\omega}{k_0} = c_s\left[1 - \frac{\left((\gamma-1)\hat{k}_{VT}\nu_{VT} + 2(\nu_{VT} + \xi\nu_R)\right)^2}{8\omega^2}\right] \tag{6.167}.$$

It is important to note that the initial vibrational–translational nonequilibrium results in amplification of acoustic waves (in contrast to the case of relaxation gas dynamics [Equation 6.165]). The amplification coefficient (increment) of the acoustic waves in the nonequilibrium gas at the high-frequency limit can be obtained from Equation 6.166 as:

$$\frac{\delta}{k_0} = \frac{(\gamma - 1)\hat{k}_{VT}\nu_{VT} + 2(\nu_{VT} + \xi\nu_R)}{2\omega} \tag{6.168}$$

3. **Acoustic wave dispersion in the presence of intensive plasma chemical reaction**. This case is of the most practical interest for plasma chemical applications. It implies that the reaction frequency greatly exceeds the frequency of vibrational relaxation; heating is mostly related to the chemical heat release ($\xi\nu_R >> \nu_{VT}$); and the steady-state T_v is balanced by vibrational excitation and chemical reaction (see Section 5.6.7). The general dispersion (Equation 6.155) can be simplified in this case to:

$$\frac{k^2 c_s^2}{\omega^2} = 1 - \frac{i\omega a\nu_R + b\nu_R^2}{\omega^2 + i\omega c\nu_R + b\nu_R^2} \tag{6.169}$$

The dimensionless parameters a, b, and c were introduced for simplicity and can be found from the following relations:

$$a = 2\xi, \quad b = -\xi\gamma \frac{c_v T_0}{c_v^v T_v} \frac{\partial \ln k_R}{\partial \ln T_v}\left(1 + \frac{\partial \ln n_e}{\partial \ln n_0}\right), \quad c = 2\xi + \gamma \frac{c_v T_0}{c_v^v T_v} \frac{\partial \ln k_R}{\partial \ln T_v} \tag{6.170}$$

As a numerical example, take typical parameters for the nonequilibrium supersonic discharge in CO_2, where the chemical process of dissociation is stimulated by vibrational excitation:

$$T_v = 3500K, T_0 = 100K, E_a = 5.5eV, \frac{\partial \ln k_R}{\partial \ln T_v} \approx \frac{E_a}{T_v} = 18,$$

$$\xi = 10^{-2}, \frac{\partial \ln n_e}{\partial \ln n_0} \approx (-3) \div (-5)$$

In this case, according to Equation 6.170, the dispersion parameters can be estimated as $a \approx b \approx 0.2$, $c \approx 0.4$. Numerical values for characteristic frequencies of chemical reaction ν_R (Equation 6.163) and VT relaxation ν_{VT} (Equation 6.162), corresponding to the example of CO_2 dissociation stimulated in plasma by vibrational excitation, are given in Table 6.3.

Different dispersion curves for acoustic waves in nonequilibrium plasma illustrating relation 6.155 are presented in Figure 6.26 through Figure 6.28. Phase velocities and wavelengths as functions of frequency are given in Figure 6.26 for the case of intensive chemical reaction (Equation 6.169). The acoustic wave amplification coefficients as a function of frequency are presented for the same case in Figure 6.27. Finally, wave length and amplification coefficient as functions of frequency are illustrated in Figure 6.28 for a case in which the heat release is mostly due to vibrational relaxation ($\nu_{VT} >> \xi\nu_R$).

Table 6.3 Characteristic Frequencies of Chemical Reaction, ν_R, sec^{-1}, and VT relaxation, ν_{VT}, sec^{-1}, at Different T_0 and T_v

$T_0 \downarrow$	$T_v = 2500$ K	$T_v = 3000$ K	$T_v = 3500$ K	$T_v = 4000$ K
100 K	$\nu_R = 7 \cdot 10^3$	$\nu_R = 3 \cdot 10^5$	$\nu_R = 4 \cdot 10^6$	$\nu_R = 3 \cdot 10^8$
	$\nu_{VT} = 3 \cdot 10^2$	$\nu_{VT} = 4 \cdot 10^2$	$\nu_{VT} = 5 \cdot 10^2$	$\nu_{VT} = 7 \cdot 10^2$
300 K	$\nu_R = 2 \cdot 10^3$	$\nu_R = 10^5$	$\nu_R = 10^6$	$\nu_R = 10^8$
	$\nu_{VT} = 1.4 \cdot 10^4$	$\nu_{VT} = 1.6 \cdot 10^4$	$\nu_{VT} = 1.9 \cdot 10^4$	$\nu_{VT} = 2.7 \cdot 10^4$
700 K	$\nu_R = 10^3$	$\nu_R = 4 \cdot 10^4$	$\nu_R = 5 \cdot 10^5$	$\nu_R = 5 \cdot 10^7$
	$\nu_{VT} = 10^5$	$\nu_{VT} = 1.2 \cdot 10^5$	$\nu_{VT} = 1.3 \cdot 10^5$	$\nu_{VT} = 1.9 \cdot 10^5$

$n_0 = 3 \cdot 10^{18}$ cm^{-3}, $E_a = 5.5$ eV, $B = 72$ K. B is the Landau–Teller coefficient (Equation 3.57) for vibrational relaxation.

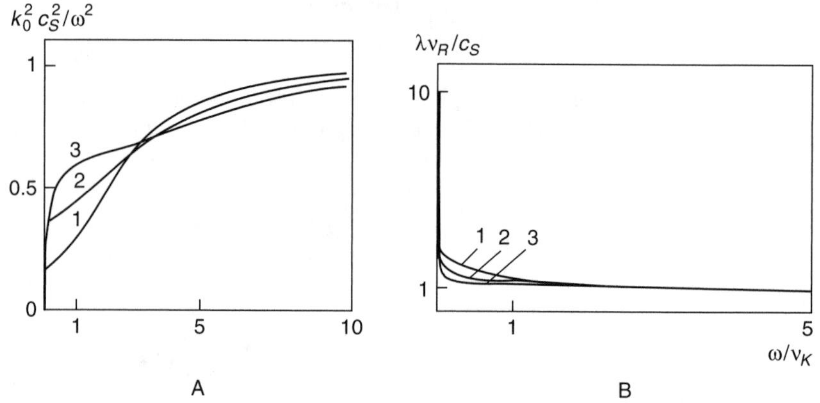

Figure 6.26 (A) Refraction index and (B) wavelength as functions of frequency; translational temperature: 1 = 100 K; 2 = 300 K; 3 = 700 K[9]

An interesting physical effect illustrating the dependence of $\delta/k_0(\omega)$ can be seen in Figure 6.27. When frequency ω is decreasing, the amplification coefficient at first grows, then passes maximum and decreases, reaching zero ($\delta = 0$) at some characteristic frequency:

$$\omega = \nu_R \gamma \frac{c_v T_0}{c_v^v T_v} \sqrt{\xi \left| 1 + \frac{\partial \ln n_e}{\partial \ln n_0} \right|} \tag{6.171}$$

This effect is due to an increase of the phase shift between oscillations of heating rate and pressure when frequency is decreasing (at the high-frequency limit this phase shift is about $2\xi\nu_R/\omega$).

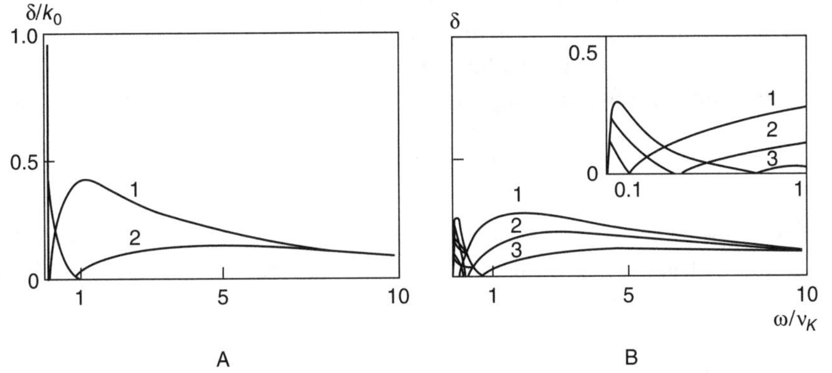

Figure 6.27 (A, B) Wave amplification coefficients as functions of frequency; translational temperature: 1A, 1B = 100 k; 2A, 3B = 700 K; 2B = 300 K

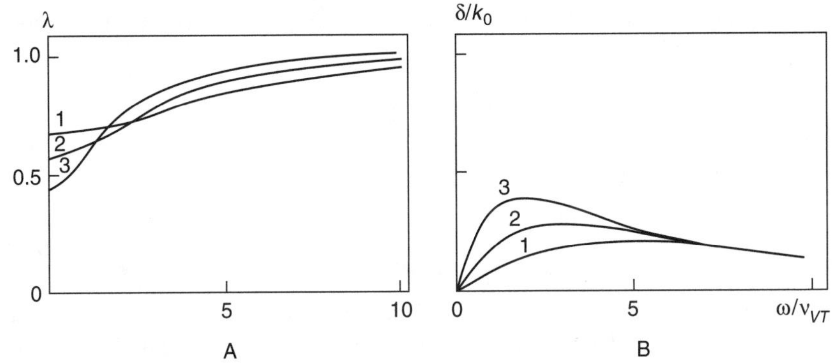

Figure 6.28 (A) Wavelength λ (in units ω/c_s) and (B) amplification coefficient δ/k_0 as functions of frequency; translational temperature: 1 = 100 K; 2 = 300 K; 3 = 700 K

It should be noted that typical values of ν_R, ν_{VT} (see Table 6.3) are in the ultrasonic range of frequencies ($\omega > 10^5$ Hz, $\lambda < 0.1$ cm). This provides possibilities to use the peculiarities of the acoustic dispersion curves (Figure 6.26 through Figure 6.28)[9] in ultrasonic diagnostics of nonequilibrium plasma chemical systems.[134,134a,b]

6.6 PROPAGATION OF ELECTROMAGNETIC WAVES IN PLASMA

6.6.1 Complex Dielectric Permittivity of Plasma in High-Frequency Electric Fields

Consider first electromagnetic waves in nonmagnetized plasma. In this case, the complex dielectric permittivity and its components—high-frequency plasma conductivity and dielectric constant—are the key concepts in describing electromagnetic wave propagation. Therefore, these should be analyzed starting with the description of an electron motion in a high-frequency electric field.

One-dimensional electron motion in the electric field $E = E_0 \cos \omega t = \mathrm{Re}(E_0 e^{i\omega t})$ of an electromagnetic wave with frequency ω can be described by the equation:

$$m\frac{du}{dt} = -eE - mu\nu_{en} \tag{6.172}$$

where ν_{en} is electron–neutral collision frequency; $u = \mathrm{Re}(u_0 e^{i\omega t})$ is the electron velocity; and E_0 and u_0 are amplitudes of the corresponding oscillations. The relation between the amplitudes of electron velocity and electric field is complex and, based on Equation 6.172, can be expressed as:

$$u_0 = -\frac{e}{m}\frac{1}{\nu_{en} + i\omega}E_0 \tag{6.173}$$

The imaginary part of the coefficient between u_0 and E_0 (complex electron mobility) reflects a phase shift between them. Alternately, the Maxwell equation:

$$curl\vec{H} = \varepsilon_0 \frac{\partial \vec{E}}{\partial t} + \vec{j} \tag{6.174}$$

allows the total current density to be presented as the sum:

$$\vec{j}_t = \varepsilon_0 \frac{\partial \vec{E}}{\partial t} + \vec{j} \tag{6.175}$$

The first current density component in Equation 6.175 is related to displacement current and its amplitude in complex form can be presented as $\varepsilon_0 i\omega E_0$. The second current density component in Equation 6.175 corresponds to the conduction current and has amplitude of $-en_e u_0$. Then the amplitude of the total current density can be taken as:

$$j_{t0} = i\omega\varepsilon_0 E_0 - en_e u_0 \tag{6.176}$$

Taking into account the complex electron mobility (Equation 6.173 and Equation 6.176) for the total current density and thus the Maxwell equation, Equation 6.174 can be rewritten as:

$$j_{t0} = i\omega\varepsilon_0\left[1 - \frac{\omega_p^2}{\omega(\omega - i\nu_{en})}\right]E_0, \quad curl\vec{H}_0 = i\omega\varepsilon_0\left[1 - \frac{\omega_p^2}{\omega(\omega - i\nu_{en})}\right]\vec{E}_0 \tag{6.177}$$

Here, ω_p is the electron plasma frequency. When considering electromagnetic wave propagation, it is always convenient to keep the Maxwell equation (Equation 6.174) for the amplitude of magnetic field H_0 in the form $curl\vec{H}_0 = i\omega\varepsilon_0\varepsilon E_0$. In this case, the conduction current is included into displacement current through introduction of the complex dielectric constant. According to Equation 6.177, this complex dielectric constant (or dielectric permittivity) in plasma can be presented as:

$$\varepsilon = 1 - \frac{\omega_p^2}{\omega(\omega - i\nu_{en})} \tag{6.178}$$

This complex relation can be used to introduce the real, high-frequency dielectric permittivity and conductivity of plasmas.

6.6.2 High-Frequency Plasma Conductivity and Dielectric Permittivity

The complex dielectric constant expressed by Equation 6.178 can be rewritten in the following form with specified real and imaginary parts:

$$\varepsilon = \varepsilon_\omega + i\frac{\sigma_\omega}{\varepsilon_0 \omega} \tag{6.179}$$

In this case, ε_ω, the real component of Equation 6.178, is the high-frequency dielectric constant of the plasma, which is given as:

$$\varepsilon_\omega = 1 - \frac{\omega_p^2}{\omega^2 + \nu_{en}^2} \tag{6.180}$$

Thus, plasma can be considered a dielectric material. However, in contrast to conventional dielectrics with $\varepsilon > 1$, a plasma is characterized by $\varepsilon < 1$. This can be explained noting that, according to Equation 6.172 and Equation 6.173, a free electron (without collisions) oscillates in phase with the electric field and in counterphase with the electric force. As a result, polarization in a plasma is negative. Obviously, negative polarization and $\varepsilon < 1$ are typically not only for plasma, but also for metals and in cases of extremely high frequencies when electrons can be considered free.

The imaginary component of Equation 6.178 corresponds to the high-frequency conductivity based on Equation 6.179, which can be expressed as:

$$\sigma_\omega = \frac{n_e e^2 \nu_{en}}{m\left(\omega^2 + \nu_{en}^2\right)} \tag{6.181}$$

Equation 6.180 and Equation 6.181 for the high-frequency dielectric permittivity and conductivity can be simplified in two cases, in the so-called collisionless plasma and in the static limit. The **collisionless plasma** limit means $\omega >> \nu_{en}$. For example, a microwave plasma can be considered collisionless at low pressures (about 3 torr and less). In this case:

$$\sigma_\omega = \frac{n_e e^2 \nu_{en}}{m\omega^2}, \quad \varepsilon_\omega = 1 - \frac{\omega_p^2}{\omega^2} \tag{6.182}$$

Thus, the conductivity in a collisionless plasma is proportional to the electron–neutral collision frequency, and dielectric constant does not depend on the frequency

ν_{en}. Note that the ratio of conduction current to polarization current (which actually corresponds to displacement current) can be estimated as:

$$\frac{j_{conduction}}{j_{polarization}} = \frac{\sigma_\omega}{\varepsilon_0 \omega |\varepsilon_\omega - 1|} = \frac{\nu_{en}}{\omega} \tag{6.183}$$

This means that the polarization current in a collisionless plasma (where $\nu_{en} << \omega$) greatly exceeds the conductivity current.

In an opposite case of the **static limit** $\nu_{en} >> \omega$, the conductivity and dielectric permittivity can be expressed as:

$$\sigma_\omega = \frac{n_e e^2}{m\nu_{en}}, \quad \varepsilon = 1 - \frac{\omega_p^2}{\nu_{en}^2} \tag{6.184}$$

In the static limit, the conductivity coincides with the conventional conductivity in DC conditions (see Equation 4.98). In this case, the dielectric permittivity also does not depend on frequency.

6.6.3 Propagation of Electromagnetic Waves in Plasma

Electromagnetic wave propagation in plasmas can be described by the conventional wave equation derived from the Maxwell equations:

$$\Delta\vec{E} - \frac{\varepsilon}{c^2}\frac{\partial^2 \vec{E}}{\partial t^2} = 0, \quad \Delta\vec{H} - \frac{\varepsilon}{c^2}\frac{\partial^2 \vec{H}}{\partial t^2} = 0 \tag{6.185}$$

Actually, all plasma peculiarities with respect to electromagnetic wave propagation are related to the complex dielectric permittivity ε (Equation 6.178 and Equation 6.179) that is present in this wave equation. Thus, the dispersion equation for electromagnetic wave propagation in dielectric medium,

$$\frac{kc}{\omega} = \sqrt{\varepsilon} \tag{6.186}$$

is valid in a plasma, again with the complex dielectric permittivity ε in form of Equation 6.178 and Equation 6.179. Here, c is the speed of light.

In the wave, electric and magnetic fields can be considered as $\vec{E},\vec{H} \propto \exp(-i\omega t + i\vec{k}\vec{r})$. Assuming the electromagnetic wave frequency ω as real, then the wave number k is complex because ε is complex in Equation 6.186. Separate the real and imaginary components of the wave number k, based on Equation 6.186, in the following way:

$$k = \frac{\omega}{c}\sqrt{\varepsilon} = \frac{\omega}{c}(n + i\kappa) \tag{6.187}$$

From this relation, the physical meaning of the parameter n is **the refractive index** of an electromagnetic wave in plasmas. The phase velocity of the wave is $v = \omega/k =$

c/n and the wavelength in the plasma is $\lambda = \lambda_0/n$ (where λ_0 is the corresponding wavelength in vacuum). The wave number κ characterizes the **attenuation of electromagnetic wave in plasma**; the wave amplitude decreases e^{κ} times on the length $\lambda_0/2\pi$.

Taking into account Equation 6.179 for ε from Equation 6.187, the relation between the refractive index and attenuation of electromagnetic wave n, κ with high-frequency dielectric permittivity ε_ω and conductivity σ_ω, is obtained:

$$n^2 - \kappa^2 = \varepsilon_\omega, \quad 2n\kappa = \frac{\sigma_\omega}{\varepsilon_0 \omega} \tag{6.188}$$

Solving this system of equations yields an explicit expression for the electromagnetic wave attenuation coefficient:

$$\kappa = \sqrt{\frac{1}{2}\left(-\varepsilon_\omega + \sqrt{\varepsilon_\omega^2 + \frac{\sigma_\omega^2}{\varepsilon_0^2 \omega^2}}\right)} \tag{6.189}$$

As is clear from Equation 6.189, the attenuation of an electromagnetic wave in a plasma is due to the conductivity if $\sigma_\omega << \varepsilon_\omega \varepsilon_0 \omega$; the electromagnetic field damping can be neglected. The explicit expression for the electromagnetic wave refractive index also can be obtained by solving the system of Equation 6.188:

$$n = \sqrt{\frac{1}{2}(\varepsilon_\omega + \sqrt{\varepsilon_\omega^2 + \frac{\sigma_\omega^2}{\varepsilon_0^2 \omega^2}})} \tag{6.190}$$

If the conductivity is negligible, the refractive index $n \approx \sqrt{\varepsilon_\omega}$. As noted earlier, the plasma polarization is negative and $\varepsilon_\omega < 1$, which means at the low conductivity limit $n < 1$. This results in an interesting conclusion: the velocity of an electromagnetic wave exceeds the speed of light $v_{ph} = c/n > c$. This is not a paradox because phase velocities are able to exceed the speed of light; the corresponding group velocity is obviously less than the speed of light.

The preceding expression for the refractive index $n \approx \sqrt{\varepsilon_\omega}$ in the case of negligible conductivity leads, together with Equation 6.182, to the well-known form of dispersion equation for electromagnetic waves in collisionless plasma:

$$\frac{k^2 c^2}{\omega^2} = 1 - \frac{\omega_p^2}{\omega^2}, \quad \omega^2 = \omega_p^2 + k^2 c^2 \tag{6.191}$$

The corresponding dispersion curve for electromagnetic waves in a plasma is compared in Figure 6.29 with the dispersion curve for electrostatic plasma waves. From Equation 6.191 and Figure 6.29, it is seen that the phase velocity in this case exceeds the speed of light as was mentioned earlier. Differentiation of the dispersion (Equation 6.191) gives the relation between the phase and group velocities of electromagnetic waves in plasma:

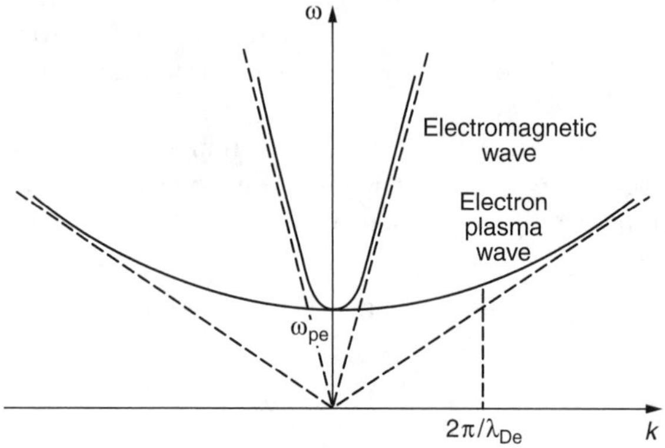

Figure 6.29 Electromagnetic and electrostatic plasma wave dispersion

$$\frac{\omega}{k} \times \frac{d\omega}{dk} = v_{ph} v_{gr} = c^2 \tag{6.192}$$

Thus, if the phase velocity exceeds the speed of light, then the group velocity is below the speed of light, which corresponds to a requirement of the theory of relativity.

6.6.4 Absorption of Electromagnetic Waves in Plasmas: Bouguer Law

The energy flux of electromagnetic waves can be described by the Pointing vector:

$$\vec{S} = \varepsilon_0 c^2 [\vec{E} \times \vec{B}] \tag{6.193}$$

where the electric and magnetic fields obviously should be averaged over the oscillation period.

On the other hand, electric and magnetic fields in the waves are related according to the Maxwell equations (when $\mu = 1$) as:

$$\varepsilon\varepsilon_0 E^2 = \mu_0 H^2 \tag{6.194}$$

Thus, damping of the electric and magnetic field oscillations is proportional to each other and can be described by the attenuation coefficient (Equation 6.187). As a result, attenuation of the electromagnetic energy flux (Equation 6.193) in plasma follows the **Bouguer law**:

$$\frac{dS}{dx} = -\mu_\omega S, \quad \mu_\omega = \frac{2\kappa\omega}{c} = \frac{\sigma_\omega}{\varepsilon_0 nc} \tag{6.195}$$

In the Bouguer law, μ_ω is called the absorption coefficient. The energy flux S (the Pointing vector) decreases by a factor e over the length $1/\mu_\omega$. The product $\mu_\omega S$ in the Bouguer law is actually the electromagnetic energy dissipated per unit volume of plasma, which obviously corresponds to energy dissipation according to the Joule heating law:

$$\mu_\omega S = \varepsilon_0 c^2 \langle EB \rangle = \sigma \langle E^2 \rangle \tag{6.196}$$

In general, the absorption coefficient μ_ω should be calculated from Equation 6.195, taking the conductivity σ_ω, refractive index n, and coefficient κ from Equation 6.181, Equation 6.189, and Equation 6.190. If the plasma degree of ionization and absorption is relatively low: $n \approx \sqrt{\varepsilon} \approx 1$, then the expression for the absorption coefficient can be simplified and, based on Equation 6.195 and Equation 6.181, expressed as:

$$\mu_\omega = \frac{n_e e^2 \nu_{en}}{\varepsilon_0 mc(\omega^2 + \nu_{en}^2)} \tag{6.197}$$

For practical calculations of electromagnetic wave absorption, it is convenient to use the following numerical formula, corresponding to Equation 6.197:

$$\mu_\omega, cm^{-1} = 0.106 n_e (cm^{-3}) \frac{\nu_{en}(\sec^{-1})}{\omega^2(\sec^{-1}) + \nu_{en}^2} \tag{6.198}$$

As can be seen from Equation 6.197 and Equation 6.198, at high frequencies $\omega >> \nu_{en}$, the absorption coefficient is proportional to the square of the wavelength $\mu_\omega \propto \omega^{-2} \propto \lambda^2$. This means that short electromagnetic waves propagate through plasma better than longer ones.

6.6.5 Total Reflection of Electromagnetic Waves from Plasma: Critical Electron Density

If the plasma conductivity is not high $\sigma_\omega << \omega\varepsilon_0|\varepsilon|$, electromagnetic wave propagates in plasma quite easily if frequency is sufficiently high. However, when frequency decreases, the dielectric permittivity $\varepsilon_\omega = 1 - \omega_p^2/\omega^2$ becomes negative and the electromagnetic wave is unable to propagate.

According to Equation 6.189 and Equation 6.190, the negative values of the dielectric permittivity make the refractive index equal to zero ($n = 0$) and attenuation coefficient $\kappa \approx \sqrt{|\varepsilon|}$. This means that the phase velocity tends to infinity and the group velocity of electromagnetic field is equal to zero. In this case, the depth of electromagnetic wave penetration in plasma:

$$l = \frac{\lambda_0}{2\pi\sqrt{|\varepsilon_\omega|}} = \frac{\lambda_0}{2\pi}\left|1 - \frac{\omega_p^2}{\omega^2}\right|^{-1/2} \tag{6.199}$$

does not depend on conductivity and is not related to energy dissipation. Such not-dissipative stopping corresponds to the phenomenon of total electromagnetic wave reflection from the plasma.

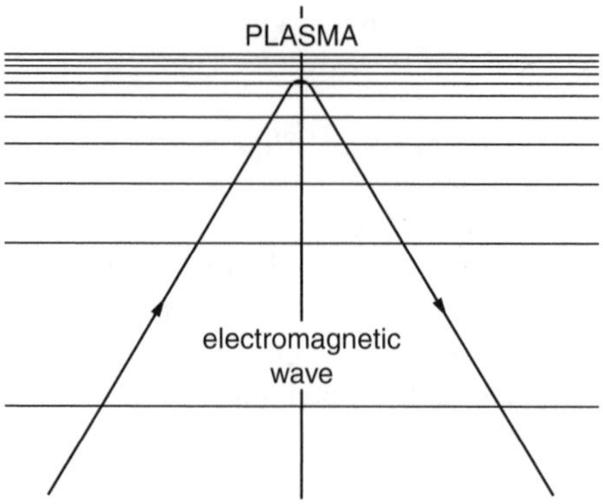

Figure 6.30 Electromagnetic wave reflection from plasma with density growing in vertical direction

To illustrate the reflection, consider an electromagnetic wave propagating in a nonuniform plasma without significant dissipative absorption (see Figure 6.30). The electromagnetic wave propagates from areas with low electron density to areas where the plasma density is increasing. The electromagnetic wave frequency is obviously fixed, but the plasma frequency increases together with the electron density, leading to decrease of ε_ω. At the point when dielectric permittivity $\varepsilon_\omega = 1 - \omega_p/\omega^2$ becomes equal to zero, the total reflection takes place. Thus, the total reflection of electromagnetic waves takes place when the electron density reaches the critical value, which can be found from $\omega = \omega_p$ as:

$$n_e^{crit} = \frac{\varepsilon_0 m\omega^2}{e^2}, \quad n_e(\text{cm}^{-3}) = 1.24 \cdot 10^4 \cdot [f(\text{MHz})]^2 \tag{6.200}$$

For example, an electromagnetic wave with frequency of 3 GHz reflects from plasma zones with electron concentrations exceeding the critical value, which, in this case, equals about 10^{11} cm^{-3}. This effect is very interesting, not only in plasma diagnostics (to measure electron concentration), but also in an important practical application such as long-distance radio signal transmission. Atmospheric air is slightly ionized by solar ultraviolet radiation at heights exceeding 100 km; electron density in the ionosphere is about $10^4 \div 10^5$ cm^{-3}. The resulting plasma reflects radio waves with frequencies of about 1 MHz, providing their long-distance transmission.

6.6.6 Electromagnetic Wave Propagation in Magnetized Plasma

If the electric field direction of a plane-polarized wave coincides with the direction of external uniform magnetic field, then the magnetic field does not influence propagation of the electromagnetic wave. However, interesting dispersion phenomena can be observed in electromagnetic wave propagation along the uniform magnetic field.

Neglecting collisions ($\nu_{en} << \omega$), an electron motion equation in the transverse wave propagating along the magnetic field B_0 is given as:

$$\frac{d\vec{v}}{dt} = -\frac{e}{m}\left(\vec{E} + \left[\vec{v} \times \vec{B}_0\right]\right) \tag{6.201}$$

If ion motion is neglected, the motion (Equation 6.201) can be rewritten in terms of the current density, $\vec{j} = -n_e e\vec{E}$, as follows:

$$\frac{d\vec{j}}{dt} = \frac{n_e e^2}{m}\vec{E} - \frac{e}{m}[\vec{j} \times \vec{B}_0] \tag{6.202}$$

To describe the electromagnetic wave dispersion in a magnetized plasma, consider Equation 6.202 together with the electromagnetic wave equation in conventional form:

$$\Delta E - \frac{1}{c^2}\frac{\partial^2 E}{\partial t^2} - \mu_0 \frac{\partial \vec{j}}{\partial t} = 0 \tag{6.203}$$

Note that, in contrast to the similar wave (Equation 6.185), in Equation 6.203 the complex dielectric permittivity is replaced by an additional term related to conduction current.

Next, look for a solution of Equation 6.202 and Equation 6.203 for a circularly polarized electromagnetic wave, when the electric vector rotates in a plane (x,y) perpendicular to direction z of the wave propagation:

$$E_x = E_0 \cos(\omega t - kz), \quad E_y = \pm E_0 \sin(\omega t - kz) \tag{6.204}$$

Here, E_0 is amplitude of electric field oscillations in the wave and signs (+) and (–) correspond to rotation of the electric field vector in opposite directions. Current density components can also be expressed in a similar way:

$$j_x = j_0 \cos(\omega t - kz), \quad j_y = \pm j_0 \sin(\omega t - kz) \tag{6.205}$$

where j_0 is amplitude of current density oscillations. Projection of Equation 6.202 and Equation 6.203 on the axis x can be presented as:

$$\frac{dj_x}{dt} = \varepsilon_0 \omega_p^2 E_x - \omega_B j_y \tag{6.206}$$

$$\frac{\partial^2 E_x}{\partial z^2} - \frac{1}{c^2}\frac{\partial^2 E_x}{\partial t^2} - \mu_0 \frac{\partial j_x}{\partial t} = 0 \tag{6.207}$$

Here, ω_p is plasma frequency; $\omega_B = eB_0/m$ is the electron cyclotron frequency, which is the frequency of an electron rotation around magnetic lines (see Section 4.3.4).

Applying Equation 6.204 and Equation 6.205 for components of the electric field and current density, rewrite Equation 6.206 and Equation 6.207 as:

$$-\omega j_0 \sin(\omega t - kz) = \varepsilon_0 \omega_p^2 E_0 \cos(\omega t - kz) \pm \omega_B j_0 \sin(\omega t - kz) \tag{6.208}$$

$$-k^2 E_0 \cos(\omega t - kz) + \frac{\omega^2}{c^2} E_0 \cos(\omega t - kz) + \mu_0 \omega j_0 \sin(\omega t - kz) = 0 \quad (6.209)$$

This system of equations has a nontrivial solution only if the following relation between wavelength and frequency is valid:

$$\frac{k^2 c^2}{\omega^2} = 1 - \frac{\omega_p^2}{\omega^2} \frac{1}{\left(1 \pm \frac{\omega_B}{\omega}\right)} \quad (6.210)$$

This is the dispersion equation for the electromagnetic wave propagation in a collisionless plasma along a magnetic field. In the absence of magnetic field, when the electron cyclotron frequency is equal to zero, $\omega_B = eB_0/m = 0$, the dispersion (Equation 6.210) obviously coincides with the conventional dispersion (Equation 6.191) for a collisionless plasma.

6.6.7 Propagation of Ordinary and Extraordinary Polarized Electromagnetic Waves in Magnetized Plasma

Two signs, (+) and (–), correspond to two different directions of rotation of vector $\vec{E}$. The (–) sign is related to the **right-hand-side circular polarization** of electromagnetic wave, when the direction of $\vec{E}$ rotation coincides with the direction of an electron gyration in the magnetic field. In optics, such a wave is usually referred to as the **extraordinary wave**. The extraordinary wave has $\omega = \omega_B$ as the resonant frequency when the denominator in the dispersion equation tends to zero. This **electron cyclotron resonance (ECR)** provides effective absorption of electromagnetic waves and is used as a principal physical effect in ECR discharges. The (+) sign in the dispersion (Equation 6.211) is related to the **left-hand-side circular polarization** of electromagnetic wave. In optics, such a wave is usually referred to as the **ordinary wave** and it is not a resonant one.

Based on the dispersion (Equation 6.210), the phase velocity of the electromagnetic waves, propagating along magnetic field, can be presented as:

$$v_{ph} = c\left[1 - \frac{\omega_p^2}{\omega^2} \frac{1}{\left(1 \pm \frac{\omega_B}{\omega}\right)}\right]^{-1/2} \quad (6.211)$$

Propagation of electromagnetic waves in the absence of a magnetic field is possible, according to Equation 6.211 and Equation 6.191, only with frequencies exceeding the plasma frequency $\omega > \omega_p$. This situation becomes very different in magnetic field. The dispersion curves corresponding to Equation 6.211 are shown in Figure 6.31. As is seen from the curves, propagation of ordinary and extraordinary waves is possible at low frequencies ($\omega < \omega_p$). Also, in contrast to the case of $B_0 = 0$, where phase velocities always exceed the speed of light, the electromagnetic waves can be "slower" than speed of light in the presence of magnetic field.

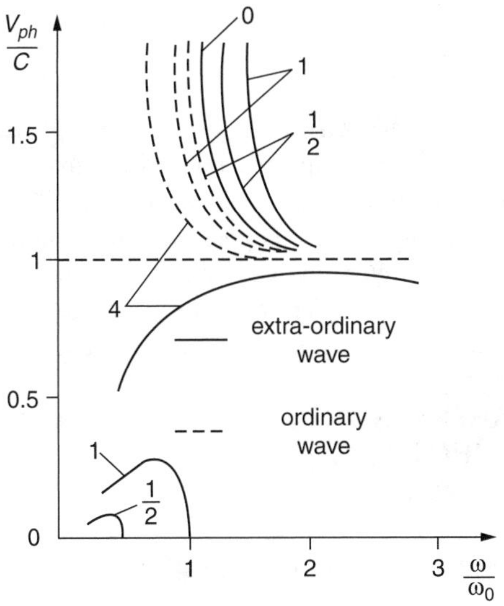

Figure 6.31 Dispersion of transverse electromagnetic wave propagating in plasma along magnetic field. Numbers correspond to values of ω_H/ω_p.

6.6.8 Influence of Ion Motion on Electromagnetic Wave Propagation in Magnetized Plasma

The extraordinary waves are able to propagate in plasma even at relatively low frequencies (see Figure 6.10). However, in this case, the ion motion can also be important and the ion cyclotron frequency should be taken into account:

$$\omega_{Bi} = \frac{e}{M} B \tag{6.212}$$

which corresponds to ions' (with mass M) gyration in magnetic field B_0.

The dispersion equation for an electromagnetic wave propagating along a uniform magnetic field B_0 can be derived taking into account also the ion motion. The derivation is actually similar to that in Section 6.6.6, leading to the following $k(\omega)$ relation:

$$\frac{k^2c^2}{\omega^2} = 1 - \frac{\omega_p^2}{\omega^2} \frac{1}{\left(1 \pm \frac{\omega_B}{\omega} - \frac{\omega_B \omega_{Bi}}{\omega^2}\right)} \tag{6.213}$$

If the electromagnetic wave frequency greatly exceeds the ion cyclotron frequency ($\omega >> \omega_B$), the influence of the term $\omega_B\omega_{Bi}/\omega^2$ in the denominator is negligible and the dispersion (Equation 6.213) coincides with Equation 6.210. Conversely, if the wave frequency is low ($\omega >> \omega_B$), then the term $\omega_B\omega_{Bi}/\omega^2$ becomes dominant in the dispersion equation. In this case, the dispersion (Equation 6.213) can be rewritten as:

$$\frac{c^2}{(\omega/k)^2} = 1 + \frac{\omega_p^2}{\omega_B \omega_{Bi}} = 1 + \frac{\mu_0 \rho}{B^2} \approx \frac{c^2}{v_A^2} \tag{6.214}$$

This introduces again the Alfven wave velocity v_A (Equation 6.139); here, $\rho = n_e M$ is the mass density in completely ionized plasma.

Only a few principal types of electromagnetic waves in magnetized plasma have been discussed. For example, numerous possible modes are related to electromagnetic wave propagation not along the magnetic field. Some of them are very important in plasma diagnostics and in high-frequency plasma heating. More details on the subject can be found in Ginsburg[215] and Ginsburg and Rukhadze.[216]

6.7 EMISSION AND ABSORPTION OF RADIATION IN PLASMA: CONTINUOUS SPECTRUM

6.7.1 Classification of Radiation Transitions

Because of its application in different lighting devices, radiation is probably the most commonly known plasma property. Radiation also plays an important role in plasma diagnostics, including plasma spectroscopy, in the propagation of some electric discharges, and sometimes even in plasma energy balance.

From a quantum mechanical point of view, radiation occurs due to transitions between different energy levels of a quantum system: transition up corresponds to absorption of a quantum, $E_f - E_i = \hbar\omega$; transition down the spectrum corresponds to emission $E_i - E_f = \hbar\omega$ (see Section 6.7.2). From the point of classical electrodynamics, radiation is related to the nonlinear change of dipole momentum, actually with the second derivative of dipole momentum (see Section 6.7.4).

It should be mentioned that neither emission nor absorption of radiation is impossible for free electrons. As was shown in Section 6.6, electron collisions are necessary in this case. It will be shown in Section 6.7.4 that electron interaction with a heavy particle, ion, or neutral is able to provide emission or absorption, but electron–electron interaction cannot.

It is convenient to classify different types of radiation according to the different types of an electron transition from one state to another. Electron energy levels in the field of an ion as well as transitions between the energy levels are illustrated in Figure 6.32. The case when both initial and final electron states are in continuum is called the **free–free transition**. A free electron in this transition loses part of its kinetic energy in the coulomb field of a positive ion or in interaction with neutrals. The emitted energy in this case is a continuum (usually infrared) called **bremsstrahlung** (direct translation: stopping radiation). The reverse process is the bremsstrahlung absorption.

Electron transition between a free state in continuum and a bound state in atom (see Figure 6.32) is usually referred to as the **free–bound transition**. These transitions correspond to processes of the radiative electron–ion recombination (see Section 2.3.5) and the reverse one of photoionization (see Section 2.2.6). These kinds

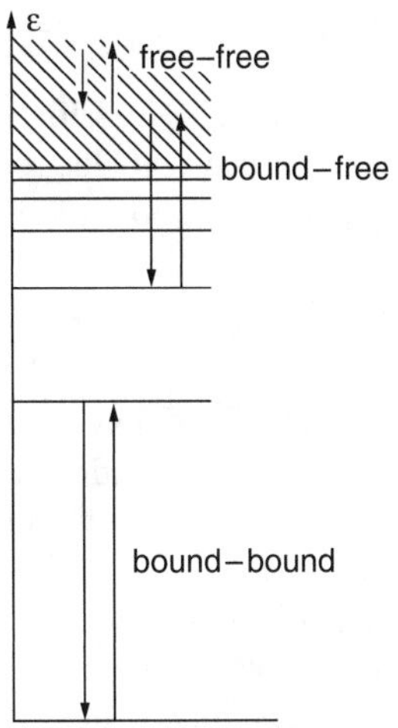

Figure 6.32 Energy levels and electron transitions induced by ion field

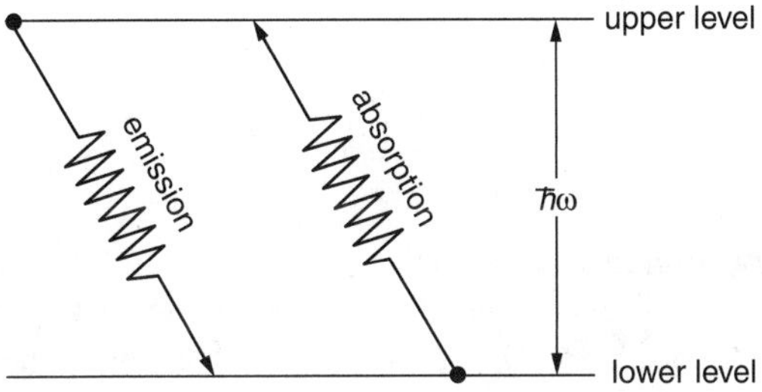

Figure 6.33 Radiative transition between two energy levels

of transitions can also take place in electron–neutral collisions. In such a case these are related to photoattachment and photodetachment processes of formation and destruction of negative ions (see Section 2.4). The free–bound transitions correspond to continuum radiation.

Finally, **the bound–bound transitions** mean transition between discrete atomic levels (see Figure 6.32) and result in emission and absorption of spectral lines. Molecular spectra are obviously much more complex than those of single atoms because of possible transitions between different vibrational and rotational levels (see Section 3.2).

6.7.2 Spontaneous and Stimulated Emission: Einstein Coefficients

Consider transitions between two states (upper u and ground 0) of an atom or molecule with emission and absorption of a photon $\hbar\omega$, which is illustrated in Figure 6.33. The probability of a photon absorption by an atom per unit time (and thus atom transition "0" → "u") can be expressed as:

$$P(\text{“}0\text{”}, n_\omega \to \text{“}u\text{”}, n_\omega - 1) = An_\omega \tag{6.215}$$

Here, n_ω is the number of photons and A is **the Einstein coefficient**, which depends on atomic parameters and does not depend on electromagnetic wave characteristics. Similarly, the probability of atomic transition with a photon emission is:

$$P(\text{“}u\text{”}, n_\omega \to \text{“}0\text{”}, n_\omega + 1) = \frac{1}{\tau} + Bn_\omega \tag{6.216}$$

Here, $1/\tau$ is the frequency of **spontaneous emission** of an excited atom or molecule, which takes place without direct relation with external fields; another Einstein coefficient, B, characterizes emission induced by an external electromagnetic field. The factors B and τ as well as A depend only on atomic parameters. Thus, if the first right-side term in Equation 6.216 corresponds to spontaneous emission, the second term is related to **stimulated emission**.

To find relations between the Einstein coefficients A and B and the spontaneous emission frequency $1/\tau$, analyze the thermodynamic equilibrium of radiation with the atomic system under consideration. In this case, the densities of atoms in lower and upper states (Figure 6.33) are related in accordance with the Boltzmann law (Equation 4.1.9) as:

$$n_u = \frac{g_u}{g_0} n_0 \exp\left(-\frac{\hbar\omega}{T}\right) \tag{6.217}$$

where $\hbar\omega$ is energy difference between the two states and g_u and g_0 are their statistical weights. According to the Planck distribution (see Section 4.1.5 and Equation 4.19), the averaged number of photons $\bar{n}_\omega$ in one state can be determined as:

$$\bar{n}_\omega = 1 \Big/ \left(\exp\frac{\hbar\omega}{T} - 1\right) \tag{6.218}$$

Then, taking into account the detailed balance of photon emission and absorption for the system illustrated in Figure 6.33,

$$n_0 P(\text{“}0\text{”}, n_\omega \to \text{“}u\text{”}, n_\omega - 1) = n_u P(\text{“}u\text{”}, n_\omega \to \text{“}0\text{”}, n_\omega + 1) \tag{6.219}$$

which can be rewritten based on Equation 6.215 and Equation 6.216 as:

$$n_0 A \bar{n}_\omega = n_u \left(\frac{1}{\tau} + B\bar{n}_\omega\right) \tag{6.220}$$

The relations between the Einstein coefficients A and B and the spontaneous emission frequency $1/\tau$ can be expressed based on Equation 6.217, Equation 6.218, and Equation 6.220 as:

$$A = \frac{g_u}{g_0}\frac{1}{\tau}, \quad B = \frac{1}{\tau} \tag{6.221}$$

Next, the final expression of emission probability of an excited atom or molecule can be rewritten from Equation 6.216 and Equation 6.221 as:

$$P(\text{“}u\text{”}, n_\omega \to \text{“}0\text{”}, n_\omega + 1) = \frac{1}{\tau} + n_\omega \frac{1}{\tau} \tag{6.222}$$

This shows that the contribution of the stimulated emission n_ω times exceeds the spontaneous times. Although the expressions (Equation 6.221) for the Einstein coefficients were derived from equilibrium thermodynamics, obviously the final relation (Equation 6.222) can be applied for the nonequilibrium number of photons n_ω.

6.7.3 General Approach to Bremsstrahlung Spontaneous Emission: Coefficients of Radiation Absorption and Stimulated Emission during Electron Collisions with Heavy Particles

The bremsstrahlung emission is related to the free–free electron transitions (see Section 6.7.1) and does not depend on external radiation in the same way as spontaneous emission. An electron slows down in a collision with a heavy particle, ion, or neutral, and loses kinetic energy, which then partially goes to radiation. It needs to be noted that a free electron cannot absorb or emit a photon because of the requirements of momentum conservation

It is convenient to describe the bremsstrahlung emission at a specific frequency ω in terms of the **differential spontaneous-emission cross section**:

$$d\sigma_\omega(v_e) = \frac{d\sigma_\omega}{d\omega} d\omega,$$

which is a function of electron velocity v_e. In this case, the emission probability per one electron of quanta with energies $\hbar\omega \div \hbar\omega + d(\hbar\omega)$ can be given by the conventional definition:

$$dP_\omega = \frac{1}{\hbar\omega} dQ_\omega = v_e n_0 d\sigma_\omega(v_e) = v_e n_0 \frac{d\sigma_\omega}{d\omega} d\omega \tag{6.223}$$

Here, dQ_ω is the emission power per one electron in the frequency interval $d\omega$; n_0 is density of heavy particles.

The differential cross section of bremsstrahlung emission $d\sigma/d\omega(v_e)$ characterizes the spontaneous emission of plasma electrons during their collisions with heavy particles in the same way as the frequency $1/\tau$ characterizes spontaneous emission of an atom (Equation 6.216). Also, similar to the Einstein coefficients A and B for atomic systems (see relation 6.215 and relation 6.216), coefficients $a_\omega(v_e)$ and $b_\omega(v_e)$ for radiation absorption and stimulated emission during the electron collision with heavy particles are introduced.

To introduce the coefficients $a_\omega(v_e)$ and $b_\omega(v_e)$, first define the radiation intensity $I(\omega)$ as the emission power in spectral interval $\omega \div \omega + d\omega$ per unit area and within the unit solid angle (Ω, measured in steradians). The product $I_\omega d\omega\, d\Omega$ actually describes the density of the radiation energy flux. Then, the radiation energy from the interval $d\omega\, d\Omega$ absorbed by electrons with velocities $v_e \div v_e + dv_e$ in unit volume per unit time can be presented as:

$$\left(I_\omega d\omega\, d\Omega\right) n_0 \left[f\left(v_e\right) dv_e\right] \times a_\omega\left(v_e\right) \tag{6.224}$$

This is the definition of the **absorption coefficient** $a_\omega(v_e)$, which is actually calculated with respect to one electron and one atom. In its definition (Equation 6.224), n_0 is the density of heavy particles and $f(v_e)$ is the electron distribution function. Similar to Equation 6.224, the stimulated emission of quanta related to electron collisions with heavy particles can be expressed as:

$$\left(I_\omega d\omega\, d\Omega\right) n_0 \left[f\left(v_e'\right) dv_e'\right] \times b_\omega\left(v_e'\right) \tag{6.225}$$

Here, the **coefficient of stimulated emission** $b_\omega(v_e')$ is introduced, which is also calculated with respect to one electron and one atom. Note that in accordance with the energy conservation law for mutually reverse processes of absorption (Equation 6.224) and stimulated emission (Equation 6.225), the electron velocities v_e and v_e' are related to each other as:

$$\frac{m(v_e')^2}{2} = \frac{mv_e^2}{2} + \hbar\omega \tag{6.226}$$

Similar to Equation 6.221, the coefficients of absorption $a_\omega(v_e)$ and stimulated emission $b_\omega(v_e')$ of an electron during its collision with a heavy particle are related to each other through the **Einstein formula**:

$$b_\omega(v_e') = \frac{v_e}{v_e'} a_\omega(v_e) = \left(\frac{\varepsilon}{\varepsilon + \hbar\omega}\right)^{1/2} a_\omega(v_e) \tag{6.227}$$

Here, ε is the electron energy (as well as $\varepsilon + \hbar\omega$). The stimulated emission coefficient is related to the spontaneous bremsstrahlung emission cross section as:

$$b_\omega(v_e') = \frac{\pi^2 c^2 v_e'}{\omega^2} \frac{d\sigma_\omega(v_e')}{d\omega} \tag{6.228}$$

6.7.4 Bremsstrahlung Emission due to Electron Collisions with Plasma Ions and Neutrals

To be correct, the Bremsstrahlung emission cross section $d\sigma_\omega/d\omega$ as well as coefficients a_ω and b_ω should be derived in the frameworks of quantum mechanics. However, classical derivation of these factors is more interesting and clear.

According to classical electrodynamics, emission of a system of electric charges is determined by the second derivative of its dipole momentum $\ddot{\vec{d}}$. In the case of an electron (characterized by a radius vector $\vec{r}$) scattering by a heavy particle at rest, $\ddot{\vec{d}} = -e\ddot{\vec{r}}$. In the case of electron–electron collisions,

$$\ddot{\vec{d}} = -e\ddot{\vec{r}}_1 - e\ddot{\vec{r}}_2 = -\frac{e}{m} \cdot \frac{d}{dt}\left(m\vec{v}_1 + m\vec{v}_2\right) = 0$$

because of momentum conservation. This relation explains the absence of bremsstrahlung emission in the electron–electron collisions. Total energy, emitted by an electron during the time of interaction with a heavy particle, can be expressed by the following relation:[217,217a]

$$E = \frac{e^2}{6\pi\varepsilon_0 c^3} \int_{-\infty}^{+\infty} \left|\ddot{\vec{r}}(t)\right|^2 dt \qquad (6.229)$$

Using the Fourier expansion of electron acceleration $\ddot{\vec{r}}(t)$ in Equation 6.229 yields the emission spectrum of the bremsstrahlung, which means the electron energy emitted per one collision in frequency interval $\omega \div \omega + d\omega$:

$$dE_\omega = \frac{e^2}{3\pi\varepsilon_0 c^3} \left| \int_{-\infty}^{+\infty} \ddot{\vec{r}}(t) \exp(-i\omega t) dt \right|^2 d\omega \qquad (6.230).$$

Taking into account that the time of effective electron interaction with a heavy particle is shorter than the electromagnetic field oscillation time, the integral in Equation 6.230 can be estimated as the electron velocity change $\Delta\vec{v}_e$ during scattering. This allows simplification of Equation 6.230:

$$dE_\omega = \frac{e^2}{6\pi^2\varepsilon_0 c^3}\left(\Delta v_e^2\right) d\omega \qquad (6.231)$$

Averaging Equation 6.232 over all electron collisions with heavy particles (frequency ν_m, electron velocity v_e) yields the formula for dQ_ω (see Equation 6.223), which is the bremsstrahlung emission power per one electron in the frequency interval $d\omega$:

$$dQ_\omega = \frac{e^2 v_e^2 \nu_m}{3\pi^2\varepsilon_0 c^3} d\omega \qquad (6.232)$$

Obviously, this formula of classical electrodynamics can only be used at relatively low frequencies $\omega < mv_e^2/2\hbar$, when electron energy exceeds the emitted quantum of radiation. At higher frequencies $\omega > mv_e^2/2\hbar$, $dQ_\omega = 0$ must be assumed to avoid the "ultraviolet catastrophe."

Based on Equation 6.232 and Equation 6.223, the cross section of the spontaneous bremsstrahlung emission of a quantum $\hbar\omega$ by an electron with velocity v_e can be expressed as:

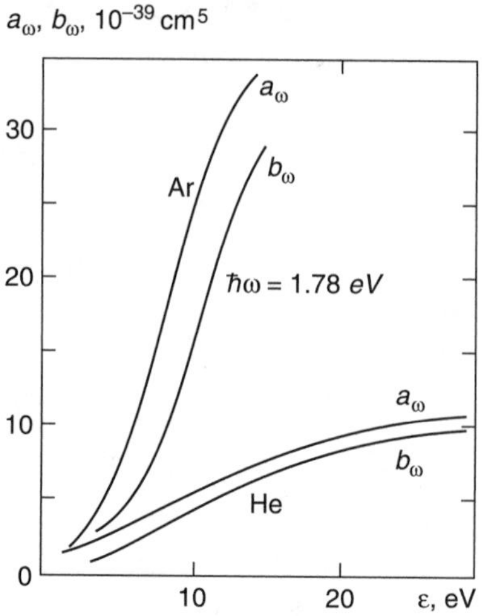

Figure 6.34 Absorption coefficients per one electron and per one Ar, He atom

$$d\sigma_\omega = \frac{e^2 v_e^2 \sigma_m}{3\pi^2 \varepsilon_0 c^3 \hbar\omega} d\omega, \quad \hbar\omega \leq \frac{mv_e^2}{2} \tag{6.233}$$

Bremsstrahlung emission is related to an electron interaction with heavy particles; therefore, its cross section (Equation 6.233) is proportional to the cross section σ_m of electron-heavy particle collisions. Equation 6.227 and Equation 6.228 then permit the devivation of formulas for the quantum coefficients of stimulated emission b_ω and absorption a_ω. For example, the absorption coefficient $a_\omega(v_e)$ can be presented in this way at $\omega >> \nu_m$ as:

$$a_\omega(v_e) = \frac{2}{3\varepsilon_o} \frac{e^2 v_e}{mc\omega^2} \frac{(\varepsilon + \hbar\omega)^2}{\varepsilon\hbar\omega} \sigma_m(\varepsilon + \hbar\omega) \tag{6.234}$$

where ε is electron energy before the absorption. Numerical examples of absorption and stimulated emission coefficients per one electron and one atom of Ar and He as a function of electron energy at a frequency of a ruby laser are presented in Figure 6.34.[101]

Plasma radiation is more significant in thermal plasma, where the major contribution in v_m and σ_m is provided by electron–ion collisions. Taking into account the coulomb nature of the collisions, the cross section of the bremsstrahlung emission (Equation 6.233) can be rewritten as:

$$d\sigma_\omega = \frac{16\pi}{3\sqrt{3}(4\pi\varepsilon_0)^3} \frac{Z^2 e^6}{m^2 c^3 v^2 \hbar\omega} \tag{6.235}$$

Here, Z is the charge of an ion; in most of our systems under consideration, $Z = 1$, but relation 6.235 in general can be applied to multicharged ions as well. Based on

Equation 6.235, the total energy emitted in unit volume per unit time by the bremsstrahlung mechanism in the spectral interval $d\omega$ can be expressed by the following integral:

$$J_{\omega}^{brems} d\omega = \int_{v_{\min}}^{\infty} \hbar\omega n_i n_e f(v_e) v_e dv_e d\sigma_{\omega}(v_e) \tag{6.236}$$

Here, n_i and n_e are concentrations of ions and electrons; $f(v_e)$ is the electron velocity distribution function, which can be taken here as Maxwellian; and $v_{\min} = \sqrt{2\hbar\omega/m}$ is the minimum electron velocity sufficient to emit a quantum $\hbar\omega$. After integration, the spectral density of bremsstrahlung emission ($Z = 1$) per unit volume (Equation 6.237) finally gives:

$$J_{\omega}^{brems} d\omega = \frac{16}{3}\left(\frac{2\pi}{3}\right)^{1/2} \frac{e^6 n_e n_i}{m^{3/2} c^3 (4\pi\varepsilon_0)^3 T^{1/2}} \exp\left(-\frac{\hbar\omega}{T}\right) d\omega = C\frac{n_i n_e}{T^{1/2}} \exp\left(-\frac{\hbar\omega}{T}\right) d\omega =$$

$$1.08 \cdot 10^{-45} W \cdot \text{cm}^3 \cdot K^{1/2} \times \frac{n_i(1/\text{cm}^3) \cdot n_e(1/\text{cm}^3)}{(T,K)^{1/2}} \exp\left(-\frac{\hbar\omega}{T}\right) d\omega \tag{6.237}$$

Obviously, this formula can only be applied to the quasi-equilibrium thermal plasmas that are able to be described by the one temperature T.

6.7.5 Recombination Emission

Recombination emission takes place during the radiative electron–ion recombination discussed in Section 2.3.5 as one of possible recombination mechanisms. According to the classification of different kinds of emission in plasma (Section 6.7.1), the recombination emission is related to free–bound transitions because, as a result of this process, a free plasma electron becomes trapped in a bound atomic state with negative discrete energy E_n. This recombination leads to emission of a quantum:

$$\hbar\omega = |E_n| + \frac{mv_e^2}{2} \tag{6.238}$$

Correct quantum mechanical derivation of the recombination emission cross section is complicated, but an approximate formula can be derived using the following quasi-classical approach. Equation 6.235 was derived for bremsstrahlung emission when the final electron is still free and has positive energy. To calculate the cross section σ_{RE} of the recombination emission, generalize the quasi-classic relation (Equation 6.235) for electron transitions into discrete bound states. Such generalization is acceptable, noting that the energy distance between high electronic levels is quite small and thus quasi classic (see Section 3.1).

The electronic levels of bound atomic states are discrete, so here the spectral density of the recombination cross section should be redefined taking into account number of levels Δn per small energy interval $\Delta E = \hbar\Delta\omega$ (around E_n):

$$\frac{d\sigma_\omega}{d\omega}\Delta\omega = \sigma_{RE}\Delta n, \quad \sigma_{RE} = \frac{d\sigma_\omega}{d\omega}\frac{1}{\hbar}\frac{\Delta E}{\Delta n} \tag{6.239}$$

For simplicity, consider the hydrogen-like atoms (where an electron moves in the field of a charge Ze); then $E_n = -I_H Z^2/n^2$. Here,

$$I_H = \frac{me^4}{2\hbar^2(4\pi\varepsilon_0)^2} \approx 13.6\,eV$$

is the hydrogen atom ionization potential and n is the principal quantum number. In this case, the ratio $\Delta E/\Delta n$ characterizing the density of energy levels can be taken as:

$$\frac{\Delta E}{\Delta n} \approx \left|\frac{dE_n}{dn}\right| = \frac{2I_H Z^2}{n^3} \tag{6.240}$$

Finally, based on Equation 6.236 and taking into account Equation 6.239 and Equation 6.240, the formula for the cross section of the recombination emission in an electron–ion collision with formation of an atom in an excited state with the principal quantum number n is obtained:

$$\sigma_{RE} = \frac{4}{3\sqrt{3}}\frac{e^{10}Z^4}{(4\pi\varepsilon_0)^5 c^3\hbar^4 m v_e^2\omega}\frac{1}{n^3} = 2.1\cdot 10^{-22}\,cm^2\cdot\frac{\left(I_H Z^2\right)^2}{\varepsilon\hbar\omega n^3} \tag{6.241}$$

Here, ε is the initial electron energy. Then, the total energy $J_{\omega n}\,d\omega$, emitted as a result of photorecombination of electrons with velocities in interval $v_e \div v_e + dv_e$ and ions with $Z = 1$ and leading to formation of an excited atom with the principal quantum number n, per unit time and per unit volume, can be expressed as:

$$J_{\omega n}d\omega = \hbar\omega n_i n_e \sigma_{RE} f(v_e) v_e dv_e = C\frac{n_i n_e}{T^{1/2}}\frac{2I_H}{Tn^3}\exp\left(\frac{I_H}{Tn^2} - \frac{\hbar\omega}{T}\right)d\omega \tag{6.242}$$

In this formula, the parameter $C = 1.08\cdot 10^{-45}\,W\cdot cm^3\cdot K^{1/2}$ is the same factor as in Equation 6.237; $f(v_e)$ is the electron velocity distribution function, which was taken as Maxwellian; and the differential relation between electron velocity and radiation frequency from Equation 6.238 was written as $mv_e dv_e = \hbar d\omega$.

It should also be taken into account that emission of quanta with the same energy $\hbar\omega$ can be provided by trapping electrons on different excitation levels with different principal quantum numbers n (obviously, the electron velocities should be relevant and determined by Equation 6.238). Thus, the total spectrum of recombination emission is the sum of similar but shifted terms:

$$J_\omega^{recomb}d\omega = \sum_{n^*(\omega)}^{\infty} J_{\omega n} \tag{6.243}$$

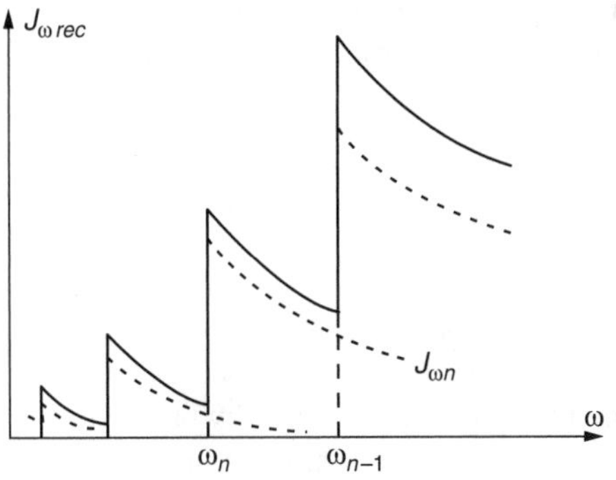

Figure 6.35 Recombination emission spectrum

Each of the individual terms in Equation 6.243, is related to the different principal quantum number n, but giving the same frequency ω, and each is proportional to $\exp(-\hbar\omega/T)$. The lowest possible principal quantum number $n^*(\omega)$ in the sum for the fixed frequency can be defined from the condition $|E_{n^*}| < \hbar\omega < |E_{n^*-1}|$. The spectrum of recombination emission corresponding to the sum (Equation 6.243) is illustrated in Figure 6.35.

6.7.6 Total Emission in Continuous Spectrum

The total spectral density of plasma emission in continuous spectrum consists of the bremsstrahlung and recombination components, Equation 6.237 and Equation 6.243. These two emission components can be combined in one approximate but reasonably accurate general formula for the total continuous emission:

$$J_\omega d\omega = C\frac{n_i n_e}{T^{1/2}}\Psi\left(\frac{\hbar\omega}{T}\right)d\omega \tag{6.244}$$

The parameter $C = 1.08\cdot 10^{-45}\, W\cdot cm^3\cdot K^{1/2}$ is the same factor as in Equation 6.237 and Equation 6.243; $\Psi(x)$ is the dimensionless function, which can be approximated as:

$$\Psi(x) = 1, \qquad \text{if } x = \frac{\hbar\omega}{T} < x_g = \frac{|E_g|}{T} \tag{6.245a}$$

$$\Psi(x) = \exp\left[-\left(x - x_g\right)\right], \qquad \text{if } x_g < x < x_1 \tag{6.245b}$$

$$\Psi(x) = \exp\left[-\left(x - x_g\right)\right] + 2x_1\exp\left[-\left(x - x_1\right)\right], \quad \text{if } x = \frac{\hbar\omega}{T} > x_1 = \frac{I}{T} \tag{6.245c}$$

Here, $|E_g|$ is the energy of the first (lowest) excited state of atom calculated with respect to transition to continuum and I is the ionization potential.

When the radiation quanta and thus frequencies are not very large ($\hbar\omega < |E_g|$), the contributions of free–free transitions (bremsstrahlung) and free–bound transitions (recombination) in the total continuous emission are related to each other as:

$$J_\omega^{recomb} \big/ J_\omega^{brems} = \exp\frac{\hbar\omega}{T} - 1 \qquad (6.246)$$

According to Equation 6.246, emission $\hbar\omega < 0.7T$ is mostly due to the bremsstrahlung mechanism, while emission of larger quanta $\hbar\omega > 0.7T$ is mostly due to the recombination mechanism. At plasma temperatures of 10,000 K, this means that only infrared radiation ($\lambda > 2\ \mu m$) is provided by bremsstrahlung; all other emission spectra are due to the electron–ion recombination.

To obtain the total radiation losses, integrate the spectral density (Equation 6.244) over all the emission spectrum (contribution of quanta $\hbar\omega > I$ can be neglected because of their intensive reabsorption). These total plasma energy losses per unit time and unit volume can be expressed after integration by the following numerical formula:

$$J, \frac{kW}{cm^3} = 1.42 \cdot 10^{-37} \sqrt{T, K}\, n_e n_i \left(\mathrm{cm}^{-3}\right) \cdot \left(1 + \frac{|E_g|}{T}\right) \qquad (6.247)$$

6.7.7 Plasma Absorption of Radiation in Continuous Spectrum: Kramers and Unsold–Kramers Formulas

The differential cross section of bremsstrahlung emission (Equation 6.235) with Equation 6.227 and Equation 6.228 permits determining the coefficient of **bremsstrahlung absorption** of a quantum $\hbar\omega$ calculated per one electron having velocity v_e and one ion:

$$a_\omega(v_e) = \frac{16\pi^3}{3\sqrt{3}} \frac{Z^2 e^6}{m^2 c (4\pi\varepsilon_0 \hbar\omega)^3 v_e} \qquad (6.248)$$

This is the so-called **Kramers formula** of the quasi-classical theory of plasma continuous radiation. Multiplying the Kramers formula by $n_e n_i$ and integrating over the Maxwellian distribution function $f(v_e)$ yields the coefficient of bremsstrahlung absorption in plasma (which actually is the reverse length of absorption):

$$\kappa_\omega^{brems} = C_1 \frac{n_e n_i}{T^{1/2} \nu^3}, \quad C_1 = \frac{2}{3}\left(\frac{2}{3\pi}\right)^{1/2} \frac{e^6}{(4\pi\varepsilon_0)^3 m^{3/2} c\hbar} \qquad (6.249)$$

To use Equation 6.249 for numerical calculations of the absorption coefficient κ_ω^{brems} (cm^{-1}), the coefficient C_1 can be taken as $C_1 = 3.69 \cdot 10^8$ cm^5/(sec$^3 \cdot$ K$^{1/2}$); here, the frequency ν $\omega/2\pi$, and $Z = 1$.

Another mechanism of plasma absorption of radiation in a continuous spectrum is **photoionization**, which was discussed in Section 2.2.6 regarding the balance of charged particles. The photoionization cross section can be easily calculated, taking into account that it is a reverse process with respect to recombination emission. Then, based on a detailed balance of the photoionization and recombination emission, Saha Equation 4.15, and Equation 6.241, the expression for the photoionization cross section for a photon $\hbar\omega$ and an atom with principal quantum number n is:

$$\sigma_{\omega n} = \frac{8\pi}{3\sqrt{3}} \frac{e^{10} m Z^4}{(4\pi\varepsilon_0)^5 c\hbar^6 \omega^3 n^5} = 7.9 \cdot 10^{-18} cm^2 \frac{n}{Z^2} \left(\frac{\omega_n}{\omega}\right)^3 \quad (6.250)$$

Here, $\omega_n = |E_n|/\hbar$ is the minimum frequency sufficient for photoionization from the electronic energy level with energy E_n and principal quantum number n. As seen from Equation 6.250, the photoionization cross section decreases with frequency as $1/\omega^3$ at $\omega > \omega_n$.

To calculate the **total plasma absorption coefficient in a continuum**, it is necessary to add to the bremsstrahlung absorption (Equation 6.248), the sum $\sum n_{0n} \sigma_{\omega n}$ related to the photoionization of atoms in different states of excitation (n) with concentrations n_{0n}. Replacing the summation by integration, the total plasma absorption coefficient in continuum can be calculated for $Z = 1$ as:

$$\kappa_\omega = C_1 \frac{n_e n_i}{T^{1/2} \nu^3} e^x \Psi(x) = 4.05 \cdot 10^{-23} cm^{-1} \frac{n_e n_i (\mathrm{cm}^{-3})}{(T,K)^{7/2}} \frac{e^x \Psi(x)}{x^3} \quad (6.251)$$

Here, factor C_1 is the same as in Equation 6.249; parameter $x = \hbar\omega/T$; and the function $\Psi(x)$ is defined by Equation 6.245. Obviously, the replacement of summation over discrete levels by integration makes the resulting approximate dependence $\kappa_\omega(\nu)$ smoother than in reality; this is illustrated in Figure 6.36.[218]

Note that the relative contribution of photoionization and bremsstrahlung mechanisms in total plasma absorption of radiation in continuum at the relatively low frequencies $\hbar\omega < |E_g|$ ($|E_g|$ is energy of the lowest excited state with respect to continuum) can be characterized by the same ratio as in case of emission (Equation 6.246). The product $n_e n_i$ in Equation 6.251 can be replaced at not very high temperatures by the gas density n_0 using the Saha Equation 4.15. At the relatively low frequencies $\hbar\omega < |E_g|$, the total plasma absorption coefficient in continuum is obtained in the following form:

$$\kappa_\omega = \frac{16\pi}{3\sqrt{3}} \frac{e^6 T n_0}{\hbar^4 c \omega^3 (4\pi\varepsilon_0)^3} \frac{g_i}{g_a} \exp\left(\frac{\hbar\omega - I}{T}\right)$$
$$= 1.95 \cdot 10^{-7} \mathrm{cm}^{-1} \frac{n_0(\mathrm{cm}^{-3})}{(T,K)^2} \frac{g_i}{g_a} \frac{e^{-(x_1 - x)}}{x^3} \quad (6.252)$$

This relation for the absorption coefficient is usually referred to as the **Unsold–Kramers formula**, where g_a, g_i are statistical weights of an atom and an ion, $x = \hbar\omega/T$, $x_1 = I/T$, and I is the ionization potential.

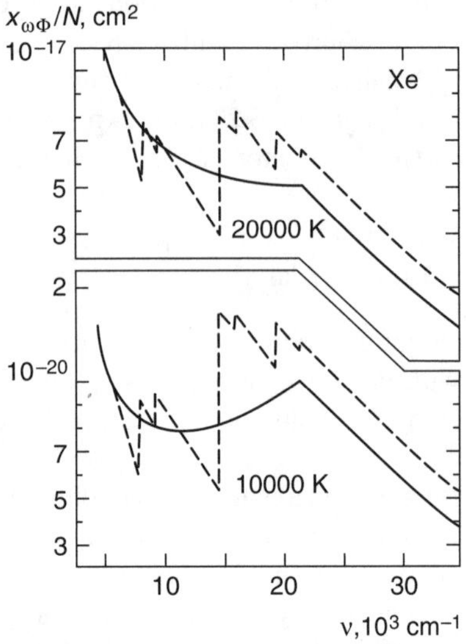

Figure 6.36 Absorption coefficient per one Xe atom, related to photoionization: dashed line = summation over actual energy levels; solid line = replacement of summation by integration[218]

As an example, the absorption length κ_ω^{-1} of red light $\lambda = 0.65\mu m$ in an atmospheric pressure hydrogen plasma at 10,000 K can be calculated from Equation 6.252 as $\kappa_\omega^{-1} \approx 180\,m$. Thus, this plasma is fairly transparent in continuum at such conditions. It is interesting to note that the Unsold–Kramers formula can be used only for quasi-equilibrium conditions, while Equation 6.251 can be applied for nonequilibrium plasma as well.

6.7.8 Radiation Transfer in Plasma

The intensity of radiation I_ω (defined in Section 6.7.3) decreases along its path s due to absorption (scattering in plasma neglected) and increases because of spontaneous and stimulated emission. The **radiation transfer equation** in a quasi-equilibrium plasma can then be given as:

$$\frac{dI_\omega}{ds} = \kappa'_\omega\left(I_{\omega e} - I_\omega\right), \quad \kappa'_\omega = \kappa_\omega[1 - \exp\left(-\frac{\hbar\omega}{T}\right)] \tag{6.253}$$

Here, the factor I_ω is the quasi-equilibrium radiation intensity:

$$I_{\omega e} = \frac{\hbar\omega^3}{4\pi^3 c^2}\frac{1}{\exp(\hbar\omega/T) - 1} \tag{6.254}$$

which can be easily obtained from the Planck formula for the spectral density of radiation (Equation 4.21). Introduce the **optical coordinate** ξ, calculated from the plasma surface $x = 0$ (with positive x directed into the plasma body):

$$\xi = \int_0^x \kappa'_\omega(x)\,dx, \quad d\xi = \kappa'_\omega(x)\,dx \tag{6.255}$$

Then, the radiation transfer equation can be rewritten in terms of the optical coordinate as:

$$\frac{dI_\omega(\xi)}{d\xi} - I_\omega(\xi) = -I_{\omega e} \tag{6.256}$$

Assuming that in the direction of a fixed ray, the plasma thickness is d; $x = 0$ corresponds to the plasma surface; and there is no source of radiation at $x > d$, then the radiation intensity on the plasma surface $I_{\omega 0}$ can be expressed by solution of the radiation transfer equation as:

$$I_{\omega 0} = \int_0^{\tau_\omega} I_{\omega e}\left[T(\xi)\right]\exp(-\xi)\,d\xi, \quad \tau_\omega = \int_0^d \kappa'_\omega dx \tag{6.257}$$

The quasi-equilibrium radiation intensity $I_{\omega e}[T(\xi)]$ is shown here as function of temperature and therefore an indirect function of the optical coordinate. In Equation 6.257 another important radiation transfer parameter is introduced: τ_ω, the **optical thickness of plasma**, obviously referred to a specific ray and specific spectral range.

6.7.9 Optically Thin Plasmas and Optically Thick Systems: Black Body Radiation

First, consider a case in which the optical thickness is small $\tau_\omega << 1$; this is usually referred to as a **transparent or optically thin plasma**. In this case the radiation intensity on the plasma surface $I_{\omega 0}$ (Equation 6.257) can be written as:

$$I_{\omega 0} = \int_0^{\tau_\omega} I_{\omega e}\left[T(\xi)\right]d\xi = \int_0^{\tau_\omega} I_{\omega e}\kappa'_\omega dx = \int_0^d j_\omega dx \tag{6.258}$$

Here, the emissivity term $j_\omega = I_{\omega e}\kappa'_\omega$ corresponds to spontaneous emission. Equation 6.258 shows that radiation of an optically thin plasma is just the result of summation of independent emission from different intervals dx along the ray. Thus, in this case all radiation generated in plasma volume is able to leave it. If, for simplicity, the plasma parameters are assumed uniform, then the radiation intensity on the plasma surface can be expressed as:

$$I_{\omega 0} = j_\omega d = I_{\omega e}\kappa'_\omega d = I_{\omega e}\tau_\omega << I_{\omega e} \tag{6.259}$$

From Equation 6.259, the radiation intensity of the optically thin plasma is seen to be much less ($\tau_\omega << 1$) than equilibrium value (Equation 6.254) corresponding to the Planck formula.

The opposite case takes place when the optical thickness is high, $\tau_\omega >> 1$. This is usually referred to as the **nontransparent or optically thick system.** If the quasi-equilibrium temperature can be considered constant in the emitting body, then for the optically thick systems, Equation 6.257 gives $I_{\omega 0} = I_{\omega e}(T)$. This means that the emission density of the entire surface in all directions is the same and equal to the equilibrium Planck value (Equation 6.254).

This is the case of the quasi-equilibrium **blackbody emission**, discussed in Section 4.1.5. Unfortunately, it cannot be directly applied to continuous plasma radiation (where the plasma is usually optically thin). The total black body emission per unit surface and unit time can be found by integration of the quasi-equilibrium radiation intensity (Equation 6.254) over all frequencies and all directions of hemisphere solid angle (2π):

$$J_e = \int d\omega \int_{2\pi} I_{\omega e} \cos\vartheta \, d\Omega = \int \pi I_{\omega e}(\omega, T) d\omega = \sigma T^4 \tag{6.260}$$

This is the **Stefan–Boltzmann law** of black body emission. In this relation, σ is the Stephan–Boltzmann coefficient (Equation 4.25); ϑ is the angle between directions of the ray (Ω) and normal vector to the emitting surface.

6.7.10 Reabsorption of Radiation, Emission of Plasma as Gray Body, Total Emissivity Coefficient

Plasmas are usually optically thin for radiation in the continuous spectrum, and the Stephan–Boltzmann law cannot be applied without special corrections. However, plasma is not absolutely transparent; the total emission can be affected by **reabsorption of radiation**. This can be illustrated by Equation 6.257 for the radiation intensity $I_{\omega 0}$, assuming fixed values of the quasi-equilibrium temperature and absorption coefficient (Equation 6.251):

$$I_{\omega 0} = I_{\omega e}\left[1 - \exp(-\tau_\omega)\right] = I_{\omega e}\left[1 - \exp(-\kappa'_\omega d)\right] = I_{\omega e}\varepsilon \tag{6.261}$$

Here, ε is the total emissivity coefficient, which characterizes the plasma as a gray body. It is interesting to note that in optically thin plasmas, the total emissivity coefficient coincides with the optical thickness: $\varepsilon = \tau_\omega$. If the total emissivity coefficient ε dependence on frequency is neglected (assumption of the **ideal gray body**), then integration similar to Equation 6.260 leads to the corrected Stefan–Boltzmann formula for emission of plasma in continuous spectrum:

$$J = \left[1 - \exp(-\kappa'_\omega d)\right]\sigma T^4 = \varepsilon \sigma T^4 \tag{6.262}$$

This obviously simplified approach is often quite convenient for qualitative description of thermal plasma emission and interpretation of experiments. As an example, the total emissivity coefficients for the air plasma layer of 1 cm are

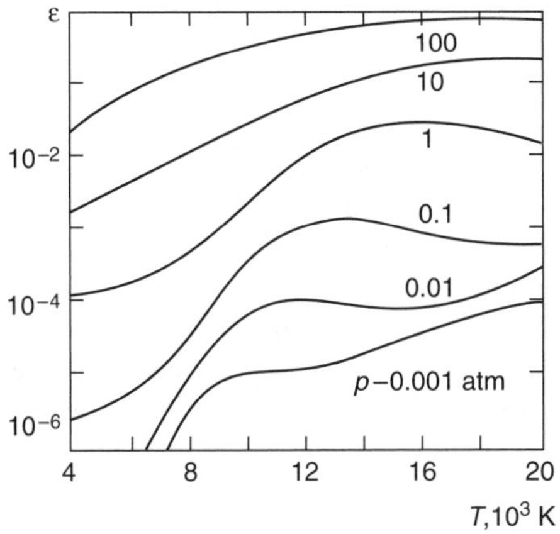

Figure 6.37 Emissivity of 1-cm air layer at different pressures[218,219]

presented in Figure 6.37 as a function of pressure and temperature.[218,219] More details on the subject can be found in references 90, 219, and 220.

6.8 SPECTRAL LINE RADIATION IN PLASMA

For discharge plasmas rather than the continuous spectrum, intensive spectral lines are often observed and make significant contributions in the total plasma emission. According to the classification given in Section 6.7.1, the spectral lines are results of the bound–bound transitions between discrete electronic states of atoms and molecules. The spectral lines determine different colors of nonthermal discharge plasmas in different gases. For example, they make glow discharge in argon dark red; in neon, red; in sodium, yellow; and in potassium and mercury, green. Also, spectral line radiation is directly related to laser generation.

6.8.1 Probabilities of Radiative Transitions and Intensity of Spectral Lines

Spontaneous discrete atomic transition from an upper energy state E_n into a lower energy state E_k results in emission of a quantum (see Figure 6.32):

$$\hbar\omega_{nk} = \frac{2\pi\hbar c}{\lambda_{mn}} = E_n - E_k \tag{6.263}$$

This radiation energy quantum determines a spectral line with frequency ω_{nk} and wavelength λ_{mn}. Intensity of the spectral line S can be characterized by the energy emitted by an exited atom or molecule per unit time in the line; this can be found in the framework of quantum mechanics as:

$$S = \hbar\omega_{nk} A_{nk} = \frac{1}{3\varepsilon_0} \frac{\omega_{nk}^4}{c^3} \left|\vec{d}_{nk}\right|^2 \tag{6.264}$$

Here, $\overrightarrow{d_{nk}}$ is the matrix element of dipole momentum of the atomic system and A_{nk} is the probability of the bound–bound transition in frequency units (A_{nk}^{-1} can be interpreted as the atomic lifetime with respect to the radiative transition $n \to k$). For the simplest atomic systems, the dipole momentum matrix elements and thus factors A_{nk} can be quantum mechanically calculated; however, in most cases the transition probabilities A_{nk} should be found experimentally. Typically, the transition probabilities for the intensive spectral lines are approximately 10^8 sec^{-1}; numerical values of the coefficients A_{nk} for some strong atomic lines are given in Table 6.4

It is interesting to note that the intensity of spectral line can also be calculated in the framework of classical electrodynamics, assuming that an electron is oscillating in an atom around equilibrium position ($\vec{r}$ is the electron radius vector). This leads to the following classical expression for the intensity of a spectral line, which is actually equivalent to the quantum-mechanical one (Equation 6.264):

$$S = \frac{1}{6\pi\varepsilon_0 c^3}\left\langle \ddot{\vec{d}}^2 \right\rangle = \frac{1}{6\pi\varepsilon_0}\frac{\omega_0^4}{c^3}\left\langle \vec{d}^2 \right\rangle = \frac{e^2\omega_0^4}{6\pi\varepsilon_0 c^3}\left\langle \vec{r}^2 \right\rangle \tag{6.265}$$

In this classical expression, c is the speed of light and ω_0 is frequency of the electron "oscillation" in an atom.

6.8.2 Natural Width and Profile of Spectral Lines

Excited states of an atom with energy E_n have a finite lifetime with respect to radiation $\tau \approx A_{nk}^{-1}$ (see Equation 6.264). Thus, according to the quantum-mechanical uncertainty principle, the energy level E_n actually is not absolutely thin, but has a characteristic width of $\Delta E \propto \hbar/\tau \approx \hbar A_{nk}$. As a result, a spectral line related to transition from this atomic energy level has a specific width $\Delta\omega \approx A_{nk}$, which is independent from external conditions and called the **natural spectral line width.** Typically, the natural width is very small with respect to the characteristic radiation frequency for electronic transitions $\Delta\omega \approx 10^8$ sec^{-1} $<< \omega_{nk} \approx 10^{15}$ sec^{-1} .

The profile of spectral lines can be determined as the photon distribution functions $F(\omega)$ over the radiation frequencies, which is usually normalized as $\int_{-\infty}^{+\infty} F(\omega)\,d\omega = 1$. To describe the natural profile of the spectral line $F(\omega)$, express the corresponding oscillations of the electric and magnetic fields in an electromagnetic wave $f(t)$, taking into account the finite lifetime of initial τ_n and final τ_k atomic states:[7]

$$f(t) \propto \exp\left(i\omega_{nk}t - \nu t\right), \quad 2\nu = \frac{1}{\tau_n} + \frac{1}{\tau_k} = \frac{1}{\tau} \approx A_{nk} \tag{6.266}$$

The frequency $\nu = 1/2\ A_{nk}$ again characterizes the photon emission probability (Equation 6.264) and at the same time describes attenuation of the electromagnetic wave because of the finite lifetime of atomic states (see Figure 6.38). Frequency

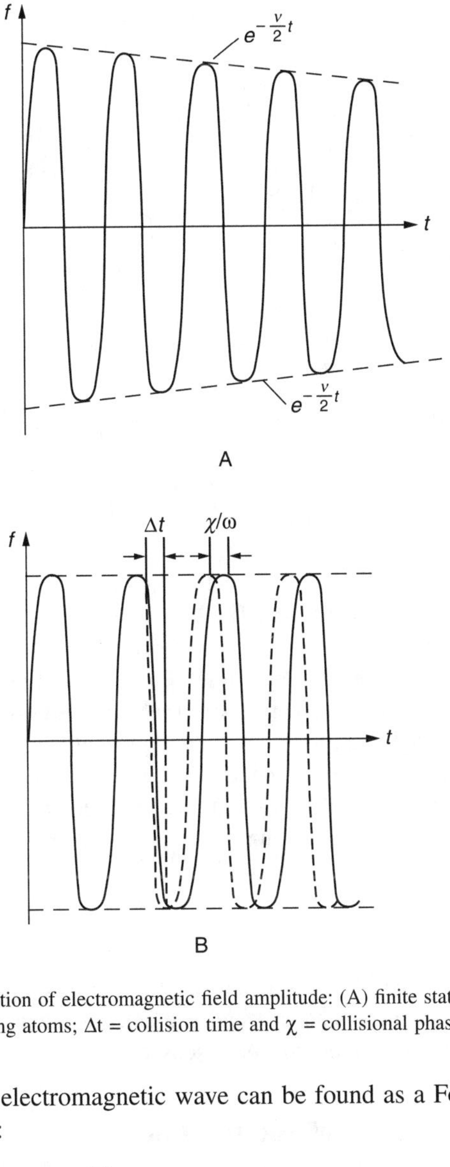

Figure 6.38 Time evolution of electromagnetic field amplitude: (A) finite state's lifetime; (B) collision of emitting and perturbing atoms; Δt = collision time and χ = collisional phase shift

distribution for the electromagnetic wave can be found as a Fourier component f_ω of the function $f(t)$:

$$f_\omega = \frac{1}{2\pi}\int_{-\infty}^{+\infty} f(t)\exp(-i\omega t)dt \propto \frac{1}{\nu + i(\omega - \omega_{nk})} \tag{6.267}$$

The radiation intensity of a spectral line at a fixed frequency ω and the photon distribution function $F(\omega)$ are proportional to the square of the Fourier component $F(\omega) \propto |f_\omega|^2$. Then, taking into account the normalization of the function $F(\omega)$, the following expression is obtained for the photon distribution function and the natural profile of a spectral line related to finite lifetime of excited states:

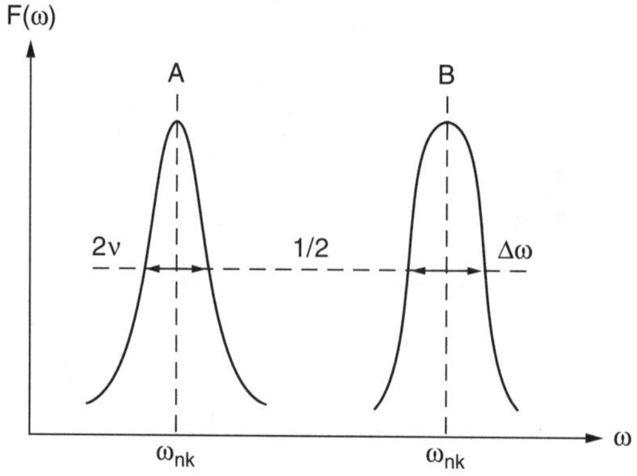

Figure 6.39 (A) "Wide wing" Lorentz and (B) "narrow wing" Gaussian profiles of spectral lines

$$F(\omega) = \frac{\nu}{\pi} \frac{1}{\nu^2 + (\omega - \omega_{nk})^2}, \quad \nu = \frac{1}{2}\left(\frac{1}{\tau_n} + \frac{1}{\tau_k}\right) = \frac{1}{2\tau} \approx \frac{1}{2} A_{nk} \tag{6.268}$$

This shape of the photon distribution function and spectral line is usually referred to as the **Lorentz spectral line profile**. The Lorentz profile decreases slowly (hyperbolically) with deviation from the principal frequency one $\omega = \omega_{nk}$. The Lorentz profile has "wide wings" (hyperbolic wings), illustrated in Figure 6.39. The parameter v in Equation 6.268 is the so-called **spectral line half-width.** In this case, the width of a line at half the maximum intensity is presented as $2v$, which is a double half-width or the entire width of a spectral line (see Figure 6.39). It is interesting to remark that the natural width of a spectral line in the wavelength-scale is independent of frequency and can be estimated as:

$$\Delta\lambda = \frac{e^2}{3\varepsilon_0 mc^2} = 1.2 \cdot 10^{-5}\,\text{nm} \tag{6.269}$$

This numerical value is very low for actual spectra. For this reason, the natural half-width of spectral lines in plasma is usually much less than those related to the Doppler and Stark effects, which will be discussed next.

6.8.3 Doppler Broadening of Spectral Lines

Real spectral lines are usually much wider than the natural one described previously. The broadening of spectral lines is due to thermal motion of the emitting particles; to collisions perturbing the emitting particles; and to the Stark effect. Consider these broadening effects separately, starting with the Doppler line broadening that is related to the thermal motion of the emitting atoms and molecules.

An electromagnetic wave emitted by a moving atom or molecule with frequency ω_{nk} is observed by a detector (at rest) at the frequency shifted according to the Doppler effect formula:

$$\omega = \omega_{nk}\left(1 + \frac{v_x}{c}\right) \tag{6.269a}$$

where v_x is the emitter velocity in the direction of wave propagation and c is the speed of light. The Doppler profile of a spectral line then can be found based on Equation 6.269a and assuming a Maxwellian distribution for atomic velocities $f(v_x)$. Taking into account that $F(\omega)\,d\omega = f(v_x)\,dv_x$, the Doppler profile is:

$$F(\omega) = \frac{1}{\omega_{mn}}\sqrt{\frac{Mc^2}{\pi T_0}}\exp\left[-\frac{Mc^2}{T_0}\frac{(\omega-\omega_{nk})^2}{\omega_{nk}^2}\right] \tag{6.270}$$

Here, M is mass of a heavy particle; T_0 is gas temperature; and c is the speed of light. In contrast to the Lorentzian hyperbolic natural profile of spectral lines, the Doppler profile decreases much faster (exponentially) with frequency deviation from the center of the line. Such shape of the curve $F(\omega)$ with relatively "narrow wings" is called the **spectral line Gaussian profile**, and it is also illustrated in Figure 6.39 to compare with the "wide wings" the Lorentz profile. The entire width (double half-width) of the Doppler profile of a spectral line can be expressed from Equation 6.270 as:

$$\frac{\Delta\omega}{\omega_{nk}} = \frac{\Delta\lambda}{\lambda_{nk}} = 7.16\cdot 10^{-7}\sqrt{T_0(K)/A} \tag{6.271}$$

where A is the atomic mass of the emitting particle. The Doppler broadening of spectral lines is most significant at high temperatures and for light atoms like hydrogen (for H_β line, $\lambda = 486.1$ nm, the Doppler width at a high gas temperature of 10,000 K is $\Delta\lambda = 0.035$ nm). However, even in this case, the broadening effect is not very strong.

6.8.4 Pressure Broadening of Spectral Lines

The influence of collisions on electromagnetic wave emission by excited atoms is illustrated in Figure 6.38. The phase of electromagnetic oscillations changes randomly during collision and therefore the oscillations can be considered as harmonic only between collisions. The effect of collisions is somewhat similar to the effect of spontaneous radiation (natural broadening) because it also restricts intervals of harmonic oscillations (compare these two effects in Figure 6.38).

For this reason, collisions with all plasma components, including heavy neutral particles, lead to this broadening effect. This effect can be described by the same Lorentz profile (Equation 6.268) as in natural broadening, although in this case with frequency v related to frequency of the emitting atom collisions with other species:[221]

$$F(\omega) = \frac{n_0\langle\sigma v\rangle}{\pi}\frac{1}{\left(n_0\langle\sigma v\rangle\right)^2 + \left(\omega-\omega_{nk}\right)^2} \tag{6.272}$$

This effect is usually referred to as the **pressure broadening (or impact broadening)** of spectral lines. In Equation 6.272, n_0 is the neutral gas density; σ is the cross section of the perturbing collision (which is usually close to the gas-kinetic cross section); and v is the velocity of the related collision partners (usually the thermal velocity of heavy particles). At high pressures, the collisional frequency obviously can exceed the frequency of spontaneous radiation; this can make the pressure broadening much more significant than the natural width of spectral lines.

6.8.5 Stark Broadening of Spectral Lines

When the density of charged particles in a plasma is sufficiently large (degree of ionization not less than about 10^{-2}), their electric microfields perturb the atomic energy levels providing the Stark broadening of spectral lines. The Stark effect is the most significant for hydrogen atoms (H_β line, 486.1 nm, is widely applied in related plasma diagnostics) and some helium levels, where the effect is strong and proportional to the first power of electric field. In most other cases, the Stark effect is proportional to square of electric field and is not strong.

The Stark broadening induced by electric microfields of ions is qualitatively different from that of electrons. For slowly moving ions, Stark broadening can be described based on the statistical distribution of the electric microfields. This approach is called the **quasi-static approximation.** In this case, the electric microfields are provided by the closest charged particles and can be estimated as $E \propto e^2/r^2 \propto n_e^{2/3}$, where $r \propto n_e^{-1/3}$ is the average distance between the charged particles. In the case of the strong linear Stark effect, the relevant width of a spectral line is also proportional to the factor $n_e^{2/3}$.

The Stark broadening induced by the electric microfields of ions in terms of wavelength $\Delta\lambda_S$ and in terms of frequencies $\Delta\omega_S$ can be estimated in the specific case of the H_β line (486.1 nm; transition from $n = 4$ to $n = 2$) by the following relation:

$$\Delta\lambda_S = \Delta\omega_S \frac{\lambda_{nk}^2}{2\pi c} \approx \frac{3e^2 a_0 n(n-1)\lambda_{nk}^2}{8\pi^2 \varepsilon_0 \hbar c} n_e^{2/3} = 5 \cdot 10^{-12} \text{nm} \cdot \left(n_e, \text{cm}^{-3}\right)^{2/3} \quad (6.273)$$

where α_0 is the Bohr radius and λ_{nk} is the spectral line principal wavelength.

The Stark broadening induced by electric microfields of the fast moving electrons can be described in the framework of **impact approximation**. In this case, the emitting atom is unperturbed most of the time and broadening is due to impacts well separated in time. The impact Stark broadening is proportional to the electron concentration and an additional factor decreasing with n_e. It is interesting that the entire Stark broadening, including effects of ions and electrons, can still be approximated as proportional to the factor $n_e^{2/3}$. In the electron temperature range of 5000 K ÷ 40,000 K and electron concentrations in the interval $10^{14} \div 10^{17}$ cm^{-3}, the total Stark broadening again of the H_β line, 486.1 nm, can be found using the following numeric formula:

$$\Delta\lambda_S \approx (1.8 \div 2.3) \cdot 10^{-11} \text{nm} \cdot \left(n_e, \text{cm}^{-3}\right)^{2/3} \quad (6.274)$$

The Stark broadening effect is quite significant in this case, and it can be effectively used in diagnostic measurements of electron concentrations in plasma (sometimes hydrogen is even specially added into discharge for such diagnostics).

The Stark broadening results in the Lorentzian profile of spectral in the same way as the pressure and natural broadenings. Obviously, different Lorentzian half-widths ($\Delta\omega_L/2$ or $\Delta\lambda_L/2$) should be used for describing different types of the Lorentzian broadenings: pressure, Stark, or natural. The total Lorentzian profile is determined then by the strongest of the three broadening effects.

6.8.6 Convolution of Lorentzian and Gaussian Profiles: Voigt Profile of Spectral Lines

The three mechanisms of spectral line broadening in a plasma (Stark, pressure, and natural) lead to the same "wide wing" Lorentzian profile; only the Doppler broadening results in the "narrow wing" Gaussian profile (see Figure 6.39). If the characteristic widths of the Lorentzian $\Delta\lambda_L$and Gaussian $\Delta\lambda_G$ profiles are of the same order of value, the resulting profile cannot be a trivial one. The central part of a spectral line can be related to Doppler broadening, while the wide wings can be provided by the Stark effect. Thus, the final shape of a spectral line in a plasma is the result of convolution of the Gaussian and Lorentzian profiles. Assuming that the Doppler and Stark broadening processes are independent, the result of the convolution is the Voigt profile of spectral lines. The Voigt profile can be presented in terms of the distribution over wavelength λ as:

$$F(\lambda) = \frac{1}{\Delta\lambda_G \sqrt{\pi}} \int_{-\infty}^{+\infty} \frac{a}{\pi} \frac{\exp(-y^2)}{(b-y)^2 + a^2} dy \tag{6.275}$$

The Voigt profile is defined by two parameters, a and b, which are related to the spectral line principal wavelength λ_{mn}, the line shift Δ, and the characteristic widths of the Lorentzian $\Delta\lambda_L$ and Gaussian $\Delta\lambda_G$ broadening effects:

$$a = \frac{\Delta\lambda_L}{\Delta\lambda_G}(\ln 2)^{1/2}, \quad b = \frac{2\ln 2}{\Delta_G}(\lambda - \lambda_{nk} - \Delta) \tag{6.276}$$

Broadening of spectral lines is quite sensitive to plasma parameters, which makes this effect especially important in plasma diagnostics. One can find many interesting details on plasma spectroscopy and its application to plasma diagnostics in Griem,[222,223] Cabanes and Chapelle,[224] and Traving.[225]

6.8.7 Spectral Emissivity of a Line: Constancy of a Spectral Line Area

Emission in a spectral line per one atom was described earlier by the spectral line intensity S. Then the total energy emitted in a spectral line, corresponding to the transition $n \to k$, in unit volume per unit time can be expressed as:

$$J_{nk} = \hbar\omega_{nk} A_{nk} n_n \tag{6.277}$$

where n_n is the concentration of atoms in the state n.

Taking into account the profile shape function $F(\omega)$, the spectral emissivity (spectral density of J_{nk} taken per unit solid angle) can be presented as:

$$j_\omega d\omega = \frac{1}{4\pi} \hbar\omega_{nk} A_{nk} n_n F(\omega)\, d\omega \tag{6.278}$$

Recall that the profile function $F(\omega)$ is normalized as

$$\int_{-\infty}^{+\infty} F(\omega)\, d\omega = 1.$$

This brings the important conclusion that *the spectral line broadening does not change the area of the spectral line*. Broadening makes a spectral line wider, but lower. This statement can be interpreted that total emission in a spectral line is determined by the internal physics of electronic transition inside an atom and does not depend on the external situation.

6.8.8 Selective Absorption of Radiation in Spectral Lines: Absorption of One Classical Oscillator

Absorption of radiation by atoms and molecules is selective in spectral lines. Efficiency of the absorption obviously depends on broadening of the spectral lines. If the broadening is provided by collisions, the cross section of electromagnetic wave absorption by one oscillator (which represents an atom in classical electrodynamics) depends on the frequency near the resonance as:

$$\sigma_\omega = \frac{e^2}{4mc\varepsilon_0} \frac{\tau^{-1} + n_0\langle\sigma v\rangle}{(\omega - \omega_{nk})^2 + \frac{1}{4}\left(\tau^{-1} + n_0\langle\sigma v\rangle\right)^2} \tag{6.279}$$

This absorption frequency curve also has the Lorentz profile (compare with Equation 6.272) completely in accordance with Kirchoff's law, which requires correlation between emission and absorption processes; $n_0 \langle\sigma v\rangle$ is the collisional frequency; τ^{-1} is frequency of the spontaneous electronic transition $n \to k$; and $\hbar\omega_{nk}$ is energy of this transition.

Equation 6.279 shows that the maximum resonance absorption cross section (Equation 6.279) in the absence of collisions ($n_0 \langle\sigma v\rangle << \tau^{-1}$, $\omega = \omega_{nk}$) has extremely high values for visible light in terms of typical atomic sizes:

$$\sigma_\omega^{\max} = \frac{3\lambda_{nk}^2}{2\pi} \approx 10^{-9}\,\text{cm}^2 \tag{6.280}$$

If collisional broadening of a spectral line is sufficiently large, the corresponding absorption cross section decreases with pressure: $\sigma_\omega \propto 1/n_0 \langle\sigma v\rangle \propto 1/p$. The area of the absorption spectral line can be found by integrating the absorption cross section σ_ω (Equation 6.279) over frequencies (traditionally v):

$$\int \sigma_\omega d\nu = \frac{1}{2\pi}\int \sigma_\omega d\omega = \frac{e^2}{4mc\varepsilon_0} = 2.64\cdot 10^{-2}\frac{\text{cm}^2}{\text{sec}} \tag{6.281}$$

This area (under the absorption curve $\sigma_\omega(v)$) describes the absorption in a spectral line of one classical oscillator, which actually represents an atom or molecule. Again, as in Section 6.8.7, the conclusion is reached that the spectral line area is constant and, in particular, does not depend on the effects of broadening.

6.8.9 Oscillator Power

The total radiation energy absorbed in a spectral line per unit time and unit volume is independent of broadening and can be expressed as:

$$\int 4\pi\kappa_\omega I_\omega d\omega = 4\pi I_{\omega nk} n_k \int \sigma_\omega d\omega \tag{6.282}$$

Here, I_ω is the radiation intensity, κ_ω is the absorption coefficient (characterizing reverse length of absorption, discussed in Section 6.7.7); and n_k is concentration of the light-absorbing atoms. Thus, according to Equation 6.282, the total radiation energy absorbed in a spectral line is determined by the area of the absorption line $\int\sigma_\omega d\omega$. It should be noted that the absorption factor $B_{kn} = 1/\omega_{nk}\int\sigma_\omega d\nu$ and A_{nk} are also usually called the Einstein coefficients (compare with those considered in Section 6.7.2 and Section 6.7.3). Based on Equation 6.278 and Kirchhoff's law, the area of the absorption line can be found as:

$$\sigma_\omega = \frac{g_n}{g_k}\frac{\pi^2 c^2}{\omega^2} A_{nk} F(\omega), \quad \int \sigma_\omega d\nu = \frac{g_n}{2g_k}\frac{\pi c^2}{\omega^2} A_{nk} \tag{6.283}$$

In this relation, g_n, g_k are statistical weights of the upper and lower electronic states and A_{nk} is the Einstein coefficient characterizing spontaneous emission of a quantum in the spectral line (see Equation 6.8.2). Obviously, the actual area of the absorption line (Equation 6.283) differs from the classical one (Equation 6.280) corresponding to absorption of one classical oscillator. It is convenient for this reason to characterize the absorption in a spectral line by the ratio of actual area of the absorption line (Equation 6.283) over the area (Equation 6.280) corresponding to one classical oscillator:

$$f = \frac{g_n}{2g_k}\frac{4m\varepsilon_0\pi c^3}{\omega^2 e^2} A_{nk} \tag{6.284}$$

Table 6.4 Parameters of Some Strong Atomic Lines in the Visible Spectrum

Atomic line	Wavelength, λ, nm	Oscillator Power, f	Coefficient A_{nk}, sec^{-1}
H_α	656.3	0.641	$4.4 \cdot 10^7$
H_β	486.1	0.119	$8.4 \cdot 10^6$
He	587.6	0.62	$7.1 \cdot 10^7$
Ar	696.5	—	$6.8 \cdot 10^6$
Na	589.0	0.98	$6.2 \cdot 10^6$
Hg	579.1	0.7	$9.0 \cdot 10^7$

This ratio is called the oscillator power and illustrates the number of classical oscillators providing the same absorption as the actual spectral line. Examples of the oscillator powers together with the Einstein coefficients A_{nk} for some spectral lines are presented in Table 6.4. From this table, it is seen that the oscillator powers for the one-electron transitions are less than unity, and they approach unity for the strongest spectral lines. It should be mentioned that situations with molecular spectra are somewhat more complicated because of the contribution of vibrational and rotational degrees of freedom (see Section 3.1 and Section 3.2).

6.8.10 Radiation Transfer in Spectral Lines: Inverse Population of Excited States and Principle of Laser Generation

The radiation transfer (Equation 6.253) for a spectral line in terms of radiative transitions $n \Leftrightarrow k$ in an atom or molecule can be presented as:

$$\frac{dI_\omega}{ds} = j_\omega + n_n \sigma_{b\omega} I_\omega - n_k \sigma_{a\omega} I_\omega \tag{6.285}$$

In this equation, n_n, n_k are concentrations of particles in the upper n and lower k states; I_ω is the radiation intensity; the coordinate s is the path along a ray propagation; j_ω is the spontaneous emissivity on the frequency ω; and cross sections of the radiation absorption $\sigma_{a\omega}$ and stimulated emission $\sigma_{b\omega}$, related to each other in accordance with the Kirchhoff and Boltzmann laws as:

$$\sigma_{b\omega} = \left(\frac{n_k}{n_n}\right)_{eq} \exp\left(-\frac{\hbar\omega}{T}\right)\sigma_{a\omega} = \frac{g_k}{g_n}\sigma_{a\omega} \tag{6.286}$$

Here, $(n_k/n_n)_{eq}$ is the equilibrium ratio of populations of lower and higher energy levels and g_k, g_n are the corresponding statistical weights. Taking into account Equation 6.286, the radiation transfer (Equation 6.285) can be rewritten as:

$$\frac{dI_\omega}{ds} = j_\omega + (N_2 - N_1) g_k \sigma_{a\omega} I_\omega \tag{6.287}$$

In this relation, $N_2 = n_n/g_n$, $N_1 = n_k/g_k$ are the concentrations of atoms in the specific quantum states related to energy levels n and k. These factors correspond in equilibrium to the exponential factors of the Boltzmann distribution.

In equilibrium gases, $N_2 = N_1 \exp(-\hbar\omega/T) < N_1$ and the radiation transfer (Equation 6.287) results in absorption corresponding to Equation 6.253. However, in nonequilibrium conditions, the population of the higher energy states can exceed the population of those with lower energy $N_2 > N_1$. Such nonequilibrium conditions are usually referred to as the **inverse population of excited states**. In the case of inverse population, according to Equation 6.287, the nonequilibrium medium does not provide absorption but rather significant amplification of radiation in a spectral line. This is the principal basis of the laser.

In homogeneous medium with $N_2 - N_1 = const > 0$, this exponential amplification of the radiation intensity along a ray can be expressed as:

$$I_\omega = I_{\omega 0} \exp\left[(N_2 - N_1) g_k \sigma_{a\omega} s\right] \tag{6.288}$$

The exponential factor $(N_2 - N_1) g_k \sigma_{a\omega}$ in Equation 6.288 is usually called the **laser amplification coefficient**.

Note that the inverse population could be quite easily achieved in nonequilibrium plasma due to intensive excitation of atoms and molecules by electron impact. This provides physical basis for gas-discharge lasers. The nonequilibrium Treanor distribution, discussed in Section 4.1.10 and Section 5.1.3, can be considered as one of the interesting examples of a plasma-induced inverse population of excited states. In particular, the Treanor effect provides the physical conditions for generation in the CO laser.

Obviously the physics of laser generation has only been briefly touched. More details on the subject can be found in numerous publications and in Bradley.[226]

6.9 NONLINEAR PHENOMENA IN PLASMA

As discussed in Section 6.5 and Section 6.6, numerous types of oscillations and waves can be stimulated in plasma. Instabilities amplify the oscillations and waves; their amplitudes grow significantly until the oscillations and waves become strongly nonlinear. The nonlinearity leads to interaction between oscillations; to their interaction with plasma; and to formation of specific structures like solitones, electric domains, shock waves, etc. that play important roles in plasma evolution. This area of plasma physics is rather complicated; probably the most complete review of the subject can be found in Kadomtsev.[178] Only the most general concepts related to plasma nonlinearity will be discussed here.

6.9.1 NonLinear Modulation Instability: Lighthill Criterion

First, analyze in general the nonlinear evolution of a perturbation in plasma. Consider a perturbation of some plasma parameter as a "wave package," which is a group of waves with different but close values of the wave vector ($\Delta k << k$). In this case, the perturbation at a point x can be presented as:

$$a(x,t) = \sum_k a(k)\exp(ikx - i\omega t) \tag{6.289}$$

where $a(k)$ is amplitude of the wave with wave vector k. To further analyze evolution of the perturbation, the following general dispersion relation $\omega(k)$ can be assumed for the wave:

$$\begin{aligned}\omega(k) &= \omega(k_0) + \frac{\partial\omega}{\partial k}(k = k_0)\cdot(k-k_0) + \frac{1}{2}\frac{\partial^2\omega}{\partial k^2}(k=k_0)\cdot(k-k_0)^2 \\ &= \omega_0 + v_{gr}(k-k_0) + \frac{1}{2}\frac{\partial v_{gr}}{\partial k}(k-k_0)^2\end{aligned} \tag{6.290}$$

Here, k_0 is the average value of the wave vector for the group of waves; ω_0 is frequency corresponding to k_0; and v_{gr} is the group velocity of the wave. Because of different group velocities for the waves with different values of wave vectors ($\partial v_{gr}/\partial k \neq 0$), the group of waves (initial perturbation) grows in size. If the waves are not interacting with each other (linear waves), the initial perturbation grows to the size $\Delta_x \approx 1/\Delta k$ during the period of time about $\tau \approx (\Delta k^2\, \partial v_{gr}/\partial k)^{-1}$ (see Equation 6.290).

The situation can be qualitatively different taking into account interaction between waves (which is the case of nonlinear waves). Take into account the non-linearity (interaction between waves) by introducing frequency dependence on the local value of amplitude $E(x)$:

$$\omega = \omega_0 - \alpha E^2 \tag{6.291}$$

where ω_0 is the wave frequency in the low amplitude limit. Using Equation 6.290 and Equation 6.291 for frequency deviations, the expression for plasma perturbation (Equation 6.289) can finally be rewritten as:

$$a(x,t) = \sum_k a(k)\exp\left[i(k-k_0)(x-x_0) - ik_0x_0 - i(k-k_0)^2\frac{\partial v_{gr}}{\partial k}t - i\alpha E^2(x)t\right] \tag{6.292}$$

Equation 6.292 shows the modulation of the group of waves, called the **modulation instability.** At some specific modes of the modulation instability, the whole "wave package" (presenting a plasma perturbation) decays into smaller groups of waves or can be "compressed" and converted into the specific lone wave called the **solitone.**

Thus, competition between the two last terms in the exponent in Equation 6.292 determines the type of evolution of the wave package. The term $i(k - k_0)^2 \, \partial v_{gr}/\partial k \, t$, as was mentioned earlier, leads to expansion of the wave package, while the nonlinear term $i\alpha E^2(x)t$ is able, in principle, to compensate the previous one. To compensate the simple expansion of a perturbation and provide an opportunity of the wave package decay into smaller groups of waves, compression, or formation of solitones, the previously mentioned two terms should at least have different signs (±). This leads to the following requirement:

$$\alpha \cdot \frac{\partial v_{gr}}{\partial k} < 0 \tag{6.293}$$

which is usually referred to as the **Lighthill criterion**. This criterion is necessary to observe the modulation instability and the mentioned nonlinear phenomena of evolution of perturbations (or "the wave packages") in plasma.

6.9.2 Korteweg–de Vries Equation

Influence of a weak nonlinearity on the wave package expansion due to dispersion $\omega(k)$ can be described using the Korteweg–de Vries equation. To derive this equation, consider propagation of the longitudinal long-wavelength oscillations (for example, the ionic sound in plasma). These oscillations at relatively long wavelength ($r_0 k << 1$) can be described by the following dispersion relation:

$$\omega = v_{gr} k\left(1 - r_0^2 k^2\right) \tag{6.294}$$

In the specific case of ionic wave propagation in plasma, the general size parameter r_0 corresponds to the Debye radius (compare the dispersion relation [Equation 6.294] with the dispersion equation for the ionic sound [Equation 6.136]). The objective now is to find a nonlinear equation that describes the propagation of waves with the dispersion. Begin with the Euler equation (see Section 4.2.3) for particle velocities in the longitudinal wave under consideration:

$$\frac{\partial v}{\partial t} + v\frac{\partial v}{\partial x} - \frac{F}{M} = 0 \tag{6.295}$$

Here, $v(x, t)$ is particle velocity in the longitudinal wave along the axis x; F is the force acting on the plasma particle; and M is mass of the particle. For the linear approximation, assume $v = v_{gr} + v'$, where $v' << v_{gr}$ is the plasma particle velocity calculated with respect to the wave. In this case, the Euler equation can be presented as a linear one:

$$\frac{\partial v'}{\partial t} + v_{gr}\frac{\partial v'}{\partial x} - \frac{F}{M} = 0 \tag{6.296}$$

with the function F/m considered as a linear operator of v'. In the harmonic approximation for the plasma particle velocities: $v' \propto \exp(-i\omega t + ikx)$, choose the linear operator $F/m\ (v')$in such a way to get from the linear equation (Equation 6.296) the dispersion relation (Equation 6.294). This leads to the linear operator $F/m\ (v') = -r_0^2 v_{gr}\ \partial^3 v'/\partial x^3$ and to the linear equations of motion in the form:

$$\frac{\partial v'}{\partial t} + v_{gr}\left(\frac{\partial v'}{\partial x} + r_0^2 \frac{\partial^3 v'}{\partial x^3}\right) = 0 \tag{6.297}$$

The last term in this linear equation describes dispersion of the long-wavelength oscillations. Now, to take into account the nonlinearity, replace v' by the plasma particle velocity v (the reverse step with respect to transition from Equation 6.295 to Equation 6.296). This results in the following nonlinear equation for plasma particles:

$$\frac{\partial v}{\partial t} + v\frac{\partial v}{\partial x} + v_{gr} r_0^2 \frac{\partial^3 v}{\partial x^3} = 0 \tag{6.298}$$

This motion equation for plasma particles is well known as the **Korteweg–de Vries equation**. This equation is especially interesting for description of nonlinear dissipative processes because it takes into account nonlinearity and dispersion of waves. It successfully describes the propagation in plasma of different long-wavelength oscillations corresponding to the dispersion equation (Equation 6.294).

6.9.3 Solitones as Solutions of Korteweg–de Vries Equation

The Korteweg–de Vries equation has a physically very interesting solution that corresponds to a lone or solitary nonharmonic wave, which is called a solitone. Solitones are particular in ability to maintain a fixed shape during propagation in a medium with dispersion. Although shorter harmonics are moving more slowly, the nonlinearity compensates this effect and the wave package does not expand but rather keeps its shape.

To describe the solitones, consider a wave propagating with velocity u. Dependence of the plasma particle velocities on coordinate and time can be presented as $v = f(x - ut)$. In this case, $\partial v/\partial t = -u\ \partial v/\partial x$ and the Korteweg–de Vries equation can be rewritten as the following third-order ordinary differential equation:

$$(v - u)\frac{dv}{dx} + v_{gr} r_0^2 \frac{d^3 v}{dx^3} = 0 \tag{6.299}$$

Integration of Equation 6.299, taking into account absence of the plasma particles' velocity at infinity ($v = 0$, $dv^2/dx^2 = 0$ at $x \to \infty$), leads to the second-order equation:

$$v_{gr} r_0^2 \frac{d^2 v}{dx^2} = uv - \frac{v^2}{2} \tag{6.300}$$

The solitone (the solitary wave) is one of the solutions of the nonlinear Equation 6.300, which can be expressed as:

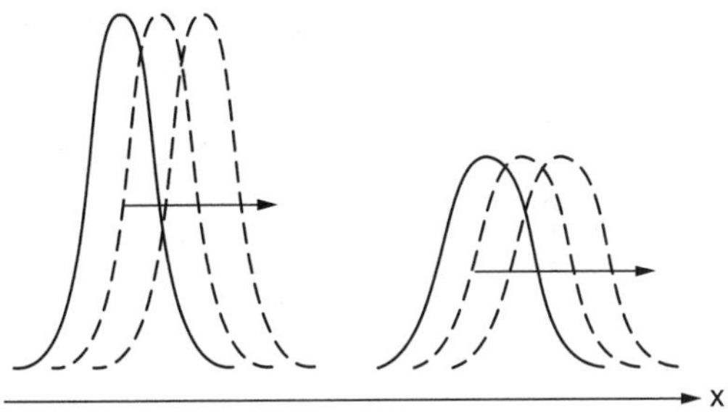

Figure 6.40 Soliton propagation

$$v = \frac{3u}{\cosh^2 \frac{x}{2r_0}\sqrt{\frac{u}{v_{gr}}}} \tag{6.301}$$

where *cosh* α is the hyperbolic cosine.

Profiles of solitones with the same fixed parameters, u, v_{gr}, and different amplitudes are illustrated in Figure 6.40. As can be seen from the figure and Equation 6.301, the solitary wave becomes narrower with increases of its amplitude. The product of the solitone's amplitude and square of its width remain constant during its evolution. If the initial nonsolitone perturbation has relatively low amplitude, then the perturbation expands with time because of dispersion until the product of its amplitude and width square corresponds to that for solitones (Equation 6.301). Then, during its evolution, the perturbation converts into a solitone. Conversely, if the amplitude of an initial perturbation is relatively high, then during evolution of the perturbation it decays to form several solitones.

6.9.4 Formation of Langmuir Solitones in Plasma

Formation of solitones in plasma is due to electric fields induced by a wave that confines a plasma perturbation in a local domain. The higher the wave amplitude is, the stronger the electric field that can be induced by this wave and the stronger the effect of perturbation compression. This effect can be illustrated using the example of plasma oscillation leading to formation of Langmuir solitones. The energy density of Langmuir oscillations at a point x can be expressed as a function of time-averaged electric field:

$$W(x) = \frac{\varepsilon_0 \langle E^2(x,t) \rangle}{2} \tag{6.302}$$

Assume the electron and ion temperatures equal to the uniform value T, and assume plasma to be completely ionized and quasi neutral. The total effective pressure is also constant in space for the long-wavelength plasma oscillations with propagation velocities much less than speed of sound:

$$2n_e(x)T + W(x) = 2n_{e0}T \tag{6.303}$$

Here, n_{e0} is electron density at infinity, where there are no plasma oscillations. The dispersion equation for plasma oscillations (Equation 6.128) can then be rewritten taking into account the relation (Equation 6.303) in a nonlinear way, including the effect of amplitude of electric field:

$$\omega^2(x) = \omega_{p0}^2\left(1 - \frac{\varepsilon_0\langle E^2\rangle}{4n_{e0}T}\right) + \frac{\gamma T}{m}k^2 \tag{6.304}$$

In this relation, ω_{p0} is plasma frequency far enough from the perturbation; T is the quasi-equilibrium plasma temperature; and γ is the specific heat ratio. From Equation 6.304, taking into account that for harmonic oscillations of the electric field, $E = E_0 \cos\omega t$, $\langle E^2\rangle = 1/2\, E_0^2$, and also introducing the Debye radius (Equation 4.129), rewrite Equation 6.304 as:

$$\omega(x) = \omega_{p0}\left(1 - \frac{\varepsilon_0 E_0^2}{16n_{e0}T} + \gamma r_D^2 k^2\right) \tag{6.305}$$

Deriving this dispersion equation for plasma oscillations, it was assumed that the second and third terms are much less than unity. This means that only the low levels of nonlinearity of the plasma oscillations are considered. The dispersion Equation 6.305 satisfies the Lighthill criterion (Equation 6.293). Differentiating Equation 6.301:

$$\alpha\frac{\partial v_{gr}}{\partial k} = \alpha\frac{\partial^2\omega}{\partial k^2} = -\frac{\gamma\varepsilon_0\omega_{p0}}{8mn_{e0}} < 0 \tag{6.306}$$

This explains the possibility for the plasma oscillations to form the solitary wave that is called, in this case, the Langmuir solitone.

6.9.5 Evolution of Strongly Nonlinear Oscillations: Nonlinear Ionic Sound

The Langmuir solitones have been considered assuming a low level of nonlinearity (Equation 6.305). Now, analyze the case of strong nonlinearity using ionic sound as an example. To describe the nonlinear ionic sound, use the system including the Euler equation of ionic motion, the continuity equation for ions, and the Poisson equation:

$$\frac{\partial v_i}{\partial t} + v_i\frac{\partial v_i}{\partial x} + \frac{e}{M}\frac{\partial\varphi}{\partial x} = 0 \tag{6.307}$$

$$\frac{\partial n_i}{\partial t} + \frac{\partial}{\partial x}(n_i v_i) = 0 \tag{6.308}$$

$$\frac{\partial^2 \varphi}{\partial x^2} = \frac{e}{\varepsilon_0}(n_e - n_i) \tag{6.309}$$

In this system of equations, v_i is the ion velocity in wave; φ is the electric field potential; n_e, n_i are the electron and ion densities; e is an electron charge; and M is an ion mass.

Consider ionic motion as a wave propagating with velocity u; then the plasma parameters (v_i, n_i, φ) depend on coordinate x and time t as $f(x - ut)$. Also take into account the high electron mobility, which results in their Boltzmann quasi equilibrium with electric field $n_e = n_{e0}\exp(e\varphi/T_e)$, where T_e is the electron temperature and n_{e0} is the average density of charged particles. Then rewrite the system of Equation 6.307 through Equation 6.309 as:

$$-u\frac{dv_i}{dx} + v_i\frac{dv_i}{dx} + \frac{e}{M}\frac{d\varphi}{dx} = 0 \tag{6.310}$$

$$\frac{d}{dx}\left[n_i(v_i - u)\right] = 0 \tag{6.311}$$

$$\frac{\partial^2 \varphi}{\partial x^2} = \frac{e}{\varepsilon_0}\left[n_{e0}\exp\left(\frac{e\varphi}{T_e}\right) - n_i\right] \tag{6.312}$$

Integrate the two first equations (Equation 6.310 and Equation 6.311) assuming that there are no perturbations far enough from the wave ($n_i = n_{e0}$, $v_i = 0$ at $x \to \infty$). This leads to the following two nondifferential relations:

$$\frac{v_i^2}{2} - uv_i + \frac{e\varphi}{M} = 0 \tag{6.313}$$

$$n_i = n_{e0}\frac{u}{u - v_i} \tag{6.314}$$

Based on these two relations, the Poisson equation (Equation 6.310) can be expressed as a second-order, nonlinear differential relation between the electric field potential and wave velocity:

$$\frac{d^2\varphi}{dx^2} = \frac{en_{e0}}{\varepsilon_0}\left[\exp\left(\frac{e\varphi}{T_e}\right) - \frac{u}{\sqrt{u^2 - \frac{2e\varphi}{M}}}\right] \tag{6.315}$$

This equation can be converted into a first-order differential equation by multiplying Equation 6.315 by $d\varphi/dx$, integrating, and assuming the electric field and potential far from the wave are both equal to zero $\varphi = 0$, $d\varphi/dx = 0$:

$$\frac{1}{2}\left(\frac{d\varphi}{dx}\right)^2 + \frac{n_{e0}T_e}{\varepsilon_0}\left[1 - \exp\left(\frac{e\varphi}{T_e}\right)\right] + \frac{n_{e0}Mu^2}{\varepsilon_0}\left(1 - \sqrt{1 - \frac{2e\varphi}{Mu^2}}\right) = 0 \tag{6.316}$$

This nonlinear differential equation describes the solitary ionic sound waves, the ionic sound solitones, without limitations on their amplitudes. Thus, this equation is able to illustrate the effects of strong nonlinearity. For example, analyze the relation between the maximum potential in the solitary wave $\varphi = \varphi_{max}$ and its propagation velocity u. To do that, assume in Equation 6.316 that $\varphi = \varphi_{max}$, $d\varphi/dx = 0$ and introduce the dimensionless variables: the dimensionless maximum potential $\xi = e\varphi_{max}/T_e$ and the dimensionless square of the wave velocity $\eta = Mu^2/2T_e$. Then, rewrite Equation 6.316 as the algebraic relation:

$$1 - \exp(\xi) + 2\eta\left(1 - \sqrt{1 - \xi/\eta}\right) \tag{6.317}$$

If the amplitude of the ionic-sound wave is small ($\xi \to 0$), Equation 6.317 gives $\eta = 1/2$ and $u = \sqrt{T_e/M}$. Obviously, this corresponds to the conventional formula for velocity of ionic sound (Equation 6.136). In the opposite limit of the maximum amplitude for a solitone, the ion energy in the solitary wave exactly corresponds to the potential energy in the wave $\xi = \eta$. In this case, the critical value of ξ can be determined, based on Equation 6.317, from:

$$1 - \exp(\xi) + 2\xi = 0 \tag{6.318}$$

Solving this equation yields $\xi = 1.26$, $e\varphi_{max} = 1.26T_e$, $u = 1.58\sqrt{T_e/M}$. At higher amplitudes than this critical one, ions are "reflected" by the wave, resulting in decay of the strongly nonlinear wave into smaller separate wave packages. Thus, the solitones only exist at some limited levels of the wave amplitude; very large amplitudes and strong nonlinearity lead to decay of solitary waves. This conclusion can be applied not only to ionic sound solitones, but to other solitary waves in plasma with strong nonlinearities.

6.9.6 Evolution of Weak Shock Waves in Plasma

Another nonlinear effect that is important in nonthermal plasma systems is related to the evolution of weak shocks and generation of strong shock waves. A weak shock wave means gas-dynamic perturbation related to a sharp change of derivatives of gas-dynamic variables, while change of these variables is continuous. The evolution of shock waves can be stimulated by VT relaxation and by chemical heat release in reactions of vibrationally excited molecules[227] and is especially important in nonequilibrium supersonic discharges.[9]

Detailed consideration of effects of nonlinearity and nonuniformity related to shock wave evolution is quite complicated and can be analyzed only with numerical methods.[227a] Some related problems were recently reviewed by Capitelli.[81] Analytical description of these phenomena can be accomplished by the transport equation for the amplitude of weak shock wave $\alpha(x)$, derived based on the method of characteristics.[228] The weak shock wave $\alpha(x)$ amplitude is introduced by the relative change of space derivatives of gas-dynamic functions (gas density n_0, velocity v_0, and pressure p) on the shock wave front:

$$n'^{(+)}_{0x} - n'^{(-)}_{0x} = \alpha(x) n_0 \tag{6.319a}$$

$$v'^{(+)}_{0x} - v'^{(-)}_{0x} = \alpha(x) c_s \tag{6.319b}$$

$$p'^{(+)}_{x} - p'^{(-)}_{x} = \alpha(x) M n_0 c_s^2 \tag{6.319c}$$

Here, c_s is the speed of sound in nonperturbed flow; n_0 is gas density in nonperturbed flow; and M is the mass of molecules. One should note that the average vibrational energy of the molecules and also its derivative are continuous across the wave front in the weak shock waves.[229] Solution of the transport equation describes the evolution of the amplitude of a weak shock wave propagating along the plasma flow:[228,230]

$$\alpha(x) = \alpha_i \xi(x) \left[1 \pm \frac{\gamma + 1}{2} \alpha_i \int_{x_0}^{x} \xi(x') \frac{dx'}{M \pm 1} \right]^{-1} \tag{6.320}$$

In this relation, α_i is initial amplitude of a weak shock wave ($x = x_0$); the sign "+" corresponds to a shock wave propagating downstream; the sign "–" corresponds to a shock wave propagating upstream; γ is the specific heat ratio; M is the Mach number; and the special function $\xi(x)$ is defined by the following exponential integral:

$$\xi(x) = \left(\frac{c_{si}}{c_s} \right)^{5/2} \left(\frac{M_i \pm 1}{M \pm 1} \right)^2 \exp \int_{x_0}^{x} \left\{ \delta_{lin}(x') + \frac{1}{c_s} \frac{\partial v}{\partial x'} \left(\frac{\pm \gamma M - 1}{M \pm 1} \right) \right\} \frac{dx'}{M \pm 1} \tag{6.321}$$

Here, c_{si}, M_i are the initial values of the speed of sound and Mach number (at $x = 0$); c_s, M are the current values of speed of sound and Mach number at any arbitrary x; v is the gas velocity; and $\delta_{lin}(x)$ is the increment of amplification of gas-dynamic perturbations in linear approximation, determined by Equation 6.168.

The nonlinear relations (Equation 6.320 and Equation 6.321) take into account nonuniformity of the medium where the weak shock wave propagates. The nonuniformity is taken into account by means of the pre-exponential factor and by the second term under the integral in Equation 6.321. If $dv/dx = 0$ and the coordinate x is close to the initial one, x_0, then the nonlinear relations (Equation 6.320 and Equation 6.321) obviously become identical to the result of linear approach Equation 6.168.

6.9.7 Transition from Weak to Strong Shock Wave

Equation 6.320 and Equation 6.321 permit describing the transition from a weak to a strong shock wave, which corresponds to $|\alpha(x)| \to \infty$. According to the considered relations, generation of a strong shock wave takes place at a critical point x_{cr}, where the denominator of Equation 6.320 becomes equal to zero:

$$\frac{\gamma+1}{2}\alpha_i \int_{x_0}^{x} \xi(x')\frac{dx'}{M \pm 1} = \mp 1 \tag{6.322}$$

As one can see from this relation, the strong shock wave can be generated only from waves of compression (for compression waves moving downstream $\alpha_i < 0$, upstream $\alpha_i > 0$) with initial amplitude exceeding the critical one:

$$|\alpha_i| > \alpha_{cr} = \left| \frac{\gamma+1}{2} \int_{x_0}^{\infty} \xi(x)\frac{dx}{M \pm 1} \right|^{-1} \tag{6.323}$$

The threshold for generation of a strong shock wave, Equation 6.323, depends on the gradient of the background flow velocity dv/dx, where the perturbation propagates (see Equation 6.321). It is especially important in vibrationally nonequilibrium plasma chemical systems ($T_v > T_0$) operating in supersonic discharges (see Section 6.4.2) in conditions close to the critical heat release. The shock wave generated in this case is supported by heat release from vibrational relaxation and can be considered a detonation wave.

PROBLEMS AND CONCEPT QUESTIONS

6.1 **Ideal and Nonideal Plasmas**. Based on Equation 6.1, calculate the minimum electron density necessary to reach conditions of the nonideal plasma at: (1) electron temperature of 1 eV and (2) electron temperature equal to room temperature.

6.2 **Derivation of Debye Radius**. To derive the formula for the Debye radius, it was assumed that $e\varphi/T \ll 1$ to expand the Boltzmann distribution in Taylor series. Show that this assumption is equivalent to the requirement of plasma ideality.

6.3 **Number of Charged Particles in Debye Sphere**. The number of particles necessary for "screening" of the electric field of one specified charged particle (Figure 6.1) can be found as the number of charged particles in a Debye sphere. Show that this number is equal to $\sqrt{T_e^3(4\pi\varepsilon_0)^3/e^6 n_e}$ and is very large in ideal plasma.

6.4 **The Bohm Sheath Criterion**. Based on Equation 6.9 for potential distribution in a sheath, show that the directed ion velocity on the steady-state plasma–sheath boundary should exceed the critical value of the Bohm velocity.

6.5 **Floating Potential**. Microparticles or aerosols are usually negatively charged in steady-state plasma and have negative floating potential (Equation 6.12) with respect to the plasma. Estimate the typical negative charge of such particles as a function of their radius. Assume that microparticles are spherical and located in nonthermal plasma with electron temperature of about 1 eV.

6.6 **Matrix and Child Law Sheaths.** Calculate the matrix and Child law sheaths for nonthermal plasma with electron temperature of 3 eV, electron concentration of 10^{12} cm^{-3}, and sheath voltage of 300 V. Compare the results obtained for the models of matrix and Child law sheaths and interpret the difference between them.

6.7 **Plasma Oscillations and Plasma Frequency.** Calculate the plasma density necessary to obtain resonance condition between plasma oscillations (Langmuir frequency [Equation 6.19]) and vibration of molecules. Is it possible to use such resonance for direct vibrational excitation of molecules in plasma without any electron impacts?

6.8 **Skin-Layer Depth as a Function of Frequency and Conductivity.** Both skin-layer depth and Debye radius reflect plasma tendency to "screen" itself from external electric field. Compare these two effects qualitatively (show the physical difference between them) as well as numerically for typical parameters of nonthermal plasma.

6.9 **Magnetic Field Frozen in Plasma.** The magnetic field becomes frozen in plasma when plasma conductivity is sufficiently high. Find the minimum value of conductivity necessary to provide the effect and calculate the corresponding level of plasma density. Use the magnetic Reynolds number criterion for calculations.

6.10 **Magnetic Pressure and Plasma Equilibrium in Magnetic Field.** Estimate the value of magnetic field necessary to provide a sufficient level of magnetic pressure to balance the hydrostatic pressure of hot confined plasma. For calculations, take typical parameters of the hot confined plasma from Table 6.2.

6.11 **The Pinch-Effect, Bennet Relation.** Derive the Bennet relation (Equation 6.40) for equilibrium of the completely ionized Z-pinch plasma, using Equation 6.36 or Equation 6.38, combined with the first Maxwell equation (Equation 6.29) for self-magnetic field of the plasma column. Calculate how the radius of a plasma cylinder in the Z-pinch depends on the discharge current.

6.12 **Two-Fluid Magneto-Hydrodynamics.** Give a physical interpretation of Equation 6.45 between the magnetic field and electron gas velocity, derived in the frameworks of the two-fluid magneto-hydrodynamics. This relation leads to the conclusion that the magnetic field is frozen specifically in the electron gas component of plasma.

6.13 **Plasma Diffusion across Magnetic Field.** Comparing Equation 6.48 and Equation 6.49, discuss the contribution of ions in plasma diffusion across the magnetic field. Consider cases of magnetized and nonmagnetized electrons and ions.

6.14 **The Larmor Radius and Diffusion of Magnetized Plasma.** Derive Equation 6.52 for the diffusion coefficient of plasma across the strong uniform magnetic field in terms of the Larmor radius. Find the criterion for the minimum magnetic field necessary to provide effective decrease of diffusion across the magnetic field.

6.15 **The Magnetic Reynolds Number**. Compare magnetic and kinematic viscosity in a plasma and discuss numerical differences between magnetic and conventional Reynolds numbers. Compare the consequences of a high magnetic Reynolds number in plasma and high Reynolds number in fluid mechanics without magnetic field.

6.16 **Thermal Instability in Monatomic Gases**. The ionization-overheating instability can be explained by the chain of causal links (Equation 6.57). Explain why the sequence of events shown in Equation 6.57 cannot take place in nonequilibrium supersonic discharges.

6.17 **Thermal and Vibrational Modes of the Ionization-Overheating Instability**. Compare Equation 6.63 and Equation 6.64 for the instability increment in thermal and vibrational modes. Explain why one of them is proportional and the other inversely proportional to the logarithmic sensitivity of the vibrational relaxation rate.

6.18 **Electron Attachment Instability**. Why does electron attachment instability affect the plasma chemical process much less than thermal (ionization-overheating) instability? Compare typical frequencies of the electron attachment instability with those of vibrational excitation by electron impact and VT relaxation at room temperature.

6.19 **Critical Heat Release in Supersonic Flows**. Estimate the maximum conversion degree for the plasma chemical process with $\Delta H \approx 1\,eV$ in supersonic flow in a constant cross section reactor with energy efficiency of about 30% and initial stagnation temperature of 300 K. Assume the maximum value of heat release in the system as a half of the critical one; assume in the calculations typical values of specific heats as those for diatomic gases.

6.20 **Profiling of Nonthermal Discharges in Supersonic Flow**. The effect of critical heat release in supersonic flow can be completely suppressed by profiling of duct and keeping the Mach number constant. Using Equation 6.82 and Equation 6.83, calculate the required increase of the reactor cross section in the discharge zone to provide the conditions of the constant Mach number for $M = 3$ in CO_2 and initial temperature $T_{00} = 300$ K.

6.21 **Dynamics of Vibrational Relaxation in Transonic Flows**. Transonic flows can be stable with respect to thermal instability of vibrational relaxation at Mach numbers less than but close to unity (see Equation 6.103). However, in this case, the critical heat release is small because of nearness of speed of sound. Estimate the maximum possible heat release for the transonic flows with Mach numbers sufficient to observe the stabilization effect.

6.22 **Space-Nonuniform Vibrational Relaxation**. Explain the qualitative physical difference between vibrational and translational modes of the nonuniform spatial vibrational relaxation. Why is the vibrational mode stable in most practical discharge conditions?

6.23 **Electrostatic Plasma Waves**. Using the dispersion (Equation 6.127), show that the product of phase and group velocities of the electrostatic plasma waves corresponds to the square of thermal electron velocities at any values of wave number. Which of the two characteristic wave velocities is larger? Discuss the difference between them.

6.24 **Ionic Sound**. Based on the dispersion (Equation 6.136), derive a relation for the group velocity of the ionic sound. Compare the group velocity with the phase velocity of the ionic sound. Discuss the result.

6.25 **Criterion of Collisionless Damping of Plasma Oscillations**. The criterion (Equation 6.145) of the collisional damping of electrostatic plasma oscillations was derived in the case in which the electron distribution function is not perturbed by interaction with waves. Prove that this criterion of the collisionless damping can be applied in general cases, even when the relative level of perturbations of plasma density is high and inequality (Equation 6.144) is not valid.

6.26 **Landau Damping**. Estimate the coefficient γ for Landau damping for electrostatic plasma oscillations (in ω_p units) in a typical range of microwave frequencies. Assume that the plasma oscillation frequency is close to the Langmuir frequency, $kr_D \approx 0.1$.

6.27 **Beam Instability**. Estimate the increment γ of Langmuir oscillations in electron beam instability (Equation 6.150) assuming plasma density of 10^{12} cm^{-3} and electron beam density of 10^9 cm^{-3}. Estimate the gas pressure at which this increment exceeds the frequency of electron–neutral collisions responsible for collisional damping of plasma oscillations (Equation 6.129).

6.28 **Amplification of Acoustic Waves in Nonequilibrium Plasma**. Based on the complex dispersion (Equation 6.169), derive a relation for coefficient of amplification of acoustic waves in nonequilibrium plasma in the presence of intensive plasma chemical reactions.

6.29 **High-Frequency Dielectric Permittivity of Plasma**. Explain why the high-frequency dielectric permittivity of plasma is less than one, while the dielectric constant is greater than one for conventional dielectric materials. What is the principal physical difference between these two cases?

6.30 **Attenuation of Electromagnetic Waves in Plasma**. Based on Equation 6.189, estimate the attenuation coefficient κ of electromagnetic waves in plasma in the case of low conductivity $\sigma_\omega << \varepsilon_\omega \varepsilon_0 \omega$. Compare the result with Relation 6.188; discuss the result.

6.31 **Electromagnetic Waves in Magnetized Plasma**. Based on the dispersion (Equation 6.213), find the resonance frequencies for electromagnetic waves propagating along the direction of the magnetic field, taking into account the ionic motion. Compare the frequencies with the electron cyclotron frequency ($\omega = \omega_B$); explain the difference.

6.32 **Emission and Absorption of Radiation by Free Electrons**. Based on the energy and momentum conservation laws for collision of an electron and photon, prove that neither emission nor absorption of radiation is possible for free electrons. Analyze the role of the third heavy collision partner, ion or neutral, in emission and absorption processes. Explain why a second free electron cannot be an effective third partner.

6.33 **Spontaneous and Stimulated Emission, the Einstein Relation**. Using the relation between the Einstein coefficients, analyze the relative contribution of spontaneous and stimulated emission in quasi-equilibrium plasma conditions. Assume the quasi-equilibrium temperature of the system as 10,000 K, and discuss the relative contribution of spontaneous and stimulated emission for different ranges of the radiation wavelengths.

6.34 **Cross Section of the Bremsstrahlung Emission**. Using Equation 6.233 and Equation 6.235, estimate numerically typical values of the bremsstrahlung emission cross sections per unit interval of radiation frequencies for typical thermal plasma parameters. Compare the relative contributions of electron–ion and electron–neutral collisions in the bremsstrahlung emission in the specific thermal plasma conditions chosen for estimations.

6.35 **Total Emission in Continuous Spectrum**. Total spectral density of plasma emission in continuous spectrum consists of bremsstrahlung and recombination components. After integration over the spectrum, these total plasma radiative energy losses per unit time and unit volume can be calculated using Equation 6.247. Calculate the typical value of this radiative power per unit volume for typical conditions of thermal plasma with $T = 10{,}000$ K and compare with the typical value of specific thermal plasma power of about 1 kW/cm.

6.36 **Total Plasma Absorption in Continuum, the Unsold–Kramers Formula**. Using the Unsold–Kramers formula, discuss the total plasma absorption in continuum as a function of temperature. Estimate the quasi-equilibrium plasma temperature when maximum absorption of red light can be achieved. Discuss the result.

6.37 **Optical Thickness and Emissivity Coefficient**. In optically thin plasmas, the total emissivity coefficient coincides with the optical thickness $\varepsilon = \tau_\omega$. Derive the relation between these two parameters in a general case. Discuss their dependence on radiation frequency.

6.38 **Probability of the Bound–Bound Transition, Intensity of Spectral Line**. Using Equation 6.264 for intensity of spectral lines, estimate the ratio of the spontaneous emission frequencies for the cases of transitions between different electronic states and transitions between different vibrational levels for the same electronic state. Estimate absolute values of the spontaneous emission frequencies, using data from Table 6.4.

6.39 **Natural Profile of a Spectral Line**. The photon distribution function and the natural profile of spectral line, related to finite lifetime of excited states of an atom or molecule, can be described by the Lorentz profile (Equation 6.268). Show that the Lorentzian profile satisfies the normalization criterion:

$$\int_{-\infty}^{+\infty} F(\omega)\, d\omega = 1.$$

6.40 **Doppler Broadening of Spectral Lines**. The Doppler broadening of spectral lines depends on gas temperature and atomic mass, while natural width does not. Estimate the minimal value of gas temperature when Doppler broadening of argon spectral line $\lambda = 696.5$ nm exceeds the natural width of the spectral line.

6.41 **Pressure Broadening of Spectral Lines**. Using Equation 6.272 for the pressure broadening profile, compare the half-widths for this case (at different pressures and temperatures) with those typical for the Doppler broadening of spectral lines in thermal and non-thermal plasmas, as well as with the natural half-width.

6.42 **The Stark Broadening of Spectral Lines**. Based on Equation 6.274 and Equation 6.272, compare the half-width for the total Stark and pressure broadening effects. For typical conditions of nonthermal plasma, determine the degree of ionization when these two broadening effects are the same order of magnitude. How does the critical ionization degree depend on pressure?

6.43 **The Voigt Profile of Spectral Lines**. The Voigt profile is a result of convolution of the Gaussian and Lorentzian profiles of spectral lines. Prove that the integral Voigt profile (Equation 6.275) becomes either Gaussian or Lorentzian in extreme cases of significant prevailing of the Doppler or Stark (or pressure) broadening effects.

6.44 **Absorption of Radiation in a Spectral Line by One Classical Oscillator**. By integrating Equation 6.281, prove that the spectral line area $\int \sigma_\omega dv$ characterizing the absorption of one classical oscillator is constant, which means it does not depend on the frequency factors $n_0 \langle \sigma v \rangle$, τ^{-1} related to broadening effects.

6.45 **The Oscillator Power**. Give a physical interpretation of the fact that the oscillator powers for the one-electron transitions are less than unity and they approach unity for the strongest spectral lines. Analyze Equation 6.284 and data presented in Table 6.4.

6.46 **Inverse Population of Excited States and the Laser Amplification Coefficient**. Based on Equation 6.288, estimate the laser amplification coefficient, assuming that the inverse population of excited states is due to the Treanor effect (see Section 4.1.10 and Section 5.1.3).

6.47 **The Korteweg–de Vries Equation and Dispersion Equation of the Ionic Sound**. Compare the dispersion relation (Equation 6.294) used for derivation of the Carteveg–De Vries equation with the dispersion equation for the ionic sound (Equation 6.136). Discuss the difference between them. Determine the criterion of the approximation. Is this criterion specific for the ionic sound or can it be generalized for consideration of nonlinear behavior of other waves?

6.48 **Solitones as Solutions of the Korteweg–de Vries Equation**. Prove that solitones (Equation 6.301) are solutions of the Korteweg–De Vries equation in the forms Equation 6.300 and Equation 6.298. Using the general expression (Equation 6.301) for solitones, analyze a relation between velocity amplitude in the solitary wave and characteristic width of the solitone. Prove that the product of the solitone's amplitude and square of the solitone's width have the constant value during its evolution.

6.49 **The Langmuir Solitones**. Analyze the Lighthill criterion for plasma oscillations, using the nonlinear dispersion equation in form Equation 6.304. Compare this criterion with Equation 6.306 derived from the dispersion Equation 6.305, where the low level of nonlinearity of plasma oscillations was directly assumed.

6.50 **Nonlinear Ionic Sound**. Analyze Equation 6.310 through Equation 6.314 describing the nonlinear ionic sound show that, in this case, the ionic velocity is always less than velocity of wave propagation ($v_i < u$). Based on the same equations, prove that electric field potential in the ionic sound wave is always positive with respect to the potential at infinity.

6.51 **Velocity of the Nonlinear Ionic-Sound Waves**. Analyzing Equation 6.317, prove that the velocity of ionic-sound waves at low amplitudes is equal to the conventional value of $c_{si} = \sqrt{T_e/M}$. During the derivation, pay attention to the fact that the second order of expansion in series is necessary to obtain the result.

6.52 **The Ionic-Sound Solitones**. Based on Equation 6.317, determine the maximum possible amplitude of the potential in the solitary ionic-sound wave. Analyze the dependence of amplitude of the wave and its velocity near the critical value of the amplitude. How does the nonlinear ionic sound wave velocity depend on amplitude of the wave?

6.53 **Evolution of Weak Shock Waves in Plasma**. As was mentioned during consideration of Equation 6.319(a, b, c), not only the average vibrational energy of molecules, but also its space derivative are continuous across the wave front in the weak shock waves in plasma. Discuss the statement and give physical interpretation of this fact.

6.54 **Comparison of Linear and Nonlinear Approaches to Evolution of Perturbations**. Analyzing Equation 6.320 and Equation 6.321, prove that the nonlinear relations (Equation 6.320 and Equation 6.321) obviously become identical to the result of linear approach (Equation 6.168), if flow can be considered as uniform $dv/dx = 0$ and the coordinate, x, is close to the initial one, x_0 (which means small shock wave amplitude).

6.55 **Generation of Strong Shock Waves and Detonation Waves in Plasma**. Analyze the behavior of plasma in conditions of strong vibrational–translational nonequilibrium sustained in supersonic flow when the heat release approaches the critical value. Apply Equation 6.323 to describe a strong shock wave generation in such a system. Is it possible to reach the conditions necessary for propagation of a detonation wave in this case?

Part II
PHYSICS AND ENGINEERING OF ELECTRIC DISCHARGES

CHAPTER

7

GLOW DISCHARGE

7.1 STRUCTURE AND PHYSICAL PARAMETERS OF GLOW DISCHARGE PLASMA, CURRENT–VOLTAGE CHARACTERISTICS: COMPARISON OF GLOW AND DARK DISCHARGES

7.1.1 General Classification of Discharges: Thermal and Nonthermal Discharges

Electric discharges in gases, in other words gas discharges, are generators of plasma. Different electric discharges provide various mechanisms and conditions of plasma formation and generate plasma with absolutely different parameters, electron temperatures, and concentrations, which can be used in numerous different applications. The term "gas discharge" initially defined the process of "discharge" of a capacitor into a circuit containing a gas gap between two electrodes. If the voltage between the electrodes is sufficiently large, breakdown occurs in the gap, the gas becomes a conductor, and the capacitor discharges. Now the term gas discharge is applied more generally to any system with ionization in a gap induced by electric field. Even electrodes are not necessary in these systems because discharges can occur simply by interaction of electromagnetic waves with gas.

All electric discharges can be classified according to their physical features and peculiarities in many different ways, for example:

- **High-pressure discharges (usually atmospheric ones such as arcs or coronas) and low-pressure discharges (10 torr and less, such as glow discharges)**. Differences between these are related mostly to how discharge walls are involved in the kinetics of charged particles, energy, and mass balance. Low-pressure discharges are usually cold and not very powerful, while high pressure discharges can be very hot and powerful (arc) as well as cold and weak (corona).
- **Electrode discharges (like glow and arc) and electrodeless discharges (like inductively coupled radio frequency [RF] and microwave)**. Differences between these are related mostly in the manner that electrodes contribute to sustaining electric current by different surface ionization mechanisms and as a result closing the electric circuit. The electrodeless discharges play an important role in "clean" technologies in which material erosion from electrodes is undesirable.
- **Direct current (DC) discharges (e.g., arc, glow, and pulsed corona) and non-DC discharges (e.g., RF, microwave, and most dielectric barrier discharges [DBD])**. The DC discharges can have constant current (arc, glow) or can be sustained in pulse-periodic regime (pulsed corona). The pulsed periodic regime permits providing higher power in cold discharges at atmospheric pressure. Non-DC discharges can be low or high frequencies (including radio frequency and microwave). Because of the skin effect, microwave discharges can be sustained continuously in a nonequilibrium way at moderate pressures at extremely high power levels up to 1 MW.
- **Self-sustained and nonself-sustained discharges**. Nonself-sustained discharges can be externally supported by electron beams and ultraviolet radiation. Probably the most important aspect of these discharges is the independence of ionization and energy input to the plasma. This permits achieving large energy inputs at high pressures without serious instabilities.
- **Thermal (quasi-equilibrium) and nonthermal (nonequilibrium) discharges**. As is clear from their names, the thermal discharges are hot (10,000 to 20,000 K), while the nonthermal discharges operate close to room temperature. The difference between these two qualitatively different types of electric discharges is primarily related to different ionization mechanisms. Ionization in nonthermal discharges is mostly provided by direct electron impact (electron collisions with "cold" non-excited atoms and molecules), in contrast to thermal discharges where ionization is due to electron collisions with preliminary excited hot atoms and molecules (see Section 2.2). Thermal discharges (the most typical example is an electric arc) are usually powerful and easily sustained at high pressures, but operate close to thermodynamic equilibrium and are not chemically selective. The nonthermal discharges (the most typical example is the glow discharge) are very selective with respect to plasma chemical reactions. The nonthermal discharges can operate very far from thermodynamic equilibrium with very high energy efficiency, but usually with limited power (see Section 5.6.1).

Classification of electric discharges into thermal and nonthermal discharges is most important from the point of view of differences in their physical parameters and properties, as well as differences in their mechanisms of stimulation of chemical

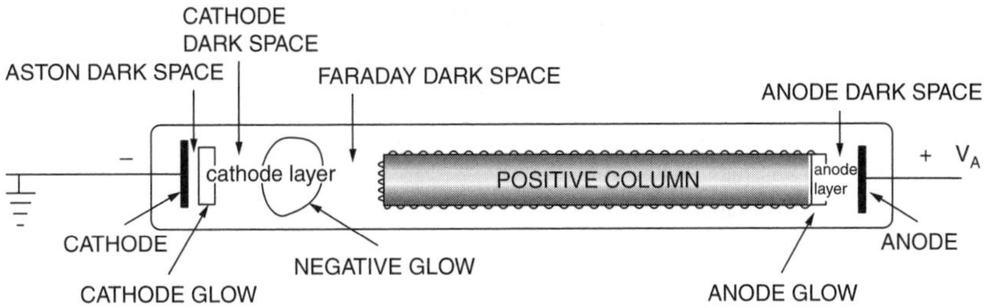

Figure 7.1 General structure of a glow discharge

processes and thus in areas of application. In this chapter, the glow discharge will be considered as the most typical and traditional example of nonthermal discharges. The most typical and traditional example of thermal discharges, the electric arc, will be considered in the following chapter.

7.1.2 Glow Discharge: General Structure and Configurations

The term "glow discharge" appeared just to point out the fact that plasma of the discharge is luminous (in contrast to the relatively low-power dark discharge, see below). According to a more descriptive physical definition: *a* **glow discharge is the self-sustained continuous DC discharge having a cold cathode, which emits electrons as a result of secondary emission mostly induced by positive ions**. A schematic drawing of a typical normal glow discharge is shown in Figure 7.1.

An important distinctive feature of the general structure of a glow discharge is the **cathode layer** (see Figure 7.1) with large positive space charge and strong electric field with a potential drop of about 100 to 500. The thickness of the cathode layer is inversely proportional to gas density and pressure. If the distance between electrodes is sufficiently large, quasi-neutral plasma with a low electric field, the so-called **positive column**, is formed between the cathode layer and anode (see Figure 7.1). The positive column of a glow discharge is the most traditional example of weakly ionized, nonequilibrium, low-pressure plasma. The positive column is separated from anode by **anode layer**. The anode layer (see Figure 7.1) is characterized by negative space charge, a slightly elevated electric field and also special potential drops.

The most conventional configuration of a glow discharge is the previously considered discharge tube with typical parameters given in Table 7.1.

This classical discharge tube was widely investigated and industrially used for several decades in fluorescent lamps as a lighting device. Other glow discharge configurations, applied in particular for thin film deposition and electron bombardment, are shown in Figure 7.2. The coplanar magnetron glow discharge convenient for plasma-assisted sputtering and deposition includes magnetic field for plasma confinement (Figure 7.2A). The configuration, optimized as an electron bombardment plasma source (Figure 7.2B), is coaxial and includes the hollow cathode ionizer as well as a diverging magnetic field.

Table 7.1 Characteristic Parameter Ranges of Conventional Glow Discharge in a Tube

Glow discharge parameter	Typical values
Discharge tube radius	0.3–3 cm
Discharge tube length	10–100 cm
Plasma volume	About 100 cm^3
Gas pressure	0.03–30 torr
Voltage between electrodes	100–1000 V.
Electrode current	10^{-4}–0.5 A
Power level	Around 100 W
Electron temperature in positive column	1–3 eV
Electron density in positive column	10^9–10^{11} cm^{-3}

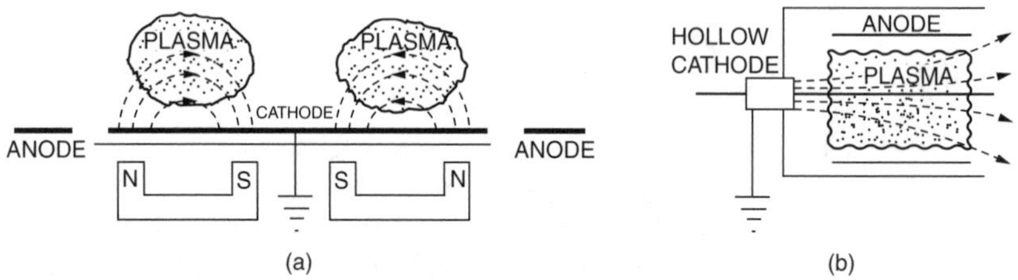

Figure 7.2 (a) Magnetron and (b) hollow cathode glow discharge configurations

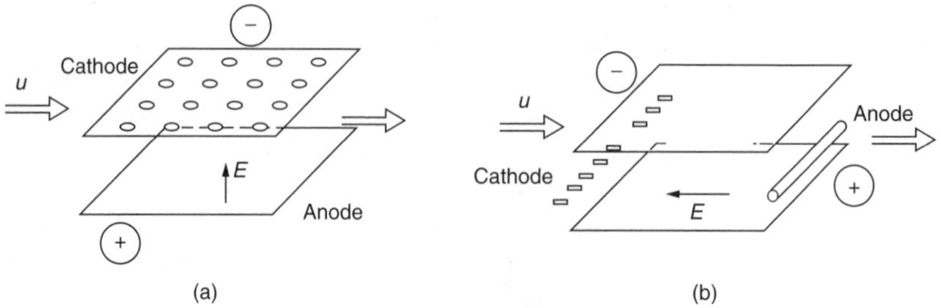

Figure 7.3 Transverse and longitudinal configurations of glow discharges in gas flow

The strongly nonequilibrium glow discharge plasma has been successfully applied at different power levels as the active medium for different gas lasers. Attempts to increase specific (per unit volume) and total power of the lasers also resulted in significant modifications of the glow discharge configuration (see Figure 7.3). Usually, at high power levels, it is a parallel plate discharge with a gas flow. The discharge can be transverse with the electric current perpendicular to the gas flow (Figure 7.3a), or longitudinal if these are parallel to each other (Figure 7.3b).

These powerful glow discharges operate at higher levels of current and voltage, reaching values of 10 to 20 A and 30 to 50 kV, respectively (compare with data presented in Table 7.1). Nevertheless, the main distinctive physical features and structure of these discharges are still similar to those of the glow discharge with conventional parameters.

7.1.3 Glow Pattern and Distribution of Plasma Parameters along Glow Discharge

Next, analyze the pattern of light emission in a classical low-pressure discharge in a tube. Along such a discharge tube, a sequence of dark and bright luminous layers is seen (Figure 7.4a). The typical size scale of glow discharge structure is usually proportional to electron mean free path $\lambda \propto 1/p$ and thus inversely proportional to pressure. For this reason, it is easier to observe the glow pattern at low pressures, when the distance between layers is sufficiently large. Thus, the layered pattern extends to centimeters when pressure is about 0.1 torr.

Special individual names were given to each layer shown in Figure 7.4(a). Immediately adjacent to the cathode is a dark layer known as the **Aston dark space** and then a relatively thin layer of the **cathode glow**. This is followed by the **dark cathode space**. The next zone is the so-called **negative glow**, which is sharply separated from the dark cathode space. The negative glow gradually decreases in brightness toward the anode, becoming the **Faraday dark space**. Only after that does the positive column begin. The **positive column** is bright (though not as bright as the negative glow), uniform, and relatively long if discharge tubes' length is sufficiently large. In the area of the anode layer, the positive column is transferred first into the **anode dark space** and finally into the narrow **anode glow** zone.

The described layered pattern of glow discharge can be interpreted based on the distribution of the discharge parameters shown in Figure 7.4(b through g). Electrons are ejected from the cathode with relatively low energy (about 1 eV) insufficient to excite atoms; this explains the Aston dark space. Then electrons obtained from the electric field with sufficient energy for electronic excitation provide the cathode glow. Further acceleration of electrons in the cathode dark space leads mostly to ionization rather than to electronic excitation (see Section 3.3.8). This explains the low level of radiation and significant increase of electron density in the cathode dark space. Slowly moving ions have relatively high concentrations in the cathode layer and provide most of the electric current.

The high electron density at the end of the cathode dark space results in a decrease of the electric field (and thus a decrease of the electron energy and ionization rate), but leads to significant intensification of radiation (see Section 3.3.8). This is the reason for the transition to the brightest layer of the negative glow. Moving further and further from the cathode, electron energy decreases, which results in transition from the negative glow into the Faraday dark space. Plasma density decreases in the Faraday dark space and electric field again grows, finally establishing the positive column.

The average electron energy in the positive column is about 1 to 2 eV, which provides light emission in this major part of the glow discharge. Notice that the cathode layer structure remains the same if electrodes are moved closer at fixed pressure, while the positive column shrinks. The positive column can be extended

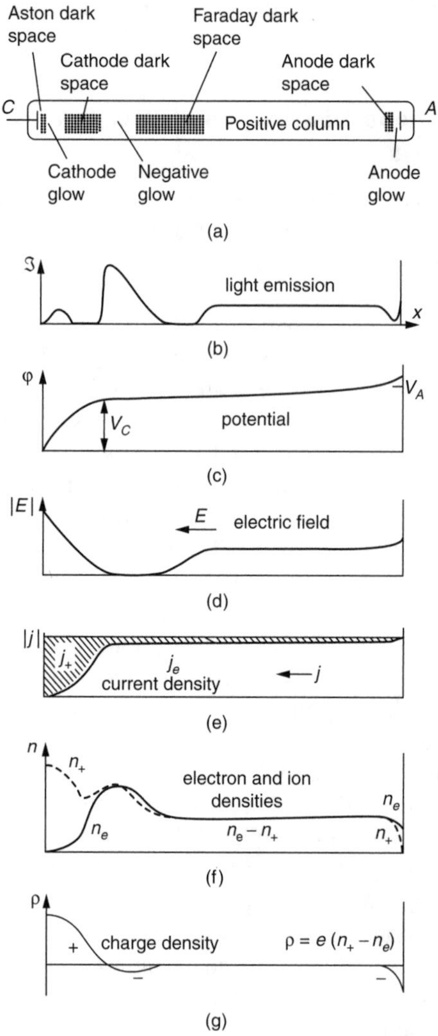

Figure 7.4 Physical parameter distribution in a glow discharge

between connecting electrodes. The anode repels ions and removes electrons from the positive column, which creates the negative space charge and leads to some increase of electric field in the anode layer. Reduction of the electron density in this zone explains the anode dark space, while the electric field increase explains the anode glow.

7.1.4 General Current–Voltage Characteristic of Continuous Self-Sustained DC Discharges between Electrodes

If the voltage between electrodes exceeds the critical threshold value V_t necessary for breakdown, a self-sustained discharge can be ignited. The current–voltage characteristic

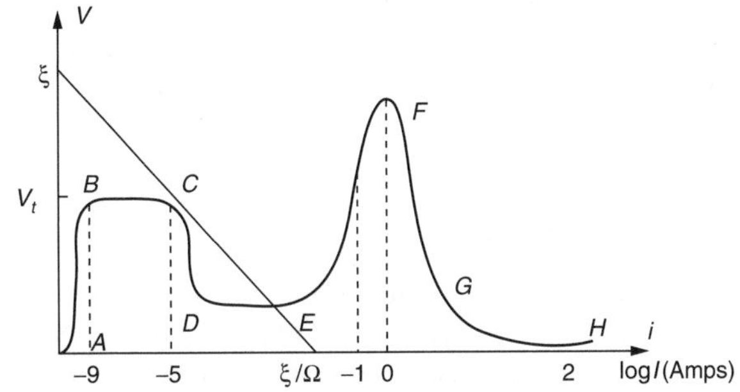

Figure 7.5 Current–voltage characteristic of DC discharges

of such a discharge is illustrated in Figure 7.5 for a wide range of currents I. The electric circuit of the discharge gap also includes an external ohmic resistance R. Then Ohm's law for the circuit can be presented as:

$$EMF = V + RI \tag{7.1}$$

where EMF is the electromotive force, V is voltage on the discharge gap. Equation 7.1 is usually referred to as the load line. This straight line corresponds to Ohm's law and is also shown in Figure 7.5. Intersection of the current–voltage characteristic and the load line gives the actual value of current and voltage in a discharge.

If the external ohmic resistance is sufficiently large and the current in circuit is very low (about 10^{-10} to 10^{-5} A), then the electron and ion densities are so negligible that perturbations of the external electric field in plasma can be neglected. Such a discharge is known as the dark Townsend discharge (obviously, this discharge can be bright at such a low electron density). The voltage necessary to sustain this discharge does not depend on current and coincides with the breakdown voltage (see Section 4.4.2, Figure 4.19). The dark Townsend discharge corresponds to the plateau BC in Figure 7.5.

An increase of the EMF or decrease of the external ohmic resistance R leads to growth of the discharge current and plasma density, which results in significant reconstruction of the electric field. This leads to reduction of voltage with current (interval CD in Figure 7.5) and to transition from a dark to a glow discharge. This still very low current version of glow discharge is called the subglow discharge. Further EMF increase or R reduction leads to the lower voltage plateau DE on the current–voltage characteristic, corresponding to the normal glow discharge. This discharge exists over a range of currents (10^{-4} to 0.1 A). This is actually the major type of glow discharges, and its general structure was discussed in Section 7.1.3.

It is important to point out that the current density on the cathode is fixed in normal glow discharges. Increases of the total discharge current occur only by growth of the so-called cathode spot through which the current flows. Only when the current is so large that no additional free surface remains on the cathode does further current growth require a voltage increase to provide higher values of current density. Such a regime is called the abnormal glow discharge and corresponds to the interval EF

on the current–voltage characteristics in Figure 7.5. Further increases of current accompanied by voltage growth in the abnormal glow regime lead to higher power levels and transition to an arc discharge, which will be discussed in the next chapter. The glow-to-arc transition usually occurs at current values of approximately 1 A.

7.1.5 Dark Discharge Physics

The distinctive feature of the dark discharge is the smallness of its current and plasma density, which keeps the external electric field almost unperturbed. In this case, the steady-state continuity equation for charged particles can be expressed, taking into account their drift in electric field and ionization, as:

$$\frac{dj_e}{dx} = \alpha j_e, \quad \frac{dj_+}{dx} = -\alpha j_e \tag{7.2}$$

In these relations, the direction from cathode to anode is chosen as the positive direction of axis x; j_e and j_+ are the electron and positive ion current densities, respectively; and α is the Townsend coefficient characterizing rate ionization per unit length (see Section 4.4.1, Equation 4.137). One should note that adding together two Equations 7.2 gives: $j_e + j_+ = j = const$, which reflects the constancy of the total current.

The boundary conditions on the cathode ($x = 0$) relate the ion and electron currents on the surface due to the secondary electron emission (with coefficient γ; see Section 8.2.5):

$$j_{eC}(x = 0) = \gamma j_{+C}(x = 0) = \frac{\gamma}{1+\gamma} j \tag{7.3}$$

Boundary conditions on the anode ($x = d_0$, d_0 is the interelectrode distance) reflects the fact of absence of ion emission from the anode:

$$j_{+A}(x = d_0) = 0, \quad j_{eA}(x = d_0) = j \tag{7.4}$$

Solution of Equation 7.2, taking into account the boundary condition on the cathode (Equation 7.3), can be expressed assuming the constancy of the Townsend coefficient α as:

$$j_e = \frac{\gamma}{1+\gamma} j \exp(\alpha x), \quad j_+ = j\left(1 - \frac{\gamma}{1+\gamma}\exp(\alpha x)\right) \tag{7.5}$$

Equation 7.5 should satisfy the anode boundary conditions. This is possible if the electric field and thus the Townsend coefficient α are sufficiently large:

$$\alpha(E) d_0 = \ln \frac{\gamma+1}{\gamma} \tag{7.6}$$

This formula describing dark discharge self-sustainment coincides with the breakdown condition in the gap (see Section 4.4.1, Equation 4.140)). Taking into account

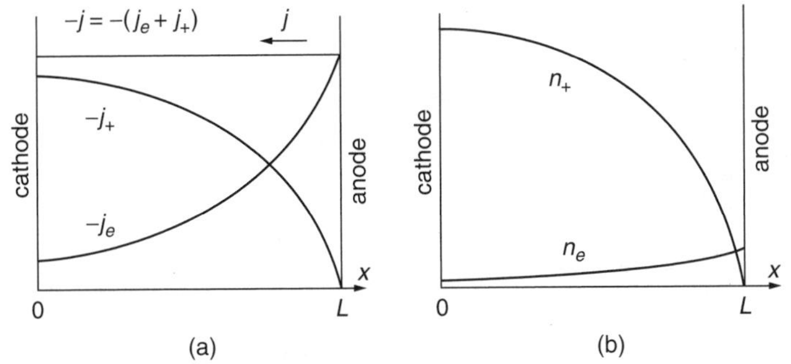

Figure 7.6 (a) Current density and (b) electron/ion density distributions in dark discharge

the discharge self-sustainment condition (Equation 7.6), the relation between currents (Equation 7.5) can be rewritten as:

$$\frac{j_e}{j} = \exp[-\alpha(d_0 - x)], \quad \frac{j_+}{j} = 1 - \exp[-\alpha(d_0 - x)], \quad \frac{j_+}{j_e} = \exp[\alpha(d_0 - x)] - 1 \quad (7.7)$$

According to Equation 7.6, $\alpha d_0 = \gamma + 1/\gamma >> 1$, because $\gamma << 1$ (numerically $\alpha d_0 = 4.6$, if $\gamma = 0.01$). Then, from Equation 7.7, the ion current exceeds the electron current over the major part of the discharge gap (Figure 7.6). The electron and ion currents become equal only near the anode ($j_e = j_+$ at $x = 0.85\ d_0$). Differences in electron and ion concentrations is even stronger because of large differences in electron and ion mobilities (μ_e, μ_+). Electron and ion concentrations become equal at a point very close to the anode (see Figure 7.6), where:

$$1 = \frac{n_+}{n_e} = \frac{\mu_e}{\mu_+}\frac{j_+}{j_e} = \frac{\mu_e}{\mu_+}\left[\exp\alpha(d_0 - x) - 1\right] \quad (7.8)$$

Assuming $\mu_e/\mu_+ \approx 100$, from Equation 7.8 the electron and ion concentrations become equal at $x = 0.998$. Thus, almost all of the gap is charged positive in a dark discharge. However, the absolute value of the positive charge is not high because of low current and thus low ion density in this discharge system.

7.1.6 Transition of Townsend Dark to Glow Discharge

The transition from the dark to the glow discharge is due to significant growth of the positive space charge and distortion of the external electric field, which results in formation of the cathode layer. To describe this transition, use the Maxwell equation for the electric field:

$$\frac{dE}{dx} = \frac{1}{\varepsilon_0} e(n_+ - n_e) \quad (7.9)$$

Taking into account that $n_+ \approx j/e\mu_+ E >> n_e$, Equation 7.9 gives the following distribution of electric field in a discharge gap:

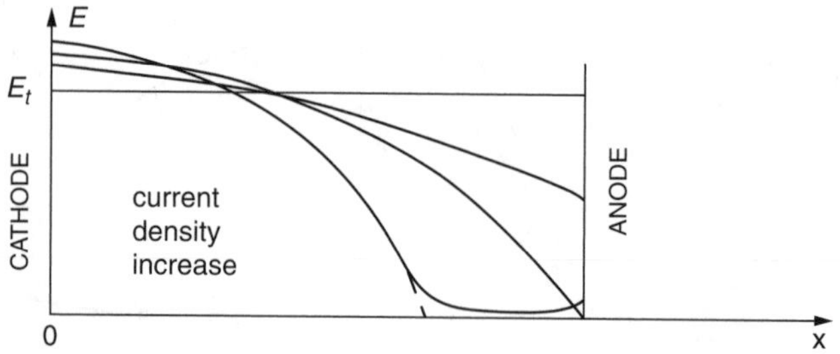

Figure 7.7 Electric field evolution in a dark discharge

$$E = E_c \sqrt{1 - \frac{x}{d}}, \quad d = \frac{\varepsilon_0 \mu_+ E_c^2}{2j} \tag{7.10}$$

Here, E_c is the electric field at the cathode. The electric field decreases near the anode with respect to the external field and grows in the vicinity of the cathode; this is illustrated in Figure 7.7. As seen from Equation 7.10 and Figure 7.7, higher values of current density lead to more distortion of the external electric field.

The parameter d in Equation 7.10 corresponds to a virtual point where the electric field becomes equal to zero. This point is located far beyond the discharge gap ($d >> d_0$) at low currents typical for dark discharges. However, when the current density becomes sufficiently high, this imaginary point of zero electric field can reach the anode ($d = d_0$). This critical current density is actually the maximum value for the dark discharge. This current density corresponds to formation of the cathode layer and to transition from the dark to glow discharge:

$$j_{\max} = \frac{\varepsilon_0 \mu_+ E_c^2}{2d_0} \tag{7.11}$$

For numerical calculations based on Equation 7.11, the electric field in the cathode's vicinity is estimated as the corresponding breakdown electric field (see Section 4.4.2, Equation 4.142). For example, in nitrogen at a pressure of 10 torr, interelectrode distance of 10 cm, electrode area of 100 cm^2, and secondary electron emission coefficient $\gamma = 10^{-2}$, the maximum dark discharge current can be estimated as $j_{\max} \approx 3 \cdot 10^{-5}\ A$.

When the current density of a dark discharge is relatively high and the electric field nonuniform, Equation 7.10 should be taken into account and the dark discharge self-sustainment condition (Equation 7.6) must be rewritten in integral form:

$$\int_0^{d_0} \alpha[E(x)]\, dx = \ln \frac{\gamma + 1}{\gamma} \tag{7.12}$$

The Townsend coefficient α strongly depends exponentially on the electric field (see Section 4.4.2, Equation 4.141). This leads to an interesting consequence: typical voltage of a glow discharge is lower than the value of a dark discharge (see Figure 7.5). Growth of the positive space charge during the dark-to-glow transition results in redistribution of the initially uniform electric field; it becomes stronger near cathode and lower near anode. However, increase of the exponential function $\alpha[E(x)]$ on the cathode side is more significant than its decrease on the anode side. Thus, electric field nonuniformity facilitates the breakdown condition (Equation 7.12), which explains why the typical voltage of a glow discharge is usually lower than in a dark discharge.

7.2 CATHODE AND ANODE LAYERS OF GLOW DISCHARGE

The cathode layer is the most distinctive zone of a glow discharge, providing its self-sustaining behavior and generating enough electrons to balance the plasma current in the positive column. Plasma chemical processes of practical interest mostly occur in the positive column of glow discharges because its volume can be much more significant. However, the most important physical processes sustaining the glow discharge take place in the cathode layer. The role of an anode layer is not as important as that of the cathode layer. It just provides total current continuity during charge transfer from the positive column to the positive electrode.

7.2.1 Engel–Steenbeck Model of Cathode Layer

When voltage is applied to a discharge gap, the uniform distribution of the electric field is not optimal to sustain the discharge. It is easier to satisfy the preceding criterion (Equation 7.12) if the sufficiently high potential drop occurs in the vicinity of the cathode, e.g., in the cathode layer. As was mentioned in Section 7.1, the required electric field nonuniformity can be provided in the discharge gap by a positive space charge formed near the cathode due to relatively low ion mobility.

The qualitative and very clear theory of a cathode layer was first developed by von Engel and Steenbeck (1934).[232,233] The electric field $E(x = d)$ on the "anode end" of a cathode layer ($x = d$, see Figure 7.7) is much less than the value near the cathode, $E(x = 0) = E_c$. Also, the ion current into a cathode layer from a positive column can be neglected due to relatively low ion mobility (usually $\mu_+/\mu_e \propto 10^{-2}$). For this reason, the Engel–Steenbeck model assumes zero electric field at the end of a cathode layer $E(x = d) = 0$ and considers the anode layer an independent system of length d (defined by Equation 7.10), where the condition (Equation 7.12) of the discharge self-sustainment should be valid. In this case, the cathode potential drop can be expressed as:

$$V_c = \int_0^d E(x)dx \tag{7.13}$$

The Engel–Steenbeck model solves the system of Equation 7.9, Equation 7.12 and Equation 7.13, taking the Townsend coefficient dependence on electric field $\alpha(E)$ in the form of Equation 4.141, and assuming a linear decrease of electric field along the cathode layer (see Figure 7.4):

$$E(x) = E_c(1 - x/d) \text{ if } 0 < x < d, \quad E = 0 \text{ if } x > d \tag{7.14}$$

Here, E_c is the electric field at the cathode. The integral (Equation 7.12) with electric field (Equation 7.14) and exponential relation $\alpha(E)$ cannot be found analytically. Simplified solutions (but with quite sufficient accuracy) can be found assuming the electric field constant over the cathode layer, $E(x) = E_c = const$, in Equation 7.12. Equation 7.12 can then be simplified to Equation 7.6 with the interelectrode distance d_0 replaced by the length d of the cathode layer; the Townsend coefficient $\alpha(E)$ can be expressed again in the form Equation 4.141; and the integral relation (Equation 7.2) can be simplified to the product $V_c = E_c d$. In this case, the Engel–Steenbeck model leads to the following relations between the electric field E_c, the cathode potential drop V_c, and the length of cathode layer pd, which are similar to those describing breakdown of a gap (Equation 4.142):

$$V_c = \frac{B(pd)}{C + \ln(pd)}, \quad \frac{E_c}{p} = \frac{B}{C + \ln(pd)} \tag{7.15}$$

Here, $C = \ln A - \ln\ln(1/\gamma + 1)$; A and B are the pre-exponential and exponential parameters of the function $\alpha(E)$ (see Equation 4.141 and Table 4.3).

The cathode potential drop V_c; electric field E_c; and similarity parameter pd depend on the discharge current density j which is close to the ion current density because $j_+ >> j_e$ near the cathode (see Figure 7.4). To determine this dependence according to the Engel–Steenbeck model, first determine the positive ion density $n_+ >> n_e$. This ion density can be found based on the Maxwell relation (Equation 7.9), taking into account the linear decrease of electric field $E(x)$ along the cathode layer from $E(x = 0) = E_c$ to $E(x = d) = 0$:

$$n_+ \approx \frac{\varepsilon_0}{e}\left|\frac{dE(x)}{dx}\right| \approx \frac{\varepsilon_0 E_c}{ed} \tag{7.16}$$

Total current density in the cathode's vicinity is close to the current density of positive ions; therefore, it can be expressed as:

$$j = en_+\mu_+ E \approx \frac{\varepsilon_0 \mu_+ E_c^2}{d} \approx \frac{\varepsilon_0 \mu_+ V_c^2}{d^3} \tag{7.17}$$

Comparing Equation 7.17 and Equation 7.10, note that accuracy of these approaches is of the order of a numerical factor. Together with Equation 7.15, this relation determines the dependence of the cathode potential drop V_c, electric field E_c, and the similarity parameter pd on the discharge current density j, which can be interpreted as the current–voltage characteristic of a cathode layer.

7.2.2 Current–Voltage Characteristic of Cathode Layer

The cathode potential drop V_c as a function of the similarity parameter pd corresponds in frameworks of the above considered Engel–Steenbeck approach to the

Paschen curve (Figure 4.19) for breakdown of a discharge gap. This function $V_c\,(pd)$ has a minimum point V_n, corresponding to the minimum value of voltage (see Equation 4.143) necessary for the breakdown.

Taking into account Equation 7.17, it is seen that the cathode potential drop V_c as a function of current density j also has the same minimum point V_n. It is convenient to express the relations among V_c, E_c, pd, and j using the following dimensionless parameters:

$$\tilde{V}=\frac{V_c}{V_n},\quad \tilde{E}=\frac{E_c/p}{E_n/p},\quad \tilde{d}=\frac{pd}{(pd)_n},\quad \tilde{j}=\frac{j}{j_n} \tag{7.18}$$

Here, electric field E_n/p and cathode layer length $(pd)_n$ correspond to the minimum point of the cathode voltage drop V_n. The subscript n stands here to denote the "normal" regime of a glow discharge, which will be discussed in the next section.

These three "normal" parameters: $\mathrm{V_c}$, E_n/p, and $(pd)_n$, as well as V_n, can be found using Equation 4.143, originally derived in Section 4.4.2 for electric breakdown as parameters of the Paschen curve. Corresponding values of the normal current density can be expressed based on Equation 7.17, using the following numeric formula constructed with similarity parameters:

$$\frac{j_n, A/\mathrm{cm}^2}{(p,\mathrm{Torr})^2}=\frac{1}{9\cdot 10^{11}}\,\frac{(\mu_+ p),\mathrm{cm}^2\mathrm{Torr}/V\sec\times(V_n,V)^2}{4\pi\left[(pd)_n,\mathrm{cmTorr}\right]^3} \tag{7.19}$$

Thus, the relations among V_c, E_c, j, and the cathode layer length pd can be expressed using the following dimensionless formulas:

$$\tilde{V}=\frac{\tilde{d}}{1+\ln\tilde{d}},\quad \tilde{E}=\frac{1}{1+\ln\tilde{d}},\quad \tilde{j}=\frac{1}{\tilde{d}(1+\ln\tilde{d})^2} \tag{7.20}$$

Dimensionless voltage $\tilde{V}$, electric field $\tilde{E}$, and cathode layer length $\tilde{d}$ are presented in Figure 7.8 as a function of dimensionless current density, which is called the cathode layer current-voltage characteristic.

7.2.3 Normal Glow Discharge: Normal Cathode Potential Drop, Layer Thickness, and Current Density

According to the current–voltage characteristic (Equation 7.20), any current densities are possible in a glow discharge. However, in reality, this discharge "prefers" to operate only at one value of current density, the normal one, j_n, which corresponds to the minimum of the cathode potential drop and can be calculated using Equation 7.19. This interesting and important effect in glow discharge physics will be discussed in the next section.

The total glow discharge current I is usually controlled by the external resistance and the load line (Equation 7.1). The current-conducting discharge channel occupies a spot with area $A = I/j_n$ on the cathode surface, which provides the required normal

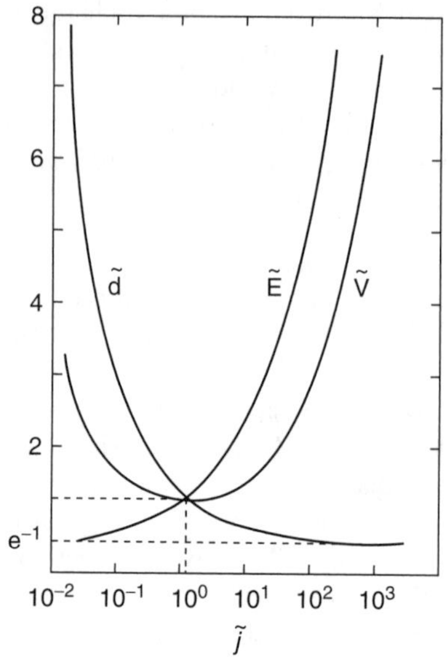

Figure 7.8 Dimensionless parameters of cathode layer

current density. This is called the cathode spot. Other current densities are unstable when the cathode surface area is sufficiently large. If, because of some perturbation $j > j_n$, the cathode spot grows until current density becomes normal; if $j < j_n$, the cathode spot decreases to reach the same normal value of current density.

Obviously, the normal current density can be reached if the cathode surface area is sufficiently large, which means $A = I/j_n$. This permits accommodating the necessary current spot on the cathode. The glow discharge with normal current density on its cathode is usually referred to as the normal glow discharge. These discharges have fixed values of current densities j_n, and corresponding fixed values of the cathode layer thickness $(pd)_n$ and voltage V_n, which only depend at room temperature on gas composition and cathode material.

Typical numerical values of these normal glow discharge parameters are presented in Table 7.2. From this table, a typical value of the normal cathode current density is about 100 μA/cm^2 at a gas pressure of about 1 torr; typical thickness of the normal cathode layer at this pressure is about 0.5 cm. A typical value of the normal cathode potential drop is about 200 V and does not depend on pressure and temperature.

7.2.4 Mechanism Sustaining Normal Cathode Current Density

When the normal glow discharge total current is changing, the current density always remains the same (see Table 7.2). This impressive effect has been investigated by numerous scientists[231,232] and explained in different ways: analyzing current stability; modeling ionization kinetics; and applying the minimum power principle.

Table 7.2 Normal Current Density,[a] Normal Cathode Layer Thickness,[b] and Normal Cathode Potential Drop[c] for Different Gases and Cathode Materials at Room Temperature

Gas	Cathode material	Normal current density	Normal thickness of cathode layer	Normal cathode potential drop
Air	Al	330	0.25	229
Air	Cu	240	0.23	370
Air	Fe	—	0.52	269
Air	Au	570	—	285
Ar	Fe	160	0.33	165
Ar	Mg	20	—	119
Ar	Pt	150	—	131
Ar	Al	—	0.29	100
He	Fe	2.2	1.30	150
He	Mg	3	1.45	125
He	Pt	5	—	165
He	Al	—	1.32	140
Ne	Fe	6	0.72	150
Ne	Mg	5	—	94
Ne	Pt	18	—	152
Ne	Al	—	0.64	120
H_2	Al	90	0.72	170
H_2	Cu	64	0.80	214
H_2	Fe	72	0.90	250
H_2	Pt	90	1.00	276
H_2	C	—	0.90	240
H_2	Ni	—	0.90	211
H_2	Pb	—	0.84	223
H_2	Zn	—	0.80	184
Hg	Al	4	0.33	245
Hg	Cu	15	0.60	447
Hg	Fe	8	0.34	298
N_2	Pt	380	—	216
N_2	Fe	400	0.42	215
N_2	Mg	—	0.35	188
N_2	Al	—	0.31	180
O_2	Pt	550	—	364
O_2	Al	—	0.24	311
O_2	Fe	—	0.31	290
O_2	Mg	—	0.25	310

[a] j_n/p^2, μA/cm²torr²

[b] $(pd)_n$ cm·torr

[c] V_n, V

Initially von Engel and Steenbeck's explanation of this effect was related to the instability of a cathode layer with $j < j_n$, where the current–voltage characteristic is "falling" (see Figure 7.8). The falling current–voltage characteristics are generally unstable in nonthermal discharges. For example, if a fluctuation results in a local increase of current density in some area of a cathode spot, the necessary voltage to sustain ionization in this area decreases. The actual voltage in this area then exceeds the required one, which leads to further increase of current until it becomes normal, $j = j_n$. If a fluctuation results in local decrease of current density, the discharge extinguishes for the same reason in this local area of a cathode spot. Such a mechanism obviously is able to explain only the instability at $j < j_n$ and the growth of current density to reach the normal one.

A more detailed model able to describe establishing the normal cathode current density $j = j_n$ starting from lower $j < j_n$ and higher $j > j_n$ current densities was developed by Vedenov[229] and Raizer and Surzhikov.[234,235] This model describes the phenomenon in terms of the charge reproduction coefficient:

$$\mu = \gamma\left\{\exp\int_0^{d(r)}\alpha[E(l)]dl - 1\right\} \tag{7.21}$$

Integration is along an electric current line l, which crosses the cathode in some point r. The charge reproduction coefficient (Equation 7.21) shows multiplication of charge particles during the only cathode layer ionization cycle. This cycle includes multiplication of the primary electron formed on a cathode in an avalanche (Section 4.4.5) moving along a current line across the cathode layer and return of positive ions formed in the layer back to cathode to produce new electrons due to secondary electron emission with the coefficient γ. To sustain the steady-state cathode layer in accordance with condition Equation 7.12 requires $\mu = 1$. Excessive ionization then corresponds to $\mu > 1$; $\mu < 1$ means extinguishing the discharge.

The curve $\mu = 1$ on the "voltage–cathode layer thickness" diagram represents the cathode layer potential drop V_c, with the near-minimum point $(pd)_n$ corresponding to the normal voltage V_n (see Figure 7.9, where the current density axis is also shown). All the area on the V–pd diagram above the cathode potential drop curve corresponds to $\mu > 1$; the area under this curve means $\mu < 1$.

If current density is less than normal $j < j_n$ ($d > d_n$, for example, point "1" in Figure 7.9), fluctuations can destroy the cathode layer by the instability mechanism described previously. At the edges, however, the cathode layer decays in this case even without any fluctuations. Positive space charge is much less at the edges and the same potential corresponds to points located further from the cathode, which can be illustrated in Figure 7.9 as moving to the right from point "1" in the area where $\mu < 1$. As a result, current disappears from the edges and the discharge voltage increases in accordance with the load line (Equation 7.1). The increase of voltage over the line $\mu = 1$ in a central part of the cathode layer leads to growth of current density until it reaches the normal value j_n.

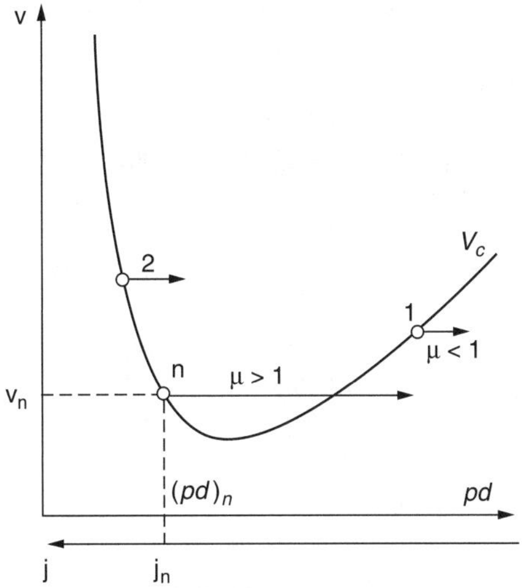

Figure 7.9 Explanation of normal current density effect

Similar consideration can be applied for a cathode layer with supernormal current density $j > j_n$, which corresponds to point "2" in Figure 7.9. In this case, the central part of the channel is stable with respect to fluctuations (compare with the Engel–Steenbeck stability analysis for $j < j_n$). At the edges of the channel, the space charge is smaller and the effective *pd* is greater for the fixed voltage (see Figure 7.9), which leads to the area of the *V–pd* diagram with the charge reproduction coefficient $\mu > 1$.

The condition $\mu > 1$ leads to breakdown at the edges of the cathode spot to an increase of total current; to a decrease of total voltage across the electrodes (in accordance with the load line Equation 7.1); and finally to a decrease of current density in the major part of a cathode layer until it reaches the normal value j_n. Thus, any deviations of current density from $j = j_n$ stimulate ionization processes at the edges of the cathode spot, which bring current density to the normal value (obviously, the cathode surface area should be large enough).

7.2.5 Steenbeck Minimum Power Principle: Application to Effect of Normal Cathode Current Density

The effect of normal current density can be illustrated using the minimum power principle that, in general, plays a special role in gas discharge physics. The total power released in the cathode layer can be found as:

$$P_c(j) = A\int_0^d jE\,dx = AjV_C(j) = IV_c(j) \tag{7.22}$$

In this relation, A is area of a cathode spot and I is the total current. The total current in glow discharge is mostly determined by external resistance (see the load line Equation 7.1) and can be considered fixed. In this case, the cathode spot area can be varied together with the current density, keeping fixed their product $j \times A = const$. According to the minimum power principle, the current density in glow discharge should minimize the power $P_c(j)$. As seen from Equation 7.22, the minimization of the power $P(j)$ at constant current $I = const$ requires minimization of $V_c(j)$, which, according to the definition, corresponds to the normal current density $j = j_n$.

The minimum power principle is very useful for illustrating different discharge phenomena, including striation and channels of thermal arcs, as well as the normal current in glow discharge. However, the minimum power principle cannot be derived from fundamental physical laws and thus should be used for illustration rather than for strict theoretical analysis.

7.2.6 Glow Discharge Regimes Different from Normal: Abnormal, Subnormal, and Obstructed Discharges

Increase of current in a normal glow discharge is provided by growth of the cathode spot area at $j = j_n = const$. As soon as the entire cathode is covered by the discharge, further current growth results in an increase of current density over the normal value. This discharge is called the **abnormal glow discharge**.

Abnormal glow discharge corresponds to the right-hand-side branches ($j > j_n$) of the dependences presented in Figure 7.8. The current–voltage characteristic of the abnormal discharge $\tilde{V}(\tilde{j})$ is growing. It corresponds to the interval EF on the general current voltage characteristic shown in Figure 7.5. According to Equation 7.18, when the current density grows further ($j \to \infty$), the cathode layer thickness decreases asymptotically to a finite value $\tilde{d} = 1/e \approx 0.37$, while the cathode potential drops and electric field grows as:

$$\tilde{V} = \frac{1}{e^{3/2}}\sqrt{\tilde{j}}, \quad \tilde{E} \approx \frac{1}{e^{1/2}}\sqrt{\tilde{j}} \tag{7.23}$$

Actual growth of the current and cathode voltage are limited by cathode overheating. Significant cathode heating at voltages of about 10 kV and current densities of 10 to 100 A/cm^2 result in transition of the abnormal glow discharge into an arc discharge.

Normal glow discharge transition to a dark discharge takes place at low currents (about 10^{-5} A) and starts with the so-called **subnormal discharge**. This discharge corresponds to the interval CD on the general current–voltage characteristic shown in Figure 7.5. The size of the cathode spot at low currents becomes large and comparable with the total cathode layer thickness. This results in electron losses with respect to normal glow discharge and thus requires higher values of voltage to sustain the discharge (Figure 7.5).

Another glow discharge regime different from the normal one takes place at low pressures and narrow gaps between electrodes, when their product pd_0 is less than normal value $(pd)_n$ for a cathode layer. This discharge mode is called the **obstructed glow discharge**. Conditions in the obstructed discharge correspond to the left-hand branch of the Paschen curve (Figure 4.19), where voltage exceeds the minimum value V_n. Because short interelectrode distance in the obstructed discharge is not sufficient for effective multiplication of electrons, to sustain the mode, the interelectrode voltage should be greater than the normal one.

7.2.7 Negative Glow Region of Cathode Layer: Hollow Cathode Discharge

Negative glow and the Faraday dark space complete the cathode layer and provide transition to the positive column. As was mentioned in Section 7.3, the negative glow region is a zone of intensive ionization and radiation (see Figure 7.4). Obviously, most electrons in the negative glow have moderate energies. However, quite a few electrons in this area are very energetic even though the electric field is relatively low. These energetic electrons are formed in the vicinity of the cathode and cross the cathode layer with only a few inelastic collisions. They provide a nonlocal ionization effect and lead to electron densities in a negative glow even exceeding those in a positive column (Figure 7.4). Details on the nonlocal ionization effect, formation and propagation of energetic electrons across the cathode into the negative glow, and the Faraday dark space can be found in References 236 through 238 in particular.

The effect of intensive "nonlocal" ionization in a negative glow can be applied to form an effective electron source, the so-called **hollow cathode discharge**. Imagine a glow discharge with a cathode arranged as two parallel plates with the anode on the side. If the distance between the cathodes gradually decreases, at some point the current grows 100 to 1000 times without a change of voltage. This effect takes place when two negative glow regions overlap, accumulating energetic electrons from both cathodes. Strong photoemission from cathodes in this geometry also contributes to intensification of ionization in the hollow cathode. More details regarding the physics of the hollow cathode discharge can be found in Moskalev.[239]

Effective accumulation of the high negative glow current can be reached if the cathode is arranged as a hollow cylinder and an anode lies further along the axis. Pressure is chosen in such a way that the cathode layer thickness is comparable with internal diameter of the hollow cylinder. The most traditional configuration of the system is the **Lidsky hollow cathode**, which is shown in Figure 7.10. The Lidsky

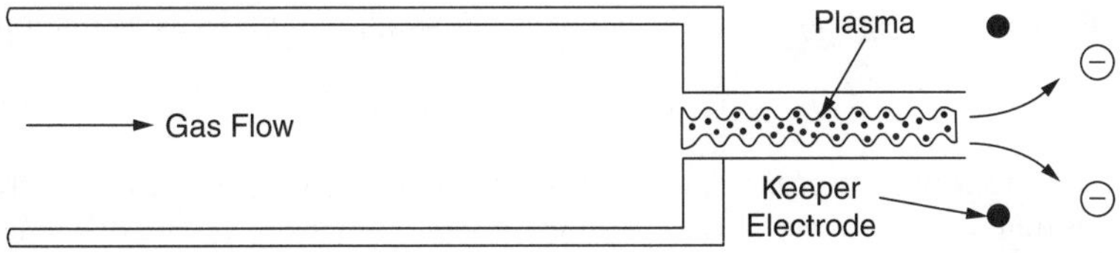

Figure 7.10 Schematic of Lidsky capillary hollow cathode

hollow cathode is a narrow capillary-like nozzle that operates with axially flowing gas. The hollow cathode is usually operated with the anode located about 1 cm downstream from the capillary nozzle; it can provide high electron currents with densities exceeding those corresponding to Child's law (see Section 6.1.5, Equation 6.14).

The Lidsky hollow cathode is hard to initiate and maintain in the steady state. For this reason, different modifications of the hollow cathode with external heating were developed to raise the cathode to incandescence and provide long-time steady operation.[240,241]

7.2.8 Anode Layer

Because positive ions are not emitted (but repelled) by the anode, their concentration at the surface of this electrode is equal to zero. Thus a negatively charged zone, called the anode layer, exists between the anode and positive column (Figure 7.4). The ionic current density in the anode layer grows from zero at the anode to the value $j_{+c} = \mu_+/\mu_e\, j$ in the positive column (here, j is the total current density and μ_+, μ_e are mobilities of ions and electrons). It can be described in terms of the Townsend coefficient α as:

$$\frac{dj_+}{dx} = \alpha j_e \approx \alpha j, \quad j_{+c} \approx j \times \int \alpha dx \tag{7.24}$$

From Equation 7.24, it is sufficient for one electron to provide only a very small number of ionization acts to establish the necessary ionic current:

$$\int \alpha dx = \frac{\mu_+}{\mu_e} << 1 \tag{7.25}$$

Number of generations of electrons produced in the anode layer (Equation 7.13) is about three orders of magnitude smaller than the corresponding number of generations of electrons produced in the cathode layer. For this reason, the anode layer's potential drop is less than the potential drop across the cathode layer (see Figure 7.4). Numerically, the value of the anode potential drop is approximately the value of the ionization potential of the gas in the discharge system. However, the anode voltage grows slightly with pressure in the range of moderate pressures around 100 torr (the dependence is a little bit stronger in electronegative gases; see Figure 7.11).

Typical values of the reduced electric field E/p in the anode layer are about $E/p \approx 200 - 600$ V/cm Torr. Thickness of the anode layer is of the order of the electron mean free path and can be estimated in simple numerical calculations as:

$$d_A(\text{cm}) \approx 0.05\ \text{cm}/p(\text{Torr}) \tag{7.26}$$

The current density j/p^2 at the anode is independent of the value of current in the same manner as for the normal cathode layer. Numerical values of the current density in anode layer are about 100 μA/cm^2 at gas pressures of about 1 torr and

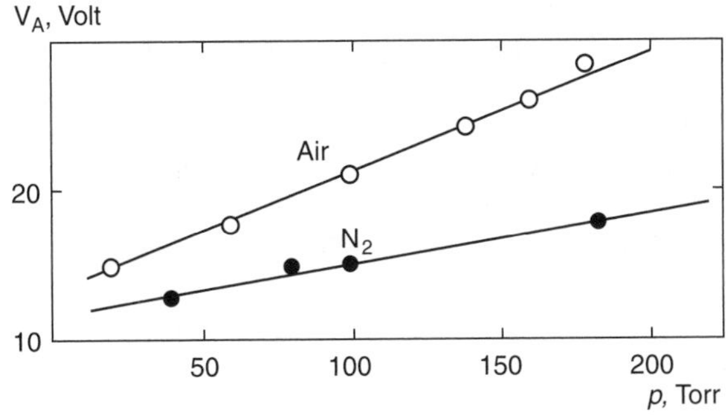

Figure 7.11 Anode fall for discharges in nitrogen and air as a function of pressure[101]

actually coincide with those of the cathode layer (see Table 7.2.). Actually, the normal cathode layer imposes the value of current density j on the entire discharge column including the anode.

Glow discharge modeling including cathode and anode layers[234,235] shows that even radial distribution of the current density $j(r)$ originating at the cathode can be repeated at the anode. Although many similarities take place between current density in the cathode and anode layers, theoretically these phenomena are quite different.[242] For example, in contrast to the cathode, the anode quite often is covered by an electric current not continuously, but in a group of spots not directly connected with each other. Also, in general, processes taking place in the anode layer can play a significant role in developing instabilities in glow discharges. Although this discussion of cathode and anode layers in glow discharges was brief, more details on the subject can be found.[101,243–245]

7.3 POSITIVE COLUMN OF GLOW DISCHARGE

7.3.1 General Features of Positive Column: Balance of Charged Particles

The positive column of glow discharge can be quite long and homogeneous, and the major part of discharge power released there. This explains the special role of the positive column in plasma chemical, plasma-engineering, and laser applications. However, the physical function of the positive column is simple: it closes the electric circuit between cathode layer and anode. In contrast to the cathode layer, a glow discharge can exist without the positive column at all.

The positive column of a normal glow discharge is probably the most conventional source of nonequilibrium nonthermal plasma. Plasma behavior and parameters in the long positive column are independent of the phenomena in cathode and anode layers. The state of plasma in the positive column is determined by local processes of charged particle formation and losses, and by the electric current, which is actually controlled by external resistance and EMF (see the load line Equation 7.1).

Losses of charged particles in the positive column due to volume and surface recombination should be balanced in a steady-state regime by ionization. The ionization rate quite sharply depends on electron temperature and reduced electric field. That is the reason the value of the electric field in the positive column is more or less fixed, which determines the longitudinal potential gradient and the voltage difference across a column of a given length.

Different modes of the charged particle balance in nonthermal discharges were considered in Section 4.5. Generation of electrons and ions in a positive column is always due to volumetric ionization processes, so classification of the steady-state regimes is determined by the dominant mechanisms of losses of charged particles. In Section 4.5, three different discharge modes were considered: those controlled by electron–ion recombination in volume (subsection 4.5.2); those controlled by electron attachment to molecules (subsection 4.5.3); and those controlled by charged particle diffusion to the walls (subsection 4.5.4).

Each mode imposes special requirements on electron temperature and, thus, on the electric field, to provide a balance for formation and losses of electrons and ions. These requirements can be interpreted as current–voltage characteristics of the discharge. Note that these three regimes can be observed in the positive column of glow discharges for different conditions.

7.3.2 General Current–Voltage Characteristics of a Positive Column and a Glow Discharge

If the balance of charged particles is controlled by their diffusion to the wall (Section 4.5.4), then the electric field E in the positive column is determined by the Engel–Steenbeck relation (Equation 4.173) and does not depend on electron density and electric current. This is due, in this case, to the fact that ionization rate and rate of diffusion losses are proportional to n_e; thus, electron density can be cancelled from the balance of electron generation and losses.

In this case, the current–voltage characteristic of a positive column and a normal glow discharge is almost a horizontal straight line (see Figure 7.5):

$$V(I) = V_n + E\,d_c \approx const \tag{7.27}$$

In this relation, d_c is the length of positive column; V_n is the normal potential drop in the cathode layer; and E is the electric field in the column (see Equation 4.173) that depends only on type of gas, pressure, and radius of discharge tube.

When the discharge current and thus the electron concentration are growing, the contribution of volumetric electron–ion recombination becomes significant. Such a discharge regime controlled by recombination was considered in Section 4.5.2. In this case, electron concentration $n_e(E/p)$ follows the formula (Equation 4.170), and then the relation between current density and reduced electric field E/p in the positive column can be expressed as:

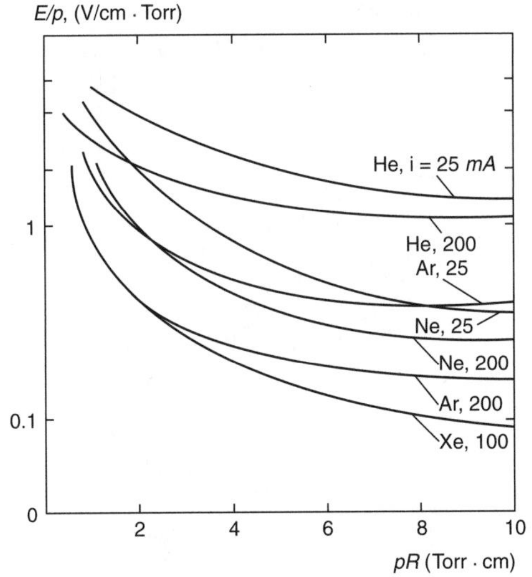

Figure 7.12 Measured values of E/p for positive column in inert gases

$$j = (\mu_e p)\frac{E}{p} e \frac{k_i(E/p)\, n_0}{(k_r^{ei} + \varsigma k_r^{ii})(1+\varsigma)} \tag{7.28}$$

Here, $\mu_e p$ is the electron mobility presented as a similarity parameter (see Table 4.1); $k_i(E/p)$ is the ionization coefficient; n_0 is the gas density; k_r^{ei}, k_r^{ii} are coefficients of electron–ion and ion–ion recombination; k_a, k_d are coefficients of electron attachment and detachment; and $\varsigma = k_a/k_d$ is the factor characterizing the balance between electron attachment and detachment.

Taking into account the sharp exponential behavior of the function $k_i(p)$, one can conclude from Equation 7.28 that electric field E grows very slowly with current density in the recombination regime. Thus, the current–voltage characteristic of positive column and of a whole normal glow discharge is almost a horizontal straight line in this case, as well as in the diffusion controlled discharge mode (see Equation 7.27 and Figure 7.5).

Some examples of the experimental dependences of reduced electric field E/p on the similarity parameter pR (p is pressure and R is radius of a discharge tube) are presented in Figure 7.12 and Figure 7.13.[101] Some of the E/p decrease with current is due to increases in the gas temperature (ionization actually depends on E/n_0; increase of temperature leads to reduction of n_0 and growth of E/n_0 at constant pressure and electric field).

Comparing Figure 7.12 and Figure 7.13, it is seen that reduced electric fields E/p are about 10 times lower in inert gases relative to molecular gases. This is obviously related to effects of inelastic collisions (mainly vibrational excitation), which significantly reduce the electron energy distribution function at the same values of reduced electric field (see Equation 4.83 and Equation 4.84). The figures

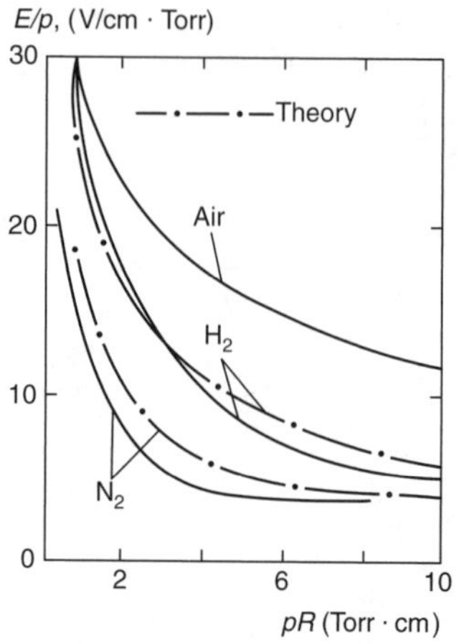

Figure 7.13 Measured values of E/p for positive column in molecular gases

also show that electric fields necessary to sustain glow discharge are essentially lower than those necessary for breakdown (see Section 4.4.2). In particular, this is due to the fact that electron losses to the walls during breakdown are provided by free diffusion, while the corresponding discharge is controlled by much slower ambipolar diffusion (see Section 4.3.9).

If electron detachment from negative ions is negligible, the discharge regime can be controlled by electron attachment. This discharge mode was considered in Section 4.5.3. Ionization rate and electron attachment rate are proportional to n_e; therefore, in this case the electron density actually can be cancelled from the balance of electron generation and losses (similar to the discharge regime controlled by diffusion). Obviously, the current–voltage characteristic in the attachment regime is again an almost horizontal straight line $V(I) = const$ (see Figure 7.5).

It is important to note that electron detachment and destruction of negative ions can be effective in steady-state discharges. For example, the reduced electric field in an air discharge controlled by electron attachment to oxygen should be equal to $E/p = 35$ V/cm·torr according to Equation 4.172. However, experimentally measured E/p values, presented in Figure 7.13, are lower: 12 to 30 V/cm·torr. This shows that the steady state discharge is able to accumulate significant amount of particles efficient for the destruction of negative ions. Another similar example, related to propagation of CO_2-discharge, was discussed in the Section 4.5.5.

7.3.3 Heat Balance and Plasma Parameters of Positive Column

The power that electrons receive from electric field and then transfer through collisions to atoms and molecules can be found as joule heating (Section 4.3.2; Equation 4.100): $jE = \sigma E^2$. If the electron energy consumption into chemical processes is neglected (Section 5.6.7), the entire joule heating should be balanced by conductive and convective energy transfer. At steady-state conditions, this is expressed as:

$$w = jE = n_0 c_p (T - T_0) \nu_T \tag{7.29}$$

where w is the discharge power per unit volume; c_p is the specific heat per one molecule; T is the average gas temperature in the discharge; T_0 is the room temperature; and ν_T is the heat removal frequency. Note that the heat removal frequency from the cylindrical discharge tube of radius R and length d_0 can be determined as:

$$\nu_T = \frac{8}{R^2} \frac{\lambda}{n_0 c_p} + \frac{2u}{d_0} \tag{7.30}$$

where λ is the coefficient of thermal conductivity and u is the gas flow velocity. The first term in Equation 7.30 is related to heat removal due to thermal conductivity; the second term describes the convective heat removal.

Based on the heat balance relations Equation 7.29 and Equation 7.30, estimate the typical plasma parameters of a positive column. If heat removal is controlled by thermal conduction, the combination of Equation 7.29 and Equation 7.30 gives the typical discharge power per unit volume, which doubles the gas temperature in the discharge ($T - T_0 = T_0$):

$$w = jE = \frac{8\lambda T_0}{R^2} \tag{7.31}$$

The thermal conductivity coefficient λ does not depend on pressure and can be estimated for most of the gases under consideration as $\lambda \approx 3 \cdot 10^{-4}$ W/cm · K. Thus, based on Equation 7.31, it can be concluded that specific discharge power also does not depend on pressure and, for tubes with radius $R = 1$ cm, it can be estimated as 0.7 W/cm^3. Higher values of specific power result in higher gas temperatures and thus in contraction of a glow discharge (see Section 6.3.1).

Typical values of current density in the positive column with the heat removal controlled by thermal conduction are inversely proportional to pressure and can be estimated based on Equation 7.31 as:

$$j = \frac{8\lambda T_0}{R^2} \frac{1}{(E/p)} \frac{1}{p} \tag{7.32}$$

Assuming the value of the reduced electric field as $E/p = 3 - 10$ V/cm·torr, Equation 7.32 gives numerically: j,mA/cm^2 $\approx$ 100/p,Torr.

Electron concentration in the positive column can be calculated from Equation 7.32, taking into account Ohm's law $j = \sigma E$ and Equation 4.98 for the electric conductivity σ. In this case, the electron concentration is easily shown to be inversely proportional to gas pressure:

$$n_e = \frac{w}{E^2} \frac{m\nu_{en}}{e^2} = \frac{w}{(E/p)^2} \frac{m k_{en}}{e^2 T_0} \frac{1}{p} \tag{7.33}$$

Here, ν_{en}, k_{en} are the frequency and rate coefficient of electron–neutral collisions. Numerically, for the previously assumed values of parameters, the electron concentration in a positive column with conductive heat removal can be estimated as n_e, $cm^{-3} = 3 \cdot 10^{11}/p$,Torr.

It was shown that the electron concentration and current densities fall linearly with pressure growth. The reduction of the plasma degree of ionization with pressure is even more significant: $n_e/n_0 \propto 1/p^2$. Thus, it can be concluded that low pressures are in general more favorable for sustaining the steady-state homogeneous nonthermal plasma.

7.3.4 Glow Discharge in Fast Gas Flows

The convective heat removal with the gas flow, (second term in Equation 7.30) promotes pressure and power increases in nonthermal discharges. Gas velocities in such systems are usually about 50 to 100 m/sec. Based on Equation 7.29 and Equation 7.30, the current density and electron concentration corresponding to doubling of the gas temperature do not depend in this case on gas pressure:

$$j = \frac{2uc_p}{d_0(E/p)}, \quad n_e = \frac{2uc_p m k_{en}}{e^2 d_0 T_0 (E/p)^2} \tag{7.34}$$

Numerically, assuming $d_0 = 10$ cm, $u = 50$ m/sec, $E/p = 10$ V/cm·torr, the typical values of current density and electron concentration in the positive column under consideration are $j \approx 40$ mA/cm^2, $n_e = 1.5 \cdot 10^{11}$ cm^{-3}.

Under the same conditions, the specific discharge power in the positive column with convective heat removal grows proportionally to pressure:

$$w = jE = c_p T_0 \frac{2u}{d_0} \cdot \frac{p}{T_o} \tag{7.35}$$

Numerically, with these values of parameters, the specific discharge power is calculated as w, W/cm^3 = $0.4 \cdot p$,Torr.

High gas flow velocity also results in voltage and reduced electric field growth (to values of about $E/p = 10 - 20$ V/cm·torr) to intensify ionization and compensate charge losses. Significant increases of charge losses in fast gas flows can be related to turbulence, which accelerates the effective charged particle diffusion to the discharge tube walls. The acceleration of charged particle losses can be estimated by replacing the ambipolar diffusion coefficient D_a by an effective one including a special turbulent term:

$$D_{eff} = D_a + 0.09Ru \tag{7.36}$$

where u is the gas flow velocity and R is the discharge tube radius (or half-distance between walls in plane geometry).

The elevated values of reduced electric field and electron temperature can be useful to improve the efficiency of several plasma chemical processes; to improve discharge stability; and to provide higher limits of the stable energy input. Therefore, to elevate the electric field, small-scale turbulence is often deliberately introduced, especially in high-power gas discharge lasers. Note that the increase of voltage and reduced electric field in fast flow discharges can be directly provided by convective charged particle losses, especially if the discharge length along the gas flow is small.[246] Also, the effect of E/p increasing in the fast flows can be explained by convective losses of active species responsible for electron detachment from negative ions.[247]

7.3.5 Heat Balance and Its Influence on Current–Voltage Characteristics of Positive Columns

The current–voltage characteristic of glow discharges controlled by diffusion (see Section 7.3.3) is slightly decreasing; a current increase leads to some reduction of voltage, an effect due to joule heating. The increase of current leads to some growth of gas temperature T_0, which at constant pressure results in decrease of gas density n_0. The ionization rate is actually a function of E/n, which is often only "expressed" as E/p, assuming room temperature (see Chapter 4 and Section 6.3.2). For this reason, the decrease of gas density at a fixed ionization rate leads to a decrease of electric field and voltage, which finally explains the decreasing current–voltage characteristics.

Analytically, this slightly decreasing current–voltage characteristic can be described based on Equation 7.29 and the similarity condition $E/n_0 \propto ET_0 \approx const$, which is valid at the approximately fixed ionization rate. In this case, the relation between current density and electric field can be expressed as:

$$\frac{j}{j_0} = \left(\frac{E_0}{E}\right)^{3/2}\left(\frac{E_0}{E} - 1\right) \tag{7.37}$$

Here, E_0 is the electric field, which is necessary to sustain low discharge current $j \to 0$ when gas heating is negligible; j_0 is the typical value of current density, which corresponds to Equation 7.32 and can be also expressed as (w_0 is a typical value of specific power; see Equation 7.14):

$$j_0 = n_0 c_p T_0 \frac{\nu_T}{E_0} = \frac{w_0}{E_0} \tag{7.38}$$

Note that this slightly decreasing current–voltage characteristic can be crossed by the load line (Equation 7.1) not in one, but in two points, a situation illustrated in Figure 7.14. In this case, only one state, namely, the lower one, is stable; the upper one is unstable. If current grows slightly in some perturbation δI with respect to the upper crossing point (Figure 7.14), the actual plasma voltage corresponding to the load line exceeds the voltage necessary to sustain the steady-state discharge. As a result of the overvoltage, the degree of plasma ionization and total current will grow further until the second (lower) crossing point is reached. If the current fluctuation from the upper crossing point is negative, $\delta I < 0$, the actual plasma voltage becomes less than required to sustain steady-state ionization and the discharge extinguishes.

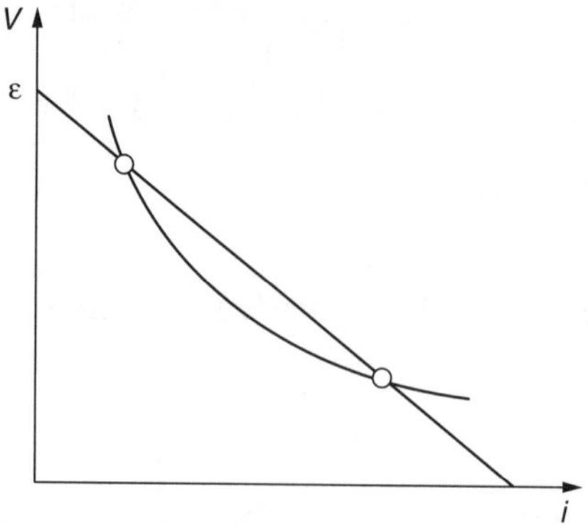

Figure 7.14 Decreasing current–voltage characteristic

Similar reasoning proves the stability of the lower crossing point (Figure 7.14). For example, current fluctuation $\delta I < 0$ results in plasma overvoltage, acceleration of ionization, and restoration of the steady-state conditions.

7.4 GLOW DISCHARGE INSTABILITIES

A general description of the main nonthermal plasma instabilities, striation and contraction, was considered in Section 6.3 together with the principal mechanisms of their inception and development. In this section, specific conditions of the instabilities in a glow discharge and some approaches to suppressing them will be discussed. Contraction is the transverse instability of a glow discharge; perturbations grow across a positive column. Striations in contrast are longitudinal. Contraction, a more serious destructive instability in nonequilibrium discharges, will be considered first.

7.4.1 Contraction of Positive Column

As was explained in Section 6.3.1, contraction is a nonthermal plasma instability related to instantaneous self-compression of a discharge column into one or several bright current filaments. This instability occurs when an attempt is made to increase pressure and current while maintaining limits on the specific energy input and specific power of a nonthermal discharge. Contraction decreases the level of nonequilibrium and, for this reason, it is very undesirable in nonthermal plasma and laser applications.

Typical parameters and peculiarities of the positive column contraction can be illustrated by the example of a glow discharge in neon at pressures of 75 to 100 torr sustained in a 2.8-cm radius tube with walls at room temperature.[248] The current–voltage characteristic of this discharge in Figure 7.15 demonstrates contraction that takes place when current exceeds a critical value of about 100 mA. Notice that the

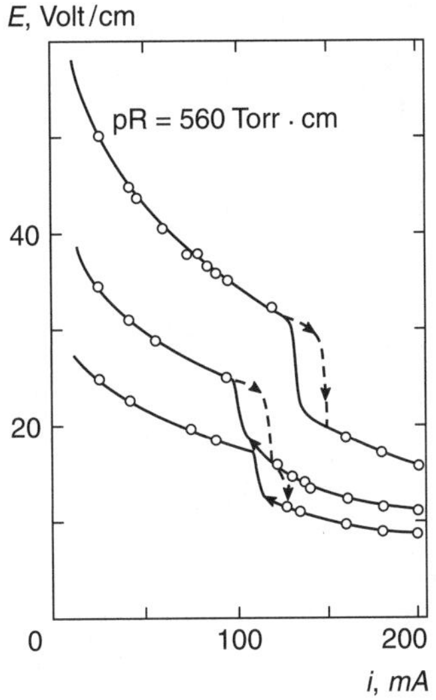

Figure 7.15 V-i characteristic of discharge in a tube containing neon in the region of transition from diffuse to contracted form. Tube radius $R = 2.8$ cm; (1) $pR = 210$ torr • cm; (2) $pR = 316$ torr • cm; (3) $pR = 560$ torr • cm. The solid curve in the region of jump was recovered as current was decreased, and the dashed curve as it was raised.[248]

current–voltage characteristic is decreasing at current values close to the critical one, which demonstrates the strong effect of joule heating (see Section 7.3.5). It is also interesting that the transition between the diffusive and contracted modes demonstrates hysteresis.

At the critical current value, the electric field in the positive column abruptly decreases with a related sharp transition of the discharge regime from the initial strongly nonequilibrium diffusive mode into the contracted mode. A brightly luminous filament appears along the axis of the discharge tube while the rest of the discharge becomes almost dark. Average current density at the transition point is 5.3 mA/cm^2; the corresponding current density on the axis of the discharge tube is 12 mA/cm^2 with an electron density on the axis of approximately 10^{11} cm^{-3}.

Radial distributions of the relative electron density before and after contraction in the same glow discharge system at a fixed pressure $p = 113$ Torr are shown in Figure 7.16. Corresponding changes of discharge parameters (electron and gas temperatures, electron concentration) during the contraction transition on the axis of the discharge tube are presented in Table 7.3.

As seen from Figure 7.16, the diameter of a filament formed as a result of contraction is almost two orders of magnitude smaller than the diameter of a tube initially completely filled with the diffusive glow discharge. Electron concentration increases about 50 times. Gas temperature increases due to localized heat release, and the electron temperature decreases because of reduction of the electric field (Figure 7.15).

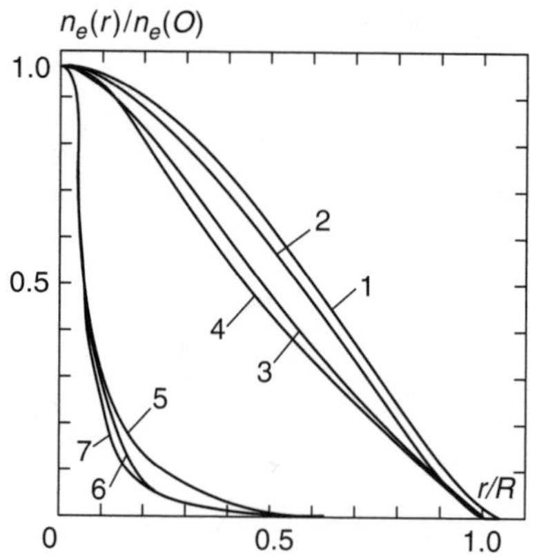

Figure 7.16 Profiles of n_e measured under conditions of Figure 7.16: (1) i/R = 4.8 mA/cm; (2) 15.4; (3) 26.8; (4) 37.5; (5) 42.9; (6) 57.2; (7) 71.5 mA/cm. The transition site occurred in the region between (4) and (5).[248]

Table 7.3 Change of Electron Temperature, Gas Temperatures, and Electron Density during Glow Discharge Contraction

Plasma parameters	Before contraction (I = 96 mA)	After contraction (I = 120 mA)
Electron temperature	3.7 eV	3.0 eV
Gas temperature	930 K	1200 K
Electron density	$1.2 \cdot 10^{11}$ cm^{-3}	$5.5 \cdot 10^{12}$ cm^{-3}

Thus, after contraction, a glow discharge is no longer strongly nonequilibrium (see Table 3.13). This is the reason the contraction phenomenon is sometimes referred to as arcing. However, this term is not completely suitable for contraction because the contracted glow discharge filaments are still not in quasi equilibrium. The filament plasma is not a thermal one in contrast to the thermal arc plasma, which will be considered in the next chapter.

7.4.2 Glow Discharge Conditions Resulting in Contraction

Instability mechanisms leading to glow discharge contraction were considered in Section 6.3. Specifically, the principal causes of the glow discharge contraction are thermal instability (subsection 6.3.2 and subsection 6.3.3); stepwise ionization instability (subsection 6.3.8b); and electron Maxwellization instability (subsection 6.3.8c). All these mechanisms provide nonlinear growth of ionization with electron density, which is the main physical cause of contraction.

All these instability mechanisms become significant at electron concentrations exceeding critical values of about 10^{11} cm^{-3} and corresponding specific powers of about 1 W/cm^3. These values correspond to the contraction conditions in glow discharges with heat removal controlled by diffusion. It should be noted that the principal mechanism of electron losses at the relatively high electron densities mentioned earlier is a volumetric one (for example, electron–ion recombination; see Section 4.5.1), which is also the necessary requirement for contraction.

Glow discharges in fast gas flows have in principle the same mechanisms of contraction. However, the heat balance and thus the gas temperature and overheating of a glow discharge in fast flow are controlled by convection, which determines the gas residence time in the discharge zone (see Section 7.3.4). For this reason, transition to the contracted mode takes place in the fast flow glow discharges, not when the specific power (discharge power per unit volume, W/cm^3), but rather the specific energy input (discharge energy released per one molecule, eV/mol or J/cm^3) and thus, discharge overheating exceeds the critical value (see Section 6.3.6, Figure 6.12).

Maximizing the specific energy input in homogeneous, nonthermal discharges is an especially important engineering problem for powerful gas lasers. Numerically, contraction of glow discharges usually takes place in the case of fast flows, when the specific energy input corresponds to a gas temperature growth of about 100 to 300 K.[249,250]

7.4.3 Comparison of Transverse and Longitudinal Instabilities: Observation of Striations in Glow Discharges

Contraction of glow discharge is the **transverse instability**. This means that plasma parameters were changed across the direction of electric field. Taking into account that the tangential component of the electric field is always continuous, a sharp decrease of the electric field in the central filament occurs as a consequence of contraction and results in an overall voltage decrease (kind of a short circuit) and loss of nonequilibrium in the discharge as a whole (see Figure 7.17). That is why the transverse instability, contraction, is so harmful for strongly nonequilibrium glow discharges.

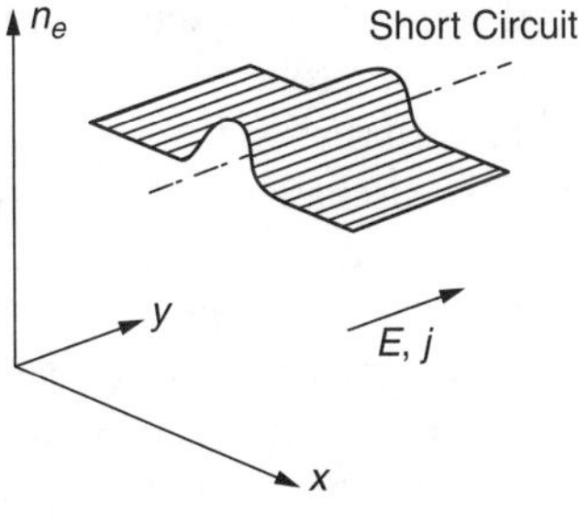

Figure 7.17 Electron density perturbation in transverse instabilities

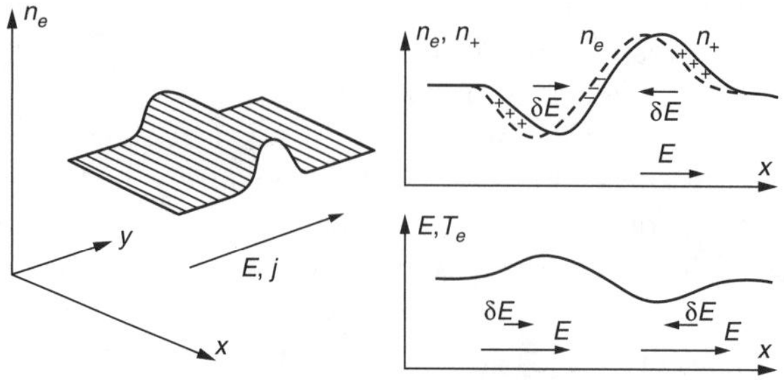

Figure 7.18 Longitudinal instability

Striations are related to longitudinal perturbations of plasma parameters, which means that changes of plasma parameters occur along a positive column. In this case, the electric current (and the current density in the one-dimensional approximation) remains fixed during a local perturbation of electron density δn_e and temperature δT_e. Local growth of electron density n_e (and electric conductivity σ) induces a local decrease of electric field E and vice versa; while the current density remains the same:

$$j = \sigma E \propto n_e E = const, \quad \frac{\delta n_e}{n_e} = -\frac{\delta E}{E} \tag{7.39}$$

The mechanism of the electric field reduction in perturbations with elevated electron density and vice versa is illustrated in Figure 7.18. Shift of electron density with respect to ion density is due to the electron drift in the electric field. As can be seen, the direction of the polarization field δE is opposite to the direction of the external electric field E if the fluctuation of electron concentration is positive. For some reason, if the electron density is reduced, the polarization field is added to the external one. Such an instability is actually unable to destroy the nonequilibrium discharge as a whole.

The definition and some general features of striations were briefly discussed in Section 6.3.1, where an illustrative picture of the phenomenon was presented in Figure 6.8. As was mentioned, striations can move fast (up to 100 m/sec from anode to cathode at pressures of 0.1 to 10 torr in inert gases) or remain at rest. The stationary striations are usually formed if some strong fixed perturbation is present in or near the positive column (e.g., an electric probe or even the cathode layer). The fixed striations build up away from the perturbation toward the anode and gradually vanish.

Plasma parameters of a glow discharge with and without striations are nearly the same. It is important to point out that this type of instability, in contrast to the previously discussed contraction, does not significantly affect the nonthermal discharge. For this reason, striations as plasma instabilities are not as detrimental for plasma chemical applications. Striations exist for a limited range of current, pressure, and radius of a discharge tube. The electric current, gas pressure, and tube radius also determine the amplitude of luminosity oscillations, the striations' wavelengths, and their propagation velocity. The striations normally behave as linear waves of low amplitude, but large amplitude nonlinear striations are possible. Additional details about observation of this phenomenon are available.[251–253]

7.4.4 Analysis of Longitudinal Perturbations Resulting in Formation of Striations

From a physical point of view, striations can be considered ionization oscillations and waves; they can be initiated by the stepwise ionization instability. In this case, an increase of electron density leads to a growth in the concentration of excited species, which accelerates stepwise ionization and results in a further increase of the electron density. When the electron concentration becomes too large, superelastic collisions deactivate the excited species. Further nonlinear growth of the ionization rate is usually due to the Maxwellization instability (see subsection 6.3.8c).

Both of these ionization instability mechanisms involved in striations—stepwise ionization and Maxwellization—are not directly related to gas overheating. Remember that overheating requires relatively high values of specific power and specific energy input. As a result, striations can be observed at less intensive plasma parameters (electron concentration, electric current, specific power) than those related to thermal (ionization overheating) instability, which finally is responsible for contraction (see Section 7.4.2).

A change of the electric field is a strong stabilizing factor for striations, which is an important peculiarity of this kind of instability. Autoacceleration of ionization and nonlinear growth of electron density in striations induces reduction of the electric field in accordance with Equation 7.39. The reduction of the electric field leads to a reduction of the effective electron temperature after a short delay, $\tau_f = 1/(\nu_{en}\delta)$, related to establishing the corresponding electron energy distribution function (here, ν_{en} is frequency of electron–neutral collisions and δ is the average fraction of electron energy transferred to a neutral particle during the collision; see Equation 4.87). The reduction of electron temperature results in an exponential decrease of ionization and finally in stabilization of the instability.

This stabilization effect actually suppresses striations if the characteristic length of the longitudinal perturbations $2\pi/k_s$ is sufficiently large (or, alternately, the wave numbers of ionization wave k_s are small enough). Thus, the striations cannot be observed if there is sufficient time for electrons to establish the new, corrected energy distribution function (corresponding to the new electric field) during the electron drift along the perturbation:

$$k_s \cdot v_d \tau_f \approx k_s \lambda / \sqrt{\delta} << 1 \tag{7.40}$$

Here, λ is the electron mean free path. In this inequality, it was taken into account that $v_d \tau_f \approx \lambda/\sqrt{\delta}$, which can be derived from the definition of τ_f given earlier (Equation 4.101 for drift velocity v_d and Equation 4.86) between the electric field and electron temperature.

In the opposite case, when perturbation wavelengths are short, electron temperature and related ionization effects do not have sufficient time to follow the changing electric field and stabilize striations. Thus, striations can be effectively generated at relatively low wavelength. However, the striations' wavelengths cannot be very small (shorter than the discharge tube radius R) because of stabilization due to electron losses in longitudinal ambipolar diffusion. The most favorable conditions for generation of striations can be expressed in terms of the striations' wave numbers as:[254]

$$k_s\lambda/\sqrt{\delta} \approx 5 \div 10, \quad k_s R \propto 1 \tag{7.41}$$

Eliminating the perturbation wavelength k_s from the Equation 7.41, rewrite the conditions favorable for generation of striations as a relation between the electron mean free path and the discharge tube radius as:

$$\frac{\lambda}{R} \approx (5 \div 10) \cdot \sqrt{\delta} \tag{7.42}$$

The factor δ is very small (Equation 4.87) in inert gases for which the conditions (Equation 7.41) were actually proposed. In this case, the condition (Equation 7.42) can be satisfied at pressures about 0.1 to 1 torr, which actually corresponds to typical pressures necessary for generation of striations. In molecular gases, the factor δ is much larger (Equation 4.87), and this is the reason it is so difficult to observe striations in molecular gases.

7.4.5 Propagation Velocity and Oscillation Frequency of Striations

Striations are usually moving in a direction from the anode to the cathode. Physical interpretation of this motion is illustrated in Figure 7.19. In the case of relatively short wavelengths typical for actual striations, the gradients of electron density in a perturbation δn_e are quite significant, and charge separation is mostly due to electron diffusion. The electric field of polarization, δE, occurring as a result of this electron diffusion actually determines the oscillation of the total electric field. Maximum of the electric field oscillations δE_{max} corresponds to the points on the wave where the electron concentration is not perturbed, $\delta n_e = 0$. As a result, the maximum of the electric field oscillations, δE_{max}, is shifted with respect to the maximum of plasma density δn_e oscillations by one quarter of a wavelength toward the cathode (see Figure 7.19). The ionization rate is fastest at the point of maximum electric field (δE_{max}), resulting in moving the point of maximum plasma density δn_e toward the cathode. Thus, the striations are propagating in the direction from anode to cathode as are the ionization waves.

To determine the striations' velocity as a velocity of the ionization wave, assume perturbations of electric field, electron concentration, and temperature change in a harmonic way: $\delta E, \delta n_e, \delta T_e \propto \exp[i(\omega t - k_s x]$. Then, based on Equation 4.98, the relation between perturbations of electric field and electron density is:

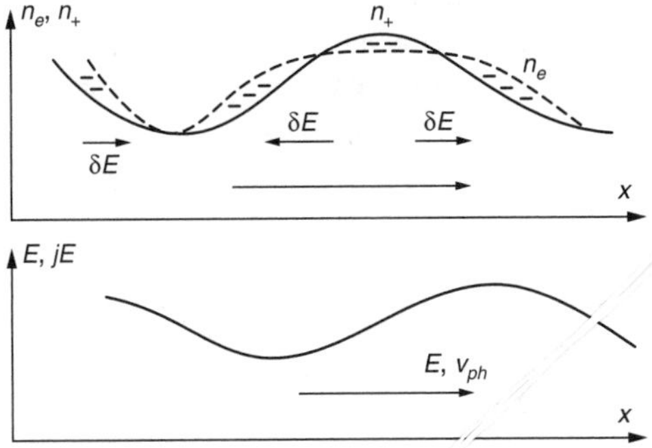

Figure 7.19 Propagation of striations

$$\delta E = ik_s \frac{T_e}{e} \frac{\delta n_e}{n_e} \tag{7.43}$$

Electrons receive a portion of energy about T_e from the electric field during their drift over the length needed to establish the electron energy distribution function ($v_d \tau_f \approx \lambda/\sqrt{\delta}$, where λ is the electron mean free path and δ is the fraction of electron energy transferred during a collision). Thus, taking into account that $eE(\lambda/\sqrt{\delta}) \approx T_e$, Equation 7.43 can be rewritten as:

$$\frac{\delta E}{E} \approx ik_s \frac{\lambda}{\sqrt{\delta}} \frac{\delta n_e}{n_e} \tag{7.44}$$

Relationships between relative perturbations of electric field and electron temperature can be derived in a similar manner from a balance of additional joule heating and electron thermal conductivity (with coefficient λ_e): $j \cdot \delta E = k_s^2 \lambda_e \, \delta T_e$:

$$\frac{\delta T_e}{T_e} \approx \frac{1}{\left(k_s \lambda / \sqrt{\delta}\right)^2} \frac{\delta E}{E} \tag{7.45}$$

Here, the Einstein relation, $\lambda_e/\mu_e = T_e/e$; the Ohm's law $j = en_e\mu_e E$; and the relation $eE(\lambda/\sqrt{\delta}) \approx T_e$ were taken into account.

Acceleration of the ionization rate $\partial n_e/\partial t$ in striations related to the electron temperature increase δT_e can be expressed as:

$$\delta\left(\frac{\partial n_e}{\partial t}\right) \approx n_e n_0 \frac{\partial k_i}{\partial T_e} \delta T_e = k_i n_e n_0 \frac{\partial \ln k_i}{\partial \ln T_e} \frac{\delta T_e}{T_e} \tag{7.46}$$

where n_e, n_0 are concentrations of electrons and neutral species in the discharge; k_i is the ionization rate coefficient; and $\partial \ln k_i/\partial \ln T_e \approx I/T_e >> 1$ is the logarithmic sensitivity of the ionization rate coefficient to the electron temperature.

As was illustrated in Figure 7.19, the electron density grows by the value of the amplitude perturbation δn_e during a quarter of a period (about $1/k_s v_{ph}$, where v_{ph} is the phase velocity of ionization wave and k_s is the corresponding wave number). This means that $\delta(\partial n_e/\partial t) \times 1/k_s v_{ph} \approx \delta n_e$. By combining this relation with Equation 7.46, a formula for the phase velocity of the ionization wave is derived as:

$$v_{ph} \approx \frac{k_i n_0}{k_s} \frac{\partial \ln k_i}{\partial \ln T_e} \left(\frac{\delta T_e}{T_e} \Big/ \frac{\delta n_e}{n_e} \right) \tag{7.47}$$

Taking into account Equation 7.44 and Equation 7.45 for the relative perturbations of electron temperature and concentration, rewrite Equation 7.47 into the final expression for the phase velocity of the ionization wave, e.g., the phase velocity of striations:

$$v_{ph} = \frac{\omega_s}{k_s} = \frac{1}{k_s^2 \lambda / \sqrt{\delta}} k_i n_0 \frac{\partial \ln k_i}{\partial \ln T_e} \tag{7.48}$$

Obviously, this is a simple derivation of the striations' phase velocity; however, detailed derivation[255,256] results in an expression similar to Equation 7.48. Interesting details on moving striations can also be found in the publications of Roth.[257,258]

Thus, the striations' velocity is proportional to the square of the wavelength; numerically, their typical value is about 100 m/sec. The frequency of oscillations of electron density, temperature, and other plasma parameters in the striations can then be found from the expression:

$$\omega = \frac{1}{k_s \lambda / \sqrt{\delta}} k_i n_0 \frac{\partial \ln k_i}{\partial \ln T_e} \tag{7.49}$$

As seen, the oscillation frequency in striations is proportional to wavelength ($2\pi/k_s$) and numerically is approximately the value of ionization frequency $10^4 \div 10^5$ sec^{-1}. It is interesting to note that according to the dispersion equation, the absolute value of group velocity of striations (v_{gr}) is equal to that of the phase velocity. However, directions of these two velocities are opposite: $v_{ph} = \omega/k = -d\omega/dk = -v_{gr}$. For this reason, some special discharge marks (e.g., bright pulsed perturbations) move to the anode, that is, in an opposite direction with respect to striations.

7.4.6 The Steenbeck Minimum Power Principle: Application to Striations

The effect of striations was explained based on physical kinetics and discharge electrodynamics. In a similar manner to the case of the normal cathode current density (see Section 7.2.5), striations can be illustrated using the Steenbeck minimum power principle.

If the discharge current is fixed, the voltage drop related to a wavelength of striations is less than the corresponding voltage of a uniform discharge. This can be explained by the strong exponential dependence of the ionization rate on the electric field value. Because of this strong exponential dependence, an oscillating electric field provides a more intensive ionization rate than an electric field fixed at the average value. Therefore, to provide the same ionization level in the discharge with striations requires less voltage and, consequently, lower power at the same current. Obviously, the Steenbeck minimum power principle is only an illustration of a phenomenon that is actually determined and controlled by the ionization instabilities of nonthermal plasma discussed earlier.

7.4.7 Some Approaches to Stabilization of Glow Discharge Instabilities

Suppression of the nonthermal discharge instabilities is the most important problem in these discharge systems at elevated currents, powers, pressures, and volumes and is very important for their applications in laser generation. As explained earlier, the most energy-intensive regimes of the glow discharge can be achieved in fast gas flows, so several approaches were developed with the objective to suppress contraction in these systems.

The most applied approach is **segmentation of the cathode**. If a high-conductivity plasma filament (contraction) occurs between two points on two large electrodes, current grows and the discharge voltage immediately drops. This can be somewhat suppressed by segmentation of an electrode, usually the cathode. Voltage is applied to each segment independently through an individual external resistance. If a filament occurs at one of the cathode segments, the discharge voltage related to other segments does not drop significantly.

Another reason for cathode segmentation is due to the relation between current density in the positive column and the normal cathode current density. This is an important question to be considered in more detail. As shown in Section 7.3.4, establishing glow discharges in fast flows provides higher specific power at relatively elevated pressures (see Equation 7.35). Thus, a pressure increase is preferable for different plasma chemical and laser applications where high levels of power are desirable.

Current density in the positive column of fast flow glow discharges does not depend on pressure (Equation 7.34) and, at typical discharge parameters, can be estimated as 40 mA/cm^2. Alternately, the normal cathode current density is proportional to the square of gas pressure and, according to the Table 7.2, can be estimated as 0.1 to 0.3 $mA/cm^2 \times p^2$ (torr). This yields a normal cathode current density of about 300 to 500 mA/cm^2 for typical pressures of 40 torr, which exceeds the positive column current density by an order of magnitude.

Glow discharges usually operate in a transition between normal and abnormal regimes when all the cathode area is covered by the electric current. Then cathode segmentation is useful to provide the previously mentioned 10-times difference in current density. Some practical ways of cathode segmentation are illustrated in Figure 7.20. In the case of transverse discharges, the cathode segments are spread over a dielectric plate; in longitudinal discharges the segments are arranged as a group of cathode rods at the gas inlet to the discharge chamber.

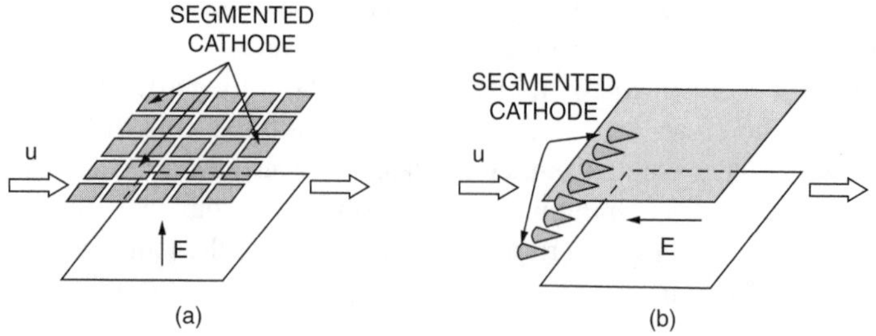

Figure 7.20 Cathode segmentation in transverse longitudinal discharge configuration

Another helpful method of suppressing the glow discharge contraction is related to the **gas flow in the discharge chamber**. Making the velocity field as uniform as possible prevents inception of instabilities. Usually, a high gas velocity also stabilizes a discharge because of reduced residence time to values insufficient to contraction. Finally, discharge stabilization can be achieved by utilizing **intensive small-scale turbulence**, which provides damping of incipient perturbations. An interesting approach of suppressing nonthermal discharge instabilities is related to using different non-self-sustained gas discharges in which plasma generation and energy input into the discharge are separated, thus preventing many instability mechanisms; these systems will be discussed in Chapter 12. Extensive details on depressing discharge instabilities in the electric-discharge gas lasers are available.[232,259]

7.5 DIFFERENT SPECIFIC GLOW DISCHARGE PLASMA SOURCES

7.5.1 Glow Discharges in Cylindrical Tubes, Parallel Plate Configuration, Fast Longitudinal and Transverse Flows, and with Hollow Cathodes

To summarize the most typical configurations of glow discharges already considered: the normal glow discharge in a cylindrical tube is the most widely used and the majority of physical features of the glow discharge previously considered were related to this configuration. Glow discharges in fast longitudinal and transverse fast gas flows especially of interest for laser applications were considered in Section 7.3.4. Segmentation of cathodes in these discharge systems to achieve higher values of specific power without contraction was discussed in Section 7.4.7. The physical basis and general principles of a hollow cathode discharge were considered in Section 7.2.7. Consider now some other configurations of practical interest.

7.5.2 Penning Glow Discharges

The special feature of the Penning discharge is its strong magnetic field (up to 0.3 torr), which permits magnetizing both electrons and ions. This discharge was proposed by

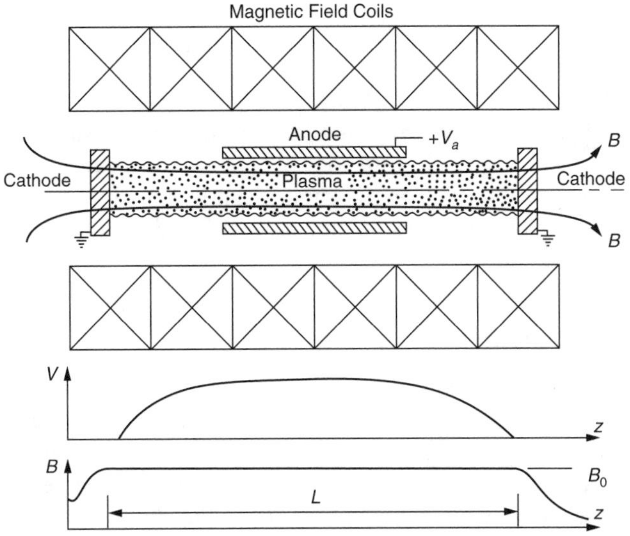

Figure 7.21 Classical Penning discharge with uniform magnetic induction and electrostatic trapping of electrons (From Roth, J. R. 2000. *Industrial Plasma Engineering*. Briston: Institute of Physics Publishing)

Penning[260,261] and further developed by Roth.[262] The classical configuration of the Penning discharge is shown in Figure 7.21. To effectively magnetize charge gas particles in this discharge requires low gas pressures, 10^{-6} to 10^{-2} torr. The two cathodes in this scheme are grounded and the cylindrical anode has voltage about 0.5 to 5 kV.

Even though the gas pressure in the Penning glow discharge is low, plasma densities in these systems can be relatively high, up to $6 \cdot 10^{12}$ cm^{-3}. The plasma is so dense because radial electron losses are reduced by the strong magnetic field and axially the electrons are trapped in an electrostatic potential well (see Figure 7.21). Although the configuration of the Penning discharge is markedly different from the traditional glow discharge in a cylindrical tube, it is still a glow discharge because the electrode current is sustained by secondary electron emission from cathode provided by energetic ions.

The ions in the Penning discharge are so energetic that they usually cause intensive sputtering from the cathode surface. Ion energy in these systems can reach several kiloelectron volts and greatly exceeds the electron energy. Plasma between the two cathodes is almost equipotential (see Figure 7.21), so it actually plays the role of a second electrode inside the cylindrical anode. As a result, the electric field inside the cylindrical anode is close to radial. Electrons and ions can be magnetized in the Penning discharge; this leads to azimuthal drift of charged particles in the crossed fields: radial electric E_r and axial magnetic B.

The tangential velocity v_{EB} of the azimuthal drift can be found from Equation 4.107, and it is the same for electrons and ions. As a result, the kinetic energy $E_K(e, i)$ of electrons and ions is proportional to their mass $M_{e,i}$:

$$E_K(e,i) = \frac{1}{2} M_{e,i} v_{EB}^2 = \frac{1}{2} M_{e,i} \frac{E_r^2}{B^2} \tag{7.50}$$

This explains why the ion temperature in the Penning discharge much exceeds the electronic temperature. Typically, the electron temperature in a Penning discharge is approximately 3 to 10 eV, while the ion temperature can be an order of magnitude higher (30 to 300 eV). Details on the subject can be found in the publications of Roth;[263] alternative configurations of the classical Penning discharge are reviewed in Roth.[264]

7.5.3 Plasma Centrifuge

The effect of azimuthal drift in the crossed electric and magnetic fields in the Penning discharge and the related fast plasma rotation allows creation of plasma centrifuges. In these systems, electrons and ions circulate around the axial magnetic field with very large velocities (Equation 4.107). This "plasma wind" is able to drag neutral particles and transfer to them the high energies of the charged particles. It is interesting to note that in the collisional regime, the gas rotation velocity is usually limited by the kinetic energy corresponding to ionization potential:

$$v_{rA} = \sqrt{2eI/M} \tag{7.51}$$

Here, I and M are the ionization potential and mass of heavy neutral particles. The maximum neutral gas rotation velocity v_{rA} in the plasma centrifuge is usually referred to as the **Alfven velocity for plasma centrifuge**.

Currently, no complete explanation for the phenomenon of the critical Alfven velocity exists. It is clear that acceleration of already very energetic ions becomes impossible in weakly ionized plasma when they have energy sufficient for ionization. Further energy transfer to ions does not go to their acceleration but rather most to ionization. The point is that ions in contrast to electrons are unable to ionize neutrals when their energy only slightly exceeds the ionization potential (see Section 2.2.7, Equation 2.37).

Effectiveness of energy transfer between a heavy ion and a light electron inside a neutral particle is very low according to the adiabatic principle (Section 2.2.7). For this reason, effective ionization by direct ion impact usually requires energies much exceeding the ionization potential (usually in the kiloelectron-volt energy range). Electrons are obviously able to ionize at energies about that of the ionization potential, but their average energies in a plasma centrifuge are usually less than those of ions (Equation 7.50). The explanation of the Alfven critical velocity is related in some way to energy transfer from energetic ions to plasma electrons responsible for ionization.

Practical gas rotation velocities in plasma centrifuges are very high and reach 2 to $3 \cdot 10^6$ cm/sec in the case of light atoms. For comparison, the maximum velocities in similar mechanical centrifuges are about $5 \cdot 10^4$ cm/sec. The fast gas rotation in a plasma centrifuge can be applied for isotope separation, which occurs as a result of a regular diffusion coefficient, D, in a field of centrifugal forces. The separation time can be estimated as:

$$\tau_S \approx \frac{T_0 R_C^2}{D \Delta M v_\varphi^2} \tag{7.52}$$

where T_0 is the gas temperature; v_φ is the maximum value of gas rotation velocity; R_C is the centrifuge radius; and $\Delta M = |M_1 - M_2|$ is the atomic mass difference of isotopes or components of gas mixture. Because of the high values of rotation velocities v_φ, the separation time in the plasma centrifuges is low. For the same reason of the high values of rotation velocities v_φ, the steady-state separation coefficient for binary mixture is significant even for isotopes with a relatively small difference in their atomic masses $M_1 - M_2$:

$$R = \frac{(n_1/n_2)_{r=r_1}}{(n_1/n_2)_{r=r_2}} = \exp\left[(M_1 - M_2)\int_{r_1}^{r_2} \frac{v_\varphi^2}{T_0}\frac{dr}{r}\right] \tag{7.53}$$

Here, (n_1/n_2) is the concentration ratio of the binary mixture components at the radius r, so obviously the heavier component moves preferentially to the centrifuge periphery.

For plasma centrifuges with weakly ionized plasma $n_e/n_0 \approx 10^{-4} - 10^{-2}$, the parameter

$$\frac{Mv_\varphi^2}{2} \Big/ \frac{3}{2}T_0$$

usually has a value of about 3.[9] As an example, the separation coefficient for mixture He–Xe in such centrifuges exceeds 300; for mixture ^{235}U to ^{238}U, it is about 1.1. Radial distribution of partial pressures of gas components for deuterium–neon separation in the plasma centrifuge is presented in Figure 7.22.[9]

It is interesting to point out that plasma centrifuges are able to provide chemical process and product separation in one system. For example, water vapor can be dissociated in the system $H_2O \rightarrow H_2 + \frac{1}{2}O_2$ with simultaneous separation of hydrogen and oxygen (the separation coefficients for this process are shown in Figure 7.23 in comparison with H_2–Ne separation[9]). Additional information on the subject can be found in Poluektov and Efremov.[265]

7.5.4 Magnetron Discharges

This glow discharge configuration is applied most for sputtering of cathode material and film deposition. A general schematic of the **magnetron discharge with parallel plate electrodes** is shown in Figure 7.24. To provide effective sputtering and film deposition, the mean free path of the sputtered atoms must be large enough and thus gas pressure should be sufficiently low (10^{-3} to 3 torr). However, in this system, because electrons are trapped in the magnetic field by the magnetic mirror, plasma density on the level of 10^{10} cm^{-3} is achieved. Ions are not supposed to be magnetized

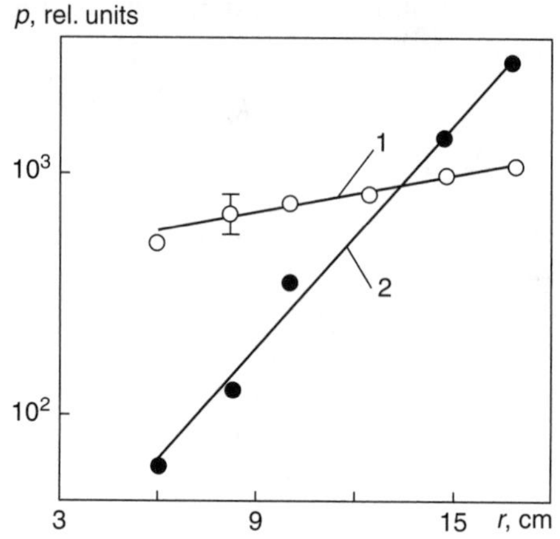

Figure 7.22 Radial distribution of partial pressures of (1) D_2 and (2) Ne in mixture

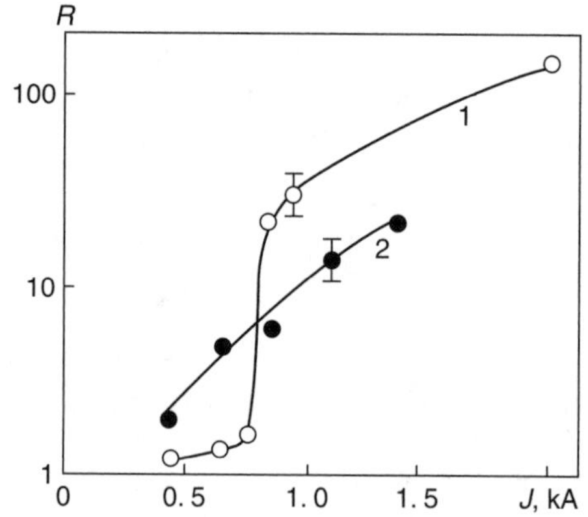

Figure 7.23 Separation coefficient for (1) water decomposition products and (2) H_2–Ne mixture (p = 0.3 torr; H = 5 kg; τ = 3 msec)

in this system to provide sputtering. Typical voltage between electrodes is several hundreds volts and magnetic induction is approximately 5 to 50 mt.

Negative glow electrons are trapped in the magnetron discharge by the magnetic mirror. The effect of the magnetic mirror causes "reflection" of electrons from areas with elevated magnetic field (see Figure 7.25). The magnetic mirror is actually one of the simplest systems for plasma confinement in a magnetic field (see Section 6.2). The magnetic mirror effect is based on the fact that, if spatial gradients of the magnetic field are small, the magnetic moment of a charged particle gyrating around the magnetic lines is an approximate constant of the motion. The particle motion in this magnetic field is said to be the adiabatic motion. The magnetic mirror effect related to the magnetron discharge will be discussed in more detail in the next section.

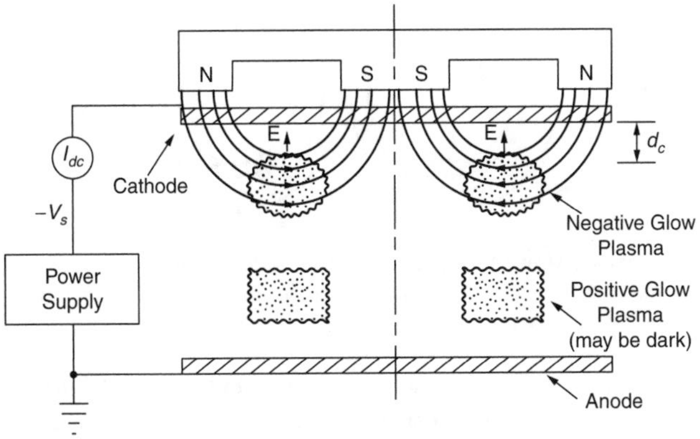

Figure 7.24 Glow discharge plasma formation in parallel plate magnetron. The negative plasma glow is trapped in the magnetic mirror formed by magnetron magnets. (From Roth, J. R. 2000. *Industrial Plasma Engineering*. Briston: Institute of Physics Publishing)

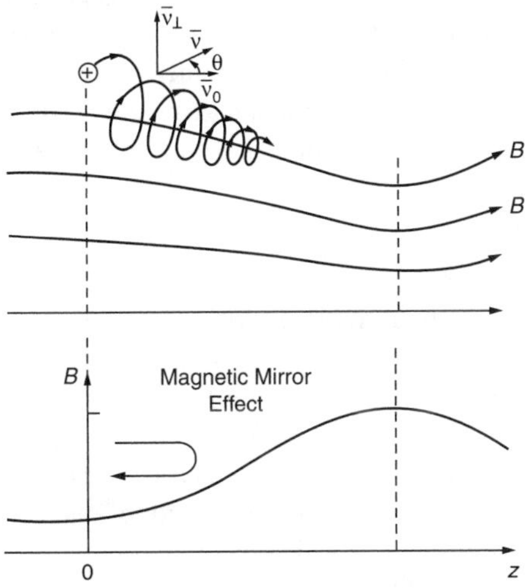

Figure 7.25 Magnetic mirror effect

The electric field between the cathode and the negative glow zone is relatively strong, providing ions with the energy necessary for effective sputtering of the cathode material. It is important that the drift in crossed electric and magnetic fields cause the plasma electrons in this system to drift around the closed plasma configuration. This makes the plasma of the magnetron discharge quite uniform, which is important for sputtering and film deposition.

The magnetron discharges can be arranged in various configurations. For example, if the cathode location prevents effective deposition, it can be relocated. This leads to the so-called **coplanar configuration of the magnetron discharge**, shown

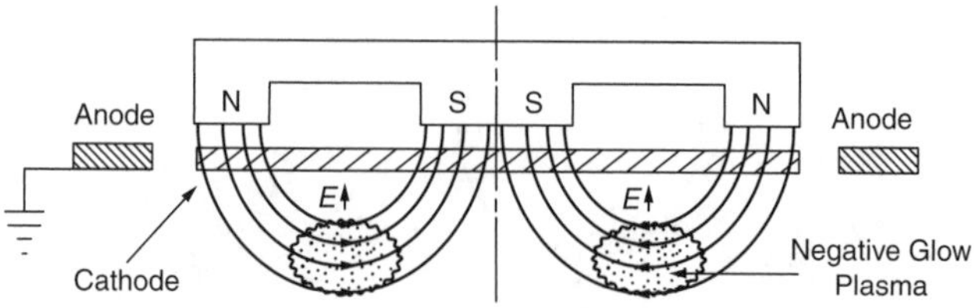

Figure 7.26 Co-planar magnetron configuration. Only the negative glow plasma, trapped in the magnetic pole pieces, is normally visible in this configuration. (From Roth, J. R. 2000. *Industrial Plasma Engineering*. Briston: Institute of Physics Publishing)

in Figure 7.26. Other possible geometries of the magnetron discharge are reviewed in Roth;[264] from the physical point of view these are almost identical.

7.5.5 Magnetic Mirror Effect in Magnetron Discharges

The magnetic mirror effect plays a key role in magnetron discharges, trapping the negative glow electrons and providing sufficient plasma density for effective sputtering at relatively low pressures of the discharge. Due to its importance, this effect will be discussed in more detail. The magnetic moment μ of a charged particle (in this case, it is an electron, mass m) gyrating in a magnetic field B is defined as the current I related to this circular motion multiplied by the enclosed area $\pi\rho_L^2$ of the orbit:

$$\mu = I \cdot \pi\rho_L^2 = \frac{mv_\perp^2}{2B} = const \tag{7.54}$$

In this relation for the magnetic moment, μ:ρ_L is the electron Larmor radius (see Equation 6.51 and Equation 6.47) and $v_\perp$ is the component of electron thermal velocity perpendicular to magnetic field (see Figure 7.25 and Equation 6.51).

The magnetic mirror effect can be understood from Figure 7.25. When a gyrating electron moves adiabatically toward higher electric fields, its normal velocity component $v_\perp$ is growing proportional to the square root of magnetic field in accordance with Equation 7.54. The growth of $v_\perp$ is obviously limited by energy conservation. For this reason, the electron drift into the zone with elevated magnetic field also should be limited, leading to reflection of the electron back to the area with lower values of magnetic field. This generally explains the magnetic mirror effect.

If the magnetic lines in the plasma zone are parallel to electrodes (see Figure 7.24), energy transfer from the electric field in this zone can be neglected in the *zero*th approximation. In this simplified model, the kinetic energy of the gyrating electron $mv^2/2$ and total electron velocity v can be considered as constant. Taking into account that $v_\perp = v \sin\theta$ (where θ is the angle between total electron velocity and magnetic field direction; see Figure 7.25), the constant magnetic momentum (Equation 7.54) can be rewritten for these simplified conditions as:

$$\frac{\sin^2\theta}{B(z)} = \frac{2\mu}{mv^2} = const \tag{7.55}$$

Thus, the angle θ between total electron velocity and magnetic field direction grows during the electron penetration into areas with higher values of magnetic field $B(z)$ until it reaches the reflection point $\theta = \pi/2$. As is clear from Equation 7.55, the electrons can be reflected if their initial angle θ_i is sufficiently large:

$$\sin\theta_i > \sqrt{B_{min}/B_{max}} \tag{7.56}$$

where B_{min}, B_{max} are the maximum and minimum values of magnetic field in the magnetic mirror (see Figure 7.25). If the initial angle θ_i is not sufficiently large and criterion (Equation 7.56) is not valid, electrons are not reflected by the mirror and are able to escape. This is why the minimum angle is usually referred as the **escape cone angle**. However, these electron losses are not crucial for practical applications of the magnetron discharge.

7.5.6 Atmospheric Pressure Glow Discharges

As was stated in Section 7.1, glow discharges in general operate at low gas pressures. If the steady-state discharge cooling is controlled by conduction (the diffusive regime), the pressure increase is limited by overheating. In Section 7.3.3 it was shown that the maximum values of current density and electron concentration proportionally decrease with pressure (see Equation 7.32 and Equation 7.33), and the ionization degree decreases as square of pressure $n_e/n_0 \propto 1/p^2$.

In Section 7.3.4, it was shown that somewhat higher operation pressures can be reached by operating the glow discharges in fast flows. However, in this case, these pressures are also limited by plasma instabilities because their very short induction times should be comparable to or longer than the gas residence times in the discharge. Thus, operating a stable, continuous nonthermal glow discharge at atmospheric pressure is a challenging task. Nevertheless, this has successfully been accomplished in some special discharge systems, e.g., glow discharges in transonic and supersonic flows of essentially electronegative gases. Application of special aerodynamic techniques permits sustaining the uniform steady-state glow discharges at atmospheric pressure and specific energy input up to 500 J/g.[204,205,266] In these investigations it was pointed out that the glow discharge decontraction at atmospheric pressure becomes possible due to suppressing the transverse diffusion influence on the temperature and current density distribution. Details regarding organization of nonthermal discharges in supersonic flows were discussed in Section 6.4.

Another interesting application of the glow discharge at atmospheric pressure is related to using special gas mixtures as working fluid and additionally elaborating some special types of electrodes. Such gas mixtures are supposed to be able to provide the necessary level of the ionization rate at relatively low values of reduced electric field E/p and are suitable to sustain glow discharge operation at atmospheric pressure. Gases for such discharges usually include different inert gas mixtures, and first of all helium.

Comparing Figure 7.12 and Figure 7.13, it is seen that the reduced electric field necessary to sustain a glow discharge in inert gases can be more than 30 times lower than that for molecular gases; this is due to the absence of electron energy losses to vibrational excitation. Mixtures of helium or neon with argon or mercury are very effective for ionization when the Penning effect takes place. In this case, hundreds of volts are sufficient to operate a glow discharge at atmospheric pressure. Use of helium is also helpful to increase heat exchange and cooling of the systems.

As was discussed earlier, the normal cathode current density is proportional to the square of pressure and becomes large at elevated pressures. For this reason, special types of electrodes should be applied, e.g., fine wires or barrier discharge type electrodes. These are necessary to avoid overheating and correlate current density on the cathode and in the discharge volume.

Although atmospheric pressure glow (APG) discharges are physically similar to such traditional nonthermal atmospheric pressure discharges as corona or dielectric barrier discharge (DBD), their voltage of approximately hundreds of volts is much less than that in a corona or DBD. More details on this type of atmospheric pressure glow discharges are available.[267–270] The key point of APG is absence of streamers, which are responsible for ionization in relatively powerful atmospheric pressure non-thermal discharges.[619] This challenging physical problem will be discussed in Section 9.3.6.

7.5.7 Some Energy Efficiency Peculiarities of Glow Discharge Application for Plasma Chemical Processes

Glow discharges are widely used as light sources; as active medium for gas lasers; for treatment of different surfaces; sputtering; film deposition; etc. Traditional glow discharges controlled by diffusion are of interest only for chemical applications in which electric energy cost-effectiveness is not an issue. Application of glow discharges as the active medium for highly energy-effective plasma chemical processes is limited by the following three major factors:

1. Typical values of specific energy input in the traditional glow discharges controlled by diffusion are of about 100 eV/mol and much exceed the optimal value of the discharge parameter $E_v \approx 1$ eV/mol (see Section 5.6.8). Taking into account that the energy necessary for one act of chemical reaction is usually of about 3 eV/mol, it is clear that the maximum possible energy efficiency in these systems is about 3% even at complete 100% conversion. These high values of the specific energy input (about 100 eV/mol) are related to relatively low gas flows passing though the discharge. In the optimal case when a molecule receives energy, $E_v \approx 1$ eV/mol $\approx 3\hbar\omega$ (see Section 5.6.8), it is supposed to leave the discharge zone. From a viewpoint of energy balance, this means that plasma cooling should be controlled by convection, which takes place only in fast flow glow discharges.
2. In Section 5.6, it was shown that the most energy effective plasma chemical processes require sufficiently high levels of ionization degree (see Equation 5.150). Taking into account that in the discharges under consideration $n_e/n_0 \propto 1/p^2$ (see Equation 7.33), the requirement of a high degree of ionization leads

to low gas pressures and thus to further growth of the specific energy input and decrease of energy efficiency. The reduction of pressure necessary for increase of the degree of ionization also results in an increase of the reduced electric field and, therefore, an increase of electron temperature (see Section 4.5.4 and the Engel–Steenbeck Equation 4.173). Such electron temperature increase is also not favorable for energy efficiency of highly energy-effective plasma chemical processes.

3. Specific power of the conventional glow discharges controlled by diffusion does not depend on pressure and is actually quite low. Numerically, the glow discharge power per unit volume is about 0.3 to 0.7 W/cm^3 (see Equation 7.31). For this reason, the specific productivity of such plasma chemical systems is also relatively low. Possible increase of the specific power and specific productivity of related plasma chemical systems can be achieved by increasing the gas pressure and applying the fast flow glow discharges with convective cooling.

Thus, energy efficiency of plasma chemical processes in conventional glow discharges is not very high with respect to other nonthermal discharges. Lower energy prices can be achieved applying nontraditional glow discharges in fast flows or those sustained by an external source of ionization, which will be discussed in Chapter 11.

PROBLEMS AND CONCEPT QUESTIONS

7.1 **Space Charges in Cathode and Anode Layers.** Space charges are formed near the cathode as well as near the anode. Explain why the space charge of the cathode layer is much more significant than that of the anode layer.

7.2 **Radiation of Plasma Layers Immediately Adjacent to Electrodes.** Explain why the plasma layer immediately adjacent to the cathode is dark (the Aston dark space), while the plasma layer immediately adjacent to the anode is bright (the anode glow).

7.3 **The Seeliger's Rule of Spectral Line Emission Sequence in Negative and Cathode Glows.** The negative glow first reveals (closer to the cathode) spectral lines emitted from higher excited atomic levels and then spectral lines related to lower excited atomic levels. This sequence is reversed with respect to the order of spectral line appearance in the cathode glow. Explain this fact, which is usually referred to as the Seeliger's rule.

7.4 **Glow Discharge in Tubes of Complicated Shapes.** Glow discharges can be maintained in tubes of very complicated shapes. This effect is widely used in luminescent lamps. Explain the mechanism of sustaining the glow discharge uniformity in such a case.

7.5 **Current–Voltage Characteristic of DC Discharges between Electrodes.** Analyze the current–voltage characteristic presented in Figure 7.5; explain why the dark discharge cannot exist at very small currents lower than some critical value (interval *AB* in the figure). Estimate this minimal current for a dark discharge.

7.6 **Space Distribution of Ion and Electron Currents in Dark Discharge.** Based on the continuity equations for electron and ions, derive Equation 7.2 for electron and ion current distributions. Estimate the accuracy of this equation related to neglecting of diffusion fluxes and recombination of charged particles in this case.

7.7 **Maximum Current of Dark Discharge.** Analyze Equation 7.11 and show that the maximum current of a dark discharge is proportional to the square of gas pressure. Derive a relation between the relevant similarity parameters: the dark discharge maximum current j_{max}/p^2, ion mobility $\mu_+ p$, reduced electric field E/p, and interelectrode distance pd_0.

7.8 **Comparison of Typical Voltages in Dark and Glow Discharges.** Typical voltage of a glow discharge is usually lower than that in a dark discharge because of the strongly exponential dependence of the Townsend coefficient on the electric field in a gap $\alpha[E(x)]$. However, it cannot be applied to very large electric fields when the dependence $\alpha(E)$ is close to saturation and not strong. Determine the maximum electric field necessary for the effect, based on Equation 4.141, for the Townsend coefficient $\alpha(E)$.

7.9 **The Engel–Steenbeck Model of a Cathode Layer.** Analyze the solution of the system of Equation 7.9, Equation 7.12, and Equation 7.13, taking the Townsend coefficient dependence on electric field $\alpha(E)$ in the form Equation 4.141 and assuming a linear decrease of electric field along the cathode layer (Equation 7.14). Find possible analytical approximations of the solution. Discuss the accuracy of replacing in Equation 7.12 the linear expression for electric field (Equation 7.14) by a constant electric field.

7.10 **Normal Cathode Potential Drop, Current Density, and Thickness of Cathode Layer.** Analyzing Equation 4.143, prove that in first approximation, the normal cathode potential drop does not depend on pressure or on gas temperature. Discuss the limits of this statement. Determine the dependence of normal current density and normal thickness of cathode layer on temperature at constant pressure and on pressure at constant temperature.

7.11 **Stability of Normal Current Density.** Prove that the central quasi-homogeneous region of cathode layer is stable even if current density exceeds the normal one, $j > j_n$. Show that the periphery of the cathode spot is unstable in the same time, which finally leads to a decrease of current density until it reaches the normal value, $j = j_n$.

7.12 **The Steenbeck Minimum Power Principle.** Analyze the minimum power principle and explain why it is not related exactly to fundamental physical principles. How can you explain the wide applicability of the minimum power principle to different specific problems in gas discharge physics? Prove this principle for the specific problem of establishing normal current density in cathode layer of a glow discharge.

7.13 **Abnormal Glow Discharge.** Determine the dependence of the average conductivity in the area of cathode layer on total current for the abnormal regime of glow discharge. Compare this average conductivity with that for cathode layer of a normal glow discharge. Estimate the total resistance of the cathode layer per unit area of cathode surface in the abnormal glow discharge; also compare it with that for the case of normal glow discharge.

7.14 **Glow Discharge with Hollow Cathode.** For the case of the Lidsky configuration of the hollow cathode (Figure 7.10), explain how the electric field becomes able to penetrate into the capillary thin hollow cathode. Take into account that plasma formed inside the capillary tube is a good conductor. Explain how to start the glow discharge inside the hollow capillary metal tube, where initially there is almost no electric field.

7.15 **Anode Layer of a Glow Discharge.** Compare typical thickness values of the anode and cathode layers and their dependences on pressure for a normal glow discharge. Discuss the result of this comparison.

7.16 **Current–Voltage Characteristics of a Glow Discharge in Recombination Regime.** Based on Equation 7.28, analyze the influence of electron attachment and detachment processes (formation and destruction of negative ions) on the current–voltage characteristic of a glow discharge in recombination regime. Analyze different effects promoting slight growing and slight decreasing of this almost horizontal current–voltage characteristic.

7.17 **Conductive and Convective Mechanisms of Heat Removal from Positive Column of a Glow Discharge.** Based on Equation 7.30, determine the critical gas velocity necessary for the convective mechanism to dominate heat removal from the positive column of a glow discharge. Consider numerical examples of gases with high and low values of heat conductivity coefficients and cases of capillary-thin and relatively thick discharge tubes.

7.18 **Glow Discharge in Fast Gas Flows.** The effect of E/p increase in the fast laminar flows can be explained not only by direct convective losses of charged particles, but also by convective losses of active species responsible for electron detachment from negative ions. Compare the effectiveness of these two mechanisms in the reduced electric field E/p increase.

7.19 **Joule Heating Influence on Current–Voltage Characteristic of Glow Discharge.** Derive Equation 7.37, describing the slightly decreasing current–voltage characteristic of a glow discharge. In the derivation, take into account the heat balance (Equation 7.29) and the similarity condition: $E/n_0 \propto ET_0 \approx const.$ Discuss the frameworks of possible applicability of this similarity condition in a glow discharge.

7.20 **Contraction of Positive Column of Glow Discharge Controlled by Diffusion.** Explain why, in addition to the nonlinear mechanism of ionization growth with electron density, the volumetric character of electron losses is necessary to provide contraction of a glow discharge. Based on relations from Section 4.5 and Section 6.3, estimate the critical electron concentrations and specific powers typical for glow discharge transition to the contracted mode.

7.21 **Contraction of Glow Discharge in Fast Gas Flow.** Explain why transition to the contracted mode takes place in fast flow glow discharges when not specific power (W/cm^3), but rather the specific energy input (eV/mol or J/cm^3)

exceeds the critical value (see Section 6.3.6, Figure 6.12). Explain why the fast flow mode of a glow discharge is preferable from the point of view of total power with respect to one controlled by diffusion.

7.22 **Limitation of the Striation Wavelength.** Derive the criterion (Equation 7.40) for suppression of striations by ionization effects that are based on the relation: $v_d\tau_f \approx \lambda/\sqrt{\delta}$. Use the definition of the time interval τ_f necessary for building up the electron energy distribution function ($\tau_f = 1/(\nu_{en}\delta)$); Equation 4.101 for drift velocity; and Equation 4.86 between electric field and electron temperature. Analyze the possible wavelength of striations in inert and molecular gases.

7.23 **Phase and Group Velocity of Striations.** Considering striations as ionization waves following the dispersion Equation 7.49, prove that their group and phase velocities have the same absolute value, but opposite signs. Give a physical interpretation of different directions of the phase and group velocities for the wave with frequencies proportional to wavelength.

7.24 **Striations from the Viewpoint of the Steenbeck Principle of Minimum Power.** Based on the strong exponential dependence of ionization rate on value of electric field $k_i\,(E/p)$, prove that if the discharge current is fixed, the voltage drop related to a wavelength of striations is less than the corresponding voltage of a uniform discharge for the same length.

7.25 **Cathode Segmentation.** Using Equation 7.34 for current density in the positive column of the fast flow glow discharge, and Table 7.2 for the normal cathode current density, determine the discharge and cathode material conditions when the cathode segmentation is necessary to correlate the current densities in cathode layer and positive column.

7.26 **The Penning Discharge.** Analyze the classical configuration of the Penning discharge and explain why the electric field in this system can be considered radial. Compare the Penning discharge with the hollow cathode glow discharge, where the electric field configuration is also quite sensitive to the plasma presence. The classical configuration of the hollow cathode glow discharge is difficult to ignite; does the Penning discharge have a similar problem?

7.27 **The Alfven Velocity in Plasma Centrifuge.** Estimate values of electric and magnetic field in plasma centrifuge when the ionic drift velocity in the crossed electric and magnetic fields reaches the critical value of Alfven velocity. Give your interpretation of the critical Alfven velocity, taking into account that direct ion impact ionization is an adiabatic process and requires much higher energy than the ionization potential.

7.28 **Isotope Separation in Plasma Centrifuge.** Using Equation 7.53 and typical parameters of a plasma centrifuge given in Section 7.5.3, estimate the value of the separation coefficient for uranium isotopes ^{235}U through ^{238}U. Compare the result of your estimations with the experimental value of the coefficient given in Section 7.5.3.

7.29 **Magnetron Discharge.** Taking into account typical values of the magnetic field induction in the magnetron discharges given in Section 7.5.4, estimate the interval of pressures when electrons are magnetized but ions are not.

Compare your estimated values of pressures with actual pressures in the magnetron discharges.

7.30 **Adiabatic Motion of Electrons in Magnetic Mirror.** Based on Equation 6.51 and Equation 6.47 for the Larmor radius and cyclotron frequency, derive Equation 7.54 for the magnetic momentum of an electron gyrating around a magnetic line in the magnetron discharge. Explain the physical basis of the magnetic moment constancy for the adiabatic motion of electrons in a magnetic mirror and its relation to plasma confinement in a magnetic field.

7.31 **The Escape Cone Angle in Magnetic Mirror.** If the initial angle θ_i between electron velocity and magnetic field direction is not sufficiently large, electrons are not reflected by the magnetic mirror and are able to escape. Derive Equation 7.56 for the escape cone angle as a function of the maximum and minimum values of magnetic field in the magnetic mirror.

7.32 **Atmospheric Pressure Glow Discharges.** Explain the main problems to be solved in order to organize glow discharges at atmospheric pressure. What is the difference between such traditional nonthermal atmospheric pressure discharges as corona and dielectric barrier discharge (DBD) and the atmospheric pressure glow discharges.

CHAPTER

8

ARC DISCHARGES

8.1 PHYSICAL FEATURES, TYPES, PARAMETERS, AND CURRENT–VOLTAGE CHARACTERISTICS OF ARC DISCHARGES

8.1.1 General Characteristic Features of Arc Discharges

In the same way as glow discharges traditionally represent nonthermal, nonequilibrium plasma, arc discharges are traditional examples of thermal quasi-equilibrium plasma (even though arcs' plasmas are quite often far from quasi equilibrium). Arc discharges have existed for a very long period (almost two centuries) and are applied in illumination devices and in metallurgy.

As a rule, the arcs are self-sustaining DC discharges, but in contrast to glow discharges, they have relatively low cathode fall voltage of about 10 eV, which corresponds to the ionization potential. Arc cathodes emit electrons by intensive **thermionic** and **field emission** mechanisms that are able to provide high cathode current already close to the total discharge current. The principal mechanisms of electron emission from the cathode will be discussed in Section 8.2. Because of the high values of the cathode current, a high cathode fall voltage is not needed to multiply electrons in the cathode layer to provide the necessary discharge current (which can be actually very large in arc discharges). Arc cathodes always receive large amounts of joule heating from the discharge current and, therefore, are able to reach very high temperatures in contrast to the glow discharges, which are actually cold. The high temperature leads to evaporation and erosion of electrodes. This is a negative feature of the arc discharges.

Table 8.1 Typical Ranges of Thermal and Nonthermal Arc Discharge Plasma Parameters

Discharge plasma parameter	Thermal arc discharge	Nonthermal arc discharge
Gas pressure	0.1–100 atm	10^{-3}–100 torr
Arc current	30 A–30 kA	1–30 A
Cathode current density	10^4–10^7 A/cm^2	10^2–10^4 A/cm^2
Voltage	10–100 V	10–100 V
Power per unit length	>1 kW/cm	<1 kW/cm
Electron density	10^{15}–10^{19} cm^{-3}	10^{14}–10^{15} cm^{-3}
Gas temperature	1–10 eV	300–6000 K
Electron temperature	1–10 eV	0.2–2 eV

The main arc discharge zone located between electrode layers is called the positive column in the same way as in glow discharges. Plasma of the arc discharge positive column can be quasi equilibrium and nonequilibrium, depending on gas pressure. Thus, a nonequilibrium DC plasma can be generated not only in glow discharges, but also in arcs at low pressures, while quasi-equilibrium DC plasma can be generated only in electric arcs. This is the reason that electric arcs are usually considered a classical example of quasi-equilibrium thermal plasma. Some interesting extreme examples can be found in publications of Helerlein,[638] Kruger,[639] and Hralovsky.[640]

8.1.2 Typical Ranges of Arc Discharge Parameters

The thermal and nonthermal regimes of arc discharges have many peculiarities and quite different parameters. In particular, the principal cathode emission mechanism is thermionic in nonthermal regimes and mostly field emission in thermal arcs. Also, the reduced electric field E/p is low in thermal arcs and relatively high in nonthermal arcs. The total voltage in any kind of arc is usually relatively low; in some special forms it can be only a couple of volts. Ranges of plasma parameters typical for the thermal and nonthermal arc discharges are outlined in Table 8.1.

As the table shows, thermal arcs operating at high pressures are much more energy intensive. They have higher currents and current densities plus higher power per unit length. For this reason, these discharges are sometimes referred to as high-intensity arcs. The preceding division of arc discharges into two groups is very simplified. The next classification, according to specific peculiarities of the cathode processes; peculiarities of the positive column; and peculiarities of the working fluid (influence of vaporized cathode material), can be more informative.

8.1.3 Classification of Arc Discharges

Actually, many quite different DC discharges with low cathode-fall voltage are considered arc discharges. As mentioned earlier, these can be usefully classified by the principal cathode and positive column process mechanisms.

- **Hot Thermionic Cathode Arcs**. In these arcs, a whole cathode has temperatures of 3000 K and greater, providing a high current due to thermionic emission. The arc in this case is stationary to a fixed and quite large cathode spot. Current is distributed over a relatively large cathode area and, therefore, its density is not extremely high—about 10^2 to 10^4 A/cm^2. Only special refractory materials like carbon, tungsten, molybdenum, zirconium, tantalum, etc. can continuously withstand such high temperatures and be used in these types of arc hot thermionic discharges. Note that the cathode can be heated to sufficiently high temperatures not only by the arc current, but also in a non-self-sustained manner from an external source of heating. Such cathodes are utilized in low-pressure arcs and, in particular, in thermionic converters. Cathodes in such arc discharges are usually activated to decrease the temperature of thermionic emission.
- **Arcs with Hot Cathode Spots.** If a cathode is made from low-melting-point metals like copper, iron, silver, or mercury, the high temperature necessary for emission obviously cannot be sustained permanently. In this case, electric current flows through hot spots that appear, move fast, and disappear on the cathode surface. Current density in the spots is extremely high—about 10^4 to 10^7 A/cm^2. This leads to localized intensive, short heating and evaporation of the cathode material, while the rest of the cathode actually stays cold. The principal mechanism of electron emission from the spots is thermionic field emission to provide a high current density at temperatures limited by melting point. Note that cathode spots appear not only in the case of the low-melting-point cathode materials, but also on refractory metals at low currents and low pressures.
- **Vacuum Arcs**. This type of low-pressure arc, operating with the cathode spots, is special because the gas phase working fluid in this discharge is provided by intensive erosion and evaporation of electrode material. Thus, the vacuum arc operates in a dense metal vapor, which is actually self-sustained in the discharge. This type of arc is of special importance in high-current electrical equipment, e.g., high-current vacuum circuit breakers and switches.
- **High-Pressure Arc Discharges**. An arc positive column plasma is a quasi-equilibrium one at pressures exceeding 0.1 to 0.5 atm. Most traditional thermal arcs obviously operate at atmospheric pressure in open air. The main parameters of such arcs, current, voltage, temperature, and electron density were shown in Table 8.1. Thermal arcs operating at very high pressures exceeding 10 atm are a special example. In this case, thermal plasma is so dense that most of the discharge power, 80 to 90%, is converted into radiation that is much greater than at atmospheric pressure. Such types of arcs in xenon and in mercury vapors are applied as special sources of radiation.
- **Low-Pressure Arc Discharges**. Positive column plasmas of arc discharges at low pressures of about 10^{-3} to 1 torr are nonequilibrium and quite similar to that in glow discharges. However, it should be pointed out that ionization degrees in nonthermal arcs are higher than in glow discharges because arc currents are much larger. Typical parameters of such nonthermal plasma, current, voltage, temperature, and electron density were presented in Table 8.1.

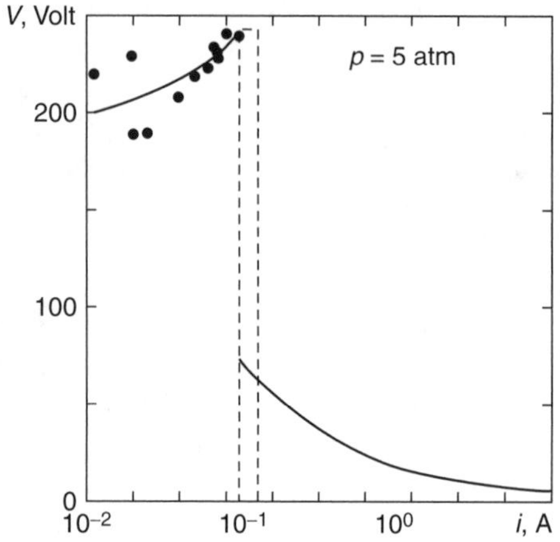

Figure 8.1 Current-voltage characteristics of a xenon lamp, p = 5 atm, in the region of transition from glow to arc discharge[620]

8.1.4 Current–Voltage Characteristics of Arc Discharges

The general current–voltage characteristic of continuous self-sustained DC discharges for a wide range of currents was discussed in Section 7.1.4 and illustrated in Figure 7.5. Transition from a glow to arc discharge corresponds to the interval FG on the current–voltage characteristic. Current density in abnormal glow discharge increases (see Section 7.2.6), resulting in cathode heating and in intensive growth of thermionic emission, which determines the glow-to-arc transition.

The glow-to-arc transition is continuous (Figure 7.5) in the case of thermionic cathodes made from refractory metals and takes place at currents of about 10 A. In contrast, cathodes made from low-melting-point metals provide the transition at lower currents of 0.1 to 1 A. This transition is sharp, unstable, and accompanied by formation of hot cathode spots. An example of the current-density characteristic for such a glow-to-arc transition is presented in Figure 8.1.

An example of the current–voltage characteristic (Figure 8.2) for an actual arc discharge corresponds to the interval GF in Figure 7.5. This example corresponds to the most classical type of arc discharges, **the voltaic arc** (Figure 8.3), which is a carbon arc in atmospheric air (arc discharges were first discovered in this form). Cathode- and anode-layer voltages in the voltaic arc are about 10 V; the balance of voltage (see Figure 8.2) corresponds to a positive column. Increase of the discharge length leads to linear growth of voltage, which means that the reduced electric field in the arc is constant at fixed values of electric current. When the discharge current grows, the electric field and voltage gradually decrease until a critical point of sharp explosive voltage reduction (see Figure 8.2); this is followed by an almost horizontal current–voltage characteristic. The transition is accompanied by specific hissing noises that are physically associated with the formation of hot anode spots with intensive evaporation.

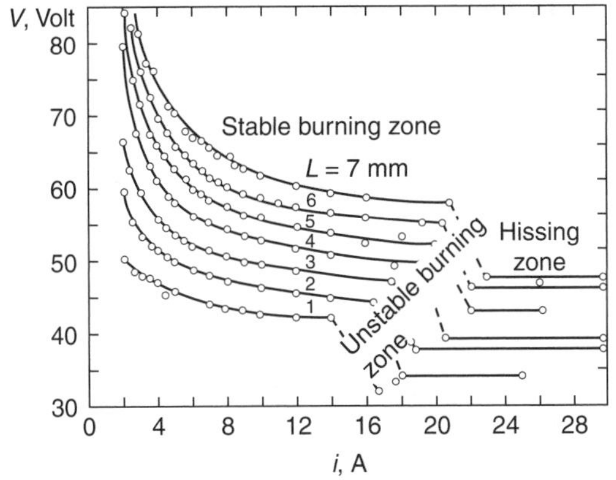

Figure 8.2 Current-voltage characteristics of carbon arc in air. Values of L indicate the distance between electrodes[620]

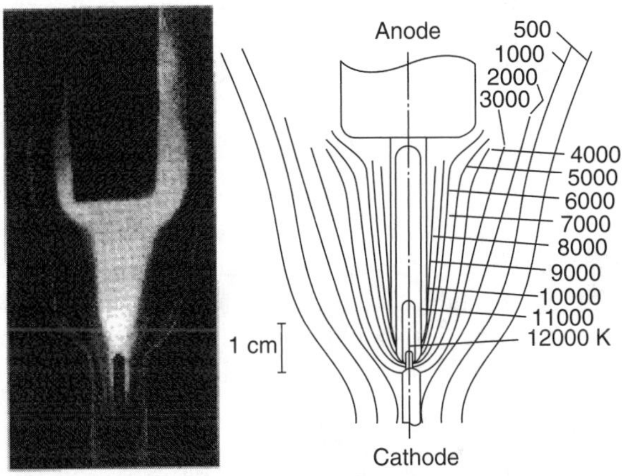

Figure 8.3 Carbon arc in air at a current of 200 A: a Toepler photograph and measured temperature field[620]

Although arc discharges may have very different configurations, in general, their structure includes cathode layer, positive column, and anode layer. Similar to the case of glow discharges, consider first the physical phenomena and characteristics of the cathode and anode layers that are responsible for forming and sustaining the discharge. Then consider the positive column that is the main zone of the arc and used for different applications. Before the physics of the cathode layer (Section 8.3) are considered, the mechanisms of electron emission that play a key role will be discussed.

8.2 MECHANISMS OF ELECTRON EMISSION FROM CATHODE

As mentioned previously, the key features of arc discharges are related to the mechanisms of electron emission from the cathode. Consider the most important of these: the mechanisms of cathode emission. This will be compared with glow discharges.

8.2.1 Thermionic Emission: Sommerfeld Formula

Thermionic emission refers to the phenomenon of electron emission from a high-temperature metal surface (e.g., hot cathode), which is due to thermal energy of electrons located in metal. Emitted electrons can remain in the surface vicinity, creating a negative space charge, which prevents further electron emission. However, the electric field in the cathode vicinity is enough to push the negative space charge out of the electrode and reach the saturation current density. This saturation current density is the main characteristic of the cathode thermionic emission.

To derive the expression for the saturation current of thermionic emission, consider the electron distribution function in metals, which is the density of electrons in velocity interval $v_x \div v_x + dv_x$, etc. The electron distribution in metals is essentially a quantum-mechanical one, and can be described by the Fermi function:[62]

$$f(v_x, v_y, v_z) = \frac{2m^3}{(2\pi\hbar)^3} \frac{1}{1 + \exp\dfrac{\varepsilon - \mu(T, n_e)}{T}} \tag{8.1}$$

Here, $\mu(T,n_e)$ is the chemical potential, which can be found from normalization of the Fermi distribution function. This means that the integral of the distribution (Equation 8.1) over all velocities should be equal to the total electron density n_e in metal. Also, in Equation 8.1, m and ε are electron mass and total energy, including kinetic and potential :

The most energetic electrons are able to leave the metal if their kinetic energy $mv_x^2/2$ in the direction x perpendicular to the metal surface exceeds the absolute value of the potential energy $|\varepsilon_p|$. The electric current density of these energetic electrons leaving the metal surface can be found by integrating the Fermi distribution function (Equation 8.1):

$$j = e\int_{-\infty}^{+\infty} dv_y \int_{-\infty}^{+\infty} dv_z \int_{\sqrt{2|\varepsilon_p|/m}}^{+\infty} v_x f(v_x, v_y, v_z) dv_x \tag{8.2}$$

The energy distance from the highest electronic level in metal (the Fermi level) to the continuum is called the **work function** W; this actually corresponds to the minimum energy necessary to extract an electron from the metal. In terms of the

Table 8.2 Work Functions of Some Cathode Materials

Material	C	Cu	Al	Mo
Work function	4.7 eV	4.4 eV	4.25 eV	4.3 eV
Material	W	Pt	Ni	W/ThO_2
Work function	4.54 eV	5.32 eV	4.5 eV	2.5 eV

work function, the integration (Equation 8.2) leads to the **Sommerfeld formula** describing the saturation current density for thermionic emission:

$$j = \frac{4\pi me}{(2\pi\hbar)^3} T^2 (1 - R) \exp\left(-\frac{W}{T}\right) \tag{8.3}$$

Here, R is a quantum mechanical coefficient describing the reflection of electrons from the potential barrier related to the metal surface. It is convenient for practical calculations to use the numerical value of the Sommerfeld constant:

$$\frac{4\pi me}{(2\pi\hbar)^3} = 120 \frac{A}{cm^2 K^2},$$

and to take into account typical values of the reflection coefficient $R = 0 \div 0.8$. Numerical values of the work function W for some cathode materials are given in Table 8.2. As the table shows, a tungsten cathode covered by thorium oxide has a lower work function than a pure tungsten cathode—almost half. As a result, the thermionic emission current from the oxide cathode exponentially exceeds that for a cathode made from pure refractory metal. In general, the complex metal-oxides cathode is very active in thermionic emission.

8.2.2 Schottky Effect of Electric Field on Work Function and Thermionic Emission Current

As shown, the thermionic current grows with electric field until the negative space charge near the cathode is eliminated and saturation is achieved. However, this saturation is rather relative. Further increase of the electric field gradually leads to an increase of the saturation current level, which is related to reduction of work function. The effect of electric field on the work function is known as the **Schottky effect.**

The work function W is actually the binding energy of an electron to a metal surface. Neglecting the external electric field, this is work $W_0 = e^2/(4\pi\varepsilon_0)4\alpha^2$ against the attractive image force $e^2/(4\pi\varepsilon_0)(2r)^2$ (between the electron and its mirror image inside metal). Here, a is the typical interatomic distance in metal. In the presence of an external electric field E, extracting electrons from cathode, the total electric field applied to the electron can be expressed as a function of its distance from the metal surface:

$$F(r) = \frac{e^2}{16\pi\varepsilon_0 r^2} - eE \tag{8.4}$$

As can be seen, the extraction takes place if electron distance from the metal exceeds the critical one: $r_{cr} = \sqrt{e/16\pi\varepsilon_0 E}$ when attraction to the surface is changed to repulsion. Thus, the work function can be calculated as the integral:

$$W = \int_a^{r_{cr}} F(r)\,dr \approx \frac{e^2}{16\pi\varepsilon_0 a} - \frac{1}{\sqrt{4\pi\varepsilon_0}} e^{3/2}\sqrt{E} \tag{8.5}$$

The Schottky relation (Equation 8.5) returns in the absence of electric field $E = 0$ to the previously mentioned expression $W = W_0$ for the work function. For practical calculations of the work function decrease in an external electric field, the Schottky relation (Equation 8.5) can be rewritten in the following numerical way:

$$W, \text{eV} = W_0 - 3.8 \cdot 10^{-4} \cdot \sqrt{E, V/\text{cm}} \tag{8.6}$$

As Equation 8.6 shows, the decrease of work function is relatively small at reasonable values of electric field. However, the Schottky effect can result in a major change of the thermionic current because of its strong exponential dependence on the work function in accordance with the Sommerfeld formula (Equation 8.3).

A relevant example of numerical dependence of the thermionic emission current density on electric field is given in Table 8.3.[271] From this table, a 4-fold change of electric field results in 800-fold increase of the thermionic current density. Table 8.3 also includes numerical examples regarding field electron emission and thermionic field emission, which will be discussed later in this section.

8.2.3 Field Electron Emission in Strong Electric Fields: Fowler–Nordheim Formula

If the external electric fields are very high (about $1 - 3 \cdot 10^6$ V/cm), they are able not only to decrease the work function but also directly and effectively to extract electrons from cold metal due to the quantum-mechanical effect of tunneling. A simplified triangular potential energy barrier for electrons inside a metal, taking into account external electric field E but neglecting the mirror forces, is presented in Figure 8.4. Electrons are able to escape from metal across the barrier due to quantum-mechanical tunneling, which is called the field emission effect. The field electron emission current density in the approximation of the triangular barrier (Figure 8.4) can be calculated by the following Fowler–Nordheim formula:

$$j = \frac{e^2}{4\pi^2\hbar}\frac{1}{(W_0 + \varepsilon_F)}\sqrt{\frac{\varepsilon_F}{W_0}}\exp\left[-\frac{4\sqrt{2m}\,W_0^{3/2}}{3e\hbar E}\right] \tag{8.7}$$

Table 8.3 Current Densities of Thermionic, Field, and Thermionic Field Emissions as a Function of Electric Field E

Electric field, 10^6 V/cm	Schottky decrease of W, V	Thermionic emission, j, A/cm^2	Field emission, j, A/cm^2	Thermionic field emission, j, A/cm^2
0	0	$0.13 \cdot 10^3$	0	0
0.8	1.07	$8.2 \cdot 10^3$	$2 \cdot 10^{-20}$	$1.2 \cdot 10^4$
1.7	1.56	$5.2 \cdot 10^4$	$2.2 \cdot 10^{-4}$	$1.0 \cdot 10^5$
2.3	1.81	$1.4 \cdot 10^5$	1.3	$2.1 \cdot 10^5$
2.8	2.01	$3.0 \cdot 10^5$	130	$8 \cdot 10^5$
3.3	2.18	$6.0 \cdot 10^5$	$4.7 \cdot 10^3$	$2.1 \cdot 10^6$

Note: Values of electrode temperature (T = 3000 K); work function (W = 4 eV); Fermi energy (ε_F = 7 eV); and pre-exponential factor ($A_0(1 - R)$ = 80 A/cm^{2}K^2) of the Sommerfeld relation are taken in the example of numerical calculations.

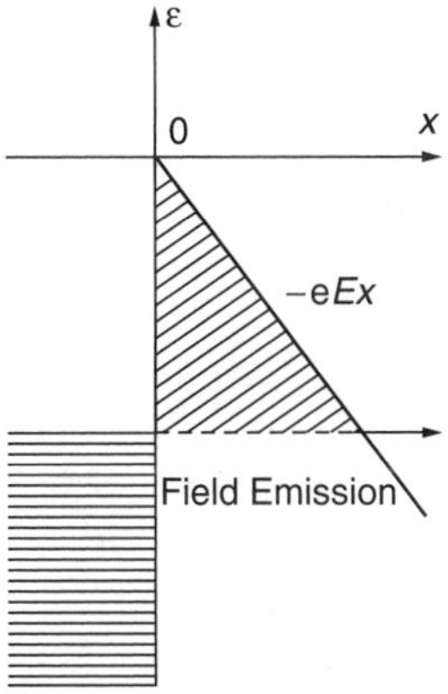

Figure 8.4 Electron potential energy on the metal surface when electric field is applied

Here, ε_F is the Fermi energy of a metal and W_0 is the work function not perturbed by external electric field (see Equation 4.54).

As seen from the Fowler–Nordheim formula, the field emission current is sensitive to even small changes in the electric field and work function, including those related to the Schottky effect (Equation 8.6). Electron tunneling across the potential barrier influenced by the Schottky effect and the corresponding field emission are illustrated in Figure 8.5. Related correction in the Fowler–Nordheim formula can be done by introducing a special numerical factor $\xi(\Delta W/W_0)$, depending on the relative Schottky decrease of work function (Equation 8.6):

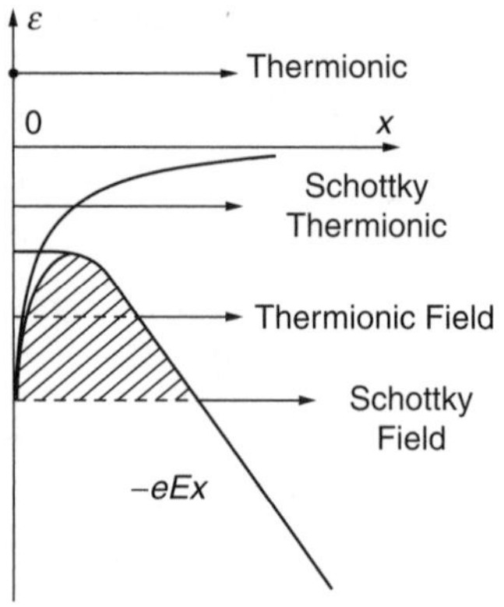

Figure 8.5 Thermionic and field emission; Schottky effect

Table 8.4 Correction Factor $\xi\Delta W/W_0$ in Fowler–Nordheim Formula for Field Emission

Relative Schottky decrease of work function $\Delta W/W_0$	0	0.2	0.3	0.4	0.5
Fowler–Nordheim correction factor $\xi(\Delta W/W_0)$	1	0.95	0.90	0.85	0.78
Relative Schottky decrease of work function $\Delta W/W_0$	0.6	0.7	0.8	0.9	1
Fowler–Nordheim correction factor $\xi(\Delta W/W_0)$	0.70	0.60	0.50	0.34	0

$$j = 6.2\cdot 10^{-6}\ \mathrm{A/cm^2} \times \frac{1}{\left(W_0,\mathrm{eV} + \varepsilon_F,\mathrm{eV}\right)}\sqrt{\frac{\varepsilon_F}{W_0}}\exp\left[-\frac{6.85\cdot 10^7 W_0^{3/2}(\mathrm{eV})\cdot \xi}{E,\mathrm{V/cm}}\right] \quad (8.8)$$

The corrective numerical factor $\xi(\Delta W/W_0)$ is tabulated for practical calculations in Table 8.4.[231]

Numerical examples of field emission calculations based on the corrected Fowler–Nordheim relation (Equation 8.8) are presented in Table 8.2 in comparison with current densities of other emission mechanisms. From the table, the field emission current density dependence on electric field is really strong in this case: a fourfold change of electric field results in an increase of the field emission current density by more than 23 orders of value. According to the corrected Fowler–Nordheim formula (Equation 8.8) and Table 8.2, the field electron emission becomes significant when the electric field exceeds 10^7 V/cm. However, the electron field emission already makes significant contribution at electric fields about $3 \cdot 10^6$ V/cm because of the field enhancement at the microscopic protrusions on metal surfaces.

8.2.4 Thermionic Field Emission

When cathode temperature and external electric field are high, thermionic and field emission make significant contributions to the current of electrons escaping the metal. This emission mechanism is usually referred to as the thermionic field emission and plays an important role in cathode spots. To compare emission mechanisms, it is convenient to subdivide electrons escaping the metal surfaces into the four groups illustrated in Figure 8.2. Electrons of the first group have energies below the Fermi level, so they are able to escape metal only through tunneling or, in other words, by the field emission mechanism. Electrons of the fourth group leave metal by the thermionic emission mechanism without any support from the electric field. These two groups of electrons present extremes in the electron emission mechanisms.

Electrons of the third group overcome the potential energy barrier because of its reduction in the external electric field. This Schottky effect of the electric field is obviously a pure classical one. The second group of electrons is able to escape the metal only quantum-mechanically by tunneling similar to that from the first group. However, in this case, the potential barrier of tunneling is not so large because of the relatively high thermal energy of the second group's electrons. These electrons escape the cathode by the mechanism of the thermionic field emission.

Because the thermionic emission is based on the synergetic effects of temperature and electric field, these two key parameters of electron emission need only to be reasonably high to provide the significant emission current. Results of calculations of the thermionic field emission[231] are also presented in Table 8.2. The thermionic field emission dominates other mechanisms at $T = 3000$ K and $E > 8 \cdot 10^6$ V/cm. Note that at high temperatures but lower electric fields $E < 5 \cdot 10^6$ V/cm, electrons of the third group usually dominate the emission, which follows, in this case, the Sommerfeld relation (Equation 8.3) with the work function diminished by the Schottky effect (Equation 8.6).

8.2.5 Secondary Electron Emission

The previously considered mechanisms of thermionic and field emissions play the most important role in the cathode processes of arc discharges. There are other mechanisms of electron emission from solids, related to surface bombardment by different particles, that are called secondary electron emissions. The secondary electron emission does not make major contributions in electrode kinetics of arc discharges, but obviously is important in other discharge systems. The most important from this group is the **secondary ion–electron emission**, which refers to the emission of electrons from a surface induced by ion impact. This secondary emission mechanism makes an important contribution to the Townsend breakdown (see Section 4.4.1) and sustaining cathode current in glow discharges (see Section 7.1). Actually, the secondary ion–electron emission is the principal distinctive feature of the glow discharges (Chapter 7).

As shown in Section 2.2.7, the direct ionization in collisions of ions with neutral atoms is not effective because of the adiabatic principle. Heavy ions are unable to transfer energy to light electrons to provide ionization. This general statement also can be applied to the direct electron emission from solid surfaces induced by ion impact. As seen from Figure 8.6,[272] the secondary electron emission coefficient γ electron yield per one ion (see Section 4.4.1) effectively starts growing with ion

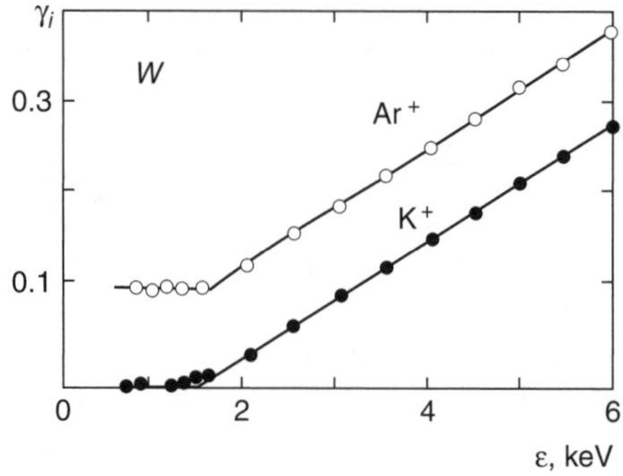

Figure 8.6 Secondary electron emission from tungsten as function of ion energy[272]

energy only at very high energies exceeding 1 keV, when the Massey parameter becomes large.

Although the secondary ion–electron emission coefficient γ is indeed lower (on the level of 0.01 to 0.1) at lower ion energies, it is not negligible (see Figure 8.6). This value of the coefficient γ stays almost constant at ion energies below the kilovolt range. This can be explained by the **Penning mechanism of the secondary ion–electron emission**. According to this mechanism, which is also called the potential mechanism, ions approaching a surface extract an electron because the ionization potential I exceeds the work function W. The defect of energy $I - W$ is usually enough ($I - W > W$) to provide escape of one more electron from the surface. Such a process is nonadiabatic (Section 2.2.7), so its probability is not negligible.

If surfaces are clean, the secondary ion–electron emission coefficient γ can be calculated with reasonable accuracy using the following empirical formula:

$$\gamma \approx 0.016(I - 2W) \tag{8.9}$$

If surfaces are not clean, the γ coefficients are usually lower and also grow with ion energy even when they are not high enough for adiabatic emission to take place.

Another secondary electron emission mechanism is related to the surface bombardment by excited meta-stable atoms with excitation energy exceeding the surface work function. This so-called **potential electron emission induced by meta-stable atoms** can have quite a high secondary emission coefficient γ. Some of these are presented in Table 8.5. Note the term "potential" is applied to this emission mechanism to contrast it with the kinetic one, related to ineffective adiabatic ionization or emission induced by kinetic energy of the heavy particles.

Secondary electron emission also can be provided by a photoeffect metallic surface. The **photoelectron emission** is usually characterized by the **quantum yield** $\gamma_{\hbar\omega}$, which is similar to the secondary emission coefficient and shows the number of emitted electrons per one quantum $\hbar\omega$ of radiation. The quantum yields as a function of photon energy $\hbar\omega$ for different metal surfaces[109] shown in Figure 8.7. From this figure, the visual light and low-energy UV radiation give the quantum yield $\gamma_{\hbar\omega} \approx 10^{-3}$, which is sensitive to the quality of the surface. High-energy UV radiation

Table 8.5 Secondary Emission Coefficient γ for Potential Electron Emission Induced by Collisions with Meta-Stable Atoms

Meta-stable atom	Surface material	Secondary emission coefficient γ
$He(2^3S)$	Pt	0.24 electron/atom
$He(2^1S)$	Pt	0.4 electron/atom
Ar*	Cs	0.4 electron/atom

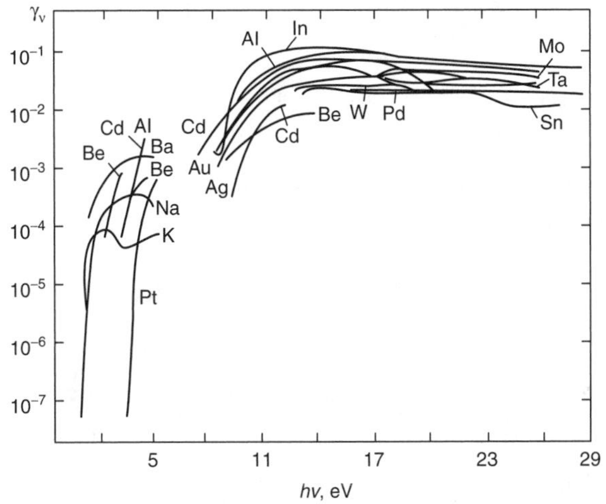

Figure 8.7 Photoelectron emission coefficients as function of photon energy[109]

provides emission with the quantum yield in the range of 0.01 to 0.1 and is less sensitive to surface characteristics.

Finally, **secondary electron–electron emission** refers to electron emission from a solid surface induced by electron impact. In general, this emission mechanism does not play any significant role in DC discharges. The plasma electrons move to the anode where electron emission is not effective in the presence of an electric field pushing them back to the anode. However, the secondary electron–electron emission can be important in the case of high-frequency breakdown of discharge gaps at very low pressures, and also in heterogeneous discharges, which will be discussed in Chapter 11.

The secondary electron–electron emission is usually characterized by the multiplication coefficient γ_e, which shows the number of emitted electrons produced by the initial one. Dependence of the multiplication coefficient γ_e on electron energy

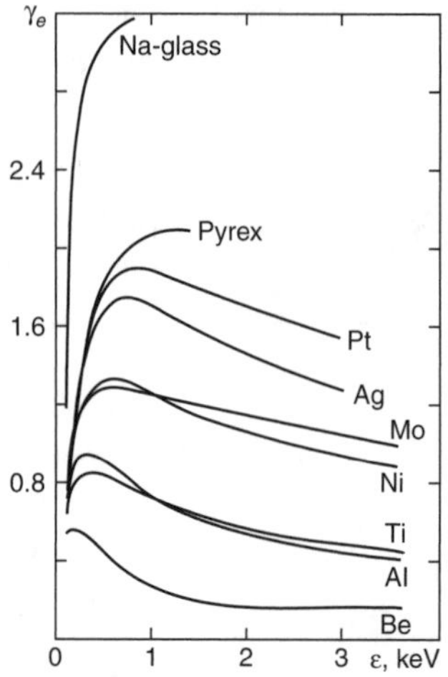

Figure 8.8 Secondary electron emission as function of bombarding electron energy[272]

for different metals and dielectrics[272,273] is shown in Figure 8.8. Additional information on the subject can be found in Modinos[274] and Komolov.[275]

8.3 CATHODE AND ANODE LAYERS IN ARC DISCHARGES

8.3.1 General Features and Structure of Cathode Layer

The general function of the cathode layer is to provide the high current necessary for electric arc operation. As discussed in the previous section, electron emission from the cathode in arcs is due to thermionic and field emission mechanisms. These are much more effective with respect to the secondary ion–electron emission that dominates in glow discharges. In the case of thermionic emission, ion bombardment provides cathode heating, which then leads to escape of electrons from the surface.

The secondary emission usually gives about $\gamma \approx 0.01$ electrons per one ion, while thermionic emission can generate $\gamma_{eff} = 2 - 9$ electrons per one ion. The fraction of electron current near the cathode in a glow discharge is very small $\gamma/(\gamma + 1) \approx 0.01$ (see Equation 7.3). The same approach gives the fraction of electron current in the cathode layer of an arc:

$$S = \frac{\gamma_{eff}}{\gamma_{eff} + 1} \approx 0.7 - 0.9 \tag{8.10}$$

Equation 8.10 shows that thermionic emission from the cathode actually provides most of the electric current in the arc discharge. On the other hand, it is known that the electric current in the positive column of arc and glow discharges is almost

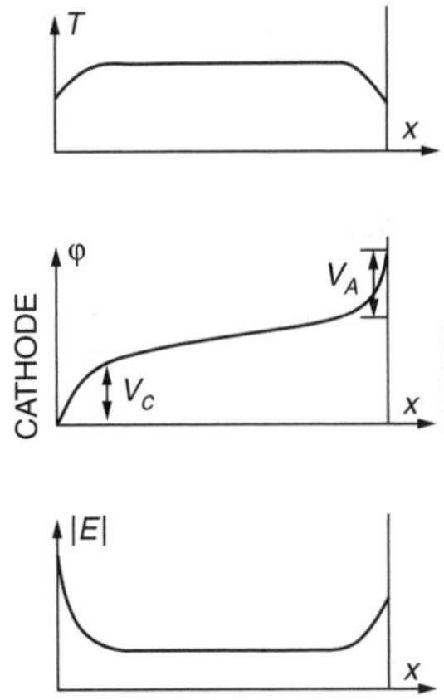

Figure 8.9 Distributions of temperature, potential, and electric field from cathode to anode

completely provided by electrons because of their very high mobility with respect to ions. In glow discharges, most of this current $(1-\gamma/(\gamma+1) \approx 99\%)$ is generated by electron-impact gas-phase ionization in the cathode layer. For this reason, the cathode layer voltage should be quite high—on the level of hundreds of volts—to provide several generations of electrons (see Section 7.2.3, Table 7.2).

In contrast, the electron-impact gas-phase ionization in the cathode layer of arc discharges should provide only a minor fraction of the total discharge current $1 - S \approx 10 - 30\%$. This means that less than one generation of electrons should be born in the arc cathode layer. The necessary cathode drop voltage in this case is relatively low, about the same as or even less than ionization potential.

Thus, the cathode layer in arc discharges has several self-consistent specific functions. First, a sufficient number density of ions should be generated there to provide the necessary cathode heating in the case of thermionic emission. Gas temperature near the cathode is the same as the cathode surface temperature and a couple of times less than the temperature in positive column (see Figure 8.9). For this reason, the thermal ionization mechanism is unable to provide the necessary degree of ionization, which requires the necessity of nonthermal ionization mechanisms (direct electron impact, etc.) and thus the necessity of elevated electric fields near the cathode (see Figure 8.9).

The elevated electric field in the cathode vicinity stimulates electron emission by a decrease of work function (the Schottky effect), as well as by contributions of field emission. On the other hand, intensive ionization in the cathode vicinity leads to a high concentration of ions in the layer and in the formation of a positive space charge, which actually provides the elevated electric field. General distribution of arc parameters, temperature, voltage, and electric field along the discharge from cathode to anode is illustrated in Figure 8.9.

Structure of the cathode layer is illustrated in Figure 8.10. A large positive space charge, with high electric fields and most of the cathode voltage drop, is located in

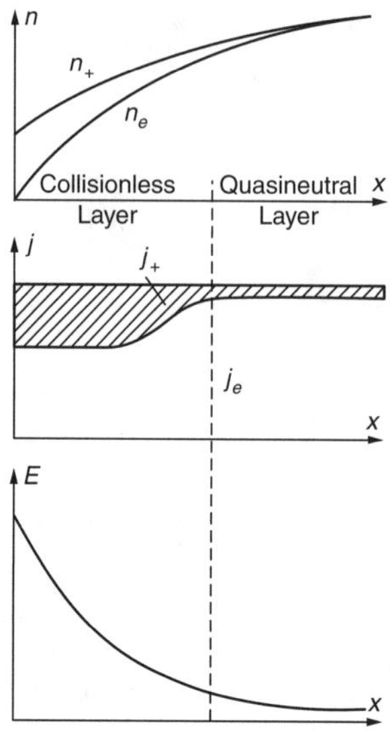

Figure 8.10 Distributions of charge density, current, and field in cathode layer

the very narrow layer near the cathode. This layer is actually even shorter than the ions' and electrons' mean free path, so it is usually referred to as the collisionless zone of cathode layer. Between the narrow collisionless layer and positive column, the longer quasi-neutral zone of cathode layer is located. While the electric field in the quasi-neutral layer is not so high, the ionization is quite intensive there because electrons retain the high energy received in the collisionless layer. Most ions carrying current and energy to the cathode are generated in this quasi-neutral zone of cathode layer.

From Figure 8.10, it is seen that the electron and ion components of the total discharge current are constant in the collisionless layer, where there are no sources of charge particles. In the following quasi-neutral zone of cathode layer, a fraction of electron current grows from $S \approx 0.7 - 0.9$ (see Equation 8.10) to almost unity in the positive column (to be more exact, to ratio of mobilities $\mu_+/(\mu_e + \mu_+)$). Plasma density $n_e \approx n_+$ in the quasi-neutral zone of cathode layer steadily grows in the direction of the positive column because of intensive formation of electrons and ions in this layer.

8.3.2 Electric Field in Cathode Vicinity

Next consider the collisionless zone of the cathode layer. Electron j_e and ion j_+ components of the total current density j are fixed in this zone and can be expressed as:

$$j_e = S \cdot j = n_e e v_e, \quad j_+ = (1 - S) j = n_+ e v_+ \tag{8.11}$$

Electron and ion velocities v_e and v_+ can then be presented as a function of voltage V, assuming $V = 0$ at the cathode and $V = V_C$ at the end of the collisionless layer (m and M are masses of electrons and positive ions, respectively):

$$v_e = \sqrt{2eV/m}, \quad v_+ = \sqrt{2e(V_C - V)/M} \tag{8.12}$$

Based on Equation 8.11 and Equation 8.12, the Poisson's equation for the voltage V in the collisionless layer can be written as:

$$-\frac{d^2V}{dx^2} = \frac{e}{\varepsilon_0}\left(n_+ - n_e\right) = \frac{j}{\varepsilon_0\sqrt{2e}}\left[\frac{(1-S)\sqrt{M}}{\sqrt{V_C - V}} - \frac{S\sqrt{m}}{\sqrt{V}}\right] \tag{8.13}$$

Taking into account that

$$\frac{d^2V}{dx^2} = \frac{1}{2}\frac{dE^2}{dV},$$

the Poisson's equation (Equation 8.13) should be integrated assuming as a boundary condition that, at the positive column side, electric field is relatively low: $E \approx 0$ at $V = V_C$. This leads to the relation between the electric field near the cathode, current density, and the cathode voltage drop (which can be represented by V_C):

$$E_C^2 = \frac{4j}{\varepsilon_0\sqrt{2e}}\left[(1-S)\sqrt{M} - S\sqrt{m}\right]\sqrt{V_C} \tag{8.14}$$

The first term in Equation 8.14 is related to the ions' contribution in formation of a positive space charge and enhancement of the electric field near the cathode. The second term is related to electrons' contribution in slight compensation of the ionic space charge. Noting from Equation 8.10 that the fraction of electron current S = 0.7 – 0.9, the second term in Equation 8.14 can be neglected and the equation can then be rewritten in the following numerical form:

$$E_c, \mathrm{V/cm} = 5 \cdot 10^3 \cdot A^{1/4}(1-S)^{1/2}(V_C, V)^{1/4}(j, \mathrm{A/cm^2})^{1/2} \tag{8.15}$$

where A is the atomic mass of ions in a.m.u. For example, for an arc discharge in nitrogen ($A = 28$) at typical values of current density for hot cathodes $j = 3 \cdot 10^3$ A/cm^2, cathode voltage drop $V_C = 10$ eV and $S = 0.8$ give, according to Equation 8.15, the electric field near the cathode: $E_c = 5.7 \cdot 10^5$ V/cm. This electric field also provides a reduction of the cathode work function (Equation 8.6) of about 0.27 eV, which permits the thermionic emission at 3000 K to triple.

Integration of the Poisson's equation (Equation 8.13), neglecting the second term related to effect of electrons on the space charge, permits finding an expression for the length of the collisionless zone of the cathode layer as:

$$\Delta l = 4V_C/3E_C \tag{8.16}$$

Numerically, this example gives the length of the collisionless layer as $\Delta l \approx 2 \cdot 10^{-5}$ cm.

8.3.3 Cathode Energy Balance and Electron Current Fraction on Cathode (*S*-factor)

The cathode layer characteristics depend on the *S*-factor showing the fraction of the electron current in the total current on the cathode. Simple numerical phenomenological estimation of the *S*-factor was given by Equation 8.10. Actually, this factor depends on a detailed energy balance on the arc discharge cathode.

Accurate calculations on the subject are quite complicated. However, reasonable qualitative estimations can be done assuming that the energy flux brought to the cathode surface by ions goes completely to provide electron emission from the cathode. Each ion brings to the surface its kinetic energy (which is of the order of cathode voltage drop V_C) and also the energy released during neutralization (which is equal to difference $I - W$ between ionization potential and work function necessary to provide an electron for the neutralization). This simplified cathode energy flux balance, neglecting conductive and radiation heat transfer components, can be expressed as:

$$j_e \cdot W = j_+(V_C + I - W) \tag{8.17}$$

The balance Equation 8.17 permits deriving a simplified but fairly accurate relation for the *S*-factor, the fraction of electron current on cathode:

$$S = \frac{j_e}{j_e + j_+} = \frac{V_C + I - W}{V_C + I} \tag{8.18}$$

For example, if $W = 4$ eV, $I = 14$ eV, and $V_C = 10$ eV, according to Equation 8.17, the fraction of electron current is $S = 0.83$, which is quite close to measurements. It is interesting to note that much more detailed calculations that take into account additionally conductive and radiative heat transfer do not essentially change the result.[276]

8.3.4 Cathode Erosion

The high energy flux to the cathode results obviously in thermionic emission (Equation 8.17), as well as in erosion of the electrode material. The erosion is very sensitive to presence of oxidizers; even 0.1% of oxygen or water vapor makes a significant effect. Numerically, the erosion effect is usually characterized by the **specific erosion**, which shows the loss of electrode mass per unit charge passed through the arc. As an example, the specific erosion of tungsten rod cathodes at moderate and high pressures of inert gases and a current of about 100 A is about 10^{-7} *g/C*.

The most intensive erosion takes place in the hot cathode spots that will be discussed next. At low pressures of about 1 torr and less, the cathode spots are formed even on refractory materials. For this reason, the rod cathodes of refractory metals are usually used only at high pressures. At low pressures, the hollow cathode configuration can be effectively used from this point of view (see Section 7.2.7). The arc is anchored to the inner surface of the hollow cathode tube, where the gas flow rate is sufficiently high. Specific erosion of this hollow cathode made from refractory metals can reach the very low level of 10^{-9} to 10^{-10} g/C.[276]

8.3.5 Cathode Spots

The cathode spots are localized current centers, which can appear on the cathode surface when significant current should be provided in the discharge, but the entire cathode cannot be heated enough to make it. The most typical cause of cathode spots is application of metals with relatively low melting points. The cathode spots can be caused as well by relatively low levels of the arc current that are able to provide the necessary electron emission only when concentrated to a small area of the cathode spot.

The cathode spots appear also at low gas pressures even in the case of cathodes made of refractory metals. At low gas pressures (usually less than 1 torr), metal vapor from cathode provides atoms to generate enough positive ions to bring their energy to the cathode to sustain electron emission. To provide the required evaporation of cathode material, the electric current should be concentrated in the cathode spots. Thus, at low pressures (less than 1 torr) and currents of 1 to 10 A, the cathode spots appear even on refractory metals (on low-melting-point metals, the cathode spots appear at any pressures and currents).

Initially, the cathode spots are formed fairly small (10^{-4} to 10^{-2} cm) and move very fast (10^{-3} to 10^{-4} cm/sec). These primary spots are nonthermal; relevant erosion is not significant and probably is related to microexplosions due to localization of current on tiny protrusions on the cathode surface. After a time interval of about 10^{-4} sec, the small primary spots merge into larger spots (10^{-3} to 10^{-2} cm). These matured cathode spots can have temperatures of 3000 K and greater, and provide conditions for intensive thermal erosion mechanisms. They also move much slower (10 to 100 cm/sec).

Typical current through an individual spot is 1 to 300 A. Growth of current leads to splitting of the cathode spots and their multiplication in this way. The minimum current through a single spot is about $I_{\min} \approx 0.1 - 1A$. The arc as a whole extinguishes at lower currents. This critical minimum current through an individual cathode spot for different nonferromagnetic materials can be numerically found using the following empirical formula:

$$I_{\min}, A \approx 2.5 \cdot 10^{-4} \cdot T_{boil}(K) \cdot \sqrt{\lambda, W/\text{cmK}} \tag{8.19}$$

In this relation, T_{boil} (in Kelvin) is the boiling temperature of the cathode material and γ (in *W*/cmK) is the heat conduction coefficient.

It is interesting to note that the cathode spots are sources of intensive jets of metal vapor. Emission of 10 electrons corresponds approximately to an erosion of one atom. The metal vapor jet velocities can be extremely high: 10^{-5} to 10^{-6} cm/sec. Numerous experimental data characterizing the cathode spots can be found in Lafferty[277] and Lyubimov and Rachovsky.[278] Some of these data are summarized in Table 8.6.

Table 8.6 Typical Characteristics of Cathode Spots

Cathode material	Cu	Hg	Fe	W	Ag	Zn
Minimum current through a spot, A	1.6	0.07	1.5	1.6	1.2	0.3
Average current through a spot, A	100	1	80	200	80	10
Current density, A/cm^2	$10^4 \div 10^8$	$10^4 \div 10^6$	10^7	$10^4 \div 10^6$	—	$3 \cdot 10^4$
Cathode voltage drop, V	18	9	18	20	14	10
Specific erosion at 100–200 A, g/C	10^{-4}	—	—	10^{-4}	10^{-4}	—
Vapor jet velocity, 10^5 cm/sec	1.5	1	0.9	3	0.9	0.4

As shown in the table, the current density in cathode spots can reach extremely high levels of 10^8 A/cm^2. Such large values of electron emission current density can be explained only by thermionic field emission (see Section 8.2.4). Contribution into initial high-current densities in a spot can also be due to the **explosive electron emission** related to localization of strong electric fields and following explosion of microprotrusions on the cathode surface. This phenomenon, which plays a key role in pulse breakdown of vacuum gaps, can influence the early stages of the cathode spot evolution.[279]

Although extensive experimental and theoretical research has been conducted,[277,278] several problems related to the cathode spot phenomenon are not completely solved. Some of them sound like paradoxes. For example, the current–voltage characteristics of vacuum arcs (where the cathode spots are usually observed) are not decreasing, as is traditional for arcs, but rather increasing. Furthermore, there is no complete explanation of a mechanism for the cathode spot motion and splitting. The most intriguing cathode spot paradox is related to the direction of its motion in the external magnetic field. If this field is applied along a cathode surface, the cathode spots move in the direction opposite to that corresponding to the magnetic force $\vec{I} \times \vec{H}$. Although quite a few hypotheses and models have been proposed on the subject, a consistent explanation of the paradox is still absent.

8.3.6 External Cathode Heating

If a cathode is externally heated, it is not necessary to provide its heating by ion current. In this case, the main function of a cathode layer is acceleration of thermal electrons to energies sufficient for ionization and sustaining the necessary level of plasma density. Losses of charged particles in such thermal plasma systems cannot be significant, resulting in low values of cathode voltage drop. The cathode voltage drop in such systems is often lower than ionization (and even electronic excitation).

If the discharge chamber is filled with a low-pressure (about 1 torr) inert gas, voltage to sustain the positive column is also low, about 1 V; anode voltage drop is also not large. The arc discharges with low total voltage of about 7 to 8 V are usually referred to as **low-voltage arcs**. For example, such low values of voltage are sufficient for the non-self-sustained arc discharge in a spherical chamber with a 5-cm radius in argon at 1- to 3-torr pressure and currents of 1 to 2 A. Such kinds of low-pressure gas discharges are used in diodes and thyratrones in particular.

8.3.7 Anode Layer

Similar to the case of a cathode, an arc can connect to the anode in two different ways: by diffuse connection or by anode spots. The diffuse connection usually occurs on large-area anodes; current density in this case is about 100 A/cm^2. The anode spots usually appear on relatively small and nonhomogeneous anodes; current density in the spots is about 10^4 to 10^5 A/cm^2. The number of spots grows with total current and pressure. Sometimes the anode spots are arranged in regular patterns and move along regular trajectories.

The anode voltage drop consists of two components. The first is related to negative space charge near the anode surface, which obviously repels ions. This small voltage drop (of about the ionization potential and less at low currents) stimulates some additional electron generation to compensate for the absence of ion current in the region. The second component of the anode voltage drop is related to the arc discharge geometry. If the anode surface area is smaller than the positive-column cross section or the arc channel is contracted on the anode surface, electric current near the electrode should be provided only by electrons. This requires higher values of electric fields near the electrode and additional anode voltage drops, sometimes exceeding the space charge voltage by a factor of two.

Each electron brings to the anode an energy of about 10 eV consisting of the kinetic energy obtained in the pre-electrode region and the work function. Thus, the energy flux in an anode spot at current density of 10^4 to 10^5 A/cm^2 is about 10^5 to 10^6 W/cm^2. Temperatures in anode spots of vacuum metal arcs are about 3000 K and, in carbon arcs, about 4000 K.

8.4 POSITIVE COLUMN OF ARC DISCHARGES

8.4.1 General Features of Positive Column of High-Pressure Arcs

The Joule heat released per unit length of positive column in the high-pressure arcs is quite significant, usually 0.2 to 0.5 kW/cm. This heat release can be balanced in three different ways that define three different manners of arc stabilization. If the joule heat is balanced by heat transfer to the cooled walls, the arc is referred to as wall stabilized. If the joule heat is balanced by intensive (often rotating) gas flow, the arc is referred to as flow stabilized. Finally, if heat transfer to electrodes balances joule heat in the short positive column, the arc is referred to as electrode stabilized.

The arc discharge plasma of molecular gases at high pressures ($p \geq 1$ atm) is always in quasi equilibrium at any values of current. In the case of inert gases, the electron–neutral energy exchange is less effective and requires relatively high values of current and electron density to reach quasi equilibrium at atmospheric pressure[231]

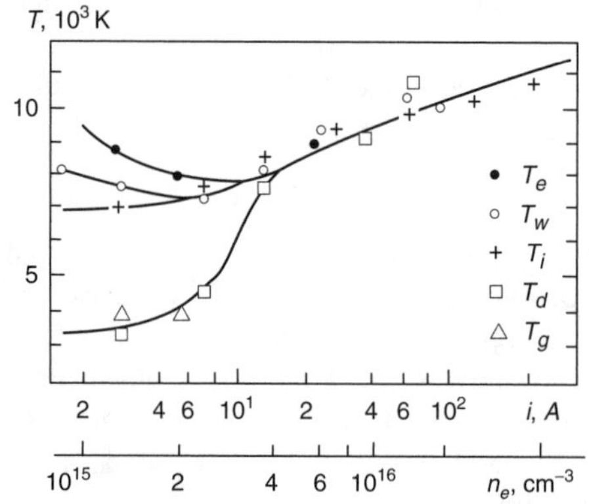

Figure 8.11 Temperature separation in the postive arc column in argon, or Ar with an admixture of H_2 at p = 1 atm as a function of the current density or electron density. T_e is the electron temperature; T_w corresponds to the population of the upper levels; the ion temperature T_i is related to n_e by the Saha formula; T_g is the gas temperature; and T_d corresponds to the population of the lower levels.[231]

(see Figure 8.11). Temperatures and electron concentrations reach their maximum on the axis of the positive column and decrease toward the walls; the electron density obviously decays exponentially faster than temperatures. Also, the temperatures of electrons and neutrals can differ considerably in arc discharges at low pressures ($p \leq 0.1$ Torr) and currents ($I \approx 1$ A).

Typical current–voltage characteristics of arc discharges are illustrated in Figure 8.12 for different pressures.[231] The electric field E is constant along the positive column, so it actually describes voltage. The current–voltage characteristics are hyperbolic, indicating that joule heat per unit length $w = EI$ does not significantly change with current I. Furthermore, the joule heat per unit length $w = EI$ grows with pressure, which is related to intensification of heat transfer mostly due to the radiation contribution of the high-density plasma.

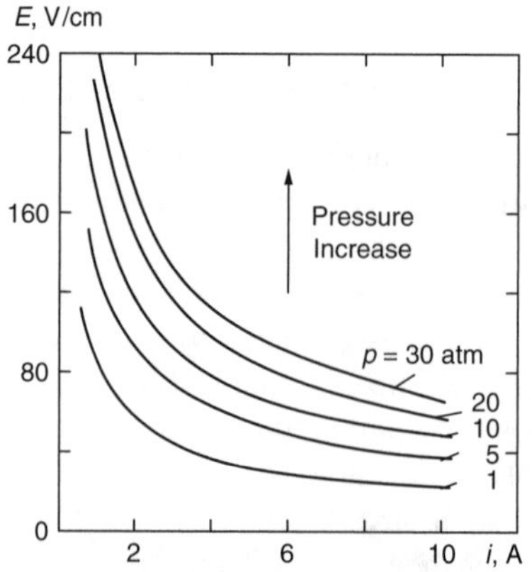

Figure 8.12 Current-voltage characteristics of positive arc columns in air at various pressures[231]

Table 8.7 Radiation Power per Unit Length of Positive Column of Arc Discharges at Different Pressures and Different Values of Joule Heat per Unit Length $w = EI$, W/cm.

Gas	Pressure, atm	Radiation power per unit length, W/cm	$w = EI$, W/cm
Hg	≥1	$0.72 \cdot (w - 10)$	—
Xe	12	$0.88 \cdot (w - 24)$	>35
Kr	12	$0.72 \cdot (w - 42)$	>70
Ar	1	$0.52 \cdot (w - 95)$	>150

The contribution of radiation increases somewhat proportionally to the square of the plasma density and thus grows with pressure. The arc radiation losses in atmospheric air are only about 1%, but become quite significant at pressures higher than 10 atm and high arc power. The highest level of radiation can be reached in Hg, Xe, and Kr. This effect is practically applied in mercury and xenon lamps. Convenient empirical formulas for calculating plasma radiation in different gases[231] are presented in Table 8.7. According to this table, the percentage of arc power conversion into radiation is very high in mercury and xenon even at relatively low values of the joule heat per unit length.

8.4.2 Thermal Ionization in Arc Discharges: Elenbaas–Heller Equation

Quasi-equilibrium plasma of arc discharges at high pressures and high currents has a wide range of applications (see, for example, Polak et al.[154] and Pfender[280]). In contrast to nonthermal plasmas, which are sensitive to the details of the discharge kinetics (see Chapter 4 and Chapter 5), characteristics of the quasi-equilibrium plasma (density of charged particles, electric and thermal conductivity, viscosity, etc.) are determined only by its temperature and pressure. This makes quasi-equilibrium thermal plasma systems much easier to describe.

Gas pressure is fixed by experimental conditions, so description of the arc positive column requires only description of temperature distribution. Such distribution can be found in particular from the Elenbaas–Heller equation, which is derived considering a long, cylindrical steady-state plasma column stabilized by walls in a tube of radius R. In the framework of the Elenbaas–Heller approach, pressure and current are supposed to be not too high and plasma temperature does not exceed 1 eV. Radiation can be neglected in this case and heat transfer across the positive column can be reduced to heat conduction with the coefficient $\lambda(T)$ (see Section 4.3.11).

According to the Maxwell equation, $curl\ E = 0$. For this reason, the electric field in a long, homogeneous arc column is constant across its cross section. Radial distributions of electric conductivity $\sigma(T)$ (see Section 4.3.2), current density $j =$

$\sigma(T)E$, and Joule heating density $w = jE = \sigma(T)E^2$ are determined only by the radial temperature distribution $T(r)$. Then, plasma energy balance can be expressed by the heat conduction equation with the joule heat source:

$$\frac{1}{r}\frac{d}{dr}\left[r\lambda(T)\frac{dT}{dr}\right] + \sigma(T)E^2 = 0 \tag{8.20}$$

This equation is known as the Elenbaas–Heller equation. Its boundary conditions are $dT/dr = 0$ at $r = 0$, and $T = T_w$ at $r = R$ (the wall temperature T_w can be considered as zero, taking into account that the plasma temperature is much larger).

The electric field E is a parameter of Equation 8.20, but the experimentally controlled parameter is not electric field but rather current, which is related to electric field and temperature as:

$$I = E\int_0^R \sigma[T(r)]\cdot 2\pi r\,dr \tag{8.21}$$

Taking into account Equation 8.21, the Elenbaas–Heller equation permits calculating the function $E(I)$, which is the current–voltage characteristic of plasma column. In this case, electric conductivity $\sigma(T)$ and thermal conductivity $\lambda(T)$ remain as two material functions that determine the current–voltage characteristic. To reduce the number of material functions to only one, it is convenient to introduce the **heat flux potential** $\Theta(T)$, instead of temperature, as an independent parameter:

$$\Theta = \int_0^T \lambda(T)\,dT, \quad \lambda(T)\frac{dT}{dr} = \frac{d}{dr}\Theta \tag{8.22}$$

Using the heat flux potential, the Elenbaas–Heller Equation 8.20 can be rewritten in the following simplified form:

$$\frac{1}{r}\frac{d}{dr}\left[r\frac{d\Theta}{dr}\right] + \sigma(\Theta)E^2 = 0 \tag{8.23}$$

In this form, the Elenbaas–Heller equation includes only the material function $\sigma(\Theta)$. The temperature dependence $\Theta(T)$ determined by Equation 8.22 is much smoother with respect to the temperature dependence of $\lambda(T)$ (see Figure 8.13[281]), which makes the material function $\mu(\Theta)$ also smooth enough.

The Elenbaas–Heller Equation 8.21 was derived for electric arcs stabilized by walls. Nevertheless, it has wider applications, including the important case of arcs stabilized by gas flow (see Section 8.4.1), because the temperature distribution in vicinity of the discharge axis is not very sensitive to external conditions.

8.4.3 The Steenbeck "Channel" Model of Positive Column of Arc Discharges

The Elenbaas–Heller equation (Equation 8.23) cannot be solved analytically because of the complicated nonlinearity of the material function $\sigma(\Theta)$. Dresvin[621] considered

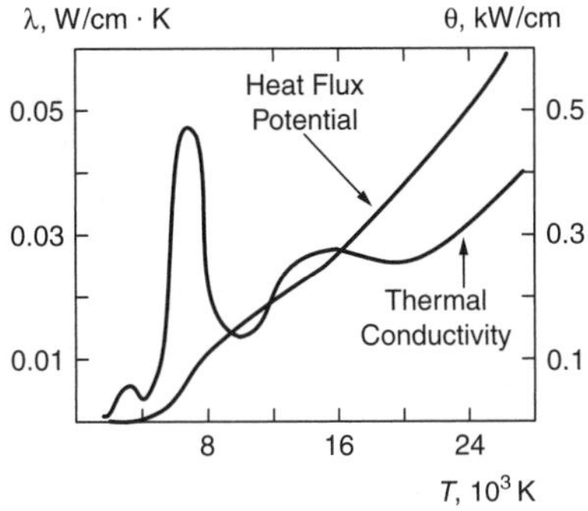

Figure 8.13 Thermal conductivity λ and heat flux potential θ in air at 1 atm[281]

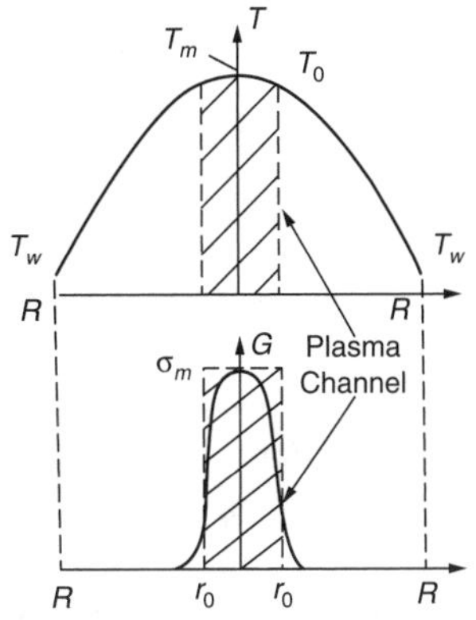

Figure 8.14 Steenbeck channel model

some simplified approaches to the problem; however, good qualitative and quantitative description of the positive column can be achieved by applying the Elenbaas–Heller equation in the analytical channel model proposed by Steenbeck.

The Steenbeck approach, illustrated in Figure 8.14, is based on the very strong exponential dependence of electric conductivity on the plasma temperature related to the Saha equation (see Section 4.1.3, Section 4.3.2, Equation 4.15, and Equation 4.98). At relatively low temperatures (less than 3000 K), the quasi-equilibrium plasma conductivity is small; however, the electric conductivity grows significantly when temperature exceeds 4000 to 6000 K. Figure 8.14 shows that the radical temperature decrease $T(r)$ from the discharge axis to the walls is quite gradual, while the electric conductivity change with radius $[\sigma T(r)]$ is very sharp. This leads to the conclusion that the arc current is located only in a channel of radius r_0—the principal physical basis of the Steenbeck channel model.

Temperature and electric conductivity can be considered as constant inside the arc channel and can be taken equal to their maximum value on the discharge axes: T_m and $\sigma(T_m)$. In this case, the total electric current of the arc can be expressed as:

$$I = E\sigma(T_m)\cdot \pi r_0^2 \tag{8.24}$$

Outside the arc channel $r > r_0$, the electric conductivity can be neglected as well as the current and the joule heat release. For this reason, the Elenbaas–Heller equation can be easily integrated outside the arc channel with boundary conditions $T = T_m$ at $r = r_0$ and $T = 0$ by the walls at $r = R$. The integration leads to the relation between the heat flux potential $\Theta_m(T_m)$ in the arc channel, related to the plasma temperature in the positive column, and the discharge power (the joule heating) per unit arc length $w = EI$:

$$\Theta_m(T_m) = \frac{w}{2\pi}\ln\frac{R}{r_0} \tag{8.25}$$

The heat flux potential $\Theta_m(T_m)$ in the arc channel and the discharge power per unit arc length $w = EI$ are defined in Equation 8.25 as:

$$w = \frac{I^2}{\pi r_0^2 \sigma_m(T_m)}, \quad \Theta_m(T_m) = \int_0^{T_m} \lambda(T)\,dT \tag{8.26}$$

The channel model of an arc includes three discharge parameters, which are assumed determined: plasma temperature T_m, arc channel radius r_0, and electric field E. Electric current I and discharge tube radius R are experimentally controlled parameters. To find the three unknown discharge parameters, T_m, r_0, and E, the Steenbeck channel model has only Equation 8.24 and Equation 8.25. Steenbeck suggested the **principle of minimum power** (see Section 7.2.5) to provide the third equation necessary to complete the system.

According to the Steenbeck principle of minimum power, the plasma temperature T_m and the arc channel radius r_0 should minimize the specific discharge power w and electric field $E = w/I$ at fixed values of current I and discharge tube radius R. Application of the minimization requirement $(dw/dr_0)_{I=const} = 0$ to the functional Equation 8.24 and Equation 8.25 gives the necessary third equation of the Steenbeck channel model in the form:

$$\left(\frac{d\sigma}{dT}\right)_{T=T_m} = \frac{4\pi\lambda_m(T_m)\sigma_m(T_m)}{w} \tag{8.27}$$

Arc discharge modeling, based on the system of Equation 8.24, Equation 8.25, and Equation 8.27, gives excellent agreement with experimental data.[282] However, the validity of the minimum power principle requires a nonequilibrium thermodynamic proof. For the case of the arc discharge, the principle has been rigorously proved.[283]

8.4.4 Raizer "Channel" Model of Positive Column

In 1972, Raizer[284] that showed the channel model does not necessarily require the minimum power principle to justify and complete the system of Equation 8.24, Equation 8.25, and Equation 8.27. The "third" equation (Equation 8.27) can be derived by analysis of the conduction heat flux J_0 from the arc channel $w = J_0 \cdot 2\pi r_0$. This flux is provided by the actual temperature difference $\Delta T = T_m - T_0$ across the arc channel (see Figure 8.14) and can be estimated as:

$$J_0 \approx \lambda_m(T_m) \cdot \frac{\Delta T}{r_0} = \lambda_m(T_m) \cdot \frac{T_m - T_0}{r_0} \tag{8.28}$$

The estimation (Equation 8.28) can be replaced by a more accurate relation by integrating the Elenbaas–Heller Equation 8.23 inside arc channel $0 < r < r_0$ and still assuming homogeneity of the joule heating σE^2:

$$4\pi\Delta\Theta = w \approx 4\pi\lambda_m \Delta T, \quad \Delta\Theta = \Theta_m - \Theta_0 \tag{8.29}$$

The key point of Raizer modification of the channel model is the definition of an arc channel as a region where electric conductivity decreases not more than e times with respect to the maximum value at the axis of the discharge. This definition permits specifying the arc channel radius r_0 and gives the "third" Equation 8.27 of the channel model. The electric conductivity of the quasi-equilibrium plasma in the arc channel can be expressed from the Saha equation (Equation 4.15 and Equation 4.98) as the following function of temperature:

$$\sigma(T) = C\exp\left(-\frac{I_i}{2T}\right) \tag{8.30}$$

where I_i is the ionization potential of gas in the arc and C is the conductivity parameter, which is approximately constant. To be accurate, it should be noted that Equation 8.30 is valid at moderate currents and temperatures, when the degree of ionization is not too high and electron–atomic collisions dominate the collisional frequency in Equation 4.98. However, Equation 8.30 can be applied even at high temperatures and currents by only replacing the ionization potential by the effective potential. For example, electric conductivity in air, nitrogen, and argon at atmospheric pressure and temperatures $T = 8000$ to 14,000 K can be expressed by the same numerical formula:

$$\sigma(T), \mathrm{Ohm}^{-1}\mathrm{cm}^{-1} = 83 \cdot \exp\left(-\frac{36{,}000}{T, K}\right) \tag{8.31}$$

that corresponds to the effective ionization potential $I_{eff} \approx 6.2$ eV.

Taking the electric conductivity of the arc discharge channel in the form of Equation 8.32 and assuming $I/2T >> 1$, the e times decrease of conductivity corresponds to the following small temperature decrease:

$$\Delta T = T_m - T_0 = \frac{2T_m^2}{I_i} \tag{8.32}$$

Combining Equation 8.29 and Equation 8.32 finally gives the required "third" equation of the arc channel model in the form:

$$w = 8\pi\lambda_m(T_m)\frac{T_m^2}{I_i} \tag{8.33}$$

The third equation of the Raizer channel model (Equation 8.33) completely coincides with that of the Steenbeck model (Equation 8.27) based on the principle of minimum power. In this case, the electric conductivity in Steenbeck Equation 8.27 obviously should be taken in the form of Equation 8.30.

8.4.5 Plasma Temperature, Specific Power, and Electric Field in Positive Column According to Channel Model

Equation 8.33 determines the plasma temperature in the arc channel as a function of specific discharge power w per unit length (with ionization potential I_i and thermal conductivity coefficient λ_m as parameters):

$$T_m = \sqrt{w \cdot \frac{I_i}{8\pi\lambda_m}} \tag{8.34}$$

It is interesting that plasma temperature does not depend directly on discharge tube radius and mechanisms of the discharge cooling outside the arc channel. The plasma temperature directly depends only on the specific discharge power w, which, in turn, depends on intensity of the arc cooling. Dependence of plasma temperature T_m on the specific power is also not very strong—less than $\sqrt{w}$ because, in general, the heat conductivity $\lambda(T)$ grows with T (see Figure 8.13).

Experimentally, the principal controlled parameter of an arc discharge is the electric current, which can vary easily by changing external resistance (see Equation 7.1). Thus, it is interesting to express the parameters of the arc channel as functions of current. Assuming, for simplicity, that $\lambda = const$, $\Theta = \lambda T$ in Equation 8.25, Equation 8.30, and Equation 8.33, the conductivity in the arc channel is almost proportional to electric current:

$$\sigma_m = I \cdot \sqrt{\frac{I_i C}{8\pi^2 R^2 \lambda_m T_m^2}} \tag{8.35}$$

Obviously, plasma temperature in the arc channel grows with electric current I, but this growth is only logarithmic:

$$T_m = \frac{I_i}{\ln\left(8\pi^2\lambda_m CT_m^2/I_i\right) - 2\ln(I/R)} \tag{8.36}$$

The fact of nearly constant thermal arc temperature reflects the strong exponential dependence of the degree of ionization on the temperature. A similar effect takes place in nonthermal discharges but with respect to electron temperature, which determines the degree of ionization in these nonequilibrium systems.

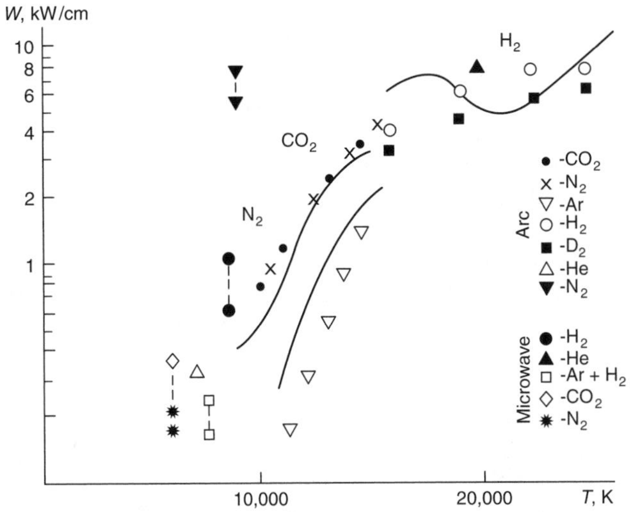

Figure 8.15 Dissipated power per unit length vs. maximum temperature for different types of discharges; solid lines correspond to numerical calculation

The weak logarithmic growth of temperature in the channel with electric current leads (according to Equation 8.33) to the similar weak logarithmic dependence on the electric current of the arc discharge power w per unit length:

$$w \approx \frac{const}{(const - \ln I)^2} \tag{8.37}$$

Experimental data demonstrating the relatively weak dependence of the arc discharge power w per unit length on arc conditions are presented in Figure 8.15. This important effect will also be discussed later regarding evolution of gliding arcs.

The previously mentioned logarithmic constancy of the arc discharge power $w = EI$ per unit length obviously results in the approximate to hyperbolic decrease of electric field with current I:

$$E = \frac{8\pi\lambda_m T_m^2}{I_i} \cdot \frac{1}{I} \approx \frac{const}{I \cdot (const - \ln I)^2} \tag{8.38}$$

This relation explains the almost hyperbolic decrease of current–voltage characteristics typical for thermal arc discharges. The radius of the arc discharge channel may be found in the framework of this model as:

$$r_0 = R\sqrt{\frac{\sigma_m}{C}} = R\sqrt{\frac{I}{R}}\sqrt[4]{\frac{I_i}{8\pi^2\lambda_m T_m^2 C}} \tag{8.39}$$

According to this equation, the arc channel radius grows as the square root of the discharge current I. This can be interpreted as $I \propto r_0^2$, which means that the increase of current primarily leads to growth of the arc channel cross section, while the current density is logarithmically fixed in the same manner as plasma temperature.

All the preceding relations include current always in combination with the discharge tube radius R (the similarity parameter I/R). Obviously, this is due to the initial assumption of wall stabilization arc in the Elenbaas–Heller equation. To generalize the results to the case of gas flow-stabilized arcs, the radius R should be replaced by an effective value describing the actual mechanism of the arc cooling. Intensive cooling in fast gas flows corresponds to small values of the effective radius; according to Equation 8.36 and Equation 8.33, this leads to an increase of plasma temperature.

8.4.6 Possible Difference between Electron and Gas Temperatures in Thermal Discharges

Previously, plasma in the channel model was always taken in quasi equilibrium, which corresponds to relatively high pressures and high currents (see Section 8.4.1). However, at lower pressures and lower currents (when the electric field E is higher; see Equation 8.38), the electron temperature T_e can exceed gas temperature T (see Figure 8.11). This effect can be described by the simplified balance equation for electron temperature (see Section 4.2.7):

$$\frac{3}{2}\frac{dT_e}{dt} = \left[\frac{e^2E^2}{m\nu_{en}^2} - \delta\left(T_e - T\right)\right]\cdot \nu_{en} \tag{8.40}$$

where ν_{en} is the frequency of electron–neutral collisions, and the factor δ characterizes the fraction of an electron energy transferred to neutrals during collisions $\delta = 2\,m/M$ in monatomic gases, and higher in molecular gases (see Equation 4.87).

The balance Equation 8.40 shows that electrons receive energy from the electric field and transfer it to neutrals if $T_e > T$. However, if the temperature difference $T_e - T$ is not large, the excited neutrals are able to transfer energy back to electrons in super-elastic collisions. Thus, Equation 8.40 gives quasi-equilibrium $T_e = T$ in the steady state and absence of electric field $E = 0$. In steady-state conditions but in presence of an electric field E, this balance equation gives the following expression for difference between temperatures:

$$\frac{T_e - T}{T} = \frac{2e^2E^2}{3\delta T_e m\nu_{en}^2} \tag{8.41}$$

Assume that gas in the discharge is mostly monatomic ($\delta = 2\,m/M$) and also replace the electron–neutral frequency ν_{en} by the electron mean free path $\lambda = v_e/\nu_{en}$ (v is the average electron velocity). Then, Equation 8.41 can be rewritten for the difference between electron and gas temperatures in the following numerical way:

$$\frac{T_e - T}{T_e} \approx 200\,A \cdot \left[\frac{E(\mathrm{V/cm}) \cdot \lambda(\mathrm{cm})}{T_e(\mathrm{eV})}\right]^2 \tag{8.42}$$

Here, A is the atomic mass of the monatomic gas in the discharge. If an amount of molecules is present in the discharge, the numerical coefficient in Equation 8.42 is obviously somewhat less than 200 and the degree of nonequilibrium is lower.

Equation 8.42 explains the essential temperature difference at low pressures when the mean free path λ is high and at low currents when electric field is elevated (see Figure 8.11). Note that the electric fields necessary to sustain thermal arc discharges are always lower than those needed to sustain nonequilibrium discharges. Electric fields in thermal discharges are responsible for providing the necessary joule heating, while electric fields in nonthermal discharges are "directly" responsible for ionization by direct electron impact.

8.4.7 Dynamic Effects in Electric Arcs

High currents in electric arc discharges can induce relatively high magnetic fields and thus discharge compression effects in the magnetic field (see "pinch effect" in Section 6.2.4). The body forces acting on axisymmetric arcs are illustrated in Figure 8.16. Assuming that the current density j in the arc channel is constant, the azimuthal magnetic field B_θ inside an arc can be expressed as:

$$B_\theta(r) = \frac{1}{2}\mu_0 j\, r, \quad r \le r_0 \tag{8.43}$$

This magnetic field results in a radial body force (force per unit volume) directed inward and tending to pinch the arc (see Figure 8.16). The magnetic body force can be found as a function of distance r from the discharge axis as:

$$\vec{F} = \vec{j} \times \vec{B}, \quad F_r(r) = -\frac{1}{2}\mu_0 j^2 r \tag{8.44}$$

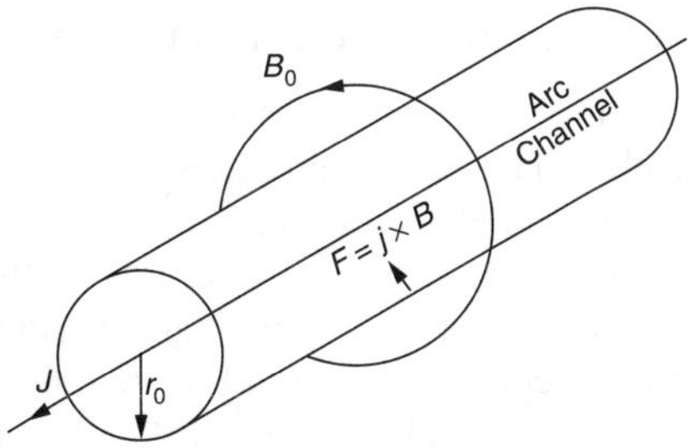

Figure 8.16 Radial body forces on cylindrical arc channel

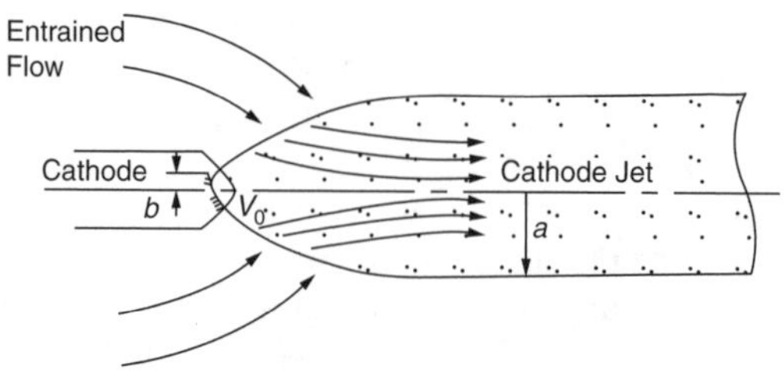

Figure 8.17 Electrode jet formation

The cylindrical arc discharge channel, in which expansionary kinetic pressure of plasma p is balanced by the inward radial magnetic body force Equation 8.44, is called the **Bennet pinch**. Equations of this balance can be expressed as:

$$\nabla p = \vec{j} \times \vec{B}, \quad \frac{dp}{dr} = -jB_\theta \tag{8.45}$$

Together with formula Equation 8.43 for the azimuthal magnetic field B_θ, this leads to the relation for pressure distribution inside the Bennet pinch of radius r_0:

$$p(r) = \frac{1}{4}\mu_0 j^2 \left(r_0^2 - r^2\right) \tag{8.46}$$

The corresponding maximum pressure (on axis of the arc channel) can be rewritten in terms of total current $I = j \cdot \pi r_0^2$ as:

$$p_a = \frac{1}{4}\mu_0 j^2 r_0^2 = \frac{\mu_0 I^2}{4\pi^2 r_0^2} \tag{8.47}$$

Numerically, according to Equation 8.47, a very high value of arc current, almost 10,000 A is necessary to provide an axial pressure of 1 atm in a channel with radius r_0 = 0.5 cm. Electric currents in most industrial arc discharges are much less than those of the Bennet pinch (defined by Equation 8.47), so the arc is stabilized primarily, not by magnetic body forces, but by other factors discussed in Section 8.4.1.

8.4.8 Bennet Pinch Effect and Electrode Jet Formation

Although the Bennet pinch effect pressure is usually small with respect to total pressure, it can initiate electrode jets—intensive gas streams that flow away from electrodes. The physical nature of electrode jet formation is illustrated in Figure 8.17.

Additional gas pressure related to the Bennet pinch effect (Equation 8.47) is inversely proportional to the square of the arc channel radius. Also, the radius of arc channel attachment to electrode ($r_0 = b$) is less than that corresponding to the positive column ($r_0 = a$; see Section 8.3 and Figure 8.17). This results in development of an axial pressure gradient that drives neutral gas along the arc axis away from electrodes:

$$\Delta p = p_b - p_a = \frac{\mu_0 I^2}{4\pi^2}\left(\frac{1}{b^2} - \frac{1}{a^2}\right) \tag{8.48}$$

Assuming that $b << a$, the jet dynamic pressure and thus the jet velocity can be found from the simple equation:

$$\Delta p \approx \frac{\mu_0 I^2}{4\pi^2 b^2} = \frac{1}{2}\rho v_{jet}^2 \tag{8.49}$$

where ρ is the plasma density. The electrode jet velocity then can be expressed as the function of arc current and radius of arc attachment to the electrode:

$$v_{jet} = \frac{I}{\pi b}\sqrt{\frac{\mu_0}{2\rho}} \tag{8.50}$$

Typical numerical range of the electrode jet velocities according to Equation 8.50 is about v_{jet} = 3 to 300 m/sec. Magnetic effects in arc discharges are discussed in many special publications.[176,285–287]

8.5 DIFFERENT CONFIGURATIONS OF ARC DISCHARGES

8.5.1 Free-Burning Linear Arcs

These simplest axisymmetric electric arcs between two electrodes can be arranged in horizontal and vertical configurations (see Figure 8.18). Sir Humphrey Davy first observed such a horizontal arc in the beginning of 19th century. Buoyancy of hot gases in the horizontal free-burning linear arc leads to bowing up or "arcing" of the plasma channel, which explains the origin of the name "arc." If the free-burning arc has a vertical configuration (see Figure 8.18b), the cathode is usually placed at the top of the discharge. In this case, buoyancy provides more intensive cathode heating, which sustains more effective thermionic emission from the electrode.

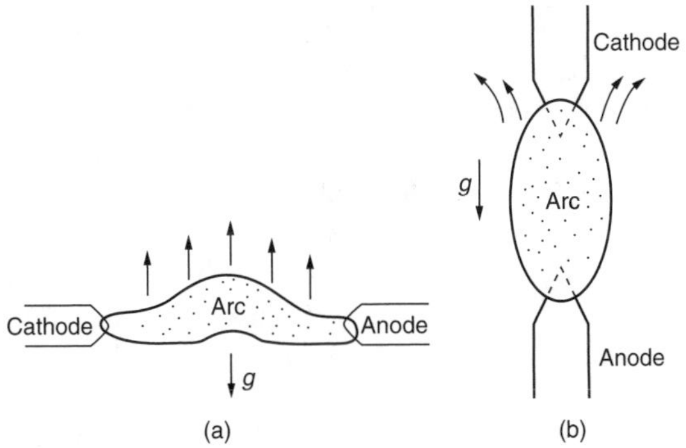

Figure 8.18 (a) Horizontal and (b) vertical free-burning arc

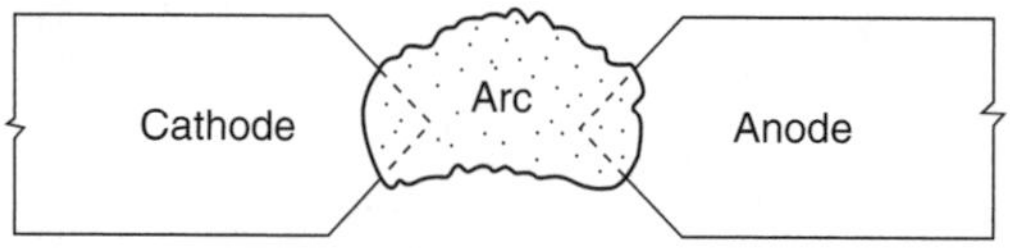

Figure 8.19 Obstructed electrode-stabilized arc

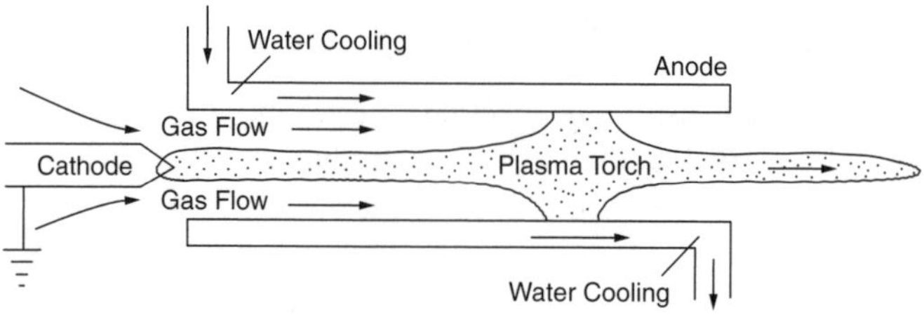

Figure 8.20 Wall-stabilized arc

The length of free-burning linear arcs typically exceeds their diameter. Sometimes, however, the arcs are shorter than their diameter (see Figure 8.19) and the "arc channel" actually does not exist. Such discharges are referred to as **obstructed arcs**. Distance between electrodes in such discharges is typically about 1 mm; nevertheless, voltage between electrodes exceeds anode and cathode drops. The obstructed arcs are usually electrode stabilized (see Section 8.4.1).

8.5.2 Wall-Stabilized Linear Arcs

This type of arc is most widely used for gas heating (see Section 8.4.1 and Section 8.4.2). A simple configuration of the **wall-stabilized arc with a unitary anode** is presented in Figure 8.20. The cathode is axial in this configuration of linear arc discharges and the unitary anode is hollow and coaxial. In this discharge, the arc channel is axisymmetric and stable with respect to asymmetric perturbations. If the arc channel is asymmetrically perturbed and approaches a coaxial anode, discharge cooling is intensified and temperature on the axis increases. The increase of temperature results in displacement of the arc channel back on the axis of the discharge tube, which stabilizes the linear arc. Then the anode of the high power arcs can be water-cooled (see Figure 8.20).

This arc discharge can attach to the anode at any point along the axis, which makes the discharge system irregular and less effective for several applications. Although the coaxial gas flow is able to push the arc-to-anode attachment point to the far end of the anode cylinder, the more regular and better-defined discharge arrangement can be achieved in the so-called **segmented wall-stabilized arc configuration**.

The wall-stabilized linear arc discharge organized with the segmented anode is illustrated in Figure 8.21. The anode walls in this system are usually water cooled, electrically segmented, and isolated. Such a configuration provides a linear decrease of axial voltage and forces the arc attachment to the wider anode segment furthest

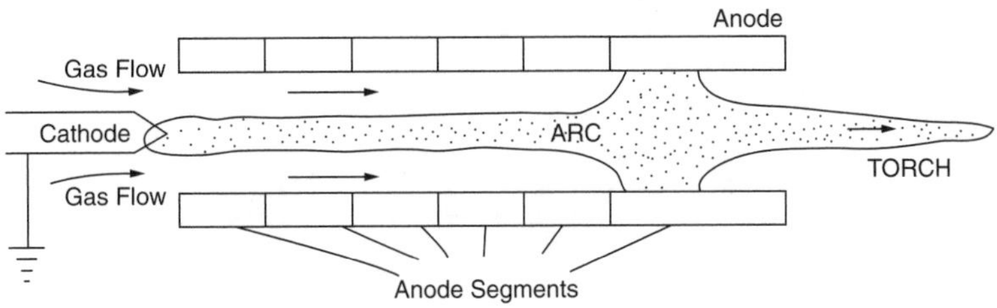

Figure 8.21 Wall-stabilized arc with segmented anode

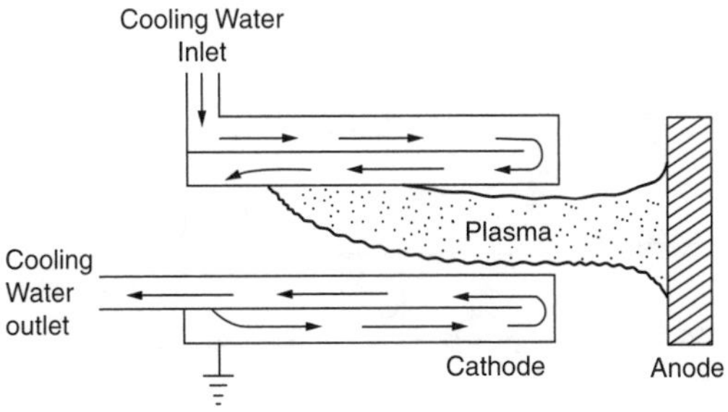

Figure 8.22 Transferred arc configuration

from the cathode (see Figure 8.21). The lengths of the other segments are taken sufficiently small to avoid breakdown between them.

8.5.3 Transferred Arcs

Linear transferred arcs with water-cooled nonconsumable cathodes are illustrated in Figure 8.22. Generation of electrons on inner walls of the hollow cathodes is provided by field emission, which permits operating the transferred arcs at multimegawatt power during thousands of hours. The electric circuit in the discharge is completed by transferring the arc to an external anode. In this case, the external anode is actually a conducting material where the high arc discharge energy is supposed to be applied. The arc "transferring" to an external anode gave the name to this arc configuration, which is quite effective in melting metal and the refining industry. The arc root in the transferred arcs is also able to move over the cathode surface, thus increasing the lifetime of the cathode and the arc in general.

8.5.4 Flow-Stabilized Linear Arcs

The arc channel can be stabilized on the axis of discharge chamber by radial inward injection of cooling water or gas. Such a configuration, illustrated in Figure 8.23, is usually referred to as the **transpiration-stabilized arc**. This discharge system is similar to the segmented wall-stabilized arc (see Section 8.5.2), but the transpiration

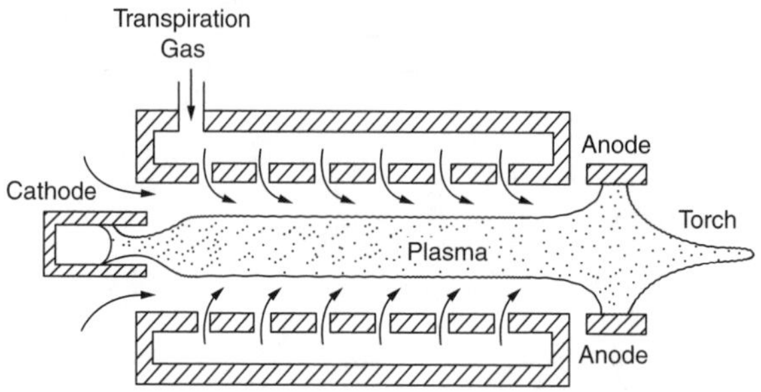

Figure 8.23 Transpiration-stabilized arc

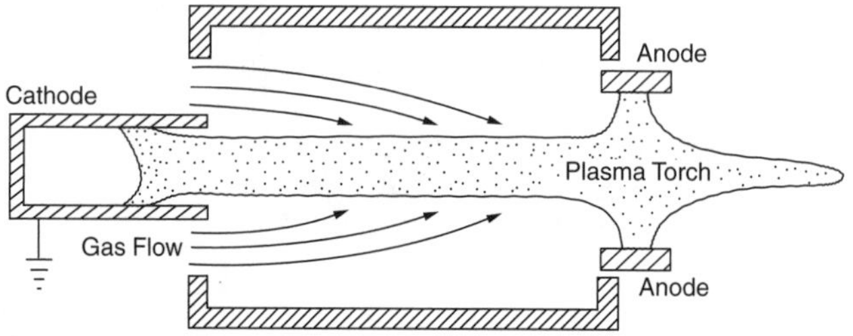

Figure 8.24 Coaxial flow stabilized arc

of a cooling fluid through annular slots between segments makes lifetime of the interior segments longer.

Another linear arc configuration providing high-power discharge is the so-called coaxial flow stabilized arc, illustrated in Figure 8.24. In this case, the anode is located too far from the main part of the plasma channel and cannot provide wall stabilization. Instead of a wall, the arc channel is stabilized by a coaxial gas flow moving along the outer surface of the arc. Such stabilization is effective without heating the discharge chamber walls if the coaxial flow is laminar. Similar arc stabilization can be achieved using a flow rotating fast around the arc column. Different configurations of vortex-stabilized arcs are shown in Figure 8.25(a, b). The arc channel is stabilized in this case by a vortex gas flow introduced from a special tangential injector. The vortex gas flow cools the edges of the arc and maintains the arc column confined to the axis of the discharge chamber. This flow is very effective in promoting the heat flux from the thermal arc column to the walls of the discharge chamber.

8.5.5 Nontransferred Arcs and Plasma Torches

A nonlinear, nontransferred, wall-stabilized arc is shown in Figure 8.26. This discharge system consists of a cylindrical hollow cathode and coaxial hollow anode located in a water-cooled chamber and separated by an insulator. Gas flow blows the arc column out of the anode opening to heat material that is supposed to be

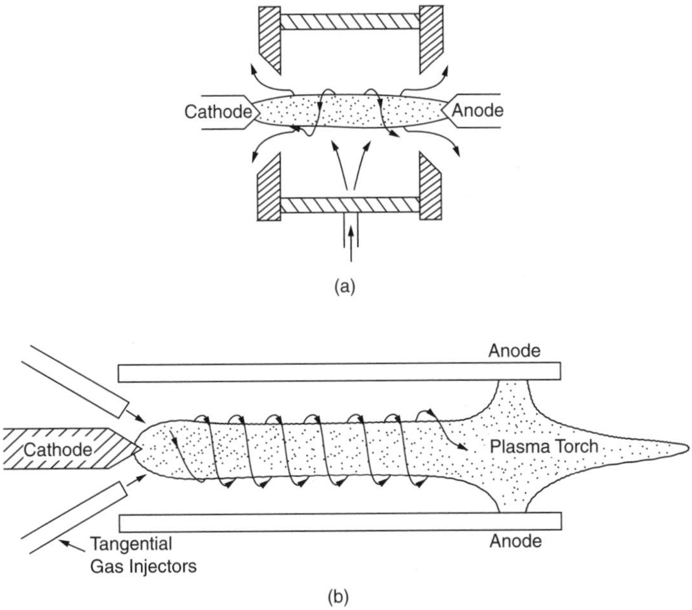

Figure 8.25 Vortex-stabilized arcs

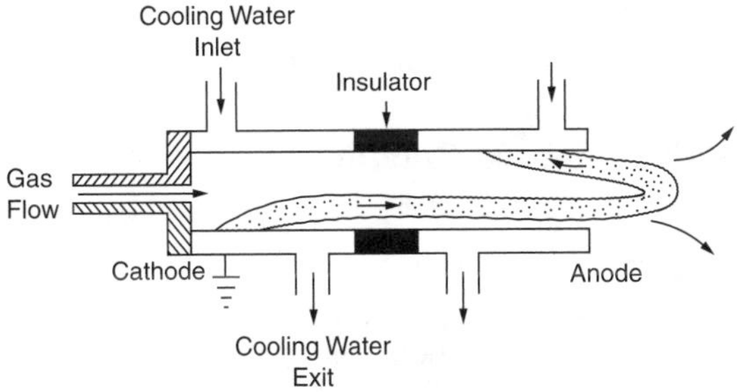

Figure 8.26 Nontransferred arc

treated. In contrast to the transferred arcs, the treated material is not supposed to operate as an anode. This explains the name "nontransferred" given to these arcs.

Magnetic $\vec{I} \times \vec{B}$ forces in these discharge systems cause the arc roots to rotate around electrodes (see Figure 8.26). This magnetic effect provides longer cathode and anode lifetime with respect to more traditional incandescent electrodes. In this case, generation of electrons on the cathode is provided completely by field emission.

An axisymmetric version of the nontransferred arc is illustrated in Figure 8.27. This discharge configuration is usually referred to as the **plasma torch** or the **arc jet**. The arc is generated in a conical gap in the anode and pushed out of this opening by the gas flow. The heated gas flow forms a very high-temperature arc jet, sometimes

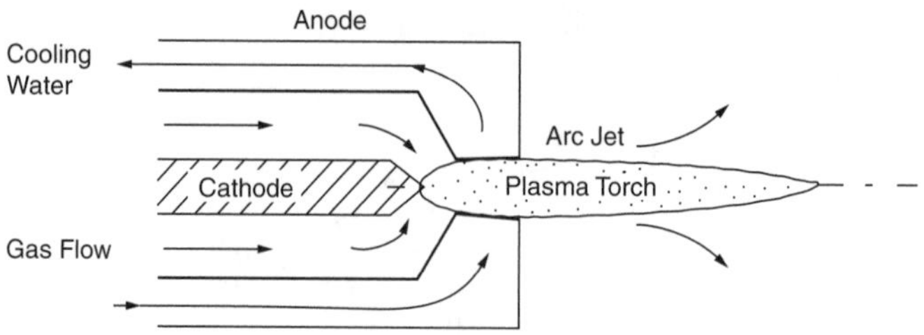

Figure 8.27 Plasma torch

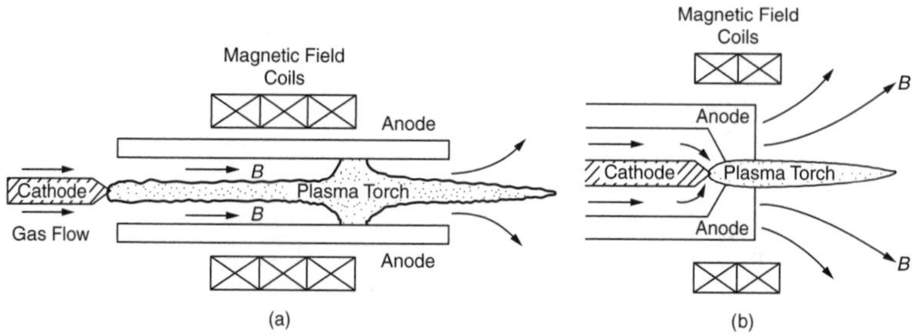

Figure 8.28 Magnetically stabilized arcs

at supersonic velocities. Today, this type of arc discharge has a quite wide area of industrial applications.

8.5.6 Magnetically Stabilized Rotating Arcs

Such configurations of arc discharges are illustrated in Figure 8.28(a, b). In this configuration, the external axial magnetic field provides $\vec{I} \times \vec{B}$ forces that cause very fast rotation of the arc discharge and protect the anode from intensive local overheating. Figure 8.28(a) shows the case of additional magnetic stabilization of a wall-stabilized arc (see Section 8.5.2). Figure 8.28(b) presents a very important configuration of the magnetically stabilized plasma torch. The effect of magnetic stabilization is an essential supplement to wall or gas flow stabilization.

8.6 GLIDING ARC DISCHARGE

8.6.1 General Features of the Gliding Arc

The gliding arc discharge is an auto-oscillating periodic phenomenon that develops between at least two diverging electrodes submerged in a laminar or turbulent gas flow. Picture and illustration of a gliding arc are shown in Figure 8.29(a, b). Self-initiated in the upstream narrowest gap, the discharge forms the plasma column connecting the electrodes. This column is further dragged by the gas flow toward the diverging downstream section. The discharge length grows with the increase of

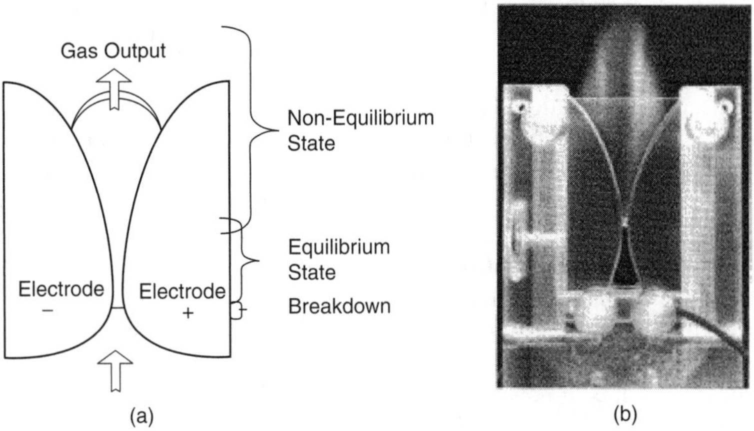

Figure 8.29 (a) Illustration and (b) picture of gliding arc discharge

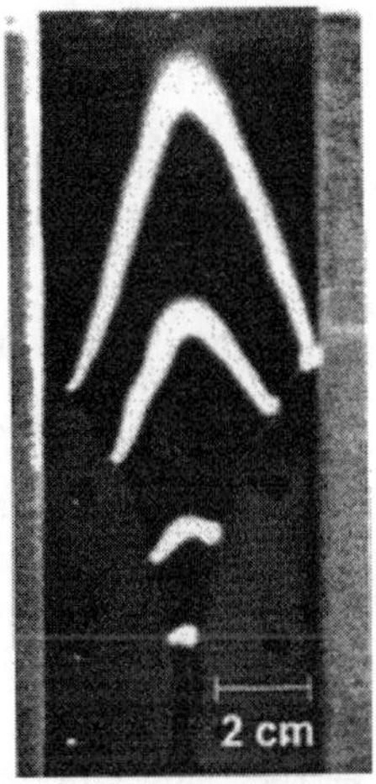

Figure 8.30 Gliding arc evolution shown with 20-ms separation between snapshots at flow rate of 110 SLM

interelectrode distance until it reaches a critical value, usually determined by power supply limits. After this point, the discharge extinguishes, but momentarily reignites at the minimum distance between the electrodes and a new cycle start. The time–space evolution of the gliding arc is illustrated by a series of photos presented in Figure 8.30.

Plasma generated by the gliding arc discharge has thermal or nonthermal properties depending on system parameters such as power input and flow rate. Along with completely thermal and completely nonthermal modes of the discharge, it is possible to define the transition regimes of the gliding arc. In this most interesting regime, the discharge starts as thermal, but during the space and time evolution becomes nonthermal. This powerful and energy-efficient transition discharge combines the benefits of equilibrium and nonequilibrium discharges in a single structure. It can provide plasma conditions typical for nonequilibrium cold plasmas but at elevated power levels.

As discussed in Section 5.6, generally two very different kinds of plasmas are used for chemical applications. Thermal plasma generators have been designed for many diverse industrial applications covering a wide range of operating power levels from less than 1 kW to over 50 MW. However, in spite of providing sufficient power levels, these appear not to be well adapted to the purposes of plasma chemistry, in which selective treatment of reactants (through the excitation of molecular vibrations or electron excitation) and high efficiency are required. An alternative approach for plasma chemical gas processing is the nonthermal one. The glow discharge in a low-pressure gas seems to be a simple and inexpensive way to achieve nonthermal plasma conditions. Indeed, cold nonequilibrium plasmas offer good selectivity and energy efficiency of chemical processes, but as Chapter 7 demonstrated, they are generally at limited to low pressures and powers.

In these two general types of plasma discharges, it is impossible to maintain a high level of nonequilibrium, high electron temperature, and high electron density simultaneously, whereas most prospective plasma chemical applications require high power for high reactor productivity and high degree of nonequilibrium to support selective chemical process at the same time. However, this can be achieved in the transition regime of the gliding arc that, for this reason, attracts interest to this type of nonequilibrium plasma generator. Gliding arcs operate at atmospheric pressure or higher; the dissipated power at nonequilibrium conditions can reach up to 40 kW per electrode pair.

The gliding arc configuration has been well known for more than 100 years. In the form of Jacob's ladder, it is often used in different science exhibits. The gliding arc was first used in chemical applications in the beginning of the 1900s for production of nitrogen-based fertilizers.[288] An important recent contribution to development of fundamental and applied aspects of the gliding arc discharge was made by Czernichowski and colleagues.[289,289a] The nonequilibrium nature of the transition regime of the gliding arc was first reported by Fridman and colleagues.[290,291]

8.6.2 Physical Phenomenon of Gliding Arc

Consider a simple case of a direct current gliding arc in air, driven by two generators. A typical electrical scheme of the circuit is shown in Figure 8.31. One generator is a high voltage (up to 5000 V) used to ignite the discharge and the second is a power generator (with voltage up to 1 kV, and a total current I up to 60 A). A variable resistor $R = 0$ to 25 Ω is in series with a self-inductance $L = 25$ mH. In principle, more advanced schemes such as an AC gliding arc, 3-phase gliding arc, configurations with several parallel or serial electrodes, etc. could be configured. However, the simplest case will be considered here for better understanding of the gliding arc phenomenon.

Initial breakdown of the processed gas begins the cycle of the gliding arc evolution. The discharge starts at the shortest distance (1 to 2 mm) between two

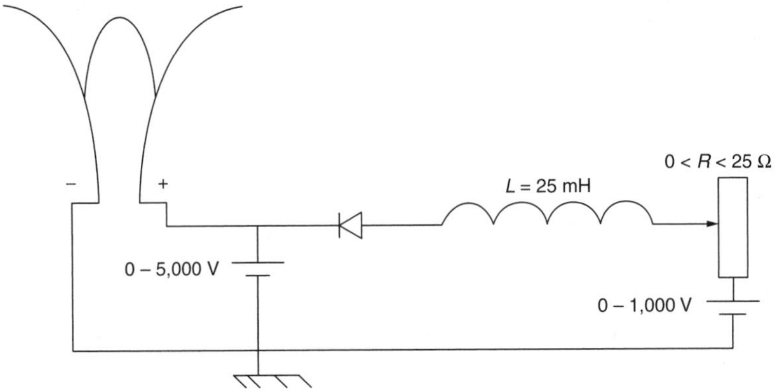

Figure 8.31 Typical gliding arc discharge electrical scheme

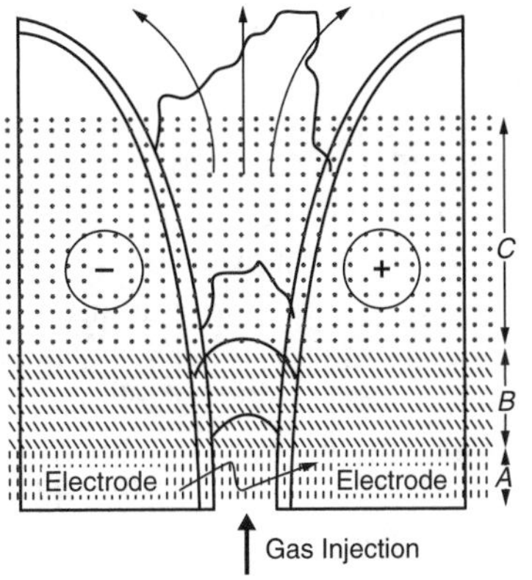

Figure 8.32 Phases of gliding arc evolution: (A) reagent gas breakdown; (B) equilibrium heating phase; (C) nonequilibrium reaction phase

electrodes (see Figure 8.32). The high-voltage generator provides the necessary electric field to break down the air between the electrodes. For atmospheric pressure air and a distance between electrodes of about 1 mm, the breakdown voltage V_b is approximately 3 kV. The characteristic time of the arc formation τ_i can be estimated from the simple kinetic equation on the electron concentration:

$$\frac{dn_e}{dt} = k_i n_e n_0 = \frac{n_e}{\tau_i} \tag{8.51}$$

where k_i is the ionization rate coefficient; n_e and n_0 are electron and gas concentrations. The estimation for τ_I in clean air and total arc electrical current $J = 1$ A give the value $\tau_i \approx 1$ μs. Within a time of about 1 μs, a low-resistance plasma is formed and the voltage between the electrodes falls.

The **equilibrium stage** takes place after formation of a stable plasma channel. The gas flow pushes the small equilibrium plasma column with a velocity of about

10 m/sec, and the length l of the arc channel increases together with the voltage. Initially, the electric current increases during formation of the quasi-equilibrium channel up to its maximum value of $I_m = V_0/R \approx 40$ A. The current time dependence during the current growth phase is sensitive to the inductance L:

$$I(t) = \left(V_0/R\right)\left(1 - e^{-t/\tau_L}\right) \tag{8.52}$$

where $\tau_L = L/R \approx 1$ msec. In this quasi-equilibrium stage of the gliding arc discharge, the equilibrium gas temperature T_0 does not change drastically and lies in the range $7000 \leq T_0 \leq 10{,}000$ K for the numerical example.[290] For experiments with lower electric current and lower power, the temperature could be different. For example, at power level 200 W and electric current 0.1 A, the gas temperature is as low as 2500 K.[292]

The quasi equilibrium arc plasma column moves with the gas flow. At a 2-kW power level, the difference in velocities grows from 1 to 10 m/s with an increase in the inlet flow rate from 830 to 2200 cm^3/sec.[293] In the reported case, the flow velocity was 30 m/sec and the arc velocity was 24 m/sec. In experiments reported by Deminsky et al.,[294] the difference in velocities was up to 30 m/sec for high flow rates (2200 cm^3/sec) and small arc lengths (5 cm). This difference decreases several meters per second when arc length increases up to 15 cm. In general, the difference in arc and flow velocities increases with increase in the arc power and absolute values of flow rates and thus flow velocities. It decreases with an increase of the arc length.

The length of the column l during this movement, and thus the power consumed by the discharge from the electrical circuit, increases up to the moment when the electrical power reaches the maximum value available from the power supply source P_{max}.

The **nonequilibrium stage** begins when the length of gliding arc exceeds its critical value l_{crit}. Heat losses from the plasma column begin to exceed the energy supplied by the source, and it is not possible to sustain the plasma in a state of thermodynamic equilibrium. As a result, the discharge plasma cools rapidly to about $T_0 = 2000$ K, and the plasma conductivity is maintained by a high value of the electron temperature $T_e = 1$ eV and step-wise ionization. After decay of the non-equilibrium discharge, a new breakdown takes place at the shortest distance between electrodes and the cycle repeats.

8.6.3 Equilibrium Phase of Gliding Arc

Physical parameters of the equilibrium phase of the gliding arc evolution are similar to those for the conventional atmospheric pressure arc discharges. After ignition and formation of a steady arc column, the energy balance can be reasonably described by the Elenbaas–Heller equation (Equation 8.20). Then, application of the Raizer "channel" model of positive column (see Section 8.4.4) leads to Equation 8.33 describing the power w dissipated per unit length of the gliding arc. As discussed in Section 8.4.5 (in particular, see Equation 8.37 and Figure 8.15), the specific power

w does not change significantly. In the temperature range of 6000 to 12,000 K, the characteristic value of the specific power w is about 50 to 70 kW/m.

Assuming a constant specific power w permits describing the evolution of current, voltage, and power during the gliding arc quasi-equilibrium phase. Neglecting the self-inductance L, Ohm's law can be written for the circuit including the plasma channel, active resistor, and power generator as:

$$V_0 = R\,I + w \cdot l/I \tag{8.53}$$

Here, V_0, R, and J are respectively the open circuit voltage of the power supply, external serial resistance, and current. The arc current can be determined from Ohm's law (Equation 8.53) and presented as a function of the arc length l, which is growing during the arc evolution:

$$\mathrm{I} = \left(\mathrm{V}_0 - \left(V_0^2 - 4\ wlR\right)^{1/2}\right)\Big/2R \tag{8.54}$$

The solution with the sign "+" describes the steady state of the gliding arc column; the solution with $J < V_0/2R$ corresponds to negative differential resistance ρ ($\rho = dV/dJ$) of the circuit and to an unstable regime of gliding arc evolution:

$$\rho = R - wl/I^2 = 2R - V_0/I < 0 \tag{8.55}$$

From Equation 8.54 the current is seen to decrease slightly during the quasi-equilibrium period. At the same time, the arc voltage is growing as wl/I, and the total arc power $P = wl$ increases almost linearly with the length l.

According to Equation 8.55, the absolute value of the differential resistance of the gliding arc grows. Arc discharges are generally stable and have descending volt–ampere characteristics. This means that the differential resistance (du/dj) of an arc as a part of the electric circuit is negative. To provide the stable regime of the circuit operation, the total differential resistance of the whole circuit should be positive. When the power dissipated by the arc achieves its maximum, the arc's differential resistance becomes equal to the differential resistance of external part of the circuit (see Equation 8.55) and the electric circuit loses its stability. This leads to a change in the electrical parameters of the circuit and significantly affects the arc parameters. The power supply cannot provide the increasing arc power, so the system transfers into a new nonequilibrium state, which will be discussed later. Numerical solutions for the current I, voltage V, and power P in the equilibrium phase are shown in Figure 8.33.

8.6.4 Critical Parameters of Gliding Arc

The quasi-equilibrium evolution of the gliding arc is terminated when the arc length approaches the critical value:

$$l_{crit} = V_0^2/(4wR) \tag{8.56}$$

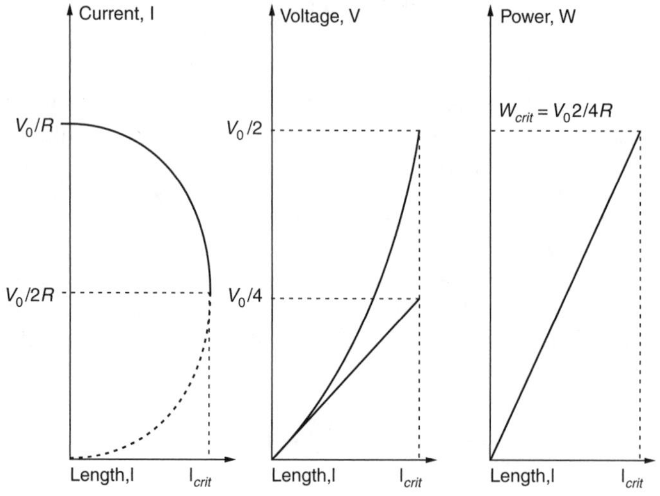

Figure 8.33 Evolution of current, voltage, and power vs. length

and the square root in Equation 8.54 (that is, the differential resistance of the whole circuit) becomes equal to zero. The current falls at this point to its minimal stable value of $I_{crit} = V_0/2R$, which is half of its initial value. Plasma voltage, electric field E, and total power at the same critical point approach their maximum values of, respectively:

$$V_{crit} = V_0/2, \quad E_{crit} = w/J_{crit}, \quad W_{crit} = V_0^2/4R \tag{8.57}$$

At the critical point (Equation 8.56 and Equation 8.57), a growing plasma resistance becomes equal to the external one and, obviously, the maximum value of discharge power corresponds to half of the maximum power of the generator. The numerical values of the critical parameters for pure air conditions are $J_{crit} = 20$ A; $V_{crit} = V_0/2 = 400$ V; $W_{crit} = 8$ kW (with a characteristic equilibrium temperature of the gliding arc discharge $7000 \leq T \leq 10{,}000$ K); and $l_{crit} = 10$ cm.

Equation 8.56 for the gliding arc critical length can be made more specific by taking into account the channel model relation (Equation 8.33) for the specific power w:

$$l_{crit} = V_0^2 I_i / 32\pi R \lambda_m T^2 \tag{8.58}$$

This relation for the discharge parameters under consideration also gives the similar estimation for the critical length of the gliding arc discharge $l_{crit} \sim 10$ cm.

8.6.5 Fast Equilibrium → Nonequilibrium Transition (FENETRe Phenomenon)

When the length of the gliding arc exceeds the critical value $l > l_{crit}$, the heat losses wl continue to grow. However, because the electrical power from the power supply cannot further increase, it is no longer possible to sustain the arc in the thermodynamic quasi equilibrium. The gas temperature falls rapidly during the conduction time (about $r^2/D \approx 10^{-4}$ sec), and the plasma conductivity can be maintained only by a high value of the electron temperature $T \approx 1$ eV and stepwise mechanism of ionization (see Section 2.2.4).

The phenomenon of the "fast equilibrium → nonequilibrium transition" (the so-called FENETRe) in the gliding arc is a type of "discharge window" from the thermal plasma zone to a relatively cold zone.[295] The arc instability is due to the slow increase of the electric field $E = w/I$ and corresponding increase of electron temperature T_e during the gliding arc evolution:

$$T_e = T_0\left(1 + E^2/E_i^2\right) \tag{8.59}$$

Here, the electric field E_i corresponds to transition from thermal to direct electron impact ionization (see Section 2.2.4).

The equilibrium arc discharge is usually stable. This thermal ionization stability is due to a peculiarity of the heat and the electric current balances in the arc discharge. Indeed, a small temperature increase in the arc discharge leads to a growth of electron concentration and electric conductivity σ. Taking into account the fixed value of current density $j = \sigma E$, the conductivity increase results in a reduction of the electric field E and the heating power σE^2, thus resulting in stabilizing the initial temperature perturbations.

In contrast to that, direct electron impact ionization (typical for nonequilibrium plasma) is usually not stable. For a fixed pressure (Pascal law) and constant electric field, a small temperature increase leads to a reduction of the gas concentration n_0 and to an increase of the specific electric field E/n_0 and the ionization rate. This results in an increase of the electron concentration n_e, the conductivity σ, the joule heating power σE^2 (and thus in an additional temperature increase), and finally in total discharge instability.

Thus, when the conductivity σ depends mainly on the gas translational temperature, the discharge is stable. When the electric field is relatively high and the conductivity σ depends mainly on the specific electric field, the ionization becomes unstable. The gliding arc passes such a critical point during its evolution and the related electric field grows. The electric field growth during the gliding arc evolution could be written based on Equation 8.54 as:

$$E = 2wR\Big/\left(V_0 + \left(V_0^2 - 4wlR\right)^{0,5}\right) \tag{8.60}$$

For a relatively high electric field, the electric conductivity $\sigma(T_0,E)$ begins to depend not only on the gas temperature as it takes place in quasi-equilibrium discharges, but also on the field E. The corresponding value of logarithmic sensitivity of the electric conductivity to the gas temperature corresponds to the quasi-equilibrium Saha ionization:

$$\sigma_T = \partial \ln \sigma\left(T_0, E\right)/\partial \ln T_0 \approx I_i/2T_0 \tag{8.61}$$

where I_i is the ionization potential. Taking Equation 8.59 into account, the logarithmic sensitivity of electric conductivity to the electric field could be written as:

$$\sigma_E = \partial \ln \sigma \left(T_0, E\right) / \partial \ln E = I_i E^2 / E_i^2 T_0 \tag{8.62}$$

To analyze the arc stability, consider Ohm's law (Equation 8.53), the logarithmic sensitivities (Equation 8.61 and Equation 8.62), and the thermal balance equation:

$$n_0 c_p \, dT_0 / dt = \sigma\left(T_0, E\right) E^2 - 8\pi\lambda T_0^2 / I_i S \tag{8.63}$$

where S is the arc cross section; n_0 and c_p are the gas concentration and specific heat. The linearization procedure permits describing the time evolution of a temperature fluctuation ΔT in the exponential form:

$$\Delta T(t) = \Delta T_0 \exp(\Omega t) \tag{8.64}$$

where ΔT_0 is an initial temperature fluctuation, and exponential frequency parameter Ω is an instability decrement that is a frequency of temperature fluctuation disappearance. In this case, the negative value of the decrement $\Omega < 0$ corresponds to discharge stability, while the positive $\Omega > 0$ corresponds to a case of instability. The linearization gives the following instability decrement for the gliding arc discharge:[295]

$$\Omega = -\omega \frac{\sigma_T}{1 + \sigma_E} \left(1 - \frac{E}{E_{crit}} \right) \tag{8.65}$$

In this relation, ω is the thermal instability frequency factor, similar to that described in Section 6.3.2 and defined by the relation:

$$\omega = \frac{\sigma E^2}{n_0 c_p T_0} \tag{8.66}$$

Development of ionization instability finally results in transition to nonequilibrium regime or in break of the discharge.

8.6.6 Gliding Arc Stability Analysis

The physical scheme of the gliding arc ionization instability is presented in Figure 8.15. It is clear that, initially, the gliding arc discharge could remain stable ($\Omega < 0$) at relatively small electric field $E < E_{crit}$ ($l < l_{crit}$), but it becomes completely unstable ($\Omega > 0$) when the electric field grows stronger (Equation 8.34). It is interesting and important that gliding arc can lose its stability even in the "theoretically stable" regime (see Figure 8.15):

$$E_i \left(T_0 / I_i\right)^{0,5} < E < E_{crit} \tag{8.67}$$

Although such an electric field is less than critical (stability in the linear approximation Equation 8.65), its influence on electric conductivity becomes dominant, which strongly perturbs stability. In this regime of the "quasi instability" (Equation 8.67), the stability factor Ω is still negative, but decreases 10 to 30 times (compared to one in the initial regime of the gliding arc) due to the influence of the logarithmic sensitivity σ_E (factor $1/(1 + \sigma_E)$). This means that the gliding arc discharge actually becomes unstable at such electric fields.

This effect of quasi instability may be better illustrated using nonlinear analysis, in which the instability decrement in addition to Equation 8.65 also depends on a level of temperature perturbation. The results of this analysis of the differential equations can be represented by a critical value of a temperature fluctuation corresponding to the gliding arc transition into an unstable form:

$$T_{cr} = \left(4T_0^2/I_i\right)\left(1 - E/E_{crit}\right)/\left(1+\sigma_E\right) \quad (8.68)$$

In a regular arc and in the initial equilibrium stage of the gliding arc, the electric field and the sensitivity of conductivity to electric field are relatively small $E \ll E_{crit}$, $\sigma_E \ll 1$. Then, the maximum permitted temperature perturbation to sustain the arc stability of the discharge is :

$$T_{cr}(\text{equil.}) = T_0^2/I_i \quad (8.69)$$

Numerically, this maximum temperature perturbation is approximately 1000 K and is high enough to guarantee the arc's stability. In the transition stage of the gliding arc (Equation 8.67), the value of the critical temperature fluctuation T decreases to:

$$T_{cr}(\text{non-eq}) = T_0^3\, E_i^2 / I_i^2\, E^2 \quad (8.70)$$

which corresponds numerically to about 100 K. In this case, a small temperature perturbation leads to arc instability and to the FENETRe phenomenon. The main qualitative results of the linear and the nonlinear analysis of the gliding arc evolution, including the instability decrement and the critical temperature fluctuation as a function of the electric field, are presented in Figure 8.34.

During the FENETRe stage, the electric field and the electron temperature T_e in the gliding arc are slightly increasing, numerically from $T_e \approx 1$ eV in the quasi-equilibrium zone to approximately $T_e \approx 1$ eV in the nonequilibrium zone. At the same time, the translational gas temperature decreases about three times from its initial equilibrium value of approximately 0.5 eV. The electron concentration falls to the value of about 10^{12} cm^{-3}.

Decrease of the electron concentration and thus conductivity σ leads to the reduction of specific discharge power σE^2. It cannot be compensated here by the electric field growth as in the case of the stable quasi-equilibrium arc because of the strong sensitivity of the conductivity to electric field (see Equation 8.62). This is the main physical reason of the quasi-unstable discharge behavior in FENETRe conditions (see Figure 8.34).

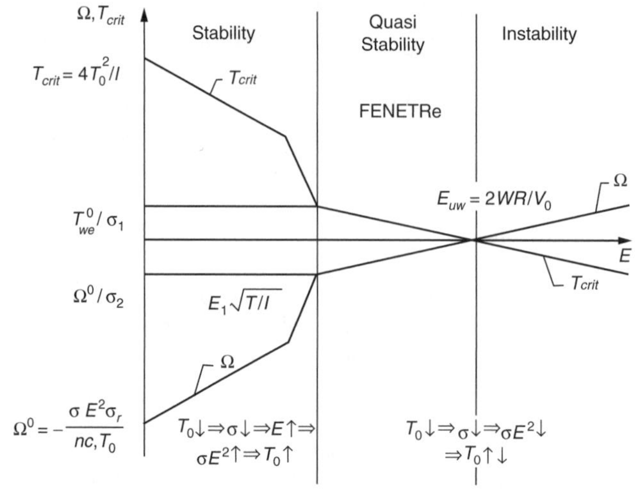

Figure 8.34 Characteristics of gliding arc instability: instability decrement and critical thermal fluctuation

During the transition phase, the specific heat losses w (Equation 8.33) also decrease (to a smaller value w_{noneq}) as well as the specific discharge power σE^2 with the reduction of the translational temperature T_0. However, according to Ohm's law ($E < E_{crit}$), the total discharge power increases due to the growth of plasma resistance. This discrepancy between the total power and specific power (per unit of length or volume) results in a possible "explosive" increase in the arc length. This increase can be easily estimated from Equation 8.56 as:

$$l_{\max}/l_{crit} = w/w_{\text{non-eq}} \approx 3 \tag{8.71}$$

This was experimentally observed in the gliding arc discharge.[296]

8.6.7 Nonequilibrium Phase of Gliding Arc

After the fast FENETRe, the gliding arc is able to continue its evolution under the nonequilibrium conditions $T_e >> T_0$.[291,296,296a,637] Up to 70 to 80% of the total gliding arc power can be dissipated after the critical transition, resulting in a plasma in which the gas temperature is approximately 1500 to 3000 K and the electron temperature is about 1 eV.

Obviously, effective gliding arc evolution in the nonequilibrium phase is possible only if the electric field during the transition is sufficiently high; this is possible when the arc current is not too large.[637] The critical gliding arc parameters before transition in atmospheric air are shown in Table 8.8 as a function of the initial electric current.

Under conditions of the decaying arc, even the relatively small (for conventional nonthermal discharges) electric field $(E/n_0)_{crit} \approx (0.5 \text{ to } 1.0) \times 10^{-16}$ V × cm^2 is sufficient to sustain the nonequilibrium phase of a gliding arc because of the influence

Table 8.8 Critical Gliding Arc Parameters before Transition in Dependence on Magnitude of Initial Current I_0

I_{crit}, A	50	40	30	20	10	5	1	0.5	0.1
I_0, A	100	80	60	40	20	10	2	1	0.2
T_0, K	10800	10300	9700	8900	7800	6900	5500	5000	4100
w_{crit}, W/cm	1600	1450	1300	1100	850	650	400	350	250
E_{crit} V/cm	33	37	43	55	85	130	420	700	2400
(E/n_0) 10^{-16} Vcm^2	0.49	0.51	0.57	0.66	0.9	1.3	3.1	4.8	14

of stepwise and, possibly, the Penning ionization mechanism. Then, taking into account data from Table 8.8, it can be concluded that:

1. If $I_0 > 5$ to 10 A, then the arc discharge extinguishes after reaching the critical values.
2. Alternatively, for $J_0 < 5$ to10 A, the value of reduced electric field $(E/n_0)_{crit}$ is sufficient for maintaining the discharge in a new nonequilibrium ionization regime.

Thus, it is possible to observe three different types of gliding arc discharge. At relatively low currents and high gas flow rates, the gliding arc discharge is nonequilibrium throughout all stages of its development. Alternately, at high currents and low gas flow rates, the discharge is thermal (quasi equilibrium) and breaks out at the critical point. Only at intermediate values of currents and flow rates does the FENETRe takes place. Experimental data and comparative analysis regarding these three types of gliding arcs are discussed in Mutaf–Yardimci et al.[297]

Detailed electron T_e, vibrational T_v, and translational T_0 temperature measurements in the nonequilibrium regimes of gliding arc have been reported.[298] Typical values of the measured temperatures were: $T_e \approx 10{,}000$ K; $T_v \approx 2000$ to 3000 K; and $T_0 \approx 800$ to 2100 K. Additional experimental data on the subject can be found in references 299, 299a, and 300.

8.6.8 Effect of Self-Inductance on Gliding Arc Evolution

Introducing a self-inductance into the gliding arc circuit (see Figure 8.31) decelerates the rate of current decrease and thereby prolongs the time of evolution of the nonequilibrium arc. Taking into account the self-inductance of the electric circuit, the Ohm's law can be written as:

$$V_0 = RI + L\,dI/dt + P/I \tag{8.72}$$

where power P in the nonequilibrium regime can be presented in the form:

$$P = P_{crit} + w_{crit} 2v\alpha\, t \tag{8.73}$$

P_{crit} and w_{crit} are the total and critical discharge power at the transition; v is the velocity of gliding arc; 2α is an average angle between the diverging electrodes; and time $t = 0$ at the transition point. Then, using the dimensionless variables:

$$i = I/I_{crit}, \qquad \tau = t/\tau_L, \qquad \lambda = 2v\tau_L\, a/l_{\max} \tag{8.74}$$

Ohm's law can be rewritten as follows:

$$(i-1)^2 + i\, di/d\tau + \lambda\, t = 0 \tag{8.75}$$

where $\tau_L = L/R$ is the characteristic time of the electric circuit. For typical conditions of the gliding arc, the magnitude of λ is rather small ($\lambda = 0.003$).

At the beginning of the process when $\tau < 0$ ($|\tau| >> 1$) and the derivative $di/d\tau$ is neglected, the approximation for i can be written as

$$i = 1 + (-\lambda\tau)^{1/2} \tag{8.76}$$

which corresponds to the current evolution shown in Figure 8.33. When $\tau > -1/\lambda^{1/3}$, Equation 8.75 can be written in the elliptic form:

$$i^2 + \lambda\, \tau^2 = 1 \tag{8.77}$$

Time evolution of electric current corresponding to solutions of Equation 8.75 through Equation 8.77 is shown in Figure 8.35. It is seen that introducing self-inductance prolongs the transition phase by the time $\tau = 1/\lambda^{1/2}$. As a result, the power dissipated throughout the transition phase can also be increased (numerically, by about 20 to 50%).

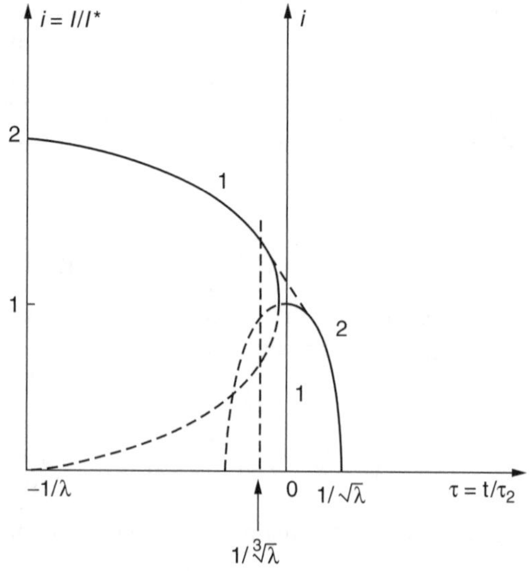

Figure 8.35 Effect of self-inductance on transient phase of gliding arc

8.6.9 Special Configurations of Gliding Arcs

The gliding arc discharges can be organized in various ways for different practical applications. Of special interest are those organized in a fluidized bed (see Figure 8.36),[632] as well as those in a fast rotation in magnetic field (see Figure 8.37).

An expanding arc configuration similar to the gliding arc is used in switchgears, as shown in Figure 8.38. The contacts start out closed, but when the contactors open, an expanding arc is formed and glides along the opening of increasing length until it extinguishes. Extinction of the gliding arc in this case is often promoted by using strongly electronegative gases like SF_6.

Gliding arc discharge can also be rotated by fast rotating gas flow. One electrode can be made spiral in this case following streamlines of the flow. The most interesting is stabilization of the gliding arc in reverse vortex flow.[622] Discharge stabilization in

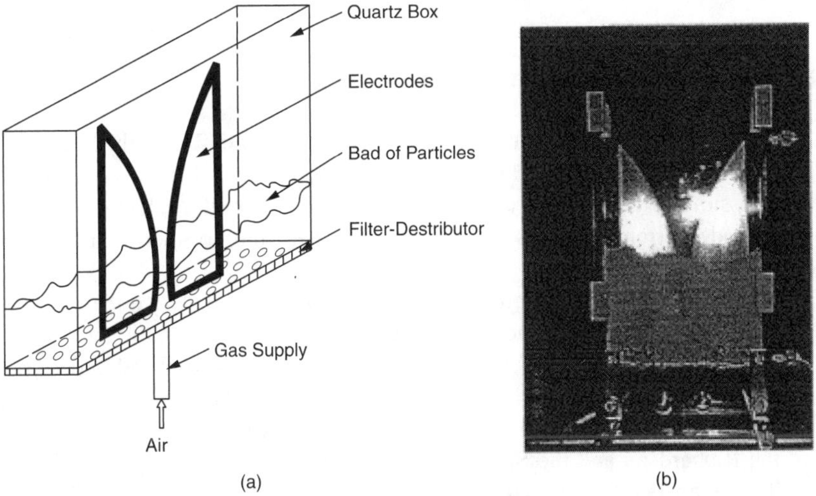

Figure 8.36 Gliding arc-fluidized bed system: (a) schematic and (b) photograph

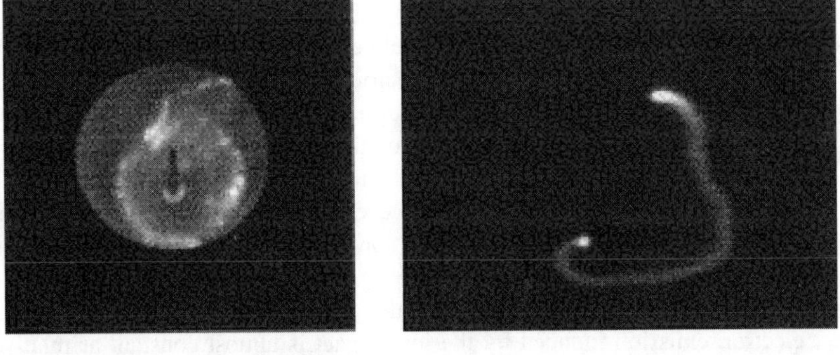

Figure 8.37 Gliding arc rotating in magnetic field

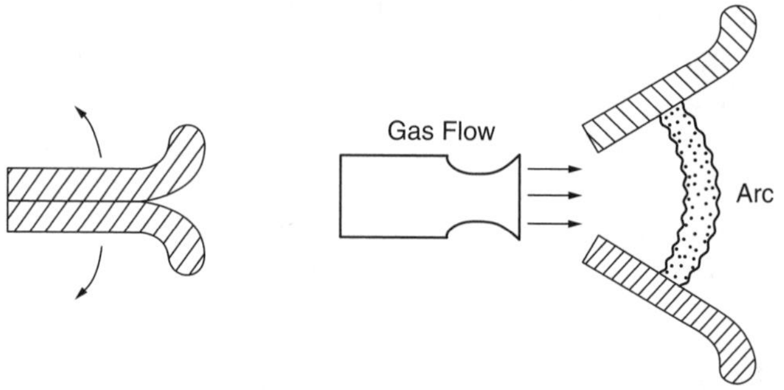

Figure 8.38 Switches as gliding arcs

the reverse vortex (TORNADO) is discussed in Section 10.1.10. A picture of the gliding arc trapped in TORNADO is shown in Figure 1.2(c).

PROBLEMS AND CONCEPT QUESTIONS

8.1 **The Sommerfeld Formula for Thermionic Emission**. Using the Sommerfeld formula (Equation 8.3), calculate the saturation current densities of thermionic emission for a tungsten cathode at 2500 K. Compare the calculated values of cathode current densities with those typical for the hot thermionic cathodes shown in Table 8.1.

8.2 **The Richardson Relation for Thermionic Emission**. Derive the thermionic current density dependence on temperature by assuming in the integration of Equation 8.2 a Maxwell distribution function for electron velocities, rather than Fermi. In 1908, Richardson first derived this relation with the pre-exponential factor proportional to T rather than to T^2. Explain the difference between the Richardson and the Sommerfeld formulas.

8.3 **The Schottky Effect of Electric Field on the Work Function**. Derive the Schottky relation by integrating Equation 8.5 with the electric force (Equation 8.4) applied to an electron. Estimate the relative accuracy of this relation. Calculate the maximum Schottky decrease of work function, corresponding to very high electric fields sufficient for field emission.

8.4 **The Field Electron Emission**. The field electron emission is characterized by very sharp current dependence on electric field (see, for example, Table 8.3). As a result of this strong dependence, one can determine the critical value of electric field when the contribution of the field electron emission becomes significant. Based on the Fowler–Nordheim formula (Equation 8.7), determine the critical value of the electric field necessary for the field electron emission and its dependence on the work function. Give physical interpretation of the dependence.

8.5 **Secondary Electron Emission**. Explain why the coefficient γ of the secondary electron emission induced by the ion impact is almost constant at relatively low energies when it is provided by the Penning mechanism. Estimate the probability of the electron emission following the Penning mechanism. Explain why the coefficient γ of the secondary electron emission induced by the ion impact grows linearly with ion energy when the ion energy is sufficiently high.

8.6 **The Secondary Ion–Electron Emission Coefficient** γ. Based on Equation 8.9, estimate the coefficient γ of secondary electron emission induced by the ion impact at relatively low ion energies for ions with different ionization potentials and surfaces with different work functions. Compare the obtained result with the experimental data presented in Figure 8.6.

8.7 **Electric Field in the Vicinity of Arc Cathode**. Integrating the Poison's equation (Equation 8.13), derive Equation 8.14 for the value of electric field near the cathode of arc discharge. Show that the first term in this formula is related to the contribution of ions in forming the positive space charge, and the second term to the effect of electrons in compensating the positive space charge. Compare contribution of these two terms numerically.

8.8 **Structure of the Arc Discharge Cathode Layer**. Integrating the Poisson's equation (Equation 8.13) twice and neglecting the second term related to effect of electrons on the space charge, derive Equation 8.16 for length of the collisionless zone of the cathode layer. Give a physical interpretation of the fact that most of the cathode voltage drop takes place in the extremely short collisionless zone of the cathode layer.

8.9 **Erosion of Hot Cathodes**. Based on the numerical example of specific erosion given in Section 8.3.4, estimate the rate of cathode mass losses per hour for the tungsten rod cathode at atmospheric pressure of inert gases and current of about 300 A.

8.10 **Cathode Spots**. The critical minimum current through an individual cathode spot for different nonferromagnetic materials can be numerically found using the empirical formula (Equation 8.19). Using this relation, calculate the minimum current density for copper and silver electrodes and compare the results with data presented in Table 8.5.

8.11 **Radiation of the Arc Positive Column**. Based on the empirical formulas presented in Table 8.6, calculate the percentage of arc power conversion into radiation for different gases at relatively low and relatively high values of the joule heat per unit length, $w = EI$, W/cm. Make your conclusion about effectiveness of different gases as the arc radiation sources at the different levels of specific discharge power.

8.12 **The Elenbaas–Heller Equation**. The Elenbaas–Heller equation in Equation 8.23 includes the only material function, $\sigma(\Theta)$, expressing dependence of the quasi-equilibrium plasma conductivity on the heat flux potential (Equation 8.22). Taking into account that thermal conductivity dependence on temperature $\lambda(T)$ is essentially nonmonotonic (see Figure 8.13), explain why the material function $\sigma(\Theta)$, nevertheless, is smooth enough.

8.13 **The Steenbeck Channel Model of Arc Discharges**. Integrating the Elenbaas–Heller equation in the framework of the Steenbeck channel model, derive the relation between the heat flux potential $\Theta_m(T_m)$ related to the plasma temperature, and the discharge power (joule heating power) per unit arc length.

8.14 **Principle of Minimum Power for Positive Column of Arc Discharges.** According to the Steenbeck principle of minimum power, the plasma temperature T_m and the arc channel radius r_0 should minimize the specific discharge power w and electric field $E = w/I$ at fixed values of current I and discharge tube radius R. Apply the minimization requirement $(dw/dr_0)_{I=const} = 0$ to the functional relations (Equation 8.24 and Equation 8.25) and derive the third equation (Equation 8.27) of the Steenbeck channel model of the positive column of electric arcs.

8.15 **The Raizer Channel Model of Arc Discharges.** Prove the equivalence of the Raizer and Steenbeck approaches to the channel model. Show that the third equation of the Raizer channel model (Equation 8.33) completely coincides with that of the Steenbeck model (Equation 8.27), taking the electric conductivity in Equation 8.30. The Steenbeck approach was based on the principle of minimum power. Is it possible to interpret the agreement between two modifications of the channel model as a proof of validation of the principle of minimum power for arc discharges?

8.16 **Arc Temperature in Frameworks of the Channel Model.** Derive Equation 8.36 for the plasma temperature in arc discharges based on the principal equations of the arc channel model. Explain why the arc temperature is close to a constant with only weak logarithmic dependence on current and radius R (characterizing cooling of the discharge). Compare this conclusion with a similar one for nonthermal discharges (for example, a glow discharge), where the range of changes of electron temperature is also not wide.

8.17 **Modifications of the Arc Channel Model for the Discharge Stabilization by Gas Flow.** Generalize the principal results of the channel model (Equation 8.36 through Equation 8.39) for the case of arc stabilization by fast gas flow, replacing the radius R by an effective one, R_{eff}, which takes into account convective arc cooling. Use the derived relations to explain increase of plasma temperature in the channel by intensive arc cooling in fast gas flows.

8.18 **Difference between Electron and Gas Temperatures in Arc Discharges.** Using the numerical relation (Equation 8.42), estimate the critical value of electric current when the difference between electron and gas temperatures in atmospheric pressure argon arc becomes essential. Compare the estimation results with data presented in Figure 8.11.

8.19 **The Bennet Pinch Pressure Distribution.** Based on the dynamic balance of expansionary kinetic pressure of plasma p and inward radial magnetic body force (Equation 8.45 together with Equation 8.43) for the azimuthal magnetic field B_θ, derive the relation for pressure distribution inside the Bennet pinch of radius r_0. Compare mechanisms of arc stabilization in the Bennet pinch and in industrial arcs with lower electric currents.

8.20 **Electrode Jet Formation.** Based on Equation 8.50, estimate typical values of the electrode jet velocity corresponding to parameters of cathode spots given in Section 8.3 for atmospheric pressure arc discharges. Discuss possible consequences of the high-speed cathode jets on the pre-electrode arc behavior.

8.21 **Stabilization of Linear Arcs near Axis of the Discharge Tube**. Based on the Elenbaas–Heller equation (see Section 8.4.2), explain why, if the arc is cooled on its edges, the thermal discharge temperature on the axis rises. Use this effect to interpret stability of linear arcs near the axis of the discharge tube.

8.22 **Critical Length of Gliding Arc Discharge**. Analyze the relation (Equation 8.54) for electric current in the quasi-equilibrium phase of gliding arc, and asymptotically simplify if in the vicinity of the critical point when the discharge power approaches its maximum value. Why is this relation unable to describe the current evolution for bigger lengths of the arc?

8.23 **Quasi-Unstable Phase of Gliding Arc Discharge**. Based on the stability diagram shown in Figure 8.34, explain the mechanism leading to instability of the formally stable regime of arc with length lower than the critical one. Compare the quasi-unstable phase of the gliding arc with totally stable and totally unstable regimes of the discharge.

8.24 **Discharge Power Distribution between Quasi-Equilibrium and Nonequilibrium Phases of Gliding Arc**. Assuming that power of the gliding arc remains on the maximum level after the FENETRe and using Equation 8.71 for the maximum equilibrium and nonequilibrium lengths, derive a formula describing the fraction of the total gliding arc energy released in the nonequilibrium phase of the discharge.

CHAPTER

9

NONEQUILIBRIUM COLD ATMOSPHERIC PRESSURE PLASMAS: CORONA, DIELECTRIC BARRIER, AND SPARK DISCHARGES

The physical principles of the discharges discussed in this chapter are closely related to the specific physics of the breakdown phenomena (avalanches, streamers, and leaders) considered in Section 4.4. In this chapter, sparks, corona, and dielectric barrier discharges as generators of nonthermal, nonequilibrium atmospheric pressure plasma will be discussed.

9.1 CONTINUOUS CORONA DISCHARGE

9.1.1 General Features of Corona Discharge

Corona is a weakly luminous discharge that usually appears at atmospheric pressure near sharp points, edges, or thin wires in which the electric field is sufficiently large. Thus, corona discharges are always nonuniform; strong electric field, ionization, and luminosity are actually located in the vicinity of one electrode. Weak electric fields simply drag charged particles to another electrode to close the electric circuit. No radiation appears from this "outer region" of the corona discharge. A corona can be observed in air around high-voltage transmission lines, lightning rods, and even masts of ships, where they are called "Saint Elmo's fire." This is the origin of the corona name, which means "crown."

Rather high voltage is required to ignite the corona discharge, which as noted earlier, mainly occupies the region around one electrode. If the voltage grows even larger, the remaining part of the discharge gap breaks down and the corona transfers into the spark (see Section 4.4). In this section, only the main physical and engineering principals of the continuous corona discharge will be discussed; more details on the subject can be found in other references.[301–303]

9.1.2 Electric Field Distribution in Different Corona Configurations

As already mentioned, corona discharges occur only if the electric field is essentially nonuniform. The electric field in the vicinity of one or both electrodes should be much stronger than in the rest of the discharge gap. Such situations usually take place if the characteristic size of an electrode r is much smaller than characteristic distance d between electrodes. For example, a corona discharge in air between parallel wires occurs only if the interwire distance is sufficiently large $d/r > 5.85$. Otherwise, an increase of the discharge system's voltage results not in a corona, but in a spark.

For this reason, engineering calculations of corona discharges require information on the electric field distribution in the system. If effects of the plasma are neglected, these distributions can be found in the framework of electrostatics. For simple discharge gap geometries (see Figure 9.1a through g), such electric field distributions can be found analytically:

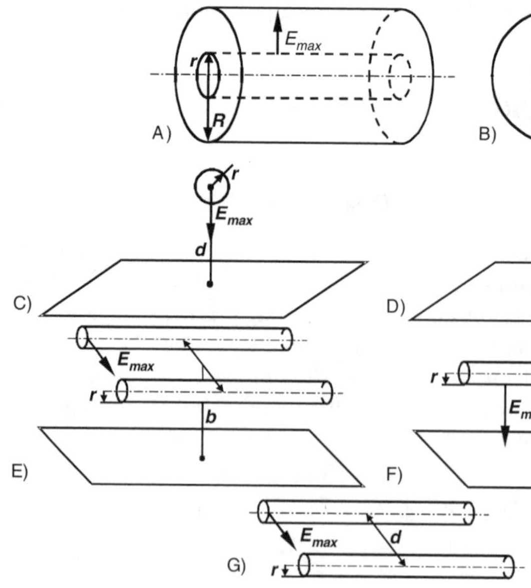

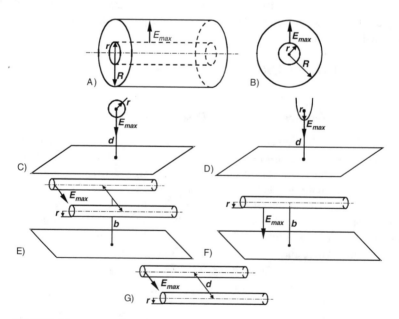

Figure 9.1 Different corona configurations

(*a*) The electric field in the space between *coaxial cylinders* of radii r (internal) and R (external) at a distance x from the axis (Figure 9.1A) can be expressed as:

$$E = \frac{V}{x \ln(R/r)}, \quad E_{max}(x = r) = \frac{V}{x \ln(R/r)} \tag{9.1}$$

where V is the voltage between two electrodes.

(*b*) The electric field in the space *between concentric spheres* also of radii r (internal) and R (external) at a distance x from the center (Figure 9.1B) is:

$$E = V\frac{rR}{x^2(R-r)}, \quad E_{max} \approx \frac{V}{r} \text{ (if } R >> r) \tag{9.2}$$

(*c*) The electric field in the space *between a sphere of radius r and a remote plane d/r* $\rightarrow \propto$ (Figure 9.1C) can be written as a function of distance x from the sphere center:

$$E \approx V\frac{r}{x^2}, \quad E_{max} = \frac{V}{r} \tag{9.3}$$

(*d*) The electric field in the space *between a parabolic tip with curvature radius r and a plane* perpendicular to it and located at a distance d from the tip (Figure 9.1D) can be expressed as a function of distance x from the tip:

$$E = \frac{2V}{(r+2x)\ln(2d/r+1)}, \quad E_{max} \approx \frac{2V}{r \ln(2d/r)} \tag{9.4}$$

(*e*) If the corona is organized *between two parallel wires* of radius r separated on distance d between each other and located on distance b from the Earth (Figure 9.1E), the maximum electric field is obviously achieved near the surface of the wires and equals to:

$$E_{max} = \frac{V}{r \ln\left[d \Big/ \left\{r \cdot \sqrt{1+(d/2b)^2}\right\}\right]} \tag{9.5}$$

(*f*) Based on Equation 9.5, the maximum electric field induced by the *single wire and a parallel plane* located at a distance b ($d \rightarrow \propto$; Figure 9.1F) can be presented as:

$$E_{max} = \frac{V}{r \ln(2b/r)} \tag{9.6}$$

(*g*) One can also derive from Equation 9.5 the maximum electric field *between two parallel wires* spaced by a distance d between each other ($b \rightarrow \propto$; Figure 9.1G):

$$E_{max} = \frac{V}{r \ln(d/r)} \tag{9.7}$$

Obviously, Equation 9.1 through Equation 9.7 describe electric fields in the corona discharge system before the generated plasma perturbs it.

9.1.3 Negative and Positive Corona Discharges

The mechanism for sustaining the continuous ionization level in a corona depends on the polarity of the electrode where the high electric field is located. If the high electric field zone is located around the cathode, such corona discharge is referred to as the **negative corona**. Conversely, if the high electric field is concentrated in the region of the anode, such a discharge is called the **positive corona.**

Ionization in the negative corona is due to multiplication of avalanches (see Section 4.4). Continuity of electric current from the cathode into the plasma is provided by secondary emission from the cathode (mostly induced by ion impact). Ignition of the negative corona actually has the same mechanism as the Townsend breakdown (see Section 4.1.1), obviously generalized taking into account nonuniformity and possible electron attachment processes:

$$\int_0^{x_{max}} [\alpha(x) - \beta(x)]dx = \ln\left(1 + \frac{1}{\gamma}\right) \tag{9.8}$$

In this equation, $\alpha(x)$, $\beta(x)$, and γ are the first, second, and third Townsend coefficients, describing ionization, electron attachment, and secondary electron emission from the cathode, respectively; x_{max} corresponds to the distance from the cathode, where the electric field becomes low enough and $\alpha(x_{max}) = \beta(x_{max})$, which means that no additional electron multiplication takes place. The equality $\alpha(x_{max}) = \beta(x_{max})$ actually corresponds to the breakdown electric field E_{break} in electronegative gases (see Section 4.4). If the gas is not electronegative ($\beta = 0$), integration of Equation 9.8 is formally not limited; however, due to the exponential decrease of $\alpha(x)$ it is.

Note that the critical distance $x = x_{max}$ determines not only the ionization, but also the electronic excitation zone and thus the zone of plasma luminosity. This means that the critical distance $x = x_{max}$ can be considered the visible size of the corona (or an active corona volume).

Ionization in the positive corona cannot be provided by the cathode phenomena because, in this case, the electric field at the cathode is low. Here, ionization processes are related to the formation of the cathode-directed (or the so-called positive) streamers (see Figure 4.25). Ignition conditions can be described for the positive corona using the criteria of cathode-directed streamer formation. In this case, generalization of the Meek breakdown criterion (Equation 4.156) is quite a good approximation, taking into account the nonuniformity of the corona and possible contributions of electron attachment:

$$\int_0^{x_{max}} [\alpha(x) - \beta(x)]dx \approx 18 - 20 \tag{9.9}$$

Comparing the similar ignition criteria (Equation 9.8 and Equation 9.9), one can see that the minimal values of the amplification coefficients should be two to three times lower to provide ignition of a negative corona (because $\ln(1/\gamma) \approx 6 - 8$). However, the critical values of the electric field for ignition of positive and negative coronas are pretty close even though these are related to very different breakdown mechanisms. This can be explained by the strong exponential dependence of the amplification coefficients on the electric field value.

9.1.4 Corona Ignition Criterion in Air: Peek Formula

According to Equation 9.8 and Equation 9.9, ignition for positive and negative coronas is mostly determined by the value of the maximum electric field in the vicinity of the electrode, where the discharge is to be initiated. The critical value of the igniting electric field (near the electrode) for the case of coaxial electrodes in air can be calculated numerically using the empirical Peek formula:

$$E_{cr}, \frac{\text{kV}}{\text{cm}} = 31\delta\left(1 + \frac{0.308}{\sqrt{\delta r(\text{cm})}}\right) \tag{9.10}$$

In the Peek formula, δ is the ratio of air density to the standard value, and r is the radius of internal electrode. The formula can be applied for pressures 0.1 to 10 atm, polished internal electrodes with radius $r \approx 0.01$ to 1 cm, with both direct current and AC with frequencies up to 1 KHz. Roughness of the electrodes decreases the critical electric field by 10 to 20%.

The Peek formula (Equation 9.10) was derived for the case of coaxial cylinders, but it can be used also for other corona configurations with slightly different values of coefficients. As an example, the critical corona-initiating electric field in the case of two parallel wires can be calculated using the following empirical formula:

$$E_{cr}, \frac{\text{kV}}{\text{cm}} = 30\delta\left(1 + \frac{0.301}{\sqrt{\delta r(\text{cm})}}\right) \tag{9.11}$$

As is seen, the empirical Equation 9.11 does not differ much from the Peek formula. Both relations correspond to simplified empirical formula for the Townsend coefficient α in air at reduced electric fields $E/p < 150$ V/cm · torr:

$$\alpha, \frac{1}{\text{cm}} = 0.14 \cdot \delta\left[\left(\frac{E, \text{kV/cm}}{31\delta}\right)^2 - 1\right] \tag{9.12}$$

Equation 9.10 and Equation 9.11 determine the critical value of the corona electric field. Voltage necessary to ignite the corona discharge then can be found based on such relations as Equation 9.1 through Equation 9.7. In this case, the critical value of the electric field is supposed to be reached in the close vicinity of an active electrode.

9.1.5 Active Corona Volume

Ionization, and thus effective generation of charged particles, takes place in corona discharges only in the vicinity of an electrode where electric field is sufficiently

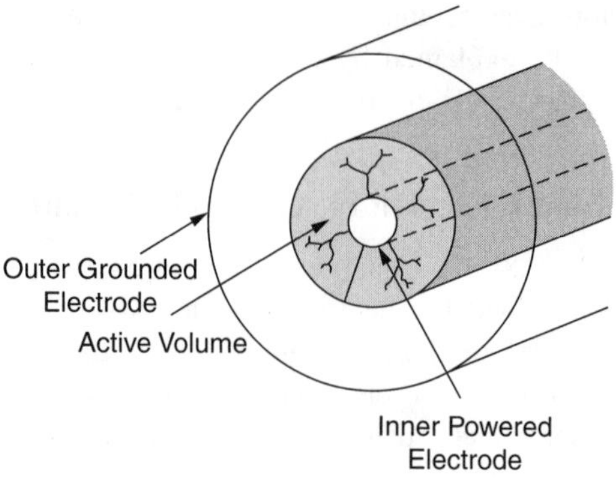

Figure 9.2 Active corona volume

high. This zone is usually referred to as the active corona volume (see Figure 9.2). The active corona volume is the most important part of the discharge from the point of view of plasma chemical applications because most excitation and reaction processes take place in this zone. External radius (or more generally, size) of the active corona volume is determined by the value of the electric field, which should correspond to breakdown value E_{break} on the boundary of the active volume (see discussion after Equation 9.8, as well as Section 4.4).

As an example, consider the generation of a corona discharge in air (at normal conditions) between a thin wire electrode of radius $r = 0.1$ cm and coaxial cylinder external electrode with radius $R = 10$ cm. According to the Peek formula, the critical igniting electric field in the vicinity of the internal electrode is about 60 kV/cm. In this case, Equation 9.1 gives the minimum value of voltage required for ignition of the corona to be about 30 kV.

At the same time, the electric field near the external electrode is relatively very low $E(R) \approx 0.6$ kV/cm, $E(R)/p \approx 0.8$ V/cm · torr and, obviously, not sufficient for ionization. Effective multiplication of charges requires breakdown of the electric field, which can be estimated as $E_{break} \approx 25$ kV/cm. This determines the external radius of the active corona volume case as $r_{AC} = rE_{cr}/\, E_{break} \approx 0.25$ cm. Thus, the active corona volume occupies the cylindrical layer $0.1 \text{ cm} < x < 0.25$ cm around the thin wire.

In general, the external radius of the active corona volume around the thin wire can be determined based on Equation 9.1 as:

$$r_{AC} = \frac{V}{E_{break} \ln(R/r)} \tag{9.13}$$

where V is the voltage applied to sustain the corona discharge. As is seen from this equation, the radius of active corona volume is increasing with applied voltage because values of R, r, and E_{break} are actually fixed by the discharge geometry and gas composition.

Similar to Equation 9.13, the external radius of the active corona volume generated around a sharp point can be expressed according to Equation 9.2 as:

$$r_{AC} \approx \sqrt{\frac{rV}{E_{break}}} \tag{9.14}$$

Based on Equation 9.13 and Equation 9.14, compare the active radii of corona around thin wire and sharp point:

$$\frac{r_{AC}(\text{wire})}{r_{AC}(\text{point})} \approx \frac{1}{\ln(R/r)}\sqrt{\frac{V}{aE_{break}}} = \frac{r_{AC}(\text{point})}{r\ln(R/r)} \tag{9.15}$$

Numerically, this ratio is typically about 3, which illustrates the advantage of corona generated around a thin wire in producing a larger volume of nonthermal atmospheric pressure plasma effective for different applications.

9.1.6 Space Charge Influence on Electric Field Distribution in Corona Discharge

The generation of charged particles takes place only in the active corona volume in the vicinity of an electrode. Thus, the electric current to the external electrode outside the active volume is provided by the drift of charged particles (generated in the active volume) in the relatively low electric field. In the positive corona, these drifting particles are positive ions and, in the negative corona, they are negative ions (or electrons, if the corona is generated in nonelectronegative gas mixtures).

The discharge current is determined by the difference between the applied voltage V and the critical one, V_{cr}, corresponding to the critical electric field E_{cr} and to ignition of the corona (see Section 9.1.4). The current value is limited by the space charge outside the active corona volume. The active corona volume is able to generate high current, but the current of charged particles is partially reflected back by the space charge formed by these particles. The phenomenon is somewhat similar to the phenomenon of current limitation by space charge in sheaths, or in vacuum diodes (see Section 6.1.4 and Section 6.1.5). However, in the case under consideration, the motion of charged particles is not collisionless, but determined by drift in electric field.

To analyze the space charge influence on the electric field distribution, consider the example of a corona generated between coaxial cylinders with radii R and r (e.g., corona around a thin wire). The electric current per unit length of the wire i is constant outside active corona volume, where there is no charge multiplication:

$$i = 2\pi x \cdot en \cdot \mu E = const \tag{9.16}$$

Here, x is the distance from corona axis; n is the number density of charged particles providing electric conductivity outside of the active volume; and μ is the mobility of the charged particles. Assuming the space charge perturbation of the electric field is not very strong, the number density distribution $n(x)$ can be found in the first approximation based on Equation 9.16 and nonperturbed electric field distribution (Equation 9.1):

$$n(x) = \frac{i}{2\pi e\mu Ex} = \frac{i\ln(R/r)}{2\pi e\mu V} = const \tag{9.17}$$

This expression for the charged particles' density distribution $n(x)$ then can be applied to find the second approximation of the electric field distribution $E(x)$, using the Maxwell equation for the case of cylindrical symmetry:

$$\frac{1}{x}\frac{d[xE(x)]}{dx} = \frac{1}{\varepsilon_0}en(x), \quad \frac{1}{x}\frac{d[xE(x)]}{dx} = \frac{i\ln(R/r)}{2\pi\varepsilon_0\mu V} \tag{9.18}$$

The Maxwell equation can be easily integrated, taking into account that the electric field distribution should follow Equation 9.1 with the critical value of voltage V_{cr} (corresponding to corona ignition) at the low current limit $i \to 0$. This integration yields the electric field distribution, which takes into account current and, thus, the space charge:

$$E(x) = \frac{V_{cr}\ln(R/r)}{x} + \frac{i\ln(R/r)}{2\pi\varepsilon_0\mu V}\cdot\frac{x^2 - r^2}{2x} \tag{9.19}$$

Obviously, in this relation:

$$\int_r^R E dx = V$$

Also, the distribution (Equation 9.19) is valid only in the case of small electric field perturbations due to the space charge outside the active corona volume. Expressions similar to Equation 9.19 describing the influence of electric current and space charge on the electric field distribution could be derived for other corona configurations.[264]

9.1.7 Current–Voltage Characteristics of Corona Discharge

Integration of the expression for the electric field Equation 9.19 over the radius x, taking into account that in most of the corona discharge gap $x^2 >> r^2$, gives the relation between current (per unit length) and voltage of the discharge, which is the current–voltage characteristic of corona generated around a thin wire:

$$i = \frac{4\pi\varepsilon_0\mu V(V - V_{cr})}{R^2\ln(R/r)} \tag{9.20}$$

As this equation shows, the corona current depends on the mobility of the main charge particles providing conductivity outside the active corona volume. Noting that mobilities of positive and negative ions are nearly equal, the electric currents in positive and negative corona discharges are also close. Negative corona in gases without electron attachment (e.g., noble gases) provides much larger currents because electrons are able to leave the discharge gap rapidly without forming a significant space charge. Even a small admixture of an electronegative gas decreases the corona current.

It is important to point out that the parabolic current–voltage characteristic (Equation 9.20) is valid not only for thin wires, but also for other corona configurations. Obviously, the coefficients before the quadratic form $V(V - V_{cr})$ are different for different geometries of corona discharges:

$$I = CV(V - V_{cr}) \tag{9.21}$$

In this relation, I is the value of total current in the corona discharge. For example, the current–voltage characteristic for the corona generated in atmospheric air between a sharp point cathode with radius $r = 3 - 50$ μm and a perpendicular flat anode located on the distance of $d = 4 - 16$ mm can be expressed as:

$$I, \mu A = \frac{52}{(d, mm)^2}(V, kV)(V - V_{cr}) \tag{9.22}$$

In this empirical relation, I is the total corona current from the sharp point cathode. The critical corona ignition voltage V_{cr} in this case can be taken as $V_{cr} \approx 2.3$ kV and does not depend on the distance d.[303]

9.1.8 Power Released in Continuous Corona Discharge

Based on the current–voltage characteristic (Equation 9.20), the electric power released in the continuous corona discharge can be determined for the case of a long thin wire as:

$$P = \frac{4\pi L\varepsilon_0 \mu V(V - V_{cr})}{R^2 \ln(R/r)} \tag{9.23}$$

where L is the length of the wire. In more general cases, the corona discharge power can be determined based on the current–voltage characteristic (Equation 9.21) as:

$$P = CV(V - V_{cr}) \tag{9.24}$$

For example, for the corona generated in atmospheric air between the sharp pointed cathode with radius $r = 3 - 50$ μm and a perpendicular flat anode located at a distance of $d = 1$ cm, the coefficient $C \approx 0.5$ if voltage is expressed in kilovolts and power is expressed in milliwatts. This yields the corona power as about 0.4 W at the voltage of about 30 kV. This power is actually very low. Similarly, corona discharges generated in atmospheric pressure air around the thin wire ($r = 0.1$ cm, $R = 10$ cm, and $V_{cr} = 30$ kV) with 40 kV voltage release power of about 0.2 W per centimeter of the discharge.

As shown, the power of the continuous corona discharges is very low and not acceptable for many possible applications. Recall that further increases of voltage and current lead to corona transition into sparks. However, these can be prevented by organizing the corona discharge in a pulse-periodic mode. Such pulsed corona discharges will be discussed in the next section. Although the corona power is relatively low per unit length of a wire, the total corona power becomes significant when the wires are very long. Such situations take place in the case of high-voltage overland transmission lines, in which coronal losses are significant. In humid and snow conditions, these can exceed the resistive losses.

One must take into account, in the case of high-voltage overland transmission lines, that the two wires generate corona discharges of opposite polarity. Electric currents outside active volumes of the opposite polarity corona discharges are provided by positive and negative ions moving in opposite directions. These positive and negative ions meet and neutralize each other between wires, which results in a decrease of the space charge and an increase of the corona current, which relates to phenomenon power losses. For example, the coronal power losses for a transmission line wire with 2.5 cm diameter and voltage of 300 kV are about 0.8 kW/km in case of fine and sunny weather. In rainy or snowy conditions, the critical voltage V_{cr} is lower, and coronal losses of the high-voltage overland transmission lines grow significantly. For this case, the same loss of about 0.8 kW/km corresponds to a voltage of about 200 kV.

9.2 PULSED CORONA DISCHARGE

9.2.1 Why the Pulsed Corona?

Nonthermal atmospheric pressure discharges, and especially corona discharges, are very attractive for different applications, such as surface treatment and cleansing of gas and liquid exhaust streams. These discharges are able to generate high concentrations of active atoms and radicals at atmospheric pressure without heating the gas as a whole. However, application of the continuous corona discharge is limited by very low currents and, thus, very low power of the discharge (see Section 9.1.8). This results in a low rate of generation of active species and low rate of treatment of materials and exhaust streams.

To increase the corona current and power, the voltage and electric field should be increased. However, as the electric field increases, the active corona volume grows until it occupies the entire discharge gap. When streamers are able to reach the opposite electrode, formation of a spark channel occurs, which subsequently results in local overheating and plasma nonuniformity that are not acceptable for applications.

Increasing corona voltage and power without spark formation becomes possible by using pulse-periodic voltages. Today, the pulsed corona is one of the most promising atmospheric pressure, nonthermal discharges. As was shown in Section 4.4.6, streamer velocity is about 10^8 cm/sec and exceeds by a factor of 10 the typical electron drift velocity in an avalanche. If the distance between electrodes is about 1 to 3 cm, the total time necessary for development of avalanches, avalanche-to-streamer transition, and streamer propagation between electrodes is about 100 to 300 nsec. This means that voltage pulses of this duration range are able to sustain streamers and effective power transfer into nonthermal plasma without streamer transformations into sparks.

For pulsed corona discharges, the key point is to make relevant pulse power supplies, which are able to generate sufficiently short voltage pulses with steep front and very short rise times. Some specific methods of generation and parameters for the pulsed corona discharges will be discussed later in this section. First, however, some important non-steady-state phenomena occurring in the continuous corona discharges, which should be taken into account in analyzing pulsed corona discharges, will be discussed.

9.2.2 Corona Ignition Delay

Numerous experimental data regarding ignition delay of the continuous corona are available. These are quite helpful in understanding pulsed corona discharge. For example, the ignition delay of the continuous negative corona strongly depends on cathode conditions and varies from one experiment to another. Such facility-specific characteristics are one reason why pulsed coronas are more often organized as positive.

Typical ignition delay in the case of a positive corona is about 100 nsec (30 to 300 nsec). This interval is much longer than streamer propagation times; thus, it is related to the time for initial electron formation and propagation of initial avalanches. Note that the initial electrons are not formed near the cathode but rather in the discharge gap.

Random electrons in the atmosphere usually exist in the form of negative ions, O_2^-; their effective detachment is due to ion–neutral collisions and effectively takes place at electric fields about 70 kV/cm. If humidity is high and the negative ions O_2^- are hydrated, the electric field necessary for detachment and formation of a free electron is slightly higher. Thus, experimental data related to the ignition delay of the continuous corona actually indicate the same limits for pulse duration in pulsed corona discharges. This indicates some advantages of the positive corona and the cathode-directed streamers, and also shows the electric fields necessary for effective release of initial free electrons by detachment from negative ions.

9.2.3 Pulse-Periodic Regime of Positive Corona Discharge Sustained by Continuous Constant Voltage: Flashing Corona

Corona discharges sometimes operate in the form of periodic current pulses even at constant voltage conditions. Frequency of these pulses can reach 10^4 Hz in the case of the positive corona and 10^6 Hz for the negative corona. This self-organized pulsed corona discharge is obviously unable to overcome the current and power limitations of the continuous corona discharges because continuous high voltage still promotes the corona-to-spark transition. However, it is an important step toward the non-steady-state coronas with higher voltages, higher currents, and higher power.

As an example, consider at first a positive corona discharge formed between a sharp point anode of radius 0.17 mm and a flat cathode located 3.1 cm apart.

Discharge ignition takes place in this system at the critical voltage $V_{cr} \approx 5$ kV. This corona operates in the pulse-periodic regime starting from the ignition voltage to a voltage $V_1 \approx 9.3$ kV.[303] Near the boundary voltage values (V_{cr}, V_1), the frequency of the current pulses is low. The frequency of the pulses reaches the maximum value of about 6.5 kHz in the middle of the interval (V_{cr}, V_1). This pulsing discharge is usually referred to as a **flashing corona**. The mean current value in this regime reaches 1 μA at the voltage $V_1 \approx 9.3$ kV.

Increasing the voltage from $V_1 \approx 9.3$ kV to $V_2 \approx 16$ kV stabilizes the corona; the discharge operates in steady state without pulses. Current grows in this regime from 1 to 10 μA. Further increase of voltage from $V_2 \approx 16$ kV up to the corona transition into a spark at $V_t \approx 29$ kV leads again to the pulse-periodic regime. Frequency grows up in this regime to about 4.5 kHz; mean value of current grows during the same time, to about 100 μA.

The flashing corona phenomenon can be explained by the effect of positive space charge, which is created when electrons formed in streamers decrease fast at the anode, but slow positive ions remain in the discharge gap. The growing positive space charge decreases electric field near anode and prevents new streamers generation. Positive corona current is suppressed until the positive space charge goes to the cathode and clears up the discharge gap. After that, a new corona ignition takes place and the cycle can be repeated again.

The flashing corona phenomenon does not occur at intermediate voltages ($V_1 < V < V_2$), when the electric field outside of the active corona volume is sufficiently high to provide effective steady-state clearance of positive ions from the discharge gap but not too high to provide intensive ionization. It is interesting to note that the electric current in the flashing corona regime does not fall to zero between pulses; some constant component of the corona current is continuously present.

From this experimental data, one can easily calculate the maximum power of a pulse-periodic corona around a sharp point to be about 3 W. This is more than 10 times higher than the maximum power in a continuous regime for the same corona system, which is 0.2 W (see also the numerical example after Equation 9.24). Thus, the pulse-periodic regime leads to a fundamental increase of corona power, although the power increase in this system is still limited by spark formation because the applied voltage is continuous.

9.2.4 Pulse-Periodic Regime of Negative Corona Discharge Sustained by Continuous Constant Voltage: Trichel Pulses

Negative corona discharges sustained by continuous voltage also can operate in the pulse-periodic regime at a relatively low value of voltages close to the ignition value. Frequency of the pulses is much higher in this case (10^5 to 10^6 Hz) relative to the positive corona; the pulse duration is short, approximately 100 nsec. If the mean corona current is 20 μA, the peak value of current in each pulse can reach 10 mA. The pulses disappear at higher voltages and, in contrast to the case of positive corona, the steady-state discharge exists until transition to spark.

The pulse-periodic regime of the negative corona discharge is usually referred to as **Trichel pulses**. General physical causes of the Trichel pulses are similar to those of the flashing corona discussed earlier, though with some peculiarities. The growth of avalanches from cathode leads to formation of two charged layers: (*a*) an internal one that is positive and consists of positive ions and (*b*) an external negative one that consists of negative ions (in air or other electronegative gases) or of electrons in the case of electropositive gases.

In electropositive gases such as nitrogen or argon, the Trichel pulses are not generated at all. Because of their high mobility, electrons reach the anode quite fast. As a result, the density of the space charge of electrons in the external layer is very low and, consequently, the electric field near the cathode is not suppressed. The positively charged internal layer even increases the electric field in the vicinity of the cathode and provides better conditions for the active corona volume. Thus, the Trichel pulses take place only in electronegative gases.

In air and other electronegative gases, negative ions are able to form significant negative space charge around the cathode. This negative space charge cannot be compensated by narrowing the layer of positive ions, which are effectively neutralized at the nearby cathode. Thus, the space charge of the negative ions suppresses the electric field near the cathode and therefore suppresses the corona current. Subsequently, when the ions leave the discharge gap and are neutralized on the electrodes, the negative corona can be reignited, and the cycle again can be repeated.

9.2.5 Pulsed Corona Discharges Sustained by Nanosecond Pulse Power Supplies

The key element of this corona discharge system is the nanosecond pulse power supply, which generates pulses with duration in the range of 100 to 300 nsec, sufficiently short to avoid the corona-to-spark transition. Also, the power supply should provide a pretty high voltage rise rate (0.5 to 3 kV/nsec), resulting in higher corona ignition voltage and higher power. As illustration of this effect, Figure 9.3 shows the corona inception voltage as a function of the voltage rise rate.[307]

The high voltage rise rates, and thus higher voltages and mean electron energies in the pulsed corona, also result in better efficiency of several plasma chemical processes requiring higher electron energies. In these processes, such as plasma cleansing of gas and liquid steams, high values of mean electron energy are necessary to decrease the fraction of the discharge power going to vibrational excitation of molecules and stimulate ionization and electronic excitation and dissociation of molecules (see Section 3.3.8).

The nanosecond pulse power supply relevant to application in pulse corona discharges can be organized in many different ways. Especially of note are the application of Marx generators; simple and rotating spark gaps; electronic lamps: thyratrones; and thyristors with possible further magnetic compression of pulses (see Pu and Woskov[304]). Application of transistors for the high voltage pulse generation is also under consideration.

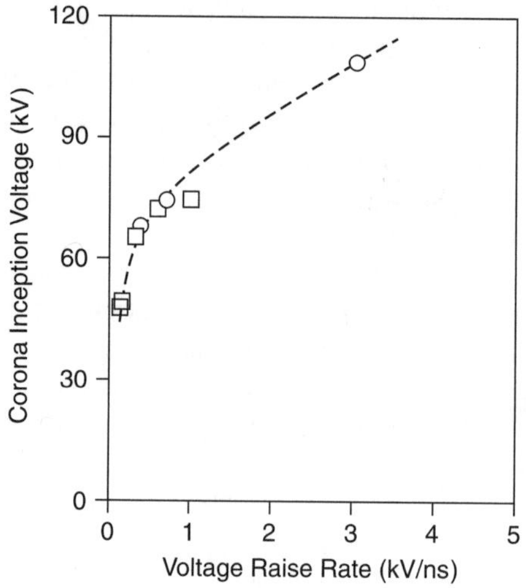

Figure 9.3 Corona inception voltage as a function of voltage raise rate[307]

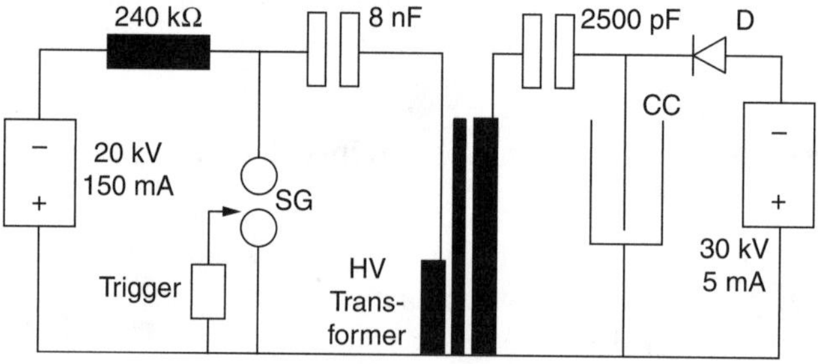

Figure 9.4 Electrical circuit for production of high-voltage pulses with DC bias[305,306]

As an example, consider the pulse power supply for a corona discharge based on the thyristor triggered spark gap, which switches the capacitor.[305,306] The general scheme of this power supply is illustrated in Figure 9.4; the generated voltage and current wave-shapes are shown in Figure 9.5. The high voltage transformer provides a 70-kV voltage pulse with a rise time of about 100 nsec and half-width of 180 nsec. The maximum pulse voltage rise rate was not very high in these experiments (0.7 kV/nsec) due to the large inductance of the transformer; this also resulted in a long and decayed oscillation on the tail of the voltage pulse. Compare this with the voltage rise rate claimed by Mattachini et al.,[307] which exceeded 3 kV/nsec. The half-widths of their corona current pulses were 140 to 150 nsec, which was sufficient to prevent the corona-to-spark transition.

To reach higher current pulses and increase the pulse energy, a DC-bias voltage can be effectively added (see Figure 9.4). At a DC-bias voltage of about 20 kV, the single pulse energy of the corona is about 0.3 to 1 J/m.[306,307] The power of a pulsed

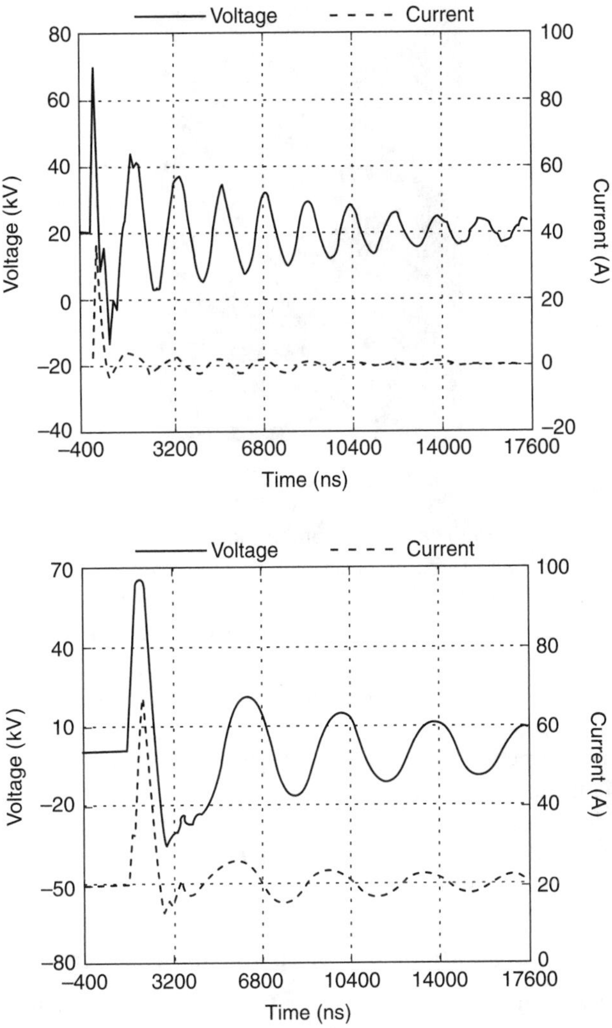

Figure 9.5 Current and voltage evolution in pulsed corona[305,306]

corona is usually varied by changing the pulse frequency. In the referenced systems, frequencies were up to 300 kHz, which yields a specific corona power of 3 W per 1 cm of wire. This specific power of the pulsed corona discharge is 15 times larger than the typical value of the specific power of continuous corona (see Section 9.1.8).

Note that the repetition frequency of high voltage pulses could be higher, thus providing the possibility of further increase of specific power of pulsed corona discharges. For example, the power supply with a magnetic pulse compression system delivers up to 35-kV, 100-nsec pulses at repetition rates up to 1.5 kHz.[308] In general, the total power of the pulsed coronas can be increased by increasing the

Figure 9.6 Illustrative picture of pulsed corona discharge

length of wire and reach high values for nonthermal atmospheric pressure discharges. Thus, the pulsed corona discharge described by Mattachini et al.[307] should operate at powers exceeding 10 kW, and that presented by Korobtsev et al.[353a] is able to operate at a power level of 30 kW. Powerful 10 kW pulsed corona created in collaboration with Drexel Plasma Institute, UIC, and Kurchatov Institute of Atomic Energy is mounted on a trailer and applied as a Mobile Plasma Laboratory for large volume air cleaning.[619]

In general, the pulsed corona can be relatively powerful, not very much but sufficiently luminous, and quite nice looking, which is more or less illustrated in Figure 9.6. More information about the physical aspects and applications of the pulsed corona discharge can be found in the publications of Penetrante and Schulteis[308,308a,309] and Masuda.

9.2.6 Specific Configurations of Pulsed Corona Discharges

The most typical configurations of the pulsed corona as well as continuous corona discharges are based on using thin wires, which maximizes the active discharge volume (see Section 9.1.5). One of these configurations is illustrated in Figure 9.7. Limitations of the wire configuration of the corona are related by the durability of the electrodes and also by nonoptimal interaction of discharge volume with incoming gas flow. The latter is important for plasma chemical applications.

From this point of view, it is useful to use another corona discharge configuration, based on multiple stages of pin-to-plate electrodes and illustrated in Figure 9.8.[310] This system is obviously more durable and, as one can see from Figure 9.9, is able to provide good interaction of the incoming gas stream with the active corona volume formed between the electrodes with pins and holes.

Combination of pulsed corona discharges with other methods of gas treatment is very practical for different applications and has been intensively investigated, with much recent development. As an example, the pulse corona was successfully combined

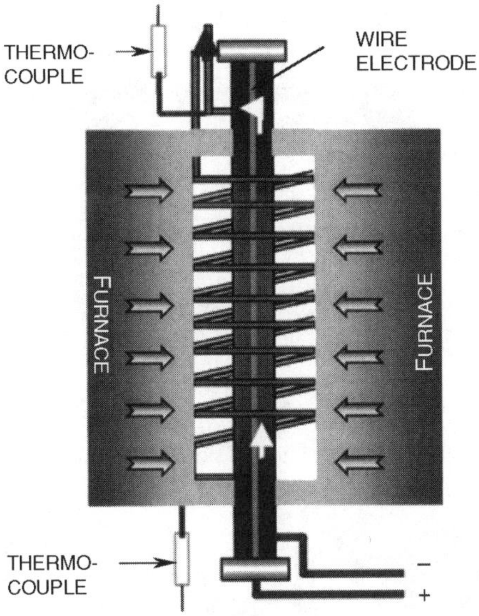

Figure 9.7 Pulsed corona discharge in wire cylinder configuration with preheating[312]

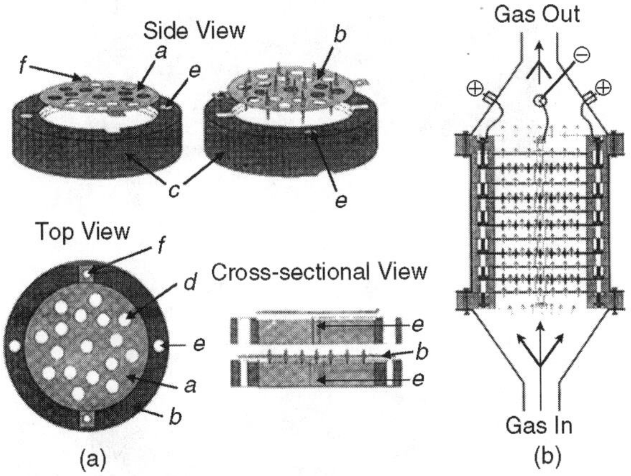

Figure 9.8 Schematic diagram of electrodes and mounting blocks (a) and cross-sectional view of the assembled plasma reactor and incoming gas stream (b); a—anode plates, b—cathode plates, c—mounting block, d—holes for gas flow, e—holes for connecting post, f—connection wings[310]

with catalysis to achieve improved results in the plasma treatment of automotive exhausts[311] and for hydrogen production from heavy hydrocarbons.[312]

Another interesting technological hybrid is related to the pulsed corona coupled with water flow. Such a system can be arranged in form of a shower, called the spray corona, or with a thin water film on the walls, which is usually referred to as the wet corona (see Figure 9.10). Such plasma scrubbers are especially effective in air cleansing processes, when plasma converts a nonsoluble pollutant into a soluble one.

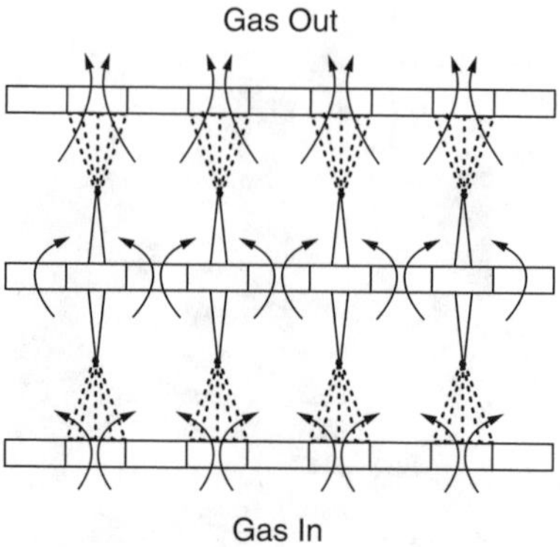

Figure 9.9 Illustration of interaction of incoming gas stream with corona discharge formed between the electrodes with pins and holes[310]

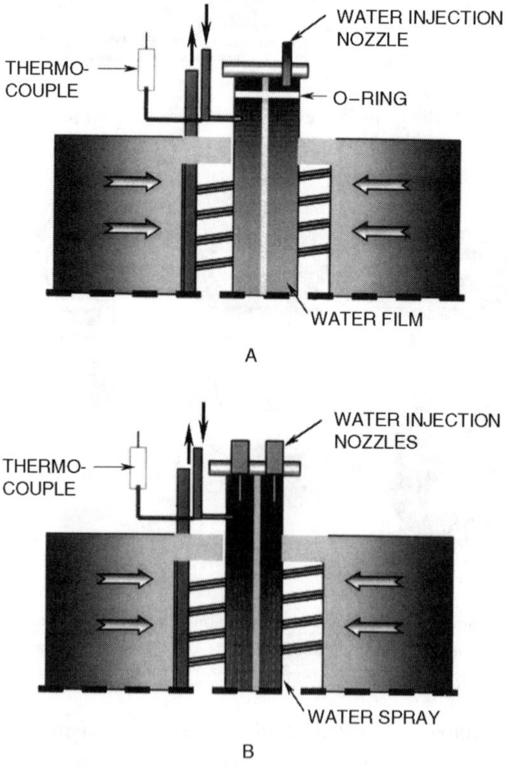

Figure 9.10 (A) Wet and (B) spray corona discharges[312]

This approach is successfully used in the above-mentioned Mobile Plasma Laboratory to clean exhaust gases from paper mills and wood plants. The technology was developed as a result of collaboration between Drexel University, UIC, Kurchatov

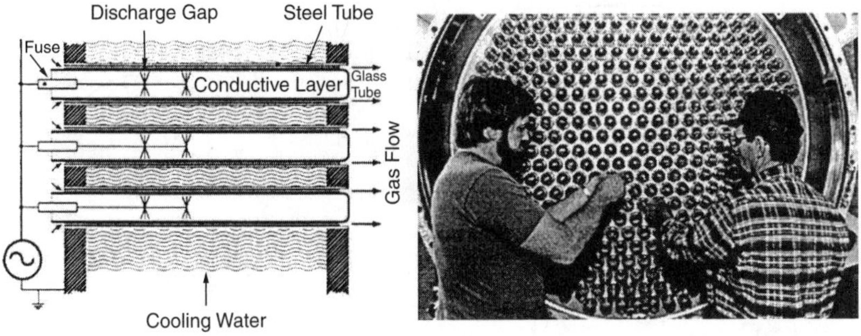

Figure 9.11 Schematic diagram of discharge tubes and photograph of large ozone generator at Los Angeles Aqueduct Filtration Plant[314,634]

Institute of Atomic Energy, Argonne National Laboratory and Pacific Northwest National Laboratory.[633]

9.3 DIELECTRIC-BARRIER DISCHARGE

9.3.1 General Features of Dielectric-Barrier Discharge

It has been shown that the corona-to-spark transition at high levels of voltage is prevented in pulsed corona discharges by employing special nanosecond pulse power supplies. An alternate approach to avoid formation of sparks and current growth in the channels formed by streamers is to place a dielectric barrier in the discharge gap. This is the principal idea of **dielectric barrier discharges** (DBD). The presence of a dielectric barrier in the discharge gap precludes DC operation of DBD, which usually operates at frequencies between 0.5 and 500 kHz. Sometimes dielectric-barrier discharges are also called **silent discharges** because of the absence of sparks that are accompanied by local overheating, generation of local shock waves, and noise.

An important advantage of the dielectric-barrier discharge is the simplicity of its operation. It can be employed in strongly nonequilibrium conditions at atmospheric pressure and at reasonably high power levels, without using sophisticated pulse power supplies. Today, the DBDs find large-scale industrial use for ozone generation[313,314,315] (also see Figure 9.11). These discharges are industrially applied as well in CO_2 lasers, and as a UV-source in excimer lamps. DBD application for pollution control and surface treatment is quite promising, but the largest expected DBD applications are related to plasma display panels for large-area flat television screens.

Important contributions in fundamental understanding and industrial applications of DBD were made recently by Kogelschatzs and Eliasson and their group at ABB.[313,314,315] However, this discharge actually has a long history. It was first introduced by Siemens in 1857[311] to create ozone; this use determined the main direction for investigations and applications of this discharge for many decades. Important steps in understanding the physical nature of the DBD were made by Bussin in

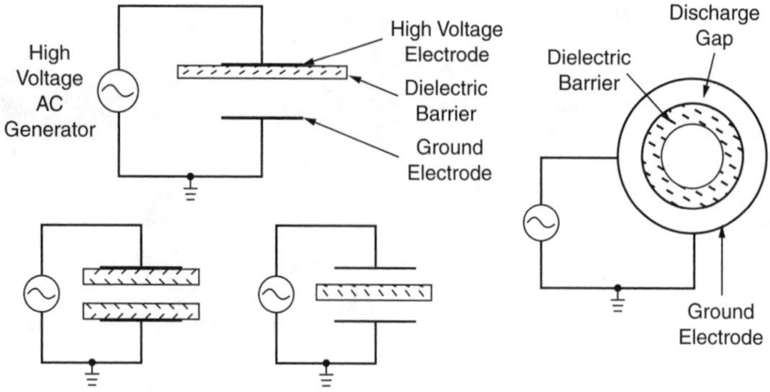

Figure 9.12 Common dielectric-barrier discharge configurations[313,314]

1932[316] and Klemenc et al. in 1937.[317] Their work showed that this discharge occurs in a number of individual tiny breakdown channels, which are now referred to as **microdischarges,** and intensively investigated their relationship with streamers.

9.3.2 General Configuration and Parameters of Dielectric Barrier Discharges

The dielectric barrier discharge gap usually includes one or more dielectric layers located in the current path between metal electrodes. Two specific DBD configurations, planar and cylindrical, are illustrated in Figure 9.12. Typical clearance in the discharge gaps varies from 0.1 mm to several centimeters. Breakdown voltages of these gaps with dielectric barriers are practically the same as those between metal electrodes. If the dielectric barrier discharge gap is a few millimeters, the required AC driving voltage with frequency of 500 Hz to 500 kHz is typically about 10 kV at atmospheric pressure.

The dielectric barrier can be made from glass, quartz, ceramics, or other materials of low dielectric loss and high breakdown strength. Then a metal electrode coating can be applied to the dielectric barrier. The barrier–electrode combination also can be arranged in the opposite manner, e.g., metal electrodes can be coated by a dielectric. As an example, steel tubes coated by an enamel layer can be effectively used in the dielectric barrier discharge.

9.3.3 Microdischarge Characteristics

As has been mentioned previously, the dielectric barrier discharge proceeds in most gases through a large number of independent current filaments usually referred to as microdischarges. From a physical point of view, these microdischarges are actually streamers that are self-organized taking into account charge accumulation on the dielectric surface. Typical characteristics of the DBD microdischarges in a 1-mm gap in atmospheric air are summarized in Table 9.1.

Table 9.1 Typical Parameters of a Microdischarge

Lifetime	1–20 nsec	Filament radius	50 ÷ 100 μm
Peak current	0.1 A	Current density	0.1 ÷ 1 kA/cm^{-2}
Electron density	10^{14} ÷ 10^{15} cm^{-3}	Electron energy	1–10 eV
Total transported charge	0.1–1 nC	Reduced electric field	$E/n = (1 \div 2)(E/n)_{Paschen}$
Total dissipated energy	5 μJ	Gas temperature	Close to average, about 300 K
Overheating	5 K		

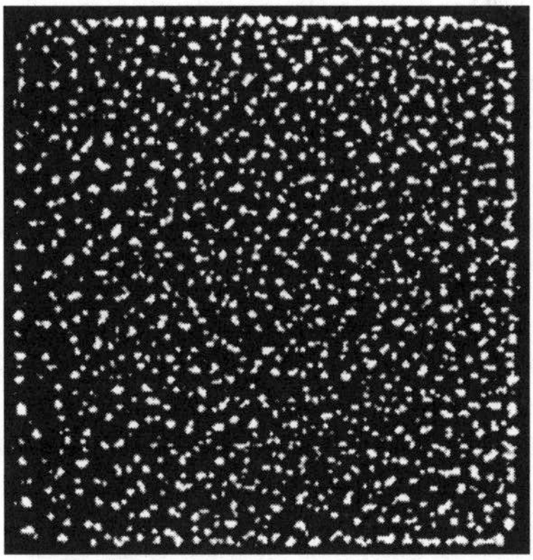

Figure 9.13 End-on view of microdischarges; original size: 6 × 6 cm; exposure time: 20 ms[314,634]

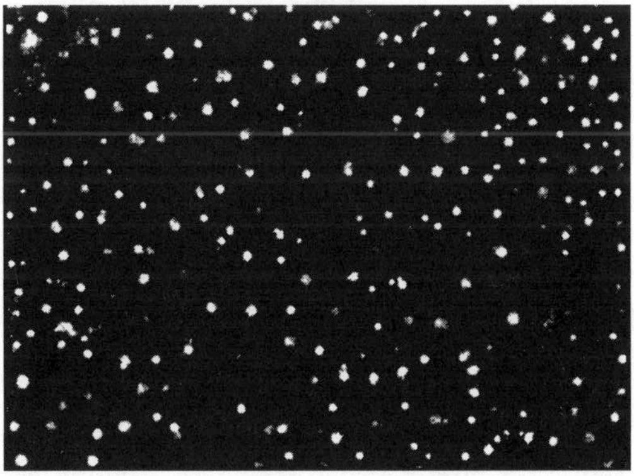

Figure 9.14 Lightenberg figure showing footprints of individual microdischarges; original size: 7 × 10 cm.[314,634]

A snapshot of the microdischarges in a 1-mm DBD air gap photographed through a transparent electrode is shown in Figure 9.13. As seen, the microdischarges are spread over the whole DBD zone quite uniformly. Footprints of the microdischarges left on a photographic plate with the emulsion facing the discharge gap and the glass plate serving as the dielectric barrier are shown in Figure 9.14.

The extinguishing voltage of microdischarges is not far below the voltage of their ignition. Charge accumulation on the surface of the dielectric barrier reduces the electric field at the location of a microdischarge. This results in current termination within just several nanoseconds after breakdown (see Table 9.1). The short duration of microdischarges leads to very low overheating of the streamer channel, and the DBD plasma remains strongly nonthermal.

New microdischarges then occur at new positions because the presence of residual charges on the dielectric barrier reduce the electric fields at the locations where microdischarges have already occurred. However, when the voltage is reversed, the next microdischarges will be formed for the same reason in the old locations. As a result, the high-voltage, low-frequency DBDs have a tendency of spreading microdischarges, while low-voltage, high-frequency DBDs tend to reignite the old microdischarge channels every half-period. The AC-plasma displays use this memory phenomenon related to deposition on the dielectric barrier.

The principal microdischarge properties for most frequencies do not depend on the characteristics of the external circuit, but only on the gas composition, pressure, and the electrode configuration. An increase of power simply leads to generation of a larger number of microdischarges per unit time, which simplifies scaling of the dielectric barrier discharges.

Modeling the microdischarges is closely related to the analysis of the avalanche-to-streamer transition and streamer propagation discussed in Section 4.4. Detailed two-dimensional modeling of formation and propagation of relevant streamers can be found in numerous publications.[318–325] More detailed two-dimensional DBD modeling of the formation and propagation of streamers includes charge accumulation on the dielectric barrier surface.[326,327] Results of such modeling show that the arrival of a cathode-directed streamer to the dielectric barrier creates, within a fraction of a nanosecond, a cathode layer with a thickness of about 200 μm and an extremely high electric field of several thousands of V/cm torr. The time and space evolution of such a streamer is presented in Figure 9.15.

An interesting phenomena can occur due to the mutual influence of streamers in a DBD. These are related to the electrical interaction of streamers themselves and with residual charge left on the dielectric barrier, and also with the influence of excited species generated in one streamer on the propagation of another streamer.[328,624]

In general, sequence of streams striking in one point form a microdischarge. These microdischarges or streamer families are actually observed experimentally. The above-mentioned interaction between streamers usually leads to repulsion of the microdischarges. If power is high enough and surface density of the microdischarges is high as well, this repulsion results in building of an organized structure of the DBD microdischarges.[625,626]

9.3.4 Surface Discharges

Closely related to the DBD are surface discharges generated at dielectric surfaces embedded by metal electrodes in a different way. The dielectric surface essentially

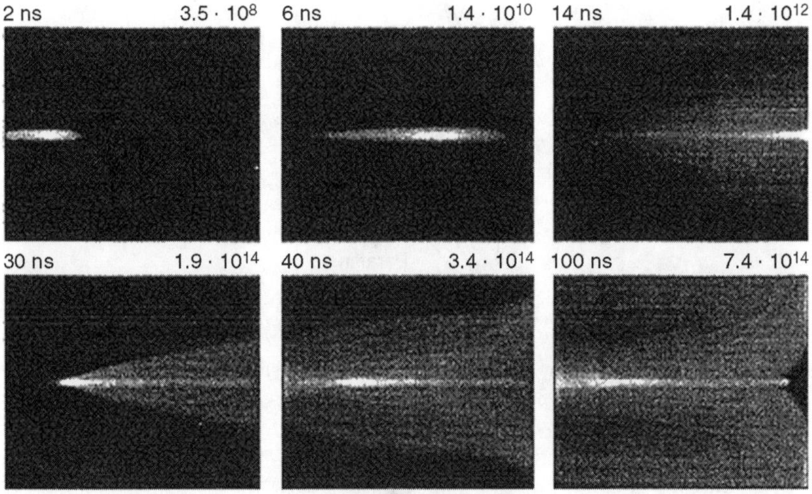

Figure 9.15 Development of microdischarge in atmospheric pressure H_2/CO_2 mixture (4/1). The 1-mm discharge gap is bounded by a plane metal cathode (left) and a 0.8-mm thick dielectric of $\varepsilon = 3$ (right). A constant voltage is applied, resulting in an initially homogeneously reduced field of 125 Td in gas space. This corresponds to an overvoltage of 90% in this mixture. The numbers in the right upper corner indicate the maximum electron density in cm^{-3} reached in that picture. The maximum current of 35 mA is reached at 40 ns.[314,634]

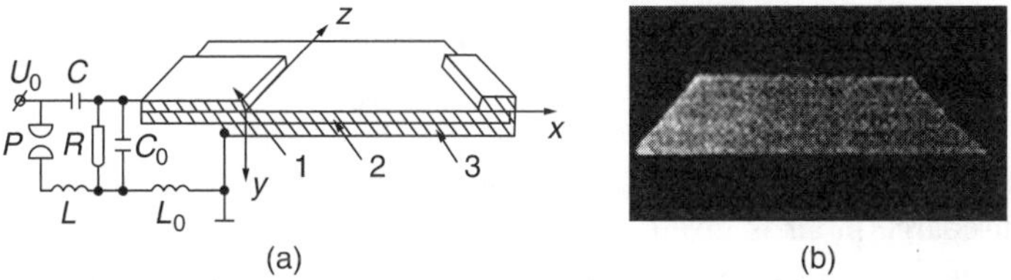

Figure 9.16 (a) Schematic and (b) emission picture of pulse surface discharge: 1 = initiating electrode; 2 = dielectric; 3 = shielding electrode[627]

decreases the breakdown voltage in such systems because of the creation of significant nonuniformities of electric field and thus creates a local over voltage. The surface discharges, as well as DBDs, can be supplied by AC or pulsed voltage.

A very effective decrease of the breakdown voltage can be reached in the surface discharge configuration in which one electrode just lies on the dielectric plate, with another one partially wrapped around it as shown in Figure 9.16(a).[329] This discharge is called **sliding discharge**. It can be fairly uniform in some regimes on the dielectric plates of high surface areas with linear sizes over 1 m at voltages not exceeding 20 kV (see Figure 9.16b).

The component of electric field E_y normal to the dielectric surface plays an important role in generating the pulse-periodic sliding discharge that does not depend essentially on the distance l between electrodes along the dielectric (axis x in Figure 9.16a). For this reason, breakdown voltages of the sliding discharge do not follow

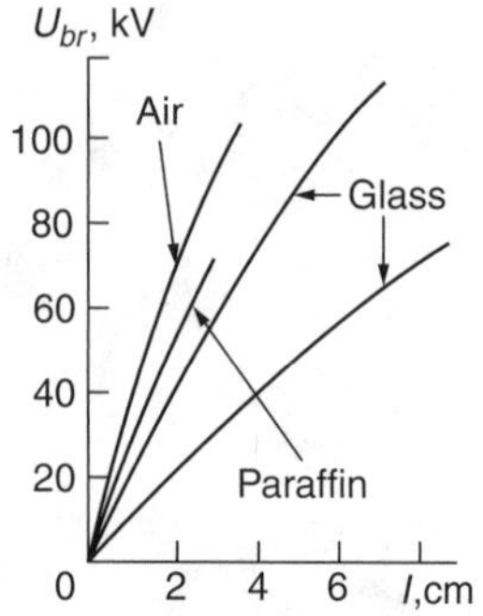

Figure 9.17 Breakdown voltage in air along different insulation cylinders (d = 50 mm, f = 50 Hz)[329]

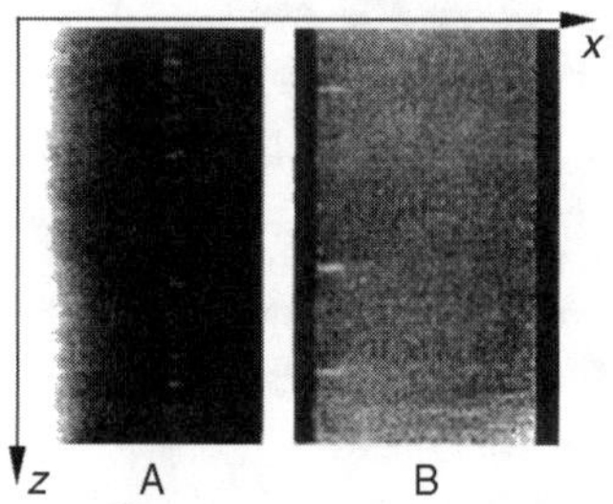

Figure 9.18 (A) Incomplete and (B) complete surface discharges (He, p = 1 atm, $\varepsilon \approx 5$, d = 0.5 mm, pulse frequency 6×10^{13} Hz)[627]

the Paschen law (see Section 4.4.2). As an example, the breakdown curve for the sliding discharge in air is given in Figure 9.17.

One can achieve two qualitatively different modes of the surface discharges by changing the applied voltage amplitude: (1) complete (sliding surface spark) and (2) incomplete (sliding surface corona). Pictures of complete and incomplete surface discharges are presented in Figure 9.18. The sliding surface corona discharge takes place at voltages below the critical breakdown value. Current in this discharge regime is low and limited by charging the dielectric capacitance. Active volume and luminosity of this discharge is localized near the igniting electrode and does not cover all the dielectric.

The sliding surface spark (or the complete surface discharge) takes place at voltages exceeding the critical one corresponding to breakdown. In this case, the formed plasma channels actually connect electrodes of the surface discharge gap. At low overvoltages, the breakdown delay is about 1 μsec. In this case, the many-step breakdown phenomenon starts with the propagation of a direct ionization wave, followed by a possibly more intense reverse wave related to the compensation of charges left on the dielectric surface. After about 0.1 μsec, the complete surface discharge covers all the electrodes of the discharge gap. The sliding spark at low overvoltage usually consists of only one or two current channels.

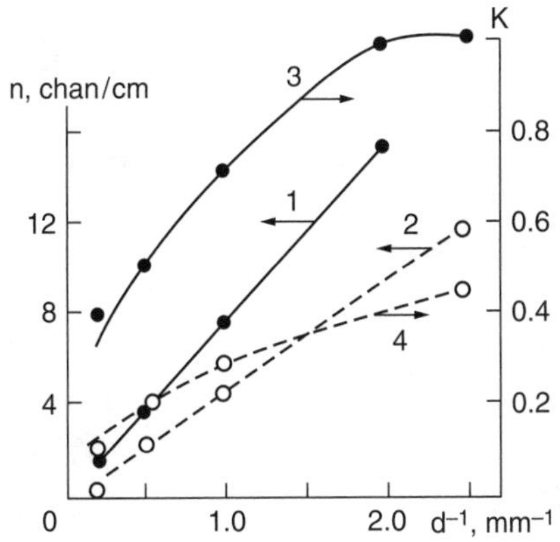

Figure 9.19 Linear density of channels n (1, 2) and surface coverage by plasma K (3, 4) as functions of d^{-1} inverse dielectric thickness: He (1,3); air (2, 4)[627]

At higher overvoltages, the breakdown delay becomes shorter, reaching the nanosecond time range. In this case, the complete discharge regime takes place immediately after the direct ionization wave reaches the opposite electrode. The surface discharge consists of many current channels in this regime. In general, the sliding spark surface discharge is able to generate the luminous current channels of very sophisticated shapes, usually referred to as the **Lichtenberg figures**. Some simple, but nontypical, examples of Lichtenberg figures are shown in Figure 9.14.

The number of the channels r depends on the capacitance factor ε/d (ratio of dielectric permittivity over thickness of dielectric layer), which determines the level of electric field on the sliding spark discharge surface. This effect is illustrated in Figure 9.19 and is important in the formation of large area surface discharges with homogeneous luminosity. Many interesting additional details related to physical principles and applications of the sliding surface discharges can be found in the book of Baselyan and Raizer.[330]

9.3.5 Packed-Bed Corona Discharge

The packed-bed corona is an interesting combination of the dielectric barrier discharge DBD and the sliding surface discharge. In this system, high AC voltage (about 15 to 30 kV) is applied to a packed bed of dielectric pellets and creates a nonequilibrium plasma in the void spaces between the pellets.[331,332] The pellets effectively refract the high-voltage electric field, making it essentially nonuniform and stronger than the externally applied field by a factor of 10 to 250 times, depending on the shape, porosity, and dielectric constant of the pellet material.

A typical scheme for organizing a packed-bed corona is shown in Figure 9.20(A); a picture of the discharge is presented in Figure 9.20(b). The discharge chamber shown on the figure is shaped as coaxial cylinders with an inner metal electrode and an outer tube made of glass. The dielectric pellets are placed in the annular gap. A metal foil or screen in contact with the outside surface of the tube

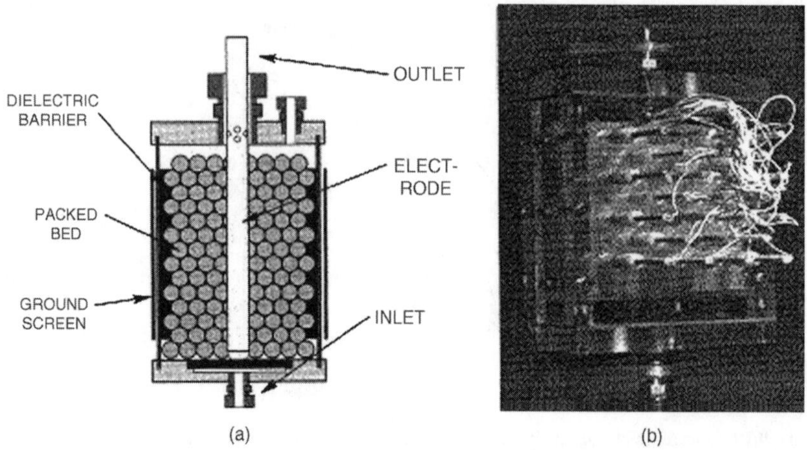

Figure 9.20 (a) Schematic and (b) photograph of packed-bed corona discharge made in Pacific Northwest National Laboratory[331,332]

serves as the ground electrode. The inner electrode is connected to a high voltage AC power supply operated on the level of 15 to 30 kV at a fixed frequency of 60 Hz or at variable frequencies.

In this discharge system the glass tube serves as a dielectric barrier to inhibit direct charge transfer between electrodes and as a plasma chemical reaction vessel. Special features of the packed-bed corona make this discharge very practical, in particular, for air purification and other environmental control processes.

9.3.6 Atmospheric Pressure Glow Modification of Dielectric Barrier Discharge

Principal concepts and physical effects regarding the atmospheric pressure glow discharges were already discussed in Section 7.5.6. Such discharges can be effectively organized in a DBD configuration. In this case, the key difference is related to using only special gases, e.g., helium. The atmospheric pressure glow DBD modification permits arranging the barrier discharge homogeneously without streamers and other spark-related phenomena. Practically, it is important that the glow modification of DBD be operated at much lower voltages (down to hundreds of volts) with respect to those of traditional DBD conditions.

A detailed explanation of the special functions of helium in the atmospheric pressure glow discharge is not known; however, it is clear that these are related mainly to the following effects. First, they are related to the high electronic excitation levels of helium and the absence of electron energy losses on vibrational excitation. This leads to high values of electron temperatures at lower levels of the reduced electric field. To see this effect, compare the reduced electric fields necessary to sustain glow discharges in inert and molecular gases, shown in Figure 7.12 and Figure 7.13.

Second, they are related to heat and mass transfer processes that are relatively fast in helium. This prevents contraction and other instability effects in the glow discharge at high pressures. One can state that streamers are overlapping in this case. Also, helium and other noble gases are chemically passive. In contrast to molecular gases, especially to electronegative gases, the noble gases do not stimulate relaxation and quenching of metastable electronically excited particles responsible for pre-ionization. Effective spreading of metastables and charged particles leads to strong pre-ionization and multiplication of a number of initial avalanches. The same processes can be important in preventing the generation of space-localized streamers and sparks.

An important role in avoiding narrow streamers is played by the "memory effect;" this is the influence of particles generated in a previous streamer on a subsequent streamer. The memory effect can be related to metastable atoms and molecules and to electrons deposited on the dielectric barrier. The memory effect provides high level of pre-ionization, which results in a larger number of initial avalanches. Overlapping of the avalanches makes the DBD uniform. Clear explanation of APG mode of DBD is probably the most challenging problem of non-thermal atmospheric pressure plasma, especially in electronegative gases. More details regarding the atmospheric pressure glow modification of the dielectric barrier discharges can be found in other references.[333–335,641]

9.3.7 Ferroelectric Discharges

Special properties of DBD of practical interest can be revealed by using ferroelectric ceramic materials of a high dielectric permittivity (ε above 1000) as the dielectric barriers.[336,337] Today, ceramics based on $BaTiO_3$ are the most employed ferroelectric material for dielectric-barrier discharges.

To illustrate the peculiarities of the ferroelectric discharge, recall that the mean power of the dielectric barrier discharges can be determined by the equation:[338]

$$P = 4f\,C_d V\left(V - V_{cr}\frac{C_d + C_g}{C_d}\right) \tag{9.25}$$

Here, f is the frequency of applied AC voltage; V is amplitude value of the voltage; V_{cr} is the critical value of the voltage corresponding to breakdown of the discharge gap; and C_d and C_g are the electric capacities of a dielectric barrier and gaseous gap, respectively. When the dielectric barrier capacity exceeds that of the gaseous gap, $C_d >> C_g$, the relation for discharge power obviously can be simplified to the following form:

$$P = 4f\,C_d V(V - V_{cr}) \tag{9.26}$$

From this equation, high values of dielectric permittivity of the ferroelectric materials are helpful in providing relatively high discharge power at relatively low values of frequency and applied voltage. For example, the ferroelectric discharge based on $BaTiO_3$, which employs a ceramic barrier with a dielectric permittivity ε

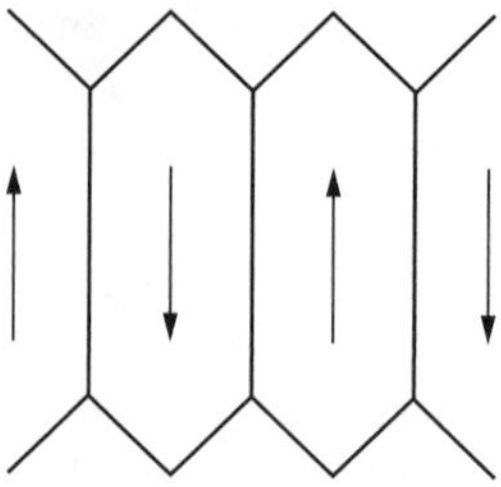

Figure 9.21 Ferro-electric domain arrangement

exceeding 3000, thickness 0.2 to 0.4 cm, and gas discharge gap 0.02 cm, operates effectively at an AC frequency of 100 Hz, voltages below 1 kV, and electric power of about 1W.[339] Such discharge parameters are of particular interest for special practical applications such as medical ones, where low voltage and frequency are very desirable for safety and simplicity reasons, and high discharge power is not required.

Important physical peculiarities of the ferroelectric discharges are related to the physical nature of the ferroelectric materials, which in a given temperature interval can be spontaneously polarized. Such spontaneous polarization means that the ferroelectric materials can have a nonzero dipole moment even in the absence of external electric field. The electric discharge phenomena accompanying contact of a gas with a ferroelectric sample were first observed in detail by Robertson and Baily.[340] The first qualitative description of this sophisticated phenomenon was developed by Kusz.[341]

The long-range correlated orientation of dipole moments can be destroyed in ferroelectrics by thermal motion. The temperature at which the spontaneous polarization vanishes is called the temperature of ferroelectric phase transition or the **ferroelectric Curie point**. When the temperature is below this point, the ferroelectric sample is divided into macroscopic uniformly polarized zones called the **ferroelectric domains** (illustrated in Figure 9.21) localized near the igniting electrode and does not cover all the dielectric.

From Figure 9.21, the directions of the polarization vectors of individual domains in the equilibrium state are set up in a way to minimize internal energy of the crystal and to make polarization of the sample, as a whole, close to zero. Application of an external AC voltage leads to overpolarization of the ferroelectric material and reveals strong local electric fields on the surface. These local surface electric fields, which stimulate the discharge on ferroelectric surfaces, can exceed 10^6 V/cm.[342]

Thus, the active volume of the ferroelectric discharge is located in the vicinity of the dielectric barrier, which is essentially the narrow interelectrode gaps typical for the discharge. In this case, scaling the ferroelectric discharge can be achieved by using some special configurations. One such special discharge configuration comprises a series of parallel thin ceramic plates. High dielectric permittivity of ferroelectric ceramics enables this multilayer sandwich to be supplied by only two edge electrodes. Another interesting configuration can be arranged by using a packed bed of the ferroelectric pellets. In the same manner described in Section 9.3.5, nonequilibrium plasma is created in such a system in the void spaces between the pellets.

9.4 SPARK DISCHARGES

9.4.1 Development of Spark Channel, Back Wave of Strong Electric Field, and Ionization

When streamers provide an electric connection between electrodes and neither a pulse power supply nor a dielectric barrier prevents further growth of current, an opportunity for development of a spark occurs. However, the initial streamer channel does not have very high conductivity and usually provides only a very low current of about 10 mA. Thus, some fast ionization phenomena take place after formation of an initial streamer channel to increase the degree of ionization and the current to an intensive spark.

As discussed in Section 4.4.8, the potential of the head of the cathode-directed streamer is close to the anode potential. This is the region of a strong electric field around the streamer's head. While the streamer approaches the cathode, this electric field is obviously growing. It stimulates intensive formation of electrons on the cathode surface and its vicinity and, subsequently, their fast multiplication in this elevated electric field. New ionization waves much more intense than the original streamer now start propagating along the streamer channel, but in an opposite direction from the cathode to anode. This is usually referred to as the **back ionization wave** and propagates back to the anode with an extremely high velocity of about 10^9 cm/sec.

The high velocity of the back ionization wave is not directly the velocity of electron motion, but rather the phase velocity of the ionization wave. The back wave is accompanied by a front of intensive ionization and the formation of a plasma channel with sufficiently high conductivity to form a channel of the intensive spark.

9.4.2 Expansion of Spark Channel and Formation of an Intensive Spark

The high-density current initially stimulated in the spark channel by the back ionization wave results in intensive joule heating, growth of the plasma temperature, and a contribution of thermal ionization. Gas temperatures in the spark channel reach 20,000 K and electron concentration rises to about 10^{17} cm^{-3}, which is already close to complete ionization.

Electric conductivity in the spark channel at such a high degree of ionization is determined by coulomb collisions and actually does not depend on electron density (see Section 4.3.5). According to Equation 4.113, the conductivity can be estimated in this case as 10^2 $ohm^{-1} \cdot cm^{-1}$. Thus, further growth of the spark current is related not to an increase of the ionization level and conductivity, but rather to expansion of the channel and increase of its cross section.

The fast temperature increase in the spark channel leads to sharp pressure growth and to generation of a cylindrical shock wave. The amplitude of the shock wave is so high (at least in the beginning) that the temperature after the wave front is sufficient for thermal ionization. As a result, the external boundary of the spark current channel at first grows together with the front of the cylindrical wave. During the initial 0.1 to 1 μsec after the breakdown point, the current channel expansion velocity is about 10^5 cm/sec. Subsequently, the cylindrical shock wave decreases in strength and expansion of the current channel becomes slower than the shock wave velocity.

Radius of the spark channel grows to about 1 cm, which corresponds to a spark current increase of 10^4 to 10^5 A at current densities of about 10^4 A/cm^2. Plasma conductivity grows relatively high and a cathode spot can be formed on the electrode surface (see Section 8.3.5). Interelectrode voltage decreases lower than the initial one, and the electric field becomes about 100 V/cm. If voltage is supplied by a capacitor, the spark current obviously starts decreasing after reaching the mentioned maximum values.

The detailed theory of electric sparks was developed by Drabkina[343] and Braginsky;[344] extensive modern experimental and simulation material on the subject are available.[330]

9.4.3 Atmospheric Phenomena Leading to Lightning

Lightning is a large-scale natural spark discharge occurring between a charged cloud and the Earth, between two clouds, or internally inside a cloud. Obviously, lightning is caused by high electric fields related to the formation and space separation in the atmosphere of positive and negative electric charges. Therefore, the consideration of lightning will begin with an analysis of the processes of formation and space separation of the charges in the atmosphere.[345,346]

In general, the formation of electric charges in the atmosphere is due mainly to ionization of molecules or microparticles by cosmic rays. Generation of electric charges can also take place during the collisional decay of water droplets. However, what is actually important for the interpretation of lightning is the fact that the negative charge in the thundercloud is located in the bottom part of the cloud and the positive charge is primarily located in the upper part of the cloud.

To understand this phenomenon, remember that polar water molecules on the water surface are mostly aligned with their positive ends oriented inward from the water surface. This effect related to the hydrogen bonds between water molecules is illustrated in Figure 9.22. Such orientation of the surface water molecules leads to formation of a double electric layer on the surface of droplets with a voltage drop experimentally determined as $\Delta\varphi = 0.26$ V. This double layer predominantly traps negative ions and reflects positive ions until their charge, Ne, compensates for the voltage drop:

$$N \approx \frac{4\pi\varepsilon_0 r\,\Delta\varphi}{e} \tag{9.27}$$

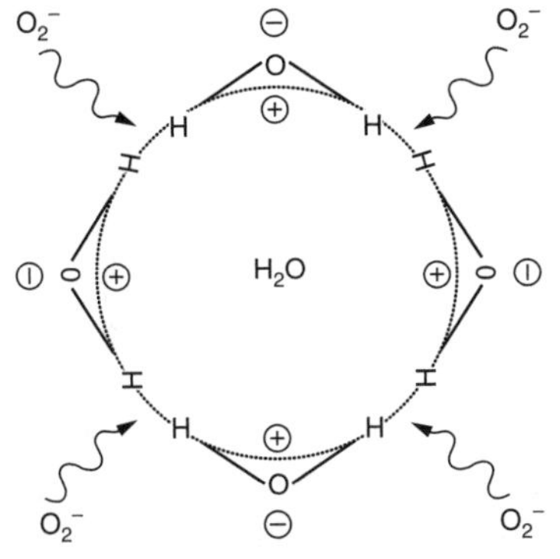

Figure 9.22 Water droplet trapping negative ions

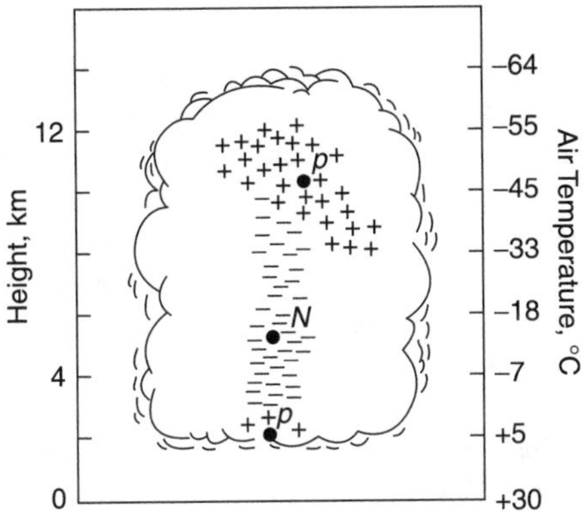

Figure 9.23 Charge distribution in thundercloud; black dots mark centroids of charge clouds. According to measurements of electric fields around clouds, $P = +40C$ (above), $p = +10C$ (below), and $N = -40C$.[346]

The simple calculation based on Equation 9.27 shows that a droplet with a radius of about $r = 10$ μm is able to absorb about 2000 negative ions effectively. Thus, the negative ions in clouds are trapped in droplets and descend, while positive ions remain in molecular or cluster form in the upper areas of a cloud. As a result, a typical charge distribution in a cloud appears as shown in Figure 9.23.[346] The charge distribution in a thundercloud explains why most lightning discharges occur inside the clouds.

9.4.4 Lightning Evolution

Typical duration of the lightning discharge is about 200 msec; the sequence of events in this natural modification of the long spark discharge is illustrated in Figure 9.24. The lightning actually consists of several pulses with duration of about 10 msec each and intervals of about 40 msec between pulses.

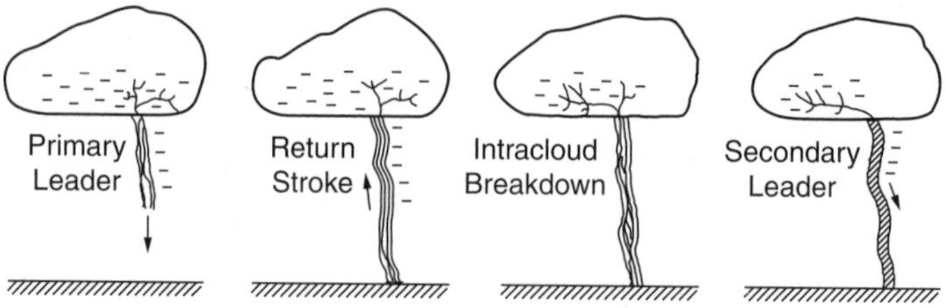

Figure 9.24 Lightning evolution

Each pulse starts with propagation of a leader channel (see Section 4.4.9) from the thundercloud to the Earth. Current in the first negative leader is relatively low—about 100 A. This leader, called the multistep leader, has a multistep structure with a step length about of 50 m and an average velocity of 1 to $2 \cdot 10^7$ cm/sec. The visible radius of the leader is about 1 m and the radius of the current-conducting channel is smaller. Leaders initiating sequential lightning pulses usually propagate along channels of the previous pulses. These leaders are called dart leaders; they are more spatially uniform than the initial one and are characterized by a current of about 250 A and higher propagation velocities of 10^8 to 10^9 cm/sec.

When the leaders approach the earth, the electric field increases between them, resulting in the formation of a strong ionization wave moving in the opposite direction back to the thundercloud. This extremely intensive ionization wave is usually referred to as the return stroke. The physical nature of the return stroke is similar to that of the back ionization wave in the laboratory spark discharges described in Section 9.4.1. The return stroke is the main phase of the lightning discharge. In this case, the velocity of the ionization wave reaches gigantic, almost relativistic, values of 0.1 to 0.3 speed of light. Maximum current reaches 100 kA, which is actually the most dangerous effect of lightning. Temperature in the lightning channel reaches 25,000 K; electron concentration approaches 1 to $5 \cdot 10^{17}$ cm^{-3}, which corresponds to complete ionization.

The electric field on the front of the return stroke is quite high, about 10 kV/cm (for comparison, electric fields in arcs in air are about 10 V/cm). Taking into account the high values of current leads to the gigantic values of specific power on the front of the return stroke, about $3 \cdot 10^5$ kW/cm (for comparison, in electric arcs in air, this is about 1 kV/cm). Intensive and fast heat release leads to strong pressure increases in the current channel of a lightning discharge and thus to shock wave generation heard as thunder.

Intensive heat release leads to fast expansion of the initial current channel (see Section 9.4.2) during propagation of the return stroke and to formation of a developed spark channel. Through this channel, some portion of the negative electric charge goes to the Earth during about 40 msec. Electric currents through the spark channel during this 40-msec period are of the order of 200 A.

Remember that negative charges are located in the lower part of the thundercloud mostly in the form of clusters or charged microparticles. These have low conductivity and cannot be evacuated quickly from the cloud through the spark channel. Thus, the effective evacuation of the negative charges takes place due to preliminary liberation of electrons from ions and microparticles under the influence of the strong electric field.

As mentioned in the beginning of this section, the lightning discharge consists of several pulses. Each pulse results in transfer to the Earth of only a part of the negative charge collected in the thundercloud. It collects negative charges only from the area close to where the return strike attacks the cloud. Then the lightning discharge temporarily extinguishes (intervals between pulses are about 40 msec) until electric charges redistribute themselves in the cloud by means of internal breakdowns.

The conductivity in the spark channel of the previous pulse has already decreased significantly when the internal charge distribution in a thundercloud is restored after about 40 msec and the system is ready for a new pulse. The new lightning pulse starts again with a leader propagating along the remains of the previous one. As previously mentioned, the new leader does not have a multistep structure; it is a dart leader. When the dart leader reaches the Earth, it stimulates formation of the second return stroke and the complete cycle repeats. The sequence of the lightning pulses continues for about 200 msec until most of the negative charge from the cloud reaches the Earth. Note that the positive charges located in the upper parts of the thundercloud (Figure 9.23) mostly stay there because the distance between these charges and the Earth is too large for breakdown. Additional detailed information about lightning discharge can be found in recent books by Baselyan and Raizer.[330,330a]

9.4.5 Mysterious Phenomenon of Ball Lightning

Even a short discussion on lightning as a natural example of long spark discharges cannot be complete without mentioning the interesting and mysterious phenomenon of ball lightning. Ball lightning is a rare natural phenomenon; the luminous plasma sphere occurs in the atmosphere, moves in unpredictable directions (sometimes against the direction of wind), and finally disappears—sometimes with an explosive release of a considerable amount of energy.

Ball lightning has some special nontrivial oddities, including the ability of the plasma sphere to move through tiny holes, possible explosion heat release exceeding all reasonable estimations of energy contained inside of the ball, etc. Difficulties in interpretation are mostly due to difficulties in observing this rare phenomenon in nature or the laboratory. Some unique pictures of ball lightning together with contradicting scientific descriptions of their observation can be found on the Web. An example of an interesting and good-quality picture of the ball lightning is shown in Figure 9.25.

Several hypotheses describe the ball lightning phenomenon, the most developed of which is the so-called chemical model first proposed by Arago in 1839.[347] Subsequently, it was developed by many researches and summarized in detail by Smirnov.[221,348] The chemical model assumes the production of some excited, ionized, or chemically active species in the channel of regular lightning followed by exo-

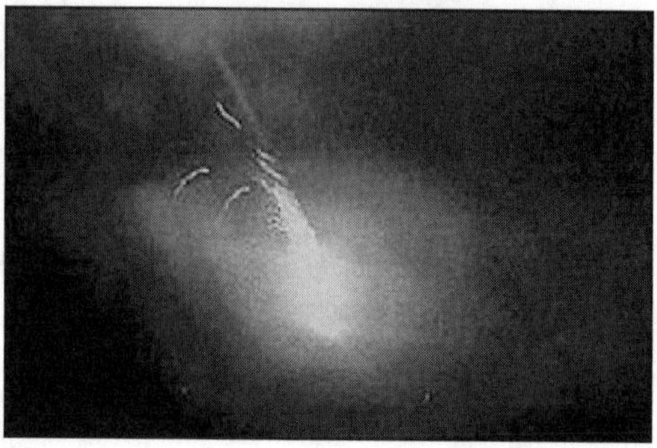

Figure 9.25 Ball lightning

thermic reaction among them, leading to formation of luminous spheres of ball lightning. There is disagreement about how energy accumulates. According to some authors, accumulation of energy is provided by positive and negative complex ions;[349,350] others hold that ozone, nitrogen oxides, hydrogen, and hydrocarbons are energy sources for the ball lightning.[221,348,351,352]

The typical ball lightning spherical shape is related in the framework of the chemical model to heat and mass balances of the process. Fuel generated by regular lightning is distributed over volumes greatly exceeding those of ball lightning. Steady-state exothermic reactions take place inside a sphere. Fresh reagents diffuse into the sphere while heat transfer provides an energy flux from the sphere to sustain the steady-state process. Analysis of the steady-state spherical wave for describing the shape of ball lightning has been done by Rusanov and Fridman.[353] This steady-state "combustion" sphere moves (in slightly nonuniform conditions) in the direction corresponding to the growth of temperature and "fuel" concentration. This explains possible observed motion of ball lightning in the opposite direction to that of the wind. The high energy release during a ball lightning explosion can be explained by noting that the explosion occupies much larger volume than the initial one related to the region of the steady-state exothermic reaction.

Several more or less successful experimental systems were developed to reproduce the ball lightning plasma experimentally in the laboatory. The latest and probably most exciting were recently presented by Egorov and Stepanov.[628] In these experiments the "man-made ball lightning" is created by a pulse discharge in humid air.

Although understanding of the ball lightning phenomenon is far from complete, numerous interesting materials have been published on the subject.[354–358]

9.4.6 Laser-Directed Spark Discharges

Modification of sparks can be done by synergetic application of high voltages with laser pulses.[359,360] A simplified scheme of this discharge system is illustrated in Figure 9.26. It is interesting that laser beams can direct spark discharges not only along

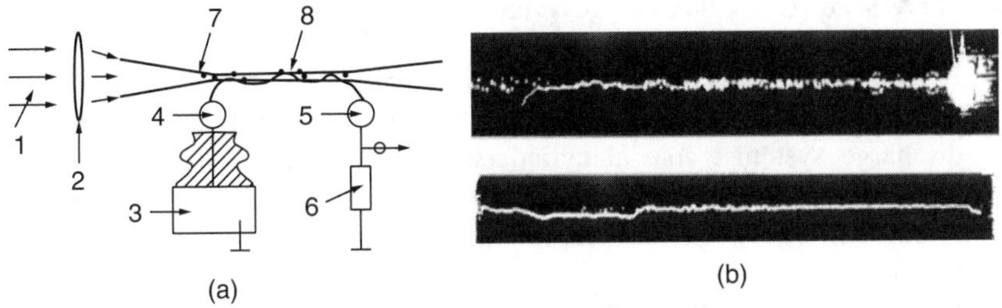

Figure 9.26 (a) Schematic and (b) pictures of laser-directed spark: 1 = laser beam; 2 = lens; 3 = voltage pulse generator; 4, 5 = powered and grounded electrodes; 6 = resistance; 7 = optical breakdown zone; 8 = discharge zone. Discharges on the pictures are 1 and 3 m long; bright right-hand spot is a high-voltage electrode overheated by the laser beam.[627]

straight lines, but also along more complicated trajectories. Laser radiation is able to stabilize and direct the spark discharge channel in space through three major effects: local preheating of the channel, local photoionization, and optical breakdown of gas.

Preheating the discharge channel creates a low gas density zone, leading to higher levels of reduced electric field E/n_0, which is favorable as a result for spark propagation. This effect works best if special additives provide the required absorption of the laser radiation. For example, if a CO_2 laser is used for preheating, a strong effect on the corona discharge can be achieved when about 15% ammonia (which effectively absorbs radiation on a wavelength of 10.6 μm) is added to air. At a laser radiation density of about 30 J/cm^2, the breakdown voltage in the presence of ammonia decreases by an order of magnitude. The maximum length of the laser-supported spark in these experiments was up to 1.5 m. Effective stabilization and direction of the spark discharges by CO_2 laser in air was also achieved by admixtures of C_2H_2, CH_3OH, and CH_2CHCN.

Photoionization by laser radiation is able to stabilize and direct a corona discharge without significantly changing the gas density by means of local preionization of the discharge channel. UV laser radiation (for example, Nd-laser or KrF-laser) should be applied in this case. Ionization usually is related to the two-step photoionization process of special organic additives with relatively low ionization potential. The UV KrF-laser with pulse energies of approximately 10 mJ and pulse duration of approximately 20 nsec is able to stimulate the directed spark discharge to lengths of 60 cm. Note that the laser photoionization effect to stabilize and direct sparks is limited in air by fast electron attachment to oxygen molecules. In this case, photodetachment of electrons from negative ions can be provided by using a second laser radiating in the infrared or visible range.

The most intensive laser effect on spark generation can be provided by the optical breakdown of the gases. The length of such a laser spark can exceed 10 m. The laser spark in pure air requires power density of an Nd-laser (λ = 1.06 μm) exceeding 10^{11} W/cm^2. Generation of these impressive discharges, as well as other discharges sustained by electromagnetic oscillations of different frequencies, will be discussed in the following chapter.

PROBLEMS AND CONCEPT QUESTIONS

9.1 **Electric Field Distribution in Corona Discharge Systems**. Derive Equation 9.1 through Equation 9.1.3 for electric field distributions for the simple corona discharge systems: coaxial cylinders, coaxial spheres, and sphere–remote plane. Interpret Equation 9.5 and then derive formulas for the maximum electric field between parallel wires as well as between a single wire and a parallel plane.

9.2 **Positive and Negative Corona Discharges**. Compare the ignition criteria of positive and negative corona discharges (Equation 9.8 and Equation 9.9) and explain why the positive corona requires slightly higher values of voltage to be initiated. Estimate the relative difference in the values of ignition voltages; show and explain why it is quite low.

9.3 **The Peek Formula for Corona Ignition**. Using the empirical formula (Equation 9.12) for the Townsend coefficient α in air and criteria of Equation 9.8 and Equation 9.9 of corona ignition, analyze and derive the Peek formula for initiating a corona discharge in air. Compare the Peek formula with a similar criterion (Equation 9.11) for the case of corona ignition between two parallel wires, estimate numerically, and explain the difference between them.

9.4 **Active Corona Volume**. Explain why most plasma chemical processes take place in the active corona volume. Explain why the active corona volume cannot be simply increased by increasing the applied voltage. Why does the formation of a corona discharge around a thin wire look more attractive for applications from the point of view of maximizing the active corona volume than a corona formed around a sharp point?

9.5 **Space Charge Influence on Electric Field Distribution in Corona**. Based on the Maxwell equation (Equation 9.18), derive the electric field distribution (Equation 9.19) in a corona discharge formed around a thin wire, taking into account the space charge effect. The Maxwell equation in Equation 9.18 was obtained assuming low perturbation of the electric field distribution by the space charge outside the active corona volume. Based on the distribution (Equation 9.19), derive the criterion for the low level of perturbation of the electric field by the space charge, which limits the corona current.

9.6 **Current–Voltage Characteristics of Corona Discharge**. Based on the electric field distribution in the corona formed around a thin wire, derive the current–voltage characteristic (Equation 9.20) of the discharge. Explain the difference in the current–voltage characteristics for positive and negative corona discharges and for electronegative and nonelectronegative gases.

9.7 **Power of Continuous Corona Discharges**. Explain the physical limitations of the power increase in the case of continuous corona discharges. According to Equation 9.23, corona power can be increased by diminishing the radius of the external electrode. What is the limitation of this approach to the corona power increase?

9.8 **Pulse-Periodic Regimes of Positive and Negative Corona Discharges**. As was observed, corona discharges are able to operate at some conditions in the form of periodic current pulses even in the constant voltage regime. The frequency of these pulses can reach 10^4 Hz in the case of a positive corona and 10^6 Hz for negative coronas. Give your interpretation of why the frequency of these pulses in a positive corona (flashing corona) is so much higher than the frequency in the case of negative pulses (the so-called Trichel pulses).

9.9 **Flashing Corona Discharge**. Continuous positive corona discharge exists in the steady-state regime only in some intermediate interval of voltages. At relatively low voltages close to corona ignition conditions, as well as at relatively high voltages close to the corona-to-spark transition, the corona exists in pulse-periodic regime. If the flashing corona phenomenon at low voltages can be explained by inefficient positive space charge drift to the cathode in the relatively weak electric field, explain the flashing corona appearance at high electric fields.

9.10 **Voltage Rise Rate in Pulse Corona Discharges**. Give your interpretation of why not only pulse duration, but also the voltage rise rates, are so important to reach high levels of voltage pulses and high efficiencies of the pulse corona discharges.

9.11 **Electric Field of Residual Charge Left by DBD–Streamer on Dielectric Barrier**. Based on the typical data presented in Table 9.1, estimate the electric field induced by the residual charge left by a streamer on a dielectric barrier. Compare the result of estimations with the corresponding modeling data given at the end of Section 9.3.3.

9.12 **Overheating of the DBD Microdischarge Channels**. As is presented in Table 9.1, the heating effect in the DBD micro-discharge channels is fairly low, about 5 K. Compare this heating effect with the general heating effect in streamer channels, which was estimated in the Section 4.4.6.

9.13 **Plasma Chemical Energy Efficiency of the Dielectric Barrier in Comparison with the Pulsed Corona Discharge**. Taking into account that pulsed corona discharges usually operate with higher voltage rise rates and, in general, with higher voltages, discuss the difference in efficiencies of ionization, electronic excitation, and vibrational excitation for these two types of nonthermal atmospheric pressure discharges.

9.14 **Sliding Surface Discharges**. The number of the electrode-connecting and current-conducting channels in the gliding spark regime of the surface discharge, as well as the plasma coverage of the discharge gap, depends on the value of electric field on the dielectric surface. Based on the simple equivalent circuit consideration, show that the electric field (and thus the number of channels and the plasma coverage of the gap) is determined by the capacitance factor ε/d, which is the ratio of dielectric permittivity over thickness of dielectric layer. Give your interpretation of the relevant experimental data presented in Figure 9.19.

9.15 **Ferroelectric Discharges**. Peculiarities of the ferroelectric discharges are related from one view to the extremely high dielectric permittivity of ferroelectric materials. This provides relatively high power at low AC frequencies and applied voltages. From another view, these are related to extremely high local electric fields on the surface of the materials. Compare the influence of these two effects on parameters of the ferroelectric discharge plasma.

9.16 **Velocity of the Back Ionization Wave**. Early stages of a spark generation are related to propagation of a back ionization wave, which actually provides a high degree of ionization in the spark channel. Give your interpretation of the experimental fact that the velocity of this wave reaches very high values of about 109 cm/sec.

9.17 **Negative Ions' Attachment to Water Droplets; Mechanism of Charge Separation in Thundercloud**. Estimate the size of a water droplet (number of water molecules in a cluster) that is able to provide effective trapping of at least one negative ion due to the surface polarization effect. Discuss proportionality of this charge to the droplet radius. Compare the negative charging of droplets due to the surface polarization effect (Equation 9.27) with possible charging of particles related to the floating potential effect of high mobility of electrons.

9.18 **Mechanism of Propagation of Ball Lightning**. Estimate the propagation rate of ball lightning in a steady-state atmosphere with slightly nonuniform spatial distributions of temperature and fuel concentration. Based on the calculation results, estimate the possibility of the ball lightning propagating opposite to the wind direction.

CHAPTER

10

PLASMA CREATED IN HIGH-FREQUENCY ELECTROMAGNETIC FIELDS: RADIO-FREQUENCY, MICROWAVE, AND OPTICAL DISCHARGES

All discharges sustained by high-frequency electromagnetic fields can be divided for plasma classification into two large groups: thermal (quasi-equilibrium) and nonthermal (nonequilibrium) discharges. Physical features of all the high-frequency thermal discharges are quite similar, although different techniques are used for plasma generation at different frequencies. Thermal discharges will be considered in the first two sections of this chapter. After that, low-pressure and moderate-pressure nonthermal discharges will be discussed in which physical phenomena are much more sensitive to frequency ranges of electromagnetic fields and modes of their excitation, such as capacitive and inductive coupling.

10.1 RADIO-FREQUENCY (RF) DISCHARGES AT HIGH PRESSURES: INDUCTIVELY COUPLED THERMAL RF DISCHARGES

10.1.1 General Features of High-Frequency Generators of Thermal Plasma

The easiest (and traditional) way of thermal plasma generation is related to electric arc discharges. The arc discharges provide high power for thermal plasma generation at atmospheric pressure, using DC power supplies with a relatively low price of about \$0.1 to \$0.5/W. The radio-frequency (RF) discharges are also able to generate plasma at high power level and atmospheric pressure, but these require more expensive RF power supplies and are characterized by a price of about \$1 to \$5/W.

Such significant difference in the prices of power supplies is especially important because modern industrial applications often require thermal plasma generation at power levels from tens of kilowatts to many megawatts. Nevertheless, different types of high-frequency discharges now are used increasingly often for thermal plasma generation. This is mostly due to the fact that direct electrode-plasma contact is not required and, in many of these discharge systems, there are no electrodes at all, as well as electrode-related problems.

High-frequency electromagnetic fields can interact with plasma in different ways. In the RF frequency range, either inductive or capacitive coupling can provide this interaction. Electromagnetic field interaction with plasma in microwave discharges is quasi optical. Specific features of these different plasma–electromagnetic field interaction modes will be considered after preliminary discussion of some general common features of the plasma energy balance.

10.1.2 General Relations for Thermal Plasma Energy Balance: Flux Integral Relation

Thermal plasma sustained by electromagnetic fields with different frequencies is always characterized by quasi-equilibrium temperature distributions determined by the absorption of energy of the electromagnetic fields. In many practical systems, gas flow is subsonic and pressure can be considered fixed. In this case, taking into account electromagnetic energy dissipation and heat transfer, the energy balance of thermal plasma can be described as:

$$\rho c_p \frac{dT}{dt} = -div\vec{J} + \sigma\left\langle \vec{E}^2 \right\rangle - \Phi, \quad \vec{J} = -\lambda \nabla T \tag{10.1}$$

In this equation, $\vec{J}$ is the heat flux; c_p is the gas specific heat at constant pressure; λ is the thermal conductivity coefficient; σ is the high-frequency conductivity (see Section 6.6.2) determined by Equation 6.181; the square of the electric field is averaged $<E^2>$ over an oscillation period that is supposed to be short; the factor Φ describes radiation heat losses that can be neglected at atmospheric pressure and $T < 11{,}000 - 12{,}000\,\text{K}$; $\rho = Mn_0$ is the gas density, related to temperature T, taking into account the constancy of pressure $p = (n_0 + n_e)T$; M is the mass of heavy particles; and n_e and n_0 are the number densities of electrons and heavy species. The material derivative dT/dt is concerned with a fixed mass of gas (the Lagrangian description of the flow) and is related to the local Eulerian derivative $\partial T/\partial t$ by the well-known convective derivative relation, $dT/dt = \partial T/\partial t + (\vec{u} \cdot \vec{\nabla})T$, where $\vec{u}$ is velocity vector of the fixed mass of gas mentioned earlier.

Neglecting the effect of gas motion on plasma temperature in the energy release zone, the energy balance (Equation 10.1) is rewritten for steady-state discharges in the following form similar to Equation 8.20 for arc discharges:

$$-div\vec{J} + \sigma\left\langle \vec{E}^2 \right\rangle = 0, \quad \vec{J} = -\lambda \nabla T \tag{10.2}$$

In contrast to the DC case, the electric field (Equation 10.2) is not constant and should be determined from the electromagnetic field energy balance. In the steady-state system, this balance can be presented in terms of the Pointing vector, which is the flux density $\vec{S}$ of electromagnetic energy (see Section 6.6.4):

$$div\left\langle \vec{S} \right\rangle = -\sigma\left\langle \vec{E}^2 \right\rangle, \quad \vec{S} = \varepsilon_0 c^2 \left[\vec{E} \times \vec{B}\right] \tag{10.3}$$

Combining of the balance (Equation 10.2 and Equation 10.3) leads to the differential relation between thermal and electromagnetic energy fluxes:

$$div\left(\vec{J} + \left\langle \vec{S} \right\rangle\right) = 0 \tag{10.4}$$

This kind of energy continuity equation illustrates that the electromagnetic energy flux coming into some volume and dissipated there is balanced by thermal energy flux going out. In other words, the total energy flux has no sources. This differential equation can be easily solved for one-dimensional systems, resulting in the so-called **integral flux relation**:

$$\vec{J} + \left\langle \vec{S} \right\rangle = \frac{const}{r^n} \tag{10.5}$$

In the flux integral, r is the one-demensional coordinate; the power $n = 0$ for plane geometry, $n = 1$ for cylindrical geometry, and $n = 2$ for spherical.

The constant of integration in Equation 10.5 can be determined from the boundary conditions. For example, radial thermal and electromagnetic fluxes are equal to zero at the axis of the cylindrical plasma column. As a result, the constant in the integral flux relation is also equal to zero in this case, thus leading to the equation of cylindrical plasma column in the following simple form: $J_r + S_r = 0$.

This simple form of the integral flux relation gives the third equation of the channel model of arc discharges (Section 8.4.3 and Section 8.4.4). It shows that thermal plasma columns sustained by DC and high-frequency electromagnetic fields have many common features.

10.1.3 Thermal Plasma Generation in Inductively Coupled RF Discharges

The general principle of plasma generation in inductively coupled discharges is illustrated in Figure 10.1. High-frequency electric current passes through a solenoid coil where the resulting high-frequency magnetic field is induced along the axis of the discharge tube. In turn, the magnetic field induces a high-frequency vortex electric field concentric with the elements of the coil, which is able to provide breakdown and sustain the inductively coupled discharge. Electric currents in this discharge are also concentric with the coil elements and the discharge is apparently

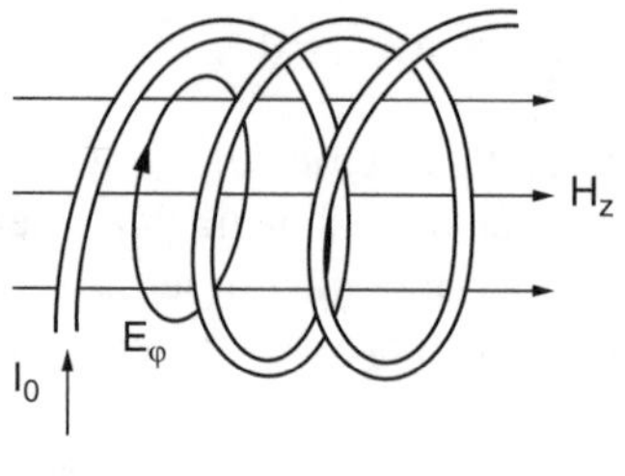

Figure 10.1 Generation of inductively coupled plasma

electrodeless. Generated in this way, **inductively coupled plasma (ICP)** can be quite powerful and effectively sustained at atmospheric and even higher pressures.

The magnetic field in the inductively coupled discharge is determined by the current in the solenoid; electric field there, according to the Maxwell equations, is also proportional to frequency of the electromagnetic fields. As a result, to achieve electric fields sufficient to sustain the ICP, high frequencies (RF) of about 0.1 to 100 MHz are usually required. Practically, a dielectric tube is usually inserted inside the solenoid coil; then ICP is sustained inside the discharge tube in the gas of interest.

It is interesting to remark that powerful radio-frequency discharges and power supplies generate noises and are able to interfere with radio communication systems. To avoid this undesirable effect, several specific frequency intervals were assigned for operation of industrial RF discharges. The most common RF frequency used in industrial plasma chemistry and plasma engineering is 13.6 MHz (corresponding wavelength is 22 m).

The inductively coupled RF discharges are quite effective in sustaining thermal plasma at atmospheric pressure. However, the relatively low values of electric fields in these systems are not sufficient for ignition of the discharges at atmospheric pressures. To ignite inductive discharges at atmospheric pressures requires special approaches. For example, an additional rod electrode can be introduced inside the coil solenoid. Then, heating the electrode by Foucault currents results in its partial evaporation, which simplifies the breakdown. After the breakdown, the additional electrode can be removed from the discharge zone.

An important specific problem of the RF discharges is the effectiveness of plasma coupling as a load with the RF generator. Electrical parameters of the plasma load, such as resistance and inductance, influence operation of the electric circuit as a whole and determine effectiveness of the coupling. For this reason, analysis of the ICP temperature distribution, which determines other parameters of quasi-equilibrium thermal plasma, is especially important in this case.

10.1.4 Metallic Cylinder Model of Long Inductively Coupled RF Discharge

Consider a dielectric discharge tube of radius R inserted inside a solenoid coil (see Figure 10.2). Plasma is sustained by joule heating induced by high-frequency AC and stabilized by heat transfer to the walls of the externally cooled discharge tube.

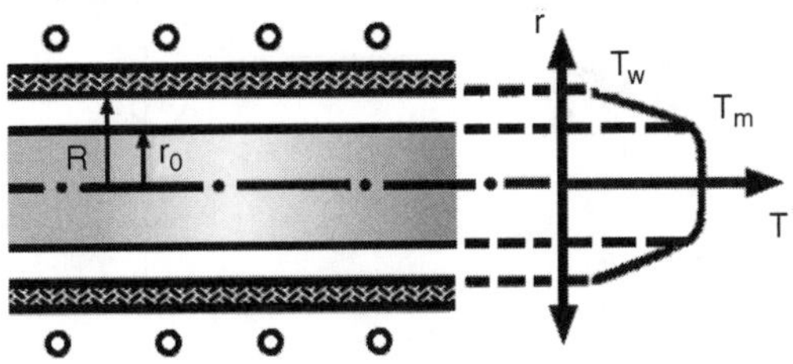

Figure 10.2 Radial temperature distribution in ICP discharge in a tube of radius R inserted inside a solenoid (r_0 is plasma radius); T_m—maximum temperature, T_w—wall temperature.

In this case, radial temperature distribution (see also Figure 10.2) can be described by the energy balance equation (Equation 10.2):

$$-\frac{1}{r}\frac{d}{dr}rJ_r + \sigma\left\langle E_\varphi^2\right\rangle = 0, \quad J_r = -\lambda\frac{dT}{dr} \tag{10.6}$$

This equation is similar to Equation 8.20 for positive column of arc discharges, although the electric field now is not axial but azimuthal and rapidly alternating. Note that in the megahertz frequency range of the RF discharges, the high-frequency conductivity (Equation 6.181) actually coincides with that of the DC case because $\omega^2 << \nu_{en}^2$; polarization and displacement currents can be neglected with respect to conductivity current; and the complex dielectric constant (Equation 6.179) is mostly imaginary.

Neglecting the displacement current and assuming $E, H \propto \exp(-i\omega t)$, the Maxwell equations (6.174 and 6.1.23) for the electric and magnetic fields in the case of cylindrical symmetry can be expressed as:

$$-\frac{dH_z}{dr} = \sigma E_\varphi, \quad \frac{1}{r}\frac{d}{dr}rE_\varphi = i\omega\mu_0 H_z \tag{10.7}$$

Together with the balance equation (10.6), the Maxwell equations complete the system of equations describing the thermal plasma column. The boundary conditions for the system of Equation 10.6 and Equation 10.7 for the discharge geometry shown in Figure 10.2—assuming low temperature on the externally cooled walls—can be taken as:

$$J_r = 0, \quad E_\varphi = 0 \quad \text{at} \quad r = 0; \quad T = T_w \approx 0 \quad \text{at} \quad r = R \tag{10.8}$$

Also, the magnetic field in nonconductive gas near the walls ($r = R$) of the discharge tube is the same as the one inside the empty solenoid:

$$H_z(r = R) \equiv H_0 = I_0 n \tag{10.9}$$

In this boundary relation, I_0 is current in the solenoid coil and n is number of the coil turns per unit of its length. Amplitudes of current and magnetic field are actually complex values, but here these can be considered real. Phase deviation between the oscillating fields H_z and E_φ can be calculated with respect to the phase of magnetic field H_0 (Equation 10.9) in the nonconductive gas near the wall of the discharge tube.

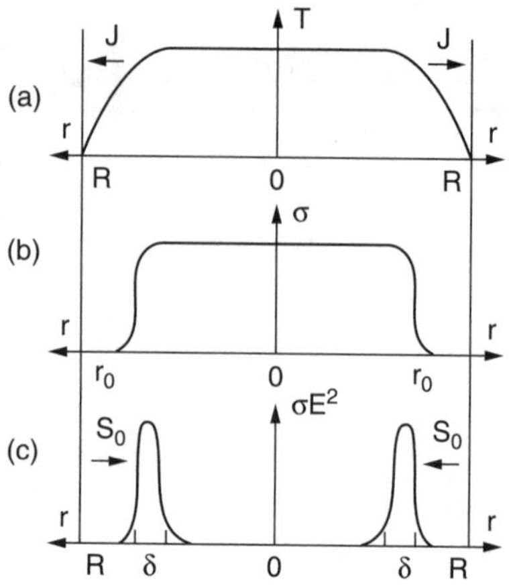

Figure 10.3 Radial distributions of (a) temperature; (b) electric conductivity; and (c) joule heating in ICP discharge

Further simplification and solution of the system of Equation 10.6 and Equation 10.7 can be done under the framework of the **metallic cylinder model**. This is actually a generalization of the channel model previously applied for arc discharges (see Section 8.4.3). According to the model, a plasma column is considered as a metallic cylinder with the conductivity fixed in the first approximation and corresponding to the maximum temperature T_m on the discharge axis.

The physical reasons underlying the metallic cylinder model are the same as those of the channel model of the arc discharge column. Thermal plasma conductivity is a very strong exponential function of the quasi-equilibrium discharge temperature. For this reason, when the temperature slightly decreases toward the discharge tube walls, conductivity becomes negligible. Note also that the plasma temperature T_m, conductivity σ, and the "metallic cylinder" (plasma column) radius r_0, are *a priori* unknown in the framework of the metallic cylinder model.

If plasma conductivity is high enough, the skin effect (see Section 6.1.7) prevents penetration of electromagnetic fields deep into the discharge column. As a result, heat release related to the inductive currents is localized in the relatively thin skin layer of the plasma column. Thermal conductivity inside the metallic cylinder provides the temperature plateau in the central part of the discharge cylinder where the inductive heating by itself is negligible. Radial distributions of plasma temperature, plasma conductivity, and the joule heating in the inductively coupled RF discharge corresponding to the metallic cylinder model are illustrated in Figure 10.3.

10.1.5 Electrodynamics of Thermal ICP Discharge in Framework of Metallic Cylinder Model

The metallic cylinder model permits separate consideration of electrodynamics and heat transfer aspects of the inductively coupled RF discharge. Analyzing first the electrodynamics of the ICP discharge, plasma conductivity σ and radius r_0 can be taken as parameters.

At the most typical RF discharge frequency, $f = 13.6\,\text{MHz}$ and high conductivity conditions of atmospheric pressure thermal plasma, the skin layer δ (see Equation 6.25 and Equation 6.26) is usually small with respect to the plasma radius $\delta << r_0$. In this case of strong skin effects, interaction of the electromagnetic field with plasma can be simplified to a one-dimensional plane geometry, which permits rewriting the Maxwell relations (10.7) as:

$$\frac{dH_z}{dx} = \sigma E_y, \quad \frac{dE_y}{dx} = -i\omega\mu_0 H_z \tag{10.10}$$

Coordinate x here is positive for the direction inward to the plasma (opposite to radius); coordinates y and z are directed tangentially to the plasma surface. Boundary conditions can be taken as $H = H_0$ at $x = 0$, and $E_y, H_z \to 0$ at $x \to \infty$. An analytical solution of this system of equations can be presented in the following form:

$$H_z = H_0 \exp\left[-i(\omega t - x/\delta) - x/\delta\right], \quad \delta = \sqrt{2/\omega\mu_0\sigma} \tag{10.11}$$

$$E_y = H_0\sqrt{\frac{\omega\mu_0}{\sigma}} \exp\left[-i\left(\omega t - \frac{x}{\delta} + \frac{\pi}{4}\right) - \frac{x}{\delta}\right] \tag{10.12}$$

The same relations for electromagnetic fields penetrating and damping in the δ skin layer of high conductivity thermal (see Equation 6.1.25 and Equation 6.1.26) can be rewritten not in complex, but in the real form:

$$H_z = H_0 \exp\left(-\frac{x}{\delta}\right)\cos\left(\omega t - \frac{x}{\delta}\right) \tag{10.13}$$

$$E_y = H_0\sqrt{\frac{\mu_0\omega}{\sigma}} \exp\left(-\frac{x}{\delta}\right)\cos\left(\omega t - \frac{x}{\delta} + \frac{\pi}{4}\right) \tag{10.14}$$

Thus, the amplitudes of electric and magnetic fields are seen to decrease exponentially inside the plasma column, with a phase shift between them of $\pi/4$.

The electromagnetic energy flux is normal to the plasma surface and directed inward to the plasma column (opposite to radius). Based on Equation 10.13 and Equation 10.14, the electromagnetic energy flux can be expressed as:

$$\langle S \rangle = S_0 \exp\left(-\frac{2x}{\delta}\right), \quad S_0 = H_0^2\sqrt{\frac{\mu_0\omega}{4\sigma}} \tag{10.15}$$

Here the flux S_0 shows the total electromagnetic energy absorbed in the unit area of the skin layer per unit time. The total power w, released per unit length of the long cylindrical plasma column, is related to this flux as $w = 2\pi r_0 \cdot S_0$.

For calculations of the electromagnetic flux S_0, it is convenient to use the additional relation (Equation 10.9) between magnetic field H_0 and current I_0 in a solenoid as well as the number, n, of turns per unit length of the coil. Together with Equation 10.15, this leads to the following convenient numerical relation for power per unit surface of the long cylindrical ICP column:

$$S_0, \frac{W}{cm^2} = 9.94 \cdot 10^{-2} \cdot \left(I_0 n, \frac{A \cdot turns}{cm} \right)^2 \sqrt{\frac{f, MHz}{\sigma, Ohm^{-1}cm^{-1}}} \tag{10.16}$$

10.1.6 Thermal Characteristics of Inductively Coupled Plasma in Framework of Metallic Cylinder Model

Integration of Equation 10.6 in the area between plasma column and discharge walls ($r_0 < r < R$, $\sigma = 0$) gives the relation of plasma temperature, radius r_0, and specific discharge power w per unit length:

$$\Theta_m(T_m) - \Theta_w(T_w) = \frac{w}{2\pi} \ln \frac{R}{r_0} \tag{10.17}$$

Quasi-equilibrium plasma temperature is expressed here in terms of the heat flux potential $\Theta(T)$ introduced by Equation 8.22; T_m is the maximum temperature in plasma column; and T_w is the temperature of discharge tube walls. Equation 10.17 is equivalent to Equation 8.25 derived in Section 8.4.3 for arc discharges, although the expressions for the specific power w are obviously different.

Noting that the interval Δr between plasma column and discharge walls is usually short $\Delta r = R - r_0 << R$, Equation 10.17 can be simplified:

$$\Theta_m(T_m) - \Theta_w(T_w) \approx \frac{w}{2\pi r_0} \cdot \Delta r = S_0 \cdot \Delta r \tag{10.18}$$

In contrast to Equation 10.17, this relation is specific for the ICP discharges, in which $\Delta r = R - r_0 << R$ and the heat release is concentrated in a cylindrical ring near the discharge walls, in contrast to arcs where the current is located within a channel near the discharge axis.

Plasma temperature in the central part of the inductive discharge is almost constant and close to the maximum value (see Figure 10.3). From the figure, the plasma temperature and conductivity are seen to decrease in the case of strong skin effect in a thin layer about $\delta/2$, where joule heating is mostly localized. The energy balance of the joule heating induced by electromagnetic fields and thermal conductivity in this layer can be expressed as:

$$\lambda_m \frac{\Delta T}{\delta/2} \approx S_0 \tag{10.19}$$

This approach is actually a modified model of the metallic cylinder and is similar to the Raizer modification of the channel model of arc discharges (see Section 8.4.4). Taking into account the relation between electromagnetic flux and specific discharge power per unit length of a column, $w = 2\pi r_0 \cdot S_0$, Equation 10.19 can be rewritten as:

$$4\pi\lambda_m \Delta T \approx w\frac{\delta}{r_0} \tag{10.20}$$

In the preceding relations, the thermal conductivity coefficient λ_m corresponds to the maximum plasma temperature T_m, and ΔT is the plasma temperature decrease related to the exponential conductivity decrease. The temperature decrease ΔT at the boundary layer of the thermal inductively coupled plasma can be determined based on the Saha equation in the same way as for arc discharges by using Equation 8.32. Using this relation for ΔT, Equation 10.15 for S_0, Equation 10.9 for H_0, and Equation 10.11 for δ, one can rewrite Equation 10.19 as:

$$2\sqrt{2}\,\lambda_m \frac{T_m^2}{I_i}\sigma_m = I_o^2 n^2 \tag{10.21}$$

This important formula relates current I_o and number of turns n per unit length in the solenoid coil with plasma conductivity σ_m and thus with ICP temperature T_m. In Equation 10.21, I_i is the effective ionization potential.

10.1.7 Temperature and Other Quasi-Equilibrium ICP Parameters in Framework of Metallic Cylinder Model

Taking into account that the plasma conductivity σ_m strongly depends on gas temperature, while T_m and λ_m are changing only slightly, one can conclude from Equation 10.21 that approximately $\sigma_m \propto (I_0 n)^2$. Then, based on the Saha equation for plasma density and conductivity, $\sigma_m(T_m)$, the plasma temperature dependence on the solenoid current and number of turns per unit length can be expressed as:

$$T_m = \frac{const}{const - \ln(I_0 n)} \tag{10.22}$$

This is not a strong logarithmic dependence $T_m(I_0,n)$ and corresponds to that derived for arc discharges (Equation 8.36). Note that in the case of a strong skin effect, plasma temperature T_m does not depend on the electromagnetic field frequency.

The following relations can illustrate the dependence of the ICP power per unit length of discharge on the solenoid current and the other parameters:

$$w = 2\pi r_0 S_0 \propto H_0^2\sqrt{\frac{\omega}{\sigma_m}} = (I_0 n)^2\sqrt{\frac{\omega}{\sigma_m}} \propto I_0 n\sqrt{\omega} \tag{10.23}$$

Thus, the specific power of the inductive discharge grows not only with the solenoid current and the number of turns, but also with the frequency of the electromagnetic field. This can be explained by an increase of the electric field with frequency (see Equation 10.14).

Taking into account that $\sigma_m \propto (I_0 n)^2$, Equation 10.23 gives $w \propto \sqrt{\sigma_m}$. This means that even a small increase of temperature T_m (which, according to the Saha, leads to an exponential increase of electric conductivity) requires a significant increase of discharge power per unit length. This effect actually limits the maximum temperature of the thermal discharges.

Using the preceding relations, one can estimate typical numerical characteristics of the thermal ICP in atmospheric pressure air on the electromagnetic field frequency $f = 13.6$ MHz. For a gas temperature of about 10,000 K, a discharge tube radius of about 3 cm, and a thermal conductivity of $\lambda_m = 1.4 \cdot 10^{-2}\ \text{W/cm} \cdot \text{K}$, the electric conductivity of the thermal plasma can be found to be $\sigma_m \approx 25\ \text{Ohm}^{-1}\text{cm}^{-1}$ and the skin layer $\delta \approx 0.27\,\text{cm}$. To sustain such a plasma, the necessary electromagnetic flux can be calculated from Equation 10.19 and Equation 8.32 as $S_0 \approx 250\ \text{W/cm}^2$.

The corresponding value of solenoid current and number of turns is $I_0 n \approx 60$ A turns/cm; magnetic field $H_0 \approx 6\,\text{kA/m}$. Then the maximum electric field on the external boundary of the plasma column is approximately 12 V/cm; the density of circular current is 300 A/cm^2; the total current per unit length of the column is approximately 100 A/cm; thermal flux potential is about 0.15 kW/cm; the distance between the effective plasma surface and discharge tube is $\Delta r \approx 0.5$ cm; and, finally, the ICP discharge power per unit length is about 4 kW/cm.

Analyzing the preceding typical numerical characteristics of a thermal ICP discharge, one can note that, according to Equation 10.23 and the following discussion, the specific discharge power should be increased from 4 kW/cm to at least 8 kW/cm to increase plasma temperature from 10,000 to12,000 K. Also, taking into account the growing radiative losses at 12,000 K requires the specific power to be even higher. As a result of such strong power requirements, the ICP temperature does not exceed 10,000 to 11,000 K.

A typical value of the circular ICP current per unit length of the RF discharge was estimated earlier as 100 A/cm. This value is about the same as and even exceeds the corresponding value of solenoid current per unit length of the coil: $I_0 n \approx 60$ A/cm. This means that inductive influence of the plasma current on the electric circuit of the RF power supply is quite significant. Thus, the RF power supply must be effectively coupled with the plasma, which can be considered a load in the circuit.

10.1.8 ICP Discharge in Weak Skin-Effect Conditions: Thermal ICP Limits

Decrease of current in the solenoid coil (from the discussed values corresponding to thermal plasma conditions) leads to a growth of the skin layer δ until it becomes about the size of the plasma ($\delta \approx r_0$, R) and the model considered previously is no longer valid. For the opposite case of low temperature and electric conductivity, assume $\delta >> r_0, R$. The magnetic field is uniform $H = H_0$ in the absence of skin effect and the electric field amplitude distribution along radius of the discharge tube can be calculated from the Maxwell equation (10.7) as:

$$E(r) = \frac{1}{2}\omega\mu_0 H_0 \cdot r \tag{10.24}$$

In this case, power per unit length of the discharge column with radius r_0 can be calculated by integrating the joule heating density:

$$w = \int_0^{r_0} \sigma \langle E^2(r) \rangle 2\pi r\,dr \approx \frac{1}{16}\pi\mu_0^2\omega^2\sigma_m H_0^2 r_0^4 \tag{10.25}$$

Taking into account the relation between the magnetic field with solenoid current and the number of turns per unit length (Equation 10.9), Equation 10.25 can be rewritten as:

$$w \approx \frac{1}{16}\pi\mu_0^2\omega^2\sigma_m I_0^2 n^2 r_0^4 \tag{10.26}$$

In contrast to the case of strong skin-effect conditions where the temperature decrease takes place mainly in the boundary layer (Equation 10.19 and Equation 10.20), here the temperature reduction is distributed over the entire plasma column radius as it was in the arc discharges. This leads to the following relation between specific power and maximum plasma temperature (Equation 8.32):

$$w \approx 4\pi r_0 \lambda_m \frac{\Delta T}{r_0} = 4\pi\lambda_m \Delta T \approx 8\pi\lambda_m \frac{T_m^2}{I_i} \tag{10.27}$$

In this relation, ΔT is the temperature decrease across the plasma column (see Equation 8.32); λ_m is the thermal conductivity corresponding to the maximum plasma temperature T_m; and I_i is the effective value of ionization potential.

Next, analyze what happens when the temperature and electric conductivity decrease in this weak skin-effect regime. The temperature cannot decrease significantly (because of the related exponential reduction of conductivity) that, according to Equation 10.27, makes the specific power w almost constant even when electric conductivity decreases. This effect of specific power stabilization is illustrated in Figure 10.4(a). Then, based on Equation 10.18, the plasma radius must also decrease

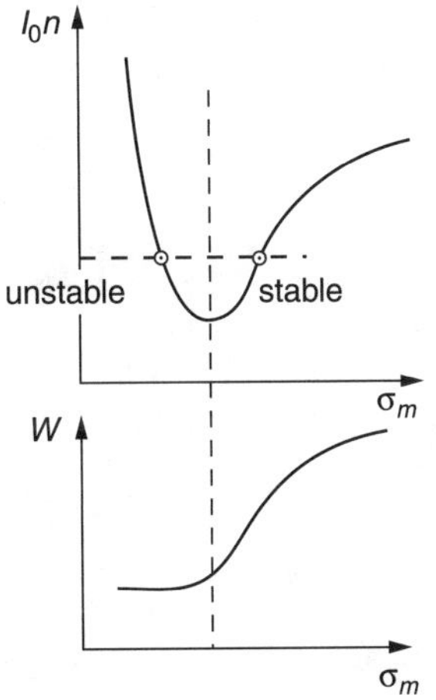

Figure 10.4 Solenoid current and specific power dependence on conductivity. Stability analysis

with temperature reduction. Taking into account Equation 10.26 leads to the interesting conclusion that current in the solenoid coil (factor $I_0 n$) in this regime is not decreasing but grows with the conductivity decrease in contrast to the strong skin-effect regime where $w \propto I_0 n \propto \sqrt{\sigma_m}$. This nonmonotonic dependence of $I_0 n(\sigma_m)$ is illustrated in Figure 10.4(b).

Thus, the dependence $I_0 n(\sigma_m)$ has a minimum corresponding to the case when the skin layer is about the same size as the discharge tube radius $\delta \approx R$. This minimum value of the solenoid current $I_0 n$, which is necessary to sustain the quasi-equilibrium thermal ICP, can be found from Equation 10.21, assuming a value of electric conductivity σ_m corresponding to the critical condition $\delta \approx R$ (see Equation 10.11). In turn, this leads to the following expression for the minimum value of the solenoid current $I_0 n$ to sustain thermal plasma column:

$$(I_0 n)_{\min} \approx \frac{2T_m}{R}\sqrt{\frac{\lambda_m \sqrt{2}}{I_i \mu_0 \omega}} \tag{10.28}$$

As an example, the minimal current necessary to provide a thermal ICP discharge in a tube with $R = 3$ cm at frequency $f = 13.6$ MHz in air, according to Equation 10.28, can be estimated as $(I_0 n)_{\min} \approx 10$ A·turns/cm; corresponding minimum values of the quasi-equilibrium temperature are approximately $T_{crit} \approx 7000$ to 8000 K.

As seen from Figure 10.4(b), when the solenoid current and number of turns per unit length exceed the critical value, in principle, two stationary states of the ICP discharge can be realized. One of these corresponds to high conductivity and strong skin-effect conditions and the other to low conductivity and no skin-effect conditions. However, only one of them, high conductivity regime, is stable. The low conductivity regime (left branch on Figure 10.4) is unstable. For example, if temperature (and thus conductivity) increases because of some fluctuation, then a current lower than the actual one is sufficient, according to Figure 10.4, to sustain the discharge. This leads to ICP plasma heating and further temperature increase until the stable high-conductivity branch is reached.

10.1.9 ICP Torches

The inductively coupled plasma torches are widely used as industrial plasma sources. They are important competitors of the DC arc jets discussed in Section 8.5. A standard configuration of an ICP torch[286] is illustrated in Figure 10.5. The heating coil is usually water cooled and is not in direct contact with plasma, thus providing the plasma purity; this is one of the most important advantages of ICP discharges.

ICP torches are difficult to start because the electric fields in these systems are relatively low (see Section 10.1.8 and Section 10.1.9). To initiate the discharge at low power levels of about 1 kW, a special graphite starting rod, shown in Figure 10.5, can be applied. At high power levels, it is convenient to use a pilot DC or RF small plasma generators to initiate the powerful ICP torch. Such hybrid plasma torches operating at power levels of 50 to 100 kW are illustrated in Figure 10.6(a, b).

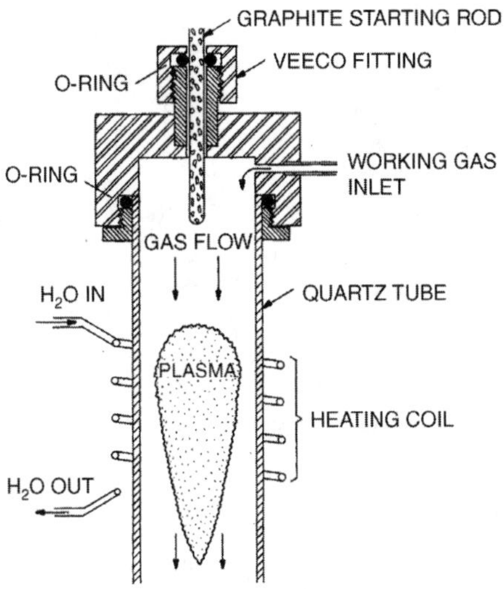

Figure 10.5 Kilowatt-level inductively coupled plasma torch[264,286]

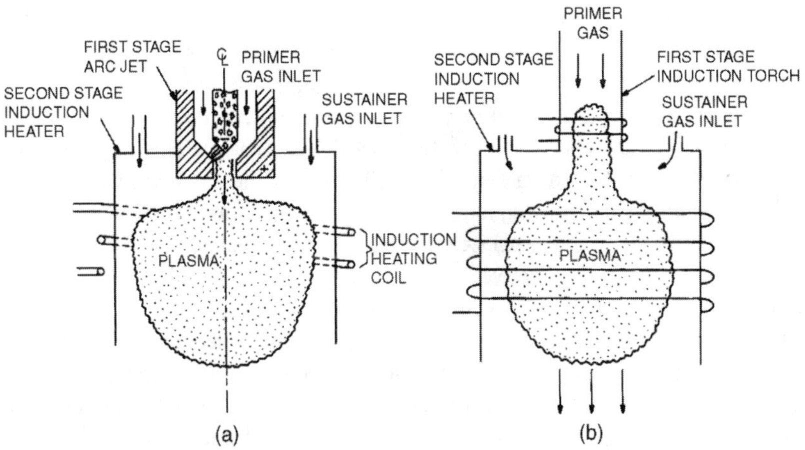

Figure 10.6 Hybrid plasma torches: (a) DC–RF; (b) RF–RF[264]

The ICP torches are often operated in transparent quartz tubes to avoid heat load on the discharge walls related to visible radiation. At power levels exceeding 5 kW, the walls should be cooled by water. Plasma stabilization and the discharge walls' insulation from direct plasma influence can be achieved in fast gas flows. Fast flowing gas flows are the most effective from this point of view. Some of these vortex methods of plasma stabilization will be discussed next.

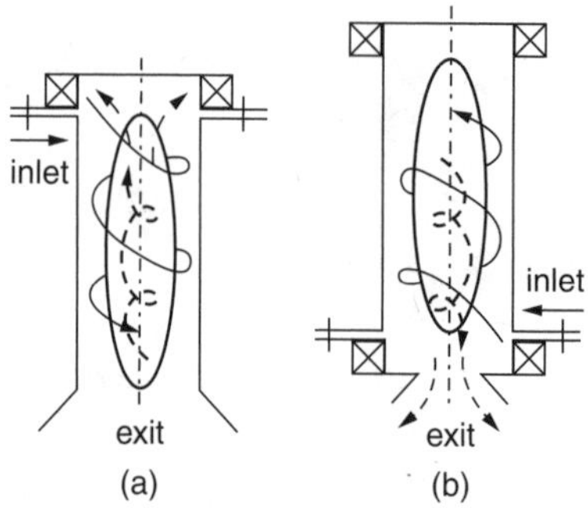

Figure 10.7 (a) Forward and (b) reverse vortex stabilization

10.1.10 ICP Torch Stabilization in Vortex Gas Flow

In the most conventional approach, the swirl generator is placed upstream with respect to the ICP discharge, and the outlet of the plasma jet is directed to the opposite side. Such configuration is usually referred to as **forward-vortex stabilization** and is illustrated in Figure 10.7(a). The rotating gas provides good protection to the walls from the plasma heat flux (see, for example, Gutsol et al.[363]). This can be explained by noting that portions of the gas change direction from an axial swirl to the swirl-radial due to viscosity and heat expansion. This radial moving gas compresses the central hot zone, decreasing the heat flux to the walls and providing effective insulation.

However, some reverse axial pressure gradient and central reverse flow appear in the case of forward-vortex stabilization (see Figure 10.7a). This effect is related to fast flow rotation and strong centrifugal effects near the gas inlet, which become a bit slower and weaker downstream. The hot reverse flow mixes with incoming cold gas and increases heat losses to the walls, which makes the discharge walls' insulation somewhat less effective.

A more effective ICP torch walls' insulation can be achieved by means of the **reverse-vortex stabilization**, developed by Gutsol and colleagues.[361,362] Flow configuration for the reverse-flow stabilization of the inductively coupled discharge is shown in Figure 10.7(b). In this case, the outlet of the plasma jet is directed along the axis to the swirl generator side. The cold incoming rotating gas in this stabilization scheme at first moves past the walls, providing their cooling and insulation, and only after that does it go to the central plasma zone and become hot. Thus, in the case of reverse-vortex stabilization, the incoming gas is entering the discharge zone from all directions except the outlet side, which makes this approach interesting for ICP heating efficiency and discharge walls' protection.

Experimental details regarding the reverse-vortex stabilization of the 60-kW thermal ICP discharge in argon can be found in Gutsol et al.[363] It should be noted that the discussed reverse-vortex stabilization could be applied not only to ICP, but to other thermal discharges and even combustion systems. Reverse-vortex stabilization of microwave discharges is discussed in Gutsol et al.[363] and application to gas burners is discussed in Kalinnikov and Gutsol.[361]

Obviously, the reverse-vortex stabilization can be applied not only to ICP torches. This "TORNADO" effect also permitted organization of very efficient configurations of gliding arcs[622] and other combustion and plasma systems.

10.1.11 Capacitively Coupled Atmospheric Pressure RF Discharges

As discussed previously, the inductively coupled plasma, where magnetic field is induced, usually has lower values of electric fields than capacitively coupled discharges, in which the electric field is the primary effect (as in a capacitor). For this reason, the ICP discharges at moderate to high pressures are usually concerned with thermal quasi-equilibrium plasma generation, in which the ionization is sustained by heating and high electric fields are not necessary. The capacitively coupled plasma (CCP) of RF discharges is able to provide high values of electric fields, which makes these discharges interesting for generating nonthermal nonequilibrium plasma. The nonthermal RF CCP discharges of moderate and low pressures (optimal for cold plasma generation) play an important role in material treatment, especially in electronics. These will be considered in detail later.

In this section the not widely used but interesting **atmospheric pressure RF glow discharge** developed recently at the University of Tennessee, Knoxville, (UTK)[364,365] will be discussed. This discharge is a member of a family of atmospheric pressure glow discharges considered in Section 7.5.6 and Section 9.3.6. A schematic drawing of the UTK radio-frequency atmospheric pressure glow discharge is shown in Figure 10.8.

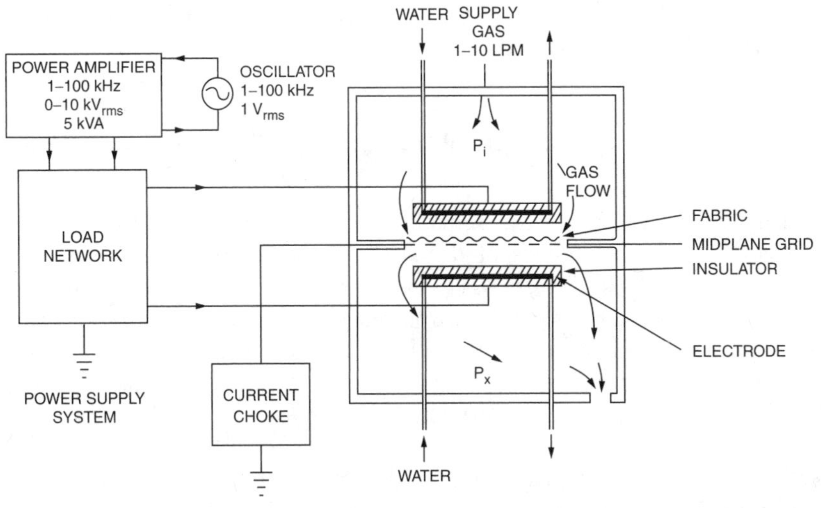

Figure 10.8 Atmospheric pressure glow discharge plasma reactor[364,365]

The discharge volume is confined by parallel electrodes across which a RF electric field is imposed. The discharge system also includes a bare metal screen located midway between the electrodes that can be grounded through a current choke (see Figure 10.8). This median screen provides a substrate surface to support the material to be treated in the atmospheric pressure glow discharge.

The electric field applied between the electrodes is on the level of kilovolts per centimeter; it should be sufficiently strong for electric breakdown and to sustain the discharge. In helium or argon, this electric field is obviously lower than in atmospheric air. Typical frequencies necessary to sustain the uniform glow regime of the discharge are in the kilohertz range (about 1 to 20 kHZ). At lower values of frequency, the discharge is difficult to initiate; at higher frequencies, the discharge is not uniform and has the filamentary structure.

The most interesting and desirable form of the discharge, the atmospheric pressure uniform glow regime, corresponds to the specific values of RF frequency, which are sufficiently high to trap the ions between the median screen and an electrode, but not high enough to trap plasma electrons as well. This frequency range provides some reduction of electron–ion recombination in the boundary layers, which promotes the ionization balance at lower electric fields.

The power density of the discharge obviously grows with an increase of the voltage amplitude and the electric field frequency. However, even at relatively high values of the voltage amplitude and the electric field frequency, the discharge power density and total power are still not high relative to, for example, pulsed corona discharges. Maximum values of the power density in the uniform regime are approximately 100 mW/cm^3; maximum total power is about 100 W.[264]

10.2 THERMAL PLASMA GENERATION IN MICROWAVE AND OPTICAL DISCHARGES

10.2.1 Optical and Quasi-Optical Interaction of Electromagnetic Waves with Plasma

Wavelengths of electromagnetic oscillations in the preceding RF discharges were much larger than typical sizes of the systems. For example, the industrial RF frequency $f = 13.6$ MHz corresponds to the wavelength of 22 m. In contrast to that, microwave plasma is sustained by electromagnetic waves in the centimeter range of wavelengths. Thus, microwave radiation has wavelengths comparable with the discharge system sizes, and the electromagnetic field interaction with plasma in microwave discharges is quasi optical. The optical discharges sustained by laser radiation are characterized by much smaller wavelengths and so the electromagnetic field interaction with plasma in this case is also optical. For this reason, thermal plasma generation in the microwave and optical discharges will be discussed together.

In the same manner as in the case of RF discharges, thermal plasma generation in microwave and optical discharges is usually related to high-pressure systems (typically atmospheric pressure). Nonthermal, nonequilibrium microwave discharges

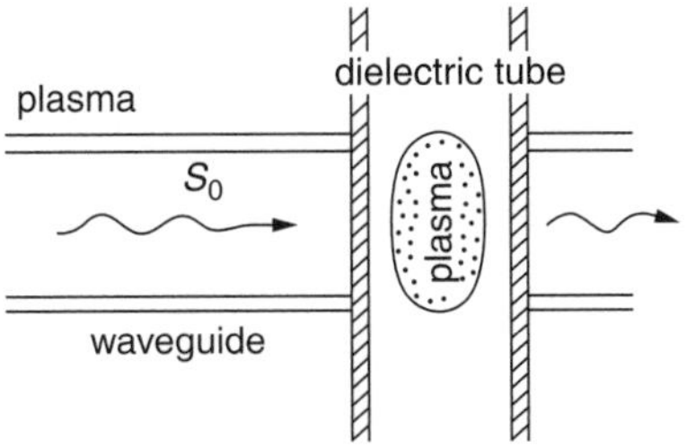

Figure 10.9 General schematic of microwave discharge in a waveguide

are usually related in a continuous mode to moderate and low pressures; these will be discussed later in this chapter.

10.2.2 Microwave Discharges in Waveguides: Modes of Electromagnetic Oscillations in Waveguides without Plasma

Microwave generators, in particular magnetrons, steadily operating with power exceeding 1 kW in gigahertz frequency range, are able to maintain the steady-state thermal microwave discharges effectively at atmospheric pressure. Electromagnetic energy in the microwave discharges can be coupled with plasma in different ways. The most typical one is related to the application of waveguides and is illustrated in Figure 10.9.

In the configuration of microwave discharges shown in this figure, the dielectric tube (usually quartz), which is transparent for the electromagnetic waves, crosses the rectangular waveguide. Plasma is ignited and maintained in the discharge tube by dissipation of electromagnetic energy. Heat balance of the thermal plasma is provided mostly by convective cooling in the gas flow. Some special features of gas flow around and across the plasma zone will be discussed later on (see also stabilization of the thermal plasma zone in Section 10.1.10).

Different modes of electromagnetic waves formed in the rectangular waveguide can be used to operate microwave discharges. The most typical one is the H_{01} mode, illustrated in Figure 10.10. In this figure, the electric field distribution in this waveguide mode can be seen without taking into account the plasma influence. The electric field in the H_{01} mode is parallel to the narrow walls of the waveguide, and its value is constant in this direction. Along the wide waveguide wall, the electric field is distributed as a sine function (in the absence of plasma) with a maximum in the center of the discharge tube and zero field on the narrow waveguide walls:

$$E_y = E_{max} \sin\left(\frac{\pi}{a_w} x\right); \quad E_x = E_z = 0 \tag{10.29}$$

In this relation, E_x is the electric field component directed along the longer wall of the waveguide, which has length a_w; E_y is the electric field component directed along the shorter wall of the waveguide, which has length b_w; and E_z is the electric field component directed along the third axis z. The maximum value of electric field E_{max}

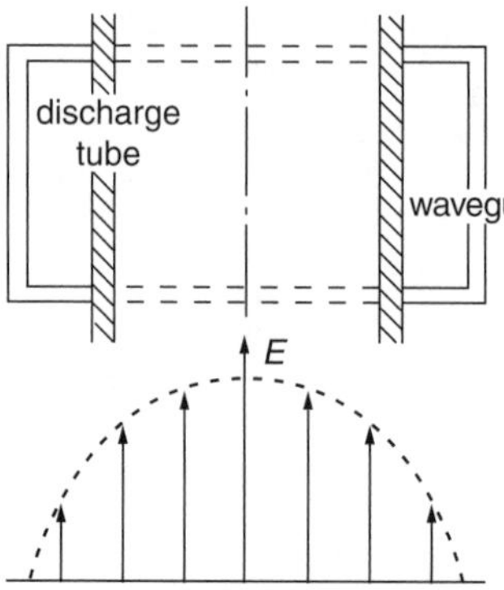

Figure 10.10 Electric field distribution for H_{01} mode in rectangular waveguide

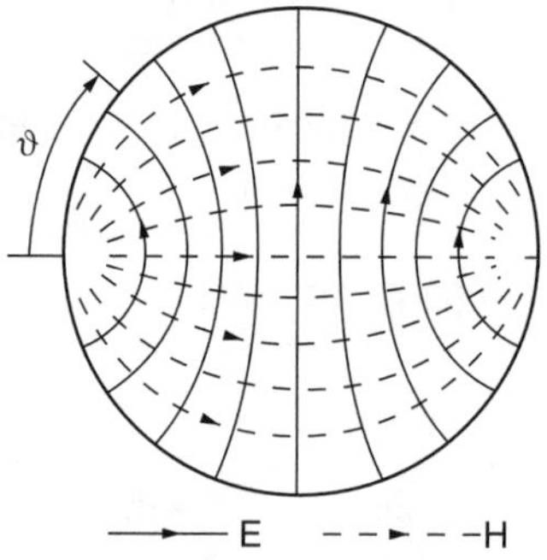

Figure 10.11 Electric and magnetic field disributions for H_{11} mode in round waveguide

(expressed in kilowatts per centimeters) in the case of the mode H_{01} is related to the microwave power P_{MW} (expressed in kilowatts) transmitted along the rectangular waveguide by the following numerical formula:

$$E_{\max}^2 = \frac{1.51 \cdot P_{MW}}{a_w b_w}\left[1-\left(\frac{\lambda}{\lambda_{crit}}\right)^2\right]^{-1/2} \tag{10.30}$$

Here, λ is the wavelength of the electromagnetic wave propagating in the waveguide; λ_{crit} is the maximum value of the wavelength when the propagation is still possible ($\lambda_{crit} = 2a_w$); and lengths of the waveguide walls a_w and b_w are expressed in centimeters.

The H_{01} mode is convenient for microwave plasma generation in the rectangular waveguides because the electric field in this case has a maximum in the center of the discharge tube (see Figure 10.9 and Figure 10.10). Microwave discharges obviously can be generated in the cylindrical waveguides with round cross section. The most typical oscillation mode in such waveguides is H_{11}. Space distribution of electric and magnetic fields in this case is illustrated in Figure 10.11 and can be expressed, without taking into account the influence of the plasma, as:

$$E_\vartheta(r,\vartheta) = E_{max} \cdot J_1'\left(\frac{1.84r}{R_w}\right)\cos\vartheta; \quad E_z = 0 \tag{10.31}$$

$$E_r(r,\vartheta) = E_{max}\frac{R_w}{1.84r} \cdot J_1\left(\frac{1.84r}{R_w}\right) \cdot \sin\vartheta \tag{10.32}$$

In these relations, R_w is radius of the round waveguide; J_1 (1.84 r/R_w) and J_1' (1.84 r/R_w) are the Bessel function of the first order and its first derivative; and E_{max} again is the maximum value of electric field (expressed in kilowatts per centimeter) now for the mode H_{11}, which is related to the microwave power P_{MW} (expressed in kilowatts) transmitted along the round waveguide by the following numerical formula:

$$E_{max}^2 = \frac{1.58P_{MW}}{\pi R_w^2}\left[1 - \left(\frac{\lambda}{\lambda_{crit}}\right)^2\right]^{-1/2} \tag{10.33}$$

Similar to Equation 10.30, the critical wavelength λ_{crit} is the maximum value of the wavelength when the propagation is still possible. For the case of H_{11} mode in the round waveguide, $\lambda_{crit} = 3.41R_w$.

10.2.3 Microwave Plasma Generation by H_{01} Electromagnetic Oscillation Mode in Waveguide

As was discussed previously and illustrated in Figure 10.10, the H_{01} mode in the rectangular waveguides is convenient for plasma generation because the electric field in this case has maximum in the center of the discharge tube. Plasma is formed along the electric field and axis of the discharge tube (see Figure 10.9). To provide plasma stabilization, gas flow is often supplied into the discharge tube tangentially, as discussed in Section 10.1.10.

From the discussion in Section 10.2.2, the waveguide dimensions are related with the frequency of the electromagnetic wave, (see the relations for critical values of wavelengths). Thus, for an electromagnetic wave frequency f = 2.5 GHz (the corresponding wavelength in vacuum λ = 12 cm), the wide waveguide wall should be longer than 6 cm and usually is equal to 7.2 cm. Typically, the narrow waveguide wall is 3.4 cm long and the dielectric discharge tube diameter is about 2 cm. The diameter of the generated microwave plasma column in such conditions is usually about 1 cm.

If power of the atmospheric pressure microwave discharge described previously is 1 to 2 kW, then typical thermal plasma temperatures in air and other molecular gases are normally in the range of 4000 to 5000 K. When temperature is not extremely high, energy exchange between electrons and heavy particles in rare gases is slower than in molecular gases, which leads to some differences between electron temperature and temperature of heavy species in such systems. For example, typical plasma temperature values in a microwave discharge in argon at similar conditions are about 6500 to 7000 K for electrons and about 4500 K for heavy particles.

In general, one can conclude that plasma temperatures in thermal microwave discharges are usually lower than temperatures in thermal RF and arc discharges under similar conditions. At lower wavelengths ($\lambda = 3$ cm) and thus at smaller waveguide cross sections, plasma temperatures are slightly higher; at similar conditions in molecular gases such as nitrogen, plasma temperatures reach about 6000 K.

The incident electromagnetic wave formed in the waveguide interacts with plasma generated in the discharge. This interaction results in partial dissipation and reflection of the electromagnetic wave. Typically, about half of the power of the incident wave can be directly dissipated in the high conductivity thermal plasma column, about a quarter of the wave is transmitted through the plasma, and the remaining is reflected. To increase the effectiveness of electromagnetic wave coupling with the thermal plasma column (fraction of the wave dissipated in plasma), the transmitted wave can be reflected back, which leads to the formation of a standing wave. Such special coupling techniques permit increasing the fraction of the electromagnetic energy absorbed in the plasma up to 90 to 95%, which is important for practical applications of the discharge system.[366,367]

A powerful microwave discharge for plasma chemical applications was developed in the Kurchatov Institute of Atomic Energy by Rusanov and co-workers.[9,134] Microwave energy was provided by four magnetrons (maximum power of each was 300 kW) and coupled with plasma in different molecular gases at moderate- and relatively high-pressure ranges. Note that the electromagnetic wave mode in this waveguide was more complicated than those described previously. The powerful microwave discharge was fast gas flow-stabilized (including the case of supersonic flow stabilization; see Section 6.4 and Section 10.1.10) and was able to operate in thermal and nonthermal regimes.

10.2.4 Microwave Plasma Generation in Resonators

The powerful atmospheric pressure thermal microwave discharge in a resonator was developed by Kapitsa and co-authors.[368] A schematic of such a cylindrical resonator with standing wave mode E_{01} is presented in Figure 10.12. A microwave generator of 175-kW power at 1.6-GHz frequency (wavelength 19 cm) was used in the Kapitsa experiments.

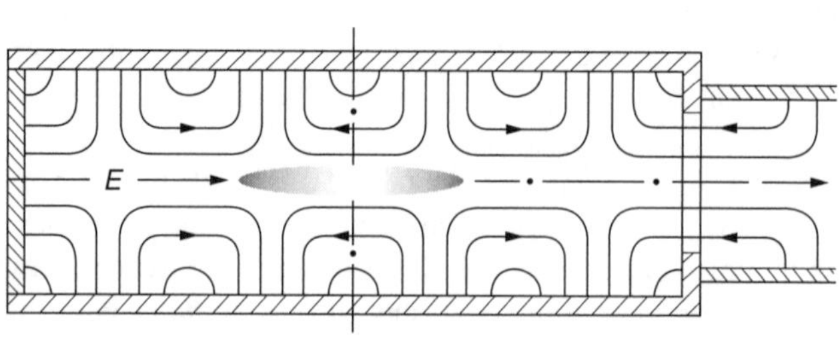

Figure 10.12 Microwave plasma in resonator

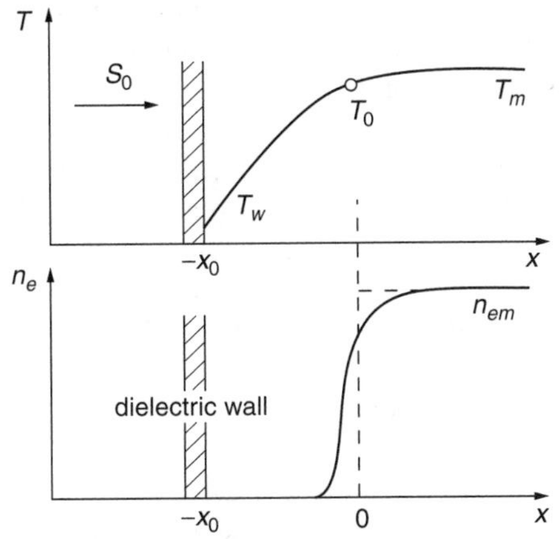

Figure 10.13 Temperature and electron density distributions in a discharge sustained by electromagnetic wave

In this case, the electric field on the axis of the cylindrical resonator is directed along the axis and varies as the cosine function with the maximum in the center of the cylinder. The electric field then decreases further from the axis. Ignition of the discharge obviously takes place in the central part of the resonator cylinder, where the electric field has its maximum value (see Figure 10.12). Microwave plasma formed in the resonator appears as a filament located along the axis of the resonator cylinder. The length of the plasma filament is about 10 cm, which corresponds to half of the wavelength; the diameter of the plasma filament is about 1 cm.

To stabilize the microwave plasma in the central part of the resonator at relatively high power levels of 20 kW in hydrogen, deuterium, and helium at atmospheric and higher pressures, gas is tangentially supplied to the discharge chamber (see Sectio n 10.1.10). Plasma temperature in this microwave discharge generally does not exceed 8000 K, which is related to intensive reflection effects at higher conductivities.

10.2.5 One-Dimensional Model of Electromagnetic Wave Interaction with Thermal Plasma

Microwave plasma generation can be qualitatively described by the simple one-dimensional model, which assumes that plane electromagnetic wave passes through a plane dielectric wall (transparent to this wave) and then meets plasma. Heat released in the plasma is subsequently transferred back to the dielectric wall, which is externally cooled, providing energy balance of the steady-state system. This simple but effective model is illustrated in Figure 10.13.

This approach actually describes local interaction of the incident electromagnetic wave with the closest plasma surface (see Figure 10.9). When the microwave power, gas pressure, and thermal plasma conductivity are relatively high, the electromagnetic wave penetration into the plasma is shallow. This makes the one-dimensional model to describe the microwave plasma generation applicable. More detailed models that take into account electromagnetic wave scattering on the microwave plasma make the system under consideration much more complicated for physical analysis but do not add important qualitative effects in the final results.

The one-dimensional model of microwave plasma generation can be described by the integral flux relation (10.5), taking into account that in the plane geometry, $n = 0$. The constant in Equation 10.5 is equal to zero ($const = 0$), because the temperature tends to the constant value T_m deep inside the plasma ($x \to \infty$). As a result, the integral flux relation can be rewritten as:

$$J + \langle S \rangle = 0, \quad J = -\lambda \frac{dT}{dx} \tag{10.34}$$

In this relation, λ is the coefficient of thermal conductivity and $\langle S \rangle$ is the mean value of the electromagnetic energy flux density, determined in Equation 10.3.

Because microwave discharges are quasi optical, such wave effects as reflection and interference should be taken into account here. In this case, description of the electromagnetic wave propagation in plasma (see Section 6.6) requires analysis of the wave equation with the complex dielectric permittivity (Equation 6.179):

$$\frac{d^2 E_y}{dx^2} + \left(\varepsilon_\omega + i \frac{\sigma_\omega}{\varepsilon_0 \omega} \right) \frac{\omega^2}{c^2} E_y = 0 \tag{10.35}$$

In this relation, E_y is a component of the electric field directed along the plasma surface; σ_w and ε_w are high-frequency plasma conductivity and dielectric permittivity determined according to Equation 6.180 and Equation 6.181.

Because the plasma temperature and degree of ionization in the thermal microwave discharges are not very high (with respect to RF and arc discharges under similar conditions), the electron–ion collisions can be neglected in calculating the plasma conductivity. From the Saha equation, the quasi-equilibrium high-frequency plasma conductivity and dielectric permittivity in the wave Equation 10.36 can be taken as:

$$\sigma_\omega \propto 1 - \varepsilon_\omega \propto n_e \propto \exp\left(\frac{I_i}{2T} \right) \tag{10.36}$$

where I_i is the ionization potential. Boundary conditions for the system of Equation 10.34 and Equation 10.35 are: the electric field is assumed to be zero deep inside the plasma ($E = 0$ at $x \to \infty$); temperature on the cooled dielectric wall is low ($T = T_w \approx 0$ at $x = -x_0$); and the electromagnetic energy flux density S_0 related to microwave power provided from the generator is given.

Solution of the system of Equation 10.34 and Equation 10.35 determines: the plasma temperature $T_m = T(x \to \infty)$; the fraction of the electromagnetic energy flux density S_1, which is dissipated in the plasma; and the microwave reflection coefficient from the plasma $\rho = (S_0 - S_1)/S_0$. A general solution of the system obviously cannot be found analytically and requires numerical approaches. Similar to the cases of thermal arc and thermal RF discharges, physically clear analytic solution of the system (Equation

10.34 and Equation 10.35) can be obtained only by assuming a sharp conductivity change on the plasma boundary and constant conductivity inside the plasma.

10.2.6 Constant Conductivity Model of Microwave Plasma Generation

To find an analytical solution of the system (Equation 10.34 and Equation 10.35), assume that the plasma has a sharp boundary at $x = 0$, where the temperature equals T_0 (see Figure 10.13). To the left from this point ($x \leq 0$, $T \leq T_0$), assume no plasma: $\sigma = 0$, $\varepsilon = 1$. To the right from the boundary point ($x > 0$, $T_0 < T < T_m$), assume a constant conductivity plasma: $\sigma = \sigma_m$, $\varepsilon = \varepsilon_m$. This **constant microwave plasma conductivity model** is somewhat similar to the channel model of arc discharges and the metallic cylinder model of the thermal ICP discharges.

In the framework of the constant conductivity model, the reflection coefficient of an incident electromagnetic wave normal to the sharp boundary of plasma, $\rho = (S_0 - S_1)/S_0$, can be found from the boundary conditions for electric and magnetic fields on the plasma surface[217] in the following form:

$$\rho = \frac{(n-1)^2 + \kappa^2}{(n+1)^2 + \kappa^2} \tag{10.37}$$

In this relation, κ is the attenuation coefficient of the electromagnetic wave, defined by Equation 6.189; and n is the electromagnetic wave refractive index, defined by Equation 6.190. Both parameters n and κ are functions of the high-frequency permittivity ε_ω and electric conductivity σ_ω, which are assumed constant.

While Equation 10.37 describes the electromagnetic wave reflection from the plasma, microwave absorption and energy flux damping in the plasma can be calculated from the Bouguer law (Equation 6.195) with the absorption coefficient μ_ω mostly dependent on the electron density (Equation 6.197). According to the Bouguer law, dissipation of the electromagnetic wave takes place in a surface layer of about $l_\omega = 1/\mu_\omega$, where temperature grows from T_0 to T_m.

The energy balance in this layer between the absorbed electromagnetic flux S_1 and the heat transfer to the wall can be expressed as $S_1 \approx \lambda_m \, \Delta T/l_\omega$. Similar to the channel model of arc discharges and the metallic cylinder model of RF discharges, one can determine $\Delta T = T_m - T_0$, assuming the exponential conductivity decrease in the plasma. As a result, the following equation to determine the plasma temperature is obtained:

$$S_0\left[1-\rho(T_m)\right] = \lambda(T_m)\cdot\frac{2T_m^2}{I_i}\cdot\mu_\omega(T_m) \tag{10.38}$$

Finally, to find the size of the gap x_0 between the plasma and the dielectric wall, one can use the equation for heat transfer across this gap:

$$\Theta_m(T_m) - \Theta_w(T_w) = S_1 \cdot |x_0| \tag{10.39}$$

where the heat flux potential $\Theta(T)$ is determined by Equation 8.22.

Table 10.1 Characteristics of Atmospheric Pressure Microwave Discharge in Air at Frequency f = 10 GHz, (λ = 3 cm) as a Function of Plasma Temperature

Plasma temperature, T_m	4500 K	5000 K	5500 K	6000 K
Electron density, n_e, 10^{13} cm^{-3}	1.6	4.8	9.3	21
Plasma conductivity, σ_m, 10^{11} sec^{-1}	0.33	0.99	1.9	4.1
Thermal conductivity, λ, 10^{-2} W/cm · K	0.95	1.1	1.3	1.55
Refractive index n of plasma surface	1.3	2.1	2.8	4.3
Plasma attenuation coefficient, κ	2.6	4.7	7.3	11
Depth of the microwave absorption layer, $l_\omega = 1/\mu_\omega$, 10^{-2} cm	9.1	5.0	3.2	2.2
Energy flux absorbed, S_1, kW/cm^2	0.23	0.35	0.56	1.06
Microwave reflection coefficient, ρ	0.4	0.65	0.76	0.81
Microwave energy flux to sustain plasma, S_0, kW/cm^2	0.38	1.0	2.3	5.6

10.2.7 Numerical Characteristics of Quasi-Equilibrium Microwave Discharge

Numerical data characterizing the generation of a thermal microwave discharge at frequency f = 10 GHz, (λ = 3 cm) in atmospheric pressure air are presented in Table 10.1. These data are based on quasi-equilibrium calculations similar to those discussed previously, but additionally taking into account conduction losses from the plasma filament of 0.3-cm radius and also some smearing of plasma boundaries, which slightly reduces the reflection coefficient.

From the table, it is seen that the depth of microwave energy absorption into plasma l_ω grows significantly with the quasi-equilibrium temperature reduction; at T = 3500 K, it reaches values close to 2 cm. This absorption growth is exponential and determines the lower temperature limit of thermal microwave discharges:

$$l_\omega \propto \frac{1}{n_e} \propto \exp\left(\frac{I_i}{2T}\right) \tag{10.40}$$

At relatively low temperature values, the microwave plasma becomes transparent to electromagnetic microwave radiation; thus, to sustain such a plasma, high power must be provided. Furthermore, a low-temperature and low-conductivity regime is unstable in the same way as with RF discharges (see Section 10.1.8).

The minimal temperature of the thermal microwave discharges can be estimated from the equation $l_\omega(T_{\min}) = R_f$, where R_f is the radius of plasma column or plasma filament. Numerically, the minimum temperature is about 4200 K for the filament radius of 3 mm. The corresponding value of the absorbed energy flux is $S_1 = 0.2$ kW/cm^2; then the minimal value of the microwave energy flux to sustain the thermal plasma is approximately S_0 = 0.2 kW/cm^2.

The maximum temperature value is limited by strong reflection of electromagnetic waves from the plasma at high conductivities. This effect is clearly seen numerically from Table 10.1, where the reflection coefficient reaches the high value of ρ = 81% at T_m = 6000 K. For this reason, the quasi-equilibrium temperature of the microwave plasma in atmospheric pressure air usually does not exceed a level of 5000 to 6000 K.

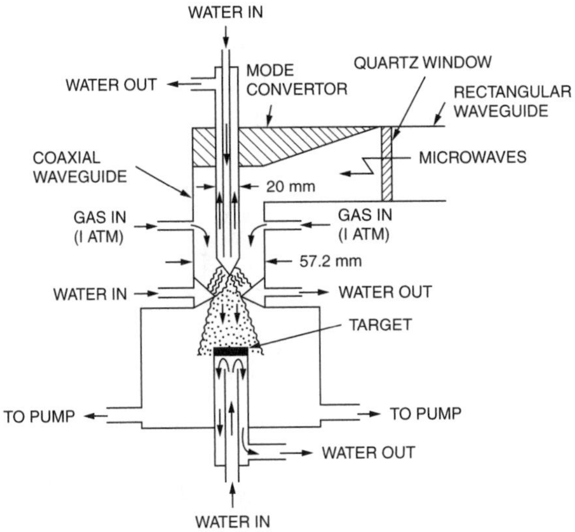

Figure 10.14 Microwave plasma torch[369]

10.2.8 Microwave Plasma Torch and Other Nonconventional Configurations of Thermal Microwave Discharges

The most common configurations of microwave discharges applied to generate high-pressure thermal plasmas, waveguide, and resonator systems have already been analyzed in Section 10.2.2 and Section 10.2.4. Other interesting and practically important concepts of microwave discharges include waveguide mode converters to provide better coupling of electromagnetic microwave radiation with plasma.

As an example, consider the microwave plasma torch developed by Mitsuda et al.[369] shown in Figure 10.14. Microwave power in this system is delivered primarily by a rectangular waveguide. Then, the electromagnetic wave passes through a quartz window into an impedance-matching mode converter, which couples the microwave power into a coaxial waveguide. The coaxial waveguide then operates somewhat as an arc jet: the center conductor of the waveguide forms one electrode and the other electrode comprises an angular flange on the outer coaxial electrode.

Another interesting configuration of the microwave discharge is based on conversion of the rectangular waveguide mode into a special circular waveguide system illustrated in Figure 10.15. This system was effectively used in generating very high-power microwave discharges. More details about different special configurations of microwave discharges can be found in Batenin et al.[367] and MacDonald and Tetenbaum.[370]

10.2.9 Continuous Optical Discharges

Thermal atmospheric pressure plasma can be generated by optical radiation somewhat similar to the way it is generated by electromagnetic waves in the RF and microwave frequency range. These not so conventional, but very interesting, thermal

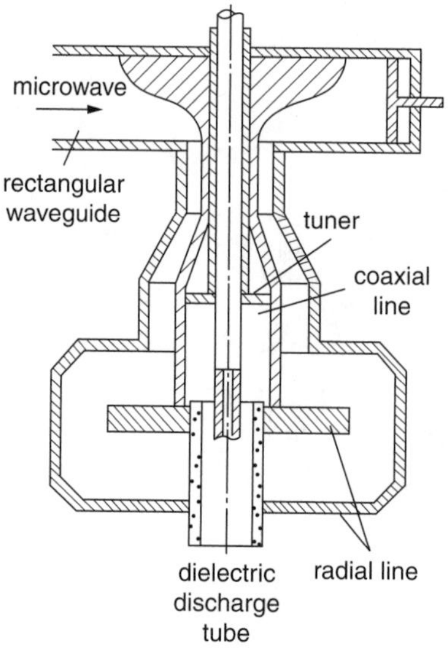

Figure 10.15 Radial and microwave plasmatron

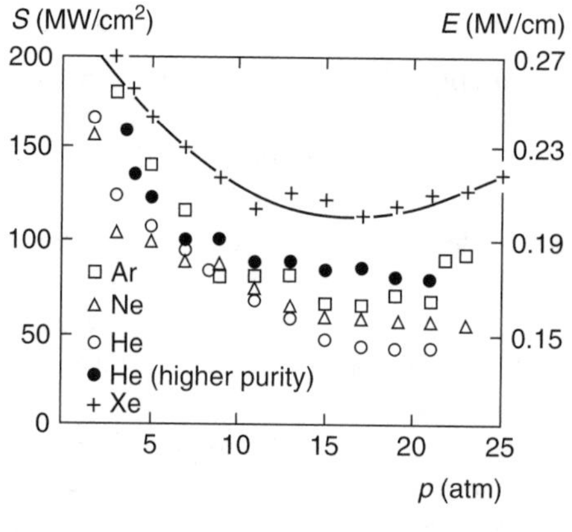

Figure 10.16 Breakdown thresholds of inert gases under CO_2 laser radiation[271]

plasma generators were first theoretically considered by Raizer[371] and then experimentally realized by Generalov et al.[372]

Optical plasma generation began before the 1970s with the discovery of the **optical breakdown** effect.[373] It became possible after development of Q-switched lasers that are able to produce extremely powerful light pulses, the so-called "giant pulses." Detailed consideration of the subject can be found in Raizer;[115] typical breakdown thresholds are presented in Figure 10.16.[271]

The continuous optical discharge is illustrated in Figure 10.17. Usually, a CO_2 laser beam is focused by a lens or mirror to sustain the discharge. Power of the CO_2 laser should be high; for continuous optical discharge in atmospheric air, it must be at least

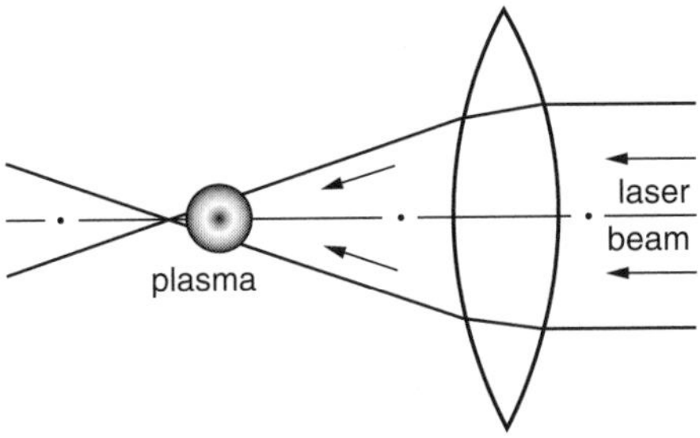

Figure 10.17 Continuous optical discharge

5 kW in the low-divergence beam. To sustain the discharge in xenon at a pressure of a couple of atmospheres, much lower power of approximately 150 W is sufficient.

Light absorption coefficient in plasma significantly decreases with growth of the electromagnetic wave frequency (see Section 6.7.7). For this reason, application of visible radiation requires 100 to 1000 times more power than that of the CO_2 lasers; this is very difficult to provide experimentally. One should also take into account that plasma directly sustained by focusing solar radiation would have temperatures less than the solar temperature, 6000 K, according to the second low of thermodynamics. Equilibrium plasma density at 6000 K is not sufficient to provide effective absorption of the radiation. To provide effective absorption of laser radiation, the plasma density should very high, close to complete ionization. As a result, plasma temperature in the continuous optical discharges is high, usually in the range of 15,000 to 20,000 K.

10.2.10 Laser Radiation Absorption in Thermal Plasma as Function of Gas Pressure and Temperature

Absorption of the CO_2 laser radiation quanta ($\hbar\omega = 0.117$ eV or in temperature units of 1360 K, which is less than the plasma temperature) is mostly provided by the bremsstrahlung absorption mechanism related to electron–ion collisions (see Section 6.7.7). The absorption coefficient $\mu_\omega(CO_2)$ then can be calculated based on Kramer's formula (Equation 6.248 and Equation 6.249) and taking into account induced emission. The resulting numerical formula, including the effect of double ionization,[271] can be presented as:

$$\mu_\omega(CO_2), cm^{-1} = \frac{2.82 \cdot 10^{-29} \cdot n_e(n_+ + 4n_{++})}{(T,K)^{3/2}} \cdot \lg\left(\frac{2700 \cdot T, K}{n_e^{1/3}}\right) \qquad (10.41)$$

All number densities are measured in this relation in cm^{-3}; n_e is the electron concentration and n_+ and n_{++} are densities of single-charged and double-charged positive ions, which can be calculated for the quasi-equilibrium plasma based on the Saha equation.

The absorption coefficient of the CO_2 laser radiation in air is shown in Figure 10.18 as a function of temperature.[101] This figure shows that this function has a maximum at constant pressure, which corresponds to about 16,000 K at atmospheric

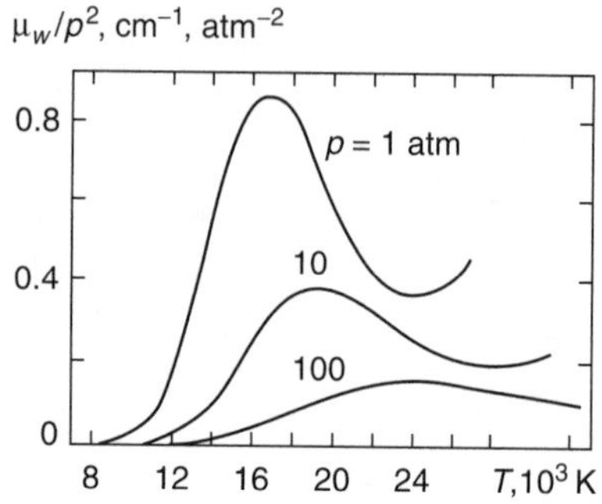

Figure 10.18 Absorption coefficients for CO_2 laser radiation in air[101]

pressure. This temperature corresponds to complete single ionization. Further increase of temperature does not lead to additional ionization (because double ionization requires significantly higher temperatures), while gas density, and thus electron concentration, decrease at fixed pressures. This explains the decrease of the absorption coefficient at higher temperatures that is shown in Figure 10.18 and therefore maximum of the dependence $\mu_\omega(T)$. Taking into account Equation 10.41, the decrease of absorption coefficient $\mu_\omega(T)$ can be estimated as:

$$\mu_\omega \propto \frac{n_e^2}{T^{3/2}} \propto \frac{1}{T^{7/2}} \qquad (10.42)$$

Obviously, at very high plasma temperatures, the absorption coefficient starts growing again because of contribution of the double ionization effect.

As one can see from Figure 10.18, the maximum value of the absorption coefficient μ_ω grows with pressure slightly slower than the square of gas pressure. The maximal absorption coefficient for the CO_2 laser radiation at atmospheric pressure is $\mu_\omega = 0.85\ \text{cm}^{-1}$, which corresponds to a light absorption length $l_\omega = 1.2$ cm. This maximum in the dependence $\mu_\omega(T)$ takes place only in the case of high electromagnetic wave frequencies $\omega^2 >> \nu_e^2$ (see Section 6.6), where ν_e is the frequency of electron collisions (at high ionization degrees typical for the optical discharges, ν_e is related to the electron–ion collisions).

In the opposite case of lower electromagnetic wave frequencies $\omega^2 << \nu_e^2$, the absorption coefficient μ_ω (which is proportional to the electric conductivity σ) continuously grows with temperature. Assuming as before, the major contribution of the electron–ion coulomb collisions into the frequency ν_e and the light absorption coefficient μ_ω dependence on temperature can be expressed as:

$$\mu_\omega \propto \sigma \propto \frac{n_e}{\nu_e} \propto \frac{n_e}{n_+ \bar{v} \sigma_{coulomb}} \propto T^{3/2} \qquad (10.43)$$

In this relation, the coulomb collision cross section $\sigma_{coulomb}$ dependence on temperature $\sigma_{coulomb} \propto 1/T^2$ (see Section 2.1.9) and the average electron velocity dependence on temperature $\bar{v} \propto \sqrt{T}$ were taken into account. It is interesting that the absorption coefficient in this case of low electromagnetic wave frequencies does not depend on gas pressure and density.

10.2.11 Energy Balance of Continuous Optical Discharges and Relation for Plasma Temperature

The energy balance of the continuous optical discharge can be analyzed considering the simplified one-dimensional spherical model of the discharge system. In the framework of this one-dimensional model, plasma is considered as a sphere of radius r_0 and constant temperature T_m (and thus also fixed absorption coefficient μ_ω). This plasma temperature is maintained by absorbing convergent spherically symmetric rays of total power P_0. If the plasma is transparent for the laser radiation, the fraction of the total radiation power absorbed in the plasma can be estimated as $P_1 = P_0\mu_\omega r_0$.

At relatively low laser power levels and not very high pressures, one can neglect radiation losses and balance absorption of the laser radiation with thermal conduction flux J_0:

$$P_1 = P_0\mu_\omega r_0 = 4\pi r_0^2 J_0 \approx 4\pi r_0^2 \frac{\Delta\Theta}{r_0} = 4\pi r_0 \Delta\Theta \tag{10.44}$$

In this relation, $\Delta\Theta = \Theta_m - \Theta_0$ is the drop of the heat flux potential in the plasma (for a definition of the heat flux potential, see Equation 8.22 in Section 8.4.2).

Taking into account that gas is cold at infinity ($\Theta(r \to \infty) = 0$), the heat balance outside the plasma sphere can be expressed as:

$$P_1 = -4\pi r^2 \frac{d\Theta}{dr}, \quad \Theta(r) = \frac{P_1}{4\pi r}, \quad \Theta_0(r = r_0) = \frac{P_1}{4\pi r_0} \tag{10.45}$$

Comparing Equation 10.44 and Equation 10.45, one can see that $\Delta\Theta = \Theta_0$, so $\Theta_m = 2\Theta_0$. From this relation between maximum plasma temperature and the one on the plasma surface, rewrite Equation 10.44 as $P_1 = 2\pi r_0 \Theta_m$. Recalling the previously mentioned formula, $P_1 = P_0\mu_\omega r_0$, the following final energy balance relation, which determines the maximum plasma temperature T_m is obtained:

$$P_0 = 2\pi \frac{\Theta_m(T_m)}{\mu_\omega(T_m)} \tag{10.46}$$

10.2.12 Plasma Temperature and Critical Power of Continuous Optical Discharges

Analyzing Equation 10.46 for plasma temperature, the important conclusion is reached that the power required to maintain the optical discharge$P_0(T_m)$ as a function of plasma temperature has a minimum, T_t (see Figure 10.19).[101] This minimum occurs because the function $\mu_\omega(T)$ has a maximum (see Figure 10.18), while the function $\Theta(T)$ grows continuously. The temperature T_t, corresponding to the minimum

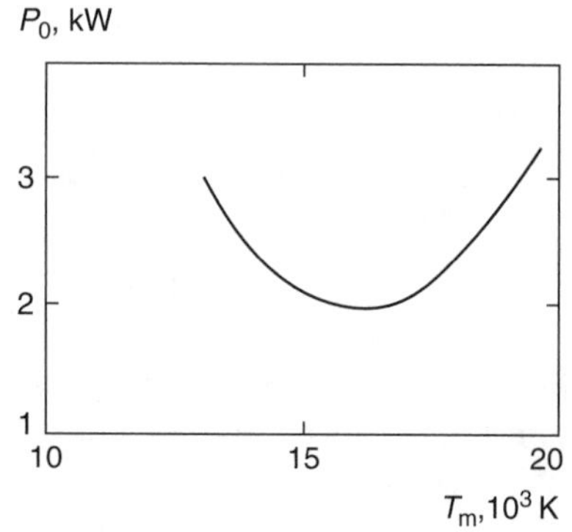

Figure 10.19 Power of spherically convergent laser beam as function of maximum plasma temperature[101]

power P_t, which is necessary to sustain the continuous optical discharge (see Figure 10.19), is close to the temperature corresponding to the maximum of the function $\mu_\omega(T)$ (see Figure 10.18).

Similar to the cases of thermal ICP and microwave discharges (see Section 10.1.8 and remarks after Equation 10.40), the left low-temperature and low-conductivity branch of the curve $P_0(T)$ in Figure 10.19 where $T_m < T_t$ is unstable. If the temperature is lower than the threshold temperature $T_m < T_t$, a small temperature increase due to any fluctuation results in a decrease of the required power necessary to sustain the stationary discharge with respect to the actual laser radiation power. This would lead to further temperature growth until it reaches the threshold temperature, $T_m = T_t$.

Equation 10.46 permits an accurate calculation of the critical minimum value of the laser radiation necessary to sustain the continuous optical discharge. Obviously, in this case, plasma temperature should be taken as the critical one, T_l:

$$P_t = P_{0\min} = 2\pi \frac{\Theta_m(T_m = T_t)}{\mu_\omega(T_m = T_t)} \tag{10.47}$$

For example, CO_2 laser radiation in atmospheric air has the maximum value of the absorption coefficient $\mu_{\omega\max} \approx 0.85\,cm^{-1}$ at the threshold temperature T_l = 18,000 K. In this case, the heat flux potential is $\Theta(T_m = T_t) \approx 0.3$ kW/cm, and the minimal threshold value of power necessary to sustain the discharge is $P_t \approx 2.2$ kW.

Again, it should be noted that the thermal plasma temperatures in continuous optical discharges are usually higher (about twice) than those of ICP and arc discharges, and significantly higher than in microwave discharges (about three to four times). This is related to the fact that $\mu_\omega \propto 1/\omega^2$ and plasma is actually transparent for the optical radiation. Only very high temperatures corresponding to almost complete ionization provide the level of absorption sufficient to sustain the discharges.

The minimal threshold value of the laser power P_t in accordance with Equation 10.47 decreases quite fast with pressure growth because of the corresponding significant growth of the absorption coefficient $\mu_{\omega\ \max}$. Also, the minimum threshold

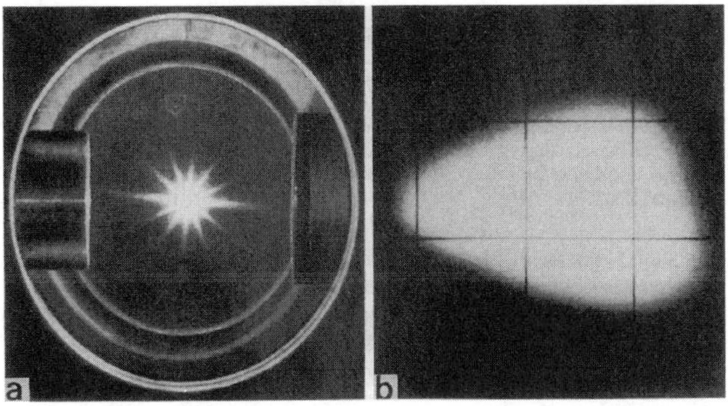

Figure 10.20 Continuous optical discharge: (a) general view; (b) large image (beam travels from right to left)[372]

value of the laser power is lower in gases with lower ionization potential (to decrease T_t) and lower thermal conductivity (to decrease Θ_m). For example, the minimum CO_2 laser power necessary to sustain the continuous optical discharge in Ar and Xe at pressures of about 3 to 4 atm is 100 to 200 W.

Additional details regarding the fundamentals of optical discharges can be found in several books by Raizer.[101,115,271] Photographs of this very interesting discharge phenomenon are shown in Figure 10.20.[372]

10.3 NONEQUILIBRIUM RF DISCHARGES: GENERAL FEATURES OF NONTHERMAL CCP DISCHARGES

10.3.1 Nonthermal RF Discharges

High-pressure continuous RF discharges were considered earlier in Section 10.1. Most of them generate quasi-equilibrium thermal plasma, in which electromagnetic fields provide only a heating effect necessary to sustain the high quasi-equilibrium temperature resulting in thermal ionization.

At lower pressures, RF plasma becomes essentially nonequilibrium. Electron–neutral collisions are less frequent at lower pressures while gas cooling by discharge walls is more intensive. Together, these lead to electron temperatures much exceeding those of neutrals somewhat similar to the case of glow discharges considered in Chapter 7. Behavior of the nonthermal RF discharges is physically determined not only by temperatures, but also by many other essentially nonequilibrium parameters, making these nonequilibrium systems more difficult to analyze. On the other hand, they have attracted significant interest recently because of large-scale applications in electronics and other technologies related mostly to high precision surface treatment. In this section and following sections, nonthermal RF discharges will be considered along the lines of (1) classification and general features of CCP discharges; (2) moderate-pressure CCP; (3) low-pressure CCP discharges; and (4) ICP discharges.

Specifying the frequency range typical for RF discharges, one should note that the upper frequency limit is related here to wavelengths close to the system sizes (electromagnetic waves with smaller wavelengths are usually referred to as microwaves). Lower RF frequency limit is related to characteristic frequency of ionization and ion transfer. Ion density in RF discharge plasmas and sheaths usually can be considered as constant during a period of electromagnetic field oscillation. Thus, RF frequencies typically exceed 1 MHz, although sometimes they can be smaller. As mentioned in Section 10.1, the most industrially used radio frequency for plasma generation is 13.6 MHz.

The nonthermal RF discharges can be subdivided into those of moderate (or intermediate) pressure and those of low pressure (possibility of organization of the RF nonthermal CCP discharges at high pressures was considered in Section 10.1.11). The discharges are usually referred to as those of moderate pressure if they are nonequilibrium, but the electron energy relaxation length is small with respect to all characteristic sizes of the discharge system. In this case, the electron energy distribution function (and therefore ionization, excitation of neutrals, and other elementary processes) is determined by local values of electric field. Usually, this occurs in the pressure range from 1 to 100 torr. The moderate pressure nonequilibrium CCP discharges can be effectively used in gas lasers, which stimulated their active research starting from the late 1960s.[374]

In the opposite case of low-pressure discharges, the electron energy relaxation length is small with respect to characteristic sizes of the discharge system. Electron energy distribution function is then determined by electric field distribution in the entire discharge zone. This discharge regime takes place at low pressures p and small characteristic discharge sizes L (for example, in inert gases it requires $p(\text{torr}) \cdot L(\text{cm}) < 1$). Low-pressure, strongly nonequilibrium RF discharges are widely applied today in electronics in etching and chemical vapor deposition (CVD) technologies. This drives the extremely high interest paid today to these discharge systems (see, for example, Lieberman and Gottscho[375]).

10.3.2 Capacitive and Inductive Coupling of the Nonthermal RF Discharge Plasmas

Similar to the case of thermal discharges (see Section 10.1), the nonthermal plasma of RF discharges can be capacitively coupled (CCP) or inductively coupled (ICP). These two ways of organization of nonthermal RF discharges are illustrated in Figure 10.21.

CCP discharges provide the electromagnetic field by means of electrodes located either inside or outside the discharge chamber, as is shown in Figure 10.21(a, b). These discharges primarily stimulate the electric field, which obviously facilitates their ignition. The inductive coil induces the electromagnetic field in ICP discharges; the discharge can be located inside the coil (Figure 10.21c) or adjacent to the plane (or quasi-plane) coil (Figure 10.21d). These discharges primarily stimulate magnetic fields, while corresponding nonconservative electric fields necessary for ionization are relatively low. For this reason, the nonthermal ICP plasma discharges are usually

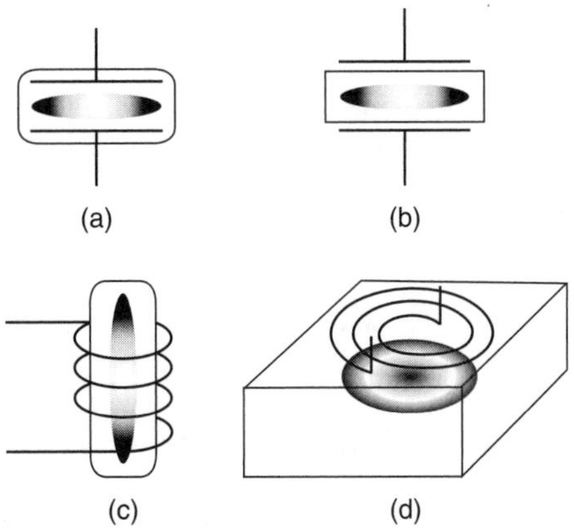

Figure 10.21 Capacitively and inductively coupled RF discharges

organized at low pressures, when the reduced electric field E/p is sufficient for ionization.

It should be noted that the inductive coils generate not only the nonconservative (vortex) electric fields, but also the conventional conservative (potential) electric field in the gap between turns (windings) of the coil. This potential electric field exceeds the nonconservative one by a fairly big factor equal to ratio of length of a turn (winding) to the distance between them. These relatively high but local electric fields (capacitive component in inductive discharge) can be quite important, especially during breakdown, unless a special electric screen is installed.

Coupling between the inductive coil and plasma inside can be interpreted as a transformer where the coil represents the primary windings and the plasma the secondary windings. A coil as primary windings consists of many turns while a plasma has only one. Thus, ICP discharges can be considered as a voltage-decreasing transformer. Effective coupling with RF power supply requires low plasma resistance. As a result, the ICP discharges are convenient to reach high currents, high electric conductivity, and high electron concentration. In contrast, the CCP discharges are more convenient to provide higher values of electric fields.

10.3.3 Electric Circuits for Inductive and Capacitive Plasma Coupling with RF Generators

RF power supplies for plasma generation typically require active load of 50 or 75 ohm. To provide effective correlation between resistance of the leading line from the RF generator and the discharge impedance, a special coupling circuit should be applied. A general schematic of such circuits for CCP and ICP discharges is shown in Figure 10.22. It is important that the RF generator not only be effectively correlated with the RF discharge during its continuous operation, but also initially provide sufficient voltage for breakdown to start the discharge. To provide the effective ignition of the nonthermal RF discharge, the coupling electric circuit should form

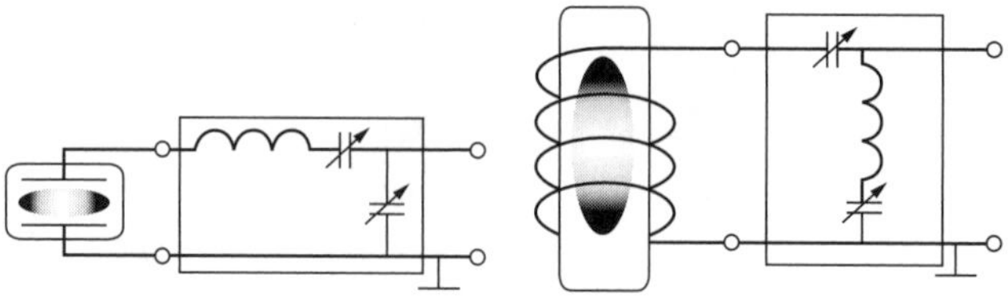

Figure 10.22 Coupling circuits for (a) CCP and (b) ICP discharges

the AC current resonance being in series with generator and discharge system during its idle operation.

For this reason, the coupling circuit in the case of CCP discharge includes inductance in series with a generator and discharge system (see Figure 10.22A). Variable capacitance is included there as well for adjustment because the design of variable inductances is more complicated. In the case of ICP discharge, the only variable capacitance, located in the electric coupling circuit in series with generator and idle discharge system, provides the necessary effect of initial breakdown.

10.3.4 Motion of Charged Particles and Electric Field Distribution in Nonthermal RF CCP Discharges

The external electric circuit is able to provide fixed voltage amplitude between electrodes or fixed current amplitude in CCP discharges. One of these regimes can be chosen, varying the relation between the discharge impedance and resistance of RF generator and elements of the previously discussed electric coupling circuit. The fixed current regime is, however, the more typical choice in practical application.

The fixed current regime can be represented simply by an electric current with density $j = -j_0 \sin(\omega t)$ flowing between two parallel plane electrodes, which implies that the electrodes' sizes exceed distance between them. For the following analysis, it can be assumed that the RF plasma density $n_e = n_i$ is high enough and the electron conductivity current in plasma exceeds the displacement current (see Section 6.6.1 and Section 6.6.2):

$$\omega << \frac{1}{\varepsilon_0}|\sigma_e| \equiv \frac{1}{\tau_e} \tag{10.48}$$

In this relation, τ_e is the Maxwell time for electrons, which characterizes (at $\omega << \nu_e$, ν_e is electron collisional frequency) the time interval necessary for the charged particles to shield the electric field; σ_e is the complex electron conductivity (see Section 6.6.1 and Section 6.6.2):

$$\sigma_e = \frac{n_e e^2}{m(\nu_e + i\omega)} \tag{10.49}$$

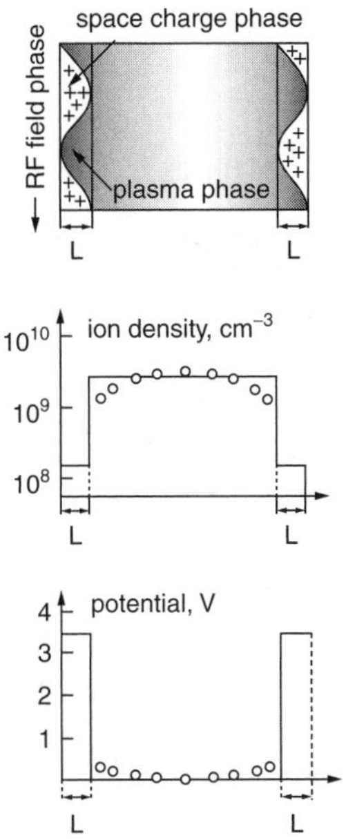

Figure 10.23 Illustration of ion density and potential distribution, and boundary layer oscillation in RF CCP discharge

An inequality opposite to the preceding equation usually takes place in the RF discharges for ions because of their much higher mass $M >> m$:

$$\omega >> \frac{1}{\varepsilon_0}\left|\sigma_i\right| \equiv \frac{1}{\tau_i}, \quad \sigma_i = \frac{n_i e^2}{M(\nu_i + i\omega)} \tag{10.50}$$

The inequality (10.50) means that the ion conductivity current can be neglected with respect to displacement current. In other words, the ion drift motion during an oscillation period of electric field can be neglected.

This leads to the following picture of electric current and the motion of charged particles in the RF discharges. Ions form the "skeleton" of plasma and can be considered at rest, while electrons oscillate between electrodes as shown in Figure 10.23. Electrons are present in the sheath of width L near an electrode only for a part of the oscillation period called the plasma phase. Another part of the oscillation period, when no electrons are in the sheath, is called the space charge phase.

The oscillating space charge shown in Figure 10.23 creates an electric field that forms the displacement current and closes the circuit. The electric field of the space charge has a constant component in addition to an oscillating component, which is directed from plasma to the electrodes. For this reason, similar to glow discharges, the quasi-neutral plasma zone is also called the positive column. However, the constant component of the space charge field provides much faster ion drift to the electrodes than in the case of ambipolar diffusion. As a result, ion density in the space charge layers near electrodes is lower with respect to their concentration in plasma (see Figure 10.23).

10.3.5 Electric Current and Voltages in Nonthermal RF CCP Discharges

Assume that the RF discharge is symmetrical and ion concentrations in plasma and sheaths are fixed and equal to n_p and n_s, respectively (see Figure 10.23). Then the one-dimensional discharge zone can be divided into three regions:

1. A plasma region with thickness L_p that is quasi neutral throughout the oscillation period. Electric current in this region is provided by electrons and conductivity is active at relatively low frequencies $\omega < \nu_e$.
2. Sheath regions in plasma phase, where the electric conductivity is also active and provided by electrons.
3. Sheath regions with space charge. Taking into account the discharge symmetry and constancy of charge concentration in the sheath, one can conclude that total thickness of the space charge region is always equal to the total thickness of the sheath region in plasma phase and equals the sheath size L (see Figure 10.23).

If total thickness of the sheath region in plasma phase is equal to L, then active resistance of the region can be expressed as:

$$R_{ps} = \frac{Lm\nu_e}{n_s e^2 S} \tag{10.51}$$

where ν_e is the frequency of electron collisions and S is the area of an electrode. Similarly, resistance of the quasi-neutral plasma region can be presented as:

$$R_p = \frac{L_p m\nu_e}{n_p e^2 S} \tag{10.52}$$

As one can see from Equation 10.49, the reactive (imaginary) resistance component of plasma itself and sheaths in the plasma phase also should be taken into account at high frequencies of electromagnetic field $\omega > \nu_e$:

$$X_{p,ps} = iL_{p,s} \frac{m\omega}{n_{p,s} e^2 S} = i\omega L^{(e)}_{p,s} \tag{10.53}$$

This impedance has the inductive nature, and factors $L^{(e)}_{p,s}$ are the effective inductance of plasma zone (p) and sheath (s) in plasma phase, respectively.

The voltage drop on a space charge sheath is proportional to the electric field near an electrode E and to the instantaneous size of the sheath $d(t)$. Assuming the constancy of ion concentration in the sheath, the voltage drop is actually proportional to the square of its instantaneous size:

$$U_r = \frac{1}{2} E_r d_r(t) = \frac{j_0}{4\varepsilon_0 \omega} L[1+\cos(\omega t)]^2 \quad (10.54)$$

Taking into account that the sheaths' half-size $L/2$ can be interpreted as the amplitude of an electron oscillation $J_0/en_s \cdot 1/\omega$, the expression (Equation 10.54) for the voltage drop on the "right-hand" sheath (subscript r) can be rewritten as:

$$U_r = \frac{j_0^2}{2\varepsilon_0 \omega^2 en_s}[1+\cos(\omega t)]^2 \quad (10.55)$$

The voltage drop on the space charge sheath near the opposite "left-hand" sheath (subscript l) is in counterphase with the voltage:

$$U_l = -\frac{j_0^2}{2\varepsilon_0 \omega^2 en_s}[1-\cos(\omega t)]^2 \quad (10.56)$$

The total voltage U_s related to the space charge sheaths obviously can be presented as a sum of voltages corresponding to both electrodes:

$$U_s = U_r + U_l = \frac{2j_0^2}{\varepsilon_0 \omega^2 en_s}\cos(\omega t) = \frac{j_0}{\varepsilon_0 \omega} L\cos(\omega t) \quad (10.57)$$

Analyzing Equation 10.57, the following two important conclusions regarding the space charge sheaths are reached:

1. Total voltage drop on the space charge sheaths includes only the principal harmonic (ω) of the applied voltage, while the voltage drop on each sheath separately contains constant component and second harmonics (2ω).
2. Taking into account that the electric current density was defined in Section 10.3.4 as $j = -j_0 \sin(\omega t)$, one can see that the phase shift between voltage and current in the space charge sheath corresponds to the capacitive resistance.

10.3.6 Equivalent Scheme of a CCP RF Discharge

As seen from the current–voltage relation (10.57), the space charge sheath can be interpreted by the equivalent capacitance:

$$C_s = \frac{\varepsilon_0 S}{L} \quad (10.58)$$

This capacitance corresponds to the capacitance of the vacuum gap with the width equal to the total size of the space charge sheaths L. If the total size of the space charge sheaths L does not depend on the discharge current, then the equivalent capacitance (Equation 10.58) is constant and the discharge circuit can be considered a linear one.

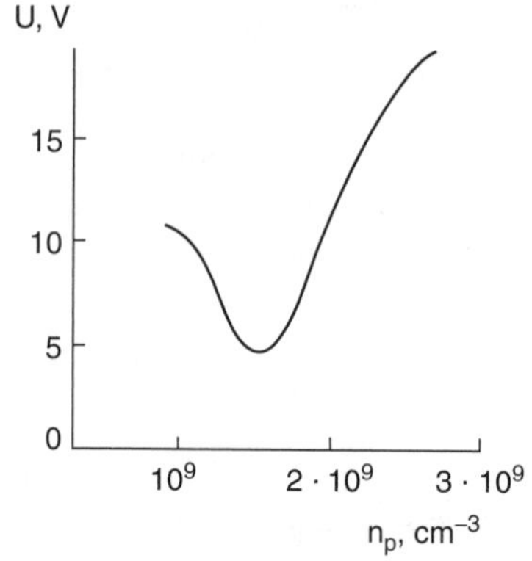

Figure 10.24 Discharge voltage (Hg, 1.2 mtorr) as a function of plasma density related to discharge current (power grows from left to right)[376]

Considering that the impedance of a plasma (width L_p) has an inductive component (Equation 10.53) and the impedance of the space charge layers (width L) is capacitive, resonance in the discharge circuit related to their connection in series is possible on the frequency:

$$\omega = \omega_p \sqrt{L/L_p} \tag{10.59}$$

In this relation, ω_p is the plasma frequency corresponding to the quasi-neutral plasma zone of the discharge. Typical discharge voltage dependence on electron concentration in the plasma zone illustrating this resonance is shown in Figure 10.24.[376]

As mentioned earlier, the constant component of the space charge field is stronger than the electric field in the plasma zone (see Figure 10.23). For this reason, even a small ionic current can lead to an energy release in the sheath exceeding that in the plasma zone. Discharge power per unit electrode area, transferred this way to the ions, can be estimated as a product of the ion current density determined at the mean electric field in the sheath:

$$j_i = en_s v_i = en_s b_i \frac{j_0}{\varepsilon_0 \omega} \tag{10.60}$$

and the constant component of the voltage (Equation 10.55). Here, v_i and b_i are the ions' drift velocity and mobility. Thus, the discharge power per unit electrode area related to ionic drift in a sheath can be expressed based on Equation 10.55 and Equation 10.60 as:

$$P_i = \frac{3}{4\varepsilon_0^2} \frac{j_0^3 b_i}{\omega^3} S \tag{10.61}$$

As seen from Equation 10.61, the ionic power P_i is proportional to the cube of the discharge current, inversely proportional to the gas pressure (because of $b_i \propto 1/p$), and proportional to the cube of frequency; it does not depend at all on the size of the sheath.

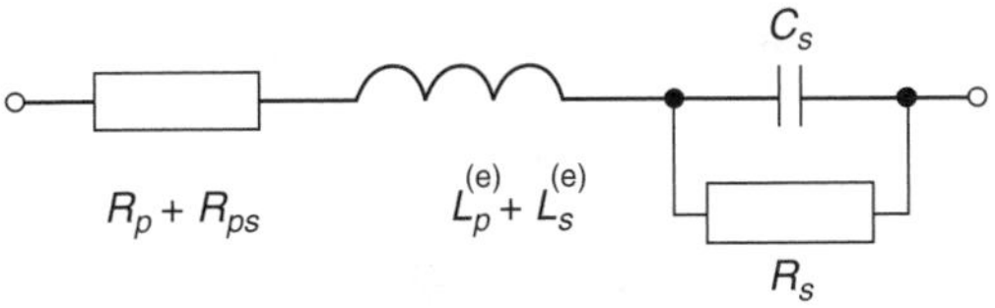

Figure 10.25 Equivalent circuit of a capacitively coupled RF discharge

The discharge power transfer to ions in a sheath can be presented in the equivalent scheme by resistance R_s connected in parallel to the capacitance of sheaths. The value of this resistance is determined by voltage drop on the capacitance and power released in both sheaths:

$$R_s = \frac{U_s^2/2}{2P_i} = \frac{1}{3}\frac{\omega L^2}{j_0 b_i S} \tag{10.62}$$

Thus, the total equivalent scheme of a capacitively coupled RF discharge is presented in Figure 10.25. The scheme includes the resistance, R_s, and the capacitance, C_s, of sheaths connected in parallel to each other and then connected in series to the active resistance of sheaths in plasma phase R_{ps} and of plasma itself R_p. At relatively low pressures, when $\omega > \nu_e$, the inductance of sheaths in plasma phase $L_p^{(e)}$ and of plasma itself $L_s^{(e)}$ (Equation 10.53) should also be taken into account in the equivalent scheme.

10.3.7 Electron and Ion Motion in CCP Discharge Sheaths

The drift oscillation of the electron gas "around ions" and thus oscillation of the boundary between plasma zone and sheath zone is harmonic only if the ion concentration in the sheath is assumed to be uniform (see Figure 10.23). Taking into account the nonuniformity of the ion concentration, the boundary motion is not harmonic if $j = -j_0 \sin(\omega t)$. For example, in the area of elevated ion density, the boundary should move more slowly to provide the same change of the space charge field and thus the same current. The equation for the plasma boundary motion can be expressed in this case as:

$$\sin z \cdot \frac{dz}{dx} = \frac{\omega e}{j_0} n_i(x) \tag{10.63}$$

where $z(x)$ is the phase of the plasma boundary in the position x (see Figure 10.23). The boundary of the zone where plasma is always quasi neutral corresponds to $z = 0$, and the electrode location corresponds in this case to phase $z = \pi$. As can be easily derived from Equation 10.63, the velocity of the plasma boundary is inversely proportional to the ion concentration $n_i(x)$. In this case, the electric field in any point of sheath in the space charge phase can be calculated by the following simple relation:

$$E(x,t) = \frac{j_0}{\varepsilon_0 \omega}\left[\cos(\omega t) - \cos z\right] \tag{10.64}$$

Time integration of $E(x,t)$ gives the analytical expression for the constant component of the space charge field in the sheath:

$$\langle E(x) \rangle = \frac{1}{\pi}\int_0^{z(x)} E(x,t)\omega dt = \frac{j_0}{\omega \pi \varepsilon_0}(\sin z - z\cos z) \tag{10.65}$$

This constant component of the electric field determines the motion of ions in the sheath, which, however, also even qualitatively depends on pressure. At relatively high pressures and low values of the reduced electric field, the ion drift velocity is proportional to the previously determined constant component of electric field $\langle E(x) \rangle$:

$$v_i = b_i \langle E(x) \rangle \tag{10.66}$$

where b_i is the ion mobility. At higher values of reduced electric field, ion energies received on the mean free path become comparable with their thermal energies. The ion drift velocity in this case becomes proportional to the square root of the electric field:

$$v_i = \sqrt{\frac{e\lambda_i\sqrt{2}}{M}\langle E(x) \rangle} \tag{10.67}$$

Here, λ_i is the mean free path of an ion in the sheath and M is the ionic mass. It is important to note that the relation for ion velocity in form (10.67) is the most typical one for the case of low-pressure CCP discharges.

Finally, in the case of very low pressures, the ionic motion can be considered completely collisionless. The ion velocities at such pressures are determined by potential drop U corresponding to the ionic motion:

$$v_i = \sqrt{2eU/M} \tag{10.68}$$

Analysis of the ion density distribution in sheaths is an important step in describing and therefore controling nonthermal radio-frequency CCP discharges. This distribution can be found in general from the continuity equation, taking into account the ions' drift and diffusion as well as ionization and recombination processes:

$$\frac{\partial n_i}{\partial t} + \nabla\left[-D_a \nabla n_i + n_i v_i\right] = I(n_i, \langle E \rangle) - R(n_i, \langle E \rangle) \tag{10.69}$$

In this equation, D_a is the coefficient of ambipolar diffusion; $I(n_i, \langle E \rangle)$ is the ionization rate; and $R(n_i, \langle E \rangle)$ is the recombination rate.

One should take into account that the constant component of the electric field of space charge in the the sheath (Equation 10.64) is quite high, and its contribution to the ionic motion is more significant than the one related to temperature (which is T_e for ambipolar diffusion). For this reason, the contribution of diffusion in the continuity equation is important only in the vicinity of the plasma zone. Otherwise, the ionic drift in the space charge field mostly determines the motion of ions in the sheath and contributes most in the continuity equation.

The continuity equation combined with the preceding relations for the ion drift velocity can be solved to determine the ion density distributions in the CCP sheaths. However, sometimes simple estimations assuming a constant ion concentration in sheaths (see Figure 10.23) are sufficient for good qualitative analysis. Further consideration requires specification of gas pressure in the discharge. As was already mentioned, the physics of CCP discharges (as well as practical applications) is qualitatively different for the cases of moderate and low pressures, which will be considered in the following two sections.

10.4 NONTHERMAL CCP DISCHARGES OF MODERATE PRESSURE

10.4.1 General Features of Moderate-Pressure CCP Discharges

Moderate-pressure CCP discharges can be defined as those in which the energy relaxation length λ_ε is less than the typical sizes of the plasma zone and sheaths:

$$\lambda_\varepsilon < L, L_p \tag{10.70}$$

The moderate-pressure range numerically corresponds to the interval 1 to 100 torr. In this pressure range:

$$\omega < \delta \nu_e \tag{10.71}$$

where δ is the average fraction of electron energy lost per one electron collision and ν_e is the frequency of the electron collisions. The inequality (Equation 10.71) means that electrons lose (and gain) energy during a time interval shorter than the period of electromagnetic RF oscillations.

Electron energy distribution functions and thus ionization and excitation rates are determined in the moderate-pressure discharges by local and instantaneous values of the electric field. In particular, this results in important contributions of ionization in the sheaths, where electric fields have maximum values and are able to provide maximum electron energies. An important peculiarity of the moderate-pressure CCP discharges is the possibility to sustain them in two qualitatively different forms, referred to as the α-discharge and the γ-discharge.[377]

10.4.2 The α- and γ-Regimes of Moderate-Pressure CCP Discharges: Luminosity and Current–Voltage Characteristics

The main experimentally observed differences between the α- and γ-regimes of the moderate-pressure CCP discharges are related to electric current density and luminosity distribution in the discharge gap. Pictures of the two regimes of a moderate-pressure

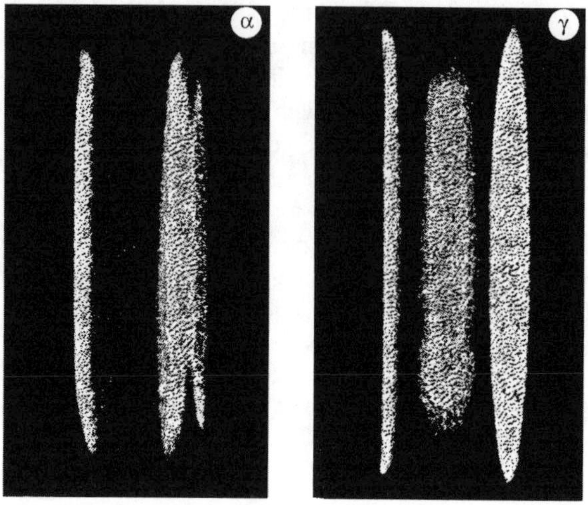

Figure 10.26 α- and γ-ICP discharges (air, 10 torr, 13.56 MHz, 2 cm). (From Raizer, Y. P. et al. *Radio Frequency Capacitive Discharges*. Moscow: Nauka Science)

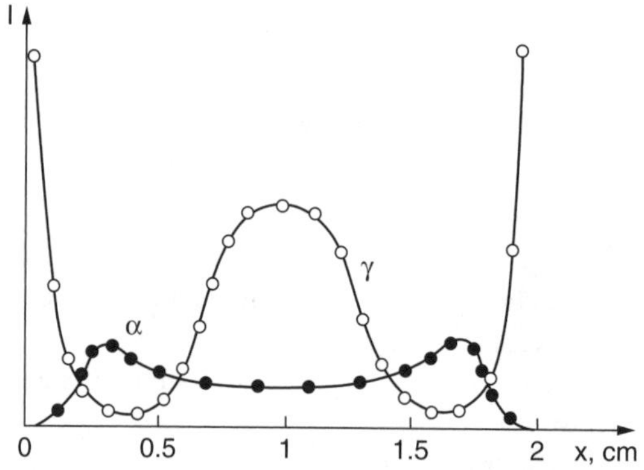

Figure 10.27 Emision intensity distribution along discharge gap (air, 10 torr, 13.56 MHz, 2 cm)[378]

CCP discharge are presented in Figure 10.26; related luminosity distribution across the discharge gap is shown in Figure 10.27.[378]

As is seen from the picture, the α-discharge can be characterized by low luminosity in the plasma volume. Brighter layers are located closer to electrodes, but layers immediately adjacent to the electrodes are dark in this case. The γ-discharge takes place at much higher values of current density. In this regime, the discharge layers immediately adjacent to the electrodes are very bright, but relatively thin. The plasma zone in also luminous and separated from the bright electrode layers by dark layers similar to the Faraday dark space in the DC glow discharges (see Section 7.1.3).

Key differences in physical characteristics of the α- and γ-discharges are related to the fact that the contribution in the γ-regime ionization processes is due to

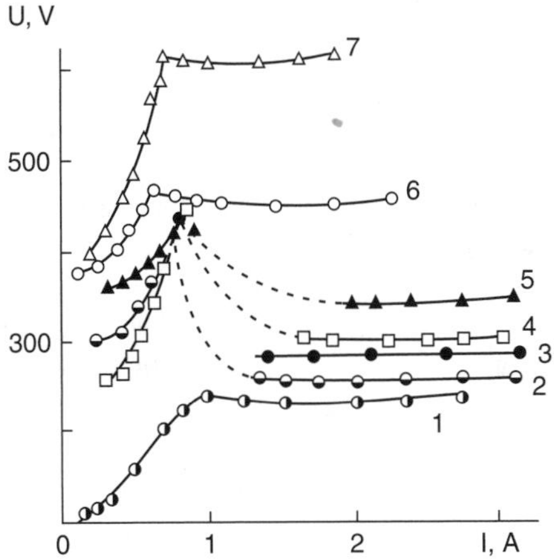

Figure 10.28 Current–voltage characteristics of CCP discharge, 13.56 MHz. 1 = He (p = 30 torr; L_0 = 0.9 cm); 2 = air (30 torr, 0.9 cm); 3 = air (30 torr, 3 cm); 4 = air (30 torr, 0.9 cm); 5 = CO_2 (15 torr, 3 cm); 6 = air (7.5 torr, 1 cm); 7 = air (7.5 torr, 1 cm).[374]

ionization provided by secondary electrons formed on the electrode by the secondary electron emission and accelerated in a sheath (see Section 7.1.3 and Section 8.2.5). This ionization mechanism is similar to that in the cathode layers of glow discharges (see Section 7.2). This is the reason why the main features of the sheaths in γ-discharges are similar to those of cathode layers.

The α- and γ-discharges operate at normal current density in the same manner as glow discharges (see Section 7.2.3). This means that an increase of the discharge current is provided by a growth of the electrode area occupied by discharge, while the current density remains constant. The normal current density in both regimes is proportional to the frequency of electromagnetic oscillations and, in the case of γ-discharges, it exceeds the current density of α-discharges more than 10 times.

Current–voltage characteristics of the moderate-pressure CCP discharge at different conditions are shown in Figure 10.28.[374] One can see the qualitative change of the current–voltage characteristics related to lower and higher levels of the RF discharge current. Currents less than 1 A correspond on this characteristic to the α-discharges; higher currents are related to the γ-discharges.

Normal α-discharge can be observed on curves 2 and 4 of the figure, where there is no essential change of voltage at low currents. Most of curves, however, show that an α-discharge is abnormal just after breakdown. The current density is so small in this regime that the discharge occupies all the electrode immediately after breakdown, which leads to voltage growth together with current (abnormal regime; see Section 7.2.6).

Transition from the α-regime to the γ-regime takes place when the current density exceeds some critical value, depending on pressure, frequency, electrode material, and type of gas. The α–γ transition is accompanied by a discharge contraction and by more than an order of magnitude growth of the current density.

Increase of current in the γ-regime after transition does not change the voltage and the discharge remains in the normal regime.

10.4.3 The α-Regime of Moderate-Pressure CCP Discharges

Two mechanisms provide ionization in sheaths. One is related to ionization in the plasma phase, when the sheath is filled with electrons; this mechanism dominates in α-discharges. The term "α-discharge" by itself reminds one of the influence of the first Townsend coefficient, α, on this ionization process (see Section 7.1.5). The second mechanism is related to ionization provided by secondary electrons produced as a result of secondary electron emission from electrodes. This mechanism dominates in γ-discharges. The term "γ-discharge" by itself shows the important contribution of the third Townsend coefficient γ (see Section 7.1.5 and Section 8.2.5) on this ionization process.

The ionization rate in the α-regime can be calculated using the approximation (Equation 4.141) for the first Townsend coefficient, and taking into account that the electric field and electron density are related in the plasma phase by Ohm's law: $j = n_e e b_e E$, where b_e is the electron mobility. Thus the ionization rate in the α-regime can be expressed as:[376]

$$I_\alpha(x,t) = Ap\frac{j_0}{e}|\sin\omega t|\exp\left(-\frac{n_e}{n_0}\frac{1}{\sin\omega t}\right) \tag{10.72}$$

where the concentration parameter $n_0 = j_0/eb_eBp$ (the concentration parameter n_0 does not depend on pressure because the electron mobility $b_e \propto 1/p$). This ionization rate grows exponentially with the instantaneous value of discharge current and also with a decrease of the charge density (this leads to a monotonic decrease of ion concentration from plasma boundary to electrode). The exponential increase of $I_\alpha(x,t)$ at lower levels of electron concentration permits neglecting the recombination effect in sheaths.

Motion of the plasma boundary and corresponding time dependence of the ionization rate in different points of the sheath are illustrated in Figure 10.29. Coordinate x is calculated in this case from the quasi-neutral plasma zone (see Figure 10.23), where $x = 0$ and phase $z = 0$, to the electrode, where $x = 2x_0$ and the phase $z = \pi$. The evolution of the ionization rate in three different points of the sheath presented in Figure 10.29 can be interpreted as follows:

1. Near the plasma zone, where the coordinate $x < x_0$ and phase $z < \pi/2$: the ionization rate of the α-discharge I_α reaches a maximum value when the current is in maximum. This takes place when $\omega t = \pi/2$ and $\omega t = 3\pi/2$.
2. In the midpoint of the sheath, where the coordinate $x = x_0$ and phase $z = \pi/2$: electrons appear only when the electric field in the plasma has already reached its maximum. That is the reason the ionization peak shown in Figure 10.29 is vertical (actual width is about the Debye radius). A similar phenomenon, but reverse in time, takes place at the phase $z = 3\pi/2$.

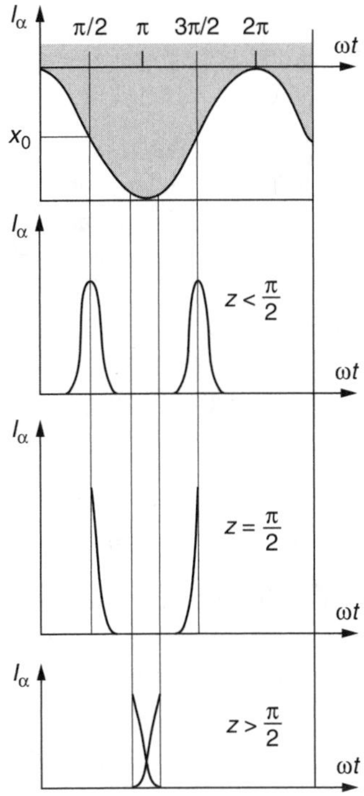

Figure 10.29 Plasma boundary motion and time dependence of ionization in different shift positions

3. Closer to the electrode, where the coordinate $x > x_0$ and phase $z > \pi/2$: electrons are absent in the moment of maximum discharge current; for this reason, the maximum of ionization is shifted until $\omega t = z$ (when electrons appear). Taking into account the strong I_α dependence on electric field, it is clear that the ionization rate significantly decreases near the electrode ($x > x_0$).

Thus, most ionization in the moderate pressure α-discharge takes place near the boundary of the plasma zone $0 < x < x_0$. This explains the picture of luminosity distribution shown in Figure 10.26 and Figure 10.27. If $n_e >> n_0$ in Equation 10.72, then the ionization maximum peak is very narrow.

10.4.4 Sheath Parameters in α-Regime of Moderate-Pressure CCP Discharges

The sheath parameters in the α-regime can be estimated based on the following simple assumptions: the ion concentration in the sheath n_s is constant; the ionization rate follows Equation 10.72 at $z < \pi/2$; and ionization is absent near the electrode $I_\alpha = 0$ at $z > \pi/2$. Balancing the ion flux to the electrode (Equation 10.60) and the ionization rate in the sheath $I_\alpha \cdot L/2$, the following relation for the ion density in the sheath can be derived:

$$n_s = n_0 \ln\left[\frac{ABp^2\varepsilon_0}{e}\sqrt{\frac{2}{\pi}\frac{b_e}{b_i}\frac{1}{n_0}\left(\frac{n_0}{n_s}\right)^{5/2}}\right] = n_0 \ln\Lambda \tag{10.73}$$

In this relation, A and B are parameters of the Townsend relation for the α-coefficient (Equation 4.141); n_0 is the pressure-independent concentration parameter introduced previously regarding Equation 10.72; p is the gas pressure; b_e and b_i are the electron and ion mobilities; and $\ln\Lambda$ is a logarithmic factor only slightly changing with other system parameters.

Taking into account that $n_0 = j_0/eb_eBp$, one can conclude that the ion concentration in the sheath is proportional to the discharge current, does not depend on frequency, and only logarithmically depends on gas pressure. Then the sheath size can be determined as:

$$L = \frac{2j_0}{\omega n_s} = \frac{2eb_eBp}{\omega\ln\Lambda} \tag{10.74}$$

From the preceding equation, it is seen that the sheath size in α-regime of the moderate-pressure CCP discharge is inversely proportional to the electromagnetic oscillation frequency and does not depend on the discharge current density.

10.4.5 The γ-Regime of Moderate-Pressure CCP Discharges

As shown previously, the electric field of the α-discharge in the space charge phase grows proportionally with current density. Also, the multiplication rate of the γ-electrons (formed as a result of the secondary electron emission from electrodes) increases exponentially with the electric field in the sheath. As a result, reaching a critical value of the current density leads to the "breakdown of the sheath" and transition to the γ-regime of the CCP discharge.

The maximum ionization in the γ-regime of the CCP discharge occurs when the electric field in the sheath has the maximum value. This takes place at the moment when the oscillating electrons are located furthest from the electrode ($z = 0$, $z = 2\pi$). This effect is illustrated in Figure 10.29 in comparison with the space and time ionization evolution in α-discharges.

To estimate the sheath parameters in the γ-regime of moderate-pressure CCP discharges, balance the averaged ion flux to electrodes:

$$\langle\Gamma_i\rangle = b_i\langle E(x,t)\rangle \tag{10.75}$$

and multiply the γ-electrons in the sheath, taking into account the time and space evolution $E(x,t)$ of the electric field:

$$\left\langle\int I_\gamma dx\right\rangle = \left\langle \gamma\Gamma_i\left[\exp\int_0^{d(t)}\alpha(E(x,t))dx - 1\right]\right\rangle \tag{10.76}$$

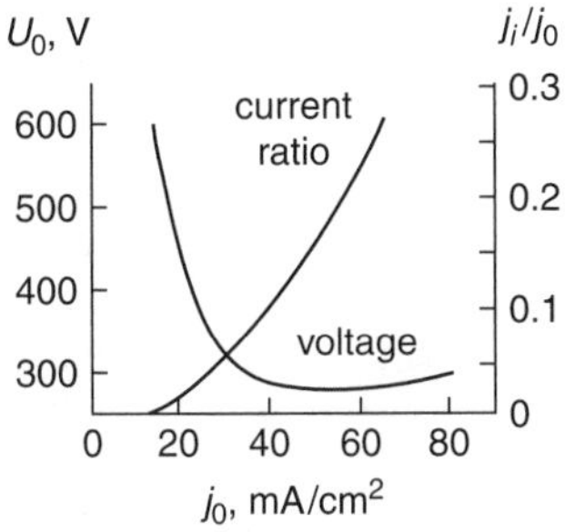

Figure 10.30 Current–voltage characteristic of γ-discharge (air, 13.56 MHz, 30 torr) and ion current ratio/discharge current ratio

In the preceding relation, $I_r(x,t)$ is the ionization rate term physically similar to Equation 10.72, but this time related to the contribution of γ-electrons; γ is the coefficient of secondary electron emission from the electrode (see Section 8.2.5); Γ_i is the ion flux on the electrode; $d(t)$ is the instantaneous size of the sheath; and $\alpha(E)$ is the first Townsend coefficient.

The preceding balance of the averaged ion flux to electrodes and the multiplication of the γ-electrons in the sheath is similar to the corresponding balance in the cathode layer of glow discharges (see Section 7.2.1). The main difference between these two cases is related to the fact that ion current in the sheath of CCP discharges is not equal to the discharge current because of a significant contribution of the displacement current.

Current–voltage characteristics of the sheath in the CCP discharge in air at 30 torr pressure calculated based on Equation 10.75 and Equation 10.76 is presented in Figure 10.30[376] together with the ratio of ion current to the total current in the discharge. As expected, the current–voltage characteristics shown in the figure are similar to the cathode voltage drop dependence on current density in glow discharges (see Section 7.2.2 and Section 7.2.3). For this reason, the effect of normal current density, discussed in detail for glow discharges (see Section 7.2.3), also takes place in the γ-regime of the moderate-pressure CCP discharges with metallic electrodes.

10.4.6 Normal Current Density of γ-Discharges

The effect of normal current density in the γ-regime of the CCP discharge with metallic electrodes can be illustrated using Figure 10.30, in which the normal current density j_n corresponds to the minimum of the current-voltage characteristics $U_0(j_o)$. At lower current density $j < j_n$, the current–voltage characteristics are "falling," which is generally an unstable condition. For example, any small increase of current density due to some fluctuation, in this case, results in lower voltages necessary to sustain the ionization. Then the actual voltage exceeds the required one, and current density grows until the normal regime $j = j_n$. The right branch ($j > j_n$) of the current–voltage characteristics $U_0(j_o)$ is stable in general, but not stable at the edge of the sheath on the boundary of current-conducting and currentless areas of the electrode. This phenomenon is similar to that described in Section 7.2.4 for glow discharges.

Thus, current growth in the γ-regime of moderate pressure CCP discharges leads to an increase of the electrode surface area covered by the discharge, while the current density and sheath voltage drop remain the same and equal to their normal values, j_n and U_n. When the discharge covers the entire electrode, the voltage and current density in principle should begin growing (similar to the abnormal regime of glow discharges).

However, as seen from Figure 10.30, the growth of voltage and current density results in an increase of the fraction of ion conductivity current in the total discharge current. As soon as the ion conductivity current and displacement current become close ($j_i \approx j_D$), the parameter, $\omega\tau_i \approx 1$, and ion motion during an oscillation period can no longer be neglected. Thus, in this case, the inequality (Equation 10.50) is no longer valid, nor is the entire previously considered approach.

The opposite limit of high ion conductivity current ($j_i > j_D$) can be interpreted as $\omega\tau_i < 1$, which means that the CCP discharge should be considered as a low-frequency one. Under these low-frequency conditions, the CCP plasma resistance is almost completely active. Time is sufficient to establish cathode and anode layers during one oscillation period, making this CCP discharge regime even closer to that of the glow discharge.

If the electrode is covered by a thin dielectric layer with capacitive resistance much less than discharge resistance, then general features of the CCP discharge remain similar to those discussed earlier. However, the distributed capacitive resistance is able to determine the discharge current on the electrode surface, which leads to the possibility of stabilizing the discharge at currents lower than the normal one, $j < j_n$.[374]

The normal values of the current density and the sheath voltage drop in the γ CCP discharges with metallic electrodes can be found by minimizing the dependence $U_0(j_0)$. This results in the following expression for the normal current density:

$$j_n = \frac{Bp\omega\varepsilon_0}{2} \cdot \psi_1(\gamma) \tag{10.77}$$

where B is the parameter of Townsend formula for α-coefficient (Equation 4.141), and the dimensionless function $\psi_1(\gamma)$ characterizes the normal current density dependence on the secondary electron emission coefficient presented in Figure 10.31.[376]

The corresponding expression for the normal value of the sheath voltage drop in the γ CCP discharges,

$$U_n = \frac{B}{A}\psi_2(\gamma) \tag{10.78}$$

also includes the A, B parameters of the Townsend formula for α-coefficient (Equation 4.141), and the dimensionless function $\psi_2(\gamma)$ characterizing the influence of the secondary electron emission coefficient (see also Figure 10.31).

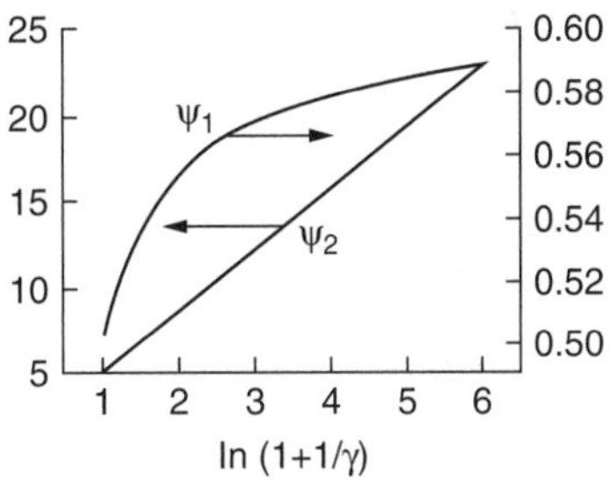

Figure 10.31 Ψ-Function dependence on secondary electron emission coefficient[376]

From Figure 10.31, the normal current density is seen to have a very weak dependence on the secondary emission coefficient γ; the normal voltage drop grows with a decrease of γ, but not significantly. Similar to the case of a DC glow discharge, the sheath voltage drop does not depend on pressure and is determined only by the gas and electrode material. The normal current density is proportional to the frequency of the electromagnetic field and to the gas pressure. The pressure growth also leads to an increase of the ion conductivity current:

$$j_i = \frac{b_i B^2 \varepsilon_0 A p^3}{4\psi_2 \psi_1^3} \tag{10.79}$$

At some critical pressure, the ion current reaches the value of the displacement current (which corresponds to $\omega\tau_i \approx 1$; see Equation 10.50 and the preceding discussion in this section), which leads to the discharge transfer in the low-frequency regime. Thus, the upper limit pressure of the RF CCP discharge in γ-regime can be expressed as:

$$p < \frac{2\omega\psi_1^2(\gamma)\psi_2(\gamma)}{(b_i p)AB} \tag{10.80}$$

10.4.7 Physical Analysis of Current–Voltage Characteristics of Moderate-Pressure CCP Discharges

An example of a numerically calculated current–voltage characteristic of a moderate pressure CCP discharge sheaths is shown in Figure 10.32.[376] The lower limit of the current densities (about 2 mA/cm^2) is related to low electron conductivity in plasma. This corresponds to the condition $\omega\tau_e \approx 1$ (see the inequality 10.48), when displacement current in the plasma zone becomes necessary. The upper limit of the current densities (200 mA/cm^2) is related to the ion density growth in the sheath with an increase of current density. The upper limit corresponds to the condition $\omega\tau_i \approx 1$ (see the inequality 10.50), when ion conductivity in sheaths becomes necessary.

The current–voltage characteristic shown in Figure 10.32 between the lower and upper limits can be subdivided into three regions:

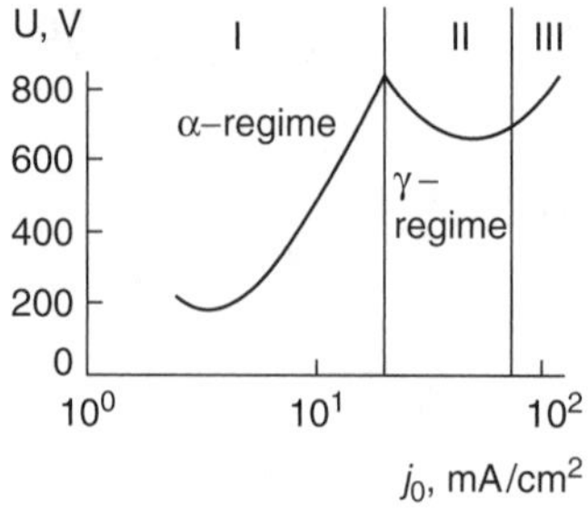

Figure 10.32 Curent–voltage characteristic of CCP discharge in nitrogen, 13.56 MHz; 15 torr[376]

1. The first region in this figure with current densities of about 2 to 20 mA/cm corresponds to the α-regime of the moderate pressure CCP discharge. In this case, the sheath voltage drop grows with current density; the sheath depth does not depend on the discharge current density; and the ion concentration decreases monotonically from the plasma zone to electrode.
2. The second region presented in the figure with current densities of about 20 to 100 mA/cm corresponds to γ-regime of the moderate-pressure CCP discharge, where most of the ionization is provided by secondary electrons formed on the surface of electrodes by ion impact during the space charge phase of the discharge. In this case, the ion density in the sheath has a maximum value exceeding the ion concentration in the plasma zone; the ion density in the sheath grows with the discharge current density; and the sheath depth decreases with current density. This results in the "falling" (and actually unstable) current–voltage characteristics in this region.
3. The sheath in the third region with current densities of about 100 to 200 mA/cm is somewhat similar to the cathode layer of the abnormal DC glow discharge. The electric field E in the space charge phase is so high that the dependence $\alpha(E)$ is no longer strong. The high values of the electric field lead to longer distances necessary to establish the electron energy distribution functions. This results in essential nonlocal effects: secondary electrons provide significant ionization in the plasma phase, and the maximum density of charged particles shifts from the sheath to the plasma zone (with possible formation of the Faraday dark space). Obviously, the minimum point on the $U_0(j_0)$ curve corresponds to the normal current density of the γ-discharge.

10.4.8 The α–γ Transition in Moderate-Pressure CCP Discharges

Here, the electric field in the sheath is not very high at low current densities corresponding to the α-regime (see Figure 10.32). For this reason, multiplication of electrons formed on the electrode surface due to secondary emission can be neglected in this regime. Growth of the current density and the related growth of the electric field lead to increasingly intense multiplication of the secondary (γ) electrons, which finally results in the phenomenon of the α–γ transition.

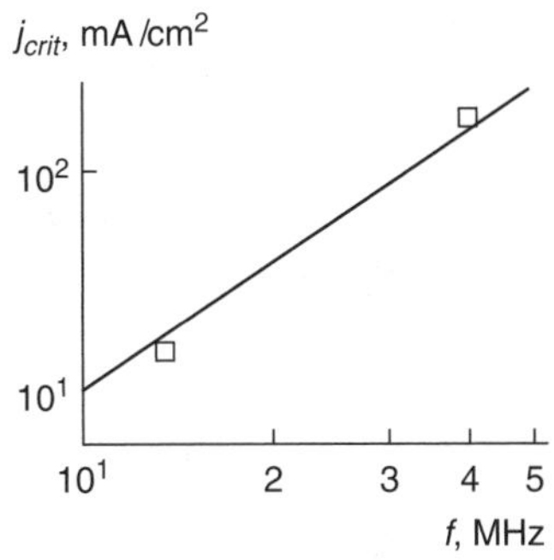

Figure 10.33 Experimental and calculated α–γ transition current dependence on frequency (air, 0.75 cm, 15 torr.[376]

At first, the increase of current density makes the multiplication of the secondary γ-electrons essential only near the electrodes, where the electric field has the maximum value. In this case, the maximum of the ionization rate function I_r is located in the vicinity of an electrode, where the drift in the electric field of the space charge determines the ions' flux and the contribution of diffusion is not significant. This can result in density profiles with a minimum.

One can conclude that the α–γ transition takes place when the generation of the ion flux due to multiplication of the secondary γ-electrons exceeds the ion flux produced in the plasma phase. This requires current densities exceeding the critical value:

$$j_{crit} = Bp\omega\varepsilon_0 \ln^{-1}\left\{\frac{ApL}{3\left[\ln\left(\left(1+\frac{1}{\gamma}\right)\sqrt{\frac{\pi ApL}{4}}\right)-\frac{Bp\omega\varepsilon_0}{j_{crit}}\right]}\right\} \tag{10.81}$$

The critical value of current density, calculated from Equation 10.81 and corresponding to the α–γ transition, is shown in Figure 10.32 by an arrow.

As seen from Equation 10.81, the critical current density of the α–γ transition in moderate-pressure CCP discharges grows with the oscillation frequency. The numerical dependence $j_{crit}(\omega)$ is presented in Figure 10.33. A similar curve describes the critical current density dependence on gas pressure.[376]

10.4.9 Some Frequency Limitations for Moderate-Pressure CCP Discharges

General frequency limitations in the radio-frequency discharges were discussed in Section 10.3.1. Now consider a frequency limitation important for the moderate-pressure CCP discharges. As was noted earlier, the sheath depth decreases with growth of the oscillation frequency. Thus, at high frequencies, the inequality $L \gg \lambda_\varepsilon$, which determines the domination of local effects in forming the electron energy distribution function, is no longer valid (here λ_ε is the length of electron energy relaxation; see a definition of low-pressure RF discharges in Section 10.3.1).

This means that the nonlocal effects typical for low-pressure discharge systems become important in these systems. During formation of the high-energy tail of the electron energy distribution function, the electrons' displacement can be longer than typical sizes of the discharge system. Characteristics of the sheaths at high frequencies become similar to those at low pressures; ionization takes place in the plasma zone and the ion flux in the sheath is constant.

Numerically, the mean free path of electrons in air at pressure of 15 torr is about 0.02 mm, and the length of electron energy relaxation and formation of the electron energy distribution function can be calculated as $\lambda_\varepsilon \approx 0.5$ mm. The sheath depth becomes shorter than this value of the length of electron energy relaxation at frequencies exceeding 120 MHz. This means that at gas pressures of 15 torr, the CCP discharges can be considered as moderate-pressure, if the oscillation frequencies do not exceed about 60 MHz.

The radio-frequency CCP discharges of moderate pressure were investigated mainly related to their application in powerful high-efficiency waveguide CO_2 lasers. It is interesting that the first $He - Ne$ and CO_2 lasers were based on the CCP radio-frequency discharges.[252,379] More details on the subject can be found in Raizer, Sneider, and Yatsenko.[374]

10.5 LOW-PRESSURE CAPACITIVELY COUPLED RF DISCHARGES

10.5.1 General Features of Low-Pressure CCP Discharges

"Low-pressure RF discharges" usually means discharges with gas pressure less than about 0.1 torr; these are widely used in electronics for etching and different kinds of chemical vapor deposition (CVD) processes. A typical scheme of a low-pressure CCP discharge installation applied for the surface treatment is illustrated in Figure 10.34. During the last decades, this large-scale application stimulated intensive

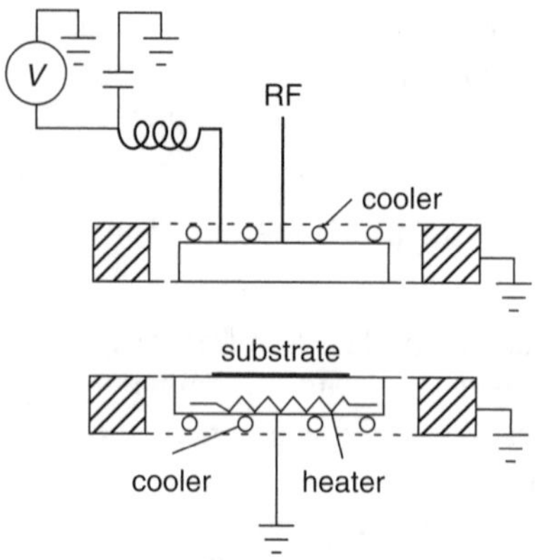

Figure 10.34 Low-pressure CCP discharge

fundamental and applied research activity (see, for example, Reference 13, Reference 380 through Reference 383, and Reference 636.

The luminosity pattern of the low-pressure CCP RF discharges is different from that of moderate-pressure discharges. In this case, the plasma zone is bright and separated from the electrodes by dark pre-electrode sheaths. The discharge usually has an asymmetrical structure: a sheath located by the electrode where the RF voltage is applied is usually about 1 cm thick, while another sheath located by the grounded electrode is only about 0.3 cm thick. Also, in contrast to moderate-pressure discharges, after breakdown plasma fills out the entire low-pressure discharge gap, the phenomenon of normal current density does not take place.

10.5.2 Plasma Electron Behavior in Low-Pressure Discharges

According to the definition given in Section 10.3.1, electron energy relaxation length, λ_ε, in the low-pressure radio-frequency CCP discharges characterizing thickness of plasma zone and sheaths, exceeds the typical sizes of the system:

$$\lambda_\varepsilon > L_p, L \tag{10.82}$$

This means that the electron energy distribution function is not local; it is determined, not by the electric field in a fixed point, but by the entire distribution of the electric fields in the zone of size λ_ε. Numerically, the inequality (Equation 10.82) for He discharges implies that $p \cdot (L_p, L) \leq 1\ \text{Torr} \cdot \text{cm}$. This inequality (10.82) usually occurs with the following:

$$\omega >> \delta \cdot \nu_e \tag{10.83}$$

where δ is the average fraction of electron energy lost per one electron collision and ν_e is the frequency of the electron collisions (compare inequalities Equation 10.82 and Equation 10.83 with the opposite requirements—Equation 10.70 and Equation 10.71).

The criterion (Equation 10.83) actually means that the characteristic time of formation of the electron energy distribution function (EEDF) is much longer than the period of RF oscillations, which permits considering the EEDF as stationary. If, at the same time $\omega < \nu_e$, then the "DC-analogy" is still valid. This means that the EEDF in the low-pressure CCP RF discharge is the same as that in the constant electric field $E_0/\sqrt{2}$ (where E_0 is amplitude of the RF oscillating field).

The electric field in the low-pressure CCP RF plasma (as well as in sheaths) can be divided into constant and oscillating components. The constant component of the electric field in the plasma (including plasma zone and sheaths in plasma phase) provides the balance of electron and ion fluxes necessary for quasi neutrality. The oscillation component of the electric field provides the electric current and heating of the plasma electrons.

Typical distribution of plasma density and electric potential φ (corresponding to the constant component of the electric field) is illustrated in Figure 10.23b, c). The electric field in the space charge sheath exceeds that in the plasma. For this reason, the sharp change of potential on the boundary of plasma and sheath is shown in Figure 10.23 as a vertical line. Thus, the plasma electrons move between the sharp potential barriers, satisfying the conservation of total energy, which can be expressed as:

$$\varepsilon = \frac{1}{2} m v_e^2 + e\phi \tag{10.84}$$

Here v_e is the electron velocity; m is an electron mass; and $\phi = -\varphi$ is the electric potential corresponding to the constant component of the electric field and taken with an opposite sign (because e is considered positive).

The electron energy distribution function depends only on total electron energy (Equation 10.84). The kinetic equation for the EEDF in the Fokker–Plank form (see Section 4.2.4 and Section 4.2.5) is similar in this case to that in the local approximation, but the diffusion coefficient in energy space should be averaged over the entire space interval in which the electron moves.

10.5.3 Two Groups of Electrons, Ionization Balance, and Electric Fields in Low-Pressure CCP Discharges

All plasma electrons in low-pressure CCP discharges can be divided into two groups. The first group consists of electrons with kinetic energy below the potential barrier on the sheath boundary (see Equation 10.84 and Figure 10.23). These electrons trapped in the plasma zone are heated by the electric field in this zone, which is determined by plasma density n_p:

$$E_p = \frac{j_0 m \nu_e}{n_p e^2} \tag{10.85}$$

The second group of electrons has kinetic energy exceeding the potential barrier $e\phi$ on the sheath boundary. These electrons spend part of the time inside the sheath, where density n_s is lower and the electric field in the plasma phase is higher:

$$E_s = \frac{j_0 m \nu_e}{n_s e^2} >> E_p \tag{10.86}$$

The energetic electrons of the second group are heated by the averaged electric field, which can be calculated as:

$$\left\langle E^2(\varepsilon)\right\rangle = \frac{L_p}{L_0} E_p^2 + \frac{L}{L_0}\left(1 - \frac{e\phi}{\varepsilon}\right) E_s^2 \tag{10.87}$$

In this relation, L_p, L, and L_0 are lengths of the plasma zone, sheaths, and total discharge gap, respectively; the square of the electric field is averaged in the equation because it determines the electron heating effect.

The fast electrons provide the main ionization in the plasma zone, where their kinetic energy has a maximum value. The "DC-analogy" can be applied in this case for the fast electrons by using the effective electric field:

$$E_{eff}^2 = \frac{1}{2}\langle E^2 \rangle \frac{\nu_e^2}{\omega^2 + \nu_e^2} \tag{10.88}$$

The effective electric field (Equation 10.88) can then be used in the empirical formula for calculation of the ionization frequency:

$$I = Ap\frac{en_s}{m\nu_e}E_{eff}\exp\left(-\frac{B}{\sqrt{E_{eff}/p}}\right) \tag{10.89}$$

where A, B are the special parameters of this relation determined by type of gas, somewhat similar to those of the Townsend formula (see Section 4.4.2).

Balancing the ionization in volume (Equation 10.89) and ion flux to the electrode $\Gamma_i = n_s v_i$, where the velocity v_i is determined by Equation 10.67, leads to the following formula for the effective electric field:

$$\frac{E_{eff}}{p} = B^2 \ln^{-2} \frac{E_{eff}ApL_pe}{m\nu_e\sqrt{\sqrt{2e\lambda_i j_0}/\varepsilon_0 M\omega}} \tag{10.90}$$

As can be seen, the effective reduced electric field is determined by the constant B and only logarithmically depends on plasma parameters. Thus, the effective reduced electric field can be actually considered as fixed in the low-pressure CCP RF discharges.

10.5.4 High- and Low-Current Density Regimes of Low-Pressure CCP Discharges

Assuming the electric field is fixed, the electron concentration in the plasma is proportional to the current density in the same manner as takes place in the DC-glow discharges (see chapter 7). This gives a possibility to consider the cases of relatively high and relatively low current densities separately.

If the current density (and thus n_p) is relatively low, then according to Equation 10.85 and Equation 10.86, the electric field in plasma is relatively high and heating of electrons takes place mostly in the plasma zone (also taking into account the larger size of this zone). The voltage drop is determined by the electron energy and its value numerically is near the lowest level of electronic excitation of neutral particles. In general, the plasma zone determines all discharge properties at low current densities. The sheaths are then arranged in a way necessary to close the discharge circuit by displacement current and to provide ion flux from the plasma zone.

If the current density (and thus the density n_p) is relatively high, then according to Equation 10.85 and Equation 10.86, the electric field in the plasma zone is much lower with respect to the electric field in sheaths in the plasma phase. Thus, most of the electron heating effect takes place in sheaths in the plasma phase. In the high-current density regime, the ionization balance determines the electric field and thus the ion density n_s in the sheath. Therefore, based on Equation 10.86 through Equation 10.88, the effective electric field can be expressed as:

$$E_{eff}^2 \approx \frac{L}{L_0} E_s^2 = \frac{L}{L_0} \frac{j_0^2 m^2 \nu_e^2}{n_s^2 e^4} \tag{10.91}$$

Taking into account that the effective electric field E_{eff} can be considered fixed (Equation 10.91) gives:

$$\frac{n_s^2}{L} \propto j_0^2 \tag{10.92}$$

Also, the Poisson equation requires the following relation between the ion concentration in sheaths and current density:

$$n_s L \propto j_0 \tag{10.93}$$

The combination of Equation 10.92 and Equation 10.93 leads to the conclusion that the sheath size L does not depend on current density j_0 in the regime when the current density is high. At the same time, the ion concentration in the sheath is proportional to the current density. Thus, in general, all discharge properties at high-current densities are determined by the sheaths. The plasma zone is then arranged in a way necessary to provide the required ion current in the sheath.

10.5.5 Electron Kinetics in Low-Pressure CCP Discharges

Heating electron gas can be described by the Fokker–Plank kinetic approach (see Section 4.2.4) as the diffusion flux along the electron energy spectrum. In this case, the electron heating is proportional to the derivative over the energy of the electron energy distribution function $\partial f(\varepsilon)/\partial\varepsilon$ and to the diffusion coefficient D_ε in the electron energy space, which depends on the electric field and the heating mechanism. In general, the diffusion coefficient D_ε in energy space can be expressed for the RF discharge systems, based on Equation 4.73, in the following form:

$$D_\varepsilon = \frac{1}{6} \frac{(eEv_e)^2}{\omega^2 + \nu_e^2} \nu_e \tag{10.94}$$

where E is the amplitude value of the electric field.

As mentioned earlier (Section 10.5.3), electrons in the systems under consideration can be divided into two groups corresponding to high and low energies relative to the potential on the sheath boundary. The diffusion coefficient in energy space (Equation 10.94) can be expressed in somewhat different ways for each of the two electron groups. Electrons of the first group are relatively "cold" and have kinetic energy lower than the potential on the sheath boundary. Thus, the cold electrons are unable to penetrate the sheath. For this reason, their diffusion coefficient in energy space is determined by the electric field in the plasma zone and can be expressed as:

$$D_{\varepsilon c}(\varepsilon) = E_p^2 \varepsilon \frac{e^2 \nu_e}{6m\left(\omega^2 + \nu_e^2\right)} \tag{10.95}$$

In this relation, E_p is the electric field in the plasma zone (Equation 10.85); ε is the electron energy; and ν_e is the frequency of their collisions.

The second group includes the "hot" electrons that have energies exceeding the potential on the sheath boundary $\varepsilon > e\phi$. These electrons are able to spend part of the time in the sheaths and, therefore, their diffusion coefficient in the energy space is determined by the averaged electric field defined by Equation 10.87 and can be presented as:

$$D_{\varepsilon h} = \left[\frac{L_p}{L_0} E_p^2 + \frac{L}{L_0}\left(1 - \frac{e\phi}{\varepsilon}\right) E_s^2 \right] \varepsilon \frac{e^2 \nu_e}{6m\left(\omega^2 + \nu_e^2\right)} \tag{10.96}$$

In this relation, E_s is the electric field in the sheath (plasma phase) and determined by Equation 10.87; L_p, L, and L_0 are lengths of the plasma zone, sheaths, and total discharge gap, respectively; and ϕ is the potential barrier between the plasma and the sheath.

For analytical estimations, it is reasonable to assume in these discharges that electrons lose all their energy in nonelastic collisions (corresponding to the lowest level of electronic excitation, ε_1) and then return to the low-energy region of the electron energy distribution function (EEDF). Then, one can conclude that the integral flux in the energy space should be constant over the electron energy interval $0 < \varepsilon < \varepsilon_1$. This means that the EEDF is a very smooth function of energy (see Section 4.2.4 and Section 4.2.5).

In the framework of this approach, the density of charge particles in the plasma zone and sheath can be presented as a product of the averaged value of the EEDF (f_c^0 for cold electrons, f_h^0 for hot electrons) and the corresponding volume in energy space. Thus, the concentration of charged particles in the plasma zone can be given as:[376]

$$n_p \approx \frac{8\pi\sqrt{2}}{3m^{3/2}} f_c^0 (e\phi)^{3/2} \tag{10.97}$$

Similarly, the electron density in sheaths in the plasma phase can be expressed by the following relation:

$$n_s \approx \frac{8\pi\sqrt{2}}{3m^{3/2}} f_h^0 \left[\varepsilon_1^{3/2} - (e\phi)^{3/2}\right] \tag{10.98}$$

Derivatives of the EEDF can be estimated in a similar manner as the averaged EEDF values divided by the relevant interval in the energy space, which permits calculating fluxes in the energy space (see Section 4.2.4). By balancing the fluxes, one can derive the equation for the potential barrier on the plasma–sheath barrier:

$$\left(\frac{\varepsilon_1}{e\phi}\right)^{3/2} - 1 = \left(\frac{e\phi}{eL_p}\frac{2\omega\varepsilon_0}{j_0}\right)^{1/2} \left(\frac{\varepsilon_1}{\varepsilon_1 - e\phi} + \frac{eL}{e\phi}\frac{j_0}{2\omega\varepsilon_0}\right) \tag{10.99}$$

As seen from Equation 10.99, the potential barrier $e\phi$ on the plasma–sheath barrier is to approximate the lowest energy level ε_1 of electronic excitation, numerically.

Combination of Equation 10.87, Equation 10.90, and Equation 10.99 allows determining the sheath size L and potential barrier $e\phi$ and, as a result, all other discharge parameters.

10.5.6 Stochastic Effect of Electron Heating

Electron heating in RF capacitive discharges of moderate pressures is due to electron–neutral collisions. Received by an electron from the electromagnetic field after a previous collision, energy of the systematic oscillations can be transferred to chaotic electron motion during the next collision. If the gas pressure is sufficiently low and $\omega^2 >> \nu_e^2$, then the electric conductivity and joule heating are proportional to the frequency of electron–neutral collisions ν_e (see Equation 6.181 and Equation 6.182). This means that the discharge power related to electron–neutral collisions also should be very low at low pressures. However, the discharge power under such conditions can be significantly higher experimentally because of the contribution from stochastic heating. Even in the collisionless case, this effect provides heating of fast electrons in RF discharges.[384–386]

The physical basis of stochastic heating can be easily explained when the pressure is so low that the mean free path of electrons exceeds the size of sheaths. Here, electrons entering the sheath are then reflected by the space charge potential of the sheath boundary. This process is similar to an elastic ball reflection from a massive wall. If the sheath boundary moves from the electrode, then the reflected electron receives energy; conversely, if the sheath boundary moves toward the electrode and a fast electron is "catching up" to the sheath, then the reflected electron loses kinetic energy. Taking into account that the electron flux to the boundary moving from the electrode exceeds that for one moving in the opposite direction to the electrode, one can conclude that energy is transferred to the fast electrons on average by this mechanism; this explains the stochastic heating effect.

The stochastic heating effect can be described by means of a special stochastic diffusion coefficient in energy space $D_{\varepsilon st}$ and an effective "stochastic" electric field E_{st}. These can be calculated noting that the frequency of the electron "collision" with the potential barrier is the inverse time of the electron motion between electrodes:

$$\Omega = \frac{\sqrt{2\varepsilon/m}}{L_0} \tag{10.100}$$

The diffusion coefficient in the energy space is determined by the square of energy change per one collision and frequency of the collisions (see Section 4.2.4). The velocity of the moving sheath boundary coincides with the electron drift velocity in the sheath; therefore, the energy change during a collision with the boundary is the same as that for a collision with neutral particles. For this reason, the diffusion coefficient in energy space, related to the stochastic heating, can be obtained from Equation 10.94 by replacing the electron–neutral collision frequency ν_e by the frequency (Equation 10.100) of "collisions" with the sheath boundary:

$$D_{\varepsilon st} = \frac{e^2 E_s^2 \varepsilon}{6m} \Omega \qquad (10.101)$$

The effective "stochastic" electric field responsible for the stochastic heating effect can be determined in a similar manner. The square of the effective stochastic electric field differs from that in the sheath in the plasma phase (Equation 10.86) by a factor of about Ω/ν_e:

$$E_{st}^2 = \frac{3\Omega}{2\nu_e} E_s^2 = \frac{3\Omega}{2\nu_e}\left(\frac{j_0 m \nu_e}{n_s e^2}\right)^2 \qquad (10.102)$$

In obtaining the relation, $\omega >> \nu_e$, the conditions when the contribution of the stochastic heating effect is essential were taken into account. Thus, the effective stochastic electric field should replace E_s in equations for potential and particle balance for calculations of the capacitive RF discharge at very low pressures.

10.5.7 Contribution of γ-Electrons in Low-Pressure CCP RF Discharges

Mean free paths of electrons with high energies of 500 to 1000 eV at gas pressures below 0.1 torr become larger than 1 cm, which can exceed the discharge gap between electrodes. Such γ-electrons accelerated near electrodes can move across the discharge gap without any collisions and not directly influence the ionization processes. However, the energetic γ-electrons can make significant contributions to physical and chemical processes on the electrode surface; this is essential in plasma chemical application of the low-pressure capacitive RF discharge.

Typical energy distributions of electrons and ions bombarding electrodes in low-pressure CCP discharges (p = 5 mtorr) are shown in Figure 10.35. As seen from this figure, the flux of the fast electrons on the grounded electrode is close to that of ions, while the energies of these electrons reach 2 keV and greatly exceed those of

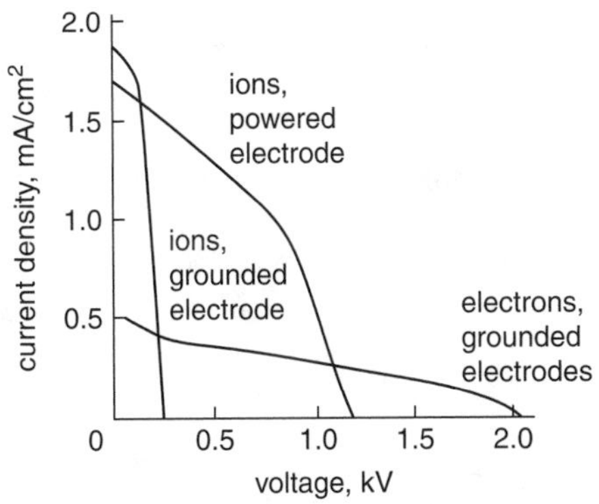

Figure 10.35 Stopping potentials for charged particles bombarding electrodes in CCP discharge

ions. Thus, the total electron energy also exceeds the total ion energy transferred to the grounded electrode.

An interesting effect can take place if the time necessary for the secondary γ-electrons to cross the discharge gap is longer than $1/\omega$. Then the secondary electrons accelerated in one sheath slow down near the opposite electrode. Such resonant electrons can be trapped in the plasma and make contributions to ionization (see the secondary-electron resonance discharge).

10.5.8 Analytic Relations for Low-Pressure RF CCP Discharge Parameters

Even after simplifications of the model, the system of equations describing low-pressure capacitive RF discharges remains too complicated for general analytical solutions. However, asymptotic solutions can be found for some extreme but practically important regimes. Consider the low- and high-current discharge regimes, assuming collisional heating ($\nu_e >> \omega$) and the proportionality of ion velocity in the plasma zone to the ambipolar electric field:

$$v_p = \frac{e}{M\nu_i}\frac{2\phi}{L_p} \tag{10.103}$$

Such assumptions are satisfied in the low-pressure RF capacitive discharge in argon if $0.5 < pL_0 < 1\ \text{Torr}\cdot\text{cm}$.

Low-Current Regime of Low-Pressure RF CCP Discharges. Heating in the sheaths is negligible at low currents ($L_pE_p^2 >> L_sE_s^2$) and the effective electric field is equal to:

$$E_{eff} = E_p/\sqrt{2} \tag{10.104}$$

Substituting this expression for the effective electric field into Equation 10.90, the following relation for the electric field in plasma is obtained:

$$E_p = \sqrt{2}B^2 p \ln^{-2}\frac{E_p ApL_p e}{m\nu_e\sqrt{2\sqrt{2}e\lambda_i j_0/\varepsilon_0 M\omega}} \tag{10.105}$$

Thus, the electric field in the plasma only logarithmically depends on the current density, which is somewhat similar to the positive column of the DC glow discharges.

The potential barrier $e\phi$ on the plasma–sheath boundary is about equal to the lowest energy level ε_1 of electronic excitation of neutral species and can be determined as:

$$e\phi = \varepsilon_1\left[1 - \left(\frac{b_i\varepsilon_1}{3eL_p}\right)^{1/2}\left(\frac{\omega\varepsilon_0 M}{j_0 e\lambda_i}\right)^{1/4}\right] \tag{10.106}$$

Numerically, this potential barrier is almost constant and can be taken for estimations in the following simple form:

$$\phi = \frac{2}{3}\frac{\varepsilon_1}{e} \tag{10.107}$$

In this case, concentration of charged particles in the plasma zone is proportional to the current density and can be expressed as:

$$n_p = \frac{j_0 m\nu_e}{e^2 E_p} \propto j_0 \tag{10.108}$$

The ions' flux is also proportional to the current density and can be given by:

$$\Gamma_i = \frac{4}{3}\frac{j_0 \varepsilon_1 m\nu_e}{E_p L_0 e^2 M\nu_i} \propto j_0 \tag{10.109}$$

where ν_e and ν_i are the electron–neutral and ion–neutral collision frequencies.

Sheaths do not make any contribution to electron heating in the low-current regime. In this case, the sheath parameters are determined by the ion flux from the plasma. Thus, balancing the ion flux in the sheath with that in plasma (Equation 10.109), one can derive the following expression for the sheath thickness, which is proportional to square root of current density:

$$L = \frac{3eE_p L_0 M\nu_i}{2\omega\varepsilon_1 m\nu_e}\sqrt{\frac{\sqrt{2}\lambda_i}{M}\frac{j_0}{\varepsilon_0\omega}} \propto \sqrt{j_0} \tag{10.110}$$

Based on Equation 10.74 between n_s and the sheath thickness in form 10.110, the ion density in the sheath can be expressed as:

$$n_s = \frac{4}{3}\frac{\varepsilon_1 m\nu_e}{E_p e^2 L_0 M\nu_i}\sqrt{\frac{M\omega\varepsilon_0 j_0}{\sqrt{2}\lambda_i}} \propto \sqrt{j_0} \tag{10.111}$$

It is also proportional to the square root of current density.

High-Current Regime of Low-Pressure RF CCP Discharges. The main electron heating in the high-current regime takes place in the sheaths. In this case, low-energy electrons are heated by the relatively low electric field in the plasma zone and their effective "temperature" decreases. For this reason, the ambipolar potential maintaining the low-energy electrons in the plasma zone is less than the excitation energy. The expression for the effective electric field can be derived in this regime based on the Poisson equation and Equation 10.86 and Equation 10.91:

$$E_{eff} = \sqrt{\frac{L}{2L_0}}E_s = \sqrt{\frac{L}{2L_0}}\frac{m\omega\nu_e L}{2e} \tag{10.112}$$

By comparing this relation with one for the effective electric field (Equation 10.90), the following formula for the sheath thickness is obtained:

$$L^{3/2} = \frac{2\sqrt{2L_0}eB^2 p}{m\omega\nu_e}\ln^{-2}\left(\frac{ApL_p^{3/2}L_p\omega}{2\sqrt{\frac{2\sqrt{2L_0}e\lambda_i}{M}\frac{j_0}{\varepsilon_0\omega}}}\right) \quad (10.113)$$

As seen from this relation, the sheath thickness is determined by a balance of ionization and ion losses and, practically, does not depend on the discharge current. The potential barrier on the sheath boundary in the high-current regime of the low-pressure discharge then can be expressed as:

$$e\phi = \varepsilon_1\left(\frac{2\varepsilon_1}{M\nu_i L}\right)^2 \frac{M}{\sqrt{2e\lambda_i}}\frac{\omega\varepsilon_0}{j_0} \propto \frac{1}{j_0} \quad (10.114)$$

As was previously mentioned, in this case the potential barrier of the sheath boundary is less than the excitation energy ε_1 and is inversely proportional to the discharge current.

The charged particles' concentration in the plasma zone increases quite strongly with the current density:

$$n_p = \frac{2j_0}{\omega eL}\frac{L_p}{L}\left(\frac{\sqrt{2e\lambda_i}}{M}\frac{j_0}{\omega\varepsilon_0}\right)^{3/2}\left(\frac{LM\nu_i}{2\varepsilon_1}\right)^3 \propto j_0^{5/2} \quad (10.115)$$

Ion concentration in the sheath is just proportional to the discharge current density and can be expressed quite simply as:

$$n_s = \frac{2j_0}{e\omega L} \propto j_0 \quad (10.116)$$

The critical value of current density j_0^{cr} dividing the previously described regimes of low and high currents can be found from Equation 10.114, assuming that $e\phi = 2/3\ \varepsilon_1$. This leads to the following formula for the critical current density:

$$j_0^{cr} = 3\sqrt{2}\left(\frac{\varepsilon_1}{LM\nu_i}\right)^2 \frac{M\omega\varepsilon_0}{e\lambda_i} \quad (10.117)$$

where the sheath thickness can be taken from Equation 10.113.

10.5.9 Numerical Values of Low-Pressure RF CCP Discharge Parameters

Experimental and numerically predicted values of parameters of the low-pressure capacitive radio-frequency discharge in argon are presented in Figure 10.36 as functions of current density.[376] Figure 10.36(a) illustrates the plasma concentration dependence on the discharge current density. The low-current branch of the dependence is

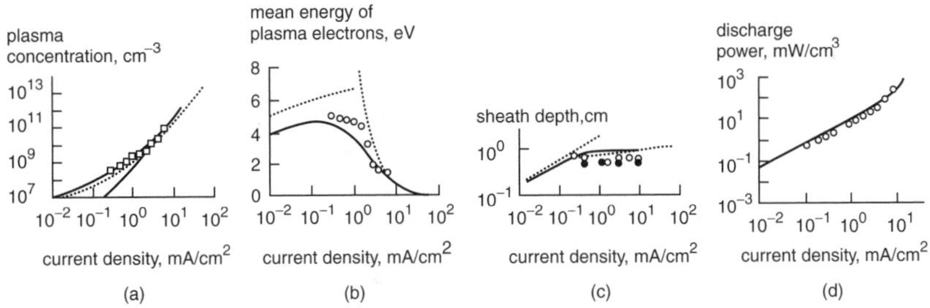

Figure 10.36 (a) Plasma concentration; (b) electron energy; (c) sheath depth; and (d) discharge power as the function of current density. Solid curve = exact calculation; dotted line = asymptotic formulas; points = experiment (f = 13.56 MHz; L_0 = 6.7 cm; p = 0.03 torr).[376]

linear in accordance with Equation 10.108, while at high currents, this dependence is stronger (see Equation 10.115).

The average energy of plasma electrons $\langle\varepsilon\rangle$ is related with the potential barrier on the plasma–sheath boundary as $\langle\varepsilon\rangle = 2/3\ e\phi$. The dependence of the average energy of plasma electrons on the discharge current density is illustrated in Figure 10.36(b). In good agreement with Equation 10.106, the average electron energy grows slightly with j_0 at low-current densities. In contrast, at high-current densities exceeding 1 mA/cm^2, the average electron energy decreases inversely proportional to the current density (see Equation 10.114).

The sheath thickness increases with current density as $\sqrt{j_0}$ at low currents (see Equation 10.110 and Figure 10.36(c) and reaches the saturation level (see Equation 10.113) close to 1 cm at current densities exceeding 1 mA/cm^2. Specific discharge power (per unit volume) grows with the current density as shown in Figure 10.36(d) and reaches values of about 1 W/cm^3 at current densities of about 10 mA/cm^2. More detailed description of experimental results related to low-pressure RF capacitive discharges can be found in the numerous publications of Godyak and co-authors.[387–390]

An especially interesting modeling approach to analyze low-pressure capacitive RF discharges is related to the so-called particle-in-cell (PIC) simulation. This approach allows following large numbers of representative particles acted upon by the basic forces and does not require many of the analytic model assumptions. The PIC approach can be considered a computer experiment, which permits determining various microscopic quantities not observable in direct experiments (see, for example, References 391 through 393).

10.6 ASYMMETRIC, MAGNETRON, AND OTHER SPECIAL FORMS OF LOW-PRESSURE CCP RF DISCHARGES

10.6.1 Asymmetric Discharges

The RF CCP discharges are usually organized in grounded metal chambers. One electrode is connected to the chamber wall and therefore also grounded; another is

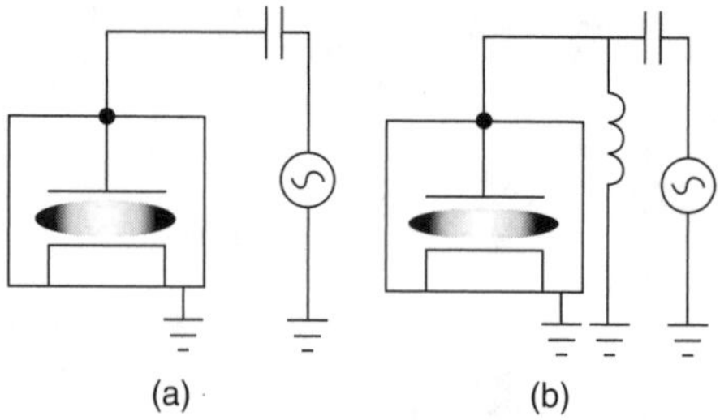

Figure 10.37 RF discharge with (a) disconnected and (b) DC-connected electrodes

obviously powered, making the discharge asymmetric in this configuration (see, for example, Lieberman[394,395]).

The current between an electrode and the grounded metallic wall in moderate-pressure discharges usually has a reactive capacitive nature and does not play any important role. However, in low-pressure discharges, the situation is different. The plasma occupies a much larger volume because of diffusion, and some fraction of the discharge current goes from the loaded electrode to the grounded walls. As a result, the current density in the sheath located near the powered (or loaded) electrode usually exceeds that in the sheath related to the grounded electrode. Also, in practical applications of low-pressure capacitive RF discharges, the surface area of the powered electrode is smaller. The treated material is located on this loaded electrode. This difference in the surface areas also results in higher values of current densities on the powered electrode.

Lower current density in this sheath corresponds to lower values of voltage. The constant component of the voltage, which is the plasma potential with respect to the electrode, is also lower at lower values of current density. For this reason, a constant potential difference occurs between the electrodes if a dielectric layer covers them or if a special blocking capacitance is installed in the electric circuit (see Figure 10.37) to avoid direct current between the electrodes. This potential difference is usually referred to as the **autodisplacement voltage.**

The plasma is charged positively with respect to the electrodes. The voltage drop in the sheath near the powered (loaded) electrode exceeds the drop near the grounded electrode. As a result, the loaded electrode has a negative potential with respect to the grounded electrode. Thus, the RF loaded electrode is sometimes called a "cathode."

10.6.2 Comparison of Parameters Related to Powered and Grounded Electrodes in Asymmetric Discharges

Consider the asymmetric low-pressure CCP RF discharge between planar electrodes that is illustrated in Figure 10.37. The ion flux to the powered electrode (cathode) can be expressed as:

$$\Gamma_i \approx n_s \sqrt{\frac{e\lambda_i}{M} \frac{j_0}{\varepsilon_0 \omega}} \tag{10.118}$$

where n_s is the ion concentration near the loaded electrode, and j_e is the current density in this pre-electrode layer. In this case, the ion flux to the grounded electrode can be expressed by a similar formula:

$$\Gamma_i' \approx n_s' \sqrt{\frac{e\lambda_i}{M} \frac{j_0'}{\varepsilon_0 \omega}} \tag{10.119}$$

where n_s' is the ion concentration near the grounded electrode, and j_e' is the current density in this pre-electrode layer.

The fluxes (Equation 10.118 and Equation 10.119) are formed in the plasma zone by means of diffusion. Taking into account that concentrations of charged particles in the sheaths are lower than those in the plasma zone, n_s, $n_s' << n_p$, and the symmetry of the system, one can conclude that the fluxes (Equation 10.118 and Equation 10.119) should be equal. This leads to the inverse proportionality of the ion densities in the sheaths and the square roots of corresponding current densities:

$$\frac{n_s}{n_s'} = \sqrt{\frac{j_0'}{j_0}} \tag{10.120}$$

Equation 10.120 shows that the charge concentration near the grounded sheath only slightly exceeds the one in the vicinity of the powered (or loaded) electrode. Then note that the electric field is determined by the space charge. For this reason, the ratio of current densities in the opposite sheaths can be expressed as:

$$\frac{j_0'}{j_0} = \frac{n_s' L'}{n_s L} \tag{10.121}$$

Based on Equation 10.120 and Equation 10.121, the relation between the thickness of the opposite sheaths and the corresponding current densities is obtained:

$$\frac{L'}{L} = \left(\frac{j_0'}{j_0}\right)^{3/2} \tag{10.122}$$

From Equation 10.122, thickness of the sheath located near the powered electrode is seen to be larger than the one in the vicinity of grounded electrode.

Finally, taking into account that voltage drop in the sheaths U is proportional to the charge density and square of the sheath thickness, one can conclude:

$$\frac{U'}{U} = \left(\frac{j_0'}{j_0}\right)^{5/2} \tag{10.123}$$

Thus, in real asymmetric discharges, the voltage drop near the grounded electrode can be neglected, and the auto-displacement voltage is close to the amplitude of the total applied RF voltage. It should be mentioned that high charge concentration, low current density, and small sheath depth at the grounded electrode result in very low intensity of electron heating in this layer. This is obviously compensated by intensive heating of electrons in the cathode sheath.

10.6.3 The Battery Effect: "Short-Circuit" Regime of Asymmetric RF Discharges

The constant potentials of electrodes can be made equal by connecting (from the DC standpoint) the powered metal electrode with the ground through an inductance. Such a scheme is illustrated in Figure 10.37(b). Here, direct current can flow between the electrodes; this means that the electron and ion fluxes to each electrode are not supposed to be equal. The average voltages on the sheaths are the same in this regime of the asymmetric CCP RF discharge, resulting in equal thickness of the sheaths. The sheath related to the grounded electrode has a lower current density, so the amplitude of the plasma boundary displacement is shorter than the thickness of the grounded electrode sheath. The main part of the sheath stays for the entire period in the phase of space charge. The plasma does not touch the electrode and the ion current permanently flows to the grounded electrode.

Here, the average positive charge per period carried out from the discharge is compensated by the electron current to the powered electrode. Thus, the direct current in this regime is determined by the ion current to the grounded electrode, as well as by the electron current to the powered one.

The asymmetric RF discharge can be considered a source of direct current. For this reason, the described phenomenon is usually referred to as the battery effect.[396]

10.6.4 Secondary-Emission Resonant Discharges

Low-pressure RF discharges can be sustained only by surface ionization processes, without any essential contribution of the gas in the discharge gap. Consider an electron emitted from one electrode and moving in the accelerating electric field toward another electrode, where it is able to produce a secondary electron during the secondary electron emission. If the electron "flight time" to the opposite electrode coincides with the half-period of the electric field oscillations, then the secondary electron also can be accelerated by the electric field, which turned around during this flight time.

Obviously, to sustain such a resonant discharge regime, the secondary electron emission coefficient should exceed unity, which can take place at energies of about 0.3 to 1 Kev (see Figure 10.38[376]). The electron flight time t_n necessary to sustain the discharge can be equal to one, or any, odd number of half-periods. Thus, for the nth order of the resonance in such discharge, it is necessary to maintain:

$$t_n = \frac{\pi}{\omega}(2n-1) \tag{10.124}$$

where ω is the frequency of the electric field oscillations in the discharge system.

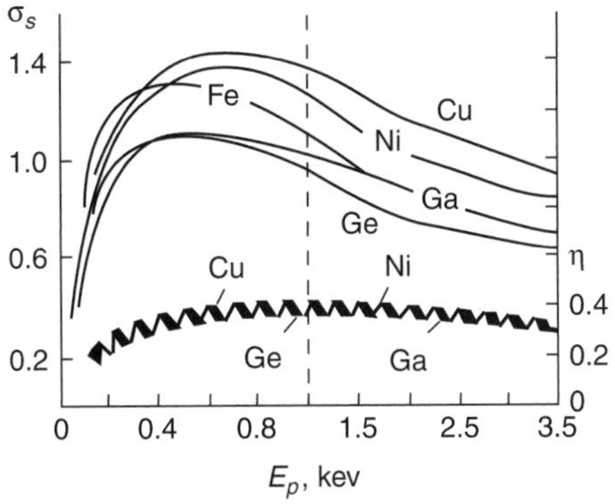

Figure 10.38 Secondary electron emission coefficient (σ_s, solid line) and inelastic reflection coefficient (η, zig-zag line) as functions of electron energy[376]

10.6.5 RF Magnetron Discharges: General Features

The RF capacitive discharge is one of the most widely used plasma sources applied to low-pressure processing of materials in electronics and other industries. However, these discharges have some important disadvantages that limit their application. First of all, the sheath voltages in RF CCP discharges are relatively high; this results at a given power level in low ion densities and ion fluxes, as well as in high ion-bombarding energies. Also in these discharges, the ion-bombarding energies cannot be varied independently of the ion flux.

The RF magnetron discharges were developed especially to make the relevant improvements; in electronics, these discharge systems are usually referred to as magnetically enhanced reactive ion etchers (MERIE). In RF magnetrons, a relatively weak DC magnetic field (about 50 to 200 G) is imposed on the low-pressure CCP discharge parallel to the powered electrode and, therefore, perpendicular to the RF electric field and current.

The RF magnetron discharge permits increasing the degree of ionization at lower RF voltages and decreasing the energy of ions bombarding the powered electrode (or a sample placed on the electrode for treatment). Also, this discharge permits increasing the ion flux from the plasma, which in turn leads to intensification of etching. The RF magnetron discharges can be effectively sustained at much lower pressures (down to 10^{-4} torr). As a result, a well-directed ion beam is actually able to penetrate the sheath without any collisions. The ratio of the ion flux to the flux of active neutral species grows, which also leads to a higher quality of etching.

A general scheme of the RF magnetron used for ion etching[269] is shown in Figure 10.39. The RF voltage is applied to the smaller lower electrode, where the sample under treatment is located. A rectangular samarium–cobalt constant magnet is placed under the powered electrode. As seen from Figure 10.39, the part of the electrode surface where the magnetic field is horizontal is quite limited and does

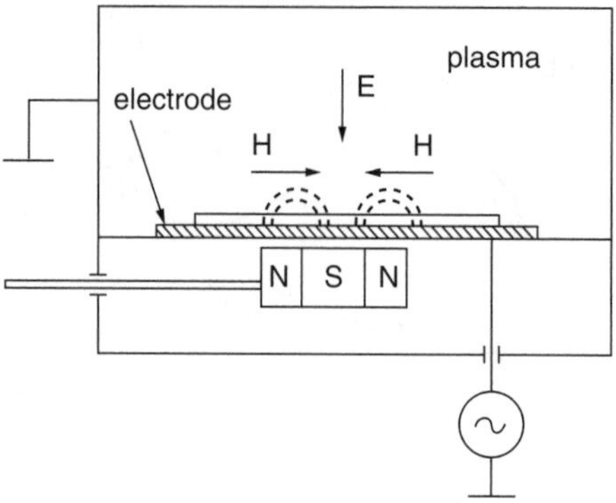

Figure 10.39 RF magnetron[269]

not cover the entire electrode. For this reason, a special scanning device is used to move the magnet along the electrode and to provide "part-time" horizontal magnetic field for the entire electrode. The RF magnetron discharges also can be organized as cylindrical systems with coaxial electrodes.[395–399] The magnetic field in these systems is directed along the cylinder axis.

The physics of the RF magnetron is based on the following effect. Electrons, oscillating together with the sheath boundary, additionally rotate around the horizontal magnetic field lines with the cyclotron frequency (Equation 4.106). When the magnetic field and the cyclotron frequency are sufficiently high, the amplitude of the electron oscillations along the RF electric field decreases significantly. The magnetic field "traps" electrons. In this case, the cyclotron frequency actually plays the same role as the frequency of electron–neutral collisions: electrons become unable to reach the amplitude of their free oscillations in the RF electric field.

The amplitude of electron oscillations determines the thickness of sheaths. Thus, a decrease of the amplitude of electron oscillations in the magnetic field results in smaller sheaths and lower sheath voltage near the powered electrode. This leads to lower values of the autodisplacement, lower ion energies, and lower voltages necessary to sustain the RF discharge.

10.6.6 Dynamics of Electrons in RF Magnetron Discharge

To analyze the RF magnetron effect, consider an electron motion in crossed electric and magnetic fields, taking into account electron–neutral collisions. Assume that the electric field is directed along the x axis perpendicular to the electrode surface $E \equiv E_x$, and the magnetic field is directed along the z axis parallel to the electrode $B = B_z$. The electron motion equation, including the Lorentz force and electron neutral collisions, can be expressed as:

$$m\frac{d\vec{v}}{dt} = -\vec{E}_a e^{i\omega t} - e\left[\vec{v} \times \vec{B}\right] - m\nu_{en}\vec{v} \qquad (10.125).$$

In this equation, $\vec{v}$ is the electron velocity; $\vec{E}_a$ is the amplitude of oscillating electric field; ν_{en} is the frequency of electron–neutral collisions. Projections of the motion equation and to the x and y axes give the following system of equations:

$$m\frac{d}{dt}v_x = -eE_a e^{i\omega t} - ev_y B - m\nu_{en} v_x \tag{10.126}$$

$$m\frac{d}{dt}v_y = ev_x B - m\nu_{en} v_y \tag{10.127}$$

To solve this system of equations with respect to velocities, it is convenient to transform variables to $v_x \pm iv_y$, multiplying Equation 10.127 by i and then adding and subtracting Equation 10.126. As a result, the forced electron oscillations in the crossed electric and magnetic fields can be described by the following complex relations:

$$v_x \pm iv_y = \frac{eE_a e^{i\omega t}}{m\left(\nu_{en} + i\left(\omega \mp \omega_B\right)\right)} \tag{10.128}$$

where ω_B is the cyclotron frequency defined by Equation 4.106. This system (Equation 10.128) gives the final expressions for the velocity of electron oscillations in the direction perpendicular to electrodes $v_x = d\xi/dt$:

$$v_x = \frac{eE_a e^{i\omega t}}{2m}\left[\frac{1}{\nu_{en} + i\left(\omega - \omega_B\right)} + \frac{1}{\nu_{en} + i\left(\omega + \omega_B\right)}\right] \tag{10.129}$$

The corresponding displacement coordinate $\xi(t)$ of the electron oscillations is determined as the integral of the normal velocity v_x:

$$\xi = -\frac{ieE_a e^{i\omega t}}{2m\omega}\left[\frac{1}{\nu_{en} + i\left(\omega - \omega_B\right)} + \frac{1}{\nu_{en} + i\left(\omega + \omega_B\right)}\right] \tag{10.130}$$

At very low pressures, the frequency of electron–neutral collisions is also low, $\nu_{en} << \omega$, and the electron oscillations can be considered collisionless. The amplitude of electron velocity in the collisionless conditions and in the absence of magnetic field is:

$$u_a = \frac{eE_a}{m\omega} \tag{10.131}$$

The electron displacement amplitude (Equation 10.130) under the same collisionless, nonmagnetized conditions can be expressed as:

$$a = \frac{eE_a}{m\omega^2} \tag{10.132}$$

The main objective of applying a magnetic field is to decrease the amplitudes of the electron velocity and displacement (Equation 10.131 and Equation 10.132), which then results in smaller sheaths and lower sheath voltage near the powered electrode. As was mentioned earlier, this leads to lower values of the autodisplacement, lower ion energies, and lower voltages necessary to sustain the RF discharge desirable for RF magnetrons applications.

If the magnetic field is sufficiently high ($\omega_B >> \omega$), then the amplitude of electron velocity in the collisionless conditions can be given as:

$$u_{aB} = \frac{eE_a\omega}{m\omega_B^2} \tag{10.133}$$

This is less than the amplitude in the absence of a magnetic field by the factor $(\omega_B/\omega)^2 >> 1$. Similarly, the amplitude of the electron oscillation displacement (Equation 10.130) in the magnetized collisionless conditions can be expressed as:

$$a_B = \frac{eE_a}{m\omega_B^2} \tag{10.134}$$

which is also less than the displacement under nonmagnetized conditions by the factor $(\omega_B/\omega)^2 >> 1$. It should be noted that heavy ion motion in the RF magnetrons is not magnetized; the ion cyclotron frequency is less than the oscillation frequency ($\omega_{Bi} << \omega$). This means that trajectories of the heavy ions are not perturbed in these systems by the magnetic field.

10.6.7 Properties of RF Magnetron Discharges

Experimental investigations and numerical simulations prove that the RF magnetron discharge retains all the typical properties of low-pressure RF CCP discharges, obviously taking into account the peculiarities related to the constant magnetic field.[400–402] The constant horizontal magnetic field perpendicular to the electric field decreases the electron flux and prevents electron losses on the powered electrode. The residence time of the electrons grows, promoting the ionization properties of the electrons.

Space charge sheaths are also created in RF magnetrons near electrodes because electrons leave the discharge gap; however, in the strong magnetic field, the effective electron mobility across the magnetic field may become lower than the ion mobility. The amplitude of the ion oscillations exceeds those of electrons under such conditions. Electron current to the electrode does not appreciably vary during the oscillation period. This is mostly due to diffusion, which is effective in spite of the magnetic field because electron temperature significantly exceeds the ion temperature. Stochastic heating of the magnetized electrons is also possible for some conditions of the oscillating sheath boundaries.[403]

The asymmetric magnetron discharge keeps all the properties, considered in Section 10.118 through section10.120, typical for the nonmagnetized asymmetric low pressure RF discharges. The auto-displacement effect can be observed in the asymmetric RF magnetrons in the presence of a blocking capacitance in the circuit. The battery effect when the external circuit is able to "generate" direct current can also be observed in the magnetron discharges.

Sheath thicknesses in RF magnetrons are much smaller than in nonmagnetized discharges. For this reason, the autodisplacement in the asymmetric magnetrons is also smaller. Energy spectrum of ions bombarding the electrode in magnetrons is similar to those in nonmagnetized discharges. However, the ion energies depend on the value of the applied magnetic field, because the magnetic field determines the sheath voltage. More details regarding ion spectra in RF magnetron discharges can be found in Knypers et al.[404] and Lukyanova et al.[405]

10.6.8 Low-Frequency RF CCP Discharges: General Features

As was discussed in Section 10.3.1, the lower limit of the RF frequency range is generally related to the characteristic frequency of ionization and ion transfer. Ion density in RF discharge plasmas and ion sheaths usually, but not always, can be considered as constant during a period of electromagnetic field oscillation. Thus, typical RF frequencies exceed 1 MHz to avoid effects of "long-distance" ion motion. In this section, however, an example and peculiarities of the low-frequency (less than 100 kHz) RF CCP discharges will be considered.

Low-frequency capacitively coupled nitrogen plasma, for example, is particularly effective in polymer surface treatment with the objective to promote adhesion of silver to polyethylene terephtalate.[406,407] Also, low-frequency CCP nitrogen plasma has been effectively applied to the surface treatment of polyester web to promote adhesion of gelatin-containing layers related to production of photographic film.[408] Low-frequency (≤100 kHz) discharges are effectively used as well for sputter deposition of metals, where the application of lower RF frequencies results in a higher fraction of input power dissipated in the sputter target.[607–612]

The block diagram of the low-frequency RF discharge system is shown in Figure 10.40.[409] As seen from this figure, the discharge is organized in the coplanar con-

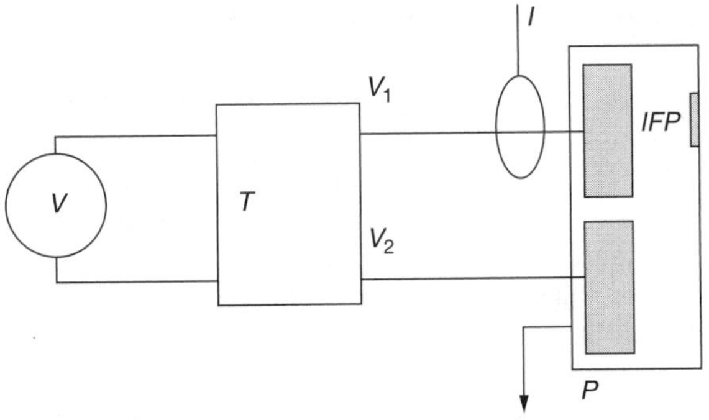

Figure 10.40 Block diagram of plasma source supply electronics, and voltage and current measurement electronics. Power supply and transformer are respectively denoted V and V_1, V_2. The position of the ion flux probe is indicated by IFP.

figuration, where the low frequency RF voltage is applied across two electrodes located in the same plane. Each electrode alternately serves as cathode and anode every half-cycle. In this case, anode conductivity is maintained, charging of insulating areas formed on the cathode is minimized, and arcing is avoided.

The discharge chamber in the experiments of Conti et al.,[409] illustrated in Figure 10.40, consists of two coplanar water-cooled aluminum electrodes (35.6 cm long, 7.6 cm wide) positioned side by side, separated by a gap of 0.32 cm, and housed in the grounded shield. The interior of the grounded enclosure has a volume of roughly 36 cm × 16 cm × 3.3 cm above the electrode pair. Typical pressure in the experiments was quite low—50 to 150 mtorr—and gas-flow rate was 180 to 800 sccm. Applied peak voltage was 900 to 1600 V, power about 300 W, and frequency 40 kHz. Typical current and voltage waveforms are shown in Figure 10.41.

10.6.9 Physical Characteristics and Parameters of Low-Frequency RF CCP Discharges

In the low-frequency discharge, the ion frequency is comparable to the driving frequency and all plasma characteristics are time dependent. The sheath voltages are high and the discharge is essentially sustained, not by bulk processes, but by secondary electron emission from the cathode, provided by ion impact (see the γ-regime of the RF discharges, Section 10.4.2 and Section 10.4.5). The discharge system is somewhat similar to DC glow discharge, with alternating cathode positions from one electrode to another each half-period of the electric field oscillations. Thus, the co-planar system can be considered as two discharges operating in counterphase modes.

Spatial averaged values of the electron and ion concentrations in the low-frequency RF CCP discharge are presented in Figure 10.42 as a function of time and in comparison with the time-evolution of voltage. As seen from this figure, the discharge is actually active only for half of the cycle, when the potential on the specified electrode is negative. During this portion of the cycle, the electron concentration increases because of ionization, while in the other half-cycle, it slowly decreases. The plasma does not completely decay after every cycle, so a new breakdown is not necessary. During the period when voltage is applied, the ion concentration is higher than the electron concentration because quasi neutrality in the sheath region is not achieved.

The corresponding average electron energy variation with time is presented in Figure 10.43, also in comparison with the time evolution of voltage. As seen from Figure10.42 and Figure 10.43, the average electron concentration and energy increase and decrease simultaneously with voltage variation. This can be attributed to intense electron avalanching in the sheath regions, where the electric field is sufficiently high to provide the intense ionization.

It should be pointed out that the presented results were obtained by using the two-dimensional particle-in cell (PIC) code with a Monte Carlo scheme for modeling collisions of charged particles and neutrals.[409] As already discussed in Section 10.6.5, the PIC simulation permits following a large number of representative particles acted

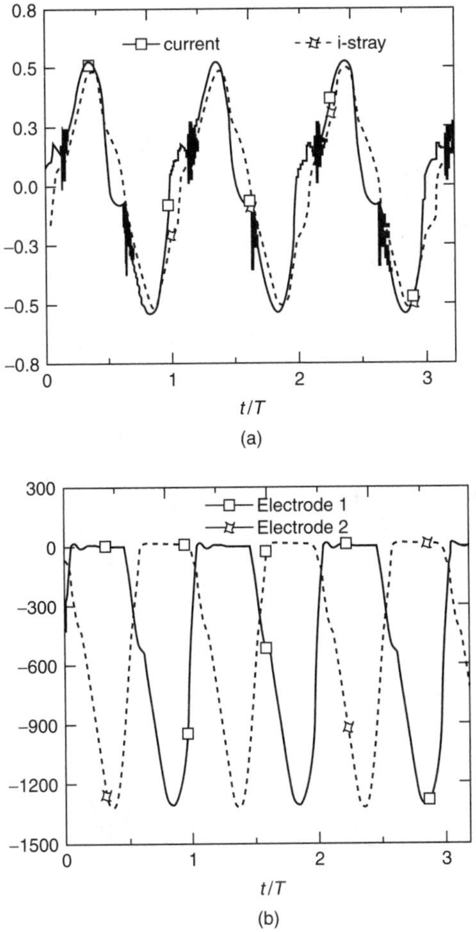

Figure 10.41 (a) Typical voltage waveforms for V_1 and V_2; (b) typical current waveform for I. I-stray denotes the current corrected for stray capacitance to ground. Plasma conditions: 150 mtorr nitrogen; 330 W[409]

upon by the basic forces. The PIC code applied for describing the low-frequency RF coplanar discharge is a modified version of the PDP2 code.[410]

The simulation of the external circuit allows evaluating the total current (displacement current plus conduction current), while the conduction current is directly obtained from the movement of charged particles. The relative contribution of the two current components is illustrated in Figure 10.44, where total current as well as electron and ion conduction currents are plotted as a function of time. Almost 50% of the calculated peak current in the circuit is due to the ion conduction current, which is typical for low-frequency RF discharges. Note that a positive current peak can be observed when the voltage increases from the negative peak value back to zero. This peak is related to a reactive current provided by parasitic capacitance.

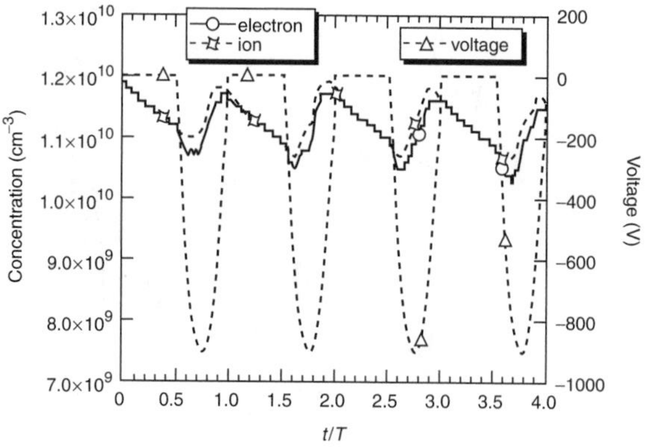

Figure 10.42 Steady-state electron and ion concentrations as function of time for 900 V and 0.15 torr[409]

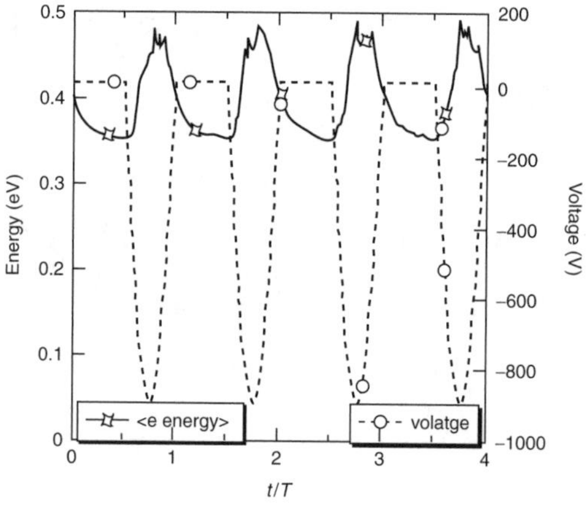

Figure 10.43 Average electron energy at steady states as function of time for 900 V and 0.15 torr[409]

Spatial distribution of charged particles, electrons, and ions between the powered electrode ($x = 0$) and the opposite grounded wall is shown in Figure 10.45 at different moments during an oscillation period. Because the maximum voltage of the powered electrode is near the ground potential, the plasma potential is not driven sufficiently high to have the grounded wall opposite to the electrode serve as a cathode. Furthermore, the high value of the driving voltage produces considerable sheath expansion by forcing the electrons away from the cathode. Consequently, the spatial distribution of charged particles is essentially asymmetric with respect to the center of the axis between the powered electrode and opposite grounded wall (see Figure 10.45). Concentrations of electrons and ions in plasma are shown in Figure 10.46 as a function of applied voltage.

Potential profile between the powered electrode ($x = 0$) and the opposite grounded wall at the moment of the peak voltage is shown in Figure 10.47. The resulting sheath decreases with increasing potential because the plasma density is

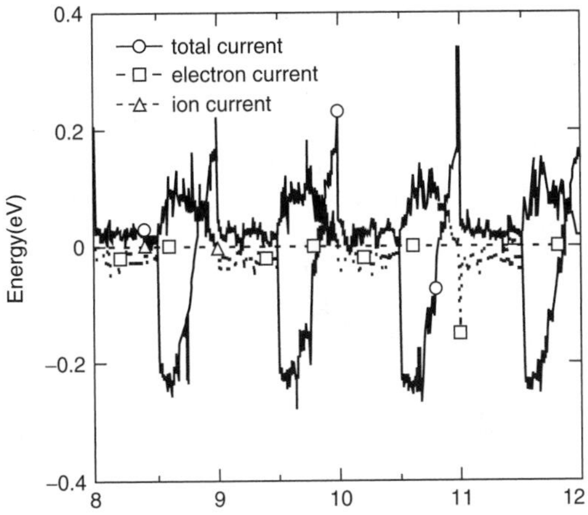

Figure 10.44 Total current and conduction current as funtion of time for 1300 V and 0.15 torr. The electron and ion current are shown separately for 1300 V and 0.15 torr.[409]

higher (and thus the screening effect of plasma is stronger) at higher values of voltage. This is consistent with general theory of γ-discharges.[411] The shape of the sheath voltage is close to parabolic because ion concentration in the sheath is approximately constant (see Figure 10.45).

10.6.10 Electron Energy Distribution Functions (EEDF) in Low-Frequency RF CCP Discharges

The EEDF for electrons in the plasma zone, calculated by the PIC code for the previously described conditions of the low-frequency RF discharge (and divided by square root of energy), is shown in Figure 10.48.[409] As depicted, the EEDF shows a Maxwellian behavior at relatively low energies, but has a long and nonthermalized tail at high energies. The "cut-energy" is well correlated with the specific values of peak voltages.

The origin of the energetic electrons in the low-frequency RF discharge is related to the sheath region near the powered electrode. In general, three groups of electrons can be clearly seen in the EEDF of this discharge. The major one is in the plasma bulk and is characterized by a low energy of about 1 eV. The second group is formed by the scattered γ-electrons and the electrons formed in the cathode sheath by ionization collisions of the γ-electrons and neutrals. Energies of these electrons are in the range of 20 to 50 eV. Finally, the third group of electrons is formed by those γ-electrons that managed to cross the sheath and the plasma bulk and reach the grounded wall retaining the high energy gained in the sheath region. The energy distribution of these highly energetic electrons is quite flat and a single specific temperature value cannot be assigned to this third group.

In other words, the three groups of electrons can be interpreted as follows. The third group of the most energetic electrons can be considered as a high-energy electron beam. The second group can be considered secondary electrons produced

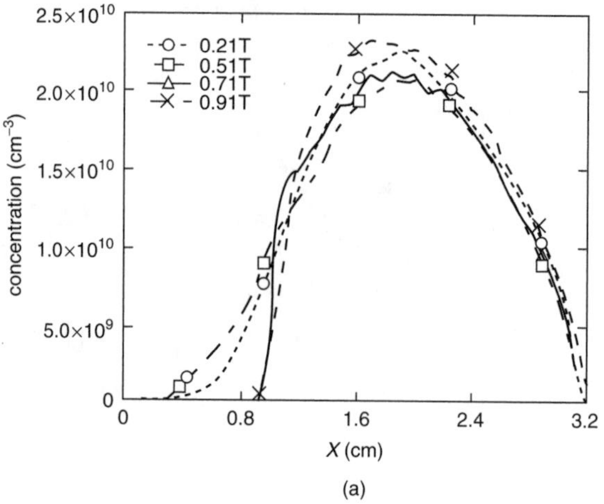

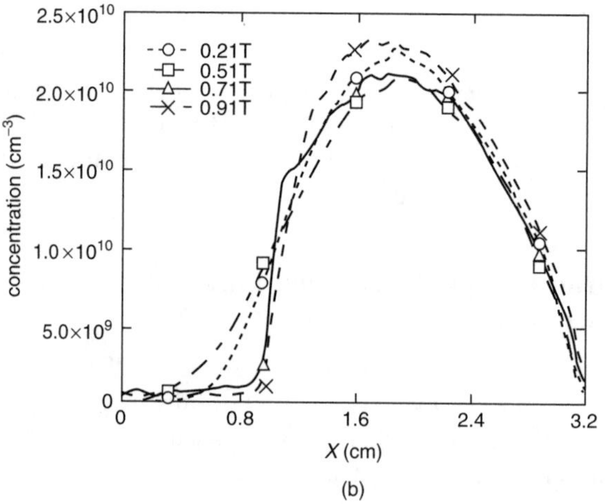

Figure 10.45 (a) Electron and (b) ion concentrations at different moments of the period for 900 V and 0.15 torr[409]

by the beam (such secondary electrons usually have energies of about two to three ionization potentials). Finally, the first group consists of numerous "tertiary" low-energy plasma electrons; such a discharge is an effective source of electrons with energies of about 30 eV. This is especially useful for N^+ atomic ion generation in a molecular nitrogen plasma, which is of interest in the application of low-frequency RF discharges in nitrogen for low-pressure treatment of polymer surfaces.

The existence of these three groups of electrons was experimentally proven by ion flux probe measurements in a low-frequency RF discharge specifically applied for polymer surface treatment to promote adhesion of silver to polyethylene terephtalate.[409] These diagnostic measurements were based on the deposition tolerant flux probe technique developed by Braithwaite and colleagues.[412]

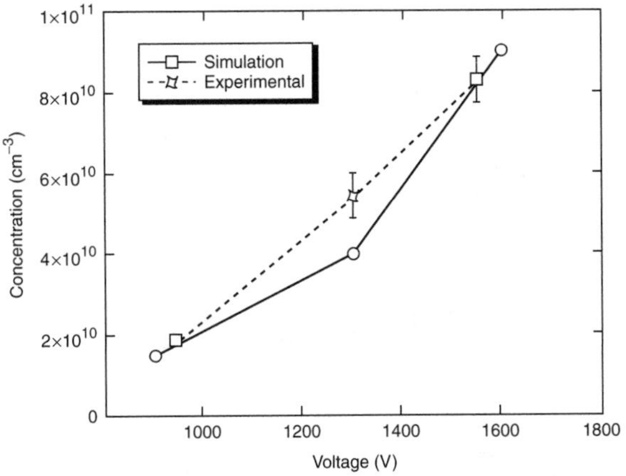

Figure 10.46 Comparison of ion and electron concentrations for different voltages and pressure of 0.15 torr[409]

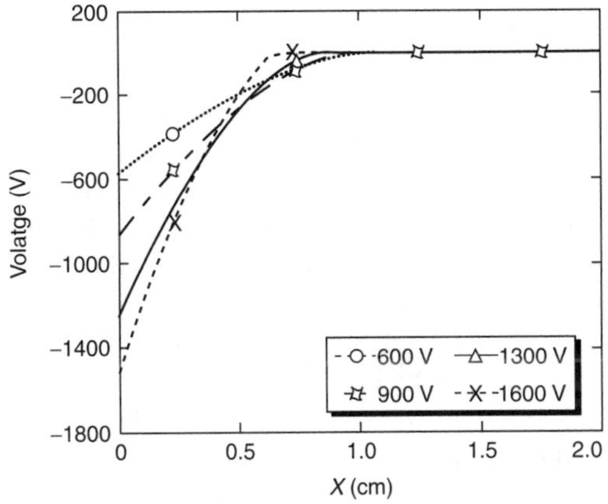

Figure 10.47 Voltage profiles at peak voltage (0.75 t/T) for different applied voltages at 0.15 torr[409]

10.7 NONTHERMAL INDUCTIVELY COUPLED (ICP) DISCHARGES

10.7.1 General Features of Nonthermal ICP Discharges

As was mentioned in Section 10.3, the electromagnetic field in ICP discharges is induced by an inductive coil (Figure 10.21c, d) in which the magnetic field is primarily stimulated and the corresponding nonconservative electric fields necessary for ionization are relatively low. Thus, nonthermal ICP plasma discharges are usually

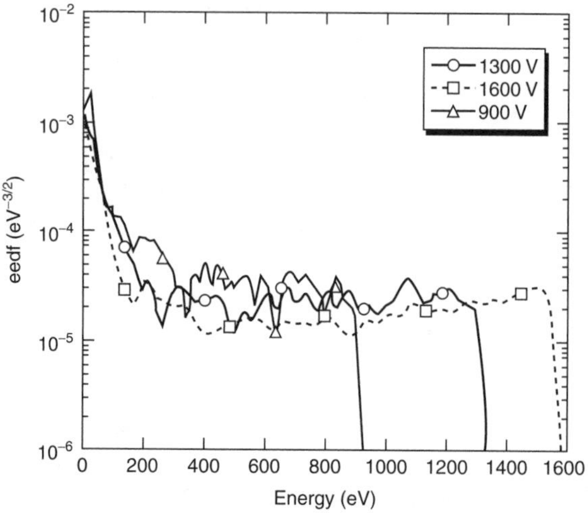

Figure 10.48 Electron energy distribution function of gamma electrons that hit the web plane. Bulk electrons are not included (1600 V, 1300 V, 900 V, 0.15 torr)

configured at low pressures to provide the reduced electric field E/p sufficient for ionization.

The coupling between the inductive coil and the plasma can be interpreted as a transformer in which the coil presents the primary windings and plasma represents the secondary ones. The coil as the primary windings consists of many turns, while plasma has only one. As a result, the ICP discharge can be considered a voltage-decreasing and current-increasing transformer. The effective coupling with the RF power supply requires a low plasma resistance. Thus, the ICP discharges are convenient to reach high currents, high electric conductivity, and high electron density at relatively low values of electric field and voltage.

For example, the low-pressure ICP discharges effectively operate at electron densities of 10^{11} to 10^{12} cm^{-3} (not more than 10^{13} cm^{-3}), which exceeds by an order of magnitude typical values of electron concentration in the capacitively coupled RF discharges. Because of the relatively high values of the electron concentration in ICP discharge systems, they are sometimes referred to as **high-density plasma (HDP)**. These features of ICP (or HDP) discharges make them very attractive in electronics and other applications related to high-precision surface treatment.

Large-scale application of low-pressure ICP discharges in electronics was stimulated by important disadvantages of the RF capacitive discharges, (see Section 10.6.5 and Section 10.6.7). The sheath voltages in the RF CCP discharges are relatively high; at a given power level, this results in low ion densities and ion fluxes, as well as in high ion-bombarding energies. Also, the ion-bombarding energies cannot be varied in the low-pressure capacitively coupled RF discharges independently of the ion flux.

Another important advantage of the ICP discharges for high precision surface treatment (in addition to high plasma density and low pressure) is that RF power is coupled to the plasma across a dielectric window or wall, rather than by direct connection to an electrode in the plasma, as occurs in the CCP discharges. Such "noncapacitive" power transfer to the plasma provides an opportunity to operate at low voltages across all sheaths at the electrode and wall surfaces. The DC plasma potential and energies of ions accelerated in the sheaths is typically 20 to 40 V, which is very good for the numerous surface treatment applications.

In this case, the ion energies can be independently controlled by an additional capacitively coupled RF source, called the RF bias, driving the electrode on which the substrate for material treatment is placed (see Figure10.21c, d). Thus, ICP discharges are able to provide independent control of the ion and radical fluxes by means of the main ICP source power and the ion-bombarding energies by means of power of the bias electrode.

In this section, a brief discussion on the main concepts of the nonthermal ICP discharges will be undertaken; more details can be found in special publications on the subject. Earlier works regarding ICP discharge in cylindrical coil geometry with pressures exceeding 20 mtorr are reviewed by Eckert.[413] More recent developments of the ICP discharges at pressures less than 50 mtorr in planar coil geometry that are especially important in electronics are reviewed by Lieberman and Gottscho.[375]

10.7.2 Inductively Coupled RF Discharges in Cylindrical Coil

ICP discharges in the planar configuration are more interesting for practical applications, but more complicated for physical analysis because of their two-dimensional symmetry. For this reason, consider the inductive RF discharge in a long cylindrical tube placed inside the cylindrical coil. The physical properties of such a discharge are similar to those of the planar one, but they are easier for analysis because of the circular symmetry of this discharge system.

The electric field $E(r)$ induced in the discharge tube, which is located inside the long coil, can be calculated from the Maxwell equation:

$$E(r) = -\frac{1}{2\pi r}\frac{d\Phi}{dt} \tag{10.135}$$

In this equation, Φ is the magnetic flux crossing the loop of radius r perpendicular to the axis of the discharge tube; r is the distance from the discharge tube axis.

The magnetic field is created in this system by the electric current in the coil $I = I_c e^{i\omega t}$ as well as by the electric current in the plasma (which mostly flows in the external plasma layers). Assuming a plasma conductivity σ (r) = constant and considering the current in plasma as a harmonic one $j(r) = j_0(r)e^{i\omega t}$, the following equation for the current density distribution along the plasma radius is obtained:

$$\frac{\partial^2 j_0}{\partial r^2} + \frac{1}{r}\frac{\partial j_0}{\partial r} - \frac{1}{r^2} j_0 = i\frac{\sigma\omega}{\varepsilon_0 c^2} j_0 \tag{10.136}$$

If the pressure is very low and the plasma can be considered as collisionless, then its conductivity is inductive and can be expressed as:

$$\sigma = -i\varepsilon_0 \frac{\omega_p^2}{\omega} \tag{10.137}$$

where ω_p is the plasma frequency. Substituting this plasma conductivity in Equation 10.136 yields the current density distribution in the ICP plasma (finite at $r = 0$) in the form of the modified Bessel function:

$$j_0(r) = j_b I_1\left(\frac{r}{\delta}\right) \tag{10.138}$$

where $I_1(x)$ is the modified Bessel function and δ is the skin-layer thickness (Equation 6.25). Note that if collisions are important, the Bessel function must have a complex argument. In such a case, phase of the current in the plasma also depends on the radius. Current density on the plasma boundary in the vicinity of the discharge tube is determined by the nonperturbed electric field by the plasma conductivity on the boundary of the plasma column:

$$j_b = \sigma \frac{\omega a N}{2\varepsilon_0 c^2 l} I_c \tag{10.139}$$

In this relation, a is the radius of the discharge tube; N is the number of turns in the coil; l is the length of the coil; c is the speed of light; and I_c is the amplitude of current in the coil.

If the skin-layer thickness exceeds the radius of the discharge tube, currents in the plasma do not perturb the electric field. Here the electric field grows linearly with the radius from the axis to the walls of the discharge tube. Such a regime is possible only if the plasma conductivity depends slightly on the electric field; this requires quite an exotic discharge situation, e.g., the application of inert gas with easy-to-ionize additives that are completely ionized.

Under more realistic conditions, plasma conductivity grows with the electric field and the skin-layer thickness is smaller than discharge radius. Then most of electric current is located in the relatively thin δ-layer on the discharge periphery. In this case, the Bessel function for the current density distribution can be simplified to a simple exponential function:

$$j_0(r) \approx j_b \exp\left(\frac{r-a}{\delta}\right) \tag{10.140}$$

10.7.3 Equivalent Scheme of ICP RF Discharges

In general, an ionization balance of charged particles determines electric fields in nonthermal discharges. For this reason, some special coupling or feedback effect is always assumed to establish the electric field on the level necessary to provide the relevant ionization balance in the steady-state discharges. Such coupling in the CCP discharges is due to shielding (or screening) of the electric field in the sheaths. For inductively coupled discharges, the coupling is provided by the external electric circuit.

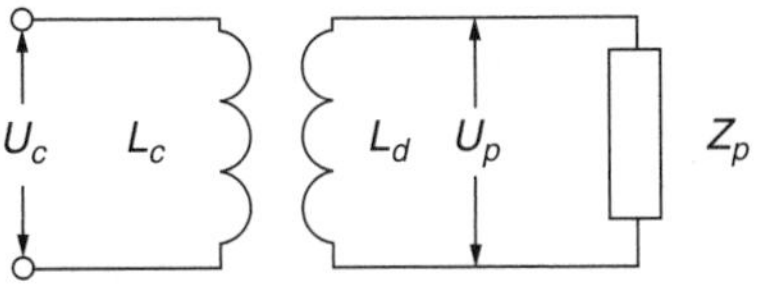

Figure 10.49 Equivalent scheme of an ICP discharge

As was mentioned earlier, the system inductor plasma can be interpreted as a transformer decreasing voltage and increasing the electric current. Thus, the discharge stabilization takes place at high currents and high values of plasma conductivity. In this case, the Bessel function (Equation 10.138) can be simplified to an exponential function (Equation 10.140) and the total current can be considered as concentrated in the skin layer. Taking into account Equation 10.139, the total discharge current then can be related to the current in the coil I_c as:

$$I_d = \sigma E_b \delta l = \sigma\delta \frac{a\omega N}{2\varepsilon_0 c^2} I_c \tag{10.141}$$

where E_b is the electric field in the thin-skin layer on the plasma boundary.

Based on this relation for the total discharge current, the complex inductively coupled plasma impedance can be expressed as:

$$Z_p = \frac{U}{I_d} = \frac{2\pi a}{\sigma\delta l} = \frac{2\pi am}{n_e e^2 \delta l} \nu_{en} + \frac{2\pi am}{n_e e^2 \delta l} i\omega \tag{10.142}$$

In this relation, n_e is the electron concentration and ν_{en} is the frequency of electron-neutral collisions. From this equation, the plasma impedance is seen to have active (first-term) and inductive (second-term) components. Analyzing the equivalent circuit, one must take into account the "geometrical" inductance, which can be attributed to the plasma considered as a conducting cylinder:

$$L_d = \frac{\mu_0 \pi a^2}{l} \tag{10.143}$$

Thus, if the skin-layer thickness is small with respect to the discharge radius, the equivalent scheme of the ICP discharge can be represented as a transformer with a load in the form of impedance (Equation 10.142; see Figure 10.49). The transformer can be characterized by the geometrical plasma inductance (Equation 10.143) and inductance of the coil:

$$L_c = \frac{\mu_0 \pi R^2}{l} N^2 \tag{10.144}$$

where R is radius of the coil; l is the coil length; and N is the number of its windings. Also, the transformer presented in Figure 10.49 should be characterized by the mutual inductance, which can be expressed as:

$$M = \frac{\mu_0 \pi a^2}{l} N \tag{10.145}$$

The system of the electric circuit equation for the equivalent scheme of the radio-frequency ICP discharge (Figure 10.49) includes two equations: one describes the amplitude of the voltage applied to the inductor (the coil):

$$U_c = i\omega L_c I_c + i\omega M I_d \tag{10.146}$$

and the other describes the amplitude of the voltage on the plasma loop:

$$U_p = -I_d Z_p = i\omega M I_c + i\omega L_d I_d \tag{10.147}$$

In these equations, I_c is the amplitude of the electric current in coil and I_d is the amplitude of electric current in plasma. The system of Equation 10.146 and Equation 10.147 describing the equivalent scheme of the ICP discharge is especially helpful in determining plasma parameters for these types of RF discharges.

10.7.4 Analytical Relations for ICP Discharge Parameters

The current in coil I_c is determined by external circuit and can be considered a given parameter. The electric field in the plasma is related to the voltage on the plasma loop as $E_p = U_p/2\pi a$, and its value only logarithmically (weakly) depends on other plasma parameters. The value of the electric field in a plasma at high currents is small relative to one in the idle regime without plasma:

$$E = \frac{\omega M I_c}{2\pi a},$$

where M is the mutual inductance (Equation 10.145). Thus, the voltage drop U_p related to the plasma can be neglected for relatively high currents. The discharge current can then be expressed as:

$$I_d = -\frac{M}{L_d} I_c = -N I_c \tag{10.148}$$

As seen from this equation, the discharge current flows in a direction opposite to the inductor current, and what is very important for practical applications is that the value of the plasma current exceeds that in the inductor.

If the electric field is known, the electron concentration in plasma can be found based on the current density from Equation 10.148:

$$n_e = j\frac{m\nu_{en}}{e^2 E_p} = \frac{N I_c}{l\delta}\frac{m\nu_{en}}{e^2 E_p} \tag{10.149}$$

Using Equation 6.25 and Ohm's law in differential form for the current conducting plasma layer on the discharge periphery, one can derive the following formula for the thickness of skin layer in the ICP discharge:

$$\delta = \frac{2E_p l}{\omega\mu_0 N I_c} \propto \frac{1}{I_c} \tag{10.150}$$

From this relation, the thickness of the skin layer, where most of the ICP current is concentrated, is inversely proportional to the electric current in the inductor coil. Using this skin-layer relation for the electron concentration in the plasma, Equation 10.149 can be rewritten as:

$$n_e = \left(\frac{NI_c}{elE_p}\right)^2 \frac{\omega\mu_0 m\nu_{en}}{2} \propto I_c^2 \tag{10.151}$$

Thus, the plasma density in the ICP discharges is proportional to the square of electric current in inductor coil.

10.7.5 Moderate- and Low-Pressure Regimes of ICP Discharges

Similar to the case of CCP discharges, regimes of the inductively coupled discharges are different at moderate and low pressures. In particular, the electric field is determined differently in these two regimes.

Moderate-Pressure Regime. The energy relaxation length in this regime is less than the thickness of the skin layer. This means that heating of electrons is determined by local values of the electric field and takes place in the skin layer. Therefore, ionization processes as well as plasma luminosity also are concentrated at moderate pressures in the skin layer, which is illustrated in Figure 10.50.

The internal volume of the discharge tube, located closer to the tube axis relative to the skin layer, is filled with the plasma only due to the radial inward plasma diffusion from the discharge periphery. If losses of charged particles in the internal volume resulting from recombination and diffusion along the axis of discharge tube are significant, plasma concentration can be lower in the central part of the discharge than in the periphery.

Balancing the ionization rate in the skin layer of the moderate-pressure discharge and the diffusion flux from the layer to the discharge tube surface, the following logarithmic relation for electric field in the skin layer is obtained:

$$E_p = Bp\sqrt{2} \cdot \ln^{-1} \frac{Apel^2 E_p^3}{\mu_0 \omega^2 m\nu_{en} D_a N^2 I_c^2} \tag{10.152}$$

In this relation, D_a is the coefficient of ambipolar diffusion; A and B are factors that determine the α-coefficient of Townsend (Equation 4.141).

Low-Pressure Regime. The energy relaxation length in this regime exceeds the thickness of the skin layer, so although heating of electrons takes place in the skin layer, ionization processes are effective in the plasma volume, where the electrons have a maximum value of kinetic energy. Distributions of plasma density, amplitude

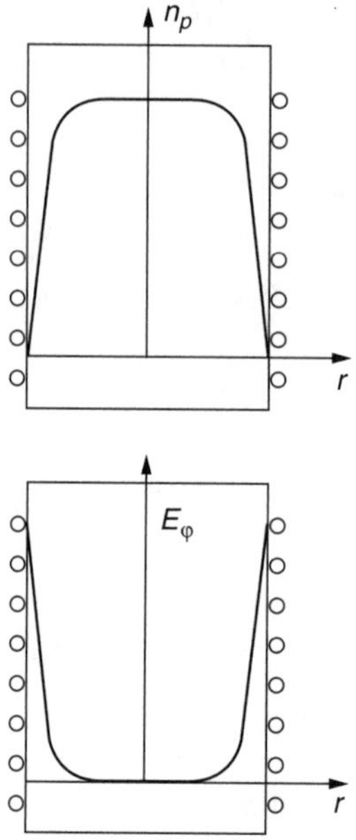

Figure 10.50 Radial distributions of plasma density and electric field in moderate-pressure ICP discharge

of the oscillating electric field, and ambipolar potential along the radius of the low-pressure discharge tube are illustrated in Figure 10.51. Thus, in the low-pressure regime of the radio-frequency ICP discharges, electron concentration on the discharge axis can significantly exceed that in the skin layer.

Electrons spend only part of their lifetime in the skin layer. For this reason, similar to the case of CCP discharges, an average or effective value of electric field can be used. If the electron energy relaxation length exceeds the radius of the discharge tube, the effective average value of the electric field can be estimated as:

$$E_{eff}^2 = \frac{2\delta}{a} E_p^2 \frac{\nu_{en}^2}{\omega^2 + \nu_{en}^2} \tag{10.153}$$

In this relation, δ is the thickness of the skin layer; a is the radius of discharge tube; and E_p is the electric field in the plasma (actually in the skin layer; see Figure 10.51). The value of this electric field also can be determined from the balance of charged particles in the low-pressure ICP discharge:[376]

$$E_p = \left(\frac{Bp}{\nu_{en}}\right)^{2/3} \sqrt[3]{\frac{\mu_0 \omega N I_c a}{l}\left(\omega^2 + \nu_{en}^2\right)} \times \ln^{-2/3}\left[\frac{pea^3}{5.8 \cdot m\nu_{en}^2 D_a} \sqrt{\frac{\mu_0 \omega N I_c a}{l} E_p\left(\omega^2 + \nu_{en}^2\right)}\right] \tag{10.154}$$

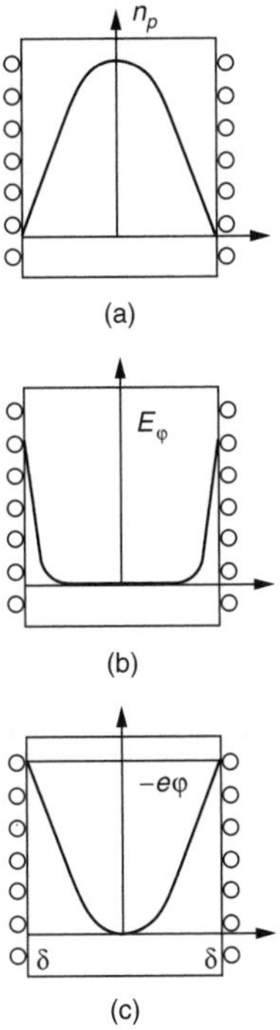

Figure 10.51 Radial distributions of plasma density, electric field, and potential in low pressure ICP-discharge

10.7.6 Abnormal Skin Effect and Stochastic Heating of Electrons

Electrons in the low-pressure discharges move through the skin layer faster than changes in the oscillation period and the following inequality is valid:

$$\frac{v_e}{\delta} > \omega, \nu_{en} \tag{10.155}$$

Here, v_e is the average thermal velocity of electrons and δ is the thickness of the skin layer determined by Equation 6.25.

Each electron receives momentum and kinetic energy while moving across the skin layer and transports them to the plasma zone outside the skin layer, resulting in formation of the electric current outside the skin layer. At the same time, chaotic electrons from the central part of the discharge come to the skin layer without any organized drift velocity, which leads to a decrease of current density and effective plasma conductivity. Thus, the effective skin-layer thickness in the low-pressure discharges under consideration exceeds the value determined by Equation 6.25. This phenomenon is known as the abnormal skin effect.

Again, plasma electrons moving across the skin layer receive not only momentum but also kinetic energy, even without any collisions with neutrals. This results in the effect of stochastic heating of electrons, somewhat similar to that considered in the low-pressure CCP discharge (see Section 10.5.6).

To analyze the thickness δ_c of the abnormal skin layer, estimate at first the drift velocity u, which the plasma electrons receive during the time interval S_c/v_e of their flight across the skin layer:

$$u = \frac{eE}{m}\frac{\delta_c}{v_e} \tag{10.156}$$

Based on this relation for the electron drift velocity, the effective value of conductivity is determined as:

$$\sigma_{eff} = \frac{n_e eu}{E} = \frac{n_e e^2}{m}\frac{\delta_c}{v_e} = \frac{n_e e^2}{m\nu_{eff}} \tag{10.157}$$

Using this expression for the electric conductivity in the general relation (6.25) for skin layer, one obtains the following equation for the abnormal skin-layer thickness δ_c:

$$\delta_c = \frac{c}{\omega_p}\sqrt{\frac{2\nu_{eff}}{\omega}} \tag{10.158}$$

In this relation, ω_p is the plasma frequency (see Equation 6.17); c is the speed of light; and ν_{eff} is the effective frequency:

$$\nu_{eff} = \frac{v_e}{\delta_c} \tag{10.159}$$

which replaces, in this collisionless case, the frequency of electron–neutral collisions. Equation 10.158 leads to the following final equation for thickness of the abnormal skin layer in the collisionless sheath of the ICP discharge:

$$\delta_c = \sqrt[3]{\frac{2c^2 v_e}{\omega\omega_p^2}} \tag{10.160}$$

Stochastic heating of electrons in the low-pressure collisionless regime can be described by using the general collisional formulas and replacing the electron–neutral collision frequency by the effective frequency and, thus, the collisional electric conductivity by the effective one (Equation 10.157). Therefore, in the low-pressure collisionless regime, the electron–neutral collisions are effectively replaced by "collisions" with the skin-layer boundaries.

10.7.7 Planar Coil Configuration of ICP Discharges

ICP discharges in the planar configuration are widely used in electronics and a general scheme is illustrated in Figure 10.52 (such a reactor was applied by Keller

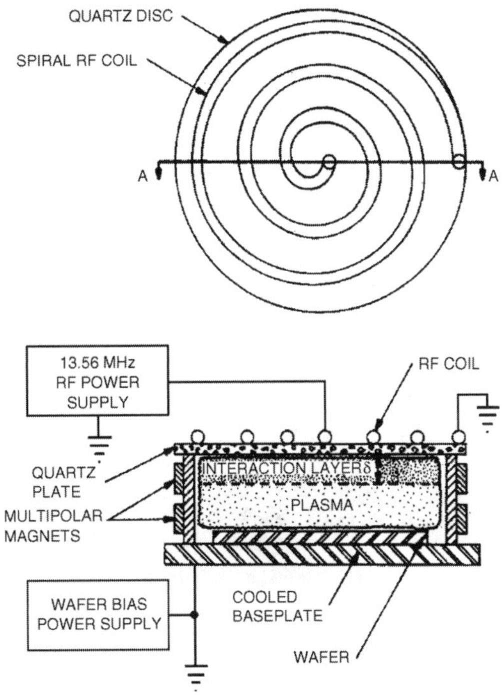

Figure 10.52 ICP parallel plate reactor[264]

and his colleagues at IBM to microelectronic plasma processing, see Roth[264]). This RF discharge scheme is quite similar geometrically to the conventional RF CCP parallel plate reactor. However, here the RF power is applied to a flat spiral inductive coil separated from the plasma by quartz or another dielectric insulating plate.

The RF currents in the spiral coil induce image currents in the upper surface of the plasma (see Figure 10.52) corresponding to the skin layer. Thus, this discharge is inductively coupled and, from the physical point of view, is similar to the simple cylindrical geometry of the inductive coils considered previously. The analytical relations derived for the low-pressure ICP discharge inside an inductive coil can also be applied qualitatively for the planar coil configuration of the inductive RF discharges at low pressures.

As one can see from Figure 10.52, the planar ICP discharge also includes two practically important elements: multipolar permanent magnets and a DC wafer bias. The multipolar permanent magnets are located around the outer circumference of the plasma to improve plasma uniformity and plasma confinement and to increase plasma density. The DC wafer bias power supply is used to control the energy of ions impinging on the wafer, which typically ranges 30 to 400 eV in the planar discharge system under consideration.

Structure of magnetic field lines in the planar coil configuration of the ICP discharges is obviously more complicated than in the case of the cylindrical inductive coil. The RF magnetic field lines in the planar coil configuration in the absence of

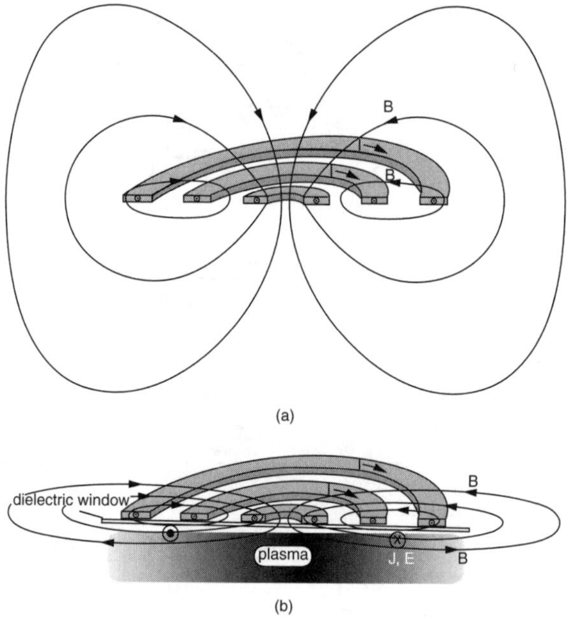

Figure 10.53 RF magnetic field near planar inductive coil: (a) without nearby plasma and (b) with nearby plasma[415a]

plasma are illustrated in Figure 10.53(a).[415a] These magnetic field lines encircle the coil and are symmetric with respect to the plane of the coil.

Deformation of the magnetic field in presence of plasma formed below the coil is shown in Figure 10.53(b). In this case, an azimuthal electric field and an associated current (in the direction opposite to that in the coil) are induced in the plasma skin layer. Both the multiturn coil current and the "single-turn" induced plasma current generate the total magnetic field. As seen from this figure, the dominant magnetic field component within the plasma is vertical near the axis of the planar coil and horizontal away from the axis. More details regarding space distribution of magnetic field, electric currents, and concentration of charged particles in the low-pressure planar ICP discharge can be found in References 414 and 415.

The planar radio-frequency ICP discharge illustrated in Figure 10.52[264] is able to produce uniform plasma and uniform plasma processing of wafers with diameters at least 20 cm. Power level of this reactor is approximately 2 kW—about an order of magnitude larger than the power of a CCP discharge under similar conditions. This results in a higher flux of ions and other active species that accelerate the surface treatment process. For example, ion flux in the system under discussion was 60 mA/cm^2 at a power input of 1600 W, corresponding to a high level of etch rate (1 to 2 μm/min for polyimide film).

The reactor has been operated at frequencies ranging from 1 to 40 MHz, but the usual operating frequency is 13.56 MHz. Operation pressure of this planar ICP discharge is in the range from 1 to 20 mtorr, far lower than typical pressure values of the corresponding CCP discharges (which is about several hundreds mtorr). Obviously, such low-pressure values are desirable for CVD (chemical vapor deposition) and etching technologies because they imply longer mean free paths and little scattering of ions and active species before they reach the wafer, thus improving the surface treatment processes.

10.7.8 Helical Resonator Discharges

The helical resonator discharge is a special case of the low-pressure RF ICP discharges. The helical resonator consists of an inductive coil (helix) located inside the cylindrical conductive screen and can be considered as a coaxial line with an internal helical electrode. A general schematic of the helical resonator plasma source is shown in Figure 10.54.

Electromagnetic wave propagates in such a coaxial line with a phase velocity much lower than the speed of light: $v_{ph} = \omega/k << c$, where k is the wavelength and c is the speed of light. This property allows the helical resonator to operate in the megahertz frequency range, which permits generating low-pressure plasma in these systems. The coaxial line of the helical discharge becomes resonant when an integral number of quarter waves of the RF field fit between the two ends of the system. The criterion of the simplest resonance can then be expressed as:

$$2\pi r_h N = \frac{\lambda}{4} \tag{10.161}$$

where r_h is the helix radius; N is the number of turns in the coil; and λ is the electromagnetic wavelength in vacuum.

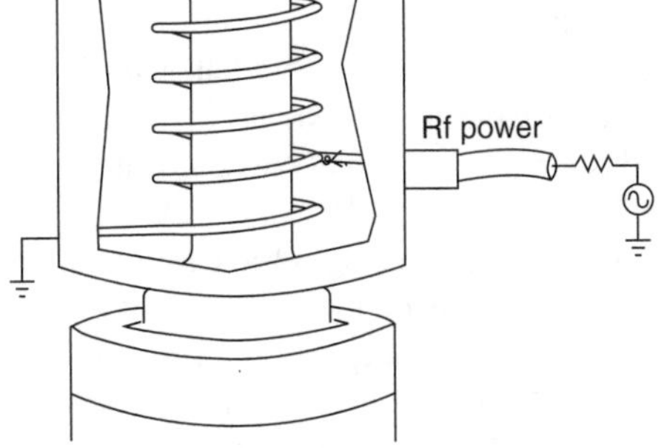

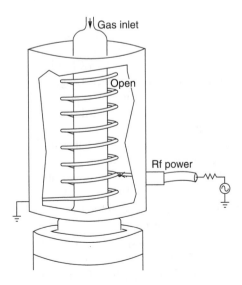

Figure 10.54 Helical resonator

Helical resonator discharges effectively operate at radio frequencies of 3 to 30 MHz with simple hardware and do not require a DC magnetic field. The resonators exhibit high Q-values (typically 600 to 1500 without plasma). This means that, in the absence of plasma, the electric fields are quite large, facilitating the initial breakdown of the system. Also, helical resonator discharges have high characteristic impedance and can be operated without a matching network.

Because of the resonance, large voltages necessarily appear between the open end of the helix and the plasma. Therefore, the electric field is not exactly azimuthal in the helical resonator, and the discharge cannot be considered purely inductively coupled. On the other hand, the discharge sizes in this system are close to the electromagnetic wavelength, which means that the helical resonator discharge in some sense is similar to microwave discharges.

10.8 NONTHERMAL LOW-PRESSURE MICROWAVE AND OTHER WAVE-HEATED DISCHARGES

10.8.1 Nonthermal Wave-Heated Discharges

Quasi-optical interaction of electromagnetic waves with plasma, electromagnetic wave propagation, and organization of microwave discharges in waveguides and resonators at different modes were already considered in some detail in Section 10.2.1 through Section 10.2.4 with regard to thermal plasma generation in wave-heated discharges. In Section 10.8 and Section 10.9, nonthermal and strongly nonequilibrium plasma generation in microwave and other wave-heated discharges will be considered. Low-pressure homogeneous discharges, used mostly for surface treatment processes, will be discussed in the present section; less uniform, moderate-pressure microwave discharges, mostly related to bulk gas treatment, will be considered in Section 10.9.

Consider the low-pressure, wave-heated plasmas and focus on the following three discharge systems: electron–cyclotron resonance (ECR) discharges, helicon discharges, and surface wave discharges. In ECR discharges, a right circularly polarized wave (usually at microwave frequencies, e.g., 2.45 GHz) propagates along the DC magnetic field (usually quite strong, 850 G at resonance) under the conditions of electron–cyclotron resonance, which provides the wave energy absorption through a collisionless heating mechanism.

In the helicon discharges, an antenna radiates the whistler wave, which is subsequently absorbed in plasma by collisional or collisionless mechanisms. The helicon wave-heated discharges are usually excited at RF frequencies (typically 13.56 MHz), and a weak magnetic field of about 20 to 200 G is required for the wave propagation and absorption. Finally, in the case of surface wave discharges, a wave propagates along the surface of the plasma and is absorbed by collisional heating of the plasma electrons near the surface. The heated electrons then diffuse from the surface into the bulk plasma. Surface wave discharges can be excited by RF or microwave sources and do not require DC magnetic field.

The plasma potential with respect to all wall surfaces for wave-heated discharges is relatively low (about five electron temperatures, $5T_e$) similar to the case of ICP discharges. This results in the effective generation of high-density plasmas at reasonable absorbed power levels, making wave-heated discharges especially useful for intensive surface treatment processes.

10.8.2 Electron Cyclotron Resonance (ECR) Microwave Discharges: General Features

A thermal discharge sustained by microwave radiation requires only a sufficient heat energy flux to provide thermal balance and thermal ionization (see Section 10.2.5 through Section 10.2.7). In contrast, nonthermal discharges require a sufficient level of electric field for heating electrons and effective nonthermal ionization. The necessary high level of electric field can be provided by applying the relatively high microwave power and power density typical for the moderate-pressure regime (see Section 10.9), or by applying resonators with a high value of Q. Another widely used approach applies a steady magnetic field and effective electron heating due to the electron cyclotron resonance (ECR).

ECR resonance between an applied electromagnetic wave frequency ω and the electron cyclotron frequency $\omega_{Be} = eB/m$ (see Equation 4.106) allows electron heating sufficient for ionization at relatively lower values of electric fields in the electromagnetic wave. For calculations of the electron cyclotron frequency, it is convenient to use the following numerical formula: f_{Be} (MHz) = $2.8 \cdot B(G)$.

This effective electron heating in the ECR resonance takes place because the gyrating electrons rotate in phase with the right-hand polarized wave, seeing a steady electric field over many gyro-orbits. Obviously, pressure in the ECR discharge system is low in order to have a low electron–neutral collision frequency, $\nu_{en} << \omega_{Be}$, and to provide electron gyration sufficiently long to obtain the energy necessary for ionization. This phenomenon determines the key effect guiding the physics of ECR microwave discharges.

Furthermore, the injection of microwave radiation along the magnetic field (with $\omega_{Be} > \omega$ at the entry into the discharge region) allows a wave to propagate to the absorption zone $\omega \approx \omega_{Be}$ even in dense plasma with $\omega_{pe} > \omega$. Electromagnetic wave propagation in nonmagnetized plasma is obviously impossible at frequencies below the plasma frequency (see Section 6.6.5) or, in other words, when the plasma density exceeds the critical value (see Equation 6.200). High plasma densities lead to the total reflection of electromagnetic waves from the nonmagnetized plasma. However, application of magnetic fields permits propagation of electromagnetic waves even at high plasma densities exceeding the critical value; this is important for practical applications favoring high densities to accelerate the surface treatment processes.

This effect can be explained by analyzing the dispersion equation (6.210) for electromagnetic wave propagation in magnetized plasma. From this dispersion equation, it is seen that the right-hand polarized wave (corresponding to the "–" sign) has a real wave number even at high densities, $\omega < \omega_{pe}$, if the magnetic field is sufficiently large and $\omega_{Be} = eB/m > \omega$.

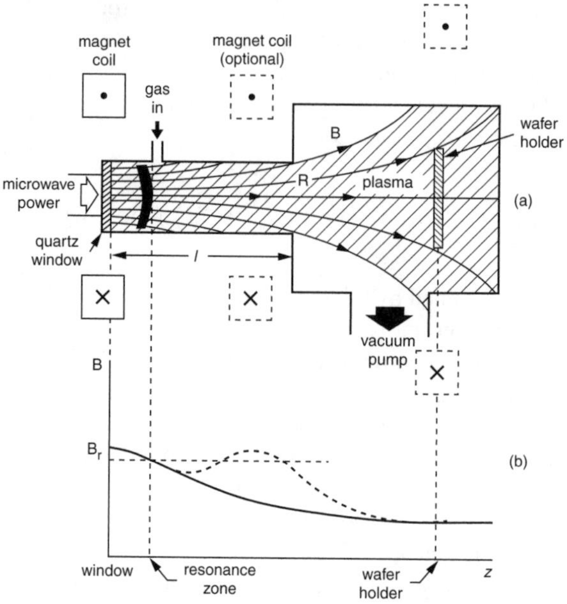

Figure 10.55 Typical high-profile ECR system: (a) geometric configuration; (b) axial magnetic filed variation, showing one or more resonance zones (From Lieberman, M. A. and Lichtenberg, A. J. 1994. *Principles of Plasma Discharges and Material Processing*. New York: John Wiley & Sons)

10.8.3 General Scheme and Main Parameters of ECR Microwave Discharges

A schematic of the ECR-microwave discharge with microwave power injected along the axial nonuniform magnetic field is shown in Figure 10.55.[13] The magnetic field profile is chosen in this discharge system to provide effective propagation of the electromagnetic wave from the quartz window to the zone of the ECR resonance without major reflections even at high plasma densities. Special magnetic field profiles can provide multiple ECR resonance positions as shown in the figure by the dashed line.

Low-pressure gas introduced into the discharge chamber forms a highly nonequilibrium plasma that streams and diffuses along the magnetic field toward a wafer holder shown in Figure 10.55. Energetic ions and free radicals generated within the entire discharge region are then able to provide the necessary surface treatment effect. Additionally, a magnetic field coil at the wafer holder can be used to modify the uniformity of the etch or deposition processes; this is the principal technological application of this discharge system.

Typical ECR microwave discharge parameters are pressure: 0.5 to 50 mtorr; power: 0.1 to 5 kW; characteristic microwave frequency: 2.45 MHz; volume: 2 to 50 L; magnetic field: about 1 kG; plasma density: 10^{10} to 10^{12} cm^{-3}; ionization degree: 10^{-4} to 10^{-1}; electron temperature: 2 to 7 eV; ion acceleration energy: 20 to 500 eV; and typical source diameter: 15 cm.

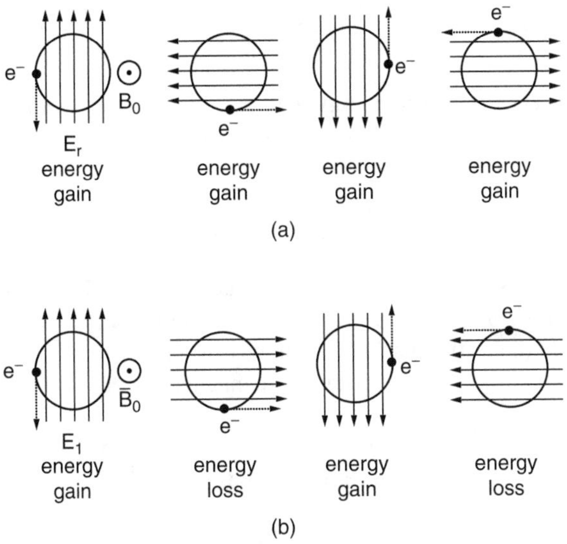

Figure 10.56 Mechanism of ECR heating: (a) continuous energy gain for right-hand polarization; (b) oscillating energy for left-hand polarization

10.8.4 Electron Heating in ECR Microwave Discharges

Consider the ERC microwave discharge sustained at low pressure by a linearly polarized electromagnetic wave. This wave can be decomposed into the sum of two counter-rotating, circularly polarized waves: right-hand polarized and left-hand polarized. The basic physical principle of the ECR heating of magnetized electrons is illustrated in Figure 10.56.

As seen from the figure, the electric field vector of the right-hand polarized wave rotates around the magnetic field at frequency ω, while an electron in the uniform magnetic field also gyrates in the same "right-hand" direction at frequency ω_{Be}. Thus, at the ECR resonance conditions, $\omega = \omega_{Be}$, the electric field continuously transfers energy to the electron providing its effective heating. In contrast, the left-hand polarized electromagnetic wave rotates in the direction opposite to the direction of the electron gyration. So, at the ECR resonance conditions $\omega = \omega_{Be}$, for a quarter of the period, the electric field accelerates the gyrating electrons, and for another quarter of the period slows them down, resulting in no average energy gain.

One should note that the described electron heating effect takes place only close to the ECR conditions, which are necessary for continuous energy transfer from microwave to electron. Because the actual magnetic field and electron cyclotron frequency are not constant along the z axis of the discharge (Figure 10.56), the actual electron heating takes place only locally, near the ECR resonance point, where the electron–cyclotron frequency can be expressed as:

$$\omega_{Be}(z) = \omega\left(1 + \frac{1}{\omega}\frac{\partial \omega_{Be}}{\partial z}\Delta z\right) = \omega\left(1 + \frac{e}{\omega m}\frac{\partial B}{\partial z}\Delta z\right) \tag{10.162}$$

Then, the average electron energy gain per one pass across the ECR resonance zone in the collisionless regime can be calculated as:

$$\varepsilon_{ECR} = \frac{\pi e E_r^2}{v_{res}\left|\partial B/\partial z\right|} \tag{10.163}$$

In this relation, E_r is the amplitude of right-hand polarized electromagnetic wave (which is half of the linearly polarized microwave amplitude); $\partial B/\partial z$ is the gradient of magnetic field along the discharge axis near the resonance point; and v_{res} is the component of electron velocity parallel to magnetic field also in the vicinity of resonance point.

The width of the resonance zone along the discharge axis z, where most of the energy is transferred to the electron, is given as:

$$\Delta z_{res} = \sqrt{\frac{2\pi m v_{res}}{e\left|\partial B/\partial z\right|}} \tag{10.164}$$

Finally, the absorbed electromagnetic wave power per unit area (or microwave energy flux) can be expressed by the following formula:

$$S_{ECR} = \frac{\pi n_e e^2 E_r^2}{e\left|\partial B/\partial z\right|} \tag{10.165}$$

where n_e is the electron density in the ECR resonance zone.

More details related to the theory and modeling of electromagnetic wave propagation and absorption under ECR conditions, as well as theory and modeling of ECR microwave discharges, can be found in Reference 416 through Reference 418. More experimental facts regarding this type of low-pressure microwave discharge can be found in Reference 419 and Reference 420.

10.8.5 Helicon Discharges: General Features

Another high-density plasma (HDP) discharge that can be used for different material processing applications is the helicon discharge. The HDP plasma generation in the helicon discharge was first investigated by Boswell.[421] The detailed theory of this discharge and the general propagation and absorption of the helicon mode in plasma was developed by Chen.[422]

Helicon discharges are sustained by electromagnetic waves propagating in magnetized plasma in the so-called helicon modes. The driving frequency in these discharges is typically in the radio-frequency range of 1 to 50 MHz (the industrial radio frequency of 13.56 MHz is commonly used for material processing discharges). It is interesting to note that, in contrast to the RF discharges considered in Section

10.3 through Section 10.7, the helicon discharges can be considered as wave heated even though they operate in the radio-frequency range. This can be explained by taking into account that the phase velocity of electromagnetic waves in magnetized plasma can be much lower than the speed of light (see Section 6.6.7 and Section 10.8.7). This provides the possibility to operate in a wave propagation regime with wavelengths comparable with the discharge system size even at radio frequencies that are much below the microwave frequency range.

The magnetic field in helicon discharges applied for material processing varies from 20 to 200 G (for fundamental plasma studies, it reaches 1000 G) and is much below the level of magnetic fields applied in ECR microwave discharges. Application of lower magnetic fields is an advantage of the helicon discharges. Plasma density in these wave-heated discharges applied for material processing is about 10^{11} to 10^{12} cm^{-3}, but in some special cases can reach very high values of about 10^{13} to 10^{14} cm^{-3}.

Excitation of the helicon wave is provided by an RF antenna that couples to the transverse mode structure across an insulating chamber wall. The electromagnetic wave mode then propagates along the plasma column in the magnetic field, and plasma electrons due to collisional or collisionless damping mechanisms absorb the mode energy.

A typical general schematic of a helicon discharge is illustrated in Figure 10.57. As one can see from the figure, the material processing chamber is located downstream from the plasma source. The plasma potentials in the helicon discharges are

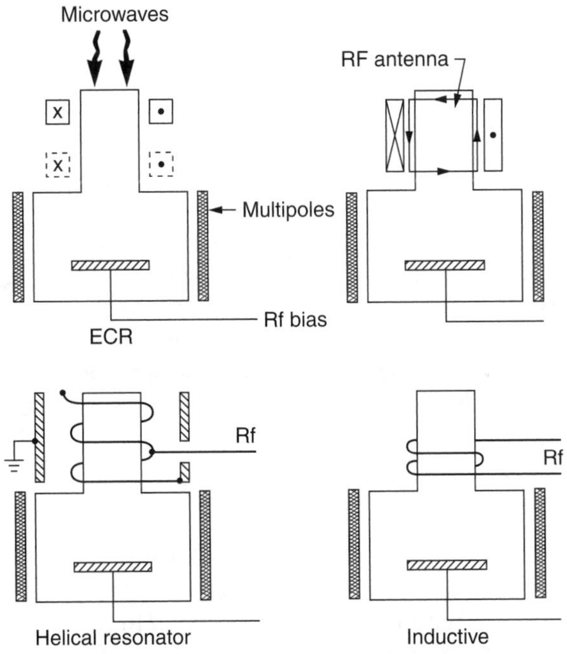

Figure 10.57 Helicon discharge

typically low, about 15 to 20 V, similar to ECR microwave discharges. Important advantages of the helicon discharges with respect to ECR-discharges are related to relatively low values of magnetic field and applied frequency. However, the resonant coupling of the helicon mode to the antenna can lead to nonsmooth variation of the plasma density with source parameters. This effect, known as "mode jumps," restricts the operating regime for a given design of plasma source.

10.8.6 Whistlers and Helicon Modes of Electromagnetic Waves Applied in Helicon Discharges

Consider helicon discharges using a simple analysis of propagation of the helicon modes of electromagnetic waves in magnetized plasma. The helicons are propagating electromagnetic "whistler" wave modes in an axially magnetized, finite diameter plasma column. The electric and magnetic fields of the helicon modes have radial, axial, and, usually, azimuthal variations. They propagate in a low-frequency, high plasma density regime with relatively low magnetic fields, which can be characterized by the following frequency limitations:

$$\omega_{LH} << \omega << \omega_{Be}, \quad \omega_{pe}^2 >> \omega\omega_{Be} \tag{10.166}$$

In these inequalities, ω_{pe} is the electron plasma frequency; ω_{Be} is the electron cyclotron frequency; and ω_{LH} is the lower hybrid frequency of the electromagnetic field in magnetized plasma, which occurs taking into account the ions' mobility (see Section 6.6.8):

$$\frac{1}{\omega_{LH}^2} \approx \frac{1}{\omega_{pi}^2} + \frac{1}{\omega_{Be}\omega_{Bi}} \tag{10.167}$$

where ω_{pi} is the plasma ion frequency (Equation 6.137) and ω_{Bi} is the ion cyclotron frequency (Equation 6.212).

The right-hand polarized electromagnetic waves in magnetized plasma with frequencies between ion and electron cyclotron frequencies $\omega_{Bi} << \omega << \omega_{Be}$ are known as the **whistler waves**. In the particular case of the right-hand polarized electromagnetic wave propagation along the magnetic field at frequencies below the frequency of the electron cyclotron resonance, the dispersion equation (6.210) can be rewritten in the following form:

$$\frac{k^2c^2}{\omega^2} = 1 + \frac{\omega_{pe}^2}{\omega\omega_{Be}} \tag{10.168}$$

As seen from the dispersion equation (10.168), propagation of the electromagnetic whistler waves is possible at frequencies below the plasma frequency $\omega < \omega_{pe}$. Taking into account the helicon frequency conditions (10.166) and introducing the wave number $k_0 = \omega/c$ corresponding to the electromagnetic wave propagation without plasma, the dispersion equation for the whistler waves (10.168) becomes:

$$\omega = \frac{k_0^2 \omega_{pe}^2}{k^2 \omega_{Be}} \tag{10.169}$$

The whistler waves can propagate at an angle to the axial magnetic field. Thus, the dispersion equation for the whistlers can be rewritten in more general form as:

$$\omega = \frac{k_0^2 \omega_{pe}^2}{k k_z \omega_{Be}} \tag{10.170}$$

where $k = \sqrt{k_\perp^2 + k_z^2}$ is the wave-vector magnitude, which takes into account not only axial k_z but also radial component $k_\perp$.

The helicon frequency condition (10.166) requires $\omega_{pe}^2 >> \omega_{Be}\omega$, which together with the dispersion equation (10.169) shows that $k^2 >> k_0^2$. This means that wavelengths of the helicon waves in the magnetized plasma are much less than those of electromagnetic waves of the same frequency without magnetic fields. For this reason, in contrast to RF CCP and RF ICP discharges, helicon discharges can be characterized by wavelengths comparable with the typical discharge size and can be considered as wave heated, even though they operate in a relatively low radio-frequency range.

10.8.7 Antenna Coupling of Helicon Modes and Their Absorption in Plasma

Helicons are a superposition of the low-frequency whistler waves propagating at a common fixed angle to the axial magnetic field. The helicon modes are mixtures of electromagnetic ($div\vec{E} \approx 0$) and quasi-static ($curl\vec{E} \approx 0$) fields, which can be presented as:

$$\vec{E}, \vec{H} \propto \exp i(\omega t - kz - m\theta) \tag{10.171}$$

where θ is the angle between the wave propagation vector and magnetic field and the integer m specifies the azimuthal mode.

Helicon sources have been developed based on excitation of the $m = 0$ and $m = 1$ modes. The $m = 0$ mode is axisymmetric and the $m = 1$ mode has a helical variation; therefore, both modes generate time-averaged axisymmetric field intensities. The transverse electric field patterns for the $m = 0$ and $m = 1$ modes and the way they propagate along the axial magnetic field are shown in Figure 10.58.[422] The quasi-static and electromagnetic axial electric field components exactly cancel in the undamped helicon modes, which means that the total $E_z = 0$. Thus, the antenna is able to couple to the transverse electric or magnetic field to excite the modes.

To design an RF antenna for efficient power coupling, its length should be correctly related to magnetic field and plasma density. The following simplified formula derived for the case $k_z >> k_\perp$ can be practically used for this purpose if electron density is not very high:[423]

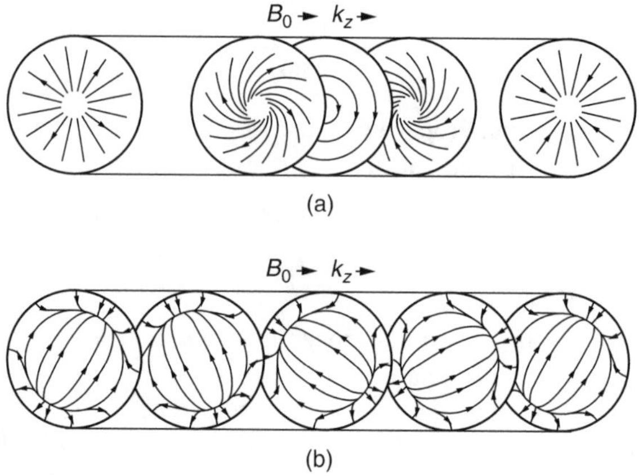

Figure 10.58 Transverse electric fields of helicon modes at five different axial positions: (a) $m = 0$; (b) $m = 1$[422]

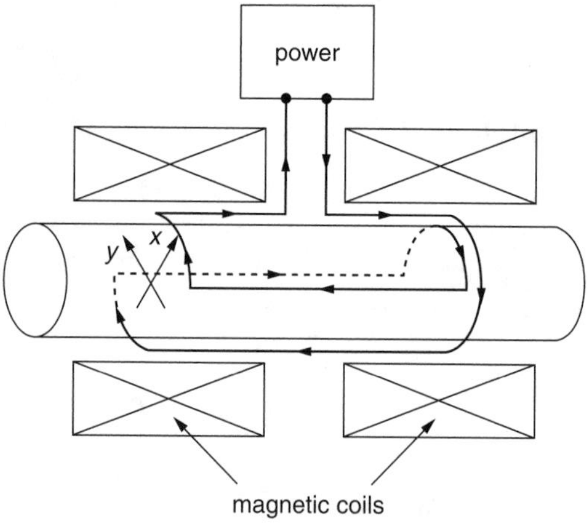

Figure 10.59 Helicon mode $m = 1$ excitation by antenna

$$\lambda_z = \frac{2\pi}{k_z} = \frac{3.83}{R} \frac{B}{e\mu_0 n_e f} \tag{10.172}$$

In this relation, R is the radius of insulating (or conducting) wall; n_e is the plasma density; and $f = \omega/2\pi$ is the electromagnetic wave frequency.

A typical schematic of a radio-frequency antenna to excite the helicon $m = 1$ mode is shown in Figure 10.59. The antenna generates a B_x radio-frequency magnetic field over an axial antenna length, which can couple to the transverse magnetic field of the helicon mode. The antenna also induces an electric current within the plasma column just beneath each horizontal wire in a direction opposite to currents shown in the figure. This current produces a charge of opposite signs at the two ends of the antenna. In turn, these charges generate a transverse quasi-static RF electric field E_y that can couple to the transverse quasi-static fields of the helicon mode (see Figure 10.58b).

The helicon mode energy can be transferred to plasma electron heating by collisional damping or collisionless Landau damping (see Section 6.5.6). The collisional damping mechanism transfers the electromagnetic wave energy to the thermal bulk electrons, while the collisionless Landau damping preferentially heats nonthermal electrons with energies much exceeding the bulk electron temperature. The collisional damping mechanism dominates at relatively higher pressures, while at lower pressures (less than 10 mtorr in the case of argon), the Landau damping dominates.

10.8.8 Electromagnetic Surface Wave Discharges: General Features

Electromagnetic waves can propagate along the surface of a plasma column and be absorbed by the plasma, sustaining the surface wave discharge. Such surface waves, which have strong fields only near the plasma surface, were first described by Smullin and Chorney[424] and Trivelpiece and Gould,[425] and then extensively investigated by Moisan and Zakrzewski.[426]

The surface wave discharges can generate high-density plasma (HDP) with diameters as large as 15 cm. The absorption lengths of the electromagnetic wave surface modes are quite long in comparison with the ECR microwave discharge (see Section 10.8.2). The surface wave discharge typically operates at frequencies in the microwave range of 1 to 10 GHz without an imposed axial magnetic field. Komachi[629] has developed planar rectangular configurations of this discharge that may be effective for large area material processing applications.

Damping in both directions away from the surface, the electromagnetic surface wave can be arranged in different configurations. One of them, a planar configuration on the plasma–dielectric interface, will be considered in more detail in the following section. In another configuration, plasma is separated from a conducting plane by a dielectric slab. This planar system also admits propagation of a surface wave that decays into the plasma region. Although this electromagnetic wave does not decay into the dielectric, it is confined within the dielectric layer by the conducting plane. Finally, a surface wave also is able to propagate in the cylindrical discharge geometry. In this case, the surface wave propagates on a nonmagnetized plasma column confined by a thick dielectric tube.

10.8.9 Electric and Magnetic Field Oscillation Amplitudes in Planar Surface Wave Discharges

The simplest planar surface wave configuration can be described in the following way. The electromagnetic surface wave is supported at an interface between a dielectric and plasma. At the interface between a semi-infinite plasma and dielectric, a solution can be found for which the wave amplitude decays in both directions away from the plasma–dielectric interface. This solution actually corresponds to the surface wave discharge.

Assuming that the semi-infinite plasma zone is located in positive semispace $x > 0$, and that the electromagnetic wave propagates in the direction of z, the amplitude of magnetic field oscillations in plasma can be presented as:

$$H_{yp} = H_{y0} \exp\left(-|\alpha_p| x - ik_z z\right) \tag{10.173}$$

In this expression, H_{y0} is the amplitude of magnetic field component directed along the plasma–dielectric interface; α_p characterizes the electromagnetic wave damping in plasma; and k_z is the wave number in the direction of electromagnetic wave propagation.

Assuming that magnetic field H_y in the electromagnetic surface wave is continuous across the interface at $x = 0$, the amplitude of the magnetic field oscillations in dielectric can be expressed similarly to Equation 10.173 as:

$$H_{yd} = H_{y0} \exp\left(|\alpha_d| x - ik_z z\right) \tag{10.174}$$

where α_d characterizes the electromagnetic wave damping in dielectric. As one can see from Equation 10.173 and Equation 10.174, the magnetic field oscillations decrease in plasma where $x > 0$ and in dielectric where $x < 0$.

From the wave equation, the damping coefficients α_d and α_p for the transverse electromagnetic surface waves can be related to the wave number k_z as:

$$-\alpha_d^2 + k_z^2 = \varepsilon_d \frac{\omega^2}{c^2} \tag{10.175}$$

$$-\alpha_p^2 + k_z^2 = \varepsilon_p \frac{\omega^2}{c^2} \tag{10.176}$$

In these relations: ε_d is the dimensionless dielectric constant of the dielectric semispace under consideration; and ε_p is the dimensionless plasma dielectric constant (6.178), which can be expressed in the collisionless regime as a function of the plasma frequency ω_{pe}:

$$\varepsilon_p = 1 - \frac{\omega_{pe}^2}{\omega^2} \tag{10.177}$$

The electric field amplitude from dielectric (d) and plasma (p) sides can then be related to magnetic field amplitude based on Maxwell equations:

$$E_{zd} = H_{y0} \frac{\alpha_d}{i\omega\varepsilon_0\varepsilon_d} \exp\left(|\alpha_d| x - ik_z z\right) \tag{10.178}$$

$$E_{zp} = -H_{y0} \frac{\alpha_p}{i\omega\varepsilon_0\varepsilon_p} \exp\left(-|\alpha_p| x - ik_z z\right) \tag{10.179}$$

As one can see from Equation 10.173 and Equation 10.174, the electric field oscillations, similar to the case of magnetic fields (see Equation 10.173 and Equation 10.174), also decrease in plasma where $x > 0$ and in dielectric where $x < 0$.

10.8.10 Electromagnetic Wave Dispersion and Resonance in Planar Surface Wave Discharges

Taking into account the continuity of E_z at the plasma–dielectric interface $x = 0$, one can derive, based on Equation 10.178 and Equation 10.179, the relation between the damping coefficients α_d and α_p in plasma and dielectric zones as:

$$\frac{\alpha_p}{\varepsilon_p} = -\frac{\alpha_d}{\varepsilon_d} \tag{10.180}$$

Substituting Equation 10.175 and Equation 10.176 into Equation 10.180, one obtains the relation, which is free of the unknown damping coefficients α_d and α_p:

$$\varepsilon_d^2\left(k_z^2 - \varepsilon_p\frac{\omega^2}{c^2}\right) = \varepsilon_p^2\left(k_z^2 - \varepsilon_d\frac{\omega^2}{c^2}\right) \tag{10.181}$$

This relation can be solved for the wave number k_z, which, taking into account Equation 10.177, leads to the final dispersion equation for the planar surface waves:

$$k_z = \sqrt{\varepsilon_d}\,\frac{\omega}{c}\sqrt{\frac{\omega_{pe}^2 - \omega^2}{\omega_{pe}^2 - (1+\varepsilon_d)\omega^2}} \tag{10.182}$$

This dispersion equation in form of dependence $k_z c/\omega_{pe}$ (ω/ω_{pe}) is illustrated in Figure 10.60. It is interesting to note that propagation of the surface waves is possible at lower electromagnetic field frequencies in contrast to the conventional case of electromagnetic wave propagation in plasma (see the dispersion equation 6.191 to compare). In the case of conventional dispersion (Equation 6.191), electromagnetic wave propagation is possible only when the plasma density is lower than the critical value (Equation 6.200). No such kind of limit for surface wave discharges exists; thus, an opportunity is provided to operate these discharges at high plasma densities in the HDP-regimes.

As one can see from the dispersion equation (10.182) and Figure 10.60, the wave number k_z of the electromagnetic surface wave is real and wave propagation is possible for frequencies below the resonant value ($\omega \le \omega_{res}$):

$$\omega_{res} = \frac{\omega_{pe}}{\sqrt{1+\varepsilon_d}} \tag{10.183}$$

In the case of low frequencies ($\omega << \omega_{res}$), $k_z \approx \frac{\omega}{c}\sqrt{\varepsilon_d}$ and the surface wave propagates as a conventional one in the dielectric. The frequency of interest for the

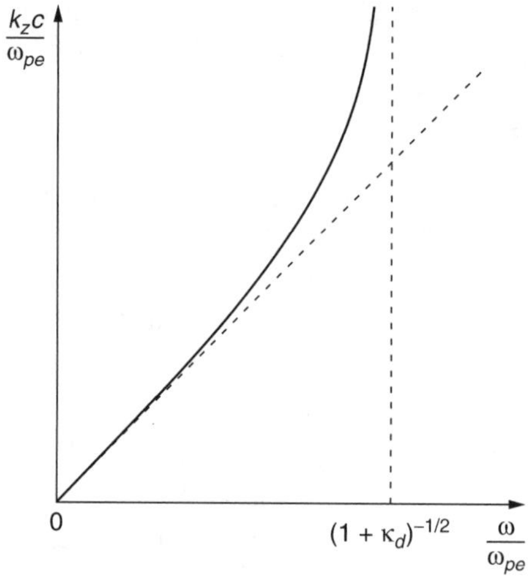

Figure 10.60 Surface wave dispersion curve $k_z(\omega)$

surface wave discharges is near, but below the resonant one, ω_{res}, which for the HDP plasma sources corresponds to microwave frequencies exceeding 1 MHz.

When the electromagnetic wave frequency ω is fixed, the surface wave propagation is possible only at plasma densities exceeding the critical one:

$$n_e \geq n_{res} = \frac{m\omega^2 \varepsilon_0 (1 + \varepsilon_d)}{e^2} \tag{10.184}$$

Thus, unlike the conventional case (Equation 6.200), propagation of surface waves and sustaining the surface wave discharge is possible when the plasma density exceeds the relevant critical value (Equation 10.184) even without external steady magnetic field.

10.9 NONEQUILIBRIUM MICROWAVE DISCHARGES OF MODERATE PRESSURE

10.9.1 Nonthermal Plasma Generation in Microwave Discharges at Moderate Pressures

The low-pressure nonthermal microwave discharges considered in Section 10.9 are mostly applied in heterogeneous processes for material surface processing. Gas conversion and gas treatment processes such as hydrogen sulfide, carbon dioxide, methane (and other organic compound) conversion, hydrogen production, etc. usually require high or moderate pressures to provide sufficient levels of technological productivity. However, powerful, steady-state atmospheric pressure discharges usually generate thermal plasmas. For the specific case of microwave discharges, this was considered in Section 10.2.

As discussed in Section 5.6.1, energy efficiency of the plasma chemical processes can be improved by organizing these processes under nonequilibrium, nonthermal conditions; this is especially true when it is possible to stimulate chemical reactions by vibrational excitation of molecules (see Section 5.6.2) Microwave discharges permit organizing such strongly nonequilibrium, plasma chemical processes with very high energy efficiency in the intermediate range of moderate pressures (usually between 30 and 200 torr).

One should note that the general schematics of moderate-pressure microwave discharges are almost identical to those considered in Section 10.2.2 through Section 10.2.4 regarding atmospheric pressure microwave discharges. Thus, the waveguide- and resonator-based configurations of the microwave plasma generators (see Figure 10.9 through Figure 10.12, and Figure 10.15) can be applied to create nonthermal and strongly nonequilibrium plasma at moderate pressures.

10.9.2 Energy Efficiency of Plasma Chemical Processes in Moderate-Pressure Microwave Discharges

As shown in Section 5.6.1 and Section 5.6.2, as well as Section 5.6.7 through Section 5.6.9, the highest plasma chemical energy efficiency can be reached under strongly nonequilibrium conditions with contributions of vibrationally excited reagents. However, such highly energy-efficient regimes require generation of plasma with specific parameters: electron temperature T_e should be about 1 eV and higher than the translational one (≤1000 K); the degree of ionization and specific energy input (energy consumption per molecule; see Section 5.6.7) should be sufficiently high, $n_e/n_0 \geq 10^{-6}$, $E_v \approx 1$ eV/mol.

Simultaneous achievement of these parameters is rather difficult, especially in steady-state uniform discharges (see Section 6.3). For example, the conventional low-pressure nonthermal discharges are characterized by values too large for the specific energy inputs of at least 30 to 100 eV/mol; the streamer-based atmospheric pressure discharges have low values of specific power and average energy input; and powerful steady-state atmospheric pressure discharges usually operate in close to quasi-equilibrium conditions. In contrast, moderate-pressure microwave discharges are able to generate nonequilibrium plasma with the previously mentioned optimal parameters,[9] thus attracting special interest to these systems if minimization of the cost of electrical energy is an important issue of a specific plasma chemical technology. Under such conditions, microwave discharges are not spatially uniform; however, this is not so important for gas treatment applications.

An important advantage of moderate-pressure microwave discharges is related to the fact that formation of an overheated plasma filament within the plasma zone does not lead to an electric field decrease because of the skin effect in the vicinity of the filament.[149] Such peculiarity of the electrodynamic structure permits sustaining strong nonequilibrium conditions $T_e > T_v >> T_0$ in the microwave discharges at relatively high values of specific energy input.

For example, a steady-state microwave discharge investigated by Krasheninnikov[434] is sustained in CO_2 at a frequency of 2.4 GHz; power of 1.5 kW;

pressure of 50 to 200 torr; and flow rate of 0.15 to 2 sl/sec. Specific energy input is in the range of 0.2 to 2 eV/mol and specific power is up to 500 W/cm^3 (compare with conventional glow discharge values of 0.1 to 3 W/cm^3). Spectral diagnostics prove that vibrational temperature in the discharge can be on the level of 3000 to 5000 K and significantly exceed rotational and translational temperatures that are about 1000 K.[134,427]

10.9.3 Microstructure and Energy Efficiency of Nonuniform Microwave Discharges

Processes taking place on the front of the propagating microwave discharge at moderate pressures make an important contribution in total plasma chemical kinetics and energy efficiency of the plasma chemical process. As was shown in Section 4.5.5 and Section 4.5.6, propagation of nonequilibrium discharges in fast gas flows is related to the processes on the plasma front, and it is only slightly sensitive to the processes in the bulk of the plasma.

The velocity of the ionization wave and thickness of the plasma front are determined by the diffusion coefficient D of heavy particles and the characteristic time τ of the limiting gas preparation process for ionization (in particular, by vibrational excitation). Reactions of vibrationally excited molecules can be characterized by the same time interval $\tau_{chem} \approx 1/k_{eV}n_e \approx \tau$ (see Section 5.6.7), where k_{eV} is the rate coefficient of vibrational excitation by electron impact. Thus, one can conclude that not only ionization but also main nonequilibrium chemical processes take place on the front of a propagating moderate-pressure microwave discharge.

If reactions are stimulated by nonequilibrium vibrational excitation (Section 5.6.7 through Section 5.6.9), the total energy efficiency can be expressed as the following function of initial gas temperature T_0^i, degree of ionization n_e/n_0, and specific energy input E_v:

$$\eta = \eta_{ex}\eta_{chem}\frac{E_v - k_{VT}\left(T_0^i\right)n_0\hbar\omega\left(\tau_{eV}+\tau_p\right) - \varepsilon_v\left(T_v^{\min}\right)}{E_v\left(1-\dfrac{\varepsilon_v\left(T_v^{\min}\right)}{\Delta H}\right)} \tag{10.185}$$

In this relation: $\tau_{eV} = E_v/k_{eV}n_e\hbar\omega$ is the total time of vibrational excitation; $\tau_p = c_v^v(T_v^{\min})^2/k_{VT}n_0E_a\hbar\omega$ is the reaction time in the passive phase of the discharge; η_{ex}, η_{chem} are the excitation and chemical components of the total energy efficiency (Section 5.6.9); k_{VT} is the rate coefficient of vibrational VT relaxation; $\varepsilon_v(T_v^{\min})$ is average vibrational energy of a molecule at the critical vibrational temperature $T_v^{\min}$ corresponding to equal rates of chemical reaction and vibrational relaxation (Equation 5.161); ΔH and E_a are the plasma chemical reaction enthalpy and activation energy; and c_v^v is the vibrational part of the specific heat per one molecule.

Thus, the total energy efficiency of the plasma chemical process can be found using Equation 10.185 as a function of the initial gas temperature T_0^i, ionization degree n_e/n_0, and specific energy input E_v, which can be identified in moderate-pressure microwave discharges with those on the plasma front. Next, determine these key plasma chemical parameters on the front of microwave discharge propagating in fast gas flow.

Propagation of the nonequilibrium discharge is determined by the value of the reduced electric field on the plasma front $(E/n_0)_f$, which depends on two external parameters: gas pressure p and electric field on the front E. Taking into account that the reduced electric field on the front is almost fixed by the ionization rate requirements, the initial gas temperature can be found by the following simple relation:[428,429]

$$T_f = \left(\frac{E}{n_0}\right)_f \frac{p}{E} \tag{10.186}$$

Following this approach, the electron concentration on the plasma front can be found from the energy balance as:

$$n_{ef} = \frac{D}{r_p^2} \frac{T_p}{T_p - T_g} \frac{p}{\mu_e e E^2} \tag{10.187}$$

In this relation, r_p is the characteristic radius of the nonuniform microwave plasma zone; T_p, T_g are translational temperatures in the plasma zone and ambient gas, respectively; and μ_e is the electron mobility.

Finally, the value of the specific energy input E_V in the discharge, which essentially determines the energy efficiency of the plasma chemical process, can be determined using the conventional relations (see Section 5.6.7 and Section 5.6.8):

$$E_v = P/Q_f \tag{10.188}$$

In this relation, P is the microwave discharge power absorbed in the plasma and Q_f is the flow rate, which in nonuniform discharges is only the portion of the flow crossing the plasma front, not the total flow rate.

10.9.4 Macrostructure and Regimes of Moderate-Pressure Microwave Discharges

Microstructure of the plasma front considered previously, together with discharge fluid mechanics and general energy balance, permits describing the macrostructure and shape of moderate-pressure microwave discharges.[430,460a] The macrostructure analysis includes consideration of transition among the three major forms of the discharges: diffusive (homogeneous), contracted, and combined.

The three major macrostructures, diffusive (homogeneous), contracted, and combined discharge forms, take place at different values of pressure p and electric field E, which is related to the electromagnetic energy flux $S = \varepsilon_0 c E^2$ (see Equation 6.193 and Equation 6.192). Critical values of pressure p and electric field E separating the three discharge forms are shown in Figure 10.61. The area above curve 1-1, for

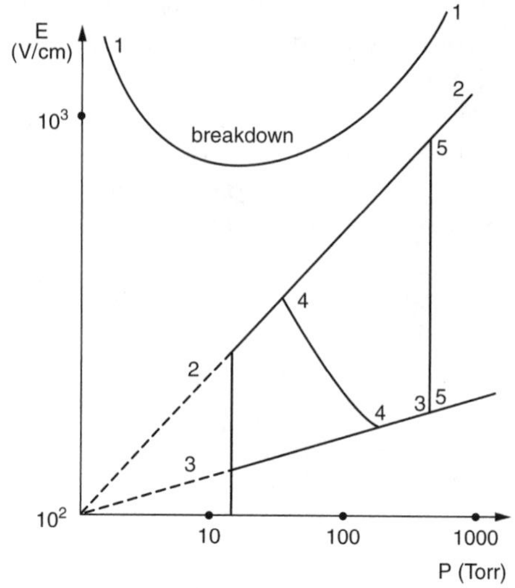

Figure 10.61 Three regimes of moderate pressure microwave discharges

example, corresponds to microwave breakdown conditions; the microwave discharges are sustained below this curve.

Critical curve 2-2 in the figure corresponds to the maximal ratio E/p of the steady-state microwave discharges, which is sufficient according to Equation 10.186 to sustain ionization on the plasma front even at room temperature. The diffusive (homogeneous) regime, illustrated in Figure 10.62(a), takes place when $E/p < (E/p)_{max}$, but pressure is relatively low (close to 20 to 50 torr), though still in the moderate pressure range. As seen from the figure, the space configuration of the discharge front in this regime is determined by the stabilization of the front in axial flow. This stabilization requires the normal velocity of the discharge propagation u_m to be equal to the component of the gas flow velocity $v \sin\theta$ perpendicular to the discharge front (see Section 4.5.5 and Section 4.5.6).

The ratio E/p decreases at higher pressures, and temperatures on the discharge front should increase (according to Equation 10.186) to provide the necessary ionization rate. Critical curve 3-3 in Figure 10.61 determines the minimal value of the reduced electric field $(E/p)_{min}$ when the microwave discharge is still nonthermal. The maximal temperature T^* on the discharge front, corresponding to $(E/p)_{min}$, is related to transition from nonthermal to thermal ionization mechanisms (see Section 2.2.4 and, especially, Equation 2.28):

$$T^* = I \ln^{-1}\left\{k_i\left[\left(\frac{E}{n_0}\right)_f\right] \times \frac{(4\pi\varepsilon_0)^5\hbar^3 T^*}{me^{10}} \frac{g_0}{g_i}\right\} \qquad (10.189)$$

In this relation, I is the ionization potential; g_i, g_0 are the statistical weights of ion and ground state neutrals; and $k_i[(E/n_0)_f]$ is the rate coefficient of nonthermal ion-

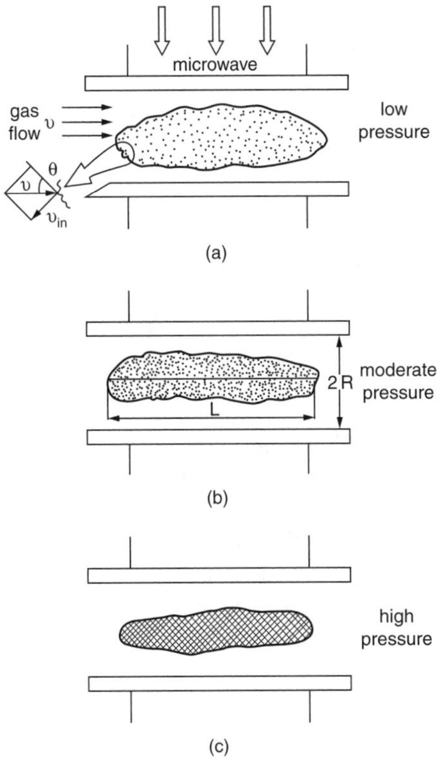

Figure 10.62 Transition of diffusive microwave discharge (a) into a contracted one (c) related to pressure increase; (b) intermediate regime

ization by direct electron impact at the reduced electric field given on the discharge front (see Equation 10.186).

It is important to point out that, at an intermediate pressure range of about 70 to 200 torr and $(E/p)_{min} < E/p < (E/p)_{max}$, the combined microwave discharge regime can be sustained when the nonthermal ionization front co-exists with the thermal one. This interesting and practically important regime is illustrated in Figure 10.62(b). In this case, a hot, thin filament of thermal plasma is formed interior to a relatively large nonthermal plasma zone. Skin effect prevents penetration of the electromagnetic wave into the hot filament and most of the energy still can be absorbed in the strongly nonequilibrium, nonthermal plasma surrounding.

The combined regime is able to sustain strongly nonequilibrium microwave plasma at relatively high values of degrees of ionization and specific energy input. This unique feature makes this regime especially interesting for plasma chemical applications, where the high-energy efficiency is the most important factor.

The combined regime requires relatively high pressures and electric fields. To derive the criterion of existence for this regime, take into account that T^* (Equation 10.189) is the minimal temperature sufficient for the contraction of the filament.

Energy balance of the thermal filament determined by the skin layer $\delta(T_m)$ can be expressed, based on Equation 8.33, as:

$$S = \varepsilon_0 c E^2 = \frac{4\lambda_m T_m^2}{I\delta(T_m)} \tag{10.190}$$

In this relation, λ_m is the thermal conductivity coefficient at maximum plasma temperature T_m in the hot filament and I is the ionization potential. Then, criterion of existence for the combined regime of microwave discharges can be expressed as:

$$E^2\sqrt{p} \geq \frac{2\lambda_m e\cdot(T^*)^2}{\varepsilon_0 cI}\sqrt{\frac{2\omega\mu_0 T^* n_e(T^*)}{mk_{en}}} \tag{10.191}$$

where k_{en} is the rate coefficient of electron–neutral collisions. The corresponding critical curve 4-4 in Figure 10.61 separates the lower pressure regime of the homogeneous discharge (Figure 10.62a) from the higher pressure regime of the combined discharge (Figure 10.62b).

At high pressures, to the right from the critical curve 5-5, radiation heat transfer becomes comparable with the molecular heat transfer. Because of reduction of the mean free path, the radiative front overheating becomes essential, and the contracted microwave discharge becomes completely thermal, which is illustrated in Figure 10.62(c).[431]

10.9.5 Radial Profiles of Vibrational $T_v(r)$ and Translational $T_0(r)$ Temperatures in Moderate-Pressure Microwave Discharges in Molecular Gases

The profiles $T_v(r)$ and $T_0(r)$ are qualitatively different in these three regimes: homogeneous (occurring at lower pressures), contracted (occurring at higher pressures), and combined, which is related to intermediate pressure range (Figure 10.61 and Figure 10.62). Results of detailed spectral measurements of vibrational and translational (rotational) temperatures are illustrated in Figure 10.63, together with radial distribution of power density.[134,427] These can be summarized as follows.

The radial profiles of vibrational and translational temperatures are obviously close to each other in the quasi-equilibrium regime at higher pressures (Figure 10.63c). At relatively low pressures, electron density is low and the skin effect can be neglected. In this case, the maximum deviation of the vibrational temperature from the translational one occurs on the axis of the discharge tube, where electron concentration and power density are maximal (Figure 10.63b).

Qualitatively, different temperature distributions can be observed at intermediate pressures, as illustrated in Figure 10.63(a). In this case, effective vibrational excitation and strong vibrational–translational nonequilibrium take place only on the nonequilibrium front of the microwave discharge, which is located at some intermediate radii. The vibrational excitation is not effective near the discharge axis because of the low electric field (which slows down the excitation) and high translational

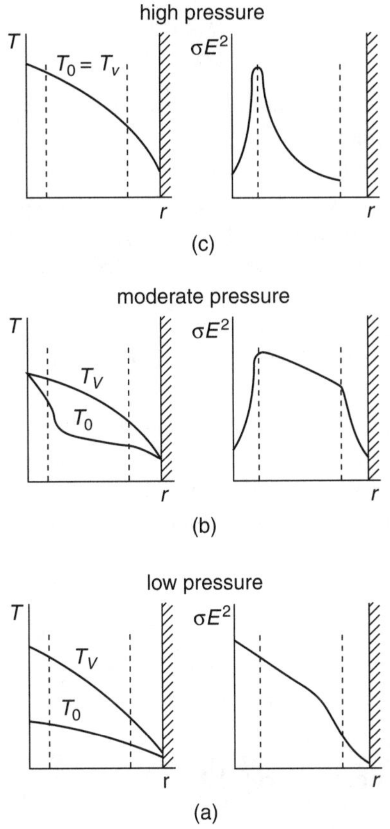

Figure 10.63 Radial distributions of vibrational and transitional temperatures (as well as specific power) at different pressures

temperature (which accelerates vibrational relaxation). Near the walls of the discharge tube, the vibrational excitation is ineffective because of low electron density.

10.9.6 Energy Efficiency of Plasma Chemical Processes in Nonuniform Microwave Discharges

The energy efficiency of nonequilibrium plasma chemical processes determined by Equation 10.185 is related only to formation of products on the discharge front. Considering nonuniform microwave discharges (first in the combined regime; see Figure 10.62b), one should take into account that part of the products effectively formed in the nonthermal zone can be destroyed in thermal inverse reactions in the hot filament. These losses, as well as other nonuniformity losses like those related to microwave absorption in the hot skin layer, can be taken into account by the additional special efficiency factor:

$$\frac{T_{\max} - T_{\min}}{T_{\max}}$$

that is related to the discharge nonuniformity and appears similar to the Carnot thermal efficiency coefficient.[195,432]

Physically, the nonuniformity factor in energy efficiency can be interpreted taking into account that large temperature differences between the cold surrounding nonthermal plasma zone and the hot filament prevent the products from having long contact with the high-temperature zone. This protects the plasma chemical products from inverse reactions and increases the energy efficiency. Also, the high-temperature $T_{\max}$ in the plasma filament leads to higher plasma density there and to a stronger skin effect, which decreases energy losses in this zone and also promotes high-energy efficiency.

In this case, the total plasma chemical energy efficiency of the nonequilibrium processes can be presented by a combination of the quasi-uniform energy efficiency η, given, for example, by Equation 10.185, and the nonuniformity factor introduced earlier. This leads to the following expression for the total energy efficiency:

$$\eta_{total} = \eta \times \frac{T_{\max} - T_{\min}}{T_{\max}} \tag{10.192}$$

In this relation, $T_{\min}$ is the gas temperature in the cold nonequilibrium discharge zone, while $T_{\max}$ is the gas temperature in the hot plasma filament.

10.9.7 Plasma Chemical Energy Efficiency of Microwave Discharges as Function of Pressure

Equation 10.192 can be applied in particular to describe pressure dependence of the energy efficiency of nonthermal microwave discharges. Taking into account Equation 10.186 between pressure and temperature on the plasma front:

$$T_f = T^* \frac{p}{p_2(E)} \tag{10.193}$$

The pressure dependence of the quasi-uniform component η of the energy efficiency in accordance with Equation 10.185 can be expressed as:

$$\eta \approx 1 - a_k \sqrt[3]{p/p_2(E)} \tag{10.194}$$

In this relation, $a_k \approx 0.3$ is the dimensionless kinetic parameter of a plasma chemical process; $p_2(E) = T^* E/(E/n_0)_f$ is the maximum value of pressure when the nonequilibrium plasma front is still possible (see Figure 10.61). Equation 10.194 actually describes the well-known tendency of growth of relaxation losses with pressure.

The energy balance of the plasma filament in the microwave discharges gives the following simplified pressure dependence of the maximum temperature:[194,433]

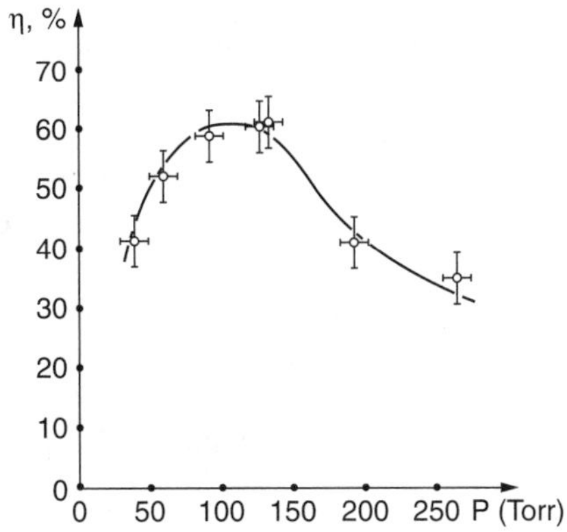

Figure 10.64 Energy efficiency as function of pressure[434]

$$T_{max} = T^* \sqrt{p/p_1} \tag{10.195}$$

where p_1 is the minimal pressure of the thermal filament appearance (see Figure 10.61). This expression for the maximum temperature in the hot plasma zone permits rewriting the energy efficiency nonuniformity factor as:

$$1 - \frac{T_{min}}{T_{max}} = 1 - \frac{T_{min}}{T^*}\sqrt{\frac{p_1}{p}} \tag{10.196}$$

As expected, the nonuniformity factor continuously increases with pressure. The total energy efficiency of plasma chemical processes stimulated by vibrational excitation (Equation 10.192) based on Equation 10.194 and Equation 10.196 can be presented as:

$$\eta_{total} = \left(1 - \frac{T_{min}}{T^*}\sqrt{\frac{p_1}{p}}\right) \times \left(1 - a_k \sqrt[3]{\frac{p}{p_2(E)}}\right) \tag{10.197}$$

According to this relation, the total energy efficiency as a function of pressure has a maximum in a good agreement with experiments of Krasheninnikov[434] (see Figure 10.64). The maximum of the $\eta_{total}(p)$ dependence takes place at the pressure:

$$p_{opt} = p_1^{3/5} p_2^{2/5} \left(\frac{T_{min}}{T^*}\frac{3}{2a_k}\right)^{6/5} \tag{10.198}$$

The temperature factor in the parentheses is of order unity and therefore the optimal pressure is in the interval $p_1 < p < p_2$, which corresponds to the combined regime of moderate-pressure microwave discharges (Figure 10.62b). In other words, the maximum value of energy efficiency can be achieved during transition from the homogeneous diffusive to contracted form of moderate-pressure microwave discharges.

This statement has a clear physical interpretation. The maximum efficiency of the reactions stimulated by vibrational excitation of molecules (see Section 5.6.7 and Section 5.6.8) requires, on one hand, that the pressure decrease to diminish the relaxation losses. On the other hand, it requires higher pressures to provide larger temperature gradients to prevent penetration of the reaction products and electromagnetic energy into the thermal plasma zone. These two requirements are satisfied simultaneously in the moderate-pressure range during the microwave discharge transition from the homogeneous diffusive into contracted form.

To give a relevant numerical example, consider experiments carried out in CO_2.[434] The microwave plasma transition from the homogeneous diffusive into the contracted form took place at $p_1 = 90$ torr; transition to the quasi-equilibrium thermal discharge regime took place at $p_1 = 90$ torr; transition $p_2 = 300$ torr; and the maximum energy efficiency of the plasma chemical process (which means minimum energy cost of CO_2 dissociation) was achieved at an optimal pressure of $p_{opt} \approx 120$ torr.

10.9.8 Power and Flow Rate Scaling of Space-Nonuniform, Moderate-Pressure Microwave Discharges

As discussed in Section 5.6.7 through Section 5.6.9, scaling of plasma chemical processes can be based on the value of the specific energy input E_v. For example, endothermic plasma chemical processes of gas conversion usually achieve the maximum value of energy efficiency at specific energy inputs of about 1 eV/mol. This means that optimization of energy efficiency requires the fixed ratio of the discharge power to the gas flow rate through the discharge on a level close to 1 kWh per standard m^3. Increase of power requires the corresponding proportional increase of the flow rate.

This simple and very practical scaling rule should be corrected for the case of spatial-nonuniform discharges (in particular, the moderate pressure microwave discharges), where some significant portion of the gas flow can avoid direct contact with the plasma zone. To make these important corrections and determine the power limitations of the discharge, consider a microwave discharge with the characteristic radial size of skin layer of about δ and length of about L, sustained in gas flow (see Figure 10.62).

If the normal velocity of discharge propagation is u_{in} and axial gas velocity is equal to v, then the angle θ between the vector of axial velocity and the quasi plane of the discharge front (see Figure 10.62) can be found as:

$$\sin\theta = \frac{u_{in}}{v} \tag{10.199}$$

If the axial gas velocity v is not very large, then the angle θ is rather large, and most of gas flow is able in principle to cross the discharge front. In this case, at a fixed value of pressure (for example, on the optimal level P_{opt} determined by Equation 10.198), the flow rate across the discharge front Q_f is close to the total gas flow rate

Q and increases with the axial gas velocity v. In this regime, the nonuniformity effects on the energy efficiency are small.

At high gas velocities, the influence of the nonuniformity effects becomes crucial. Further increase of the gas flow across the discharge front becomes impossible when the axial gas velocity reaches the maximum value:

$$v_{\max} = \frac{u_{in}}{\sin\theta} \approx u_{in}\frac{L}{\delta} \tag{10.200}$$

If the axial gas velocities exceed the critical value determined by Equation 10.200, $v > v_{\max}$, the fast flow compresses the discharge (see Equation 10.199). In this case, the effective discharge cross section decreases with gas velocity as $1/v^2$, and the gas flow across the discharge front decreases linearly with gas velocity as $1/v$.

Thus, there is maximum value of the gas flow rate across the discharge front that can be expressed, taking into account Equation 10.200, as:

$$Q_{f\max} \approx u_{in}\pi n_0 \delta L \tag{10.201}$$

where n_0 is the neutral gas density.

The relation between the gas flow rate across the discharge front and total flow rate, assuming that at low velocities, discharge radius is correlated with radius of the discharge tube, can be expressed as:

$$Q_f \approx Q, \quad \text{if} \quad Q < Q_{f\max} \tag{10.202a}$$

$$Q_f \approx \frac{Q_{f\max}^2}{Q}, \quad \text{if} \quad Q \geq Q_{f\max} \tag{10.202b}$$

Thus, scaling of the plasma chemical process by proportional increase of the gas flow rate Q and discharge power P, keeping the specific energy input $E_v = P/Q$ constant, is possible only at flow rate values below the critical one of $Q < Q_{f\max}$. Proportional increase of the gas flow rate and the discharge power at higher flow rates leads to an actual increase of the specific energy input proportionally to the square of power: $E_v \propto P^2$. In this case of higher flow rates $Q > Q_{f\max}$, the energy efficiency of plasma chemical processes decreases as $\eta \propto 1/P^2$ (see Section 5.6.8).

Therefore, based on Equation 10.202, the maximum power level, where high plasma chemical energy efficiency is still possible in nonuniform microwave discharges of moderate pressure, is:

$$P_{\max} = \pi E_v u_{in} \delta L n_0 \tag{10.203}$$

At the following typical values of the nonthermal microwave plasma and gas flow parameters, $E_v \approx 1$ eV/mol; $u_{in} \approx 300$ cm/sec; $n_0 \approx 5 \cdot 10^{18}$ cm^{-3}; $\delta \approx 1$ cm; and $L \approx 10 - 30$ cm, the power limit (Equation 10.203) gives $P_{\max} = 10$ to 30 kW, which is in a good agreement with the experiment.[201]

One should note that further increase of microwave power, still with high-energy efficiency, becomes possible by special changes of the discharge geometry and by organizing the nonthermal microwave discharges in supersonic gas flows (see Section 6.4).

PROBLEMS AND CONCEPT QUESTIONS

10.1 **Integral Flux Relation and the Channel Model of Arc Discharges.** Use the integral flux relation in the simple form, $J_r + S_r = 0$, which describes the cylindrical plasma column, together with the Maxwell relation for magnetic field ($curl\vec{H}$) to derive the third equation closing the channel model of arc discharges.

10.2 **Equations Describing a Long Inductively Coupled RF Discharge.** Derive the boundary condition (Equation 10.9) for the system of equations (Equation 10.6 and Equation 10.7) describing the inductively coupled RF discharge. In other words, show that the magnetic field in a nonconductive gas near the walls of the discharge tube is the same as the field inside an empty solenoid.

10.3 **Damping of Electromagnetic Fields in Skin Layer of ICP.** Based on the Maxwell equations (Equation 10.10), derive the distribution of electric and magnetic fields inside ICP (Equation 10.13 and Equation 10.14) in the framework of the metallic cylinder model, assuming constant conductivity σ inside the thermal plasma.

10.4 **ICP Temperature as a Function of Solenoid Current and Other Parameters.** Combining Equation 10.21 with the formula for plasma conductivity and the Saha equation, derive Equation 10.22 for the ICP temperature with an explicit expression for constants. Analyze this relation and show that the ICP temperature in the case of strong skin effect does not depend on the frequency of the electromagnetic field.

10.5 **Temperature Limits in Thermal ICP Discharges.** Typical values of temperature in the ICP discharges are usually limited to the interval between 7000 and 11,000 K. Analyzing Equation 10.23 for the specific ICP discharge power, explain why the thermal plasma temperature usually does not exceed 11,000 K. Taking into account the stability problem of the thermal ICP discharges at low conductivity conditions (see Equation 10.28), give your interpretation of the lower temperature limit of the discharge on the level of about 7000 K.

10.6 **Critical Solenoid Current to Sustain Thermal ICP Discharge.** Derive Equation 10.28 for the minimum value of the solenoid current $I_0 n$, which is necessary to sustain the quasi-equilibrium thermal ICP. For the derivation, use Equation 10.21, assuming a value of electric conductivity σ_m corresponding to the critical condition $\delta \approx R$ (see Equation 10.11). Give your interpretation of the fact that high solenoid current is necessary to sustain the thermal ICP in a smaller discharge tube.

10.7 **Stability of Thermal ICP Discharge at High Conductivities.** In the manner of the conclusion of Section 10.1.8 where it was shown that the low conductivity regime of the thermal ICP discharge is unstable (left branch in Figure 10.4), prove that the ICP regime with high conductivity and strong skin effect (left branch in Figure 10.4) is stable with respect to temperature, and thus conductivity fluctuations.

10.8 **Capacitively Coupled Atmospheric Pressure RF Discharges.** Give your interpretation why the power density of the atmospheric pressure CCP discharge grows with an increase of the voltage amplitude and the electric field frequency. Explain why the power densities of the uniform atmospheric pressure CCP discharges (see Section 10.1.11) are usually lower than those of the pulsed corona discharges (see Section 9.2.5).

10.9 **Microwave Discharge in H_{01} Mode of Rectangular Waveguide.** Calculate the maximum electric field for an electromagnetic wave in the H_{01} mode of a rectangular waveguide at the frequency f = 2.5 GHz (the corresponding wavelength in vacuum λ = 12 cm). Assume the microwave discharge power is approximately 1 kW; the wide waveguide wall is equal to 7.2 cm; and the narrow one is 3.4 cm long. Calculate the value of the reduced electric field E/p in this system without the plasma influence at atmospheric pressure; compare the result with typical E/p values in nonthermal glow discharges and thermal arc discharges.

10.10 **Microwave Discharge in H_{11} Mode of Round Waveguide.** Based on Equation 10.31 and Equation 10.32, analyze the electric field distribution along the radius (but in different directions) of the round waveguide operating in H_{11} mode without a plasma. Assume the electromagnetic wave frequency f = 2.5 GHz (the corresponding wavelength in vacuum λ = 12 cm); the microwave discharge power is approximately 1 kW; and the round waveguide radius is equal to 4 cm. Calculate the maximum electric field in the case. Compare the calculated maximum electric field with the one calculated in the previous problem under similar conditions but for a rectangular waveguide.

10.11 **Constant Conductivity Model of Microwave Plasma Generation.** In the framework of the constant conductivity model, derive the reflection coefficient $\rho = (S_0 - S_1)/S_0$ of an incident electromagnetic wave normal to the sharp boundary of plasma. This can be derived (see Section 10.2.6) based on boundary conditions for electric and magnetic fields on the plasma surface.

10.12 **Energy Balance of the Thermal Microwave Discharge.** Derive the energy balance equation (10.38), which permits determining the microwave plasma temperature. Based on this equation, analyze dependence of the microwave plasma temperature on power, pressure, and other discharge system parameters.

10.13 **Laser Radiation Absorption in Thermal Plasma.** Based on Equation 10.41 for the absorption coefficient of laser radiation, give your interpretation of the absorption dependence on pressure illustrated in Figure 10.18. Explain in particular why the maximum of the temperature dependence μ_ω/p^2 (T) shifts to the right in this figure (to the higher temperature levels) when gas pressure increases from 1 to 100 atm.

10.14 **Geometry of the Continuous Optical Discharge.** Analyzing the discharge pictures presented in Figure 10.20, clarify the physical factors, which determine the size of the continuous optical discharge. Experimentally, the minimum plasma radius corresponds in this case to the critical threshold conditions (see Equation 10.47) and numerically is about $r_t \approx 0.1$ cm. Estimate the radiation flux density for such conditions.

10.15 **Stable and Unstable Regimes of Continuous Optical Discharges.** As was shown in Section 10.2.12, the left low-temperature and low-conductivity branch of the curve $P_0(T)$ in Figure 10.19 where $T_m < T_t$ corresponds to unstable regime. Prove in a similar way that the opposite right high-temperature and high-conductivity branch of the curve $P_0(T)$ in Figure 10.19 where $T_m > T_t$ is stable. Explain the similarity from this point of view (stability only at high-conductivity conditions) of the continuous thermal ICP, microwave, and optical discharges.

10.16 **Comparative Analysis of Temperatures in the Thermal ICP, Microwave, and Optical Discharges.** As was shown, the thermal plasma temperatures in the continuous optical discharges are usually about double those of ICP and arc discharges, and about three to four times larger than in microwave discharges. Give your interpretation why the plasma temperature changes nonmonotonically with frequency of the electromagnetic fields.

10.17 **Reactive Component of Capacitively Coupled RF Plasma Resistance.** Derive Equation 10.53 for the reactive (imaginary) resistance component of the plasma and sheaths in plasma phase at high frequencies of electromagnetic field $\omega > \nu_e$. Explain why the reactive resistance component has the inductive nature.

10.18 **Voltage Drop on Space Charge Sheaths of CCP Discharges.** Derive Equation 10.57 for the total voltage U_s related to the space charge, and explain why it includes only the principal harmonic (ω) of the applied voltage, while the voltage drop on each sheath separately contains a constant component and second harmonics (2ω). Prove that the phase shift between voltage and current in the space charge sheath corresponds to the capacitive resistance.

10.19 **CCP Discharge Power Transferred to Ions in Sheaths.** Derive Equation 10.61 for power transferred per unit electrode area to ions in a sheath of the ICP discharge. Give your interpretation of the inverse proportionality of this power to the cube of electromagnetic oscillation frequency.

10.20 **Equivalent Scheme of a Capacitively Coupled RF Discharge.** Based on the equivalent scheme of the ICP discharges presented in Figure 10.25, calculate the resonant frequency of electromagnetic oscillations for this circuit. Compare the obtained result with Equation 10.59 for the resonance frequency. Discuss peculiarities of the result in the case of relatively high and relatively low gas pressures.

10.21 **Motion of the Plasma–Sheath Boundary in CCP Discharges, Taking into Account Nonuniformity of Ion Density in the Sheath.** Motion of the plasma–sheath boundary should be considered nonharmonic even in the case of $j = -j_0 \sin(\omega t)$, taking into account the nonuniformity of ion density in the sheath zone. Based on Equation 10.63 for phase evolution across the sheath zone, prove that velocity of the plasma boundary is inversely proportional to ion concentration $n_i(x)$. Explain why the duration of the space charge phase is longer in the vicinity of the ICP discharge zone, where plasma is always quasi neutral.

10.22 **Ion Concentration in Sheaths of Moderate-Pressure CCP Discharges in α-Regime.** Balancing the ion flux to the electrode (Equation 10.60) and the ionization rate in the sheath $I\alpha \cdot L/2$, derive Equation 10.73 for the ion density in the sheath of a moderate pressure CCP discharge. Estimate typical numerical values for pressure and current density dependence of the weakly changing logarithmic factor $\ln\Lambda$ in this relation.

10.23 **Sheath Size of Moderate-Pressure CCP Discharges in α-Regime.** Based on Equation 10.74 and Equation 10.73, estimate typical sizes of sheaths in moderate-pressure α-discharges. Compare the sheath sizes in this discharge with Debye radius at relevant plasma parameters. Give your interpretation of the comparison.

10.24 **Ion Current of Moderate-Pressure CCP Discharges in γ-Regime.** Analyze Equation 10.79 and show that the ion current in the normal regime of the γ-discharge is proportional to the square of gas pressure: $j_i \propto p^2$. Give your interpretation of the discharge transition to the low frequency regime with the gas pressure increase.

10.25 **High-Pressure Limit of the γ-CCP Discharge.** Prove that the upper pressure limit of the γ-CCP discharges (Equation 10.80) is equivalent to the following requirement: number of ions reaching the electrode during one oscillation period is less than total number of ions in the sheath. Give numerical estimation of the upper pressure limit (Equation 10.80).

10.26 **Critical Current of the α–γ Transition.** Analyze the dependence of the critical current of the α–γ transition on gas pressure (using Equation 10.81). Simplify Equation 10.81 to make it easier for numerical calculations.

10.27 **Effective Electric Field in Low-Pressure CCP RF Discharges.** Derive Equation 10.90 for the effective reduced electric field in low-pressure CCP discharges. Analyze numerical value of the logarithmic factor in this relation and explain why the electric field in this case can be considered constant. Compare this effect with the similar one in the case of positive column of DC glow discharges.

10.28 **Potential Barrier on the Plasma–Sheath Boundary.** Analyze numerically the kinetic flux balance equation (Equation 10.99) for low-pressure capacitive RF discharges, and show that the potential barrier on the plasma–sheath boundary in this system can be estimated as the lowest energy level of electronic excitation of neutral species.

10.29 **Stochastic Heating Effect.** Analyze reflection of electrons from the sheath boundaries moving to and from electrodes and calculate the energy transfer to and from an electron in this process. Take into account that the electron flux to the boundary moving from the electrode exceeds that of the one moving in an opposite direction to the electrode; estimate the energy transferred to the fast electrons in average by the stochastic heating effect.

10.30 **Potential Barrier on the Plasma Boundary at Low and High Current Limits.** Analyze numerically Equation 10.106 and Equation 10.114 for the potential barrier on the plasma boundary at the low and high current limits. Compare values of these potential barriers between themselves and with the electronic excitation energy ε_1. Discuss evolution of the potential barriers (and relation between them) with discharge current density.

10.31 **Plasma Density and Sheath Thickness Dependence on Current Density in Low-Pressure CCP RF Discharges.** Based on relations considered in Section 10.5.8, describe the plasma concentration and sheath thickness dependence on the current density, including regimes of high and low current. Give your interpretation of the dependences, and illustrate your conclusions using data presented in Figure 10.36.

10.32 **Critical Current Density of Transition between Low- and High-Current Regimes in Capacitive Discharges.** Using Equation 10.117, estimate numerically the critical value of the discharge current density corresponding to the transition between low- and high-current regimes of the low-pressure CCP RF discharges. Compare the results of your estimations with the relevant data presented in Figure 10.36.

10.33 **Asymmetric Effects in Low-Pressure Capacitive RF Discharges.** Assuming that the surface area of the powered electrode is 10 times the area of the grounded one and neglecting current losses on the walls of the grounded discharge chamber, calculate the ratio of voltages on the corresponding sheaths. Give your interpretation why the plasma-treated material is usually placed in the sheath related to the powered electrode.

10.34 **Secondary-Emission Resonant RF Discharge.** The secondary-emission resonant discharge requires that the electron "flight time" to the opposite electrode coincide with a half-period (or odd number of half-periods) of the electric field oscillations. Show that such discharges can be stable if the electron emission phase shift with respect to moment $E = 0$ is also small enough to make electrons emitted in different but close phases group together.

10.35 **Magnetron RF CCP Discharge.** Velocity of electron oscillations in the direction perpendicular to electrodes and the corresponding coordinate of this oscillation can be described in the magnetron RF discharges by Equation 10.129 and Equation 10.130. Derive these relations based on Equation 10.125 of electron motion in the crossed magnetic and oscillating electric fields. Analyze the resonant characteristics of Equation 10.129 and Equation 10.130.

10.36 **Current Density Distribution in ICP Discharges.** For the simple case of an ICP discharge arranged inside an inductor coil, analyze the current density distribution (Equation 10.138) along radius of the discharge tube. Determine conditions when the Bessel function describing this distribution can be simplified to the exponential form (Equation 10.140), which corresponds to plasma current localizing in the narrow layer on the discharge periphery.

10.37 **Equivalent Scheme of an ICP Discharge.** Analyze the equivalent scheme of the ICP discharge shown in Figure 10.49, and explain why the discharge current exceeds the current in the inductor coil exactly N times, where N is the number of windings in the coil (Equation 10.148). This result is derived from Equation 10.147 by neglecting the voltage drop on the plasma loop, which is correct only at high current levels. Determine a criterion required to have some simple relation (Equation 10.148) between currents in plasma and in inductive coil.

10.38 **Plasma Density in ICP Discharges.** Low-pressure ICP discharges are able to operate effectively at electron densities of 10^{11} to 10^{12} cm^{-3} (even up to 10^{13} cm^{-3}), which is more than 10 times the typical values of electron concentration in the capacitively coupled RF discharges and provides an important advantage for practical application of the discharges. Analyzing Equation 10.149 and Equation 10.151, give your interpretation of this important feature of ICP discharges. Explain also the dependence of the plasma density on the electromagnetic oscillation frequency and electric current in the inductor coil.

10.39 **Abnormal Skin Effect in Low-Pressure ICP Discharges.** Based on the expression (Equation 10.157) for effective plasma conductivity in the collisionless sheath of a low-pressure ICP discharge, derive Equation 10.160 for thickness of the abnormal skin layer. Compare Equation 6.25 and Equation 10.160 for normal collisional and abnormal skin layers, and determine the ICP discharge parameters corresponding to transition between them.

10.40 **Helical Resonator Discharge.** Analyze the resonance condition (Equation 10.161) and estimate the typical number of helix turns necessary to operate the discharge at an electromagnetic wave frequency of 13.6 MHz. Taking into account the Q-values of about 600 to 1500 in the helical resonator, estimate typical values of the electric fields in the discharge in the absence of plasma and compare these electric fields with those necessary for breakdown at low pressures.

10.41 **Propagation of Electromagnetic Waves in ECR Microwave Discharge.** Based on the dispersion equation (Equation 6.210), determine the phase and group velocities of the right-hand polarized electromagnetic waves in an ECR microwave discharge. Assume that the electromagnetic wave frequency is below the plasma frequency and electron–cyclotron frequency. Analyze the peculiarity of the wave propagation near the ECR-resonance condition.

10.42 **ECR Microwave Absorption Zone.** Using Equation 10.164 and Equation 10.165, estimate the typical width of the effective ECR-resonance zone and of the absorbed microwave power per unit area. Compare these energy flux levels with those of radio-frequency CCP and ICP discharges applied for material treatment.

10.43 **Whistler Waves and Helicon Discharges.** Analyzing the dispersion equation (Equation 6.210) for electromagnetic wave propagation in magnetized plasma at helicon conditions (Equation 10.166), derive the relevant dispersion equation for whistler waves, superposition of which forms the helicon modes. Derive a relation for phase velocity of the whistlers, compare this phase velocity with the speed of light, and explain how the helicon discharge can operate as the wave-heated one even at a relatively low level of radio frequency.

10.44 **Landau Damping of Helicon Modes.** The antenna length l_a for the helicon modes' excitation can be chosen related to the electromagnetic wave vector as $k_z \approx \pi/l_a$. By means of the Landau damping, the excited helicon modes are able to heat plasma electrons, whose kinetic energies ε correspond to the wave phase velocity $\varepsilon = 1/2m\,(\omega/k)^2$. Estimate the antenna length necessary to provide heating of electrons with energies $\varepsilon \approx 30$ to 50 eV, the most effective for ionization. Assume the excitation radio frequency $f = 13.56$ MHz.

10.45 **Planar Surface Wave Discharges.** Analyze the dispersion equation (Equation 10.182) of electromagnetic surface waves and give physical interpretation of the possibility of surface waves to propagate at lower frequencies in contrast to the conventional case of electromagnetic wave propagation in nonmagnetized plasma.

10.46 **Critical Plasma Density for Surface Wave Propagation.** Based on Equation 10.184, estimate the minimal plasma density necessary for surface wave propagation with frequency of 2.45 GHz. Estimate the gas pressure range, which permits considering this plasma as collisionless. Analyze the effect of dielectric constant on the critical plasma density.

10.47 **Combined Regime of Moderate-Pressure Microwave Discharges.** Derive the criterion (Equation 10.191) of formation of the hot filament of thermal plasma inside a nonequilibrium one in moderate-pressure microwave discharges. Based on the derived formula, make numerical estimations of the critical value of $E^2\sqrt{p}$ to form the hot filament and compare the result with data presented in Figure 10.61. Give your interpretation why the filament formation requires relatively high values of pressure and electric field. Compare microwave and radio-frequency discharges from this point of view.

10.48 **Nonuniformity Factor of Energy Efficiency of Microwave Discharges at Moderate Pressures.** Give your interpretation and conceptual derivation of the nonuniformity factor in the energy efficiency relation (Equation 10.192), describing in particular the energy losses related to inverse reactions of products in the hot plasma filament zone. Estimate this nonuniformity factor numerically for the combined microwave discharge regime and the more uniform, relatively low-pressure regimes of the microwave plasma.

10.49 **Pressure Dependence of Energy Efficiency of Microwave Discharges.** The energy efficiency of plasma chemical processes determined by Equation 10.197 reaches its maximum value at the pressure defined by Equation 10.198. Derive a formula for calculating the maximum value of the energy efficiency that can be reached in this case. Analyze the dependence of the energy efficiency on the ratio T_{min}/T^* of the nonthermal zone temperature and the temperature corresponding to transition from nonthermal to thermal ionization mechanisms.

10.50 **Power and Flow Rate Scaling of Moderate-Pressure Microwave Discharges.** Proportional increase of the total gas flow rate Q and the discharge power P at high flow rates $Q > Q_{f\max}$ leads to an actual increase of the specific energy input proportional to the square of the power $E_v \propto P^2$. Show that the energy efficiency of plasma chemical processes in this case decreases as $\eta \propto 1/P^2$.

CHAPTER

11

DISCHARGES IN AEROSOLS, AND DUSTY PLASMAS

Conventional plasma systems consist of electrons, atomic and molecular ions, and neutrals (see Section 2.1). Interesting and qualitatively different plasma properties can be observed if macroparticles of nanometer or micrometer scale (clusters and aerosols) are also actively present in discharge systems. In particular, macroparticles are easier to ionize because the ionization potential (work function) for them is lower. On the other hand, aerosols are usually more effective in electron attachment than macrosurfaces and individual neutral molecules—the former because of their large specific surface area per unit of system volume and the latter due to the absence of quantum-mechanical limitations. Micro and nanoparticles are very chemically and catalytically active in plasma, which determines many specific features of their application.

Photo- and thermoionization processes in aerosols will be considered in the first two sections of this chapter (effective ionization of macroparticles by electron impact usually requires high electron energies and will be discussed in the next chapter). The following two sections are concerned with breakdown of the aerosols and different electric discharges in these heterogeneous systems. The last sections of the chapter are related to processes of cluster and macroparticle formation in plasma and to some special features of the dusty plasmas.

11.1 PHOTOIONIZATION OF AEROSOLS

11.1.1 General Remarks on Macroparticle Photoionization

The photoionization of aerosols can play an important role in increasing electron concentration and thus electric conductivity of dusty gases.[435] These processes also make important contributions in astrophysics to understanding some of the characteristics of cosmic plasma because significant amounts of interstellar matter are in the form of macroparticles.[436]

The photoionization of macroparticles makes the most important contribution in generation of a dusty plasma if the radiation quantum energy exceeds the work function of macroparticles, but is below the ionization potential of neutral gas species. In this case, the steady-state electron density is determined by photoionization of the aerosols and electron attachment to the macroparticles. Here, the positive ions in this case are replaced by charged aerosol particulates with some specific distribution of charges.[437–439]

11.1.2 Work Function of Small and Charged Aerosol Particles Related to Photoionization by Monochromatic Radiation

Calculate the steady-state electron concentration n_e provided by photoionization of spherical monodispersed macroparticles of radius r_a and concentration n_a.[440,441] The aerosol concentration n_m distribution over different charges me should take into account that the work function of the aerosol surface depends on their charge before ionization:

$$\varphi_m = \varphi_0 + \frac{me^2}{4\pi r_a} \tag{11.1}$$

and thus there are upper n_+ and lower n_- limits of the macroparticle charge.

The upper charge limit is due to the fixed value of monochromatic photon energy $E = \hbar\omega$ and increasing value of the work function with a particle charge:

$$n_+ = \frac{4\pi\varepsilon_0 (E - \varphi_0) r_a}{e^2} \tag{11.2}$$

In these relations, φ_0 is the work function of a noncharged macroparticle ($m = 0$), which slightly exceeds the work function A_0 of the same material with a flat surface:[442]

$$\varphi_0 = A_0 \left(1 + \frac{5}{2}\frac{x_0}{r_a}\right) \tag{11.3}$$

$x_0 \approx 0.2$ nm is the numerical parameter of the relation.

The lower aerosol charge limit n_e ($n_- < 0$) is related to the charge value when the work function, and thus the electron affinity to a macroparticle, become equal to zero and attachment becomes noneffective:

$$n_- = -\frac{4\pi\varepsilon_0 \varphi_0 r_a}{e^2} \tag{11.4}$$

11.1.3 Equations Describing Photoionization of Monodispersed Aerosols by Monochromatic Radiation

The system of equations describing the concentration of aerosol particles with different values of electric charges, generated as a result of the photoionization process, can be expressed as:

$$\frac{dn_{n-}}{dt} = -\gamma p n_{n-} + \alpha n_e n_{n-+1} \tag{11.5a}$$

$$\frac{dn_i}{dt} = \gamma p n_{i-1} - \alpha n_e n_i - \gamma p n_i + \alpha n_e n_{i+1}, \quad n_- + 1 \le i \le n_+ - 1 \tag{11.5b}$$

$$\frac{dn_{n+}}{dt} = \gamma p n_{n+-1} - \alpha n_e n_{n+} \tag{11.5c}$$

In this system of equations, $\alpha \approx \sigma_a v_e$ is the coefficient of attachment of electrons with averaged thermal velocity v_e to the macroparticles of cross section σ_a, which is assumed here to be constant at $i > n_-$; γ is the photoionization cross section of the macroparticles, which is also supposed here to be constant at $i < n_+$; n_i is the concentration of macroparticles carrying charge ie ($i < 0$ for negatively charged particles); and p is the flux density of monochromatic photons.

Considering steady-state conditions $d/dt = 0$, and adding sequentially the equations of the system Equation 11.5, the balance equation for the charged species in the heterogeneous system under consideration is expressed as:

$$\alpha n_e n_{i+1} = \gamma p n_i, \quad q = \frac{n_{i+1}}{n_i} = \frac{\gamma p}{\alpha n_e} \tag{11.6}$$

Here, $q = n_{i+1}/n_i$ is an important factor describing the distribution of aerosol particles over electric charges. Taking into account the electroneutrality of the system as well as the total mass balance, the balance equation for electron density can be presented as:

$$\frac{n_e}{n_a} = \frac{\sum_{k=n-}^{k=n+} k q^k}{\sum_{k=n-}^{k=n+} q^k} \approx \frac{\int_{n-}^{n+} x q^x dx}{\int_{n-}^{n+} q^x dx} \tag{11.7}$$

Integrating Equation 11.7 leads to the final equation, which describes the density of electron gas generated by the monochromatic photoionization of monodispersed aerosols as:[440]

$$\frac{n_e}{n_a} = \frac{n_+ q^{n+} - n_- q^{n-}}{q^{n+} - q^{n-}} - \frac{1}{\ln q} \tag{11.8}$$

11.1.4 Asymptotic Approximations of Monochromatic Photoionization

This general aerosol photoionization equation determines the relation between electron concentration n_e; photon energy $E = \hbar\omega$; flux p; and aerosol parameters: radius r_a, work function A_0, and concentration n_a. In general, the dependence $n_e = f(E, p, r_a, n_a, A_0)$ cannot be expressed explicitly from Equation 11.8, but it can be analyzed asymptotically in the opposite extremes of low and high fluxes of monochromatic photons.

1. **The high photon flux extreme $q >> 1$.** According to definition of the q-factor (Equation 11.6), this regime requires $p >> \alpha n_e/\gamma$. In this case, the electron concentration based on Equation 11.8 can be found as:

$$\frac{n_e}{n_a} = \frac{n_+ q^{n+}}{q^{n+}} = n_+ \tag{11.9}$$

   Thus, if the photon flux is sufficiently high to satisfy the inequality:

$$p >> \frac{\alpha n_a n_+}{\gamma} \tag{11.10}$$

   the photoionization of aerosols is so intensive that each macroparticle receives the maximum possible electric charge $n_+ e$, which leads to the electron density (Equation 11.9).
2. **The low photon flux case ($p << \alpha n_a/\gamma$).** A decrease of the photon flux p leads to a reduction of the q factor as well, until at the extreme case $p << \alpha n_a/\gamma$, the electron concentration becomes relatively low $n_e/n_a << 1$ and the q factor reaches its minimal value. Then, based on Equation 11.8, this can be determined from the equation:

$$\frac{n_+ q^{n+} - n_- q^{n-}}{q^{n+} - q^{n-}} = \frac{1}{\ln q} \tag{11.11}$$

   As seen from this equation, the minimal value of the q-factor at low photon fluxes is close to unity. This means that the electron density at low photon fluxes can be estimated as $q = 1$ (see Equation 11.6).

Thus, the total asymptotic expression for the electron density provided by monochromatic photoionization of aerosols can be given as:

$$n_e = n_a n_+, \quad \text{if} \quad p >> \alpha n_a n_+/\gamma \tag{11.12a}$$

$$n_e = \gamma p/\alpha, \quad \text{if} \quad p << \alpha n_a/\gamma \tag{11.12b}$$

A convenient chart for calculating the electron density n_e as a function of $\log p$ and $\log(n_a n_+)$ at monochromatic photoionization based on Equation 11.12 is presented in Figure 11.1. In this chart, it is assumed that $\alpha/\gamma = 10^8$ cm/sec. Figure 11.1 also shows the relative accuracy of calculations based on Equation 11.12. It is

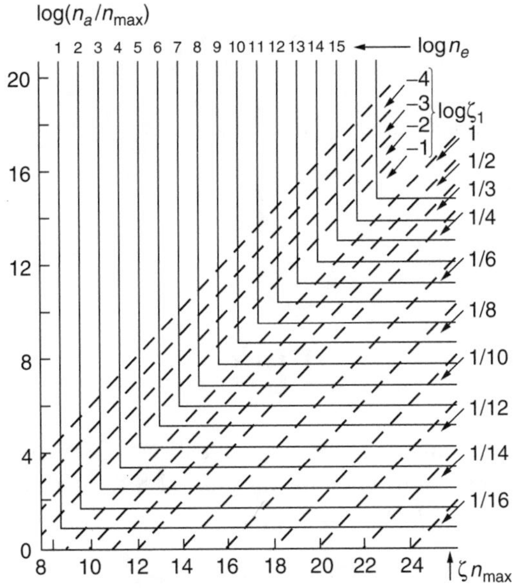

Figure 11.1 Calculation of electron concentration provided by photoionization of aerosols[440]

presented in a form of simple relative accuracy, $\eta = \Delta n_e / n_e$, as well as in a form of relative accuracy of calculations of $\ln n_e$:

$$\eta_1 \approx \frac{\Delta n_e}{2.3 \cdot n_e \log n_e} \tag{11.13}$$

11.1.5 Photoionization of Aerosols by Continuous Spectrum Radiation

Generalization of steady-state relations for macroparticle density distribution over charges n_i (Equation 11.6) for the case of radiation with a continuous spectrum can be expressed as:

$$\gamma p_i n_i = \alpha n_e \tag{11.14}$$

In this case, p_i is the flux density of photons with energy sufficient for the $(n + 1)$ – *st* photoionization. Using the energy distribution of photons in the form: $\xi_i = p_i/p_0$, $(\xi_0 = 1)$, the concentration of macroparticles with charge *ke* is:

$$n_1 = n_0 \left(\frac{\gamma p_0}{\alpha n_e} \right) \xi_0 \tag{11.15a}$$

$$n_2 = n_1 \left(\frac{\gamma p_0}{\alpha n_e} \right) \xi_1 = n_0 \left(\frac{\gamma p_0}{\alpha n_e} \right)^2 \xi_0 \xi_1 \tag{11.15b}$$

$$n_k = n_0 \left(\frac{\gamma p_0}{\alpha n_e} \right)^k \prod_{i=0}^{k-1} \xi_i \tag{11.15c}$$

In general, these relations can be applied for $k < 0$ as well. In this case, the product

$$\prod_{i=0}^{k} \xi_i$$

implies

$$\prod_{i=0}^{-k} (1/\xi_{-i})$$

Considering electroneutrality and the material balance of the aerosol system, Equation 11.15 leads to a generalization of Equation 11.8 for the case of radiation in a continuous spectrum:

$$\frac{n_e}{n_a} = \frac{\sum_{k=n-}^{\infty} k n_k}{\sum_{n-}^{\infty} n_k} = \frac{\sum_{k=n-}^{\infty} k q_0^k \prod_{i=0}^{k-1} \xi_i}{\sum_{k=n-}^{\infty} q_0^k \prod_{i=0}^{k-1} \xi_i} \tag{11.16}$$

In this equation, the q_0 factor for the nonmonochromatic radiation is defined similar to Equation 11.6 as $q_0 = \gamma p_0 / \alpha n_e$.

11.1.6 Photoionization of Aerosols by Radiation with Exponential Spectrum

Equation 11.16 obviously cannot be analytically solved, but it can be significantly simplified, assuming an exponential behavior of the photon energy distribution along the spectrum: $p(E) = p\exp(-\delta E)$. Such a distribution is valid for the tail of the thermal radiation spectrum. In this case, parameters of the photon flux density p_0 and ξ_i relevant to the photoionization problem can be expressed as:

$$p_0 = \int_{\varphi_0}^{\infty} p(E)dE = \frac{p}{\delta} \exp(-\delta\varphi_0) \tag{11.17}$$

$$\xi_i = \frac{p_i}{p_0} = \frac{1}{p_0} \int_{\varphi_i}^{\infty} p(E)dE = \exp\left(-\delta i \frac{e^2}{4\pi\varepsilon_0 r_a} \right) \tag{11.18}$$

Substituting p and ξ from Equation 11.17 and Equation 11.18 into Equation 11.16 and assuming $n_- \to -\infty$, Equation 11.16 can be rewritten as:

$$\frac{n_e}{n_a} = \frac{\sum_{-\infty}^{\infty} k q_0^k \exp\left(-\lambda \frac{k(k-1)}{2}\right)}{\sum_{-\infty}^{\infty} q_0^k \exp\left(-\lambda \frac{k(k-1)}{2}\right)} \tag{11.19}$$

where $\lambda = \delta e^2/4\pi\varepsilon_0 r_a$. This equation can be solved, and the electron density provided by the nonmonochromatic radiation can be expressed in the following differential form using elliptical functions:[443]

$$\frac{n_e}{n_a} = y + \frac{\rho}{2\pi}\frac{d}{dy}\ln\theta_3(y,\rho) \tag{11.20}$$

In this relation, $\theta_3(y,\rho)$ is the elliptical θ function depending on two new variables, $y = 1/\lambda \ln q_0 + 1/2$, $\rho = 2\pi/\lambda$. In the extreme case of $\rho >> 1$, Equation 11.10 can be simplified, and the relation for electron concentration can be given as:

$$\frac{n_e}{n_a} = \frac{1}{\lambda}\ln q_0 + \frac{1}{2} \tag{11.21}$$

In the opposite extreme of $\rho << 1$, the asymptotic expression for the elliptical function gives the electron concentration in the form of:

$$\frac{n_e}{n_a} = \frac{q_0}{1+q_0} \tag{11.22}$$

11.1.7 Kinetics of Establishment of Steady-State Aerosol Photoionization Degree

Non-steady-state evolution of the electron concentration provided by photoionization of aerosol particles can be described by the following kinetic equation:

$$\frac{dn_e}{dt} = \gamma\sum_{i=n-}^{\infty} p_i n_i - \alpha n_e (n_a - n_{n-}) \tag{11.23}$$

In this equation, the electron attachment is effective for all macroparticles except those with the minimal charge n_-. The simple case of monochromatic radiation permits assuming that the photon flux $p_i = p$ for $n_- \le i \le n_+ - 1$ and $p_i = 0$ outside this aerosol charge interval. Then, the kinetic equation (Equation 11.23) can be rewritten as:

$$\frac{dn_e}{dt} = \gamma p(n_a - n_{n+}) - \alpha n_e (n_a - n_{n-}) \tag{11.24}$$

This equation cannot be solved because it requires analysis of the evolution of partial concentration of macroparticles with all possible charges $n_- \le i \le n_+$. However, the characteristic time for establishing steady-state electron concentrations can be found for the extreme cases of low and high photon flux densities.

As discussed earlier, at relatively low photon flux densities, $p \ll \alpha n_a/\gamma$; as discussed previously, $q \approx 1$. In this case, one can conclude that $n_{n-}, n_{n+} \ll n_a$. This permits finding a solution of the kinetic equation (Equation 11.24) in the form:

$$n_e(t) = \frac{\gamma p}{\alpha}\left[1 - \exp(-\alpha n_a t)\right] \tag{11.25}$$

Thus, the characteristic time establishing the steady-state electron concentration in the case of low photon flux density is:

$$\tau = 1/\alpha n_a \tag{11.26}$$

In the opposite case of high photon flux densities $p \gg \alpha n_a n_+/\gamma$, most macroparticles have the same maximum electric charge: $Z_a e = n_+ e$. While aerosol particles are obtaining this charge, the ionization process obviously slows down. Taking into account that, for each particle in this case, $dZ_a/dt \approx \gamma p$ $(Z_a < n_+)$, the characteristic time for establishing the steady-state electron concentration can then be expressed as:

$$\tau = n_+/\gamma p \tag{11.27}$$

Equation 11.26 and Equation 11.27 can be generalized to describe the characteristic photoionization time at any values of photon flux:

$$\tau = \frac{n_+}{\alpha n_a n_+ + \gamma p} \tag{11.28}$$

At low and high photon fluxes, respectively, this formula obviously corresponds to the extreme asymptotic relations (Equation 11.26 and Equation 11.27).

11.2 THERMAL IONIZATION OF AEROSOLS

11.2.1 General Aspects of Thermal Ionization of Aerosol Particles

The energy necessary for ionization of macroparticles is related to their work function, which is usually lower than ionization potential of atoms and molecules. Taking into account that ionization potentials determine the exponential growth of ionization rate with temperature, one can conclude that thermal ionization of aerosol particles can be very effective and provide high electron density and conductivity at relatively low temperatures. This factor points out the main physical features and application areas of macroparticle thermal ionization.

Thermal ionization of aerosol particles was applied to increase the electron concentration and conductivity in MHD (magneto-hydrodynamic) generators.[444] The conductivity increase in this case is more significant with respect to the alternative approach using alkaline metal additives. The effect of thermal macroparticle ionization plays a special role in rocket engine torches.[445] Absorption and reflection of radio waves by the plasma of rocket engine torches affects and complicates control of the rocket trajectory. Electron density in the flame plasma can be abnormally high because of thermal ionization of macroparticles, (see, for example, Sholin et al.[446]).

Numerous publications address the thermal ionization of aerosol particles. For example, simple relations for electron concentration in aerosol systems with macroparticle charge less than the elementary one can be based on the Saha equation.[447–451] If the temperature is relatively high, the average charge of a macroparticle can exceed that of an electron, and the electric field of an aerosol particle affects the level of thermal ionization. This important case was first considered by Einbinder;[452] the Einbinder formula was then analyzed in detail by Arshinov and Musin.[447,448,453,454] The space distribution of electrons around a thermally ionized particle was first analyzed by Samuilov[455,456] and electric conductivity of thermally ionized macroparticles has been considered in Samuilov.[449,450,457]

In this section, only main features of the phenomenon will be considered, starting with ionization of the heterogeneous system stimulated by its radiative heating. The discussion will then move to space distribution of electrons around a thermally ionized aerosol particle and, finally, electric conductivity of thermally ionized aerosols.

11.2.2 Photoheating of Aerosol Particles

To analyze the thermal ionization of aerosol particles provided by their photoheating, first calculate the photoheating effect of aerosols by itself. Consider the heating of particles by a monochromatic radiation flux pE, assuming that the photon energy $E = \hbar\omega$ is not sufficient for direct ionization. The energy balance equation for the average temperature T of aerosol particles (considered as black body) can be expressed as:

$$c\rho\frac{4}{3}\pi r_a^3\frac{\partial T}{\partial t} = pE\pi r_a^2 + 4\pi r_a^2\lambda\frac{\partial T}{\partial r}\bigg|_{r=r_a} - \sigma T^4 4\pi r_a^2 \tag{11.29}$$

In this equation, c, ρ, and r_a are specific heat, density, and radius of aerosol particles; λ is the effective gas conduction coefficient; and σ is the Stephan–Boltzmann coefficient. In terms of the quasi-steady-state heat transfer coefficient α from the macroparticle surface, the energy balance equation (Equation 11.29) can be rewritten as:

$$\frac{1}{3}c\rho r_a\frac{\partial T}{\partial t} = \frac{1}{4}pE - \alpha(T - T_\infty) - \sigma T^4 \tag{11.30}$$

where T_∞ is the ambient gas temperature. In this case, the steady-state temperature of aerosol particles can be expressed as:

$$T_{st} = T_\infty + \frac{pE}{4\alpha}, \quad \text{if} \quad \gamma = \frac{4\sigma}{pE}\left(T_\infty + \frac{pE}{4\alpha}\right)^4 << 1 \tag{11.31a}$$

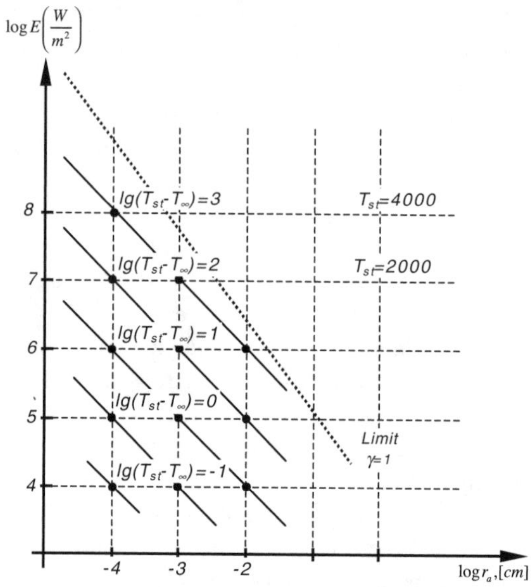

Figure 11.2 Heating of microparticles by irradiation (temperature in Kelvin)

$$T_{st} = \left(\frac{pE}{4\sigma}\right)^{1/4}, \quad \text{if} \quad \gamma = \frac{4\sigma}{pE}\left(T_\infty + \frac{pE}{4\alpha}\right)^4 >> 1 \tag{11.31b}$$

As can be seen, the case of $\gamma << 1$ corresponds to conductive cooling of the macroparticles, while $\gamma >> 1$ corresponds to a radiative cooling mechanism.

The numerical dependence of the steady-state aerosol particle temperature on the logarithm of radiation power flux log(pE, W/cm^2) and logarithm of the macroparticle radius log(r_a, cm) is presented in Figure 11.2. This dependence is related to the case of photoheating of aerosol particles in air, assuming the Nusselt number $Nu = 2\alpha r_a/\lambda \to 2$. The cooling mode separation line $\gamma = 1$ is also shown in Figure 11.2.

11.2.3 Ionization of Aerosol Particles Due to Their Photoheating: Einbinder Formula

If the photon energy is insufficient for the photoeffect, the ionization of macroparticles is mostly provided by the thermal ionization mechanism. Assume that the average distance between macroparticles greatly exceeds their radius: $n_a r_a^3 << 1$ (n_a is the concentration of aerosols). In this case, if the average charge of a macroparticle is less than the elementary one ($n_e << n_a$), the electron density can be found by the following relation similar to the Saha equation (see Section 4.1.3):

$$n_e = K(T) = \frac{2}{(2\pi\hbar)^3}(2\pi mT)^{3/2}\exp\left(-\frac{A_0}{T}\right) \tag{11.32}$$

From a practical viewpoint, a more interesting case corresponds to the average macroparticle charge exceeding the elementary one and, therefore, to the electron concentration exceeding that of the aerosol particles. In this case, ($n_e/n_a >> 1/2$) and the electron concentration can be found by the **Einbinder formula,**[447,452] taking into account the work function dependence on the macroparticle charge:

$$\frac{n_e}{n_a} = \frac{4\pi\varepsilon_0 r_a T}{e^2} \ln\frac{K(T)}{n_e} + \frac{1}{2} \tag{11.33}$$

For numerical calculation of the electron density in the typical range of parameters: $n_a \approx 10^4 \div 10^8 \,\text{cm}^{-3}$ with work functions of about 3 eV, Equation 11.32 and Equation 11.33 can be given in the following form:

$$n_e = 5\cdot 10^{24}\,\text{cm}^{-3}\cdot \exp\left(-\frac{A_0}{T}\right), \quad \text{if} \quad T < 1000\,\text{K} \tag{11.34a}$$

$$n_e = n_a \frac{4\pi\varepsilon_0 r_a T}{e^2}\left(20 - \frac{A_0}{T}\right), \quad \text{if} \quad T > 1000\,\text{K} \tag{11.34b}$$

These relations describing thermal ionization of aerosol particles imply thermodynamic quasi equilibrium and thus equality of temperatures of macroparticles T_a, electrons T_e, and neutrals. However, in the case of relatively large distances between macroparticles $n_a r_a^3 << 1$, which takes place in the photoheating, aerosol temperatures can exceed that of neutral gas. Electron temperatures are determined not only by the macroparticle temperature, but also by the energy transfer to relatively cold gas, and by possible direct radiative heating. As a result of these processes, $T_e \neq T_a$. The nonequilibrium effects influence the electron density provided by thermal ionization and should be taken into account.

To take these nonequilibrium effects into account, note that $K(T)$ in Equation 11.32 corresponds to the electron concentration near the surface. Then the Boltzmann electron energy distribution function permits generalizing the Einbinder formula to the nonequilibrium case of $T_e \neq T_a$:[458–460]

$$\frac{n_e}{n_a} = \frac{4\pi\varepsilon_0 r_a T_e}{e^2} \ln\frac{K(T_a)}{n_e} + \frac{1}{2} \tag{11.35}$$

For numerical calculation, this formula can be expressed in the following form similar to the quasi-equilibrium relation (Equation 11.34b):

$$n_e = n_a \frac{4\pi\varepsilon_0 r_a T_e}{e^2}\left(20 - \frac{A_0}{T_a}\right) \tag{11.36}$$

but distinguishing the temperature values of aerosol particles and electron gas.

11.2.4 Space Distribution of Electrons around a Thermally Ionized Macroparticle: Case of High Aerosol Concentration

The Einbinder formula assumes that electrons move from the macroparticle surface at a distance much exceeding their radius r_a and, for this reason, cannot be applied at high aerosol concentrations $n_a r_a^3 \approx 1$. The space distribution of electrons around thermally ionized macroparticles can be qualitatively modeled as the one-dimensional space distribution $n_e(x)$ of electrons between two plane cathodes heated to the high temperature T.

Combining the Boltzmann distribution function of the thermally emitted electrons $n_e(x) = K(T)\exp[e\varphi(x)/T]$ with the Poisson's equation gives the following differential equation for the electric potential $\varphi(x)$ distribution in this case:

$$\frac{d^2y}{dx^2} + b\exp y = 0 \tag{11.37}$$

Here, $y = e\varphi(x)/T$ characterizes the potential distribution and parameter $b = e^2K(T)/\varepsilon_0 T$ is related to the electron concentration $K(T)$ near the macroparticle (cathode) surface, which can be found from Equation 11.32. The second-order differential equation (Equation 11.37) can be simplified to a first-order one:

$$\frac{dy}{dx} = -\sqrt{2|b|(C_1 + \exp y)} \tag{11.38}$$

The constant C_1 can be found from the boundary conditions on the cathode: $\varphi = 0$, $\exp y = 1$, $|y'| = eE_0/T$, where $E_0 = 1/2\varepsilon_0\, \bar{n}_e ea$ is the electric field near the cathode; $\bar{n}_e$ is the average electron concentration; and a is the distance between cathodes. The boundary conditions determine the relation between the constant C_1 and the electric field energy density:

$$1 + C_1 = \frac{\varepsilon_0 E_0^2/2}{K(T)T} = \alpha \tag{11.39}$$

Taking into account the symmetry of the potential distribution between cathodes, one can conclude that $C_1 < 0$ and $\alpha < 1$. This means, according to Equation 11.39, that the thermal energy density of electrons near the cathode exceeds the density of the electric field energy.

Solution of Equation 11.38 gives the electric potential and thus the electron density distribution between two cathodes:[461]

$$n_e(x) = \frac{K(T)\cdot(1-\alpha)}{\cos^2\left(\sqrt{\frac{1-\alpha}{2}}\cdot\frac{x}{r_D^0}\right)} \tag{11.40}$$

In this relation, $K(T)$ is the electron density on the cathode surface (Equation 11.32); coordinate $x = 0$ is assumed in the center of the interelectrode gap; and $r_D^0 = \sqrt{\varepsilon_0 T/K(T)e^2}$ is the Debye radius corresponding to the electron density near the surface. The boundary conditions on the cathodes $n_e(x = a/2) = K(T)$ permit finding the electric field energy density parameter α:

$$\alpha = \sin^2\left(\sqrt{\frac{1-\alpha}{2}}\,\frac{a}{2r_D^0}\right) \tag{11.41}$$

If $r_D^0 << a$, the electron density distribution $n_e\,(x)$ is essentially nonuniform; most electrons are located in the close vicinity of a macroparticle. The electric field energy density parameter α can be presented in this case as:

$$\alpha \approx 1 - 4\pi^2\left(\frac{r_D^0}{a}\right)^2 + 32\pi^2\left(\frac{r_D^0}{a}\right)^3 \tag{11.42}$$

which leads to the following simple expression for the average electron concentration between cathodes (aerosol particles):

$$\overline{n}_e = \sqrt{\frac{\varepsilon_0 K(T)\cdot T}{a^2 e^2}} \approx K(T)\frac{r_D^0}{a} \tag{11.43}$$

The electric field distribution between the cathodes can be found based on the distribution of electron density (Equation 11.40) and the Boltzmann distribution:

$$E(x) = \frac{T}{er_D^0}\sqrt{2(1-\alpha)}\tan\left(\sqrt{\frac{1-\alpha}{2}}\cdot\frac{x}{r_D^0}\right) \tag{11.44}$$

As seen from Equation 11.44, the electric field is equal to zero in the middle of the intercathode interval and has a maximum value on the cathode surface.

Characteristics of the electron density and electric field distributions (Equation 11.40, Equation 11.43, and Equation 11.44) can be generalized for the case of relatively compact aerosol particles ($n_a r_a^3 \approx 1$), assuming that $r_a \approx a$. The electron density distribution (Equation 11.40) in the macroparticle corresponds to the **Langmuir relation** for the electron density distribution near a plane cathode:[462]

$$n_e(z) = \frac{K(T)}{\left(1 + \frac{z}{\sqrt{2}r_D^0}\right)^2} \tag{11.45}$$

In the Langmuir relation, z is distance from the plane cathode surface and $K(T)$ is the electron density just near the cathode (Equation 11.32).

11.2.5 Space Distribution of Electrons around a Thermally Ionized Macroparticle: Case of Low Concentration of Aerosol Particles

The case of relatively low aerosol concentration $n_a r_a^3 << 1$ is the most typical one in plasma chemistry. In this case, distances between aerosol particles greatly exceed their radius, and description of the electron density and electric field space distributions first requires consideration of a lone macroparticle. Combining the Poisson equation with the Boltzmann distribution leads to the following equation for electric potential, $\varphi(r)$, around a spherical, thermally ionized aerosol particle:

$$\frac{d^2\varphi}{dr^2}+\frac{2}{r}\frac{d\varphi}{dr}=\frac{e}{\varepsilon_0}K(T)\exp\left(\frac{e\varphi}{T}\right) \tag{11.46}$$

This equation has no solution in elementary functions; a detailed numerical analysis was made by Samuilov.[456] To analyze Equation 11.46, it is convenient to rewrite it with respect to electron concentration, taking into account the Boltzmann distribution:

$$\frac{d}{dr}\left(\frac{r^2}{n_e}\frac{dn_e}{dr}\right)=\frac{e^2}{\varepsilon_0 T}n_e r^2 \tag{11.47}$$

For the case of spherical symmetry, the Boltzmann relation $n_e(x) = K(T)\exp[e\varphi(x)/T]$ can be rewritten in the following differential form:

$$E(r)=-\frac{d\varphi}{dr}=-\frac{T}{e}\frac{1}{n_e}\frac{dn_e}{dr} \tag{11.48}$$

As a boundary condition for Equation 11.47, take $dn_e/dr\ (r = r_{av}) = 0$, where $r_{av} \approx n_a^{-1/3}$ is the average distance between aerosol particles. Taking into account this boundary condition, integrate the right- and left-hand sides of Equation 11.47 to obtain the following interesting integral relation:

$$N(r)=\int_r^{r_{av}} n_e(r)4\pi r^2 dr=-\frac{4\pi\varepsilon_0 T r^2}{e^2}\frac{1}{n_e(r)}\frac{dn_e(r)}{dr} \tag{11.49}$$

Here, $N(r)$ is number of electrons located in the radius interval from $r\,(r \geq r_a)$ to r_{av}. Combining Equation 11.49 and Equation 11.48 gives:

$$N(r)=\frac{E(r)}{e/4\pi\varepsilon_0 r^2} \tag{11.50}$$

In the close vicinity of a macroparticle ($z = r - r_a << r_a$), electron density distribution is expressed using the Langmuir relation (Equation 11.45). Then, the total charge of an aerosol particle can be calculated based on Equation 10.30 as:

$$Z_a = N(r_a)=\frac{4\pi\varepsilon_0 r_a T}{e^2}\left(\frac{r_a}{r_D^0}\right) \tag{11.51}$$

Integrating the Langmuir relation (Equation 11.45), one finds that most of the electrons are located in a thin layer about $r_D^0 << r_a$ around the macroparticle. These electrons are confined in the Debye layer by the strong electric field of the aerosol particle. This field can be expressed, based on Equation 11.49, as:

$$E(r) = \frac{T}{e\left(\frac{r_D^0}{\sqrt{2}} + \frac{r - r_a}{2}\right)} \tag{11.52}$$

11.2.6 Number of Electrons Participating in Electric Conductivity of Thermally Ionized Aerosols

Because thermally ionized aerosol particles have high values of the inherent electric field (Equation 11.52) when their charge Z_a is sufficiently high, the electric conductivity of such an aerosol system at $n_a r_a^3 << 1$ depends on the value of external electric field E_e. This phenomenon occurs because, at higher external electric fields, more electrons can be released from being trapped by the inherent electric field of a macroparticle.

The electric conductivity of the free space between macroparticles under consideration can be presented as:

$$\sigma_0 = en_a bN(r_e) \tag{11.53}$$

where b is the electron mobility in heterogeneous medium, and

$$N(r_e) = \int_{r_e}^{r_{av}} n_e(r) 4\pi r^2 dr$$

determines how many electrons participate in the electric conductivity in the external electric field E_e with respect to one macroparticle.

If the external electric field is relatively low, the concentration of electrons n_e * participating in electric conductivity can be found from the Einbinder formula:

$$N(r_{e1}) = \frac{n_e *}{n_a} = \frac{4\pi\varepsilon_0 r_a T}{e^2} \ln \frac{K(T)}{n_e *} >> 1 \tag{11.54}$$

This expression for $N(r_{e1})$, together with Equation 11.50 and Equation 11.52, permits specifying a virtual sphere of radius r_{e1} around a macroparticle. Electrons located outside this sphere are not "trapped" by the inherent electric field of the aerosol particle and provide the Einbinder value of electron density (Equation 11.33):

$$r_{e1} - r_a \approx r_a \ln^{-1} \frac{K(T)}{n_e *} < r_a \tag{11.55}$$

The following relation can then express the electric field of a macroparticle on the virtual sphere of radius r_e:

$$E(r_{e1}) = \frac{T}{r_a e} \ln \frac{K(T)}{n_e *} \tag{11.56}$$

If the external electric field is less than the critical value determined by Equation 11.56, $E_e < E(r_{e1})$, then the "free" electron concentration follows the Einbinder formula and does not depend on the value of the external electric field. As a consequence, the electric conductivity of aerosols also does not depend on the external electric field in this case.

When the external electric field exceeds the critical value, the number of electrons per one macroparticle able to participate in the electric conductivity $N(r_{e1})$ becomes dependent on the external electric field. To find this dependence, assume that only electrons not trapped in the vicinity of the surface by the inherent electric field of a macroparticle ($E(r) \le E_e$) can participate in the electric conductivity.

The electric field of a macroparticle $E(r)$ decreases with the radius r (distance from the center of a macroparticle). Therefore, if $E(r_e) > E(r_{e1})$, then, taking into account Equation 11.55, one can conclude that $r_e - r_a < r_a$. In this case, based on Equation 11.50, the number of electrons $N(r_e)$ (per one macroparticle) participating in the electric conductivity in the external electric fields in the interval $E(r_{e1}) < E_e < E(r_a)$ is:

$$N(r_e) = \frac{4\pi\varepsilon_0 E_e r_a^2}{e} \tag{11.57}$$

Here, $E(r_a)$ is the maximum value of the inherent electric field of a macroparticle, which is obviously realized on its surface (see Equation 11.52). If the external electric field exceeds the maximum value of the inherent electric field $E_e > E(r_a)$, then all thermally emitted electrons are able to participate in the electric conductivity $N(r_e) = N(r_a)$ (see Equation 11.51).

11.2.7 Electric Conductivity of Thermally Ionized Aerosols as Function of External Electric Field

Combining the expressions for the number of electrons participating in electric conductivity $N(r_e)$, Equation 11.54, Equation 11.57, and Equation 11.51, together with Equation 11.53, give the final dependence of the electric conductivity of the thermally ionized aerosols on the value of the external electric field, which can be expressed as:

$$\sigma_0 = \frac{4\pi\varepsilon_0 r_a T}{e^2}\left(\frac{r_a}{r_D^0}\right) n_a eb \quad \text{if} \quad E_e \ge \frac{T}{e r_D^0} \tag{11.58a}$$

$$\sigma_0 = 4\pi\varepsilon_0 E_e r_a^2 n_a b \quad \text{if} \quad \frac{T}{e r_a}\ln\frac{K(T)}{n_e *} < E_e < \frac{T}{e r_D^0} \tag{11.58b}$$

$$\sigma_0 = \frac{4\pi\varepsilon_0 r_a T}{e^2}\left(\ln\frac{K(T)}{n_e *}\right) n_a eb \quad \text{if} \quad E_e \le \frac{T}{e r_a}\ln\frac{K(T)}{n_e *} \tag{11.58c}$$

From Equation 11.58, the electric conductivity of aerosols begins depending on the external electric field explicitly when $E_e > T/er_a \ln K(T)/n_e{}^*$. Numerically, this means that the external electric field should be greater than about 200 V/cm, if $T = 1700$ K, $r_a = 10$ μm, and $A_0 = 3$ eV. The maximum number of electrons participate in electric conductivity if $E_e \geq T/er_D^0$, which numerically means that, at the same values of parameters, the external electric field should exceed about 1000 V/cm.

To calculate the total electric conductivity of the heterogeneous medium, it is necessary to take into account also the conductivity σ_a of the macroparticle material:[463]

$$\sigma = \sigma_0\left(1 - \frac{4}{3}\pi r_a^3 n_a \frac{\sigma_0 - \sigma_a}{2\sigma_0 + \sigma_a}\right) \tag{11.59}$$

This electric conductivity is based only on thermal ionization of the aerosol and neglects any breakdown effects.

11.3 ELECTRIC BREAKDOWN OF AEROSOLS

11.3.1 Influence of Macroparticles on Electric Breakdown Conditions

Macroparticles can be present in a discharge gap for different reasons, for example, air impurities or detachment of micropikes from the electrode surface.[464,465] Also, aerosol particles can be injected into a discharge gap to increase its breakdown resistance or, conversely, to initiate the breakdown.[46] Three major physical factors determine the decrease of the breakdown resistance related to the presence of macroparticles in a discharge gap:[466,467]

- Moving macroparticles are somewhat similar to electrode surface irregularities and induce local intensification of electric field, which stimulates breakdown.
- When a macroparticle approaches the electrode surface, its electric potential changes quickly. The macroparticle collision with the surface leads to a local change of the electric field structure and also to possible local intensification of the electric field.
- Decrease of the breakdown resistance also occurs due to microbreakdowns between macroparticles, and between macroparticles and electrodes.

These factors are able to decrease the breakdown voltage at atmospheric pressure air by a factor of two if the size of the aerosol particles is rather large (up to 1 mm).[46] Details regarding experimental and theoretical consideration of this effect in different heterogeneous discharge systems can be found in Raizer[468] and Bunkin and Savransky.[469] However, in many cases, admixture of the aerosol particles leads to an increase, rather than a decrease, of the breakdown resistance and breakdown voltage (see, for example, Walker[470] and Stengach.[471,472] This effect is mostly due to intensive attachment of electrons to the surface of aerosol particles; taking into account the practical importance of this effect, it will be considered later in more detail.[473]

11.3.2 General Equations of Electric Breakdown in Aerosols

Neglecting the nonuniformity effects discussed in the previous section, the kinetic equations describing the evolution of electron and positive ion densities in the aerosol system can be given:

$$\frac{dn_e}{dt} = An_e + Bn_i \tag{11.60a}$$

$$\frac{dn_i}{dt} = Cn_e - Dn_i \tag{11.60b}$$

The effective frequencies *A*, *B*, *C*, and *D* describing the breakdown of the aerosol system can be written in the following simplified form:

$$A = \alpha v_d^e - \beta v_d^e - \frac{D_e}{d^2} + K - \frac{v_d^e}{d} \tag{11.61a}$$

$$B = \gamma_{ia}\sigma_{ia}v^i n_a + \gamma_{iw}\frac{v_d^i}{d} \tag{11.61b}$$

$$C = k_i n_0 \tag{11.61c}$$

$$D = k_a n_a + \frac{D_i}{d^2} + \frac{v_d^i}{d} \tag{11.61d}$$

$$K = \gamma_{pa}\mu\nu n_0 + \gamma_{pw}\mu(1-\nu)n_0 \tag{11.61e}$$

In these relations, n_0 and n_a are concentrations of neutral gas and aerosol particles; v_d^i, v^i, v_d^e, v^e are thermal and drift (with subscript *d*) velocities of electrons (superscript *e*) and ions (superscript *i*); α and β are the first and second Townsend coefficients (see Section 4.4.1 and Section 4.4.3) that determine the formation and attachment of electrons per unit length along the electric field during electron drift; D_e and D_i are the diffusion coefficients of electrons and ions (either free or ambipolar values, depending on system parameters); *d* is the characteristic size of a discharge gap; σ_{ia} is the cross section of ion collisions with the aerosol particles; γ_{ia}, γ_{pa}, γ_{pw}, and γ_{iw}, are probabilities of an electron formation in an ion collision with a macroparticle–photon interaction with a macroparticle, photon interaction with a wall, and ion collision with a wall; μ is the rate coefficient of a photon formation calculated with respect to an electron–neutral collision; ν is the probability for a photon to reach an aerosol particle; k_i is the rate coefficient of neutral gas ionization by electron impact; and k_a is the rate coefficient of ion losses related to collisions with aerosol particles.

In the system of the kinetic equation (Equation 11.60), electron formation due to neutral gas ionization; secondary electron emission from the surface of macroparticles; ion collisions with aerosols; photoemission from macroparticles; and discharge walls are taken into account. Electron losses are due to attachment to gas molecules and macroparticles plus diffusion and drift along the electric field. Ions are formed in electron–neutral collisions, and their losses are due to collisions with macroparticles and the discharge walls. Because of relatively low concentration of the charged particles, their volume recombination can usually be neglected in these systems.

Solution of Equation 11.60 for the electron concentration n_e with the initial conditions $n_e(t=0)=n_e^0$ and $n_i(t=0)=n_i^0$ can be given as:

$$n_e(t)=\frac{(An_e^0+Bn_i^0)-\lambda_2 n_e^0}{\lambda_1-\lambda_2}\exp(\lambda_1 t)+\frac{\lambda_1 n_e^0-(An_e^0+Bn_i^0)}{\lambda_1-\lambda_2}\exp(\lambda_2 t) \quad (11.62)$$

Here, $\lambda_{1,2}$ are solutions of the characteristic equation related to the system (Equation 11.60) and can be expressed as:

$$\lambda_{1,2}=\frac{A-D\pm\sqrt{(A+D)^2+4BC}}{2} \quad (11.63)$$

11.3.3 Aerosol System Parameters Related to Breakdown

To apply Equation 11.62 for specific types of electric breakdown, consider first the main parameters of the aerosol system related to the relations in Equation 11.61. The mean free paths of electrons with respect to all collisions, collisions with neutrals, and collisions with macroparticles can be respectively expressed as:

$$\lambda_\Sigma=\frac{1}{n_0\sigma_{e0}+n_a\sigma_a},\quad \lambda_0=\frac{1}{n_0\sigma_{e0}},\quad \lambda_a=\frac{1}{n_a\sigma_a} \quad (11.64)$$

where σ_{e0} and σ_a are cross sections of collisions of electrons with neutral species and macroparticles. Then, write down the expressions for electron temperature T_e; thermal and drift velocities of electrons v^e and v_d^e; electron diffusion coefficient D_e; and the first and second Townsend coefficients α and β in aerosol systems, using the corresponding values of $T_e^0, v_0^e, v_{d0}^e, D_e^0, \alpha_0, \beta_0$ for neutral gas in the same conditions but without macroparticles. Thus, at first, the electron temperature in the aerosol system can be considered as:

$$T_e=\frac{eE}{\delta_1/\lambda_0+\delta_2/\lambda_a}=\left[\frac{1}{T_e^0(E)}+\frac{\delta_2}{eE\lambda_a}\right]^{-1} \quad (11.65)$$

In this relation, δ_1 and δ_2 are the fraction of electron energy lost in collision with a molecule or macroparticle, respectively. Similarly, the electron thermal velocity in aerosol can be expressed, based on Equation 11.65, as:

$$v^e = \sqrt{\frac{2eE}{m(\delta_1/\lambda_0 + \delta_2/\lambda_a)}} = \left[\left(\frac{1}{v_0^e}\right)^2 + \frac{\delta_2 m}{2\lambda_a eE}\right]^{-1/2} \tag{11.66}$$

The diffusion coefficient of electrons in the aerosol medium can be written using the additional parameter $\xi = n_a\sigma_a/n_0\sigma_{e0}$ as:

$$D_e = \frac{\lambda_\Sigma}{3}\sqrt{\frac{2eE}{m(\delta_1/\lambda_0 + \delta_2/\lambda_a)}} = \frac{D_e^0}{1+\xi}\left(1 + \frac{\delta_2 T_e^0}{eE\lambda_a}\right)^{-1/2} \tag{11.67}$$

The electron drift velocity in aerosols can be presented as:

$$v_d^e = \lambda_\Sigma\sqrt{\frac{eE}{2m}\left(\frac{\delta_1}{\lambda_0} + \frac{\delta_2}{\lambda_a}\right)} = \frac{v_{d0}^e}{1+\xi}\left(1 + \frac{\delta_2 T_e^0}{eE\lambda_a}\right)^{1/2} \tag{11.68}$$

Note that relations for thermal velocity, drift velocity, and diffusion coefficient for positive ions are obviously similar to those for electrons (Equation 11.66 through Equation 11.68). The first Townsend coefficient shows ionization per unit length of electron drift and, in the aerosol system, can be expressed as:

$$\begin{aligned}\alpha &= \frac{2}{\lambda_\Sigma(\delta_1/\lambda_0 + \delta_2/\lambda_a)}\left[\frac{1}{\lambda_0}\exp\left(-\frac{I_i}{T_e}\right) + \frac{1}{\lambda_a}\exp\left(-\frac{I_a}{T_e}\right)\right] \\ &= \frac{2}{\lambda_\Sigma(eE/T_e^0 + \delta_2/\lambda_a)}\left[\frac{\alpha_0\lambda_0}{2}\frac{eE}{T_e^0} + \frac{1}{\lambda_a}\exp\left(-\frac{I_a}{T_e}\right)\right]\end{aligned} \tag{11.69}$$

In this relation, I_i is the ionization potential of molecules and I_a is the critical electron energy, when the secondary electron emission coefficient from macroparticle reaches unity. The second Townsend coefficient describing electron attachment to molecules and aerosol particles can be presented similarly to Equation 10.57 as:

$$\begin{aligned}\beta &= \frac{2}{\lambda_\Sigma(\delta_1/\lambda_0 + \delta_2/\lambda_a)} \times \left[\frac{1}{\lambda_0}w^0 + \frac{1}{\lambda_a}w^a\right] \\ &= \frac{2}{\lambda_\Sigma(eE/T_e^0 + \delta_2/\lambda_a)} \times \left[\frac{\beta_0\lambda_0}{2}\frac{eE}{T_e^0} + \frac{1}{\lambda_a}w^a\right]\end{aligned} \tag{11.70}$$

where w^0 and w^a are the electron attachment probabilities in a collision with a molecule or macroparticle, respectively.

11.3.4 Pulse Breakdown of Aerosols

The plasma chemical kinetic factors of the aerosol system, calculated in Section 11.3.3, permit analyzing breakdown conditions for different specific discharges. Begin by considering the pulse breakdown, where it is assumed that $v_d^e \cdot \tau \leq d$, $d \leq c\tau$, $\lambda_\Sigma \leq d$ (Raether;[106] τ is the voltage pulse duration and c is the speed of light in the aerosol medium). Considering the pulse breakdown, neglect the terms related to diffusive and drift losses in Equation 11.61, assuming $1/d \to 0$. Also, it is usually assumed that the contribution of ion–macroparticle collisions in electron balance in relatively low: $B << (A = D)^2/C$. Then, taking into account that the conventional numerical pulse breakdown criterion[106] is:

$$n_e(t = \tau) = n_b = 10^8 \times n_e^0 \tag{11.71}$$

the pulse breakdown condition in an aerosol system based on Equation 11.62 can be given in the following form:

$$\tau = \frac{\ln \dfrac{n_b}{n_e^0} + \dfrac{BC}{(A+D)^2}}{A + \dfrac{BC}{A+D}} \tag{11.72}$$

This pulse breakdown condition can be specified based on Equation 11.61. It then permits estimating the maximum possible increase of the pulse breakdown voltage related to electron attachment to aerosol particles, by assuming the attachment probability as $w^a = 1 - \exp(-I_a/T_e)$. This effect of the maximum breakdown voltage increase can be calculated using the following relation:[473]

$$\frac{\alpha_0 \lambda_0 eE}{2T_e^0} + \frac{2\exp\left(-\dfrac{I_a}{T_e}\right) - 1}{\lambda_a} = \frac{eE\lambda_0 \ln \dfrac{n_b}{n_0}}{2T_e^0 v_{d0}^e \tau} \sqrt{1 + \frac{\delta_2 T_e^0}{eE\lambda_a}} \tag{11.73}$$

For example, if $\tau = 10$ nsec, $p = 10$ torr, $d = 10$ cm, $I_a = 100$ eV, and $\delta_2 = 1$, the pulse breakdown voltage in nitrogen without aerosol particles is $E_0 \approx 2$ kV/cm ($\xi = 0$). Addition of aerosol particles characterized by $\xi = n_a\sigma_a/n_0\sigma_{e0} = 0.1$ increases the breakdown voltage by factor of about 1.2 (20%).

11.3.5 Breakdown of Aerosols in High-Frequency Electromagnetic Fields

The electric breakdown conditions of aerosols in accordance with Equation 11.62 correspond to $\max(\lambda_1, \lambda_2) = 0$, which can be written as:

$$AD + BC = 0 \tag{11.74}$$

In the case of breakdown in high-frequency electromagnetic fields (where drift losses can usually be neglected), the general condition (Equation 11.74) can be given as:[473]

$$\alpha = \beta + \frac{1}{d^2\left(\frac{eE}{T_e^0} + \frac{\delta_2}{\lambda_a}\right)} - \gamma_{ia}\sigma_{ia}n_a\frac{v^i}{v_d^e}\frac{k_i n_0}{\left(k_a n_a + \frac{D_i}{d^2}\right)} - \frac{K}{v_d^e} \tag{11.75}$$

The electric field in this and the following relations should be taken as an effective one (see Equation 4.78), taking into account the high frequency of the electromagnetic field. Similar to Equation 11.73, the maximum effect of breakdown voltage increase related to electron attachment by aerosol particles can be calculated, taking into account formulas from Section 11.3.3, as:

$$\lambda_0\frac{eE}{T_e^0}\alpha_0 + \frac{2\xi}{\lambda_0}\left[2\exp\left(-\frac{I_a}{T_e}\right) - 1\right] = \frac{\lambda_0}{d^2(1+\xi)} \tag{11.76}$$

Consider a numerical example applying Equation 11.76 similar to that in Section 11.3.4: p = 10 torr, d = 10 cm, I_a = 100 eV, and δ_2 = 1. The addition of aerosol particles, characterized by $\xi = n_a\sigma_a/n_0\sigma_{e0} = 0.1$ in nitrogen, gives a possibility to increase the breakdown voltage by a factor of about 1.3 (30%) with respect to the same system without macroparticles ($\xi = 0$).

The breakdown of aerosols in radio-frequency and microwave electromagnetic fields is quite sensitive to the macroparticle heating in these fields. These heating effects are related not only to thermal emission (see Section 11.3.2) but also to evaporation and thermal explosion of macroparticles, which affects the breakdown conditions.[468]

11.3.6 Townsend Breakdown of Aerosols

The Townsend breakdown of aerosols can be described by the relation similar to that for pure gas (see Section 4.4.1):

$$\gamma_{eff}\left\{\exp(\alpha-\beta)d - 1 + \frac{\beta}{\alpha-\beta}[\exp(\alpha-\beta)d - 1]\right\} = 1 \tag{11.77}$$

but with Townsend coefficients α and β calculated according to Equation 11.69 and Equation 11.70, taking into account all effects related to macroparticles. The third Townsend coefficient γ_{eff} is equal to the total number of generated secondary electrons with respect to one ion formed in the volume. In this case, the coefficient γ_{eff} is an effective value and should take into account ion and photon interaction, not only with the cathode but also with aerosol particles.

The probability for an ion formed in the volume to reach the cathode without being trapped by aerosol particles can be estimated as:

$$w = \exp\left(-n_a\sigma_{ia}d\frac{v^i}{2v_d^i}\right) \tag{11.78}$$

Then, the effective value of the third Townsend coefficient γ_{eff} can be calculated by the following cumulative relation:

$$\gamma_{eff} = \gamma_{iw} \exp\left(-n_a \sigma_{ia} d \frac{v^i}{2v_d^i}\right) + \gamma_{ia}\left[1 - \exp\left(-n_a \sigma_{ia} d \frac{v^i}{2v_d^i}\right)\right] + \gamma_{pw} \frac{\mu(1-\nu)}{k_i} + \gamma_{pa} \frac{\mu\nu}{k_i} \quad (11.79)$$

To estimate the maximum increase of the Townsend breakdown voltage due to the electron attachment by macroparticles similar to Equation 11.73, one can use the equation:

$$\frac{2d}{\lambda_\Sigma\left(eE/T_e^0 + \delta_2/\lambda_a\right)}\left[\frac{\alpha_0 \lambda_0 eE}{2T_e^0} + \frac{2\exp\left(-I_a/T_e\right) - 1}{\lambda_a}\right] = n_a \sigma_{ia} d \frac{v^i}{2v_d^i} - \ln \gamma_{ia} \quad (11.80)$$

As an example, the Townsend breakdown voltage of nitrogen at 10 torr with tantalum electrodes (d = 1 cm) can have a 10% increase by adding macroparticles with $\xi = n_a\sigma_d/n_0\sigma_{e0} = 0.1$; $\delta_2 = 1$; and $I_a = 100$ eV.

11.3.7 Effect of Macroparticles on Vacuum Breakdown

In the preceding examples, the relatively low addition of macroparticles $\xi = n_a\sigma_d/n_0\sigma_{e0} < 1$ was assumed when electron collisions with molecules occur more often than collisions with aerosol particles. As was shown in the numerical examples of Section 11.3.4 through Section 11.3.6, the maximum increase of the breakdown voltage related to electron attachment to macroparticles is about 10 to 30%. Although this effect is not very strong at low aerosol concentration $\xi = n_a\sigma_d/n_0\sigma_{e0} < 1$, it is of interest for different applications, in particular for electric filters.[630]

Much larger increases of breakdown voltage can be reached when $\xi = n_a\sigma_d/n_0\sigma_{e0} > 1$ and most electron collisions are related to aerosol particles. In this case, the breakdown conditions actually do not depend on the gas characteristics, so the breakdown can be interpreted as the vacuum values. Thus, the pulse breakdown conditions in aerosols at $\xi >> 1$ and $d > 1/2\ \lambda_a \ln n_b/n_e^0$ can be expressed from Equation 11.61 and Equation 11.62 as:

$$2\exp\left(-\frac{I_a}{eE\lambda_a}\right) - 1 = \frac{\lambda_a}{2\tau}\sqrt{\frac{2m}{eE\lambda_a}} \ln \frac{n_b}{n_e^0} \quad (11.81)$$

In the specific case of $d > \tau\sqrt{I_a/2m} >> \lambda_a \ln n_b/n_e^0$, the most physically clear pulse breakdown condition can be obtained from Equation 11.81:

$$E = \frac{I_a}{e\lambda_a \ln 2} \quad (11.82)$$

According to Equation 11.82, if I_a = 100 eV; $n_a = 10^7$ cm^{-3}; $\sigma_a = 10^{-6}$ cm^2; d = 10 cm; τ = 10 nsec; and p = 0.1 torr, the pulse breakdown electric field should be numerically equal to E = 1000 V/cm.

The breakdown of relatively dense aerosols ($\xi >> 1$) in the high-frequency electromagnetic fields can be described by the simplified relation of Equation 11.76 as:

$$2\exp\left(-\frac{I_a}{eE\lambda_a}\right) - 1 = \frac{\lambda_a^2}{d^2} \tag{11.83}$$

The effective breakdown electric field (see Equation 4.78) in the radio-frequency or microwave range can be found from Equation 11.83 as:

$$E_{eff} = \frac{I_a}{e\lambda_a \ln\left[2/\left(1+\lambda_a^2/d^2\right)\right]} \tag{11.84}$$

This solution for the breakdown in high frequency electromagnetic field corresponds to the similar condition (Equation 11.82) for the pulse breakdown, if $\lambda_a^2 << d^2$.

11.3.8 Initiation of Electric Breakdown in Aerosols

Aerosol particles can be applied to increase a discharge gap resistance to electric breakdown. At the same time, the presence of macroparticles provides controlled initiation of breakdown, in particular by pulsed photoionization.[46,474]

Photoionization of aerosols (see Section 11.3.1) can be applied for initiation (commutation) of the pulsed nanosecond breakdown, as well as for initiation of the breakdown in a constant electric field. In the case of pulsed nanosecond breakdown, the photoinitiation is provided by fast generation of the necessary initial electron concentration, n_e^0, due to photoionization. In the case of breakdown in constant electric field, the initiation effect is related to the induced local nonuniformities of electric field. Electrons formed during photoionization move from the discharge gap to cathode, leaving the charged macroparticles in the gap. Positive space charge of the macroparticles provides the nonuniformity of electric field, resulting in the breakdown initiation.

11.4 STEADY-STATE DC ELECTRIC DISCHARGE IN HETEROGENEOUS MEDIUM

11.4.1 Two Regimes of Steady-State Discharges in Heterogeneous Medium

Consider the steady-state DC discharge in a heterogeneous medium, neglecting the effects of thermionic and field emission (see Section 8.2.1 through Section 8.2.4) from the surface of macroparticles. The main mechanisms of electron formation in this case are ionization of the neutral gas by direct electron impact and secondary electron emission from the aerosol particles. The electroneutrality condition of the aerosol system, including electrons, positive ions, and macroparticles, can then be written as:

$$n_e = n_i + Z_a n_a \tag{11.85}$$

where n_e, n_i, and n_a are concentrations of electrons, positive ions, and macroparticles, respectively, and Z_a is the average charge of a macroparticle (in elementary charges).

If the concentration and radius of aerosol particles in the discharge system are relatively low, then formation of electrons is mostly due to neutral gas ionization. In this case, usually $n_e >> Z_a n_a$, and, according to Equation 11.85, the concentrations of electrons and positive ions are nearly equal $n_e \approx n_i$. This regime of discharge in aerosols is referred to as the quasi-neutral one, which points out the quasi neutrality of the system in the gas phase.

Another heterogeneous discharge regime takes place if the macroparticles' number density and sizes are sufficiently high and their ionization is through secondary electron emission. In this case, $n_i << n_e \approx Z_a n_a$, and the heterogeneous plasma actually consists of electrons and positively charged macroparticles (instead of ions). This regime is usually referred to as the **electron–aerosol plasma** regime to emphasize the contrast with conventional electron–ion plasma.

11.4.2 Quasi-Neutral Regime of Steady-State DC Discharge in Aerosols

The quasi-neutral aerosol discharge $n_e \approx n_i >> Z_a n_a$ was investigated in detail by Musin,[445] Konenko et al.,[475] and Konenko and Musin.[476,477] In the positive column of such a discharge, electron formation is mostly due to ionization of neutral species in the gas phase, while scattering and recombination of charged particles take place mostly on the surfaces of macroparticles and the discharge tube. Macroparticles in this regime are negatively charged and their potential is close to the floating value.

The quasi-neutral regime of a steady-state DC discharge in aerosols can be described by the following system of equations:[445]

$$\left(4\pi r_a^2 n_a + \frac{2}{R}\right)\sqrt{\frac{m}{M}} = n_0 \sigma_{e0} \exp\left(-\frac{I_i}{T_e}\right) \tag{11.86}$$

$$E = I_i n_S \left[1 + \frac{1}{n_S}\left(n_0 \sigma_{e0} + n_e \sigma_{ei}\right)\right] \cdot \Phi^2(T_e) \tag{11.87}$$

$$n_e = \frac{j_e}{e}\sqrt{\frac{m}{2eI_i}} \cdot \Phi(T_e) \tag{11.88}$$

In this system, R is the radius of discharge tube; m and M are masses of an electron and a positive ion; n_0 is the gas density; T_e is the electron temperature; I_i is the ionization potential; E is the electric field; σ_{e0} and σ_{ei} are the characteristic cross sections of electron–neutral and electron–ion collisions corresponding to average electron energy; and j_e is the electron current density. The function $\Phi(T_e)$ is determined by the expression:

$$\Phi(T_e) = n_S \sqrt{\frac{m}{2M} \cdot \frac{T_e}{I_i}} \times \frac{\left(2 + \frac{1}{2}\ln\frac{M}{m}\right)\frac{T_e}{I_i} + 1}{n_0 \sigma_{e0} + n_e \sigma_{ei}} \tag{11.89}$$

Another concept applied in the preceding relations is n_S "density of surfaces," which represents the surface area of macroparticles and discharge tube walls per unit volume. In the case of a long cylindrical discharge tube, this can be expressed as:

$$n_S = 4\pi r_a^2 n_a + \frac{2}{R} \tag{11.90}$$

The system of Equation 11.86 through Equation 11.88 shows that addition of aerosol particles in the discharge at the fixed value of current density leads to an essential increase of electron temperature, while growth of electron concentration is relatively small. If the surface area of macroparticles is less than that of the discharge tube walls, the system of Equation 11.86 through Equation 11.88 corresponds to the Langmuir–Klarfeld theory.[478,479] Equation 11.86 shows that an increase of the surface area density n_S leads to a logarithmic increase of electron temperature. If macroparticles make the major contribution to the total surface area $2/R << 4\pi r_a^2 n_a$, this dependence can be presented as:

$$T_e = I_i \ln^{-1} \frac{n_{a,cr}}{n_a} \tag{11.91}$$

where $n_{a,cr}$ is the critical value of aerosol concentration when the electron temperature tends to infinity $T_e \to \infty$. Thus, the considered quasi-neutral regime can exist only at relatively low concentrations of aerosol particles:

$$n_a < n_{a,cr} = n_0 \frac{\sigma_{e0}}{4\pi r_a^2} \sqrt{\frac{M}{m}} \tag{11.92}$$

11.4.3 Electron–Aerosol Plasma Regime of Steady-State DC Discharge, Main Equations Relating Electric Field, Electron Concentration, and Current Density

At aerosol particle concentrations exceeding this critical value (Equation 11.92), an additional ionization mechanism is required to compensate the electron losses on the macroparticles. This becomes possible in the electron–aerosol plasma regime, where additional formation of electrons is related to the secondary electron emission from the macroparticles.[474]

Consider this regime of predominant ionization of aerosols $n_i << n_e \approx Z_a n_a$ and relatively low gas concentration. Because the mass of macroparticles is relatively high, the ambipolar diffusion to the discharge walls can be neglected and, therefore, the steady-state balance of charged processes should be given by volume processes. This means that the total coefficient of the secondary electron–electron emission (see Section 8.2.5) from the surface of macroparticles should be equal to unity, if thermionic and field emissions of electrons are neglected.

The total coefficient δ of the secondary electron–electron emission from aerosol particles depends on the work function $\varphi(Z_a)$ that is related to the macroparticle charge according to Equation 11.1, and average energy $\varepsilon \approx eE\lambda_a$ of electrons bombarding the macroparticles. Thus, the charged particle balance equation can be expressed as:

$$\delta\left(\varepsilon = eE\lambda_a, \varphi(Z_a) = \varphi_0 + \frac{Z_a e^2}{4\pi\varepsilon_0 r_a}\right) = 1 \tag{11.93}$$

This dependence permits finding the macroparticle charge Z_a as a function of the electric field E in the following form:

$$Z_a = Z_0 \frac{E - E_0}{E_0} \tag{11.94}$$

In this relation, E_0 is the breakdown electric field in the heterogeneous medium; the aerosol charge parameter Z_0 is determined as:

$$Z_0 = -\varepsilon_{00}\left(\frac{\partial\delta}{\partial\varepsilon}\right)_{\varphi=\varphi_0} \Bigg/ \frac{e^2}{4\pi\varepsilon_0 r_a}\left(\frac{\partial\delta}{\partial\varphi}\right)_{\varepsilon=\varepsilon_{00}} \tag{11.95}$$

where the energy parameter $\varepsilon_{00} = E_0\lambda_a e$. For numerical calculations of Z_0 at the following parametric values: ε_0 = 100 eV, φ_0 = 3 eV, $(\partial\delta/\partial\varphi)_{\varepsilon=\varepsilon_{00}}$ = –0.05 eV^{-1}, $(\partial\delta/\partial\varepsilon)_{\varphi=\varphi_0}$ = 0.01 eV^{-1}, it is convenient to use the relation:

$$\log Z_0 = 8 + \log r_a(\text{cm}) \tag{11.96}$$

Equation 11.94 for the electric field should be combined with the equations for current density and electron concentration:

$$j_e = n_e e\sqrt{\frac{eE\lambda_a}{2m}}, \quad n_e = Z_a n_a \tag{11.97}$$

to form the system of equations sufficient to determine the electric field and electron concentration in the heterogeneous discharge as a function of current density.

11.4.4 Electron–Aerosol Plasma Parameters as Function of Current Density

The system of Equation 11.94 and Equation 11.97 permits finding the electric field and electron concentration as a function of current density in the electron–aerosol plasma. This function can be determined in terms of the special dimensionless factor B characterizing the current density in the heterogeneous discharge:[474]

$$B = \frac{j_e}{Z_0 n_a e}\sqrt{\frac{2m}{\varepsilon_0}} \tag{11.98}$$

For numerical calculations of the current density factor B at the following values of parameters: $\varepsilon_0 = 100$ eV, $\varphi_0 = 3$ eV, $(\partial\delta/\partial\varphi)_{\varepsilon=\varepsilon_{00}} = -0.05$ eV^{-1}, $(\partial\delta/\partial\varepsilon)_{\varphi=\varphi_0} = 0.01$ eV^{-1}, it is convenient to use the relation:

$$\log B = 2 + \log\frac{j_e\left(A/\text{cm}^2\right)}{n_a\left(\text{cm}^{-3}\right)\cdot r_a(\text{cm})} \tag{11.99}$$

Thus, according to the system of Equation 11.94 and Equation 11.97, at low current densities when the factor $B < 1$, the electric field in the heterogeneous discharge is slowly growing with current density as:

$$E = E_0(1 + B) \tag{11.100}$$

Electron concentration and thus average charge of macroparticles at low current densities ($B < 1$) grows up proportionally to j_e:

$$n_e = n_a Z_0 B, \quad Z_a = Z_0 B \tag{11.101}$$

In the case of high current densities ($B > 1$), the electron concentration and average charge of macroparticles reach the maximum values:

$$n_e = n_a Z_0, \quad Z_a = Z_0 \tag{11.102}$$

These saturation values of electron concentration and average macroparticle charge are larger at larger values of the aerosol particle radius (see Equation 11.96). At the same time, the electric field growth with j_e becomes much stronger in the high-current regime ($B > 1$), which can be expressed by the relation:

$$E \approx E_0 B^2 \tag{11.103}$$

The dependence of the electric field lg $(E - E_0)/E_0$ and electron concentration log n_e/n_a on the current density

$$\log\frac{j_e(A/cm^2)}{n_a(cm^{-3})\cdot r_a(cm)}$$

in the electron–aerosol plasma at the preceding parametric values of $\varepsilon_0 = 100$ eV, $\varphi_0 = 3$ eV, $(\partial\delta/\partial\varphi)_{\varepsilon=\varepsilon_{00}} = -0.05$ eV^{-1}, $(\partial\delta/\partial\varepsilon)_{\varphi=\varphi_0} = 0.01$ eV^{-1} is presented in Figure 11.3.

From Figure 11.3, the dependence of the average charge of aerosol particles on the current density appears as a saturation curve. The average charge of aerosol particles of radius 10 μm and concentration $n_a = 10^6$ cm^{-3} reaches its maximum value at electron current densities exceeding 10 μA/cm^2. The maximum values of the average charge of the aerosol particles depend on their size and can be quite large, about 10^5 for the 10-μm macroparticles.

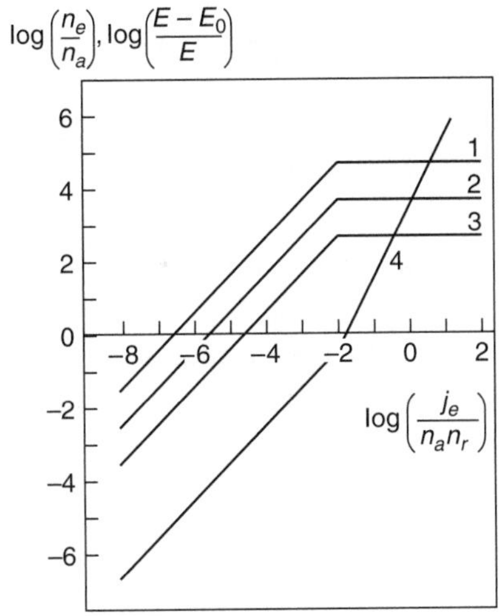

Figure 11.3 Electric discharge in aerosols: 1, 2, 3: $\log(n_e/n_a)$; 4: $\log[(E - E_0)/E]$; 1: $r_a = 10^{-3}$ cm; 2: 10^{-4} cm; 3: 10^{-3} cm; j_e [A/cm^2]; n_a [cm^{-3}][474]

11.4.5 Effect of Molecular Gas on Electron–Aerosol Plasma

In the preceding consideration of the electron–aerosol plasma, the contribution of molecular gas in the balance of charged particles and ionization kinetics was neglected. In this case, such an approach is quite acceptable because most of the electron generation in an electron–aerosol plasma is related to the secondary electron emission from the surface of aerosol particles. Nevertheless, the effect of a molecular gas is important even in this regime, because it determines the concentration of positive ions and kinetic limits of the regime $n_i << n_e$.

The molecular gas effect is revealed in the kinetic equation for positive ions in the heterogeneous discharge system. The balance of positive ions should take into account their formation provided by ionization of molecules and their losses on the surfaces of macroparticles and discharge tube; this can be expressed as:

$$\frac{n_S j_{is}}{e} = n_e n_0 \sigma_{e0} \sqrt{\frac{2T_e}{m}} \exp\left(-\frac{I_i}{T_e}\right) \tag{11.104}$$

In this relation, n_S is the "density of surfaces" defined by Equation 11.90; n_e and n_0 are the concentrations of electrons and neutral gas, respectively; σ_{e0} is the characteristic cross section of electron-neutral collisions; I_i is the ionization potential of molecules; m and T_e are the electron mass and temperature, respectively; and j_{is} is the ion current density on the surfaces of macroparticles and discharge walls, which can be found from the Bohm formula (see, for example, Huddlestone and Leonard[480] and also Section 6.1.4):

$$j_{is} \approx 0.5 n_i e \sqrt{\frac{T_e}{M}} \tag{11.105}$$

Combining Equation 11.104 and Equation 11.105 gives the qualitative ion balance equation in the following form:

$$n_s n_i \sqrt{\frac{m}{M}} = n_e n_0 \sigma_{e0} \exp(-\frac{I_i}{T_e}) \tag{11.106}$$

Taking into account that, in the electron–aerosol plasma regime, the ions' contribution to charge balance is relatively low, $n_i << n_e$, and, therefore, $n_e \approx Z_a n_a$, the concentration of the positive ions can be found as:

$$n_i = \frac{Z_a n_a n_0 \sigma_{e0}}{\frac{2}{R} + 4\pi r_a^2 n_a} \sqrt{\frac{M}{m}} \exp\left(-\frac{I_i}{T_e}\right) \tag{11.107}$$

If most of the surface area in the heterogeneous discharge is related to the macroparticles ($2/R << 4\pi r_a^2 n_a$; R is the radius of the long discharge tube), then the ratio of ion and electron densities can be given as:

$$\frac{n_i}{n_e} = \frac{n_0 \sigma_{e0}}{4\pi r_a^2 n_a} \sqrt{\frac{M}{m}} \exp\left(-\frac{I_e}{T_e}\right) \tag{11.108}$$

The requirement of relatively low value of ion concentration at any levels of electron temperature in the heterogeneous system is the main limit of the electron–aerosol plasma regime. This can be expressed, based on Equation 11.108, as:

$$n_0 << n_a \frac{4\pi r_a^2}{\sigma_{e0}} \sqrt{\frac{m}{M}} \tag{11.109}$$

Obviously, this criterion of the electron–aerosol plasma regime is opposite to the criterion (Equation 11.92) of the quasi-neutral regime of discharges in aerosols. Numerically, the criterion (Equation 11.109) shows that the electron–aerosol plasma regime takes place at $n_a = 10^8$ cm^{-3}, $r_a = 10$ μm if the molecular gas concentration is less than 10^{17} cm^{-3}.

11.5 DUSTY PLASMA FORMATION: EVOLUTION OF NANOPARTICLES IN PLASMA

11.5.1 General Aspects of Dusty Plasma Kinetics

In the aerosol plasmas discussed in Section 11.1 through Section 11.4, the macroparticles were considered of fixed size and shape, and as not chemically active. However, numerous fundamental and applied problems are related to formation and growth of molecular clusters, and nano and microscale particles in plasma. Plasma physics and plasma chemistry are essentially coupled in the so-called dusty plasma systems to be discussed in this section.[26,481–483]

Applied aspects of dusty plasmas are mostly related with electronics environment control, powder chemistry, and metallurgy. Active research on particle formation and behavior has been generated by contamination phenomena of industrial plasma reactors used for etching, sputtering, and PECVD. For this reason, the process of particle formation in low pressure glow or RF discharges in silane (SiH_4, or mixture SiH_4–Ar) has been extensively studied.[484–490] These systems clearly illustrate the dusty plasma formation and behavior to be considered next.

Many efforts have dealt with the detection and dynamics of relatively large size (more than 10 nm) nanoparticles in silane discharge. Such particles are electrostatically trapped in the plasma bulk and significantly affect the discharge behavior.[491–496] The most fundamentally interesting phenomenon takes place in the early phases of the process, starting from an initial dust-free silane discharge. During this period, small deviations in plasma and gas parameters (gas temperature, pressure, electron concentration) can alter cluster growth and the subsequent discharge behavior.[497,498,499,500]

Experimental data concerning super-small particle growth (particle diameter as small as 2 nm) and negative ion kinetics (up to 500 or 1000 amu) has demonstrated the main role of negative ion clusters in the particle growth process.[27,500–504] Thus, the initial phase of formation and growth of the dust particles in a low-pressure silane discharge is a homogeneous process, stimulated by fast silane molecule and radical reactions with negative ion clusters.

The first particle generation appears as a monodispersed one with a crystallite size of about 2 nm. Due to a selective trapping effect, the concentration of these crystallites increases up to a critical value, where a fast (such as a phase transition) coagulation process takes place leading to large size (50 nm) dust particles. During the coagulation step, when the aggregate size becomes larger than some specific value of approximately 6 nm, another critical phenomenon, the so-called "α–γ transition," takes place with a strong decrease of electron concentration and a significant increase of their energy. A small increase of the gas temperature increases the induction period required to achieve these critical phenomena of crystallite formation: agglomeration and the α–γ transition.

11.5.2 Experimental Observations of Dusty Plasma Formation in Low-Pressure Silane Discharge

The low-pressure RF discharge for observing dusty plasma formation has been conducted in the Boufendi–Bouchoule experiments in a grounded cylindrical box (13 cm inner diameter) equipped with a shower type RF-powered electrode. A grid was used as the bottom of the chamber to allow a vertical laminar flow in the discharge box. The typical Boufendi–Bouchole experimental conditions are: argon flow 30 sccm; silane flow 1.2 sccm; total pressure 117 mtorr (so the total gas concentration is about 4 10^{15} cm^{-3}, silane 1.6 10^{14} cm^{-3}); neutral gas residence time in the discharge is approximately 150 msec; and RF power is10 W. A cylindrical oven to vary gas temperature from the ambient up to 200°C surrounds the discharge structure.

The Boufendi–Bouchoule experiments show that the first particle size distribution is monodispersed with the diameter about 2 nm and is practically independent of temperature. Appearance time of the first-generation particles is less than 5 msec. The measurements give a clear indication that the particle growth proceeds through the successive steps of fast, super-small 2-nm particle formation and the growth of their concentration up to the critical value of about 10^{10} to 10^{11} cm^{-3}, when the new particle formation terminates and formation of aggregates with diameters of up to 50 nm begins by means of coagulation.

During the initial discharge phase (until the α–γ transition, up to 0.5 sec for room temperature and up to several seconds for 400 K), the electron temperature remains about 2 eV; the electron concentration is about $3 \cdot 10^9$ cm^{-3}; the positive ion concentration is approximately $4 \cdot 10^9$ cm^{-3}; and the negative ion concentration is about 10^9 cm^{-3}. After the α–γ transition, the electron temperature increases up to 8 eV while the electron concentration decreases 10 times and the positive ion concentration increases 2 times. These concentrations are correlated with the negative volume charge density of the charged particles through the plasma neutrality relation.[505,506]

The Boufendi–Bouchoule experimental data show that the critical value of super-small particle concentration before coagulation practically is independent of temperature. The induction time observed before coagulation is a highly sensitive function of the temperature; ranging from about 150 msec at 300 K, it increases more than 10 times when heated to only 400 K. For a temperature of 400 K, the time required to increase the super-small neutral particle concentration is much longer (10 times) than the gas residence time in the discharge, which can be explained by the neutral particle trapping phenomenon. According to the Boufendi–Bouchoule experiments, the particle concentration during the coagulation period decreases and the average radius grows. The total mass of dust in plasma remains almost constant during the coagulation.

11.5.3 Dust Particle Formation: a Story of Birth and Catastrophic Life

The previously described dust particle formation and growth in a SiH_4–Ar low-pressure discharge can be subdivided into four steps. These include the growth of super-small particles from molecular species, and three successive catastrophic events: selective trapping, fast coagulation, and finally strong modification of discharge parameters, e.g., the α–γ transition.[28]

A scheme for formation of the first **super-small particle generation** is presented in Figure 11.4(a). It begins mainly with SiH_3^- negative ion formation by dissociative attachment. The occurrence of the nondissociative three-body attachment to SiH_3 radicals could occur in a complementary way ($e + SiH_3 = (SiH_3^-)^*$, $(SiH_3^-)^* + M = SiH_3^- + M$). Then the negatively charged cluster growth is due to ion–molecular reactions such as: $SiH_3^- + SiH_4 = Si_2H_5^- + H_2$, $Si2H_5^- + SiH_4 = Si_3H_7^- + H_2$ (also see Section 2.6.6 about these reactions). This chain of reactions could be accelerated

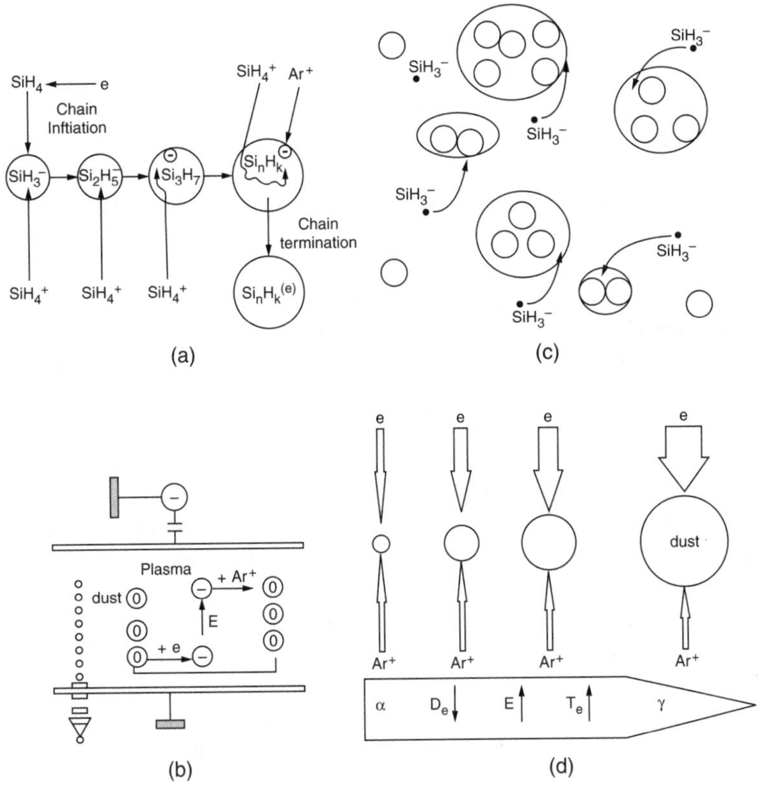

Figure 11.4 Major phases of dust particle formation: (a) Physical scheme of first generation of super-small particle growth; (b) physical scheme of electrical trapping of neutral particles; (c) mechanism of particle coagulation; (d) physical scheme of α–γ transition

by silane molecules' vibrational excitation in the plasma. Typical reaction time in this case is about 0.1 msec and, in this way, becomes much faster than ion–ion recombination. This last process (typical time: 1 to 3 msec) is the main one by which this chain reaction of cluster growth is terminated.

As the negative cluster size increases, the probability of their reactions with vibrationally excited molecules decreases because of a strong effect of vibrational VT relaxation on the cluster surface. When the particle size reaches a critical value (about 2 nm at room temperature), the chain reaction of cluster growth becomes much slower and is finally terminated by the ion–ion recombination process. The typical time of 2-nm particle formation by this mechanism is about 1 msec at room temperature.

To understand the critical temperature effect on particle growth, one must take into account that silane vibrational temperatures in the discharge are determined by VT relaxation in the plasma volume and on the walls, depending exponentially on the translational gas temperature according to the Landau–Teller effect (see Section 3.4). For this reason, a small increase of gas temperature results in a reduction of the vibrational excitation level, making all the cluster growth reactions much slower.

The **selective trapping effect** of neutral particles is schematically described in Figure 11.4(b). Trapping negatively charged particles in any discharges is related to the repelling forces exerted on such particles in the electrostatic sheaths when they reach the plasma boundary. For the super-small particles under consideration here, the electron attachment time is about 100 msec and almost two orders of magnitude longer than the fast ion–ion recombination; thus, most of the particles are neutral during this period. The effect of trapping neutral particles in electric field should occur here to allow the concentration of particles to reach the critical value sufficient for effective coagulation.

The trapping of neutral particles becomes clear, noting that for 2-nm particles, the electron attachment time is shorter than the residence time and each neutral particle is charged at least once during the residence time and quickly becomes trapped by the strong electric field before recombination. It is very important to note that the rate coefficient of two-body electron nondissociative attachment grows strongly with the particle size. For this reason, the attachment time for particles smaller than 2 nm is much longer than their residence time, and there are no possibilities to be charged even once—thus, no possibilities to be trapped in the plasma and survive. Only "large" particles exceeding the 2-nm size can survive; this is the size-selective trapping effect.

This first catastrophe associated with small particles explains why the first particle generation appears with well-defined sizes of crystallites. This could also explain the strong temperature effect on dust production. A small temperature increase results in reduction of the cluster growth velocity and, thus, of the initial cluster size; this leads to a loss of the main part of the initial neutral particles with the gas flow and, in general, determines the observed long delay of coagulation, α–γ transition and dust production process.

The **fast coagulation phenomenon** occurs when the increasing concentration of surviving, monodispersed 2-nm particles reaches a critical value of about 10^{10} to 10^{11} cm^{-3} (see Figure 11.4c). At such levels of concentrations, the attachment of small negative ions like SiH_3^- to 2-nm particles becomes faster than their chain reaction to generate new particles. New chains of dust formation almost become suppressed and the total particle mass remains almost constant. In this fast coagulation process, it was shown experimentally that the mass increase by "surface deposition" process is negligible.

Moreover, when such high particle concentrations are reached, the probability of multibody interaction increases. For this reason, the aggregate formation rate constant grows drastically, and the coagulation appears as a critical phenomenon of phase transition. In this case, because of the very small probability of aggregate decomposition (inverse with respect to coagulation), the critical particle concentration value does not depend on temperature. Besides electrostatic effects, this is one of the main differences between the critical phenomenon under consideration and the conventional phenomenon of gas condensation.

The typical induction time before coagulation is about 200 msec for room temperature and much longer for 400 K. This could be explained taking into account the selective trapping effect discussed before. The trapped-particle production rate for 400 K is much slower than that for room temperature, but the critical particle concentration value remains the same.

The **critical phenomenon of fast changing of discharge parameters, the so-called α–γ transition**, takes place during the process of coagulation when the particle size increases and the concentration decreases. Before this critical moment, the electron temperature and other plasma parameters are mainly determined by the balance of volume ionization and electron losses on the walls. The α–γ transition occurs when the electron losses on the particle surfaces become larger than on the reactor walls. The electron temperature increases to support the plasma balance and the electron concentration dramatically diminishes. This fourth step of the particle and dusty plasma formation story is the "plasma electron catastrophe" (Figure 15.4d).

During the coagulation period, the total mass of the particles remains almost constant and so the overall particle surface in the plasma decreases during coagulation. Thus, one has an interesting phenomenon: the influence of particle surface becomes more significant when the specific surface value decreases.

This effect can be explained if the probability of electron attachment to the particles growing exponentially with the particle size (see Figure 11.4d) is taken into account. For this reason, the effective particle surface grows during the coagulation period.

When the particle size exceeds a critical value of about 6 nm, an essential change of the heterogeneous discharge behavior takes place with significant reduction of the free electron concentration. Taking into account the increase of the electron attachment rate on these size-growing particles, it becomes clear that most of them become negatively charged soon after the α–γ transition.

The typical induction time before the α–γ transition is about 500 msec for room temperature and more than an order of magnitude longer for 400 K. This strong temperature effect is due to the threshold character of the α–γ transition, which takes place only when the particle size exceeds the critical value; for this reason, it is determined by the strongly temperature-dependent time of the beginning of coagulation.

The preceding discussion outlined the physical nature of the main critical phenomena accompanying the initial period of particle formation in a low-pressure silane plasma from super-small particle formation to the plasma crisis of α–γ transition. All these catastrophic phenomena are illustrated in Figure 11.4(a through d). The kinetic peculiarities of these critical phenomena in a dusty plasma will be considered next. Additional and more detailed information regarding formation, growth, and coagulation of particles in different dusty plasmas can be found in References 482 and 507 through 511.

11.6 CRITICAL PHENOMENA IN DUSTY PLASMA KINETICS

The critical phenomena in dusty plasma kinetics will be examined using the specific example of the low-pressure RF silane plasma. The general characteristics of the discharge system as well as the main steps of particle formation and growth were discussed in Section 11.5.

11.6.1 Growth Kinetics of First Generation of Negative Ion Clusters

Cluster formation in the low-pressure RF silane plasma begins from formation of negative ions SiH_3^- and their derivatives. For the discharge parameters described previously, negative ion concentration (SiH_3^- and their derivatives) could be found from the balance of dissociative attachment and positive ion–negative ion recombination:

$$\frac{d[SiH_3^-]}{dt} = k_{ad} n_e [SiH_4] - k_r^{ii} n_i [SiH_3^-] \tag{11.110}$$

where the rate coefficient of dissociative attachment $k_{ad} = 10^{-12}$ cm^3/sec for $T_e = 2$ eV; the ion–ion recombination rate coefficient $k_r^{ii} = 2\ 10^{-7}$ cm^3/sec; and n_e and n_i are the electron and positive ion concentrations. One can see from this equation that, in the absence of strong particle influence (before the α–γ transition), the electron and positive ion concentrations are nearly equal and the concentration of SiH3$^-$ (and their derivatives, other small negative ions) are about 10^9 cm^{-3}, completely in accordance with experimental data presented in the previous section. The time to establish the steady-state ion balance according to Equation 11.110 is approximately 1 msec.

The other mechanisms of initial SiH3$^-$ production, including nondissociative electron attachment to SiH_3, are discussed in Perrin et al.[501] However, in the SiH_4–Ar discharge regime under consideration, this mechanism probably is not the principal one. Indeed, in this case, the SiH_3 radical production is due to dissociation by direct electron impact with the rate constant $k_d = 10^{-11}$ cm^{-3}, and their losses are due to diffusion to the walls (the diffusion coefficient $D = 3 \times 10^3$ cm^2/sec; characteristic distance to the walls $R = 3$ cm):

$$\frac{d[SiH_3]}{dt} = k_d n_e [SiH_4] - \frac{D}{R^2}[SiH_3] \tag{11.111}$$

The establishment time for the steady-state regime according to this equation is about 3 msec, and the steady-state radical concentration is less than 0.1% from SiH_4. As a result, the dissociative attachment is the main initial source of negative ions in the system.

After the initial SiH_3^- formation, the main pathway of cluster growth is the chain of ion–molecular reactions (Equation 2.108 through Equation 2.112), discussed under the Winchester mechanism of ion–molecular processes. In the specific case of silane cluster growth, the Winchester mechanism is not fast relative to ion–ion recombination. For example, the first reaction (Equation 2.6.29) rate coefficient is of the order of 10^{-12} cm^3/sec. Such reaction rates are due to the thermoneutral and even the endothermic character of some reactions from the Winchester chain; these result in an intermolecular energy barrier for such reactions[512,513] and even in a bottleneck effect in their kinetics.[514]

11.6.2 Contribution of Vibrational Excitation in Kinetics of Negative Ion Cluster Growth

The vibrational energy of polyatomic molecules is very effective in overcoming the intermolecular energy barrier in the case of thermoneutral and endoergic reactions.[11,515–517]

To estimate the influence of vibrational excitation on the cluster growth rate (Equation 2.108 through Equation 2.112), it is necessary to analyze the vibrational energy balance (see Section 5.6.7), taking into account SiH_4 vibrational excitation by electron impact ($k_{ev} = 10^{-7}$ cm^3/sec) and vibrational VT relaxation on molecules (rate coefficient $k_{VT,silane}$); on argon atoms ($k_{VT,Ar}$); and on the walls with the accommodation coefficient P_{VT} (see Chesnokov and Panfilov[518]):

$$k_{ev} n_e [SiH_4] \cdot \hbar\omega = \left(k_{VT,silane} [SiH_4]^2 + k_{VT,Ar} n_0 [SiH_4] + P_{VT} [SiH_4] \frac{D}{R^2} \right) \times \left[\frac{\hbar\omega}{\exp(\hbar\omega/T_v) - 1} - \frac{\hbar\omega}{\exp(\hbar\omega/T_0) - 1} \right] \quad (11.112)$$

Here, n_e and n_0 are the electron and neutral gas (which is mostly argon) concentrations; T_v and T_0 are the vibrational and translational gas temperatures; and $\hbar\omega$ is the vibrational quantum approximating the only mode excitation (see Section 3.4.6 and Section 3.5.4). Vibrational relaxation on the cluster surface will be considered later.

Taking into account the Landau–Teller formula for VT relaxation (see Section 3.4), and activation energy E_a (N) as a function of the number N of silicon atoms in a negatively charged cluster, the growth rate of the initial very small clusters (Equation 2.108 through Equation 2.112) can be expressed as:

$$k_{i0}^{initial}(T_0, N) = k_0 N^{2/3} \exp\left(-\frac{\Lambda + B \dfrac{\Delta T}{3T_{00}^{4/3}}}{\hbar\omega} E_a(N) \right) \quad (11.113)$$

Here T_{00} is the room temperature; $\Delta T = T_0 - T_{00}$ is a temperature increase; k_0 is the gas-kinetic rate coefficient; B is the constant Landau–Teller parameter; and Λ is a slightly changing logarithmic factor describing vibrational relaxation:

$$\Lambda = \ln \frac{k_{VT,silane}[SiH_4] + k_{VT,Ar} n_0 + P_{VT} D / R^2}{k_{eV} n_e} \quad (11.114)$$

The initial cluster growth rate coefficient (Equation 4.9) increases with the cluster size. Later, when the number of atoms N exceeds a value of about 300, the relaxation on the cluster surface becomes significant and the rate of negative cluster growth decreases. Taking into account the probability of VT relaxation on the cluster surface, $P_{VT} \approx 0.01$, and the number of the relaxation-active spots on the cluster surface, $s = (N^{1/3})^2 = N^{2/3}$, the probability of chemical reaction on the cluster surface according to the Poisson distribution must be multiplied by the factor:[96]

$$\left(1-P_{VT}\right)^{s}=\exp\left(-P_{VT}N^{2/3}\right) \tag{11.115}$$

Taking into account Equation 11.113 with the kinetic restriction (Equation 11.115), the final expression for the negatively charged cluster growth rate constant can be presented as:

$$k_{i0}(T_0,N)=k_0N^{2/3}\exp\left[-\frac{\Lambda+\dfrac{B\Delta T}{3T_{00}^{4/3}}}{\hbar\omega}E_a(N)-P_{VT}N^{2/3}\right] \tag{11.116}$$

This formula demonstrates the initial, relatively strong temperature dependence of the cluster growth as well as the limitation of growth when the number of atoms in the cluster becomes of the order of $N = P_{VT}^{-2/3}$. Numerically, the critical N value is about 1000 at room temperature, which corresponds to the cluster radius of about 1 nm. At higher temperatures, this cluster size decreases.

11.6.3 Critical Size of Primary Nanoparticles

Without taking into account the ion–ion recombination involved in the chain termination, the particle size growth in the series of ion-molecular reactions (Equation 2.108 through Equation 2.112) can be described, based on Equation 11.116, by the equation:

$$\frac{dN}{dt}=k^*_{i0}\,[SiH_4]N^{2/3}\exp\left(-P_{VT}N^{2/3}\right) \tag{11.117}$$

where $k_{io}^* = k_{io}$ $(T_0,\ N = 1)$ is the rate coefficient of the ion–molecular reaction (Equation 2.108), stimulated by vibrational excitation. It is convenient to present the solution of this equation using the critical particle size $N_{cr} = P_{VT}{}^{-3/2}$, and the reaction (Equation 2.108) characteristic time $\tau = 1/k_{i0}^*\ [SiH_4]$, which is numerically about 0.1 to 0.3 msec at room temperature:

$$t=\tau N_{cr}^{1/3}\frac{\exp\left(N/N_{cr}\right)^{2/3}-1}{\left(N/N_{cr}\right)^{1/3}} \tag{11.118}$$

Equation 11.118 shows that, initially (for number of atoms N or cluster size much less than critical values of 1000 atoms or 1 nm), the function $N^{1/3}$ (not N) is increasing linearly with time. This means that, rather than particle mass and volume, the particle radius grows linearly with time, and the time of formation of a 1000-atom cluster is only 10 times longer than the characteristic time of the first ion–molecular reaction.

Equation 11.118 also shows that super-small particle growth is limited by the critical size of 2 nm, i.e., by the critical number of atoms N_{cr} = 1000. The time of such particle formation is about 1 to 3 msec at room temperature. Then, this process of particle growth is terminated by fast ion–ion recombination. A small change of

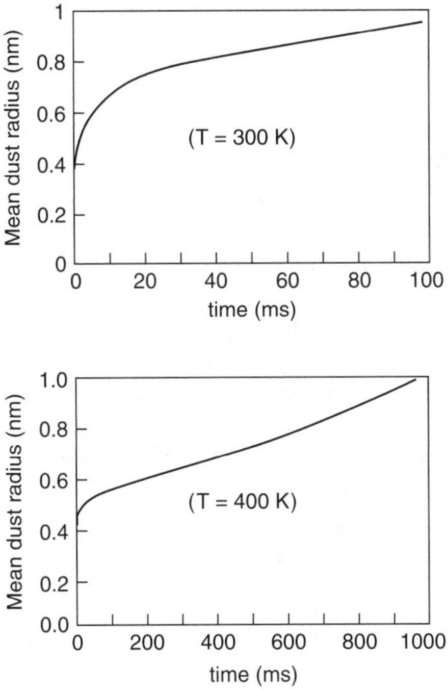

Figure 11.5 Numerical modeling of initial stage of particle growth: first curve—room temperature; second curve—$T = 400$ K[28]

gas temperature accelerates vibrational relaxation; diminishes the level of vibrational excitation and the chain reaction rate; and thus makes the process of particle growth much slower. Figure 11.5 presents the results of numerical calculations that demonstrate the saturation of super-small particle growth and the strong effect of temperature on the particle growth time.[28,519]

11.6.4 Critical Phenomenon of Neutral Particle Trapping in Plasma

The mechanism of super-small cluster growth described before leads to a continuous production of super-small particles (2 nm at room temperature) with the reaction rate corresponding to initial formation of SiH_3^- by dissociative attachment. Most of these particles are neutral according to the Boufendi–Bouchoule experiments. For such particles, the ion–ion recombination rate ($k_r^{ii} = 3 \cdot 10^{-7}$ cm³/sec) is much higher than the rate of electron attachment ($k_a = 3 \cdot 10^{-9}$ cm³/sec). A simple kinetic balance shows in this case that the percentage of negatively charged particles is approximately $k_a/k_r^{ii} = 1\%$. Nevertheless, in experiments, the concentration of such neutral particles grows during a period longer than the neutral gas residence time in the plasma volume. This means that the trapping effect, well known for negatively charged particles (particle repelling from the walls by electrostatic sheaths when they reach the edge of the plasma), takes place here for these nanometer-size neutral particles.

The neutral particle lifetime in a plasma volume is determined by its diffusion and drift with gas flow to the walls. The particle lifetime due to the drift to the wall, the so-called residence time, obviously does not depend on the super-small particle size (in experimental conditions, it is about 150 msec. On the contrary, the diffusion time R^2/D is about 3 msec for molecules and radicals and increases with the particle size proportionally to $N^{7/6}$ (due to particle section increase and their velocity reduction). These two processes become comparable for N about 60, and thus the main losses of the 2-nm particle are due to the gas flow.

To explain neutral particle trapping, it is necessary to take into account that for 2-nm particles, the electron attachment time ($1/k_a\ n_e$, about 100 msec) becomes shorter than the residence time (150 msec). Thus, each neutral particle has a possibility to be charged at least once during the residence time. The simple estimation shows that having a negative charge before recombination even for a short time interval (which happens after about 1 msec) is sufficient for particle trapping by the discharge electric field repelling the negatively charged particles back into the RF plasma volume.

11.6.5 Size-Selective Neutral Particle Trapping Effect in Plasma

The critical phenomenon of size-selective trapping is due to the fact that the rate coefficient of two-body electron nondissociative attachment to a particle grows strongly with the particle size. For this reason, when clusters are smaller than 2 nm, the attachment time is longer than their residence time. In this case, chances for such clusters to get a charge even once are very slight and thus no possibilities exist to be trapped in the RF plasma and survive.

To describe this effect, consider the probability and rate constant of the two-body electron nondissociative attachment. The small probability of this process for relatively small particles is due to the adiabatic character of the process related to energy transfer. An electron must transfer its entire energy through polarization to molecular vibrations of a super-small cluster (characteristic quantum $\hbar\omega$). The electron attachment rate can be found using the Massey parameter (see Section 2.2.7). Taking into account that the cluster polarizability is about $\alpha \approx r^3$ and the total polarization energy of an electron interaction with a cluster of radius r is:

$$\Delta E \approx 4\pi\alpha\varepsilon_0 E^2 \approx 4\pi r^3 \varepsilon_0 \left(\frac{e}{4\pi\varepsilon_0 r^2} \right)^2 = \frac{e^2}{4\pi\varepsilon_0 r} \tag{11.119}$$

leads to the following formula for coefficient of electron attachment to cluster of radius r:

$$k_a = \pi r^2 \sqrt{\frac{8T_e}{\pi m}} \exp\left(-\frac{e^2}{4\pi\varepsilon_0 \hbar\omega r} \right) \tag{11.120}$$

The fact that the rate coefficient of two-body electron nondissociative attachment grows strongly with particle size can be illustrated by comparing the electron mean free path in a cluster with a particle size: when the particle size is relatively small, the probability of direct attachment is negligible. From Equation 11.120, a 5-nm particle, $k_a = 10^{-7}$ cm^3/sec, can be obtained for a 2-nm particle, $k_a = 3 \cdot 10^{-9}$ cm^3/sec. The rate coefficient of two-body electron nondissociative and dissociative attachment becomes the same order of magnitude when the particle size is slightly less than 1 nm. The condition of a neutral particle trapping requires at least one electron attachment during the residence time τ_R:

$$k_a(r)n_e\tau_R = n_e\tau_R\pi r^2\sqrt{\frac{8T_e}{\pi m}}\exp\left(-\frac{e^2}{4\pi\varepsilon_0\hbar\omega r}\right) = 1 \tag{11.121}$$

According to this equation, the critical radius of trapping R_{cr} is about 1 nm (the critical particle diameter is 2 nm). It is known that particle growth at room temperature is limited by approximately the same value of about 2 nm. For this reason, most of the particles initially produced at room temperature are trapped; on the other hand, their size distribution is monodispersed because of losses of smaller clusters to the gas flow.

The losses of relatively small particles with a gas flow (the selective trapping effect) explain the strong temperature effect on dust production. A small temperature increase results in reduction of cluster growth velocity (Equation 11.116). Thus, the initial cluster size R_{in} before recombination is reduced. According to Equation 11.121, this leads to significant losses of the initial neutral particles with a gas flow, and to delay of coagulation, α–γ transition, and dust production in general at relatively high temperatures.

Taking into account the particle production (initiated by dissociative attachment) whose size is limited by the temperature-dependent R_{in} value, and their losses with gas flow, the concentration of relatively small particles having the radius R_{in} can be expressed as:

$$n(R_{in}) = k_{ad}n_e[SiH_4]\tau_R \tag{11.122}$$

Formation of 2-nm particles that will be effectively trapped in the plasma is determined at temperatures higher than 300 K by the relatively slow process of electron attachment to the neutral particles with the radius R_{in} less than R_{cr}, and concentration defined by Equation 5.3. The production rate of the 2-nm particles can be expressed in this case as:

$$\begin{aligned} W_p &= k_a(R_{in})n_e n(R_{in}) = k_a(R_{in})n_e k_{ad}n_e[SiH_4]\tau_R \\ &= [k_a(R_{cr})n_e\tau_R]k_{ad}n_e[SiH_4][k_a(R_{in})/k_a(R_{cr})] \end{aligned} \tag{11.123}$$

Taking into account the trapping condition (Equation 11.121) and the formula for electron attachment coefficient (Equation 11.120), Equation 11.123 can be rewritten as the relation between the rate of particle production W_p and the initial negative ion SiH_3^- production by dissociative attachment W_{ad}:

$$W_p = W_{ad} \exp\left[\frac{e^2}{4\pi\varepsilon_0 \hbar\omega}\left(\frac{1}{R_{cr}} - \frac{1}{R_{in}}\right)\right] \tag{11.124}$$

This formula illustrates the radius-dependent catastrophe of particle production. One can see that, if the rate of particle growth is sufficiently large and the particle size during the initial period (the so-called first generation) also becomes large enough for trapping $R_{in} = R_{cr}$, the rate of particle production is near the rate of SiH_3^- production. Such conditions take place for room temperature. For higher temperatures, the first-generation initial radius R_{in} is less than the critical one R_{cr} for selective trapping, and the particle production rate becomes much slower than dissociative attachment.

11.6.6 Temperature Effect on Selective Trapping and Particle Production Rate

To rewrite Equation 11.124 for nanoparticle production directly as a function of temperature, it is necessary to take into account the linear law of radius increase (Equation 11.118) and the exponential dependence of the growth rate on temperature (Equation 11.116):

$$W_p = W_{ad} \exp\left\{\frac{e^2}{4\pi\varepsilon_0 \hbar\omega R_{cr}}\left[\exp\left(-\frac{B\Delta T}{3T_{00}^{4/3}}\right) - 1\right]\right\} \tag{11.125}$$

This double exponential law shows an extremely strong dust production-dependence on temperature. The double exponential relation can be presented (for room temperature and higher) as a simple and general criterion for fast dust production in a silane plasma, taking into account the trapping criterion (Equation 11.121) and the expression (Equation 11.120) for electron attachment coefficient:

$$\exp\left(-\frac{\Delta TB}{3T_{00}^{4/3}}\right) \cdot \ln\frac{kn_e R}{v} > 1 \tag{11.126}$$

Here, v is the gas flow linear velocity; R is the distance between electrodes; and k is the pre-exponential factor in the expression Equation 11.120 for electron attachment coefficient.

This criterion of fast dust production in a silane plasma is presented numerically in Figure 11.6. In this figure, the area of effective particle formation is placed over the critical curve and ΔT is the temperature in Celsius. One can see from the figure that dust production in the plasma can be stimulated by increasing the electron concentration and neutral gas residence time (for example, by increasing the distance between electrodes) and can be restricted even by small gas heating.

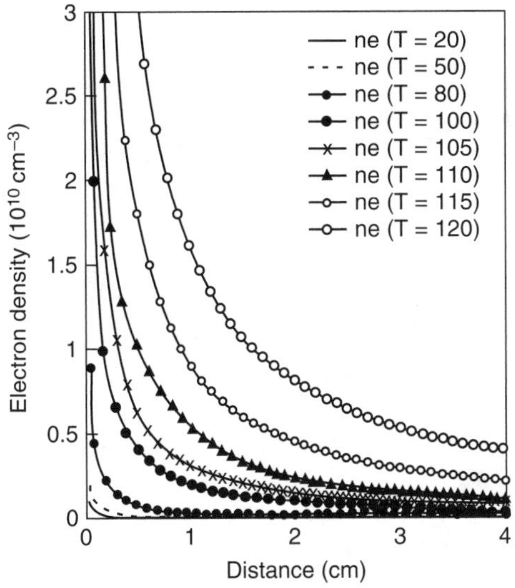

Figure 11.6 Critical conditions of particle formation: gas temperature, electron concentration, and distance between electrodes. Dust production area is above the presented curves. Gas temperature in Celsius. Gas velocity in the discharge is 20 cm/sec.[28]

11.6.7 Critical Phenomenon of Super-Small Particle Coagulation

When the growing concentration of surviving, monodispersed 2-nm particles reaches a critical value of about 10^{10} to 10^{11} cm^{-3}, the fast coagulation process begins. Experimentally, the beginning of the coagulation process has a threshold character and proves the critical character of this phenomenon. It is important to note that the critical value of particle concentration for coagulation experimentally does not depend on temperature; this is not typical for phase transition processes.

The simple estimations as well as detailed modeling show that when the cluster concentration reaches a critical value of about 10^{10} to 10^{11} cm^{-3}, the attachment of small negative ions (like SiH_3^-) to 2-nm neutral particles becomes faster than the chain reaction of new particle growth. Thus, new dust particle formation becomes much slower, taking into account that cluster mass growth from an initial negative ion SiH_3^- has an essential acceleration with particle radius (Equation 11.118). The total particle mass remains almost constant during the coagulation. The deposition of silane radicals on neutral surfaces can be neglected in the fast coagulation phase.

To describe the critical coagulation phenomenon, consider consequently the probability of two, three, and many particle interactions, assuming that the mean radius R of particle interaction is proportional to the physical radius and all direct collisions result in aggregation. The reaction rate and rate coefficient for binary particle collision are:

$$W(2) = \sigma v N^2, \quad K(2) = \sigma v \tag{11.127}$$

where N is the particle concentration and σv is the mean product of their cross section and velocity. From Equation 11.127, the steady-state concentration of the two-particle complexes $N(2)$ can be estimated, taking into account their characteristic lifetime R/v, as:

$$N(2) = (\sigma v N^2)\frac{R}{v} = N(\sigma R N) \tag{11.128}$$

Based on Equation 11.128, the reaction rate and rate coefficient of the agglomerative three-particle collision can be written as:

$$W(3) = \sigma v \cdot (\sigma R N)N^2, \quad K(3) = \sigma v(\sigma R N) \tag{11.129}$$

Repeating this procedure[48] and taking into account that σR is proportional to the particle mass, for simple estimations the rate coefficient of $(k + 2)$ body collision can be given in the following form:

$$K(k) = \sigma v \cdot (\sigma R N)^k \cdot k! \tag{11.130}$$

The total rate constant of the coagulation process can be calculated as the sum of the partial rates (Equation 11.130), which, in principal is divergent, but can be asymptotically found by a special integration procedure:[520]

$$K_c = \sum \sigma v \cdot (\sigma R N)^k \cdot k! = \sigma v \left[1 + \frac{1}{\ln(N_{cr}/N)}\right] \tag{11.131}$$

This relation shows that when the initial particle concentration N is less than critical N_{cr} (which is proportional to $1/\sigma R$, and numerically is about 10^{10} to 10^{11} cm^{-3}), the coagulation rate coefficient is the conventional one related to binary collisions. However, when the initial particle concentration N approaches the critical value N_{cr}, the coagulation rate sharply increases, demonstrating typical features of a phase transition critical phenomenon.

From Equation 11.131, it is seen that because of a very small probability of aggregate decomposition (the process reversed with respect to coagulation), the critical particle concentration value does not depend on temperature. This is the principal difference between the critical phenomenon under consideration and conventional gas condensation. Numerical results describing time evolution of the processes under consideration are presented in Figure 11.7 in a comparison with experimental data. The typical induction time before coagulation is about 100 to 200 msec for room temperature and much longer for 400 K. This temperature effect is due to the selective trapping. The particle production rate for 400 K is much slower than for room temperature, but the critical particle concentration value remains almost the same and thus the induction period becomes much longer.

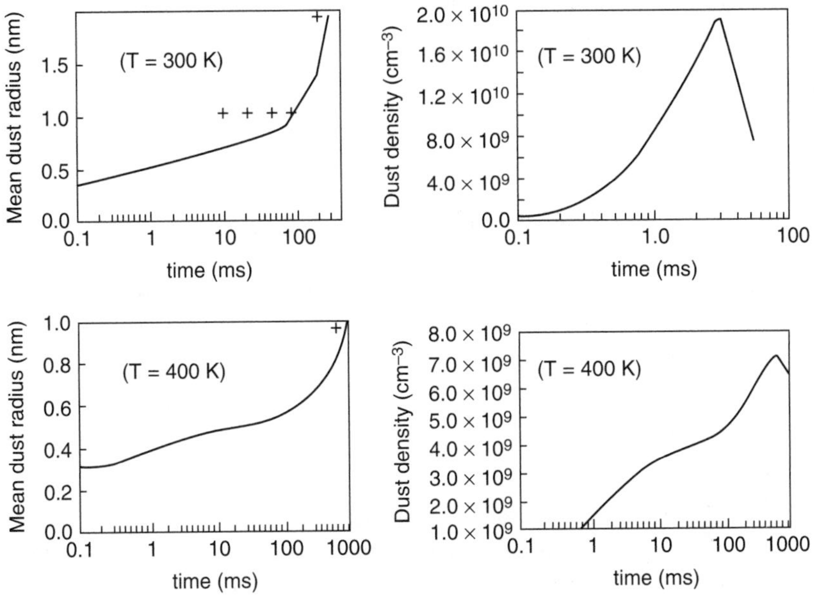

Figure 11.7 Time evolution of dust density and radius for different gas temperatures (minus modeling results, plus experimental data)[28]

11.6.8 Critical Change of Plasma Parameters during Dust Formation (α–γ Transition)

When particle size reaches a second critical value during the coagulation process, another change of plasma parameters takes place: electron temperature increases and electron concentration decreases dramatically. To analyze this critical phenomenon, the balance equation of electrons (n_e), positive ions (n_i), and negatively charged particles (N_n) should be considered. The main processes taken into account are: ionization ($k_i(T_e)$); electron and positive ion losses on the walls by ambipolar diffusion ($n_e D_a/R^2$); the electron attachment to neutral particles ($k_a(r_a)$), which is the strong function of the particle radius r according to Equation 11.120; and "ion–ion" recombination k_r of positive ions and negatively charged particles. Taking into account that before the α–γ transition, negatively charged particles have only one electron, the balance equations can be presented as:

$$\frac{dn_e}{dt} = k_i(T_e)n_e n_0 - n_e \frac{D_a}{R^2} - k_a(r)n_e N \qquad (11.132)$$

$$\frac{dn_i}{dt} = k_i(T_e)n_e n_0 - n_i \frac{D_a}{R^2} - k_r N_n n_i \qquad (11.133)$$

$$\frac{dN_n}{dt} = k_a(r)n_e N - k_r n_i N_n \qquad (11.134)$$

Here, N and n_0 are the neutral particle and gas concentrations, respectively. For the steady-state discharge regime, the derivatives on the left side can be neglected. The threshold of the $\alpha - \gamma$ transition can be calculated from Equation 11.132, assuming for the ionization coefficient: $k_i(T_e) = k_{0i} \exp(-I / T_e)$ (I is the ionization potential), and for electron attachment coefficient (see Equation 11.120):

$$k_a(r) = k_{0a} \left(\frac{r}{R_0} \right)^2 \exp\left(-\frac{e^2}{4\pi\varepsilon_0 \hbar\omega r} \right) \tag{11.135}$$

Thus, the critical phenomenon of the α–γ transition can be seen from the electron concentration balance equation (Equation 11.132) that, taking into account Equation 11.135, looks like:

$$k_{0i} \exp\left(-\frac{I}{T_e} \right) n_0 = \frac{D_a}{R^2} + N k_{0a} \left(\frac{r}{R_0} \right)^2 \exp\left(-\frac{e^2}{4\pi\varepsilon_0 \hbar\omega r} \right) \tag{11.136}$$

Before the critical moment of α–γ transition, when particles are relatively small as is any attachment to them, the electron balance is determined by ionization and electron losses to the walls. The α–γ transition is the moment when the electron losses on the particles sharply grow with their size and become more important than those on the walls. The electron temperature then increases to support the plasma balance.[521]

As shown in Section 11.6.7, the total mass and volume of the particles remain almost constant during the coagulation period, so the specific particle surface (the total surface per unit of volume) decreases with the growth of mean particle radius. Thus, the influence of particle surface becomes more significant when the specific surface area decreases. Equation 11.136 gives an explanation of this phenomenon: the exponential part of the electron attachment dependence on particle radius is much more important than the pre-exponential one. So the comparison of the first and the second terms in the right side of Equation 11.136 gives a critical value of particle size R_c, such that α–γ transition begins only when the particle radius becomes larger than the critical one during coagulation:

$$R_c = \frac{e^2}{4\pi\varepsilon_0 \hbar\omega \ln\left(k_{0a} N R^2 / D_a \right)} \tag{11.137}$$

Numerically, this critical particle radius for the α–γ transition is about 3 nm and practically does not depend on temperature.

11.6.9 Electron Temperature Evolution in α–γ Transition

Equation 11.136 also describes the electron temperature evolution as a function of aggregate radius r during the α–γ transition. There are two exponents in Equation 11.136 that play the major role in the electron balance. So, neglecting the ambipolar diffusion term leads to the simple relation for the electron temperature evolution during the α–γ transition:

$$\frac{1}{T_e} - \frac{1}{T_{e0}} = \frac{e^2}{4\pi\varepsilon_0 \hbar\omega}\left(\frac{1}{r} - \frac{1}{R_c}\right) \tag{11.138}$$

Here, T_{e0} (about 2 eV) is the electron temperature just before the α–γ transition, when the aggregate radius is near its critical value $r = R_c$. One can see from Equation 11.138 that during the particle growth after α–γ transition, the electron temperature increases with the saturation on the level of $e^2/4\pi\varepsilon_0\hbar\omega R_c$, which is numerically about 5 to 7 eV. Thus, as a result of the α–γ transition, the electron temperature can grow about three times. The electron temperature increase as a function of the relative difference of positive ion and electron concentrations $\Delta = (n_i - n_e)/n_i$ could be derived from the positive ion balance equation (Equation 11.133), taking into account plasma electroneutrality ($n_i = n_e + N_n$):

$$k_{0i} \exp\left(-\frac{I}{T_e}\right) n_0 = \frac{D_a}{R^2} + k_r n_i \frac{\Delta}{1-\Delta} \tag{11.139}$$

One can see that electron temperature is essentially related to the relative difference Δ between positive ion and electron concentrations. When, during the α–γ transition, electron temperature grows, electron concentration decreases and the Δ factor approaches 1.

As was mentioned previously, the electron concentration before α–γ transition is almost constant at about $n_{eo} = 3 \cdot 10^9$ cm^{-3}. In the experimental conditions, current density $j_e = n_e e b_e E$ is proportional to $n_e T_e^2$ and remains fixed during α–γ transition. Thus, the electron and positive ion concentrations during this period can be derived from the balance equation as:

$$n_i = \sqrt{\frac{k_{0i} n_{e0} T_{e0}^2 n_0 \exp(-I/T_e)}{k_r T_e^2}}, \quad n_e = n_{e0} \frac{T_{e0}^2}{T_e^2} \tag{11.140}$$

According to this relation, electron concentration decreases during α–γ transition approximately 10 times, and positive ion concentration slightly increases. Modeling and experimental data describing evolution of electron density and temperature during the α–γ transition are shown in Figure 11.8.[28]

11.7 NONEQUILIBRIUM CLUSTERIZATION IN CENTRIFUGAL FIELD

11.7.1 Centrifugal Clusterization in Plasma Chemistry

Quite a few high-pressure, gas-phase plasma chemical processes form products that are liquid or solid at normal conditions. Examples of such processes include dissociation

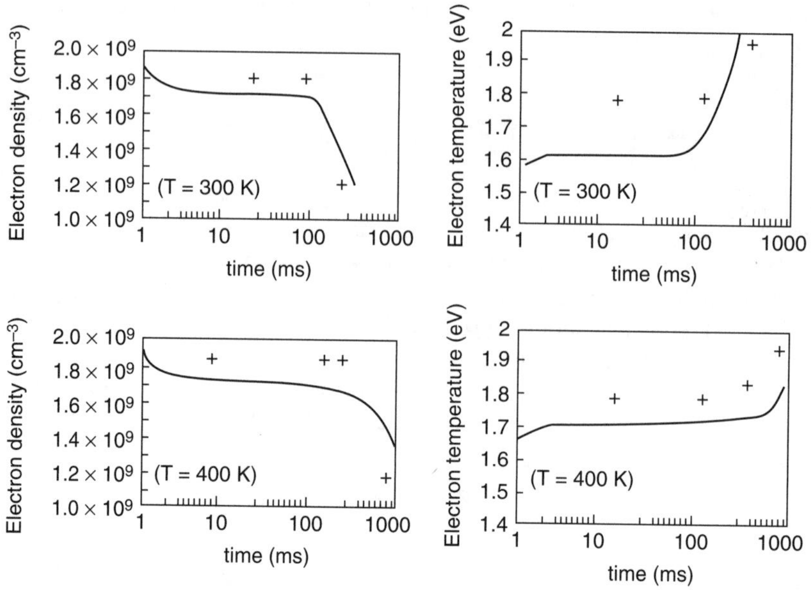

Figure 11.8 Time evolution of electron density and temperature, including α–γ transition, for different gas temperatures (minus modeling results, plus experimental data)[28]

of hydrogen sulfide and metal halogenides. Products of these processes—sulfur, metals, etc.—are usually formed as atoms in the high-temperature zone, and then move to cold areas, where they become clusters and condense phase particles (see, for example, Givotov et al.[522]).

The clusterization process can be affected by fast rotation of the plasma, which in RF and microwave discharges (see Section 10.1 and Section 10.2) often can be characterized by tangential velocities v_φ close to the speed of sound. Centrifugal forces proportional to the cluster mass push the particles to the discharge periphery. The clusters can be transferred to the discharge periphery faster than heat transfer takes place. This can result in a shift of chemical equilibrium in the direction of product formation. This effect provides product separation in the course of the process and decreases its energy cost.

The classic condensation theories of Becker and Doring,[523] Zeldovich,[524] and Frenkel[525] consider formation and growth of the quasi-equilibrium clusters at the condensation temperature and below. Cluster formation in fast rotating plasma can take place at temperatures exceeding the condensation point, which will be considered next.

11.7.2 Clusterization Kinetics as Diffusion in Space of Cluster Sizes

The clusters A_n, which consist of n molecules A, can be formed and destroyed in chemical processes:

$$A_{n1} + A_{n-n1} \Leftrightarrow A_n, \quad A_{n-k} + A_{m+k} \Leftrightarrow A_n + A_m \tag{11.141}$$

These clusters can also diffuse and drift in the centrifugal field. Consider the cluster size n as a continuous coordinate (varying from 1 to ∞). Another coordinate is the radial one x, varying from 0 to the radius R of the discharge tube. Then, clusterization in the centrifugal field can be considered in terms of the evolution of the distribution function $f(n,x,t)$ in the space (n,x) of cluster sizes. Plasma chemical reaction of dissociation of the initial compound can be taken into account as a source of molecules A located in the point $x = 0$. Thus, evolution of the cluster distribution function can be described by the continuity equation in the space (n,x) of cluster sizes:[526]

$$\frac{\partial f(n,x,t)}{\partial t} + \frac{\partial j_n}{\partial n} + \frac{\partial j_x}{\partial x} = 0 \tag{11.142}$$

In this equation, j_n is the flux of clusters along the axis of their sizes n, and j_x is the radial flux of the particles. The radial flux j_x, which takes into account nonisothermal diffusion and centrifugal drift, can be given as:

$$j_x = -D_x\left(\frac{\partial f}{\partial x} + \frac{f}{T}\frac{\partial T}{\partial x} - \frac{nmv_\varphi^2}{xT}f\right) \tag{11.143}$$

where D_x is the diffusion coefficient and m is the mass of a molecule A. Number of clusters is less than the number of molecules at temperatures above that of condensation. In this case, diffusion in the space of cluster sizes is mostly due to atom attachment and detachment processes:

$$A_{n-1} + A \Leftrightarrow A_n \tag{11.144}$$

which permits describing the flux of clusters along the axis of their sizes n in the following linear Fokker–Planck form:[526]

$$j_n = -D_n\left(\frac{\partial f}{\partial n} - \frac{\partial \ln f_0}{\partial n} f\right) \tag{11.145}$$

In this expression, D_n is the diffusion coefficient along the coordinate n; $f_0(n)$ is the distribution function in thermodynamic equilibrium conditions, which influences the nonequilibrium flux (Equation 11.145) and will be discussed in more detail.

11.7.3 Quasi-Equilibrium Cluster Distribution over Sizes

The quasi-equilibrium number density of clusters A_n is determined at fixed temperature T by the change of thermodynamic Gibbs potential ΔG_n related to A_n formation from n molecules A:

$$f_0(n) = const \cdot \exp\left[-\frac{\Delta G(n)}{T}\right] \tag{11.146}$$

The change in the Gibbs potential ΔG depends, in turn, on changes of enthalpy and entropy in the reaction $nA \rightarrow A_n$:

$$\Delta G(n) = n\Delta h(n) - n\Delta s(n) \tag{11.147}$$

In this relation, $\Delta h(n)$ and $\Delta s(n)$ are the specific enthalpy and entropy changes calculated per one A molecule. The entropy change Δs is mostly determined by change of volume per one molecule A during the clusterization process from V_2 in molecule to V_{cl} in clusters:[527]

$$\Delta s \approx \ln \frac{V_{cl}}{V_2} \approx \ln \frac{pm}{T\rho_c} \tag{11.148}$$

where p and T are the gas pressure and temperature and ρ_c is the condense phase density. Thus, as can be seen from Equation 11.148, the entropy change Δs actually does not depend on the cluster size n.

One can assume that the decomposition energy of a molecule A from the cluster A_n does not depend on n, and equals to the evaporation enthalpy λ (for example, the bonding energy of S_2 in sulfur cluster S_{2n} is almost fixed: $\lambda \approx 30$ kcal/mol at any n). In this case, the specific enthalpy change in the clusterization process can be expressed as:

$$\Delta h(n) = -\lambda\left(1 - \frac{1}{n}\right) \tag{11.149}$$

For larger spherical aggregates ($n >> 1$) that should be considered not as clusters but as condense phase particulates:

$$\Delta h(n) = -\lambda\left(1 - const/\sqrt[3]{n}\right) \tag{11.150}$$

The second term on the right-hand side is related to the surface tension energy. The corresponding second term $1/n$ in Equation 11.149 can be interpreted as the one dimensional analog of the surface tension energy. Thus, Equation 11.146 through Equation 11.149 lead to the almost exponential distribution of clusters over their sizes:

$$f_0(n) \propto \exp\left(-\frac{n}{n_0}\right) \tag{11.151}$$

where the exponential parameter n_0 of the distribution function decreases or increases depending on temperature and can be presented using the condensation temperature $T_c = -\lambda/\Delta s$ as:

$$-\frac{\partial \ln f_0}{\partial n} = \frac{1}{n_0} = -\Delta s - \frac{\lambda}{T} = \lambda\left(\frac{1}{T_c} - \frac{1}{T}\right) \tag{11.152}$$

As seen from Equation 11.151 and Equation 11.152, the quasi-equilibrium distribution function $f_0(n)$ is exponentially decreasing at temperatures exceeding the condensation point ($T > T_c$). As soon as the temperature reaches the condensation point, the parameter $n_0 \to \infty$ and clusters grow to large-scale condense phase particulates.

11.7.4 Magic Clusters

Deviation of the clusterization enthalpy from Equation 11.149 leads to nonexponential distributions $f_0(n)$ over cluster sizes. Bonding energy in clusters of some specific sizes can be stronger, resulting in relatively high concentrations of clusters with these "magic" sizes.[528] In the particular example of sulfur, clusters S_6 and S_8 are more stable than other S_{2n} at temperatures exceeding the condensation point.

Taking into account a magic cluster size n_m, the specific enthalpy change (Equation 11.149) can be rewritten in the form:

$$\Delta h(n) = -\lambda\left(1-\frac{1}{n}\right)-\frac{1}{n_m}\cdot\delta Q(n) \qquad (11.153)$$

where $\delta Q(n)$ is the sharp peak at $n = n_m$ with the characteristic width about unity and maximum value $\Delta Q_{\max}$. The corresponding addition to the logarithm of quasi-equilibrium distribution function in the magic point is equal to:

$$\Delta \ln f_0 = \frac{\delta Q(n)}{T} \qquad (11.154)$$

Thus, the effect of magic clusters is negligible at very high temperatures and becomes significant only at temperatures exceeding the condensation point:

$$\Delta T \approx T_c \frac{\Delta Q_{\max}}{\lambda n_m} \qquad (11.155)$$

11.7.5 Quasi-Steady-State Equation for Cluster Distribution Function *f*(*n*,*x*)

The general equation for the quasi-steady-state cluster distribution function over cluster sizes $f(n,x)$ can be given based on Equation 11.142 as:

$$\frac{\partial j_n}{\partial n}+\frac{\partial j_x}{\partial x}=0 \qquad (11.156)$$

The simplest solution of this equation is for quasi equilibrium, which can be obtained assuming that fluxes j_n and j_x are equal to zero:

$$f(n,x) \propto \frac{T(x=0)}{T(x)} f_0(x)\exp\left[n\int\frac{mv_\varphi^2 dx}{xT(x)}\right] \qquad (11.157)$$

However, this distribution in plasma chemical systems does not satisfy the typical boundary conditions in the space of cluster sizes. Also, the characteristic establishment time of the distribution $f(n,x)$ along n (about n^2/D_n) is much shorter for medium clusters of interest ($n < 10^2 \div 10^3$) than characteristic time along $x(R^2/D_x)$. For this reason, it is better to assume in Equation 11.156 that the distribution $f(n,x)$ is determined by fast diffusion along the n axis, and along the radius x only by the centrifugal drift.

In the preceding expressions for characteristic times, R is the discharge radius; D_x is the conventional diffusion coefficient; and D_n is the diffusion coefficient in the space of cluster sizes, which can be taken as:

$$D_n = k_0[A]\exp\left(-\frac{E_a}{T}\right) \tag{11.158}$$

k_0 is rate coefficient of gas-kinetic collisions; $[A]$ is the concentration of molecules A; and E_a is activation energy of the direct processes (Equation 11.144).

Assuming that the activation energy E_a does not depend on n, the diffusion coefficient D_n also can be considered independent of the cluster sizes. If the conventional diffusion coefficient D_x is assumed independent on n and x for the conditions of interest, the derivative of the centrifugal drift can be simply expressed as:

$$\frac{\partial}{\partial x}\left[D_x \frac{mv_\varphi^2}{xT} nf\right] \approx \frac{D_x}{R^2}\frac{mv_\varphi^2}{T} nf \tag{11.159}$$

In this case, the quasi-steady-state kinetic equation for the nonequilibrium cluster distribution function $f(n,x)$ can be derived, taking into account diffusion along the n axis and centrifugal drift along the radius x as:

$$\frac{\partial^2 f}{\partial n^2} - \frac{\partial \ln f_0}{\partial n}\frac{\partial f}{\partial n} - \left(\frac{\partial^2 \ln f_0}{\partial n^2} + \frac{D_x}{D_n R^2}\frac{mv_\varphi^2}{T} n\right) f = 0 \tag{11.160}$$

Finally, the linear kinetic equation (Equation 11.160) for the nonequilibrium cluster distribution function $f(n,x)$ can be simplified by introducing the modified cluster distribution function: $y(n) = f(n)/\sqrt{f_0(n)}$, which gives:[529]

$$y'' - \left[\frac{1}{4}\left(\frac{\partial \ln f_0}{\partial n}\right)^2 + \frac{1}{2}\frac{\partial^2 \ln f_0}{\partial n^2} + \frac{D_x}{D_n R^2}\frac{mv_\varphi^2}{T} n\right] y = 0 \tag{11.161}$$

11.7.6 Nonequilibrium Distribution Functions $f(n, x)$ of Clusters without Magic Numbers in Centrifugal Field

In the case of a monotonic equilibrium distribution function $f_0(n,x)$ of clusters without magic numbers, the kinetic equation (Equation 11.161) can be rewritten considering Equation 11.151 and Equation 11.152 as:

$$y'' + \left(\frac{1}{4n_0^2} + bn\right) y = 0 \tag{11.162}$$

In this equation, the parameter

$$b = \frac{D_x}{D_n R^2} \frac{m v_\varphi^2}{T}$$

reflects the frequency competition between centrifugal drift and diffusion along the axis of cluster sizes in formation of the distribution function $f(n,x)$; the factor n_0 (Equation 11.151 and Equation 11.152) characterizes the average cluster size at temperatures exceeding the condensation point.

The differential equation (Equation 11.162) is a Bessel type of equation and its solution can be expressed using the Hankel function $H_{1/3}^{(1)}(z)$.[530] A solution at $n \to \infty$ can be obtained as:

$$y(n) \propto \sqrt{\frac{1}{4n_0^2} + bn} \cdot H_{1/3}^{(1)}\left[i \frac{2}{3b}\left(\frac{1}{4n_0^2} + bn \right)^{3/2} \right] \tag{11.163}$$

As mentioned previously, diffusion along the axis of cluster sizes in the systems under consideration is usually faster than centrifugal drift, and:

$$b\left|n_0^3\right| = \frac{D_x / R^2}{D_n / n_0^2} \frac{\left|n_0\right| m v_\varphi^2}{T} << 1 \tag{11.164}$$

In this case, an asymptote of the Hankel function at large argument values can be applied. This permits rewriting Equation 11.163 to obtain the nonequilibrium clusterization distribution function in the following form:

$$f(n) = \frac{const}{\sqrt[4]{1 + 4bn_0^2 n}} \exp\left[-\frac{n}{2n_0} - \frac{\left(1 + 4bn_0^2 n\right)^{3/2}}{12bn_0^3} \right] \tag{11.165}$$

As seen from Equation 11.165, the distribution $f(n)$ is close to the quasi-equilibrium value (Equation 11.151) at relatively small cluster sizes $n < 1/4bn_0^2$. At relatively large cluster sizes $n > 1/4bn_0^2$, the distribution function decreases because of intensive centrifugal losses:

$$f(n) \propto \exp\left(-\frac{2\sqrt{b}}{3} n^{3/2} \right) \tag{11.166}$$

In this case, the most probable cluster size, taking into account the centrifugal effect, can be found as follows:

$$\langle n \rangle = n_0, \quad \text{if} \quad n_0 > 0 \tag{11.167a}$$

$$\langle n \rangle = 1/4bn_0^2, \quad \text{if} \quad n_0 < 0 \tag{11.167b}$$

11.7.7 Nonequilibrium Distribution Functions $f(n,x)$ of Clusters in Centrifugal Field (Taking into Account Magic Cluster Effect)

If the quasi-equilibrium distribution is not exponential, the solution of the kinetic equation (Equation 11.161) becomes much more complicated. Nevertheless, it can be done, taking into account the similarity of this equation to the Schrodinger equation for one-dimensional motion of a particle in a potential field. In this case, the factor $-1/4n_0^2$ corresponds to the energy E of a particle in the Schrodinger equation, and factor bn corresponds to potential $U(x) = bx$.

In this quantum-mechanical analogy, one is looking for solutions of Equation 11.161 at $n > 0$, which corresponds to particle penetration in the classically forbidden zone where $E < U(x)$. Then, the quasi-classical approximation can be applied far from the return point, when the following criterion is valid:

$$\frac{d}{dn}\frac{1}{\sqrt{\varphi(n)}} << 1 \tag{11.168}$$

In this criterion of the quasi-classical approximation, another function, $\varphi(n)$, related to the quasi-equilibrium cluster distribution is introduced:

$$\varphi(n) = \frac{1}{4}\left(\frac{\partial \ln f_0}{\partial \ln n}\right)^2 + \frac{1}{2}\frac{\partial^2 \ln f_0}{\partial n^2} + bn \tag{11.169}$$

The criterion of the quasi-classical approximation (Equation 11.168) coincides with the criterion (Equation 11.164) of kinetic domination of clusterization over the centrifugal drift and asymptotic simplification of the special Hankel function. Solution of the kinetic equation (Equation 11.161) for the cluster distribution function in the centrifugal field in the framework of this analogy with the Schrodinger equation can be found in the quasi-classical approximation as:

$$f(n) \propto \frac{\sqrt{f_0(n)}}{\sqrt[4]{\varphi(n)}}\exp\left[-\int \sqrt{\varphi(n)}dn\right] \tag{11.170}$$

In particular, for the quasi-equilibrium distribution with the magic number correction (Equation 11.154), $\Delta Q_{max}/T > 1$ in the vicinity of $n = n_m$, one can derive the following expression for the actual cluster distribution function in the centrifugal field:

$$f_m(n) \approx f_1(n)\exp\left[\frac{\delta Q(n)}{2T} + \sqrt{\frac{1}{4n_0^2} + bn}\left(\sqrt{1 + \frac{\frac{1}{4}\left(\frac{\delta Q(n)}{T}\right)^2}{\frac{1}{4n_0^2} + bn}} - 1\right)\right] \tag{11.171}$$

In this relation, $f_1(n)$ is the cluster distribution in a centrifugal field without magic cluster effect, determined by Equation 11.165. Note that if

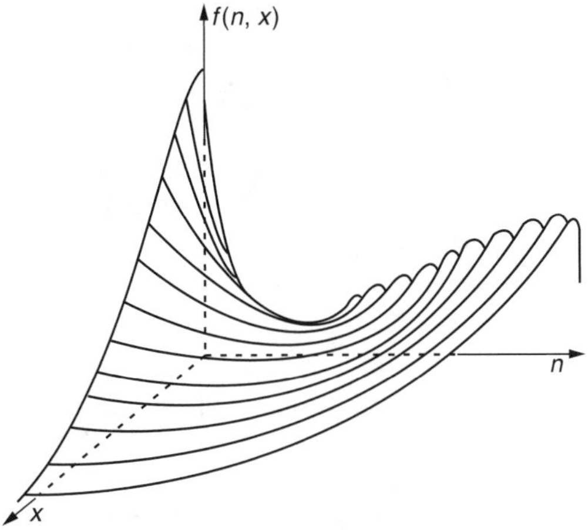

Figure 11.9 Typical cluster distribution function in (n, x) space

$$n_0\left(\frac{\Delta Q_{\max}}{T}\right)^2 >> 1 + 4bn_0^2 n_m \tag{11.172}$$

which usually takes place, then correction related to magic clusters in the vicinity of $n = n_m$ is the same as in equilibrium:

$$f_m(n) \approx f_1(n)\exp\left(\frac{\delta Q(n)}{T}\right) \tag{11.173}$$

An example of the nonequilibrium cluster size distribution function $f(n,x)$ in the centrifugal field is shown in Figure 11.9.

11.7.8 Radial Distribution of Cluster Density

Based on the distribution function $f(n,x)$ described previously, the cluster density $\rho(x)$ can be determined as:

$$\rho(x) = \int_0^\infty nf(n, x)dn \tag{11.174}$$

Total flux of clusters J_0 to the discharge periphery (calculated in number of molecules in the clusters) can be found by multiplying Equation 11.143 by n and integrating from 0 to ∞. Taking into account the definition of the cluster density yields the following equation:

$$-D_x\left[\frac{\partial\rho}{\partial x}+\frac{\rho}{R}\left(\frac{R}{T}\frac{\partial T}{\partial x}-\frac{mv_\varphi^2}{T}\frac{\langle n^2\rangle}{\langle n\rangle}\right)\right]=J_0 \tag{11.175}$$

Assume formation of the substance A on the axis $x = 0$, and absorption of the substance on the discharge walls. This gives the boundary conditions for Equation 11.175: $\rho(x = 0) = \rho_0$, $\rho(x = a) = 0$, where a is the discharge tube radius. Solution of the homogeneous part of this linear differential equation can be given as:

$$\rho^{(0)}(x)=\rho_0\frac{T_h}{T(x)}\exp\left[\frac{mv_\varphi^2}{R}\int_0^x\frac{\langle n^2\rangle}{\langle n\rangle}\frac{dx}{T(x)}\right] \tag{11.176}$$

In this relation, $T_h = T(x = 0)$ is the temperature of the hot zone on the discharge axis. Solution of the nonhomogeneous equation (Equation 11.175) can be found by the method of variation of constant in the form: $\rho(x) = C(x)\cdot\rho^{(0)}(x)$. Taking boundary conditions into account leads first to the following expression for the total flux:

$$J_0=\rho_0 D_x\Bigg/\int_0^a\frac{T(x)}{T_h}\exp\left[-\frac{mv_\varphi^2}{R}\int_0^x\frac{\langle n^2\rangle}{\langle n\rangle}\frac{dx}{T(x)}\right]dx \tag{11.177}$$

Then the radial cluster density distribution $\rho(x)$ in centrifugal field can be expressed in the following integral form:

$$\rho(x)=\rho_0\frac{T_h}{T(x)}\exp\left[\frac{mv_\varphi^2}{R}\int_0^x\frac{\langle n^2\rangle}{\langle n\rangle}\frac{dx}{T(x)}\right]\times \tag{11.178}$$

$$\left\{1-\int_0^x T(x)\exp\left[-\frac{mv_\varphi^2}{R}\int_0^x\frac{\langle n^2\rangle}{\langle n\rangle}\frac{dx}{T(x)}\right]dx\Bigg/\int_0^a T(x)\exp\left[-\frac{mv_\varphi^2}{R}\int_0^x\frac{\langle n^2\rangle}{\langle n\rangle}\frac{dx}{T(x)}\right]dx\right\}$$

Taking into account the peculiarities of the functions under the integral, the cluster flux (Equation 11.177) can be simplified for estimations to the form:

$$J_0=\rho_0 D_x\frac{T_h}{\tilde{T}}\frac{\partial\ln T}{\partial x} \tag{11.179}$$

where $\tilde{T}$ is the average temperature in the discharge zone. Equation 11.179 shows that cluster flux to the discharge periphery is limited by diffusion of molecules A into a zone where sufficiently large clusters can be formed.

11.7.9 Average Cluster Sizes

The average size of clusters moving to the discharge periphery from a given radius x can be determined based on the flux $j(x,n)$ in the combined radius–cluster size space as:

$$\langle n \rangle_f = \frac{\int_0^\infty n \cdot j(x,n)dn}{\int_0^\infty j(x,n)dn} \tag{11.180}$$

Application of the preceding relations for any cluster sizes, including small ones ($n_0 << 1$), requires replacement of the equilibrium average size parameter n_0 by an effective one $\tilde{n}_0$, determined as:

$$\frac{1}{\tilde{n}_0} = 1 - \exp\left(\frac{\partial \ln f_0}{\partial n}\right), \quad \frac{1}{n_0} = -\frac{\partial \ln f_0}{\partial n} \tag{11.181}$$

For larger clusters ($n_0 >> 1$), as Equation 11.181 indicates, $\tilde{n}_0 \approx n_0$. The replacement $n_0 \to \tilde{n}_0$ permits using the Fokker–Planck diffusion approach for small clusters, where the mean free path in the cluster space is comparable with the characteristic system size. Taking the preceding remark into account, the integrals (Equation 11.180) can be calculated with the flux (Equation 11.143) to find the average cluster size in the centrifugal field:

$$\langle n \rangle_f = \frac{\tilde{n}_0 + \alpha n_m + \frac{mv_\varphi^2}{T}\left(2\tilde{n}_0^2 + \alpha n_m^2\right)}{1 + \frac{mv_\varphi^2}{T}\left(1 + \alpha n_m\right)} \tag{11.182}$$

In this relation, $\alpha = [A_{nm}]/[A]$ is the ratio of concentration of magic clusters ($n = n_m$) to the concentration of molecules A.

11.7.10 Influence of Centrifugal Field on Average Cluster Sizes

Average sizes of clusters moving from the radius x with temperature $T(x)$ in the absence of centrifugal forces ($mv_\varphi^2/T \to 0$) can be found from Equation 11.182 as:

$$\langle n \rangle_f = \tilde{n}_0(T) + \alpha(T) n_m \tag{11.183}$$

At relatively high temperatures, $\alpha(T) << n_m^{-1}$ and $< n >_f \approx \tilde{n}_0 \approx 1$, which means that most molecules move to the discharge periphery. The fraction of magic clusters grows with a temperature decrease. Large clusters with sizes $n \geq n_m$ begin dominating in the flux to the discharge periphery only when temperature exceeds the condensation point T_c on :

$$\Delta T = T_c \frac{\Delta Q_{max}}{\lambda n_m} \tag{11.184}$$

Centrifugal forces stimulate domination of large clusters in the flux of particles to the discharge periphery even at temperatures exceeding the condensation point. If the centrifugal effect is sufficiently strong:

$$\frac{mv_\varphi^2}{T} \cdot \alpha(T) \cdot n_m^2 > 1 \tag{11.185}$$

then, according to the general relation (Equation 11.182), the average cluster sizes become relatively large and can be determined as:

$$\langle n \rangle_f \approx \frac{mv_\varphi^2}{T} \cdot \alpha(T) \cdot n_m^2 >> 1, \quad \text{if} \quad \frac{mv_\varphi^2}{T} \cdot \alpha(T) \cdot n_m < 1 \tag{11.186a}$$

$$\langle n \rangle_f \approx n_m, \quad \text{if} \quad \frac{mv_\varphi^2}{T} \cdot \alpha(T) \cdot n_m > 1 \tag{11.186b}$$

The criterion (Equation 11.185) determines the temperature when more large clusters than molecules are moving to the discharge periphery. The larger the value of the centrifugal factor mv_φ^2/T, the more this critical temperature exceeds the condensation point. For example, the condensation temperature of sulfur is $T_c \approx 550$ K at 0.1 atm. At large values of the centrifugal factor mv_φ^2/T, when the tangential velocity is close to the speed of sound, the effective clusterization temperature reaches the high value of 850 K. The magic clusters in this case are sulfur compounds S_6 and S_8.

11.7.11 Nonequilibrium Energy Efficiency Effect Provided by Selectivity of Transfer Processes in Centrifugal Field

As was shown in Section 5.7.7, the transfer phenomenon does not affect the maximum energy efficiency of plasma chemical processes when there are no external forces and the Lewis number is close to unity. However, strong increases in energy efficiency can be achieved in the centrifugal field if the molecular mass of products exceeds that for other components. In this case, the fraction of products moving from the discharge zone can exceed the relevant fraction of heat. It can result in a decrease of the product energy cost with respect to the minimum value for quasi-equilibrium thermal systems (see Section 5.7). As an example, consider the practically important plasma chemical process of hydrogen sulfide decomposition in thermal plasma with production of hydrogen and elemental sulfur:[531–533]

$$H_2S \rightarrow H_2 + S_{solid}, \quad \Delta H = 0.2 \text{ eV} \tag{11.187}$$

This process begins in the high-temperature thermal plasma zone with hydrogen sulfide decomposition forming sulfur dimers:

$$H_2S \rightarrow H_2 + S_{solid}, \quad \Delta H = 0.2 \text{ eV} \tag{11.188}$$

followed by clusterization and condensation of sulfur in the lower temperature zones on the discharge periphery:

$$S_2 \rightarrow S_4, S_6, S_8 \rightarrow S_{solid} \tag{11.189}$$

Minimum energy cost of the process in quasi-equilibrium systems with ideal quenching (see Section 5.7.2) is 1.8 eV/mol. This cost can be significantly decreased when tangential gas velocity in the discharge zone is sufficiently high and criterion (Equation 11.185) is satisfied. In this case, selective transfer of the sulfur clusters to the discharge periphery permits producing hydrogen and sulfur with a minimal energy cost of 0.5 eV/mol. Special experiments in thermal microwave and RF discharges with strong centrifugal effects gave a minimum value for the energy cost of approximately about 0.7 to 0.8 eV/mol.[530,531,531a] The optimal reaction temperature in this case is about 1150 K and the effective clusterization temperature is 850 K.

One should note that the lowest energy cost discussed here can be achieved only if the strong centrifugal-effect criterion (Equation 11.185) is satisfied in the discharge zone where the dissociation process takes place. If the necessary high gas rotation velocities exist only in the relatively lower temperature clusterization zones, the minimum energy cost is somewhat higher: 1.15 eV/mol. Experimental and modeling results illustrating the plasma chemical process of hydrogen sulfide decomposition producing hydrogen and elemental sulfur, and centrifugal effect on its efficiency, are presented in Figure 11.10.

11.8 DUSTY PLASMA STRUCTURES: PHASE TRANSITIONS, COULOMB CRYSTALS, AND SPECIAL OSCILLATIONS

11.8.1 Interaction of Particles; Structures in Dusty Plasmas

The 10- to 500-nm dust particles may acquire a very large charge $Z_d e = 10^2 - 10^5\, e$. As a result, the mean energy of coulomb interaction between them is proportional to Z_d^2 and can exceed the particle thermal energy. Thus, the dusty plasma can be highly nonideal with the charged particles playing the role of multiply charged heavy ions.[534]

The strong coulomb interaction between particles results in the formation of ordered special structures in dusty plasma similar to those in liquids and solids. Critical phenomena of phase transitions between gas–liquid and liquid–solid structures can be observed in dusty plasma as well.[613] The crystalline structures formed by charged particles in dusty plasma are usually referred to as **coulomb crystals**.[535–537] Interaction of charged particles in a dusty plasma can provide not only space, but also time–space structures. This leads to modification of wave and oscillation

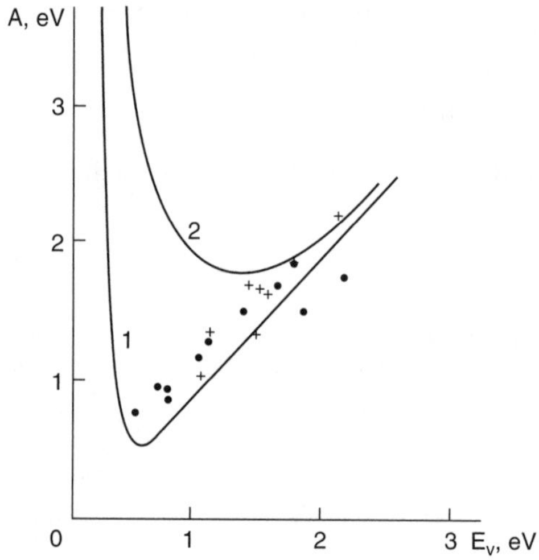

Figure 11.10 Energy cost of H_2S dissociation as function of specific energy input (both per one molecule): 1-nonequilibrium process, modeling; 2-minimal energy cost in quasi-equilibrium process, modeling. Experiments: • corresponds to microwave discharge, 50 kW; + RF discharge, 4 kW[530,531]

modes existing in nondusty plasmas, and also to the appearance of new modes typical only for dusty plasmas. The different structures of dusty plasmas will be discussed next, starting with the concept of nonideality of dusty plasma.

11.8.2 Nonideality of Dusty Plasmas

Ideal and nonideal plasmas were briefly discussed in Section 6.1.1. In general, the systems of multiple interacting particles can be characterized, by the **nonideality parameter**; this is determined as the ratio of the average potential energy of interaction between particles' neighbors and their average kinetic energy. For dust particles with concentration n_d, charge $Z_d e$, and temperature T_d, the nonideality parameter (which is also called the **coulomb coupling parameter**) can be presented as:

$$\Gamma_d = \frac{Z_d^2 e^2 n_d^{1/3}}{4\pi\varepsilon_0 T_d} \tag{11.190}$$

In general, for a system of multiple interacting particles, if the nonideality parameter is low ($\Gamma << 1$), such a system is ideal (see Section 6.1.1); if $\Gamma > 1$, the system of interacting particles is referred to as nonideal. The nonideality criterion of dusty plasma (Equation 11.194) also requires taking into account the shielding of electrostatic interaction between dust particles provided by plasma electrons and ions.

Similar to the definition (Equation 11.190), the nonideality parameter can be introduced for plasma electrons and ions, taking into account that their charge is equal to the elementary one:[538]

$$\Gamma_{e(i)} = \frac{e^2 n_{e(i)}^{1/3}}{4\pi\varepsilon_0 T_{e(i)}} \tag{11.191}$$

The nonideality parameter Γ is related to number of particles in Debye sphere N^D, which for electrons and ions can be expressed as:

$$N_{e(i)}^D = n_{e(i)} \frac{4}{3}\pi r_{De(i)}^3 \propto \frac{1}{\Gamma_{e(i)}^{3/2}} \tag{11.192}$$

where r_D is the Debye radius determined by Equation 6.4.

The number of electrons and ions in Debye sphere (Equation 11.192) is usually fairly large in dusty plasma conditions (and in most plasmas). For this reason, the nonideality parameter is small for electrons and ions ($\Gamma << 1$) and thus the electron–ion subsystem in a dusty plasma can be considered the ideal one. The dust subsystem is also ideal if $N_d^D >> 1$, which means that many dust particles are in the total Debye sphere. The dust particles can be considered an additional plasma component that participates in electrostatic shielding and makes a contribution to the value of the total Debye radius r_D:

$$\frac{1}{r_D^2} = \frac{1}{r_{De}^2} + \frac{1}{r_{Di}^2} + \frac{1}{r_{Dd}^2} \tag{11.193}$$

In this relation, r_{De}, r_{Di}, and r_{Dd} are the Debye radii, calculated separately for electrons, ions, and dust particles (see Equation 6.4).

In the opposite case of $N_d^D << 1$, the subsystem of dust particles is not necessarily the nonideal one. The dust particles cannot be considered as a plasma component under these conditions, and the Debye radius is determined only by electrons and ions. The distance between dust particles can exceed their Debye radius r_{Dd}, but interaction between them is not necessarily strong because of possible shielding provided by electrons and ions. Taking into account this shielding (or screening) effect, the modified nonideality parameter for the dust particles can be presented as:

$$\Gamma_{ds} = \frac{Z_d^2 e^e n_d^{1/3}}{4\pi\varepsilon_0 T_d} \exp\left(-\frac{1}{n_d^{1/3} r_D}\right) \tag{11.194}$$

where the Debye radius is determined by electrons and ions: $r_D^{-2} = r_{De}^{-2} + r_{Di}^{-2}$.

Thus, the degree of nonideality of the dust particle subsystem is determined by combining two dimensionless parameters: Γ_d (Equation 11.190) related to the number of dusty particles in their Debye sphere, and the parameter:

$$K = 1/n_d^{1/3} r_D \tag{11.195}$$

showing the ratio of the interparticle distance to the length of electrostatic shielding (screening), provided by plasma electrons and ions. The factor exp(−K) in the nonideality parameter (Equation 11.194) describes the weakening of the coulomb interaction between particles due to the shielding.

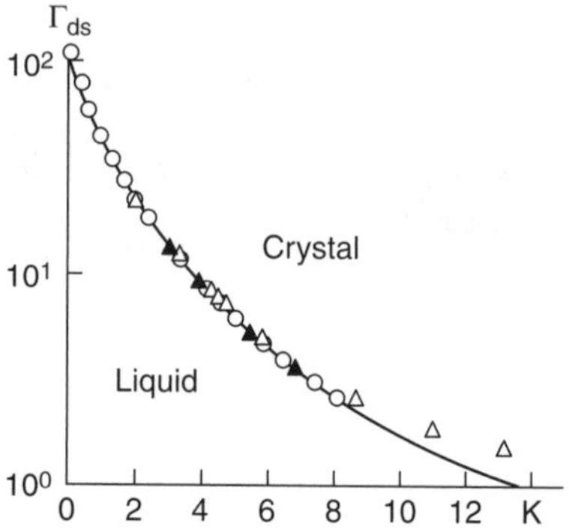

Figure 11.11 Crystallization curve calculated in Yukawa model[539–541]

11.8.3 Phase Transitions in Dusty Plasma

When interaction between the dust particles is strong and the nonideality parameter is high, the strong coupling of particles leads to an organized structure called the **coulomb crystal**. The coulomb crystallization can be qualitatively described by the relatively simple **one-component plasma model (OCP)**. The dusty plasma is treated in the framework of the OCP model as an idealized quasi-neutral system of ions and dust particles interacting according to the binary coulomb potential. This model does not directly include the electrostatic shielding effect, so $K \to 0$ in the relations Equation 11.194 and Equation 11.195.

The OCP model obviously describes the dusty plasma in terms of the nonideality parameter Γ_d (Equation 11.190) because $K \to 0$ (see, for example, Fortov and Yakubov[538]). According to the OCP model, the dusty plasma is not structured and can be considered as "gas" at low values of the parameter $\Gamma \leq 4$. At higher nonidealities Γ, some particle coupling and ordering take place, which can be interpreted as a "liquid" phase. Finally, when the nonideality parameter exceeds the critical value $\Gamma \geq \Gamma_c = 171$, the three-dimensional regular crystalline structure is formed according to the qualitative one-component plasma model.

The more detailed **Yukawa model** takes into account the effect of shielding the electrostatic interaction between charged particles by plasma electrons and ions (see Section 11.8.2). This model describes the dusty plasma nonideality, coupling, and ordering of dust particles in terms of parameters Γ_{ds} and K. Interaction between dust particles in the framework of the Yukawa model is described by the Debye–Huckel potential. Numerical calculations give the phase transition to coulomb crystals at critical values of the nonideality parameters Γ_{ds} and K, illustrated in Figure 11.11.[539–541] Yukawa modeling results shown in Figure 11.11 can be summarized by the following empirical criterion of coulomb crystallization:[542]

$$\Gamma_{ds}\left(1+K+K^2/2\right) \geq 106 \tag{11.196}$$

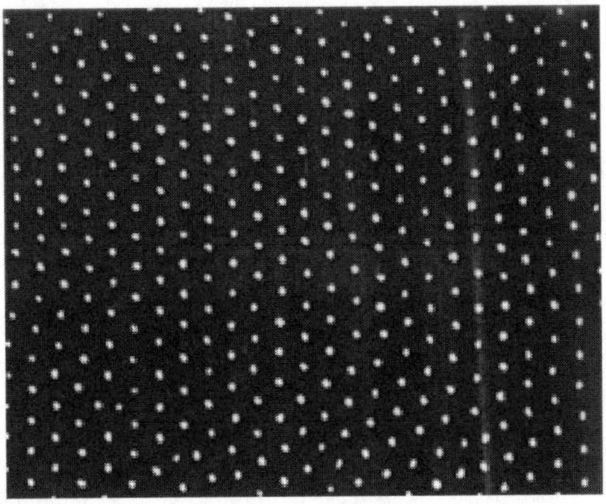

Figure 11.12 Horizontal structure of macroparticles achieved near RF discharge electrode[544]

11.8.4 Coulomb Crystal Observation in Dusty Plasma of Capacitively Coupled RF Discharge

Detailed observation of the coulomb crystals is related to RF–CCP discharges. The RF discharge was sustained in argon, and electrodes were located horizontally, with the lower one powered and operated as an effective cathode (see Section 10.6.1). Frequency was about 14 MHz; pressure was in the range of torrs; and particle size was about a few microns. The particles are negatively charged in such a discharge (see Section 11.5 and Section 11.6) and trapped in the sheath space charge near the lower electrode, which on average is an effective cathode.

The particles suspended in the sheath of the RF discharge are able to form the coulomb crystal structure.[543] Crystallization of the RF-dusty plasma is usually observed at electron and ion densities of about 10^8 to 10^9 cm^{-3}, and electron temperature on the level of electronvolts. One should note that in the sheath, where the coulomb crystal is formed, the electron concentration exceeds that of ions.

A typical example of such coulomb crystals is shown in Figure 11.12.[544] The observed space-organized structures can be characterized by the correlation function $g(r)$ showing the probability for two particles to be found on the distance r, one from another. An example of the correlation function $g(r)$ for the coulomb crystal structure is presented in Figure 11.13.[543] The correlation function on the figure substantiates the organized structure for at least five coordination spheres.

Analysis of this structure shows that dust particles form a hexagonal two-dimensional crystal grid in horizontal layers. In the vertical direction, dust particles position themselves exactly one under another and form a cubic grid between the crystal planes. The number of layers in the vertical direction is limited because of the requirement of balance between gravitational and electrostatic forces necessary for suspension. When the average distance between particles is in the order of hundreds of microns, a typical number of layers is about 10 to 30. Taking into account that in horizontal plane, particles are located inside a several-cm circle, the

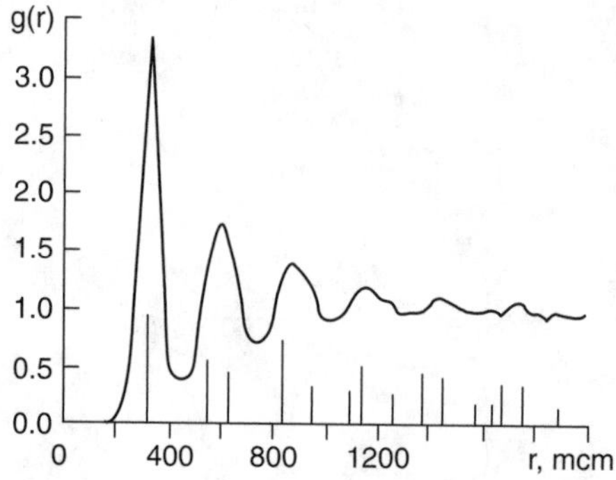

Figure 11.13 Correlation function $g(r)$ showing organized structure of coulomb crystals[543]

coulomb crystal in the RF CCP discharge can be interpreted as a 2.5-dimensional structure. "Melting" of the coulomb crystals can be initiated by reducing the neutral gas pressure or by increasing the discharge power. The crystal–liquid phase transition in both cases is due to a decrease of the nonideality parameter.

11.8.5 Three-Dimensional Coulomb Crystals in Dusty Plasmas of DC Glow Discharges

Three-dimensional quasi-crystal structures were observed in the positive column of DC glow discharge. Similar to the discussed structures in RF–CCP discharges, here horizontal layers make a two-dimensional crystal grid and, in a vertical direction, dust particles are positioned exactly one under another, forming a cubic grid between the crystal planes. In contrast to the case of RF–CCP discharges, the vertical size of the three-dimensional crystal can reach several centimeters in DC glow discharges.[545]

The coulomb crystals in glow discharges are stabilized in the standing striations (see Section 7.4.4 through Section 7.4.6), where the electric field is sufficiently high (about 15 V/cm) to balance gravitation and to provide suspension of dust particles. The relatively large thickness of the striations (up to several centimeters) results in the quite large vertical size of the coulomb clusters. This offers the possibility of considering them as three-dimensional structures.

Coulomb structures have also been observed in RF–ICP discharges and in nuclear-induced dusty plasma[545] in atmospheric-pressure thermal plasma at temperatures of about 1700 K and even in hydrocarbon flames and other systems.[546] It is interesting that higher nonideality of dusty plasma and stronger coupling effects can also be reached by decreasing temperature (see Equation 11.190 and Equation 11.194); this was demonstrated in experiments with cryogenic plasmas of DC glow and RF discharges.[547,548] Formation of ordered coulomb structures in dusty plasma of particles charged by UV radiation was also studied in microgravity experiments carried out on board the MIR space station.[549,550]

11.8.6 Oscillations and Waves in Dusty Plasmas: Dispersion Equation

Presence of dust particles in plasma leads to additional characteristic space and time scales even at low levels of nonideality (see Section 11.8.2). This makes some changes in the dispersion equations of traditional plasma oscillations and waves, and also qualitatively creates new modes specific only for dusty plasmas. An example is the modification of the ionic sound dispersion equation in dusty plasmas (Section 11.8.6) and also the appearance in these heterogeneous systems of a new low-frequency oscillation branch, called the dust sound.

To analyze wave and oscillations in nonmagnetized dusty plasma, the following dispersion equation is used (see Section 6.5):

$$1=\frac{\omega_{pe}^2}{(\omega-ku_e)^2-\gamma_e k^2 v_{Te}^2}+\frac{\omega_{pi}^2}{(\omega-ku_i)^2-\gamma_i k^2 v_{Ti}^2}+\frac{\omega_{pd}^2}{(\omega-ku_d)^2-\gamma_d k^2 v_{Td}^2} \quad (11.197)$$

In this equation, ω_{pe} is the plasma–electron frequency, determined by Equation 6.17 and Equation 6.19; ω_{pi} is the plasma–ion frequency, determined by Equation 6.137; ω_{pd} is the plasma–dust frequency, determined similar to Equation 6.137 and taking into account the dust particle mass m_d and charge $Z_d e$; u_e, u_i, and u_d are the directed velocities of electron, ions, and dust particles; v_{Te}, v_{Ti}, and v_{Td} are the average chaotic thermal velocities of electrons, ions, and dust particles; and $\gamma_j(e,i,d)$ are factors related to the equation of state for electrons, ions, and dust particles:

$$p_j = const \cdot n_j^{\gamma_i} \quad (11.198)$$

$\gamma_j = 1$ corresponds to isothermal oscillations of the j component; $\gamma_j = 5/3$ corresponds to the adiabatic ones.

In the absence of direct motion of all three components, the dispersion equation of nonmagnetized dusty plasma (Equation 11.197) can be expressed as:

$$1=\frac{\omega_{pe}^2}{\omega^2-\gamma_e k^2 v_{Te}^2}+\frac{\omega_{pi}^2}{\omega^2-\gamma_i k^2 v_{Ti}^2}+\frac{\omega_{pd}^2}{\omega^2-\gamma_d k^2 v_{Td}^2} \quad (11.199)$$

This dispersion equation can be analyzed in different frequency ranges to describe the major oscillation branches of dusty plasma. For example, at high frequencies $\omega \gg kv_{Te} \gg kv_{Ti} \gg kv_{Td}$, the dispersion equation (Equation 11.199) can be simplified to the form:

$$1=\frac{\omega_{pe}^2}{\omega^2}+\frac{\omega_{pi}^2}{\omega^2}+\frac{\omega_{pd}^2}{\omega^2} \quad (11.200)$$

Taking into account: $\omega_{pe} \gg \omega_{pi} \gg \omega_{pd}$ (because of the very high mass of dust particles with respect to that of electrons and ions), one can conclude that the high-frequency range is actually not affected by dust particles and gives the conventional electrostatic Langmuir oscillations $\omega \approx \omega_{pe}$ (see Section 6.5.1). Essential effects of the dust particles on plasma oscillation modes can be observed at lower frequencies.

11.8.7 Ionic Sound Mode in Dusty Plasma: Dust Sound

At low frequencies of electrostatic plasma oscillations $kv_{Te} >> \omega >> kv_{Ti} >> kv_{Td}$, the dispersion equation (Equation 11.8.9) can be rewritten in the following form ($\gamma_e = 1$):

$$\omega^2 \approx \omega_{pi}^2 \frac{k^2 r_{De}^2}{1 + k^2 r_{De}^2} \tag{11.201}$$

where r_{De} is the electron Debye radius. When the corresponding wavelengths of the oscillations are shorter than the electron Debye radius ($kr_{De} >> 1$), the dispersion equation (Equation 11.201) gives the conventional ion–plasma oscillations $\omega \approx \omega_{pi}$ (see Equation 6.137).

In the opposite case of longer wavelengths ($kr_{De} << 1$), the ionic sound waves $\omega \approx kc_{si}$ take place according to the dispersion equation (Equation 11.201). Here, the speed of ionic sound can be expressed in the following form:

$$c_{si} = \omega_{pi} r_{De} = \sqrt{\frac{T_e}{m_i} \cdot \frac{n_i}{n_e}} = v_{Ti}\sqrt{\tau(1 + zP)} \tag{11.202}$$

Here, m_i is the ionic mass; factor $\gamma = T_e/T_i$ characterizes the ratio of electron and ion temperatures; factor $z = |Z_d| e^2/4\pi\varepsilon_0 r_d T_e$ is the dimensionless charge of a dust particle of radius r_d; and factor

$$P = \frac{4\pi\varepsilon_0 r_d T_e}{e^2} \frac{n_d}{n_e}$$

is the density parameter of dust particles and is approximately equal to the ratio of the charge density in the dust component to the charge density in electron component. Comparing Equation 11.202 and Equation 6.137 reveals that the main influence of dust particles on the ionic sound is related to the difference in concentration of electrons and ions in the presence of charged dust particles when the dimensionless parameter $zP > 1$.

At even lower frequencies of electrostatic plasma oscillations $kv_{Te} >> kv_{Ti} >> \omega >> kv_{Td}$, the dispersion equation (Equation 11.199) can be rewritten ($\gamma_e = 1$, $\gamma_i = 1$) as:

$$\omega^2 \approx \omega_{pd}^2 \frac{k^2 r_D^2}{1 + k^2 r_D^2} \tag{11.203}$$

where the Debye radius is determined by electrons and ions $r_D^{-2} = r_{De}^{-2} + r_{Di}^{-2}$, similar to Equation 11.194. When the wavelengths of the oscillations are shorter than the Debye radius ($kr_D >> 1$), the dispersion equation (Equation 11.201) gives oscillations with plasma–dust frequency $\omega \approx \omega_{pd}$.

In the opposite case of longer wavelengths ($kr_d << 1$), the dispersion equation (Equation 11.203) leads to a qualitatively new type of wave, the **dust sound** $\omega \approx kc_{sd}$, which propagates in the dusty plasmas with the velocity:

$$c_{sd} = \omega_{pd} r_D = v_{Td} \sqrt{\frac{|Z_d| T_i}{T_d}} \sqrt{\frac{\tau z P}{1 + \tau(1 + zP)}} \tag{11.204}$$

As one can see, the speed of the dust sound can exceed the thermal velocity of dust particles (which is actually a criterion of its existence), mostly due to the large dust charges Z_d (concurrently, the dimensionless dust density should not be very low).

PROBLEMS AND CONCEPT QUESTIONS

11.1 **Work Function Dependence on Radius and Charge of Aerosol Particles.** Give your physical interpretation of Equation 11.1 and Equation 11.3 describing the work function dependence on radius and charge of aerosol particles. Consider separately the cases of conductive and nonconductive materials of the macroparticles.

11.2 **Equation of Monochromatic Photoionization of Aerosols.** Derive Equation 11.8 describing the steady-state electron concentration during the monochromatic photoionization of monodispersed aerosol particles. The derivation assumes substitution of summation by integration; estimate accuracy of this procedure.

11.3 **Calculation of Electron Concentration at Monochromatic Photoionization of Aerosols.** Using the chart in Figure 11.1, calculate the electron density provided by monochromatic photoionization (photon flux 10^{14} 1/cm^2 sec, photon energy $E = 2$ eV) of aerosol particles with radius of 10 nm and work function $\varphi_0 \approx A_0 = 1$ eV. Consider two different values of the aerosol particle concentration: $n_a = 10^3$ cm^{-3} and $n_a = 10^7$ cm^{-3}. Using the same chart, estimate accuracy of the electron density calculations in both cases.

11.4 **Equation of Aerosol Photoionization by Thermal Radiation Tail.** Analyze the differential equation (Equation 11.20) describing nonmonochromatic photoionization of aerosols in terms of elliptical functions and prove that it is identical to Equation 11.19, expressed in the form of series. Taking into account analytical properties of the elliptical θ-functions, derive the asymptotic forms of the differential equation (Equation 11.20).

11.5 **Calculation of Electron Concentration at Nonmonochromatic Photoionization of Aerosols.** Calculate the electron concentration provided by photoionization of aerosols by thermal radiation. Assume the ratio of attachment and photoionization coefficients as constant, the same as in the case of monochromatic radiation: $\alpha/\gamma = 10^8$ cm/sec. Consider monodispersed aerosol particles with 10-nm radius; work function $\varphi_0 \approx A_0 = 1$ eV; and the aerosol particle concentration $n_a = 10^7$ cm^{-3}. For the nonmonochromatic radiation of the total power density 10 W/cm^2, assume that its spectrum corresponds to that of a black body with temperature $T = 3000$ K.

11.6 **Characteristic Monochromatic Photoionization Time**. Calculate the characteristic time for establishing the steady-state electron concentration in the process of monochromatic radiation of monodispersed aerosol particles. Assume the photon flux 10^{17} 1/cm^2 sec; photon energy $E = 2$ eV; an aerosol particle radius = 10 nm; a work function $\varphi_0 \approx A_0 = 1$ eV; and an aerosol concentration $n_a = 10^7$ cm^{-3}, $\alpha/\gamma = 10^8$ cm/sec.

11.7 **The Einbinder Formula**. Analyze the Einbinder formula (Equation 11.33), which describes the electron concentration at the thermal ionization of aerosols, taking into account the influence of the macroparticle charge on the work function. Discuss the applicability of the Einbinder formula in the case of low average values of the macroparticle electric charge.

11.8 **Thermal Ionization of Aerosols Stimulated by Their Photoheating**. Using Figure 11.2, calculate the steady-state photoheating temperature of 10-μm aerosol particles under radiation with power density $pE = 10^7$ W/m^2. Estimate the characteristic time for establishing the steady-state temperature. Calculate the average electron concentration provided by such photoheating, if the concentration of the aerosol particles is $n_a = 10^5$ cm^{-3}.

11.9 **Space Distribution of Electron Density and Electric Field around an Aerosol Particle**. Consider thermal ionization of a macroparticle of 10-μm radius; work function of 3 eV; and temperature 1500 K. Find the electron concentration near the surface of the aerosol particle and the corresponding value of the Debye radius r_D^0. Calculate the total electric charge of the macroparticle electric field on its surface and on a distance 3 μm from the surface.

11.10 **Electric Conductivity of Thermally Ionized Aerosols**. Consider thermal ionization of an aerosol system with T = 1700 K; $r_a = 10$ μm; $n_a = 10^4$ cm^{-3}; and $A_0 = 3$ eV. Assuming the electron mobility in this heterogeneous system as that in atmospheric air, calculate the electric conductivity of aerosols as a function of the external electric field. Consider the external electric field E_e in all three intervals of interest; corresponding the Einbinder relation; explicit dependence on E_e; and range of maximum value of the electric conductivity.

11.11 **Macroparticles Kinetic Parameters**. Relations for thermal velocity, drift velocity, and diffusion coefficient for positive ions are similar to those for electrons (Equation 11.66 through Equation 11.68). Derive these relations for ions in aerosol systems, replacing collisional cross sections, mean free paths, and collisional energy losses of electrons by those for ions. Analyze the contribution of macroparticles in these relations with respect to the contribution of the neutral gas.

11.12 **Vacuum Breakdown of Aerosols**. Based on Equation 11.81 and Equation 11.83, estimate typical values of breakdown voltages of aerosol systems. Take the characteristic parameters of the aerosol system from the example in Section 11.3.7. Compare the result with typical breakdown voltages of discharge gaps without aerosols.

11.13 **Photoinitiation of Pulse Breakdown in Aerosol Systems**. Assuming that the initial electron concentration sufficient for initiating a pulse breakdown is about $n_e^0 = 10^4$ cm^{-3}, estimate typical values of the radiation power density, concentration, and radius of the aerosol particles necessary for the breakdown initiation. For the numerical estimations, use the aerosol photoionization data presented in Figure 11.1.

11.14 **Quasi-Neutral Regime of Steady-State DC Discharge in Aerosols**. Determine typical values of aerosol particle sizes and concentration when the contribution to the total surface area density n_S exceeds that of the long cylindrical discharge wall surface. Estimate the maximum critical concentration of macroparticles with a 10-μm radius when the quasi-neutral regime of steady-state DC discharge is still possible.

11.15 **Electron–Aerosol Plasma Regime of Heterogeneous Discharges**. The maximum value of the average charge of macroparticles is determined by the parameter Z_0 (Equation 11.95 and Equation 11.96). Explain the dependence of this average macroparticle charge on its size. How does the charge density on the surface of aerosol particles depend on radius of macroparticles?

11.16 **Electric Field of Aerosol Particles in Electron–Aerosol Plasma**. Using data presented in Figure 11.3, calculate the inherent electric field on the surface of aerosol particles as a function of current density in the discharge and radius of macroparticles. Compare values of the inherent electric field on the aerosol surface at different currents and macroparticle radii with the external electric field in heterogeneous discharge.

11.17 **Effect of Vibrational Excitation on Kinetics of Negative Ion-Cluster Growth**. The contribution of vibrational excitation of polyatomic silane molecules on the kinetics of negative ion-cluster growth (Equation 2.108 through Equation 2.112) is determined by the logarithmic factor Λ (Equation 11.114) describing the competition between vibrational excitation and relaxation. Discuss the sign of this factor Λ in general, and determine its relation to the effectiveness of vibrational excitation in the cluster growth process. Make a numerical estimation of the logarithmic factor.

11.18 **Kinetics of Super-Small Particle Growth in Dusty Plasma**. Give your interpretation of the fact that the cubic root of number of atoms $N^{1/3}$ (not N) increases linearly with time (see Equation 11.118). Thus the particle radius, not the mass, grows linearly with time. As a result, the time for formation of the critical size 1000-atom cluster is only 10 times longer than the characteristic time of the first ion–molecular reaction.

11.19 **Neutral Particle Trapping in RF Plasma**. Using Equation 11.120, show that the electron attachment time ($1/k_a n_e$) for 2-nm particles becomes shorter than the residence time. This means that each neutral particle has a possibility to be charged at least once during the residence time. This shows that having a negative charge before recombination even for a short time interval (about 1 msec) is sufficient for particle trapping by the RF discharge sheath electric field, which explains the "neutral particle trapping" effect.

11.20 **Coagulation of Neutral Nanoparticles in Plasma**. Analyze the summation (Equation 11.131) of partial recombination rates, which describes the total coagulation rate of nanoparticles in plasma. Demonstrate the divergence of this sum that provides the critical nature of the phenomenon, and explain the physical limitation of the divergence. Replacing the summation by integration, estimate the coagulation rate of neutral nanoparticles (Equation 11.131) and compare the result with conventional relations describing kinetics of condensation.

11.21 **The α–γ Transition during Dusty Plasma Formation**. The α–γ transition, including the sharp growth of electron temperature and decrease of electron density, is related to the contribution of electron attachment to particle surfaces and occurs during the fast coagulation stage. It is interesting that the total surface of particles decreases during the coagulation process, because, while their diameters grow, the total mass remains almost fixed. Give your interpretation of the phenomenon; does it depend on the particle material?

11.22 **Clusterization Process as Diffusion in Space of Cluster Sizes**. At temperatures exceeding that of condensation, evolution of clusters is mostly due to atom attachment and detachment processes: $A_{n-1} + A \Leftrightarrow A_n$. Show that the diffusive flux of clusters along the axis of their sizes n can be expressed in the linear Fokker–Plank form (Equation 11.145). Analyze the criteria of applicability of the linear flux to the condensation processes.

11.23 **Magic Clusters**. The effect of magic clusters, which means relatively high concentration of clusters with some specific sizes, is negligible at very high temperatures. Show that this effect becomes significant only at temperatures slightly exceeding the condensation point determined by Equation 11.155.

11.24 **Kinetic Equation for Nonequilibrium Clusterization Process in Centrifugal Field**. Based on Equation 11.160, derive the Fokker–Plank-type kinetic equation (Equation 11.161) for the modified distribution function $y(n) = f(n/\sqrt{f_0}(n)$, taking into account the diffusion along the n axis of cluster sizes and centrifugal drift along the radius x. In general, analyze types of solutions that can be obtained from the kinetic equation.

11.25 **Cluster Flux to Discharge Periphery**. Simplifying integration in the general relation Equation 11.177, derive the simple formula (Equation 11.179) for the cluster flux. Give a physical interpretation of the formula, taking into account that the cluster flux to the discharge periphery is limited by diffusion of molecules into a zone where sufficiently large clusters can be formed. Using Equation 11.179, estimate typical numerical values of the cluster flux.

11.26 **Effective Clusterization Temperature as a Function of Centrifugal Factor mv_φ^2/T**. At a pressure of 0.1 atm and high values of the centrifugal factor $mv_\varphi^2/T \approx 1$, the effective clusterization temperature in sulfur (formation of S_6 and S_8) is about 850 K. Taking into account that the condensation temperature of sulfur is $T_c \approx 550$ K at 0.1 atm, calculate the effective clusterization temperature for sulfur as a function of the centrifugal factor at tangential velocities below the speed of sound.

11.27 **Nonideality Criterion of Dusty Plasmas**. The nonideality of a dusty plasma (strong coupling of dust particles) takes place when the parameter Γ_{ds} (Equation 11.194), including plasma shielding of interaction between particles, is sufficiently large. Explain why the Debye radius describing the shielding combines the effect of electrons, ions, and dust particles (Equation 11.193) in the case of ideal dusty plasma and does not include a contribution of the particles in the nonideal case (Equation 11.194).

11.28 **"Melting" of Coulomb Crystals**. "Melting" of two-dimensional coulomb crystals in RF CCP discharges can be initiated by reducing neutral gas pressure or by increasing discharge power. Explain how the change of pressure and power lead to a decrease of the nonideality parameter and to the crystal–liquid phase transition.

11.29 **Plasma–Dust Frequency**. Using the one-component plasma model (OCP; see Section 11.8.3), derive a formula for the plasma–dust frequency ω_{pd}. Give a physical interpretation of this frequency and analyze its relation to the Debye radius. From this point of view, compare the cases of ideal and nonideal dusty plasmas.

11.30 **Ionic Sound in Dusty Plasma**. Based on the dispersion equation (Equation 11.201), derive Equation 11.202 for the speed of the ionic sound in a dusty plasma. Make the derivation in terms of difference of electron and ion concentrations in dusty plasma, and also in terms of the product zP of dimensionless charge of dust particles and their dimensionless concentration. Give numerical examples of calculations of c_{si} using Equation 11.202. Compare the speed of ionic sound under similar plasma conditions with and without dust particles; consider cases of predominantly positive and negative charging of the dust particles.

11.31 **Dust Sound Wave in Plasma**. The dust sound is the propagation of low-frequency waves in a dusty plasma at wavelengths exceeding the Debye radius r_d with the speed determined by Equation 11.204. Show that the dust sound can exist only if its velocity exceeds the thermal one for dust particles. Derive the existence criterion of the dust sound and determine the minimum concentration of dust particles necessary for this wave.

CHAPTER

12

ELECTRON BEAM PLASMAS

Numerous types of electron beams employ a wide range of currents and energies, including high-current relativistic electron beams. These are found in applications ranging from initiation of super-powerful gas lasers and controlled thermonuclear reactions to processes of surface treatment and depollution of high-volume exhaust streams.[551] General properties of electron beams and their propagation and degradation will be discussed in the first two sections of this chapter. The following two sections will be devoted to special discharges sustained by electron beams at low and high pressures. The two final sections will deal with plasma generation by nuclear debris and electron beam interaction with aerosols.

12.1 GENERATION AND PROPERTIES OF ELECTRON-BEAM PLASMAS

12.1.1 Electron-Beam Plasma Generation

This plasma is formed by injecting a cylindrical or plane electron beam with electron energies usually ranging from 10 keV to 1 MeV into a neutral gas. A general scheme of electron beam plasma generation is illustrated in Figure 12.1. Here, a thin cylindrical electron beam is formed by an electron gun located in a high-vacuum chamber. Then the beam is injected through a special window into a gas-filled discharge chamber in which the high-energy electrons generate plasma and transfer their energy into the gas by various mechanisms, depending on pressure, beam current density, electron energy, and plasma parameters.

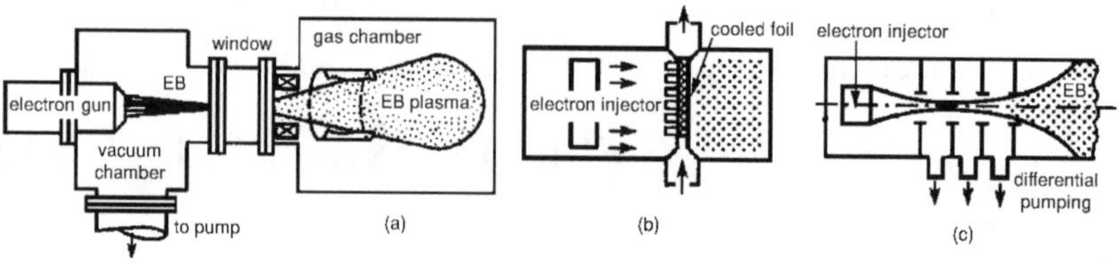

Figure 12.1 Electron beam generator: (a) general schematic; (b) foil window; (c) differential pumping window

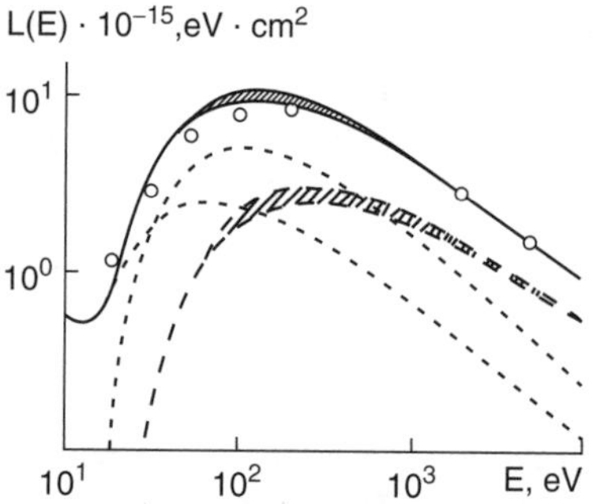

Figure 12.2 Different calculations for electron energy losses' function in nitrogen[552]

The electron beam losses related to neutral gas ionization (and thus creation of electron-beam plasma) were discussed in Section 2.2.5, as well as the effective ionization rate coefficients for the high-energy beam electrons (see Equation 2.34). Total energy losses of a fast electron beam in a dense gas due to interaction with neutrals along its trajectory (axis z) can be simply described as:

$$-\frac{1}{n_0}\frac{dE}{dz} = L(E) \tag{12.1}$$

where n_0 is the neutral gas density; E is the electron energy; and $L(E)$ is the electron energy loss function, which is well known for different gases and mixtures. As an example, the $L(E)$ function for nitrogen is shown in Figure 12.2.[552]

For calculations of the electron energy loss function $L(E)$, it is convenient to use the numerical **Bethe formula**, which can be applied even for relativistic electron beams (although not for the strong relativistic case):

$$L(E) = \frac{A}{E}\frac{\left(1+E/mc^2\right)}{\left(1+E/2mc^2\right)}\ln\frac{1.17\cdot E}{I_{ex}} \tag{12.2}$$

In this relation, m is an electron mass; c is the speed of light; and I_{ex} is the characteristic excitation energy (equal to 87 eV for air; 18 eV for hydrogen; 41 eV

Table 12.1 Ionization Energy Cost at Neutral Gas Irradiation by an Electron Beam with Energy of 0.1 to 1 MeV

Gas	Ionization cost	Gas	Ionization cost
Air	32.3 eV	Carbon Dioxide (CO_2)	31.0 eV
Nitrogen (N_2)	35.3 eV	Oxygen (O_2)	32.0 eV
Hydrogen (H_2)	36.0 eV	Helium (He)	27.8 eV
Argon (Ar)	25.4 eV	Krypton (Kr)	22.8 eV
Xenon (Xe)	20.8 eV	—	—

for helium; 87 eV for nitrogen; 102 eV for oxygen; 190 eV for argon; 85 eV for carbon dioxide; and 72 eV for water vapor). The factor A depends only on gas composition; for example, in air $A = 1.3 \cdot 10^{-12}$ eV^2 cm^2.

If kinetic energy of electrons exceeds mc^2 (numerically, about 0.5 MeV), such electrons are called relativistic, and their beam is usually referred to as the **relativistic electron beam.** Dynamic analysis of such beams definitely requires taking into account special effects of relativistic mechanics.

12.1.2 Ionization Rate and Ionization Energy Cost at Gas Irradiation by High-Energy and Relativistic Electrons

The rate of ionization (and thus plasma generation) provided by an electron beam with current density j_b in a unit volume per unit time was discussed in Section 2.2.5 in terms of an effective ionization rate coefficient. Now, the same ionization rate can be determined using the fast electron energy losses dE/dz as:

$$q_e = -\frac{dE}{dz}\frac{j_b}{eU_i} \tag{12.3}$$

In this equation, U_i is the energy cost of an electron–ion pair formation (the ionization energy cost), an important parameter describing plasma formation by an electron beam. If the beam electron energy is in the range of 0.01 to 1 MeV, the ionization energy cost has an almost constant value, which is given in Table 12.1.

Electron beam plasma obviously has a complicated composition, including charged, excited atomized, and other species. To analyze the concentration of these species, it is convenient to consider the electron beam as their source and specify the energy cost of their formation. The ionization energy cost for an electron beam irradiation of gases, as well as energy cost of formation of different kinds of excited species, can be calculated using the degradation spectrum kinetic approach, which will be considered in Section 12.2.

12.1.3 Classification of Electron Beam Plasmas According to Beam Current and Gas Pressure

Independent changes in the current density of electron beam and neutral gas pressure results in qualitatively different regimes of electron beam plasmas. These regimes can be classified into five principal groups:

1. **Powerful Electron Beam at Low-Pressure Gas.** Here, electron beam energy losses are mostly due to beam instability or the Langmuir paradox (see Section 6.5.7), when the electron beam energy can be transferred by nonlinear beam–plasma interaction into the energy of Langmuir oscillations. This regime usually takes place when the electron beam current and beam current density are sufficiently high ($I_b > 1$ A, $j_b > 100$ A/cm^2), while pressure is relatively low (less than a few torreys). This system is called the plasma–beam discharge and will be considered in Section 12.3. Plasma in this discharge can be characterized by high degrees of ionization ($\alpha = n_e/n_0 \geq 10^{-3}$), and electron temperatures of several electronvolts.
2. **Low-Current Electron Beam in Rarefied Gas.** This regime is related to current densities $j_b < 0.1$ A/cm^2 and pressures below 1 torr. In this case, the beam instability is ineffective and the plasma–beam discharge cannot be sustained. The degree of ionization is relatively low in this regime ($\alpha = n_e/n_0 < 10^{-7}$), and the temperature of plasma electrons is also low and close to the neutral gas temperature.
3. **Moderate-Current Electron Beam in a Moderate-Pressure Gas.** Current density of an electron beam in this regime is in the range of 0.1 A/cm^2 and 100 A/cm^2, with pressures from 1 to 100 torr. The degree of ionization depends on pressure and beam density and can vary in the wide range between 10^{-7} and 10^{-3}. In this case, the electron temperature is determined by the beam degradation in nonelastic and then elastic collisions; this will be discussed in Section 12.2. Specific energy input in these systems is usually relatively low, and translational temperature does not grow significantly during the electron beam interaction with a portion of neutral gas (see Section 5.6.7). Thus, the discharge can be considered a nonequilibrium discharge.
4. **High-Power Electron Beam in a High-Pressure Gas.** This regime is related to high current densities $j_b > 100$ A/cm^2 and pressures above 100 torr. The degree of ionization and electron temperatures are similar to those in the case of moderate pressures and currents. However, high values of beam power and specific energy input result in more intensive relaxation and neutral gas heating. In this regime, the gas temperature distribution is determined not only by pressure and beam density, but also by the heat exchange between plasma zone and environment. Thus, in the stationary situation, this electron beam plasma is thermal and local temperatures can exceed 10,000 K.

12.1.4 Electron Beam Plasma Generation Technique

The electron beam injectors for electron beam plasma generation can be arranged using direct electrostatic accelerators, although other types of accelerators can be applied as well.[553,554] To generate beams of electrons with energies not exceeding 200 to 300 keV, the injectors can be based on electron guns with thermionic, plasma, or field-emission cathodes. Power can be supplied by special transformers with AC–DC conversion.

Electron beams are formed in a vacuum and then should be transmitted across the inlet system (window) into the dense gas chamber to generate plasma. These inlet system windows are usually arranged in one of two ways: foil windows and lock systems with differential pumping. Figure 12.2 shows a general schematic of both inlet systems. The application of foil windows for nonrelativistic, continuous, concentrated electron beams has complications related to possible high thermal flux; thinness of the foil for electron energies below 300 keV; and chemical aggressiveness of gases in plasma chemical systems.

12.1.5 Transportation of Electron Beams

An important advantage of plasma chemical applications of the high-energy and relativistic electron beams is related to the long distance of their decay even in high-pressure systems. Transportation length of the high-energy and relativistic electron beams in dense gases (including atmospheric air) is determined by energy losses due to ionization and excitation of atoms and molecules, which was described by Equation 12.1 and Equation 12.2.

For practical calculations, it is convenient to use the following numeric relation, describing the total length L (in meters) for stopping a high-energy electron along its trajectory:

$$L = AE_{b0}^{1.7}\frac{T_0}{p} \qquad (12.4)$$

In this relation, T_0 is gas temperature in Kelvins; p is gas pressure in torreys; E_{b0} is initial electron energy; and A is a constant that depends only on gas composition for air $A = 1.1 \cdot 10^{-4}$.

Electron beam transportation can also be limited by collisionless effects related to beam instability, electrostatic repulsion of the beam electrons, and magnetic self-contraction. The first of these effects implies that the electron beam energy transfers to electrostatic Langmuir oscillations. This high-current–low-pressure effect was mentioned earlier as the first regime of electron beam plasma generation and will be considered in detail in Section 12.3.

The second and third effects are related to the electric and magnetic fields of the beam. The total force pushing a periphery beam electron in a radial direction away from the axis (combining electric and magnetic components) can be expressed as:

$$F = \frac{n_{eb}r_b e^2}{2\varepsilon_0}(1-\beta^2 - f_e) \qquad (12.5)$$

In this relation, r_b is radius of the cylindrical beam; n_{eb} is the electron density of the beam (or beam plasma); $\beta = u/c$ is the ratio of beam velocity to the speed of light;

and $f_e = n_i/n_{eb}$ is the beam space charge neutralization degree (n_i is the ion density). If $f_e > 1 - \beta^2$, which is usually valid in dense gases, the generated plasma neutralizes the electron beam space charge and the magnetic field self-focuses the beam.

If the high-current electron beam is not neutralized, it can be destroyed by its own electric field. To provide effective beam transportation in such conditions, special focusing systems, in particular electrostatic lenses, can be applied.[554] When an electron beam is neutralized ($f_e = 1$), its current cannot exceed a critical value known as the **Alfven current**:

$$I_A = \frac{mc^3}{4\pi\varepsilon_0 e\beta\gamma}, \quad I_A(\text{kA}) = 17\beta\gamma \tag{12.6}$$

where $\gamma = 1/\sqrt{1-\beta^2}$ is the relativistic factor. The Alfven current gives the inherent magnetic field of the beam, which makes the Larmor radius (Equation 6.51) less than half the beam radius and stops the electron beam propagation. More details regarding the transportation of relativistic electron beams with very high current can be found in Mesyats et al.[555]

12.2 KINETICS OF DEGRADATION PROCESSES: DEGRADATION SPECTRUM

12.2.1 Kinetics of Electrons in Degradation Processes

Chapter 4 considered physical kinetics of electrons in gas-discharge plasmas in terms of electron energy distribution functions, EEDF $f(E)$, determined by the Boltzmann or Fokker–Planck kinetic equations. Such an approach is very fruitful if the general shape of the electron energy distribution function is not very far from the quasi-equilibrium one. The situation is different for analyzing stopping of high-energy electron beams, where the energy is degrading from large, sometimes relativistic, values down to the level of ionization and excitation of neutrals and below.

In this case of electron beam energy degradation, it is more convenient to operate, not in terms of probability for any electron to have some given energy E, which is $f(E)dE$, but rather in terms of the degradation spectrum, which is the number of electrons $Z(E)dE$ with some given energy E during the degradation of one initial high-energy electron. This approach is called the **degradation spectrum** method and has been successfully applied to describing the kinetics of stopping processes of electrons and also energetic ions, photons, and neutrals.[144,446,556–560]

12.2.2 Energy Transfer Differential Cross Sections and Probabilities during Beam Degradation Process

The degradation process of an energetic beam is determined by a group of elastic and inelastic collisional processes of an electron (in principle, it can be generalized to energetic ions, photons, and neutrals) with background particles, which can be considered at rest. Each of the degradation processes k is characterized by the total cross section $\sigma_k(E)$, and the differential cross section $\sigma_k(E, \Delta E)$, where E is the electron energy during degradation, and ΔE is the electron energy loss during the collision.

It is convenient to normalize the differential cross section $\sigma_k(E, \Delta E)$ to the number of electrons formed after collision.[561] This means that degradation processes that do not change the number of electrons (for example, excitation processes) have the conventional normalization equation:

$$\int_{-\infty}^{+\infty} \frac{\sigma_k(E,\Delta E)}{\sigma_k(E)} d(\Delta E) = 1 \tag{12.7}$$

At the same time, ionization processes that double the number of electrons have the corrected normalization equation:

$$\int_{-\infty}^{+\infty} \frac{\sigma_k(E,\Delta E)}{\sigma_k(E)} d(\Delta E) = 2 \tag{12.8}$$

and the electron–ion recombination and electron attachment processes, where the degrading electron disappears, have the normalization:

$$\int_{-\infty}^{+\infty} \frac{\sigma_k(E,\Delta E)}{\sigma_k(E)} d(\Delta E) = 0 \tag{12.9}$$

Then the differential probability for an electron with energy E to lose the energy portion ΔE in the collisional process k can be determined as:

$$p_k(E,\Delta E) = \frac{\sigma_k(E,\Delta E)}{\sum \sigma_m(E)} \tag{12.10}$$

The corresponding value of the total probability for an electron with energy E to participate in the collisional process k can be found as:

$$p_k(E) = \frac{\sigma_k(E)}{\sum \sigma_m(E)} \tag{12.11}$$

Obviously, the probabilities (Equation 12.10 and Equation 12.11) can be determined only if $\sum\sigma_m(E) > 0$, e.g., above the energy threshold of the degradation processes; otherwise, the preceding probabilities should be considered as zero.

12.2.3 Degradation Spectrum Kinetic Equation

The degradation spectrum $Z(E)$ can be determined as the average number of particles (electrons, in this case) that appears during the whole degradation process in the energy interval $E - E + dE$. Then, the kinetic equation for the degradation spectrum can be presented in general as:

$$Z(E) = \int_0^\infty p(W, W - E) \cdot Z(W)dW + \chi(E) \tag{12.12}$$

with boundary condition $Z(E \to \infty) = 0$. In this equation, $\chi(E)$ is the source function of the high-energy electrons. If the degradation process is started by the only electron with energy E_0, the source function is determined by the δ-function, $\chi(E) = \delta(E - E_0)$. Finally, $p(W, W - E)$ in Equation 12.12 is the total probability of formation of an electron with energy E by collisions of an electron with initial energy W and the energy loss $\Delta E = W - E$:

$$p(W, W - E) = \sum p_m(W, \Delta E = W - E) \tag{12.13}$$

It is interesting to note that, at low energies (electron energies below the lowest excitation threshold) when the preceding collisional cross sections can be neglected, the degradation spectrum coincides with the change of the electron energy distribution function at final and initial moments of the degradation process:[561]

$$Z(E) = f(E, t \to \infty) - f(E, t \to 0) \tag{12.14}$$

Thus, assuming an absence of the initial EEDF, $f(E, t \to 0) \to 0$, one can conclude the equality of EEDF and degradation spectrum of electron beam under the excitation threshold.

12.2.4 Integral Characteristics of Degradation Spectrum: Energy Cost of Particle

The degradation spectrum $Z(E)$ is helpful in calculating many important integral characteristics of high-energy electron interaction with materials. For example, integration of the degradation spectrum gives, based on Equation 12.11, the total number of specific collisions k (ionization, excitation, dissociation, etc.) per one initial high-energy degrading electron:

$$N_k = \int_0^\infty p_k(E) \cdot Z(E)dE = \int_0^\infty Z(E) \frac{\sigma_k(E) \cdot dE}{\sum \sigma_m(E)} \tag{12.15}$$

Taking into account that the initial energy of a beam electron is equal to E_0, the energy cost of the specific collision k (ionization, excitation, dissociation, etc.) is:

$$U_k = \frac{E_0}{N_k} = E_0 \Big/ \int_0^\infty p_k(E)Z(E)dE \tag{12.16}$$

This approach is very useful in calculations of ionization energy cost (see Section 12.1.2); energy cost of specific excited species; and energy cost of dissociation and formation of specific atoms and radicals during the electron beam degradation process.

12.2.5 Alkhazov's Equation for Degradation Spectrum of High-Energy Beam Electrons

Different variations of the degradation kinetic equations were developed for different special degradation problems. The Fano–Inokuti approach is close to the one discussed earlier, but includes integration over energy losses instead of electron energies, and is referred to as the **differential electron free path approach**.[562,562a–568]

Another approach, called the **Green model**, was successfully applied to describe degradation of high-energy electrons in the upper atmosphere. This approach divides electrons into three degradation cascades: primary, secondary, and tertiary; it then considers degradation in each cascade separately.[569–571]

The simplest and most effective approach to degradation kinetics of high-energy electron beams was proposed by Alkhazov[556] and developed by Nikerov and Sholin.[561] In this case, the general degradation spectrum equation (Equation 12.12) is specified in the following form, known as **Alkhazov's equation**:

$$Z(E) = \sum_k p_{ex,k}(E+U_k)\cdot Z(E+U_k) + \int_{E+I}^{\infty} p_i(W, W-E)\cdot Z(W)dW + \delta(E-E_0) \tag{12.17}$$

The boundary condition for the equation is again $Z(E\to\infty)=0$. In Alkhazov's equation, $p_{ex,k}(E+U_k)$ is the probability of excitation of an atom into the kth excited state in a collision with an electron having energy $E+U_k$; $p_i(W, W-E)$ is the differential probability of ionization of an atom in collision with an electron having initially kinetic energy W and during the collision losing energy $\Delta E = W-E$; I and U_k are, respectively, the ionization potential and the excitation energy of the kth state; and $\delta(E-E_0)$ is the δ-function describing the source of degradation, the initial electron with high energy E_0.

Simple solution of Alkhazov's equation can be obtained in the Bethe–Born approximation, taking into account only ionization by primary electrons:[556]

$$Z(E) = \frac{\sigma(E)}{L(E)} = \frac{\aleph^2 \ln(cE)}{4R\ln\left(\sqrt{\frac{e}{2}}\frac{E}{I^*}\right)} \tag{12.18}$$

In this relation, $L(E) \approx (16R^2/E)\ln(\sqrt{e/2}E/I^*)$ is the function characterizing electron energy losses; $\sigma(E) \approx (4R/E)\,\aleph^2 \ln(cE)$ is the total cross section of inelastic collisions for an electron with energy E; R = 13.595 eV is the Ridberg constant; $\aleph^2 = 0.7525$

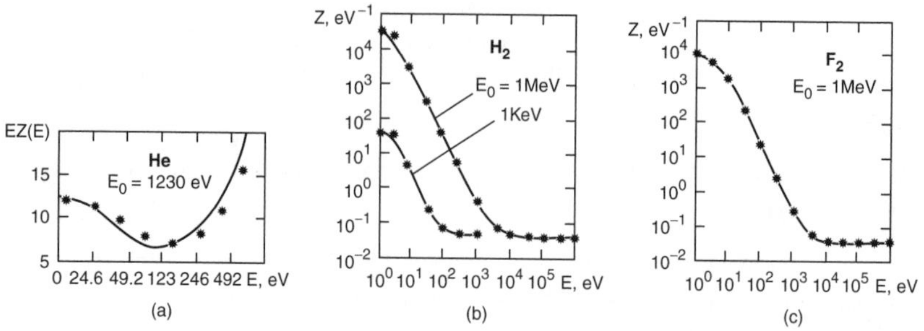

Figure 12.3 Degradation spectra in (a) helium; (b) hydrogen; (c) fluorine; stars correspond to numerical calculations; curves correspond to analytical calculations[556,558]

is square of the matrix element for the sum of transitions into continuous and discrete spectra; factor $c = 0.18$ eV; and $I^* = 42$ eV is the characteristic excitation potential (numerical data are given for helium).

12.2.6 Solutions of Alkhazov's Equation

The analytical solution of Alkhazov's equation gives only a qualitative description of degradation spectrum. Numerical solutions of Alkhazov's equation for electron beam degradation in helium, hydrogen, and fluorine are presented in Figure 12.3.[556,558] Alkhazov's equation can also be solved analytically with sufficient accuracy by using special analytical approximations for cross sections of the collisional processes. These results are also presented in Figure 12.3 in comparison with numerical solutions. Accurate estimations of the degradation spectrum above the ionization threshold can be achieved by using the following formula:[561]

$$Z(E) = \frac{0.6}{I} + \frac{0.5 \cdot E_0}{(E+I)^2}\left[1 + \frac{3I^2}{(E+I)^2}\ln\left(\frac{E}{I}+2\right)\right] \tag{12.19}$$

For numerical estimations of the degradation spectrum below the ionization threshold, one can use the alternate formula:

$$Z(E) = \frac{2}{(1+E/I)^4 I}\ln\left(\frac{E}{I}+2\right) + \frac{1}{I(1+E/I)^2} \tag{12.20}$$

This degradation spectrum, according to Equation 12.14, can be applied for estimating the electron energy distribution function (EEDF).

Once determined, the degradation spectra for the electron beam stopping can then be applied for calculating the ionization and excitation energy costs and total number of ions and excited species produced by one high-energy electron (Equation 12.15 and Equation 12.16). Results of such calculations in hydrogen are presented in Table 12.2.

Table 12.2 Integral Production of Ions and Excited Molecules by Degradation of an Electron with Initial Energy of 1 MeV and 1 keV in Molecular Hydrogen

State of molecule	E_0 = 1 MeV	E_0 = 1 keV
Ionized	$3.06 \cdot 10^4$	29.6
H_2 (Σ_u^+)	$1.10 \cdot 10^4$	10.7
H_2 (B^1 Σ_u^+)	$1.43 \cdot 10^4$	13.9
H_2 * (vibr. exc.)	$1.84 \cdot 10^5$	187

12.3 PLASMA–BEAM DISCHARGE

12.3.1 General Features of Plasma–Beam Discharges

This discharge corresponds to the first type of electron beam plasmas according to classification given in Section 12.1.3. In this case, the electron beam does not directly interact with the neutral gas. At first it interacts with Langmuir plasma oscillations, which subsequently transfer energy through electrons to neutrals. Thus, here the electron beam can be considered as a source of microwave oscillations, which then sustain the "microwave discharge" (see Section 10.8). In contrast to conventional microwave discharges, the source of electrostatic oscillations is present inside plasma, which excludes the skin-effect problems.

As mentioned previously, the plasma–beam discharge usually requires high beam currents and current densities ($I_b > 1$ A, $j_b > 100$ A/cm^2), and pressures below a few torreys. The discharge is stable and efficient over a wide range of pressures (10^{-4} to 3 torr) and powers (0.5 to 10 kW). The degree of ionization varies over the wide range of 10^{-4} to 1, as well as electron temperature of 1 to 100 eV.[572–577]

Physical basis of the beam instability is related to the collisionless interaction of electrostatic plasma waves with electrons discussed in Section 6.5.5. Injection of an electron beam in a plasma creates the distribution function illustrated in Figure 6.24, where the derivative Equation 6.145 is positive. As was shown in Section 6.5.5, this corresponds to energy transfer from the electron beam and exponential amplification of electrostatic plasma oscillations. The electron beam transfers energy to Langmuir oscillations until the EEDF becomes continuously decreasing. The increment of the beam instability describing the exponential growth of the Langmuir oscillation can be calculated from Equation 6.150.

12.3.2 Operation Conditions of Plasma–Beam Discharges

Plasma–beam discharges are based on a complicated self-consistent series of physical processes: (1) beam instability and energy transfer from the electron beam to Langmuir oscillations; (2) then energy transfer from electrostatic oscillations to plasma electrons; and finally (3) energy transfer from plasma electrons to neutral components, which determines ionization and therefore the effectiveness of beam instability. As a result, effective operation of plasma–beam discharges requires very delicate and precise choice of system parameters. The principal operation condition of plasma–beam discharges requires beam energy dissipation into Langmuir oscillations to exceed beam energy losses in electron–neutral collisions. In molecular gases, this requirement can be expressed as:[576]

$$\frac{6\pi^2\hbar\omega\nu_{eV}}{\bar{\varepsilon}_b\nu_{bn}}\left(\frac{\omega_{pb}}{\nu_{eV}}\right)>1 \tag{12.21}$$

In this relation, ν_{eV} is the vibrational excitation frequency; $\hbar\omega$ is a vibrational quantum; ν_{bn} is the frequency of beam electrons collisions with neutral particles; $\bar{\varepsilon}_b$ is the average energy loss of a beam electron during its collision with a neutral particle; n_b is density of the beam electrons; and ω_{pb} is plasma frequency corresponding to the beam electron density:

$$\omega_{pb}=\sqrt{\frac{n_be^2}{\varepsilon_0m}} \tag{12.22}$$

Actually, the criterion (Equation 12.21) of collisionless losses' domination restricts the operational pressure of the plasma–beam discharge below the level of a few torreys.

The second condition of effective plasma–beam discharge operation is related to energy transfer from Langmuir oscillations to plasma electrons. The dissipation of Langmuir oscillations can be provided (especially at low pressures) by the modulation instability. This instability leads to a collapse of the Langmuir oscillations, and energy transfer along the spectrum into the range of short wavelengths. Energy of short-wavelength oscillations can then be transferred to plasma electrons due to the Landau damping (see Section 6.5.6).

However, this dissipation mechanism of Langmuir oscillations based on the modulation instability leads to significant energy transfer to "too energetic" plasma electrons from the far tail of the EEDF. With energies exceeding ionization potential, such electrons are not very effective in plasma chemical processes, especially in those related to vibrational excitation. This makes the modulation instability not desirable.

A more effective dissipation mechanism of Langmuir oscillations for plasma chemical processes is related to joule heating, which results in uniform energy transfer to plasma electrons along the energy spectrum. Electron energy transfer in coulomb collisions from high to lower energies takes place only at a degree of ionization exceeding $\alpha \approx 0.1$. Then, the domination of the joule dissipation over the dissipation related to the modulation instability requires the electron–neutral collision frequency to be higher than the increment of modulation instability. This leads to the following criterion of effective energy transfer from Langmuir oscillations to low-energy (1 to 3 eV) plasma electrons:[576]

$$\left(\frac{\omega_{pb}}{\nu_{en}}\right)^2 \sqrt{\frac{4\pi^2}{3}\frac{\hbar\omega}{T_e}\frac{m}{M}\frac{\nu_{eV}}{\nu_{en}}} < 1 \tag{12.23}$$

Here, m and M are masses of electrons and heavy particles, respectively. It is important that, in contrast to the requirement (Equation 12.21), the criterion (Equation 12.23) determines the low limit of gas pressure for effective plasma chemical operation of plasma–beam discharges.

12.3.3 Plasma–Beam Discharge Conditions Effective for Plasma Chemical Processes

At low pressures (high values of the parameter ω/ν_{en}), when the criterion (Equation 12.23) is not valid, the electron beam effectively transfers energy into Langmuir oscillations. However, in this case, the Langmuir oscillations dissipate most of their energy only to high-energy electrons, which results in high energy-cost radiation–chemical effects.

A similar situation takes place at relatively high pressures (low values of the parameter ω/ν_{en}), when the criterion (Equation 12.21) is not valid. In this case, beam instability is suppressed by direct electron–neutral collisions, and the chemical effect is provided only by relatively low energy efficiency radiation–chemical processes. However, these two regimes could be applied for stimulating chemical processes in which high-energy efficiency is not a first priority, e.g., metal reduction from their halogenides. One example is the direct process of $TiCl_4$ dissociation with the production of metallic titanium and chlorine, which has been successfully demonstrated in the plasma–beam discharge.[578]

The optimal conditions for plasma chemical applications of plasma–beam discharges in molecular gases require two criteria (Equation 12.21 and Equation 12.23) to be valid. This leads to the relatively moderate pressures of about 1 torr. Even in the quite narrow pressure range, stimulation of chemical reactions through vibrational excitation of molecules requires sufficient specific power of the plasma–beam discharges

$$\sigma E^2 > \nu_{eV} n_e \hbar\omega \tag{12.24}$$

To satisfy this requirement, a relatively high level of Langmuir noise W should be achieved in the plasma–beam discharge:

$$\frac{W}{n_e T_e} = \frac{\varepsilon_0 E^2}{2 n_e T_e} \geq \frac{\nu_{eV}}{\nu_{en}}\frac{\hbar\omega}{T_e} \tag{12.25}$$

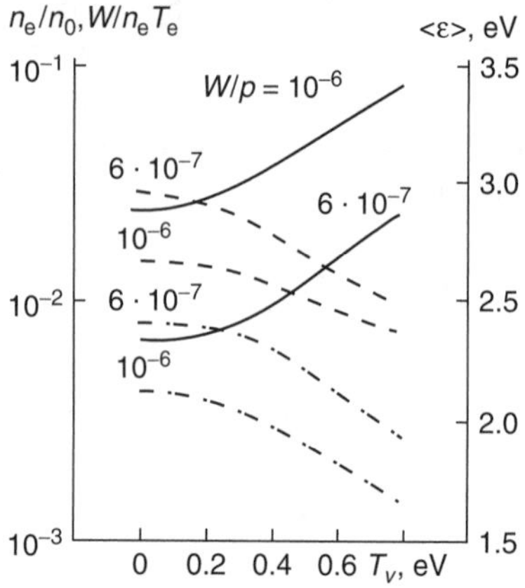

Figure 12.4 Dependence of ionization degree n_e/n_0 (—); relative Langmuir noise W/n_eT_e (–•–); and average electron energy ⟨ε⟩ (– – –) on vibrational temperature T_v.[579,580]

Numerically, the relative level of Langmuir noise W/n_eT_e should exceed the value of about 10^{-3} to provide effective vibrational excitation of molecules.

The dependence of degree of ionization; level of Langmuir noise W/n_eT_e; and average electron energy in the plasma–beam discharge is shown in Figure 12.4 for hydrogen as a function of vibrational temperature for two different values of the parameter W/p, where p is the gas pressure.[579,580] From this figure, it is seen that higher vibrational temperatures obviously correspond to higher electron concentrations (degree of ionization) that, on the other hand, correspond to lower values of W/n_eT_e and average electron energy.

12.3.4 Plasma–Beam Discharge Technique

A general schematic of a plasma chemical reactor based on the plasma–beam discharge is shown in Figure 12.5.[575] The discharge chamber is quite large; its internal diameter is 50 cm and length is 150 cm. At 10 mtorr, the flow rate is about 30 cm^3/sec; pressure in the electron gun chamber is about 10^{-5} torr and sustained by differential pumping. An electron beam with a maximum power of 40 kW is transported into the discharge chamber shown in this figure along the magnetic field of 300 kA/m; beam electrons' energy is 13 keV. In this system, the electron concentration sustained in the plasma–beam discharge is about $5 \cdot 10^{13}$ cm^{-3} at a discharge power of 3 kW.

Experiments with plasma–beam discharge show that the effectiveness of the electron beam dissipation in a plasma at fixed beam power strongly depends on the electron concentration n_b and energy E_b in the electron beam. An increase of the electron energy E_b in the beam leads to lower effectiveness of beam relaxation. Conversely, an increase of the electron concentration n_b results in higher effectiveness of beam relaxation. The maximum fraction of the beam power absorbed in plasma in these experiments reached 71%.[581]

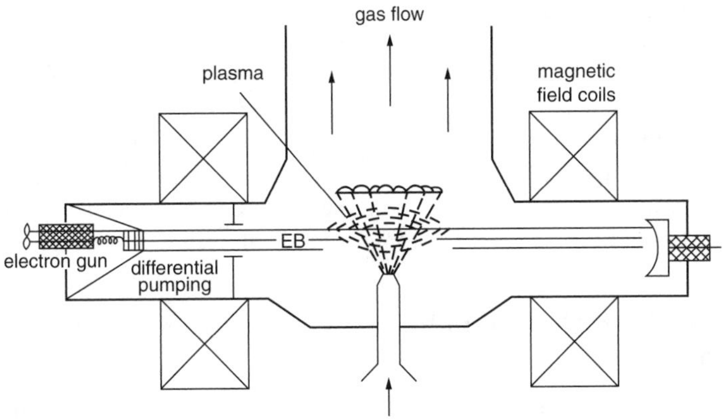

Figure 12.5 General schematic of a plasma beam discharge[575]

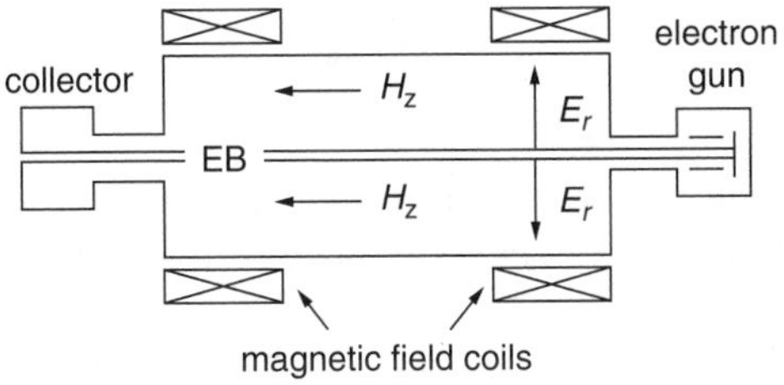

Figure 12.6 Electron beam in a magnetic field as the plasma centrifuge[582,583]

12.3.5 Plasma–Beam Discharge in Crossed Electric and Magnetic Fields: Plasma–Beam Centrifuge

General aspects of the plasma centrifuges, in which ionized gas is rotating fast in crossed electric and magnetic fields, and effects of gas separation were discussed in Section 7.5.3. A specific feature of the plasma centrifuge, based on the plasma–beam discharge, is the radial electric field provided directly by the electron beam.[582,582a,583] A schematic of this plasma centrifuge sustained by an electron beam is shown in Figure 12.6.

The plasma–beam discharge in this system is stable in inert gases at pressures of 0.1 to 1 mtorr and fills the entire volume of the discharge chamber. The maximal magnetic field in the coil center is about 800 kA/m; in the middle of the discharge chamber, it is about 500 kA/m. This magnetic field is sufficiently large to magnetize not only electrons but also ions. This means that the ion–cyclotron frequency much exceeds the frequency of the ion–ion coulomb collisions.

Electron energy in the beam is about 10 keV and the beam current is 1.5 A. Voltage between the electron beam and the external wall of the discharge chamber can be chosen to provide the ion current from center to the periphery. Radial current in plasma varies from 0.1 to 6 A. Plasma in this discharge system is completely ionized with density of about 10^{13} cm^{-3}; electron temperature is quite high on the level of 10 eV; and ion temperature is also very high on the level of 2 to 4 eV.

The velocity of the plasma rotation in the crossed electric and magnetic fields in this system is very high. According to the Doppler shift of argon ions' spectral lines, this circulation drift velocity is about $5 \cdot 10^5$ cm/sec, which is in good agreement with Equation 4.107 for drift in the crossed electric and magnetic fields. Also, the plasma rotation velocity is very high; the related binary mixture separation coefficient is not significant (see Equation 7.53) because the ions' temperature in the centrifuge is very high. In this case, separation can be achieved due to the difference in radial ion currents for components with different masses.

12.3.6 Radial Ion Current in Plasma Centrifuges and Related Separation Effect

In these plasma centrifuges, ions' displacement in radial direction at the moment of their formation is due to combined effects of the electric and magnetic fields. This radial displacement Δ_r of an ion can be found as:

$$\Delta_r = \frac{v_{EB}}{\omega_{Bi}} = \frac{1}{\omega_{Bi}} \frac{E}{B} \tag{12.26}$$

where $v_{EB} = E/B$ is velocity of the drift in the crossed electric E and magnetic B fields (Equation 4.107); and $\omega_{Bi} = eB/M_i$ is the ion–cyclotron frequency (see Section 4.3.4).

The radial displacement Δ_r of an ion, which requires a time interval of about $1/\omega_{Bi}$, takes place once per ion lifetime; this can be estimated as the ionization time $\tau_{ion} = 1/\nu_{ion} >> 1/\omega_{Bi}$. Thus, the average ion radial velocity $\bar{v}_{ri}$ can be estimated based on Equation 12.27 as:

$$\bar{v}_{ri} = \frac{E}{B} \frac{1}{\omega_{Bi}\tau_{ion}} = \frac{E}{B} \frac{\nu_{ion}}{\omega_{Bi}} = \left(\frac{E\nu_{en}}{eB^2} \right) \cdot M_i \tag{12.27}$$

Because the ionization frequency does not depend on the mass of ions, it can be concluded from Equation 12.27 that the average ion radial velocity is proportional to its mass. Thus, the ion current effect, similar to the centrifugal effect, results in a preferential flux of heavier particles to the periphery. The binary mixture separation coefficient R (Equation 7.53) can be expressed in this case as:[580]

$$\ln R = \frac{M_{iH} - M_{iL}}{\overline{M}_i} \frac{\nu_{ion}}{\nu_c} \sqrt{2} \int_{r1}^{r2} \frac{E/B \, dr}{\bar{v}_{Ti} \, \bar{r}_{Li}} \tag{12.28}$$

In this relation, M_{iH}, and M_{iL} are the masses of a heavier and lighter ion, respectively; $\overline{M}_i$ is their reduced mass; ν_c is the frequency of ion–ion coulomb collisions, which provide a mutual diffusion of the ions of different component and reduce the separation coefficient; and $\bar{v}_{Ti}$ and $\bar{r}_{Li}$ are the ionic thermal velocity and Larmor radius, calculated for the reduced ionic mass.

Typical values of the separation coefficients achieved in this type of plasma centrifuge are: for an Ar–He mixture, $R = 6.6$; for a Kr–Ar mixture, $R = 2.5$; for a Xe–Kr mixture, $R = 2.0$; and for an isotopic mixture, ^{20}Ne–^{22}Ne, the coefficient $R = 1.3$.[580]

12.4 NONEQUILIBRIUM HIGH-PRESSURE DISCHARGES SUSTAINED BY HIGH-ENERGY ELECTRON BEAMS

12.4.1 Non-Self-Sustained High-Pressure Discharges

According to the classification given in Section 12.1.3, high-current electron beam propagation in high-pressure gas (the fourth group of beam discharges) results in generation of thermal plasma. Nonthermal plasma can be generated in such systems only if the short-pulse electron beam is applied and the specific energy input does not exceed the critical value of about 0.1 to 0.3 eV/mol that leads to gas overheating.

However, most electron beam energy in these nonequilibrium systems is spent for ionization and electron excitation of neutral species (see the degradation spectrum analysis discussed in Section 12.2). To provide the necessary energy input into vibrational excitation of molecules, additional energy can be transferred to the plasma electrons from an external electric field. This electric field can be chosen sufficiently low to avoid ionization, but optimal for vibrational excitation or other processes requiring electron temperature at the relatively low level of 1 to 2 eV.

Such non-self-sustained discharges with ionization provided by an electron beam and most energy input due to an external electric field are especially important for gas lasers (in particular, powerful pulsed CO_2 lasers) and high-efficiency plasma chemical systems (see Section 5.6.2). Examples of these systems are given in References 584 through 589. Discharge aspects of the system have been reviewed in detail by Velikhov et al.[250] and the plasma chemical processes by Rusanov and Fridman.[9]

The non-self-sustained discharges are interesting for plasma chemical and laser applications first of all because they can provide a high level of nonequilibrium uniformly in very large volumes at high pressures (up to several atmospheres). An important advantage of these systems is the mutual independence of ionization and energy input in the molecular gas, which permits choosing the optimal values of the reduced the electric field E/n_0 (for example, for vibrational excitation) without care to sustain the discharge. It is also important that ionization (see Section 6.3.3) instabilities cannot destroy the discharge, simply because ionization is independent and cannot be affected by local overheating.

Ionization in such non-self-sustained, high-pressure, nonequilibrium discharges can be provided not only by electron beams, but also by using UV radiation, x-rays, etc. The important advantage of electron beams with respect to these other external sources of ionization is the relatively high intensity of ionization at the same power consumption.

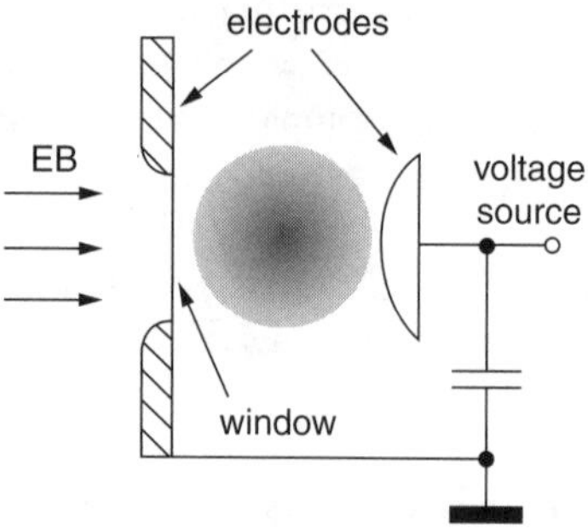

Figure 12.7 Discharge sustained by an electron beam

12.4.2 Plasma Parameters of Non-Self-Sustained Discharges

A general schematic of the discharges sustained by electron beams is illustrated in Figure 12.7. The electron beam is injected into the gas usually normal to electrodes across the accelerator's window closed by a metallic foil. Voltage is applied to the further electrode by the capacitance. Thus, plasma generation is due to electron beam ionization and the electric current is provided by the applied external electric field. Such discharge systems can be arranged in stationary and in short-pulse regimes.

In the short pulse regime, the pulse duration is usually about 1000 nsec; the electron beam energy about 300 keV; and the beam current density can exceed 100 A/cm^2. According to Equation 2.35, in atmospheric pressure air, the ionization rate provided by such an electron beam reaches $q_e = 10^{22}$ 1/cm^3 sec.

Losses of plasma electrons in this pulse-beam system are due to the electron–ion recombination and electron attachment to electronegative air components followed by the ion–ion recombination. These losses can be described by the effective recombination coefficient, determined based on the electron balance (Equation 4.1) as:

$$k_r^{eff} = k_r^{ei} + \varsigma k_r^{ii} \tag{12.29}$$

where k_r^{ei}, k_r^{ii} are rate coefficients of electron–ion and ion–ion recombination processes; and $\varsigma = k_a/k_d$ is ratio of the attachment and detachment coefficients. For calculating the electron beam degradation in atmospheric air, the effective empirical recombination coefficient can be taken as $k_r^{eff} \approx 10^{-6}$ cm^3/sec. The concentration of plasma electrons n_e can then be estimated as a simple function of the electron beam ionization rate q_e:

$$n_e = \sqrt{q_e / k_r^{eff}} \tag{12.30}$$

Numerically, for the example given previously, the concentration of plasma electrons generated by the electron beam is about 10^{14} cm^{-3}. Thus, even at atmospheric pressure, electron beams are able to produce a degree of ionization about $3 \cdot 10^{-6}$; this is high enough to sustain a high level of vibrational nonequilibrium (see Equation 5.150) and provide effective plasma chemical and laser processes.

One should note that the discharges sustained by electron beams are effective only if the power and energy input from the electron beam is much below those provided by the external electric field. For example, at the previously considered values of parameters, the specific power of the pulse electron beam can be calculated as:

$$P_b = q_e U_i \approx 0.05\ \mathrm{MW/cm^3} \tag{12.31}$$

where U_i is the energy price of ionization (see Table 12.1). At the same time, the specific power transferred from the external electric field to the atmospheric-pressure, nonthermal pulse discharge can be estimated as follows:

$$P_d = \sigma E^2 \approx k_{eV} n_e n_0 \hbar\omega \approx 1\ \mathrm{MW/cm^3} \tag{12.32}$$

where σ is the discharge plasma conductivity; n_e and n_0 are plasma electrons and neutral gas densities, respectively; and k_{eV} is the rate coefficient of vibrational excitation of molecules with relevant vibrational quantum $\hbar\omega$.

If the pulse electron beam duration is 1000 nsec, the specific energy input from the beam, according to Equation 12.31, is approximately 0.05 J/cm^3 at normal conditions. The specific energy input from the external electric field into the discharge, according to Equation 12.32, is about 1 J/cm^3, which is 20 times higher than that for the electron beam.

These discharges are sustained by the short-pulse electron beams and can obviously be arranged without essential modifications in pulse-periodic regimes.[9,250] The non-self-sustained discharges also can be organized in a continuous way. High-pressure discharges sustained by continuous electron beams are discussed in Velikhov et al.[587,588] Application of such discharges for stimulation of lasers and high-efficiency plasma chemical reactions related to vibrational excitation of molecules (see Section 5.6.2) is limited because of relatively low beam current density that results, according to Equation 12.30, in a low degree of ionization.

12.4.3 Maximum Specific Energy Input and Stability of Discharges Sustained by Short-Pulse Electron Beams

As discussed in Section 5.6.8, the energy efficiency of plasma chemical processes (in particular those stimulated by vibrational excitation) depends on the specific energy input, which is the energy consumed in the discharge per one molecule (or per unit volume at normal conditions). The optimal values of the specific energy input necessary to stimulate endothermic plasma chemical processes are approximately 1 eV/mol, or about 5 J/cm^3 at normal conditions. The minimum threshold values of the specific energy input in these systems, according to Equation 5.6.15, are about 0.1 to 0.2 eV/mol or 0.5 to 1 J/cm^3 at normal conditions.

Reaching the previously mentioned values of specific energy input in nonthermal homogeneous atmospheric pressure discharges is limited by the numerous instabilities described in Section 6.3. Although ionization instabilities are suppressed in non-self-sustained discharges because ionization is provided independently by electron beams, the fast overheating instabilities put serious limits on the specific energy

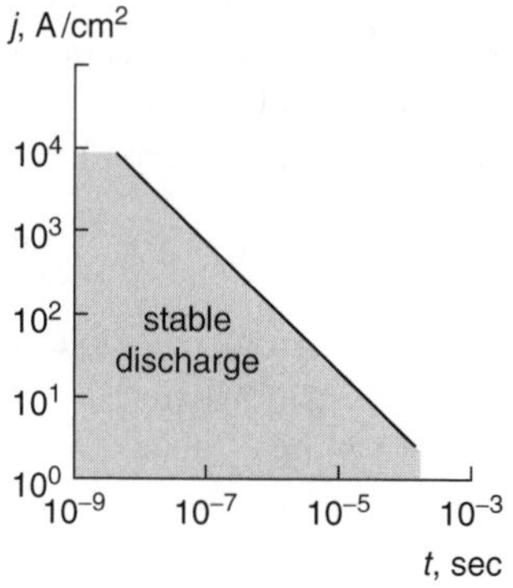

Figure 12.8 Current densities and pulse durations of a stable discharge sustained by an electron beam

input. Especially severe instability effects are related to the fast mechanisms of chemical and VV relaxation heat release, described in Section 6.3.6 (see Figure 6.12).

The non-self-sustained pulse discharges can be arranged for much longer duration than the previously discussed nanosecond range. The pulse duration can be increased to 10 to 100 μsec and more. However, at relatively high levels of the discharge specific power, duration of the discharge stability is limited. Typical stable values of current densities and pulse durations of non-self-sustained discharges are shown in Figure 12.8.

From this figure, one can calculate that specific energy inputs exceeding 0.5 J/cm^3 require special consideration of the discharge stability. It is interesting that the preceding restriction actually divides the effective plasma chemical and effective laser regimes of the discharges. Chemical conversion of laser mixtures (for example, CO_2 dissociation in CO_2 lasers), which takes place at high specific energy inputs, is not acceptable for effective laser generation.

12.4.4 Electric Field Uniformity in Non-Self-Sustained Discharges

If ionization in the non-self-sustained discharge is provided by an electron beam, then high values of the external electric field can result in decrease of electron concentration near the cathode. The elevated electric field in the cathode layer due to the formed space charge can lead to additional ionization exceeding that of the beam. This effect converts the discharge into a self-sustained one; these self-sustained pulse discharges are of interest for special gas laser applications.[590,591] The electric field uniformity criterion to avoid the ionization effects in the cathode layer can be written as:

$$\lambda_c / d << E_0 / E_c \qquad (12.33)$$

where d is the distance between electrodes; λ_c is the characteristic thickness of the space charge in the cathode layer; E_c is electric field in the cathode layer; and E_0 is the quasi-uniform electric field in the discharge gap.

The cathode layer thickness λ_c can be estimated from the steady-state condition for charged particles in the layer:

$$eq_e = \frac{1}{\lambda_c} j_e = \frac{1}{\lambda_c} n_e b_e e E_0 \tag{12.34}$$

where q_e is the ionization rate provided by an electron beam; j_e is the plasma current density; and b_e is the electron mobility. Taking into account Equation 12.30 for electron density n_e, the cathode layer thickness λ_c can be found from Equation 12.34 as:

$$\lambda_c = \frac{b_e E_0}{\sqrt{q_e k_r^{eff}}} \tag{12.35}$$

The Gauss equation for the cathode layer where the ions' concentration n_{ic} prevails over that of electrons can be presented for estimations as:

$$E_c / \lambda_c = e n_{ic} / \varepsilon_0 \tag{12.36}$$

Then, the ions' concentration n_{ic} in the cathode layer can be expressed from $j_e = n_{ic} b_i e E_c$ (b_i is the ion mobility) in the following form:

$$n_{ic} = \sqrt{q_e \varepsilon_0 / e b_i} \tag{12.37}$$

Finally, taking E_c from Equation 12.36 and λ_c from Equation 12.36 and substituting them in the inequality (Equation 12.33) yields the electric field uniformity criterion for the non-self-sustained discharges as:[459]

$$E_0 << \frac{b_i d}{b_e^2} k_r^{eff} \sqrt{\frac{q_e \varepsilon_0}{e b_i}} \tag{12.38}$$

Numerically, Equation 12.38 requires E_0 << 10 kV/cm at atmospheric pressure; electron beam current density of 10^3 A/cm^2; and d = 100 cm.

12.4.5 Nonstationary Effects of Electric Field Uniformity in Non-Self-Sustained Discharges

The total electric field in the non-self-sustained discharges can be quasi uniform even if the preceding criterion (Equation 2.59) is not valid and the external electric field is very high. This can be achieved if the pulse duration is too short to make a significant perturbation. The time interval necessary for establishing the steady-state electric field distribution can be found from the electric current continuity equation including the displacement current:

$$\varepsilon_0 \frac{\partial E_c}{\partial t} + e n_{ic} b_i E_c = e n_e b_e E_0 \tag{12.39}$$

In the continuity equation, it was taken into account that major changes of the electric field take place in the cathode layer. Then, the assumption (for simplicity) that the ion concentration in the cathode layer is in the form of Equation 10.78 and constant leads to the following expression for the establishment time of the electric field distribution in the discharge:

$$\tau_E = \frac{1}{\sqrt{q_e e b_i \varepsilon_0}} = \frac{1}{\sqrt{k_r^{eff} e b_i \varepsilon_0}} \frac{1}{n_e} \tag{12.40}$$

Numerically, if electron concentration in the atmospheric pressure plasma is $n_e = 10^{15}$ cm^{-3}, then the establishment time of the electric field distribution is about 30 nsec. Thus, the pulse discharge can be considered as quasi uniform for shorter pulse durations even at high values of external electric fields and if the requirement (Equation 2.59) is not satisfied.

12.5 PLASMA IN TRACKS OF NUCLEAR FISSION FRAGMENTS: PLASMA RADIOLYSIS

12.5.1 Plasma Induced by Nuclear Fission Fragments

As discussed in Section 5.6, energy efficiency of endothermic chemical reactions in plasma can be very high, up to 80 to 90%. This means that almost all discharge energy can be selectively directed into some chosen channels of chemical reactions, in particular through vibrational excitation of reactants (see Section 5.6.2). Although the energy cost of the processes is low, usually plasma chemical systems consume electricity, which is relatively expensive. For this reason, it is always interesting to consider plasma generation based on nonelectric energy sources. For example, such plasma systems can be arranged in the stopping process of nuclear fission fragments.

The nuclear fission process of interaction between a slow neutron with uranium-235 can be presented as:

$$_{92}U^{235} + {}_0n^1 \rightarrow {}_{57}La^{147} + {}_{35}Ba^{87} + 2{}_0n^1 \tag{12.41}$$

The total fission energy is 195 MeV. Most of this energy, 162 MeV, goes to kinetic energy of the fission fragments, barium and lanthanum (see Glasstone and Edlung[614]). Plasma chemical effects can be induced by the nuclear fission fragment, if they are transmitted into gas phase before thermalization. The fission fragment appears in the gas as multiply charged ions with $Z = 20 - 22$ and initial energy of approximately $E_0 \approx 100$ MeV.[592]

Degradation of these fission fragment ions in a gas is related to formation of the so-called δ-electrons moving perpendicularly to the fission fragment with kinetic energy of approximately $\varepsilon_0 \approx 100$ eV. These δ-electrons form a radial high-energy electron beam, which then generates plasma in the long cylinder surrounding the trajectory of the fission fragment. This plasma cylinder is usually referred to as the **track of nuclear fission fragments**.

If the plasma density in the track of nuclear fission fragments is relatively low, then degradation of each δ-electron in the radial beam is quasi independent. The chemical effect of the independent energetic electrons' (or other particles') interaction with molecule gas is called radiolysis. The radiolysis induced by high-energy electrons in different molecular gases has been investigated in detail (see, for example, Aulsloos[596]). Mostly related to primary ionization and electronic excitation (see Section 12.4.2), this process is not very energy effective (see Section 5.6), but is of interest for many special applications related to gas and surface treatment.

More interesting physical and chemical effects can be observed if the plasma density in the tracks of nuclear fission fragments is relatively high, and collective effects of electron–electron interaction, vibrational excitation, etc. take place.[593] These effects, usually referred to as **plasma radiolysis,** require special conditions of nuclear fragment degradation. To determine the nuclear fragment degradation conditions necessary for plasma radiolysis, first the phenomenon of plasma radiolysis is analyzed by itself. In particular, the required degree of ionization in the fragment's track for realization of the effect should be specified, so it is necessary to begin with analysis of plasma radiolysis using specific examples of irradiation of water vapor and carbon dioxide.

12.5.2 Plasma Radiolysis of Water Vapor

The energy efficiency of radiolysis is usually characterized by the so-called **G-factor**, which shows the number of chemical processes of interest (in this case, the number of dissociated water molecules) per 100 eV of radiation energy spent. Value of the G-factor of water vapor radiolysis by high-energy electrons (in particular, by δ-electrons in the tracks of nuclear fission fragments) can be found as:[594]

$$\frac{G}{100} = \left[\frac{1}{U_i(H_2O)} + \frac{U_i(H_2O) - I_i(H_2O)}{U_i(H_2O) \cdot I_{ex}(H_2O)}\right] + \frac{\mu}{U_i(H_2O)} \tag{12.42}$$

In this relation, $U_i(H_2O) \approx 30$ eV is the ionization energy cost for water molecules (see Section 12.1.2); $I_i(H_2O) = 12.6$ eV, $I_{ex}(H_2O) = 7.5$ eV are energy thresholds of ionization and electronic excitation of water molecules; and the factor μ shows how many times one electron with an energy below the threshold of electronic excitation $\varepsilon < I_{ex}(H_2O)$ can be used in the water molecule destruction processes.

The first two terms in Equation 12.42 are mostly related to dissociation through electronic excitation and give the radiation yield $G \approx 12$ mol/100 eV,[595] which describes water dissociation by β-radiation and low current density electron beams. The third term in the relation (Equation 12.42) describes the contribution of H_2O destruction due to the dissociative attachment with a following ion–molecular reaction:

$$e + H_2O \rightarrow H^- + OH, \quad H^- + H_2O \rightarrow H_2 + OH \tag{12.43}$$

The contribution of this mechanism to water dissociation during the conventional radiolysis is relatively low ($\mu = 0.1$) because of the resonance character of the dissociative attachment cross section dependence on energy and low ionization degree.[158] The factor μ can be significantly larger at higher degrees of ionization in plasma because of effective energy exchange (Maxwellization) between plasma electrons with energies below the threshold of electronic excitation $\varepsilon < I_{ex}$ (H_2O). Thus, the main plasma effect in water vapor electrolysis is related to the Maxwellization of plasma electrons; this results in growth of the G-factor and an increase of energy efficiency of water dissociation and hydrogen production in the system.

12.5.3 Maxwellization of Plasma Electrons and Plasma Effect in Water Vapor Radiolysis

The critical value of the degree of ionization in the tracks of nuclear fragments necessary for effective Maxwellization of plasma electrons and realization of the plasma radiolysis effect in water vapor can be found from the following kinetic equation for electron velocity distribution function f_e:[158]

$$0 = L\cdot\left(f_e^2 - \frac{\partial^2 f_e}{\partial v_i \partial v_k}\frac{\partial^2 \Psi}{\partial v_i \partial v_k}\right) + f_e\cdot\sigma_a(v)\cdot v\cdot\left[\mathrm{H_2O}\right] + \qquad (12.44)$$

$$Z_d(v)\cdot k_d n_e\left[\mathrm{H^-}\right] + Z(v)\cdot q_e - f_e\cdot\sigma_r^{ei}\cdot v\cdot n_i$$

In this equation, the factor $L = \Lambda(e^2/m\varepsilon_0)^2$; Λ is the Coulomb logarithm; $Z_d(v)$ is the normalized electron velocity distribution function after its detachment from a negative ion H^-; $Z(v)$ is the normalized electron distribution function related to total rate q_e of electrons' appearance in the energy range below the threshold of electronic excitation (a function directly related to the degradation spectrum; see Section 12.2); $\sigma_a(v)$, $\sigma_r^{ei}(v)$ are cross sections of dissociative attachment and electron–ion recombination; n_i is the positive ion concentration; and the function $\Psi_e(\vec{v})$ is determined through the electron velocity distribution function f_e:

$$\Psi_e(\vec{v}) = \frac{1}{8\pi}\int \left|\vec{v} - \vec{v}'\right|\cdot f_e(\vec{v}')\cdot d\vec{v}' \qquad (12.45)$$

Analysis of Equation 12.44 shows that effective Maxwellization of electrons under the threshold of electronic excitation $\varepsilon < I_{ex}$ (H_2O) takes place when the degree of ionization exceeds the following critical value:

$$\frac{n_e}{\left[\mathrm{H_2O}\right]} >> \sigma_a^{\max}\frac{T_e^2\varepsilon_0^2}{e^4\Lambda} \qquad (12.46)$$

In this criterion, $\sigma_a^{\max} \approx 6\cdot 10^{-18}\,\mathrm{cm}^2$ is the maximum value of the dissociative attachment cross section of electrons to water molecules. Numerically, at an electron temperature of $T_e = 2$ eV, the criterion (Equation 12.46) requires $n_e/[H_2O] >> 10^{-6}$.

Thus, if ionization in a track of nuclear fission fragment is sufficiently high (Equation 12.46), the effective maxwellization of electrons under the threshold of electronic excitation permits the μ-factor (Equation 12.42) to increase from about $\mu = 0.1$ (typical for conventional radiolysis) to higher value determined by the degradation spectrum $Z(\varepsilon)$:

$$\mu = \frac{\int_0^I \varepsilon Z(\varepsilon)\, d\varepsilon}{\varepsilon_a \int_0^I Z(\varepsilon)\, d\varepsilon} \tag{12.47}$$

that can be achieved in the plasma radiolysis. In this relation, I is the ionization potential; $Z(\varepsilon)$ is the degradation spectrum under the threshold of electronic excitation determined by Equation 12.20; and ε_a is energy necessary for dissociative attachment (see Figure 2.7).

Integrating Equation 12.47 with the degradation spectrum (Equation 12.20) gives $\mu \approx 1$. This result means that intensive Maxwellization of electrons under the electronic excitation threshold in water vapor leads almost all of the electrons to dissociative attachment. Thus, the first two terms in Equation 12.42, which are related to dissociation through electronic excitation, give the radiation yield $G \approx 12$ mol/100 eV. The third term in Equation 12.42 specific for plasma radiolysis gives an additional $G \approx 7$ mol/100 eV, which is due to dissociative attachment. This brings the total radiation yield of water dissociation in plasma radiolysis to $G \approx 19$ mol/100 eV, which corresponds to a high water decomposition energy efficiency of approximately 55% (see Section 5.6).

12.5.4 Effect of Plasma Radiolysis on Radiation Yield of Hydrogen Production

The main final saturated products of water radiolysis are hydrogen and oxygen. However, the radiation yield of water dissociation discussed earlier does not directly correspond to the radiation yield of hydrogen production because of inverse reactions. If the radiation yield of conventional water radiolysis reaches $G \approx 12$ mol/100 eV without plasma effects, the corresponding radiation yield of hydrogen production is only $G \approx 6$ mol/100 eV.[596] Plasma radiolysis can bring the total radiation yield of water dissociation to $G \approx 19$ mol/100 eV; the corresponding radiation yield of hydrogen production grows in this case to $G \approx 13$ mol/100 eV, which means an energy efficiency of about 40%.[593]

The energy efficiency of hydrogen production in the process of water vapor dissociation during radiolysis strongly depends on gas temperature and intensity of the process.[597] This is related to the fact that the primary products of water dissociation related to mechanisms of electronic excitation and electron–ion recombination are atomic hydrogen H and the hydroxyl radical OH. Formation of molecular hydrogen is then related to the following endothermic bimolecular reaction:

$$H + H_2O \rightarrow H_2 + OH, \quad E_a = 0.6 \text{ eV} \tag{12.48}$$

The inverse reactions leading back to forming water molecules are mostly three-molecular recombination processes:

$$H + OH + M \rightarrow H_2O + M \qquad (12.49)$$

that are about 100 times faster than molecular hydrogen formation in the three-molecular recombination processes:

$$H + H + M \rightarrow H_2 + M \qquad (12.50)$$

High intensity of radiolysis leads to higher concentration of radicals and therefore to a larger contribution of the three-molecular processes and lower radiation yield of hydrogen production. The radiation yield can be increased with temperature, which intensifies the endothermic reaction (Equation 12.48). It is important that hydrogen production in the specific reactions of plasma radiolysis (Equation 12.43) has no energy barrier and proceeds faster than inverse processes (Equation 12.49). This makes the plasma radiolysis effect especially important at the high process intensities that can be achieved in the tracks of nuclear fission fragments.

12.5.5 Plasma Radiolysis of Carbon Dioxide

Radiation yield of CO_2 dissociation, similar to the case of water vapor radiolysis, is determined by the conventional term G_{rad} typical for radiation chemistry, and by a special plasma radiolysis term related to the contribution of electrons under the threshold of electronic excitation.

The first conventional radiolysis term, similar to Equation 12.42, is due to electronic excitation and dissociative recombination and can reach a maximum $G \approx$ 8 mol/100 eV.[596] In this case, the main contribution of the low-energy electrons (under the threshold of electronic excitation) is due to vibrational excitation of CO_2 molecules, which is proportional to the degree of ionization n_e/n_0 in the track of nuclear fission fragment:

$$\Delta T_v \approx \frac{n_e}{n_0} \left[\frac{\int_0^I \varepsilon Z(\varepsilon)\, d\varepsilon}{c_v \int_0^I Z(\varepsilon)\, d\varepsilon} \right] \qquad (12.51)$$

In this relation, c_v is the specific heat related to the vibrational degrees of freedom of CO_2 molecules; the factor in parentheses is determined by the degradation spectrum of δ-electrons in tracks and numerically is several electronvolts. Plasma effects in CO_2 radiolysis are related to stimulation of the reaction:

$$O + CO_2 \rightarrow CO + O_2 \qquad (12.52)$$

by vibrational excitation of CO_2 molecules. This reaction prevails over three-molecular recombination of oxygen molecules:

$$O + O + M \rightarrow O_2 + M \tag{12.53}$$

at vibrational temperatures corresponding, in accordance with Equation 12.51, to the degree of ionization $n_e/n_0 >> 10^{-3}$. In this case, the radiation yield of CO_2 dissociation becomes twice as large as in conventional radiolysis and reaches $G \approx$ 16 mol/100 eV. This corresponds to an energy efficiency of about 50% (see Section 5.6).

It should be noted that the radiation yield of CO in conventional radiolysis is suppressed by inverse reactions:

$$O + CO + M \rightarrow CO_2 + M \tag{12.54}$$

that are much faster at standard conditions than recombination (Equation 12.53) because CO concentrations usually greatly exceed the concentration of atomic oxygen. As a result, the radiation yield of conventional CO_2 dissociation $G \approx 8$ mol/100 eV leads to a radiation yield of CO production of only $G \approx 0.1$ mol/100 eV.[598]

Only high-intensity radiolysis, which provides large concentrations of oxygen atoms, permits suppressing the inverse reactions (Equation 12.54) in favor of Equation 12.53 and increasing the radiation yield.[599] This can take place in tracks of nuclear fission fragments and also in the radiolysis provided by relativistic electron beams. It is interesting that high intensity of radiolysis increases the radiation yield of CO production in contrast to the case of hydrogen production (Section 12.5.4). This is because the rate coefficient of Equation 12.53 exceeds that of Equation 12.54.

12.5.6 Plasma Formation in Tracks of Nuclear Fission Fragments

As demonstrated earlier, the energy efficiency of endothermic chemical processes can be increased by a factor of two when the degree of ionization in the tracks of nuclear fission fragments exceeds some critical value (10^{-4} for H_2O and 10^{-3} for CO_2). For this reason, the degree of ionization in the tracks of fragments provided by the degradation of δ-electrons with kinetic energy $\varepsilon_\delta \approx 100$ eV formed by the fragments should be analyzed. At first it is necessary to take into account that degradation length of the fragments can be presented as:

$$\lambda_0 = \frac{4\pi\varepsilon_0^2 E_0 m v^2}{n_0 Z^2 Z_0 e^4} \ln^{-1}\left(\frac{2mv^2}{I}\right) \tag{12.55}$$

In this relation, v, Z_e, and E_0 are velocity, charge, and kinetic energy of the nuclear fission fragments; m and e are mass and charge of an electron; Z_0 is the sum of charge numbers of atoms forming molecules that are irradiated by the nuclear fragments; n_0 is the neutral gas density; and I is the effective value of ionization potential.

The total number of energetic δ-electrons formed by a nuclear fission fragment per unit length can be calculated by the relation:

$$N_\delta = \frac{n_0 Z^2 Z_0 e^4}{4\pi\varepsilon_0^2 m v^2 \varepsilon_\delta} \ln\left(\frac{2mv^2}{I}\right) \quad (12.56)$$

For example, generation of δ-electrons in water vapor at atmospheric pressure is characterized by $N_\delta = 10^6$ cm^{-1} and that in carbon dioxide is $N_\delta = 3 \cdot 10^6$ cm^{-1}.

The tracks start with very narrow channels of complete ionization. The radius of these initial, completely ionized channels can be expressed as:

$$r_0 = \frac{Ze^2}{2\pi\varepsilon_0 v\sqrt{m\varepsilon_\delta}} \ln\left(\frac{2mv^2}{I}\right) \approx 3 \cdot 10^{-8}\text{cm} \quad (12.57)$$

After the initial track (Equation 12.57), expansion is at first free and collisionless when radius r of the channel is less than length λ_i necessary for ionization. Then, the expansion mechanism becomes diffusive.

12.5.7 Collisionless Expansion of Tracks of Nuclear Fission Fragments

Analyzing the collisionless track expansion, two different regimes can be noted depending on the linear density of δ-electrons N_δ. If the linear density is sufficiently high:

$$N_\delta > 4\pi\varepsilon_0 \frac{\varepsilon_\delta}{e^2} \ln^{-1}\left(\frac{\lambda_i}{r_0}\right) \approx 10^8 \text{cm}^{-1} \quad (12.58)$$

then the plasma remains quasi neutral during the collisionless expansion. In the opposite case of low linear density, when $N_\delta < 10^8$ cm^{-1}, electrons and ions are actually independent during the collisionless expansion of the initial tracks. In the case of high linear density of δ-electrons (Equation 12.58) and thus quasi-neutral plasma, the collisionless expansion of the plasma channel can be described by the following system of equations, which includes a dynamic equation (here, T_e is the effective electron temperature):

$$nM\frac{d\vec{v}}{dt} = -T_e \, grad\, n \quad (12.59)$$

as well as the continuity equation for ion concentration n_i and ion velocity v_i:

$$\frac{\partial n}{\partial t} + div(n\vec{v}) = 0 \quad (12.60)$$

Taking into account the axial symmetry of the system, the preceding system of equations can be reduced to an ordinary differential equation, using the automodel variable $\xi = r/t$:

$$\frac{M}{T_e}(v-\xi)^2\frac{\partial v}{\partial \xi}=\frac{\partial v}{\partial \xi}+\frac{v}{\xi} \tag{12.61}$$

This equation can be solved and its solution can be presented as the following function of the automodel variable:

$$v=\xi+c_{si}\sqrt{1+\Omega(\xi)} \tag{12.62}$$

In this expression, $c_{si}=\sqrt{T_e/M}$ is the ionic sound velocity (see Section 6.5.3), and the special function $\Omega(\xi)$ can be estimated as:

$$\Omega(\xi)\approx 1-\frac{c_{si}}{2\xi}+\sqrt{\left(\frac{c_{si}}{2\xi}\right)^2+\frac{2c_s}{\xi}} \tag{12.63}$$

In this case, the distribution of ion concentration in the expanding plasma channel also can be expressed as a function of the automodel variable:

$$n(\xi)\propto \exp\left[-\int\sqrt{1+\Omega(\xi)}\frac{d\xi}{c_{si}}\right] \tag{12.64}$$

Equation 12.62 and Equation 12.64 permit determination of the average value of the square of the ion velocity as:

$$\langle v^2\rangle=\int n(\xi)\cdot v^2(\xi)\cdot d\xi \tag{12.65}$$

This integration leads to the important conclusion that about two thirds of the total initial energy of δ-electrons is going into directed radial motion of ions.

12.5.8 Energy Efficiency of Plasma Radiolysis in Tracks of Nuclear Fission Fragments

As discussed earlier, two thirds of δ-electrons' energy (about 100 eV) is transferred at $N_\delta > 10^8$ cm^{-1} into radial motion of ions. On the other hand, the efficiency of the 70-eV ions in ionization processes and in chemical reactions is very low. Thus, it can be concluded that high linear density of δ-electrons ($N_\delta > 10^8$ cm^{-1}) is not effective in plasma radiolysis.

Improved energy efficiency can be achieved during the expansion of the initial track when the linear density of δ-electrons is relatively low ($N_\delta < 10^8$ cm^{-1}); electrons do not draw ions with them and transfer most of their energy into ionization and excitation of the neutral gas. This is quite useful in plasma chemistry. In this case, the radius of the tracks can be determined as $\lambda_i \approx 1/n_0\sigma_i$, where σ_i is a cross section of ionization provided by δ-electrons. The degree of ionization in the tracks of nuclear fragments can then be calculated as:

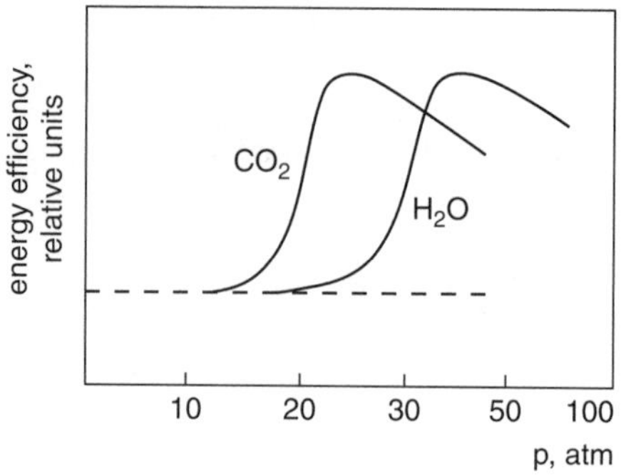

Figure 12.9 Energy efficiency of water vapor and carbon dioxide dissociation as a function of pressure

$$\frac{n_e}{n_0} = \left(\frac{3Z_0 e^4 \sigma_i^2}{4\pi\varepsilon_0^2 m\varepsilon_\delta} \right) \cdot \left(\frac{Z}{v} \right)^2 n_0^2 \tag{12.66}$$

It is important that the degree of ionization in the plasma channel of the nuclear tracks is proportional to the square of the neutral gas density and pressure. Numerically, at atmospheric pressure, the degree of ionization in the tracks of nuclear fragments in water vapor is $n_e/n_0 \approx 10^{-6}$; in carbon dioxide it is about $n_e/n_0 \approx 3 \cdot 10^{-6}$.

The necessary values of the degree of ionization for significant plasma effects in radiolysis (which can be estimated for water and carbon dioxide as $10^{-4} \div 10^{-3}$) cannot be achieved at atmospheric pressure. However, taking into account that $n_e/n_0 \propto n_0^2$, the sufficient degree of the ionization can be achieved by increasing pressure to 10 to 20 atm. Further pressure increase is limited by the criterion opposite to Equation 12.58.

At high pressures, the linear densities N_δ are also high, and the main l portion of electron energy is ineffectively lost in favor of ions. Thus, ion heating and transfer to thermal plasma in the tracks takes place in water vapor at pressures of approximately 100 atm, and in carbon dioxide at pressures of about 30 atm. Calculated dependence of the energy efficiency of water and carbon dioxide decomposition in the tracks of nuclear fission fragments is shown in Figure 12.9[9]; this is in qualitative agreement with experiments of Hartech and Doudes.[598] In principle, the highest values of energy efficiency of water and carbon dioxide decomposition in the tracks of nuclear fission fragments can be achieved if δ-electrons transfer their energy, not directly to neutral gas by different degradation processes (see Section 12.2), but rather through intermediate transfer of most of the energy to plasma formed in the tracks.

In this case, the major part of the energy of a nuclear fragment can be transferred to plasma electrons by mechanisms similar to those in the plasma–beam discharges described in Section 12.3. Then, the energy of plasma electrons can be transferred to the effective channels of stimulation of plasma chemical reactions, in particular to vibrational excitation.

Thus, nuclear fission fragments can transfer their energy in such systems mostly into formation of chemical products (in particular, in hydrogen) rather than heating.

The chemo-nuclear reactor based on such principles actually can be "cold" because only a small portion of its energy generation is related to heating. Such a nuclear reactor was considered in detail in Bochin et al.[600]

It is interesting that, in the tracks of α-particles with energies of about 1 MeV, the factor Z / v in Equation 12.66 becomes lower with respect to the case of nuclear fission fragments. Then, the pressure interval optimal for plasma radiolysis provided by α-particles is shifted to a higher-pressure range and becomes wider.

12.6 DUSTY PLASMA GENERATION BY RELATIVISTIC ELECTRON BEAM

12.6.1 General Features of Dusty Plasma Generated by Relativistic Electron Beam Propagation in Aerosols

Ionization of aerosols by electron beams is somewhat similar to photoionization of aerosols considered in Section 11.1. The key difference between these two cases is that high-energy electrons are able to ionize neutral gas and aerosols, while the energies of photons are sufficient only for ionization of dust particles with relatively low work functions.

The charge of macroparticles irradiated by an electron beam is determined by two main effects. One is related to neutral gas ionization and plasma formation, which lead to negative charging of macroparticles because electron mobility prevails over that of ions.[601,602] The second effect is related to secondary electron emission by electron impact from the aerosol surfaces, which tends toward positive charging of aerosols. Actually, competition between these two ionization effects determines the composition and characteristic of the dusty plasma irradiated by electrons.

Special interest is directed toward interaction of β-radiation (which can be considered as a low-current high-energy electron beam) with aerosols, and to plasma formation and charging related to β-active hot aerosols. Considering the interaction of β-radiation with aerosols in air, it should be taken into account that low electron fluxes usually lead to fast attachment of plasma electrons to electronegative molecules and to electron conversion into negative ions. This makes the first previously mentioned effect of negative charging less strong.[603]

The β-active hot aerosols present an interesting example of dusty plasma.[604–606] In this case, the electron density in the volume is relatively low, and the macroparticle charge becomes positive. At β-radiation rates exceeding one electron per second per particle, the total positive charge of the dust particles usually exceeds hundreds of elementary charges. For this case, analysis of macroparticles' heating and evaporation was presented by Vdovin in 1975.[631]

12.6.2 Charging Kinetics of Macroparticles Irradiated by Relativistic Electron Beam

In this system, the average charge of macroparticles is determined by electron and ion fluxes to the surface of the particles, and by secondary electron emission induced by the electron beam. Steady-state kinetics of charged particles can then be described by the following system of equations:

$$n_{eb} v_{eb} n_0 \sigma_i = \sigma_{rec}^{ei} \bar{v}_e n_e n_i + f_i(n_i, Z_a) n_a \tag{12.67}$$

$$n_{eb} v_{eb} \pi r_a^2 (\delta - 1) + f_i(n_i, Z_a) = f_e(n_e, Z_a) \tag{12.68}$$

$$n_e = n_i + Z_a n_a \tag{12.69}$$

In this system, n_0, n_a, n_i, n_e, and n_{eb} are the concentrations of gas, macroparticles, ions, plasma, and beam electrons, respectively; Z_a is the average charge of an aerosol particle; $v_{eb}, \bar{v}_e$ are the velocities of beam electrons and plasma electrons; $\sigma_i, \sigma_{rec}^{ei}$ are averaged effective cross sections of gas ionization by the beam electrons and the electron–ion recombination (effectively taking into account possible electron attachment in electronegative gases; see Section 4.5.2); δ is the coefficient of the secondary electron emission from the aerosol surfaces; $f_i(n_i, Z_a)$, $f_e(n_e, Z_a)$ are the fluxes of plasma ions and electrons on the aerosol surfaces (see Rosen[601] and Dimick and Soo[602]); and r_a is the average radius of macroparticles. Equation 12.67 describes the balance of ions; Equation 12.68 describes the macroparticles' charge; and Equation 12.69 is the quasi-neutrality condition.

12.6.3 Conditions of Mostly Negative Charging of Aerosol Particles

Most aerosols are negatively charged if the secondary electron emission from their surfaces can be neglected. Based on the preceding system of equations, this means:

$$n_{eb} v_{eb} \pi r_a^2 (\delta - 1) << f_i(n_i, Z_a) \tag{12.70}$$

Assuming a Bohm flux of ions on the aerosol surfaces (see Section 6.1.4 and Equation 6.10), and that electron–ion recombination mostly follows volume mechanisms (which takes place at moderate electron beam densities $j_{eb} > 100\ \mu A/cm^2$), the inequality (Equation 12.70) can be rewritten as a requirement of relatively low electron concentration in the beam and thus a low value of the electron beam current density:[460]

$$n_{eb} << n_0 \frac{\sigma_i}{\sigma_{rec}^{ei}} \frac{\bar{v}_e}{v_{eb}} \frac{1}{(\delta - 1)^2} \frac{m}{M} \tag{12.71}$$

where m and M are, respectively, electron and ion masses. Numerically, the criterion (Equation 12.71) of negative charging of the aerosol particles means $n_{eb} < 3 \cdot 10^7\ cm^{-3}$, if: $\bar{v}_e = 10^8$ cm/sec; $\sigma_i = 3 \cdot 10^{-18}\ cm^2$; $v_{eb} = 3 \cdot 10^{10}$ cm/sec, $\sigma_{rec}^{ei} = 10^{-16} cm^2$, $\delta = 100$ (which corresponds to a particle radius $r_a = 10\ \mu m$); $m/M = 10^{-4}$, $n_0 = 3 \cdot 10^{19}\ cm^{-3}$.

According to Equation 12.71, the electron beam current density should be relatively low ($j_{eb} < 0.1\ A/cm^2$) to have mostly negative charging of the aerosol particles and to have the possibility to neglect the secondary emission from the macroparticle surfaces.

12.6.4 Conditions of Balance between Negative Charging of Aerosol Particles by Plasma Electrons and Secondary Electron Emission

At higher electron beam current densities, the contribution of secondary electron emission becomes more important. At the critical value of electron concentration in the beam:

$$n_{eb0} = n_0 \frac{\sigma_i}{\sigma_{rec}^{ei}} \frac{\overline{v}_e}{v_{eb}(\delta - 1)^2} \left[1 - \frac{n_a}{n_0} \frac{\pi r_a^2}{\sigma_i} \frac{\overline{v}_i}{\overline{v}_e} (\delta - 1) \right] \tag{12.72}$$

and corresponding electron beam current density, the average charge of aerosol particles becomes equal to zero $Z_a = 0$. In Equation 12.72, the average ion velocity $\overline{v}_i \approx 10^5$ cm/sec, and the aerosol particle concentration n_a is to be less than 10^9 cm^{-3}. Expressions for the electron and ion fluxes $f_i(n_i, Z_a)$, $f_e(n_e, Z_a)$ become simple if macroparticles are not charged; this makes solving Equation 12.67 through Equation 12.69 not difficult. At the preceding values of parameters, Equation 12.72 gives $n_{eb0} = 3 \cdot 10^{11}$ cm^{-3}. This means that the contribution of secondary electron emission becomes significant and the average charge of macroparticles becomes equal to zero, when the current density of the relativistic electron beam j_{eb} increases to $j_{eb0} = 10^3$ A/cm^2.

In this case of uncharged aerosol particles, plasma density in the heterogeneous system can be given by:

$$n_{e0} = n_0 \frac{\sigma_i}{\sigma_{rec}^{ei}} \frac{1}{\delta - 1} \tag{12.73}$$

which numerically gives $n_{e0} \approx 10^{16}$ cm^{-3}.

12.6.5 Regime of Intensive Secondary Electron Emission: Conditions of Mostly Positive Charging of Aerosol Particles

If $j_{eb} < j_{eb0}$, aerosol particles are negatively charged. When the electron beam current density exceeds the critical value j_{eb0} determined by Equation 12.72, the contribution of the secondary electron emission becomes so significant that the aerosol particles become positively charged. Consider this case, assuming that the total charge of aerosol particles does not significantly affect the balance of electron and ion concentrations:

$$Z_a n_a << n_e \approx n_i \tag{12.74}$$

When macroparticles are positively charged, they attract electrons. The electrons flux to the aerosol surface can be estimated as:

$$f_e\left(n_e, Z_a > 0\right) \approx n_e \overline{v}_e r_a^2 \left(1 + \frac{Z_a e^2}{4\pi\varepsilon_0 r_a T_e} \right) \tag{12.75}$$

where T_e is the temperature of plasma electrons. Equation 12.67 through Equation 12.69 can be rewritten in this case as:

$$n_{eb}v_{eb}n_0\sigma_i = \sigma_{rec}^{ei} n_e^2 \bar{v}_e \tag{12.76}$$

$$n_{eb}v_{eb}(\delta - 1) = n_e\bar{v}_e\left(1 + \frac{Z_a e^2}{4\pi\varepsilon_0 r_a T_e}\right) \tag{12.77}$$

This system of equations permits finding an expression for the average positive charge of an aerosol particle in the following form:

$$Z_a = \frac{4\pi\varepsilon_0 r_a T_e}{e^2}\left(\sqrt{\frac{j_{eb}}{j_{eb0}}} - 1\right) \tag{12.78}$$

Obviously, $Z_a = 0$ when electron beam current density is equal to the critical one, $j_{eb} = j_{eb0}$. The concentration of plasma electrons and ions in this regime can be expressed as follows:

$$n_e = n_{e0}\sqrt{\frac{j_{eb}}{j_{eb0}}} \tag{12.79}$$

The sufficient condition of equality of electron and ion densities (Equation 12.74) now can be rewritten in the following simple form:

$$\frac{4\pi\varepsilon_0 r_a T_e}{e^2}\frac{n_a}{n_{e0}} < 1 \tag{12.80}$$

which is valid at the previously considered values of parameters.

One should note that the time necessary for establishing these steady-state conditions of a relativistic electron beam interaction with aerosols can be found as:

$$\tau = \frac{n_{e0}}{n_{eb0}n_0\sigma_i v_{eb}} = \frac{1}{n_{eb0}v_{eb}\sigma_{rec}^{ei}(\delta - 1)} \tag{12.81}$$

This time interval can be estimated as $\tau \approx 10\,n\,\mathrm{sec}$. Thus, these results for steady-state systems also can be applied for the pulsed and pulse-periodic relativistic electron beams, where a typical pulse duration is about 100 nsec.

12.6.6 Electron Beam Irradiation of Aerosols in Low-Pressure Gas

In the case of low pressure ($n_0 \to 0$), the positive ions in the volume are actually absent and the aerosol particles are positively charged at any values of the electron beam current density. Then, the general system of equations (Equation 12.67 through Equation 12.69) can be simplified to :

$$n_{eb}v_{eb}\pi r_a^2(\delta - 1) = f_e(n_e, Z_a) \tag{12.82}$$

$$n_e = n_a Z_a \tag{12.83}$$

Because macroparticles are positively charged in these conditions, they attract electrons; the electrons' flux to the aerosol surface $f_e(n_e, Z_a > 0)$ can also be estimated in this case using Equation 12.75. Then, the average positive charge of an aerosol particle Z_a in a low-pressure gas can be calculated from the following formula:

$$Z_a = \frac{n_e}{n_a} = \min\left\{\frac{n_{eb}}{n_a}\frac{v_{eb}}{\bar{v}_e}(\delta - 1);\ \sqrt{\frac{n_{eb}}{n_a}\frac{v_{eb}}{\bar{v}_e}(\delta - 1)\frac{4\pi\varepsilon_0 r_a T_e}{e^2}}\right\} \tag{12.84}$$

The approach in general ($n_0 \to 0$) and Equation 12.84 are valid only if $n_0 << Z_a n_a$, where the average macroparticle charge Z_a is determined from Equation 12.74.

PROBLEMS AND CONCEPT QUESTIONS

12.1 **Relativistic Effects in Electron Energy Losses Function *L*(*E*).** Using the Bethe formula (Equation 12.2), analyze the relativistic effect on electron energy losses related to stopping a relativistic electron beam in a neutral gas. Give your interpretation and estimation of the effect, and calculate values of the energy losses function $L(E)$ in air for nonrelativistic 10-keV electrons and relativistic 1-MeV electrons.

12.2 **Ionization Energy Cost at Gas Irradiation by High-Energy Electrons.** Using Table 12.1, calculate the fraction of the electron beam energy going to excitation of neutral species in different molecular and atomic gases. Analyze Table 12.1 and explain why the ionization energy cost in molecular gases exceeds that for inert gases. Why is the ionization energy cost in helium the largest one between inert gases?

12.3 **Stopping Length of Relativistic Electron Beams in Atmospheric Air.** Using Equation 12.4, calculate the initial energy of a relativistic electron necessary to provide its stopping length along a trajectory exceeding 1 m. Calculate the relativistic factor for this electron. Estimate the stopping length along the beam axis in this case.

12.4 **Critical Alfven Current for Electron Beam Propagation.** Derive Equation 12.6 for the limiting case of a relativistic electron beam propagation related to self-focusing in its own magnetic field. Take into account that the Alfven current gives the inherent magnetic field of the beam, which makes the Larmor radius (Equation 6.51) less than half the beam radius.

12.5 **Bethe–Born Approximation of Degradation Spectrum.** Derive the solution (Equation 12.18) of Alkhazov's equation in the Bethe–Born approximation, taking into account only ionization by primary electrons. Analyze the derived expression of the degradation spectrum. Explain why this simple formula gives only qualitative features of the degradation spectrum.

12.6 **Analytical Solutions of Alkhazov's Equation.** Compare the analytical solutions of the Alkhazov's equation (12.19) and (12.20) for electron energies above and below the excitation threshold. Using the degradation spectrum

(12.20) and its relation with electron energy distribution function, calculate the average electron energy under the excitation threshold.

12.7 **Energy Cost of Ionization and Excitation at Electron Beam Stopping in Gases**. Using the data presented in Table 12.2, calculate the energy cost of ionization and excitation of electronically and vibrational excited states in hydrogen irradiated by electron beam with an energy of 1 MeV and 1 keV. Compare the energy costs of ionization and vibrational excitation for 1-MeV and 1-keV electron beams and interpret the result.

12.8 **Operation Conditions of Plasma–Beam Discharges**. The principal operational condition (Equation 12.21) of plasma–beam discharges requires the beam energy dissipation into Langmuir oscillations to exceed the beam energy losses in electron–neutral collisions. Actually, the criterion (Equation 12.21) determines the upper limit of effective operation of the plasma–beam discharges. Give a numerical estimation of the pressure limit, taking the characteristic beam parameters from Section 12.21.

12.9 **Limitation of Langmuir Noises for Efficient Vibrational Excitation in Plasma–Beam Discharges**. Vibrational excitation of molecules requires sufficient values of specific power of plasma–beam discharges, which leads to the limitation (Equation 12.25) of the minimal level of Langmuir noise W/n_eT_e. Derive the criterion (Equation 12.25) and analyze the numerical value of the minimal level of Langmuir noise W/n_eT_e.

12.10 **Plasma–Beam Discharge Parameters in Molecular Gases**. Analyze and explain the dependence of the degrees of ionization, level of Langmuir noise W/n_eT_e, and average electron energy in the plasma–beam discharge in hydrogen as a function of vibrational temperature; this is shown in Figure 12.4 for two different values of the parameter W/p. Explain why higher electron concentrations (degrees of ionization) correspond to lower values of W/n_eT_e and to a lower level of average electron energy in plasma–beam discharges.

12.11 **Plasma–Beam Discharge in Crossed Electric and Magnetic Fields**. In the plasma centrifuge based on the plasma–beam discharge, the radial electric field is provided directly by the electron beam. Derive a relation for the drift velocity in the crossed electric and magnetic fields in this system as a function of magnetic field and the electron beam current density. Use the derived relation to calculate the plasma rotation velocity for the plasma–beam discharge parameters given in Section 12.3.5. Compare the result with experimentally measured values.

12.12 **Magnetized Ions' Separation Effect in Plasma Centrifuge Based on Plasma–Beam Discharge**. Analyze the ions' displacement in the radial direction at the moment of their formation due to the combined effect of the electric and magnetic fields. Derive Equation 12.26, and explain why the velocity in this relation corresponds, not to the thermal one, but rather to the velocity of the centrifugal drift, in the crossed electric and magnetic fields.

12.13 **Stable Regimes of Non-Self-Sustained Discharges with Ionization Provided by High-Energy Electron Beams**. The relatively high levels of the discharge specific power given by Equation 12.32 can be stable only at limited values of the discharge durations. The stable values of current densities and pulse durations of the non-self-sustained discharges are shown in Figure 12.8. Using data presented in this figure, estimate the maximum values of the specific energy input for the pulse discharge sustained by a relativistic electron beam.

12.14 **Uniformity of Non-Self-Sustained Discharges with Ionization Provided by High-Energy Electron Beams**. High external electric field can result in a decrease of electron concentration near the cathode and thus leads to elevated electric field in the layer. Using Equation 12.38, calculate the maximum external electric field (at atmospheric pressure, electron beam current density = 10^3 A/cm^2 and $d = 100$ cm), which does not lead in the steady-state approximation to essential additional ionization in the cathode layer.

12.15 **Plasma Radiolysis of Carbon Dioxide**. The plasma effect in CO_2 radiolysis is related to stimulation of the reaction $O + CO_2 \rightarrow CO + O_2$ by vibrational excitation of CO_2 molecules, which then prevails over three-molecular recombination $O + O + M \rightarrow O_2 + M$. Determine the vibrational temperature of CO_2 molecules necessary for the plasma effect, taking into account that the necessary degrees of ionization should satisfy the requirement $n_e/n_0 >> 10^{-3}$.

12.16 **Initial Tracks of Nuclear Fission Fragments**. Derive Equation 12.57 for the radius of the initial completely ionized plasma channel formed by the tracks of the nuclear fission fragments. Explain why the plasma in these channels is completely ionized. Using the derived equation, calculate the radius of the initial tracks and compare it with the mean free path of electrons and the length necessary for ionization and electronic excitation provided by the δ-electrons.

12.17 **Tracks and Plasma Effects in Radiolysis Provided by α-Particles**. Using Equation 12.66 and Equation 12.58, estimate the pressure range optimal for plasma radiolysis in water vapor and carbon dioxide provided by α-particles with energies of about 1 MeV. Compare the physical and chemical characteristics of the tracks made by α-particles and by nuclear fission fragments.

12.18 **Relativistic Electron Beam in Aerosols**. Analyze the system of equations (Equation 12.67 through Equation 12.69) describing the charging of aerosol particles in a plasma generated by a relativistic electron beam for the case when the predominant electron flux to the aerosols is balanced by secondary electron emission and the average charge of macroparticles is equal to zero. Derive the expressions for electron beam current density, which are related to Equation 12.72 and to plasma density (Equation 12.73). Take into account that expressions for the electron and ion fluxes $f_i(n_i, Z_a), f_e(n_e, Z_a)$ can be taken in this case in a simple way without considering electrostatic effects related to aerosol particles.

References

1. Kondratiev, V. N., ed. 1974. *Energy of Chemical Bonds. Ionization Potential and Electron Affinity*. Moscow: Nauka Science.
2. Smirnov, B. M. 2001. *Physics of Weakly Ionized Gases. Physics of Ionized Gases*. New York: John Wiley & Sons.
3. McDaniel, E. W. 1964. *Collision Phenomena in Ionized Gases*. New York: Wiley.
4. McDaniel, E. W. 1989. *Atomic Collisions: Electron and Photon Projectiles*. New York: Wiley.
5. Massey, H. 1976. *Negative Ions*. Cambridge: Cambridge University Press.

5a. Massey, H., Burhop, E., and Gilbody, H. B. 1974. *Electron and Ion Impact Phenomena*. Oxford: Clarendon Press.

6. Landau, L. D. and Lifshitz, E. M. 1997. *The Classical Theory of Fields*. Oxford: Butterworth–Heinemann.

6a. Tomson, J. J. 1912. *Philos. Mag*. 23:449.

6b. Tomson, J. J. 1924. *Philos. Mag*. 47:337.

7. Smirnov, B. M. 1974. *Ions and Excited Atoms in Plasma*. Moscow: Atomizdat.
8. Biondi, M. A. 1976. In *Principles of Laser Plasma*, ed. G. Bekefi. New York: John Wiley & Sons.
9. Rusanov, V. D. and Fridman, A. 1984. *Physics of Chemically Active Plasma*, Moscow: Nauka.
10. Virin, L., Dgagaspanian, R., Karachevtsev, G., Potapov, V., and Talrose, V. 1978. *Ion–Molecular Reactions in Gases*, Moscow: Nauka.
11. Talrose, V. L., Vinogradov, P. S., and Larin, I. K. 1979. *Gas Phase Ion Chemistry*, vol. 1, ed. M. Bowers. Academic Press, 305.
12. Zeldovich, Ya. B. and Raizer, Yu. P. 1966. *Physics of Shock Waves and High Temperature Hydrodynamic Phenomena*, New York: Academic Press.

12a. Adamovich, I., Saupe, S., Grassi, M. J., Schulz, O., Macheret, S., and Rich, J. W. 1993. *Chem. Phys*. 174:219.

13. Lieberman, M. A. and Lichtenberg, A. J. 1994. *Principles of Plasma Discharges and Material Processing*, New York: John Wiley & Sons.
14. Bloch, F. and Bradbury, N. 1935. *Phys. Rev*. 48:689–696.
15. Alexandrov, N. 1981. In *Plasma Chemistry—8*, ed. B. M.Smirnov. Moscow: Energoizdat.
16. Smirnov, B. M. 1982. *Negative Ions*, New York: McGraw–Hill.
17. Inokuti, M., Kim, Y., and Platzman, R. L. 1967. *Phys.Rev*. 164:55–61.
18. Natanson, G. 1959. *Sov. Phys. J. Tech. Phys*. 29:1373–1378.
19. Kirillov, I. A., Polak, L. S., and Fridman, A. 1984. *IYth USSR Symp. Plasma Chem*. Dnepropetrovsk. 1:101.
20. Talrose, V. L. 1952. Ph.D. Dissertation, N. N. Semenov Institute of Chemical Physics, Moscow; also, 1952. *Sov. Phys. Doklady*. 86:909.

20a. Plonjes, E., Palm, P., Rich, J. W., and Adamovich, I. 2001. 32nd AIAA Plasmadynamics and Lasers Conference, AIAA-2001–3008, Anaheim, CA.

21. Su, T. and Bowers, M. T. 1975. *Int. J. Mass Spectrom. Ion Phys*. 17:221.
22. Daniel, J. and Jacob, J. 1986. *J. Geophys. Res*. 91:9807.
23. Potapkin, B. V., Deminsky, M., Fridman, A., and Rusanov, V. D. 1995. *Radiat. Phys. Chem*. 45:1081–1088.
24. Potapkin, B. V., Rusanov, V. D., and Fridman, A. 1989. *Sov. Phys. Doklady*. 308:897.
25. Mutaf–Yardimci, O., Saveliev, A., Fridman, A., and Kennedy, L. 1998. *Int. J. Hydrogen Energy* 2312;1109.
26. Garscadden, A. 1994. *Pure Appl. Chem*. 66:1319.
27. Howling, A. A., Dorier, J. L., and Hollenstein, Ch. 1993. *Appl. Phys. Lett*. 62:1341.
28. Fridman, A., Boufendi, L., Bouchoule, A., Hbid, T., and Potapkin, B. 1996. *J. Appl. Phys*. 79:1303–1314.
29. Eyring, H., Lin, S. H., and Lin, S. M. 1980. *Basic Chemical Kinetics*, New York: John Wiley & Sons.

30. Kondratiev, V. N. and Nikitin, E. E. 1981. *Chemical Processes in Gases*, New York: John Wiley & Sons.
31. Landau, L. D. 1997. *Quantum Mechanics*, 3rd. ed., Oxford: Butterworth–Heinemann.
33. Huber, K. P. and Herzberg, G. 1979. *Molecular Spectra and Molecular Structure IV, Constants of Diatomic Molecules*, New York: Van Nostrand Reinhold Company.
34. Radzig, A. A. and Smirnov, B. M. 1980. *Handbook on Atomic and Molecular Physics*, Moscow: Atomizdat.
35. Herzenberg, A. 1968. *J. Phys. B: Atomic and Molecular Physics*, 1:548–553.
36. Rusanov, V. D., Fridman, A., and Sholin, G. 1986. Vibrational kinetics and reactions of polyatomic molecules. In *Topics of Current Physics*—39, ed. M. Capitelli, Berlin: Springer–Verlag.
37. Rusanov, V. D., Fridman, A., and Sholin, G. 1981. *Adv. Phys. Sci., Uspehi Phys. Nauk*. 134:185.
38. Zaslavskii, G. and Chirikov, B. 1971. *Adv. Phys. Sci., Uspehi Phys. Nauk*. 105:3.
39. Platonenko, V. and Sukhareva, N. 1980. *Sov. Phys. JETP, J. Exp. Theor. Phys*. 78: 2126.
40. Makarov, A. and Tyakht, V. 1982. Sov. Phys. JETP, *J. Exp. Theor. Phys*. 83:502.
41. Schultz, G. J. 1976. In *Principles of Laser Plasma*, ed. G. Bekefi, New York: John Wiley & Sons.
42. Crowford, O. H. 1967. Mol. Phys. 13:181–185.
43. Gerjoy, E. and Stein, S. 1955. Phys. Rev. 95:1971–1976.
44. Drawin, H. W. 1968. *Ztschr.*, Bd. 211, S:404–408.
45. Drawin, H. W. 1969. *Ztschr.*, Bd. 255, S:470–475.
46. Slivkov, I. N. 1966. *Electric Breakdown and Discharge in Vacuum*, Moscow: AtomIzdat.
47. Kochetov, I. V., Pevgov, V. G., Polak, L. S., and Slovetsky, D. I. 1979. In *Plasma Chemical Reactions*, ed. L. S. Polak, Moscow: Russian Academy of Science.
48. Nikitin, E. E. 1970. *Theory of Elementary Atom–Molecular Processes in Gases*, Moscow: Chimia Chemistry.
49. Gorgiets, B. F., Osipov, A. I., and Shelepin, L. A. 1980. *Kinetic Processes in Gases and Molecular Lasers*, Moscow: Nauka Science; B. Gordiets. 1988. Gordon and Breach Science Pub.
50. Landau, L. D. and Teller, E. 1936. *Phys. Ztschr. Sow.*, Bd. 10, S:34–39.
51. Schwartz, R. N., Slawsky, Z. I., and Herzfeld, K. F. 1952. *J. Chem. Phys*. 20:1591.
52. Billing, G.D. 1986. Vibrational kinetics and reactions of polyatomic molecules. In *Topics of Current Physics—39*, ed. M. Capitelli, Berlin: Springer–Verlag.
53. Lifshitz, A. 1974. *J. Chem. Phys*. 61:2478–2483.
54. Millican, R. S. and White, D. R. 1963. *J. Chem. Phys*. 39:3209–3215.
55. Andreev, E. A. and Nikitin, E. E. 1976. In *Plasma Chemistry—3*, ed. B. M. Smirnov, Moscow: Atomizdat.
56. Gershenson, Yu., Rosenstein, V., and Umansky, S. 1977. In *Plasma Chemistry—4*, ed. B. M. Smirnov, Moscow: Atomizdat.
57. Rockwood, S. D., Brau, J. E., Proctor, W. A., and Canavan, G. H. 1973. *IEEE J. Quantum Electron*. 9:120.
58. Geffers, W. Q. and Kelley, J. D. 1971. *J. Chem. Phys*. 55:4433.
59. Rusanov, V. D., Fridman, A., and Sholin, G. 1979. *Sov. Phys. J. Tech. Phys*. 49:2169.
60. Rapp, D. 1965. *J. Chem. Phys*. 43:316.
60a. Rapp, D. and Francis, W. E. 1962. *J. Chem. Phys*. 37:2631.
61. Makarov, A., Puretzky, A., and Tyakht, V. 1980. *J. Appl. Phys*. 23:391
62. Landau, L. D. and Lifshitz, E. M. 1980. *Statistical Physics*, Part 1, Oxford: Butterworth–Heinemann.
63. Parker, J. G. 1959. *Phys. Fluids* 2:449.
64. Bray, C. A. and Jonkman, R. M. 1970. J. *Chem.Phys*. 52:477.
65. Bjerre, A. and Nikitin, E. E. 1967. *Chem. Phys. Lett*. 1:179.
66. Levine, R. D. and Bernstein, R. 1978. In *Dynamics of Molecular Collisions*, ed. W. Miller, New York: Plenum Press.
67. Le Roy, R. L. 1969. *J. Phys. Chem*. 73:4338.
68. Baulch, D. L. 1992–1994. Evaluated kinetic data..., series of publications in *J. Phys. Chem. Ref. Data*.
69. Kondratiev, V. N. 1971. *Rate Constants of Gas Phase Reactions*, Moscow: Nauka Science.
70. Sato, S. J. 1955. *J. Chem. Phys*. 23:2465.
71. Sabo, Z. G. 1966. *Chemical Kinetics and Chain Reactions*, Moscow: Nauka Science.

72. Tihomirova, N. and Voevodsky, V. V. 1949. *Sov. Phys. Doklady* 79:993.
73. Kagija, N. 1969. *Bull. Chem. Soc. Jpn.* 42:1812.
74. Polishchuk, A., Rusanov, V. D., and Fridman, A. 1980. *Theor. Exp. Chem.* 16:232.
75. Rusanov, V. D., Fridman, A., and Sholin, G. V. 1978. In *Plasma Chemistry—5*, ed. B. M. Smirnov, Moscow: Atomizdat.
76. Levitsky, A. A., Macheret, S. O., and Fridman, A. 1983. *High Energy Chem.* 17:625.
77. Macheret, S. O., Rusanov, V. D., and Fridman, A. 1984. *Sov. Phys. Doklady* 276:1420.
78. Secrest, D. 1973. *Annu. Rev. Phys. Chem.* 24:379.
79. Smirnov, B. M. 1968. *Atomic Collisions and Elementary Processes in Plasma*, Mo ow: Atomizdat.
80. Engelgardt, A. G., Felps, A. V., and Risk, G. G. 1964. *Phys. Rev.* 135:1566.
81. Capitelli, M. 1986., ed. Non-equilibrium vibrational kinetics. In *Topics in Current Physics—39*, Berlin: Springer–Verlag.
81a. Capitelli, M. 2000. ed.,Plasma kinetics in atmospheric gases. In *Springer Series on Atomic, Optical and Plasma Physics*, vol. 31, Berlin: Springer–Verlag.
82. Givotov, V. K., Rusanov, V. D., and Fridman, A. 1982, 1984. In *Plasma Chemistry—9* and *Plasma Chemistry—11*, ed. B. M. Smirnov, Moscow: Atom–Izdat.
83. Losev, S. A., Sergievska, A. L., Rusanov, V. D., Fridman, A., and Macheret, S. O. 1996. *Sov. Phys. Doklady* 346:192.
84. Macheret, S. O., Fridman, A., Adamovich, I. V., Rich, J. W., and Treanor, C. E. 1994. Mechanism of non-equilibrium dissociation of diatomic molecules, AIAA-Paper # 94-1984.
85. Park, C. 1987. Assessment of two-temperature kinetic model for ionizing air, AIAA-Paper #87-1574.
86. Losev, S. A. and Generalov, N. A. 1961. *Sov. Phys. Doklady* 141:1072.
87. Marrone, P. V. and Treanor, C. E. 1963. *Phys. Fluids* 6:1215.
88. Sergievska, A. L., Kovach, E. A., and Losev, S. A. 1995. *Mathematical Modeling in Physical Chemical Kinetics*, Moscow State University, Institute of Mechanics.
89. Nester, S., Demura, A. V., and Fridman, A. 1983. Disproportioning of the vibrationally excited CO-molecules, Kurchatov Institute of Atomic Energy, #3518/6, Moscow.
90. Boulos, M. I., Fauchais, P., and Pfender, E. 1994. *Thermal Plasmas. Fundamentals and Applications*, New York, London : Plenum Press.
91. Mayer, J. E. and Mayer, G. M. 1966. *Statistical Mechanics*, 11th ed., New York: Wiley.
92. Drawin, H. W. 1972. The thermodynamic properties of the equilibrium and non-equilibrium states of plasma. In *Reactions under Plasma Conditions*, ed. M. Venugopalan, New York: Wiley.
93. Nester, S., Potapkin, B., Levitsky, A., Rusanov, V., Trusov, B., and Fridman, A. 1988. *Kinetic and Statistic Modeling of Chemical Reactions in Gas Discharges*, Moscow: CNII ATOM INFORM.
94. Slovetsky, D. I. 1980. *Mechanisms of Chemical Reactions in Non-Equilibrium Plasma*, Moscow: Nauka.
95. Veprek, S. 1972. *J. Chem. Phys.* 57:952, and *J. Crystal Growth* 17:101.
96. Legasov, V. A., Rusanov, V. D., and Fridman, A. 1978. In *Plasma Chemistry—5*, ed. B. M. Smirnov, Moscow: Atomizdat.
96a. Legasov, V. A., Asisov, R. I., and Butylkin, Yu. P. 1983. In *Nuclear–Hydrogen Energy and Technology—5*, ed. V. A. Legasov, 71, Moscow: Energo–Atom–Izdat.
97. Kuznetzov, N. M. 1971. *J. Tech. Exp. Chem. Tekh. Eksp. Khim.* 7:22.
97a. Kuznetsov, N. M. and Raizer, Yu. P. 1965. *Sov. Phys.J. Appl. Mech. Tech. Phys.* 4:10.
98. Treanor, C. E., Rich, I. W., and Rehm, R. G. 1968. *J. Chem. Phys.* 48:798.
99. Lifshitz, E. M. and Pitaevsky, L. P. 1979. *Physical Kinetics, Theoretical Physics*, vol. 10, eds. L. D. Landau and E. M. Lifshitz, Moscow: Nauka Science.
100. Krall, N. A. and Trivelpiece, A. W. 1973. *Principles of Plasma Physics*, New York: McGraw–Hill.
101. Raizer, Yu. P. 1991. *Gas Discharge Physics*, Berlin, Heidelberg, New York : Springer–Verlag.
102. Uman, M. A. 1964. *Introduction to Plasma Physics*, New York: McGraw–Hill.
103. McDaniel, E. W. and Mason, E. A. 1973. *The Mobility and Diffusion of Ions in Gases*, New York: Wiley.
104. Huxley, L. G. H. and Crompton, R. W. 1974. *The Diffusion and Drift of Electron in Gases*, New York: Wiley.

105. Eletsky, A. V., Palkina, L. A., and Smirnov, B. M. 1975. *Transfer Phenomena in Weakly Ionized Plasma*, Moscow: Atomizdat.
106. Raether, H. 1964. *Electron Avalanches and Breakdown in Gases*, London: Butterworth & Co.
107. Loeb, L. B. 1960. Basic Processes of Gaseous Electronics, Berkley: University of California Press.
108. Meek, J. M. and Craggs, J. D. 1978. *Electrical Breakdown of Gases*, New York: Wiley.
109. Lozansky, E. D. and Firsov, O. B. 1975. *Theory of Sparks*, Moscow: Atomizdat.
110. Dawson, G. A. and Winn, W. P. 1965. *Zs. Phys.* 183:159.
111. Gallimberti, I. 1972. *J. Phys. D., Applied Phys.* 5:2179.
112. Klingbeil, R. D., Tidman, A., and Fernsler, R. F. 1972. *Phys. Fluids* 15:1969.
113. Baselyan, E. M. and Goryunov, A. Yu. 1986. *Electrichestvo J. Electricity* 11:27.
114. Gallimberti, I. 1977. *Electra* 76:5799.
115. Raizer, Yu. P. 1974. *Laser Spark and Discharge Propagation*, Moscow: Nauka Science.
116. Baranov, V. Yu., Vedenov, A. A., and Niziev, V. G. 1972. *Sov. Phys. Thermophys. High Temp.* 10:1156.
117. Munt, R., Ong, R. S. B., and Turcotte, D. L. 1969. *J. Plasma Phys.*11:739.
118. Liventsov, V., Rusanov, V., Fridman, A., and Sholin, G. 1981. *Sov. Phys. J. Tech. Phys. Lett.* 9:74.
119. Boutylkin, Yu. P., Givotov, V. K., Krasheninnikov, E. G., Rusanov, V. D., and Fridman, A. 1981. Sov. Phys.J. Tech. Phys. 51:925.
120. Nikitin, E. E. and Osipov, A. I. 1977. *Vibrational Relaxation in Gases*, Moscow: VINITI.
121. Likalter, A. A. and Naidis, G. V. 1981. In *Plasma Chemistry—8*, ed. B. M. Smirnov, Moscow: Energoizdat.
122. Gordiets, B. and Zhdanok, S. 1986. Non-equilibrium vibrational kinetics, In *Topics in Current Physics—39*, ed. M. Capitelli, Berlin: Springer–Verlag.
123. Rusanov, V. D., Fridman, A., and Sholin, G. 1979. *Sov. Phys. J. Tech. Phys.* 49:554.
124. Demura, A. V., Mahceret, S. O., and Fridman, A. 1984. *Sov. Phys. Doklady* 275:603.
125. Rusanov, V. D. and Fridman, A. 1976. *Sov. Phys. Doklady* 231:1109.
126. Gordiets, B. F., Osipov, A. I., Stupochenko, E. B., and Shelepin, L. A. 1972. *Adv. Phys. Sci. Uspehi Phys. Nauk* 108:655.
127. Macheret, S .O., Rusanov, V. D., Fridman, A., and Sholin, G. V. 1979. *Plasma Chemistry 1979, III Symposim on Plasma Chemistry*, 101, Moscow: Nauka Science.
128. Sergeev, P. A. and Slovetsky, D. I. 1979. *Plasma Chemistry 1979, III Symposium on Plasma Chemistry*, 132, Moscow: Nauka Science.
129. Likalter, A. A. 1975. *Sov. Phys. Prikl. Mech. Tech. Phys. Appl. Mech. Tech. Phys.* 3:8.
130. Likalter, A. A. 1976. *Sov. Phys. Prikl. Mech. Tech. Phys. Appl. Mech. Tech. Phys.* 4:3.
131. Likalter, A. A. 1975. *Sov. Phys. Kvantovaya Electronica, Quantum Electron.* 2:2399.
132. Zhdanok, S. A., Napartovich, A. P., and Starostin, A. N. 1979. *Sov. Phys. J. Exp. Theor. Phys.* 76:130.
133. Fridman, A. and Rusanov, V.D. 1994. *Pure Appl. Chem.* 66(6):1267.
134. Givotov, V. K., Rusanov, V. D., and Fridman, A. 1985. *Diagnostics of Non-Equilibrium Chemically Active Plasma*, Moscow: Energo–Atom–Izdat.
134a. Potapkin, B. V., Rusanov, V. D., Samarin, A .E., and Fridman, A. 1980. *Sov. Phys. High Energy Chem.* 14:547.
134b. Rusanov, V. D., Fridman, A., and Sholin, G. V. 1977. *Sov. Phys. Doklady* 237:1338.
135. Losev, S. A., Shatalov, O. P., and Yalovik, M. S. 1970. *Sov. Phys. Doklady* 195:585.
136. Macheret, S. O., Rusanov, V. D., Fridman, A., and Sholin, G. V. 1980. *Sov. Phys. J. Tech. Phys.* 50:705.
137. Rich, J. W. and Bergman, R. C. 1986. Isotope separation by vibration–vibration pumping. In *Topics of Current Physics—39*, ed. M. Capitelli, Berlin: Springer–Verlag.
138. Bergman, R. C., Homicz, G. F., Rich, J. W., and Wolk, G.L. 1983. *J. Chem. Phys.* 78:1281.
139. Belenov, E. M., Markin, E. P., Oraevsky, A. N., and Romanenko, V. I.1973. *Sov. Phys. J. Exp. Theor. Phys. Lett.* 18:116.
140. Akulintsev, V. M., Gorshunov, V. M., and Neschimenko, Y. P. 1977. *Sov. Phys. J. Appl. Mech. Tech. Phys.* 18:593.
141. Eletsky, A. V. and Zaretsky, N. P. 1981. *Sov. Phys. Doklady* 260:591.

142. Margolin, A. D., Mishchenko, A. V., and Shmelev, V. M. 1980. *Sov. Phys. High Energy Chem.* 14:162.
143. Biberman, L. M., Vorobiev, V. S., and Yakubov, I. T. 1982. *Kinetics of the Non-Equilibrium Low-Temperature Plasma*, Moscow: Nauka Science.
144. Gudzenko, L .I. and Yakovlenko, S. I. 1978. *Plasma Lasers*, Moscow: Atomizdat.
145. Beliaev, S. T. and Budker, G. I. 1958. *Plasma Physics and Problems of the Controlled Thermo-Nuclear Reactions*, vol. 3, Moscow: Academy of Sciences of the USSR.
146. Biberman, L. M., Vorobiev, V. S., and Yakubov, I. T. 1979. *Adv. Phys. Sci. Uspehi Phys. Nauk* 128:233.
147. Safarian, M. N. and Stupochenko, E. V. 1964. *Sov. Phys. J. Appl. Mech. Tech. Phys.* 7:29.
148. Anderson, H. C., Oppenheim, I., and Shuler, K. E. 1964. *J. Math. Phys.* 5:522.
149. Vakar, A. K., Givotov, V. K., Krasheninnikov, E. G., and Fridman, A. 1981. *Sov. Phys. J. Tech. Phys. Lett.* 7:996.
150. Ahmedganov, R. A., Bykov, Yu. V., Kim, A. V., and Fridman, A. 1986. *J. Inst. Appl. Phys.* (Gorky, USSR) 147:20.
151. Grigorieva, T. A., Levitsky, A. A., Macheret, S. O., and Fridman, A. 1984. *Sov. Phys, High Energy Chem.* 18:336.
152. Piley, M. E. and Matzen, M. K. 1975. *J. Chem. Phys.* 63:4787.
153. Baranchicov, E. I., Denisenko, V. P., and Rusanov, V. D. 1990. *Sov. Phys. Doklady* 339:1081.
154. Polak, L. S., Ovsiannikov, A. A., Slovetsky, D. I., and Vursel, F. B. 1975. *Theoretical and Applied Plasma Chemistry*, Moscow: Nauka Science.
155. Givotov, V. K., Krotov, M. F., Rusanov, V. D., and Fridman, A. 1981. *Int. J. Hydrogen Energy* 6:441.
156. Asisov, R. I., Vakar, A., Fridman, A., and Givotov, V. K. 1985. *Int. J. Hydrogen Energy* 10:475.
156a. Asisov, R. I., Krotov, M. F., and Potapkin, B. V. 1983. *Sov. Phys. Doklady* 271:94.
157. Conti, S., Fridman, A., and Raoux, S. 1999. *14th International Symposium on Plasma Chemistry* ISPC 14.
158. Bochin, P., Legasov, V. A., Rusanov, V. D., and Fridman, A. 1978. In *Nuclear–Hydrogen Energy and Technology* vol.1, ed. V. A. Legasov, 183, Moscow: Atom–Izdat.
158a. Bochin, V. P., Legasov, V. A., Rusanov, V. D., and Fridman, A. 1979. In *Nuclear–Hydrogen Energy and Technology* vol. 2, ed. V. A. Legasov, 206, Moscow: Atom–Izdat.
159. Losev, S. A. 1977. *Gas-Dynamic Lasers*, Moscow: Nauka Science.
160. Kurochkin, Yu. V., Polak, L. S., and Pustogarov, A. V. 1978. *Sov. Phys. Thermal Phys. High Temp.* 16:1167.
161. Liventsov, V. V., Rusanov, V. D., and Fridman, A. 1983. *Sov. Phys. J. Tech. Phys. Lett.* 7:163.
162. Liventsov, V. V., Rusanov, V. D., and Fridman, A. 1984. *Sov. Phys. Doklady* 275:1392.
163. Rusanov, V. D., Fridman, A., and Sholin, G. V. 1982. In *Heat and Mass Transfer in Plasma-Chemical Systems*, vol. 1, 137, Minsk: Nauka Science.
164. Asisov, R. I., Givotov, V. K., Rusanov, V. D., and Fridman, A. 1980. *Sov. Phys. High Energy Chem.* 14:366.
165. Levitsky, A. A., Polak, L. S., Rytova, I. M., and Slovetsy, D. I. 1981. *Sov. Phys. High Energy Chem.* 15:276.
166. Polak, L. S., Slovetsky, D. I., and Butylkin, Yu. P. 1977. *Carbon Dioxide Dissociation in Electric Discharges*, Institute of Petrol-Chemical Synthesis, Moscow: Nauka Science.
167. Potapkin, B. V., Rusanov, V. D., and Fridman, A. 1984. *Sov. Phys. High Energy Chem.* 18:252.
168. Givotov, V. K., Malkov, S. Yu., Rusanov, V. D., and Fridman, A. 1983. *J. Nucl. Sci. Technol. Nucl.-Hydrogen Energy* 114:52.
169. Givotov, V. K., Malkov, S. Yu., Fridman, A., and Potapkin, B. V. 1984. Sov. Phys. High Energy Chem. 18:252.
170. Williams, F.A. 1985. *Combustion Theory*, 2nd ed., Redwood City, CA: Addison–Wesley.
171. Potapkin, B. V., Rusanov, V. D., and Fridman, A. 1985. *Non-Equilibrium Effects in Plasma, Provided by Selectivity of Transfer Processes*. Kurchatov Institute of Atomic Energy, vol. 4219/6, Moscow.
172. Alfven, H. 1950. *Cosmical Electrodynamics*, Oxford: Clarendon Press.

173. Kruscal, M. D. and Kulsrud, R. M. 1958. Phys. Fluids 1:265.
174. Kadomtsev, B. B. 1958. *Plasma Physics and Problem of Controlled Thermonuclear Reactions*, vol. 3, Moscow: Academy of Science of the USSR.
175. Allis, W. P. 1960. *Nuclear Fusion*, Princeton, NJ: D.Van Nostrand Company.
176. Bennett, W. H. 1934. Magnetically self-focusing streams. *Phys. Rev.* 45:890.
177. Thompson, W. B. 1962. *An Introduction to Plasma Physics*, Oxford: Pergamon Press, Addison–Wesley Publishing Company.
178. Kadomtsev, B. B. 1976. *The Collective Phenomena in Plasma*, Moscow: Nauka Science.
179. Braginsky, S. I. 1963. *Problems of Plasma Theory*, vol. 1, ed. M. A.Leontovich, 183–272, Moscow: Atomizdat.
180. D'Angelo, N. and Rynn, N. 1961. *Phys. Fluids* 4:275.
181. Lundquist, S. 1952. *Ark. f. Fys.* 5:297.
182. Chen, F. F. 1984. *Introduction to Plasma Physics and Controlled Fusion: Plasma Physics*, 2nd ed., New York: Plenum Press.
183. Rutherford, P. H. and Goldston, R. J. 1995. *Introduction to Plasma Physics*, Institute of Physics Publishing, Bristol, U.K.
184. Melrose, D. B. 1989. *Instabilities in Space and Laboratory Plasmas*, Cambridge: Cambridge University Press.
185. Mikhailovskii, A. B. 1998. *Theory of Plasma Instabilities*, Institute of Physics Publishing.
186. Mikhailovskii, A. B. 1998. *Instabilities in a Confined Plasma, Plasma Physics Series*. Institute of Physics Publishing.
187. Givotov, V. K., Kalachev, I. A., Krasheninnikov, E. G. 1983. *Spectral Diagnostics of a Non-Homogeneous Plasma-Chemical Discharge*, Kurchatov Institute of Atomic Energy, vol. 3704/7, Moscow.
188. Nedospasov, A. V. 1968. *Adv. Phys. Sci., Uspehi Phys. Nauk* 94:439.
189. Landa, P.S., Miskinova, N. A., and Ponomarev, Yu. V. 1978. *Adv. Phys. Sci. Uspehi Phys. Nauk* 126:13.
190. Nighan, W. L. 1976. In *Principles of Laser Plasma*, ed. G. Bekefi, New York: John Wiley & Sons.
191. Haas, R. 1973. *Phys. Rev., A – General Phys.* 8:1017.
192. Nighan, W. L. and Wiegand, W. G. 1974. *Appl. Phys. Lett.* 25:633.
193. Vedenov, A. A. 1973. *Ionization Explosion in Glow Discharge*, 108, ICPIG-XI, Prague.
194. Kirillov, I. A., Rusanov, V. D., and Fridman, A. 1984. In Kinetic and Gas-Dynamic Processes in Non-Equilibrium Medium, 97, Moscow: Moscow State University.
195. Kirillov, I. A., Rusanov, V. D., and Fridman, A. 1984. *Sov. Phys. J. Tech. Phys.* 54:2158.
196. Kirillov, I. A., Potapkin, B. V., Fridman, A., and Strelkova, M. I. 1984. Sov. Phys. J. Appl. Mech. Tech. Phys. 6:77–80.
197. Meyer, J. 1969. *J. Phys. D: Appl. Phys.* 2:221.
198. Kekez, M., Barrault, M. R., and Craggs, J. D. 1970. *J. Phys. D: Appl. Phys.* 3:1886.
199. Chalmers, J. D. 1972. *Proc. R. Soc., A, London* 369:171.
200. Gildenburg, V. B. 1981. In Non-Linear Waves, 15, ed. A. V. Gaponov–Grekov, Moscow: Nauka Science.
201. Ageeva, N. P., Novikov, G. I., and Raddatis, V. K. 1986. Sov. Phys. High Energy Chem. 20:284.
202. Smirnov, B. M. 1977. *Introduction to Plasma Physics*, Moscow: Mir; 1982. Moscow: Nauka.
203. Provorov, A. S. and Chebotaev, V. P. 1977. In *Gas Lasers*, 174, ed. R. I. Soloukhin, Novosibirsk: Nauka Science.
204. Hill, A. E. 1971. *Appl. Phys. Lett.* 18:194.
205. Gibbs, W. E. and McLeary, R. 1971. *Phys. Lett. A* 37:229.
206. Zyrichev, N. A., Kulish, S. M., and Rusanov, V. D. 1984. CO_2 *Dissociation in Supersonic Plasma-Chemical Reactor*, Kurchatov Institute of Atomic Energy, vol. 4045/6, Moscow.
207. Potapkin, B. V., Rusanov, V. D., and Fridman, A. 1983. *Sov. Phys. High Energy Chem.*17:528.
208. Kirillov, I. A., Potapkin, B. V., and Fridman, A. 1984. *Sov. Phys. High Energy Chem.* 18:151.
209. Kirillov, I. A., Potapkin, B. V., and Strelkova, M. I. 1984. *Dynamics of Space-Non-Uniform Vibrational Relaxation in Chemically Active Plasma*, Moscow: Kurchatov Institute of Atomic Energy, vol. 3608/6.
210. Stix, T. H. 1992. *Waves in Plasmas*, Berlin: Springer–Verlag.

211. O'Neil, T. M. 1965. *Phys. Fluids* 8:2255.
212. Ivanov, A. A. 1977. Physics of Strongly Non-equilibrium Plasma, Moscow: Atom–Izdat.
212a. Kadomtsev, B. B. 1968. *Adv. Phys.Sci. Uspehi Phys. Nauk* 95:111.
213. Rayleigh, J. W. S. 1878. *The Theory of Sound*, vol. 2, later printed: 1945. New York: Dover Publications.
214. Jacob, J. M. and Mani, S. A. 1975. *Appl. Phys. Lett.* 26:53.
214a. Kirillov, I. A., Potapkin, B. V., Rusanov, V. D., and Fridman, A. 1983. *Sov. Phys. High Energy Chem.* 17:519.
215. Ginsburg, V. L. 1960. *Propagation of Electromagnetic Waves in Plasma, Physics and Mathematics*, Moscow: Giz.
216. Ginsburg, V. L. and Rukhadze, A. A. 1970. *Waves in Magneto-Active Plasma*, Moscow: Nauka Science.
217. Landau, L. D. 1982. *Quantum Electrodynamics*, 2nd ed., Oxford: Butterworth-Heinemann.
217a. Landau, L. D., Pitaevsky, L. P., and Lifshitz, E. M. 1957. *Electrodynamics of Continuous Media*, Moscow: Nauka.
218. Biberman, L. M. and Norman, G. E. 1967. *Adv. Phys. Sci. Uspehi Phys. Nauk* 91:193.
219. Biberman, L. M. 1970. *Optical Properties of Hot Air*, Moscow: Nauka Science.
220. Modest, M. F. 1993. *Radiative Heat Transfer*, New York: McGraw–Hill.
221. Smirnov, B. M. 1977. *Plasma Chemistry–4*, 191, ed. B. M. Smirnov, Moscow: Atomizdat.
222. Griem, H. R. 1964. *Plasma Spectroscopy*, New York: McGraw–Hill.
223. Griem, H. R. 1974. *Spectral Broadening by Plasma*, New York: Academic Press.
224. Cabannes, F. and Chapelle, J. 1971. *Reactions under Plasma Conditions*, vol. 1, New York: Wiley Interscience.
225. Traving, G. 1968. In *Plasma Diagnostics*, ed. Lochte–Holtgreven, New York: Wiley.
226. Bradley, E. B. *Molecules and Molecular Lasers for Electrical Engineers,* series in electrical engineering, Washington, D.C.: Hemisphere.
227. Parity, B. S. 1967. *J. Fluid Mech.* 27:49.
227a. Sydney, R. 1970. in: *Non-Equilibrium Flows*, vol. 2, 160, ed. P. P. Wegener, New York: Marcel Dekker.
228. Kirillov, I. A., Rusanov, V. D., and Fridman, A. 1983. *Sov. Phys. J. Phys. Chem.* 5:280.
229. Vedenov, A. A. 1982. *Physics of Electric Discharge CO_2 Lasers*, Moscow: Energo–Atom–Izdat.
230. Rogdestvensky, B. L. and Yanenko, N. N. 1978. *Systems of Quasi-Linear Equations*, Moscow: Nauka Science.
231. Granovsky, V. L. 1971. *Electric Current in Gas, Steady Current*, Moscow: Nauka Science.
232. von Engel, A. 1965. *Ionized Gases*, Oxford: Clarendon Press; 1994. In *American Vacuum Society Classics*, Heidelberg: Springer–Verlag.
233. von Engel, A., Steenbeck, M. 1934. *Elektrische Gasentladungen Ihze Physik und Technik,* vol. 2, Berlin: Springer.
234. Raizer, Yu. P. and Surzhikov, S. T. 1987. *Sov. Phys. J. Tech. Phys. Lett.* 13:452.
235. Raizer, Yu. P. and Surzhikov, S. T. 1988. *Sov. Phys. Thermal Phys. High Temp.* 26:428.
236. Gill, P. and Webb, C. E. 1977. *J. Phys. D Appl. Phys.* 10:229.
237. Boeuf, J. P. and Marode, E. 1982. *J. Phys. D Appl. Phys.* 15:2169.
238. Bronin, S. Ya. and Kolobov, V. M. 1983. *Sov. Phys. Plasma Phys.* 9:1088.
239. Moskalev, B. I. 1969. *The Hollow Cathode Discharge*, Moscow: Energy.
240. Poeschel, R. L., Beattle, J. R., Robinson, P. A., and Ward, J. W. 1979. 14th Int. Electric Propulsion Conf., Princeton, NJ, paper 79-2052.
241. Forrester, A. T. 1988. *Large Ion Beams—Fundamentals of Generation and Propagation*, New York: John Wiley & Sons.
242. Sabo, Z. G. 1966. *Chemical Kinetics and Chain Reactions*, Moscow: Nauka Science.
243. Brown, S. C. 1959. *Basic Data of Plasma Physics*, Cambridge, MA: MIT Press.
244. Brown, S. C. 1966. *Introduction to Electrical Discharges in Gases*, New York: John Wiley & Sons.
245. Howatson, A. M. 1976. *An Introduction to Gas Discharges*, 2nd ed., Oxford: Pergamon Press.
246. Pashkin, S. V. and Peretyatko, P. I. 1978. Sov. Phys. Quantum Electron. 5:1159.
247. Velikhov, E. P., Golubev, V. S., and Pashkin, S. V. 1982. *Adv. Phys.Sci. Uspehi Phys. Nauk* 137:117.

248. Golubovsky, Yu. B., Zinchenko, A. K., and Kagan, Yu. M. 1977. *Sov. Phys. J. Tech. Phys*. 47:1478.
249. Generalov, N. A., Kosynkin, V. D., Zimakov, V. P., and Raizer, Yu. P. 1980. *Sov. Phys. Plasma Phys*. 6:1152.
250. Velikhov, E. P., Kovalev, A. S., and Rakhimov, A. T. 1987. *Physical Phenomena in Gas-Discharge Plasma*, Moscow: Nauka Science.
251. Francis, G. 1956. The glow discharge at low pressure. In *Encyclopedia of Physics*, ed. S. Flugge, *Handbuch der Physik*, Bd. XXII, 55–208, Berlin.
252. Patel, C. K. N. 1964. *Phys. Rev. Lett*. 13:617.
253. Garscadden, A. 1978. Ionization waves. In *Gaseous Electronics – I, Electrical Discharges*, ed. M. N. Hirsh and H. J. Oskam, New York: Academic Press.
254. Nedospasov, A. V. and Khait, V. V. 1979. *Oscillations and Instabilities of Low-Temperature Plasma*, Moscow: Nauka Science.
255. Nedospasov, A. V. and Ponomarenko, Yu. B. 1965. *Sov. Phys. Thermal Phys. High Temp*. 3:17.
256. Tsendin, L. D. 1970. *Sov. Phys. J. Tech. Phys*. 40:1600.
257. Roth, J. R. 1967. *Phys. Fluids* 10:2712.
258. Roth, J. R. 1969. *Plasma Phys*. 11:763.
259. Abilsiitov, G. A., Velikhov, E. P., and Golubev, V. S., 1984. *High Power CO_2 Lasers and Their Applications in Technology*, Moscow: Nauka Science.
260. Penning, F. M. 1936. *Physica* 3:873.
261. Penning, F. M. 1937. *Physica* 4:71.
262. Roth, J. R. 1966. *Rev. Sci. Instrum*. 37:1100.
263. Roth, J. R. 1973. *IEEE Trans. Plasma Sci*. 1:34.
264. Roth, J. R. 2000. *Industrial Plasma Engineering*, Bristol: Institute of Physics Publishing.
265. Poluektov, N. P. and Efremov, N. P. 1998. *J. Phys. D* 31:988.
266. Chebotaev, V. 1972. *Sov. Phys. Doklady* 206:334.
267. Kanazava, S., Kogoma, M., Moriwaki, T., and Okazaki, S. 1987. *International Symposium on Plasma Chemistry*, ISPC-8, Tokyo.
268. Yokayama, T., Kogoma, M., and Okazaki, S. 1990. *J. Phys. D Appl. Phys*. 23:1125.
269. Okano, H., Yamazaki, T., and Horiike, Y. 1982. *Solid State Technol*. 25:166.
270. Babukutty, Y., Prat, R., Endo, K., Kogoma, M., Okazaki, S., and Kodama, M. 1999. *Langmuir* 15:7055.
271. Raizer, Yu. P. 1977. *Laser-Induced Discharge Phenomena*, New York: Consultants Bureau.
272. Dobretsov, L. N. and Gomounova, M. V. 1966. *Emission Electronics*, Moscow: Nauka Science.
273. Fransis, G. 1960. *Ionization Phenomena in Gases*, London: Butterworths.
274. Modinos, A. 1984. Field, *Thermionic and Secondary Electron Emission Spectroscopy*, New York: Plenum Press.
275. Komolov, S. A. 1992. *Total Current Spectroscopy of Surfaces*, Gordon & Breach Science Pubishers.
276. Zhukov, M. F. 1982. Electrode Processes in Arc Discharges, Novosibirsk: Nauka Science.
277. Lafferty, J. M. 1980. ed., *Vacuum Arcs, Theory and Applications*, New York: Wiley.
278. Lyubimov, G. A. and Rachovsky, V. I. 1978. Cathode spot of vacuum arc, *Adv. Phys. Sci. Uspehi Phys. Nauk* 125:665.
279. Korolev, Yu. D. and Mesiatz, G. A. 1982. Field Emission and Explosive Processes in Gas Discharge, Novosibirsk: Nauka Science.
280. Pfender, E. 1978. Electric arcs and arc gas heaters. In *Gaseous Electronics*, vol. 1, eds. M. N. Hirsh and H. J. Oscam, chap. 5, New York: Academic Press.
281. Penski, V. 1968. Theoretical calculations of transport properties in nitrogen plasma. *IV Symp. Thermophys. Properties*, College Park, MD: University of Maryland.
282. Finkelburg, W. and Maecker, H. 1956. Elektrische Bogen und Thermisches Plasma. *Handbuch Physik*. 22:S.254.
283. Rozovsky, M. O. 1972. *Sov. Phys. J. Appl. Mech. Tech. Phys*. 6:176.
284. Raizer, Yu. P. 1972. *Sov. Phys. Thermal Phys. High Temp*. 10:1152.
285. Tonks, L. and Langmuir, I. 1929. *Phys. Rev*. 34:876.
286. Gross, B., Grycz, B., and Miklossy, K. 1969. *Plasma Technology*, New York: Elsevier.

287. Hirsh, M. N. and Oscam, H. J., eds. 1978. *Gaseous Electronic*, vol. 1, *Electrical Discharges*, New York: Academic Press.
288. Naville, A. A. and Guye, C. E. 1904. French Patent No. 350 120.
289. Lesueur, H., Czernichowski, A., and Chapelle, J. 1990. *J. Phys.* Colloque C5:51.
289a. Czernichowski, A. 1994. *Pure Appl. Chem.* 66,(6):1301.
290. Fridman, A., Czernichowski, A., Chapelle, J., Cormier, J.-M., Lesueur, H., and Stevefelt, J. 1993. International Symposium on Plasma Chemistry, ISPC-11, Loughborough, U.K.
291. Rusanov, V. D., Petrusev, A. S., Potapkin, B. V., Fridman, A., Czernichowski, A., and Chapelle, J. 1993. *Sov. Phys. Doklady* 332(6):306.
292. Fridman, A., Nester, S., Yardimci, O., Saveliev, A., and Kennedy, L. A. 1997. 13th International Symposium on Plasma Chemistry, ISPC-13, Beijing, China.
293. Richard, F., Cormier, M., Pellerin, S., and Chapelle, J. 1996. *J. Appl. Phys.* 79(5):2245.
294. Deminsky, M. A., Potapkin, B. V., Cormier, J. M., Richard, F., Bouchoule, A., and Rusanov, V. D. 1977. *Sov. Phys. Doklady* 42:337.
295. Fridman, A., Nester, S., Kennedy, L., Saveliev, A., and Mutaf–Yardimci, O. 1999. *Prog. Energy Combustion Sci.* 25:211.
296. Cormier, J.-M., Richard, F., Chapelle, J., and Dudemaine, M. 1993. *Proc. 2nd Int. Conf. Elect. Contacts, Arcs, Apparatus Appl.* 40–42.
296a. Kennedy, L., Fridman, A., Saveliev, A., and Nester, S. 1997. *APS Bull.* 419:1828.
297. Mutaf–Yardimci, O., Saveliev, A., Fridman, A., and Kennedy, L. 1999. *J. Appl. Phys.* 87:1632.
298. Czernichowski, A., Nassar, H., Ranaivosoloarimanana, A., Fridman, A., Simek, M., Musiol, K., Pawelec, E., and Dittrichova, L. 1996. *Acta Phys. Pol.* A89:595.
299. Dalaine, V., Cormier, J.-M., Pellerin, S., and Lefaucheux, P. 1998. *J. Appl. Phys.* 84:1215.
299a. Pellerin, S., Cormier, J.-M., Richard, F., Musiol, K., and Chapelle, J. 1996. *J. Phys. D Appl. Phys.* 29:726.
300. Mutaf–Yardimci, O., Kennedy, L., Saveliev, A., and Fridman, A. 1998. Plasma exhaust after treatment, SAE, SP-1395, 1.
301. Loeb, L.B. 1965. Electrical Coronas—Their Basic Physical Mechanisms, Berkeley: University of California Press.
302. Bartnikas, R. and McMahon, E. J., eds. 1979. *Engineering Dielectrics*, vol. 1, Corona Measurement and Interpretation, ASTM Special Technical Publication, No. 669, Philadelphia, PA.
303. Goldman, M. and Goldman, N. 1978. Corona Discharges. In *Gaseous Electronics*, vol. 1, *Electrical Discharges*, eds. M. N. Hirsh and H. J. Oscam, New York: Academic Press.
304. Pu, Y. K. and Woskov, P. P. 1996. eds., *Proc. Int. Workshop Plasma Technol. Pollution Control Waste Treat.*, MIT, Cambridge, MA.
305. van Veldhuizen, E. M., Rutgers, W. R., and Bityurin, V. A. 1996. Plasma Chem. Plasma Process. 16:227.
306. Zhou, L. M. and van Veldhuizen, E. M. 1996. Eindhoven University of Technology Report 96-E-302, ISBN 90-6144-302-4.
307. Mattachini, F., Sani, E., and Trebbi, G. 1996. In *Proc. Int. Workshop Plasma Technol. Pollut. Control Waste Treat.*, MIT, Cambridge, MA.
308. Penetrante, B. M., Hsiao, M. C., Bardsley, J. N., Meritt, B. T., Vogtlin, G. E., Wallman, P. H., Kuthi, A., Burkhart, C. P., and Bayless, J. R. In *Proc. Int. Workshop Plasma Technol. Pollut. Control Waste Treat.*, MIT, Cambridge, MA.
308a. Penetrante, B. M., Brusasco, R. M., Meritt, B. T., Pitz, W. J., Vogtlin, G. E., Kung, M. C., Kung, H. H., Wan, C. Z., and Voss, K. E. 1998. In *Plasma Exhaust Aftertreatment*, Warrendale, PA: Society of Automotive Engineers.
309. Penetrante, B. and Schulteis, S., eds., 1993. *NATO ASI Series*, vol. G 34B, Berlin: Springer–Verlag.
310. Park, M., Chang, D., Woo, M., Nam, G., and Lee, S. 1998. In *Plasma Exhaust Aftertreatment*, Warrendale, PA: Society of Automotive Engineers.
311. Siemens, W. 1857. Poggendorfs *Ann. Phys. Chem.* 102:66.
312. Sobacchi, M., Saveliev, A. Fridman, A., Gutsol, A., and Kennedy, L. 2003. *Plasma Chemistry and Plasma Processing*. 23:347.

313. Kogelschatz, U. 1988. Advance ozone generation. In *Process Technologies for Water Treatment*, ed. S. Stucki, 87, New York: Plenum Press.
314. Kogelschatzs, U. and Eliasson, B. 1995. Ozone Generation and Applications. In *Handbook of Electrostatic Processes*, eds. J. S. Chang, A. J. Kelly, and J. M. Crowley, 581, New York: Marcel Dekker.
315. Kogelschatzs, U., Eliasson, B., and Egli, W. 1997. J. Phys. IV, 7, Colloque C4, 4–47.
316. Buss, K. 1932. *Arch. Elektrotech.* 26:261.
317. Klemenc, A., Hinterberger, H., and Hofer, H. 1937. *Z. Elektrochem.* 43:261.
318. Kunchardt, E. E. and Tzeng, Y. 1988. *Phys. Rev.* A38:1410.
319. Marode, E. 1995. *Gas Dicharges Their Appl.* GD-95, 11–484, Tokyo.
320. Dali, S. K. and Williams, P. F. 1987. *Phys. Rev.* A31:1219.
321. Yoshida, K. and Tagashira, H. 1979. *J. Phys. D Appl. Phys.* 12:3.
322. Kulikovsky, A. A. 1994. *J. Phys. D Appl. Phys.* 27:2556.
323. Babaeba, N. and Naidis, G. 1996. *J. Phys. D Appl. Phys.* 29:2423.
324. Morrow, R. and Lowke, J. J. 1997. *J. Phys. D Appl. Phys.* 30:614.
325. Vitello, P. A., Penetrante, B. M., and Bardsley, J. N. 1994. Phys. Rev., vol. E49, pg. 5574.
326. Eliasson, B. and Kogelschatz, U. 1991. *IEEE Trans. Plasma Sci.* 19:309.
327. Pietsch, G. J., Braun, D., and Gibalov, V. I. 1993. Modeling of dielectric barrier discharges, NATO ASI series, vol. G34, part A, *Nonthermal Plasma Techniques for Pollution Control*, eds. B. M. Penetrante and S. E. Schultheis, 273, Berlin: Springer–Verlag.
328. Xu, X. P. and Kushner, M. J. 1998. *J. Appl. Phys.* 84:4153.
329. Borisov, B. M. and Khristoforov, O. F. 2001. In *Encyclopedia of Low-Temperature Plasma*, vol. 2, ed. V. E. Fortov, 350, Moscow: Nauka Science.
330. Baselyan, E. M. and Raizer, Yu. P. 1997. *Spark Discharge*, Edition of the Moscow Institute of Physics and Technology, Moscow.
330a. Baselyan, E. M. and Raizer, Yu. P. 2001. *Physics of Lightning and Lightning Protection*, Moscow: Nauka Science.
331. Heath, W. O. and Birmingham, J. G. 1995. PNL-SA-25844, Pacific Northwest Laboratory, Annual Meeting of American Nuclear Society, Philadelphia, Pennsylvania.
332. Birmingham, J. G. and Moore, R. R. 1990. Reactive Bed Plasma Air Purification, United States Patent #4, 954, 320.
333. Kanazawa, S., Kogoma, M., Moriwaki, T., and Okazaki, S. 1988. *J. Phys. D* 21:838.
334. Lacour, B. and Vannier, C. 1987. *J. Appl. Phys.* 38:5244.
335. Honda, Y., Tochikubo, F., and Watanabe, T. 2001. *25th Int. Conf. Phenomena Ionized Gases*, ICPIG-25, 4:37, Nagoya, Japan.
336. Szymanski, A. 1985. *Beitr. Plasmaphys.* 2:133.
337. Levitsky, A., Macheret, S., and Fridman, A. 1983. In *Chemical Reactions in Non-Equilibrium Plasma*, ed. L. S. Polak, Moscow: Nauka Science.
338. Samoylovich, V. G., Gibalov, V. I., and Kozlov, K. V. 1989. *Physical Chemistry of Barrier Discharge*, Edition of the Moscow State University, Moscow.
339. Opalinska, T. and Szymanski, A. 1996. *Contrib. Plasma Phys.* 36:63.
340. Robertson, G. D. and Baily, N. A. 1965. *Bull. Am. Phys. Soc.* 2:709.
341. Kusz, J. 1978. Plasma Generation on Ferroelectric Surface, ed. PWN, Poland: Warsaw–Wroclaw.
342. Hinazumi, H., Hosoya, M., and Mitsui, T. 1973. *J. Phys. D Appl. Phys.* 1973:21.
343. Drabkina, S. I. 1951. *Sov. Phys. J. Theor. Exp. Phys.* 21:473.
344. Braginsky, S. I. 1958. *Sov. Phys. J. Theor. Exp. Phys.* 34:1548.
345. Frenkel, Ya. I. 1949. *Theory of Atmospheric Electricity Phenomena*, Moscow: Gos. Tech. Izdat.
346. Uman, M. 1969. *Lightning*, New York: McGraw–Hill.
347. Arago, D. F. 1859. *Thunder and Lightning*, Paris: Ecole Politechnique.
348. Smirnov, B. M. 1975. *Adv. Phys. Sci. Uspehi Phys. Nauk* 116:111.
349. Stakhanov, I. P. 1973. *Sov. Phys. J. Theor. Exp. Phys. Lett.* 18:193.
350. Stakhanov, I. P. 1974. Sov. Phys. J. Tech. Phys. 44:1373.
351. Dmitriev, M. T. 1967. *Priroda Nature* 56:98.
352. Dmitriev, M. T. 1969. *Sov. Phys. J. Tech. Phys.* 39:387.

353. Rusanov, V. D. and Fridman, A. 1976. *Sov. Phys. Doklady* 230:809.

353a. Korobtsev, S., Medvedev, D., Rusanov, V., and Shiryaevsky, V. 1997. *13th Int. Symp. Plasma Chem.* p. 755, Beijing, China.

354. Leonov, R. 1965. *Ball Lightning Problems*, Moscow: Nauka Science.

355. Stakhanov, I. P. 1976. *Physical Nature of Ball Lightning*, Moscow: Nauka Science.

356. Singer, S. 1973. *The Nature of Ball Lightning*, ASIN 0306304945.

357. Barry, J. D. 1980. *Ball Lightning and Bead Lightning: Extreme Forms of Atmospheric Electricity*, New York: Plenum Press.

358. Stenhoff, M. 2000. *Ball Lightning: An Unsolved Problem in Atmospheric Physics*, Dordrecht: Kluwer Academic Publishers.

359. Vasilyak, L. M., Kostuchenko, S. V., Kurdyavtsev, N. N., and Filugin, I. V. 1994. *Adv. Phys. Sci., Uspehi Phys. Nauk* 164:263.

360. Asinovsky, E. I. and Vasilyak, L. M. 2001. In *Encyclopedia of Low-Temperature Plasma*, vol. 2, ed. V. E. Fortov, 234, Moscow: Nauka Science.

361. Kalinnikov, V. T. and Gutsol, A. 1997. *Sov. Phys. Doklady* 35342:469179.

362. Gutsol, A. and Fridman, A. 2001. *XXVth Int. Conf. Phenomena Ionized Gases*, ICPIG-25, Nagoya, Japan, 1:55.

362a. Gutsol, A., Fridman, A., Chirokov, A., Kennedy, L. A., and Worek, W. 2001. *2nd Int. Conf. Computational Heat Mass Transfer*, Rio de Janeiro, Brazil, 65.

363. Gutsol, A., Larjo, J., and Hernberg, R. 1999. *XIV Int. Symp. Plasma Chem.*, ISPC-14, Prague, 1:227.

364. Roth, J. R., Laroussi, M., and Liu, C. 1992. *Proc.19th IEEE, Int. Conf. Plasma Sci.*, Tampa, FL.

365. Kanda, N., Kogoma, M., Jinno, H., Uchiyama, H., and Okazaki, S. 1991. *X Int. Symp. Plasma Chem.*, ISPC-10, 3:3.2–20.

366. Blinov, L. M., Volod'ko, V. V., Gontarev, G. G., Lysov, G. V., and Polak, L. S. 1969. In *Low-Temperature Plasma Generators*, ed. L. S. Polak, Moscow: Energia Energy.

367. Batenin, V., Klimovsky, I. I., Lysov, G. V., and Troizky, V. N. 1988. *Microwave Plasma Generators: Physics, Engineering, Applications*, Moscow: Energo AtomIzdat.

368. Kapitsa, P. L. 1969. *Sov. Phys. J. Exp. Theor. Phys.* 57:1801.

369. Mitsuda, Y., Yoshida, T., and Akashi, K. 1989. *Rev. Sci. Instrum.* 60:249.

370. MacDonald, A. D. and Tetenbaum, S. J. 1978. In *Gaseous Electronics*, vol. 1, eds. M. N. Hirsh and H. J. Oskam, *Electrical Discharges*, New York: Academic Press.

371. Raizer, Yu. P. 1970. Sov. Phys. Lett. J. Exp. Theor. Phys. 11:195.

372. Generalov, N. A., Zimakov, V. P., Kozlov, G. I., Masyukov, V. A., and Raizer, Yu. P. 1971. *Sov. Phys. J. Exp. Theor. Phys.* 61:1444.

373. Maker, P. D., Terhune, R. W., and Savage, C. M. 1964. In *Quantum Electronics III*, eds. P. Grivet and N. Bloembergen, New York: Columbia University Press.

374. Raizer, Yu. P., Shneider, M. N., and Yatsenko, N. A. 1995. *Radio-Frequency Capacitive Discharges*, Moscow: Nauka Science and Boca Raton, FL: CRC Press.

375. Lieberman, M. A. and Gottscho, R. A. 1994. In *Physics of Thin Films*, vol. 18, eds. M. H. Francombe and J. L. Vossen, New York: Academic Press.

376. Smirnov, A. S. 2000. In Encyclopedia of Low-Temperature Plasma, vol. 2, ed. V. E. Fortov, 67, Moscow: Nauka Science.

377. Levitsky, S. M. 1957. *Sov. Phys. J. Tech. Phys.* 27:970, 1001.

378. Yatsenko, N. A. 1981. Sov. Phys. J. Tech. Phys. 51:1195.

379. Javan, A., Bennett, W. R., and Herriot, D. R. 1961. *Phys. Rev. Lett.* 6:106.

380. Manos, D. M. and Flamm, D. L. 1989. Plasma Etching: An Introduction, New York: Academic Press.

381. Konuma, M. 1992. Film Deposition by Plasma Technologies, New York: Springer–Verlag.

382. Vossen, J. L. and Kern, W. 1978. eds., Thin Film Processes, New York: Academic Press.

383. Vossen, J. L. and Kern, W., eds., 1991. *Thin Film Processes II*, New York: Academic Press.

384. Godyak, V. A. 1971. Sov. Phys. *J. Tech. Phys.* 41:1361.

385. Godyak, V. A. 1971. *Sov. Phys. Plasma Phys.* 2:141.

386. Lieberman, M. A. 1988. *IEEE Trans. Plasma Sci.* 16:1988.

387. Godyak, V.A. 1986. *Soviet Radio Frequency Discharge Research*, Falls Church, VA: Delphic Associates.

388. Godyak, V. A., Piejak, R. B., and Alexandrovich, B. M. 1991. *IEEE Trans. Plasma Sci.* 19:660.
389. Godyak, V. A. and Piejak, R. B. 1990. *Phys. Rev. Lett.* 65:996.
390. Godyak, V. A. and Piejak, R. B. 1990. *J. Vac. Sci. Technol.* A8:3833.
391. Vahedi, V., Birdsall, C. K., Lieberman, M. A., DiPeso, G., and Rognlien, T. D. 1994. *Plasma Sources Sci. Technol.* 2:273.
392. Vender, D. and Boswell, R. W. 1990. *IEEE Trans. Plasma Sci.* 18:725.
393. Wood, B. P. 1991. Sheath Heating in Low Pressure Capacitive Radio Frequency Discharges, Thesis, University of California, Berkley.
394. Lieberman, M. A. 1989. *IEEE Trans. Plasma Sci.* 17:338.
395. Lieberman, M. A. 1989. *J. Appl. Phys.* 65:4168.
396. Alexandrov, A. F., Godyak, V. A., Kuzovnikov, A. A., and Sammani, A. Y. 1967. VIII Int. Conf. Phenomena Ionized Gases, ICPIG-8, Vienna, 165.
397. Yeom, G. Y., Thornton, J. A., and Kushner, M. J. 1989. *J. Appl. Phys.* 65:3816.
398. Knypers, A. D. and Hopman, H. J. 1990. *J. Appl. Phys.* 67:1229.
399. Lin, I. 1985. *J. Appl. Phys.* 58:2981.
400. Porteous, R. K. and Graves, D. B. 1991. *IEEE Trans. Plasma Sci.* 19:204.
401. Gurin, A. A. and Chernova, N. I. 1985. *Sov. Phys. Plasma Phys.* 11:244.
402. Lukyanova, A. B., Rahimov, A. T., and Suetin, N. B. 1990. *Sov. Phys. Plasma Phys.* 16:1367.
403. Lieberman, M. A., Lichtenberg, A. J., and Savas, S. E. 1991. *IEEE Trans. Plasma Sci.* 19:189.
404. Knypers, A. D., Granneman, E. H. A., and Hopman, H. J. 1988. *J. Appl. Phys.* 63:1899.
405. Lukyanova, A. B., Rahimov, A. T., and Suetin, N. B. 1991. *Sov. Phys. Plasma Phys.* 17:1012.
406. Spahn, R. G. and Gerenser, L. J. 1994. U.S. Patent No. 5,324,414.
407. Gerenser, L. J., Grace, J. M., Apai, G., and Thompson, P. M. 2000. *Surface Interface Anal.* 29:12.
408. Grace, J. M. 1995. U.S. Patent No. 5,425,980.
409. Conti, S., Porshnev, P. I., Fridman, A., Kennedy, L. A., Grace, J. M., Sieber, K. D, Freeman, D. R., and Robinson, K. S. 2001. *Exp. Thermal Fluid Sci.* 24:79
410. Vahedi, V., Birdsall, C. K., Lieberman, M. A., DiPeso, G., and Rognlien, T. D. 1993. *Phys. Fluids* B57:2719.
411. Godyak, V. A. and Khanneh, A. S. 1986. *IEEE Trans. Plasma Sci.* PS-14 2:112.
412. Braithwaite, N., Booth, J. P., and Cunge, G. 1996. *Plasma Sources Sci. Technol.* 5:677
413. Eckert, H. U. 1986. *Proc. II Int. Conf. Plasma Chem. Technol.*, ed. H. Boening, Lancaster, PA: Technomic Publishing.
414. Hopwood, J., Guarnieri, C. R., Whitehair, S. J., and Cuomo, J. J. 1993. *J. Vac. Sci. Technol.* A11:147.
415. Hopwood, J., Guarnieri, C. R., Whitehair, S. J., and Cuomo, J. J. 1993. *J. Vac. Sci. Technol.* A11:152.
415a. Wendt, A. E. and Lieberman, M. A. 1993. *2nd Workshop on High Density Plasmas and Applications*, AVS Topical Conference, San Francisco, CA.
416. Budden, K. G. 1966. *Radio Waves in Ionosphere*, Cambridge: Cambridge University Press.
417. Williamson, M. C., Lichtenberg, A. J., and Lieberman, M. A. 1992. *J. Appl. Phys.* 72:3924.
418. Wu, H.-M., Graves, D. B., and Porteous, R. K. 1994. *Plasma Sources Sci. Technol.* p. 345.
419. Stevens, J. E., Huang, Y. C., Larecki, R. L., and Cecci, J. L. 1992. *J. Vac. Sci. Technol.* A10:1270.
420. Matsuoka, M. and Ono, K. 1988. *J. Vac. Sci. Technol.* A6:25.
421. Boswell, R.W. 1970. *Plasma Phys. Controlled Fusion,* vol. 26, pg. 1147.
422. Chen, F. F. 1991. *Plasma Phys. Controlled Fusion* 33:339.
423. Komori, A., Shoji, T., Miyamoto, K., Kawai, J., and Kawai, Y. 1991. *Phys. Fluids* B3:893.
424. Smullin, L. D. and Chorney, P. 1958. *Proc. IRE* 46:360.
425. Trivelpiece, A. W. and Gould, R. W. 1959. *J. Appl. Phys.* 30:1784.
426. Moisan, M. and Zakrzewski, Z. 1991. *J. Phys. D Appl. Phys.* 24:1025.
427. Vakar, A. K., Krasheninnikov, E. G, and Tischenko, E. A. 1984. *4th Symp. Plasma Chem.*, Dnepropetrovsk, 35.
428. Kirillov, I. A., Rusanov, V. D., and Fridman, A. 1986. *Sov. Phys. Doklady* 284:1352.
429. Kirillov, I. A., Rusanov, V. D., and Fridman, A. 1987. *Sov. Phys. High Energy Chem.* 21:262.
430. Kirillov, I. A., Potapkin, B. V., and Fridman, A., 1984. *Sov. Phys. High Energy Chem.* 18:151.
431. Alexandrov, A. F. and Rukhadze, A. A. 1976. *Physics of the High-Current Electric Discharge Light Sources*, Moscow: Atomizdat.

432. Liventsov, V. V., Rusanov, V. D., and Fridman, A. 1988. *Sov. Phys. High Energy Chem.* 22:67.
433. Mayerovich, B. E. 1972. *Sov. Phys. J. Exp. Theor. Phys.* 63:549.
434. Krasheninnikov, E.G., 1981. Experimental Investigations of Non-Equilibrium Plasma Chemical Processes in Microwave Discharges at Moderate Pressures, Moscow: Kurhcatov Institute of Atomic Energy.
435. Guha, S. and Kaw, P. K. 1968. *Brit. J. Appl. Phys.* 1:193.
436. Sodha, M. S. and Guha, S. 1971. Adv. Plasma Phys. 4:219.
437. Spitzer, L. 1941. Astrophys. J. 93:369.
438. Spitzer, L. 1944. Astrophys. J. 107:6.
439. Sodha, M. S. 1963. Brit. J. Appl. Phys. 14:172.
440. Karachevtsev, G.V. and Fridman, A. 1974. Sov. Phys. J. Tech. Phys. 44:2388.
441. Karachevtsev, G.V. and Fridman, A. 1975. IV SU Conf. Phys. Low Temp. Plasma, Kiev, 2:26.
442. Maksimenko, A. P. and Tverdokhlebov, V. P. 1964. *Sov. Phys. News Higher Educ. Phys.* 1:84.
443. Sayasov, Yu. S. 1958. Sov. Phys. Doklady 122:848.
444. Kirillin, V. A. and Sheindlin, A. E. 1971. *MHD—Generators as a Method of Electric Energy Production*, Moscow: Energia (Energy).
445. Musin, A. K. 1974. *Ionization Processes in Heterogeneous Gas Plasma*, ed. Moscow State University, Moscow.
446. Sholin, G. V., Nikerov, V. A., and Rusanov, V. D. 1980. *J. Phys.* 41, C9:305.
447. Arshinov, A. A. and Musin, A. K. 1958. *Sov. Phys. Doklady* 120:747.
448. Arshinov, A. A. and Musin, A. K. 1958. *Sov. Phys. Doklady* 118:461.
449. Samuilov, E. V. 1966. *Sov. Phys. Doklady* 166:1397.
450. Samuilov, E. V. 1966. *Sov. Phys. Thermal Phys. High Temp.* 4:143.
451. Samuilov, E. V. 1966. *Sov. Phys. Thermal Physics High Temp.* 4:753.
452. Einbinder, H. 1957. *J. Chem. Phys.* 26:948.
453. Arshinov, A. A. and Musin, A. K. 1959. *Sov. Phys. J. Phys. Chem.* 33:2241.
454. Arshinov, A. A. and Musin, A. K. 1962. Sov. Phys. Radio-Tech. Electron. 7:890
455. Samuilov, E. V. 1968. In *Physical Gas Dynamics of Ionized and Chemically Reacting Gases*, 3, Moscow: Nauka Science.
456. Samuilov, E. V. 1973. In *Thermo-Physical Properties of Gases*, 153, Moscow: Nauka Science.
457. Samuilov, E.V. 1967. In *Property of Gases at High Temperatures*, 3, Moscow: Nauka Science.
458. Fridman, A., Czernichowski, A., Chapelle, J., Cormier, J.-M., Lesuer, H., and Stevefelt, J. 1994. *J. Phys.* 1449.
459. Fridman, A. 1976. *Ionization Processes in Heterogeneous Media*, ed. Moscow Institute of Physics and Technology, MIPT, Moscow.
460. Fridman, A. 1976. *Plasma-Chemical Processes in Non-Self-Sustained Discharges with Ionization Provided by High-Current Relativistic Electron Beams*, ed. Moscow Institute of Physics and Technology, MIPT, Moscow.
460a. Kirillov, I. A., Rusanov, V. D., and Fridman, A. 1981. In *Physical Methods of Investigations of Biological and Chemical Systems*, 53, Moscow: Moscow Institute of Physics and Technology.
461. Karachevtsev, G. V. and Fridman, A. 1977. *Sov. Phys. Thermal Phys. High Temp.* 15:922.
462. Langmuir, I. 1961. *The Collected Works*, vol. 3, 107, Oxford: Pergamon Press.
463. Mezdrikov, O. A. 1968. *Electrical Methods of Volumetric Granulometry*, Leningrad: Energia Energy.
464. Berger, S. 1975. XII Int. Conf. Phenomena Ionized Gases, ICPIG-12. 1:154, Eindhoven, Netherlands.
465. Crichton, B. H. and Lee, B. 1975. XII Int. Conf. Phenomena Ionized Gases, ICPIG-12. 1:155, Eindhoven, Netherlands.
466. Berger, S. 1974. III Int. Conf. Gas Discharge, London, 380.
467. Cookson, A. and Wotton, R. 1974. III Int. Conf. Gas Discharge, London, 385.
468. Raizer, Yu. P. 1972. *Sov. Phys. Adv. Phys. Sci. Uspehi Phys. Nauk* 108:429.
469. Bunkin, F. V. and Savransky, V. V. 1973. *Sov. Phys. J. Exp. Theor. Phys.* 65:2185.
470. Walker, A. B. 1974. US Patent # 3568400 cl.55-5).
471. Stengach, V. V. 1972. *Sov. Phys. J. Appl. Mech. Tech. Phys.*1:128.

472. Stengach, V. V. 1975. *Sov. Phys. J. Appl. Mech. Tech. Phys*. 2:159.
473. Karachevtsev, G. V. and Fridman, A. 1976. *Sov. Phys. J. Tech. Phys*. 46:2355.
474. Karachevtsev, G. V. and Fridman, A. 1979. *Sov. Phys. Izvesia VUZov Higher Educ. News. Phys*. 4:23.
475. Konenko, O. R., Musin, A. K., and Utenkova, S. F. 1973. *Sov. Phys. J. Tech. Phys*. 43:1685.
476. Konenko, O. R. and Musin, A. K. 1972. *Sov. Phys. J. Tech. Phys*. 42:782.
477. Konenko, O. R. and Musin, A. K. 1973. *Sov. Phys. J. Exp. Theor. Phys*. 43:2075.
478. Langmuir, I. and Tonks, L. 1929. *Phys. Rev*. 34:876.
479. Klarfeld, B. N. 1940. Works of the All-Union Electro-Technical Institute VEI. vol. 41, 165.
480. Huddlestone, R. H. and Leonard, S. L., eds. 1965. *Plasma Diagnostic Techniques*, New York: Academic Press.
481. Bouchoule, A. 1993. *Phys. World* 47, August issue.
482. Bouchoule, A. 1999., ed. Dusty Plasmas: Physics, Chemistry and Technological Impacts in Plasma Processing, New York: John Willey & Sons.
483. Watanabe, Y., Shiratani, M., Kubo, Y., Ogava, I., and Ogi, S. 1988. *Appl. Phys. Lett*. 53:1263.
484. Spears, K., Kampf, R., and Robinson, T. 1988. *J. Phys. Chem*. 92:5297.
485. Selwyn, G. S., Heidenreich, J. E., and Haller, K. L. 1990. *Appl. Phys. Lett*. 57:1876.
486. Howling, A. A., Hollenstein, Ch., and Paris, P. J. 1991. *Appl. Phys. Lett*. 59:1409.
487. Bouchoule, A., Boufendi, L., Blondeau, J. Ph., Plain, A., and Laure, C. 1991. *J. Appl. Phys*. 70:1991.
488. Bouchoule, A., Boufendi, L., Blondeau, J. Ph., Plain, A., and Laure, C. 1992. *Appl. Phys. Lett*. 60:169.
489. Bouchoule, A., Boufendi, L., Blondeau, J. Ph., Plain, A., and Laure, C. 1993. *J. Appl. Phys*. 73:2160.
490. Jellum, G., Graves, D., and Daugherty, J. E. 1991. *J. Appl. Phys*. 69:6923.
491. Boeuf, J. P. 1992. *Phys. Rev. A* 46:7910
492. Sommerer, T. J., Barnes, M. S., Keller, J. H., McCaughey, M. J., and Kushner, M. 1991. *Appl. Phys. Lett*. 59:638.
493. Barnes, M. S., Keller, J. H., Forster, J. C., O'Neil, J. A., and Coutlas, D. K. 1992. *Phys. Rev. Lett*. 68:313.
494. Daugherty, J. E., Porteuos, R. K., Kilgore, M. D., and Graves, D. B. 1992. *J. Appl. Phys*. 72:3934.
495. Daugherty, J. E., Porteuos, R. K., Kilgore, M. D., and Graves, D. B. 1993. *J. Appl. Phys*. 73:1619.
496. Daugherty, J. E., Porteuos, R. K., Kilgore, M. D., and Graves, D. B. 1993. *J. Appl. Phys*. 73:7195.
497. Bouchoule, A. and Boufendi, L. 1993. *Plasma Sources Sci. Technol*. 2:204.
498. Bohm, C. and Perrin, J. 1991. *J. Phys. D Appl. Phys*. 24:865
499. Perrin, J. 1991. *J. Non-Cryst. Solids* 137/138:639.
500. Roth, J. R. 1973. *Plasma Phys*. 15:995.
501. Perrin, J., Bohm, C., Etemadi, R., and Lloret, A. 1993. NATO Advanced Research Workshop Formation, Transport and Consequences of Particles in Plasma, France.
502. Perrin, J. 1993. *J. Phys. D: Appl. Phys*. 26:1662.
503. Boufendi, L. and Bouchoule, A. 1994. *Plasma Sources Sci. Technol*. 3:204.
504. Hollenstein, Ch., Howling, A. A., Dorier, J. L., Dutta, J., and Sansonnens, L. 1993. NATO Advanced Research Workshop Formation, Transport and Consequences of Particles in Plasma, France.
505. Boufendi, L., Hermann, J., Stoffels, E., Stoffels, W., De Giorgi, M. L., and Bouchoule, A. 1994. *J. Appl. Phys*. 76:148.
506. Boufendi, L., Ph.D. Thesis, Faculte de Science, GREMI, University of Orleans, 1994.
507. Watanabe, Y., Shiratani, M., and Koga, K. 2001. XXV- Int. Symposium on Phenomena in Ionized Gases, ICPIG-25, vol. 2, pg. 15
508. Watanabe, Y., Shiratani, M., Fukuzava, T., and Koga, K. 2000. *J. Tech. Phys*. 41:505.
509. Koga, K., Matsuoka, Y., Tanaka, K., Shiratani, M., and Watanabe, Y. 2000. *Appl. Phys. Lett*. 77:196.
510. Girshick, S., Kortshagen, U. R., Bhandarkar, U. V., and Swihart, M. T. 1999. *Pure Appl. Chem*. 71:1871.
511. Girshick, S., Bhandarkar, U. V., Swihart, M. T., and Kortshagen, U. R. 2000. *J. Phys. D Appl. Phys*. 33:2731.
512. Reents, W. D. Jr. and Mandich, M. L. 1992. *J. Chem. Phys*. 96:4449.
513. Raghavachari, K. 1992. *J. Chem. Phys*. 96:4440.
514. Mandich, M. L. and Reents, W. D. Jr. 1992. *J. Chem. Phys*. 96:4233.

515. Veprek, S. and Veprek–Heijman, M.G. 1990. *Appl. Phys. Lett.* 56:1766.
516. Veprek, S. and Veprek–Heijman, M.G. 1991. *Plasma Chem. Plasma Process.* 11:323.
517. Baronov, G. S., Bronnikov, D. K., Fridman, A. A., Potapkin, B. V., Rusanov, V. D., Varfolomeev, A. A., and Zasavitsky, A. A. 1989. *J. Phys. B Atom. Mol. Opt. Phys.* 22:2903.
518. Chesnokov, E.N. and Panfilov, V. N. 1981. *J. Theor. Exp. Chem.* 17:699.
519. Porteous, R. K., Hbid, T., Boufendi, L., Fridman, A., Potapkin, B. V., and Bouchoule, A. 1994. ESCAMPIG-12, Euro-Physics Conference Abs., Noordwijkerhout, Netherlands, 18E, pg. 83.
520. Migdal, A. B. 1981. *Qualitative Methods in Quantum Mechanics*, Moscow: Nauka Science.
521. Belenguer, Ph., Blondeau, J. P., Boufendi, L., Toogood, M., Plain, A., Laure, C., Bouchoule, A., and Boeuf, J. P. 1992. *Phys. Rev. A* 46:7923.
522. Givotov, V. K., Kalachev, I. A., Mukhametshina, Z. B., Rusanov, V. D., Fridman, A., and Chekmarev, A. M. 1986. *High Energy Chem.* 20:354.
523. Becker, R. and Doring, W. 1935. Condensation theory, *Ann. Phys.* 24:719.
524. Zeldovich, Ya. B. 1942. *Sov. Phys. J. Exp. Theor. Phys.* 12:525.
525. Frenkel, Ya. I. 1945. *Kinetic Theory of Liquids*, ed. the Academy of Sciences of USSR, Moscow.
526. Macheret, S. O., Rusanov, V. D., and Fridman, A. 1985. *Non-Equilibrium Clusterization in the Field of Centrifugal Forces and Dissociation of Molecules in Plasma*, ed. Kurchatov Institute of Atomic Energy, IAE-4220/6, Moscow.
527. Vostrikov, A. A. and Dubov, D. Yu. 1984. *Real Cluster Properties and Condensation Model*, ed. Institute of Thermal Physics of the Siberian Branch of Academy of Sciences of USSR, vol. 112/84, Novosibirsk.
528. Sattler, K. 1982. *13th Int. Symp. Rarefied Gas Dynamics*, vol. 1, 252, Novosibirsk.
529. Rusanov, V. D., Fridman, A., and Macheret, S. O. 1985. *Sov. Phys.* Doklady 283:590.
530. Krasheninnikov, E. G., Rusanov, V. D., Saniuk, S. V., and Fridman, A. 1986. *Sov. Phys. J. Tech. Phys.* 56:1104.
531. Balebanov, A. V., Butylin, B. A., Givitov, V. K., Matolich, R. M., Macheret, S. O., Novikov, G. I., Potapkin, B. V., Rusanov, V. D., and Fridman, A. 1985. *J. Nucl. Sci. Technol. Nucl.-Hydrogen Energy* 3:46.
531a. Balebanov, A. V., Butylin, B. A., Givitov, V. K., Matolich, R. M., Macheret, S. O., Novikov, G. I., Potapkin, B. V., Rusanov, V. D., and Fridman, A. 1985 *Sov. Phys. Doklady* 283:657.
532. Nester, S. A., Rusanov, V. D., and Fridman, A. 1985. *Dissociation of Hydrogen Sulfide in Plasma with Additives*, ed. Kurchatov Institute of Atomic Energy, vol. 4223/6, Moscow.
533. Harkness, J. B. L. and Doctor, R. 1993. *Plasma-Chemical Treatment of Hydrogen Sulfide in Natural Gas Processing*, Gas Research Institute, GRI-93/0118.
534. Ichimaru, S. 1982. *Rev. Mod. Phys.* 54:1017.
535. Ikezi, H. 1986. *Phys. Fluids* 29:1764.
536. Chu, J. H. and Lin, I. 1994. *Phys. Rev. Lett.* 72:4009.
537. Thomas, H., Morfill, G. E., and Demmel, V. 1994. *Phys. Rev. Lett.* 73:652
538. Fortov, V. E. and Yakubov, I. T. 1994. *Non-Ideal Plasma*, Moscow: Energo—Atom–Izdat.
539. Meijer, E. J. and Frenkel, D. 1991. *J. Chem. Phys.* 94:2269.
540. Stevens, M. J. and Robbins, M. O. 1993. *J. Chem. Phys.* 98:2319.
541. Hamaguchi, S., Farouki, R. T., and Doubin, D. H. E. 1997. *Phys. Rev., E* 56:4671.
542. Molotkov, V. I., Nefedov, A. P., Petrov, O. F., Khrapak, A. G., and Khrapak, S. A. 2000. In *Encyclopedia of Low-Temperature Plasma*, vol. 3, ed. V. E. Fortov, 160, Moscow: Nauka Science.
543. Trottenberg, T., Melzer, A., and Piel, A. 1995. *Plasma Sources Sci. Technol.* 4:450.
544. Morfill, G. E. and Thomas, H. 1996. *J. Vacuum Sci. Technol.* A 14:490.
545. Nefedov, A. P. 2001. *XXV Int. Conf. Phenomena Ionized Gases*, ICPIG-25, Nagoya, Japan, 2:1.
546. Fortov, V. E., Molotkov, V. I., and Nefedov, A. P. 1999. *Phys. Plasmas* 6:1779.
547. Vasilyak, L. M., Vetchinin, S. P., Zimnukov, V. S., Nefedov, A. P., Polyakov, D. N., and Fortov, V. E. 2001. *XXV Int. Conf. Phenomena Ionized Gases*, ICPIG-25, Nagoya, Japan, 3:55.
548. Balabanov, V. V., Vasilyak, L. M., Vetchinin, S. P., Nefedov, A. P., Polyakov, D .N., and Fortov, V.E. 2001. *Sov. Phys. J. Exp. Theor. Phys.* 92:86.
549. Nefedov, A. P., Petrov, O. F., and Fortov, V. E. 1997. *Adv. Phys. Sci. Uspehi Phys. Nauk* 167:1215; *Phys.Usp.* 40:1163.

550. Fortov, V. E., Nefedov, A. P., and Vaulina, O. S. 1998. *J. Exp. Theor. Phys.* 114:2004 *JETF* 87:1087.
551. Bykov, V. L., Vasiliev, M. N., and Koroteev, A. S. 1993. Electron-Beam Plasma. Generation, Properties, Application, ed. Moscow State University, Moscow.
552. Vasiliev, M. N. 2000. In *Encyclopedia of Low-Temperature Plasma*, ed. V. E. Fortov, vol. 4, 436, Moscow: Nauka Science.
553. Humphries, S. 1990. *Charge Particle Beams*, New York: John Wiley & Sons.
554. Fridman, A. and Rudakov, L. I. 1973. *Electrostatic Lenses for Focusing of Relativistic Electron Beams*, Moscow: Moscow Institute of Physics and Technology.
555. Mesyats, G. A., Mkheidze, G. P., and Savin, A. A. 2000. In *Encyclopedia of Low-Temperature Plasma*, vol. 4, ed. V. E. Fortov, 108, Moscow: Nauka Science.
556. Alkhazov, G. D. 1971. *Sov. Phys. J. Techn. Phys.* 41:2513.
557. Nikerov, V. A. and Sholin, G. V. 1978. *Sov. Phys. Plasma Phys.* 4:1256.
558. Nikerov, V. A. and Sholin, G. V. 1978. *Fast Electrons Stopping in Gas, Degradation Spectrum Approach*, ed. Kurchatov Inst. of Atomic Energy, 2985.
559. Vysotsky, Yu. P. and Soshnikov, V. N. 1980. *Sov. Phys. J. Techn. Phys.* 50:1682.
560. Vysotsky, Yu. P. and Soshnikov, V. N. 1981. *Sov. Phys. J. Techn. Phys.* 51:996.
561. Nikerov, V. A. and Sholin, G.V. 1985. *Kinetics of Degradation Processes*, Moscow: Energo–Atom–Izdat.
562. Choi, S. and Kushner, M. 1993. *J. Appl. Phys.* 74:853.
562a. Fano, U. 1953. *Phys. Rev.* 92:328.
563. Spencer, L. V. and Fano, U. 1954. *Phys. Rev.* 93:1172.
564. Inokuti, M. 1971. *Rev. Mod. Phys.* 43:297.
565. Inokuti, M. 1974. *Radiation Res.* 59:343.
566. Inokuti, M. 1975. *Radiation Res.* 64:6.
567. Rau, A. R. P., Inokuti, M., and Douthart, D. A. 1978. *Phys. Rev.* 18:971.
568. Douthart, D. A. 1979. *J. Phys. B: Atomic Molecular Phys.* 12:663.
569. Green, A. E. S. and Barth, C. A. 1965. *J. Geophys. Res.* 70:1083.
570. Green, A. E. S. and Barth, C. A. 1967. *J. Geophys. Res.* 72:3975.
571. Stolarsky, R. S.and Green, A. E. S. 1967. *J. Geophys. Res.* 62:3967.
572. Ivanov, A. A. 1975. *Sov. Phys. Plasma Phys.* 1:47.
573. Ivanov, A. A. 1982. In *Plasma Physics*, ed. V. D. Shafranov, 105, Moscow: VINITI.
574. Ivanov, A. A. and Soboleva, T. K. 1978. *Non-Equilibrium Plasma Chemistry*, Moscow: Atom–Izdat.
575. Ivanov, A. A. and Nikiforov, V.A. 1978. In *Plasma Chemistry—5*, ed. B. M. Smirnov, Moscow: Atom–Izdat.
576. Krasheninnikov, S. I. and Nikiforov, V.A. 1982. In *Plasma Chemistry—9*, ed. B. M. Smirnov, Moscow: Atom–Izdat.
577. Berezin, A. K., Lifshitz, E. V., and Fainberg, Ya. B. 1995. *Plasma Phys.* 21:226.
578. Atamanov, V. M., Ivanov, A. A., and Nikiforov, V. A. 1979. *Sov. Phys. J. Tech. Phys.* 49:2311.
579. Krasheninnikov, S. I. 1980. Electron beam interaction with chemically active plasma, Ph.D. thesis, Moscow Institute of Physics and Technology, Moscow.
580. Ivanov, A. A. 2000. In *Encyclopedia of Low Temperature Plasma*, vol. 4, ed. V. E. Fortov, 428, Moscow: Nauka Science.
581. Alekseew, A. M., Ivanov, A. A., and Starikh, V. V. 1979. *IV Int. Symp. Plasma Chem.* ISPC-4, 2:427, Zurich.
582. Babaritsky, A. I., Ivanov, A. A., Severny, V. V., and Shapkin, V. V. 1975. *II Int. Symp. Plasma Chem.* ISPC-2, 1: Rome.
582a. Babaritsky, A. I., Ivanov, A. A., Severny, V. V., Sokolova, T. I., and Shapkin, V. V. 1975. *II SU Symp. Plasma Chem.* 2:35, Riga.
583. Zhuzhunashvili, A. I., Ivanov, A. A., Shapkin, V. V., and Cherkasova, E. K. 1975. *II SU Symp. Plasma Chem.* 2:35, Riga.
584. Hoad, E., Rease, H., Stall, J., and Zav, J. 1973. *J. Quantum Electron.*, 9:652.
585. Basov, N. G., Belenov, E. M., and Danilychev, V. A. 1973. *Sov. Phys. J. Exp. Theor. Phys.* 64:108.
586. Panteleev, V. I. 1978. Electro-ionization synthesis of chemical compounds, Ph.D. thesis, Physical Institute of Academy of Sciences, Moscow.

587. Velikhov, E. P., Golubev, S. A., and Rakhimov, A. T. 1973. *Sov. Phys. J. Exp. Theor. Phys*. 65:543.
588. Velikhov, E. P., Golubev, S. A., and Rakhimov, A. T. 1975. *Sov. Phys. Plasma Phys*. 1:847.
589. Velikhov, E. P., Pismennyi, V. D., Rakhimov, A. T. 1977. *Adv. Phys. Sci*., Uspehi Phys. Nauk 122:419.
590. Basov, N. G. 1974. *Adv. Phys. Sci. Uspehi Phys. Nauk* 114:213.
591. Baranov, V. Yu., Napartovich, A. P., and Starostin, A. N. 1984. In *Plasma Physics*, ed. V. D. Shafranov, vol. 5, Moscow: VINITI.
592. Hassan, H. A. and Deese, J. E. 1976. *Phys. Fluids* 19:2005.
593. Belousov, I. G., Krasnoshtanov, V. F., and Rusanov, V. D. 1979. *J. Nucl. Sci. Technol. Nucl.–Hydrogen Energy* 15:43.
594. Pikaev, A. K. 1965. *Pulse Radiolysis of Water and Water Solutions*, Moscow: Nauka Science.
595. Firestone, R. F. 1957. *J. Am. Chem. Soc*. 79:5593.
596. Aulsloos, P., ed. 1968. *Fundamental Processes in Radiation Chemistry*, Danville, N.Y.: Interscience Publishers.
597. Dzantiev, B. G., Ermakov, A. N., and Popov, V. N. 1979. *J. Nucl. Sci. Technol. Nucl.–Hydrogen* Energy 15:89.
598. Hartech, P. and Doodes, S. 1957. *J. Chem. Phys*. 26:1727.
599. Kummler, R. 1977. *J. Phys. Chem*. 81:2451.
600. Bochin, V. P., Krasnoshtanov, V. F., Rusanov, V. D., and Fridman, A. 1983. *J. Nucl. Sci. Technol. Nucl.–Hydrogen* Energy 215:12.
601. Rosen, G. 1962. *Phys.Fluids* 5:737.
602. Dimick, R. C. and Soo, S. L. 1964. *Phys. Fluids* 7:1638.
603. Fuks, N. A. 1964. *Sov. Phys. News USSR Acad. Sci. Geophys*. 4:579.
604. Kirichenko, V. N. 1972. *Sov. Phys. Doklady* 205:78.
605. Ivanov, V. D. 1969. *Sov. Phys. Doklady* 188:65.
606. Ivanov, V. D. 1972. *Sov. Phys. Doklady* 203:806.
607. Este, G. and Westwood, W. D. 1984. *J. Vac. Sci. Technol*. A63:1845.
608. Este, G. and Westwood, W. D. 1988. *J. Vac. Sci. Technol*. A2:1238.
609. Butterbaugh, J. W., Baston, L. D., and Sawin, H. H. 1990. *J. Vac. Sci. Technol*. A82:916.
610. Schiller, S., Beister, G., Buedke, E., Becker, H. J., and Schmidt, H. 1982. *Thin Solid Films* 96:113.
611. Ridge, M. I. and Howson, R. P. 1982. *Thin Solid Films* 96:213.
612. Affinito, J. and Parson, R. R. 1984. *J. Vac. Sci. Technol*. A2:1275.
613. Fortov, V. E. 2001. *XXV Int. Conf. Phenomena Ionized Gases*, ICPIG-25, Nagoya, Japan, 2:13.
614. Glasstone, S. and Edlung, M. C. 1952. *Nuclear Reactor Theory*, Princeton, N.J.: D. Van Nostrand Co.
615. Su, T. and Bowers, M. T. 1973. *Int. J. Mass Spectrom. Ion Phys*. 12:347.
616. Shchuryak, E. 1976. *Sov. Phys.,* JETP, Journal of Experimental and Theoretical Physics. 71:2039.
617. Rusanov, V. D., Fridman, A. A., Sholin, G. V., Potapkin, B. V. 1985. Kurchatov Institute of Atomic Energy, Preprint #4201, Moscow.
618. Dobkin, S. V., Son, E. E. 1982. Sov. Phys. Thermal Physics of High Temperatures. 20:1081.
619. Fridman, A. 2003. Non-thermal Atmospheric Pressure Discharges: Physics and Applications. ISPC-16, pg. 8, UOPAC, Italy.
620. Finkelnburg, W., Maecker, H. 1956. Elektrishe Bögen und Thermisches Plasma, Handbuch der Physik, Bd. XXII.
621. Dresvin, S. V. 1977. Physics and Technology of Low Temperature Plasmas, Iowa State University Press, Ames, IA.
622. Calra, C. S., Kossitsyn, M., Iskenderova, K., Chirokov, A., Cho, Y. I., Gutsol, A., Fridman, A. ISPC-16, UOPAC, 2003, p. 218, Italy.
623. Masuda, S., ed. Chang, T. S., Oda, T. 2001. Applied electrostatic studies. IEJ, Institute of Electrostatics of Japan, Tokyo, 589 pg.
624. Iskenderova, K., Chirokov, A., Gutsol, A., Fridman, A., Sieler, K., Grace, J. 2001. ISPC-15. *UOPAC*. Orleans, France.
625. Iskenderova, K., Chirokov, A., Gutsol, A., Fridman, A., Sieber, K., Grace, J. 2003. ISPC-16, UOPAC, Italy.

626. Chirokov, A., Gutsol, A., Fridman, A., Sieber, K., Grace, J. 2003. Plasma Chemistry and Plasma Processing. vol. 23.
627. Encyclopedia of Low Temperature Plasma. 2000. Fortov, V. E., Ed. Moscow: Nauka.
628. Egorov, A. I., Stepanov, S. I. 2002. Sov. Phys., Journal of Technical Physics. 72:102.
629. Komachi, K. 1992. J. Vac. Sci. Technol. A11:164.
630. Uzhov, V. N. 1967. Industrial Gas Cleaning by Electric Filters. Khimia (Chemistry), Moscow.
631. Vdovin, Yu. A. 1975. Sov. Phys., Journal of Technical Physics, 45:630.
632. Fridman, A., Skop, H., Nester, S., Saveliev, A., Kennedy, L. 1977. ISPC-13, pg. 802–807, UOPAC, Beijing, China.
633. Fridman, A., Harkness, J. 2000. The Technical and Economic Feasibility of Using Low-Temperature Plasmas to Treat Gaseous Emissions from Pulp Mills and Weed Product Plants. NCASI Technical Journal. vol. 795, pg. 1–50.
634. Rakness, K. L. 1977. J. Phys. IV. vol. 7. Collogue, C4.
635. Barnett, C. F. 1989. In *A Physicist's Desk Reference,* Ed. Anderson, H. L., American Institute of Physics.
636. Kouprine, A., Gitzhofer, F., Booulos, M., Fridman, A. 2003. Polymerlike C-H Thin Film Coating of Nano-Powders in CCP-RF Discharge. Plasma Chemistry and Plasma Processing. vol. 23. Kluwer Academic Publ., NY.
637. Kuznetsova, I., Kalashnikov, N., Gutsor, A., Fridman, A., Kennedy, L. A. 2002. Journal of Applied Physics. 92:4231.
638. Heberlein, J. 2002. Pure and Applied Chemistry. 74:327.
639. Kruger, C. H., Laux, C. O., Packan, D. M., Yu, L., Pierrot, L. 2002. Pure and Applied Chemistry. 74:337.
640. Hrabovsky, M. 2002. Pure and Applied Chemistry. 74:429.
641. Okazaki, K., Nozaki, T. 2002. Pure and Applied Chemistry. 74:447.
642. Bray, K. N. 1968. J. Physics B: Atomic and Molecular Physics. 1:705.
643. Brau, C. A. 1972. Physica. 58:533.

INDEX

A

B

C

D

E

F

G

H

I

J

K

L

M

N

O

P

Q

R

S

U

V

W

X

Y

Z